Physiology of Cotton

James McD. Stewart · Derrick Oosterhuis
James J. Heitholt · Jack Mauney

Editors

Physiology of Cotton

 Springer

Editors

Prof. James McD. Stewart
University of Arkansas
Dept. Crop, Soil, &
Environmental Sciences
115 Plant Science Building
Fayetteville AR 72704
USA

Prof. James J. Heitholt
Texas A&M University
Commerce, TX 75429
USA

Prof. Derrick M. Oosterhuis
University of Arkansas
Dept. Crop, Soil &
Environmental Sciences
1366 Altheimer Drive
Fayetteville AR 72704
USA

Dr. Jackson R. Mauney
161 E. 1st Street, Suite 4
Mesa, AZ 85201
USA

Cover legend: Cotton production today is not to be undertaken frivolously if one expects to profit by its production. If cotton production is to be sustainable and produced profitably, it is essential to be knowledgeable about the growth and development of the cotton plant and to the adaption of cultivars to the region as well as the technology available. In addition, those individuals involved in growing cotton should be familiar with the use of management aids to know the most profitable time to irrigate, apply plant growth regulators, herbicides, foliar fertilizers, insecticides, defoliants, etc. The chapters in this book were assembled to provide those dealing with the production of cotton with the basic knowledge of the physiology of the plant required to manage to crop in a profitable manner.

ISBN 978-90-481-3194-5   e-ISBN 978-90-481-3195-2
DOI 10.1007/978-90-481-3195-2
Springer Dordrecht Heidelberg London New York

Library of Congress Control Number: 2009932906

Printed on acid-free paper

Springer is part of Springer Science+Business Media (www.springer.com)

Physiology of Cotton

FOREWORD

As a result of the success of the first *Cotton Physiology* book, edited by Jack R. Mauney and James McD. Stewart and published by the Cotton Foundation in 1985, discussion began for a continuation or an update of that book. During ensuing discussions we decided that rather than update *Cotton Physiology*, the new book would attempt to cover deficiencies and areas in which cotton research results were lacking in the first book, or in which research advances were being made at an accelerated rate since the publication of *Cotton Physiology*, e.g., plant response to the environment, crop modeling and biotechnology, among others. The current book *Physiology of Cotton* covers the literature and developments between 1985 and 1999. In many cases the chapters provide the definitive word, but in other areas, such as biotechnology, research has continued at a rapid pace. Unlike this book, whose conception originated with the first two editors, subsequent books are being planned as a series of smaller volumes under the auspices of the Cotton Agronomy and Physiology Conference. The planned books will each cover a more limited number of select topics of cotton physiology and will compile results and references that have ensued since 2000.

PREFACE

Cotton production today is not to be undertaken frivolously if one expects to profit by its production. If cotton production is to be sustainable and produced profitably, it is essential to be knowledgeable about the growth and development of the cotton plant and in the adaptation of cultivars to the region as well as the technology available. In addition, those individuals involved in growing cotton should be familiar with the use of management aids to know the most profitable time to irrigate, apply plant growth regulators, herbicides, foliar fertilizers, insecticides, defoliants, etc. The chapters in this book were assembled to provide those dealing with the production of cotton with the basic knowledge of the physiology of the plant required to manage the cotton crop in a profitable manner.

Research and promotion of cotton has been very successful in recent years, so that now the cotton textile industry ranks ahead of synthetic fibers (although these remain a looming threat to the industry). The main threat to the US producer currently is from foreign production of cotton, especially since foreign produced textiles have such a large share of the US market. Improved understanding of cotton fiber is essential if the quality parameters are to be improved in traditional and biotechnological genetic improvement

The new technologies of BT and herbicide tolerance and the boll weevil eradication program in the US allowed the resurgence of profitable cotton production in the Southeast where production had fallen to essentially to nil. Now the state of Georgia ranks second behind Texas in the number of acres planted to cotton and acreage is on the rise in Virginia and Florida. The BT and herbicide tolerance technologies have been adopted very rapidly by producers where they have been introduced. This includes China and India, countries which now rank #1 and #2 in bales of cotton produced. With the rapid adoption of biotech cotton came new problems (e.g., herbicide tolerance by weeds) that require a basic understanding of cotton physiology. The intent of this book is to provide this understanding for individuals dealing with all aspects of cotton production.

The Editors

ACKNOWLEDGEMENTS

The editors and contributors are indebted to Mr. Buddy Formby, previously of Microflo Chemical Company, for generously donating $50,000 towards the publication of the book in order to keep the final price affordable by a large number of cotton workers. We also thank Marci Milus for preparation and production of the "camera-ready" copy of the manuscripts.

TABLE OF CONTENTS

Chapter 1

THE ORIGIN AND EVOLUTION OF *GOSSYPIUM*

Jonathan F. Wendel[1], Curt L. Brubaker[2], and Tosak Seelanan[3]

[1]*Department of Botany, Iowa State University, Ames, IA 50011;* [2]*Present address: Centre for Plant Biodiversity Research, CSIRO Plant Industry, GPO 1600, Canberra ACT 2601, Australia*

I. INTRODUCTION

The genus *Gossypium* has a long history of taxonomic and evolutionary study. Much of this attention has been stimulated by the fact that the genus includes four domesticated species, the New World allopolyploids *G. hirsutum* and *G. barbadense* (2n = 52), and the Old World diploids *G. arboreum* and *G. herbaceum* (2n = 26). These cultivated species embody considerable genetic diversity, but this diversity is dwarfed by that included in the genus as a whole, whose 50 species have an aggregate geographic range that encompasses most tropical and subtropical regions of the world.

A remarkable morphological diversification accompanied the global radiation of *Gossypium* in response to the demands of particular ecological settings and selective environments. Plant habit, for example, ranges from fire-adapted, herbaceous perennials in northwest Australia to small trees in southwest Mexico that "escape" the dry season by dropping their leaves. Corolla colors embrace a rainbow of mauves and pinks ("Sturt"s Desert Rose", *G. sturtianum*, is the official floral emblem of the Northern Territory, Australia), whites and pale yellows (Mexico, Africa-Arabia) and even a deep sulphur-yellow (*G. tomentosum* from Hawaii). Seed coverings range from nearly glabrous (*e.g.*, *G. klotzschianum* and *G. davidsonii*), to short, stiff, dense brown hairs that aid in wind-dispersal (*G. australe, G. nelsonii*), to long, fine white fibers that characterize highly improved forms of the four cultivated species. There are even seeds that produce fat bodies to facilitate ant-dispersal (section *Grandicalyx* cottons from NW Australia). Much of this morphological diversity is lucidly detailed in Fryxell

(1979), and need not be belabored here. Perhaps it is worthwhile, though, to express the truism that the morphological and ecological breadth must have parallels in physiological and chemical diversity. The wild species of cotton, consequently, represent an ample genetic repository for exploitation. Although these wild species remain a largely untapped genetic resource, examples abound of their productive inclusion in breeding programs (*e.g.,* Meyer, 1974, 1975; Fryxell, 1976; Narayanan *et al.,* 1984; Niles and Feaster, 1984; Meredith, 1991). Further utilization of the many wild relatives of the cultivated cottons requires first that we understand their biology and relationships. This understanding grows from a combination of basic plant exploration, detailed taxonomic investigations, and phylogenetic studies designed to incorporate what is known about the biology of species into an evolutionary perspective.

Against this backdrop, it seems entirely appropriate to start a book on agronomic *Gossypium* with a review of what is known about wild and agronomically primitive *Gossypium*. This is the intent of the present chapter. Specifically, we first discuss the evolutionary origin of the cotton genus, and then focus on the taxonomy and diversification of *Gossypium* itself. This is followed by a synopsis of the biogeography of the genus and what is known about the origin of its many species. Particular attention is focused on the evolution of the New World allopolyploids, including the lineage to which *G. hirsutum* and *G. barbadense* belong. Finally, we discuss the development of the cultivated cottons, from their original domestication by aboriginal cultivators through the various stages of their progressive agronomic refinement.

J.McD. Stewart et al. (eds.), *Physiology of Cotton*,
DOI 10.1007/978-90-481-3195-2_1, © Springer Science+Business Media B.V. 2010

2. EMERGENCE AND DIVERSIFICATION OF THE COTTON TRIBE

Our taxonomic understanding of the cotton tribe developed from more than a century of study involving traditional taxonomic methods as well as modern tools such as comparative analysis of DNA sequences. This accumulating synthesis has led to a reasonably coherent taxonomic concept of a group of genera that are aligned into a single tribe, the *Gossypieae*. This relatively small tribe, which includes only eight genera, has traditionally been distinguished from other Malvaceae (Fryxell, 1968, 1979) on the basis of morphological features of the embryo, wood and seed coat anatomy, and by the presence of the punctae or lysigenous cavities ("gossypol glands") that are widely distributed throughout the plant body. More recently, the naturalness of the tribe has been investigated using tools from molecular biology, including comparative analyses of variation in chloroplast DNA restriction sites (LaDuke and Doebley, 1995) and in DNA sequence (Seelanan *et al.*, 1997). This latter study employed DNA sequences from both the chloroplast genome (cpDNA) and the nuclear genome to develop a phylogenetic framework for relationships among seven of the eight genera in the tribe, lacking information only for the rare genus *Cephalohibiscus*, which is represented by a single species from New Guinea and the Solomon Islands.

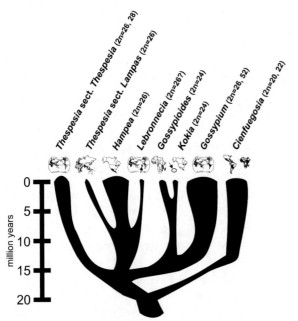

Figure 1-1. Phylogenetic relationships in the cotton tribe (*Gossypieae*), as inferred from molecular sequence data (Seelanan *et al.,* 1997). The sizes of the branches are scaled approximately to the number of species within genera, and numbers following generic names indicate somatic chromosome counts (not available for *Lebronnecia kokioides*). Ambiguities regarding branch orders are shown as trichotomies. The best available evidence suggests that the tribe is approximately 20 million years old, and that *Gossypium* emerged roughly 12.5 million years ago.

A simplified version of the molecular phylogeny from Seelanan *et al.* (1997) is presented in Figure 1-1, which in addition to displaying the evolutionary branching pattern among genera, provides a biogeographic, chromosomal, and temporal context. As shown, one of the two branches resulting from the earliest divergence in the tribe led to the evolution of *Cienfuegosia*, a genus of 26 species from the neotropics and parts of Africa and the Arabian peninsula. This phylogenetically basal position, based on DNA sequence data, is consistent with the relative morphological distinctiveness of the genus as well as its chromosome numbers (2n = 20, 22), which are not found elsewhere in the tribe. The phylogeny also shows a close relationship among *Hampea*, with 21 neotropical species, *Lebronnecia*, which consists of only a single rare species from the Marquesas Islands, and one portion of the genus *Thespesia* (the taxonomic section *Lampas*). The remainder of the 17 species of the pantropically distributed genus *Thespesia* are placed elsewhere in the molecular phylogeny (Fig. 1-1, as *Thespesia* sect. *Thespesia*). One of the more remarkable revelations is that the East African - Madagascan genus *Gossypioides* (with only two species) is the closest living relative of the Hawaiian endemic genus *Kokia* (with three extant and one extinct species). These two genera uniquely share a somatic chromosome number of 24, which, given the context of the phylogenetic relationships and chromosome numbers shown in Figure 1-1, may have been derived from a single ancestral aneuploid reduction from an ancestor with 26 chromosomes. In this respect Hutchinson's (1943) observation of an unusually long pair of chromosomes in *Gossypioides brevilanatum* (as *Gossypium brevilanatum*) is tantalizing, in that it suggests chromosome fusion as the underlying mechanism. Hutchinson also noted that successful grafts could be made between *Kokia rockii* [= *Kokia drynarioides*] and *Gossypioides kirkii*, providing additional support for the close relationships between the two genera. Finally, the molecular phylogenetic data indicate that the genus *Gossypium* is reasonably isolated from all other genera in the tribe, but that its closest living relatives lie in the *Gossypioides/Kokia* lineage.

The oldest Malvacean pollen is from the Eocene (38 - 45 million years before present - mybp) in South America and Australia and from the Oligocene (25 - 38 mybp) in Africa (Muller, 1981, 1984; Macphail and Truswell, 1989). This information suggests that the Malvaceae originated during the first third of the Tertiary and that by approximately 30 mybp it had achieved a world-wide distribution. Beyond this limited information, no clues regarding the origin of the *Gossypieae* or of *Gossypium* are available from the fossil record. Divergence in DNA sequence among species provides an alternative approach to estimate divergence times, given a "molecular clock" and an appropriate clock calibration. The time scale in Figure 1-1, for example, is based on sequence divergence percentages among taxa for the approximately 2100 base-pair cpDNA gene *ndhF* (Seelanan *et al.*, 1997), and a clock calibration based on an average divergence rate for single copy cpDNA genes of 5×10^{-10} nucleotide substitutions per site per year (Palmer, 1991). There are several potential sources of error in these calculations (Hillis *et al.,* 1996), so we regard the values in Figure 1-1 only as useful "ballpark" estimates, which nonetheless contribute to our understanding of the history of the tribe.

Within the tribe, the mean *ndhF* sequence divergence between the two basal lineages (*Cienfuegosia* and other genera) translates into an initial divergence of 19 mybp. This estimate, to the extent that the molecular clock has operated and that we have calibrated it correctly, functions as a lower bound on the age of the tribe. In a similar fashion, other divergence estimates are summarized in Figure 1-1. The *Lebronnecia/Hampea/*Thespesia sect. *Lampas* lineage, for example, separated from the rest of the tribe approximately 15 mybp, with divergence into its three sub-lineages occurring approximately half as long ago. *Gossypium* is inferred to have branched off from its closest relatives (*Kokia* and *Gossypioides*) approximately 12.5 mybp, with the latter two genera becoming separated relatively recently, circa 3 mybp. This recent a separation between genera now geographically isolated from one another by many thousands of kilometers of open ocean (*Kokia* from Hawaii and *Gossypioides* from Madagascar and East Africa) implies that trans-oceanic dispersal was involved in the evolution of one or both genera. In this respect the *Gossypioides - Kokia* floristic relationship represents the latest in a series of examples of long-distance, salt-water dispersal in the tribe (Stephens, 1958, 1966; Fryxell, 1979; Wendel and Percival, 1990; Wendel and Percy, 1990; DeJoode and Wendel, 1992; Wendel and Albert, 1992). These many examples serve to underscore the importance of oceanic dispersal as a factor in the evolution of the tribe.

3. EMERGENCE AND DIVERSIFICATION OF THE COTTON GENUS

3.1 A Global Pattern of Diversity

The framework of Figure 1-1 suggests that the cotton genus has a history that extends back at least 12.5 million years, and according to other molecular data (Wendel and Albert, 1992) the lineage may be twice this old. Since its emergence, the genus has achieved pantropical distribution with three primary centers of diversity. These are in Australia, especially the Kimberley region, the Horn of Africa and southern part of the Arabian Peninsula, and the western part of central and southern Mexico.

Because of the economic importance of the cultivated cottons, the genus has long attracted the attention of taxonomists, whose work has been summarized in several useful volumes (Watt, 1907; Hutchinson *et al.,* 1947; Saunders, 1961; Fryxell, 1979, 1992). The most widely followed taxonomic treatments are those of Fryxell (Fryxell, 1979;

Fryxell *et al.,* 1992), in which species are grouped into four subgenera and eight sections. This classification system is based primarily on morphological and geographical evidence, although most infrageneric alignments are congruent with cytogenetic and molecular data sets as well. At present, *Gossypium* includes 49 species (Table 1-1), including some recently recognized from Africa/Arabia (Vollesen, 1987) and Australia (Fryxell *et al.,* 1992). Although the classification given in Table 1-1 is unlikely to change dramatically in the coming years, the taxonomic status of a number of species is still uncertain. This is especially the case for taxa from Africa/Arabia, many of which are poorly represented in collections and for whom information is largely lacking on such central taxonomic issues as natural patterns of diversity and geographic distribution. Similarly, the wealth of diversity in taxa from the Kimberley region of Australia is presently being studied in detail from both morphological and molecular perspectives (by C. Brubaker, L. Craven, T. Seelanan, J. Stewart, and J. Wendel). This work will undoubtedly lead to some changes in our taxonomic concepts within the genus (Table 1-1).

Global radiation of the genus was accompanied not only by an impressive diversification in morphology and ecology, but also by extensive chromosomal evolution (Beasley, 1940, 1942; Phillips and Strickland, 1966; Edwards and Mirza, 1979; reviewed by Endrizzi *et al.,* 1985; also see Stewart, 1994). Genomes typically are reasonably similar among close relatives, and this is reflected in the ability of related species to form hybrids that display normal meiotic pairing and high F_1 fertility. In contrast, wider crosses are often difficult or impossible to effect, and those that are successful are usually characterized by meiotic abnormalities. The collective observations of pairing behavior, chromosome sizes, and relative fertility in interspecific hybrids have led to the recognition of eight diploid "genome groups" (designated A through G, plus K). This cytogenetic partition of the genus is largely congruent with taxonomic and phylogenetic divisions, as discussed below.

A summary cytogeographic depiction of the diploid genome groups is shown in Figure 1-2. Although all diploid *Gossypium* species have n = 13 chromosomes, DNA content per genome varies more than three-fold (Edwards *et al.,* 1974; Kadir, 1976; Bennett *et al.,* 1982; Michaelson *et al.,* 1991). The 2C contents range from approximately 2 picograms per cell in the New World D genome diploids to approximately 7 pg per cell in Australian K genome species (J. McD. Stewart, pers. comm.). The variation in DNA content is probably caused by modification of the repetitive DNA fraction, with relatively little difference in the absolute amounts of single-copy DNA (Geever *et al.,* 1989).

3.2 Taxonomy and Phylogeny

Speculation regarding the time and place of origin of *Gossypium* has a long history (Hutchinson *et al.,* 1947; Saunders, 1961; Fryxell, 1965a; Johnson and Thein, 1970;

Table 1-1. Taxonomy of *Gossypium*[1] and some notable features of its species.

Gossypium L.	Genome[2]	Comments	References
Subgenus *Sturtia* (R. Brown) Todaro			
Section *Sturtia*	C	This subgenus consists of all the indigenous Australian species.	Fryxell, 1992
		The species of this section, as well as those in section *Hibiscoidea*, are the only *Gossypium* species that do not deposit terpernoid aldehydes ("gossypol") in the seed Flower mauve.	Brubaker *et al.*, 1996; Fryxell, 1992
G. sturtianum J. H. Willis	C₁	"Sturt's Desert Rose", the floral emblem of the Northern Territory, is distributed widely across the Australian continent in the temperate arid zone.	Craven *et al.*, 1994
G. robinsonii F. Mueller	C₂	Some phylogenetic analyses place this species basal in the Australian *Gossypium* lineage.	Wendel and Albert, 1992
Section *Grandicalyx* Fryxell	K	This section includes the unusual herbaceous perennials from the Kimberley region of NW Australia. These species have a thick root-stock from which they resprout following fire or seasonal drought, and eliosomes on nearly hairless seeds to facilitate ant-dispersal. This section has the largest genome in the genus.	Fryxell, 1992; Fryxell *et al.*, 1992; Stewart, 1994
G. costulatum Todaro	K	One of the first Australian *Gossypium* species to be collected, along with *G. cunninghamii* and *G. populifolium*, by Alan Cunningham between 1818 and 1820 where each occurs near coastal waters accessible by ship.	Craven *et al.*, 1994
G. cunninghamii Todaro	K	The only sessile or subsessile species in *Gossypium*. This species may have originated from an ancient hybridization in which one parent (maternal) was a species similar to present-day *G. sturtianum*. A similar cytoplasm is also found in *G. bickii* (see below). The paternal parent, however, is located in the Northern Territory.	Wendel and Albert, 1992
G. exiguum Fryxell, Craven & Stewart *G. pilosum* Fryxell *G. rotundifolium* Fryxell, Craven & Stewart	K	These mainly prostrate species are more widely distributed than the other section *Grandicalyx* species, may be difficult to distinguish in the field, and may have imprecise taxonomic descriptions.	Fryxell *et al.*, 1992; Stewart, Craven, Brubaker, and Wendel (personal observations)
G. enthyle Fryxell, Craven & Stewart *G. nobile* Fryxell, Craven & Stewart *G. pulchellum* (C. A. Gardner) Fryxell *G. londonderriense* Fryxell, Craven & Stewart *G. marchantii* Fryxell, Craven & Stewart *G. populifolium* (Bentham) F. Mueller ex Todaro *G. pulchellum* (C. A. Gardner) Fryxell *G. anapoides*	K	These suberect and erect species are more narrowly distributed than the former three species, with some being known from only a few populations. The newest species *G. anapoides* has yet to be formally named.	Fryxell *et al.*, 1992; Stewart, Craven, Brubaker, and Wendel (unpublished)
Section *Hibiscoidea* Todaro	G	These three species do not deposit terpernoid aldehydes in the seeds (see section *Sturia* above).	Brubaker *et al.*, 1996
G. australe F. Mueller *G. nelsonii* Fryxell	G	These morphologically similar species possess stiff spreading seed hairs that allow the seed to "climb" out of the capsule, and are the only species that are wind dispersed.	Stewart *et al.*, 1987; Fryxell, 1992

continued

Table 1-1. Continued.

	Genome[2]	Comments	References
G. bickii Prokhanov	G_1	The origin of this species involved hybridization in which the maternal parent was similar to present-day G. sturtianum and the paternal parent was like modern G. australe or G. nelsonii.	Wendel et al., 1992
Subgenus Houzingenia (Fryxell) Fryxell			
Section Houzingenia			
Subsection Houzingenia	D	Large shrubs and small trees from the New World, primarily Mexico	Fryxell, 1992
	D	The two species in this subsection are morphologically similar and interfertile.	Fryxell, 1965b, 1967, 1979
G. thurberi Todaro	D_1	Northern most species which tolerates mild frost via defoliation. The D-genome species employed by J.O. Beasley to create the triple hybrid that was used to introgress high fiber strength into G. hirsutum.	Fryxell, 1976
G. trilobum (DC.) Skovsted	D_8	Sister species to G. thurberi. Source of male sterile cytoplasm and restorer factor.	Fryxell, 1965b, 1967; Stewart, 1992b
Subsection Integrifolia (Todaro) Todaro	D	Interspecific hybrids between either species in this subsection and several other species except B,C,G are embryo lethal.	Phillips, 1977; Lee, 1981, 1986
G. davidsonii Kellogg	D_{3-d}	These two species represent a progenitor-derivative species-pair, whereby the latter taxon, from the Galapagos Islands, was derived from the former following long-distance dispersal from Baja California.	Wendel and Percival, 1990
G. klotzschianum Andersson	D_{3-k}		
Subsection Caducibracteolata Mauer	D	The three species of this subsection are caliciphiles typically found in arid habitats around the Gulf of California. They have reduced leaves with thick cuticles and a double palisade layer and the largest seeds of the diploid species. The floral bracts are caduceus, abscising well before anthesis in G. armourianum, and shortly before to just after anthesis in the other species.	Phillips and Clement, 1967; Fryxell, 1992
G. armourianum Kearney	D_{2-1}	Germplasm pool for bacterial blight resistance gene	Endrizzi et al., 1985
G. harknessii Brandegee	D_{2-2}	Source of cytoplasmic male sterility and restorer factors.	Meyer, 1975
G. turneri Fryxell	D_{10}	Sister species to or derivative from G. harknessii.	Fryxell, 1978
Section Erioxylum (Rose & Standley) Prokhanov	D		
Subsection Erioxylum	D	This group of species have a unique flowering phenology. At the height of the dry season, while leafless, the plants flower and fruit. After the fruits mature, the plants remain dormant until returning rains stimulate new vegetative growth.	
G. aridum (Rose & Standley ex Rose) Skovsted	D_4	There is evidence of cytoplasmic introgression from subsection Integrifolia into populations of G. aridum from the State of Colima, Mexico. Other populations of this widely distributed species are normal in this respect.	Wendel et al., 1995a
G. lobatum H. Gentry	D_7	Leaves nearly disticous	Fryxell, 1979, 1992

continued

Table 1-1. Continued.

	Genome[2]	Comments	References
G. laxum Phillips	D₉		DeJoode, 1992; Wendel and Albert, 1992
G. schwendimanii Fryxell & S. Koch	D₁₁	The most recently described species among the New World diploids.	Fryxell
Subsection Selera (Ulbrich) Fryxell			
G. gossypioides (Ulbrich) Standley	D₆	The only diploid species that shows evidence of the original A X D hybridization that gave rise to the allotetraploids. This species may have arisen via introgressive speciation.	Fryxell and Koch, 1987
Subsection Austroamericana Fryxell			
G. raimondii Ulbrich	D₅	This species, a relatively recent immigrant to Peru, and *G. gossypioides* are the two taxa whose genomes are most similar to the D- subgenome of the allopolyploids, and hence serve as models of the D-genome diploid parent.	Endrizzi et al. 1985; Wendel et al., 1995a
Subgenus Gossypium	A, B, E, F	This subgenus includes the African, Arabian, and Asian diploid species.	Fryxell, 1979, 1992; Vollesen, 1987
Section Gossypium			
Subsection Gossypium	A, B, E, F	An A-genome progenitor served as the female parent in the hybridization event that resulted in allopolyploid formation	Wendel, 1989
G. herbaceum L.	A₁	Still cultivated on a small scale. Germplasm pool for several agronomically desirable traits.	Endrizzi et al.. 1985; Stewart, 1994
G. arboreum L.	A₂	Type species of the genus. Still cultivated on a small scale; Germplasm pool for several agronomically desirable traits	Endrizzi et al. 1985; Stewart, 1994
Subsection Anomala Todaro	B		
G. anomalum Wawra & Peyritsch	B₁	Germplasm pool for bacterial blight resistance	Endrizzi et al. 1985
G. triphyllum (Harvey & Sonder) Hochreutiner	B₂	Fryxell (1992) placed it in Section *Triphylla*, but molecular data demonstrate that its affinities are with other B-genome taxa.	Wendel and Albert, 1992; Seelanan et al. 1997
G. capitis-viridis Mauer	B₃	This island endemic is similar to *G. anomalum*, with which it may be conspecific.	Vollesen, 1987
Subsection Longiloba Fryxell	F		
G. longicalyx J. B. Hutchinson & Lee	F₁	This cytologically unique *Gossypium* species is also unusual for its ecological adaptation to mesic environments	Phillips and Strickland, 1966; Fryxell, 1979
Subsection Pseudopambak (Prokhanov) Fryxell	E	Seven species adapted to the extremely arid habitats of eastern Africa, the southeastern tip of Arabia, and the Sind in Pakistan	Vollesen, 1987; Fryxell, 1992
G. benadirense Mattei	E	These two species were only recently reinstated from limited materials and hence are incompletely known.	Vollesen, 1987
G. bricchettii (Ulbrich) Vollesen	E		
G. vollesenii Fryxell	E	This species was given as *Gossypium* sp. A by Vollesen (1987). Fryxell recognized it as a new species distinct from *G. incanum* and *G. somalense*.	Fryxell, 1992
G. stocksii Masters in Hooker	E₁		
G. somalense (Gurke) J. B. Hutchinson	E₂	May contain a complementary lethal factor when combined with *G. hirsutum*.	Stewart, per. comm..
G. areysianum Deflers	E₃		
G. incanum (Schwartz) Hillcoat	E₄	Foliage has a putrid odor	

continued

Table 1-1. Continued.

	Genome[2]	Comments	References
Section *Serrata* Fryxell	?		
G. trifurcatum Vollesen	?	This poorly known species may be distinguished from other *Gossypium* species by its serrate leaves.	Vollesen, 1987; Fryxell, 1992
Subgenus *Karpas* Rafinesque	AD	The allopolyploid cottons were formed by hybridization between A- and D-genome diploids, probably during the Pleistocene.	Wendel, 1989; Wendel and Albert, 1992; Seelanan *et al.*, 1997
G. hirsutum L.	AD_1	A geographically narrow range was greatly expanded following domestication. Great morphological diversity developed among the wild, semi-wild and domesticated stocks, which may provide many agronomically desirable genes.	Bell, 1984a; Meredith, 1991; Brubaker and Wnedel, 1993, 1994
G. barbadense L.	AD_2	Domesticated in NW South America. Modern cultivars are highly introgressed with *G. hirsutum*. Landraces provide disease resistance and fiber quality genes.	Shepherd, 1974; Meredith, 1991; Percy and Wendel, 1990; Wang *et al.* 1995
G. tomentosum Nuttall ex Seemann	AD_3	Hawaiian Island endemic. Source of the nectariless trait.	Meyer and Meyer, 1961; DeJoode and Wendel, 1992
G. mustelinum Miers ex Watt	AD_4	This species, which consists of widely scattered populations in NE Brazil, represents one of three modern lineages of allopolyploid *Gossypium*.	Wendel *et al.* 1994
G. darwinii Watt	AD_5	Galapagos Islands relative of *G. barbadense*. Source of resistance to *Fusarium* and *Verticillium*.	Bell, 1984a; Wendel and Percy, 1990

[1] Modified from Fryxell (1992) to accommodate recent molecular evidence (Wendel and Albert, 1992; Cromn *et al.*, 1996; Seelanan *et al.*, 1997)

[2] From Stewart (1994).

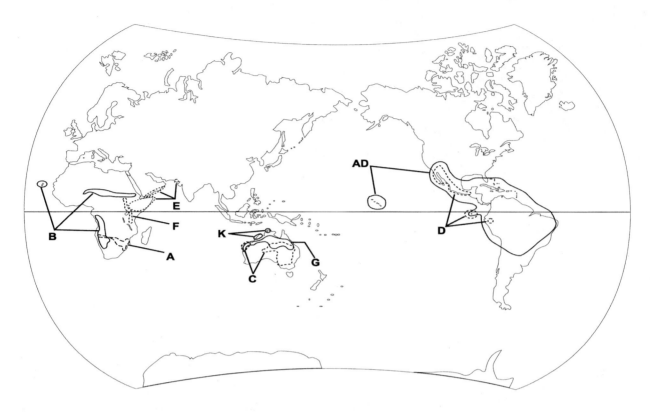

Figure 1-2. Genome biogeography of *Gossypium*. Genome designations
and geographic distributions are shown. See text for additional detail.

Edwards *et al.,* 1974; Valiček, 1978). Much of this literature has been critically evaluated elsewhere (Wendel, 1989; Wendel and Albert, 1992). The essential point is that in the absence of a clear fossil record, the most rigorous method for reconstructing the history of the genus is formal phylogenetic analysis. This approach becomes especially powerful when combined with measurements of DNA sequence divergence, thereby allowing ages of particular lineages to be estimated.

Phylogenetic investigations in *Gossypium* based on molecular data include cpDNA restriction sites (Wendel and Albert, 1992), and DNA sequences for the 5S ribosomal genes and spacers (Cronn *et al.,* 1996), for the chloroplast gene *ndhF*, and for the nuclear 5.8S gene and its flanking internal transcribed spacers (Seelanan *et al.,* 1997). Based on these analyses, genealogical lineages of species are largely congruent with genome designations and geographical distributions. Accordingly, each genome group corresponds to a single natural lineage, and in most cases, these lineages are also geographically cohesive. This information has been embodied in the classification presented in Table 1-1 and the synthesis depicted in Figure 1-3.

As shown, there are three major lineages of diploid species, corresponding to three continents: Australia (C, G, K genomes), the Americas (D genome), and Africa/Arabia (A, B, E, and F genomes). Not illustrated is the uncertainty regarding some of the earliest branch points, which exists because the phylogenies inferred from the different molecular data sets differ in some details. To a certain extent, this is not

surprising, given that divergences among the major genomic groups appear, on the basis of DNA sequence divergence (Seelanan *et al.,* 1997), to have been closely spaced, with the major lineages having been established by approximately 11.5 mybp, shortly after *Gossypium* originated and diverged from the *Kokia-Gossypioides* clade. The evolutionary picture envisioned is one of rapid radiation early in the history of the genus, leading ultimately to the emergence of the modern cytogenetic groups. Relationships among species within each genome group are in many cases solidly established, whereas in others there is continuing uncertainty.

3.2.1　Australian Species

Some uncertainty resides with the Australian cottons (subgenus *Sturtia*), which include 16 named species as well as a new species yet to be named (Stewart, Craven, and Wendel, unpublished). Collectively, these taxa are comprised in the C-, G-, and K-genome groups. These three groups of species are implicated by DNA sequence data (Seelanan *et al.,* 1997) to be natural lineages, consistent with their formal alignments (Table 1-1) into the taxonomic sections *Sturtia* (C genome), *Hibiscoidea* (G genome), and *Grandicalyx* (K genome). Relationships among the three groups, however, are still not clear. Wendel and Albert (1992) place *G. robinsonii* basally within the entire assemblage of Australian species (suggesting that radiation of *Gossypium* in Australia proceeded eastward from the westernmost portion of the continent. However, recent analyses by Seelanan *et al.* (1997) are equivocal in this regard.

With respect to the taxonomy within each of the three Australian genome groups, there is little uncertainty for the C and G genome groups, as these are well represented in collections and have been thoroughly studied (Wendel *et al.,* 1991; Wendel and Albert, 1992; Seelanan *et al.,* 1997; and references therein). Much less certain is the taxonomy of the K genome species, which are all placed in section *Grandicalyx.* These unusual species have a distinctive geography, morphology and ecology, and exhibit a syndrome of features characteristic of "fire-adapted" plants. In particular, they are herbaceous perennials with a bi-seasonal growth pattern whereby vegetative growth dies back during the dry season, or as a result of fire, to underground rootstocks that initiate a new cycle of growth with the onset of the next wet season. In addition, they are distinctive in the genus in having upright flowers whose pedicels recurve in fruit, the latter releasing onto the ground lintless seeds with a fleshy elaisome that functions in ant dispersal. Recent expeditions to the Kimberley area have enhanced our understanding of the group and have resulted in the discovery of at least seven new species, six of which have been formally described (Fryxell *et al.,* 1992), however circumscriptions of the species are poorly understood. Our ongoing investigations, in collaboration with J. Stewart and L. Craven, are

expected to expand our knowledge of the group and lead to a more stable and justified taxonomy.

3.2.2 African-Asian Species

Subgenus *Gossypium* comprises fourteen species from Africa and Arabia in the most recent taxonomic treatment of the genus (Fryxell, 1992). In Table 1-1 these species are divided into two sections, one (section *Gossypium*) with four subsections and the other (section *Serrata*) with a single included species, *G. trifurcatum.* This latter species, from deserts in eastern Somalia, is poorly understood taxonomically and cytogenetically. The unusual feature of dentate leaves raises the possibility that it may not belong in *Gossypium*, and may instead be better referred to *Cienfuegosia* (Fryxell, 1992), a possibility that requires future evaluation.

This latter example underscores the provisional nature of much of the taxonomy of the African-Arabian species of *Gossypium*. Although there are recent valuable contributions to our knowledge of the group (Vollesen, 1987; Holubec, 1990; Fryxell, 1992), the need remains for basic plant exploration for nursery material and additional taxonomic study. In addition to the uncertain position of *G. trifurcatum*, the specific status of some taxa within subsection *Anomala* and especially subsection *Pseudopambak* is unsure. Within the former, for example, Fryxell differenti-

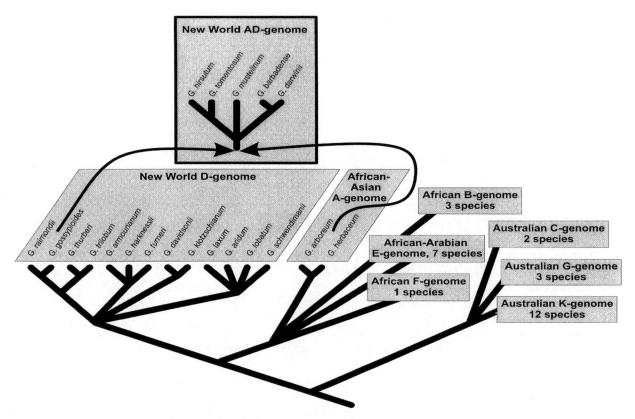

Figure 1-3. Evolutionary history of *Gossypium.* Relationships among diploid groups are shown, as is the origin of the allopolyploids following inter-genomic hybridization (top). Ambiguities regarding branch orders are shown as trichotomies. See text for additional detail.

ates *G. triphyllum* at the sectional level (as the sole species of section *Triphylla*) based on some unusual morphological characteristics and similarities to some section *Hibiscoidea* taxa. Other taxonomic opinion, however (Vollesen, 1987) and molecular data (Wendel and Albert, 1992; Seelanan *et al.*, 1997) place this SW African species squarely in section *Anomala* with the other two B genome species. Within section *Pseudopambak*, species recognition and definition are in some cases based on rather limited herbarium material (*e.g., G. benadirense, G. bricchettii, G. vollesenii*) without the benefit of analyses typically conducted in *Gossypium* to increase taxonomic confidence, such as "common garden", cytogenetic, or molecular experiments. Thus, this tentative classification must suffice until additional material becomes available for comparative study.

From a cytogenetic standpoint, the African-Arabian species exhibit considerable diversity, collectively accounting for four of the eight genome groups. The A genome, comprising the two cultivated cottons *G. arboreum* and *G. herbaceum* of subsection *Gossypium*, has been extensively studied (reviewed in Wendel *et al.*, 1989; see also discussion below). The three African species in subsection *Anomala* make up the B genome, as discussed above. The sole F-genome species, *G. longicalyx*, is cytogenetically distinct (Phillips and Strickland, 1966), morphologically isolated (Fryxell, 1971; Valiček,1978; Vollesen, 1987), and is perhaps adapted to more mesic conditions than any other diploid *Gossypium* species. The remaining seven species (subsection *Pseudopambak*) are considered to possess E genomes, although three of these have not been examined cytogenetically.

Efforts to resolve relationships within the entire subgenus *Gossypium* and to reconstruct its evolutionary history have met with mixed success. On the whole, taxonomic subdivisions within the subgenus (as discussed above; see Table 1-1) appear to correspond to natural lineages that are congruent with cytogenetic designations. Molecular data (Wendel and Albert, 1992; Seelanan *et al.*, 1997) uniformly support the recognition of four evolutionary lines, corresponding to the A, B, E, and F genomic groups. Different data sets differ in branch order among these groups, however, so in the interest of conservatism we have shown the relationships among subsections as unresolved in Figure 1-3. To a certain extent, this phylogenetic ambiguity is expected, given the DNA sequence data (Seelanan *et al.*, 1997) implicating an evolutionary history of rapid and early diversification of the primary lineages (and hence, short "interior branches" that are difficult to detect and discern). Notwithstanding this potential roadblock to gathering definitive phylogenetic evidence, additional study of this group is critical, not only for understanding the history of the African-Arabian lineages, but for evaluating earlier speculations that the genus originated in Africa (*e.g.,* Hutchinson *et al.,* 1947; Saunders, 1961: Johnson and Thein, 1970).

3.2.3   American Diploid Species

Subgenus *Houzingenia* contains two sections and six subsections, whose species collectively represent the New World D-genome diploids (Table 1-1). Given their proximity to American taxonomists and agriculturalists, these species have been more thoroughly collected and studied. Consequently their taxonomy is reasonably well understood. The subgenus has also received considerable phylogenetic attention (Wendel and Albert, 1992; Wendel *et al.,* 1995a; Cronn *et al.,* 1996; Seelanan *et al.,* 1997), which provides strong support for the naturalness of most of the recognized subsections. Specifically, the 13 species are divided into lineages corresponding precisely to the six taxonomic subsections. Evolutionary relationships among these subsections are less evident, however, and so a conservative representation is shown in Figure 1-3. Illustrated is the apparent lack of correspondence between phylogeny and the alignment of subsections into the two sections *Houzingenia* and *Erioxylum*, which Fryxell (1979, 1992) has long maintained as natural. Although this alignment is not directly contradicted by molecular data, different data sets lead to various reconstructions, with a consensus view yet to emerge. An additional revelation from molecular work (Wendel and Albert, 1992; Wendel *et al.,* 1995a; Cronn *et al.,* 1996) is that *G. gossypioides*, a species with some unusual morphological features and the sole representative of subsection *Selera*, is the closest relative of *G. raimondii*, a geographically isolated species (with a narrow distribution in Peru) and the exclusive member of subsection *Austroamericana*. This relationship has implications for the origin of allopolyploid cottons, as discussed below.

Available evidence indicates that the D-genome lineage is no more than 10 million years old, which implies an origin following long-distance dispersal from Africa. Because the center of diversity of this assemblage of 13 species is western Mexico, it is likely that the lineage became established and initially diversified in this region. Later range extensions are inferred to have arisen from relatively recent (probably Pleistocene) long-distance dispersals, leading to the evolution of endemics in Peru (*G. raimondii*) and the Galapagos Islands (*G. klotzschianum*; Wendel and Percival, 1990).

4.   BIOGEOGRAPHY AND SPECIATION

4.1   Speciation Mechanisms in the Diploids

4.1.1   Dispersal and Vicariance

As discussed above and as illustrated in Figure 1-1, *Gossypium* is estimated to have originated 12.5 (Seelanan *et al.,* 1997) to 25 (Wendel and Albert, 1992) mybp. Because paleocontinental reconstructions of the Earth at this time show the African and Australian plates to be separated by several thousand km of ocean, disjunction of the Australian and African lineages probably occurred via long-distance dispersal rather than range fragmentation caused by the emergence of a natural barrier such as a mountain range or desert (vicariance). Similar arguments for long-distance dispersal apply to later divergences among the major lineages, including D-genome radiation from Africa into the Americas, and a second colonization of the New World by the A-genome ancestor of the AD-genome allopolyploids (discussed below). Indeed, the propensity for long-distance dispersal appears characteristic of the entire cotton tribe, given the lineage ages and geographic framework of Figure 1-1.

Long-distance dispersal clearly played an important role not only in diversification of major evolutionary lines but also in speciation within *Gossypium* lineages. Examples include dispersals from southern Mexico to Peru (*G. raimondii*), from northern Mexico to the Galapagos Islands (*G. klotzschianum*; Wendel and Percival, 1990), from western South America to the Galapagos Islands (*G. darwinii*; Wendel and Percy, 1990), from Africa to the Cape Verde Islands (*G. capitis-viridis*), and from the neotropics to the Hawaiian Islands (*G. tomentosum*; DeJoode and Wendel, 1992). These latter examples, as well as those cited above for the origin of genera within the tribe *Gossypieae*, suggest a common dispersal mechanism of oceanic drift. In this respect, seeds of many species of *Gossypium* are tolerant of prolonged periods of immersion in salt water (Stephens, 1958, 1966). Remarkably, seeds of the Hawaiian endemic cotton, *G. tomentosum*, are capable of germination after three years immersion in artificial sea water (Fryxell, pers. comm.).

Many *Gossypium* species have relatively small and isolated geographic distributions, raising the possibility that their disjunction from other taxa arose via range fragmentation accompanying climatic fluctuations (Fryxell, 1965a, 1979) or other forms of vicariance. In the Kimberley region of Australia, for example, speciation may have been promoted during periods of desertification by progressive range restriction of one or more widespread species into pockets of suitable habitat. The accompanying geographic isolation is envisioned to have permitted genetic and morphological divergence, and ultimately the genesis of novel taxa. Similar processes may have been instrumental in generating one or more of the many species-pairs that are common in *Gossypium,* for example, the morphologically similar and interfertile, yet geographically disjunct *G. thurberi* (Sonora to Chihuahua, Mexico and Arizona) and *G. trilobum* (Sinaloa to Morelos, Mexico). Fryxell (1965a, 1979) suggested that this disjunction, as well as several others, arose as a consequence of range restrictions caused by increasing aridity during the Pleistocene-to-Recent transition (*ca.* 11,000 years ago).

4.1.2 Cytoplasmic Introgression and Recombinational Speciation

In addition to the conventional phenomena of divergence mediated by dispersal and vicariance, speciation in *Gossypium* has involved evolutionary processes that are less well understood. This was first discovered when molecular markers from the plastid and nuclear genomes were used to document the evolutionary history of *G. bickii* (Wendel *et al.*, 1991). This species is one of three morphologically similar G-genome cottons (along with *G. australe* and *G. nelsonii*) in section *Hibiscoidea*. In contrast to expectations based on this taxonomy, the maternally inherited chloroplast genome of *G. bickii* was nearly identical to the plastid genome of *G. sturtianum*, a morphologically distant C-genome species in a different taxonomic section (*Sturtia*). In contrast, nuclear markers showed the expected relationship, *i.e.*, that *G. bickii* shares a more recent common ancestor with its close morphological allies, *G. australe* and *G. nelsonii*, than it does with *G. sturtianum*. This discrepancy was explained by invoking a bi-phyletic ancestry for *G. bickii*, whereby *G. sturtianum*, or a similar species, served as the maternal parent in an ancient hybridization with a paternal donor from the lineage leading to *G. australe* and *G. nelsonii*. Interestingly, no *G. sturtianum* nuclear genes were detected in *G. bickii*, suggesting that the nuclear genomic contribution of the maternal parent was eliminated from the hybrid or its descendent maternal lineage.

This phenomenon of "cytoplasmic capture" or "recombinational speciation" has subsequently been implicated elsewhere in the genus. A likely example concerns the K-genome species *G. cunninghamii*, which perhaps not coincidentally has an unusual morphology and is geographically widely disjunct from its close relatives. This species is restricted to the Cobourg Peninsula of the Northern Territory approximately 500 km distant from the Kimberley region, where all other K-genome taxa are found. Analogous to *G. bickii*, the chloroplast genome of *G. cunninghamii* appears to have been donated by the *G. sturtianum* lineage, although in this case the hybridization event appears to have been more ancient (Wendel and Albert, 1992).

A final example involves *G. aridum*, one of four species of Mexican cottons that are comprised in subsection *Erioxylum*. These four species form a morphologically coherent and distinctive group whose shared common ancestry is supported, with one exception, by all molecular data (DeJoode, 1992; Wendel *et al.*, 1995a; Cronn *et al.*, 1996; Seelanan *et al.*, 1997). The exception relates to populations from the Mexican state of Colima that have a chloroplast genome that is strikingly divergent from that found in the remainder of the species (DeJoode, 1992; Wendel and Albert, 1992). This alien cytoplasm is inferred to have originated through an ancient hybridization with a member of the *Integrifolia* subsection, whose two extant species (*G. davidsonii* and *G. klotzschianum*) have geographic ranges (Baja California and the Galapagos Islands, respectively) that are distant from the range of *G. aridum*. As was the case with *G. bickii*, the nuclear genome of *G. aridum*, including the Colima populations, exhibits no evidence of this introgression event.

The foregoing examples highlight the possibility of cytoplasmic introgression between species whose present ranges give no indication that hybridization is even possible, and apparently without long-term survival of nuclear genes from the maternal progenitor. Possibly one or more of these species (or others) actually originated through an evolutionary process that was "seeded" by a hybridization event. In the case of *G. aridum*, however, cytoplasmic introgression without speciation is implicated, in that only populations from a small part of the range contain the aberrant cytoplasm. Possible mechanisms of cytoplasm transfer include repeated backcrossing of the hybrid as female into the paternal donor lineage, cytoplasmic male sterility, selection against recombinant nuclear genomes and a form of apomixis known as semigamy (Wendel *et al.*, 1991; Rieseberg and Wendel, 1993).

4.1.3   Nuclear Introgression and Recombinational Speciation

The most recently discovered "unusual" speciation mechanism in *Gossypium* is one involving recombination between diverged nuclear genomes. The species in question is *G. gossypioides*, the exclusive member of subsection *Selera* and a taxon that persists in several small isolated populations in a single river drainage in Oaxaca, Mexico.

Until recently, there was no indication that *G. gossypioides* had a noteworthy evolutionary history, in that the inferred relationships between *G. gossypioides* and the other D-genome species, based on comparative gross morphology (Fryxell, 1979, 1992), cytogenetic data (Brown and Menzel, 1952a), interfertility relationships (Brown and Menzel, 1952b), and allozyme analysis (Wendel, unpublished data) were congruent. Wendel *et al.* (1995b), surprisingly, have demonstrated that nuclear ribosomal DNA sequences from *G. gossypioides* are unlike those of all other D-genome taxa. In fact, the DNA sequence data implicate extensive recombination with rDNA sequences from A-genome cottons. Subsequent to this finding, other repetitive DNAs from *G. gossypioides* have been found to be similarly introgressant (Zhao *et al.*, in prep).

Wendel *et al.* (1995b) attributed these data to an ancient hybridization event, whereby *G. gossypioides* experienced contact with an A genome, either at the diploid level, or at the triploid level as a consequence of hybridization with a New World allopolyploid, followed by repeated backcrossing of the hybrid into the *G. gossypioides* lineage, thereby restoring the single-copy component of the D nuclear genome. Possibly the *G. gossypioides* lineage was spawned by this process. To the extent that this is true, this example highlights an additional mechanism of recombinational speciation in the genus *Gossypium*. As discussed below, these

data also raise the possibility that *G. gossypioides*, rather than *G. raimondii*, is the closest living descendant of the ancestral D-genome parent of the allopolyploids.

4.2   Origin and Diversification of the Allopolyploids

The previous section highlighted the potential significance of hybridization on species formation at the *diploid* level in *Gossypium*. Long before these examples were discovered, however, the role hybridization played in polyploid speciation in *Gossypium* had been well documented.

Ample evidence establishes that the New World tetraploid cottons are allopolyploids containing an A genome similar to those found in the Old World cultivated diploids and a D genome similar to those found in the New World diploids species (reviewed in Endrizzi *et al.*, 1985; also see Wendel, 1989). This implies that the two genomes must have established physical proximity, at least ephemerally, at some time in the past. Presently, however, the two parental genomic groups exist in diploid species with geographical distributions that are half a world apart. This mystery of allopolyploid formation has led to considerable speculation regarding the identity of the progenitor diploid species, the question of how many times polyploids were formed, and the time of allopolyploidization (Endrizzi *et al.*, 1985).

Molecular data indicate that all allopolyploids in *Gossypium* share a common ancestry, lending support to the hypothesis that polyploid formation occurred only once (*contra* suggestions of multiple origins: Johnson, 1975; Parks *et al.*, 1975). In addition, all allopolyploids contain an Old World (A genome) chloroplast genome, indicating that the seed parent in the initial hybridization event was an African or Asian A-genome taxon (Wendel, 1989).

Views concerning the time of polyploid formation have varied widely, from proposals of a Cretaceous origin with subsequent separation of genomic groups arising from tectonic separation of the South American and African continents, to suggestions of very recent origins in prehistoric times involving transoceanic human transport. Molecular data suggest a geologically recent (Pleistocene) origin of the allopolyploids, perhaps within the last 1 to 2 million years, consistent with earlier suggestions based on cytogenetic (Phillips, 1963) and ecological considerations (Fryxell, 1965a, 1979).

A Pleistocene origin for allopolyploid cotton has several evolutionary implications. First, morphological diversification and spread of allopolyploid taxa subsequent to polyploidization must have been relatively rapid. This is apparent in published data (Wendel, 1989; Wendel and Albert, 1992) as well as in recently generated DNA sequence data (Wendel, unpubl.) where few differences among the allopolyploid species were discovered in a survey of over 5000 base pairs of chloroplast DNA. Collectively, the data implicate an early radiation into three lineages (Fig. 1-3). *Gossypium mustelinum*, the sole descendant of one branch of the earliest polyploid radiation, is restricted to a relatively small region of northeast Brazil (Wendel *et al.,* 1994). Each of the other two lineages is represented by two species, one of which is cultivated (*G. barbadense* and *G. hirsutum*), while the other is an island endemic that originated from long-distance dispersals: *G. barbadense* with *G. darwinii* (Galapagos Islands; Wendel and Percy, 1990) and *G. hirsutum* with *G. tomentosum* (Hawaiian Islands; DeJoode and Wendel, 1992). The two cultivated species have large indigenous ranges in Central and South America, the Caribbean, and even reach distant islands in the Pacific (Solomon Islands, Marquesas etc.). Some have recognized a sixth allopolyploid species (*e.g.,* Fryxell, 1979), *G. lanceolatum* (= *G. hirsutum* "race palmeri"), which is known only as a cultigen. Brubaker and Wendel (1993) reviewed the evidence that bears on the specific status of this taxon and concluded that *G. lanceolatum* is more properly considered a variant of *G. hirsutum.*

A second implication of a Pleistocene origin of the allopolyploids is the possibility of identifying modern diploid lineages that closely approximate the original genome donors. Many investigators have considered a species similar to *G. raimondii*, from Peru, as representing the closest living model of the D-genome, paternal donor to the allopolyploids, although other species have been suggested (Endrizzi *et al.,* 1985; Wendel, 1989). *Gossypium raimondii* belongs to an evolutionary lineage that is otherwise Mexican and its closest relative is *G. gossypioides*, from Oaxaca, suggesting a recent dispersal to South America. The identity of the A-genome donor is also uncertain. Genomes of the only two A-genome species, *G. arboreum* and *G. herbaceum*, differ from the A sub-genome of allopolyploid cotton by three and two reciprocal chromosomal arm translocations, respectively (Brown and Menzel, 1950; Gerstel, 1953; Menzel and Brown, 1954), suggesting that *G. herbaceum* more closely resembles the A-genome donor than *G. arboreum.*

A third implication of a Pleistocene polyploid origin concerns the biogeography of their formation. Cytogenetic data, combined with the observation that the only known wild A-genome cotton is African (*G. herbaceum* subsp. *africanum*), has been used to support the suggestion that polyploidization occurred following a trans-Atlantic introduction to the New World of a species similar to *G. herbaceum*. While this theory is plausible, *G. herbaceum* is clearly not the actual maternal parent, as indicated by its cytogenetic and molecular differentiation from the A subgenome of the allopolyploids (Brown and Menzel, 1950; Gerstel, 1953; Menzel and Brown, 1954; Wendel, 1989; Brubaker *et al.,* 1999). Wendel and Albert (1992) raised the possibility of a pre-agricultural radiation of the A genome into Asia followed by a trans-Pacific, rather than trans-Atlantic, dispersal to the Americas. This possibility is supported by the biogeography of the D-genome species, which are hypothesized to have originated in western Mexico. The relatively recent arrival of *G. raimondii* in Peru also suggests that the initial hybridization event may have taken place in Mesoamerica rather than South America.

The evolution of the AD genome allopolyploids is conceptually intertwined with the phenomenon of A genome introgression into the *G. gossypioides* lineage, as discussed in the section *Nuclear Introgression and Recombinational Speciation*, above. Because *G. gossypioides* is the only D-genome diploid that exhibits evidence of genetic "contact" with an A-genome plant, the long-distance dispersal event that led to an ephemeral presence of an A-genome entity in the New World is suggested to have occurred after *G. gossypioides* diverged from its closest relative, *G. raimondii*, consistent with other indications of a Pleistocene allopolyploid origin (Phillips, 1963; Fryxell, 1979; Wendel, 1989). This evolutionary history raises the possibility (Wendel *et al.,* 1995a) that the *G. gossypioides* lineage was involved in the origin of allopolyploid cotton rather than *G. raimondii*, as is the prevailing opinion (Endrizzi *et al.,* 1985). Indeed, A-genome introgression into *G. gossypioides* and initial allopolyploid formation may have been spatially and temporally associated events, and may even have occurred in the same population (Wendel *et al.,* 1995a). The proposal that *G. gossypioides*, or more precisely its extinct ancestor or close relative, is the D-genome parent of the allopolyploids is intriguing and worthy of further study. At present, this hypothesis provides a simple scenario that accounts for all biogeographic, cytogenetic, and molecular data.

5.   DOMESTICATION OF THE CULTIVATED COTTONS

All *Gossypium* species have seeds with hairs that are elongated epidermal cells (Fryxell, 1963). During development, the seed hairs of most wild *Gossypium* species deposit thick secondary walls, and when these hairs desiccate after the capsules open, the hairs retain their circular aspect in cross-section. The lint-bearing species, from which the cultivated cottons were domesticated, have a "fuzz" of short cylindrical hairs (*ca.* 1 to 3 mm) similar to the hairs of the wild *Gossypium* species[1], but they also have a second "fiber" or "lint" layer of longer hairs (*ca.* 10 to 25 mm) that have thinner secondary cell walls (Fryxell, 1963). When these lint hairs desiccate at maturity, they form a flattened ribbon, and because the cellulose strands of the secondary walls are laid down in periodically reversing spirals, the entire hair convolutes and twists. It is this characteristic that allows the fibers to be spun into a yarn (Hutchinson and Stephens, 1947).

The layer of elongated lint is restricted in its distribution to the A-genome and AD-genome species of *Gossypium*, and perhaps the rare Madagascan species, *Gossypioides brevilanatum* (Hutchinson and Stephens, 1947). Four of these species have be domesticated: the Old World diploids, *G. arboreum* and *G. herbaceum,* and the New World allotetraploids,

---

[1]   A few New World cultigens may lack the fuzz layer (Fryxell, 1963),

G. barbadense and *G. hirsutum*. Hutchinson *et al.* (1947) suggested that the evolution of the lint occurred under domestication. This scenario, however, rests upon a number of implausible events and is no longer accepted (Hutchinson, 1954). More likely, lint evolved by natural selection and that the original domesticators recognized the utility of this naturally occurring characteristic (Fryxell, 1979). In this regard, all archeobotanical materials of *Gossypium* recovered to date can be attributed to the four domesticated lint-bearing species (Gulati and Turner, 1928; Chowdhury and Buth, 1971; Smith and Stephens, 1971; Stephens and Moseley, 1974; Damp and Pearsall, 1994). Lint-bearing seeds probably evolved only once, in the ancestor of the A-genome lineage, prior to allopolyploid formation and long before human intervention (Saunders, 1961).

Regardless of the timing of the evolution of the lint, accumulated evidence suggests that each of the four cultivated species existed as a distinct wild species prior to domestication (Wendel *et al.*, 1989; Percy and Wendel, 1990; Wendel *et al.*, 1992; Brubaker and Wendel, 1994). Apparently early in the evolution of the A genome, a mutation that gave rise to longer, thinner-walled seed hairs in addition to a layer of shorter cylindrical seed hairs became fixed either because of some selective advantage or by stochastic processes (Hutchinson, 1962). The species that later evolved from this now extinct A-genome entity, either via divergence (*G. arboreum* and *G. herbaceum*) or allopolyploid speciation (*G. barbadense* and *G. hirsutum*), possessed a useful characteristic that independently attracted the attention of aboriginal domesticators in four different pre-historical cultures. In this regard, the genus *Gossypium* is unique among crop plants.

5.1   Evolution Under Domestication

The effects of human manipulation on the four cultivated species have been so extensive that the evolution of these species can only be understood within the context of domestication. Starting with the original domestication events, the genetic structures, morphologies, and geographical ranges of *G. arboreum, G. herbaceum, G. barbadense*, and *G. hirsutum* have been modified by parallel changes due to human selection and human-mediated influences. The wild progenitors most likely had restricted and widely separated geographic ranges, typical of most wild *Gossypium* species. Following domestication within these geographically restricted indigenous ranges, low levels of human-mediated germplasm diffusion modestly extended the native ranges. This was followed by more intensive agronomic development which in turn led to more extensive germplasm diffusion. The advent of global travel produced additional rounds of agronomic improvement and germplasm diffusion.

The end result of these overlapping series of agronomic development and germplasm diffusion is that all four cultivated species exhibit greater levels of morphological diversity and encompass much larger geographic regions than most of their wild counterparts (Hutchinson *et al.*, 1947). The four cultivated species, however, are not exceptionally diverse genetically (Wendel *et al.*, 1989; Percy and Wendel, 1990; Wendel *et al.*, 1992; Brubaker and Wendel, 1994), a phenomenon that reflects the genetic bottlenecks through which each species has passed during various stages of domestication. Some of the genetic diversity that exists may actually derive from interspecific introgression, as human-mediated germplasm diffusion brought formerly separated species (*G. arboreum-G. herbaceum* and *G. barbadense-G. hirsutum*) into contact. Throughout their current ranges, all four species are found primarily as cultivars with varying levels of development, or as feral derivatives that have established self-perpetuating populations in human-modified environments (*e.g.*, roadsides, field edges, dooryards). The wild progenitors exist in relic populations that occupy small and sometimes peripheral parts of the indigenous ranges.

For all four cultivated species, human selection has converged to a similar phenotype: short, compact, day-length-neutral shrubs with large fruits and large seeds with permeable seed coats and a marked differentiation between the lint and fuzz layers (Fryxell, 1979). In contrast, their antecedents are perennial, sprawling or upright shrubs or small trees, with small fruits and seeds with impermeable seed coats and poor differentiation between the lint and fuzz fiber layers. The transformation from one end of this spectrum to the other occurred gradually during the past four millennia and in concert with changing agricultural technology. The first selection probably was unintentional and was for genotypes that could grow well in human-modified environments. Subsequent morphological modifications can all be attributed to changes in technology (Fryxell, 1979). The development of weaving would have led to selection for lint quality characteristics. The transport of cotton into higher latitudes led to the development of day-length neutral varieties. The requirement of hand-ginning favored smooth-seeded varieties that had easily detached lint fibers (*e.g.*, the *G. barbadense* kidney cottons). The invention of mechanical gins allowed smooth-seeded varieties to be replaced by fuzzy-seeded variants with higher quality lint and lint percentages.

The role of introgression in the development of the cultivated species is worthy of special mention. Few of the wild species grow together in the wild, and not surprisingly, therefore, documented cases of hybridization and introgression are rare (Wendel *et al.*, 1991). Among the cultivated species, both intentional and unintentional hybridizations have been more common, the latter occurring in many parts of the indigenous ranges prior to the 20th century. This can be attributed to human-mediated range expansion that brought formerly isolated species into contact. While this led to overlapping morphological attributes between *G. arboreum* and *G. herbaceum* and between *G. barbadense* and *G. hirsutum* (Stephens, 1967; Boulanger and Pinheiro, 1971; Stephens, 1974; Stanton *et al.*, 1994), species distinction has not been obscured (Wendel *et al.,* 1989; Brubaker *et al.,* 1993).

The maintenance of species boundaries can be attributed to genetic or cytological barriers between the species. When *G. barbadense* and *G. hirsutum,* or *G. arboreum* and *G. herbaceum,* hybridize, the F$_1$ typically is vigorous and fertile, but numerous depauperate progeny occur in the F$_2$ and later generation progenies (Harland, 1939; Silow, 1944; Knight, 1945; Stephens, 1949, 1950b; Hutchinson, 1954; Phillips, 1960). Typically the most vigorous and fertile plants recovered from such interspecific pairings resemble one of the parents. In addition, the two A-genome diploids differ from each other by a reciprocal chromosome arm translocation (Menzel and Brown, 1954), and in areas of sympatry, *G. barbadense* and *G. hirsutum* populations have high frequencies of alternate alleles for the corky locus (Stephens, 1946, 1950a; Stephens and Phillips, 1972); interspecific hybrids heterozygous for these alleles are stunted and female sterile.

Despite these isolating mechanisms, introgression between *G. barbadense* and *G. hirsutum* and between *G. arboreum* and *G. herbaceum* has occurred within their indigenous ranges. Patterns of introgression between *G. arboreum* and *G. herbaceum* have received only preliminary study, but each species shows evidence of ancient nuclear introgression from the other species (Wendel *et al.,* 1989). Among the indigenous New World cottons, nuclear and cytoplasmic introgressions are not symmetrical (Brubaker *et al.,* 1993). Nuclear introgression is geographically more widespread and more frequently detected than is cytoplasmic introgression. Nuclear introgression is rarely detected in *G. barbadense,* but has been detected in accessions from west of the Andes as well as from Paraguay and Argentina (Percy and Wendel, 1990). In contrast, populations of *G. hirsutum* frequently contain *G. barbadense* alleles, especially in accessions from sympatric regions. The highest frequency of *G. barbadense* alleles is detected in *G. hirsutum* race *mariegalante* (Brubaker and Wendel, 1994). This day-length sensitive arborescent cotton is found primarily where the two species co-occur (the Caribbean, northern South American, and Central America) and is morphologically intermediate between *G. barbadense* and *G. hirsutum* (Stephens, 1967; Boulanger and Pinheiro, 1971; Stephens, 1974).

In contrast to the higher levels of "indigenous" introgression observed in *G. hirsutum* than in *G. barbadense,* the deliberate interspecific introgression that has characterized modern breeding programs of both species has had a much greater impact on *G. barbadense* than on *G. hirsutum.* Transpecific alleles are only rarely detected in modern *G. hirsutum* cultivars (Brubaker *et al.,* 1993), but on average 8.9% of RFLP alleles in modern improved *G. barbadense* cultivars come from *G. hirsutum* (Wang *et al.,* 1995). This may, in part, explain why modern, highly improved *G. barbadense* cultivars, as a group, exhibit levels of genetic diversity greater than accessions from all indigenous regions except the west Andean region, the center of genetic diversity for the species (Percy and Wendel, 1990).

Despite the low levels of transpecific alleles detected in indigenous *G. barbadense,* there is evidence that some of these *G. hirsutum* alleles were captured prior to the development of the first modern improved *G. barbadense* cultivars (Sea Island). Stephens (1975) hypothesized that the successful introduction of *G. barbadense* into the United States depended on the introgressive transfer of day-length-neutrality from *G. hirsutum* to the short-day indigenous *G. barbadense* cultivars. Support for this idea comes from the observation that most indigenous day-neutral forms of *G. barbadense* show evidence of introgression from *G. hirsutum.* Wang *et al.* (1995) detected *G. hirsutum* RFLP alleles in the agronomically primitive Peruvian accessions, Tanguis 45 and Mollendo, and in all Sea Island accessions they examined, and Percy and Wendel (1990) found the highest percentages of introgressed *G. hirsutum* alleles in indigenous cultivars from Argentina and Paraguay. Introgression of this trait probably occurred before the introduction of *G. barbadense* into the United States, because day-length sensitive accessions would not have flowered could not have persisted long enough for introgression to occur following their introduction (Hutchinson and Manning, 1945).

5.2 Domestication of Cotton in the Old World

Little is known about the time and place of domestication of *G. arboreum* and *G. herbaceum* in the Old World. Cloth fragments and yarn that most likely were derived from *G. arboreum,* dated to 4300 years B.P., were recovered from archeological sites in India and Pakistan (Gulati and Turner, 1928). No clearly identified archeological remains of *G. herbaceum* have been recovered, but the stronger geographic discrimination among subtypes and its wide distribution prior to the development of industrial textile manufacturing imply a history of domestication at least as long as that of *G. arboreum.*

Gossypium herbaceum is known primarily as a crop plant (grown from Ethiopia to western India), with the exception of an endemic form from southern Africa, *G. herbaceum* ssp. *africanum.* This morphologically distinct entity, which occurs in regions far removed from historical or present diploid cotton cultivation, has a unique ecological status in that it is fully established in natural vegetation in open forests and grasslands. Its small fruit, thick, impervious seed coats, sparse lint, and absence of sympatric cultivated *G. herbaceum* suggest that *G. herbaceum* ssp. *africanum* is a wild plant. Consistent with the expectation that the site of original domestication lies within the range of the wild progenitors, this is generally accepted as the source of the original *G. herbaceum* cultigens (Hutchinson, 1954). The most agronomically primitive *G. herbaceum* cultivars, constituting the perennial race *acerifolium* forms, are distributed along the coasts boarding the Indian Ocean trade routes. This suggests that the primary dispersion involved the diffusion of *G. herbaceum* northward into northern

Africa, Arabia and Persia. Hutchinson (1954) suggests that secondary agronomic development and diffusion led to expansion into western Africa and the development of annualized forms in more northerly temperate climates. The agronomic success of the annualized *G. herbaceum* races fostered a later dispersal into peninsular India that replaced perennial *G. arboreum* cultigens.

Gossypium arboreum is the only cultivated species for which a clearly identified wild progenitor is not known. This assertion warrants further investigation because it is certainly biased by the assumption that lint evolved under domestication, hence the dismissal of self-perpetuating forms of linted *G. arboreum* in indigenous vegetation as feral derivates of domesticated forms (Hutchinson *et al.*, 1947). Two geographic regions can be nominated as possible regions for the location of the original wild progenitor populations and the original domesticated forms. On Madagascar two forms of *G. arboreum* have been described. The first is a primitive arborescent form found only in xerophytic woodlands; the second is a primitive cultigen found only in association with human settlements (Hutchinson, 1954). The ecological discrimination between these geographically adjacent *G. arboreum* forms is consistent with a progenitor-derivative relationship, but without access to these accessions, this hypothesis cannot be tested. The Indus valley (Mohenjo Daro) has also been nominated as a possible site of original domestication or agronomic development on the basis that it represents the center of diversity for *G. arboreum*. The first documented archeological evidence for Old World cultivated cotton was found there (Gulati and Turner, 1928). This may be true, but as a cautionary note, current centers of diversity do not necessarily correspond to original geographic points of origin. Instead, they may have developed during later stages of domestication and human-mediated germplasm diffusion. Hutchinson (1954) considers the Indus Valley cottons (Gulati and Turner, 1928) to be more similar to northern, more agronomically advanced *G. arboreum* cultivars, suggesting that the Indus Valley may represent a site of secondary development and diffusion.

Without a clearly identified geographical point of origin, one can only comment on more recent stages of agronomic development. Within *G. arboreum* two series of developments appear to have occurred (Hutchinson, 1954). The first involves the development of a primitive perennial domesticate, race *indicum* in western India, that subsequently was dispersed into peninsular India and along the east coast of Africa via the Indian Ocean trade routes and perhaps into East Asia (Hutchinson, 1954, 1962). Race *indicum* cottons represent the most agronomically primitive form of *G. arboreum* and, thus, may exemplify the remnants of the first agronomic dispersal. A second flowering of development is centered around the Indus Valley, from which a number of dispersions occurred. One pathway led west through Egypt into Western Africa, while eastern diffusion encompassed China and East Asia. Much of this dispersion occurred with the advent of the modern textile industry, and it is this north-

ern group that came to dominate Old World cotton cultivation prior to the introduction of New World cottons.

The morphological and genetic similarities (Silow, 1944) between the primitive *G. arboreum* race *indicum* and *G. herbaceum* race *acerifolium* cultivars has been cited as evidence of a progenitor-derivative relationship (Hutchinson, 1954, 1962). Molecular data, however, support the alternative that the two species were independently domesticated from divergent wild progenitors (Wendel *et al.*, 1989). In light of this observation, morphological similarity may be attributed to parallel retention of characteristics from a common ancestor or from post-domestication introgression between *G. arboreum* and *G. herbaceum* as they came into contact along the Indian Ocean trade routes (Wendel *et al.*, 1989). One might expect that the retention of ancestral characteristics would be highest in the least developed cultigens, consistent with the observations noted above. To the extent that introgression has contributed to the observed similarities, it would parallel the evolution of *G. hirsutum* race *marie-galante*, which probably originated from interspecific introgression between *G. hirsutum* and *G. barbadense* as their indigenous ranges jointly expanded into the Caribbean under domestication (Stephens, 1967; Boulanger and Pinheiro, 1971; Stephens, 1974).

5.3   Domestication of Cotton in the New World

Although *G. arboreum* is still a significant crop plant in India and Pakistan, and *G. herbaceum* is cultivated on a small scale in several regions of Africa and Asia, New World tetraploid cultivars presently dominate worldwide cotton production. Primary production areas for *G. barbadense* (Pima and Egyptian cotton) include several regions of Central Asia, Egypt, Sudan, India, the United States, and China. *Gossypium barbadense* is favored for some purposes because of its long, strong, and fine fibers, but its relatively low yield has limited its importance to less than 10% of total world production. The bulk (>90%) of the world's cotton is supplied by modern cultivars of *G. hirsutum*, or "Upland" cotton. Upland cultivars currently are grown in more than 40 nations in both tropical and temperate latitudes, from 47°N in the Ukraine and 37°N in the United States to 32°S in South America and Australia (Niles and Feaster, 1984).

Gossypium barbadense was most likely domesticated in northwest South America, perhaps along the coast or inland near watercourses. The earliest archeobotanical remains (seed, fiber, fruit, yarn, fishing nets, and fabrics) of *G. barbadense* were recovered from central coastal Peru, dating to 5500 years B.P. (Vreeland, pers. comm.). The primitive agronomic characteristics of the remains support the general belief they were derived from early domesticated forms and that the original domestication must have occurred somewhere in this region (Hutchinson *et al.*, 1947). This hypothesis is supported by molecular evidence (Percy and Wendel, 1990), which reveals a center of genetic diversity that is geographically congruent with the distribution of wild populations. Following a primary domestication west of the Andes, the primary dispersal appears to have been a trans-Andean expansion into northern South America. A secondary stage of agronomic development and dispersal expanded the range of *G. barbadense* into Central America, the Caribbean and the Pacific. A post-Columbian dispersal from the West Andean gene pool into Argentina and Paraguay further expanded the range. The trans-Andean dispersal of *G. barbadense* in northern South America and then into Central America, the Caribbean, and the Pacific was accompanied by sequential reductions of genetic diversity and allelic richness (Percy and Wendel, 1990).

The modern elite *G. barbadense* cultivars trace their origins to the Sea Island cottons developed on the coastal islands of Georgia and South Carolina. Historical data regarding the development of the first Sea Island cottons are limited and contradictory. Stephens (1976) suggested that they originated from the West Indian Sea Island cottons, but Hutchinson and Manning (1945) suggest that the original introductions came from west Andean Peruvian stocks. Multivariate analyses of allozyme data (Percy and Wendel, 1990) support the latter hypothesis. If the day-length neutrality and the extra-long staple fiber of the Sea Island cottons did arise via introgression from *G. hirsutum* (Stephens, 1975), the most likely progenitors would be those from areas where introgression is highest, viz. west of the Andes, Paraguay, and Argentina (see above). The Sea Island industry of the United States eventually collapsed under boll weevil pressure by 1920 (Niles and Feaster, 1984), but the Sea Island lineage contributed to the development of the Egyptian cottons, which were later reintroduced into United States as a part of the Pima genepool.

Gossypium hirsutum has a large indigenous range encompassing most of Mesoamerica and the Caribbean, where it exhibits a diverse array of morphological forms spanning the wild-to-domesticated continuum. The oldest archeobotanical remains are from the Tehuacan Valley of Mexico, dating from 4000 to 5000 years B.P. (this estimate should be considered tentative until additional stratigraphic and carbon-14 dating information become available; P. Fryxell and J. Vreeland, pers. comm.). A genetic survey of over 500 accessions collected from throughout the range of *G. hirsutum* reveals two centers of genetic diversity: southern Mexico-Guatemala and the Caribbean (Wendel *et al.,* 1992). The Mesoamerican center of diversity corresponds to the center of morphological diversity. Its correspondence with the original geographic point of domestication suggests that it may represent the primary center of diversity. A high frequency of *G. barbadense* alleles in Caribbean *G. hirsutum* suggests that the Caribbean represents a secondary center of diversity that developed as primitive *G. hirsutum* cultivars spread into the Caribbean and hybridized with *G. barbadense.*

The most likely site of original domestication is the Yucatan peninsula in Mesoamerica (Brubaker and Wendel,

1994). It is here that most wild forms of *G. hirsutum*, race *yucatanense*, are found, and it is only here that *G. hirsutum* is a dominant constituent of indigenous vegetation (Standley, 1930; Hutchinson, 1951; Stephens, 1958; Sauer, 1967; Stewart, personal comm.). Race *yucatanense* is phylogenetically and phenetically allied with geographically adjacent populations of race *punctatum* with which it intergrades morphologically (Brubaker and Wendel, 1994). Race *punctatum* can be nominated as the earliest domesticated form of *G. hirsutum*. It is a sprawling perennial shrub with agronomic characteristics that are intermediate between the wild race *yucatanense* and other indigenous forms that show greater evidence of human manipulation (*e.g.,* races *latifolium* and *palmeri*; Hutchinson, 1951). Its morphological, genetic, and geographic association with wild *G. hirsutum* populations is consistent with the inference of a progenitor-derivative relationship (Zohary and Hopf, 1988).

Following its development, race *punctatum* was dispersed throughout the Yucatan peninsula and along the Mexican gulf coast, reaching as far as Florida and some of the Caribbean islands (Hutchinson, 1951). From these original *punctatum* cultivars several localized derivatives were developed: *richmondi* (south coast of the Isthmus of Tehuantepec), *morilli* (central Mexico highlands), *palmeri* (the Mexican States of Guerrero and Oaxaca), and *latifolium* (Hutchinson, 1951). Race *marie-galante*, which is widespread throughout southern Central America, northern South America, and the Caribbean most likely arose from some form of Mesoamerican *G. hirsutum,* but the details have not been satisfactorily elucidated. The superior agronomic characteristics of the southern Mexican-Guatemalan *latifolium* cultivars resulted in a second diffusion that dispersed *latifolium* throughout Mesoamerica.

The development of the modern elite *G. hirsutum* cultivars, or Upland cottons, started with the introduction of various indigenous Caribbean and Mesoamerican cultivars, including *G. barbadense* accessions (Ware, 1951; Ramey, 1966; Niles and Feaster, 1984; Meredith, 1991). From this diverse genepool two categories of cultivars arose: "greenseed" and "black seed". As the textile industry in the United States developed, however, these cultivars proved inadequate and from 1806 forward they were replaced in part by cultivars developed from *latifolium* accessions collected from the Mexican highlands (Niles and Feaster, 1984). RFLP and allozyme evidence suggest that these Mexican highland stocks were refined *latifolium* cultivars that had been transported northward during some earlier, unknown time from southern Mexico and Guatemala (Wendel *et al.,* 1992; Brubaker and Wendel, 1994). Further augmentation of the modern Upland gene pool involved a series of additional, deliberate introductions, beginning in the early 1900s in response to the devastation brought on by the boll weevil. Eventually selection for locally adapted cultivars led to the development of four basic categories of Upland cultivars (Acala, Delta, Plains, Eastern), whose modern derivatives account for the majority of Upland cotton grown worldwide (Niles and Feaster, 1984; Meredith, 1991).

6.   SUMMARY

This chapter presents an overview of the diversity in the cotton genus and summarizes recent contributions to our understanding of its evolutionary development. An appreciation of the taxonomic and phylogenetic relationships in *Gossypium* provides the necessary underpinnings for understanding the physiology of the cultivated crop. Cotton improvement programs have exploited diploid species for genes for fiber strength, disease resistance, cytoplasmic male sterility and fertility restoration, whereas genes for disease resistance, nectariless, and glandless cotton have been deliberately introduced from wild and feral tetraploids (Meyer and Meyer, 1961; Meyer, 1974, 1975; Fryxell, 1976; Bell, 1984a; Narayanan *et al.*, 1984; Niles and Feaster, 1984; Meredith, 1991; Zhu and Li, 1993). These genetic enhancements, involving intentional interspecific introgression from a minimum of two allopolyploid and four diploid *Gossypium* species, were obtained through classical genetic and plant breeding approaches. Unfortunately , the wealth of diversity has seen very little exploitation for physiological traits. Further exploitation of wild *Gossypium* and more phylogenetically distant sources of germplasm will employ traditional methods (Stewart, 1994) as well as genetic engineering. Understanding the relationships among species and their evolutionary development will continue to provide insights into the biology of cotton (Reinisch *et al.*, 1994; Wendel *et al.*, 1995b; Brubaker *et al.*, 1999), which in turn will increase the effectiveness of improvement efforts.

7.   ACKNOWLEDGMENTS

We wish to express our thanks to those whose shoulders we stand on, for their collective insights and inspiration: G.S. Zaitzev, J.O. Beasley, P.A. Fryxell, J.B. Hutchinson, M.Y. Menzel, L.L. Phillips, A. Skovsted, and S.G. Stephens. Much of the authors' work has been funded by grants from the National Science Foundation, to whom we also are grateful.

Chapter 2

GERMPLASM RESOURCES FOR PHYSIOLOGICAL RESEARCH AND DEVELOPMENT

James McD. Stewart[1]

[1]Department of Crop, Soil, and Environmental Sciences, University of Arkansas, Fayetteville, AR 72701

1. INTRODUCTION

Any research that attempts to explain the mechanisms responsible for a particular trait must use germplasm accessions that differ in expression of the particular trait being examined. The ideal situation is to select accessions that are positive and negative for the trait, *i.e.*, a plus-minus situation. In this way the researcher may look for responses associated with the positive accession that are absent from the negative accession. This chapter presents information concerning the locations of and diversity within the most important germplasm collections and provides examples where germplasm diversity has been used in research to achieve specified objectives.

2. GERMPLASM COLLECTIONS

Most cotton producing countries have limited germplasm collections which are readily available to the researchers of that country. Several cotton germplasm banks exist in the world, but the availability of the accessions are generally quite limited, or knowledge of the diversity is limited so that the end result is limited utilization of the genetic resources that are available. In the past, breeding lines of cotton tended to be informally exchanged by the scientist, but with the entry of "biotech" cotton there has been increased emphasis on "breeders' rights," and with increasing acceptance of the "Cartagena Accord" (Grajal, 1999) the flow of germplasm between countries, and even between institutions within a country, has become quite restricted. A few countries, by nature of the scientific enlightenment of the political leaders and or scientists, have made germplasm available to other countries, either through exchange agreements, or direct donations. (The major cotton germplasm banks are discussed below.) Originally the FAO dealt with international germplasm issues, but in the latter part of the 20th century an independent organization, International Plant Genetic Resources Institute (IPGRI), was spun off from this organization and became the main organization to promote germplasm in the world.

Although the IPGRI emphasizes food crops, it has designated the *Gossypium* collections in India and the US as the world cotton germplasm collections. Theoretically the collections are supposed to be duplicated at each site. But the inclination and logistics must be in place for this to happen in reality. The USDA collection, being close to the center of origin of *G. hirsutum* (Mexico) has many unique accessions of that species, whereas India has many more accessions of *G. arboreum*. The two germplasm banks partially corrected the imbalances by exchanging approximately 1000 accessions in the 1990's (*i.e.*, *G. hisutum* accessions for *G. arboreum* accessions).

2.1 The USDA Cotton Collection

The most complete cotton collection is maintained by the USDA at College Station, TX. This is a public collection that represents most of the phenotypic diversity that has been discovered in the world. The "cotton collection" is actually a series of collections that include 1) obsolete varieties, 2) *Gossypium barbadense*, 3) Texas race stock, 4) Asiatic cotton, and 5) wild species. The original sources of the germplasm in the collections have accumulated over the years from planned collecting trips to various parts of the world, from donations by individual collectors, and from exchange of materials from other germplasm collections throughout the world (Percival, 1987).

J.McD. Stewart et al. (eds.), *Physiology of Cotton*,
DOI 10.1007/978-90-481-3195-2_2, © Springer Science+Business Media B.V. 2010

In addition to the main *Gossypium* collection there are two other collections that are maintained by individuals outside of the public collection maintained by USDA. These are 1) the Genetic Marker collection of *G. hirsutum* maintained by Dr. Russel Kohel of the USDA, ASR at College Station, Texas and 2) the Cytogenetical collection maintained by Dr. David Stelly of Texas A& M University, College Station, TX. The former collection, as the name implies, contains the various mutations and genetic markers of *G. hirsutum*, while the latter contains the monosomic, telosomic, translocation, and duplication-deficiency stocks that were developed primarily by J. Endrizzi at the University of Arizona and the various cytogeneticists who have been in charge of the Beasley Laboratory at Texas A&M University since the 1940's [*i.e.,* M. Brown and student M. Menzel (Florida State University), A. Edwards, and D. Stelly].

Accessions in the first five collections are freely available in lots of 25 seeds to all *bona fide* requesters. A *bona fide* requester is a person with a legitimate reason for requesting the seed, such as a cotton breeder or other cotton researcher. Usually the requester must request specific accessions rather than make a blanket request for a specific trait, *e.g.,* drought tolerance, since the curator of the collection does not have the responsibility to obtain this type of information. The information that is available on each accession can be obtained through the Germplasm Resource Information Network (GRIN) at http://www.ars-grin.gov/npgs/searchgrin.html. Unfortunately the complete cotton collection (particularly the later accessions acquired) has yet to be entered into GRIN and evaluation data on many of the accessions already in the GRIN system have not been obtained. Evaluation data is considered to be the greatest deficiency with the germplasm collections. Nevertheless, the GRIN system is a good place to start when looking for specific traits that may be present in the collection.

As the name implies, the wild species collection contains many of the 45 known wild *Gossypium* species (3 AD tetraploid and 42 diploid species) including all of those from the Western Hemisphere, the arid zone of Australia, and many of the diploids from Africa and Southern Asia. A few of the species of the E genomic group known to occur in the Horn of Africa have never been collected, and a few species (K genome) from the north Kimberley region of Western Australia apparently have been lost to the germplasm bank.

The Asiatic cotton collection comprises the two diploid cultivated cottons, *G. herbaceum* (A_1) and *G. arboreum* (A_2). For the most part this collection contains cultivars of the two species but it also contains accessions of wild representatives of *G. herbaceum* ssp. a*fricanum* from South Africa and Namibia. The wild populations of *G. herbaceum* are known only for this species while no wild populations of *G. arboreum* are known to exist.

The "race stock" collect is somewhat of a mis-nomer because all *G. hirsutum* collected from any region of the world goes into this collection. The race stock designations were created by Hutchinson (1951) (latifolium, punctatum, marie galante, richmondi, palmeri, morilli, and yucatanense) based on plant types observed when all of the accessions were grown in a common garden at Shambat, Sudan. These are not taxonomic divisions and it is likely that no one currently working with cotton can distinguish among the various races. In the "race stock" collection a race designation is available only for the early entries. Most later entries do not have a race designation. All accessions of *G. hirsutum* collected from the Americas and various other parts of the world in the last 30 years are included in this collection.

The *G. barbadens*e collection includes all accessions of this species, including both the obsolete cultivars and the wild and unimproved accessions from South America. This collection comprises the original Arizona AS, B, and K collections maintained by the USDA Pima breeding program at Tempe, AZ, plus new accessions of *G. barbadense* that have been obtained since the collections were moved from Arizona to College Station, TX.

The Obsolete Cultivar collection, as the name implies, comprises those accessions of *G. hirsutum* that were once popular among growers but which are no longer cultivated, being replaced by cultivars with increased yield or quality or both. This collection was originally maintained at Stoneville, MS, but is now a part of the *Gossypium* collection at College Station, TX.

2.2   The India Cotton Collection

Indian scientists have conducted a number of expeditions throughout India to collect the diploid Asiatic cottons as well as the cultivated tetraploids. These expeditions have resulted in a large collection of *G. arboreum* and *G. herbaceum.* The curator of the US collection visited Nagpur, India in 1993 and an agreement was reached to exchange germplasm. Upon his return to the US the curator shipped more than 2000 accessions of *G. hirsutum* to Nagpur. He returned to Nagpur in 1996, and managed to acquire 1,147 accessions of mostly *G. arboreum* from the Indian Collection (Hays, 1996). This acquisition more than tripled the number of accessions available in the USDA Asiatic cotton collection. At this time, the number of accessions in the Indian collection is unknown, but a working figure of approximately 3,000 accessions is a reasonable estimate.

2.3   The Uzbekistan Cotton Collection

The most famous plant collector in history is N.I. Vavilov. Through his efforts and students who followed him, an extensive collection of cotton germplasm was acquired from Central America and preserved at St. Petersburg at the N.I. Vavilov Institute (VIR) with working collections maintained primarily in Uzbekistan. With the dissolution of the former Soviet Union the VIR Institute has struggled financially to maintain the collection and has sent a number of accessions to the US (Hays, 1996). The working collections in Uzbekistan (maintained by the Uzbek Research Institute of Plant Industry, and the Institute of Genetics and Experimental Biology of Plants) have taken on more importance because of the decline in VIR. Unfortunately, the collections apparently are maintained by yearly growout, so genetic drift and possible technical errors no doubt are rampant in those collections. Nevertheless, valuable germplasm is available in these collections (James Olvey, personal communication) including the four cultivated species as well as diploid wild species and hybrids among these. They support approximately 6,000 *Gossypium* accessions (Hays, 1996).

2.4   The CIRAD Cotton Collection

The French have acquired much cotton germplasm through formal collection efforts in the Americas and through their efforts in former colonies in Africa and SE Asia. This germplasm is maintained at Montpellier by the Centre de coopération internationale en recherche agronomique pour le développement (CIRAD) at 4C. Accessions are increased by growout every 12 to 15 years. For many years the French maintained a cotton nursery on the island of Guadeloupe in the West Indies. According to A.E. Percival (personal communication) the accessions are suspect because of outcrossing, that is, the individual accessions were not maintained by controlled self-pollination. In recent years, with movement of their nursery to Costa Rica, the French have improved maintenance of the collection. Approximately 3,000 accessions are maintained in this collection, which comprises the four cultivated species as well as numerous accessions of the wild species. Through contacts in Vietnam and Laos the CIRAD cotton collection contains Asiatic cottons from outside of India that should possess unique diversity. Since CIRAD's main function is to improve the agriculture of developing nations (especially the francophone countries of Africa) this germplasm is readily available to those countries,

3   DIVERSITY WITHIN *GOSSYPIUM*

The germplasm available for cotton (*G. hirsutum*) was placed into germplasm pools by Stewart (1995) based on the ease with which one could expect to obtain introgression of a particular trait. The primary pool comprises all of the tetraploid AD species, because they have the same genomic chromosome constituency and will form fertile hybrids with cotton. The secondary germplasm pool comprises the species of the A, F, B, and D genomes because these are relatives of the ancestral parents that gave rise to the AD genome. Once a fertile hybrid with cotton is obtained, one can expect that recombination and gene transfer can occur. The tertiary germplasm pool consists of the species in the C, G, K, and E genomes. Because the chromosome of these

species are quite divergent from the A or D genomes, much effort must be devoted to transferring a particular trait from the wild diploid to the cultivated tetraploid. Even when a fertile hybrid is obtained, recombination of genetic material is difficult because of the divergence of the chromosomes.

3.1   The Primary Germplasm Pool

The primary germplasm pool consists of all the accessions in the Obsolete Cultivar collection, Texas Race Stock collection, the *G. barbadense* collection, and the three tetraploid species (*G. tomentosum, G. mustelinum,* and *G. darwinii*) in the Wild *Gossypium* Collection (Stewart, 1995). Although these collections have not been thoroughly characterized, obviously there is extensive diversity within the primary germplasm pool.

For breeders whose charge is to develop cultivars, the first choice of germplasm is that which will introduce the fewest deleterious genes. In most cases this will be elite breeding lines and obsolete cultivars since most deleterious genes have already been eliminated. This is an essential component for the breeder where the pressure is to develop a cultivar rapidly. Because of the pressure to release cultivars felt by most cotton breeders, there is obviously a niche and need for pre-breeders, that is, individuals who will enhance exotic germplasm to the point that a cotton breeder can utilize the germplasm directly in his/her breeding program.

If the desired trait is not in the germplasm of first choice, subsequent choices in order of acceptability would be accessions of the Race Stock Collection, the *G. barbadense* Collection, then the wild tetraploid species. Each of these germplasm resources present special problems that must be over come. For the race stocks if the breeder is developing cultivars for a temperate zone, the problem of photoperiodicity must be overcome, and deleterious genes must be eliminated in subsequent generations. Crosses between *G. hirsutum* and *G. barbadense* are notorious for genetic "breakdown" in the F2 (Stephens, 1950b; Phillips, 1961) and beyond generations and recovery of plants that look like one or the other parents with few intermediates. To transfer a trait from an AD_2 plant to an AD_1 plant requires a large population of plants from which to make selections. Utilization of the three wild tetraploids presents the problem of photoperiodicity and numerous deleterious genes. The question of F2 breakdown has not been specifically addressed with the wild species, but based on the phylogenetic relationships among the species (Wendel and Albert, 1992), such breakdown would be expected in $AD_1 \times AD_5$ hybrids but not necessarily in $AD_1 \times AD_3$ hybrids. In the absence of data, any F_2 breakdown in $AD_1 \times AD_4$ hybrids would only be speculative.

3.2   The Secondary Germplasm Pool

The secondary germplasm pool comprises the A-genome cottons of the Asiatic collection, the D-genome species of the Americas, and the species in the B and F genomes of Africa (Stewart, 1995). The two Asiatic species chromosomes are closely related to the A subgenome chromosomes of the tetraploid species so that recombination in hybrids occurs readily. Although the diploid species of the Americas all belong to the D genome and their chromosomes have reasonably close homoeology with the D subgenome of the tetraploids, some divergence has occurred. Nevertheless , the degree of divergence does not prevent them from establishing fertile trispecies hybrids with cotton. Exceptions to this are the species *G. davidsonii* and *G. klotzschianum* which possess complementary lethal genes with many other *Gossypium* species. However, these two species will form viable hybrids with *G. anomalum* of the B genome. The B- and F-genome chromosomes are closely related to the A-genome chromosomes (Phillips and Strickland, 1966) and can substitute for these in interspecific hybrids with D-genome species. Stewart (1995) suggested that any of the species in the secondary germplasm pool can recombine genetically with upland cotton if it is incorporated into a fertile synthetic allotetraploid bridge combination and then crossed with cotton. This concept has been proven with *G. herbaceum, G. arboreum, G. anomalum, G. armourianum,* and *G. trilobum* (Stewart, personal communication).

3.3   The Tertiary Germplasm Pool

The tertiary germplasm pool consists of the species in the C, G, and K genomes of Australia and the E genome of Africa and Southern Asia. Although some of the species form hybrids easily, generally the genes residing in this germplasm pool are difficult to transfer to *G. hirsutum* because of the divergence that has occurred among the homoeologous chromosomes. This germplasm pool is usually utilized only when a trait is unavailable in the primary or secondary germplasm pools.

4.   DIVERSITY IN THE GERMPLASM COLLECTIONS

The genetic diversity within a germplasm pool may be a morphological trait of interest, in which case it may be introgressed into cotton via hybridization and selection using a visual screen. An example of this would be the nectariless trait of *G. tomentosum* (Meyer and Meredith, 1978). Alternatively, the trait may be visually unseen and require a more extensive selection method, for example a chemical test. The terpenoid aldehyde, raimondal (Stipanovic *et al.,* 1980), provides an excellent example of this type of trait. The technology exists now for traits to be associated with molecular markers so that selections can be made based on the marker rather than a phenotypic trait. This would be particularly useful when the screen for the trait is time consuming and laborious. As an example, current screens

for reniform nematode resistance require a minimum of 60 days and do not provide the precision required to make single plant selections in a segregating population. For a practical breeding program a molecular marker closely linked to the trait is essential before the trait will be utilized.

Based on the habitat of the wild species (or on particular morphological traits observed to be in the wild germplasm) one can surmise that a particular trait of interest might be present in the germplasm. The primary germplasm pool would be expected to have a high degree of salt tolerance because their original mode of dispersal to various parts of the world appears to have been by ocean currents (Fryxell, 1979). Certainly, *G. tomentosum* and *G. darwinii,* endemic to the Hawaiian and Galapagos Islands, respectively, would have been distributed to those islands by ocean currents. Cotton is considered to be relatively salt tolerant, but some genotypes are more salt-resistant than others. This has allowed a number of comparative tests to be conducted (Gossett *et al.,* 1994a, 1994b; Ashraf and Ahmad, 2000) on the nature of the tolerance.

Another example where one might infer tolerance to water deficit would be in *G. anomalum* and *G. davidsonii,* both endemic to relatively dry areas and both with broad leaf structure (Fryxell, 1979). One could speculate that a hybrid of these two species should provide a good source of genes for physiological drought tolerance (as opposed to morphological tolerance). The hybrid of these two species, when the chromosomes are doubled in number (2 DB), should be compatible with an upland cotton possessing a D3-d compatibility gene from AD_2, *e.g.,* Paymaster 464.

In any physiological test for a specific feature where a number of genotypes are examined one intuitively recognizes that diversity is necessary to draw a conclusion concerning the feature. Several cotton fiber mutants (Nadarajan and Rangasamy, 1988; Narbuth and Kohel, 1990; Kohel and McMichael, 1990; Ruan and Chourey, 1998) have been used (for example, see Triplett, 1989, 1990) and are essential to draw conclusions concerning fiber development. Other mutations that may be useful in physiological work include okra leaf, frego bract, and smooth leaf, all traits that impart non-preference insect resistance to cotton (Wilson and George, 1982).

Diversity in cotton germplasm has proven to be useful in physiological work, especial in relation to water-deficit stress tolerance. Scientists at the Plant Stress Laboratory in Lubbock, TX, examined several accessions in the primary germplasm pool for tolerance to water-deficit stress (Quisenberry *et al.,* 1981; Rosenow *et al.,* 1983) and published methods to assess tolerance (Quisenberry *et al.,* 1982, Quisenberry and McMichael, 1982; Gausman *et al.,* 1984, Quisenberry and McMichael, 1991). Leidi *et al.* (1999) used δ-13C discrimination to select for drought tolerance among more than 20 genotypes of *G. hirsutum* cultivars from around the world. Other physiologists have sought correlations between drought tolerance and other physiological traits (Hampton *et al.,* 1987). Rooting potential of different types of germplasm has also received attention (McMichael *et al.*, 1985; Oosterhuis *et al.*, 1988; McMichael and Quisenberry, 1991; Quisenberry and McMichael, 1996). In addition to the roots of different germplasm accessions, the crop canopy relative to leaf shape also has received much attention from physiologists. Parameters such as leaf morphology (Wells *et al.*, 1986; Heitholt and Meredith, 1998), water relations (Karami *et al.*, 1980), temperature (Hatfield *et al.*, 1987), and photosynthesis (Rosenthal and Gerik, 1991; Quisenberry *et al.*, 1994) have been compared among different genotypes of cotton.

Tolerance to water-deficit is often correlated to tolerance to heat stress, so screens of germplasm for tolerance to heat have also been conducted. Often heat tolerance is related to stomatal conductance (Lu *et al.*, 1997, 1998; Radin *et al.*, 1994). Most of the work on tolerance to low temperature was done in the middle of the 20th century, but some screens were done in the last part of the 20th century. For example, Bradow (1991) examined three cultvars and found that one developed on the High Plains of Texas was more tolerant to low temperature at germination than Delta cultivars.

A number of specific traits have been identified as being limited to a species or a few species that would be useful if transferred to upland cotton. Specific examples include reniform nematode resistance from *G. longicalyx* (F₁) (Yik and Birchfield, 1984) or from *G. arboreum* and *G. herbaceum* (Stewart and Robbins, 1996). Also, the leaf content of terpenoid aldehydes (TA) differs both qualitatively and quantitatively among *Gossypium* species (Magboul *et al.*, 1978; Khan *et al.*, 1999). The TA, raimondal, is found only in *G. raimondii* (Stipanovic *et al.*, 1980). Another species-specific expression is the glandless seed trait of the C- and G-genome species of the arid zone of Australia. The closely related K-genome species have lysigenous glands in the seeds (Brubaker, 1996).

The distribution pattern of TAs seems to be related to the area from which the accessions being examined were collected. This suggests that the TA sequestered in the leaves of each species is the one that is most effective against the predominant insect herbivore of the area. For example, all species from Baja California accumulate gossypol also exclusively, whereas *G. hirsutum* (and species endemic to other areas) has only a low level of this TA in the leaves. Most of the leaf TAs in *G. hirsutum* are the heliocides (Khan *et al.*, 1999), where as the TA in the embryo is almost exclusively gossypol. The TAs are sequestered in the lysigenous glands that are characteristic of the genus (Fryxell, 1979). Diversity exists among the species in the size and occurrence of the glands and considerable comparative work as been done on the diversity. Generally the higher the number of glands the more TA the plant will accumulate (McCarty *et al.*, 1996). High TA content is correlated to resistance to herbivorous insects (Parrott *et al.*, 1989; Vilkova *et al.*, 1989; Percy *et al.*, 1996).

In addition to the traits discussed above, diversity among cotton genotypes has been observed for numerous other traits that might be useful in physiological studies, including trichome number (Navasero and Ramaswamy, 1991), interaction with *Agrobacterium* (Velten *et al.*, 1998), semigamy expression (Stelley, 1992), regeneration ability (see Chapter 33), earliness (Godoy and Palomo, 1999), herbicide tolerance (Molin and Khan, 1996), and the response to insects (Rummel and Quisenberry, 1979; Mehetre and Thombe, 1982).

Most of the diversity screens have involved the cultivated cottons and most involve only accessions of *G. hirsutum*. Obviously there is tremendous morphological diversity within the wild germplasm. From this one can assume that the same level of diversity exists for less visible traits such as pest resistance. Totally unknown is the amount of diversity that may be present within a wild species since most screens that include a wild species usually include only one accession as representing the species.

5. SUMMARY

The largest cotton germplasm collection in the world is maintained by the US Department of Agriculture at College Station, TX. Accessions for which seeds are available are distributed at no cost to any *bona fide* person making a request. Other sizable germplasm collections are maintained in India, Uzbekistan, and France. Smaller collections are maintained in most cotton- growing countries, or at cotton research institutions within a country, but the purity of line and pertinent passport data is often suspect. For cotton (Goss*ypium hirsutum*) improvement, the available germplasm can be grouped into 1°, 2°, and 3° pools based on the level of expectation that recombination will occur in F₁ hybrids. For cotton there is extensive diversity available in the primary and secondary pools, but generally pre-breeding is necessary to enhance the germplasm to the point that a breeder can utilize it for cotton improvement. Although there are numerous potentially useful traits in the tertiary germplasm pool, a pre-breeder can devote several years to capturing one trait. With molecular markers, it may be possible to more efficiently transfer traits from the germplasm pools to cotton. Diversity in trait express is available in almost all screens than have utilized germplasm accessions. For many tests the diverse genotypes are selected to provide a range of responses (*e.g.*, okra leaf vs. normal leaf).

CHAPTER 3

MORPHOLOGICAL ALTERATIONS IN RESPONSE TO MANAGEMENT AND ENVIRONMENT

Randy Wells[1] and Alexander M. Stewart[2]
[1]*North Carolina State University, Raleigh, NC;* [2]*Louisiana Agricultural Center, Alexandria, LA*

1. INTRODUCTION

The morphology of the cotton plant serves two main functions. The first is that it provides photosynthate sources (*i.e.,* leaves) and sinks (*i.e.,* squares, flowers, and bolls). The second function is the plant superstructure (*i.e.,* roots, stems, branches, petioles) upon which the sources and sinks reside, and through which they are interconnected via vascular tissues. Excellent reviews of the morphology of the cotton plant have been published earlier by numerous authors, including Mauney (1984) and Oosterhuis and Jernstedt (1999).

Cotton has no theoretical limit to plant development due to its indeterminate growth habit (Hearn and Constable, 1984). Plant stature, however, will be determined via the interaction of genotype and environment. Production of both leaves and fruiting sites is necessarily tied together with the production of each new main-stem node. Fruiting will continue until active boll load equals the supply of assimilate produced by photosynthesis (Guinn, 1985). The demand for assimilates by reproductive growth increases exponentially, while the increase in assimilate supply rises asymptotically (Hearn, 1995). The result is a continual increase in proportional assimilate demand by the fruit and a slowing and eventual cessation of vegetative morphological development.

The rate of nodal development is largely dependent on temperature, however, it is important to recognize that any limiting factor (*i.e.,* water, light, nitrogen) can determine the ultimate growth potential and final stature of the plant. A theory proposed well over a century ago by Liebig (Carter, 1980) addresses this phenomenon. He proposed that '*The yield of a crop is limited by the plant nutrient element present in the smallest quantity relative to the crop's requirements, all other being present in adequate amounts*'. While this principle was originally directed towards nutrient supply, it also applies to any crop growth requirement. If all crop requirements are present in the environment in non-limiting supply except, for example, water, growth and yield will be determined by water supply. If irrigation is available and moisture deficits are removed, the next most limiting growth factor will determine maximal yielding potential. While this premise is a very simplistic view of a rather complex situation, it helps one recognize the importance of environmental factors and their manipulation through management decisions in determining growth and development of a crop. Furthermore, crop managers can control or influence some environmental requirements for plant growth, including water, nutrition, pH, soil tilth, and pest control. On the other hand, little or no control is exercised over other factors, such as temperature and light. Regardless, an understanding of how these many factors influence plant growth is important if future yield advances are to be realized.

The following discussion will examine how, and to what extent, environmental and crop managerial factors influence plant growth and development, with emphasis given to the attainment of plant stature. Genetic control of leaf and plant morphology will be addressed where appropriate.

2. EARLY GROWTH EVENTS

2.1 Leaf Area, Light Interception, and the Compound Interest Law

The ability to intercept light is one of, if not the most, important factor determining growth of plants prior to sig-

J.McD. Stewart et al. (eds.), *Physiology of Cotton*,
DOI 10.1007/978-90-481-3195-2_3, © Springer Science+Business Media B.V. 2010

nificant inter-plant competition. Therefore, leaf area development is a critical determinant of plant growth (Potter and Jones, 1977). Muramoto *et al.* (1965) reported that the rate of leaf area development among cotton genotypes was directly associated with differences in dry matter production. Relative leaf area growth rates (RLAGR) ranged from a high of 0.12 $cm^2/cm^2/d$ for a interspecific hybrid versus 0.10 $cm^2/cm^2/d$ for the mid-parent mean. Concurrently, the respective plant dry weights of the hybrid and the mid-parent mean were 69 and 21 g/plant. Leaf photosynthetic rates did not exhibit significant differences, thereby implying that the greater leaf area growth rate of the hybrid was allowing greater light interception per plant. In another study, intraspecific hybrids possessed greater leaf area due to a greater RLAGR than observed in their parents (Wells and Meredith, 1986a). There was a linear relationship between RLAGR and relative growth rate in both the hybrids and their parents (Fig. 3-1).

Greater early growth will provide a greater base upon which further growth will depend. Blackman (1919) stated that growth of a plant is more nearly approximated by money growth at compound interest which is added continuously. The equation proposed is referred to today as relative growth rate (RGR) and is calculated as follows

$$RGR = [\ln W_2 - \ln w_1]/t \qquad (3.1)$$

where W_1 is the initial dry weight, W_2 is the dry weight after the growth period in question, RGR is relative growth rate, and t is the growth period examined, usually in days. Essentially, growth will be dependent on the rate of accumulation, how much plant mass was there initially, and the duration of growth. For this reason, even if RGR is similar between treatments, the advantage of a larger plant at an earlier point in ontogeny will be maintained. This response can also be important if leaf area is mutually affected by generalized growth responses, thus resulting in greater light interception.

Intraspecific hybrids were found to have greater leaf area per plant and correspondingly greater rates of photosynthesis per plant prior to heightened plant-to-plant competition, characteristic of plants which are large enough to mutually shade their neighbors (Wells *et al.,* 1988). By early reproductive growth, the advantage of greater leaf area per plant is abated as critical leaf area indices (LAI) are approached, indicating that light interception is no longer a limiting factor. By definition, a critical LAI is one which intercepts 95% of available PAR (photosynthetically active radiation, 400-700 nm wavelengths). Critical LAI for cotton is approximately 3 m^2 leaves/m^2 ground (Hearn, 1969b; Ludwig *et al.,* 1965), although leaf areas in excess of critical levels approaching 6 and higher have been reported (Hearn and Constable, 1984; Wells and Meredith, 1984a). It is impossible for a plant canopy which does not close "over the middles" to reach its maximal growth potential because interception of solar radiation will not be maximized.

Cotton leaves are generally heliotropic. Lang (1973) reported that cotton leaves turn throughout the day to stay at a more perpendicular angle to the incoming solar radiation. By turning, the cotton leaves intercept a greater intensity of radiation as explained by Lambert's Cosine Law which states:

$$I = I_o \cos \theta \qquad (3.2)$$

where I is the realized radiation flux density, I_o is the flux density on a surface that is perpendicular to the beam, and θ is the angle that the actual beam is from perpendicular (Jones, 1992).

Genotypes possessing okra and super-okra leaf morphologies have lower LAI values than their normal leaf isogeneic counterparts (Wells *et al.,* 1986). This trait can lead to reduced light intercepting capacity and canopy photosynthetic rates when growth conditions are limiting to canopy development and canopy closure is not reached. On the other hand, increased PAR penetration to lower canopy leaves could greatly increase their photosynthetic contribution to the total assimilate supply (Guinn, 1974). Heitholt *et al.,* (1992) found that okra-leaf cotton increased its PAR interception in narrow row (0.5 m) when compared with wide row (1.0 m) spacing. Furthermore, optimal plant density was greater for okra-leaf cotton than normal-leaf (Heitholt, 1994a). Okra-leaf canopies required a LAI of 4-5 to reach 95% PAR interception. There are also reduced levels of boll disease in closed okra-leaf canopies as compared to normal leaf, apparently due to increased air movement in the canopy (Andries *et al.,* 1969). As suggested by Cothren (1994), the challenge is reaching a balance between light interception capability by the canopy, greater air movement at deeper canopy strata, and increased light penetration to leaves which would otherwise be shaded, or nearly so.

Modern cultivars have smaller plant canopies than their ancestors (Wells and Meredith, 1984a). This observation may reflect an earlier transition into reproductive development and, subsequently, an earlier cessation of new vegetative growth related to cutout (cessation of new vegetative and reproductive growth). Modern cultivars also possess larger reproductive-to-vegetative dry weight ratios

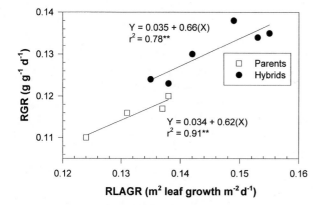

Figure 3.1. Relationship of relative growth rate (RGR) to relative leaf area growth rate (RLAGR) during vegetative growth of hybrids and their parents (adapted from Wells and Meredith, 1986).

throughout reproductive development. Canopies larger than that required for maximal light interception may represent a poor expenditure of growth resources when increased fiber production is the objective.

2.2   Plant Stature and Plant Population Density

Stature of the cotton plant's superstructure is dependent on three factors: 1) main-stem height, 2) number and length of vegetative branches, and 3) the number and length of reproductive branches. Before reproductive development, plant height is positively associated with the plant population density (Buxton et al., 1977; Leffler, 1983). Buxton et al. (1977) reported that cotton plants grown in populations of 21 and 14 plants/m² were 30 and 15% taller than plants grown in a population of 7 plants/m² at 49 days after planting (DAP). As the plants grow larger, however, reciprocal associations between plant height and plant population are evident. Fowler and Ray (1977) found that individual plant stature, measured as mature plant dry weight, plant height, main-stem node number, vegetative branch number, reproductive branch number, and leaf area per plant increased as the equidistant plant spacing increased in two cultivars. 'Paymaster 101A' grown at plant spacings of 51 cm had a plant height, stem diameter, and main-stem node number that was 21, 73, and 31% greater than plants grown at 13 cm spacings. The greater stature on a per-plant basis of expression may reflect the later maturity of plants grown in reduced plant populations. When grown at 2 plants/m², there was an average delay of 16 days in peak flower production when compared to a population of 12 plants/m² over two years (Jones and Wells, 1997). Later maturity, and subsequently later cutout, would lead to continued vegetative development in more widely spaced plants and larger plants would result. This expectation is supported by the production of a greater number of main-stem nodes in response to wider plant spacings (Buxton et al., 1977; Jones and Wells, 1997).

The stature of the plant community is reduced in low plant population densities, indicating that the increased size of the individual plants will not compensate for populations reduced beyond a particular critical plant density (Fowler and Ray, 1977; Jones and Wells, 1997). Fowler and Ray (1977) reported LAI of 5.4, 4.0, 3.0, 2.5, and 1.8 m² leaves/m² ground at 80 DAP for plants with equidistant spacings of 13, 18, 25, 36, and 51 cm, respectively. Based on a critical LAI (LAI required to reach 95% PAR interception) of approximately three, the two smallest plant populations would be unable to fully intercept available light (Ludwig et al., 1965). Furthermore, dry weight of stems, leaves, reproductive branches, and the total plant per unit ground area all declined as the spacing between plants increased.

2.3   Ultra-Narrow Row Cotton

Due to rising production costs and static or declining cotton prices, alternative methods of producing cotton more economically are always being investigated. One potential method involves *ultra-narrow row cotton* (UNRC) in which cotton is grown in drastically reduced row spacings (<25 cm) and high plant populations. Equipment costs in UNRC are reduced compared to conventional systems primarily because the crop is harvested using a finger stripper instead of a more expensive spindle-picker. Early research into the UNRC system demonstrated the potential for increasing earliness and yield compared with wide-row cotton (Hawkins and Peacock, 1973; Lewis, 1971). However, UNRC was not a practical alternative to wide-row cotton for producers because of challenges in weed control. The UNRC system does not allow the use of cultivation or post-directed herbicide options for weed control. The advent of the over-the-top broadleaf herbicide pyrthiobac {sodium salt of 2-chloro-6-[(4,6-dimthoxy-2-pyrimidinyl)thio]benzoic acid}, and transgenic cottons tolerant of glyphosate [N-(phosphonomethyl)glycine] and bromoxynil (3,5-dibromo-4-hydroxybenzonitrile) have enabled weeds to be controlled adequately in the UNRC system (Culpepper and York, 2000). Thus, there is renewed interest in UNRC.

Plant density in UNRC is markedly increased compared with wide-rows and cotton plants exhibit morphological differences in the two systems. Increasing plant population results in earlier peak flower production, and thus earlier maturity (Jones and Wells, 1997). Cawley (1999) found populations of 34.7 plants/m² in 19-cm rows resulted in peak bloom occurring 11 days earlier than a population of 11.3 plants/m² in 97-cm rows. Cotton plants grown in wide rows have been shown to be taller when plant density is increased from 7 to 14 to 21 plants/m². However, Jost and Cothren (2000) noted that decreasing row width from 76- to 19-cm with populations of 13.1 to 13.6 and 39.4 to 45.8 plants m², respectively, resulted in a 17% reduction in final plant height and a 14% reduction in main-stem nodes. This reduction in plant stature can be attributed to earlier fruit-set and maturity in UNRC which would decrease the assimilate supply available for nodal development and stem elongation compared to wide-row cotton.

UNRC results in more first-position bolls and more bolls set on lower nodes than wide-row cotton (Jost and Cothren, 2000). Jost and Cothren (2000) noted that the number of bolls per plant decreased from 6.5 to 3.1 when row spacing was reduced from 76 to 19 cm. However, yields were equal between the two row spacings in a wet year and increased in 19-cm rows in a dry year (Jost and Cothren, 2000). Cawley (1999), Gerik et al.(1998), Gwathmey et al. (1999), and Kerby (1998) have reported lint yields in UNRC to be equal to or greater than wide-row production.

2.4   Light Spectral Balance and Plant Morphology

As seen in the preceding section, plants can perceive their spatial environment sufficiently to alter their growth patterns in three dimensions. How do plants succeed in doing so and with considerable sensitivity?

Kasperbauer (1971, 1988) concluded that plants perceived their spatial environment through red to far-red light ratios (R/FR) via the phytochrome system. Phytochrome is a chromoprotein whose form is modified by red (R) and far-red (FR) signals (Sage, 1992). The two forms, Pfr and Pr, are photo-convertible and can be measured using spectrophotometric techniques (Fig. 3-2). The proportion of total phytochrome (P_{total}) existing as Pfr, the biologically active form of phytochrome, is a function of R/FR ratio. Greater R/FR ratios produce higher amounts of Pfr. Ballaré *et al.* (1989), using fiber optic probes inserted into the stems of *Datura ferox* L. seedlings, found a negative linear relationship between the estimated Pfr/P_{total} inside the stem and the final internode length. These changes were induced in plant spacings too sparse to cause mutual shading between plants. Similarly, plants growing in full sunlight and exposed low supplemental amounts of FR modified their morphology accordingly. Greater proportions of FR induced greater internode elongation rates in Chinese thornapple (*Datura ferox* L.), mustard (*Sinapis alba* L.), and lambsquarters (*Chenopodium album*) seedlings (Ballaré *et al.*, 1987). Smith *et al.* (1990) used artificial canopies of mustard and tobacco (*Nicotiana tabacum* L.) to measure spectral distributions, R/FR ratios, and phytochrome photoequilibria (Pfr/P_{total}) at various distances away. The R/FR and Pfr/P_{total} showed increases as one moved away from the artificial canopies up to 20 to 30 cm away (Fig. 3-3). The smoothness of these curves, the authors point out, indicates that plants may not only detect the presence of neighboring plants but also their proximity. Many plant species appear to have the capacity to alter their morphological development in response to physical changes in their surroundings, even before a direct shading effect is realized.

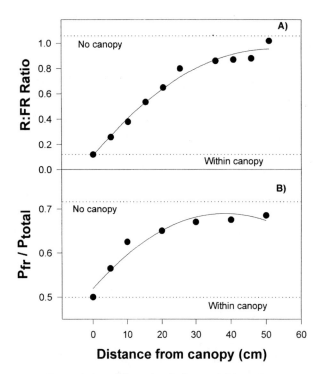

Figure 3.3. Red:far-red ratio (A) and Pfr/Ptotal (B) as a function of the distance from an artificial canopy of tobbaco. Spectroradiometer sensor was held at a 180° angle to the solar beam for the R:FR ratio measurements. Phytochrome cuvette was placed at a 0° angle to the solar beam for 15 min (adapted from Smith *et al.*, 1990).

Cotton plants exhibited similar alterations in morphology as the light spectral quality was changed through use of different colored mulches and soils (Kasperbauer and Hunt, 1992). Reduced flux of upwardly directed R/FR caused longer stems, leaves with lower specific leaf weight, less root dry matter allocation, and greater shoot/root dry weight ratio. Shoot lengths were 40 and 31% longer when grown with green and red mulches when compared with white mulch. Similarly, responses to different soil colors were similar to variations induced by mulch color.

Taller young cotton plants in response to larger plant population densities (Buxton *et al.*, 1977; Leffler, 1983) are consistent with the observations of Kasperbauer and Hunt (1992). Higher plant population densities exhibit lower R/FR due to a higher proportion of reflected and transmitted light caused by the closer proximity of neighboring plants. The failure of these high population plants to exhibit taller plants later in the season may be an indication of the earlier maturity of plants in high population densities as compared with more widely spaced plants (Jones and Wells, 1997). Earlier dry matter allocation into reproductive organs in high plant populations would decrease the assimilate supply available for further nodal development, thus causing stem elongation to cease.

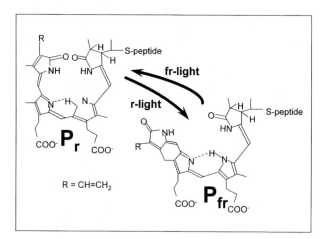

Figure 3.2. Structures of red (P_r) and far-red (P_{fr}) absorbing forms of phytochrome (adapted from Sage, 1992)

3.   TEMPERATURE

3.1   Temperature and Plant Development

Temperature is the paramount environmental factor which influences cotton growth. Temperature effects during germination have been studied intensively and reviews of germination and temperature have been completed by Gipson (1986) and Christensen and Rowland (1986). This section will not attempt to address temperature's effect on germination and subsequent seedling growth. Instead, this section will address vegetative development, plant stature, and temporal relationships as influenced by temperature differentials during growth.

Days between growth events (*i.e.,* appearance of leaves, floral buds, and flowers) are decreased as average growth temperature is increased from 13 to 30°C (17/11 and 32/29°C day/night temperature regimes)(Hesketh *et al.,* 1972). Reddy *et al.* (1992) examined early growth (through 56 days after emergence) at five temperature regimes in growth chamber studies ranging from 20/12 to 40/32°C. The greatest stem elongation, leaf area expansion, and dry weight accumulation occurred at 30/22°C, with less growth at both lower and higher temperature regimes. The leaf area per plant at 30/22°C was approximately 530, 67, 26, and 88% greater than found at 20/12, 25/17, 35/27, and 40/32°C, respectively. The accumulation of biomass was similar with the 30/22°C treatment having approximately 550, 76, 11, and 95% greater mass per plant than found at the same respective temperatures. Moraghan *et al.* (1968) observed that leaf area per plant at squaring was small for plants developing at low night and day temperatures, low night temperatures, and if low night temperatures were imposed only during seedling growth.

Few field studies of high temperature stress response in cotton exist due to the variable and inconsistent nature of temperatures in the field. Vegetative growth in field situations is seldom inhibited by high temperatures as evidenced by positive relationships between heat unit accumulation and time between growth events (Guthrie, 1991; Young *et al.,* 1980). Reproductive development, however, can be negatively impacted by high temperatures. Meyer (1969) found the flower sterility was increased by maximum daily temperatures above 38°C, occurring 15 to 16 days before flowering. Taha *et al.* (1981) found genotypic differences in anther dehiscence in response to high temperatures. Over three dates in July, genotype ST3 had only 7% non-dehiscent anthers, while 83 % were non-dehiscent for genotype B557. ST3 developed bolls from over 28% of its flowers in response to high temperatures. In contrast, genotype B557 matured only 11% of its flowers into bolls. Average maximum temperatures during the period of May through August ranged from 32.9 to 42.7°C. These losses of fruiting forms can have large responses in vegetative growth through compensatory plant growth resulting from continued production of reproductive structures and delayed crop maturity (Sadras, 1995).

Rooting depth increases are also related to the rate of soil warming (Bland, 1993). Downward root extension rates were greatest (1.7 cm/d) when the entire soil profile was warmed in a manner similar to the near-surface soil segment. In contrast, when the sub-soil regions were warmed more like normal cotton-growing soil profiles, the growth rates were 1.4 cm/d. An even slower warming soil profile pattern resulted in root extension rate of 0.7 cm/d. McMichael and Burke (1994) found a lower root metabolic (2,3,5-triphenyltetrazolium reduction) sensitivity to temperature changes during early seedling development, presumedly related to period of seed reserve remobilization. After seed reserves were depleted, the range of metabolic sensitivity increased. Hydraulic conductivity decreases in roots as soil temperatures decline (Bolger *et al.,* 1992). Conductance at 20, 18, and 7°C was 57, 43, and 18 % of that seen at 30°C. No acclimation to an 18°C root zone temperature was observed.

The number of fruiting sites increased linearly as the temperature increased to 40/32°C, however, the retention of squares and bolls was maximal at the 30/22°C temperature (Reddy *et al.,* 1991). This response is indicative of the repressive effect of high temperatures, especially high night temperatures, on reproductive development (Gipson and Joham, 1968; Reddy *et al.,* 1991). Mauney (1966) reported that the lowest main-stem node of the first floral branch was found when warm day temperatures (28 to 32°C) were combined with cool night temperatures (20 to 22°C). Initiation and timing of anthesis were affected in a more complex manner since vegetative growth rates were also affected by temperature differences (Hesketh *et al.,* 1972; Mauney, 1966). Anthesis was delayed by high day temperatures when night temperatures were also high. Delays in floral initiation due to high night temperatures were somewhat abated by low day temperatures, but not completely. In terms of management concerns, moderate to warm temperatures would be favorable during vegetative growth, however, extreme temperatures during reproductive development can greatly reduce fruiting potential.

Effects of temperature on plant growth have been expressed in biochemical terms through the concept of thermal kinetic windows (TKW) (Burke *et al.,* 1988). The TKW is the range of temperatures within which the Michaelis constant, K_m, is at or below 200% of the maximal observed rate. For example, the activity of glyoxylate reductase exhibits a TKW that is 23.5 to 32°C. The relation between biomass production and the growth time within the TKW is linear. The proposed existence of TKW, with a mid-point of 27.5, was further developed via the use of the crop-specific biochemical temperature optimum as the baseline for a thermal stress index (TSI). The TSI was calculated as follows: TSI $= [(T_f > T_b) - T_b]/T_b$, where $T_f =$ temperature of the foliage and $T_b =$ temperature of the baseline which has been determined through biochemical experimentation (Burke *et al.,* 1990).

3.2   Planting Date and Growth

Plant growth is greatly influenced by variable dates of planting. Normal and okra-leaf genotypes were planted one month apart, 27 April and 27 May (Wells and Meredith, 1986b). By approximately 70 DAP, the plants sown at 27 May had accumulated approximately three times the total dry weight than plants sown on 27 April. Similarly, a study that examined genotypes at an optimal (26 April) and a later (12 May) planting date found that plants sown in April had less than 80% of the maximum dry weight found in the later planting (Wells and Meredith, 1984a). In addition, planting a month after the optimum date resulted in a peak white flower production that was over three weeks earlier (relative to the date of planting) than found in the late April planting (Wells and Meredith, 1986b).

Many of these differences in growth and development are attributable to greater cumulative heat units in response to later planting dates. In North Carolina, Guthrie (1991) found heat units (15.6°C base) during the first five days after planting were over 200% greater in response to planting during late May as opposed to early May in one year. Similarly, Young *et al.*(1980) found that the number of days from planting to squaring dropped from 74 to 50 as the date of planting was changed from 1 April to 27 May. Heat units (12.8°C base) accumulated over this period was 31% greater in response to the late planting. Intermediate dates of planting displayed temporal and heat unit differences that were consistent with the trends of the extremes. Despite the greater vegetative growth associated with later planting dates, planting considerably later than the optimal date for a region will often result in decreased fiber yields (Guthrie, 1991; Wells and Meredith, 1984b).

4.   WATER

4.1   Water Deficits

Water relations in plants is an extremely complex subject and certainly greater than the scope of the present discussion. However, a basic premise of crop management is that biomass production and the cumulative seasonal evapotranspiration (ET) are closely associated in environments where other growth factors are not limiting (Ritchie, 1983). Evapotranspiration can be determined by the amount of solar radiation intercepted by the crop canopy and the availability of water (Hearn, 1979). Water-deficit stress imposed via numerous treatments caused progressively greater reductions in ET depending on the severity and timing of stress (Grimes and Yamada, 1982). Hearn (1969c) found that final plant dry weight was linearly related to water usage. Furthermore, main-stem node number was curvilinearly associated with available water. In a desert environment, new node development ceased at earlier dates as water supply was reduced (Hearn, 1969a). Regardless of whether water was supplied frequently or in a single application, nodes developed at the same rate prior to differential cessation points, indicating that nodes were produced independently of water supply until some threshold was reached. Grimes and Yamada (1982) described a linear decrease in stem elongation as leaf water potential (LWP) declined. No further growth was noted when LWP was -2.4 or -2.5 MPa. Crop growth rates (CGR) are lower in environments with low water supply (Hearn, 1969b; Marani and Levi, 1973). In three environments, Constable and Hearn (1981) found that most of the variability in CGR due to varying water supply was attributable to differences in LAI and solar radiation. The CGR per unit leaf area (net assimilation rate) was not affected by these same treatments and reflected a reduced rate of leaf area expansion.

Leaf area production is reduced with limiting water supply (Marani and Levi, 1973; Hearn 1969b) causing a reduced LAI. Thomas *et al.* (1973) found field-grown plants exhibiting a reduction in leaf area per plant of 77 % from the control in response to a single stress cycle. Marani and Levi (1973) found that with moderate and high water supply levels, leaf area increases were greatest from early flowering and continued so for approximately a month. In low water treatments, however, LAI remained at values of 1.2 or less during the same period and was the characteristic that was most persistently affected by water supply.

Marani *et al.* (1985) suggested that two different mechanisms may be responsible for negative effects of water stress on canopy photosynthetic rates. The first is an effect on individual leaves through closed stomata and lessening of related leaf activities. The second is a combination of reduced leaf area, increased average leaf age in the canopy, and hastened leaf senescence due to limited water supply. Leaf expansion is more sensitive than stomatal closure to water supply. Expansion of leaves declines parallel to the decreasing water supply during a drying cycle, while stomatal closure occurs after some threshold water deficit is attained (Hearn, 1979). The actual water deficit levels at which cessation of leaf expansion (Cutler and Rains, 1977) and stomatal closure (Thomas *et al.,* 1976) occur are dynamic and depend on the water supply history of the crop. Frequently irrigated plants displayed greater reductions in leaf elongation rates at greater predawn LWP than plants watered less frequently (Cutler and Rains, 1977). The effect of water supply on leaf growth and expansion was addressed in detail by Radin and Mauney (1986).

During drying cycles, root growth appears closely associated with shoot growth, with elongation ceasing concurrently with cessation of growth in the shoot (Taylor and Klepper, 1974). Root elongation ceased when the soil water content was 0.06 to 0.07 cm^3/cm^3 (-0.1 MPa). Ball *et al.* (1994) found that leaf expansion was more sensitive to water stress than root elongation, with the former ceasing two days into a six day drying cycle. Root elongation did not cease elongation until the sixth day.

Root depth patterns change during a drying cycle (Klepper *et al.,* 1973). Depth of new root proliferation is greater as drying continues, which is a trend not observed in well-watered control plants. If plants have a history of frequent soil-surface wetting, they lack the ability to adjust to conditions of progressive drying by rapidly exploring wetter, deeper soil profiles (Carmi *et al.,* 1993). These root systems lack a strong growing taproot, and they have instead, a preponderance of thin lateral roots near the soil surface (Carmi *et al.,* 1992). Depth of rooting should be taken into account when designing irrigation regimes.

For excellent reviews of water relations in cotton, refer to Hearn (1979, 1995).

4.2   Flooding

Excess water can inhibit plant development just as too little water. Two major phenomena occur in the soil during flooding. The first is a large reduction in soil oxygen partial pressure (Meyer *et al.,* 1987). The second is decreased nitrogen uptake by the roots (Hocking *et al.,* 1985). The latter case may be explained by either a reduced soil nitrogen concentrations, due to denitrification, or a reduced capacity for active N uptake (Meyer *et al.,* 1987). In either case, reduced availability of oxygen manifests itself in a reduced nitrogen supply for plant growth and maintenance.

Meyer *et al.* (1987) reported that waterlogging cotton root systems caused reductions in leaf photosynthesis rates after 2 d and continued to decrease until a difference of 39% existed between flooded and control plants. Concomitant reductions in leaf growth and stomatal conductance occurred at 6 d after flooding. Reductions in leaf growth due to waterlogging are common among different studies (Meyer *et al.,* 1987; Reicosky *et al.,* 1985; Soomro and Waring, 1987). Changes in leaf water potential were not consistently associated with the observed reductions in leaf growth and indicates that immediate changes in growth are not directly mediated through plant water status (Reicosky *et al.,* 1985). Later flooding events do not result in as severe effects on growth and adaptive processes appear to be at present. After a second flooding, nitrogen transport has partially recovered and no visible symptoms of stress appear (Reicosky *et al.,* 1985).

5.   NITROGEN

How large a cotton plant will become is greatly determined by the amount of N it has available for growth. In rain-fed areas of the world, where irrigation is limited, N is the essential factor of growth that the grower can most easily control. It is also a factor that can often be present in either limiting or excess supply. In either case, plant productivity is reduced. Gerik *et al.* (1998) provide an excellent review of N and its management.

Plant height (Malik *et al.,* 1978; Gardner and Tucker, 1967), stem diameter (Gardner and Tucker, 1967), mainstem node number (Jackson and Gerik, 1990; Thompson *et al.,* 1976), total plant dry weight (Wullschleger and Oosterhuis, 1990b), leaf area expansion rate (Wullschleger and Oosterhuis, 1990b), LAI (Constable and Hearn, 1981; Jackson and Gerik, 1990; Thompson *et al.,* 1976; Wullschleger and Oosterhuis, 1990b), vegetative branching (Gardner and Tucker, 1967), and the number of fruiting sites (Thompson *et al.,* 1976; Malik, *et al.,* 1978) are all positively related to N supply. In addition, the time required for growth events is reduced when N is in ample supply. Thompson *et al.* (1976) found that greater N supply reduced the plastochrons of leaf emergence (days/leaf) at nodes 5 through 10. In addition, several accounts of delayed reproductive development have been reported due to low levels of N fertility. Malik *et al.* (1978) found that the time to first flower was increased an average of 30% in two cultivars if only 17% of the non-limiting rate of N was available. Non-limiting N supply actually lengthened the reproductive period because more bolls are produced over a longer period. Hearn (1975) found that high N and frequent irrigations caused a slower boll set but resulted in an extended reproductive period and greater boll loads. The longer reproductive period prolonged the need for pest control three weeks later in the season, possibly resulting in less economic return for the producer. Excessive nitrogen can lead to extravagant vegetative growth and crop canopies which promote boll disease, low micronaire, and late maturity (Hearn, 1975).

Fiber yield is closely related to the number of bolls set by a crop of a particular cultivar and crop history (Wells and Meredith, 1984b). Further, integrated seasonal canopy photosynthesis is positively correlated to yield (Wells *et al.,* 1986), indicating that the supply of photosynthate may determine the carrying capacity of a particular crop. Hearn and Constable (1984), on the other hand, suggest that N supply is also an important determinant of yield and carrying capac-

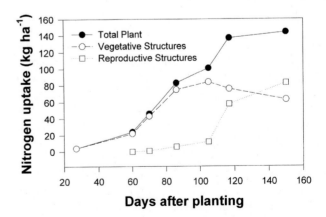

Figure 3.4. Changes in vegetative, reproductive, and whole shoot N content as a function of days after planting (DAP) for cotton given 60 kg N/ha at planting and 60 kg N/ha at 70 DAP (adapted from Oosterhuis *et al.,* 1983).

ity. They proposed that the capacity of a crop to initially take up, store in vegetative organs, and subsequently remobilize N, without excessive vegetative development (rank growth), may be as important as photosynthesis in determining boll carrying capacity. Oosterhuis *et al.* (1983) reported that about 42% of the shoot N resides in the seed and fiber and found that vegetative and reproductive structures exhibit inverse trends during reproductive growth (Fig. 3-4). Further, Zhu and Oosterhuis (1992) found that at 28, 42, and 62 days after main-stem leaf initiation the sympodial leaves of the tenth node sympodial branch contained 75, 52, and 20% of the total branch N. Contrastingly, the bolls contained 11, 38, and 73% of the branch N at these same respective dates. Despite these inverse patterns of vegetative and reproductive change, not all N requirements could be fulfilled by the subtending sympodial leaves and other sources of N would be required (*i.e.*, main-stem leaf, upper-canopy leaves, and leaves further out the sympodium). Therefore, N is required to produce the photosynthetic factory and much of this N will be remobilized for boll development as reproductive development proceeds.

6. GROWTH REGULATORS

Plant growth regulators (PGRs) influence a multitude of specific physiological processes in cotton. These responses are an indication of complex interactions of cultivar, management, and environment. Therefore, growth responses are often not predictable and somewhat inconsistent. The following section will only touch on the wealth of knowledge concerning PGRs and their influence on plant growth and development. For a more complete discussion refer to Cothren (1994) and Cothren and Oosterhuis (this book, chapter 26).

6.1 Mepiquat Chloride

Mepiquat chloride (MC), an anti-gibberellin, is probably the most widely used growth regulator in cotton production. Numerous alterations in plant stature are caused by MC, including smaller more compact plants with fewer nodes (Bader and Niles, 1986; Reddy *et al.*, 1990; Reddy *et al.*, 1992; York, 1983a), narrower plant canopies (Walter *et al.*, 1980), reduced leaf area expansion (Fernandez *et al.*, 1992), altered light interception patterns at the canopy mid-point (Bader and Niles, 1986), reduced LAI (Reddy *et al.*, 1990; Stuart *et al.*, 1984), and increased specific leaf weights (Reddy *et al.*, 1990; Xu and Taylor, 1992). The impact of MC on stem elongation and general height differences are temperature dependent (Reddy *et al.*, 1992) and help explain the observed responses of growth to MC in combination with different planting dates (Cathy and Meredith, 1988). Mepiquat chloride treated plants exhibited heights that were 9, 17, and 20% less than controls when

planted at mid-April, early-May, and mid-May, respectively. Beneficial effects of MC are evident in late-planted cotton which, as discussed earlier, generate greater vegetative development.

Earlier crop maturity is induced by MC. Kerby *et al.* (1986) reported that MC-treated plants set more bolls at lower plant positions and fewer at higher positions. Late season boll load was limited by reduced boll retention and not by a decline in the number of fruiting sites. In addition, MC-treated plants had greater first-position bolls up to the fifteenth node than the untreated control. Bader and Niles (1986) found four of five maturity measures indicating that MC caused earlier crop maturity. They also indicated that full-season cultivars are more responsive to MC-induced maturity effects than a short-season, determinate cultivar such as 'TAMCOT CAMD-E'. Constable (1994), on the other hand, found cultivar responses that were unexpected. Early maturing cultivars 'Siokra S324' and 'CS 8S' exhibited greater positive yield responses than the later maturing 'Siokra L23' and Siokra V-15'. Others have found little effect of cultivar on the observed responses to MC (Cathey and Meredith, 1988; York, 1983a).

Constable (1994) indicated that thirty years ago it was common to stress a cotton crop at early flowering to promote early boll retention and set the plant on the course of increasing boll assimilate demand, concurrent with declining leaf initiation. This practice has been replaced by use of MC to control excessive vegetative development. The key for attaining maximal yield responses to MC has been determining the timing and amount of MC to apply to mitigate excessive vegetative growth while promoting maximal fruiting potential. A number of monitoring methods have been utilized to more effectively determine MC needs and timing (Edmisten, 1994; Fletcher *et al.*, 1994). Edmisten (1994) utilized a point system, where points were assessed based on the moisture, plant height, height/node ratio and other crop history factors. Constable (1994), in a five year study, found that yield responses to MC occurred when internodes were longer than 6.5 cm/node. This observation is consistent with responses usually observed when environmental conditions are non-limiting to growth and prone to excessive vegetative development. Measuring internodal length development is a promising monitoring technique since adjustment to plant vigor must be made during its manifestation and not after the plants are already too large (Constable, 1994; Fletcher *et al.*, 1994). Landivar *et al.* (1996) proposed using the average length of the top five internodes (ALT5) of the main-stem as an indicator of the stem elongation rate. ALT 5 is measured using a special ruler (MEPRT: **MEP**iquat Chloride **R**ate and **T**iming) calibrated to estimate the potential height of the plants at the cessation of vegetative growth, or cutout (Landivar, 1998). The goal is to maintain the ALT5 values between 3.6 and 4.1 cm (1.4 to 1.6 in.).

6.2 Gibberellins, Auxins, and Cytokinins

In contrast to the effects of MC, many growth regulators have been used to stimulate early vegetative and reproductive growth of cotton. Gibberellic acid (GA) was applied by either soaking seed with 10 to 100 ppm GA or by applying the same concentrations at squaring by El-Fouly and Moustafa (1969). Both application methods resulted in increased plant height, while time to reproductive growth was unaffected. More recently, GA_3 (RyzUp) has been applied to plants with three to seven true leaves in both greenhouse and field (Hansen *et al.*, 1996). Increases in leaf area per leaf, plant height, and photosynthetic rate were observed in response to GA_3 application. Cytokinins or similar compounds have also been studied under a number of trade names (Burst, Cytozyme, Cytokin, Triggrr). Positive yield responses to cytokinin compounds have been reported, however, responses are generally inconsistent (Cothren, 1994). Cytokinin effects on growth and plant stature are poorly understood. Some PGRs are mixtures of different compound types. The best known product of this group is probably PGR-IV, consisting mainly of GA, indolebutyric acid, and a proprietary fermentation broth. Positive effects of PGR-IV have been reported to manifest themselves as increased root growth, enhanced nutrient uptake, greater leaf area, increased shoot growth, enhanced photosynthesis, and higher yields than in untreated controls (Oosterhuis, 1995b). Responses indicate that PGR-IV may alleviate some of detrimental effects of mild stresses on plant processes. Zhao and Oosterhuis (1997) reported significantly greater photosynthetic rate, root dry weight, and floral bud dry weight in PGR-IV treated water-stressed plants than untreated water-stressed plants.

Growth and yield responses from PGRs are generally inconsistent. Much of the variability comes from the fact that even small differences in management, genotype and environment can greatly alter plant response to PGRs.

7. OZONE

Ozone is the most commonly occurring atmospheric pollutant that negatively affects cotton growth and yield. Yield response is very dose dependent and is variable in occurrence due to varying climatic conditions and nearness of urban centers (Olszyk *et al.*, 1993). As urban areas encroach into agricultural regions, ozone will continue to expand its importance as a yield-limiting factor. Yield decreases of 10 to 20% are common in response to ambient ozone concentrations (Heagle *et al.*, 1986; Heagle *et al.*, 1988; Olszyk *et al.*, 1993; Temple *et al.*, 1985). Under well-watered conditions several alterations to plant growth and morphology are noteworthy. The major alteration in response to ozone is increased abscission of main-stem leaves (Temple, 1990b) and reduced leaf area duration (Miller *et al.*, 1988). Associative reductions in net assimilation rates are also apparent (Miller *et al.*, 1988; Oshima *et*

al., 1979), indicating that plant assimilatory capacity is impaired. Ozone-exposed plants tend to be more 'leafy' early in growth; a trait measurable by greater leaf area ratios (leaf weight per unit plant weight)(Miller *et al.*, 1988; Oshima *et al.*, 1979), leaf weight ratio (leaf weight per unit plant weight), and specific leaf area ratios (leaf area per unit leaf weight) than observed in low ozone controls (Miller *et al.*, 1988). Later in growth these increases in leaf partitioning are reversed since damaged leaves are then abscised from the plant (Miller *et al.*, 1988). Plant height does not appear reduced due to ozone, however, internodal lengths are decreased by ozone exposure (Temple, 1990b). Observed growth effects are much greater in well-watered than water-stressed plants and reflect a need for open stomata as a point of entry to internal leaf tissues (Miller *et al.*, 1988; Heagle *et al.*, 1988; Temple *et al.*, 1985). Internally, carboxylation efficiency is the initial cause of decline in photosynthesis following ozone exposure (Farage *et al.*, 1991). Inhibition of electron transport is not involved in initial events and stomatal closure is predominantly a secondary effect occurring later in leaf ontogeny (Pell *et al.*, 1992).

There are significant cultivar x ozone interactions concerning growth and yield, however, these differences appear more closely tied to the degree of determinacy rather than to some physiological mechanism (Temple, 1990b). Determinate cultivars enter reproductive development when ozone concentrations are high. Similarly, indeterminate cultivars can respond to leaf loss by altering growth patterns, such as producing more branches. An excellent review of the effects of gaseous pollutants on cotton growth can be found in Chapter 15 (Temple and Grantz, this book).

8. SUMMARY

The number of fruiting sites produced by the plant is dependent on the stature and branching pattern. Fruiting site quantity, combined with the capacity of the plant to supply assimilates, will determine the final fiber production. Environmental factors, some influenced through management inputs and some not, will determine the plant stature. Further, the assimilatory capacity of the crop is determined through both plant stature (*e.g.*, LAI, light interception patterns) and the influence of the environment on the physiological mechanisms (*e.g.*, photosynthesis, respiration, translocation). For these reasons, crop managers must "stack the deck" in the crop's favor, allowing maximal yield potential regardless of what uncontrolled environmental conditions (*e.g.*, temperature, solar radiation, water in rain-fed regions, air pollution) exist. All controllable growth factors (*e.g.*, fertility, pH, pests, growth regulators) must be present in sufficient amounts and must be applied in response to individual field needs as they arise during the growth cycle. For these reasons, an intimate knowledge of the crop and its morphological development is perhaps a crop manager's most powerful tool.

Chapter 4

PHYSIOLOGICAL AND ANATOMICAL FACTORS DETERMINING FIBER STRUCTURE AND UTILITY

C.H. Haigler[1]

[1]*Departments of Plant Biology and of Crop Science, NC State University, Raleigh, NC 27695.*

1. INTRODUCTION

The yarn and textile industries rely on the cotton plant to produce desirable fibers, which for the purposes of manufacturing are essentially a material with particular physical properties. To some extent, current expectations of the industry and current fiber uses are dictated by what nature and human domestication of cotton gave the 19th and 20th centuries as a starting material. More than five millennia ago, humans realized the potential for spinning and weaving of the natural fibers that presumably aided cottonseed dispersal (Arthur, 1990). Hence, agriculturalists have long selected for higher-yielding, longer, and perhaps even stronger cotton fibers that were better suited for making yarns for cloth. Continuing since the industrial revolution, changes in manufacturing technology, fiber analysis potential, and consumer demands for higher quality end-products have led to more rapid improvement in the basic properties of the fiber (or combinations of desirable properties) through the efforts of cotton breeders. Fiber improvement is likely to accelerate in an increasingly sophisticated cotton industry and market that demands and rewards particular fiber properties. For example, cotton fiber strength has improved greatly in just the past decade after increasing use of cheaper rotor spinning put an increasing premium on strength to maintain high-quality yarn production, and automated HVI testing allowed strength to be monitored and at least partially factored into pricing at the gin (Chewning, 1992; Sasser, 1992; Lewis and Benedict, 1994). Also, genetic engineering offers the prospect of expanding the rather narrow potential of the domesticated cotton genomes (Wendel *et al.*, 1992) by allowing targeted improvement in traditional fiber properties and addition of completely novel properties to fibers through changes in genetic composition or regulation (John, 1994).

Nature has given us a few cotton mutants with novel and useful properties, such as light brown and green color due to integration of suberin into the walls in the case of the green cotton (Schmutz *et al.*, 1993). These mutants have been exploited in niche markets for organically-grown, undyed cotton products (Lee, 1996). Progress by genetic engineering toward adding other novel properties to fibers, for example by including a thermoplastic polyester compound in the lumen (John *et al.*, 1996), is occurring faster than modification of traditional fiber properties because the necessary genes can be identified from other organisms (John, 1994). In contrast, modification of traditional fiber properties by genetic engineering has not yet occurred because we lack knowledge about which are the critical controlling genes (John, 1992). Work to identify the genes that are expressed in cotton fibers represents a beginning toward reaching that goal (John and Crow, 1992; Delmer *et al.*, 1995; Ma *et al.*, 1995; Pear *et al.*, 1996; Song and Allen, 1997), but extensive physiological and biochemical work, often coupled with production of transgenic plants with altered gene expression, will be required to understand which of these genes code for proteins with critical regulatory roles. Understanding the control mechanisms of the basic agronomic traits is an important goal if we are to improve these traits or even maintain them under stressful production conditions and with sustainable farming practices. Although cotton with special properties will occupy important niches in the future, it is probable that traditional cotton will continue to meet many of the bulk demands for this natural fiber. Identification of controlling genes is predicated on a basic understanding of cellular and biochemical phenomena relevant to particular fiber traits, and even that understanding at a mechanistic level is in its infancy.

J.McD. Stewart et al. (eds.), *Physiology of Cotton*,
DOI 10.1007/978-90-481-3195-2_4, © Springer Science+Business Media B.V. 2010

The purpose of this chapter is to discuss the current state of knowledge about physiological and anatomical factors that determine fiber structure and utility. Emphasis will be placed on research accomplished since the mid-1980s because excellent comprehensive reviews of fiber development were published near that time (Basra and Malik, 1984; Ryser, 1985; DeLanghe, 1986; Kosmidou-Dimitropoulou, 1986). Hormonal controls of fiber development will not be covered extensively because this has not been a particularly active area of recent research since those reviews. The organizing principle will be to consider the major fiber traits now valued by the textile industry (fineness, length, maturity, and strength) in the context of their value to the industry and their control at the level of fiber development. Consideration will be given to four levels of structural organization that dictate the properties of natural fibrous polymers (Rebenfield, 1990): (a) molecular structure of the predominant homopolymer in cotton fiber, cellulose; (b) macromolecular structure, including cellulose chain length; (c) supramolecular organization, or aggregation of cellulose molecules into microfibrils; and (d) aggregation of microfibrils into the fibrous composite of the cell wall under genetic control.

Uniformity of fiber properties will also be discussed because of its value relative to: (a) more efficient design and use of textile machinery; (b) more uniform and predictable product properties; and (c) generation of higher prices paid to the producer and even long-term sale contracts to particular end-users (Booth, 1968; Lands' End Inc., 1997). Fiber yield will be discussed since no particular quality trait, or even uniformity, is valued unless it can be optimized along with other traits in a high-yield crop. Finally, attempts to modify fiber properties will be discussed. Because of the broad scope of this chapter, it does not represent exhaustive coverage of any fiber property; readers interested in particular topics are encouraged to consult other reviews and the recent primary literature for more information. This book also contains more detailed discussions of cotton molecular biology, biochemistry, production systems, and environmental interactions that have bearing on fiber quality.

2.   OVERVIEW OF FIBER DEVELOPMENT

Traditionally, fiber development has been divided into four stages: (1) initiation; (2) elongation; (3) secondary wall thickening; and (4) maturation (Jasdanwala *et al.*, 1996). The terminology of stages (3) and (4) has recently become confusing, since the industry would like to replace the traditionally measured fiber property of micronaire, which is an indirect composite measurement dependent on fiber perimeter (fineness) and wall thickness (maturity), with separately quantified measurements of perimeter and maturity (Bradow *et al.*, 1996). It is the process of secondary wall thickening that leads to maturity; hence, confusion arises

if stage (4) is referred to as maturation. Stage (4) is really the stage of death and desiccation of the fiber, which results in collapse of the formerly round lumen and twisting of the fiber. Therefore, this chapter will refer to the four stages of fiber development as: (1) initiation; (2) elongation; (3) thickening; and (4) death/dessication. The changes that occur in stage (4) are critical to the physical properties and use of cotton fibers (DeLanghe, 1986); for example, twisting (or formation of convolutions) of the fiber upon drying increases elongation to break and aids spinning into composite yarns. However, the occurrence and periodicity of the twists are determined by the fiber structure that was formed by active cellular processes in the first three stages. Therefore, only the physiological and anatomical controls of the first three stages and their impact on fiber properties will be discussed here.

Stage (1), initiation, refers to the ballooning out of the fiber initial above the seed epidermal surface on the day of flowering, or anthesis. Therefore, customarily and in this chapter, fiber age is described by days post-anthesis, DPA. Fiber age is indicated by positive DPA numbers, whereas differentiation stages of fiber development (before initiation) are indicated by negative DPA numbers. The fiber initial increases in length and perimeter relative to the foot of the fiber that remains embedded in the outer layer of the epidermis (Ryser, 1985). The stage of initiation continues over several days across the whole epidermal surface. Although it has been said that fibers initiate in particular patterns over the ovule surface (Tiwari and Wilkins, 1995), there can be extensive variability between ovules so that no particular pattern exists (Tiwari and Wilkins, 1995). Each individual fiber remains in the initiation stage with a bulbous tip for about two days. The property of fiber perimeter (or diameter when the initial is circular) is determined during or shortly after initiation.

Stage (2), elongation, can be defined as beginning when the individual fiber develops a sharply tapered tip. This stage is characterized by rapid primary cell wall synthesis as the single-celled fiber attains lengths that can be greater than 2.25 inches (Booth, 1968). Elongation continues until 14 to 40 DPA, with the duration dependent upon genotype and environment (Thaker *et al.*, 1989; Ryser, 1985; Haigler *et al.*, 1991; Xie *et al.*, 1993). The latter point emphasizes that days of starting or duration cannot be rigidly assigned to any of these fiber developmental stages in all cultivars or growing regimes; any particular study that depends on identification of exact developmental stages should include data to establish internal reference points.

Stage (3), thickening, begins when the cell wall starts to thicken. The times of initiation and duration of this phase also depend on genotype and environment (Thaker *et al.*, 1989; Haigler *et al.*, 1991). Generally, thickening begins between 12 and 20 DPA while elongation continues (DeLanghe, 1986). Therefore, elongation and thickening are overlapping phases for a variable period that is as long as 21 days in *G. barbadense* (Schubert *et al.*, 1976). Cell wall thickening begins with deposition of a thicker primary wall (Meinert and Delmer, 1977), but soon the deposition of a cellulose-rich secondary wall begins. The cellulose-rich secondary wall forms the bulk of the mature fiber, and its deposition is completed by 35 to 55 DPA (Schubert *et al.*, 1976; Thaker *et al.*, 1989; Haigler *et al.*, 1991). The secondary wall of the cotton fiber is the purest cellulose structure produced in bulk by higher plants, containing more than 95% cellulose (Meinert and Delmer, 1977).

3. FINENESS

Fineness is a relative property arising from the diameter, cross-sectional area, weight per unit length, perimeter, and specific surface area of the fiber (Arthur, 1990). The textile industry is placing increasing value on fine fibers, or those for which fiber mass per unit length is lower, because rotor spinning places more value on fineness and strength than length (Deussen, 1992). Fine fibers produce the strongest and most uniform yarn for a given fiber length because more fibers fit into the yarn cross-section, which increases strength and tends to minimize the effect of individual fiber irregularity (Booth, 1968; Basra and Malik, 1984). Also, finer fibers are less rigid, promoting ease of spinning and better fabric drape (Booth, 1968; Deussen, 1992). Since mature fibers with thick walls are also desirable to enhance single fiber strength and dyeability (up to the limit of suboptimum fiber rigidity), fineness is best reduced by lower fiber perimeter. Only control of fiber perimeter will be discussed in this section, with wall thickness (maturity) being discussed in a following section.

3.1 Fiber Property Measurements Relevant to Fiber Fineness

3.1.1 Micronaire

Micronaire is a comparative value with units arising from measurement of air-flow through a fiber mass of standard weight. The air flow depends on packing of the fibers, which in turn depends on a composite of fiber properties including large contributions by fineness and maturity. A mass of large perimeter (coarse) fibers with thin walls (immature) could have the same micronaire value as a mass of small perimeter (fine) fibers with thick walls (mature).

The premium micronaire range is 3.5 to 4.9 (Culp, 1992).

3.1.2 Linear Density (mtex = mg/km)

Linear density depends on both the perimeter of the fiber and the thickness of the secondary wall. If the fibers were perfectly circular with uniform thickness, only diameter would have to be considered; using mass/unit length accounts for deviations in these parameters (Booth, 1968). Linear density is determined from the number of fibers and the weight of fibers in a 1.5 cm bundle.

A typical linear density value for upland cotton is 130 mtex.

3.2 Control at Fiber Initiation

Fiber perimeter is related to the extent to which the fiber initial expands in diameter as it balloons above the epidermal surface (usually between 0 and 2 DPA)(Wilkins, 1992) and becomes wider than the fiber foot (Berlin, 1986). A positive correlation has been shown between initial perimeter and final fiber perimeter (DeLanghe, 1986). However, perimeter is also under regional control in older fibers. Lint fibers become wider until about 10 mm distance from the ovule surface (Boylston *et al.*, 1993), with a region near the tip of older fibers being narrower than this maximum. The taper was restricted to the terminal 1000 to 1500 μm of the fiber tip in 2 cultivars of *G. hirsutum* L., with the shorter taper region occurring during primary wall synthesis (R.W. Seagull, personal communication). Final fiber perimeter typically varies between 40 and 70 μm (Vincke *et al.*, 1985), indicating substantial variation in this trait between cultivars. Within one cultivar, fiber perimeter values show a normal distribution with a population standard deviation between 10 and 30% of the value, although some of this variation is likely accounted for by differences in point of cross-section along the fiber length (Petkar *et al.*, 1986; Boylston *et al.*, 1993). Fiber perimeter is not affected to an appreciable extent by environment; its variation is controlled genotypically (Kloth, 1992).

Cells swell in perimeter if cell wall properties are such that the turgor pressure of the cell can cause expansion. (It is possible for a wall to be so inextensible that no expansion occurs even though turgor is a constant feature of unstressed, vacuolated plant cells.) Therefore, the initial swelling of particular fiber cells, as contrasted with non-swelling neighbors, must imply an increase in turgor and/or a loosening of the cell wall in those particular cells (Wilkins, 1992). The physical properties of the cell wall at any time imply that there exists a particular yield threshold of pressure required to induce any expansion, and once that threshold is exceeded the magnitude of turgor pressure can affect the rate of cell expansion as long as the wall remains extensible (Cosgrove and Knievel, 1987; Cosgrove, 1997). Therefore, reduction of wall extensibility, perhaps through increased synthesis of enzymes or polymers that can participate in cross-linking, is a possible target for decreasing fiber perimeter.

The direction of reinforcement of the primary cell wall by the innermost layers of cellulose microfibrils determines the direction of cell expansion (Richmond, 1983). The orientation of cellulose microfibrils is determined by the orientation of cytoskeletal microtubules in many plant cells (Seagull, 1989a; 1991), and when the fiber initial first swells as an approximate spherical balloon the orientation of microtubules in these initials is random (Seagull, 1992a). However, within 2 to 3 days, the microtubules reorient into hoops or a flat helix parallel to the ovule surface, and consequently the innermost wall microfibrils begin to be synthesized around the circumference of the fiber (Seagull, 1992b). This constrains circumferential expansion, and the fiber begins to lengthen unidirectionally. We do not understand why a tapered tip develops at this stage, although it could indicate the addition of a tip-growth mechanism (see below) to the intercalary growth already proceeding.

Fiber perimeter could readily change as long as synthesis of an extensible cell wall continues, implying that this property is under active cellular control. Evidence that microtubules are important in control of perimeter over most of the fiber length during primary wall synthesis comes from experiments with microtubule polymerization inhibitors (oryzalin) or microtubule stabilizers (taxol), both of which cause fiber swelling (Seagull, 1995). These experiments suggest that the presence of microtubules in their normal state of dynamic function is required to restrict excessive fiber expansion, probably because of their effect on determining orientation of the cellulose microfibrils (Seagull, 1989a, b). In regions of intercalary growth with an expanding cell surface, newly arranged microtubules would be required to orient the newly synthesized microfibrils, which explains why even stabilized microtubules allow fiber swelling (Seagull, 1995). Contrary to the common idea that perimeter cannot change after the thick, less-extensible secondary wall begins to be deposited, increases in fiber perimeter have been observed through 30 DPA in MD51 and DP50 cultivars of *G. hirsutum* (R.W. Seagull, personal communication). In the latter stages of fiber thickening, further perimeter increases may cease; there is a tendency for a 10 to 20% decline in average perimeter of air-dried *G. hirsutum* fibers between 24 and 63 DPA (Petkar *et al.*, 1986; Boylston *et al.*, 1993).

4.   LENGTH

Desirable commercial cotton cultivars generally produce fibers with length (standardized as UHM = upper half mean; the mean length of the longest one/half of the fiber population) in the range of ≤0.99 inch (short) to ≥1.26 inches (extra long). Staple length is an estimate of the average length of fibers in a bale, which varies between 0.87 and 1.5 inches in commercial cultivars (Rebenfield, 1990). Span length (*e.g.*, 50% span length) is a measurement from clamped fibers indicating that 50% of the fibers are greater than or equal to the stated length; this measurement is often viewed as more realistic of fiber processing since the fibers start out randomly arrayed (Booth, 1968).

Both the rate (for example up to 2.9 mm/day; Thaker *et al.*, 1986) and duration of fiber elongation (14 to 38 days in a representative experiment) are positively correlated with final fiber length (Thaker *et al.*, 1989). Long fibers are most highly valued for fine textiles because long fibers can be spun into stronger, finer yarns (Arthur, 1990). Therefore, improving length without sacrificing other properties is a major goal for cotton improvement. Individual cotton fibers can be more than 2.25 inches long in *G. barbadense* cv. Sea Island, but fibers are much shorter in *G. hirsutum* (typically not more than 1.1 inches long) (Booth, 1968). Even in *G. barbadense*, there is no reason to believe that an inherent limit of length has been reached since elongation commonly stops before the fiber dies (Schubert *et al.*, 1976) and much longer plant cells do exist (extraxillary fibers of flax and ramie are 2.7 inches and up to 21.6 inches long, respectively; Fahn, 1990).

Efforts to improve fiber length are limited because we do not know of one or a few genes that control length. If elongation were to occur faster or if its period were to be extended, at least some of a whole suite of genes (for example those establishing sufficient turgor to exceed the yield threshold of the cell wall and those required for plasma membrane and wall synthesis) would have to be up-regulated and/or continue to be expressed. Such a goal would be easier to accomplish if there were developmental timing genes (Seagull, 1992a; Wilkins, 1992; Haigler *et al.*, 1994) that regulate the elongation process such that manipulation of these would allow the process to be extended. Alternatively, one or a few enzymes could represent the major limitations of length, with excess capacity for elongation built into other aspects of the system. Recent research that has yielded additional information about cellular mechanisms related to the fiber elongation process is discussed below.

4.1   Timing of Fiber Initiation

The long lint fibers, which initiate *in planta* between 0 and 2 DPA (Berlin, 1986), are used in the spinning of yarns to make textiles. There is another population of short fibers, called fuzz fibers (or sometimes linters) that arise from fiber initials that did not achieve their length potential. Since the fuzz fibers are the last to initiate between 6 and 8 DPA (Berlin, 1986), the timing of initiation may be one controller of fiber length. It can be speculated that there is an inherent stimulus of the suite of genes controlling extensive elongation, for example a particular hormone or combination of hormones that are present in the ovule only for a certain time. Once this time is passed, fiber initials that form later cannot be stimulated to elongate extensively. However, the complexity of control of elongation is indicated by the fact that ovules cultured constantly with exogenous hormones require a longer time to initiate the same number of fibers as occur on plant-grown ovules (R.W. Seagull, personal communication). Hormones affecting fiber initiation are likely to modulate the cytoskeleton; it has recently been shown that lint fiber number can be increased through stabilizing microtubules with taxol (Seagull, 1998).

Coincidentally with fiber initiation on 0 DPA, both sucrose synthase mRNA and protein are upregulated in cotton ovules (Nolte *et al.*, 1995). Sucrose synthase (EC 2.4.1.13) catalyzes the reversible reaction: sucrose <-----> fructose + UDP-glucose, although it often works preferentially in the degradative direction under physiological conditions (Kruger, 1990). Immunolocalization showed that the enzyme amount increased during the earliest stages of fiber initiation, but not in adjacent non-expanding epidermal cells, and that it remained abundant through early stages of fiber elongation. This increase, as well as the beginning of fiber expansion, occurred on the day of flowering with or without pollination occurring (Nolte *et al.*, 1995). Sucrose synthase operating degradatively could have diverse roles in the expanding fiber, including support of respiration, production of osmolytes, and providing UDP-glucose for cell wall biosynthesis (Nolte *et al.*, 1995).

Recent research has indicated that the formation and properties of the vacuole(s) are critical to fiber initiation and elongation. Electron microscopy accomplished by rapid freeze-fixation (which reveals more accurate views of membrane compartments than traditional chemical fixation) has shown that a large central vacuole is created in fiber initials after anthesis. At -1 DPA, differentiated fiber cells that have not yet expanded have only a few small spherical vacuoles (Tiwari and Wilkins, 1995) and a second unique, highly polymorphic, anastomosed vacuolar network (Tiwari and Wilkins, personal communication). The second unique type of vacuole differentiates by -3 DPA only in incipient fibers, not in their neighboring ovule epidermal cells. It contains an electron dense material that depletes from the vacuole between -1 and 3 DPA in normal fibers, but the material never depletes in epidermal cells of the Ligon lintless mutant that lacks long fibers (Kohel *et al.*, 1993). [It should be noted that previous electron microscopic work suggested that this electron dense material was released from the vacuole into the cytoplasm of incipient fibers (Berlin, 1986), but this was likely an artifact of chemical fixation (Wilkins and Tiwari, 1994).] Therefore, this electron dense material could function as an elongation signal or a metabolic precursor supply that is required to be mobilized after fiber cell differentiation to facilitate fiber initiation and early elongation (Tiwari and Wilkins, personal communication).

4.2 Cellular Requirements for Elongation

4.2.1 Generation of Intracellular Turgor Pressure

For fiber elongation to occur by intercalary growth (see section 4.2.2), cell turgor is essential to provide the force that expands the primary cell wall. Generation of turgor is dependent on the influx of water that follows import of osmotically active molecules, predominantly K^+ and malate, into the large central vacuole (Wilkins, 1992). Rate of fiber elongation and fiber length are inversely related to the K^+/malate ratio (Basra and Malik, 1983). The specific activity of malate dehydrogenase, which is required to synthesize malate from oxaloacetate, did not vary greatly over 24 inbred cultivars of *G. hirsutum* varying from 11.4 to 15.5 cm 50% span fiber length. These results suggest that either the activity of this enzyme to synthesize malate is not a limitation to length or that it has already been optimized in commercial cultivars of upland cotton (Kloth, 1992). As a further indication of outstanding questions about the possible regulatory role of this enzyme in determining fiber properties, there are conflicting reports about the relative activity of malate dehydrogenase during primary and secondary wall development including reports of more protein abundance and activity during the secondary wall stage (Ferguson *et al.*, 1996).

From 2 to 3 DPA after the fiber initial has ballooned above the surface, many dispersed provacuoles appear, and these are proposed to fuse to form the large central vacuole that persists for the life of the fiber (Joshi *et al.*, 1988). The small vacuoles have heavier cytochemical deposits indicating ATPase activity (Joshi *et al.*, 1988), suggesting that the extra ATPase activity needed to facilitate import of K^+ and malate activity into the central vacuole is added by fusion of these small vacuoles. Vacuolar H+-ATPases generate an electrochemical gradient to which ion and solute transport is coupled (Joshi *et al.*, 1988; Wilkins, 1992). The H+-ATPases, which are large multi-subunit proteins that are virtually ubiquitous in plant cells, have been the focus of recent cytochemical and molecular biological analyses in cotton fiber. Their function can be regulated at the level of gene transcription and at the level of subunit assembly/disassembly (Wilkins, 1992; Wilkins *et al.*, 1996). ATPase activity has been detected cytochemically in the tonoplasts of expanding cells, but not in adjacent non-expanding epidermal cells or in any epidermal cells of a lintless mutant (Joshi *et al.*, 1988). Positive correlations with fiber expansion rate have been shown for amounts of mRNA corresponding to three H+-ATPase subunits in separated fibers and for levels of H+-ATPase protein and enzyme activity in vacuolar or microsomal membranes from combined ovules and fibers (Wilkins, 1996). These data indicate that there are housekeeping and expansion forms of the H+-ATPase and multiple levels of control of the enzyme activity (translational and post-translational). The genes are organized into two families found in both diploid and tetraploid cotton (Wilkins, 1992). The Ligon lintless-2 mutant lacks the expansion form (Wilkins *et al.*, personal communication).

4.2.2 Cell Wall Synthesis and Expansion

Primary walls are synthesized rapidly in elongating cotton fibers, and cellular activity must increase to support this increased demand for deposition of cellulose microfibrils at newly synthesized plasma membrane and for export of structural matrix polysaccharides and proteins of the primary wall by vesicles of the Golgi apparatus (Haigler,

1985; Bolwell, 1993). Matrix polysaccharides are important in control of elongation, as indicated by the change in organization of methyl-esterified pectin in the primary wall during fiber initiation and the onset of polarized elongation (Wilkins and Tiwari, 1994). The number of Golgi stacks increases at least 3X during fiber initiation (2 DPA) compared to -1 DPA epidermal cells, although the number per unit area actually decreases. Similarly, the number of cisternae per Golgi stack increases from 5 to 6 (in 50% of the stacks) between -1 DPA and 2 DPA (Tiwari and Wilkins, 1995). Seagull (personal communication) also observed an increase in the number of Golgi cisternae per stack. Comparison of the two studies suggests that the number of cisternae per stack decreased between 2 and 8 DPA, only to increase again by 15 DPA; such changes could reflect quantitative or qualitative differences in Golgi function.

The mechanism of primary wall expansion is relevant to the control of fiber length. Two major mechanisms of plant cell wall extension occur: (a) intercalary growth, where new wall material is added all along the cell surface and cellulose microfibrils in the older wall become more net-like to allow surface area expansion; and (b) tip growth, where new wall is synthesized only at the elongating tip and older parts of the wall do not change. The activity of β-glycerophosphatase has been cytochemically localized to expanding parts of fiber walls (Joshi *et al.,* 1985), and activity of covalently-bound cell wall glycosidases shows a close correlation with the rate of fiber elongation activity (Thaker *et al.,* 1987). Similarly, genes coding for endo-β-1,4-glucanase, which could solubilize cell wall xyloglucan to facilitate cell expansion, and expansin, which can loosen cell walls by an unidentified non-hydrolytic mechanism (Cosgrove, 1997), both have elevated expression during primary wall synthesis (Shimizu *et al.,* 1997). These types of proteins are likely to be involved in wall loosening to allow expansion by intercalary growth. However, tip-growth probably also contributes to fiber elongation. Cells engaged in tip vs. intercalary growth have differently oriented cytoskeletons, differently distributed Golgi stacks and vesicles, and different responses to alterations of the cytoskeleton (reviewed in Seagull, 1995; Tiwari and Wilkins, 1995; Wilkins and Tiwari, 1994). Two groups have recently investigated these phenomena in cotton fibers using different ages of fibers. A synthesis of these results follows.

4.3   Phases and Mechanisms of Elongation

4.3.1   Onset of Polarized Eelongation, *e.g.,* 2 DPA, Intercalary Growth Mechanism

Numerous cellular features or responses are indicative of intercalary growth on 2 DPA, although many fibers have already developed a tapered tip. Golgi stacks and vesicles are dispersed along the cell surface and no other organelle zonation is detected. Wall microfibrils are synthesized in

transverse hoops paralleled by microtubules, and there is no actin meshwork at the tips as is found in tip-growing cells. The microtubule antagonist colchicine prevents polarized expansion and elongation of the fiber initials, whereas the actin antagonist cytochalasin D does not (Tiwari and Wilkins, 1995).

4.3.2   Early Elongation Phase, *e.g.,* 3-7 DPA, Dual Intercalary and Tip-Growth Mechanism?

Fibers at this stage of development are in a period of exponentially increasing elongation rate since the maximum rate is reached by 10 to 12 DPA in many fibers (Schubert *et al.,* 1976; Thaker *et al.,* 1986). Golgi stacks and mitochondria are dispersed along the cell surface, except for a void volume within 1 μm of the tip, indicative of intercalary growth. However, Golgi vesicles bind more abundantly at the tip, indicative of tip growth (Seagull, 1995). Cell wall microfibrils are transverse on the innermost wall layer and net-like on the outside (Seagull, 1995; Willison, 1983), indicative of expansion of older wall layers as part of intercalary growth. Microtubules are transverse to the longitudinal cell axis along most of the fiber length, which is consistent with elongation by intercalary growth but not tip growth (Ryser, 1985; Seagull, 1989b, 1992b). Whereas tip-growing cells have randomly oriented microtubules and microfibrils at the tip, cotton fibers have been alternatively reported to match these feature (*G. arboreum* L.; Ryser, 1985)) or to have oriented microtubules and microfibrils at the tip (*G. hirsutum*; Seagull, 1995). Microtubule depolymerizing agents cause fiber swelling along most of the fiber length, but not at the tapering tip (Seagull, 1995), consistent with both intercalary and tip growth occurring regionally.

4.3.3   Intermediate Elongation Phase, *e.g.,* 5 DPA, Dual Intercalary and Tip-Growth Mechanism, Shifting to Predominance of Tip Growth?

Fibers at this stage under typical growing conditions have initiated wall thickening and just passed their most rapid elongation rate, up to 2.4 to 2.9 mm/day, after which the rate will continuously decline for the remainder of the elongation period (Schubert, 1976; Thaker *et al.,* 1986). At 15 DPA, Golgi stacks are still found all along the fiber surface, although they have become more enriched at the tapering tip (by 2-fold) when compared to 8 DPA with no organelle zonation (R.W. Seagull, personal communication). These results are consistent with autoradiography, which indicated growth all along the surface with more intense label incorporation at the tip (Ryser, 1985). This stage of fiber development may correspond to thickening of the primary wall or synthesis of the winding layer, which probably occur sequentially at the primary to secondary wall transition

(Meinert and Delmer, 1977). The winding layer is the first layer of cellulose microfibrils oriented in a steeper helix relative to the longitudinal axis of the fiber, and the microfibrils are also more widely spaced than in the true secondary wall (Seagull, 1992a). It is possible that the synthesis of the winding layer represents a change in patterns of wall reinforcement so that elongation by intercalary growth with its accompanying wall expansion is no longer possible.

4.3.4   Late Elongation, Tip Growth Only?

Older cotton fibers have thick secondary walls that do not allow swelling in the presence of microtubule-altering drugs. They are also notoriously difficult to fix for electron microscopy by chemical methods because of the penetration barrier of the thick secondary wall. Therefore, similar reliable morphological analyses to those described above have not been completed during secondary wall thickening (DeLanghe, 1986), and we can currently only make hypotheses about the mechanism of continuing elongation during that time. (Recently, we have achieved successful fixation of late secondary wall stage, plant-grown, cotton fibers by freeze substitution, and analysis of the results by electron microscopy is in progress, C.H. Haigler and R.W. Seagull and coworkers, unpublished). Once the secondary wall begins to be deposited, further elongation in those zones is unlikely. Therefore, the phase of elongation that continues after secondary wall thickening begins may well occur exclusively by tip growth (Delmer *et al.*, 1992; Meinert and Delmer, 1977). It has been suggested that secondary wall thickening begins in the middle of the fiber (Ryser, 1985), which would leave the tip free to elongate. There is intracellular spatial control over secondary wall deposition within one fiber because the foot of lint fibers extending into the epidermis is not extensively thickened (Berlin, 1986), a factor that also explains the ability of the lint fibers to break off during ginning. The possible exclusive use of tip growth for later stages of fiber elongation is consistent with the increased sensitivity of later stages of elongation (excluding initiation) to cool temperatures (Stewart, 1986).

Final elongation by tip growth leads to some of the longest plant cells known, specifically the primary extraxillary fibers of ramie that can be up to 21.65 inches long (6.5 inches on average). These fibers elongate first by intercalary growth (as the cells expand along with the whole tissue) then finally by only intrusive tip growth, which can continue for months (Fahn, 1990). Since cotton fiber elongation ceases long before cell death, there must be a genotypically-controlled length limit that might be overcome, for example by blocking the signal to end elongation by putative late-stage tip growth. It has been suggested that the extent of late tip-growth may be variable between varieties and represent a key to length determination (Delmer *et al.*, 1992; Meinert and Delmer, 1977). Even in long-staple Pima with a prolonged elongation period, elongation only ended when 90% of fiber weight was attained (Schubert *et*

al., 1976). The power of this genotypically-controlled signal for the end of elongation is illustrated by the fact that, in *in vitro* fibers under 34/22°C cycling, the elongation period is prolonged until the same length is attained as fibers growing at the *in vitro* optimum of constant 34°C (Haigler *et al.*, 1991). The same results have also been obtained in a more prolonged experiment for 34/15°C cycling (C. H. Haigler and J.-Y. Huang, unpublished). These *in vitro* findings parallel results from field studies showing prolonged elongation periods under lower average temperatures (Thaker, 1989). Interestingly, exceptionally long primary phloem fibers are multinucleate (Fahn, 1990) whereas cotton fibers are mononucleate, so the limit to length could arise by several mechanisms.

4.4   Signals to End Elongation

It has been suggested, based on studying 3 cotton cultivars (two *G. hirsutum* and one *G. barbadense*), that a delay in onset of secondary wall deposition may lead to longer fibers (Beasley, 1979), possibly because the dual intercalary/tip-growth mechanisms can operate longer. Very similar results for *G. barbadense* (Schubert *et al.*, 1976) and *G. hirsutum* (Jasdanwala *et al.*, 1977) indicate that about 80% of fiber length was attained before secondary wall thickening commenced between 18 and 21 DPA (as indicated by the beginning of rapid increase in fiber weight/length). Therefore, the onset of the wall thickening phase is correlated with a decrease in elongation rate in currently available cultivars, perhaps because intercalary growth ceases at that time. There is no mutant available lacking all wall thickening to disprove the idea that the onset of wall thickening limits elongation, since even the *imim* immature fiber mutant undergoes substantial wall thickening before the process shuts down prematurely, with rapid elongation ending just as wall thickening starts (Kohel *et al.*, 1974).

Peroxidases can have multiple cellular roles that impact fiber length, including oxidation of free auxin, which is required for elongation (Rama Rao *et al.*, 1982b), mediation of cross-linking leading to wall rigidification that might hinder elongation by intercalary growth (Rama Rao *et al.*, 1982b; Thaker *et al.*, 1986; John and Stewart, 1992), and generation of hydrogen peroxide (via NADH-linked malate dehydrogenase) that could be an intracellular signal for onset of secondary wall deposition (Delmer *et al.*, 1995). As elongation ceases, there is an increase in both cytoplasmic and wall-bound peroxidases (Rama Rao *et al.*, 1982b; Thaker *et al.*, 1986), but we do not know what, if any, causal role these enzymes play in slowing elongation.

Even in short growing seasons, there is usually sufficient time for the apparent elongation potential of current cultivars to be achieved. However, it remains possible that these fibers would become longer if the thickening phase were delayed so that the highest rate elongation (by dual intercalary and tip growth?) could continue for a longer period. Supporting this possibility, ramie primary phloem fibers that elongate both by intercalary and tip growth achieve average lengths of 6.5 inches, whereas secondary phloem fibers that elongate only by tip growth achieve average lengths of only 0.6 inches (Fahn, 1990). A goal to prolong rapid elongation would have to be balanced with allowing sufficient time to achieve a mature fiber before the frost-induced end to the growing season. Near-isogenic lines have been identified that vary by 4 to 5 days in time of onset of secondary wall deposition under the same environmental conditions, and these can be analyzed to determine if fiber length increases and may allow identification of critical "developmental timing" genes (Triplett, 1992).

5.   MATURITY

Secondary wall thickness, or maturity, of the cotton fiber contributes to single fiber strength of dried fibers (see section 4.1) and to greater dyeability. At present, the only routine method of assessing maturity is through the composite index of micronaire (described in section 3), but an independent assessment method that can be applied in routine cotton grading is needed (Bradow *et al.*, 1996). Increased maturity is generally desirable up to the limit when an overly thick wall would limit the fiber flexibility that confers valuable textile properties. The micronaire range of current commercial cultivars falls between 3.5 and 5.0, with the premium price range established for micronaire of 3.6 to 4.2 (Chewning, 1992). An optimal fiber for modern spinning systems would have fineness of 120 to 140 mtex with micronaire between 3.2 and 3.9 (Deussen, 1992).

Fiber thickening is accomplished by deposition of cellulose inside the primary wall so that the secondary wall progressively fills the lumen. This phase begins about 12 to 22 DPA and continues until 35 to 55 DPA depending on cultivar and environment (Schubert *et al.*, 1976, Thaker *et al.*, 1986; Haigler, 1991) (see also below). The mature fiber wall is composed of about 95% cellulose (Meinert and Delmer, 1977; Benedict *et al.*, 1992), and the dilution effect of the cellulose-poor primary wall indicates that the secondary wall must be almost pure cellulose. [The dilution effect of the secondary wall on primary wall properties can be documented by sequential analysis over time of the concentration of Ca^{++} by X-ray fluorescence spectroscopy since Ca^{++} is extensively cross-linked only in the primary wall (Wartelle *et al.*, 1995).] Although the percent protein in the wall decreases dramatically with the onset of wall thickening, available data suggest the possibility of continued deposition of protein in the wall during the thickening phase since the percentage (about 2%) and amount of protein (0.6 ng/mm) did not decrease between 18 and 20 DPA (the last day of analysis) (Meinert and Delmer, 1977). The mRNA coding for a proline-rich protein that may belong to the arabinogalactan protein group has been identified, and it has been hypothesized that the protein could participate in secondary wall assembly (John and Keller, 1995).

The cellulose is in the form of structural microfibrils crystallized from high molecular weight β-1,4 glucan chains in extended, parallel chain conformation (predominantly the cellulose I allomorph). As is typical for higher plants, 60 to 80% of cotton cellulose is in a crystallographic sub-class of cellulose I called cellulose Iβ (Atalla and VanderHart, 1984; O'Sullivan, 1997). The deviation of this value from 100% may, in part, relate to the lower crystalline order of microfibrils in the primary wall (Chanzy *et al.*, 1978, Herbert, 1992). The synthesizing apparatus is localized at the plasma membrane/cell wall interface, and the cellular organization of the enzymes modulates the size and crystallinity of the microfibrils that form. One crystalline microfibril may form from one synthetic complex, or from two or more in a close group since the temporal gap between chain polymerization and crystallization gives chains from closely spaced synthetic sites the potential to mingle before crystallization (Haigler, 1985, 1991, 1992; Blanton and Haigler, 1996). As will be discussed further in the section on fiber strength, the crystallization process is moderated by the density of synthetic sites and the abundance of matrix components. Individual crystalline microfibrils may associate further by hydrogen bonding along their surfaces into larger bundles, and this occurs during fiber primary and secondary wall synthesis so that loose bundles of microfibrils are visible (Willison and Brown, 1977).

5.1   Regulation of the Transition to Fiber Thickening

We are only now obtaining evidence about the signals that regulate the transition to the maturation phase. The signals operate on a seed basis, since onset of fiber secondary wall synthesis correlates both with timing of embryo expansion (DeLanghe, 1986) and with onset of secondary wall deposition in non-fiber ovule epidermal cells (Berlin, 1986; Stewart, 1986). As mentioned earlier, it has been hypothesized that an oxidative burst and associated synthesis of H_2O_2 could regulate the transition (Delmer *et al.*, 1995). Two genes for small GTPase Rac proteins are up-regulated in fibers between 14 and 24 DPA (Delmer *et al.*, 1995). In addition to possible roles in binding actin to mediate the change to a helical microtubule array at the primary to secondary wall transition, Rac proteins can activate NADPH oxidases, resulting in production of H_2O_2 (Delmer *et al.*, 1995). Evidence has recently been obtained that H_2O_2 (a diffusible signal that could have pleiotropic effects in the seed) is produced in cotton fibers coincidently with onset of secondary wall deposition. Furthermore, antagonism of H_2O_2 synthesis or its premature addition have the effects of delaying or prematurely inducing fiber secondary wall synthesis, respectively (T. Potikha *et al.*, 1999). The activity of malate dehydrogenase in fibers correlated negatively (r=-0.47) with micronaire, wall thickness, and maturity in a study of 24 cotton cultivars (Kloth, 1992). Does this result suggest that the higher activity of this enzyme, which is involved in generation of turgor pressure to drive elongation, is associated with delayed onset of secondary wall deposition? Environmental conditions also factor into the signaling system for onset of fiber thickening, since cool night temperatures will cause a delay in its onset (*e.g.*, from 12 to 22 DPA) (Thaker *et al.*, 1986; Haigler *et al.*, 1991).

The developmental signals are translated into expression of particular fiber genes as required to accomplish secondary wall synthesis. The genes may code for unique proteins (or unique isoforms) not required previously in fiber development or mediate up-regulated synthesis of proteins already in the fiber. A tubulin gene has been reported to be up-regulated at the primary to secondary wall transition (Delmer *et al.*, 1992), which correlates with the unique appearance of two β-tubulin isotypes in secondary wall stage fibers (Dixon *et al.*, 1994). A probable cellulose synthase gene is up-regulated between 14 and 17 DPA and remains strongly expressed throughout the duration of fiber development (Pear *et al.*, 1996). The amount of a membrane-bound sucrose synthase also increases during fiber development (Amor *et al.*, 1995) (see section 5.2 for further discussion). Finally, a gene with unknown function, FbL2A, is up-regulated at the transition between primary and secondary wall synthesis (Rinehart *et al.*, 1996).

5.2   Two Phases of Fiber Thickening

The onset of secondary wall deposition is often interpreted as the time when fiber weight per unit length begins to increase. However, detailed analysis of fiber wall composition has indicated that the first period of wall thickening (12 to 16 DPA in one experiment) is accomplished by continued synthesis in the same proportions of all the primary wall components, including cellulose, pectin, xyloglucan, and protein (Meinert and Delmer, 1977). These data are consistent with increasing wall birefingence while the cellulose microfibrils remain transversely oriented in the primary wall pattern (Seagull, 1986). Next, a more cellulose-rich wall begins to be deposited (Meinert and Delmer, 1977), probably corresponding to the winding layer laid down between approximately 16 and 22 DPA (Maltby *et al.*, 1979). True secondary wall deposition and the highest rate of cellulose synthesis commences around 24 DPA after all elongation has ceased (Meinert and Delmer, 1977). An immature cotton fiber mutant (*imim*) has been identified with a recessive, single-locus genetic lesion that prevents the last phase of cellulose synthesis so that cell wall thickening (as seen by electron microscopy, Benedict, personal communication) is initiated but cannot be completed (Benedict *et al.*, 1994, Kohel *et al.*, 1974). Final fiber dry weight in *imim* was 66.7% of the TM-1 parent and crystalline cellulose content was reduced by 77.5% (Benedict *et al.*, 1994).

The biochemical and molecular regulation of cellulose synthesis are at the heart of a full understanding of control of fiber maturity. Research in these areas has just reached a critical stage where new findings should allow rapid future progress (Delmer and Amor *et al.*, 1995; Blanton and Haigler, 1996; Brown *et al.*, 1996; Kawagoe and Delmer, 1997; Delmer, 1999). Evidence has been obtained to suggest that membrane-bound sucrose synthase (see section 4.1 for an introduction to this enzyme) is required to channel UDP-glucose to the cellulose synthase to support the polymerization of cellulose during secondary wall synthesis in cotton fibers (Delmer, 1994; Amor *et al.*, 1995). This hypothesis is consistent with previous experiments based on feeding exogenous substrates including sucrose to cotton fibers (Pillonel *et al.*, 1980; Buchala and Meier, 1985) and with previous evidence that sucrose synthase had a possible role in cell wall biosynthesis (Rollit and Maclachlan, 1974; Chourey *et al.*, 1991). It is also consistent with the demonstration of a functional symplastic continuum passing through plasmodesmata at the fiber foot between the phloem-unloading region of the cottonseed and the fibers (Ryser, 1992; Ruan *et al.*, 1997). Production of UDP-glucose through cleavage of sucrose-by-sucrose synthase rather than through pyrophosphorylase reactions would be energetically and biochemically favorable (Delmer, 1994; Amor *et al.*, 1995; Delmer, 1999). Strong candidates for cellulose synthase genes have also been identified in cotton fibers (Pear *et al.*, 1996). Differential gene expression and/or changes in enzyme activity may account for different rates of cellulose synthesis that occur within the thickening period (Meinert and Delmer, 1977; Pear *et al.*, 1996).

5.3   Relationship to Fiber Absorptivity and Dyeability

The increasing thickness of the wall increases the mass available to absorb water and bind cellulosic dyes, but the increased crystallinity of the secondary wall cellulose (see section 6.3) implies that absorptivity and the binding of dye per unit area of wall is reduced. Crystallite size and water and dye absorption are inversely related, since higher crystallinity decreases the number of accessible glucan chain hydroxyl groups on microfibril surfaces (Rowland and Bertoniere, 1985). Raw cotton has 95% of its pores (spaces between microfibrils and on the surfaces of crystalline microfibrils that can hold water and dyes) with dimensions less than 10 nm, and the larger pores are preferentially eliminated upon processing (scouring) in alkali although total pore volume increases (Zahn, 1988). Mercerization of cotton fiber also results in increased dyeing capacity and absorption because the native cellulose I crystallites are disrupted, then replaced with recrystallized cellulose II with a lower degree of order (Rowland and Bertoniere, 1985).

5.4   Effect of Temperature on Fiber Thickening

Fiber maturity is affected greatly by environment, particularly cool night temperature (Haigler, 1992; El-Zik and Thaxton, 1994; Haigler *et al.*, 1994) because cellulose synthesis is greatly hindered by cool temperatures. Cellulose synthesis in the seed and fiber responds more severely than respiration to cool temperature stress in *G. hirsutum* cv. Acala SJ-1; cellulose synthesis has an apparent $Q_{10} = 4$ between 18 and 28°C, whereas respiration has an apparent $Q_{10} = 3$ (Roberts *et al.*, 1992). Similarly, in *G. arboreum*, cellulose synthesis in the fiber has an apparent $Q_{10} = 6$ between 15 and 25°C, whereas callose synthesis has an apparent $Q_{10} = 2.3$ (Pillonel and Meier, 1985). Many plant enzymatic processes such as respiration and the dark reactions of photosynthesis have $Q_{10} = 2$ to 3 in the physiological temperature range (Fitter and Hay, 1987); therefore, cellulose synthesis appears more temperature sensitive than many other processes.

Fibers maturing under cool night temperatures (*e.g.*, 22°C) have rings in the secondary wall that are revealed in cross-sections subjected to chemical swelling. Each ring indicates a slower period of cellulose deposition. For old and new commercial cotton cultivars, cellulose synthesis at 15°C (a typical night temperature in northerly growing regions) occurs at only 12 to 26% of the control rate at 34°C (a typical day temperature) (Haigler *et al.*, 1994). This diurnal change in rate of cellulose synthesis occurs without any effect of the light/dark cycle (reviewed in Haigler *et al.*, 1991). The rings appear in the fiber wall after swelling because the cellulose synthesized during the cool night swells differently than the cellulose produced during a warm day (DeLanghe, 1986), suggesting an as yet undefined change in the cellulose physical properties. There is no definitive evidence to indicate whether or not the presence of rings reduces individual fiber strength, but bundle strength based on an equal mass of fiber is not reduced (Haigler *et al.*, 1991). The presence of rings does indicate an increased chance of fiber immaturity in a short growing season, since only part of each 24 hour period has been used efficiently to thicken the fiber wall (Gipson, 1986).

The temperature effect observed in field-grown fibers is also observed in ovules with attached fibers cultured with constantly available substrate (glucose, which is preferred by *in vitro* cultures) in incubators under cycling temperatures, indicating that a large part of the cool temperature hindrance of cellulose deposition occurs in the ovule and/or fiber itself (Haigler *et al.*, 1991). Build-up of glucose-6-P at 15°C, which is likely part of a partially adaptive mechanism as observed in other plants (Labate and Leegood, 1989), occurs inside the fibers of ovules cultured on glucose (Haigler *et al.*, 1994). These data also suggest the existence of a major temperature block in the cellulose biosynthesis pathway beyond the formation of glucose-6-P in cultured fibers. The cellulose biosynthetic pathway is likely to represent a point for future improvement in resistance of fiber development

to cool temperature stress, and work is in progress through biochemical and molecular testing methods to identify which steps in the pathway are most cool temperature sensitive (C.H. Haigler and coworkers, unpublished).

The problem of potential fiber immaturity in a short growing season is often made worse by chemical defoliation of cotton plants prior to harvest so that boll development is abruptly ended by photosynthate deprivation (Lewis, 1992). In contrast, fiber developing under excessive heat can develop a wall that is too thick (has high micronaire), leading to excessive rigidity, poor spinability, and discounts in the selling price. This problem could be helped by chemical defoliation at the right time, but there is no reliable way to choose the time (Lewis, 1992), especially given the variability in boll development across the whole plant (Bradow *et al.*, 1996).

5.5   Signals to End Fiber Thickening

The period of fiber thickening is flexible within a genetically constrained range within and between cultivars. For example, cool night temperatures result in a prolonged period of secondary wall thickening (up to 60 DPA if the growing season is long enough) until the apparent genetic potential for maturity of a cultivar is reached (Thaker *et al.*, 1989; Haigler *et al.*, 1991). Since cool night temperatures also prolong the elongation period, it is possible that the ultimate developmental signals are inherent in the fiber, being a mass- or size-based system (Schubert *et al.*, 1976; Basra and Malik, 1983). This speculation is supported by the observation that direct attempts to determine fiber expansion mechanisms by attaching small beads to the surface fail because the fiber stops growing (Seagull, 1995), perhaps because the extra mass is a signal to end fiber development. Also *G. barbadense*, with its prolonged elongation period and longer overlap between elongation and thickening compared to *G. hirsutum*, has finer fibers and thinner secondary walls (Schubert *et al.*, 1976). Finally, we do not know whether a true autolytic event occurs to end fiber development (see discussion in Delmer, 1999) or whether it stops due to fiber dehydration when the boll opens. This question is of obvious importance; if an autolytic event occurs and could be delayed, fiber maturation might be enhanced.

6.   STRENGTH

Fiber strength refers to the ability of the fiber to withstand a load before breaking (Rebenfield, 1990). Fiber bundle strength (as indicated by yarn strength) has a strong genetic component with 71% of observed variability depending on variety (Meredith, 1992a, b). Furthermore, there is evidence from breeding research that one or a few genes are major controllers of strength (Meredith, 1992b). In numerous past studies, fiber strength has been correlated with fiber length, fiber fineness, cellulose molecular weight, microfibril orientation, frequency and distribution of reversals in the microfibril helix, fiber convolution frequency, and fiber surface properties (Meredith, 1992; Triplett, 1993). However, we do not know how the cell controls any of these parameters at the level of gene regulation, enzyme activity, or cellular processes (Lewis, 1992; Lewis and Benedict, 1994). Research is active on the cellular and genetic determinants of fiber strength because further improvement in strength is highly desired by the textile industry, especially to maintain the competitiveness of cotton compared to stronger synthetic fibers such as polyester (McGovern, 1990). This demand is driven by factors such as: (a) more extensive use of high-speed rotor spinning, which demands higher single fiber strength to produce stronger yarns; (b) consumer preferences for fine fabrics without neps arising from tangled and broken fibers; and (c) demand for wrinkle-free cotton fabric, which at present can only be produced through by treatments that weaken the yarn (Arthur, 1990; Deussen, 1992). The discussion below will emphasize secondary wall properties that impact fiber strength.

It is important to clarify whether bundle strength or single fiber strength is being discussed in any particular case; each of these has importance to the textile industry, but high bundle strength and high individual fiber strength reflect at least partially different combinations of fiber characteristics. For example, high bundle strength (breaking force of a known weight of fibers) can be obtained from a bundle of more numerous immature fibers, each of which could be shown to be weak on an individual fiber basis. It is the weak individual fibers that cause the breakage and tangling leading to nep formation. It is also important to clarify which measurement of strength is being used when different studies are compared. Some of the most common strength measurements (Sasser, 1992) and their synonyms are summarized below:

- single fiber breaking force (g)
  - The force required to break a single clamped fiber.
  - synonyms: breaking stress, fiber breaking load, force-to-break
  - typical values: 4-8 g at 1.8 inches (3.2 mm) gage (or clamping) distance
- bundle strength (g/tex)
  - The force required to break a bundle of fibers that is one tex unit in size. (See section 3 for explanation of the tex unit.) Use of this unit allows comparison of strength between varieties with different fiber fineness (Booth, 1968).
  - synonyms: tenacity, breaking strength, tensile strength, specific strength
  - typical values: 25 to 50 g/tex
- breaking extension (%)
  - The extension of the fiber as a percentage of its original length at the breaking point (extension/original length x 100). This unit is useful for comparing strength of different fiber structures, for example single fibers and yarns (Booth, 1968).
  - synonyms: breaking elongation, ultimate elongation, elongation-at-break
  - typical values: 5 to 15%

6.1 Relationship to Stages of Fiber Development

Understanding how each stage of fiber development contributes to strength is critical for future efforts to improve strength by targeted genetic engineering. The strength of the fiber is derived from its cell wall, which becomes thicker during development, and, as will be discussed further below, comes to contain cellulose in higher percentage and with increasing microfibrillar order, molecular weight, fibril size, % crystallinity, and orientation parallel to the longitudinal fiber axis.

Superficially and based on older data (reviewed in Lewis and Benedict, 1994), the thin primary wall and winding layer (about 0.1 to 0.5 μm thick) would seem to be such small mass components of the final fiber wall (up to 6 μm thick) that they might be disregarded in terms of the origin of strength. However, recent work demonstrates that these early stages are important for the development of bundle strength, but not single fiber breaking force.

When micronaire and bundle strength (HVI) were compared in developing fiber hand-harvested from green bolls, it was shown that micronaire development lagged strength development, suggesting that about 50% of bundle strength was determined by the end of the primary to secondary wall transition before extensive fiber wall thickening (Lewis and Benedict, 1994). Similarly, the linear density (dependent on the number and weight of fibers in a 1.5 cm bundle), single fiber breaking force, and bundle strength of 14 DPA fibers were 28%, 16%, and 57%, respectively, of values for mature fibers (Herbert, 1993). Since the number of fibers leading to the linear density values would not change appreciably between 14 DPA and maturity, these results indicate that 57% of the bundle strength had been obtained when the wall was less than 30% of its final thickness (based on 28% linear density value). In contrast, by 19 DPA (when fibers were assumed to still contain mostly primary wall) only 12.5% of the single fiber breaking force of mature fibers had been attained (Tsuji *et al.*, 1992).

In a study where bundle strength and single fiber breaking force were directly compared, bundle strength reached its maximum earlier (by 21 DPA, when cell walls are still quite thin with no obvious secondary wall deposition) than single fiber breaking force (Hseih, 1994; Hseih *et al.*, 1995). The single fiber breaking force of dried fibers increased the most rapidly (from 1.5 to 3g) between 27 and 29 DPA (coincident with the sharpest rise in tex and onset of rapid secondary wall thickening as shown by electron microscopy) and continued to increase more slowly until the end of fiber development (to 4g). (Onset of rapid secondary wall thickening was late in this study due to unknown reasons since greenhouse conditions were not specified.) In the same study, the single fiber breaking force of wet fibers went up sharply after 25 DPA to reach its maximum at 27 DPA (from 2 to 5 g). The different observations for wet and dried fibers may relate to the lower strength of dried fibers, which cannot absorb strain as well as wet fibers because intermicrofibrillar slippage is hindered by more extensive hydrogen bonding (Hearle, 1985).

Therefore, the deposition of the winding layer at the primary to secondary wall transition may contribute most of the single fiber breaking force of wet fibers, with increased thickness of the cellulosic secondary wall (fiber maturity) contributing significantly to single fiber breaking force in dried fibers. (Properties of dried fibers are more important to consider for practical purposes since these are spun into yarn.) This hypothesis is consistent with the fact that bolls on lower branch positions, which were in the shade, had lower single fiber breaking force, probably because they were less mature (Hseih *et al.*, 1995). It is also consistent with a 60.2% reduction in single fiber breaking force of immature secondary walls of the *imim* mutant (Benedict *et al.*, 1994).

6.2   Microfibril Orientation

The orientation of cellulose microfibrils impacts the mechanical deformation properties and strength of the fiber. The orientation of microfibrils relative to the long fiber axis (described by an angle of deviation from that axis) changes throughout fiber development as shown in Fig. 4-1: (a) during primary wall synthesis they are approximately transverse (70° angle); (b) in the winding layer synthesized at the primary to secondary cell wall transition they shift slightly toward longitudinal (45 to 55° angle); and (c) during secondary wall synthesis they become increasingly longitudinal (45° shifting to about 20° final angle) (Arthur, 1990; Rebenfield, 1990). The microfibrils also show reversals at intervals in the handedness of their helical winding around the fiber circumference. Reversals occur up to about 100 times over the length of the fiber, with the absolute number determined by interaction of genotype and environment (Rebenfield, 1990). Finally, when the hollow fiber dries, it twists about its own axis, forming convolutions. The convolutions must arise due to crystalline microfibrils in their native state, since they disappear upon mercerization (Rebenfield, 1990).

The net microfibril orientation in a dried cotton fiber is affected by the convolutions; (Rebenfield, 1990) and is quantitated by average X-ray fibril angle, which is determined from the angle between half maximum intensity of the cellulose 002 diffraction arc arising from an oriented fiber bundle and the equator of the diffraction pattern (Hebert, 1992). Lower fibril angle in cotton correlates with higher elastic modulus (implying less extensibility or extension to break) and higher tenacity. Conversely, higher fibril angle indicates less spring-like extension potential and correlates with lower elastic modulus (implying less extensibility) and

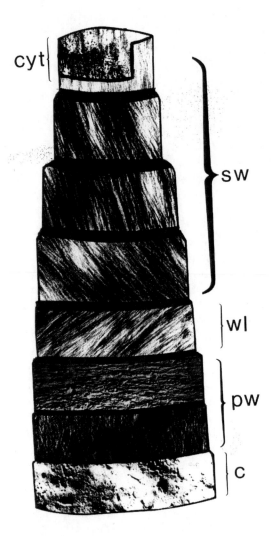

Figure 4-1. Cartoon of the layers of a mature cotton fiber showing direction of cellulose microfibril orientation. Cyt = cytoplasm; sw = secondary wall; wl = winding layer; pw = primary wall; and c = cuticle.

higher tenacity (Rebenfield, 1990). Therefore, fiber maturation is correlated with increasing elastic modulus (less extensibility) and strength. Others have argued that differences in tenacity between cotton varieties are not specifically due to changes in X-ray angle, but rather to differences in the frequency and handedness of the convolutions since the first event upon stretching is straightening of the convolutions (Hearle, 1985). Other physical properties correlated with fibril angle include resilience, work, recovery, and permanent set. Cotton fiber shows the same correlations between fibrillar orientation and strength as man-made fibers, but cotton has the highest orientation (Rebenfield, 1990).

6.2.1   What Determines Microfibril Orientation?

There is substantial evidence that cortical microtubules, which are fibrous cytoskeletal elements underlying the plasma membrane, establish the overall orientation of the microfibrils (and crystallites) in cotton fibers, even paralleling microfibrils through reversal points (Seagull, 1986, 1989b, 1990a, 1992a, b, 1993; Yatsu and Jacks, 1981; Quader, 1987). This finding for cotton fibers is similar to results from many other plant cells (Seagull, 1989a, 1991). The correspondence between microtubule and microfibril angle is close, but not exact (Seagull, 1993), emphasizing that we do not yet understand the mechanism that links microtubule and microfibril orientation (Giddings and Staehelin, 1991; Seagull, 1989b). Nonetheless, the organization of microtubules has a substantial impact on fiber strength. In experiments in which the microtubule antagonist colchicine was injected into developing bolls, fibers with disorganized secondary walls formed near the injection site and these fibers showed reduction in bundle strength (tenacity) to 39.4 g/tex compared to 57.1 g/tex in the control (Yatsu, 1983). Colchicine caused substantial disruption of cotton microtubules with corresponding disruption in the patterns of microfibril deposition (Seagull, 1990a; Yatsu and Jacks, 1981).

Tubulin synthesis and microtubule assembly are regulated dynamically in parallel with cotton fiber development. Among numerous isotypes of tubulin (the protein monomers of microtubules) in cotton fibers, four are unique to fibers with two α-tubulin isotypes and two β-tubulin isotypes appearing uniquely in 10 and 20 DPA fibers, respectively (Dixon *et al.*, 1994). The overall abundance of tubulin also increases about 3 times between 10 and 20 DPA in close approximation to the increasing length of the cell over that time (Kloth, 1989). Between primary wall synthesis (2 to 19 DPA) and secondary wall synthesis (30 to 36 DPA), the number of microtubules (indicated by number of microtubule ends observed in thin sections) increased 4 times and the microtubules also became about 4 times longer (Seagull, 1992b). The existence of more numerous and longer microtubules at the secondary wall stage of fiber development correlates with increased cellulose biosynthesis activity (Meinert and Delmer, 1977).

Furthermore, evidence has been obtained that an undisturbed actin network is required to maintain the microtubules in the transverse orientation during primary wall synthesis; if actin antagonists are applied for brief periods, the microtubules reorient prematurely into an angled helix. Over longer periods of application to relatively young fibers, actin antagonists cause completely randomized microtubules followed by disordered microfibrils. Older fibers are more resistant, indicating that the effect of actin on microtubules may be transient in fiber development (Seagull, 1990a). This possibility correlates with the maximum expression of genes for proteins known to associate with actin (Racs) at the transition between primary and secondary wall synthesis (17 DPA) (Delmer *et al.*, 1995). The actin networks visualized so far do not mirror microtubule orientation, since actin filaments exist as a longitudinal network during secondary wall deposition along with a finer 3-D network of actin (Quader *et al.*, 1987; Seagull, 1990a,

1993). Cytoskeletal organization changes dramatically coincidentally with major shifts in fiber development that are in turn critical determinants of fiber properties; thus, manipulation of this regulation could be a future target for genetic engineering to change fiber properties (Seagull, 1992a).

6.3 Cellulose Crystallite Size and Percent Crystallinity

The size and degree of perfection of cellulose crystals determine the crystallite size and percent crystallinity, respectively. Both cellulose crystallite size and percent crystallinity depend on the extent of aggregation and crystallization of cellulose with itself, factors that ultimately depend on how many cellulose chains are synthesized in close proximity and the abundance of other molecules that could interfere with their co-crystallization (Haigler, 1991). The ribbon-like conformation of β-1,4-linked cellulose predisposes chains synthesized in close proximity and in the absence of interfering molecules to crystallize in parallel orientation because of hydrophobic interactions between the flat faces of pyranose rings that are then stabilized by intra- and intermolecular hydrogen bonds (Nevell and Zeronian, 1985).

One study by freeze fracture electron microscopy of the putative synthetic sites for cotton fiber cellulose microfibrils showed that circles (called rosettes) of 6 intramembrane protein aggregates moderately increased their density at 20 to 22 DPA (0.8 to 1.7 μm^2) compared to 10 DPA (0.05 to 0.2/ μm^2). Another membrane structure that is perhaps related to cellulose synthesis, the microfibril terminal globule that appears in the opposite half of the fractured plasma membrane bilayer from the rosette, showed a greater density of 5 to 10/μm^2 during secondary wall synthesis (Willison, 1983). (For reviews of membrane structures associated with cellulose biosynthesis, see Emons, 1991; Delmer and Amor, 1995; Blanton and Haigler, 1996). However, either of these densities is far lower than in tracheary elements that are rapidly depositing a wall composed of only about 50% cellulose (190 rosettes/μm^2) (Herth, 1985). It is possible that the stresses of sampling and preparing the long cotton fibers for examination by freeze fracture resulted in disappearance of synthetic sites, which are known to be quite labile to cell disturbance (Herth, 1989).

The cotton primary wall contains less than 24% cellulose embedded in a matrix of other polysaccharides and protein (Meinert and Delmer, 1997). Widely spaced rosettes together with the abundant matrix components probably limit microfibril and crystallite size to being less than or equal to the glucan chain bundle produced by one rosette-based synthetic site; very small cellulose microfibrils with low crystallinity have been observed in cotton fiber primary walls (Chanzy *et al.*, 1978). In contrast, during cellulose-rich secondary wall deposition rosettes are likely to be more dense and matrix molecules more sparse (<5% of wall weight; Meinert and Delmer, 1977) so that extensive glucan chain co-crystallization is promoted. However, a layer of β-1,3-glucan between the plasma membrane and the wall into which microfibrils are first spun could still hinder extensive inter-microfibrillar aggregation and co-crystallization without the β-1,3-glucan becoming incorporated into the wall since its molecular conformation is different from cellulose (Waterkeyn, 1981; Haigler, 1992).

The overall ordering of cellulose microfibrils within the wall increases during the secondary wall stage of development (Hsieh *et al.*, 1997; Hu and Hsieh, 1996), a finding that is consistent with the developmental change to a cellulose-rich wall. Cotton cellulose crystallite size (reflecting the dimensions of the perfect crystalline lattice, not including possible poorly-organized surface chains) has been determined to be as high as 5.5 nm by small angle X-ray scattering and X-ray line broadening (Ryser, 1985). In a more recent study of *G. hirsutum* cv. Acala SJ-2, minimum crystallite size in the lattice planes of the microfibril surfaces (L101 and L101) were estimated by the line broadening method to reach 3.5 nm at fiber maturity (Hu and Hsieh, 1997). Using current data on cellulose I crystalline packing (Sugiyama, 1985) and models suggesting that rosettes are associated with the synthesis of 36 glucan chains (Brown *et al.*, 1996), the 3.5 nm crystallite size might be consistent with a crystallite formed by only one rosette-based synthetic site (3.5 nm predicts 30 to 42 chains within the crystallite), but the larger 5.5 nm estimate would require co-crystallization of at least 80 to 100 glucan chains, probably arising from 3 rosettes. More reliable freeze fracture views of cotton fiber membranes engaged in secondary wall synthesis will be required before we know whether there is an apparent biological basis for co-crystallization of chains from multiple, closely-packed rosettes (*e.g.,* regular occurrence of groups of rosettes).

Both crystallite size and percent crystallinity increased between the primary and secondary wall stages of fiber development (Tsuji *et al.*, 1992; Hu and Hsieh, 1996), making a contribution to increased fiber strength. In mature cotton fibers, percent crystallinity is 50 to 60%, a range that was established by 19 DPA in one study and did not change appreciably upon drying (Hirai *et al.*, 1990; Tsuji *et al.*, 1992). [Note that the apparent % crystallinity of cotton fiber cellulose ranges from 58 to 95% depending on the measuring method used (Benedict *et al.*, 1992).] Another study showed that crystallite size and % crystallinity of the cellulose increased rapidly from 21 to 34 DPA (including the first half of secondary wall thickening) then remained approximately constant for the duration of the thickening phase (Hu and Hsieh, 1996). However, since the whole fiber was analyzed, these results could reflect the increasing predominance of the secondary wall material (diluting out the primary wall) rather than a significant change in the character of the cellulose as thickening progresses. Measurements of single fiber breaking force of different ages of fibers between 21 and 28 DPA showed a positive correlation with increasing average crystallite size in *G. hirsutum* cultivars SJ-2 and Maxxa. A less positive correlation exists for overall crystallinity (Hsieh *et al.*, 1997). Therefore, the addition of secondary wall cellulose of higher crystallite size does contribute to higher fiber strength.

6.4   Cellulose Molecular Weight

The molecular weight of the cellulose chains within the microfibrils is also a potential contributor to strength. A non-degradative method was developed for solvating all of the polymers in the cotton fiber wall before molecular weight analysis (Timpa, 1991, 1992). Since the cotton secondary wall is almost pure cellulose, this method gives an accurate measure of molecular weight of secondary wall cellulose and a degree of polymerization (DP) can be calculated for the homopolymer. In contrast, the method establishes an approximate value for the molecular weight of cellulose but a DP cannot be calculated because cellulose is solvated along with the other matrix polysaccharides. These data showed that the molecular weight of cellulose increases from $\leq 1.4 \times 10^5$ daltons at the primary wall stage to about 2.5×10^6 daltons at the secondary wall stage of fiber development, although some higher molecular weight material appears as early as 8 DPA (Timpa and Triplett, 1993). The secondary wall cellulose would then have a calculated DP = 15,000, which corresponds to a chain length of 7.2 µm (Timpa, 1992) or about 1/2 to 1/4 the perimeter of the cotton fiber. (It is important to note that microfibrils are likely longer than 7.2 µm because of the random ends of individual glucan chains within them.) Comparing three cotton cultivars, increasing HVI bundle strength correlated positively with increasing cellulose peak molecular weight (Timpa and Ramey, 1989). Other comparisons indicate that weight average molecular weight accounts for at least 38% of the variability in strength (Timpa, 1992). When eight cultivars were compared in an independent study, HVI bundle strength and the weight average molecular weight of the cellulose in the crystalline regions of the microfibrils were correlated with $r^2 = 0.94$ (Benedict *et al.*, 1992).

How cellulose molecular weight is controlled is unknown, but could relate to the turn-over time of individual cellulose-synthesizing complexes (Delmer, 1990), barring stress effects that might lead to premature chain termination. In a study in which fibers from dryland and irrigated cotton were compared, dryland production (presumably with greater water and temperature stress) caused 18 to 45% decrease in cellulose average molecular weight, but no changes in fiber bundle strength were observed. However, the authors speculated that there might be an unanalyzed difference in individual fiber strength leading to inconsistent textile processing (Timpa and Wanjura, 1989). Similarly, sub-optimal temperatures have been reported to induce more diversity in cellulose molecular weight (Timpa, 1992).

7.   UNIFORMITY

Even on one seed or among seeds of one locule, fiber properties including length, strength, perimeter, and maturity are not uniform (DeLanghe, 1986; Davidonis and Hinojosa, 1994). This most likely reflects the complex metabolic control mechanisms and signals that regulate the activities of each individual cell. There is also evidence that competition for assimilate among seeds and even regions on a seed may be a factor promoting non-uniformity of at least some fiber properties. For example, micronaire of fibers from field plants was inversely related to number of seeds per boll (Stewart, 1986). Similarly, partial defruiting led to an increase of 8% in micronaire and 6% in maturity, whereas partial deleafing and shading led to a decrease in micronaire of 11% (note that temperature would also decrease with shading) (Pettigrew, 1995). However, another study comparing extent of fiber secondary wall deposition across different positions on the same seed and among differently positioned seeds in the same locule both *in planta* and *in vitro* (cultured ovule system) showed that positional diversity in fiber maturity was preserved even when substrate was unlimited *in vitro*, suggesting that competition for photosynthate is not the cause of this diversity. In the same study, positional differences in fiber length distributions observed *in planta* did not exist *in vitro*, suggesting that differences in length are primarily controlled by resource competition *in planta* (Davidonis and Hinojosa, 1994). Fiber properties such as maturity are also greatly affected by adverse weather, soil, plant disease, or pests (Booth, 1968), indicating that growing conditions often do not allow realization of the full genetic potential. It has been inferred from delayed impact of earlier drug treatments of cultured fibers that stress at certain critical stages of fiber development (*e.g.*, 1 to 11 DPA) may result in less uniform or poorer quality fiber at harvest, including diminished secondary wall properties (Davidonis, 1993b).

8.   YIELD

Fiber yield (weight) on one seed is determined by fiber number, fiber length, and fiber maturity. Controls of length and maturity have already been dealt with in this chapter. Increased length of 1/32 of an inch or 0.1 micronaire unit correlate with about 3% and 2% increases, respectively, in yield potential (Lewis, 1992), which is a meaningful difference to a producer paid primarily according to pounds sold. Lint fiber number is determined by the number of fibers that initiate during the optimum period from the day of anthesis until 3 DPA. Usually about one of every 3 to 4 ovule epidermal cells initiates as a lint fiber (Stewart, 1975; Basra and Malik, 1984), leading to 10 to 12,000 fibers/seed in upland cotton (Lewis, 1992). In culture, the period of lint fiber initiation is prolonged until about 6 DPA. Antagonists of microtubules and actin cause a 17 to 33% reduction in fiber number in culture, whereas microtubule stabilization by taxol causes a 10 to 15% increase in numbers of fibers initiated (Seagull, 1998). In this century, yields have been steadily increasing, and previously negative correlations with important fiber properties such as strength have

been broken by cotton breeders (examples can be found in Cooper, 1992; Culp, 1992; El-Zik and Thaxton; 1992, 1994; Gannaway and Dever, 1992; Smith and Golladay, 1994).

9.   ATTEMPTS TO MODIFY FIBER PROPERTIES

Through biotechnology, we now have the potential to relate controlling elements of fiber development to particular physical properties (John and Stewart, 1992; Triplett, 1993). One strategy has been to identify genes that are uniquely expressed or up-regulated during certain phases of fiber elongation and then to manipulate their expression through genetic engineering. As an example, the quantity of E6, a cytoplasmic fiber protein with unknown function and maximum abundance during rapid elongation (5 to 15 DPA), was manipulated through anti-sense technology. The protein abundance was reduced to 1.7% of the wild-type level, but there was no change in fiber development or traditionally-measured fiber properties (John and Crow; 1992; John, 1996). This lack of change could be due to a non-critical role for E6 or the presence of other genes coding for proteins with redundant functions. The lack of effect on fibers is perhaps not surprising since this gene is conserved in fiber-less *Gossypium* species and other plants of the family Malvaceae (John and Crow, 1992). Similarly, up-regulation of auxin and cytokinin synthesis or anti-sense reduction of expression of a gene for a proline-rich protein (H6; John and Keller, 1995) and a gene with an unknown protein product (FbL2A; Rinehart *et al.*, 1996) that is expressed only during secondary wall development had no effect on fiber length, strength, or micronaire (M.E. John, personal communication). Therefore, it is clear that modification of the traditional properties of cotton fiber that currently confer value to the textile industry will require more understanding of the critical control points and the genetic regulation and redundancy at these points. Gaining this understanding represents a substantial challenge for the future, and the necessary research would be greatly facilitated by more efficient cotton transformation to allow more extensive testing of possible control mechanisms by genetic engineering.

10.   ACKNOWLEDGMENTS

Appreciation is extended to D.P. Delmer, Y.L. Hsieh, R.W. Seagull, and T.A. Wilkins for providing preprints of articles prior to publication and to R.W. Seagull for critical reading of the manuscript. The authors research on cotton has been supported by Cotton Incorporated, the Texas Advanced Research and Technology Programs, and the Texas Tech University Institute for Research in Plant Stress.

Chapter 5

GERMINATION AND SEEDLING DEVELOPMENT

Judith M. Bradow[1] and Philip J. Bauer[2]
USDA, ARS, [1]Southern Regional Research Center, New Orleans, La. 70179 and [2]Coastal Plains, Soil, Water, and Plant Conservation Research Center, Florence, S.C. 29502

1. INTRODUCTION

The responses of cotton (*Gossypium* spp.) seeds to the germination environment depend upon (1) the point in the germination-through-emergence sequence at which conditions cease to promote germination and seedling development, (2) the magnitude and duration of the deviations from conditions promotive of germination, and (3) seedling development 'success' potentials determined by the genetic and seed vigor of a particular cotton seed lot and genotype. Suboptimal environmental factors, both abiotic and biotic, modulate, delay, or terminate cotton seed germination and seedling development during any of the four universal phases of seed germination: (1) imbibition, (2) mobilization of seed reserves (cotyledonary lipids and proteins in cotton), (3) radicle protrusion and elongation through resumption of cell division, and (4) hypocotyls and cotyledon emergence above the soil with the shift from metabolic dependence on seed storage compounds to photosynthetic autotrophy. In cotton and other oilseeds, cotyledonary lipid mobilization depends upon subcellular organelle-cooperativity and membrane-transport phenomena elucidated as the gluconeogenic glyoxylate cycle of oilseed species.

Among the environmental factors that affect cotton seed germination and seedling establishment are temperature, water availability, soil conditions such as compaction, rhizosphere gases, seed and seedling pathogens, and interactions among theses and other biotic and abiotic factors that are present in the seed bed and post-emergence micro-environments. This chapter refers to earlier reviews of cotton seed germination and seedling establishment and provides a guide to recent investigations of the two essential physiological processes, seed germination and seedling establishment, that ultimately determine both the yield and the quality of a crop.

2. PHYSIOLOGY OF GERMINATING COTTON SEEDS

Under promotive environmental condtions, the four sequential phases of cotton seed germination and seedling emergence occur during a relatively brief period (*ca.* four to six days) in the physiological progression from fertilized ovule to the mature plant that produces the next crop of seeds and fiber. When a quiescent, but viable, seed is planted (Baskin *et al.,* 1986; Delouche, 1986; Association of Official Seed Analysts, 1988; McCarty and Baskin, 1997), the return of the embryo and the sustaining seed storage tissues to active metabolism is initiated by water imbibition, the first step in seed germination (Ching, 1972; Bewley and Black, 1978; Pradet, 1982; Simon, 1984; Christiansen and Rowland, 1986). However, in 'hard" seeds of some cotton species and varieties, this chalazal pore is plugged with water-insoluble parenchymatous material (Tran and Cavanaugh, 1984). The presence and persistence of the plug can produce 'hardseed' or 'seed-coat' dormancy, a form of dormancy in which there is no or minimal water uptake (Christiansen and Moore, 1959; Benedict, 1984; Christiansen and Rowland, 1986, Delouche *et al.,* 1995). Seed coat impermeability can also be induced in cotton when seed water content is reduced to ≤10% before planting or germination testing (Delouche, 1986; Delouche *et al.,* 1995).

2.1 Early Imbibition

In the presence of adequate water and oxygen, viable, non-dormant cotton seeds, depending on the ambient temperature, require four to six hours for full hydra-

J.McD. Stewart et al. (eds.), *Physiology of Cotton*,
DOI 10.1007/978-90-481-3195-2_5, © Springer Science+Business Media B.V. 2010

tion (Benedict, 1984; Christiansen and Rowland, 1986). Initially, seed rehydration is a consequence of the matric potential (Ψ_m) of the cell walls and cell contents of the seed (Bewley and Black, 1978). Thus, the earliest phase of imbibitional water can occur in both in dead and viable seeds.

Exposure of imbibing Upland cotton seeds to temperatures below 5°C during the initial phase of imbibition results in seedling death (Christiansen, 1967; Christiansen, 1968). Depending on the duration of chilling exposure, imbibitional temperatures below 10°C cause radicle abortions or, in cotton seedlings that survive chilling injury, necrosis of the tap root tip and abnormal lateral root proliferation (Christiansen, 1963). Germination of Pima cotton seeds is inhibited by exposure to temperatures of 5 to 10°C at the beginning of the imbibition period (Buxton *et al.,* 1976). Pima seeds have been reported to be resistant to chilling damage after four hours of warm imbibition.

Significant chilling injury is induced by exposure of cotton seeds to cold water during the initial hours of imbibition. Chilling during earliest seed imbibition has also been associated with increased leakage of solutes from seeds (Simon, 1979, 1984). Both chilling injury and solute leakage reduce seedling vigor and increase seed and seedling susceptibility to pathogens. Both processes are manifestations of events during the first few hours of imbibition, and the severity of both chilling injury and solute leakage may be reduced if the seeds are preconditioned under warm, germination-promotive conditions (Christiansen, 1968; Simon, 1979).

2.2 Post-Imbibitional Periods of Sensitivity to Chilling Temperatures

In viable seeds only, the initial phase of imbibition is followed by an apparent 'lag phase' characterized by reduction in the rate of water uptake, rapid increases in metabolic activity, *e.g.,* protein and mRNA synthesis, and reactivation of preexisting organelles and macromolecules (Ching, 1972; Bewley and Black, 1978; Pradet, 1982; Simon, 1984). Significant water uptake resumes when protrusion of the radicle through the seed coast signals 'true' germination with the concomitant resumption of cell division in embryonic axis coupled with rapid mobilization of seed storage reserves (Ching, 1972; Simon, 1984).

During the germination process, seed respiration rates follow a triphasic curve similar to the cubic rate of imbibition (Ching, 1972; Simon, 1984). The initial period of high respiration overlaps the rapid initial stage of imbibition and the second germination phase characterized by the reactivation of preexisting macromolecules and organelles. The post-imbibitional phase in seed germination represents a 'steady state' for both water uptake and respiration during which preexisting metabolic systems synthesize the substrates needed for biogenesis of new proteins, mRNA, membranes, and organelles. Upland cotton seeds showed increased sensitivity to chilling between 18 and 30 hours after exposure to initial germination temperatures of 31°C

(Christiansen, 1967). A similar period of increased sensitivity to chilling was observed at 28 to 32 hours when Pima seeds were germinated at 35°C and 40 to 56 hours when germination was at 25°C (Buxton *et al.,* 1976). Respiration and water uptake both increase rapidly after radicle protrusion and the resumption of cell division in embryo tissues. These processes occur in Upland cotton seeds after a rehydration/germination period of approximately 48 hours at 30 to 31°C, the temperature considered optimal for cotton seed germination (McCarty and Baskin, 1997). One field study also identified a third, later period of chilling sensitivity at *ca.* 140 to 170 h after planting (Steiner and Jacobsen, 1992). Under the conditions of that study, the third period of sensitivity corresponded to the 'early crook' stage of development for the chilling-stressed seedlings when the hypocotyls were near the soil surface and ready to emerge.

2.3 Glyoxylate Cycle and Storage Lipid Metabolism

In the lipid-storage tissue of cotton cotyledons, mitochondrial respiration is intergrated with glyoxysomal gluconeogensis (Trelease, 1984; Trelease and Doman, 1984). Thus, lipid mobilization during cotton seed germination involves four subcellular compartments, *i.e.,* lipid bodies, glyosomes, mitochondria, and cytosol. As germination and seedling development progress, lipases associated with the lipid bodies liberate fatty acids stored in the cotyledons as triaclyglycerides during seed development. The free fatty acids are transported across the membranes of the lipid body and glyoxysome into the glyoxysomal matrix. Within the single unit membrane of the glyoxysome are the enzymes necessary for the β-oxidation of the fatty acids to acetyl-CoA and the specialized glyoxylate cycle by which the acetyl-CoA is metabolized with the result of net synthesis of succinate within the glyosome (Goodwin and Mercer, 1983; Trelease and Doman, 1984).

Enzymes needed for further metabolism of succinate synthesized during the glyoxylate cycle are not located in the glyoxysomes, and succinate must be transported across the glyoxysomal and mitochondrial membranes for conversion the oxaloacetate in the tricarboxylic acid cycle of mitochondrial respiration (Goodwin and Mercer, 1983; Benedict, 1984; Trelease, 1984; Trelease and Doman, 1984). Additional information on the biochemistry of embryogenesis and the development of glyoxylate cycle organelles and enzymes during seed maturation are discussed in Chapter 25 of this book.

Nearly all lipid reserves in oil-rich seeds like cotton are mobilized *after* radicle protrusion (Trelease and Doman, 1984). Once lipid mobilization is initiated during the imbibition period, lipid utilization is rapidly completed over a relatively brief period during the first week of cotton seedling development (Smith *et al.,* 1974, Trelease and Doman, 1984). The activity of isocitrate lyase (ICL, EC 4.1.3.1), a marker enzyme unique to glyoxysomes and the glyoxylate cycle, peaked after two-days germination in the dark, and ICL activity slowly declined to undetectable levels after eight days (Smith *et al.,* 1974). Exposure to light accelerated the decline in ICL activity, which was no longer detectable in illuminated cotton cotyledons after four days.

Radicle protrusion through the seed coat marks the completion of seed *germination.* Thus, lipid mobilization and the associated membrane transport and organelle cooperativity of the glyoxylate cycle are more precisely treated as seedling *emergence* phenomena. Glyoxylate cycle activity peaks when seedling survival and development are dependent on cotyledonary reserves and photosynthesis is beginning in the greening cotyledons and hypocotyl. When true leaves appear seven to ten days after emergence, cotyledonary lipid reserves have been metabolized and the transiently essential glyoxysome-mitochondria complex has disappeared. However, during the period between radicle protrusion (*ca.* 48 h post-planting) and full photosynthetic autotrophy, cotton seedling growth and emergence depend on the organellar enzymatic complex composed of the mitochondria and glyoxysomes. Studies of temperature and other factors that affect seed germination must, therefore, include consideration of the mechanisms by which environmental factors modulate both the primarily physical phenomena of imbibition and the biochemical and physiological phenomena of metabolic reserve mobilization and seedling development.

3.   TEMPERATURE

Of the many abiotic factors that influence seed germination, seedling emergence, and stand establishment, temperature, which the cotton producer can monitor but not modulate, is the most important (Waddle, 1984; Stichler, 1996; Spears, 1997). Insufficient rainfall before or after planting can be augmented by irrigation, but the producer can neither schedule nor accelerate the rise in spring air temperatures to 16°C (60°F), the temperature below which cotton ceases to grow (Tharp, 1960; Munro, 1987; Edminsten, 1997a). This significant relationship between cotton production practices and temperature is usually quantified by the Degree-Day-60°F (DD-60) or Degree-Day-16°C (DD-16) heat unit frequently cited in cotton production guides and research reports (Bradow and Bauer, 1997; Edminsten, 1997a; 1997b).

3.1   Planting Date Criteria

Each spring, cotton producers in areas where cotton is grown as an annual must select the 'most favorable planting date' based on current and historic mean temperatures from analyses of long-term weather patterns (Waddle, 1985; Norfleet *et al.,* 1997). Several temperature criteria are included in this selection process. During the first five to ten days after planting, soil temperatures <10°C cause

significant chilling injury (Gipson, 1986; Edmisten, 1997c). Therefore, production consultants in the short-season areas of the U.S. Cotton Belt east of the Mississippi River advise growers to delay planting until (1) the *soil* temperature at a depth of 7.6 cm (3 inches)has reached 18°C at 1000 hours standard time, and (2) more than 25 DD-60 heat units are predicted with the temperature to be >10°C for the first two nights after planting (Ferguson, 1991; Edmisten, 1997c). In Califorina, Kerby and coauthors (1989) concluded that cotton should not be planted when fewer than 10 DD-60 heat units per day were expected for the five days after planting. Similar 'rule of thumb' recommendations have been developed for the Texas High Plains and Mid-South regions of the U.S. Cotton Belt (Gipson, 1986). The relationship between yield and the number of degree days below 16°C after planting is seen in Figure 5-1 on which yield date were regressed on the number of days after planting for which the minimum temperature was <60°F (<15.6°C). The regression line plotted in Figure 5-1 was derived from yield data of PD-3 from 1991 through 1994 in South Carolina (Camp *et al.,* 1997).

Planting date selection and risk management techniques based on historical weather records and current meteorological predictions are often inadequate for cotton, a highly temperature-sensitive plant for which both early and late planting reduces yield (Kittock *et al.,* 1987; Bauer and Bradow, 1996; Edminsten, 1997c). Further, producers prefer to plant cotton as early as possible to maximize growing season length and yield, reduce post-emergence insect infestations, and avoid fall storms that lower fiber quality and interfere with crop defoliation and harvest (Edminsten, 1997c, Steiner and Jacobsen, 1992; Bird, 1997). When early planting exposes cotton seeds to cool, wet conditions, germination is delayed or fails to occur and seedlings are damaged or killed (Bird, 1986; Christiansen and Rowland, 1986; Wanjura, 1986; McCarty and Baskin, 1997; Spears,

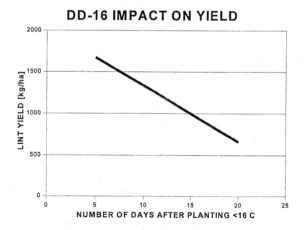

Figure 5-1. Impact of number of days after planting on which minimum temperatures were <60°F (15.6°C) on lint yield of trickle-irrigated 'PD-3' cotton in Florence, S.C. Regression line based on 1991, 1992, 1993, and 1994 treatment means (Camp *et al.,* 1997).

1997). Stand establishment in cool wet soil is often so poor that replanting, which entails significant costs for additional seeds and pesticides, becomes necessary (Edmisten, 1997a). There are no universal guidelines for cotton replanting decisions (Jones and Wells, 1997). The advantages of a more uniform stand must be balanced by the reduced maturity of late-planted cotton, and no guarantees for the uniform emergence of a second planting.

3.2 Seed Germination and Seed Vigor Testing

Growers regularly attempt to compensate for uncertain and suboptimal planting conditions by sowing extra seeds and/or using high quality, *i.e*, high vigor, seed lots (Kerby *et al.*, 1989; Bird, 1997). High vigor seeds emerge faster (Quinsenberry and Gipson, 1974) and from lower soil depths (Wanjura *et al.*, 1967). The potential of a seed to germinate or a seedling to survive in a chilling environment depends on the vigor of the seed (Bird, 1986; 1997; Spears, 1997). Thus, the well-characterized sensitivity of cotton seeds to chilling stress has become the basis for the 'cool germination test' (CGT) of cotton seed vigor (Smith and Varvil, 1984; McCarty and Baskin, 1997; Spears, 1997; Tolliver *et al.*, 1997).

The CGT is conducted in addition to the standard germination test (SGT), which is performed under temperature and moisture conditions that are highly favorable for germination and seedling development (McCarty and Baskin, 1997; Spears, 1997; Tolliver *et al.*, 1997). In both the SGT and CGT, cotton seeds are planted on special moistened towels scrolled to hold the seeds in place and to reduce drying (McCarty and Baskin, 1997). In the SGT, the scrolls are placed in a germinator operated at constant 30°C or in a 20/30°C cycle (16 h at 20°C and 8 h at 30°C). The SGT consists of four replicates of 50 or 100 seeds each.

The first SGT evaluation is made four days after planting. The scrolls are unscrolled, the normal seedlings are counted and removed from the towels, and the count data are recorded. The towels containing the ungerminated seeds are rescrolled and returned to the germinator. The second SGT evaluation is made eight days after planting. If no additional *normal* seedlings have developed, the SGT is terminated. If the SGT is not terminated at eight days, the towels are rescrolled and returned to the germinator until the final SGT evaluations are combined, and the 'standard' germination percentage is calculated and printed on the tag attached to the bag of seeds (seed lot) from which the SGT subsamples were drawn.

Cotton seeds tested under the SGT protocol are tested under laboratory conditions that are much more promotive of seed germination than normal field conditions are. In the CGT, seed scrolls and four replicates of 50 seeds each are also used; but the CGT germinator is held at a constant 18°C (Smith and Varvil, 1984; McCarty and Baskin, 1997). The contents of chilled CGT scrolls are evaluated once at seven days after planting. Only strong, vigorous seedlings that reach a combined root-hypocotyl length of 4.0 cm (1.5 in.) are counted. The percentage of high-vigor seedlings is calculated and reported as 'percentage cool test germination.'

There remains considerable variability among laboratories performing the CGT, and CGT results are not printed on the official seed-lot tag (Tolliver *et al.*, 1997). Further, the CGT ratings do not predict the success of either field germination or stand establishment (Spears, 1997). However, producers should obtain CGT information from seed dealers since there is a significant difference between the vigor levels of cotton seed lots with 85 percent and 60 percent CGT ratings. A seed lot with a high CGT percentage is more likely to perform well under chilling conditions, *i.e.*, at 18±1°C, the temperature of the CGT protocol (Smith and Varvil, 1984; Bird, 1997). Indeed, the *minimum* recommended soil temperature for planting cotton is the temperature used in the CGT (McCarty and Baskin, 1997). When field temperature are similar to those used in the SGT, lower vigor seed lots will usually perform as well as high vigor seed lots.

3.3 Seed Vigor and Emergence Prediction

Incubation periods in the SGT and CGT protocols approximate the lengths of time required for completion of the imbibition-to-emergence sequence of seed germination and seedling establishment. Thus, SGT and CGT data should be useful for predicting cotton seedling emergence (Buxton *et al.*, 1977; Smith and Varvil, 1984; Kerby *et al.*, 1989). However, seed vigor, quantified as the ratio of seedling axis weight to total seedling weight (percent transfer), was a poor predictor of field emergence (Buxton *et al.*, 1977). Further, combining percent transfer with percent germination into a 'germination index' (Buxton *et al.*, 1997) did not consistently improve emergence prediction, compared to predictions based on percent germination alone.

Combining SGT and CGT percentages improved the prediction of cotton seedling emergence under fluctuating soil temperature conditions. Kerby and coauthors (1989) used multiple regression analyses of (SGT% + CGT%) and DD-60 heat unit accumulations at ten days after planting to explain >64% of the variation in cotton seedling emergence. When interaction terms from the multiple regressions were included in the predictive equations, the combination of (SGT% + CGT%) DD-60 heat units at five days after planting, and the interaction term explained >68% of the variation of seed vigor and potential yield capacity (Wanjura *et al.*, 1969).

More recently, Steiner and Jacobsen (1992) examined the effects of planting time of day and cool-temperature stress from naturally varying soil temperatures by following seedling emergence and seedling rate of development. Final seedling emergence percentages were not affected by planting-day heat units, although seeds planted in the morning (0800 hours) were more sensitive to soil temperature than were seeds planted in the afternoon (1600 hours). There was no defined relationship between emergence and seedling rate of development. However, genotype-related differences in cotton response to cool soil conditions at planting were reported. The authors suggested that high levels of cotton emergence under cool soil conditions could be achieved if chilling-tolerant genotypes were selected and planting was done in the afternoon to take advantage of diurnal warming of the seed-zone environment.

An examination of the relationships between seed quality (vigor) and preconditioning at 50°C and 100% relative humidity indicated that warm, moist preconditioning decreased cotton seed resistance to chilling (Bird, 1997). Pre-conditioning for ≤2.5 d increased germination percentages at 18°C, but field emergence was reduced by seed preconditioning. The reduced stands obtained from preconditioned seeds were associated with infection by pathogenic (damping-off) fungi.

3.4 Temperature and Post-Emergent Cotton Seedling Physiology

Most studies of responses of cotton seeds and emerging seedlings to chilling temperatures have concentrated on germination *per se* (Christiansen, 1963; 1967; 1968; Guinn, 1971; Cole and Wheeler, 1974; Buxton *et al.,* 1976) or low temperature effects upon seedling emergence (Pearson *et al.,* 1970; Wanjura and Buxton, 1972a, 1972b; Fowler, 1979; Wanjura and Minton, 1981). Relatively little is know about the effects of suboptimal temperatures on photosynthetic seedlings that have successfully emerged from the soil (Bradow, 1990a). Even less is know about seedling physiology during recovery from exposure to suboptimal temperatures (Clowes and Stewart, 1967; Bagnall *et al.,* 1983; Bradow, 1990a; 1990b). A few cotton genotypes have been screened for chilling resistance (Anderson, 1971; Krieg and Carroll, 1978; Bradow, 1991; Bauer and Bradow, 1996; Schulze *et al.,* 1997), but most such studies have concentrated on germination and heterotrophic seedling growth.

Without a simple assay for suboptimal-temperature sensitivity in photosynthetic seedlings, growers have relied on personal experience and anecdotal information when selecting cotton genotypes that might survive the effects of cold weather fronts which arrive after seedlings have emerged from the soil. This problem was addressed by adapting the SGT protocol to provide uniform seedling populations 48 hours after planting (Bradow, 1990a, 1991). Using a modified CGT protocol, morphologically homogeneous populations of photosynthetic seedlings of three cotton genotypes were exposed to light (14-h day) and growth temperatures of 10, 15, 20, 30, or 35°C for seven days (ten days total from imbibition to seedling harvest and evaluation) (Bradow, 1991). The effects of suboptimal temperatures on seedling root and shoot growth were evaluated by separate measurements of root and shoot lengths, fresh weight, dry weights,

and relative water contents (Weatherley, 1950; Bradow, 1991). Significant genotype differences in seed vigor, inhibition of root and shoot elongations, fresh weight, and relative water contents were reported (Bradow, 1991).

The most marked differences between genotypes were in the capacities to resume growth, and to reestablish normal root and shoot water relations, after a five-day exposure to suboptimal temperatures (≤20° for roots and ≤25°C for shoots) (Bradow, 1991). The capacities of cotton seedling roots and shoots for recovering rapidly and fully from exposure to moderate suboptimal temperatures (>15°C) can be a strong determinant of stand establishment and subsequent yields in years in which post-emergent chilling occurs (Kittock *et al.,* 1987, Bauer and Bradow, 1996). Genotype recovery capacity in seedlings exposed for five days to temperature below 30°C could be gauged by changes in root relative water contents after a 48-h 'recover' period at 30°C (Fig. 5-2).

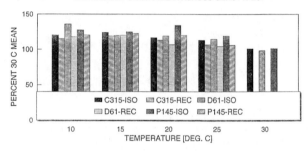

ROOT RELATIVE WATER CONTENTS
NORMALIZED ON 30 C MEAN ACROSS GENOTYPES

Figure 5-2. Root relative water contents of 10-day seedling roots of 'Coker 315' (C315), 'DPL 61' (D61), and 'Paymaster 145' (P145) exposed to 10, 15, 20, 25, or 30°D for seven days before evaluation (ISO treatment) or to 10, 15, 20, 25, or 30°C for five days, followed by two-day recovery at 30°C (REC treatment). Treatment relative water content percentages were normalized on the isothermal 30°C mean across all genotypes (adapted from Bradow, 1991).

Moderate chilling stress alters the water relations of both roots of shoots of photosynthesizing cotton seedlings by decreasing the shoot water content and increasing root water content in a cultivar-specfic manner (Bradow, 1991). However, the changes in root and shoot steady-state water relationships after a chilling-stress period differ from the shoot dehydration [wilting] observed under cold-shock (≤10°C) conditions (Bradow, 1990a). Recovery from chilling is related to the capacity of a seedling to resume normal water translocation from the roots and the capacity of shoot tissue to rehydrate upon restoration to growth-promoting temperatures (Fig. 5-3). At temperatures < 30°C, the root relative water contents of the three genotypes in Figure 5-2 increased after 48 h at 30°C, seedling root relative water contents of the three genotypes were lower (Fig. 5-2) and the shoot relative water contents of chilled 'Coker 315" seedlings were higher (Fig. 5-3). An increase in shoot

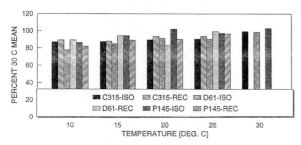

Figure 5-3. Shoot relative water contents of 'Coker 3115' (C315), 'DPL 61' (D61), and 'Paymaster 145' (P145) 10-day seedlings exposed to 10, 15, 20, 25, or 30°C for seven days before evaluation (ISO treatment) or to 10, 15, 20, 25, or 30°C for five days, followed by two days at 30°C (REC treatment). The treatment relative water content percentages were normalized on the isothermal 30°C mean across all genotypes (adapted from Bradow, 1991).

relative water content after a 48-h recovery period was also observed in 'DPL 61' seedlings that had previously been chilled at 10, 15, and 25°C. No post-chilling shoot rehydration was observed in 'PM 145' seedlings. Instead, 'PM 145' shoot relative water contents decreased during the 48-h recovery period (Bradow, 1991).

Seedling root elongation studies of 'DPL 20', 'PL 50', 'PL 5690', and 'DPL Acala 90' revealed genotype differences in response to temperatures <30°C in a controlled environment (Bauer and Bradow, 1996). 'DPL 20' root elongation was most inhibited by moderate chilling, and in the year in which planting was followed by cold weather, 'DPL 20' lint yields were lower than the yields of the three genotypes that were less sensitive to chilling stress.

3.5 Early-Season Heat Unit Accumulations, Yield, and Fiber Quality

Temperatures and heat unit accumulations during the first 50 days after planting (DAP) affect both fiber yield (Camp *et al.,* 1997) and fiber maturity (Bradow and Bauer, 1997). The impact of the number of heat units during the first 50 DAP is seen in the regression of lint yield for four years on the cumulative heat units at 50 DAP in those years (Fig. 5-4).

The effects of temperature, treated as heat unit accumulation, are equally clear in the relationship of fiber maturity to DD-16 accumulations (Bradow and Bauer, 1997). In Figure 5-5 is described the dependence of Immature Fiber Fraction (IFF) (Bradow *et al.,* 1996a) on DD-16 heat unit accumulations during the first 50 DAP in a two-year planting-date study of 'DPL 20', 'DPL 50', 'DPL 5690', and 'DPL Acala 90' grown in Florence, South Carolina. Overall, 1991 was the shorter, hotter, drier crop year, but

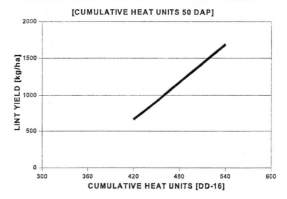

Figure 5-4. Impact of the number of heat units during the first 50 days after planting on lint yield in trickle irrigated cotton at Florence, S.C. The genotype was PD-3, and heat units were calculated as $\sum$ ((daily high temperature + daily low temperature)·0.5) - 15.6°C.

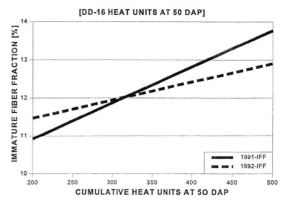

Figure 5-5. Impact of temperature during first 50 days fter planting on fiber maturity. Fiber maturity was quantified as Immature Fiber Fraction (Bradow *et al.,* 1996), and data were pooled across four genotypes within years (1991 and 1992).

the differences in the 1991 and 1992 thermal environments derived mainly from the higher spring temperatures during the first 50 DAP in 1991. When data were pooled across genotypes and IFF data were regressed on DD-16, the 1991 maturation rate (rate of decreasing IFF) was 1.6 times higher than the maturation rate in 1992, the cooler year. The same 1991:1992 maturation rate ration (1991 rate= 1.6 x 1992 rate) was found when fiber maturity was quantified as micronaire (Bradow and Bauer, 1997).

Temperature modulated and controls cotton seed germination, stand establishment, and post-emergence seedling growth and development. The effects of suboptimal temperatures extend beyond mere decrease in seed germination and stand percentages to significant reductions in the yield from cotton plants that survive exposure to chilling temperatures and consequential decrease in fiber quality, particularly fiber maturity. Indeed, just as soil and air temperatures before planting can be used to estimate the probabilities of germination success, temperatures (as DD-60 or DD-16 heat unit accumulations) during the first 50 days after planting are valid indicators of yield, fiber quality and, therefore, the value of the cotton crop to both producers and processors.

4.   WATER, LIGHT AND COTTON SEEDLING PHYSIOLOGY

4.1   Soil Moisture and Planting Date

Selection of the optimum planting date can depend as much upon the quantity and timing of the water supply as upon soil and air temperatures (Munro, 1987). In cotton-growing regions where the rainfall pattern is unimodal (a single wet period lasting up to six months and followed by a significantly drier period during the rest of the growing season), the best yields are obtained from cotton sown as early as possible in the wet period. In temperate cotton-growing areas, planting date selection must be governed by both water supply and temperature to minimize exposure of seeds and seedlings to cool, wet soil. When soil moisture content was at field capacity (-0.033 MPa), Upland and Pima genotypes attained 50% emergence 36 h earlier when soil temperatures were 31 to 36°C than when soil temperature ranges were 26 to 29°C or 32 to 45°C (Gonzales *et al.,* 1979). Low soil moisture and elevated temperatures (32 to 45°C) either prevented seedling emergence or increased the time required for 50% emergence by more than 100 hours.

Generally, *ca.* 50 mm of soil-penetrating rainfall should be recorded before cotton is planted (Munro, 1987). High initial water contents in clay and sandy soils accelerated cotton emergence and counteracted the negative effects of soil compaction or high soil impedance (Gemtos and Lellis, 1997). High soil physical impedance (1.12 to 3.36 kg cm^{-1}) reduced both hypocotyl and radicle elongation in cotton seedlings grown at 32°C and -0.03 MPa (Wanjura and Buxton, 1972a). Decreased hypocotyl elongation was consistently noted with lower soil moisture contents, but decreasing soil moisture increased root elongation as roots sought water at lower depths.

4.2   Light and Cotton Seeds and Seedlings

Seeds of commercial cotton genotypes do not require light for germination and the far-red/red (FR/R) light response is not a factor in cottonseed germination. Thus, depending on soil type and condition, a sowing depth of 2 to 4 cm is recommended (Munro, 1987). Shallow sowing decreases the amount of moisture available to the germinating seeds. Deeper sowing increases soil impedance and reduces

seedling emergence. In a good stand of cotton seedlings the cotyledons are 4 to 7 cm above the soil surface.

Unlike the germinating seed, cotton seedlings are morphologically responsive to FR/R light ratios from the time of emergence (Kasperbauer and Hunt, 1992; Kasperbauer, 1994). Cotton seedlings were highly sensitive to FR/R ratios at the end of the day, and the responses were photoreversible so that the plants responded to the color received last (Kasperbauer and Hunt, 1992). Seedlings that received a high FR/R ratio last on each day developed lower specific weights, longer and heavier stems, less massive roots, and higher shoot/root biomass ratios. When the responses of cotton seedlings to FR/R ratios reflected from the soil surface were characterized by growing the plants over red, green, or white soil covers (mulches) in the field, stems of the five-week-old seedlings grown in sunlight over green or red surfaces were 130% longer than the stems of similar seedlings grown over white surfaces (Kasperbauer, 1994).

5   IMPACTS OF TILLAGE AND OTHER SOIL FACTORS ON COTTON SEEDLING EMERGENCE

5.1   Soil Impedance, Crusting, and Compaction

Soil physical impedance inhibits the elongation of both radicles and hypocotyls (Wanjura and Buxton, 1972a, Wanjura, 1973). Soil compaction and crusting are also important factors in cotton seedling emergence (Bennett *et al.,* 1964; Wanjura and Buxton, 1972a, Wanjura, 1973; Munro, 1987; Tupper, 1995). Emerging cotton hypocotyls are particularly sensitive to soil-surface crusting (Bennett *et al.,* 1964; Wanjura, 1973; Stichler, 1996). In addition to soil 'caps' formed when the surface dries, soil crusting due to the impact of rain or aerial irrigation water droplets mechanically impedes the emergence of seedlings (Arndt, 1965; Munro, 1987). When the soil crust forms a seal or cap above the germinating seed, the U-bend of the emerging hypocotyl may fracture before the emerging cotyledons are pulled free of the soil (Munro, 1987). If the soil type or prevailing weather conditions at planting increase the probability of surface crusting, producers are advised to sow seeds more thickly since the combined emergence pressure of three or four closely-spaced cotton seedlings may break through a soil cap that a single seedling could not penetrate. Soil sealing and crusting also increase runoff and erosion; and cover-crop residue management practices or soil-surface treatments have been developed to reduce soil-crust formation and the related inhibitions of seedling emergence (Bradford *et al.,* 1988; Pikul and Zuzel, 1994; Zhang and Miller, 1996).

5.2 Impacts of Tillage Methods and Equipment Traffic on Soil Aeration and Seedling Growth

Inhibition of seedling radicle elongation by soil impedance and the related slowing of emergence and early-season growth (Wanjura and Buxton, 1972a; Burmester *et al.,* 1995) have been related to tillage methods (Burmester *et al.,* 1995; Busscher and Bauer, 1995), equipment traffic patterns (Khalilian *et al.,* 1995; Gemtos and Lellis, 1997), and soil strength differences among soil horizons (Box and Langdale, 1984; Busscher and Bauer, 1995). The growth of seedling radicles is also restricted by soil compaction or 'sealing' that results when knife openers are used on overly moist soils (Stichler, 1996). When cotton seedlings in conservation-tillage plots were dug up two weeks after emergence, many plants had developed lateral taproots that ran along the top of a compacted zone 5 to 10 cm below the soil surface (Burmester *et al.,* 1995). Cotton growth and yield depend on efficient root penetration to reach nutrients and water in the subsoil horizons. Therefore, equipment traffic patterns must be controlled to limit soil compaction in the root zone and to reduce the formation of the high-impedance subsoil zones associated with some equipment and cultural methods. Use of some seed planters and tillage methods may lead to hypoxia (<3.5 mmol $O_2 \cdot mol^{-1}$) or increases in soil CO_2 levels (>20 mmol $CO_2 \cdot mol^{-1}$) that result in inhibition of cotton root elongation (Leonard and Pinckard, 1946; Minaei and Coble, 1989; 1990).

Cotton seed germination is also affected by the O_2:CO_2 ratios. Germination was suppressed at concentrations <5 mmol $O_2 \cdot mol^{-1}$, regardless of CO_2 concentration (Minaei and Coble, 1990). Higher concentrations (20 to 30 mmol $CO_2 \cdot mol^{-1}$) promoted germination but inhibited root elongation. Higher oxygen levels, combined with CO_2 concentrations of 1.5 to 2 mmol $CO_2 \cdot mol^{-1}$, promoted radicle elongation.

The low tolerance of cotton roots for poor soil aeration is also a factor in reduced seedling emergence and root elongation associated with flooding and water-logging of the soil (Jackson *et al.,* 1982; Hodgson and Chan, 1982; Lehle *et al.,* 1991). Ethanolic fermentation was induced immediately after introduction of imbibed cottonseeds to N_2 or CO_2 atmospheres at 28°C (Lehle *et al.,* 1991). During a 2-h hypoxic stress period, cotton radicle elongation was briefly halted before growth resumed at a reduced rate. Upon restoration of aerobic conditions, radicle growth recovered fully; and ethanol produced by hypoxic fermentation was assimilated rapidly once hypoxic stress was relieved.

5.3 Conservation Tillage and Cotton Seedling Growth

Conservation tillage systems represent one of most common methods for reduction of soil erosion on cotton acreage (Valco and McClelland, 1995). However, slower early-season growth has been reported for cotton planted into no-till cotton or no-till wheat residues (Burmester *et al.,* 1995). In comparison to conventional tillage, use of conservation tillage also reduced cotton stand populations at four weeks after planting (Colyer and Vernon, 1993). Other researchers have also reported reduced plant populations under reduced tillage (Brown *et al.,* 1985; Rickerl *et al.,* 1984; 1986; Chambers, 1990).

Conservation-tillage and the sometimes negative effects of cover crop residues on cotton stand establishment have been attributed to increased seedling disease (Rickerl *et al.,* 1988; Chambers, 1995b); ammonia toxicity (Megie *et al.,* 1967), seedling growth inhibition by volatile organic compounds released by cover crop residues decomposing in the root zone (Bradow and Connick, 1988; Bradow and Bauer, 1992; Bradow, 1993; Bauer and Bradow, 1993), and seedbed moisture depletion by the cover crop (Bauer and Bradow, 1993). When 'Coker 315' seedlings were grown for two weeks in soil collected immediately after zero (Day 0) or seven days (Day 7) after cover crop (fallow weeds or crimson clover) residue incorporation, cotton seedling root elongation was inhibited 50%, compared to root growth in a sterile control soil of similar impedance (Bradow and Bauer, 1992; Bauer and Bradow, 1993). The Day 0 soil samples containing decomposing residues also inhibited shoot elongation (>25%) and cotyledon expansion (>20%). The inhibitory activity of decomposing plant residues was increased by soil crusting but disappeared with time (*ca.* 14 days after residue incorporation). Seedling growth inhibition by cover crop volatile emissions was minimized when cotton planting was delayed two weeks after residue incorporation or until no recognizable plant residues remained in the soil.

6. SEEDLING DISEASE COMPLEX AND OTHER BIOTIC STRESS FACTORS THAT AFFECT COTTON SEEDLING EMERGENCE

6.1 Seedling Disease and Seedbed Environment

Conservation tillage, specifically no-till, increases the severity of seedling disease, particularly in early plantings (Chambers, 1995b). Even when warm, dry soil conditions are present at planting, stand counts and plant vigor were lower in no-till than in conventional tillage. Seedling disease complex, which causes an estimated 2.8 to 5% annual loss, is the most important disease problem of cotton in the United States (Rothrock, 1996; Bailey and Koenning, 1997). For example, in controlled-environment studies of cotton black root rot caused by *Thielaviopsis basicola,* the weights of infected seedlings were reduced 22 to 31% at 20°C and 13 to 19% at 24°C. The effects of seedling infection persisted as reductions in yield. Crop rotation or summer flooding reduced *T. basicola* frequency in areas where *T. basicola* was found in 100% of fields planted to continuous cotton (Holtz *et al.,* 1994). However, rotation with peanut did not reduce the severity of seedling disease caused by *Rhizoctonia solani* (Sumner, 1995).

Seedling diseases are caused by several soil-borne fungi that thrive under cool, wet conditions and seem more prevalent in sandy, low-organic matter soils (Ferguson, 1991; Bailey and Koenning, 1997). Other factors that increase seedling disease frequencies are planting too deep, poor seedbed conditions, compacted soil, nematode infestation, and misuse of soil-applied herbicides such as the dinitroanilines. The primary agents of seedling disease are the fungi *Rhizoctonia solani, Pythium* spp., and *Fusarium* spp. The same fungi may cause seed decay, seedling root rot, or both. *Pythium* and *Fusarium* spp. usually attack the seed and below-ground parts of the young seedlings. *Rhizoctonia* spp. infections appear as reddish brown, sunken lesions at or below ground level ('sore-shin') and may occur at anytime from emergence until the seedlings are six inches tall.

The control of seed and seedling diseases is preventive, rather than remedial (Garber, 1994; Bailey and Koenning, 1997). Fungicides (as seed treatments, in-furrow applications and hopper-box treatments) are the primary control program components. In most years, seed treatment fungicides are sufficient unless the seed is of low quality or the weather is unfavorable for germination. In some years, protection from the seed-applied treatment may not persist until the cotton seedlings have grown to a stage of lower disease susceptibility; and in-furrow fungicide application is recommended for early plantings or when cool, wet weather is expected after planting (McLean *et al.,* 1994; Bailey and Koenning, 1997). The hopper-box method is less expensive and less effective than the in-furrow application.

Cultural practices that reduce the severity of seedling diseases are use of high-quality seed that grows more rapidly and produces more vigorous seedlings, delay of planting until the soil has reached 18°C at a three-inch depth; crop rotation; proper fertilization and liming to promote seedling growth, avoidance of excess rates and deep incorporation of herbicides, early cutting and shredding of stalks to reduce the level of inoculum carried over from one year to the next, use of raised beds for early plantings and control of nematodes through crop rotation, planting resistant

genotypes, and use of nematicides (Bailey and Koenning, 1997). Recently, several biological fungicides have been developed for use with cotton (EI-Zik *et al.,* 1993; Bauske *et al.,* 1994; Brannen and Backman, 1994a; Howell, 1994; Kenney and Arthur, 1994; Howell and Stipanovic, 1995). The effectiveness of biocontrol fungicides, however, can be unpredictable and sometimes significantly lower than the commercial seed-treatment fungicides (Davis *et al.,* 1994).

Although the weather cannot be controlled, cotton producers can and should choose those cultural practices that create the most favorable conditions for seedling germination and emergence (Garber, 1994). Choice of planting dates, seed bed depths, and cultural practices that create an environment favorable for 'friendly' microorganisms antagonistic to seedling pathogens allow the producer to modulate the temperature-related factors so that a good stand of vigorous cotton seedlings is obtained under suboptimal and inhibitory conditions.

7.   SUMMARY

Cottonseed germination and seedling development are highly sensitive to the environment at planting and for several weeks after that. The environmental impacts on germinating cotton seeds and emerging cotton seedlings depend on the point during the germination-through-emergence sequence at which conditions cease to promote germination and seedling development and the magnitude and duration of the deviations from conditions promotive of germination. Further, the stand establishment 'success' potential of a particular cotton seed lot and genotype is determined by genetic factors and seed vigor of that seed lot or genotype. Suboptimal environmental factors, both abiotic and biotic, modulate, delay, or may terminate cotton seed germination and seedling development at any point from seed imbibition to photosynthetic autotrophy. Among the environmental factors that affect cotton seed germination and seedling establishment are temperature, water availability, soil conditions such as compaction, rhizosphere gases, seed and seedling pathogens, and interactions among these and other biotic and abiotic factors that are present in the seed bed and post-emergence micro-environments. This chapter provides a guide the impacts of the seedbed environment on two essential physiological processes, seed germination and seedling establishment, that ultimately determine both the yield and the quality of a cotton crop.

Chapter 6

GROWTH AND DEVELOPMENT OF ROOT SYSTEMS

B.L. McMichael, D.M. Oosterhuis, J.C. Zak, and C.A. Beyrouty

1. INTRODUCTION

The importance of the root system to productivity has been acknowledged for nearly a century (Weaver, 1926). It is common knowledge that roots serve as an anchor to the plant and act as the means by which the plant takes up water and nutrients that are necessary for plant survival and growth. In recent years, the use of modern technology has revealed such things as genetic diversity in cotton root systems and the role of root signals in impacting the overall growth and development of the plant.

In the chapter on root growth in the book *Cotton Physiology* (McMichael,1986), a brief overview of the development of the cotton root system was given, which included some factors that affect root growth and aspects of root-shoot interactions. Also presented were some methods that could be used in measuring the growth of roots both in the field and in plants grown in pot culture. While certainly some aspect of root development have not changed since that first chapter was written (*i.e.,* root anatomy and root morphological development), new insights have been gained in areas dealing with root-shoot relationships, the impact of genetic diversity of root systems and the interaction of this diversity with both the soil and above-ground environment to influence plant productivity. The present chapter will, therefore, briefly touch on the areas of root anatomy and morphology and concentrate in greater detail on some factors affecting root growth and the impact of genetic variability to future directions for improvement of cotton productivity.

2. ANATOMY OF COTTON ROOTS

As described by McMichael (1986), the cotton plant consists of a primary or 'tap' root from which branch or secondary, tertiary, etc. roots emanate (Spieth,1933; Hayward,1938; Brown and Ware,1958). The cotton root, whether primary or branch root, has a single layer of epidermal cells surrounding the root cortex. The endodermis, another single cell layer, surrounds the stele which contains the xylem and phloem vascular elements and a cell layer called the pericycle. The xylem elements are arranged in either a tetrarch (four distinct xylem bundles, the most common) or in a pentarch or greater bundle arrangement (McMichael *et al.*,1985). This difference in arrangement is apparently genetically controlled and may result in significant genetic diversity in root branching.

The pericycle eventually becomes the protective layer for the root as the root ages and the endodermis and cortex are sloughed. In the mature root, the sieve tubes and companion cells of the phloem are formed from the protophloem located between the vascular bundles. As the root further matures the sieve tubes and xylem elements are arranged along the outer edge of the stele (Speith,1933).

3. DEVELOPMENT OF THE COTTON ROOT SYSTEM

To briefly summarize the information presented in *Cotton Physiology* (McMichael, 1986), the development of the cotton root system has been described by a number of researchers over the years under a number of different growth conditions (Balls,1919a; Collings and Warner, 1927; Brown and Ware, 1958; Taylor and Klepper, 1978).

J.McD. Stewart et al. (eds.), *Physiology of Cotton*,
DOI 10.1007/978-90-481-3195-2_6, © Springer Science+Business Media B.V. 2010

Since cotton is a taproot crop, the shape of the root system, the volume of soil explored by the roots and the overall root density is dependent on the development of secondary or lateral roots. These roots can extend outward from the taproot to a distance of over two meters (Taylor and Klepper, 1974). These roots also remain fairly shallow (less than one meter deep, Hayward, 1938). The lateral roots are formed from the cambial layer of the taproot and are arranged in a row according to the number of vascular bundles present in the primary root. If the number of vascular bundles increase above the normally occurring four, then the potential for a greater number of lateral roots increases. McMichael *et al.* (1987) observed a direct correlation between the number of vascular bundles in the taproots of cotton seedlings and the production of lateral roots. The implications of these findings will be discussed in more detail in the section dealing with genetic variability in root systems.

The depth of root penetration depends on a number of factors, but in general the taproot can reach depths of over three meters and can elongate at a rate from less than one to over six centimeters per day. The elongation rate of the lateral or secondary roots would generally fall within the same range.

In general, the root system continues to grow and increase in length until young bolls begin to form (Taylor and Klepper, 1974) at which time root length declines as older roots die. New roots continue to be formed past this point but the net result is a decline in total length (Hons and McMichael, 1986).

4. FACTORS AFFECTING COTTON ROOT GROWTH

The major factors influencing cotton root growth were briefly described previously by McMichael (1986). These included soil temperature, soil strength, soil aeration, and soil water. In this chapter these factors are discussed to include more recent information. Additional factors such as fertility, growth regulators, pathogens, and osmotic adjustment are covered. Also information concerning genetic diversity is discussed as a factor influencing root development.

4.1 PHYSICAL FACTORS

4.1.1 Temperature

The temperature of both the soil and air can have a significant influence on the growth of cotton root systems. Most research has shown that in general, the growth of cotton roots increases with increasing soil temperature until an optimal temperature is reached beyond which growth declines. Early work suggested that the optimal soil temperature for the growth cotton roots was approximately 35°C (Bloodworth, 1960; Lety *et al.*,1961; Pearson *et al.*,1970; Taylor *et al.*, 1972). Pearson *et al.* (1970) showed that root

elongation increased to a maximum of 32°C and then declined sharply as soil temperature increased in 80-hour-old seedlings. Research by Bland (1993) in controlled environment experiments showed that the rate of cotton root growth increased with the rate at which the soil warmed. His experiments indicated that the root system grew at progressively lower rates of elongation as the rate of soil warming was reduced from isothermal conditions.

In research on the growth of roots of cotton seedlings at various soil temperatures, McMichael and Burke (1994) showed that the optimal temperature for root elongation may depend on the level of available substrate or stored seed reserves. They suggested that the measured root length at 10 DAP (Days After Planting), for example, represented a composite of both narrow and broad metabolic temperature responses. Analysis of mitochondrial electron transport showed that the temperature optimum for root metabolism at 10 DAP for example, was lower than that obtained from the measure of accumulated root growth during the same time period.

Kaspar and Bland (1992) indicted that changes in soil temperature can affect growth of a number of root system components. For example, low temperatures generally reduced cotton root branching (Brower and Hoagland, 1964), while higher temperatures approaching the optimum tend to increase branching (Nielsen, 1974). The uptake of water by roots is reduced at low temperature (Nielsen, 1974) while higher temperatures result in increased uptake. Bolger *et al.* (1992) demonstrated that the hydraulic conductance of cotton roots declined as the root zone temperature decreased below 30°C and that conductance at 18°C averaged 43% of that at 30°C.

Differences in the response of different root types to temperature were also apparent. Research conducted by Arndt (1945) indicated that the cotton taproot may be more adapted to adverse soil temperatures than subsequent branch roots at least until the taproot had developed to approximately 10 cm in length. Later work on seedling development of a number of exotic cotton strains grown in hydroponics showed similar results (McMichael and Burke, unpublished data; Fig. 6-1). Steiner and Jacobsen (1992) also noted differences between two cotton cultivars in their sensitivity to soil temperature.

The interaction between soil and air temperature in influencing cotton root development is shown in Table 6-1 for 10-day-old cotton seedlings (McMichael and Burke, 1994). When the root temperature was low (20°C), root growth was reduced regardless of the temperature of the air (shoot). The root-shoot interaction in response to temperature may be related to changes in source-sink relationships. Guinn and Hunter (1968), for example, showed changes in carbohydrate levels in shoots and roots in response to temperature with a build-up of sugars occurring at low root temperatures.

The successful emergence and initial growth of cotton seedlings is important for the establishment of healthy plants

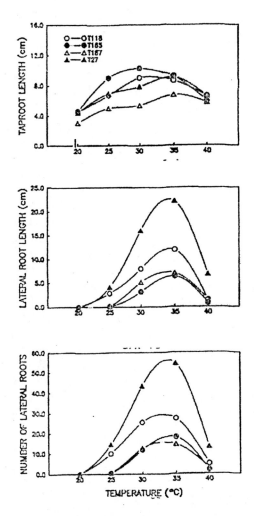

Figure 6-1. The influence of temperature on the growth of primary (tap) roots and lateral roots of 10-day-old cotton seedlings. From McMichael (unpublished data).

Table 6-1. Root development of 10-day-old cotton seedlings grown in growth pouch system at four temperature regimes. From McMichael and Burke, 1994.

Treatment		Taproot length	Lateral root length	Total root length	No. laterals
Shoot temp.	Root temp.				
----(°C)----		----------------(cm)---------------			
28	20	3.05 ± 0.24 a^z	0.0 a	3.08 ± 0.24 a	0.0 a
20	28	5.06 ± 0.28 b	1.83 ± 3.5 b	6.89 ± 0.54 b	8.00 ± 8 b
28	28	7.71 ± 59.00 c	5.82 ± 1.15 c	13.53 ± 1.07 c	8.38 ± 0.9 c
20	20	4.71 ± 0.35	0.0 a	4.71 ± 3.50 a	0.0 a

z Means followed by the same letter are not significantly different at the 0.05 probability level based on a Duncan's Multiple Range Test.

and improved productivity. Wanjura and Buxton, (1972 a,b) showed that when the minimum soil temperature at planting depth dropped from approximately 20°C to 12°C, the hours required for initial seedling emergence increased from 100 to approximately 425 hours. In many cotton-growing areas the soil temperature can be significantly lower than the optimum when seeds are planted. Therefore the development of cultivars that possess a root system that can grow and function at low temperatures could improve plant performance. However, since the exact mechanism(s) of the response of cotton roots to temperature are not known, further research, perhaps in the molecular area, is needed to elucidate the nature of the response.

4.1.2 Soil Strength

The presence of compacted soil layers, whether natural or as a result of artificial hardpans, can significantly restrict root growth in cotton and ultimately impact productivity. Taylor (1983) provides an excellent overview of the influence of soil strength on the growth of cotton root systems.

In general, root growth is restricted as the soil strength is increased (Fig. 6-2, Gerard *et al.*, 1982). Taylor and Ratliff (1969) observed that once the taproot entered a compacted zone, the root length measured at various intervals varied inversely with the soil resistance as measured by penetrometers. They reported that an increase in the penetrometer resistance 100 MPa reduced cotton root elongation rates 62% from the rate at 0 MPa. Grimes *et al.* (1975) also noted restricted root growth of cotton in high strength zones under field conditions. Wanjura and Buxton (1972) showed that the growth of both shoots and roots of young cotton plants were reduced by increased soil strength. Gerard *et al.* (1982) indicated that the critical strength at which root growth stopped ranged from 600 to 700 MPa in coarse textured soil and was approximately 250 MPa in clay soil.

Morphological changes in the cotton root system may occur as a result of changes in soil strength. Glinski and Lipiec (1990) observed that the responses of most roots grown in mechanically impeded soil have reduced root size, diminished elongation rates, uneven root distribution, thickening of roots, greater lateral branching, and a reduction of water and nutrient uptake. Camp and Lund (1964) observed that the cells of cotton roots grown in loose soil were elongated as compared to similar type root cells grown in compacted soils. The cells in the latter case were much thicker (thick cell walls). Similar observations of root thick-

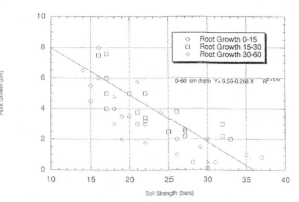

Figure 6-2. The overall relationship between
soil strength of Abilene clay loam soil and root growth
in cm. Root growth in cm for each dept is represented
about the regression line. From Gerard *et al.,* 1982.

ening were noted by Bennie (1996) who also suggested that
the decrease in root growth in a compacted zone is the same
for most species, but that the main differences are related
to the ability of the plants to produce branch roots in the
compacted layers.

Interactions between soil strength and soil water status
have been shown to be important in the response of cotton
root systems to mechanical impedance. Taylor and Gardner
(1963) observed a positive relationship between soil moisture content and root penetration in cotton but a more significant relationship occurred between root penetration and
soil strength. In other studies, Taylor and Ratliff (1969) indicated that at any penetrometer resistance there was little effect of water content *per se* on cotton root elongation within
a compacted layer. Taylor and Gardner (1963) concluded
that soil strength, and not bulk density changes, was the
critical factor controlling root penetration, particularly in
sandy soils. Camp and Lund (1964) however, showed that
as moisture decreased, the effects of soil density became
more critical and that no root penetration of cotton roots
occurred at densities of 1.7 to 1.8 g/cc. Bennie (1996) also
showed that impedance at a specific bulk density increases
with soil drying and that soil wetting decreased interparticle
attraction and decreased the pressure needed for deformation of the soil. Glinski and Lipiec (1990) also suggested
that the critical impedance of soil for root growth changes
with differences in soil moisture. Regardless of the magnitude of the interaction of soil strength and moisture content,
mechanical impedance has been shown to be a significant
factor in reducing cotton yields (Carmi and Shalavet,1983)
probably due to a restricted water supply as a result of the
limited rooting volume. Studies by Carmi (1986) indicate
that a restricted root volume can significantly affect the
above ground growth of the cotton plant.

4.1.3   Soil Aeration

The composition of the soil air is approximately 79%
nitrogen, 20% oxygen, and 0.5 to 1% carbon dioxide at
depths of 15 to 20 cm (Stolzy, 1974). Oxygen levels can
range from as low as 5% and carbon dioxide levels can increase to approximately 20% depending on such factors as
soil temperature and soil water content (Cannon, 1925).

Cotton roots appear to be very sensitive to changes in
oxygen concentrations relative to changes in carbon dioxide concentrations (Kramer, 1969). Leonard (1945) showed
that an increased water table under Houston clay prevented
cotton roots from growing deeper that 24 cm until late in
the growing season. Excess water due to temporary saturation can deplete dissolved oxygen from the bulk soil by root
respiration and soil microorganisms resulting in depressed
root growth (Drew and Stolzy,1991). Huck (1970) observed
that cotton can survive under low oxygen conditions but
for only a very short while (0.5 to 3 hours). He also indicated that the top roots died after reaching the anoxic layers
followed by stimulation of additional lateral root growth.
Camp and Lund (1964) showed that when a subsoil layer
was aerated with mixtures of oxygen and carbon dioxide
root growth was depressed at low bulk densities when either oxygen was decreased below 19% or the carbon dioxide levels were increased. They also observed that mechanical impedance was the most important factor reducing
root growth at the higher soil densities. Whitney (1941)
observed similar results when cotton roots were exposed to
toxic levels of carbon dioxide or reduced levels of oxygen.

4.1.4   Soil Water

The water content of the soil can have a significant influence on the growth and function of cotton roots. Rooting
depth and density can change as a result of soil water content (Fig. 6-3, Klepper *et al.,* 1973) and root activity can also
change as the soil dries since root proliferation may occur at
lower depths to maintain water uptake rates (Klepper *et al.,*
1973). In general, soils with a small water-holding capacity
have deeper roots while those with a larger capacity have
shallow roots (Glinski and Lipiec, 1990). McMichael (unpublished data) showed that rooting densities of cotton increased significantly at lower depths and decreased in upper
soil layers in several commercial cotton cultivars when the
upper soil profile dried. Klepper *et al.* (1973) also observed
that the rooting patterns of cotton in a drying soil shifted as
the soil dried. Initially more roots were in the upper layers,
but as a result of the death of the older roots in the top due
to the soil drying and production of new roots at the lower
depths, the rooting density increased with depth. Cotton
plants grown in uniformly moist soil did not show this reversal. Malik *et al.* (1979) also showed that emergence of
cotton roots from soil cores of different water contents into
a soil zone where water was freely available to the roots increased as the soil dried. The root/shoot ratios also increased
as the water content increased due to an absolute increase in
root weight with shoot weight not being affected.

Changes in water distribution as a result of irrigation
practices can also impact the growth of cotton roots. Radin

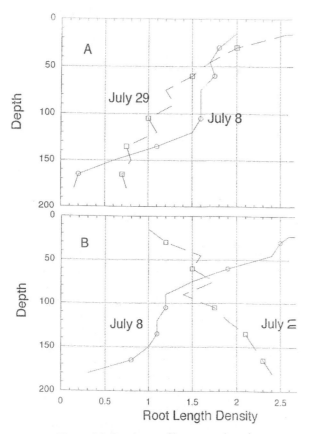

Figure 6-3. Rooting profiles at two dates in a soil profile that was maintained moist (A) and allowed to dry (B): densities refer to length of root per cm³ of the soil. From Klepper *et al.*, 1973.

et al. (1989) noted that long irrigation cycles tended to trigger more rapid deterioration of the root system during periods of heavy fruiting above the normal net reduction in root growth as fruit develops. This trend was slow to be reversed. Carmi *et al.* (1992) observed that in cotton irrigated with a drip system a shallow root system with a high percentage of the roots less than one millimeter in diameter were concentrated around the drippers which resulted in a strong dependence on a frequent supply of water for continued growth. In other studies, Carmi *et al.* (1993) also showed that the capability of more mature cotton plants to adjust rooting patterns to large changes in water distribution was slow and that preferential root growth relative to shoot development did not occur in response to progressive soil drying in their case. Carmi and Shalhevet (1983) also observed that dry matter production by cotton roots was less severely inhibited than shoots under decreasing soil moisture. This implies that changes in the root dry weight/ root length relationships can change in response to changes in soil moisture.

In terms of water extraction, Taylor and Klepper (1975) observed that water uptake in cotton was proportional to the rooting density as well as the difference in water potential between the root xylem and the bulk soil. Jordan (1983) showed that rooting densities may decrease to as low as 0.2

cm/cm³ and still extract water. Taylor and Klepper (1974) showed that root length did not increase in a soil layer when the water content fell below 0.06 cm³/cm³ which was equivalent to a soil water potential of -0.1 MPa. In other work, Taylor and Klepper (1971) also observed that water extraction per unit length of root was greater in wet soil and decreased exponentially with soil water potential. In general, they found that deep roots were as effective as shallow roots in extracting water.

Interactions between soil water status and soil temperature can also influence the function of cotton roots. Radin (1990) showed that the hydraulic conductance of cotton roots declined at cooler temperatures which would affect water uptake. Bolger *et al.* (1992) also showed that conductance decreased when the root temperatures were lowered from 30°C to 18°C. These results would suggest that under certain conditions the water uptake by cotton roots may decrease as a result of low soil temperatures even though water was not a limiting factor.

The importance of the water relations of cotton roots *per se* (*i.e.*, axial vs. radial water flow and cell water relations) is certainly not to be overlooked in any discussion of the impact of water on root development. Oertli (1968) has provided an excellent review of water transport through the root systems of plants and soil-root interactions. Since much of this information is directly related to other factors mentioned in this chapter, a more comprehensive rendering is included in the later discussion on osmotic adjustment.

4.1.5   Tillage

A number of experiments have been conducted to determine the impact of tillage on crop yields (Taylor, 1983). Studies have addressed issues such as the types of plows to use and the depth of tillage that is most effective in improving crop performance. In recent years conservation tillage practices where crops such as cotton are planted into the residue of previous crops have also been shown to improve cotton production and reduce production costs (Segarra *et al.*, 1991). Other studies have also shown that pruning of shallow roots can occur as a result of tillage. There has not been a clear correlation however, between the loss of roots in this manner and reduction of crop yields. Since there is a strong interaction between tillage and changes in other soil properties such as soil aeration and soil strength (Taylor *et al.*, 1972) the direct impact of tillage on root development is difficult to ascertain. Therefore, for brevity, these aspects have been discussed in previous sections and will not be repeated.

4.2   CHEMICAL FACTORS

4.2.1   Fertilizer Placement

There is sufficient evidence that fertilizer placement in the soil can influence root growth and rooting patterns. The distribution pattern of plant roots in the soil is largely determined by the plant's genetic makeup. However, the actual proliferation of plant roots is substantially modified by the environment in which the plant grows. Roots tend to concentrate in soil layers with higher fertility levels (Weaver, 1926; Kapur and Sekhon, 1985) and high plant available water (Klepper *et al.,* 1973). Earlier work with cotton by Rios and Pearson (1964) demonstrated that the presence or absence of certain ions in the upper soil profile influenced root growth deeper into the soil profile. They related this to the mobility of the ion and the ability of the plant to translocate the specific ion to the developing root. For example, in a soil where the subsurface zone was well limed and fertilized, cotton roots failed to develop into the lower zone due to an inability to move calcium to the growing root tip. Whereas the absence of phosphorus (P) in the subsoil did not prevent root growth due to the adequate translocation of P from surface soil layers. Nitrate and phosphate have a major influence on root growth and branching (Wiersum, 1958; Price *et al.,* 1989) as does vesicular-arbuscular mycorrhizas (Price *et al.,* 1989; see Chapter 19). Mullins (1993) showed that P uptake and root growth of cotton seedlings were affected by P placement, particularly in low P status soils. Root length in the P-treated soil volume increased as the fraction of soil fertilized with P increased and that this response was influenced by soil type.

Brouder and Cassman (1994) demonstrated that K acquisition and root proliferation by cotton is strongly influenced by the quantity and distribution of nitrate nitrogen in the root zone. Outbreaks of K deficiency in the U.S. Cotton Belt in recent years have been attributed to a number of causes including poor root growth, low soil-K status, excessive K-soil fixation, and the introduction of higher-yielding, faster-fruiting, earlier-maturing cultivars (Oosterhuis, 1995b). Suggested remedies have included foliar fertilization (Oosterhuis, 1995b) and deep placement of K fertilizer (Tupper *et al.,* 1988). However, the yield response to added K fertilizer has been very variable and unpredictable due to differing seasonal growing conditions, genotypic differences and soil type. Brouder and Cassman (1990) demonstrated that significant genotypic differences in K uptake were related to determinacy in root growth after the peak flowering period.

Because cotton is a taprooted crop that poorly exploits the topsoil layer (Gulick *et al.,* 1989; Brouder and Cassman, 1990), placement of fertilizer within the soil profile can be expected to influence root growth and nutrient uptake. This is also because root growth is strongly influenced by temperature (McMichael and Quisenberry, 1993) and soil-water status. Cotton root penetration is also highly influenced by soil pH and Al concentration (Adams and Lund, 1966). However, Carmi and Shalhevet (1983) presented evidence to show that differences in cotton growth rate were caused by differences in root system development as affected by restrictions to root growth rather than by N, P, or K deficiencies or water stress.

Current N-fertilizer placement techniques minimize early-season contact with cotton roots. In most cases, N is banded at a spacing and depth that require cotton roots to proliferate into the fertilizer zone for uptake to occur (Varco, 1997). Obviously, fertilizer placement too close to the seed will have minimal or undesirable effects on root growth (Walker and Brooks, 1964). Use of an encapsulated slow-release N or K fertilizer has permitted all the fertilizer (100 lb N/A or 60 lb K/A) to be placed with the seed in-furrow at planting with no detrimental affect on cotton growth or yield (Howard *et al.,* 1998). More work is needed on the actual nature of cotton root response to fertilizer placement, including information on root branching, root length, and root dry matter.

4.2.2   Species (Ionic) and Salinity

Cotton is a relatively salt tolerant species, but growth can still decline when the plant is exposed to saline stress. Germination and emergence (El-Zahab, 1971b) and seedling growth (Zhong and Lauchli, 1993a) are particularly salt-sensitive. See the review in this book on plant responses to salinity by Gorham *et al.* (2007). Salinity generally reduces root growth (Silberbush and Ben-Asher, 1987), but there have been reports of mild salinity enhancing root growth (Jafri and Ahrnad, 1994; Leidi, 1994). The ions Na^+, K^+, Ca^{2+}, Mg^{2+} and Cl^- are the common constituents involved in high salinity and altered plant growth and root expression. Primary root growth of cotton seedlings was severely inhibited by high concentrations of NaCl in the growing medium, but supplemental Ca reduced Na influx and improved root growth (Cramer *et al.,* 1987; Zhong and Lauchli, 1993). The protective effect of supplemental Ca on root growth under high salinity has been associated with improved Ca status and maintenance of K/Na selectivity (Cramer *et al.,* 1987) and improved cell production (Kurth *et al.,* 1986). Obviously high soil salinity can cause effects similar to water-deficit stress on plant growth (Kramer and Boyer, 1995). The degree of salinity influences the plant's ability to osmotically adjust to the altered water potential gradient between the soil solution and the plant root.

4.2.3   Timing of Application

In cotton production fertilizer has traditionally been applied at, or prior to planting. However, during the 1990's, production practices have moved towards applying less fertilizer at planting and more during the season when plant demand increases and early-season leaching losses decrease. Nitrogen, for example, was for many years incorporated prior to planting. But currently, less than 10% of U.S. cotton acreage receives N at planting, with most N applied prior to planting or as post-planting applications at early squaring (Gerik *et al.,* 1997). Foliar fertilization during boll development, particularly for N, K, and B, is sometimes used to further supplement plant nutrient requirements when tissue tests indicate potential deficiencies (Oosterhuis, 1995b).

Fertilizer placement and timing should take into account the temporal patterns of root growth and the nutrient requirements of the developing crop. Cotton root growth generally increases up to flowering and then decreases slightly after flowering (Cappy, 1979; McMichael, 1990) as the competition for available assimilate from the boll load increases. However, Basset *et al.* (1970) reported that cotton root activity intensified at lower depths as the season progressed. Nutrient uptake by cotton generally follows a pattern similar to dry weight accumulation, except that dry matter continued to increase until maturity, and nutrient accumulation peaks during boll development. For example, K is absorbed more rapidly than dry matter is produced, as evidenced by the higher concentration of K in young plants (Bassett *et al.,* 1970), and maximum K accumulation is reached in about 112 days after which there was a decrease (Halevy, 1976). Oosterhuis (1995b) speculated that the decrease in root growth after flowering may be one of the main reasons for the appearance of late-season K deficiency in the U.S. Cotton Belt. The rate of plant uptake of K depends on root length density and total root surface area (Brouder and Cassman, 1990). The high requirement for K in cotton coupled with an inherent low root length density relative to other major row crops (Gerik *et al.,* 1987) and the immobile nature of the element (Barber, 1984), means that K uptake is particularly sensitive to poor root growth and deficiencies may appear even in soils with a relatively high K content. Thus the relative sensitivity of cotton to the soil K supply may reflect in part the low density root system of the cotton plant.

Another application has been the use of starter fertilizer, which has been successfully applied to upland cotton (Funderberger, 1988; Guthrie, 1991), corn (*Zea mays* L.) (Ketcheson, 1968), and grain sorghum [*Sorghum bicolor* (L.) Moench] (Touchton and Hargrove, 1983) to stimulate early-season crop development. Unfortunately, most of these studies did not measure root growth, but the assumption was made that the improved shoot growth was the result of enhanced root growth which led to increased nutrient and water uptake. In corn, yield response to row-applied starter fertilizer in northern production areas has been attributed to compensation for reduced root growth (Nielsen *et al.,* 1961) and nutrient availability (Ching and Barber, 1979) in cool soils encountered with early planting dates or in reduced tillage. Temperature has a dramatic affect on the rate and proliferation of root growth (McMichael and Quisenberry, 1993) and thus, will have an affect on plant response to starter fertilizers. Kaspar and Bland (1992) suggested that rooting depth may follow the downward progression of a particular isotherm. There have also been some reports of minimal responses of cotton to starter fertilizer usually on soils high in P (Maples and Keough, 1973), or when the starter fertilizer was placed too close to the seed (Walker and Brooks, 1964). Indirect evidence from Radin (1990) showed that P deficiency decreased hydraulic conductivity. Therefore, we can assume that P in the starter fertilizer will increase conductivity and this should translate into improved growth.

4.2.4   Salinity

The growth of most crop plants is affected by saline conditions. According to Zhong and Lauchli (1993), cotton is a relatively salt tolerant plant, but can be very sensitive to salt conditions in the seedling stage. Water stress and ion toxicity are most likely the result of high salt conditions that reduce plant growth. Cramer *et al.* (1987) observed that the growth of the taproot of cotton seedlings was reduced in the presence of NaCl but that the effects could be countered somewhat by the addition of calcium to the growing media. Zhong and Lauchli (1993) found that the elongation of the taproot of cotton seedlings was reduced by 60% over the control plants when the roots were exposed to 150 mol/m^3 NaCl. The addition of calcium increased the elongation rate to within 80% of the controls. They also observed that the growth zone (the region of root cell elongation) of the taproot was shortened by the increased salt content of the media. Kurth *et al.* (1986) showed that the rate of cell production declined in cotton roots in the presence of high salt and that the shape of the cortical cells were affected.

Reinhardt and Rost (1995d) also observed that high salt reduced the width and length of metaxylem vessels in cotton seedlings which increase with plant age. These changes in root morphology along with changes in osmotic relationships as a result of high salt, can result in a significant reduction in root growth and root activity to reduce plant productivity.

4.2.5   Plant Growth Regulators

Interest in root-produced plant hormones or plant growth regulators (PGRs) has stemmed mainly from observations that roots affect shoots beyond the influence of their supply of water and minerals. Since most plant growth and development processes are regulated by natural plant hormones, these processes may be manipulated either by altering the plant hormone level, or by changing the capacity of the plant to respond to its endogenous hormones. In recent years, PGRs, which are synthetic hormones, have been investigated for their ability to alter cotton growth and development in an attempt to improve production. These compounds are diverse in their chemistry and in their uses. There are numerous reviews on plant growth regulation (*e.g.*, Davies *et al.,* 1987) and plant growth regulators in cotton have been recently reviewed by Cothren and Oosterhuis (2006). It is not the purpose of this review to cover all the hormonal relationships in roots, but rather to mention those relevant to cotton.

The synthesis of phytohormones by roots has been reviewed by Itai and Birnbaum (1996) who concluded that roots are capable of synthesising all known PGRs. Recently, Jasmonic acid has been added to the list of the five accepted phytohormones since evidence of its presence and the possibility that it is synthesized by roots has been established (Gross and Partheir, 1994).

Despite its chemical simplicity and gaseous nature, ethylene is a potent PGR involved in the regulation of many physiological processes in plant growth and development. In roots, ethylene is involved in adventitious root formation, gravitropism, hook formation, respiratory changes, RNA synthesis, storage product hydrolysis, and growth inhibition. Typically, root growth is stimulated at low concentrations of ethylene, and inhibited at higher concentrations (Jackson, 1985). This is important because soils can contain high concentrations of ethylene (Jackson, 1982). Root anatomy may also be affected by ethylene, *e.g.* the development of aerenchyma in flooded maize (Jackson, 1982). Ethylene can also increase the production of root hairs (Reid, 1985). However, not much has been published about the specific effects of ethylene on cotton roots.

Although the literature is full of references to plant responses to auxins, there are relatively few reports on auxin activity in cotton (*e.g.*, Rodgers, 1981; Guinn, 1986), and even fewer reports on the effects of auxins on cotton root growth. Higher concentrations of auxins are inhibitory to root growth which has important ramifications in hormonal control of plant growth. IAA and indole butyric acid (IBA) have been used to induce root formation in cuttings (Hartmann and Kester, 1984). Urwiler and Oosterhuis (1986) showed that IBA stimulated root growth of cotton seedlings. Also gibberellins (GAs) are involved in almost all phases of growth and development from germination to senescence, and in particular in the stimulation of both cell division and elongation.

It is well established that roots tips are a primary source of endogenous cytokinin, which are transported via the xylem to the shoot where they exert a major influence on growth, photosynthesis, and senescence (*e.g.*, Kulaeva, 1962; Kende, 1964). Root responses to cytokinins include enhanced root development and root hair formation, and inhibition of root elongation. Sattelmacher and Marschner (1978) reported that nitrogen deficiency decreased the cytokinin activity in root exudates of potato (*Solanum tuberosum* L.), and Radin and Ackerson (1981) showed that nitrogen deficiency in cotton increased abscisic acid (ABA) content of cotton leaves.

Since the discovery of ABA in cotton (Ohkuma *et al.*, 1963), the presence of ABA has been recorded in roots and their exudates. Like all other plant hormones, ABA has multiple physiological effects influencing plant growth and development. Among the physiological effects of ABA, the role of ABA in stomatal closure and regulating water balance in plants has been studied the most. In roots ABA is likely involved in promoting sencsence, regulating turgor pressure, and modification of both enzyme activity and RNA synthesis. Again, there are only a limited number of references to ABA and root growth in cotton. Both ABA (and cytokinin) contents of roots are clearly affected by the root environment. Cornish and Zeevart (1985) showed that roots respond to water stress by increasing their ABA levels more rapidly and with greater sensitivity than leaves.

Recently, ABA has been shown to be intimately involved in signal transduction between roots and shoots (Davis and Zhang, 1991). Root signals are thought to reflect soil water, nutrients, and mechanical attributes as sensed by roots, and the transduction involves ABA, nitrate flux, cytokinin, and hydraulic changes for regulation of whole plant growth (Aiken and Smucker, 1996). The role of ABA in signal transduction has not been extensively studied in cotton but there is no reason to think that the response would be different.

Plant growth is regulated by hormonal balance which can affect and be affected by competition for organic and inorganic nutrients. As Guinn (1986) mentioned in *Cotton Physiology*, hormonal balance could be affected by limitations in root growth because of competition by developing bolls. There are probably many complex interactions between hormones and competition for inorganic nutrients (Guinn, 1986). Auxins, cytokinin, and gibberellins promote growth whereas abscisic acid and ethylene inhibit growth; however, it is rather the balance between growth promoting and growth inhibiting that probably mediates growth in plants. More research is needed on system analysis, *i.e.*, an integrated understanding of all the known hormones for the growth process or plant response being studied. Also, information is needed on the molecular basis of the mode of action of each PGR (Itai and Birnbaum, 1996). Microbial biosynthesis of all the major plant hormones in the soil (Frankenberger and Arshad, 1995) could potentially have a physiological role in plant growth and development and in particular on root growth.

4.2.6   Herbicides

The use of herbicides to control weeds in cotton is a wide-spread practice across the Cotton Belt. The herbicide is generally either incorporated into the soil or applied as a non-incorporated surface spray. One common herbicide in use for weed control is triflualin. Anderson *et al.* (1997) reported that the soil incorporated application of trifluralin from 0.25 to 1.0 lb/A caused stunting in height and prevented lateral root production in cotton seedlings. Similar results have been reported by other researchers. They also observed that lateral root production was more affected by depth of incorporation of the herbicide than by the dosage. Hess and Bayer (1974) showed that the absence of microtubules in the root meristem of cotton disrupted normal cell mitosis. Kleifeld *et al.* (1978) however, reported that the presence of trifluralin appeared to protect cotton roots from the damage to roots caused by other herbicides such as flometuron, diuron, and prometryne.

4.3   BIOLOGICAL FACTORS

4.3.1   Pathogens

The presence of soil-borne pathogens can impact the growth and function of cotton root systems. Pathogens such as *Phymatotrichopsis omnivera* are common agents that cause root rot in cotton (Rogers, 1937). Domsch *et al.* (1980) have indicated that cotton seedlings may be more resistant to attack by this organism than older plants due to a reduced carbon content of the root bark. An increase in the carbon content of the roots due to loss of branches and fruit tends to reverse this effect. King and Presley (1942) reported that a disease of cotton that was characterized by a swollen taproot and internal black rot of the vascular tissue was found in Arizona in 1922. The organism was identified as *Thielaviopsis basicola* and was found to be most damaging to the cotton root system in the seedling stage. Rothrock (1992) later showed an interaction of this organism with soil temperature, soil water, and soil texture on the infection of cotton roots. Burke and Upchurch (1995, unpublished data) observed that cotton plants grown at low temperatures in the absence of pathogens had increased lateral root production even at the low temperatures (13°C).

Other studies have shown that infection of cotton roots by nematodes may impact the growth and development of the plant (Kirkpatrick *et al.,* 1991). These authors indicated that the effects of the infection were similar to water stress. The hydraulic conductivity was reduced and drought resistance was increased.

4.4   COTTON-MYCORRHIZAL INTERACTIONS

The interaction of cotton with mycorrhizal fungi is covered in greater detail in Chapter 19.

Several studies have shown that the association of root systems of crop plants, including cotton, with arbuscular mycorrhizae (AM) can increase the ability of the plant to absorb water and nutrients and can result in increased biomass production and yields (Afek *et al.,* 1991; Bethlenfalvay and Linderman 1992; Robson *et al.,* 1994). Arbuscular mycorrhizae can produce an extensive mycelial system outside the root, which in effect, extends the root system so that a larger soil volume may be explored. This can be especially beneficial in the case of cotton since the root system has a low density per unit soil volume (McMichael, 1990). Cotton may be more dependent on mycorrhizal associations than some other crops with denser root systems (Manjunath and Habte, 1991). Cotton has been shown to be mycorrhizal and to benefit from this symbiosis in more mesic regions (*e.g.,* Pugh *et al.,* 1981; Price et el., 1989). Rich and Bird (1974) reported that arbuscular mycorrhizae were detected in Georgia cotton 5 days after seedling emergence and that the infection was predominately arbuscular. They observed that AM development subsequently increased logarithmi-cally between 5 and 25 days. The increase in mycorrhizal roots was followed by an increase in the growth and development of the cotton and increased yield. Zak *et al.* (1998) showed that the mycorrhizal status of cotton in the semi-arid environment of the Southern High Plains of West Texas may be crucial for seedling survival.

Agricultural systems that are heavily managed, such as cotton, have been shown to have a low species richness of AM fungi (see review by Johnson and Pfleger, 1992). Moreover, those species that do occur under these highly disturbed conditions may not be very efficient symbionts, but may be those that can tolerate the stressful conditions imposed by agricultural practices. Tillage practices in particular have been shown to negatively impact the AM development of various crop species. The magnitude of tillage effects is dependent upon the degree to which the soil has been disturbed (*e.g.,* McGonigle *et al.,* 1990). Much of what occurs in most crop production systems has been shown to be detrimental to the survival of AM inoculum and the maintenance of an intact mycelial network (Kurle and Pfleger, 1994). The maintenance of an intact AM mycelial network appears to be crucial for ensuring high colonization levels for a cropping system (McGonigle *et al.,* 1990). However, on the Southern High Plains, fields are plowed to reduce wind erosion thereby destroying the AM mycelial network and encouraging decomposition of root fragments that are also important to future colonization in this semi-arid environment (Friese and Allen, 1991). Any practice that can maintain or enhance the AM inoculum potential of fields before planting cotton and at the same time reduce soil erosion from wind should have significant positive effects on the cotton production system in this semi-arid region. Laboratory experiments have demonstrated that disturbances, such as tillage, reduce AM development and effectiveness (*e.g.,* Jasper, 1989 a&b). Disruption of the mycorrhizal network of hyphae (Read, 1992) that radiates from mycorrhizal roots appears to be the mechanism for the observed reduction in AM development and P absorption by crop plants (Jasper *et al.,* 1989 a, b; Evans and Miller, 1990). Tillage may also influence the rates of AM development by altering the distribution and abundance of AM propagules in the soil (Jasper *et al.,* 1992; Johnson and Pfleger, 1992).

Mycorrhizal inoculum can exist either as asexual spores, as hyphae comprising the hyphal network, or as infected root fragments. In semi-arid and arid systems, mycorrhizal hyphae associated with root pieces may be the most important source of inoculum (Friese and Allen, 1991). It is very likely that crops in semi-arid and arid regions depend on infective mycorrhizal roots and on the AM network associated with companion plants than on mycorrhizal spores. Examinations of the cotton fields in the West Texas area have indicated that AM spore numbers are either very low or are below detectable numbers prior to and immediately after cotton planting (Zak *et al.,* unpublished data).

One of the potential benefits that can occur from the mycorrhizal colonization of agricultural plants is that these symbiotic associations can affect the incidence and severity of root diseases. Much of the literature suggests that AM fungi can reduce soilborne diseases or the effects that the disease causes (Dehne, 1982). However, other reports indicate mixed results with either no effects (*e.g.*, Baath and Hayman, 1984) or increased disease severity (Davis and Menge, 1981). Lindermann (1992) argues that mycorrhizae can contribute to disease suppression by either increasing P uptake, which would result in more vigorously growing plants that are either able to ward off or tolerate root diseases, or by reducing drought stress which predisposes the plants to pathogen attack. Davis (1980) showed that citrus trees colonized by AM fungi and subsequently infected by *Thielaviopsis basicola* were larger then non-AM colonized plants unless the latter were fertilized with phosphorus. For mycorrhizae to suppress root diseases they must be established and functioning before pathogen attack begins. This was the case for *Phytophthora* root rot (*Phytophthora parasitica*) of citrus (Graham and Egel, 1988). Wick and Moore (1984) found that mycorrhizal holly plants showed increased wood barrier formation that inhibited *Thielaviopsis* infection. Even infections by root pathogenic nematodes are generally less for mycorrhizal plants (e. g., Ingham, 1988). If mycorrhizae are to be effective in reducing diseases then agricultural practices that decrease AM colonization and inoculum should be avoided (Johnson and Pfleger, 1992).

4.4.1   Root Tissue Water Relations

Most of the studies on the water relations of cotton have focused on the whole plant (*e.g.*, Ackerson *et al.*, 1977). Field research using mini-rhizotrons has shown that non-irrigated cotton had a deeper root length than irrigated cotton (McMichael, 1990; Keino *et al.*, 1994). Furthermore, only non-irrigated cotton showed cultivar differences in root length density (Keino *et al.*, 1994). These results suggested that cotton cultivars express large differences in root length distribution under water stress, and therefore, deep rooting cultivars should be selected within environments where water is limiting. Carmi *et al.* (1992) showed that a shallow and restricted root system resulted in strong dependence of the plants on frequent and sufficient water supply, such that temporary minor changes in irrigation affected plant water status and productivity. However, a shallow root system allowed maximum flexibility for using irrigation to quickly and efficiently affect plant water status and influence processes which determine productivity.

In the last ten years there have been a number of new methods introduced to measure the water relations of roots. In cotton, thermocouple psychrometers have been used to measure root water potential (Oosterhuis, 1987; Yamauchi *et al.*, 1995) and osmotic potential (Oosterhuis, 1987), and the vapor pressure osmometer has also been used to record osmotic potential (Ball *et al.*, 1990) in excised roots. There are few reports on the nature of the osmotica in cotton and the importance of proline (McMichael and Elmore, 1977) and glycine betaine.

Root resistance accounts for a significant fraction of the hydraulic resistance in most plants (*e.g.*, Fiscus, 1983). Radial root resistance is usually substantially higher than the axial resistance (Yamauchi *et al.*, 1995). Hydraulic conductivity in cotton roots is reduced under conditions of water-deficit stress (Oosterhuis and Wiebe, 1980). Methods to measure cotton root hydraulic conductance were compared by Yang and Grantz (1996) with the reverse flow and transpirational methods appearing to have more physiological validity than the root exudation method. There have been reports of oscillations of 30 to 50 minutes in apparent hydraulic conductance in cotton plants (Passioura and Tanner, 1985), which is similar to the oscillations in stomatal conductance of cotton leaves (Barrs, 1971). Water deficit decreased cotton root pressure by 51% compared to a well-watered control, but had no effect on the exponential pressure-flux relationship (Oosterhuis and Wiebe, 1986).

4.2.3   Osmotic Adjustment and Drought Tolerance in Cotton Roots

Osmotic adjustment, or osmoregulation, is a plant mechanism for drought tolerance and the maintenance of water (Ψ_w) potential gradients (Wyn Jones and Gorham, 1983). Osmotic adjustment involves the active accumulation of osmotica (*e.g.* sugars, organic acids and mineral ions) in the cytosol during periods of water deficit or salt stress to lower the osmotic potential (Ψ_s) (Munns and Termaat, 1986). The lowered Ψ_s in response to decreasing Ψ_w, allows for the maintenance of pressure potential (Ψ_p) for turgor (Hsiao, 1973). Turgor maintenance under water stress allows continuation of growth, although at a reduced rate in comparison to optimal conditions (Sharp and Davies, 1979). Osmotic adjustment may be an important mechanism in plant tolerance although some crops do not undergo adjustment (Morgan, 1980; Oosterhuis and Wullschleger, 1988). Osmotic adjustment is a well accepted phenomenon in higher plants (Morgan, 1984). The occurrence of osmotic adjustment, however, is not universal. Varying degrees of adjustment will depend on the nature of the applied stress, and also on the crop or species, cultivar, organ, and developmental age of the organ (Morgan, 1984; Turner and Jones, 1980).

In cotton, as in most other crops, research on osmotic adjustment has focused on the leaves (Ackerson, 1981; Ackerson and Herbert, 1981; Cutler and Rains, 1977, 1979), and there are few reports of adjustments in the water relations of cotton roots in response to water stress (Oosterhuis and Wullshleger, 1987a). Cotton appears to have a greater ability to osmotically adjust to water stress than most other major row crops (Table 2) (Oosterhuis and Wullschleger, 1988). The magnitude of osmotic adjustment in cotton was greater in leaves (0.41 MPa) than roots (0.19 MPa), although the percentage change was greater in roots (46%) than leaves (22%) (Fig. 6-4) (Oosterhuis and Wullschleger, 1987a). The authors related this to the drought tolerance and survival capabilities of cotton. There is only one report-

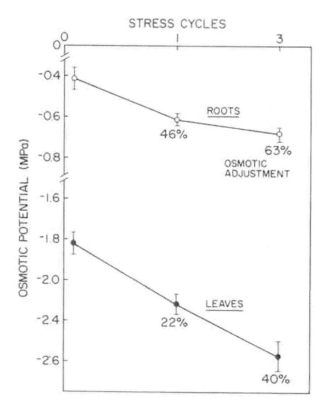

Figure 6-4. OSAD in the leaves and roots of cotton. From Oosterhuis and Wullschleger, 1987a.

Table 6-2. Magnitude and percentage osmotic adjustment in response to water stress by various crop plants. From Oosterhuis and Wullschleger, 1995.

	Osmotic adjustment			
	Magnitude		Percentage[z]	
Crop	Leaves	Roots	Leaves	Roots
	---------(MPa)--------		---------- (%) -----------	
Cotton	0.41 a[y]	0.21 a	22.4	46.3
Sorghum	0.31 a	0.19 a	25.1	37.1
Sunflower	0.17 b	0.16 a	13.9	25.2
Wheat	0.08 c	0.03 b	6.6	4.4
Soybean	0.05 c	0.00 b	4.0	-0.8

[z] Percentage osmotic adjustment refers to the percentage decrease in osmotic potential compared to the well-watered control
[y] Means within columns followed by the same letter are not significantly different at the 5% level of probability.

ed study of the role of osmotic adjustment with the growth of a root system in droughted field plants (Ball, 1992). This study showed only a small, limited amount of osmotic adjustment in the roots of field-grown cotton and a substantial adjustment in the leaves in agreement with Oosterhuis and Wullschleger (1987a). Osmotic potential of leaves varies diurnally (Hsiao, 1973), independently of daily cycles of leaf hydration. Therefore, leaves can maintain turgor during the daytime at the same level as during the night (Acevedo *et al.,* 1979). Radin *et al.* (1989) interpreted the diurnal cycling of osmotic potential in cotton as an indication of a "sink-limited" condition within the plant during the boll development period. However, there have not been any similar studies on cotton roots.

There is only a small range of genetic diversity of this trait in commercial cotton cultivars (Oosterhuis *et al.,* 1987), although Nepumeceno *et al.* (1997) recently reported significant drought tolerance in an Australian commercial cultivar, CS-50. However, a more substantial range of osmotic adjustment exists in the primitive landraces and wildtypes of cotton (Oosterhuis *et al.,* 1987). However, the role of osmotic adjustment in a cultivar bred for production as an annual crop may be quite different from that of osmotic adjustment in a perennial wildtype. Osmotic adjustment has been favored as a trait offering potential for manipulation in the breeding of drought resistant crops (Sharp and Davies, 1979; Morgan and Condon, 1986; Turner, 1986). Work in Australia on wheat (Morgan and Condon, 1986) and sor-

ghum genotypes (Ludlow and Muchow, 1988; Ludlow *et al.,*1989) has shown increased yield in high osmotic adjusting phenotypes. The yield increase in sorghum of nearly 30% over low adjusting phenotypes was related to deeper rooting resulting in more carbon fixation and increased harvest index. A clear yield advantage from osmotic adjustment in cotton has not been demonstrated. The role of osmotic adjustment in maintaining root growth, allowing water uptake longer in drying soil, has been emphasized by Acevedo and Hsiao (1974). The premise that osmotic adjustment allows for turgor maintenance and increased root growth at low water potentials implies that the plant will be able to exploit a greater and deeper soil volume for water. The role of the root system during drought is receiving current research attention as a possible sensing organ and in root-to-shoot ratios. Jones and Turner (1978) cautioned that the capacity to tolerate drought may be attributed to factors other plant water relations, such as rooting habit, conductance of water through the xylem, and desiccation tolerance.

4.2.4   Genetic Potential

The growth of the root system of cotton is under genetic control (McMichael *et al.,* 1987) but may be modified by the environment as discussed in previous sections of this chapter. McMichael (1990) has shown variability for root weight in a number of exotic cotton accessions. Variability in root/shoot ratios was also observed in these studies. Earlier, McMichael *et al.* (1985) showed genetic differences in the number of vascular (xylem) bundles in cotton taproots and suggested that variability in lateral root production was associated with the differences in vascular arrangement. Later research indicated this to be the case (McMichael *et al.,* 1987; Quisenberry *et al.,* 1981). McMichael *et al.* (unpublished data) also found genetic differences in the response of cotton seedlings to changes in temperature. Quisenberry *et al.* (1981) found differences in older plants in lateral root production as well as taproot growth. It was further suggested by McMichael *et al.* (1985) that the observed increase in the vascular system and enhanced lateral root production could lead to improved water status of the plant in drought conditions since the potential for additional water uptake and utilization might be possible. Work by Cook and El-Zik (1992) suggested that cotton genotypes having deep roots and increased lateral root production would be more drought resistant based on the variability in root traits. Oosterhuis and Wullschleger (1987b), however, were unable to show significant improvement in hydraulic properties of the plants with the increased vascular arrangement.

In field studies, Hons and McMichael (1986) showed that water extraction patterns from fallow rows of a 2x2 skip row pattern were significantly less that cotton planted in every row. This suggested that there was not sufficient rooting density in the cultivar used to use the additional water in the fallow rows. This led Quisenberry and McMichael (1996) to use a more extensive skip-row planting technique to show significant differences in rooting potential in a number of cotton genotypes by measuring differences in yield as a function of the ability of the plant root systems to extract water. This approach can be utilized to rapidly evaluate genetic differences in root development under field conditions.

5.   ENHANCEMENT OF COTTON ROOT GROWTH

5.1   Foliar Fertilization

Adverse environmental conditions during seedling development have prompted interest in foliar fertilization of seedlings to enhance growth, roots in particular, during this critical stage. Although foliar application of urea-N or boron to cotton during flowering is a widely used practice to enhance boll development, there has been relatively few studies on the benefits of foliar fertilization of cotton seedlings and none of these recorded the effect on root growth. Field research at five sites in Alabama showed that cotton yield was not influenced by the foliar application of N-P$_2$O-K$_2$O (12-48-8) fertilizer made in one to three applications at 10- to 14-day intervals starting at the 2- to 3-leaf stage (Edmisten *et al.,* 1993). However, root growth was not measured in these studies. It has been speculated that foliar fertilization could have a positive effect on droughted cotton seedlings. In growth chamber studies in Arkansas, foliar-applied urea, potassium nitrate, or Bayfolan did not improve the drought tolerance of seedlings, *i.e.,* plant-water relations were not improved for continued growth during the stress (Holman and Oosterhuis, 1992). In related studies, foliar fertilization of vegetative cotton with the commercial fertilizers Bayfolan or Solu-spray in waterlogged soil or under cool temperatures had no positive effect on growth (Holman and Oosterhuis, 1992). Again, root growth was not measured in these studies. Although plants partition a greater percent-

age of N to the roots compared to the shoots under water deficit (Hake and Kerby, 1988), the overall concentration of N is reduced and the plant is unlikely to benefit from the small quantity of foliar N (Holman, 1993). Recent research has suggested that applying foliar fertilizers after the relief of drought stress to stimulate recovery and enhance growth may be beneficial (E.M. Holman, unpublished data). It is generally accepted that the use of foliar fertilizers on seedling cotton has no significant effect on yield or root growth although there may be some growth advantage after relief of a stress.

5.1.1   Genetic Variability

Genetic variability in a number of root parameters in cotton has been shown in a previous section of this chapter, to occur across a range of environmental conditions. Quisenberry and McMichael (1996) indicated that genetic differences in rooting potential was related to plant productivity and that an increase in potential (primarily increases in root branching and distribution) could result in increases in yield of cotton under conditions of a drying soil profile. Greenhouse studies conducted using twenty-five cotton genotypes ranging from exotic accessions to commercial cultivars showed significant variability in the dry weights of root systems of sixty day-old plants. The variability was greater in the exotic accessions than in the commercial varieties (Table 6-3; McMichael and Quisenberry, 1991). McMichael *et al.* (1985) showed that the increased root xylem (vascular bundle) arrangements in the taproot of some of the exotic cotton accessions resulted in a significant increase in total vessel cross-sectional area and an increased number of lateral roots. This increase suggested an overall decrease in axial resistance to water flow in the root system which may be associated with characteristics of drought tolerance in plants with the increased xylem vessels. Oosterhuis and Wullschleger (1987a) supported the finding that increased water flux was associated with increased xylem cross sectional area. However, an increased number of vessel elements in the xylem of the primary root did not result in any apparent decrease in axial resistance to water flow. The increased number of lateral roots associated with increased vascular bundles resulting in increased xylem vessels may be important characteristics associated with drought tolerance in plants with the increased xylem vessels.

5.1.2   Plant Growth Regulators

Plant growth regulators (PGRs) have long been used by researchers and producers in an attempt to control or enhance cotton growth. Most of this attention has focused on the above ground plant parts with little attention to the root system. However, there have been a number of recent reports of synthetic PGRs enhancing root growth. IBA and mepiquat chloride (Pix™, 1,1-dimethylpiperidinium chloride) plus IBA stimulated cotton root growth, but Pix alone

Table 6-3. Mean root dry weights averaged over experiments for 25 cotton genotypes grown in the greenhouse. Plants were 60 days old at time of harvest. From McMichael and Quisenberry, 1992.

Genotype	Root dry weight
	(g)
T184	3.95
T141	3.86
T252	3.30
T283	3.12
T1	3.11
T171	3.08
T256	3.07
T461	2.83
T25	2.80
T115	2.79
T15	2.74
T1236	2.73
T185	2.47
T80	2.45
T45	2.36
Paymaster 145	2.16
Deltapine 61	2.15
G. herbaceum	2.15
Coker 5110	2.05
T151	2.03
T50	2.01
Tamcot CAMD-E	1.91
T169	1.87
Pima S-5 (*G. barbadense*)	1.77
Lubbock dwarf	1.63
LSD (0.05)	0.47

did not stimulate root growth (Urwiler and Oosterhuis, 1986). There have been reports of Pix™ decreasing root growth and increasing the root/shoot ratio (*e.g.*, Oosterhuis *et al.,* 1991a). Pix™ and Burst™ (a mixture of cytokinins) increased the terpenoid aldehyde content of cotton roots, whereas indole acetic acid (IAA), kinetin and gibberellic acid (GA) either had no significant effect or increased them (Khoshkhoo *et al.,* 1993). Oosterhuis and Zhao (1994) showed that soil-applied PGR-IV (Microflo Company, Memphis, TN) in-furrow at planting increased root length (+47%), total number of lateral roots (+23%), and root dry weight (+20%) of cotton grown in pots in a growth chamber. A number of commercially available PGRs have been shown to have beneficial affects on root growth (Oosterhuis, 1996) and seedling development of field-grown cotton, but these have not shown similar advantages in the field. While there have been numerous exogenous PGRs applied to cotton, few of these reports have paid any attention to root growth. The enhancement and stimulation of root growth in the field with PGRs is a relatively new and exciting practice, but additional research is needed to understand and confirm the benefits and to determine the best way to use this practice.

6. METHODOLOGY OF ROOT INVESTIGATIONS

There are numerous methods available to measure root growth of plants (McMichael, 1986). The selection of an appropriate method depends upon the objectives of the study as well as availability of resources such as time, labor, and money. Certain techniques are most suited for controlled environment studies, while others are adaptable to field investigations. Destructive techniques, although generally more time consuming and labor intensive, are essential if direct measurement of root surface area or root dry weight is desired, especially if the data are to be used for purposes of modeling growth or uptake of water and nutrients. Nondestructive techniques, however, can be quite satisfactory if more qualitative measurements of root growth are used to compare treatment responses.

6.1 Destructive Techniques

Destructive techniques may result in removal of a portion or all of the root system of a single or group of plants from the origin of growth or exposure of a root system such that serious alteration in plant growth occurs. The intent of destructive methods is to obtain direct access to the root system so that the roots can be analyzed directly for specific characteristics such as length, weight, volume, anatomical and morphological features, and microbial relationships. It is often assumed that direct measurements of root growth from destructive sampling are more accurate than indirect, nondestructive methods. However, destructive sampling is often more time consuming and laborious than nondestructive techniques and does not allow repeated measurements on the same plant. There is also a serious limitation on the number of samples that can reasonably be taken during a growing season because of the time required to sample and process roots and the limitations in storage space for the unprocessed samples.

One of the most common destructive techniques for measuring root growth is the soil core. Root length data obtained from this method are often used as a baseline to compare the accuracy of measuring root length by other methods such as the minirhizotron (Sanders and Brown, 1978; Parker *et al.,* 1991). This method allows direct measurement of the length of roots within a known volume of soil, often by the line intersect method (Tennant, 1975). Root length density (length of roots per unit volume of soil) can be calculated and is often reported to describe root distribution within a soil profile. However, inaccuracies in measurement of roots may result from variations in fine root recovery from soil and inability to distinguish between live and dead roots. Brouder and Cassman (1990) and Mullins *et al.* (1994) used the core method to evaluate the effect of in-row subsoiling and plant growth in a vermiculitic soil, respectively on root length densities of cotton. In addition to root length and root length density, Prior *et al.* (1994)

used the soil core method to measure cotton root volumes and dry weights in response to CO_2 enrichment.

Other destructive methods for studying root growth include the monolith and trench profile techniques. The monolith technique provides a three dimensional representation of either part or all of a root system. With this technique, large vertical sections of soil are removed with spade, probe or in boxes and soil is removed by washing with water. The size of the soil section removed depends upon the plant species studied. Needle boards are sometimes driven into the soil-root sample prior to washing so that the natural original configuration of roots remains following removal of soil. This technique allows for detailed studies of root morphology (Nelson and Allmaras, 1969) and has been an important tool to assess the impact of soil characteristics such as compaction on root growth in the field (Vepraskas *et al.*, 1986). In order to gain an appreciation for the morphological characteristics of roots without removing huge monoliths, the trench profile technique can be employed. With this technique, a vertical trench is excavated adjacent to a root system of interest. The wall face is smoothed with a spade, blade, or knife and 3- to 5-mm of soil is removed with a gentle spray of water, air, or by gently brushing away soil. Root distribution, root number, and root length can be determined with this technique and is especially valuable on row crops (Bohm, 1979).

6.2   Non-Destructive Techniques

Non-destructive techniques do not allow direct sampling and analysis of the root system of an actively growing plant. Since these methods do not allow intimate contact with plant roots, there are limitations in the type of data that can be obtained. Root dry weight can not be determined by these methods, thus precluding the calculation of root to shoot ratio on a weight basis, which is an important parameter to evaluate carbon allocation within a plant. The advantages of non-destructive sampling, however, are that measurements are usually more rapid than with destructive sampling techniques, repeated measurements can be made on the same set of plants throughout the growing season, a greater frequency of sampling of roots can often be made during a season, and root turnover rates can be measured to assess root death, root initiation, and possibly root contribution to soil carbon (Cheng *et al.*, 1991).

The most common of these techniques are the rhizotron and minirhizotron. The rhizotron is an underground laboratory lined with transparent walls that make up the exposed face of lined chambers or cells. These cells are often lined with concrete and open at the surface. Soil is placed in the cells to an appropriate bulk density and water content and planted. Throughout the growth of the plants, roots that come in contact with the transparent wall are traced and root length calculated. New roots are traced with different colored pencils each day in order to identify the most recent growth. Sometimes roots are exposed to ultraviolet light and those that fluoresce are identified as roots that are active in at least nutrient uptake (Dyer and Brown, 1980). Root branching and root color can also be determined with this technique. The rhizotron is the only technique that allows the scientist to co-exist at the same soil depth as the actively growing roots of a plant. Direct visual observations can be made without the aid of a camera or mirror. However, several disadvantages of this technique have limited adoption of the rhizotron by most root researchers. These include cost of building the structure, time and effort required to empty and refill each cell with soil, and the limited number of cells in a rhizotron which artificially restricts the number of treatments that can be evaluated.

A miniaturized version of the rhizotron, the minirhizotron, has become a focus of product development and root scientists over the last two decades of the 20th century because of it's portability, accessability, and lower labor and financial investment. The minirhizotron, like the rhizotron, is a non-destructive technique that employs the use of a transparent wall below ground. But unlike the rhizotron, the minirhizotron does not require the construction of a large, permanent facility in which soil must be prepared and carefully placed into a limited number of cells prior to making root measurements. The minirhizotron refers to the hollow, transparent tube that is placed in excavated holes in which a camera is usually inserted sometime during the growing season to record roots of plants at the soil-tube interface. Scientists have used tubes made of rigid materials such as glass, plexiglass, and poly butyrate, or inflatable materials (Gijsman *et al.*, 1991) that are shaped into cylindrical or rectangular dimensions and placed vertically, horizontally, or at an angle to the soil surface. Historically, the root image has been obtained with a mirror (Meyer and Barrs, 1985), 35-mm single lens reflex camera (Sanders and Brown, 1978), black and white video camera (Dyer and Brown, 1980), and with a color micro-video camera recording system (Upchurch and Ritchie, 1984). A disadvantage of the minirhizotron is that it appears to underpredict root length near the soil surface in comparison to the soil-core technique. Suggestions for this under prediction have been light leakage at the soil surface into the tube (Levan *et al.*, 1987), compaction of the soil surface during tube placement, and differential drying near the soil-tube interface at the soil surface. However, Merrill and Upchurch (1994) found a predicable relationship between the numbers of cotton roots measured with the minirhizotron and the bulk soil root length density. Although this technique has been used to study the root growth of a number of agronomic and horticultural crops, only recently has it been reported on cotton. Bland and Dugas (1989) used horizontally placed minirhizotrons to measure cotton root growth and soil water extraction from lysimeters during a 2-yr period. Keino *et al.* (1994) compared root growth of six cotton cultivars representing three maturities under irrigated and non-irrigated conditions. As with the rhizotron, the minirhizotron is a valuable tool for comparison of growth characteristics. In addition to having many of the advantages as the rhizotron, the minirhizotron is portable, has a lower startup investment, and is less labor intensive.

6.3 Other Techniques

There are a number of other techniques besides those mentioned that have been used to study root growth of plants. Several excellent papers have provided overviews of many of these techniques (Bohm, 1979; Upchurch and Taylor, 1990). Often one technique is not satisfactory to obtain all the desired information about root growth. Root scientists should be familiar with several techniques and use the appropriate ones as the need arises. Considerable information about root growth is needed to provide a more complete understanding about whole plant growth and to develop strategies to increase the efficiency of production practices. Techniques to study root growth are tools to be used by the scientist. They should not drive the type of research conducted. Rather, they should provide the means by which the desired information is to be obtained.

7. CONCLUSIONS AND FUTURE DIRECTIONS

The growth and development of the root system of cotton has been shown to be genetically controlled, but sub-

ject to modifications by a wide range of both above- and below-ground environmental conditions. The overall productivity of the plant is, therefore, influenced by the integrated response of the roots to environmental stimuli. In this chapter we have briefly touched on how the cotton root system initiates and grows as well as a number of physical, chemical, and biological factors that influence root development. We have also presented some strategies for enhancing root growth in cotton such as taking advantage of genetic variability and the use of plant growth regulators. Finally, we have discussed some current methods for investigating root growth both in the field and in plants grown in the greenhouse or growth chamber. Since these techniques are readily available and can be incorporated into most cotton research programs, future work should not neglect the importance of taking into account the development of the root system in evaluating cotton growth and productivity. As molecular biology continues to make inroads into our understanding of plant development and presents the possibilities for genetic engineering of plant growth processes, the opportunity also exists for manipulating the growth and development of the root system. These advances coupled with the new concepts of precision farming for example, may provide the means for maximizing cotton root system for and function for maximum plant productivity.

Chapter 7

TEMPORAL DYNAMICS OF COTTON LEAVES AND CANOPIES

G.A. Constable[1] and D.M. Oosterhuis[2]
[1]*Australian Cotton Cooperative Research Centre, CSIRO Cotton Research Unit, Locked Bag 59, Narrabri NSW, 2390, Australia.*
[2]*Department of Crop, Soil, and Environmental Sciences, University of Arkansas, Fayetteville, AR 72701, USA.*

1.   INTRODUCTION

Photosynthesis and its carbohydrate products are the fundamental building blocks of most crops. That situation is especially the case with cotton: the primary product for harvest and profit is cellulose (lint) which is 99% carbohydrate.

The temporal dynamics of cotton leaf photosynthesis are affected by many factors. Internal (plant) factors include leaf age and competition between organs within the plant for nutrients. External (environmental) factors affecting photosynthesis include light, temperature, water status, soil fertility, pests, diseases, and competition from weeds or neighbors. Although many factors are important, this chapter concentrates on leaf age and light because of their importance in cotton's canopy photosynthesis profile. The age of a leaf, its light history and competition history are strongly correlated because as a leaf ages, it is more shaded by leaves subsequently formed and also has fruit developing nearby.

The use of simulation models and carbon budgets as research tools has expanded in the past two decades at many levels of plant organization. Single leaf studies enable calculations of export (this chapter; Ho *et al.*, 1984; Shishido *et al.*, 1990); whole plant studies allow us to refine understanding of the use of storage for grain filling (Hall *et al.*, 1990) and the relative energy requirements for assimilation of NH_4 or NO_3 (Jeschke and Pate, 1992; Cramer *et al.*, 1993). Whole tree budgets for apples (DeJong and Goudriaan, 1989; Wibbe *et al.*, 1993) or wood production (Deleuze and Houllier, 1995) have been vital in research where the size of plants prohibits intensive sampling. Carbon budgets for forests (Singh *et al.*, 1991; Lugo and Brown, 1992; Kurz and Apps, 1994) or forest systems (Delcourt and Harris, 1980)

are the only way to evaluate the role of different systems or activities in cycling and as sinks for carbon when the size ($x10^3$ ha) and time scale (decades) also prevent direct measurement.

Carbon budgets will be utilized in this chapter to illustrate the consequences of photosynthetic patterns for cotton leaves and canopies. In *Cotton Physiology* (Mauney and Stewart, 1986) some chapters dealt with important background material. Jensen (1986) presented the biochemistry of photosynthesis and Mauney (1986b) discussed carbohydrate production and distribution in cotton canopies. Krizek (1986) covered photosynthesis light response of cotton leaves. Mauney (1986b) compared the measured rates of leaf net photosynthesis in cotton at about 25 $\mu molesCO_2/m^2/sec$ (maximum 31) to canopy net photosynthesis rates of up to 73 g $CO_2/m^2/day$ to produce measured rates of crop dry matter growth rates up to 30 g/m^2/day. Baker *et al.* (1972) used measurements such as these to calculate the potential cotton lint yield to be over 3000 kg/ha.

2.   PLANT-CROP MORPHOLOGY

Mauney (1986a) stated that "The cotton plant has perhaps the most complex structure of any major field crop. Its indeterminate growth habit and sympodial fruiting branches cause it to develop a four-dimensional occupation of space and time which often defies analysis." Understanding the growth habit of cotton has been a special challenge for crop physiologists for many years. The ultimate objective is to apply this knowledge to manage the crop to improve yield, earliness, quality, and other characters. Some examples of the application of physiology to cotton crop management are presented later in this chapter.

J.McD. Stewart et al. (eds.), *Physiology of Cotton*,
DOI 10.1007/978-90-481-3195-2_7, © Springer Science+Business Media B.V. 2010

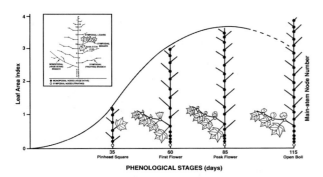

Figure 7-1. Progressive development of plant development and leaf area showing the main phenological stages and main-stem nodes, with the leaf and boll development at main-stem node five shown in more detail. Note the branches are depicted diagrammatically in an alternate arrangement but actually arise in a three-eights phyllotaxis. Inset shows the general structure of the cotton plant.

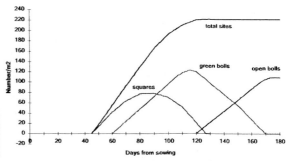

Figure 7-2. Seasonal pattern of fruiting dynamics showing the production of fruiting sites and the development of squares to green bolls then open bolls. In this example, 49% of total sites survived to the open boll stage.

The morphology of cotton is described by Oosterhuis and Jernstedt (1999) and the fruiting branch structure is shown in Figure 7-1. A typical pattern of fruiting in a cotton canopy is shown in Figure 7-2. Hearn and da Roza (1985) and Brook *et al.* (1992a) present examples of actual patterns. In the Figure 7-2 example, the crop begins to produce squares on day 45 and by day 60 these earliest squares have reached the flower stage, so the number of green bolls in the canopy begin to increase. The earliest bolls begin to open by day 120 and green boll numbers decline as bolls open progressively towards the top of the plant.

The dynamics of fruiting have been well documented by Hearn and da Roza (1985), who present the concepts of "feed forward": where there is a rapid increase in fruiting site initiation/production for larger plants; then "feedback" where boll load reduces the rate of site production and fruit retention (new branch and main-stem nodes). "Carrying capacity" is an important concept consistent with the nutritional hypothesis of fruit survival in cotton (Ehlig and Lemert, 1973; Mauney, 1986b; Turner *et al.*, 1986). These concepts are shown in Figure 7-2 as:

- A decreased rate of site production once a significant boll load is on the plant after day 90.
- A high rate of square and boll retention in the early part of the season (lower fruiting branches), then a reduced retention for later fruit (Ehlig and Lemert 1973; Turner *et al.*, 1986). The change in retention pattern is evident in Figure 7-2 as a reduced slope of the successive sites, squares, green bolls and open boll curves.

Fruiting dynamics are also popularly analyzed in plant mapping where the horizontal axis of Figure 7-2 becomes the vertical axis (stem) of a cotton plant. The change in fruiting site production and retention at different positions on a plant are shown by Kerby *et al.* (1987) and Constable (1991). The reduced retention of later fruit in those studies is clearly shown to be from upper nodes or outer fruiting branch positions. This type of mapping data is now in common use to benchmark field crop growth. Hormones are clearly involved in morphology, fruiting and shedding in cotton (Guinn, 1986), either directly or as the message carrier during growth and development (Hearn and Constable, 1984).

Cotton is definitely unique in its perennial nature and indeterminate growth habit; many leaves and fruit are produced - particularly with developing fruit(s) present at the bottom of the leaf canopy and adjacent to older and more heavily shaded leaves. This is in contrast to the growth habit with other field crop plants such as wheat, grain sorghum, or sunflower: the developing fruit(s) are at the top of a canopy, near to youngest leaves and in full light. The consequences of that difference in growth habit on the need for assimilate movement and dependence of fruit on local leaves for assimilates will be presented and discussed in the following sections. Fundamental to the understanding of cotton and its unique growth habit is the historical fact that cotton is a tropical perennial shrub grown as an annual in many temperate locations and in modern times usually as a row crop.

3. LEAF MORPHOLOGY

Results from measurements of leaf size profiles in unmodified cotton canopies show that the largest leaves are produced in the lower to central regions of the plant (McClelland, 1916; Hearn, 1969b; Constable and Rawson, 1980b; Mutsaers, 1983; Constable, 1986). Leaves on fruiting branches are smaller than the corresponding main-stem leaf at the same node position by a factor of about 0.55, 0.4, and 0.3 for the first three positions on a fruiting branch, respectively (Hearn, 1969b; Constable and Rawson, 1980b; Wullschleger and Oosterhuis, 1987a). A typical profile of final leaf size on intensively managed upland cotton plants is shown in Figure 7-3. In this example, the largest leaves (140 cm^2) are on the main-stem at node 8. Largest leaves on the fruiting branch at that position in the canopy are only 77 cm^2.

Mutsaers (1983) presented a detailed physiological explanation for these leaf size patterns. For lower leaves during early growth, competition exists for assimilates

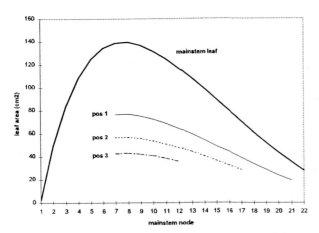

Figure 7-3. Typical profile of final area of individual main-stem and fruiting branch leaves on a cotton plant (pos 1, pos 2, and pos 3 refer to the first three leaves on each fruiting branch). This example has no vegetative branches and the first fruiting branch at node 7. The total area of leaves is 3500 cm² if plant density was 10/m² giving a maximum leaf area index of 3.5.

between the leaf primordia within the growing point. For upper leaves, the competition is between leaves and developing fruit. As a consequence of these two stages of competition, the middle region of a plant has the largest leaf size because the competition in young seedlings has decreased and competition with fruit has yet to begin.

Individual leaf growth follows a typical sigmoid pattern, with the duration of leaf area expansion and specific leaf weight increase being relatively constant at about 18 days between unfolding and final leaf area at normal temperatures (Mutsaers, 1983; Constable, 1986).

4. LIGHT INTERCEPTION

Light penetration through a crop canopy is usually related to leaf area. Leaf size, arrangement, angle, and crop density/row space will influence the relationship. Saeki (1963) analyzed light interception data as:

$$I/Io = exp(-k * LAI) \qquad (7\text{-}1)$$

where I is light at ground level; Io is light above the crop canopy; LAI is leaf area index; and k is the light extinction coefficient. The value of k varies from 0.4 in canopies of erect grass leaves to more than 1.0 in canopies of broad leaf crops. Measurements of k for cotton canopies range from 0.5 to 1.0 (Hearn, 1969b; Mutsaers, 1980; Constable, 1986; Heitholt *et al.*, 1992; Sadras, 1996b). For a cotton canopy with LAI of 3, about 95% of incident light is intercepted by the plants but light penetration through a canopy can be affected by plant spacing (Constable, 1986; Heitholt *et al.*, 1992). Cotton stems and bolls can intercept up to 20% of light (Constable, 1986), so equation 7-1 may not be linear (Fasheun and Dennett, 1982). In addition, within a plant, a cotton leaf's diaheliotropic movement (Lang, 1973;

Ehleringer and Hammond, 1987) may increase access to light for leaves on the end of branches, giving a calculated daily photosynthesis increase of 9% when compared with a horizontal leaf (Constable, 1986).

An understanding and description of the light interception/penetration in cotton canopies is required to calculate photosynthetic rates of individual leaves. Light receipt by an individual leaf will be determined by the total leaf area above that leaf and the extinction coefficient of the canopy above that leaf, as in equation 7-1. The frequency distribution of light in a crop canopy is such that a leaf is either in the sun or in the shade, not at half light. In fact, the same leaf can alternate from shade and sun during the day, particularly up to mid morning and after mid afternoon (Whitfield and Connor, 1980).

During early growth especially, a leaf will not stay at the top of the canopy for very long because new leaves are produced about every three days (Munro, 1971; Hesketh *et al.*, 1975; Constable, 1986). It is possible that six new leaves have appeared above a leaf in the time it has reached 18 days of age. Using leaf sizes shown in Figure 7-3 with equation 7-1, a leaf at node 18 has 300 cm² of leaf area (LAI of 0.3) above it, meaning that approximately 25% of total solar radiation is intercepted by the leaves on nodes 19 to 22. For leaf positions lower in the canopy, that degree of shading is greater because the leaves above that position are larger (Fig. 7-3).

5. PHOTOSYNTHETIC LIGHT RESPONSE

A typical light response curve for net photosynthesis of a cotton leaf is shown in Figure 7-4. This example using parameters measured by Constable and Rawson (1980a), is for a leaf near the top of a canopy at its peak of photosynthate production where 95% of peak net photosynthesis occurs at about 50% of full sunlight. Instantaneous dark respiration is less than 5% of instantaneous light saturated photosynthesis and this leaf requires only 0.5 % of full sunlight to be a net carbon exporter. This general pattern of photosynthesis light response in cotton has been measured by many authors (Burnside and Bohning, 1957; El-Sharkawy *et al.*, 1965; Pasternak and Wilson, 1973; Patterson *et al.*, 1977; Constable and Rawson, 1980a; Krizek, 1986).

Figure 7-5a shows the diurnal pattern of net photosynthesis for a young and old leaf with the light response curve of Figure 7-4. The curve for the young leaf shows a sharp response to light after sunrise and before sunset. In addition the saturation of net photosynthesis above 1500 μmoles/m2/sec is evident as a plateau in the diurnal pattern. The curve shape for an old leaf emphasizes the shading which occurs at the bottom of a canopy, plus the reduction in net photosynthesis for an old leaf (discussed below and Figure 7-6). In this example, the older leaf exhibits less than 20% of the net photosynthesis rate of a young, upper leaf. This difference

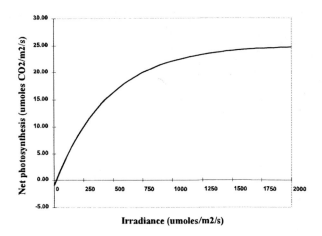

Figure 7-4. A typical response to irradiance of net photosynthesis of a young fully-expanded cotton leaf.

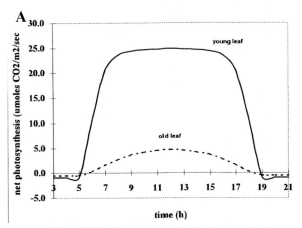

Figure 7-5a. Time course of net photosynthesis for a young fully-expanded cotton leaf at the top of the crop canopy and a 50 day old leaf at the bottom of the crop canopy. This day had daylength of 13 h, maximum irradiance of 2300 μmoles/m²/sec and 15% of light transmission to the lower leaf.

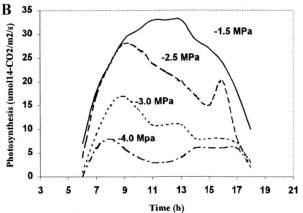

Figure 7-5b. Time course of net photosynthesis in response to water deficit (minimum leaf water potential shown) for a young fully-expanded cotton leaf at the top of the crop canopy (from Turner *et al.*, 1986).

in photosynthetic production has consequences on local assimilate supply and necessitates longer distance transport of assimilates to bolls at this position in the canopy.

Pallas *et al.* (1967), Parsons *et al.* (1979), Turner *et al.* (1986) and Puench-Suanzes *et al.* (1989) also demonstrated the strong relationship between cotton crop water status and leaf or canopy photosynthesis. Water stress has a strong effect on net photosynthesis, particularly below 50% plant available moisture (Constable and Rawson, 1982): as a drying cycle progresses, the reduction in net photosynthesis is first evident in the mid-afternoon when temperature and vapour pressure deficit are highest (Turner *et al.*, 1986). The decline in response of photosynthesis to light between morning and afternoon found by Baker (1965) is also possibly due to water stress although Pettigrew and Turley (1998) found mid-afternoon reductions in net photosynthesis even in well-watered plants. This was proposed to be inhibition of the photosynthetic apparatus to dissipate heat. These reductions in mid-afternoon net photosynthesis are shown in Figure 5b.

6. PHOTOSYNTHESIS PATTERN WITH LEAF AGE

The pattern of light saturated net photosynthesis for a cotton leaf through its development and aging (Fig. 7-6) has been measured in cotton by a number of research groups (Elmore *et al.*, 1967; Brown, 1973b; Nagarajah, 1975; Constable and Rawson, 1980a; Cornish, 1988; Krieg, 1988; Wullschleger and Oosterhuis, 1990a). Maximum net photosynthesis rates occur for a leaf about 16 days after unfolding when the leaf is about 75 to 90% of final leaf area. After a short plateau around 24 days of age, the rate of net photosynthesis declines slowly, reaching near zero by leaf senescence at about 70 days. Concomitant with this reduction in photosynthesis, light saturation occurs at about 25% of full sunlight for older leaves (Constable and Rawson, 1980a).

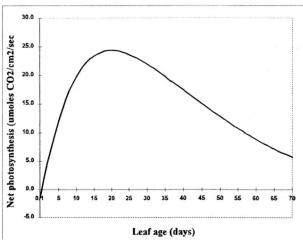

Figure 7-6. The typical pattern of light saturated net photosynthesis of a cotton leaf for the 70 days following leaf unfolding.

At least two research laboratories have studied photosynthesis patterns of leaves during development and ageing in a range of species. In Australia, Rawson found similar ageing patterns for photosynthesis in a range of crop species including tobacco (*Nicotiana tabacum*, Rawson and Hackett, 1974), soybean (*Glycine max*, Woodward and Rawson, 1976), cotton (*Gossypium hirsutum*, Constable and Rawson, 1980a), sunflower (*Helianthus annuus*, Rawson and Constable, 1980a), pigeonpea (*Cajanus cajan*, Rawson and Constable, 1980b) and wheat (*Triticum aestivum*, Rawson *et al.*, 1983). In Czechoslovakia, Catsky *et al.* (1976) studied ontogenic changes in photosynthesis of bean (*Phaseolus vulgaris*). Generally for both research groups, all leaf positions in the crop canopy (either along a branch, or different positions on the main-stem), had a similar overall pattern of aging as described above.

What are the reasons for changes in photosynthesis as a leaf ages? The photosynthetic characteristics of leaves as affected by age and insertion level on the plant have been renewed by Sestak and co-workers in a series of articles from 1977 to 1983. The reviews covered numerous plant species including cotton. Separate reviews covered chlorophyll (Sestak, 1977a), electron transport (Sestak, 1977b), carotenoids (Sestak, 1978), enzymes (Zima and Sestak, 1979), CO_2 compensation concentration (Ticha and Catsky, 1981), intercellular conductance (Catsky and Ticha, 1982), stomata (Ticha, 1982) and stomatal reaction (Solarova and Pospisilova, 1983).

From these reviews it was clear that all components of the photosynthetic apparatus change in parallel with leaf ontogeny. Compared with a fully-expanded leaf, a developing leaf has less chlorophyll, less enzymes, and less developed stomata; an aging leaf also has reduced chlorophyll and enzyme activity (also shown in cotton by Wells, 1988). In general there was a slightly greater change in mesophyll conductance than in stomatal conductance during leaf ageing, probably because more than 75% of total resistance in a leaf is in the mesophyll (Catsky and Ticha, 1982). Ticha and Catsky (1981) surmised that CO_2 compensation concentration, net photosynthesis, gross photosynthesis, light respiration, dark respiration, mesophyll resistance, and photosynthetic enzyme activity were all highly correlated during leaf ontogeny. In other words, all the changes to photosynthetic characteristics of a leaf were associated or at least changed in parallel through their lifetime. Such balance in developmental sequence makes sense as there is little evolutionary advantage in a leaf investing in protein resources in a leaf with low potential/capacity for photosynthesis due to shading. Furthermore, withdrawal of nutrients from a leaf in the time leading up to senescence allows for these nutrients to be redistributed.

The pattern of photosynthetic activity for cotton leaves is typical of most other plants. Accelerated reduction of leaf photosynthesis with lower leaves of cotton was attributed to light history or nitrogen status (Wullschleger and Oosterhuis, 1990b). Also, low light induced photosynthesis reduction has been measured in cotton (Nagarajah, 1976)

and it is reasonable to expect such an effect, especially in dense leaf canopies. Resumption of upper leaf growth once boll growth has ceased also shows the role of competition from bolls in affecting leaf ageing (Wells, 1988).

7. CARBON BUDGET - LEAF

The pattern of carbon assimilation by a leaf can be used to calculate potential export from leaves within the crop canopy. Information at that level is important for our understanding of cotton physiology because of the indeterminate growth habit. Labelling studies have shown that the leaves adjacent to a boll are the most important in providing assimilates to that boll (Ashley, 1972; Horrocks *et al.*, 1978); although other leaves above a boll also provide assimilates (Constable and Rawson, 1982; Brown, 1973a; see below). The distribution of leaf area in the canopy in relation to boll weight has also been described (Oosterhuis and Wullschleger, 1988). The pattern of boll set and growth is therefore affected by the pattern of assimilate production and distribution within the plant. Crop management to manipulate yield and earliness of cotton relies on accurate knowledge and exploitation of this physiology.

The values of dark respiration, leaf growth, and net photosynthesis can be integrated to estimate daily net potential carbon export for a leaf (Constable and Rawson, 1980b). Figure 7-7 shows the results of those calculations for a single cotton leaf with a final leaf area of 100 cm². Daily demands for leaf growth peak at 20 mg C on day eight; the leaf is exporting a similar amount of carbon by that age. Daily (24h) leaf respiration of 5 mg C peaks at about day 15; the leaf is exporting more than 20 times that amount at this age. Daily export of carbon from a leaf peaks at about 100 mg C at age 23 days, equivalent to about 1 mg C/cm²/day.

Similar calculations can be made for minerals. Measurements of nitrogen (N) concentration of cotton leaves as they emerged, expanded and aged show that a leaf imports N up to about 23 days of age (Thompson *et al.*, 1976). Comparison of the carbon and nitrogen budgets at a leaf level show one phase of leaf development from age 7 to 23 days, where a leaf exports C and imports N. Younger leaves import both C and N; older leaves export both C and N.

For modelling purposes, potential photosynthesis is determined by developmental age using the types of patterns illustrated in Figure 7-6. Puech-Suanzes *et al.* (1989) estimated canopy photosynthesis from leaf photosynthesis using similar relationships. Wullschleger and Oosterhuis (1992) used age class dynamics to calculate leaf area and carbon production profiles in cotton canopies. Peng and Krieg (1991) showed that leaf photosynthesis decreased as a cotton crop aged from 70 to 115 days, canopy photosynthesis reached a maximum at 80 to 90 days, then decreased. Similarly, Bourland *et al.* (1992) showed canopy photosynthesis in cotton was maximized at 60 to 70 days and then decreased sharply. These measured patterns are consistent with results from classical growth analysis (Hearn, 1969b; Constable and Gleeson, 1977) where Net Assimilation Rate decreased with canopy age even at similar total leaf areas.

The pattern and rates of development, photosynthesis and respiration are very similar for different leaves on the fruiting branch and insertion levels on the main-stem (Constable and Rawson, 1980a; Ticha and Catsky, 1981; Wullschleger and Oosterhuis, 1989b). Thus, the pattern presented in Figure 7-6 is repeated at regular intervals along a fruiting branch and up the main-stem. Additional assimilates are required for stem growth and respiration associated with plant structure. Leaf-stem ratios for different positions within the cotton plant have been measured (Constable and Rawson, 1980b) and allow the assimilate requirement for stems to be estimated at the same time as their associated leaves.

8. CARBON BUDGET - BOLL

Figure 7-7 shows potential export for a leaf. How do those values compare with requirements of growing fruit? Growth analysis can measure fruit growth (eg. Hearn, 1969b) and estimates or measurements of respiration can allow similar carbon budgets for fruit for comparison with leaf export (Inamdar, 1925; Baker, 1965; Constable and Rawson, 1980c; Wullschleger and Oosterhuis, 1992).

Bracts and boll walls are capable of photosynthesis, at about 10% of the rate of leaf tissue (Morris, 1965; Brown, 1968; Ashley, 1972; Elmore, 1973; Benedict and Kohel, 1975; Constable and Rawson, 1980c; Wullschleger and Oosterhuis, 1990c). Reduced numbers of stomata and less chlorophyll, particularly in the boll wall, have been associated with the low rates of photosynthesis (Brown, 1968; Elmore, 1973; Patterson *et al.*, 1977; van Volkenburgh and Davies, 1977; Nagarajah, 1978; Bondada *et al.*, 1994).

Estimates show the fruit can provide between 4 and 10% of their own carbon requirements (Benedict and Kohel, 1975; Constable and Rawson, 1980c; Wullschleger and Oosterhuis, 1990d). Some of the carbon respired by fruit is recycled through photosynthesis in cotton (Wullschleger *et al,.* 1991) and soybean (Sambo *et al.*, 1977).

The net import requirements for a square increase from about 2 mg C/day to 15 mg C/day between appearance and the day before anthesis. A high rate of respiration on the day of anthesis means 25 mg C is required on that day. The high rate of respiration on the day of anthesis is notable and has been measured in both cotton (Baker and Hesketh, 1969; Constable and Rawson, 1980c) and sunflower (Rawson and Constable, 1980a). The calculated daily photosynthesis, respiration and growth for a cotton boll is shown in Figure 7-8. The pattern shows the small positive photosynthesis from bracts and boll wall during daylight up to 10 days past anthesis, with increasing dark respiration reaching 20 mg C/day by day 40. Growth requirements for a boll peak at about 90 mg C/day on day 20. Combining growth and respiration requirements for a boll gives a peak import requirement of about 100 mg C/day at 25 days after anthesis.

Combining the leaf area profile (Fig. 7-3) with the carbon budget calculations (Fig. 7-7) will produce a profile of potential carbon export for a main-stem segment of leaf, stem, branch and fruit. These patterns are shown in Figure 7-9 for main-stem nodes 5, 7 and 13. These data show the importance of leaf size and leaf ageing on assimilate production profiles. The peak daily demand by a cotton boll of about 100 mg C, is strikingly similar to the production by a single leaf (Fig. 7-7). However the timing of the supply and demand are not synchronised (Hearn, 1969b; Constable and Rawson, 1980b; Wullschleger and Oosterhuis, 1990d). This difference is particularly evident for the bottom (lower) fruit positions on the cotton plant: peak production of carbon by bottom leaves is about 25 days before peak demand of bottom bolls, so at the time of peak boll growth in that region, the boll must obtain carbon from other (upper) leaves and/

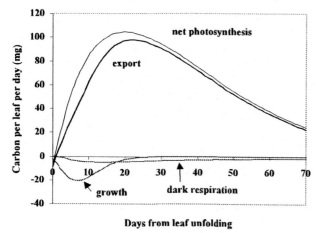

Figure 7-7. The calculated carbon budget for a cotton leaf with a final area of 100 cm². Export of carbon is leaf growth and respiration requirements subtracted from integrated daily net photosynthesis for a day of full sun (from Constable and Rawson, 1980b).

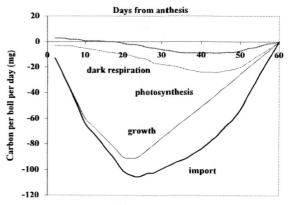

Figure 7-8. Calculated carbon budget for a cotton boll, showing daily photosynthesis, dark respiration, growth, and total import required. This example is for a boll with final weight of bracts, boll wall, seed and lint of 7 g (from Constable and Rawson, 1980c).

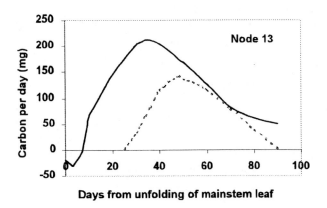

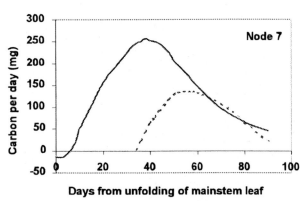

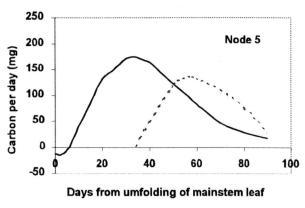

Figure 7-9. Calculated carbon budget for three main-stem nodes comprising main-stem leaf, stem, fruiting branch, with three leaves and one boll. These values are for full sun (from Constable and Rawson, 1980b).

or have a reduced growth (Fig. 7-9). This is illustrated for main-stem node 5 at 60 days where all the leaves at that node only provide half the carbon requirements for a boll at that location. Longer-distance movement of assimilates from upper leaves will be required to support those lower bolls.

Redistribution of assimilates must be a fundamental process within the cotton plant. During vegetative growth the largest leaves are providing assimilates for new growth, including new leaves. During boll growth these same upper leaves are now required to provide supplementary assimilates to lower bolls because leaves in that zone are old and shaded: [14]C label applied to pre-squaring plants has been

subsequently found in bolls on those plants, confirming redistribution (Constable and Rawson, 1982).

9.   SIGNIFICANCE FOR MANAGEMENT OF COTTON

Boll size profiles measured in cotton canopies are similar to the leaf size profile shown in Figure 7-3 (Kerby *et al.*, 1987; Constable, 1991), emphasizing the importance of local assimilate supply in determining boll growth rates and confirming labeling studies mentioned above. Largest bolls are found near the bottom of a cotton plant, on about the third fruiting branch and with bolls closest to the main-stem on each fruiting branch. Loss of fruit from insect, disease, or stress will create an opportunity for compensation where other fruit on the same branch, or elsewhere on the plant will survive instead of being shed; or grow larger (Hearn and Room, 1979; Constable, 1991; Sadras, 1995). The loss of fruit from the first position means bolls on the second or third position will grow larger (Constable, 1991). Compensation on a whole canopy basis is an important aspect of damage thresholds in cotton pest management (Brook *et al.*, 1992a; Sadras, 1995) and up to 40% of early fruit can be lost before economic damage can be measured.

Fibre quality profiles are similar to boll size profiles (Kerby and Ruppenicker, 1989). Lower and upper bolls have reduced fibre quality compared with middle bolls. Lower bolls generally have better temperature conditions and less competition, although rank crops may have reduced fibre properties from lower bolls (Hearn, 1976). Upper bolls may be maturing during cooler temperatures and are competing with other bolls at a time of reduced assimilate supply from older leaves (Fig. 7-9).

The morphology of a cotton plant and these patterns of photosynthate production emphasize many aspects which assist in understanding field management and crop response to climate. The close association between leaf size patterns on a plant and subsequent boll size patterns confirms the heavy reliance of bolls on local photosynthate. The carbon budgets presented here also indicate longer distance movement of assimilates is required. On this basis alone a good balance between vegetative and reproductive growth is very important – it would be desirable to have larger leaves, but to do so would create considerable shading of lower leaves. Landivar *et al.* (1983) demonstrated from modeling that maintaining leaf photosynthesis instead of aging would improve yields. Management of fertilizer, irrigation, and growth regulators are aimed at assisting crop agronomists to maintain the required balance. Growth regulators have become a common management tool in cotton. The ability to restrict unnecessary vegetative growth can increase yield and earliness under some conditions (Kerby, 1985; Cothren, 1994). Guidelines on decisions for growth regulator use have been based on optimizing vegetative growth during early flowering (Constable, 1994).

Cotton is normally grown on rows about 100 cm wide with a plant density about 10/m². This spacing is to facilitate many aspects of management, especially harvest. Ultra-narrow row (UNR) cotton is a production system where row spacing maybe 25 cm, with a plant density of 25/m². Finger-stripper harvesters are required. The UNR system was researched more than 30 years ago (Briggs *et al.*, 1967) and adoption is now being facilitated by developments in growth regulators, insect- and herbicide-tolerant transgenics, as well as harvesting equipment. A higher plant density means fewer bolls are required per plant, potentially reducing the time to set the same yield. Applying the morphology and photosynthesis patterns presented here to UNR systems will emphasize the importance of balancing vegetative growth. Heavy shading of lower bolls will prevent or inhibit their development (Brown, 1971) negating the intention of UNR to achieve earliness.

Cotton's morphology and physiology may be more complicated than some other field crops, but the package works well under ideal conditions. There are examples in Australia of commercial yields near the 3000 kg lint/ha potential calculated by Baker *et al.* (1972). Crop management and plant breeding have both played a role in the progress over 30 years. The challenge for scientists and managers is to increase the proportion of cotton fields which reach their potential in a sustainable way.

10. SUMMARY

The morphology of cotton plants and canopies and the pattern of photosynthate production of leaves are key characteristics of cotton which make it very different than most other field crops. An indeterminate growth habit and relatively long development periods for fruit are important components of fruiting dynamics. There is now considerable literature to show that the nutritional hypothesis of fruit survival can be used to understand fruiting dynamics, with concepts such as feed forward, feedback, and carrying capacity explaining fruiting site generation and fruit shedding.

The largest leaves on a cotton plant are produced on the main-stem in the lower to central regions of a plant. Typical large leaves are about 140 cm². Leaves on fruiting branches are usually half the area or less than the main-stem leaf on the same node. Competition with other leaves, and particularly with fruit, has been proposed as the mechanisms behind these leaf size patterns. Leaves appear at about three day intervals and each leaf takes about 18 days to reach full size. As a result, by the time a leaf reaches full size, there may be six new leaves produced above it, intercepting about 25% of solar radiation – a significant degree of shading.

Light-saturated net photosynthesis of cotton leaf peaks when the leaf nears full size, then declines, reaching near zero by leaf senescence. All components of the photosynthetic apparatus change in parallel with leaf ontogeny. Compared with a fully expanded leaf, a developing leaf has less chlorophyll, less enzymes, and less developed stomata; an aging leaf also has reduced chlorophyll and enzyme activity. Withdrawal of nutrients from senescing leaves to better placed leaves or to fruit is a strategy to best utilize plant resources.

It was calculated that a young fully expanded leaf could produce about 1 mg carbon/cm²/day and that the peak carbon demand by a cotton boll was about equal to the leaf production. However, the timing of photosynthate supply from leaves and demand from fruit are not synchronized: at lower levels on a cotton plant, peak leaf photosynthate is about 25 days before peak demand by adjacent bolls. Therefore, longer distant transport of assimilates is required to support lower bolls. It was concluded that a good balance between vegetative and reproductive growth is required to optimize fruit initiation, boll setting, and yield.

Chapter 8

COTTON SOURCE/SINK RELATIONSHIPS

Donald N. Baker[1] and Jeffery T. Baker[2]

[1]Baker Consulting,1230 Morningside Drive, Starkville, MS, 39759; and [2] Remote Sensing and Modeling Laboratory, USDA, ARS, 007, Rm.008, 10300 Baltimore Ave., Beltsville, MD 20705.

1. INTRODUCTION

Metabolite source/sink relationships govern assimilate partitioning, developmental rates, and fruit abscission in cotton. This subject is, therefore, of primary importance in the improvement of cotton plant types and in cotton culture. Here, we focus on research which has led to an understanding of metabolite source/sink interactions and secondary physiological effects resulting from those interactions. Much of this research has been done in controlled environments and some of it has been aimed at the development and testing of crop simulation models.

In nature, variations in source/sink relations result from stresses. A convenient definition of 'stress' is any factor that reduces organ growth below its genetic potential at a given temperature. Here growth is defined as dry matter accretion. In general, stresses should be thought of as syndromes in which various physiological processes and even various organs are affected at different stages of stress development. For example, as drought stress becomes increasingly more severe, photosynthesis is reduced from its maximum rate before leaf growth is affected (Boyer, 1970) and leaf growth is reduced before that of root growth. Moreover, within a tissue, cell elongation is far more sensitive to drought than cell division and the developmental period of the cell may be extended, unless significant osmotic adjustment occurs (Meyer and Boyer, 1972). Thus, on rewetting, an extraordinarily large number of unexpanded cells may exist which can cause a source/sink imbalance and may result in fruit shed. Or, where osmotic adjustment has occurred (*c.f.,* Kirkham *et al.,* 1972; Terry *et al.,* 1971), cell enlargement may be affected at a later, more severe stage of drought than cell division. Drought stress may affect transpiration and

the consequent entrainment and supply of mineral nutrients differently than photosynthesis (Baker *et al.,* 1983).

We will explore these and other plant stresses in this chapter, but with this simple discussion it is clear that source/sink relations can become a very complicated subject. Thus, special tools and facilities are needed to unravel these interconnected effects into their component parts and to reassemble them into an understandable form whose validity can be tested against field observations.

1.1 Tools and Facilities for the Study of Source/Sink Relations

F.W. Went (1963) pioneered the development of phytotrons as facilities for the complete and reproducible control of plant environments, and he pointed out their capability for independent control of most of the relevant environmental variables affecting crop physiological processes. The work of J.D. Hesketh and the others in the Canberra Phytotron and in the Southeastern Plant Environment Laboratory (SEPEL) at Duke University will be reviewed here because of its great contribution to the early database on cotton growth and development.

Phytotrons, however, have a number of very severe limitations for the study of plant source/sink relations, and of the many that were built, very few are still operational. One problem is the cost of operations. Another problem is that neither the light quality in phytotrons nor the restricted root environments of potted plants are similar to that found in the field. For example, in a carbon dioxide enrichment study with cotton, Thomas and Strain (1991) demonstrated that inadequate rooting volume was clearly associated with reduced photosynthetic capacity. They found that the reduced sink strength caused by restricting root growth in pots resulted in reductions in photosynthetic rate that were not associated with decreased stomatal conductance, but rather due to a down regulation of ribulose-1,5 bisphosphate carboxylase (Rubisco) activity. Arp (1991) also reported a similar sink-limited feedback inhibition of photosynthesis resulting from the restricted root growth of potted plants. As another example, Baker *et al.,* (1983) found that the time intervals between main-stem nodes were thirty percent longer in the field than in the phytotron. Phytotrons have some value in studies of temperature effects on growth (sink capacity) and developmental rates, but perhaps their greatest shortcoming is their inability to provide photosynthesis (source), respiration and transpiration data. Not only are data on photosynthesis under natural light lacking, but the capability to systematically manipulate source, via atmospheric CO_2 relative to sink, is lacking, and this hampers understanding of source/sink relations.

The **S**oil-**P**lant-**A**tmosphere-**R**esearch (SPAR) system, (Phene *et al.,* 1978) was developed to overcome these and other problems in phytotrons. These outdoor, naturally sunlit, computer controlled plant growth chambers (Fig. 8-1) were equipped with a soil lysimeter, 1 m deep and 2 x 0.5 m in cross section. The lysimeters have wire reinforced glass fronts for root observation and measurement, and are topped with transparent Plexiglas boxes 1.5 m in height. To prevent soil respiration from complicating measurements of canopy gas exchange, the soil surface is sealed around plant stems with an inert plastic film. Apparent photosynthesis is calculated and recorded every 15 min from a set of mass balance equations that depend mainly on the amount of CO_2 injected into each chamber to replace CO_2 taken up by the growing crop canopy in order to maintain a desired CO_2 concentration set point. Respiration rate is measured after sunset at the daytime air temperature (Acock and Acock, 1989) and this is used as a surrogate for daytime respiration for the estimation of the preceding daytime gross photosynthetic rates. Photosynthetic rates can be manipulated by controlling the chamber atmospheric CO_2 concentration over a wide range. Transpiration is measured every 15 min as the amount of condensate from the cooling coils used to control chamber humidity. A 10-unit SPAR installation costs a small fraction of what is required to build and operate a phytotron.

Controlled environment facilities are necessary sources of data for the development of physiological process rate equations for the analysis and understanding of source/sink interactions. The assembly of these data into a coherent mathematical system is the primary goal in the development of process-level crop simulation models. This assembly of the dynamic physical/physiological process-level simulation model from such data bases and the subsequent meticulous validation of the resulting models against field observations is one of the most powerful means available to develop and confirm our understanding of the interrelationships among soil, plant and atmospheric processes. The SPAR system was conceived expressly for the purpose of crop simulation modeling.

Figure 8-1. A Soil Plant Atmosphere Research (SPAR) unit.

1.2 Dynamic, Process-Level, Crop Simulation Models

Most scientists studying source/sink interactions in plants have some sort of mental image (model) of how metabolites are produced and distributed. Photosynthesis, growth, and development must all be considered, and these, along with underlying physiological processes, are all treated explicitly in materials balance, dynamic simulation models such as GOSSYM (Baker *et al.,* 1983). Materials balance crop simulation models are typically built up from modules representing the plant and soil processes within the constraints of the available database. The modules are arranged so that driving variables for a particular process are calculated, as needed, in modules executed earlier in the computations. For example, the cotton simulation model GOSSYM is arranged as shown in Fig. 8-2 (McKinion and Baker, 1982).

In GOSSYM, the weather data is processed before the soil subroutines. The soil routines provide the plant model with estimates of soil water potential in the rooted portion of the soil profile (used in estimating plant turgor), estimates of the nitrogen entrained in the transpiration stream and available for growth, and estimates of the sink strength of the root system. The below ground processes are treated in a two-dimensional grid, 1 cm thick. Material balances of water, nitrate, ammonium, organic matter, and three age classes of root dry matter are maintained and updated several times per day. Roots grow and water and nitrates flow

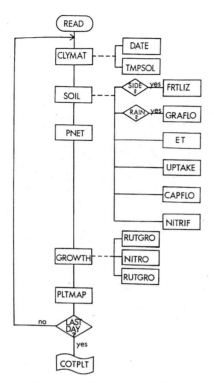

Figure 8-2. Flow diagram of
GOSSYM. Rectangular boxes represent
various modules consisting of one or more subroutines.

2.   THE SOURCE

2.1   Photosynthesis

The genetic makeup of the plant determines its un-stressed capacity to carry on photosynthesis. Genetic make-up also determines the plant's responses to various stresses, many of which operate by altering stomatal aperture and rate of leaf senescence. Thus, we may think of photosynthesis as some stress-modulated fraction of its genetic potential. The amount of assimilate actually produced by the plant depends in part on the size of the plant and the physical arrangement of its leaves. The architectural character of the plant, including size and heliotropism are, in part, genetically determined.

Pettigrew *et al.,* (1993) compared photosynthesis, and leaf anatomy in field plots of super okra, okra, and normal leaf-type isolines differing in size and degree of lobing. The okra and super okra isolines averaged 22 and 24% greater leaf P_n, respectively, than the normal leaf isoline. Leaves of the super okra and okra isolines were 42% thicker than the normal leaves. They observed greater stomatal diffusion resistances in the super okra and okra leaves than in the normal leaves and water use efficiency was greater for the okra and super okra than for the normal leaves. They found that Rubisco concentrations per unit stroma area were similar among chloroplasts of the different genotypes.

Photosynthesis was also affected by plant morphological and environmental factors. El Sharkawy and Hesketh (1965) found no difference in leaf photosynthesis among five *Gossypium* species with leaf thickness ranging from 115 to 187 µm. However, Hesketh (1968) reported lower leaf apparent or net photosynthesis (Pn) rates for cotton grown at low irradiances. Benedict (1984) speculated that this lower Pn is probably due to the lower-irradiance leaves being thinner. Patterson *et al.* (1977) found field-grown cotton had a two-fold higher leaf Pn than that of plants grown in growth chambers under artificial light then compared stomatal diffusive resistance, leaf anatomy, and chloroplast lamellar characteristics in search of the reasons for these differences. They found that light saturated stomatal diffusive resistances did not differ in leaves of the same age class. The leaves of plants grown in the field were thicker with smaller photosynthetic units than those from chamber-grown plants. Because leaves of field-grown plants also contained more chlorophyll per unit area, they also had many more photosynthetic units per unit leaf area. When leaf Pn values were expressed on mesophyll volume or per unit of chlorophyll, differences between field-grown and chamber-grown leaves were in substantial agreement except at very high light levels. Both Patterson *et al.* (1977) and Pettigrew *et al.* (1993) concluded that genotypic differences in Pn are likely due to a greater concentration of photosynthetic apparatus per unit leaf area caused by leaf thickness differences.

along water concentration gradients. Root growth increments are modeled as functions of soil temperature and soil water content at the location of the growing root. Here, root growth increments are proportional to the biomass in an age category capable of growth. Root growth increments are calculated separately for three categories of root tissue age, with young and middle aged roots representing the strongest sinks (g of assimilate required per g of growing tissue).

In the subroutine PNET, leaf water potential, canopy light interception, and canopy photosynthesis and respiration are all calculated. In the subroutine GROWTH, potential dry matter accretion of each organ is calculated from organ temperatures generated by a set of energy balance equations. This potential growth is then adjusted for turgor and nitrogen availability. Photosynthate and any reserve carbohydrates are partitioned among various plant organs in proportion to each organs contribution to the total demand. This partitioning factor is the carbohydrate supply:demand ratio. In PLTMAP, fruit loss and developmental delays are added to plastochron intervals calculated as functions of air temperature. The program cycles through these modules one day at a time to the end of the season when selected plant descriptors can be printed out.

Our point in this brief description of a materials-balance, dynamic simulation model is to illustrate that it is feasible to assemble, in a useful and comprehensive structure, knowledge and data about all the processes involved in source/sink interactions. Our purpose here is to review the literature pertaining to these processes in cotton.

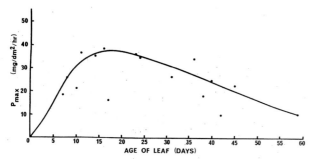

Figure 8-3. Maximum apparent photosynthetic rates of cotton leaves of various stages. From Reddy *et al.* (1991).

The photosynthetic efficiency of cotton leaves varies with leaf age as shown for field-grown cotton in Fig.8-3. For cotton, leaf Pn increases until leaf expansion is complete at 16 to 20 days and then declines with increasing leaf age and senescence (Reddy *et al.,* 1991). Constable and Rawson (1980a) and Wullschleger and Oosterhuis (1990a, d) reported similar declines in leaf Pn measured at near light saturation (P_{max}) (Fig. 8-3).

2.2   The Role of Leaf Senescence in Source Reduction

An indeterminate woody perennial, such as cotton, has a completely different strategy with regard to senescence than does an annual plant. Even extremely determinate varieties of cotton do not invest so much photosynthate in fruit, to the exclusion of leaves and roots, that the latter become ineffective. In fact, depending on the cultivar and environmental conditions, whole canopy gross phtosynthesis (Pg) may continue with little reduction in efficiency through cutout and re-growth.

Kornish (1988) reviewed all of the factors associated with (but not necessarily causing) leaf senescence. He noted that any response to a plant growth substance (*e.g.,* hormones) is really a response to the balance of the plant growth substances in the tissue.

Peng and Krieg (1991) measured canopy and leaf photosynthesis along with canopy light interception in a highly determinate variety of cotton in order to investigate the effects of plant age on senescence during the fruiting period. The leaf area index (LAI) in their 1 m rows remained virtually constant, while crop canopy light interception increased from 55 to 77% during the fruiting period. This indicates that the crop was growing new leaves at about the same rate that senescing leaves were being shed. The increase in canopy light capture over this period indicates an improvement in leaf display, probably via increased stem extension. They recorded a decline in leaf and canopy Pn between day 103 and day 111 of their experiment. But on further examination of the data they found that if the data were corrected for increased respiration associated with the increase in biomass over this time period, the Pg remained constant (Kreig, personal communication). These findings are similar to those of Baker *et al.* (1972).

2.3   Temperature Effects on Photosynthesis

Baker *et al.* (1972) measured Pg in field-grown cotton immediately after enclosing 2 m² segments of two crop canopies in a naturally sunlit, portable, temperature controlled chamber. The second setup contained older plants with a greater biomass. These results are presented in Fig. 8-4. When the canopy Pn data were adjusted for respiration losses, all of the data fell on the same response curve regardless of temperature or biomass. It is important to note that the data in Fig. 8-4 are whole canopy Pg and not single leaf photosynthesis. Single leaf photosynthesis typically displays a distinct temperature optimum (Percy and Björkman, 1983; Long, 1991). Lack of sensitivity of whole canopy photosynthesis to a rather broad range of season-long air temperature treatments has also been reported for other species including soybean (Jones *et al.*, 1985) and rice (Baker and Allen, 1993).

Reddy *et al.* (1991) measured cotton canopy Pg in 5 sunlit SPAR units over a 37-day period during fruiting. Day/night temperature treatments ranged from 20/10° to 40/30°C in 5° increments. They also recorded dry matter partitioning to stems, leaves, roots, and fruit. They found a strong temperature optimum for boll growth at 30/20°C (Fig. 8-5). As in the case of the earlier work (Baker *et al.,* 1972), initially there was very little temperature effect on photosynthesis.

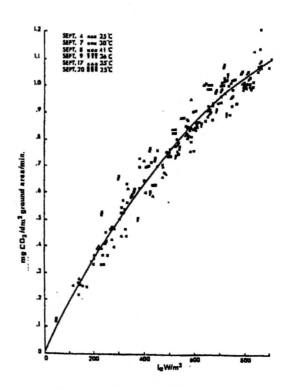

Figure 8-4. Gross canopy photosynthesis vs. light intensity in cotton crops of different weights, and at various air temperatures and vapor pressure deficits. From Baker *et al.* (1972).

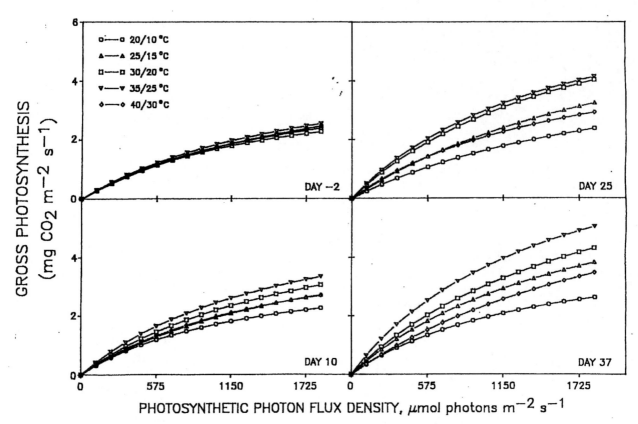

Figure 8-5. Canopy photosynthetic light response curves for cotton growing in
different air temperature treatments during the fruiting period. From Reddy *et al.* (1991).

Reddy *et al.* (1991) concluded that there was an association between boll growth rate and photosynthetic efficiency. The 40/30°C plants shed virtually all of their fruit in the square stage and partitioned nearly all their photosynthate to growth of new vegetative structures including leaves. Thus, the temperature optimum plants at 30/20°C had high photosynthetic rates in spite of the fact that their leaves were much older than in the 40/30°C treatment.

On the other hand, chilling temperatures reduce growth in cotton (Gipson, 1986; Burke *et al.,* 1988; Winter and Koniger, 1991). Cotton seedling growth is completely stopped at temperatures below 16°C (Munro, 1987). More specifically, temperatures below 20°C have been shown to reduce both starch utilization and subsequent daytime Pn (Warner and Burke, 1993). Three of the temperature treatments in the Reddy *et al.* (1991) study had nighttime temperature treatments at or below 20°C, and this may have been a factor in the subsequent reduction in canopy Pg, since measured light interception was unaffected by the temperature treatments.

2.4   Drought Stress and Photosynthesis

Another factor affecting photosynthesis in cotton is drought. Linear or near linear declines in leaf and whole canopy Pn with decreasing leaf water potential (Ψ_l) have been reported (Ackerson *et al.,* 1977a; Parsons *et al.,* 1979;

Marani *et al.,* 1985). In all cases, the decline in Pn was well established by the time water potential had dropped to -1.2 MPa. After declining to -2.5 MPa, leaf water potential was allowed to rise again to -1.5 MPa (Fig. 8-6). Very little photosynthetic recovery was observed after re-watering. Marani *et al.* (1985) attributed this to senescence of the leaf canopies during the 3-week drying cycle. In a growth chamber experiment, Bielorai and Hopmans (1975) put cotton plants through a drying cycle and measured leaf Pn and transpiration rates as well as stomatal conductance for several days after re-watering. They found that neither leaf Pn, stomatal conductance, nor transpiration fully returned to pre-stress conditions after irrigation, and even the partial recovery that was observed required several days.

Drought stress is, perhaps, the best example of a stress with both direct and indirect effects on photosynthesis. Drought can directly affect several plant processes, which in turn can lead to indirect effects on photosynthesis. The challenge to plant physiologists has long been to sort through these primary, secondary, and tertiary effects and then to elucidate the various feedback mechanisms. Boyer (1964) cited numerous investigations showing that as water deficits increase, stomata close and as a result, photosynthesis is reduced. He also noted (Boyer, 1970) that leaf growth is affected by falling leaf water potential (Ψ_l) at the first sign of turgor loss, but that a measurable reduction in photosynthesis begins later at about -1.2 MPa. Hsiao *et al.* (1982)

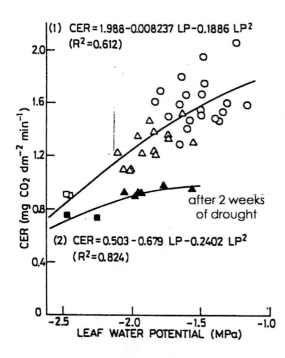

(1) $CER = 1.988 - 0.008237\ LP - 0.1886\ LP^2$
$(R^2 = 0.612)$

(2) $CER = 0.503 - 0.679\ LP - 0.2402\ LP^2$
$(R^2 = 0.824)$

after 2 weeks of drought

Figure 8-6. The effect of midday leaf water potential on canopy net photosynthesis at a radiation level of 698 W m^{-2}. Filled symbols are observations made after 2 weeks of drought. From Marani *et al.* (1985).

reported experiments in which the first measurable sign of a developing drought stress was a suppression of leaf expansive growth while photosynthesis was not affected at that leaf water potential. He found that osmotic adjustment (defined as a gain in cell solute concentration sufficient to reduce turgor loss with falling Ψ_l) in mature leaves was negligible initially and became substantial only when expansive growth was markedly suppressed by stress. Moreover, the maximum osmotic adjustment observed was only 10 bars. Hsaio *et al.* (1982) found that osmotic adjustment, though partly maintaining cell turgor in spite of declining tissue water potential, did not enable leaves to maintain photosynthesis and that on re-watering, assimilation rates remained depressed and required several days for recovery. After re-watering, the osmotically adjusted leaves lost much of their extra solutes and their osmotic adjustment.

Examining Ψ_l, osmotic potential and relative water content data from several years' irrigation experiments, Girma and Krieg (1985) found that drought stress reduced both seasonal and diurnal osmotic potential. However, they observed no difference in the osmotic potential of fully expanded leaves when corrected to 100 percent relative water content. Water content per unit leaf area and per unit dry matter were much lower under drought stress conditions and this accounted for the major difference in osmotic potential. They concluded that increased solute concentration under drought stress is due to tissue dehydration and that it is a passive phenomenon of limited extent. Furthermore, Sinclair and Ludlow (1985) reviewed the inherent difficul-

ties in relating thermodynamic properties such as either osmotic pressure or plant water potential to plant physiological responses and persuasively argued for the use of tissue relative water content as a much more relevant measure for the study of plant physiological responses to water deficit. Many authors agree, however, that a major effect of drought stress on photosynthesis in cotton has to do with canopy light capture. Light capture under drought stress is reduced by reduced rates of leaf initiation, stem and leaf growth and by premature leaf senescence and abscission.

2.5 Nitrogen Deficiency and Photosynthesis

Nitrogen deficiency is another syndrome in cotton, with secondary and tertiary effects on photosynthesis being uppermost. In our previous reference to Reddy *et al.* (1991), we noted that reduced boll growth rates, (which can be caused by nitrogen shortage) may reduce photosynthesis. Nitrogen deficiency also reduces stem extension and leaf expansion resulting in a smaller plant, which captures less light and carries on less photosynthesis (Wullschleger and Oosterhuis, 1990). Fernandez *et al.* (1993) exposed cotton seedlings in growth chambers to N limitations which resulted in reduction of leaf N concentration to nearly half that of the well fertilized control. They found that the N limitation did not decrease gross carbon uptake on a per unit leaf area basis, but after 11 days the N-limited plants had less than half the leaf area of the unstressed plants and a correspondingly reduced rate of whole plant photosynthesis.

Radin *et al.* (1982) demonstrated a secondary effect of N stress on photosynthesis. In controlled environment experiments at 35/22°C and 42/28°C day/night air temperature treatments, they found an interaction between effects of temperature, leaf water potential and leaf nitrogen concentration on stomatal conductance. These results are shown in Figs.8-7 and 8-8. The suboptimal N treatment raised the leaf water potential required for stomatal closure by about 0.4 MPa. Photosynthesis was reduced at higher leaf water potential than that required for stomatal closure (Figs. 8-7 and 8-8), but as drought proceeded stomatal closure further (in addition to mesophyll resistance) reduced photosynthesis. They demonstrated two components to this low N induced change in the leaf water potential for stomatal closure; first, an increase in ABA concentration, and second, increased stomatal sensitivity to ABA. Furthermore, they found that kinetin applied to the low N leaves decreased the stomatal response to ABA relative to that of the high N leaves. In the high N leaves, kinetin by itself had little effect on the stomatal response to ABA. Radin *et al.* (1982) concluded that a cytokinin-ABA balance is altered by suboptimal nitrogen nutrition, which promotes stomatal closure during stress. Subsequently, Radin (1983) showed that nitrogen deficiency in cotton reduced leaf Pn by 31% and leaf expansion rate 56%. In those experiments, N deficiency inhibited leaf expansion during the day, when high rates of transpiration

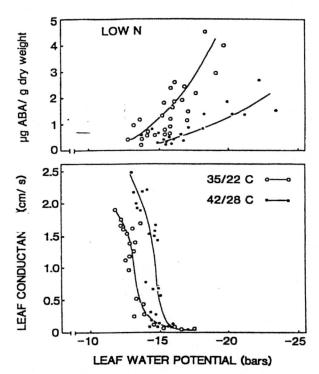

Figure 8-7. Stomatal conductances and ABA concentrations of leaves of low nitrogen plants at two temperatures. From Radin *et al.* (1982).

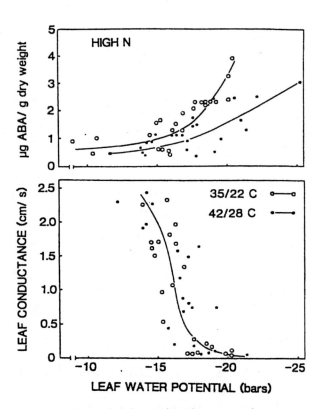

Figure 8-8. Stomatal conductances and ABA concentrations of leaves of high nitrogen plants at two temperatures. From Radin *et al.* (1982).

were occurring, but had little or no effect at night. These results support the hypothesis that N stress inhibits leaf cell expansion by reducing the hydraulic conductivity of the root system.

In other experiments, Radin and Eidenbock (1986) found that phosphorous deficiency on other plant processes were similar to those of a N deficiency. Two comments may be made about the implications of this in regard to the agronomic management of this crop. First, nitrogen is comparatively mobile in the soil, while phosphorous is relatively immobile. This means that the issue of nutritional stress and optimum fertilizer placement will differ in regard to these two elements. If N and P are improperly managed, the effects will most likely be manifested during lint growth. In terms of yield reduction, this is the worst possible developmental stage for the crop to encounter a nutritional stress. Secondly, the combination of tissue nutrient monitoring, plant mapping and crop simulation can predict crop nutrient supply and demand and thus avoid a nutrient stress by the timely scheduling of foliar sprays to remedy N shortages during this economically critical growth period.

In addition to internal leaf conditions, the geometric arrangement of leaf elements of the crop canopy, day length, and solar elevation angle determine canopy light interception and photosynthesis. Baker and Meyer (1966) showed that canopy photosynthesis is proportional to canopy light interception when solar elevation angle, row orientation, or row spacing are used to vary photosynthesis (Fig. 8-9).

Baker *et al.* (1978a) reviewed modeling efforts beginning with the growth analysis concepts of Watson (1947) and Nichiporovich (1954) followed by the computerized models of de Wit (1965) and Duncan *et al.* (1967) used to calculate crop canopy photosynthesis on a leaf element or a leaf area index (LAI) basis. In comparing rates of leaf area development, photosynthesis, and dry matter production in cultivated cottons, Muramoto *et al.* (1965) found that while differences among cultivars in leaf photosynthetic rates were not detectable, variability within any one plant was large. They also found that net assimilation rates did not differ greatly among cultivars, but there were differences in dry matter production associated with differences in rates of leaf area development and thus light interception. Baker *et al.* (1978a) demonstrated that the relationship between LAI and canopy light capture is affected by plant water status (Fig. 8-10). Here, the amount of light captured by a compact, drought-stressed crop is considerably lower than that of the taller, well-watered crop when the two canopies are compared at the same LAI.

Kharche (1984) identified another condition under which height alone is not a valid predictor of canopy light interception. In experiments involving irrigation and leaf shape (okra *vs.* normal) he found that leaf shed caused by canopy senescence, whether induced by drought stress or heavy boll load, could proceed until LAI fell below 3.1 before canopy light interception was reduced. A correction has been added to GOSSYM to correct light interception estimates from height and row width in defoliating canopies.

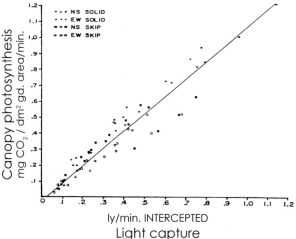

Figure 8-9. Canopy photosynthesis vs. light intensity in cotton crops planted in rows orientated north-south and east-west. In fig. legend, symbols labeled as "Solid" are for 1 m rows and "Skip" are for pairs of 1 m rows with every other pair unplanted. From Baker and Meyer (1978).

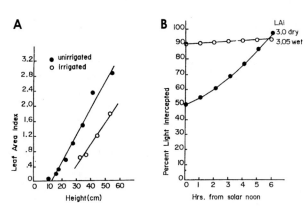

Figure 8-10. A. Plant height vs. leaf area index for cotton in 1 m rows, with and without irrigation. B. Percent canopy light interception vs. hours from solar noon in cotton crops at leaf area indices of 3 with and without irrigation. From Baker *et al.* (1978a).

Other important failures of simple leaf area based models of canopy photosynthesis include the use of a single light response curve to represent all leaves in the canopy. As described earlier, leaf light response curves can be expected to vary greatly depending on their temperature experience during the growth period and their previous exposure to stress. Other causes of failure of the LAI approach to estimating canopy photosynthesis are assumptions that only leaves intercept light, that heliotropism is nonexistent and that the leaves are distributed in a spatially uniform manner over the ground surface. The latter is, obviously, untrue in a row crop.

2.6 Additional Factors Affecting Photosynthesis

Aside from physical environmental factors, including the geometry of the plant, a number of external chemical factors, *e.g.,* atmospheric CO_2 and ozone, insecticides, and plant growth regulators can influence photosynthesis.

Atmospheric carbon dioxide concentration has increased from about 280 µL L^{-1} to over 350 µL L^{-1} in the last 200 years (Friedli *et al.* 1986; Keeling, 1995). Due mainly to the continued burning of fossil fuels, this trend is expected to continue and current atmospheric CO_2 concentration (near 360 µL L^{-1}) are projected to increase to about 670 to 760 µL L^{-1} by the year 2075 (Rotty and Marland, 1986; Trabalka *et al.* 1986; Watson, 1990). A primary direct effect of elevated CO_2 on plants such as cotton with the C_3 carbon fixation pathway is an increase in photosynthetic rate. Due mainly to enhanced photosynthesis, Mauney *et al.* (1994, see also Chapter 16) and Reddy *et al.* (1997) reported 37 and 40% increases in cotton biomass, respectively, with carbon dioxide enrichment. Baker (1965) measured the effects of atmospheric CO_2 concentration from 100 to 600 µL L^{-1} on cotton canopy CER over a range of light intensities and temperatures in an intact, field grown cotton crop canopy. These results are presented in Fig. 8-11. GOSSYM contains an algorithm, based on the data Fig. 8-11, to make an annual adjustment in photosynthesis calculations due to projected higher levels of atmospheric CO_2.

Adverse effects of ozone on cotton yield have been reported (Heagle and Heck, 1980; Heck *et al.,* 1982; Heggestad and Christiansen, 1983; Temple *et al.,* 1985, 1988). Reddy *et al.* (1989) fitted dry matter accumulation data over a range of atmospheric ozone concentrations from 0.027 to 0.107

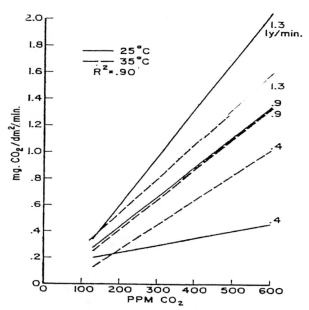

Figure 8-11. Apparent photosynthesis vs. atmospheric CO_2 concentration at three light intensities and two air temperatures. From Baker (1965).

μL L⁻¹, of Miller *et al.* (1988), to the following quadratic form (where OZONE is concentration in μL L⁻¹):

$$OZFTR = 1.01 + 0.7168 \times OZONE - 43.385 \times OZONE^2.$$

to provide a Pn adjustment factor (OZFTR) for GOSSYM. This equation provides estimates of reductions in canopy Pn of 1.0 and 5.9%, respectively, at ozone concentrations of 0.03 and 0.107 μL L⁻¹. Reddy *et al.* (1989) present April to September averages of ozone concentrations from US-EPA at 5 locations in the American Cotton Belt for the years 1963 to 1985 ranging from 0.04 to 0.09 μL L⁻¹. Within these years and locations, ozone concentrations on individual days undoubtedly caused significant reductions in photosynthesis.

Insecticides are the most common biologically active chemicals routinely applied to cotton in the field. Very few experiments and none that we are aware of in the past 18 years have been conducted to measure the effects of insecticides on stomate physiology or on photosynthesis. Baker (1966) measured Pn and transpiration rates in an intact field grown cotton canopy after treatment at recommended rates with six commercial insecticides and insecticide combinations. Air temperature was maintained at 30°C, vapor pressure deficits were at 0-1.0 MPa, and soil moisture was maintained at -0.3MPa. There were no significant effects of insecticide treatment on stomatal aperture, transpiration, or photosynthesis under these conditions.

Another class of chemicals routinely applied to cotton is plant growth regulators. Mepiquat chloride (MC) is commonly used to reduce plant height in order to improve canopy ventilation and avoid boll rot in the crop as well as facilitate pest management. However, Hodges *et al.* (1991) found some enhancement of single-leaf or whole plant photosynthesis with MC application with day/night temperatures at and above 25/15°C. We have found that height: width ratios change with applications of MC. MC seems to reduce internodal growth of main-stems more than that of sympodia (unpublished data). A mixture of gibberellic acid and indolebutyric acid (trade name PGR-IV) was found by Oosterhuis and Zhao (1993) to increase leaf Pn along with growth and branching of roots in cotton seedlings.

3.   THE SINKS

3.1   Translocation Capacity Does Not Limit Growth in Cotton

In some species translocation capacity should be considered in a discussion of source/sink relations. Indeed, some crop simulation models deal with it explicitly on the basis of metabolite concentration gradients. However, there are two lines of evidence which lead us to conclude that translocation does not limit photosynthesis or growth in cotton. Firstly, cotton can carry on photosynthesis at very high rates under elevated CO_2 and high radiance (PPFD)

levels for long periods with no apparent feedback inhibition due to starch loading either in the phloem or chloroplasts. Under these conditions leaf starch levels may reach 50% of leaf dry weight. The high starch levels occur during the daytime, but the plant has a vascular system which seems to be able to move most of the photosynthate to growing points by the end of the night. Secondly, Ashley (1972), showed with labeled CO_2 that most photosynthate in a boll comes from the subtending leaf, with some coming from an adjacent leaf on the same sympodium. However, Wullschleger and Oosterhuis (1990) concluded that because of the lack of synchrony between boll growth and photosynthate production by the subtending leaf, substantial carbon must be imported from other leaves. They found that only at main-stem node 12 were leaves capable of supplying the carbon needs of their associated bolls. Carbon import requirements, beyond that supplied by the subtending leaf, for the first three fruiting positions of node 10 were 50, 37, and 21%, respectively, of the total translocated to the developing boll.

3.2   Root Growth as a Sink

Huck (1983) summarized the data of Pearson *et al.* (1970) showing the time course of cotton taproot extension in a sandy soil at several temperatures (see also Chapter 6). He stated that root extension remained at a fairly constant rate until competition for food reserves began among organs within the seedling. He reported an extension rate of 8 cm d⁻¹ at the optimum temperature. Bland (1993) conducted rhizotron experiments under artificial lights and ambient CO_2 to address the role of seasonal patterns of soil temperature on root system development. He compared rates of root extension in three temperature regimes, two of which were typical soil temperature profiles, warming as the season progressed. He commented that the application of his root extension model assumed no limitation of structural substrate and was characterized only by the capability of the root tips for growth. His model predicted a maximum extension rate of 2.6 cm d⁻¹ at a day/night temperature of 32/15°C and less than 2.0 cm d⁻¹ at 22/16°C. Kimball and Mauney (1993) compared end-of-season cotton biomass and root/shoot ratios in open top chambers in Arizona at ambient and elevated (650 μL L⁻¹) CO_2 with irrigated and dryland treatments and with and without added N fertilizer. Due to the natural rapid decomposition of root biomass over the season and due to the fact that at harvest the plants were simply pulled out of the ground, their biomass data represents mainly structural material. That fraction of the photsynthate invested in small roots, which decompose rapidly during the season, was not included and therefore their data cannot be representative of the true sink capacity of the root system. Nevertheless, comparisons of the influences of drought stress and assimilate and nitrogen supply on root growth are possible. Under well-watered and well-fertilized conditions, root/shoot ratios were fairly constant. With CO_2 enrichment, the enhanced assimilate supply was apparently distributed between roots and shoots equally.

Taylor and Ratliff (1969) found that root elongation rate was reduced, as soil resistance increased. They stated that at any particular penetration resistance there was no effect of soil water content *per se* on root elongation within a compacted layer. However, soil strength is a function of water content. Moreover, Grimes *et al.* (1975) reported that less soil strength was required to restrict root growth at greater soil depths. Browning *et al.* (1975), working in a different rhizotron from the above recorded seasonal total root lengths and root lengths in soil wetter than –1 bar. The soils were not temperature controlled and the plants were not grown in a crop canopy configuration, although, they were grown outdoors. Nevertheless, the seasonal time courses are similar to those found in field-grown plants. The data in Table 8-1 for their Bin No.2 (Cahaba loamy fine sand) are typical. Bulk density of this soil was 1.3g cm^{-3}. Soil drying began in mid-August (Taylor and Klepper, 1974). This crop was planted on May 2 and by August 11 total root-length and root-length in soil with greater than –1.0 bar matric potential were at a maximum. The August 11 maximums for root-length would have been about 85 days after emergence and approximately two weeks after first bloom. After August 11, the rate of root decomposition greatly exceeded further root extension.

Descriptions of factors affecting root elongation and function are fairly common in the literature. Unfortunately, due to the technical difficulty of recovering roots, especially small roots, from soil, root growth data in terms of the accretion of biomass is extremely rare in the literature. The controlled environment data needed to calculate sink strength is virtually nonexistent. Phene *et al.* (1978) conducted one such experiment. Whisler *et al.* (1986), while developing the root growth component of GOSSYM, began with the assumption that root sink strength would be the same as that of bolls. In the effort to make the model mimic the data Phene *et al.* (1986) and Whisler *et al.* (1986) had to increase the initial estimate of root sink strength by 10-fold. Subsequent analyses with GOSSYM by Fye *et al.* (1984) showed that the root sink strength must be at least 6 times that of rapidly growing bolls. Thus, any estimate of total

sink strength in cotton roots must be regarded as extremely tentative. Clearly, research aimed at describing root sink strength, as a function of a wide array of environmental factors as well as soil conditions is greatly needed.

3.3 Stem and Leaf Sinks

Baker *et al.* (1983) used the phytotron data of Hesketh and Low (1968) to construct the leaf growth module of GOSSYM. Shown in Fig. 8-12 are leaf area growth rate and specific leaf weight vs. temperature from that experiment. In GOSSYM these two leaf characteristics determine the leaf contribution to the total plant sink strength. Leaf area growth showed a pronounced maximum near 30°C after which it declined rapidly with further increases in temperature. Specific leaf weight, on the other hand, increased throughout the range from 15 to 40°C.

Reddy *et al.* (1991) moved plants grown outdoors in pots to SPAR units 5 days prior to flowering. In each SPAR unit a dense canopy was created which intercepted 98% of the incident solar radiation in the chamber atmosphere. CO_2 was controlled to 350 µL L^{-1}. Day/night air temperatures ranged from 20/10 to 40/30°C in 5°C increments were maintained for 49 days. Leaf and stem growth as a function of temperature from this experiment are shown in Figs. 8-13 and 8-14. Although the leaf growth data in Fig. 8-13c have the units of cm^2 d^{-1} while the data in Fig. 8-12 are presented in terms of dm^2 dm^{-2}d^{-1}, there is a pronounced peak at 30/20°C in both cases. At 30/20°C, main-stem node 17 produced a relatively large leaf. However, because these plants were fruiting rapidly and maintained at a CO_2 concentration of 350 µL L^{-1} there was likely some substrate limitation. Therefore, these data represent the genetic potential for growth, at that temperature (demand) reduced according to the existing supply: demand ratio. As an example of CO_2 and supply:demand effects on leaf growth, Mauney *et al.* (1978) measured 153 dm^2 of leaf area at 330 µL L^{-1} and

Table 8-1. Total root length and root length in soil with water potential greater than -1 bar (-0.1 MPa). Each value is the root length of 2 plants ± S.E. in a population density of 2.8 plants m^2 (Browning *et al.,* 1969).

Date	Total length	Length in soil >-1 bar
	---------------- (km) ---------------------	
June 26	1.54	1.54
July 3	2.66	2.66
July 10	5.30	5.30
July 21	10.67	10.67
July 31	17.12	17.10
August 11	22.85	22.85
August 21	19.73	19.73
Sept. 1	7.91	7.91
Sept. 5	4.49	4.49

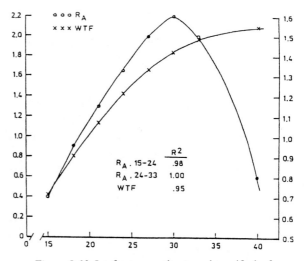

Figure 8-12. Leaf area growth rate and specific leaf weight vs. air temperature. From Hesketh and Low (1968).

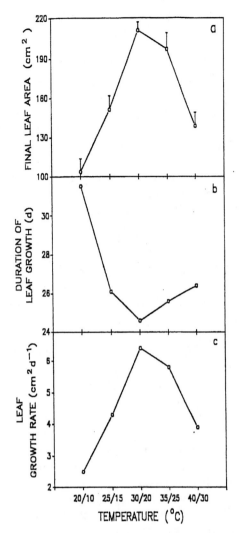

Figure 8-13. The effect of air temperature on final leaf size, duration of leaf growth and rate of leaf area growth at main-stem node 17 in cotton. From Reddy *et al.* (1991).

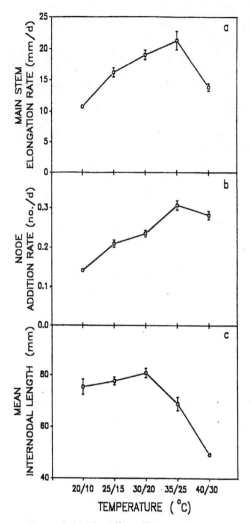

Figure 8-14. The effect of temperature on stem growth and development. From Reddy *et al.* (1991).

292 dm² at 630 µL L⁻¹ after 12 weeks of growth for cotton. At 40/30°C (Fig. 8-13) limitations due to higher respiration rates may have contributed to growth reduction. Duration of leaf growth (Fig. 8-13b) was significantly longer at very low temperatures and this does not seem to have offset the effect of growth rate on final leaf size very much. In subsequent work with seedlings, Reddy *et al.* (1992) found the same temperature response for leaf growth beginning at about 27 day after emergence. Prior to that, leaf growth was very slow with little temperature effect (Fig. 8-14). They attributed the slow early growth of leaves and stems to a very strong sink in the young root system. They also attributed the later seedling growth restriction at temperatures above 30/22°C to assimilate limitations.

Assimilate limitation factors should also be considered in the stem elongation rate data. For example, Reddy *et al.* (1994) measured stem elongation rates of 14.5 and 18 mm d⁻¹ at 350 and 700 µL L⁻¹, respectively, at 30°C. They observed a nearly identical pattern in the time course of main-

stem node initiation. Reddy *et al.* (1991 and 1992b) noted that at 30/20°C and at 30/22°C the main-stem growth was exponential at first, became linear as the number of fruiting sites increased and then decreased somewhat. They attributed this decrease in mean internode length and node initiation rate to decreased substrate supply. They further concluded that intraplant competition for the available carbohydrate supply demonstrated the greater sink strength of developing fruit compared to vegetative structures.

Besides temperature and substrate supply, other factors affecting growth of leaves and stems include tissue turgor and plant growth regulators. The SPAR data of Marani *et al.* (1985) illustrate the effect of leaf water potential (Ψ_l) on growth of leaves and stems (Figs. 8-15 and 8-16). These data clearly indicate that by the time midday Ψ_l has fallen to -1.2 MPa, large reductions in growth are already underway. We noted earlier that Boyer (1970) found that growth was reduced at the first sign of turgor loss and Hsiao *et al.* (1982) showed that reduction in leaf expansive growth is the first measurable sign of developing drought stress. Plant height

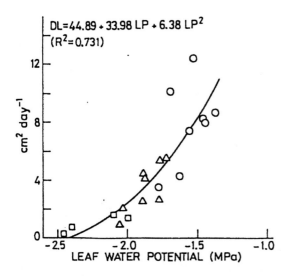

$$DL = 44.89 + 33.98\ LP + 6.38\ LP^2$$
$$(R^2 = 0.731)$$

Figure 8-15. The effect of midday leaf water potential on leaf growth in cotton. From Marani *et al.* (1985).

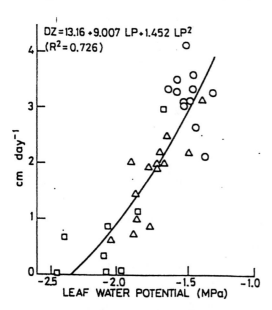

$$DZ = 13.16 + 9.007\ LP + 1.452\ LP^2$$
$$(R^2 = 0.726)$$

Figure 8-16. The effect of mid-day leaf water potential on stem growth in cotton. From Marani *et al.* (1985).

and leaf area are both reduced if Ψ_l does not rise about -0.8 MPa for a portion of the day (Jordan, 1970).

In GOSSYM, Baker *et al.* (1983) simulate the impact of turgor loss on a day and nighttime stem and leaf growth by using the fraction of that time period during which Ψ_l becomes increasingly more negative, cell elongation is reduced first, followed sequentially by reduction in cell division, stomatal aperture, and photosynthesis. Reduced photosynthesis further limits growth and in turn the reduced growth results in reduced light capture and further reduction in photosynthate supply. Other variations may include a reduction in mineral nutrient uptake with drying of the soil and restricted water uptake for areas of the soil containing the nutrients. As an example of this effect, we point to Radin (1983) who found that nitrogen deficiency reduced

leaf stomatal conductance and hydraulic conductivity of the root system.

Reddy *et al.* (1990) applied mepiquat chloride (MC) at 49 g ha^{-1} at first bloom and then maintained the plants at five temperature regimes in well-watered and fertilized SPAR units. Mepiquat chloride treatment reduced main-stem leaf area and this effect persisted for at least 21 days. Similarly, main-stem node initiation rate, main-stem elongation were also reduced by MC treatment (Reddy *et al.*, 1990). Reddy *et al.* (1992) has modeled these effects on the basis of falling tissue concentrations of MC since plant dry matter continued to increase subsequent to application of MC.

3.4 Reproductive Sinks

MacArthur *et al.* (1975) published the data in Fig. 8-17 describing boll growth rate *vs.* temperature. These data were collected by J.D. Hesketh in a glasshouse in Arizona at high atmospheric CO_2 in an attempt to ensure no assimilate limitation. As was the case with leaf (Fig. 8-13) and stem growth (Fig. 8-14) at 30/20 and 32/20°C noted previously, the data (Fig. 8-17) show a rapid increase in boll growth with temperature up to 27°, and then a rapid decline to zero growth at 33°C. Temperature effects on the time interval between flowering and open boll are shown in Fig. 8-18. It is interesting that several distinctly different types of cotton are represented in these data, all displaying a similar temperature response. The boll fill period ranged from 105 days at 22°C to 30 days at 36°C (Fig. 8-18). In applying these data in GOSSYM, Baker *et al.* (1983) made an adjustment to boll temperature based on leaf water potential to account for higher boll temperatures under water deficits. Reddy *et al.* (1991) reported a similar very pronounced boll growth

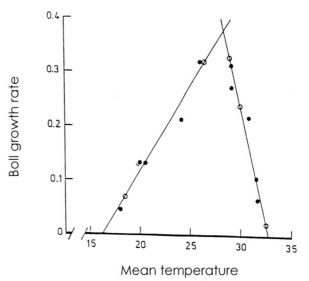

Figure 8-17. The effect of temperature on boll growth rate. Values at 27°C and above were determined for plants in air enriched with CO_2 and every other flower removed. From MacArthur *et al.* (1975).

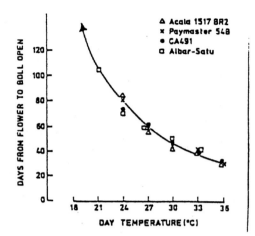

Figure 8-18. The effect of temperature on
boll fill period. From Hesketh and Low (1968).

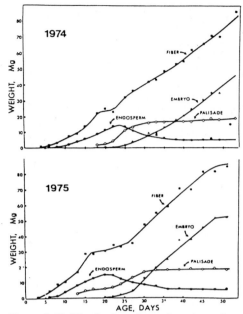

Figure 8-19. Distribution of mass into the various
parts of cotton (cv, 'Coker 310') seed during development.
Fiber includes fuzz fiber and endosperm includes inner
integument and nucleus. Each point is the average
of the contents of 15 bolls. From Stewart (1986).

temperature optimum at 30/20°C. This was a reflection of a dramatic shift in partitioning from stems and roots to bolls. Reddy *et al.* (1991) also observed very high boll abscission rates at 35/25°C and essentially no bolls were set at 40/30°C.

While some information is available to establish the effect of temperature on sink strength in whole bolls, practically none is available to break this down into boll components. However, there are a few reports from field experiments which provide information on relative time courses of reproductive events measured in terms of days post anthesis (DPA). Marani (1979) measured boll and capsule dry matter accretion at several locations in Israel. He expressed the results in terms of physiological time units as defined by McKinion *et al.* (1975) in the cotton simulation model SIMCOT II. This physiological time scale is calculated using a base temperature of 12°C and defines any temperature above 30°C as equal to 30°C. Marani (1979) found that capsule dry weight increased linearly for 20 DPA. After that there was a small decline of dry weight for 10 days, after which dry weight accretion resumed.

Growth of individual components of the developing boll over time are shown in Fig. 8-19 (after Stewart, 1986). Fiber growth begins at the very earliest stages of boll growth and is evident by 3 DPA (Fig. 8-19). Stewart (1986) noted that fiber initiation (fiber number) is related more to the time of anthesis than to fertilization and that unfertilized seeds and bolls enlarge at the same rate as fertilized seeds and bolls until 3 DPA. Further, from -1 DPA (*e.g.,* one day before anthesis) to 4 to 6 DPA, nearly all the dry matter partitioned to the seed goes into the outer integument with its newly forming fibers; so this is the tissue that is responding to changes in nutrition and the environment. Fiber weight increased for about 16 DPA, remained nearly constant for the next 5 to 10 days and then resumed growth (Fig. 8-19). The outer integument tripled in weight between 15 and 20 DPA. Then, as in the case of the capsule, it decreased in weight for about two weeks and then resumed a slow increase in weight until opening. The endosperm grew until

20 DPA and then it declined in weight as the embryo began growth. After 30 DPA, essentially all dry matter increase was in fibers and embryo with the former accumulating cellulose and the latter accumulating oil and protein. From this description of boll growth it appears that the external part of the seed receives the greatest portion of photosynthate during the first few DPA. After 4 to 6 DPA the internal weight increases somewhat faster to about 20 DPA. After 4 to 6 DPA photosynthate is distributed about evenly between external and internal seed parts. Stewart (1986) summarized the time course of boll development in Table 8-2. In reference to the fibercellulose entry in Table 8-2, Schubert *et al.* (1973) depicted this as a bell shaped curve of daily increments of weight gains, with the peak at about 30 days.

3.5   Phenological Development of Sinks

The previous section dealt with the assessment of sink strength at the organ level. Here we consider plant development for the purpose of gaining an inventory of growing organs. The hormone systems which mediate fruit set and are driven, in part, by source/sink imbalance, seem to operate at the level of the whole plant. Plant simulation models (*e.g.* GOSSYM) must make numerical estimates of source/sink ratios in order to simulate "natural" fruit shed.

In the late 1960's and early 1970's Hesketh conducted a series of phytotron (Canberra and SEPEL Duke) experiments (*c.f.,* Hesketh and Low, 1968; Low *et al.,* 1969; Moraghan *et al.,* 1968; Hesketh *et al.,* 1972) which provided a data base for the calculation of sink strength in cotton.

Table 8-2. Developmental periods and developmental events particularly sensitive to competition for assimilate or to environmental factors. Negative and positive numbers for period represent time in days prior to or after anthesis, respectively. Factor abbreviations are for air temperature (Temp.), soil nitrogen supply (N), soil water (H_2O), relative humidity (R.H.), internal plant carbohydrate supply (CHO), and soil potassium supply (K). From Stewart, 1986.

Period	Event	Factor
(days)		
-40 to -35	Initiation of floral buds	Temp., N, H_2O
-35 to -30	Carpel number (maybe)	
	Anther number	CHO
-25 to -22	Ovules/ovary	CHO
	Anther number	R.H.
-19 to -15	Pollen viability	High temp., R.H.
-2 to 12	Fiber density (f/mm²)	Temp., CHO
0	Anther dehiscence	Temp., R.H., rain
0 to 3	Rate of fiber initiation	Temp., K
	Pollen tube growth,	
	fertilization	Temp., R.H.
1 to 14	Boll abscission	CHO, H_2O
3 to 25	Fiber length, seed volume	Temp., K
15 to 45	Fiber cellulose	Temp.
25 to 50	Protein and oil accumulation;	
	oil/protein ratio	Temp., H_2O
49 to 50	Boll opening	Temp., R.H.

These controlled temperature experiments together with experiments measuring canopy light interception, canopy photosynthesis, and respiration also underway at that time set the stage for the physiological process level of simulation modeling of cotton. Hesketh's experiments provided rate functions for the following time intervals: emergence to first square, square to bloom, bloom to boll open, and plastochron intervals for the main-stem and branches. Recent SPAR experiments by Reddy and others (Reddy *et al.* 1991, 1992a, b, 1993) have updated these rate functions for modern cultivars of G. *hirsutum* and G. *barbadense*, and have added cultivar specific functions for the duration of stem internode elongation and leaf expansion.

Reddy (1994a), referring to SPAR experiments at CO_2 concentrations of 350 and 700 µL L⁻¹, stated that assimilate supply did not influence the time required to initiate the first square, but the production of all subsequent squares and bolls was sensitive to assimilate supply. Indeed, the modeling work of Baker *et al.* (1983) had shown that the plant lengthens both main-stem and sympodial plastochrons in response to source/sink imbalance. Reddy (1994) also observed that main-stem plastochrons for nodes below the first fruiting branch were considerably longer than for subsequent nodes. Reddy *et al.* (1994) was convinced that this was caused by assimilate shortage resulting from rapid partitioning to the seedling root system. Reddy *et al.* (1994) collected canopy Pn data which showed no source limitation even at 40°C in their experiments. Moreover, the rate functions presented below represent developmental events

only up through the first 17 main-stem nodes or the beginning of the boll fill period. They (Reddy *et al.*, 1993) presented data showing that at 26.4°C, CO_2 ranging from 350 to 700 µL L⁻¹ had no effect on main-stem plastochron intervals up through node 17.

Figures 8-20 and 8-21 contain the main-stem and fruiting branch plastochrons from the controlled temperature experiments of Hesketh *et al.* (1972) and Reddy *et al.* (1993), respectively. The Reddy *et al.* (1993) data are in the from of daily progress increments. This is the form used by simulation models such as GOSSYM. Hesketh *et al.* (1972) reported a main-stem plastochron of 2.4 days leaf⁻¹ at 27°C. At 27°C, main-stem plastochron for the modern cultivars (Reddy *et al.*, 1993) was 2.6 days leaf⁻¹. Sympodial plastochrons at 27°C were 7 days leaf⁻¹ in the Hesketh *et al.* (1972) experiment and 5.3 days leaf⁻¹ in the Reddy *et al.* (1993) experiments. From this it appears that the modern

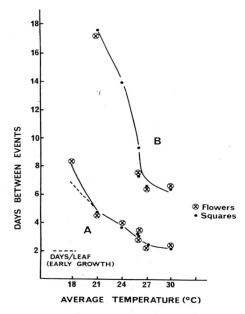

Figure 8-20. The effect of temperature on main-stem (A) and fruiting branch (B) plastochrons. From Hesketh *et al.* (1972).

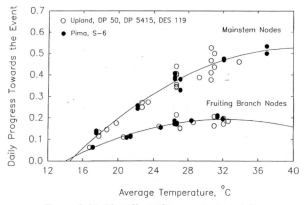

Figure 8-21. The effect of temperature on daily increments of progress toward new main-stem and fruiting branch node formation. From Reddy (1993).

cultivars studied by Reddy *et al.* (1993) have faster rates of vegetative development than the older cultivars from the experiments of Hesketh *et al.* (1972). Reddy *et al.* (1993) presented the data in Fig. 8-22 describing rates of main-stem internode extension and leaf expansion as functions of temperature. The rates in G. *barbadense* and G. *hirsutum* varieties were not different and the data were pooled to obtain the curves in Fig. 8-22.

The advent of flowering marks the appearance of reproductive structures that compete intensively with vegetative organs for assimilates. This in turn results in source/sink imbalance, which slows the developmental rate of the stem. Both main-stem and sympodial plastochrons are lengthened after flowering, as the imbalance becomes more severe. Characterizing the relationships between these developmental delays and the source/sink imbalance is an analytical problem which has been approached by Baker *et al.* (1983) in several steps, beginning with the assumption that the relevant point of imbalance is the growing boll. This procedure using the data of Bruce and Römkens (1965), was approached as follows. Daily net photosynthesis was expressed on a per plant basis to represent the carbohydrate supply available for growth. Next, using plant maps as an inventory of organs for an "average" plant, potential growth rate of each organ was calculated for each day and night period as a function of temperature. Organ sink strength was summed to represent total demand. An age vector for each organ, including roots, was maintained, and only organs in the age category for growth were considered in estimating this demand. Then developmental delays were calculated by subtracting the plastochron estimated from the functions of Hesketh *et al.* (1972) from those observed by Bruce and Römkens (1965) in their seasonal time courses of plant development. These results are presented in Fig. 8-23. Turgor was maintained at a high level in the Bruce and Römkens (1965) well-irrigated treatment, so no adjustments for turgor loss were needed, and nitrogen supplies were not limiting.

In subsequent model validation experiments Bruce and Römkens (unpublished) found it necessary to assume that under nitrogen stress conditions, the plant has some way to preferentially partition N to the developing bolls, with any excess being made available for vegetative growth. With the high N fertilizer rates in the Bruce and Römkens (1965) experiments, FSTRES in Fig. 8-23 is equivalent to the carbon source/sink ratios in the bolls. Fig. 8-23 shows that the sympodia are more sensitive to source/sink imbalance than monopodia and delays begin when 70% of the assimilate requirement is available. On the main-stem delays begin when about 55% of the assimilate requirement is available.

The SPAR facilities were designed to provide far better methods to derive the relationships in Fig. 8-23 since canopy Pn can be measured directly. Both CO_2 and nitrogen availability to the crop can be precisely controlled and varied systematically. This experimental capacity permits investigations into the possibility that the relationships in Fig. 8-23 are, themselves, functions of temperature or other variables.

3.6   Fruit Sinks

The effects of temperature on the time from emergence to first square has been investigated in a few studies. At an air temperature of 27°C, Moraghan *et al.* (1968) reported that cotton required 38 days from emergence to first square, while 34 days were reported by Hesketh *et al.* (1972) at this temperature. Daily progress increments towards first square reported by Reddy *et al.* (1993) for three modern cultivars of cotton are shown in Fig. 8-24. Here, 27 days at 27°C were required from emergence to first square.

From square to bloom, cotton required 20 and 26 days at 27°C in experiments conducted by Hesketh and Low (1968) and Hesketh *et al.* (1972), respectively. Here again, for modern cultivars, Reddy *et al.* (1993) reported a shorter time interval of 24 days from square to first bloom. From bloom to open boll at 27°C required 61 days in the study of Hesketh and Low (1968) while Reddy *et al.* (1993) reported 48 days. Thus, it appears that modern cultivars not only square earlier than obsolete cultivars but the time interval from square initiation to open boll is 7 to 12 days shorter. Interestingly, the temperature optimum for all of these developmental processes has remained consistent at 27°C between both the older and more modern cultivars.

By varying atmospheric CO_2, Reddy *et al.* (1993) found that time from emergence to first square was not influenced by carbohydrate supply. Reddy (1993) also found that assimilate supply does not influence time from square to bloom, or bloom to open boll. However, he did observe that assimilate supply did influence final boll size. Further, Reddy *et al.* (1993) found that assimilate supply did not influence main-stem or sympodial branch plastochron intervals until after first bloom (*i.e.*, after main-stem node 17).

The second process determining the fruit component of the metabolite sink is abscission, or what is commonly called "natural" shed. The plant normally initiates more fruit than can adequately be supplied with metabolites, and

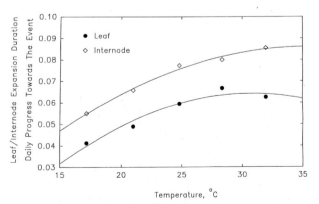

Figure 8-22. The effect of temperature on daily increments of progress in leaf expansion and stem internode extension. From Reddy (1993).

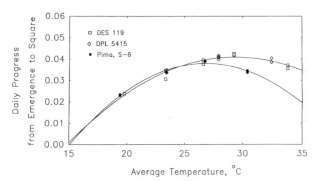

Figure 8-24. The effect of temperature on daily increments of progress from emergence to first square in modern cottons. From Reddy (1993).

the result is abscission. Beginning with the work of Mason (1922), a "nutritional theory" of boll shedding was developed which stated that the cotton plant would seek to balance demands by the bolls for carbohydrate and other nutrients against supply by abscising fruit. Wadleigh (1944) and many others (Eaton, 1955, Johnson and Addicott, 1967) found evidence supporting this theory, but for many years this theory was not conclusively proven. For example, Eaton and Ergle (1953) found no correlation between boll shedding and carbohydrate or nitrogen levels in various plant tissues. Further, they observed that an early planting and a late planting had similar concentrations of carbohydrate and nitrogen in mid-August and that the early planting was shedding fruit rapidly but the later planting was not. This stimulated an interest in hormone research to control fruit loss (Hall, 1952; Morgan, 1967; Morgan and Hall, 1964; Morgan and Gausman, 1966). Because gibberellic acid application did not increase yields, and highly determinate cotton lines are lower yielding, we propose that the source-sink balance between fruiting and vegetative meristems must be carefully considered if yields are to be increased. For example, a rapid and complete transition from the vegetative to the fruiting mode may permit rapid deterioration of the plant's photosynthetic system before the fruit can mature.

Beginning with SIMCOTT II and later with GOSSYM, Baker *et al.* (1973) and Baker *et al.* (1983) took the position that the large amount of data supporting the nutritional hypothesis could not be ignored, nor could the excellent data of Eaton and Ergle (1953) be discarded. Therefore, the modelers steered a mid-course, proposing that the plant adjusts to a disparity between real and potential growth, and abscises fruit on some basis involving both source and sink strength in the boll rather than on the basis of the carbohydrate supply *per se*. These models suggested that the source/sink trigger for fruit abscission activates the hormonal mechanisms involved in abscission. GOSSYM was eventually validated against more than 100 field crop data sets. This may have been the beginning of the development of consensus, now current among cotton physiologists (*c.f.*, Constable, 1991) that the primary driving force in natural fruit shed is a source/sink imbalance which determines the

relative rates of production and transport of various hormones in the plant which, in turn, mediate both developmental delays and fruit shed.

The reader will note that in these models, carbohydrate supply/demand ratio explicitly and separately affects three processes; photosynthate partitioning (growth), development (plastochron intervals), and fruit shed. The fruit loss function from GOSSYM is presented in Fig. 8-25. The shape of this curve varies somewhat to reflect cultivar differences in sensitivity to source/sink imbalance *c.f.* Heitholt (1993). Examining this hypothesis, Guinn (1985) manipulated source strength by varying plant spacing in the field and he manipulated sink strength by selectively defruiting the plants. His results confirmed that growth, flowering, and boll retention decrease when the photosynthate demand exceeds the supply. In earlier experiments Guinn (1984) had shown that the abscisic acid (ABA) concentration did not increase with boll load as the plant approached cutout and did not change with partial defruiting. In other experiments Guinn (1976, 1982) had shown that nutritional stress caused boll shed through an increase in ethylene production. Abscission of a given fruit may be simulated by GOSSYM until the fruit exceeds 16 days age. Guinn (1982) found that fruit abscission declines rapidly after a boll age of 10 days and that it is near zero by 15 days.

The metabolite pools are highly mobile and the abscission zones are very sensitive to the relative concentrations of the hormones. Those building simulation models have found that to simulate abscission and the developmental delays, it is necessary to assume that the carbohydrate reserves (up to 30 percent of leaf dry weight) are all available within a 24 hour period. These simulations have shown that with a heavy fruit load, demand always greatly exceeds supply. In other words, the plant is living on a single day's photosynthate production during much of the fruiting period. The trigger to abort a fruit may be reached over a very short time period, and this, of course, is irreversible.

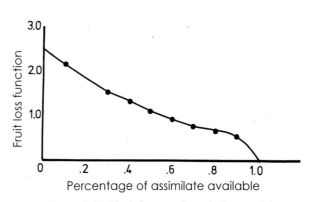

Figure 8-25. The influence of metabolite supply/demand ratio (FSTRES) on daily fruit loss (FLOSS).

4.   SUMMARY

Phytotron research laid the ground work for the process level simulation modeling of cotton. SPAR installations were built specifically for the purpose of manipulating source/sink relations and for characterizing the effects of external and internal factors on plant growth and development. These installations and the discipline of modeling have provided insight into interrelationships among plant processes and a reasonably complete coherent picture has emerged – so much so that plant physiological research results which once were primarily of academic interest are now being applied to cotton crop management. Terminology and usage have been clarified. For example, the term "stress" which has often referred to loss of turgor has come to be thought of as any factor reducing growth below its genetic potential at a given temperature. Growth is generally considered to be the accretion of dry matter and it is no longer as likely to be confused with developmental processes.

We have seen a reconciling of the "nutritional" and "hormal" theories of fruit shed. An understanding of the range of physiological factors which set in place the photosynthetic efficiency of each leaf quickly convinces the modeler of the futility of attempts to calculate crop canopy dry matter production with leaf element models. There is no unique relationship, even among varieties of the same leaf shape, between LAI and canopy light capture. Fortunately, simpler methods treating the crop canopy as an optical surface are available for that purpose. Starch buildup in the chlorplasts and vein loading appear to be of little consequence in cotton photosynthesis. A well-designed vascular system seems to assure rapid movement of assimilates to sinks and the boll is not obligated to obtain its photosynthate from the subtending leaf. Osmotic adjustment appears to be much less important than originally thought, and cotton can be modeled successfully without specific reference to translocation or osmotic adjustment.

A gram of growing root tissue has 6 to 10 times the sink strength of a gram of boll tissue. Roots are particularly effective competitors for photosynthate in seedlings. The reduced photosynthate supply further limits growth and finally, a smaller plant may capture less light. Additionally, drought stress enhances ABA production which may stimulate ethylene production and leaf senescence. The stress does not have to be very severe or long lasting for the effects to be irreversible and recovery must await the growth of new leaves.

The temperature optimum for cotton photosynthesis, growth and development has remained at about 27°C. Boll growth virtually ceases above 33°C. Any factor, (*e.g.*, temperature, turgor loss, photosynthate supply, or mechanical damage) which reduces boll growth rate causes increased ethylene production, and reduced IAA supply, which, in turn, causes fruit shed and delays stem development. Sympodial delays are greater than main-stem delays.

Modern cultivars are faster in vegetative development and in squaring than the varieties of 30 years ago. Modern cultivars are 7 to 12 days faster from square initiation to open boll. Source/sink imbalance does not influence the time to first square, square to bloom or bloom to open boll. There is a consensus that the primary driving force in natural shed is the source/sink imbalance which determines the relative rates of production and transport of various hormones in the plant which, in turn mediate these developmental delays and fruit shed. Ultimately, fruit shed is related to the growth rate of the fruit prior to and during the first 10 to 15 days after anthesis.

Chapter 9

RELATION OF GROWTH AND DEVELOPMENT TO MINERAL NUTRITION

G.L. Mullins[1] and C.H. Burmester[2]
[1]Virginia Tech, Blacksburg, VA and [2]Auburn University, Auburn University, AL

1. INTRODUCTION

Plants are unique organisms that have the capacity to absorb inorganic elements and water through their root systems and carbon dioxide from the atmosphere and combine these into cellular constituents using energy from sunlight. Managing the supply of inorganic elements (nutrients) is a fundamental component of all plant production systems. If nutrient deficiencies are occurring, achievement of optimal yields requires a producer to supply fertilizer nutrients to supplement the pool of available nutrients in the soil to meet the nutritional needs of the specific plant. In addition, the timing of fertilizer applications should ensure that high availability of the applied nutrient(s) corresponds to the peak nutrient requirements of the developing root system. A basic understanding of the growth pattern and nutrient uptake with time by the cotton plant is essential in making wise nutrient management decisions, especially in production systems where nutrients may be deficient (Gerik et al., 1998). This chapter will focus on the elements considered essential for cotton plants, and their uptake and distribution within the plant.

2. ESSENTIAL NUTRIENTS

Research in plant nutrition during the past century has greatly increased our understanding of the nutrient requirements of plants. Three criteria must be met before an element can be considered essential for plant growth: 1) a plant is unable to complete its life cycle in the absence of the element, 2) the function of the element cannot be replaced by another element, and 3) the element in question must be

directly involved in the nutrition of the plant (Mengel and Kirkby, 1987). Based on these criteria there are 18 elements that are now considered essential for plant growth (Table 9-1). Some elements (i.e. Na, I, V, Si, F) may enhance the growth of some plants but are not required by all plants (Mengel and Kirkby, 1987; Marschner, 1995).

Carbon, hydrogen, and oxygen are the major components of organic compounds and since they are obtained from water and the atmosphere, they typically are not considered in soil fertility evaluations. The remaining mineral elements can be subdivided into two major groups: macronutrients and micronutrients (Mengel and Kirkby, 1987; Marschner, 1995). Macronutrients (N, P, K, S, Mg, and Ca) are nutrients needed by plants in relatively large amounts. Micronutrients (Cu, Fe, Mo, B, Mn, Zn, Cl, Ni, and Co) are nutrients that are needed in much smaller quantities as compared to the macronutrients. Macronutrients and micronutrients are equally important to the metabolism of the plant, the difference is that macronutrients are needed in larger quantities. Discussion of the physiological functions of the essential elements can be found in the work of Hearn (1981), Mengel and Kerby (1987), Cassman (1993), and Marschner (1995) and other citations.

3. NUTRIENT UPTAKE AND DISTRIBUTION

Dry matter accumulation and nutrient uptake by cotton has been the subject of several studies. A majority of the work has focused primarily on the accumulation of N, P, and K and most of this work was conducted prior to the mid-1940s (McBryde and Beal, 1896; White, 1914, 1915;

J.McD. Stewart et al. (eds.), *Physiology of Cotton*,
DOI 10.1007/978-90-481-3195-2_9, © Springer Science+Business Media B.V. 2010

Table 9-1. Essential elements for plant growth (Brady and Weil, 1996).

Macronutrients	Micronutrients
Carbon (C)[z]	Iron (Fe)
Hydrogen (H)[z]	Manganese (Mn)
Oxygen (O)[z]	Boron (B)
Nitrogen (N)	Molybdenum (Mo)
Phosphorus (P)	Copper (Cu)
Potassium (K)	Zinc (Zn)
Calcium (Ca)	Chlorine (Cl)
Magnesium (Mg)	Nickel (Ni)
Sulfur (S)	Cobalt (Co)

[z] Primary components of organic compounds. Obtained mostly from air and water and typically not addressed in nutrient management. The remaining nutrients are obtained primarily from the soil.

Fraps, 1919; McHargue, 1926; Armstrong and Albert, 1931; Olson and Bledsoe, 1942). Most of the earlier work consisted of harvesting mature plants and determining the dry matter production and nutrient content of the various plant parts. White (1914) presented the first attempt in looking at nutrient uptake at various times (growth stages) during the growing season, however, he used a very low plant population and results were expressed in terms of nutrient uptake per plant. In the 1930s investigators looked at the effects of N rate on N uptake by cotton (Armstrong and Albert, 1931; Crowther, 1934). Christidis and Harrison (1955) provided a summary of available nutrient uptake data collected up to the mid-1950s for dryland cotton. The work of Olson and Beldsoe (1942) still remains a primary reference for evaluating nutrient uptake by cotton. They used typical production systems on three soils in Georgia to look at uptake and distribution of N, P, K, Ca, and Mg in cotton plants throughout the growing season. Later work has focused on nutrient uptake under irrigation (Bassett *et al.*, 1970; Halevy, 1976; Halevy *et al.*, 1987), including the uptake of selected micronutrients, and the effects of cultivars on nutrient uptake (Bhatt and Appukuttan, 1971; Cassman *et al.*, 1989a; Mullins and Burmester, 1991).

Lint yields and nutrient uptake can be highly dependent on growing conditions and soil fertility. Thus, as noted by Hodges (1991), a majority of studies relate nutrient uptake and removal to yields using a nutrient uptake index. This index is based on the amount of nutrient taken up or the amount of nutrient required to produce a certain amount of lint. In this chapter nutrient uptake indexes will be expressed as kg or g of nutrient per 100 kg of lint.

Growth and dry matter production by the cotton plant follows a sigmoidal curve (Oosterhuis, 1990) with the most rapid rate of production occurring after flowering (Fig. 9-1). Nutrient uptake follows a similar pattern (Figs. 9-1, 9-2, and 9-3). Maximum or peak average daily accumulation rates for most nutrients occur during the period of early to peak flowering (Fig. 9-4), which corresponds not only to a period of high demand for nutrients by the plant but also to the time period where the rate of root growth is at a maximum (Schwab, 1996; Huber, 1999; Schwab *et al.*, 2000).

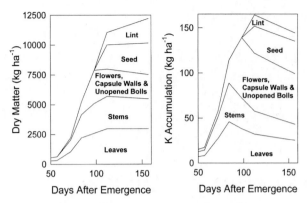

Figure 9-1. Dry matter production and potassium accumulation by irrigated cotton (Halevy, 1976).

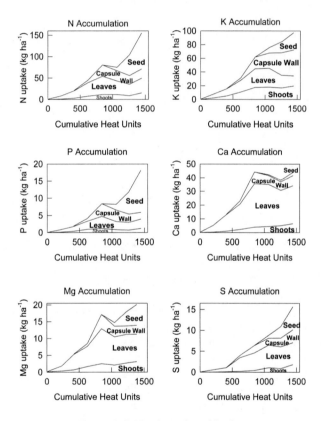

Figure 9-2. Uptake and partitioning of macronutrients by non-irrigated cotton (Mullins and Burmester, 1991, 1992, 1993b). Capsule walls include squares, flowers, and unopened bolls.

A close comparison of cotton growth curves and nutrient uptake curves (Figs. 9-1 and 9-2) shows that the uptake of most nutrients precedes the production of dry matter. These data demonstrate that an adequate supply of nutrients will be needed towards the middle of the growing season to utilize photosynthates and to sustain the production of dry matter by the cotton plant.

During the course of a growing season there appears to be some redistribution of some of the nutrients between the various plant tissues (Figs. 9-1, 9-2, and 9-3).

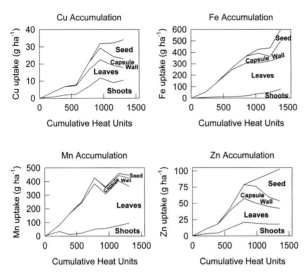

Figure 9-3. Accumulation and partitioning of selected micronutrients by non-irrigated cotton (Mullins and Burmester, 1993a). Capsule walls include squares, flowers, and unopened bolls.

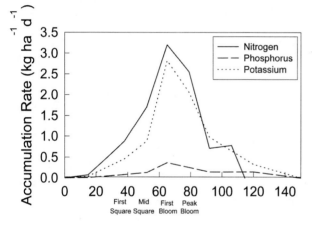

Figure 9-4. Average daily uptake rates for N, P, and K by non-irrigated cotton (Mullins and Burmester, 1991).

Redistributions occur after the onset of flowering as nutrients are transported from the leaves and/or shoots into the reproductive tissues. This phenomenon is particularly evident for K (Figs. 9-1 and 9-2), and research has shown that the fruit, especially the capsule walls, are a major sink for K (Leffler and Tubertini, 1976).

From a historical perspective there appears to be some major differences between more modern and the older, obsolete cultivars in terms of the amounts of nutrients accumulated early in the season. Meredith and Wells (1989) demonstrated that modern cultivars partition more of their dry matter into reproductive tissues. In one of the older studies, White (1914) reported that more than 60% of the N and K, 80% of the P, and less than half of the dry matter had been accumulated by first flower. Even by the mid 1950s,

Christidis and Harrison (1955) stated that a majority of the N is accumulated by first flower. In contrast, Oosterhuis *et al.* (1983) showed that 60% of the dry matter was produced and 40% of the N was taken up between 10 and 16 weeks after sowing in a long-season cultivar. Other studies show that lower amounts of nutrients and dry matter are accumulated prior to first flower. Halevy (1976) reported that irrigated, high yielding cotton accumulated an average of 29, 22, and 21% of the N, P, and K, respectively, and produced 16.8% of the dry matter by full flower. Bassett *et al.* (1970) who worked with irrigated cotton and Mullins and Burmester (1991) who worked with rain-fed cotton had similar results. Thus, modern cultivars accumulate higher proportions of their nutrients after the onset of flowering.

3.1 Specific Nutrients

3.1.1 Nitrogen

As noted previously, N is the nutrient accumulated in the largest quantities by cotton. Total seasonal-N accumulation has been reported to range from about 53 to 301 kg ha^{-1} (Table 9-2). As evident from the data in Table 9-2, total N uptake is dependent on the growing conditions and yield potential. Nitrogen removal as lint and seed range from 43 to 60% of the total plant N. In a non-irrigated study, N in mature cotton plants was distributed as 15.9% stems, 27.6% leaves, 14.1% capsule wall, and 42.4% seed (Mullins and Burmester, 1991). During fruit development (Figs. 9-1 and 9-2), the fruit becomes a primary sink for N in the plant, and redistribution of N from the leaves and stems to the developing fruit can occur (Zhu and Oosterhuis, 1992). The fruit becomes a greater sink for N under N stress (Halevy *et al.*, 1987).

Nitrogen uptake indexes range from 8 to 51 kg N per 100 kg lint. Older studies have higher values due to lower yields and the use of older cotton cultivars which have lower percentage lint and higher vegetative to reproductive ratios. For irrigated cotton primarily in desert regions, N uptake indexes range from 8 to 21 kg N (typically $\cong$11) per 100 kg lint (Bassett *et al.*, 1970; Halevy, 1976; Halevy *et al.*, 1987; Unruh and Silvertooth, 1996b). Under rain-fed conditions, values are usually higher as demonstrated by a value of 19.9 kg N per 100 kg lint reported by Mullins and Burmester (1991) in Alabama. Higher values can be attributed to greater stress under rain-fed conditions due a combination of factors including short-term drought stress, shortterm water excesses, and variable solar radiation. If available, cotton can absorb more N than it needs (Hearn, 1981). High N and water can result in rank vegetative growth, shedding of young bolls, and can delay the attainment of a full fruit load and maturity (Scarsbrook *et al.*, 1959; Thompson *et al.*, 1976; Hearn, 1981).

Peak daily uptake/accumulation rates range from 0.6 to 5.7 kg N ha^{-1} d^{-1} (Table 9-3) in rainfed studies and from 1.5 to 4.6 kg N ha^{-1} d^{-1} in irrigated studies. Daily accumula-

Table 9-2. Lint yields, total nitrogen, phosphorus, and potassium uptake by mature cotton plants, nutrient removal (seed and lint), and nutrient uptake indexes (kg nutrient accumulated per 100 kg lint produced).

Lint yield	Irrig	Nitrogen uptake			Phosphorus uptake			Potassium uptake			Reference
(kg ha⁻¹)		Total	Removal	Index	Total	Removal	Index	Total	Removal	Index	
		----(kg ha⁻¹)----		(kg)	----(kg ha⁻¹)----		(kg)	----(kg ha⁻¹)----		(kg)	
--	N	--	--	20.7	--	--	3.51	--	--	12.1	McBryde and Beal (1896)
62-432	N	71	--	36.0	8.2	--	3.80	52	--	27.0	Fraps (1919)
336	N	53	24	15.7	10	5.6	2.98	48	3.6	14.3	McHargue (1926)
428-893	N	60-217	--	--	--	--	--	--	--	--	Armstrong and Albert (1931)
183-740	N	94-150	--	20-51	11.5-29	--	3.9-7.6	54-112	--	15.1-27	Olson and Bledsoe (1942)
560	N	105	40	18.7	18	6.9	3.2	66	14	11.8	Christidis and Harrison (1955)
490-1650	Y	--	--	--	--	--	--	56-393	--	--	Bennett et al. (1965)
1178-1628	Y	142	70	9.0-13.1	19	9-12	1.3-1.7	127	16-24	7.6-11.4	Bassett et al. (1970)
996-1040	Y	130-217	70-93	--	44.2-72.3	26-34	--	136-227	62-73	--	Bhatt and Appukuttan (1971)
1700	Y	224-235	98-109	13.5	44-46	19-21	2.65	164-185	43-47	10.2	Halevy (1976)
1369-2367	Y	110-301	66-166	8.0-13.9	31-46	16-29	1.61-2.26	120-251	35-65	8.1-10.8	Halevy et al. (1987)
693-1089	y	--	--	--	--	--	--	53-122	--	--	Cassman et al. (1989a,b)
839	N	128	58	19.9	17.3	2.5	2.5	106	20	15.3	Mullins and Burmester (1991)
965-1328	Y	201	--	15-21	32	--	2.3-3.3	254-276	--	19-23	Unruh and Silvertooth (1996)

3.1.2 Potassium

Total K accumulation ranged from 52 to 112 kg ha⁻¹ in non-irrigated tests and from 53 to 393 kg ha⁻¹ in irrigated tests (Table 9-2) and potassium removal as seedcotton ranged from 7.5 to 46% of the total plant K. In one non-irrigated study (Mullins and Burmester, 1991), K in mature plants was distributed as 24.8% shoots, 20.3% leaves, 36.5% capsule wall, and 18.4% seed. Potassium uptake indexes ranged from 12.1 to 27 kg K per 100 kg lint in non-irrigated tests and from 7.6 to 23 kg K per 100 kg lint in irrigated tests. Kerby and Adams (1985) concluded that if the K uptake index is greater than 13 kg K per 100 kg lint then K has been in abundant supply and the cotton crop has indulged in luxury consumption.

Peak uptake rates for K range from 2.1 to 4.6 kg K ha⁻¹ d⁻¹ (Table 9-3). Halevy (1976) observed maximum K uptake in cotton at 72 to 84 days after emergence during which time one-third of the total K was accumulated. Bassett *et al.* (1970) found that for irrigated cotton in California, 67% of the total K was accumulated during a six-week period from mid- to late-bloom. Mullins and Burmester (1991) reported that between 31 to 41% of the total K was accumulated during the peak 2 week uptake intervals (approximately mid-bloom). Cotton is considered to be less efficient at obtaining K from the soil compared to many other plant species, and K deficiency occurs more frequently and with greater intensity compared to many agronomic crops (Kerby and Adams, 1985). Gulick *et al.* (1989) concluded that cotton under greenhouse conditions was less efficient at obtaining soil K compared to barley (*Hordeum vulgare* L.).

Cotton plants accumulate a substantial proportion of K after flowering. Most of the K accumulated during flowering goes into developing fruit, with the capsule wall serving as a major sink for K. Cotton plants have been reported to contain from 40 to 70% (typical values of about 55 to 60%) of the total K in the bolls at maturity (Olson and Bledsoe, 1942; Bassett *et al.*, 1970; Halevy, 1976; Mullins and Burmester, 1991; Unruh and Silvertooth, 1996b).

During the season there is some translocation of K from the leaves to the developing bolls. In Mississippi, Leffler and Tubertini (1976) sampled cotton bolls from 7 days after flowering to maturity and found that boll K during the sampling period increased from 0.17 to 1.1 mg boll⁻¹. Approximately two-thirds of the boll K was in the capsule wall. Leffler and Tubertini (1976) concluded that K in the capsule wall serves as a reserve for the developing seed and fiber. In contrast, Rosolem and Mikkelsen (1991) have suggested that K is required most by the capsule wall itself and is not translocated to seeds or fiber.

3.1.3 Phosphorus

Reported total P uptake ranged from 8.2 to 72.3 kg ha⁻¹ (Table 9-2). Phosphorus removal from the seed and lint ranges from 43 to 59%. Cotton grown in nutrient solution

tion rates typically reach a peak during early-mid flower. In some of the more recent irrigated studies as much as 40 to 53% of the seasonal N was accumulated during the peak 2 to 6 week period (early-mid flower), whereas in the most recent rain fed study 23 to 39% of the seasonal total was accumulated during the peak uptake period (Table 9-3). Armstrong and Albert (1931) observed an increase in the peak N uptake rate with increasing N fertilizer rate.

Table 9-3. Published peak average daily nutrient uptake rates and the percentage of the seasonal total that was accumulated during the peak uptake period (depending on the study, sampling interval ranged from 12 to 42 days).

Nutrient	Peak uptake rate	% of total uptake[z]	Reference
Nitrogen	0.6-5.7 kg ha^{-1}d^{-1}	--	Armstrong and Albert (1931)
Nitrogen	4.2 kg ha^{-1}d^{-1}	41	Olson and Bledsoe (1942)
Nitrogen	1.5-2.0 kg ha^{-1}d^{-1}	45	Bassett *et al.* (1970)
Nitrogen	4.5-4.6 kg ha^{-1}d^{-1}	27-29	Halevy (1976)
Nitrogen	2.4-3.7 kg ha^{-1}d^{-1}	--	Halevy *et al.* (1987)
Nitrogen	2.54-3.87 kg ha^{-1}d^{-1}	23-39	Mullins and Burmester (1991)
Nitrogen	3.3 kg ha^{-1}d^{-1}	40	Oosterhuis *et al.* (1983)
Phosphorus	0.74 kg ha^{-1}d^{-1}	38	Olson and Bledsoe (1942)
Phosphorus	0.17-0.34 kg ha^{-1}d^{-1}	45	Bassett *et al.* (1970)
Phosphorus	0.7-0.72 kg ha^{-1}d^{-1}	21-23	Halevy (1976)
Phosphorus	0.31-0.48 kg ha^{-1}d^{-1}	21-36%	Mullins and Burmester (1991)
Potassium	4.2 kg ha^{-1}d^{-1}	39%	Olson and Bledsoe (1942)
Potassium	2.1-3.4 kg ha^{-1}d^{-1}	67%	Bassett *et al.* (1970)
Potassium	4.6 kg ha^{-1}d^{-1}	30%	Halevy (1976)
Potassium	2.2-3.5 kg ha^{-1}d^{-1}	35%	Mullins and Burmester (1991)
Magnesium	1.4 kg ha^{-1}d^{-1}	52.8	Olson and Bledsoe (1942)
Magnesium	0.3-0.8 kg ha^{-1}d^{-1}	30-52	Mullins and Burmester (1992)
Calcium	3.1 kg ha^{-1}d^{-1}	35.7	Olson and Bledsoe (1990)
Calcium	1.5-3.1 kg ha^{-1}d^{-1}	46-49	Mullins and Burmester (1992)
Sulfur	0.34-0.49 kg ha^{-1}d^{-1}	30	Mullins and Burmester (1993b)
Zinc	1.9-4.1 g ha^{-1}d^{-1}	25-45	Mullins and Burmester (1993a)
Manganese	8.2-14.4 g ha^{-1}d^{-1}	35-47	Mullins and Burmester (1993a)
Copper	0.34-1.33 g ha^{-1}d^{-1}	29-58	Mullins and Burmester (1993a)
Iron	23-27 g ha^{-1}d^{-1}	41-60	Mullins and Burmester (1993a)

[z] % of seasonal total that was accumulated during the peak uptake period. For most nutrients the peak uptake period occurred during early to mid-bloom. In a few instances secondary peaks were observed near maturity.

contained 79% of the total P in the bolls and buds at maturity (Ergle and Eaton, 1957). In studies of irrigated and rain-fed cotton, P removal at harvest accounted for about 52% of the total P. Phosphorus in mature, non-irrigated cotton plants in Alabama (Mullins and Burmester, 1991) was distributed as 11.7 % shoots, 19.5% leaves, 16.0% capsule walls, and 52.8% seed. Phosphorus uptake indexes ranged from 1.3 to 7.6 kg P per 100 kg lint. In the more recent studies, irrigated and rain-fed cotton had similar P uptake indexes. In irrigated studies, P uptake indexes ranged from 1.3 to 3.3 kg P per 100 kg lint while in the rain fed study (Mullins and Burmester, 1991) the P uptake index was 2.5 kg P per 100 kg lint.

Peak daily P uptake rates (early to mid flower) ranged from 0.17 to 0.72 kg P ha^{-1} d^{-1} (Table 9-3). Mullins and Burmester (1991) reported that 21 to 36% of the seasonal total was accumulated during the peak 2-week uptake period for rain-fed cotton in Alabama. Bassett *et al.* (1970) reported that irrigated cotton in California accumulated 45% of the seasonal total during a 6-week peak uptake period from early to mid flower.

3.1.4   Magnesium

Magnesium uptake in non-irrigated tests conducted in the southern U.S. ranged from 18 to 41 kg ha^{-1}, whereas in an irrigated test in California total uptake was much higher ranging from 35 to 104 kg ha^{-1} (Table 9-4). Removal by

harvested seedcotton has been reported for only two non-irrigated tests, and in these reports 27 to 31% of the total Mg was removed at harvest. Mullins and Burmester (1992) reported that Mg in mature non-irrigated cotton plants was distributed as 17% stems, 38% leaves, 14% capsule wall, and 31% seed. Magnesium uptake indexes for non-irrigated cotton ranged from 2.5 to 13.6 kg Mg per 100 kg lint. A value of 2.6 was noted by (Mullins and Burmester, 1992) with the modern cultivar. Irrigated cotton in California had uptake indexes of 3 to 6.4 kg of Mg per 100 kg lint. Olson and Bledsoe (1942) reported a peak daily uptake rate of 1.4 kg of Mg ha^{-1} d^{-1} during a 15 day period at early square to early boll formation (Table 9-3). With modern cultivars, Mullins and Burmester (1992) reported peak uptake rates of 0.3 to 0.8 kg of Mg ha^{-1} d^{-1} during the two week peak period observed during early to mid flower. During the peak uptake periods (early- to mid-flower) of these two studies, 52.8% and 30-52% of the seasonal total was accumulated, respectively.

3.1.5   Calcium

Calcium accumulation by non-irrigated cotton has been reported to range from 50 to 129 kg ha^{-1} (Table 9-4). In the study of Mullins and Burmester (1992) an uptake of 64 kg Ca ha^{-1} was reported. They also reported that Ca in mature plants was distributed as 19% stems, 64% leaves, 14% capsule wall, and 3% seed. Calcium uptake indexes for non-irrigated cotton ranged from 6 to 77 kg Ca per 100 kg lint (Table 9-4) with the highest index of 77 being observed on a field site with extremely low lint yield. A Ca uptake index of 9.3 kg Ca per 100 kg lint was reported by Mullins and Burmester (1992). Maximum daily uptake rates have ranged from 1.5 to 3.1 kg ha^{-1} d^{-1} (Table 9-3). In the study of Mullins and Burmester (1992), 46 to 49% of the total Ca was accumulated during the peak two week uptake period at early- to mid-flower.

3.1.6   Sulfur

Sulfur uptake by cotton has been reported in only a limited number of studies. McHargue (1926) reported a S uptake of 23 kg ha^{-1} for mature cotton plants in Mississippi. Kamprath *et al.* (1957) conducted a three-year study (1953 to 1955) on two soils in North Carolina. At maturity total S uptake increased with S fertilizer rate and ranged from 5.2 to 34 kg ha^{-1}. Stanford and Jordon (1966) reported total S uptake values for mature cotton plants ranging from 10 to 29 kg ha^{-1}. The study of Mullins and Burmester (1993b) is the only field study where S uptake by the cotton plant at various growth stages has been evaluated. They reported total uptake values of 15.6 to 25.1 kg S ha^{-1} for mature, non-irrigated cotton plants in Alabama. Sulfur in the mature plants was distributed as 14.2% stems, 39.0% leaves, 22.2% capsule walls, and 24.6% seed. Sulfur uptake indexes have ranged from 1 to 6.8 kg S per 100 kg lint. In the 1993 study, Mullins and Burmester (1993b) reported an index of 2.7 kg S per 100 kg lint. Maximum daily uptake rates ranged from 0.34 to 0.49 kg S ha^{-1} d^{-1} (Table 9-3). An average of 30% of the total S was accumulated during the peak uptake period (boll formation).

3.1.7   Zinc

Total Zn uptake by mature cotton in Alabama (Mullins and Burmester, 1993a) was reported to be 103 ± 38 g ha^{-1} (Table 9-5). Zinc in mature plants was distributed as 23% in leaves, 18% in stems, 4% in capsule walls, and 48% in seed. In Mississippi (McHargue, 1926) mature cotton accumulated 184 g Zn ha^{-1} and mature cotton in Australia (Constable *et al.*, 1988) accumulated an average of 60 g Zn ha^{-1} in 35 commercial fields. For three cotton cultivars (Alimov and Ibragimov, 1976) total Zn uptake needed to produce 1 Mg seedcotton ranged from 25-38 g ha^{-1}. Constable *et al.* (1988) reported a an average removal of 5.7 g Zn in seedcotton per 100 kg lint produced with a range from 5.7 to 358 g ha^{-1}. Donald (1964) reported that the harvest of one bale of lint removes 358 g Zn ha^{-1} in the seedcotton. Other researchers (Table 9-5) have reported that from 47 to 55% of the total uptake is removed at harvest. Zinc uptake indexes ranged from 55 g Zn per 100 kg lint for cotton produced in Mississippi (McHargue, 1926) during the 1920s to 15 g Zn per 100 kg lint for cotton produced during the late

1980s in Alabama (Mullins and Burmester, 1993a). Mullins and Burmester (1993a) observed peak daily uptake rates of 1.9 to 4.1 g Zn ha^{-1} d^{-1} (~mid-bloom; Table 9-3). During the peak two-week uptake period (~mid-flower), 25 to 45% of the seasonal total was accumulated.

3.1.8   Manganese

Manganese uptake by mature cotton plants in Mississippi (McHargue, 1926) was reported to be 101 g ha^{-1} (Table 9-5), while cotton in Australia accumulated an average of 450 g ha^{-1} (Constable, 1988). Three cotton cultivars accumulated 86 to 163 g Mn ha^{-1} to produce 1 Mg seedcotton (Alimov and Ibragimov, 1976). Mature cotton in Alabama accumulated 451 ± 175 g ha^{-1} (Mullins and Burmester, 1993a) and was distributed as 56% in leaves, 20% in stems, 18% in capsule walls, and 6% in seed. Constable *et al.* (1988) reported an average removal of 5.2 g Mn in seedcotton for every 100 kg lint produced. Donald (1964) reported that for one bale of lint 123 g Mn ha^{-1} is removed as harvested seedcotton. Other researchers (Table 9-5) have reported that 4.5 to 5.9% of the total plant Mn is removed at harvest. Manganese uptake indexes ranged from 30 g Mn per 100 kg lint for cotton produced in the 1920s to 15 g Mn per 100 kg lint in the 1993 study (Mullins and Burmester, 1993a). Peak average daily influx rates were reported to occur at early-mid flower for non-irrigated cotton in Alabama and ranged from 8.2 to 14.4 g Mn ha^{-1} d^{-1} (Table 9-3). The peak, two-week uptake period near mid flower accounted for 35 to 47% of the seasonal total.

3.1.9   Copper

Copper accumulation by mature cotton was reported to be 41 g ha^{-1} in Mississippi (McHargue, 1926) and 20 g ha^{-1} in Australia (Constable *et al.*, 1988; Table 9-5). Three cotton cultivars accumulated 19-26 g Cu ha^{-1} (Alimov and Ibragimov, 1976). Mature cotton in Alabama accumulated 28±14 g Cu ha^{-1} (Mullins and Burmester, 1993a) and was distributed as 26% in leaves, 29% in stems, 17% in capsule walls, and 28% in seed. Constable *et al.* (1988) sampled 35 commercial cotton fields in Australia and reported that an average removal of 1.4 g Cu in seedcotton for every 100 kg lint. Donald (1964) reported for one bale of lint 67 g Cu ha^{-1} was removed as harvested seedcotton. In other studies, (Table 9-5) 17.9 to 41% of the total plant Cu was removed at harvest. Copper uptake indexes ranged from 12.3 g Cu for every 100 kg lint for cotton grown in the 1920s to 4.0 g Cu per every 100 kg lint in the 1993 study (Mullins and Burmester, 1993a). Peak average daily influx rates of 0.34 to 1.33 g Cu ha^{-1} (Table 9-3) were reported for non-irrigated cotton in Alabama (Mullins and Burmester, 1993a). During the peak two-week uptake period (~ mid-flower), 29 to 58% of the seasonal total was accumulated.

Table 9-4. Lint yields, total magnesium, calcium, and sulfur uptake by mature cotton plants, nutrient removal (seed and lint), and nutrient uptake indexes (kg nutrient accumulated per 100 kg lint produced).

Lint yield	Irrig	Magnesium uptake			Calcium uptake			Sulfur uptake			Reference
		Total	Removal	Index	Total	Removal	Index	Total	Removal	Index	
(kg ha⁻¹)		----- (kg ha⁻¹) -----		(kg)	----- (kg ha⁻¹) -----		(kg)	----- (kg ha⁻¹) -----		(kg)	
62-432	N	18	--z	2.9	70	--	8.9	--	--	--	Fraps (1919)
336	N	--	--	2.9	49.8	0.6	14.8	23	--	6.8	McHargue (1926)
183-740	N	20-41	--	5.5-13.6	84-129	--	17.4-77	--	--	--	Olson and Bledsoe (1942)
560	N	16.1	4.3	2.9	57	2.9	10.2	--	--	--	Christidis and Harrison (1955)
205-583	N	--	--	--	--	--	--	5.3-34	--	1.0-5.0	Kamprath et al. (1957)
560	N	--	--	2.5	--	--	6.0	--	--	--	Donald (1964)
560	N	--	--	--	--	--	--	10-29	--	1.9-3.3	Stanford and Jordan (1966)
1178-1628	Y	35-105	--	3-6.4	--	--	--	--	--	--	Bassett et al. (1970)
839	N	18	5.6	2.6	64	1.92	9.3	--	--	--	Mullins and Burmester (1992)
839	N	--	--	--	--	--	--	15.6-25.1	5.0	2.7±0.7	Mullins and Burmester (1993b)

z Not reported.

Table 9-5. Lint yields, total zinc, manganese, and copper uptake by mature cotton plants, nutrient removal (seed and lint), and nutrient uptake indexes (g nutrient accumulated per 100 kg lint produced).

Lint yield	Irrig	Zinc uptake			Manganese uptake			Copper uptake			Reference
		Total	Removal	Index	Total	Removal	Index	Total	Removal	Index	
(kg ha⁻¹)		----- (kg ha⁻¹) -----		(g)	----- (kg ha⁻¹) -----		(g)	----- (kg ha⁻¹) -----		(g)	
62-432	N	18	--z	2.9	70	--	8.9	--	--	--	Fraps (1919)
336	N	184	101	55	101	4.5	30	41	16.8	12.3	McHargue (1926)
560	N	--z	358	--	--	123	--	--	67	--	Donald (1964)
330	--	25-38	--	--	86-163	--	--	19-26	--	--	Alimov and Ibragimov (1976)
1575	Y	62	5.7	--	450	5.2	--	20	1.4	--	Constable (1988)
--	N	103	49	15	388	23	15	28	5.0	4.0	Mullins and Burmester (1993a)

z Not reported.

Table 9-6. Lint yields, total iron, boron, and molybdenum uptake by mature cotton plants, nutrient removal (seed and lint), and nutrient uptake indexes (g nutrient accumulated per 100 kg lint produced).

Lint yield	Irrig	Iron uptake			Boron uptake			Molybdenum uptake			Reference
		Total	Removal	Index	Total	Removal	Index	Total	Removal	Index	
(kg ha⁻¹)		----- (kg ha⁻¹) -----		(g)	----- (kg ha⁻¹) -----		(g)	----- (kg ha⁻¹) -----		(g)	
336	N	814	47	242	--z	--	--	--	--	--	McHargue (1926)
333	--	--	--	--	66-107	--	--	1.97-4.03	--	--	Alimov and Ibragimov (1976)
1575	Y	600	29	--	200	9.3	--	--	--	--	Constable (1988)
--	N	626	88	90	--	--	--	--	--	--	Mullins and Burmester (1993a)

z Not reported.

3.1.10   Iron

Total Fe uptake w reported to range from 600 to 814 g ha (Table 9-6). Iron in mature, non-irrigated cotton plants in Alabama (Mullins and Burmester, 1993b) was distributed as 43% in leaves, 20% in stems, 23% in capsule walls, and 14% in seed. Constable *et al.* (1988) determined that for irrigated cotton in Australia, an average (35 commercial fields) of 29 g Fe was removed in seedcotton for every 100 kg lint. Other researchers have reported that 5.8 to 14.1% of the total plant Fe was removed in the seedcotton. Iron absorption indexes have ranged from 242 g Fe per 100 kg lint in the oldest reference (McHargue, 1926) to 90 g Fe per 100 kg lint in the 1993 study (Mullins and Burmester, 1993a; Table 9-6). Mullins and Burmester (1993a) reported peak average daily influx rates (early- to mid-flower) of 23-27 g Fe ha^{-1} d^{-1} (Table 3). During the peak two-week period at ~ mid-flower 41 to 60% of the seasonal total was accumulated.

3.1.11   Other Nutrients

Limited data are available for the remaining micronutrients. Constable *et al.* (1988) reported an average total uptake of 200 g boron ha^{-1} for 35 irrigated cotton fields in Australia and an average removal of 9.3 g B as seedcotton for every 100 kg lint (Table 9-6). Alimov and Ibragimov (1976) determined that the total plant uptake of B, Mo, and Co to produce 1 Mg seedcotton by 3 cotton cultivars ranged from 66 to 107, 1.97 to 4.03, and 2.44 to 4.35 g ha^{-1}, respectively.

McHargue (1926) reported that mature cotton accumulated 814 g F ha^{-1} and that 186 g F ha^{-1} were removed in the harvested seedcotton. Irrigated cotton in California (Bassett *et al.*, 1970) was reported to accumulate 4.3 to 17.1 kg Na ha^{-1}. No uptake data were found for Ni or any of the so-called beneficial elements.

4.   EFFECTS OF CULTIVARS ON NUTRIENT UPTAKE

Available data suggests that cultivar differences in some instances may affect nutrient uptake enough to be considered in nutrient management decisions. Unfortunately, in the earliest study with cotton where more than one cultivar was included (Fraps, 1919), cultivar differences could not be evaluated since all of the five cultivars were not grown in the same location. More recently, Mullins and Burmester (1991, 1992, 1993a,b) found no differences in nutrient uptake or distribution by four modern cultivars when grown under non-irrigated conditions in Alabama. In contrast, Bhatt and Appakuttan (1971) found that a bushy cotton cultivar and a short branch cultivar produced the same cotton yields, but the short branch cultivar took up considerably less N, P, and K. Halevy (1976) compared two Acala cultivars and found differences in total K uptake and distribu-

tion in the plant. The authors concluded that one cultivar had difficulties in accumulating K because of a smaller root system. Cassman *et al.* (1989a) on a high vermiculitic soil also observed significant differences in K uptake between a K efficient and a K inefficient cultivar. Under K limited conditions, yield differences were not related to differences in K partitioning between vegetative and fruiting structures. Cultivar differences have been attributed to differences in plant demand and differences in rooting. Cultivar differences may also be related to the susceptibility to verticillium wilt (Kerby and Adams, 1985) and K fertilization can decrease the severity of verticillium wilt (Hafez *et al.*, 1975). Thus, in some situations the cultivar selected may have a significant impact on nutrient management decisions.

5.   BOLL LOAD/NUTRIENT STATUS ON UPTAKE AND DISTRIBUTION

In addition to cultivar differences, a number of environmental factors including soil differences, fertility levels, and water availability can affect nutrient uptake and distribution. Effects of fertilizer rates on nutrient uptake by cotton have been limited primarily to N and K, with the uptake of both nutrients increasing with increasing fertilizer rate. For N, Halevy *et al.* (1987) showed that total N uptake increased from 110 to 322 kg ha^{-1} as the N rate increased from 0 to 180 kg N ha^{-1}. Others (Scarsbrook *et al.*, 1959; Oosterhuis *et al.*, 1983) have also reported increased N uptake with increasing rate. Bennett *et al.* (1965) reported that for irrigated cotton in Alabama, total K uptake increased from 85 to 390 kg K ha^{-1} as the K fertilizer rate increased from 0 to 560 kg ha^{-1}. Work with irrigated cotton in vermiculitic soils in California (Cassman *et al.*, 1989a, b) has also demonstrated that total K uptake increases with K fertilization. Mullins *et al.* (1994) reported that K uptake by cotton may not be increased even with K fertilization unless adverse soil physical conditions (*i.e.* root-restricting hardpan) are alleviated.

As mentioned previously, the fruit can be a major sink for several nutrients (Figs. 9-4 and 9-5) and there is evidence of an apparent redistribution of some elements from vegetative tissues into the fruit as the plant matures. For those nutrients where the fruit is a major sink (*i.e.* N, K, P, etc.), these observations lead one to speculate that under limited availability, there would be a greater partitioning of these nutrients into the fruit as compared to vegetative tissues. One would also speculate that the requirements for these nutrients would increase under production conditions favoring a high boll load. Evidence to support this hypothesis has been provided for N and K. Halevy *et al.* (1987) reported that the proportion of N in the bolls decreased from 67 to 55% as the N rate increased from 0 to 180 kg ha^{-1}. Thus, a greater proportion of the N was partitioned to the developing fruit to support the developing boll load under N limiting conditions. Rosolem and Mikkelsen (1989) re-

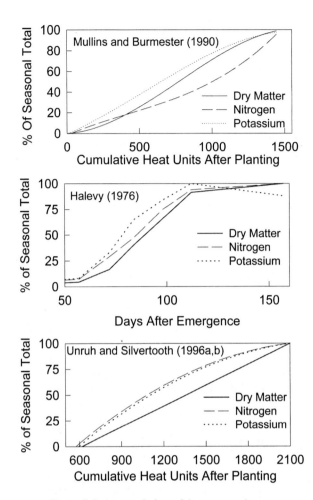

Figure 9-5. Accumulation of dry matter, nitrogen and potassium by cotton, expressed as a percentage of the seasonal total accumulation (Mullins and Burmester, 1991; Halevy, 1976; Unruh and Silvertooth, 1996a,b).

ported that the stems and capsule walls act as temporary storage organs, remobilizing N to the seeds late in the growing season.

Similar findings have been reported for K uptake by irrigated cotton in California (Cassman *et al.*, 1989a, b). Cassman al. (1989a) found that the proportion of K in the fruit increased from 70% on a K sufficient soil to 90% on a severely K deficient soil. Recently, K deficiency has been a major nutritional concern in cotton. Potassium deficiency symptoms appear first in the upper portion of the plant canopy (Maples *et al.*, 1988) and increased incidences of K deficiency have been attributed in part to modern cotton cultivars having a larger boll load than older cultivars (Oosterhuis, 1995b). A higher boll load places greater demand on the plant during fruit development.

6.   SUMMARY

This review has provided a basic understanding of the growth pattern and nutrient uptake with time by the cotton plant that is essential in making wise nutrient-management decisions. A description has also been provided of the elements considered essential by cotton plants, and their uptake and distribution within the plant. Achievement of optimum yields requires a producer to supply fertilizer nutrients at rates to supplement the pool of available nutrients in the soil to meet the nutritional needs of the specific plant. In addition, the timing of fertilizer applications should ensure that high availability of the applied nutrients corresponds to the peak nutrient requirements of the developing root system. Cultiar differences may affect nutrient uptake enough to be considered in nutrient-management decisions.

Chapter 10

CYCLES AND RHYTHMS IN COTTON

William A. Cress[1] and James McD. Stewart[2]

[1]Centro De Investigación Científica De Yucatán, A. C., Calle 43 No.130, Col. Chuburna De Hidalgo, 97200, Mérida, Yucatán, México; and [2]Department of Crop, Soil, and Environmental Sciences, University of Arkansas, Fayetteville, AR 72701 U.S.A.

1. INTRODUCTION TO BIOLOGICAL RHYTHMS

The natural physical environment in which plants grow is a dynamic and changing habitat. Many fluctuations in the environment occur at regular and predictable intervals. During each 24 hour period the earth completes one light/dark cycle, and plants growing in their natural habitat are subjected to alternating periods of sunlight and darkness. With the change of the seasons, the duration of each light/dark period changes in a constant predictable fashion in relation to the earth's yearly seasonal changes.

In close relationship to the physical cycles of our planet are the natural rhythms of biological organisms on the earth. The interruption of developmental processes and alteration of growth often result when plants are maintained under constant environmental conditions. The effects of a constant environment on plant growth and development suggest the extent of the interdependence between internal rhythms and external cycles.

Rhythmic phenomena, until 40 years ago, were generally overlooked (Palmer, 1976). The early workers in biological rhythms asserted that Claude Bernard's straight-line homeostatic constancy needed to be revised to a rhythmic stasis (Palmer, 1976). Today biological rhythmicity is recognized as an integral part of the physiology of a plant (Hammer, 1960; Cummings and Wagner, 1968; Hillman, 1976; Hastings and Schweiger, 1976; Koukkari and Warde, 1985).

Rhythmic processes are inseparable from the factors which regulate growth and development. In a strict sense, growth in plants can be defined as an irreversible increase in volume, involving cell enlargement which may or may not be accompanied by cell division (Thiamin, 1969; Koukkari and Warde, 1985; Salisbury and Ross, 1992). At times growth processes are also described as increases in mass, size, and number. Irrespective, growth and cell division have been demonstrated to be rhythmic in some organisms. However, a rhythmic interpretation of cell division or mitosis in higher plants is not straightforward because experimental designs have not always been sufficient to determine whether the processes are rhythmic or not (Koukkari and Warde, 1985).

Flowering is perhaps the classic example of differentiation that is rhythm-regulated (under the control of photoperiod). Due to the change in photoperiod there is a transition from the vegetative state to the reproductive state. In addition to the photoperiodicity of flowering in some plants, there are a number of other examples that demonstrate that morphogenesis and development can be regulated by the temporal organization of rhythms (Sweeney, 1974; Koukkari and Warde, 1985).

Biological rhythms in plants are divided into three time or free running periods: ultradian, circadian, and infradian rhythms (Koukkari and Warde, 1985). Ultradian rhythms are those which have more than a single cycle in a 20 hour period. Circadian rhythms are those which have a cycle of 20 to 25 hours while infradian rhythms are those in which the cycle is greater than 28 hours. These rhythms must have certain characteristics if they are to be termed true biological rhythms. For example, the rhythm must be self-sustaining when isolated from its synchronizer for a period of time under conditions of constant light (Sweeney, 1969; Koukkari and Warde, 1985) and the rhythm must shift its phase when the cycle of the synchronizer is shifted (Bunning, 1973; Koukkari et al., 1974; Lumsden, 1991).

J.McD. Stewart et al. (eds.), Physiology of Cotton,
DOI 10.1007/978-90-481-3195-2_10, © Springer Science+Business Media B.V. 2010

1.1 Ultradian Rhythms

High frequency rhythms or ultradian rhythms are those which complete more than one complete cycle in less than 20 hours (Koukkari and Warde, 1985). Many ultradian rhythms have periods of 20 minutes to one hour, and are easily overlooked unless continuous observations are taken. Henson *et al.* (1986) have described ultradian rhythmic oscillations in starch concentrations and the activities of amylolytic enzymes as well as rhythmic oscillations of invertase in nodules of *Medicago sativa*.

1.2 Circadian Rhythms

Circadian rhythms maintain a free running period of 20 to 28 hours, with an average of 24 hours.

1.2.1 Classical Rhythm Characteristics

A rhythm is a change or event that recurs systematically with a specifiable pattern, probability, and frequency (Koukkari *et al.*, 1974). Even though an event may possess a rhythmic pattern and be a part of a living system, it is not necessarily a biological rhythm. It is quite possible that a periodic biological event could occur solely in response to an environmental oscillation (*e.g.*, change in light intensity or temperature). Therefore, during rhythm studies, it is important that external factors and synchronizers (Koukkari *et al.*, 1974) are rigorously controlled and manipulated. With the advent of modern studies in molecular biology, extensive advances have been made in our understanding of the mechanism of the basic circadian clock.

Time keeping by circadian clocks enable organisms to respond physiologically to the passage of time and to adjust their response to daily fluctuations in the environment. The best characterized rhythms are those that display a 24-hour period length, with a resetting of the clock in response to light-dark transitions and with a temperature regulated adjustment of the period of the clock.

More recent molecular studies have re-emphasized the three elements of circadian clocks (Fig. 10-1: a. an input pathway(s) which relays environmental information to the circadian pacemaker; b. a pacemaker (oscillator) which generates the circadian oscillation; and c. a signal transduction pathway(s) through which the pacemaker regulates the output of the various rhythms). These input and output signal pathways of the circadian system appear to be specific for each organism while the core mechanism of the pacemaker appears to be the same in all organisms. More than a single clock may be present in an individual cell.

During the last 40 years, rapid advances have been made in the physiology of circadian biology and more recently advances have been made at the level of the gene transcription. A large body of evidence indicates that the mechanism of circadian rhythms is not regulated at the level of the whole cell but involves time dependent gene expression at the level of input to the signal transduction pathway.

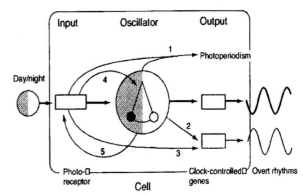

Figure 10-1. The basic circadian model. The basic model of the clock is illustrated by the central bold arrows in the figure. This basic model has been established by many physiological studies on circadian rhythms and consists of an input, an oscillator, and an output. Oscillators are located in individual cells and generate the circadian oscillation. The input transmits a light-derived signal to the oscillator to reset it. The oscillator controls the output which drives the circadian rhythm. The rhythms and the oscillator are assumed to be independent. Possible additions to the basic model to make it realistic in the context of plants are also shown: 1 photoperiodic time measurement; 2 multiple output control; 3 dual control by both the light and the clock signal; 4 alteration of clock properties such as the period; 5 modulation of sensitivity by the clock. An additional property which should be added is dual clock systems within single cells (Kondo and Ishiura, 1999).

1.2.2 Light Sets the Molecular Controls of Circadian Rhythm

Light sets the circadian rhythm by eliminating a key protein needed for the molecular mechanism of the circadian clock (Lee *et al.*, 1996, Myers *et al.*, 1996). The results from fruit fly studies (Lee *et al.*, 1996) demonstrate how this environmental parameter sets the circadian clock.

Living organisms adjust their intrinsic circadian cycles, which intrinsically range from about 23 to 25 hours, to the 24-hour solar day. Work by Lee *et al.* (1996) and Myers *et al.* (1996) identified a molecular key to light's circadian control in *Drosophila* and have suggested that there is a similar mechanism(s) in humans. These results may explain how humans are able to adjust their body clocks after traveling across time zones (Lee *et al.*, 1996). Both *Drosophila* and humans have activity rhythms that adapt well to a 24-hour light-dark cycle. This cycle can be set to new time zones by light. The influence of light is quite strong. When fruit flies are raised in total darkness they maintain an activity rhythm of about 23.5 hours, however, even a brief exposure to light can either delay or advance this activity cycle.

Molecular models for the circadian oscillator have been developed in *Drosophila*, *Neurospora*, mice, and Cyanobacteria. While all share a similar feedback system, the key proteins in each oscillator are different. A clock model has not yet been proposed for plants although the molecular system for the model plant Arabidopsis is expected to be available soon. The current information suggests that there may be two types of output in plant cells. The first is a direct regulation which creates a circadian rhythm, while the second has an indirect effect which can modulate the response to a non-rhythmic signal (Miller, 1998). Miller suggests that it would be important to determine whether this "circadian gating" extends to signaling pathways in addition to phototransduction.

The first oscillator components were identified by genetic screening in *Drosophila* and *Neurospora*. The period (*per*) and timeless (*tim*) genes have been isolated and cloned from *Drosophila* and the frequency (*frq*) gene has been isolated and cloned from *Neurospora*. In the *Drosophila*, the rhythm is set by the action of two proteins, PER and TIM, which are encoded by the *per* and *tim* genes. While all cells of the fly have *per* and *tim* genes, only cells located in the eye set the body clock. The PER and TIM proteins are accumulated in the nuclei of the eye cells which are sensitive to light (photoreceptors), as well as pacemaker cells of the central brain.

The *Drosophila* circadian cycle starts at noon when the *per* and *tim* genes are transcribed into mRNA to be translated into the PER and TIM proteins, however only after sunset is there an accumulation of the PER and TIM proteins. At night, the two proteins pair and are transported into the nucleus. About four hours before sunrise, the level of the PER/TIM protein complex reaches its greatest concentration, this signals the *per* and *tim* genes to stop transcription of the mRNA thereby stopping the formation of the protein complexes. At dawn, light induces the PER/TIM protein complexes to disintegrate. When the complexes are sufficiently depleted, the *per* and *tim* genes are again induced to make additional mRNA (around mid-day). TIM and PER proteins must complex with each other in order to be transported into the nucleus. When *Drosophila* is exposed to light the TIM protein is rapidly degraded. This blocks the transport of the remaining PER protein into the nucleus. In the normal day-night cycle, even though mRNA levels begin to increase by mid-day, sunlight degrades the TIM protein and prevents it from accumulating until nightfall. This delay prevents the binding and nuclear activity of the PER and TIM proteins until the dark part of the cycle.

Their research has also demonstrated how exposure to light at different times of day can shift the rhythm. For example, Lee *et al.* (1996) demonstrated that when flies were exposed to one hour of daylight in the evening, around 10 p.m., their normal night time level of accumulation of the TIM protein was delayed, and this postponed the behavioral cycle by four to five hours. The flies behaved as if they were adapting to westward travel. The TIM proteins did accumulate but at a later time, which is in accord with the observed behavioral delay. If the flies were exposed to daylight an hour before dawn they lost their TIM proteins earlier, and the proteins did not reappear until the following afternoon.

The early light exposure advanced their behavioral rhythm by one to two hours, which would be expected of flies traveling in an eastward direction.

In *Neurospora*, significant progress has been made in elucidating the molecular nature of the *frq* gene. Expression of the *frq* gene is influenced by light and is subject to negative feedback control by the Frq protein. Experimental evidence suggests there are distinct differences between *per* and *frq*, though a common feature of both circadian clocks seems to be a transcription-translation autoregulatory feedback loop.

For the first time a molecular explanation of the mechanism of rhymicity at the level of the gene is available which can account for all of the properties of an entrainable molecular clock. While these models have withstood a number of tests with regards to the *frq* gene in *Neurospera* and the *tim* and *per* genes in *Drosophilia* all of the predictions have not yet been tested (Miller, 1998). A comparison of the *per* and *tim* cycling between *Drosophilia* and the silk moth suggest that the current model may not be general model (Hall, 1996).

The eukaryotic clock model developed from *Drosophilia*, *Neurospora*, and mice is shown in Figure 10-2.

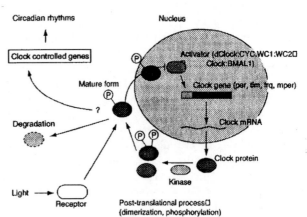

Figure 10-2. The current eukaryotic clock model. The components in the model are from the *Drasophila*, *Neurospora*, and mouse models. The translocation of the clock proteins into the nucleus provide the time delay needed for stable oscillations (Kondo and Ishiura, 1999).

1.3   Infradian Rhythms

Infradian rhythms are those in which the period is greater than 28 hours but typically take periods of days to years. These rhythms are cycles that occur less frequently than 20 to 28 hours and are subdivided into circannual (about once a year), circatrigentan (about 30 days), and circaseptan (weekly). These and other rhythms are currently under study with renewed interest on their molecular mechanisms.

2. BIOLOGICAL RHYTHMS IN COTTON

Studies on rhythmic phenomena in cotton are very limited. The current studies are limited to the rhythmic changes in ethylene production, the effect of low temperature and relative humidity on chilling resistance in cotton seedlings, the rhythmic changes in sensitivity of cotton to herbicides and the perturbation of this phenomenon by abscisic acid, and lastly the affect of fusaric acid and gibberellic acid on leaf movement rhythms.

The rhythmic production of ethylene has been studied in cotton seedlings (Rikin *et al.,* 1984). In the biosynthetic pathway of ethylene from methionine to ethylene, the conversion of [3,4-^{14}C] methionine into ethylene follows a rhythmical pattern in ethylene evolution. The conversion is low when ethylene evolution is low and the conversion is high when ethylene evolution is high. The same rhythm was maintained after a 12-hour light and 12-hour dark entrainment period, when the light-dark periods were reversed. When exposed to continuous light the rhythm of ethylene production was shortened to less than a 24-hour period. The conversion of ACC into ethylene was not affected by the light-dark cycle, however, the evolution of ethylene was always decreased by light and increased by darkness, implying that the concentration of ACC fluctuated in a rhythmic manner. Thus, in the biosynthetic pathway of ethylene production the steps from methionine to ACC are controlled by an endogenous rhythm while the step from ACC to ethylene is regulated by the immediate conditions of light. Treatment of *G. hirsutum* with silver ions stimulates the rhythmic production of ethylene in both the normal light-dark cycle and when the light-dark period is reversed (Rikin *et al.,* 1985). The rate of conversion of [3-4-^{14}C]-methionine into ethylene also is stimulated by silver ions in the same manner, while treatment with the translational inhibitor aminoethoxyvinylglycine (AVG) decreases this stimulation. The conversion of ACC to ethylene was not affected by the presence of silver ions. Cotyledon movement, phenlyalanine ammonia lyase (PAL) activity and resistance to bentazon in cotton, all of which demonstrate rhythmic changes that have been correlated with rhythmic ethylene production, were not affected by silver ions or AVG. This suggests that these processes are not dependent on or controlled by the rhythmicity in ethylene production.

Low environmental temperatures down to 15°C at 85% relative humidity (RH) will induce chilling resistance (CR) or hardening in cotton seedlings. A six hour exposure to the hardening temperature will induce a maximum level of CR (McMillan and Rikin, 1990; Rikin, 1991). If the seedling growth temperature and RH are decreased from a standard growing temperature of 33°C and 85% RH to 26°C or, 19°C at 85% RH for 6 to 12 hours, CR is induced and the seedlings will remain resistant through the light-dark cycle irrespective of the circadian cycle. Additional studies (Rikin,

1992) on the temporal organization of chilling resistance in cotton seedlings demonstrated the effect of temperature and RH on changing the time course of the endogenous rhythm of chilling resistance when the seedlings were grown under light-dark cycles of 12:12 h. The chilling resistance at 5°C and 85% RH lasted most of the dark period while exposure at 5°C and 100% RH lengthened the CR period and extended it into the last half of the light period due to a transient phase advance. The circadian rhythm of CR in cotton allows chilling resistance to develop at the end of the day when the temperatures begin to fall. If the night is cold then the plants will acquire CR toward the end of the night and the beginning of the day. The resistant phase is retained if the lower temperatures are reached at the beginning or the middle of the night and extend into the day. Additionally Rikin *et al.,* (1993) demonstrated that the CR that develops rhythmically, as well as low-temperature-induced CR, coincides with increased levels of polyunsaturated fatty acids.

An oscillating sensitivity of cotton seedlings to the herbicides acifluorfen, fluazifop, and bentazon was demonstrated by Rikin *et al.* (1984). When cotton seedlings were grown under a 12:12 hour light-dark photoperiod they observed that the seedling had oscillating sensitivity to these three herbicides. The sensitivity of the seedlings to the herbicides, expressed as areas of necrosis on the cotyledons or decreased shoot growth, was least at the beginning until the middle of the light period. It increased from the middle of the light period reaching a maximum between the beginning and the middle of the dark period, then subsequently declined. Seedlings germinated and grown under continuous light demonstrated very little oscillation in their sensitivity to the herbicides.

Plant movement has been studied for over a hundred years (Darwin and Darwin, 1897; Cummings and Wagner, 1968; Anderson and Koukkari, 1979). The circadian rhythm of leaf movement in cotton was demonstrated by the work of Sundararajan (1980). Two additional studies (Sundararajan *et al.,* 1978) demonstrated that the phase of the leaf movement shifted when seedlings were exposed to fusaric or gibberellic acid. Fusaric acid, a toxin produced during the pathogenesis of *Fusarium oxysporum f.sp vasinfectum* Akt. (Fusarium wilt disease), was demonstrated to produce a phase shift in the circadian leaf movement rhythms of cotton seedlings. The rhythms were shown to vary in degree and magnitude as a function of the phase at which treatment was applied. Phase advances occurred if treatment with fusaric acid was applied during the subjective day (90-270°) while phase delays occurred if treated during the subjective night (270-90°). Gibberellic acid acted to advance the phase of circadian leaf-movement rhythm in cotton when applied at various phases of the rhythm (Viswanathan and Subbaraj, 1983). Background illumination did not affect the quality or the magnitude of the shift. Thus, apparently no interaction between gibberellic acid and light was present.

3.   FUTURE DIRECTIONS

While the studies of rhythms in cotton have been limited to: leaf movements and their interaction with fusaric and gibberellic acid; chilling resistance and the involvement of fatty acids; herbicide resistance; and ethylene production and sap flux, there are a number of additional areas that could be productive. Studies on phytochrome and its interactions with rhythms are lacking in cotton and could prove of particular interest in elucidating the molecular clock in plants. Also, any effects of pathogens on circadian movements have not been addressed in cotton. Additional areas of interest would be leaf movement in regard to pH, K^+ flux and age of the tissue. The sole examples of non-circadian rhythms in cotton are sap flux and some leaf movements. Non-circadian rhythms such as protein synthesis and translational regulation have been reported in other plant species and should be explored in cotton. With a molecular model of a circadian clock for plants imminent, an integration of the current knowledge of rhythms in cotton with other plant species would be helpful for understanding of the physiology, growth and metabolism in cotton.

Chapter 11

PHYSIOLOGY OF SEED AND FIBER DEVELOPMENT

Reiner H. Kloth[1] and Rickie B. Turley[1]
[1]*Cotton Physiology and Genetics Research Unit, USDA-ARS, P.O. Box 345, Stoneville, MS 38776*

1. INTRODUCTION

The ovules of cotton are composed of the immature seed (embryo and seed coat) and developing fiber. By the coincidence of their proximity, fiber and seed are competing sinks fed through a common funiculus. Partitioning of photosynthate between fiber and seed has been altered by plant breeding. Selection for high yield has increased the thickness of the fiber wall, as indicated by increased micronaire readings (Bridge and Meredith, 1983; Wells and Meredith, 1984). Concomitantly, boll and seed size has decreased (Bridge *et al.*, 1971). These changes are not without an agronomic cost. Small seeds have poorer germination and lower seedling survival. Post-harvest problems are caused by small seed as well: small seed will often pass through the gin with the fiber. The agronomic impact of small seed and the economic impact of fiber make a compelling argument for understanding the physiological relationships between fiber and seed development and the physiology associated with specific events of fiber development.

Fiber and seed share similar, synchronous phases of development. Within the first 25 days post anthesis (DPA) there is rapid cell growth: expansion in the fiber (Berlin, 1986) and division in the embryo (Reeves and Beasley, 1935). Macromolecular constituents are deposited for the next 20 days. During this period, cellulose microfibrils are added to the wall of the fiber (Chapter 32) and oil, protein, and carbohydrate (Chapter 29) are deposited in the embryo. The final phase begins with the disintegration of the funiculus at approximately 45 DPA (Benedict *et al.*, 1976). Without a vascular connection the seed and fiber desiccate.

In this chapter we have attempted to review physiological events of fiber and seed development. One goal of this review was to discuss interactions between these two parts of the ovule. We have tried to limit the review to findings from *Gossypium* species, but we have introduced the physiology of other plants to bridge any gap in the existing information from cotton, or provide an illuminating comparison. We have written about the processes of development as the interaction between anatomy, enzymes, hormones, and metabolites, and thereby hope to avoid a biochemical review of fiber and seed development.

2. FIBER INITIATION

The chronological development of epidermal cells of the ovule has been discussed comprehensively by Berlin (1986). To facilitate the inclusion of new information into the vast collection of morphological data, the authors have taken liberties to report many of the original observations of Berlin (1986) intertwined with the physiological findings of the last decade.

At 23 days before anthesis (DBA), when the flower bud is visible to the naked eye, the ovules are sheathed with approximately 1,000 cuboidal, epidermal cells (8 to 9 μm in diameter). These epidermal cells pass through 5 to 6 mitotic divisions reaching 50,000 to 60,000 cells by anthesis. Preanthesis epidermal cells appear unchanged cytologically until 3 DBA. Major cellular transformations become apparent between 4 to 1 DBA: the nucleus in a few cells enlarges; the nucleolus doubles to 2 μm in diameter; the cytoplasmic vacuoles begin coalescing, becoming larger; and the staining of cells with acid toluidine blue, periodic acid-Schiff, ferric chloride, safranin-fast green, or osmic green indicates the accumulation of phenolic-like compounds in the vacuoles of epidermal cells (Berlin, 1986). Nucleolar enlargement and accumulation of phenolics were early indicators

J.McD. Stewart et al. (eds.), *Physiology of Cotton*,
DOI 10.1007/978-90-481-3195-2_11, © Springer Science+Business Media B.V. 2010

of which epidermal cells eventually differentiated into fiber cells. Fiber initials are epidermal cells that will become fiber cells, but have not yet started elongation. The transformation of epidermal cells into pre-fiber initials appears to be a random process, occurring in nondistinct patterns across the ovule epidermis between 4 and 1 DBA (Stewart, 1975).

Ovule culture can be used to experimentally manipulate the development of fiber initials; ovules harvested from pre- and post-anthesis flowers produce comparable quantities of fiber (Graves and Stewart, 1988a). Ovules taken from flower buds 3, 2, and 1 DBA were placed on culture media containing IAA and GA. Approximately 22% of the ovules from flower buds collected 3 DBA, 78% of the ovules from flower buds collected 2 DBA, and 96% of the ovules from flower buds collected 1 DBA produced fiber (Graves and Stewart, 1988a). The number of ovules producing fiber in the 3 and 2 DBA groups can be increased by withholding IAA and GA from the media for two days. Under this regime, 40% of the ovules from 3 DBA flower buds and 90% of the ovules from 2 DBA flower buds produced fiber (Graves and Stewart, 1988a). Graves and Stewart (1988a) concluded that fiber initials were preprogrammed to remain latent until a stimulus occurred at anthesis. Plant growth regulators inducing previously selected cells to differentiate into trichomes may reflect a theme in trichome development. Plant growth regulators promoted root trichomes (hairs) to initiate during the later stages of epidermal cell differentiation in *Arabidopsis* (Masucci and Schiefelbein, 1996). The cellular trigger which determines the selection of initials remains unknown, and appears to be transitory as well: fiber initials did not remain competent when unfertilized ovules were kept on hormone-free media past the theoretical day of anthesis (Graves and Stewart, 1988a).

At 16 hours before anthesis, Ramsey and Berlin (1976) reported the disappearance of phenolic-like compounds from vacuoles of fiber initials. Loss of the phenolic-like compounds consistently coincided with fiber initiation. The disappearance of these phenol-like compounds has produced much speculation about their roles in fiber initiation. An attractive hypothesis proposed monophenols as activators of IAA oxidase and diphenols as inactivators of IAA oxidase activity (Goldacre *et al.*, 1953). Activated IAA oxidase would theoretically metabolize IAA and prevent cellular elongation and fiber initiation. The reverse would hold true for the diphenolic compounds in fiber initials. Questions should be asked about the presence of phenol-like compounds in fiber and whether they actually modify activity of IAA oxidase. However, a more appropriate question is whether IAA oxidase activity exists. The existence of an IAA oxidase (peroxidase) activity is challenged because products of IAA oxidase do not occur in significant amounts in plant tissue, and transgenic plants with a 10-fold increase in peroxidase activity show no change in IAA levels (Normanly *et al.*, 1995).

Much of the work investigating the role of phenolic compounds during fiber initiation has concentrated on mea-suring the enzyme polyphenol oxidase (PPO; Naithani *et al.*, 1981; Naithani, 1987), which is localized in the thylakoids of plastids (Mueller and Beckman, 1978; Vaughn *et al.*, 1988). Naithani (1987) reported that PPO activity, measured with substrates catechol and DOPA, increased after 0 DPA in ovules. This increase does not match the disappearance of phenol-like compounds from vacuoles at 1 DBA. It does, however, coincide with fiber elongation. Though the number of plastids in fiber cells is not recorded, chloroplasts in elongating leaf cells increase in number (Boffey *et al.*, 1979; Scott and Possingham, 1980). Lanker *et al.* (1987) studied PPO protein during leaf development. They found the greatest amount of immunoprecipitable PPO (from *in vitro* translation of mRNA) to be present during the mid stage of leaf development when plastids are actively developing and dividing. Naithani *et al.* (1981, 1987) fail to account for the changes in PPO activity caused by plastid development and replication. Any study of PPO is complicated by, and must account for the latency of the enzyme: amount of activity may not reflect the total amount of PPO protein (see Vaughn *et al.*, 1988).

At 8 h before anthesis, fiber initials appeared "darker" in electron micrographs than undifferentiated epidermal cells. Ramsey and Berlin (1976) attributed the "darker" appearance to an increase in cytoplasmic ribosomes needed for escalated protein synthesis during fiber initiation and development. Recently, cDNA encoding five different cytoplasmic ribosomal proteins have been characterized from cotton, including: S4e (Turley *et al.*, 1995); S16 (Turley *et al.*, 1994a); L37a (Kloth and Turley, 1997); L41 (Turley *et al.*, 1994b); and L44 (Hood *et al.*, 1996). With these clones, *in situ* hybridization may allow the selection of fiber initials to be visualized days before anthesis.

2.1 POSTANTHESIS DEVELOPMENT OF COTTON SEED EPIDERMAL CELLS

Fiber initiation and pollination occur independently from each other. Fiber initiation is not dependent upon pollination nor is pollination dependent upon initiation. Elimination of either event results in the loss of fiber or boll. An interaction may occur 1 DPA between fiber initiation, elongation, and pollination. By 1 DPA the majority of fiber initials protrude from the ovule epidermis, but a few initials will begin elongating at any time during the next 2 days. At 1 DPA the flower has entered into a period of senescence, easily noticed by the reddening and collapse of the petals. With senescence the flower produces ethylene. Lipe and Morgan (1973) reported that at 1 DPA, greater than 50% of the ethylene produced by the flower was in the style, stigma, and stamen. The ovule at 1 DPA also produced low quantities of ethylene, however, this may have resulted from mechanical wounding during their removal from the boll. Ethylene production, therefore, occurs in close proximity to ovules. In ovule culture, ethylene was reported to have negative effects on fiber initiation and elongation (Davidonis, 1993; Hsu and Stewart, 1976). These groups used ethephon (2-chloroethylphosphonic acid) which degrades in the cytoplasm of cells to ethylene (Warner and Leopold, 1969). Hsu and Stewart (1976) found that ethephon concentrations of 10^{-5} M or above resulted in inhibition of fiber elongation and development, and induced the production of callus at the micropylar end of ovules. Lower quantities of ethylene (10^{-5} M or lower) had no effect on fiber production (Davidonis, 1993a). It is unknown whether ethylene produced by senescent flowers affects fiber initiation and elongation *in vivo*. In contrast to data from cotton ovule culture experiments, ethylene enhances the formation of root hairs (trichomes) in pea, faba bean, lupine (Abeles *et al.*, 1992). Abeles *et al.* (1992) also reported that tulip roots, which usually lack root hairs, can be induced to form hairs by ethylene treatments. The mechanisms which induce ovular and root trichomes clearly differ in their response to ethylene.

Numerous similarities, however, do occur between *Arabidopsis* and ovular trichomes of cotton (Marks and Esch, 1992). *Arabidopsis* trichomes are single cells which pass through three growth phases – initiation, elongation, and secondary wall thickening. Trichome development occurs in a basipetal fashion in *Arabidopsis* leaves (Marks and Esch, 1992) maturing first at the tips of the leaves, then progressing to the base of the leaves. Stewart (1975) reported that trichomes initiated at the chalazal end of the ovule and moved progressively toward the micropyle. Also, *Arabidopsis* root hair initials can be identified by their vacuolation and cytoplasmic density, similar to cotton fiber initials (Masucci and Schiefelbein, 1996).

In *Arabidopsis* four loci are known which influence the production of trichomes on the plant: *GLABRA2 (GL2;* Masucci *et al.*, 1996); *TRANSPARENT TESTA GLABRA (TTG; Galway et al., 1994); AUXIN RESISTENT2 (AXR2;* Wilson *et al.*, 1990); and *CONSTITUTIVE TRIPLE RESPONSE (CTR;* Dolan *et al.*, 1993). Mutations in either *GL2* or *TTG* essentially eliminate all trichomes from *Arabidopsis* shoots (Koomeef *et al.*, 1982). *GL2* appears to specifically affect trichome development, whereas *TTG* affects trichome development and has a role in controlling pigmentation and formation of mucilage in the seed coat (Marks and Esch, 1992). All the effects of a mutation to *TTG* can be reversed by the maize regulatory factor R (Lloyd *et al.*, 1994). Therefore, the maize R gene is believed to be a homologue of *TTG* (Larkins *et al.*, 1996). The possible correlation between specific phenotypic variations in cotton are easily seen: *TTG* controls pigmentation of seed coat, whereas, cotton fiber initials selectively lose compounds involved in development of pigmentation of the seed (Ramsey and Berlin, 1976).

Indole acetic acid and GA are essential plant growth regulators for production of fiber in ovule culture (Beasley

and Ting, 1973, 1974). In fertilized ovules, GA induced a marked stimulation, whereas, IAA provided only a positive trend of fiber production (Beasley and Ting, 1973). In unfertilized ovules, GA and IAA were required for ovule growth and fiber development (Beasley and Ting, 1974). Both kinetin and ABA had inhibitory effects on fiber development. Gibberillic acid is an essential plant growth regulator for the stimulation of trichome production. Gibberillic acid is required in ovule culture to produce a maximum quantity of fiber (Beasley and Ting, 1974), and induces trichome development in *Arabidopsis* leaves (Chein and Sussex, 1996).

Indole acetic acid and GA induce enzymes in the developing fiber of cultured ovules. The activities of three enzymes considered important in producing the fiber osmoticum malate (counterion for K^+) during later stages of fiber elongation (Basra and Malik, 1984), phosphoenolpyruvate carboxylase (PEPC), malate dehydrogenase (MDH), and glutamic oxaloacetic transaminase (GOT), increased after addition of IAA and GA to the culture media of unfertilized (0 DPA) ovules (Dhindsa, 1978a). Activities of enzymes not directly associated with fiber elongation, such as glucose-6-phosphate dehydrogenase, isocitrate dehydrogenase, and catalase, were only slightly stimulated (Dhindsa, 1978a). Gibberillic acid is the most critical hormone for the induction GOT, MDH, and PEPC. Gibberillic acid also induces an *Arabidopsis* tonoplast intrinsic protein (γ-TIP; Phillips and Huttley, 1994), which have been correlated with cell elongation. Ferguson *et al.* (1997) have identified and characterized a δ-TIP in cotton, but no relationship between TIP and fiber elongation has been established in cotton fiber. In another study, Thaker *et al.* (1986) reported on peroxidase and esterase activity; these activities were measurable during the first three days of fiber initiation but became prominent during later stages.

Cytochemical and immunocytochemical staining has been used to determine expression of proteins in specific cell types. These methods are most useful when the age of the fiber makes it impossible to separate it from the other cells of the ovule. Nolte *et al.* (1995) used immunolocalization in their study of sucrose synthase (SuSy) in developing ovules. Sucrose synthase has recently been postulated to provide precursors for cell wall formation (Amor *et al.*, 1995). No sucrose synthase was detected in fiber initials at 1 DBA. Joshi *et al.* (1985, 1988) compared ovules from a linted and a lintless line during fiber initiation. Cytochemical staining of β-glycerophosphatase and ATPase activity was not detected in the lintless line at any age (1 DBA to 3 DPA). Staining was easily identified in the fiber-producing line postanthesis, increasing substantially with fiber development. These studies demonstrated that the SuSy, β-glycerophosphatase, and ATPase in fiber had increased activities after anthesis, but were not the critical switch that induced epidermal cells to become fiber initials (Joshi *et al.*, 1985, 1988).

Two-dimensional polyacrylamide gel electrophoresis (2-D PAGE) offers numerous benefits to the analysis of mutants and study of plant developmental biology that are distinct from molecular techniques (Santoni *et al.,* 1994); 2-D PAGE can serve to identify and quantify proteins, and to make apparent the modification and degradation of proteins. Databases which compile information on micro sequencing of proteins from electorphoretically separated proteins are now being produced (Appel *et al.,* 1993). In yeast, major proteins have been identified that were previously overlooked with molecular techniques (Garrels *et al.,* 1994).

Turley and Ferguson (1996) used 2-D PAGE to study fiber development in a fiber-producing line (DPL 5690) and a fiberless line (SL 1-7-1) from 3 DBA to 4 DPA. Of the 37 proteins developmentally expressed, five were expressed solely in the fiber producing line after anthesis. Sixteen proteins showed developmental differences. The data on the five proteins (designated D7, D10, D12, D13, and D14) indicate that they are involved in early events of fiber development. Preliminary evidence indicates that D14 (shown in Fig. 11-1) is correlated with elongation of fiber. It is first seen at or soon after anthesis, is present at 14 DP A, but declines by 21 DPA when fiber elongation ceases (Ferguson *et al.,* 1996). Protein D14 is not present at any age in the 2-D PAGE protein profiles of the fiberless line. Graves and Stewart (1988b) also used 2-D PAGE to identify protein changes between 6 DBA to 20 DPA in developing ovules.

The study of fiber initiation at the protein and molecular level is complicated by the numerous changes occurring in ovules. As mentioned earlier, the presence of ethylene near ovules could be regulating specific ovular genes. Fertilization occurs by 2 DPA. Therefore, changes in the structure of the megasporocyte (egg to zygote) should result in changes to both in mRNA and protein composition of the ovule. Visualization of these changes in 2-D PAGE would likely be difficult due to the small size of the zygote

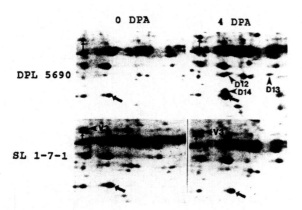

Figure 11-1. Separation of low to medium molecular mass proteins (<39 kD) of ovules at 0 and 4 days post anthesis (DPA) from a fiber-producing line (DPL 5690) and a fiberless line (SL 1-7-1). Large and small arrows (upper left corner in each panel) makr proteins that are constant between lines. Protein V3 (29.8 kD, 5.6 pI) is consistent between SL 1-7-1 panels. Proteins D12 (25.9 kD, pI 5.8) and D13 (25.7 kD, 5.98 pI) are involved in early events of fiber development. Protein D14 (24 kD, 5.79 pI) is correlated with elongation.

and egg sac in comparison to the ovule. The zygote remains relatively small for the first week of development. The zygote does not divide for 2 to 3 days after fertilization (3 to 5 DPA; Jensen, 1968), remaining relatively small during the peak period of fiber elongation (1 to 10 DPA), after which elongation slows and embryo expansion occurs at a maximum rate between 14 and 26 DPA (Mauney, 1961).

3. PHYSIOLOGY OF FIBER DEVELOPMENT

3.1 Comparison of Results

In the discussions that follow, there is some attempt to compare results of enzyme assays or metabolite concentrations between investigators. However, there is great difficulty in doing so; some results seem to be in direct conflict – compare Wäfler and Meier (1994) and Basra and Malik (1983) – without apparent explanation. The generic candidates for discrepancies are cultivar differences and environment. Variation in fiber biochemistry between species of *Gossypium,* or cultivars within a species is poorly studied, but there are real effects such as dry matter accumulation over the course of fiber development between G. *hirsutum* L. and *herbaceum* L. (Naithani *et al.,* 1982), or different malate dehydrogenase (MDH) activities between upland cotton cultivars (Kloth, 1992). Environment is always a good choice. It has a significant effect on the amount of low molecular weight sugars in the capsule (Conner *et al.,* 1972), or MDH activity extracted from developing fiber (Kloth, 1992). The range of environments is essentially global. Enzymes and metabolites are extracted from *Gossypium* plants grown in the greenhouses and growth chambers of Switzerland [see Buchala (1987) or Wäfler and Meier (1994)], or the fields of the India's Punjab region [see Basra and Malik (1983)].

The method of protein extraction complicates the comparison of assay results. More soluble protein is extracted from cotton tissues with buffers that contain reagents which protect proteins from chemical modification by secondary compounds (Schmidt and Wells, 1986). MDH extracted from fiber between 7 and 14 DPA can have 20% less activity if proteins are not protected from modification, though after 14 DPA protectants are not as critical (Kloth, unpublished). No two groups of investigators use the same reagent cocktail—even when the same enzyme is sought. Compare MDH extractions by Basra and Malik (1983), Kloth (1992), and Wäfler and Meier (1994). Indeed, Wäfler and Meier (1994) choose not to chemically protect their proteins, but rely on glandless cotton, which has approximately 50-fold less gossypol than glanded types (Schmidt and Wells, 1990). However, gossypol is apparently not the sole compound capable of modifying protein in young fiber (Kloth, unpublished). Therefore, some caution interpreting the range of enzyme activity is warranted; values for enzyme activity that occur past 14 DPA may be inflated relative to early stages of fiber development.

Standardization between experiments is possible. Charts of fiber dry matter accumulation or fiber length measured over several days postanthesis present physiological data that can be used to standardize results from plants grown in California, India, and Switzerland. The comparisons are crude. For example, no mention of enzyme activity can be made at the start of secondary cell wall deposition; secondary cell wall deposition is inferred from the gain in fiber weight, not from the determination of cellulose.

3.2   Flow and Utilization of Photosynthate

Developing fruits are supplied the bulk of their photosynthate through the subtending leaf (Ashley, 1972). Fruits developing without a subtending leaf produce fiber with reduced micronaire, an indirect measure of fiber thickness (Pettigrew, 1995). Carpel walls are also important in meeting the nutritional needs of the embryo and fiber (Bondada et al., 1994; Brown, 1968), though in comparison to leaves and bracts they are photosynthetically the least efficient (Bondada et al., 1994). Sucrose produced through photosynthesis can be transported directly to the ovule through a complex vasculature of pholem that permeates the epidermis of the fruit wall and connects to the funiculus (van Iersel et al., 1995).

The transfer of nutrients from the ovule to the fiber would appear to be symplastic. Fiber cells have an enlarged base riddled with plasmodesmata located in pits along the periclinal wall, but only sparsely pitted along the anticlinal wall (Ryser, 1992). Ryser (1992) calculated that the flux (1.33 pg fiber^{-1} sec^{-1}) through the symplast would be adequate to support the carbohydrate needs of developing fiber. How much carbon assimilate is moving through the symplast, and what proportion of the carbohydrate is diverted from the embryo to the developing fiber is unclear. Potentially, the fiber could receive little sucrose from symplastic transport after 20 DPA: analysis of sugars from immature ovules separated into embryonic and non-embryonic ("seed coat") tissue from which the fiber was removed and discarded reveals that the embryo has three times more sucrose than the "seed coat" tissue (Hendrix, 1990). However, the seed coat tissue has much more hexose sugars than the embryo (Hendrix, 1990), and this may be supplied to the fiber via the symplast. Therefore the importance of nutrient transport from the carpel wall via the apoplast can not be discounted – particularly for sucrose. Radiolabelled monosaccharides are routinely detected within fibers when these compounds are incubated with intact seed clusters or cultured ovules. Asymmetrically labeled sucrose or ^{14}C-glucosyl-1'-flurosucrose, a compound difficult for invertase to cleave, can enter the fiber intact (Buchala, 1987), indicating a disaccharide transport mechanism in the wall.

Sucrose, once inside the fiber cell, is degraded by either sucrose synthase (EC 2.4.1.13) or invertase (EC 3.2.1.26). Invertase activity is easily detected in a developing fiber (Basra et al., 1990; Buchala, 1987; Wäfler and Meier, 1994), and at least three have been found. They are divided into two groups by the pH optimum for the catalyzed reaction (Basra et al., 1990). Those forms with reaction optima at acidic pH are further partitioned by their sub-cellular location (Basra et al., 1990; Buchala, 1987). Acidic invertases can be found in the cytoplasm or associated with the cell wall (Basra et al., 1990; Buchala, 1987), but this latter group – also known as insoluble invertases – is most likely an invertase that is secreted into the apoplast (Buchala, 1987). The activity of all invertase forms from G. hirsutum decline precipitously, and with a startling synchrony (Basra et al., 1990; Wäfler and Meier, 1994), as the fiber elongation concludes and secondary cell wall synthesis accelerates. Gossypium arboreum L. does not have as precipitous a decline in activity (Buchala, 1987).

Sucrose synthase (SuSy) activity can be detected at the earliest stages of fiber growth (Nolte et al., 1995) and it increases far into the period of secondary cell wall development (Basra et al., 1990; Amor et al., 1995), but its deployment within the cell is of most interest. Sucrose synthase immunolabel is seen in the cytoplasm and in association with the plasma membrane at 1 DPA when the trichomes have already elongated 3 to 5 fold (Nolte et al., 1995). Older fibers allow a more precise quantification of membrane-bound and cytoplasmic SuSy: the proportion of SuSy protein associated with the membranes increased from 9 to 20 DPA, and represents approximately 75% of the total sucrose synthase at 20 DPA (Amor et al., 1995). Sucrose synthase and invertase are jointly and carefully regulated by plant cells (Koch et al., 1996). Developmental and environmental stresses cause changes in the distribution and the isoform of SuSy and invertase (Koch et al., 1996). Indeed, too much or too little invertase has morphological (Dickinson et al., 1991; Klann et al., 1996) and even dire developmental consequences (Miller and Chourey, 1992). The alterations in SuSy location and the decline in invertase activity need to be further discussed in the context of fiber development, and particularly the need for and use of sucrose. Two growth processes, occurring simultaneously, are in need of the metabolites that can be wrought from sucrose: fiber elongation, which is nearing its completion, and secondary cell wall synthesis, which is accelerating. The changes in SuSy and invertase activities are made to meet the joint demands of elongation and secondary cell wall synthesis and reflect changes in both the concentration and use of sucrose.

Sucrose accumulates in tissues when invertase activity is reduced (Klann et al., 1996). This has also been observed in G. hirsutum; as fiber invertase declines, sucrose increases (Jaquet et al., 1982). When the decline of invertase is delayed and not as great (relative to G. hirsutum), as found with G. arboreum fibers (Buchala, 1987), the rise in sucrose is delayed and not as large (Jaquet *et al.,* 1982). The loss of the apoplastic invertase would increase the amount of sucrose that enters from the apoplast, and, most likely, create a dramatic increase in the concentration of sucrose at the plasma membrane.

The increased amount of membrane-bound SuSy would utilize the additional sucrose to catalyze the production of UDP-glucose in the vicinity of the glucan synthases. UDP-glucose produced from SuSy seems to be preferred during secondary cell wall synthesis. Incorporation of labeled UDP-glucose into glucans rises only 1.6- fold at this time, whereas incorporation of labeled sucrose into glucans jumps 138-fold (Pillonel *et al.,* 1980). Detached, permeabilized fibers synthesizing secondary cell walls produce more β-l,4 glucan than β-1,3 glucan when fed sucrose (Amor *et al.,* 1995). However, directly feeding UDP-glucose to fibers favors β-1,3 glucan synthesis over β-1,4 glucan (Amor *et al.,* 1995). The UDP-glucose produced by SuSy would be preferable on the basis of energy conservation (Amor *et al.,* 1995). UDP-glucose made by UDPG-pyrophosphorylase requires two ATP, whereas production of UDP-glucose from sucrose by SuSy would help conserve the energy in the sucrose linkage.

Plant metabolism also uses the enzyme pyrophosphate:fructose 6-P l-phosphotransferase (PFP, EC 2.7.1.90) to deal with developmental and environmental stress (Black *et al.,* 1987). Pyrophosphate-dependent phosphotransferase activity is present in the developing fiber at a nearly steady state, until it drops 40%, when secondary cell wall synthesis reaches mid-point (Wäfler and Meier, 1994). This enzyme catalyzes the reversible phosphorylation of fructose-6phosphate (F-6-P) to fructose -1,6-bisphosphate ($F-l,6-P_2$) in the presence of pyrophosphate and UTP. Reversible phosphorylation of F-6-P is a unique feature of PFP – other enzymes capable of phosphorylating F-6-P or dephosporylating $F-l,6-P_2$ catalyze irreversible reactions – so PFP was thought to control the flow of carbohydrate between glycolysis and gluconogeneis (Black *et al.,* 1987); however PFP also functions in pyrophosphate regulation during photosynthetic sucrose synthesis (Neuhaus *et al.,* 1990) and phosphate starvation (Theodorou *et al.,* 1992). Because no single physiological role can be assigned to PFP, the enzyme's function in developing fiber is unclear.

The difficulty in discerning the function of PFP stems from the flexibility of the enzyme; nutritional status of the cell governs the function of PFP. Phosphate-starved cells of *Brassica nigra* have a 1:1 ratio of regulatory and catalytic subunits, whereas PFP isolated from well-fed cells is composed solely of catalytic subunits (Theodorou and Plaxton, 1996). The enzyme is activated by fructose-2,6-bisphosphate ($F-2,6-P_2$), but the affinity for $F-2,6-P_2$ is dependent on subunit composition, the presence of F-6-P and F-l,6-P_2, and the ratio of pyrophosphate to inorganic phosphate (Theodorou and Plaxton, 1996). Phosphate starved cells can catalyze the reaction in either direction, but the net flux will most likely be glycolytic because inorganic phosphate can increase the affinity of PFP for $F-2,6-P_2$ (Theodorou

and Plaxton, 1996). Activation by $F-2,6-P_2$ is most likely inconsequential in cells that are not undergoing a stress (Theodorou and Plaxton, 1996). In such cells, the PFP may catalyze the reaction toward gluconeogensis, since the V_{max} for this direction is an order of magnitude greater than the V_{max} for glycolysis (Theodorou *et al.,* 1992).

The tempting hypothesis for PFP in the developing fiber is that this enzyme redirects F-1,6-P 2 from glycolysis to UDP-glucose synthesis, even producing UTP for the UDPG pyrophosphorylase reaction. The time course for PFP is suitable for this role; the PFP activity declines (Wäfler and Meier, 1994) when UDP-glucose for glucan synthesis is not as desirable a substrate (Pillonel *et al.,* 1980). But elements necessary to make a compelling argument are missing. Evidence for an isomerase, such as glucose phosphate isomerase (EC 5.3.1.9), does not exist. Buchala (1987) fed seed cluster with ([^{14}C]fructosyl) sucrose at a time when the primary wall would be made. The radiolabeled fructose was not found in the glucans, and the results interpreted to mean that isomerases were absent (Buchala, 1987).

3.3   Fiber Extension

Taiz (1984, 1994) describes cell expansion as a cycle. Hydrostatic pressure increases until the wall is distended to its elastic limit. The cell releases mostly unidentified "wall loosening factors" which change the mechanical properties of the wall to allow the reversible change produced by the hydrostatic pressure to become a permanent (irreversible) extension. Hydrostatic pressure increases because turgor pressure, the difference between osmotic potential and the water potential of the cell, has increased. As would be predicted by this model of cell expansion, turgor pressure declined to a minimal value when fiber cells slowed and completed their expansion (Dhindsa *et al.,* 1975). In cotton fibers, the solutes largely responsible for producing turgor pressure, K^+ and malate, were found to increase and decrease in step with the growth rate of the fiber (Dhindsa *et al.,* 1975; Basra and Malik, 1983).

Cotton is not unique in using malate to develop turgor pressure. This organic acid has long been considered responsible for developing turgor pressure in plant cells. Malate is found in a wide range of concentrations within plant cells. Specialized cells such as *Vida faba* L. guard cells can accumulate malate to staggering concentrations: 100 to 200 mM (Raschke, 1979). Leaves have a broad range from 74 mM in *Spinacia oleracea* (Speer and Kaiser, 1991) to 3 to 4 mM in *Pisum sativum* L. (Speer and Kaiser, 1991). The roots of *Zea mays* L. have malate concentrations similar to pea leaves (Chang and Roberts, 1989). The fiber of Upland cotton *(Gossypium hirsutum)* is in the high end of the range with 60 mM malate (Dhindsa *et al.,* 1975). The synthesis of such quantities of malate depends on the production of oxaloacetate, its metabolic precursor. Oxaloacetate is made by either the carboxylation of phosphoenolpyruvate (PEP), or by the deamination of glutamate.

Carboxylation of PEP to oxalacetate is catalyzed by *posphoenolpyruvate* carboxylase (PEPC, EC 4.1.1.31). The evidence of PEPC activity in developing cotton fiber was first indirect (Dhindsa *et al.*, 1975). Cotton ovules cultured *in vitro*, and in the absence light, produced less fiber when deprived of atmospheric CO_2 When the ovules were fed $H^{14}CO_3$-, the preponderance of the radiolabeled carbon was fixed into malate. Wullshleger and Oosterhuis (1990c) found seed and fiber highly efficient at CO_2 fixation in the absence of light. Dhindsa (1978a) detected PEPC activity in developing fiber, and Basra and Malik (1983) studied the development of PEPC activity. They found PEPC activity greatest prior to and during the linear phase of fiber elongation, followed by a slow decline to half of its initial value once elongation slowed and ended (Basra and Malik, 1983). Corcoran and Zeiher (1995) have re-examined the developmental regulation of PEPC and found that the enzyme increased 4-fold from 5 DPA, reaching its maximum value when elongation ceased. Details of Corcoran and Zeiher (1995) are not available, so any reconciliation of results is not yet possible. Corcoran and coworkers (1993, 1995) have established that the PEPC activity of fiber is the result of fiber-specific isozymes. Auxin and gibberellic acid supplements to *in vitro* cultured ovules increase PEPC activity (Dhindsa, 1978a) by inducing two isoforms (Ramsey *et al.*, 1996).

The second pathway for producing oxaloacetate is glutamate-oxaloacetate transaminase (GOT EC 2.6.1.1), and this activity has been found in developing cotton fibers (Basra and Malik, 1983; Wäfler and Meier, 1994). But there is a discrepancy between the results. Wäfler and Meier (1994) show the GOT activity essentially stable, whereas Basra and Malik (1983) found GOT increasing and decreasing with the malate concentration in the fiber – including a 60% decline as elongation ceased. The contribution of GOT to fiber elongation is difficult to discern; GOT may seize the oxaloacetate and transaminate it into aspartate. Experimental results indicate that GOT activity is a secondary source of oxaloacetate; when Dhindsa *et al.* (1975) reduced the availability of atmospheric CO_2, GOT activity could not compensate and revive fiber elongation. However, GOT activity increased greatly when unfertilized ovules were treated with auxin and gibberellic acid, while enzyme activities not associated with elongation were not induced (Dhindsa, 1978a). Thus it would seem unreasonable to dismiss GOT from any role in fiber elongation.

The final step, reduction of oxaloacetate to malate by malate dehydrogenase (MDH, EC 1.1.1.37), is the most studied step in the process of making this solute. Basra and Malik (1983) found total MDH activity rising and falling coincident with the concentration of malate in the fiber, and snychronous with elongation. Wäfler and Meier (1994) found total MDH activity increasing from 8 to 35 DPA, but did not measure malate concentration. They account for the rise in activity as an increase in cytoplasmic MDH. A wall-bound MDH activity was measured and found to reach a peak at 15 DPA, before declining (Wäfler and Meier, 1994). Additionally, they argue that mitochondrial MDH remains unchanged because isocitrate dehydrogenase (EC 1.1.1.42) activity, a marker for mitochondria, does not change during fiber development (Wäfler and Meier, 1994). There is other data supportive of Wäfler and Meier's (1994) hypothesis. A polypeptide from developing fiber, which increased from 14 to 21 DPA, was tentatively identified as a cytoplasmic MDH by partial amino acid sequencing (Ferguson *et al.*, 1996). As in the case of SuSy and invertase, careful analysis ofthe subcellular location ofMDH during fiber development would help clarify the contribution of each MDH enzyme to fiber development.

For malate to accumulate in large amounts it must be sequestered, and the site used is the vacuole (Buser and Matile, 1977). The process of malate deposition works against a concentration gradient, and is thereby ATP-dependent (Martinoia *et al.*, 1985). Two components are known: a poorly understood permease (Martinoia and Rentsch, 1994), and vacuolar proton pumps (Sze, 1985). The later component of the system is easily demonstrated; when ATPases are inhibited, malate uptake by vacuoles is inhibited (Martinoia *et al.*, 1985; White and Smith, 1989). However, direct evidence for the involvement of vacuolar ATPases in cell expansion comes form Gogarten *et al.* (1992). They transformed carrot root cells with antisense constructs of small portions of the 5' non-coding and coding region of vacuolar ATPase A subunit (Gogarten *et al.*, 1992). Tonoplast fractions had reduced expression of the A subunit (but not the Golgi fraction), and reduced ATPase activity. Regenerated plants showed reduced cell expansion. In Upland cotton, a K^+ stimulated ATPase has been convincingly localized on the tonoplast of developing fibers (Joshi *et al.*, 1988), and Wilkins *et al.* (1994) have cloned members from two different families of the vacuolar ATPase subunit. The results of Gogarten *et al.* (1992) indicate that alterations in the ratio of isoforms of ATPase can have morphological effects, an observation that has immediate implications for the control of fiber traits. Potassium is also accumulated, but no direct experimental evidence from cotton is available.

For cell expansion to occur, water must also enter and exit in a regulated manner. In animal cells the flow of water is controlled by a group of membrane-spanning proteins known as aquaporins (Chrispeels and Maurel, 1994). Plants have proteins with high identities to aquaporins, but their role as mediators of water flow is still controversial. This controversy flows from the nature of the functional test applied to these proteins. Plant mRNA is injected into *Xenopus* oocytes and the flow of water across the membrane is measured (Daniels *et al.*, 1996). The origin of the first aquaporin-like proteins of plants was the tonoplast membrane, and this has sealed their name as tonoplast intrinsic proteins (TIP; Chrispeels and Maurel, 1994). A protein similar to the γ form of TIP was found in greater quantity in elongating and mature regions of hypocotyls than the regions composed of dividing cells (Maeshima, 1990; 1992). When cell elongation is stimulated in dwarf *Arabidopsis* with GA₃, transcripts of γ-TIP increase (Phillips and Huttly, 1994). In cotton, members of the δ-TIP family are expressed in the developing fiber (Ferguson *et al.*, 1997).

Accumulated malate is recycled when elongation has ended (Basra and Malik, 1983). One possibility is to run the MDH reaction backwards to produce oxaloacetate; MDH activity is present in the fiber as malate declines (Basra and Malik, 1983; Wäfler and Meier, 1994). Indeed, this may be the function of the cytoplasmic MDH observed by Wäfler and Meier (1994). Additionally, malic enzyme, which catalyzed the breakdown of malate to CO_2 and pyruvate with the cofactor NADP, increased dramatically after malate levels reach their peak and elongation ceases (Basra and Malik, 1983). There are two types of malic enzyme: one enyme has a preference for NADP and the capability of decarboxylating oxaloacetate (EC 1.1.1.40), and the other has a preference for NAD and an inability to decarboxylate oxaloacetate (EC 1.1.1.39). The latter is unique to plants, the former is ubiquitous among plants and animals (Wedding, 1989). The cofactor preferences are not absolute, and in crude extracts assays for the NADP preferring enzyme will be inflated by activity from the NAD preferring enzyme (Wedding, 1989). So most likely both forms of malic enzyme are present in the fiber, but their relative amounts can not be deduced from Basra and Malik's (1983) data. Cellular location of malic enzyme is also important. The NADP preferring enzyme is primarily present in cytoplasm, and the other is a mitochondrial enzyme (Wedding, 1989). Because neither malic enzyme nor MDH are associated with a vacuole, a means for active or passive transfer of malate out of the vacuole must be devised.

Turgor pressure is necessary for expansion, but recent experiments show that it is not the only requirement. Tobacco cells adapted to osmotic stress (428 mM NaCI) have a turgor pressure several-fold higher than unstressed cells, but expand to only one-eighth the size of the unstressed controls (Iraki *et al.*, 1989). Zhu and Boyer (1992) developed a "turgor clamp" which allowed them to change the turgor pressure of a cell without altering the external conditions. A threshold for turgor pressure was found, below which growth did not occur; however, growth rate was not dependent on turgor pressure. Growth was quickly halted – without changing the turgor pressure – by poisoning with inhibitors of energy metabolism (Zhu and Boyer, 1992). Inhibitors of polysaccharide synthesis reduced the growth rate of rice coleoptiles without significantly affecting the mechanical properties of the wall (Hoson and Masuda, 1992).

An actively expanding or extending cell wall would logically require an equally active metabolism. An extensible wall requires biochemical action to loosen the matrix and separate the microfibrils, then synthesize and insert polymers into the wall. Reduced extensibility would lead to an excess of wall material with respect to the size of the cell; whereas extension without coordinated synthesis and inser-

tion would lead to the wall becoming too thin and rupturing (McCann and Roberts, 1994). In what is assumed to be the loosening of the wall matrix, pea stems release soluble pectins and xyloglucans during auxin induced elongation (Terry *et al.*, 1981). Messenger RNA for an endoxyloglucan transferase–an enzyme capable of breaking xyloglucan chains and reforming the glycosidic linkage–was found in developing cotton fiber (Shimizu *et al.*, 1997). When elongation ceased, the amount of endoxyloglucan transferase message plummeted (Shimizu *et al.*, 1997).

Glycosidases with the potential to participate in wall loosening have also been studied in developing fiber (Thaker *et al.*, 1987). Thaker *et al.* (1987) separated fiber proteins into three fractions (cytoplasmic, ionically wall-bound, and covalently wall-bound) by washing and centrifuging fibers with solutions of different ionic strength. By this method, a β-galactosidase that remained bound to the wall after extensive washing with high salt solutions was correlated with fiber extension growth (Thaker *et al.*, 1987). However, the rise and fall of cytoplasmic and ionically wall bound forms during secondary cell wall synthesis or maturation make it difficult to assign β-galactosidase a particular role with any conviction. All forms of β-glucosidase activity appear to increase and decrease in concert with the elongation rate (Thaker *et al.*, 1987), but these activities also increase as secondary cell wall is laid down. The function of β-galactosidase and β-glucosidase in elongation will remain unclear until isoforms are characterized and the regulation of these isozymes confirmed. β-1,3-glucanase activity is also found in the fiber (Bucheli *et al.*, 1985; Thaker *et al.*, 1987). Thaker *et al.* (1987) do not specify whether an endo- or exo-β-1,3-glucanase activity was measured; however, by the nature of their substrate (laminaran) and assay, an exoglucanase activity was probably assayed. Two wall bound – an ionically and a covalently bound type – β-1,3-glucanase activities were identified. These enzymes had similar developmental patterns: a small, but significant, peak in activity coinciding with elongation and a second peak during secondary cell wall synthesis with approximately twice the activity as the first peak (Thaker *et al.*, 1987). Bucheli *et al.* (1985) also found exo-β-1,3-glucanase activity in the wall fractions of developing fiber, but the activity rose slowly and reached a peak at 40 DPA, when callose dropped to its lowest level. Bucheli *et al.* (1985) may have missed the early peak because the interval between samples was too widely spaced: Thaker *et al.* (1987) found that the two wallbound β-1,3-glucanases rapidly reached their peak values after 15 DPA, but fell to the 15 DPA level by 20 DPA. Bucheli *et al.* (1985) also measured endo-β-1,3-glucanase activity during the development of the fiber, but found no change in activity over time. This is in contrast to an increase in messenger RNA for endo-β-1,3-glucanase at the start of secondary cell wall deposition (Shimizu *et al.*, 1997).

Though the glycosidases, gluconases, and endoxyloglucan transferases are presumed necessary for elongation by loosening the matrix of xyloglucan chains, cellulose microfibrils, and pectic polysaccharides (albeit that their mechanisms and exact functions are not understood), these enzymes do not explain acid induced growth, nor why the walls of plants are not progressively weakened during wall loosening (see Taiz, 1994). Expansin is the wall protein thought to break and reform hydrogen bonds during extension (McQueen-Mason and Cosgrove, 1994) and forms the final component in the enzymology of elongation. In the developing cotton fiber expansin is only known by its messenger RNA, and the decline of message as fiber's growth slows is indicative of an enzyme important in cell elongation (Shimizu *et al.*, 1997).

3.4 Regulation of Cellulose Synthesis

Cellulose can be synthesized *in vitro* with a preparation made by treating a plasma membrane-enriched fraction with digitonin (Kudlicka *et al.*, 1995; Li and Brown, 1993; Okuda *et al.*,1993). The original methods (Li and Brown, 1993; Okuda *et al.*, 1993) produced a small quantity of cellulose (4% of total glucan) relative to callose, which led Delmer and Amor (1995) to question the validity of the claim. However, improvements in the preparation of the plasma membrane fraction and digitonin extraction of proteins increased the cellulose, which proved to be the "native" or fibrillar cellulose type, to 32.1 % (Kudilicka, *et al.*, 1995).

The activators for the production of fibrillar cellulose *in vitro* are not agreed upon. Amor *et al.* (1995) synthesized a mixture of cellulose and callose with fibers that were detached from the ovule and treated with digitonin. Fibers incubated with sucrose and 25 mM EGTA produced a glucan fraction which was 55 to 65 % cellulose – the most cellulose produced by any treatment (Amor *et al.*, 1995). If the EGTA is replaced with 1 mM $CaCl_2$ and 10 mM cellobiose, cellulose is no longer the major glucan produced; cellulose falls to 30 to 40 % of the glucans. Altering the source of carbon to UDP-glucose, but retaining the $CaCl_2$ and cellobiose, decreases the cellulose content to 10 to 15%. These results (Amor *et al.*, 1995) agree with many observations and hypotheses concerning cellulose synthesis: Ca^{2+} and β-glucosides favor callose synthesis (Delmer and Amor, 1995) and cellulose synthesis prefers UDP-glucose transferred directly from SuSy (Amor *et al.*, 1995). However, Amor *et al.* (1995) did not completely test the reaction conditions optimized by Li and Brown (1993); di-cyclic GMP and Mg^{2+} were not used in conjunction with Ca^{2+} and cellobiose. This is unfortunate, because Li and Brown (1993) did not choose these compounds on an arbitrary basis. Each compound was found necessary for *in vitro* synthesis of cellulose in *Dictyostelium* (Blanton and Northcote, 1990) or *Saproiengia* (Fevre and Rougier, 1981). Though these effectors were discovered in non-plant species, their value in improving *in vitro* synthesis of cellulose in cotton were shown by Li and Brown (1993). Neither Amor *et al.* (1995) nor Brown and coworkers (Kudilicka *et al.*, 1995; Li and Brown, 1993; Okuda *et al.*, 1993) have a complete answer

for the requirements and activation of cellulose biosynthesis; fibrillar cellulose as the sole product has yet to be made *in vitro.*

Though calcium's importance in cellulose synthesis is disputed, it's involvement in regulating callose and cellulose synthesis is not (Delmer and Amor, 1995). Calcium removed with EDTA releases annexins–multifunctional proteins involved in Ca^{2+} signal transduction (Clark and Roux, 1995)–from plasma membranes isolated from developing fiber (Andrawis *et al.*, 1993). These proteins – a family of three polypeptides with molecular mass of 34 kDa – were identified as annexins by their size, sequence homology, Ca^{2+}-dependent interaction with the plasma membrane and their phosphorylation by endogenous kinases which were also bound to the membrane in Ca^{2+} dependent manner. The cotton annexins (collectively called p34) bind callose synthase; purified p34 immobilized on nitrocellulose will bind substantial amounts of callose synthase activity when Ca^{2+} is present. The p34 also inhibits callose synthase, but not completely. Because p34 inhibition is incomplete, the role of p34 as a regulator of callose synthase is arguable: it may not be relevant *in vivo.* Alternatively, the p34 may serve to localize the callose synthase on the plasma membrane.

A regulatory protein of cellulose synthesis may also be known (Delmer *et al.*, 1987). Cellulose synthesis in cotton fibers is inhibited *in vivo* by the herbicide 2,6-dichlorobenzonitrile (DCB) and two of its analogs (Montezinos and Delmer, 1980). A photo affinity analog of DCB can compete with DCB binding in cotton extracts, and labels an 18 kD polypeptide (Cooper *et al.*, 1986). This polypeptide proved to be the only polypeptide distinctly labeled in the extract, though there is considerable non-specific labeling due to the hydrophobicity of the analog and its preference to partition into the membranes (Delmer *et al.*, 1987). The 18 kD polypeptide is loosely associated with membranes – 90% of the protein was found in the 100,000 g supernatant liquid (Delmer *et al.*, 1987). The DCB-binding polypeptide's synthesis is apparently not constitutive; small amounts of the DCB analog are bound during the expansion phase of fiber development, but the amount bound increases dramatically shortly before the start of secondary cell wall synthesis and continues to rise well into secondary cell wall synthesis (Delmer *et al.*, 1987). The specificity of the labeling by the DCB analog and the developmental pattern tie the 18 kD polypeptide to cellulose synthesis, and Delmer *et al.* (1987) argue that the role of the protein is most likely regulatory because of the loose association of the polypeptide to the membrane make it unlikely that it serves as a catalytic subunit of cellulose synthase.

3.5 Interaction of Microfibrils, Microtubules, and Microfilaments

The development of normal cotton fibers depends on the correct interaction of microfibrils and microtubules (Yatsu and Jacks, 1981). Indeed, most plants require an interaction between microfibrils and microtubules to build a normal wall (Amor and Delmer, 1994). Microfibrils of the primary wall are deposited in a helical pattern with a shallow pitch and are oriented transverse to the axis of elongation (Roelofsen, 1951). Microfibrils are reorienting as the fiber elongates, as indicated by the small amount of birefringence that is seen from the primary wall (reviewed by Seagull, 1993). Reorientation of the microfibrils is taken as evidence for intercalary growth of the primary wall (Roelofsen, 1951). Microfibrils on the outermost layer of the 10 DPA wall retain their axial organization, and as fiber elongation slows wall birefringence increases.

As elongation decreases, a layer of microfibrils that have a steeply pitched spiral (with primary and secondary wall characteristics) is produced on the inside of the primary wall (see also Chapter 14). This is the winding layer (Anderson and Kerr, 1938), or inner sheath (Rollins, 1945). The winding layer retains features of the primary and secondary wall: it is attached to the primary wall, but is assumed to be chemically similar to the secondary wall because its synthesis coincides with the onset of secondary wall deposition (Seagull, 1993). The microfibrils of the winding layer have a larger diameter and an opposite helical gyre than the microfibril of the secondary wall (Anderson and Kerr, 1938; Flint, 1950; Rollins, 1945).

The secondary wall contains microfibrils organized in helical arrays with increasing pitch. A lamellar pattern is revealed when the walls are viewed in a cross section (Flint, 1950) and this pattern results from different amounts of cellulose accumulating during temperature fluctuations (Haigler *et al.,* 1991). Another feature of the secondary wall is the presence of reversals. Reversals can be seen as an abrupt change in mircofibril orientation ("Z" reversal) or as a gradual change ("S" reversal). Frequency of reversals is greater in the third of the fiber nearest the tip (Seagull and Timpa, 1990).

Ultrastructural studies of developing fiber have found cortical microtubules oriented parallel to the innermost layer of microfibrils (reviewed by Seagull, 1993). Proof of a relationship between the microtubules and microfibrils is partly observational: the orientation of cortical microtubules next to the plasma membrane is reflected by the microfibrils (Quader *et al.,* 1987; Seagull, 1986). Cytoskeletons disrupted with chemical agents further supports the interrelationship between microtubules and microfibrils (Seagull, 1990). Widely spaced microtubule arrays remain after treatment with cholchicine, oryzalin, or trifluralin. Newly deposited microfibrils are found in association with these isolated arrays. Taxol stabilizes microtubules. When developing fibers are treated with this drug, the number of microtubules in cortical cytoplasm rapidly increases. Taxol also affects cell wall patterns because normal shifts in microtubule orientation are prevented (Seagull, 1990).

Heterogeniety of tubulins is implied by the lack of sensitivity of some arrays to chemical disruption and supported by the discovery of several isotypes of α- and β-tubulin – three of which were unique to different stages of fiber development (Dixon *et al.,* 1994). Tubulin dimers assembled from specific isoforms of α- and β-tubulin polymerize into microtubules at different rates in vivo (Panda *et al.,* 1995). Heterogeneity of microtubules can also be introduced by post- translational modification. Gibberellic acid induced microtubule reorientation coincided with loss of the terminal tyrosine residue of the α-tubulin (Duckett and Lloyd, 1994). The γ-tubulin isoform has not yet been identified in cotton, so the role of this form of tubulin in stabilizing the microtubules (Liu *et al.,* 1995) of developing fibers cannot be assessed.

Microtubules associated proteins (MAP) are being identified in plants with some regularity and have been shown to stabilize against cold-induced depolymerization (Cyr and Palevitz, 1989) or bundle tubulin *in vitro* (Chang-lie and Sonobe, 1993; Schellenbaum *et al.,* 1993; Vantard *et al.,* 1991; 1994). Phosphorylation may prove to be of importance. Cold- and gibberellic acid-induced depolymerization of microtubules was prevented when kinase inhibitors were present (Mizuno, 1993; 1994). Complementing this observation, Sonobe (1990) found protoplast ghosts loose their cortical microtubule arrays when extracts made from lysed protoplasts laced with ATP were added. Tubulin (Koontz and Choi, 1993) is as likely a substrate for phosphorylation as a MAP (Lee, 1993; Vantard *et al.,* 1994).

Microfilaments are not as well described as the microtubules or microfibrils; this is partly due to the lack of appropriate techniques (Seagull, 1990). In fibers the microfilaments can be seen as filaments running parallel to the axis of elongation and as a three dimensional network in the cytoplasm. Microfilaments are associated with microtubules, but many microtubules are without apparent connection (Seagull, 1993). More curious, Seagull (1990) found microtubules were redistributed when the microfilaments were disrupted with cytochalasin, but disruption was not total: some filaments remain after prolonged incubation. When fiber younger than 16 DPA is treated with cytochalasin, microtubules are reoriented to steeply pitched helices. The reorientated microtubules appear to be continuous with older arrays that have a shallower pitch (Seagull, 1990). This indicates reorientation, but not synthesis. Microtubules in 18 DPA fiber, when treated with cytochalasin, become randomly oriented. In each treatment, as the microtubules change orientation, the microfibrils follow, so microfilaments do not directly influence the pattern of microfibril deposition. Microtubule reorientation can occur independently of the reorientation of the microfilaments in pea roots (Hush and Overall, 1992), so the two systems are not always linked dependently. The true spatial regulator may be a protein or complex of proteins that rests astride the microtubule and the microfilament, and is present only in cells that require dynamic microtubules, such as elongating cotton fiber and differentiating *Zinnia* tracheary elements (Kobayashi *et al.,* 1989). Actins probably serve a larger role in regulating fiber development. Kinases for the inositol phospholipids–compounds involved in signal transduction in plants (Einspahr and Thompson, 1990)–are found attached to F-actin (Tan and Boss, 1991; Xu *et al.,* 1992).

Messenger RNA from developing cotton fiber was the source of two cDNA clones (Delmer *et al.,* 1995) which code for proteins with high identity to animal proteins that are involved in actin organization (Downward, 1992). These animal proteins, known as Rac, belong to the *rho* subfamily of *ras* – proteins known to bind GTP and be involved in signal transduction (Bourne *et al.,* 1991). The two cotton clones, Rac 9 and 13, show differential expression in fiber. Rac 13 shows enhanced levels over Rac 9 and reaches a maximum at the transition between secondary and primary wall synthesis, a time when the cytoskeleton is reorganized (Delmer *et al.,* 1995).

4. HORMONES AND EMBRYO DEVELOPMENT

Accumulation of malate is not specific to elongating fiber. At 12 to 14 DPA, malate is not only the predominant organic acid in the fiber, but also in the liquid endosperm surrounding the embryo (Mauney *et al.,* 1967). The embryo is in the heart stage, undergoing rapid cell division. The malate concentration was 52 to 75 mM (7 to 10 mg/ml), which is very similar to the fiber. With the high concentration of malate in the embryo, activities of PEPC, MDH, and GOT are likely to be relatively high in the structures near the endosperm or the endosperm itself. Unlike the fiber, IAA and gibberellic acid inhibited growth of the embryo in culture. IAA has been reported to peak between 12 to 15 DPA in fiber (Naithani *et al.*, 1982).

Storage protein and oil deposition occurs between 20 and 45 DPA in cottonseeds. During deposition, the embryo becomes fully capable of precocious germination, *i.e.,* vivipary (Dure, 1975). Removal of embryos from the seed coat and germinating in medium or on ABA-free media has been the method of choice for measuring percent vivipary. Hendrix and Radin (1984) reported that vivipary of embryos was 0% at 34 DPA, but quickly rose to 100% by 40 DPA. A collateral increase in cottonseed ABA levels was also measured (Davis and Addicott, 1972; Hendrix and Radin, 1984; Galau *et al.,* 1986). Vivipary of excised ovules was inhibited by the addition of 3.8 μM ABA in culture media (Choinski *et al.,* 1981). Excised ovules grown in the presence of ABA were similar to embryos matured in vivo, in that they continued to accumulate protein, neutral lipid, and dry weight (Choinski *et al.*, 1981). This was not unique to cotton, storage protein deposition continued as vivipary was suppressed by ABA in culture grown wheat (Triplett and Quatrano, 1982), soybean (Ackerson, 1984; Eisenberg and Mascarenhaus, 1985), and pea (Barratt *et al.,* 1989). Also, associated with the increase in ABA levels of cottonseeds a period of desiccation begins after 35 DPA (Reeves and Beasley, 1935). Protein deposition continued until maximum dry weight was obtained 40 to 45 DPA (Reeves and Beasley, 1935) and the funiculus degenerated (Benedict *et al.,* 1976). ABA has also been reported to prevent germination of mature cottonseeds (Halloin, 1976).

The understanding of ABA synthesis in cottonseeds has progressively changed during the last two decades. Ihle and Dure (1972) deduced that ABA was synthesized by the tissues surrounding the seed. In an examination of this hypothesis, Hendrix and Radin (1984) measured ABA in both seed coat and embryo, and found two concentration peaks (approx. 28 and 38 DPA). This biphasic profile of ABA concentration was subsequently confirmed by Galau *et al.* (1987). Hendrix and Radin (1984), however, consistently found lower levels of ABA in the cottonseed coats when compared to embryos. They reported that the embryo has a strong polarity for ABA. ABA is postulated to move from the acidic seed coat toward the more alkaline embryo (Hendrix and Radin, 1984). The seed coat of cotton has been reported as being as much as 1.4 pH units more acidic than the embryo (Hendrix *et al.,* 1987). Movement of ABA into *Phaseolus* embryos also appears to occur due to pH gradients between embryo and seed coat (LePage-Degivry *et al.,* 1989).

The mode of action of ABA in preventing germination in maturing and mature seeds is unknown. Analysis of two-dimensional polyacrylamide gel electrophoresis of protein profiles of maturing cottonseeds indicated a set of proteins which increased in abundance during desiccation of the embryos (Dure *et al.,* 1981). These proteins were designated late-embryogenesis abundant proteins or *Lea* protein. Of the 18 proteins identified, 6 were demonstrated to have ABA associated expression and these 6 proteins were redesignated *LeaA* (Hughes and Galau, 1989; 1991). The functions of the *LeaA* proteins during desiccation are still in question: however, mounting evidence is accumulating on putative roles for these proteins during desiccation (Dure, 1993, 1994).

5. CONCLUSION

Understanding the basic physiological processes of fiber and seed development are far from complete. The rudiments of elongation and almost every aspect of the primary and secondary wall physiology are not understood. These are areas of physiology which effect fiber length and micronaire, fiber properties with financial sway. For example, no experiment examining K^+ channels with a *Gossypium* species is recorded, and this ion is a major component (with malate) of the solutes that produce turgor pressure. Our knowledge of the primary walls of the cotton fiber has not kept pace with the research in the field, and much of the existing results are aging: the last analysis of the primary wall was published in 1977 (Meinert and Delmer, 1977).

Research creates its own muddle. Basra and Malik (1983) observed a relationship between total MDH activity, malate levels, and fiber length. Kloth (1992), surveying total MDH activity in modern cultivars, does not find any correlation between fiber length and total MDH. But neither Basra and Malik (1983) nor Kloth (1992) consider the subcellular localization of MDH, which proved invaluable to Thaker *et al.* (1987) and maybe important for MDH as indicated by Wäfler and Meier's (1994) results. The confusion over MDH shows that an enzyme assay is only the prelude. Subcellular location of enzymes, metabolites, and developmental expression of isoforms are essential in uncovering the physiological role of an enzyme. Understanding the developmental physiology of embryo and fiber has an economic reward in not only helping to solve the problem of small seeds mentioned in the introduction, but understanding yield and fiber quality.

Chapter 12

PLANT RESPONSES TO TEMPERATURE EXTREMES

John J. Burke[1] and Donald F. Wanjura[1]
[1]USDA Cropping Systems Research Laboratory, Lubbock, Texas

1.   INTRODUCTION

Ancestors of commercial cotton varieties are of tropical origin and are naturally adapted to growth in warm environments. Today, the derived cultivars retain the high optimal temperatures for growth that are characteristic of their progenitors. It is valuable to review the reported optimal temperature(s) for cotton before examining plant responses to temperature extremes. Despite many studies evaluating cotton temperature responses during the past century, a clear picture has yet to emerge whether there is a single temperature optimum for cotton metabolism. Some reports suggest distinct temperature responses within different anatomical structures, or at diverse developmental times. Discrepancies between reported temperature responses can be related to the measure of temperature itself. Some studies describing optimal temperatures for cotton refer to air temperature, while others refer to plant temperature. The discussion below will show that air and plant temperature measurements cannot be used interchangeably. This chapter will provide brief overviews of temperature measurement techniques, reported temperature optima, and the effects of low and high temperature stresses on cotton metabolism.

2.   TEMPERATURE MEASUREMENTS

Temperature is an environmental factor that regulates the rate of phenological development and biomass accumulation in cotton. Measuring the above ground temperature of plants or of the air is complicated by temperature variations that occur in time and space. There is a diurnal temperature cycle that is driven by solar radiation. There is also a temperature variation that exists due to distance above the soil surface. Until the development of infrared thermometry, the measurement of the temperature environment of field crops was limited to air temperature and point measurements of plant temperature with thermocouples.

Automated weather stations which are located near or in agricultural fields have the capability to position air temperature sensors at variable height locations. The standard height for measuring air temperature is in the interval from 1.5 to 2 m (Doorenbos, 1976; Ley et al., 1994). However, field studies are conducted where air temperature is measured at other heights and experimental procedures need to be reviewed to determine where air temperature was measured.

Cotton plant temperature is rarely at the measured air temperature. Differences between plant and air temperature are influenced by many factors including the diurnal cycle of radiation, plant water status, moisture content of the air, crop size, and wind speed. The pattern of canopy and air temperatures follows a diurnal cycle that is driven by the cyclical pattern of solar radiation (Fig. 12-1). Temperatures typically rise and then fall during the day time as the energy flux from solar radiation increases and then decreases. Canopy temperatures of the well-watered cotton canopy and the water-stressed canopy both change as the energy input changes (Fig. 12-1). During the daytime air temperature is usually different from the temperatures of either well-watered or water-stressed canopies.

Even when cotton is irrigated and well-watered, the difference between the plant and air temperature can vary when the moisture content of the air is different as normally exists between humid and arid climates. Plant temperature is normally below air temperature during the middle of the day; however, the ability of leaves to cool by transpiration

J.McD. Stewart et al. (eds.), Physiology of Cotton,
DOI 10.1007/978-90-481-3195-2_12, © Springer Science+Business Media B.V. 2010

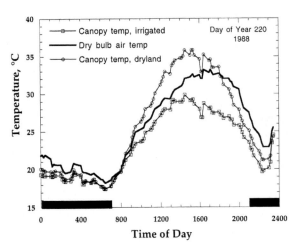

Figure 12-1. Changes in canopy and air
temperatures throughout a 24 h period in 1988.
Canopy temperatures of irrigated (□) and dryland (○)
cotton are presented with the dry bulb air
temperature (–) measured at 2 m above the canopy.

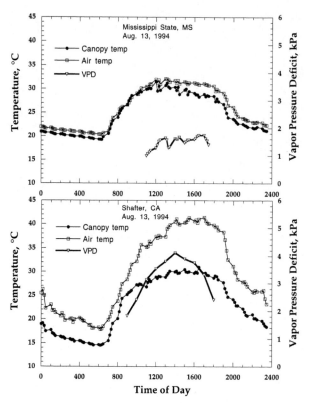

Figure 12-2. Comparison of cotton canopy
and air temperatures throughout a 24 h period in
1994 in two climates with differing vapor pressure deficits.

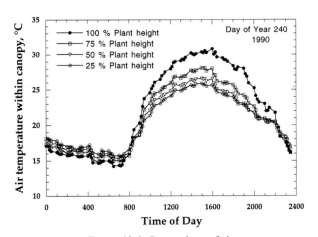

Figure 12-3. Comparison of air
temperature changes within an irrigated
cotton canopy throughout a 24 h period in 1990.

is greatly influenced by the vapor pressure gradient (Fig. 12-2). The effect of the diurnal radiation is apparent whether the comparisons are for different plant water status or between different vapor pressure deficits (VPDs).

The plant temperatures for both climates in Figure 12-2 are surface canopy temperatures. Depending on the degree of ground cover there can be a large temperature gradient from the top surface of the canopy down into the lower canopy. Air temperatures near the main-stem of a complete ground cover cotton canopy are 5°C higher at the top of the canopy compared to the temperatures at 25% of plant height (Fig. 12-3). Daily average temperatures and the range of temperatures at different vertical positions vary which changes the environment for boll development and other plant physiological processes.

Infrared thermometers measure surface radiometric temperatures and give an average temperature of the surface in its field of view (Fuchs, 1990). This report analyzed the factors that affect the canopy temperature viewed by infrared thermometers and concluded that pointing the infrared thermometer towards the canopy in the same direction as that of the sun, and at an angle of incidence close to the solar zenith angle, improves the ability to detect water stress. Gardner *et al.* (1992) reviewed the proper technique of using infrared thermometry for detecting water stress from canopy temperature measurements and provides an extensive bibliography of published research studies. The first infrared thermometers were hand-held instruments designed for intermittent use. Currently available infrared thermometers can be operated continuously which allows continuous measurement of plant temperature. Infrared scanners can provide surface temperature measurement with high spatial resolution and generate temperature frequency distributions or thermal maps.

Other sensors like thermocouples are readily available but only provide a temperature measurement for the im-mediate contact surface. Thermocouples are available in a range of wire sizes and can be inserted into plant tissue for measuring internal temperatures.

The temperature environment of the plant root system also varies temporally and spatially. Unlike the aerial environment of the plant where the plant and air temperatures are usually different, the root and soil temperatures are very similar. The diurnal range of soil temperatures varies with depth but its magnitude is changed by the amount of

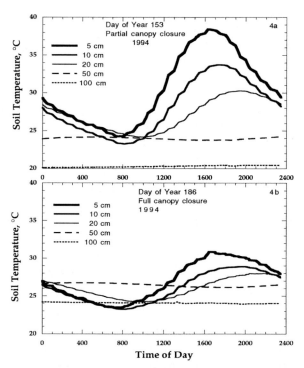

Figure 12-4. Comparison of diurnal soil temperature changes with depth throughout a 24 h period before (DOY 153) and after (DOY 186) canopy closure.

crop ground cover as illustrated in Figure 12-4. Early in the season, Figure 12-4a, the daytime rise and fall of soil temperature at 5, 10, and 20 cm is much greater than later in the season, Figure 12-4b, when crop cover is near 100%. Soil temperatures at 50 cm and 100 cm below the soil surface display little change during a diurnal cycle. Early in the season, during germination and emergence, the cotton hypocotyl and radicle are exposed to higher temperatures and with larger variations than later in the season when root activity occurs farther below the soil surface.

3. TEMPERATURE RESPONSES OF COTTON METABOLISM

The reported temperature optima for cotton enzyme function, germination, seedling growth, root development, shoot development, flowering, and lint production provide a range of "optimum" temperatures centered around 28°C ± 3°C.

3.1 Thermal Dependence of Enzyme Function

The thermal dependence of enzyme parameters have been used to delineate optimal temperatures in many plant and animal species (see review, Burke, 1994). Analysis of the thermal dependence of the apparent Km of cotton

(*Gossypium hirsutum* L.) glyoxylate reductase revealed a thermal kinetic window (TKW) of optimal enzyme function between 23.5 and 32°C, with a minimum apparent Km value at 27.5°C (Burke *et al.*, 1988). The 23.5 to 32°C TKW delineated the temperatures within which apparent Km values were within 200% of the minimum apparent Km value. Previous studies have shown that enzymes could function optimally with Km values within 200% of the minimum value (Somero and Low, 1976; Teeri and Peet, 1978). Having identified the optimal temperatures for enzyme function, a linear relationship was observed between the time that upland cotton foliage temperatures were within the TKW and plant biomass production throughout the season (Burke *et al.*, 1988).

The temperature dependence of cotton photosystem II variable fluorescence reappearance following illumination was evaluated as an alternative metabolic parameter for identification of optimal temperature(s) (Burke, 1990). Cotton leaves measured in 5°C increments between 10 and 45°C exhibited maximum reappearance of variable fluorescence at 30°C, with values declining with temperatures on either side of the optimum. The rate at which maximal reappearance occurred also increased up to 30°C. Temperatures above 30°C did not increase the rate of reappearance of variable fluorescence, but did result in a reduction in the maximum variable fluorescence reached by the leaf. More detailed analyses of cotton leaves using 2°C intervals showed that 28°C provided the highest level of photosystem II variable fluorescence and fastest rate of fluorescence reappearance.

Similar optimal temperatures in upland cotton were predicted by the temperatures providing the minimum apparent Km value (27.5°C) and the maximum reappearance of photosystem II variable fluorescence following illumination (28°C). The 28°C optimum identified by these metabolic parameters formed the basis for a successful, biologically-based irrigation system for cotton (Wanjura *et al.*, 1992).

3.2 Thermal Dependence of Germination

Germination and emergence is the first critical stage in the growth of a cotton plant. Christiansen and Rowland (1986) reviewed environmental effects on germination in Volume I of this series outlining the kinds of temperature regimes that cause injury, the nature of the injury, and the methods for preventing or ameliorating injury.

Arndt (1945) investigated the temperature characteristics of cotton germination and reported the optimum to be 33 to 36°C, and the minimum to be below 18 °C. Ludwig (1932) reported the minimum temperature for germination to be about 12°C. A soil temperature of 15.6°C (60°F) is generally considered to be the minimum temperature for planting. Low temperature exposure during cottonseed germination, emergence, and early seedling growth reduces stand establishment in addition to the physiological and morphological changes that reduce lint yields (Dennis and Briggs, 1969; Larsen and Cannon, 1966; Muller, 1968). Kittock *et al.* (1987) concluded that physiological and morphological effects of low temperatures early in the cotton planting season may often contribute as much to reduced yield as to reduced stands. The deleterious effects chilling treatments on cotton germination and seedling vigor have been described in detail by Buxton *et al.* (1976), Christiansen (1963, 1967, 1968), Christiansen and Thomas (1969), Cole and Wheeler (1974), Fowler (1979), Guinn (1971), Pearson *et al.* (1970), St. John and Christiansen (1976), Tharp (1960), Thomas and Christiansen (1971), Wanjura and Buxton (1972b), and Wanjura and Minton (1981).

3.3 Thermal Dependence of Root Growth

The optimum temperature for the growth of cotton roots has been reported to be 36°C (Arndt, 1945), 30 to 33°C (Pearson *et al.*, 1970), and 35°C (McMichael and Burke, 1994). These studies monitored the temperature dependence of root length formation and lateral root numbers during the first 2 weeks of seedlings growth. McMichael and Burke (1994) suggested that the observed temperature optima for cotton root growth drops from 35°C to 30°C as the seedling depletes its stored reserves and becomes photosynthetically competent.

McMichael and Burke (1994) compared the metabolic temperature characteristics of intact roots of cotton seedlings three and seven days after planting (DAP) by monitoring the in vivo reduction of triphenyltetrazolium chloride (TTC). The rate of TTC reduction in root tips at seven DAP was similar to the activity in root tips of seedlings three DAP at 25 and 30°C. The efficiency of the TTC reduction declined with temperatures above and below the 25 to 30°C range. These results are consistent with the observation that optimum enzyme efficiency occurs within the TKW (23.5 to 32°C) previously established for cotton. The observed broadening of the temperature characteristics of root development beyond the TKW during the period of rapid mobilization of seed reserves provides an evolutionary benefit for the establishment of the cotton root system during early seedling growth across a range of possible thermal environments.

Throughout the United States cotton (*Gossypium hirsutum* L.) is irrigated to avoid the deleterious effects of water stress on growth and development. Irrigation, while optimizing water status, results in a concomitant cooling of soil temperatures. Previous field studies have focussed on soil water status effects on root growth independent of the soil temperature changes following irrigation. Burke and Upchurch (1995) evaluated the soil temperature characteristics of irrigated and non-irrigated field plots of cotton to determine the time that temperatures were spatially and temporally within the range for optimal cotton root metabolism. Soil temperatures in the non-irrigated treatment were within cotton's thermal kinetic window (23.5 to 32°C) for

72% more time than the irrigated treatment. Predicted rooting fronts fell between the 24 and 26°C isotherms in the irrigated treatment and the 26 and 28°C isotherms in the non-irrigated treatment. Greater root length density of the cotton in the non-irrigated treatment determined from soil cores was associated with increased time at optimal temperatures. Near the soil surface, root length density decreased in the non-irrigated treatment because of reduced soil water status, not because of reduced time within the optimal temperature range. Soil temperatures cooled below the optimal range upon canopy closure in the irrigated treatment. The cooling trend was observed to a lesser extent in the non-irrigated treatments that never realized full canopy closure. This study showed that greater cotton root development occurred across time and with depth in soils exhibiting temperatures within cotton's thermal kinetic window.

The effect of soil temperature on the downward rate of cotton and soybean (*Glycine max* L.) root system extension in controlled environments has been reported (Bland, 1993). Bland (1993) evaluated a simple model for root system extension, in which daily increase in rooting depth depended on the deepest roots. Results showed that increase of rooting depth may be predicted for much of the crop life from temperature-dependent growth response of seedling roots and local soil temperatures (Bland, 1993; Kaspar and Bland,1992). Soil temperature-derived predictions of cotton root density have recently been reported (McMichael *et al.*, 1996). Linking knowledge of root growth under controlled environments with field soil temperature data, they were able to predict the development of cotton root density in the field.

Temperature not only influences root proliferation, but also impacts root function. Changes in hydraulic conductivity associated with changes in root temperatures have been reported (Bolger *et al.*, 1992). A dramatic reduction in hydraulic conductance of cotton roots was observed when the temperature was lowered from 30°C to 18°C. Root conductance at 18°C averaged 43% and at 7°C averaged only 18% of that at 30°C. Prolonged exposure of roots to 18°C did not alter the temperature-conductance relationship, indicating that there was no apparent acclimation of conductance to the 18°C temperature. The changes in hydraulic conductivity may explain the observation that chilling temperatures between 10°C and 25°C increased the relative water content of roots but not shoots (Bradow, 1991).

3.4 Thermal Dependence of Shoot Growth

The development of the cotton shoot requires emergence of the hypocotyl from the soil and the development of the photosynthetic machinery. Soil temperature determines the rate and maximum time that the cotton hypocotyl can elongate in the dark under the soil surface. The relationship between maximum emergence force and temperature was described by a quadratic relationship (Chu *et al.*, 1991). The greatest emergence force was calculated to occur at 28.5°C and was significantly lower at 22°C and 37°C. Critical time, defined as the time when changes in the hypocotyl tissue that allow increases in emergence force reach full expression, decreased linearly with increased temperature from 22°C to 37°C.

Significant stresses during the early growth phase can permanently reduce plant vigor and productivity. A field study with multiple planting dates and two cultivars was used to study diurnal soil temperature effects on emergence, the first true leaf stage of development, and seedling rate of emergence index (SRDI), computed as the time to reach four stages of development ranging from the hypocotyl crook stage to the first true leaf stage (Steiner and Jacobsen, 1992). During the first 100 h after planting, emergence was more sensitive to cool temperature stress when planting occurred at 0800 h than at 1600 h. The most sensitive periods occurred at 53 h and 143 h (early hypocotyl crook stage) after planting. Both planting times were sensitive to cool temperatures at 160 h after planting. The cultivar GC-510 was more sensitive to cool temperatures during the first 100 h than was the cultivar SJ-2. The SRDI of GC-510 was less tolerant of cool temperature than SJ-2 during the initial 30 h but afterwards it was relatively unaffected by soil temperature. SJ-2 was sensitive to cool soil temperature for 160 h after planting.

Cotton phenological and vegetative development following emergence is strongly temperature dependent. During the first 22 d after emergence of pima cotton cv. S-6, about 8 d were required to produce a node on the main-stem at 21°C, whereas only about 3.5 d were required to produce a node at the optimum day/night temperature (30°C/22°C; Reddy *et al.*, 1995). Node development at 30°C was shown to be developmentally regulated, with the time between nodes becoming shorter as the plants became older. Reddy *et al.* (1995) suggest that this is the result of intraplant competition among the plant organs, with roots having priority for available carbohydrates during early plant development. By the time the sixth node was formed, this priority for root growth was met and only 2 to 3 d were required to produce a leaf.

Temperature effects on shoot development rates reflects temperature dependent alterations in cotton metabolism. Throughout the cotton growing season on the High Plains of Texas, days provide optimal sunlight and temperature for metabolism followed by nights with temperatures of 15 to 20°C, well below the optimum. Although the low temperature sensitivity of cotton has been reported previously, few studies are available on specific metabolic response to low temperature stress. Warner and Burke (1993) evaluated the relationship between starch levels and photosystem II function in order to determine if low temperature stress at night altered these metabolic processes during the day when temperatures were optimal. Starch levels increased and photosystem II chlorophyll fluorescence levels decreased in fully expanded leaves of cotton grown under night temperatures

of 20°C compared with those grown with night temperatures of 28°C. Cotton plants grown at a constant temperature of 28°C exhibited increases in starch content in mature leaves after exposure to one night at 20°C. Cotton plants grown with a night temperature of 20°C exhibited decreases in starch content in mature leaves after one night at 28°C. Chlorophyll fluorescence responses to changes in night temperatures were reciprocal to that of leaf starch. Shading to reduce daily photosynthesis and ultimately starch levels in mature leaves of plants grown with 20°C night temperatures resulted in increases in PS II chlorophyll fluorescence. The results of this study show an accumulation of leaf starch in response to low night temperatures that is associated with a reduction in photosystem II function even when plants experience optimal temperatures during the day.

Further investigation compared the response of specific components of carbon metabolism in vegetative cotton grown with a 28°C day/28°C night and a 28°C day/20°C night (Warner *et al.*, 1995). Photosynthesis for cool-night (20°C) plants measured at 28°C the following day was only 77% of 28°C night plants. Less starch accumulation occurred during the day in the cool-night plants, yet their predawn starch levels were approximately 2.5-fold higher than the 28°C plants. Pools of triose phosphate and fructose 1,6-bisphosphate were lower at night at 20°C than at 28°C. However, the glucose 6-phosphate/fructose 6-phosphate ratio was higher for the cool night plants indicative of an apparent limitation in sucrose synthesis subsequent to cytosolic fructose 1,6-bisphosphatase. The most interesting observation is the maintenance of equal sucrose pools in both treatments even though the amount of starch catabolized at night was different. This study shows that cool night temperatures alone alter cotton carbon metabolism throughout each 24-h period.

High temperature stress can occur in moderate climates during periods of drought. Burke *et al.* (1985) showed reductions in leaf area index, plant height, and dry matter accumulation in cotton grown under dryland conditions. Canopy temperatures of the dryland cotton increased to 40°C during midday, while temperatures of the irrigated cotton remained at 30°C. Leaves of the dryland plants were shown to accumulate heat shock proteins following repeated cycles of the elevated temperatures. This study was the first to identify the heat shock response in field-grown materials, and provided the impetus for additional research on high temperature stress in crops.

3.5 Thermal Dependence of Reproductive Organ Growth

The development of fruiting branches in pima cv S-6 was temperature dependent (Reddy *et al.*, 1995). The number of fruiting branches at 64 d after emergence increased from 4 to 5 at 18°C to 14 to 16 at 27°C. Higher temperatures resulted in a decline in the number of fruiting branches, with no fruiting branches produced at 36°C.

Hodges *et al.* (1993) showed that the minimum time required to produce first square for the Delta type cotton cv. DES 119 was at a temperature of about 82°F (28°C) [average daily air temperature]. The time from first square to flower reached a minimum at 27°C and then remained constant up to an average temperature of 34°C (Reddy *et al.*, 1993). Development from square to flower was projected to stop at 14°C. The time for boll maturation declined as temperature increased but boll size was reduced at temperatures above or below 26°C. Reddy *et al.* (1995) showed that the developmental rate of pima cv. S-6 from plant emergence to the time of the first flower bud appearance was best at 27°C. About 43 d were required to form squares at 19°C, whereas only 26 d were required at 27°C. This study also showed that temperatures above 27°C caused an increase in time required to produce the first square. Reduced boll set and flower sterility also have been reported at temperatures in excess of 38°C (Hoffman and Rawlins, 1970; Taha *et al.*, 1981).

Fibers from field-grown cotton exposed to cool night temperatures have prolonged periods and reduced rates of elongation and thickening (Gipson and Joham, 1969; Gipson and Ray, 1969a, Hawkins, 1930; Thaker *et al.*, 1989) and growth rings in their secondary walls (Balls, 1919b; Kerr, 1936). Cotton fiber growth rings are revealed after swelling of whole fibers in cross-sections. The rings represent periods of reduced cellulose accumulation as a result of the cool night temperatures (Gipson and Ray, 1969a). The temperature-induced growth rings have been re-evaluated in cultured ovules across a range of temperatures (Haigler *et al.*, 1991; see also Chapter 4). They showed that in vitro fibers exhibited a growth ring for each time the temperature cycled to 22 or 15°C. Rings were rarely detected when the low point was 28°C. The rings appeared to correspond to alternating regions of high and low cellulose accumulation. These findings demonstrated that cool temperature effects on fiber development are at least partly fiber/ovule specific events independent of the physiology of the rest of the plant.

4.   SUMMARY

This chapter has provided brief overviews of temperature measurement techniques, reported temperature optima, and the effects of low and high temperature stresses on cotton metabolism. Interpretation of temperature responses of cotton growth and development requires detailed information on the methods used to measure temperature and the developmental stage of the plant. In general, cotton exhibits a plant temperature optimum of 28°C ± 3°C. Moderately low temperatures (20 to 23°C) reduce metabolic activity and chilling temperatures (below 15°C) impact membrane function thereby resulting in yield losses. High temperatures (>34°C) reduce the production of squares and may induce flower sterility.

Chapter 13

PLANT RESPONSES TO SALINITY

John Gorham[1], Andre Läuchli[2], and Eduardo O. Leidi[3]
[1]University of Wales, Bangor, U.K., [2]University of California, Davis, CA, and [3]IRNASE-CSIC, Seville, Spain

1.   INTRODUCTION

Cotton is one of the most salt-tolerant of the major annual crops, but the variable nature of both salinity and the plant's response to it make salinity a far from simple problem. Here we consider the response of cotton to salinity, and possible means of improving cotton yields on saline land or with saline water. This review will cover most of the research published in the last 30 years, but space does not permit a more extensive review of the older literature.

2.   SPECIFIC EXAMPLES OF SALINITY PROBLEMS

2.1   U.S.A.

About 23% of the world's cultivated lands are saline (Tanji, 1990), many of them being located in semiarid to arid regions. In the U.S.A. the majority of salt-affected soils are located in the western part of the country, with irrigated agriculture being prominent in the Southwest. In California about 30% of non-federal land is saline (EC 4dS/m) and much of the salt-affected area is located in the central San Joaquin Valley with extensive irrigation agriculture (Tanji, 1990) and where cotton is grown widely. The extent of the salinity problem in the San Joaquin Valley is clearly increasing with time and may approach 50% in a few years. Remote sensing has been used to estimate the effects of salinity on cotton crops in the U.S.A. (Wiegand et al., 1992, 1994) and in Russia (Golovina et al., 1992).

The management and control of soil salinity in irrigated lands of the southwestern U.S.A. is a challenge; it has been partly based on disposal of saline drainage waters.

However, particularly in the San Joaquin Valley, the disposal of saline drainage waters has experienced severe environmental constraints due to accumulation of toxic trace elements such as selenium in drainage waters and soils (Tanji et al., 1986). Reuse of saline drainage water for irrigation of relatively salt-tolerant crops is, therefore, important as an alternative to disposal. Field studies in the San Joaquin Valley have demonstrated that good yields of cotton can be obtained with saline water irrigation, if non-saline water is used to irrigate until seedling establishment (Ayars and Schoneman, 1986; Ayars et al., 1986; Rhoades, 1987; Rains et al., 1987). A nine-year field experiment was conducted from 1984 to 1992 to study the feasibility of long-term reuse of saline drainage water for crop irrigation in a cotton - cotton - safflower rotation system (Rains et al., 1987; Goyal et al., 1999a). The results showed that drainage waters with a salinity level up to 3000 ppm (4.7 dS/m) can be used for four years to irrigate cotton, provided that some soil leaching occurs by means of preplant irrigation with good quality water and the cotton is grown in rotation with an unirrigated crop such as safflower or one that is irrigated with good quality water (Goyal et al., 1999a,b). Special management procedures must be developed to sustain long-term maximum yields of cotton irrigated with saline water.

2.2   Asia

One of the best documented man-made ecological disasters of recent times is that of the Aral Sea and its catchment area (Micklin, 1988, 1994; Brown, 1991; Precoda, 1991; Aladin and Potts, 1992; Levintanus, 1992; Ghassemi et al., 1995; Pearce, 1995; Minashina, 1996; Spoor, 1998). The Aral Sea Project was inaugurated around 1960, with the intention of using the waters of two rivers, the Amu

J.McD. Stewart et al. (eds.), *Physiology of Cotton*,
DOI 10.1007/978-90-481-3195-2_13, © Springer Science+Business Media B.V. 2010

Dar'ya and the Syr Dar'ya to create a vast irrigated cotton belt from a former desert in what was then the USSR. Half of the irrigated area, at 68,000 km^2 twice the size of the state of California, was devoted to cotton production. Problems began with the lowering of the level of the Aral Sea and reduced catches of fish, from a peak of over 50,000 metric tonnes before the project started to eventual collapse caused by a three-fold increase in salinity and a reduction in the volume of the Aral Sea by two-thirds. A fishing industry supporting 60,000 people was reduced to processing Baltic fish transported more than 2,000 miles over land. However, the problem extends beyond the former Aral Sea in the form of wind-blown salt and salty rain, polluted water resources, and the classic problems of poorly-managed irrigation, *i.e.* rising water tables and salinization of agricultural land. The disaster was further compounded by the extensive use of agricultural chemicals which, since there is no outlet, accumulated within the Aral Sea Basin. Cotton production has declined both as a result of salinization and a reduction in the buffering capacity of the Aral Sea on the climate within the Basin. Summers are hotter, winters cooler, and the growing season shorter. Although steps are being taken to alleviate these problems (Pearce, 1994; Aladin *et al.*, 1995; Dukhovny, 2000), the importance of cotton exports to the economies of newly independent states such as Uzbekistan presents new challenges (Glantz *et al.*, 1993; Van Atta, 1993).

While the Aral Sea crisis is one of the most dramatic and recent examples of salinization of irrigated lands in Asia, it is not unique. In what is now Pakistan, British Army engineers started the development of existing irrigation systems in the Indus valley into what became the largest single irrigation system in the world (Ahmad and Chaudhry, 1988; Ghassemi *et al.*, 1995). The area suffers from both primary (residues of a former shallow sea) and secondary (from insufficient leaching and drainage in irrigated land) salinization, and from both salinity and sodicity (alkaline soil with cation exchange sites dominated by sodium). Again cotton is the major export crop of the region. A variety of local, national, and international programs have alleviated the problem to some extent, the most important being the Salinity Control and Reclamation Projects (SCARPs) which featured increased leaching, better drainage, and tube-wells to lower the water table. Another important alleviation project in Pakistan is the Left Bank Outfall Drain in the lower Indus valley, designed to remove saline drainage and groundwater. Mauderli *et al.* (2000) pointed to the need to consider social issues when operating such schemes. Even minor improvements to irrigation systems may not, however, be economically viable in the short term, especially when cotton prices are low (Kijne, 1998). Given these problems, and the economic importance of cotton to Pakistan, it is fortunate that cotton is one of the most salt-tolerant crops.

2.3   Spain

The area covered with salt-affected soils in Spain is one of the most extensive in Europe, with approximately 840,000 ha (Szabolcs, 1989). The cotton-producing area is mainly in the Guadalquivir River Valley, where the crop is usually grown under irrigation (about 100,000 ha). Problems with salinity affecting cotton production are evident in the soils of reclaimed areas of the Guadalquivir marshes (Marismas del Guadalquivir). In this reclaimed area (about 1/10th of the 150,000 ha of marshland), cotton and sugar beet are grown because of their higher tolerance to salinity. The hydrology and movement of salts in this area have been studied in detail by Andreu *et al.* (1994) and Moreno *et al.* (1995, 2000). Cultivation of irrigated cotton for more than 10 years reduced salinity and ESP in the top 50 cm of soil, while reduced irrigation allowed the capillary rise of saline ground water. Nutritional disorders and symptoms of salt toxicity problems have been detected in other areas (*e.g.*, Arahal, Lantejuela, Puebla de Cazalla) because of the use of wells with brackish water.

3.   PLANT RESPONSES TO SALINITY

3.1   Germination and Emergence

Salinity generally delays and reduces the germination of cotton seeds (Ashour and Abd El-Hamid, 1970; El-Zahab, 1971a; Abul-Naas and Omran, 1974; Soliman *et al.*, 1980; Kent and Läuchli, 1985; Ahmad *et al.*, 1990; Ahmad and Makhdum, 1992; Ye and Liu, 1994; Nawar *et al.*, 1995; Vargese *et al.*, 1995; Tort, 1996; Oliviera *et al.*, 1998). There are some varietal differences in the response to salinity, with Acala 1517 (Wadleigh *et al.*, 1947), NIAB-78 (Malik and Makhdum, 1987; Nawaz *et al.*, 1986; Chaudhry *et al.*, 1989; Khan *et al.*, 1995a; Qadir and Shams, 1997), and MNH-93 (Malik and Makhdum, 1987) being more tolerant at germination. Da Silva *et al.* (1992) obtained better germination of Allen 33-57 than Acala del Cerro or SU 0450-8909 over three selection cycles at 0.9 MPa with a 1:1 (w/w) mixture of NaCl and CaCl$_2$. Abul-Naas and Omran (1974) reported that *G. barbadense* L. cultivars were more tolerant than *G. hirsutum* L. or *G. arboreum* L. cottons, and that seedling growth was more sensitive to salt than germination.

Ferreira and Reboucas (1992) used pre-sowing hydration/dehydration cycles to improve the germination of three cultivars of cotton in the presence of NaCl. In another study pre-soaking of cotton seeds with 0.5 % solutions of CaSO$_4$ for eight hours prior to sowing increased the shoot and root growth of NIAB 78 after 56 days at various salinities (Rauf *et al.*, 1990), but no germination data were reported. Presoaking with ZnSO$_4$ or CaCl$_2$ was less effective.

Bozcuk (1981) found that kinetin at 10 or 20 ppm improved the germination of cotton and alleviated the effect of salt concentrations up to 150 mol m^{-3}. Shannon and Francois (1977) found that pre-soaking *G. barbadense* seeds in distilled water hastened germination, while soaking in NaCl, CaCl$_2$, or IAA reduced the percentage germination after seven days. Soaking cotton seed in water for several hours before hand-sowing is used to improve germination in India and Pakistan, but the practice is decreasing with the use of tractor-powered drills which damage the soft pre-soaked seeds.

The emergence force exerted by germinating cotton seedlings was measured by Sexton and Gerard (1982) using a transducer. Increasing salinity reduced the emergence force and increased the time required to develop the maximum force at each salinity. They suggested that such tests could be used to determine the salt tolerance of emerging seedlings. Germination of pollen grains has also been used to screen for salt tolerance (Shen *et al.*, 1997).

3.2   Root Growth

Although root growth and distribution in relation to water deficit have been studied in the field, there are no corresponding detailed studies of the effect of salinity. Thus most of the work described here relates to seedling root growth in hydroponics or soil-less systems, and may have limited relevance to field conditions. Moreover the use of different artificial substrates can have profoundly different effects on root growth and its response to salinity.

While salinity generally reduces root growth (*e.g.*, Silberbush and Ben-Asher, 1987), there are cultivars in which mild salinity can enhance the growth of roots. Jafri and Ahmad (1994) reported growth stimulation of roots, but not of shoots, in NIAB 78 and Qalandari. Leidi (1994) found longer primary roots in two cotton genotypes with 100 mol m^{-3} NaCl than without salt, while 200 mol m^{-3} inhibited root length. The number of secondary roots was inhibited at both salt levels. However, genotypic variability for primary root length and number of secondary roots has been found (Leidi, unpublished). These authors reported that root length was increased or diminished by salinity depending on the genotype, and different degrees of inhibition in secondary root growth were observed in different genotypes. The findings of Soliman *et al.* (1980) that germination and root growth of *G. barbadense* at a range of salinities were poorer in sand than in loam or clay may be attributed to the low Ca^{2+} content and cation exchange capacity of the sand. Gerard and Hinojosa (1973) reported that salinity reduced Ca^{2+} uptake and concentrations in cotton roots.

Calcium is generally thought to alleviate the effects of Na$^+$ salinity, but a variety of studies indicate complex interactions between the two cations (see also the section on calcium below). The effect of Ca^{2+} on germination in the presence of Na$^+$ was studied in detail by Kent and Läuchli

(1985). CaSO$_4$ at 10 mol m^{-3} did not enhance the germination of seven cotton cultivars in the presence of 200 mol m^{-3} NaCl, but did promote root growth. In 0.1 strength modified Hoagland's solution, length and weight of Acala SJ2 roots were slightly increased at 25 mol m^{-3} NaCl (Cramer *et al.*, 1986). The beneficial effects of Ca^{2+} were even more apparent at higher salt concentrations. Addition of 10 mol m^{-3} CaCl$_2$ resulted in stimulation of root length by salt up to 100 mol m^{-3} NaCl, and increases in fresh and dry weights up to 50 mol m^{-3} NaCl. The volumes of the cortical cells of these longer roots were similar to those of the controls, but cells were longer and narrower (Kurth *et al.*, 1986). The rate of cell production decreased with increasing salinity at 0.4 mol m^{-3} CaCl$_2$, but was unaffected by salt at 10 mol m^{-3} CaCl$_2$. In a more detailed analysis, Zhong and Läuchli (1993a) found that salinity shortened the length of the growing zone. Increasing the CaCl$_2$ concentration from 1 to 10 mol m^{-3} decreased the time for a cell to pass through the growing zone, but did not affect the length of the zone. Salinity reduced the incorporation of ^{14}C-glucose into cell wall polysaccharides (Zhong and Läuchli, 1988). At 150 mol m^{-3} NaCl, low Ca^{2+} (1 mol m^{-3}) resulted in increased wall uronic acids and intermediate-sized polysaccharides, and decreased cellulose and small-sized hemicellulose 1 fraction polymers (Zhong and Läuchli, 1993b). Polysaccharide contents were not affected by salinity when the Ca^{2+} concentration was increased to 10 mol m^{-3}.

Reinhardt and Rost (1995a,b,c,d) studied the morphology and development of roots of Acala SJ2 grown in 0.1 strength Hoagland solution or vermiculite with 1 mol m^{-3} Ca^{2+} and various concentrations of NaCl. In nine-day old, hydroponically-grown seedlings, exposed to salt for five days, salinity reduced the root growth rate and resulted in maturation of cells (except late metaxylem) nearer to the root tip. Exposure to salinity also resulted in increased vacuolation, particularly of the cortex and metaxylem, premature differentiation of protophloem sieve tube elements, dense staining of the endodermis, compression of the zone of maturation of proto- and early xylem parenchyma, and lateral initiation closer to the tip (Reinhardt and Rost, 1995d). When saline treatments were applied at the germination stage, primary root growth was delayed and peak elongation rates were reduced, but the general pattern of primary root growth was not affected. Initiation and emergence of laterals was inhibited by salinity to a smaller extent than lateral root elongation (Reinhardt and Rost, 1995c). Salinity affected the growth of roots grown in hydroponics to a greater extent than those grown in vermiculite. For plants grown in vermiculite for seven days with up to 150 mol m^{-3} NaCl, the start of secondary wall deposition in protoxylem tracheary elements was closer to the tip in roots of 50 to 150 mm, but unaffected in longer or shorter roots. Salinity and plant age interacted to reduce the length and width of metaxylem vessel members. Maturation of large, central metaxylem vessel members was delayed in salt-stressed plants (Reinhardt and Rost, 1995b). For plants growing in vermiculite for up to 10 days, salinity resulted in the development of Casparian bands and suberin lamellae in the endodermis nearer to the tip, whereas in older plants salinity had no effect on the position of suberinization of the endodermis (Reinhardt and Rost, 1995a). In plants from 5 to 28 days old, salinity induced the formation of an exodermis, complete with Casparian bands and suberin lamellae – a structure not seen in control roots – at the proximal end of the root and in the transition zone to the hypocotyl. Older (48 days) plants with secondary differentiation in the stele did not develop an exodermis in response to salinity. According to Ivonvina and Ladonina (1976) root segments with suberized exodermis can still absorb salts. Silberbush and Ben-Asher (1987) found that high salinity (>282 mol m^{-3} NaCl) resulted in very poor root growth, but increased the diameter of the roots. Even with 5 mol m^{-3} Ca^{2+}, salinity delayed the formation of thick cell walls in the stele of Stoneville 213 roots (Gerard and Hinojosa, 1973).

These detailed studies illustrate the variety of responses which can be induced at different times, and with different experimental protocols, in response to salinity, and point to the dangers of extrapolating experimental results to different tissues, times, and conditions.

3.3   Shoot Growth

Cotton is regarded as one of the most salt- (Maas, 1990; Chen *et al.*, 1996) and sodicity- (Chang and Dregne, 1955; Pearson, 1960; Ali *et al.*, 1992) tolerant crop plants. *G. hirsutum* and *G. barbadense* are also moderately (Wilcox, 1960) to very (Eaton, 1944) tolerant of boron, a toxic element when present in excess that is sometimes associated with salinity. Although in most cases increasing salinity reduces vegetative growth of cotton, there are instances of growth promotion by low concentrations of salt (Afzal, 1964; Strogonov, 1964; Twersky and Pasternak, 1972; Pasternak *et al.*, 1977; Pessarakli, 1995; Gorham, 1996a; Fig. 13-1). As with the promotion of root growth described above, this may be a nutrient-sparing effect, or the result of beneficial micronutrients present as impurities in the salt. Khalil *et al.* (1967) apparently found increasing yield of *G. barbadense* with increasing salinity in the absence of nitrogen fertilizer in soil culture. Calahan and Joham (1974) found that 10 mol m^{-3} NaCl inhibited the growth of cotton grown with 10 mol m^{-3} CaCl$_2$, but not of plants grown with only 1 mol m^{-3} CaCl$_2$. Both plant height (internode elongation) and leaf expansion were affected, and at higher salinities the differentiation of nodes was suppressed (Ahmed, 1994; Gossett *et al.*, 1994a). Vegetative node numbers may increase at the expense of fruiting nodes (Munk and Roberts, 1995). Shoot/root ratio decreases under salinity because of higher sensitivity of shoots than roots to salt stress (Leidi *et al.*, 1991; Brugnoli and Björkman, 1992). In summary, one of the main effects of salinity is a reduction in total leaf area, and consequently reduced photosynthetic output per plant.

For the same decrease in soil water potential, Shalhevet (1993) and Shalhevet and Hsiao (1986) concluded that salt

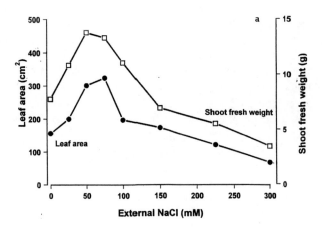

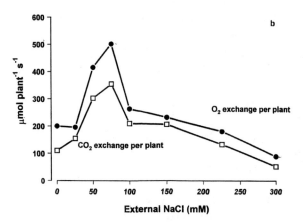

Figure 13-1. a). Shoot fresh weight and leaf area of plants of Acala SJ2 grown in hydroponic culture with or without salt (300 mol m^{-3} NaCl + 15 mol m^{-3} CaCl$_2$). b). Rates of photosynthetic gas exchange per plant. O$_2$ exchange measured in an oxygen electrode at saturating CO$_2$ supply (no stomatal limitation), and CO$_2$ exchange measured with an infra-red gas analyzer *in situ*. (Gorham and Läuchli, 1986)

stress was less detrimental to growth than the equivalent water-deficit stress. Chen and Kreeb (1989) reported greater inhibition of growth with combined salt and water stresses than with each stress alone, but the inhibition was less than the additive effects of the two stresses. In one of the few studies of combined salt and water-deficit stress in the field, Thomas (1980) found that both factors affected yield, and that from 115 days after planting plant size was correlated with the osmotic potential of the root zone and leaf Na$^+$ concentrations. Howell *et al.* (1984) examined the Crop Water Stress Index (CWSI, derived from measurements of leaf temperatures and water vapour pressure deficits) of cotton plants in well-irrigated fields of different salinities and observed that salinity reduced the vapour pressure deficit at which the plants appeared to be stressed.

At moderate salinity, and with adequate nutrition, this reduction in growth was not accompanied by symptoms of salt toxicity (Rathert, 1983; Brugnoli and Björkman, 1992), but at higher salt concentrations premature leaf senescence and leaf shedding were observed. Increasing leaf thickness (succulence) has been observed with increasing salinity (Longstreth and Nobel, 1979a; Qadir and Shams, 1997). This was the result of a slight increase in epidermal thickness and a considerable increase in mesophyll thickness resulting from increased palisade cell length and spongy mesophyll cell diameter, and additional layers of spongy cells. Effects of salinity on the structure of mitochondria and chloroplasts were reported by Gausman *et al.* (1972).

Infection with vescicular-arbuscular mycorrhizae has been reported to alleviate the effects of salinity on shoot growth of cotton in saline conditions (Jalaluddin, 1993; Feng *et al.*, 1999).

Goyal *et al.* (1999a) reported that, in a long-term study of cotton irrigated with saline drainage water, shoot height and biomass were reduced earlier, and at lower salinities, than lint yields (see below).

4. BOLL YIELD AND FIBER QUALITY

Although resources might be diverted from vegetative to reproductive development at low salinities (resulting in an increase in yield), at higher salinities boll numbers and seed and lint quality are reduced in a variety of ways (Ahmed, 1994; Babu *et al.*, 1987).

As with mild water-deficit stress (see Chapter 23), low salinity may increase the yield of seedcotton (Barakat *et al.*, 1971; Calahan and Joham, 1974; Pasternak *et al.*, 1977; Bouzaidi and Amami, 1980; El-Gharib and Kadry, 1983; Salih and Abdul-Halim, 1985; Abdullah and Ahmad, 1986), especially with adequate or supra-optimal supply of nutrients, particularly nitrogen. In some cases yields remained unaffected by salinity even though plant height and biomass were reduced (*e.g.,* Moreno *et al.*, 2000). This can be attributed to a shift in the balance of sink activity away from vegetative growth and towards reproductive structures (Vulcan-Levy *et al.*, 1998). Seed weight per boll was much less affected by salinity than the number of bolls (Abdullah and Ahmad, 1986). The reduction in the number of mature bolls as salinity increased was a function of both decreasing numbers of fruiting positions and an increase in the proportion of bolls shed, as reported earlier by Longenecker (1973, 1974). Lint weight per boll decreased more than seed weight in the study of Abdullah and Ahmad (1986), in contrast to that of Razzouk and Whittington (1991), where lint percentage increased. In the latter study attachment strength, micronaire, maturity ratio, and maturity percentage decreased with increasing salinity. Abdullah and Ahmad (1986) found an increase in mean and upper quartile fiber length with salinity, but also an increase in the percentage of short fibers. Fiber length, strength, and micronaire were reduced by salinity in both *G. hirsutum* (Acala 1517D) and *G. barbadense* (Pima S-2) grown in lysimeters (Longenecker, 1973, 1974). In *G. barbadense* cvs. Giza 75, Giza 77, and Giza 80, Nawar *et al.* (1994) found

that salinity (NaCl alone or mixtures of NaCl and $CaCl_2$ or NaCl and KCl) delayed the appearance of the first flower and reduced the numbers of fruiting branches, flowers, and bolls and reduced boll and seed size and lint percentage. NaCl alone was more detrimental than mixtures of salts.

Muhammed and Makhdum (1973) and Ahmed (1994) reported a decrease in the oil content of cotton seed with increasing soil salinity. In contrast, Abdullah and Ahmad (1986) reported increased oil contents when salinity was increased gradually, but decreasing contents after a salt shock (*i.e.* a sudden increase in salt concentration). In both diploid and tetraploid cottons, Ahmad and Abdullah (1980) found increased oil contents at low concentrations of seawater salts (4,000 to 12,000 ppm), but decreases at higher salinities (>16,000 ppm).

5. PHYSIOLOGY

The possible causes of salt-induced growth reduction have classically been identified as osmotic effects (water-deficit stress as a result of inadequate osmotic adjustment), nutritional imbalance and toxicity of Na^+ or Cl^- to metabolism. More recent evidence points to other possibilities such as alteration of cell wall extensibility (Pritchard *et al.*, 1991) and accumulation of salts in the apoplast (Flowers *et al.*, 1991; but see also Mühling and Läuchli, 2002), and imposes a more rigorous time frame on these theories (Munns, 1993). Brugnoli and Björkman (1992) reported that a large part of the reduction in shoot growth of cotton at moderate salinity was caused by increased assimilate allocation to the roots at the expense of leaf growth.

5.1 Photosynthesis

In cotton it is clear that, as in kenaf (Curtis and Läuchli, 1986), reduced leaf area resulting from impaired cell extensibility is initially more important than lower photosynthetic capacity per unit surface area (Fig. 13-2). Rehab and Wallace (1979) found no decrease in photosynthesis or transpiration in *G. hirsutum* and *G. barbadense* in greenhouse experiments when salt was added to Yolo loam soil, but in other experiments at higher salinities, photosynthetic rates were reduced (Brugnoli and Lauter, 1991). The reduction in photosynthetic rate can largely be attributed to reduced stomatal opening since sub-stomatal CO_2 partial pressures were only slightly reduced (from 277 μbar in the controls to 212 μbar at 55% seawater) and carbon isotope discrimination decreased with salinity (Brugnoli and Lauteri, 1991; Brugnoli and Björkman, 1992). Gale *et al.* (1967) reported much larger effects of salinity on stomatal aperture than on net photosynthesis, but their transpiration measurements were not consistent with their stomatal apertures. Plaut (1989) also found stomatal conductance to be more sensitive to salt, up to a certain level, than CO_2 fixation, but the reverse was true for water-deficit stress.

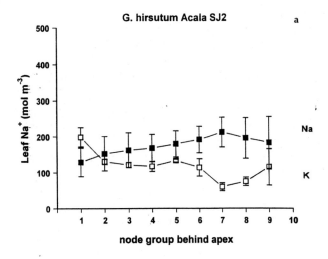

G. hirsutum Acala SJ2

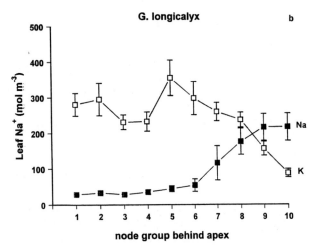

G. longicalyx

Figure 13-2. Leaf Na+ concentration profiles
along the stems of a) *G. hirsutum* cv. Acala SJ2, and
b) *G. longicalyx* at 200 mol m^{-3} NaCl + 10 mol m^{-3} CaCl$_2$.
The main-stems, which were 0.6 and 1.7 metres long
with 11 and 70 nodes, respectively, were divided
into 10 segments, starting at the apex. The leaves
nearest the apical end of each segment were harvested
in each species. (Gorham and Bridges, unpublished)

Maintenance of quantum yield, measured either as O$_2$
evolution per mol of light (Brugnoli and Lauteri, 1991;
Brugnoli and Björkman, 1992) or chlorophyll fluorescence
transients in illuminated leaves (Gorham, unpublished),
suggests that there is little impact of salinity up to about
250 mol m^{-3} NaCl on the photosynthetic machinery *per
se*. Chlorophyll fluorescence Fv/Fm ratios in dark-adapted
leaves and O$_2$ evolution in CO$_2$-saturated leaf discs were
similarly unaffected by salinity. Longstreth and Nobel
(1979a) found that up to 100 mol m^{-3} NaCl had little effect on stomatal or mesophyll resistance, but there were
considerable increases in resistance at 200 and 300 mol m^{-3} NaCl. Furthermore, mesophyll cell surface area per unit
leaf surface area increased considerably with salinity, and
hence there was an even greater increase in resistance to
CO$_2$ diffusion per unit mesophyll cell surface area. Jafri and
Ahmad (1995) reported that salt stress decreased stomatal

density, but the size of the stomata and the mesophyll surface area increased.

Brugnoli and Björkman (1992) observed a decrease in
leaf nitrogen content (on either a leaf area or dry weight basis) in response to salinity, and ascribed the increase in mesophyll resistance to CO$_2$ fixation to decreased allocation to
enzymes of carbon fixation. In contrast, Plaut and Federman
(1991) reported higher protein contents per unit leaf area in
salt-adapted Acala SJ2, and attributed higher CO$_2$ fixation
rates in acclimated than non-acclimated plants to higher ribulose bisphosphate carboxylase activity. Specific activity
of ribulose bisphosphate carboxylase decreased, however,
above -0.3 MPa external salinity. Russian workers have
found that two lines of cotton resistant to chloride salinity had higher phosphoenolpyruvate-carboxylase activities
(Shin *et al.*, 1991). Changes in photorespiration in salt-
stressed cotton have received less attention (Mert, 1989).

El-Sharkawi *et al.* (1986) reported that osmotic stress
had a greater effect than sodicity (irrigation water of 20%
sodium absorption ratio, SAR) on chlorophyll a and b contents. In both *G. arboreum* and *G. hirsutum*, Ahmad and
Abdullah (1980) found that low concentrations of seawater
salts increased chlorophyll contents, but that there was a
reduction above a salt concentration of 16,000 ppm. In contrast, Brugnoli and Björkman (1992) found that salinity increased pigment concentrations on a leaf area basis, mainly
as a result of increased mesophyll thickness. Increased chlorophyll contents per m^2 of leaf were also reported by Leidi
et al. (1992) and Plaut and Federman (1991). Other studies also show increased leaf thickness as a result of salinity
(Longstreth and Nobel, 1979a). In the study of Brugnoli
and Björkman (1992) salinity had little effect on the epoxidation state of the xanthophyll cycle.

Whereas Brugnoli and Björkman (1992) observed improved instantaneous water use efficiency and increased
stable carbon isotope discrimination in Acala SJC-1 grown
in diluted (26% or 55%) seawater, Hoffman and Phene
(1971) reported increased respiration and decreased instantaneous water use efficiency in Acala SJ-1 subjected to
NaCl-salinity (-0.64 or -1.24 MPa). The differences may
be attributed in part to the different salt mixtures used, although Plaut and Federman (1991) also reported improved
instantaneous water use efficiency with NaCl salinity. In
the study of Hoffman and Phene (1971), osmotic adjustment was observed at -0.64 MPa osmotic potential in the
medium, but was incomplete at -1.24 PMa. Brugnoli and
Björkman (1992) found that leaf Na$^+$ concentrations increased to match the increase in external Na$^+$ at -0.58 MPa
(26% seawater), but the increase in leaf Na$^+$ was more than
that in external Na$^+$ at -1.56 MPa (55% seawater). In the
case of seawater salinity (-2.2 to -2.8 MPa), Ca^{2+} is also
likely to contribute to osmotic adjustment (see below).

The importance of acclimation to salt stress was examined by Plaut and Federman (1991). They subjected cotton
plants to either slow or rapid salinization and found that
plants adapted (acclimated) to slowly increased salinity had

higher CO_2 fixation rates than those subjected to rapid (1 to 2 days) salinization, especially at -0.6 to -0.9 MPa. Some over-adjustment of leaf osmotic potentials was observed in acclimated plants, while non-acclimated plants did not achieve full osmotic adjustment and lost turgor above -0.3 MPa. Na^+ and Cl^- accounted for most of the osmotic adjustment in these plants, grown without additional Ca^{2+}.

5.2   Inorganic Ions

Changes in inorganic ion concentrations in cotton leaves in response to salinity have been widely reported (Lashin and Atanasiu, 1972: Läuchli and Stelter, 1982; Thomas, 1980; Rathert, 1982a; Silberbush and Ben-Asher, 1987; Lauter *et al.*, 1988; Plaut and Federman, 1991; Shimose and Sekiya, 1991; Leidi *et al.*, 1991, 1992; Martinez and Läuchli, 1991,1994; Brugnoli and Björkman, 1992; Gouia *et al.*, 1994; Jafri and Ahmad, 1994). Where such data are reported on a dry weight basis it is difficult to relate ion accumulation to osmotic adjustment since the water content is not known, and can often change with salinity (Brugnoli and Björkman, 1992). Leidi and de Castro (1997) found that osmotic adjustment was mainly related to Na^+ accumulation in a salt-tolerant cotton genotype.

What is clear from the published data and our own observations is that the responses of shoot inorganic ion concentrations to salinity are very varied, and apparently determined in part by cultural conditions (balance and concentrations of salts, temperature, humidity, light etc.). The responses range from massive increases in leaf concentrations of Na^+ and Cl^- and decreases in K^+ , Ca^{2+} and Mg^{2+} (Rathert, 1982a, 1983) to slight increases in K^+ and modest accumulation of Na^+ (Ahmad and Abdullah, 1980; Jafri and Ahmad, 1994). In moderately saline (up to 15 dS m^{-1}) field conditions the accumulation of Na^+ in young cotton leaves can be quite modest, with Ca^{2+}, Mg^{2+}, and K^+ concentrations higher than those of Na^+, with Ca^{2+} dominant (Gorham, unpublished observations; Thomas, 1980). Long-term field studies on the use of saline drainage water for irrigation of cotton showed that saline treatments resulted in Na^+ accumulation in the leaves, but decreases in leaf K^+ and Ca^{2+} concentrations (Goyal *et al.*, 1999a). In older leaves, at least within the vacuole, considerable substitution of one cation for another can be tolerated (see below). Joham (1986) reviewed the evidence that Na^+ stimulated the yield of K^+-deficient cotton. For a detailed review of mineral acquisition in relation to salinity see Grattan and Grieve (1992). An example of the changes in leaf solute concentrations with salt treatment is given in Table 13-1. The role of mineral nutrition in cotton is discussed in Chapters 9 and 24.

5.3.1   K/Na Ratios

Stelter *et al.* (1979) and Läuchli and Stelter (1982) reported that the most salt-tolerant cultivar they tested, Tamcot SP 37, had lower K^+/Na^+ ratios than the salt-sensi-

Table 13-1. Leaf sap concentrations of Acala SJ2 grown in hydroponic culture with or without salt (300 mol m^{-3} NaCl + 15 mol m^{-3} CaCl$_2$).

Parameter	Control	Salt
(mol m^{-3} leaf sap)	(0 mol m^{-3} NaCl)	(300 mol m^{-3} NaCl)
Sodium	1.9	161.2
Potassium	197.6	166
Magnesium	33.4	30.5
Calcium	35.5	100.7
Chloride	33.9	551.0
Nitrate	125.0	28.5
Phosphate	25.5	27.7
Sulphate	39.7	25.8
Glycine betaine	13.1	44.9

tive cultivar Deltapine 16. Similar observations were made by Leidi and Saiz (1997) who found higher accumulation of Na than K in seedlings of a salt-tolerant cotton cultivar. In contrast, Nawaz *et al.* (1986) obtained a reasonably good correlation (r=0.76) between dry matter yield and shoot K^+/Na^+ ratio in 14 cultivars of cultivated cotton. Rathert (1982a) also reported that the salt-tolerant cultivar of *G. barbadense*, Giza 45 (El-Zahab, 1971b), had less Na^+ and more K^+ in its leaves than the more sensitive Dandara. The exclusion of Na^+ was associated with salt tolerance in four Indian cotton cultivars (Janardhan *et al.*, 1976), but Jafri and Ahmad (1994) reported no clear correlation between ion accumulation and relative salt tolerance of four upland cotton cultivars. Genotypic variability detected in K+ and Na+ concentration in seedlings did not show a strong correlation with salt-stress tolerance (Leidi *et al.*, 2000).

High K/Na ratios and low Cl^- concentrations found in cotton seeds on salt-grown plants (Abdullah and Ahmad, 1986) are an indication of low retranslocation of Na^+ and Cl^- in the phloem. Eaton and Bernardin (1964) obtained very different K/Na ratios in shoots of cotton grown in soil or water culture. In soil the accumulation of Na^+ was less than in water culture, and shoot K^+ concentrations were higher.

5.3.2   Potassium Uptake

Uptake of K^+ from nutrient solution is generally reduced at all external K^+ concentrations by salinity (*e.g.*, Silberbush and Ben-Asher, 1987). Gouia *et al.* (1994) concluded that salinity induced a general decrease in all fluxes (uptake, xylem and phloem movement) involved in the partitioning of K^+ and other inorganic ions in cotton. Läuchli (1999) has emphasized the crucial role that K^+ acquisition plays in salt tolerance of crop plants.

5.3.3   Calcium

Na^+ displaces Ca^{2+} from the plasma membrane of cotton root cells (Cramer *et al.*, 1985) and allows greater efflux of K^+. Increasing the Ca^{2+}:Na^+ ratio maintains membrane integrity and K^+/Na^+ discrimination. Increasing salinity up to 100 mol m^{-3} NaCl decreased the rate of influx of Ca^{2+} into Acala SJ-2 seedlings (Cramer *et al.*, 1987), but at higher salt concentrations Ca^{2+} influx was increased. The effect was greater at 10 than at 0.4 mol m^{-3} CaCl$_2$ and Ca^{2+} influx was restored to control values at 250 mol m^{-3} NaCl at the higher CaCl$_2$ concentration. Zhong and Läuchli (1994) found that root tissue osmotic potential, although lowered by NaCl, was uniform throughout the growing zone of roots with or without supplementary Ca^{2+}. Salt reduced the rates of deposition of K$^+$ and Ca^{2+} throughout the growing zone, and 10 mol m^{-3} CaCl$_2$ mitigated the effect on K$^+$ deposition only in the apical 2.5 mm, and enhanced K/Na selectivity in the apical 2 mm. This is consistent with Ca^{2+} affecting plasma membrane K/Na selectivity, maintaining higher K$^+$ concentrations in meristematic (*i.e.* mainly cytoplasmic) regions of the root. Thus one possible mechanism by which supplemental Ca^{2+} alleviates the inhibitory effects of high Na$^+$ on root growth of cotton is by maintaining the plasma membrane selectivity of K$^+$ over Na$^+$ (Zhong and Läuchli, 1994). While Ca^{2+} uptake from solutions of low (<1 mol m^{-3}) Ca^{2+} concentration is inhibited by Na$^+$, many soils contain higher concentrations, and the extent of Ca^{2+} influx is determined by the absolute amount of Ca^{2+} as well as the Na$^+$: Ca^{2+} ratio and other factors such as transpiration rates.

The concentration of external Ca^{2+} used in different studies varies considerably, making comparisons of ion concentrations in response to salinity difficult. In solution culture experiments where plants were grown with 100 mol m^{-3} NaCl and increasing concentrations of CaCl$_2$, tissue Ca^{2+} was inversely related to Na$^+$ concentrations (Joham and Calahan, 1978). Increasing CaCl$_2$ reduces the effect of salinity on Na$^+$ and K$^+$ concentrations, but at higher concentrations results in reduced growth (Calahan and Joham, 1974) and excessive accumulation of Ca^{2+} and Cl$^-$ (Gorham and Bridges, 1995). The reduction in shoot fresh weight of the cultivar NIAB-78 with salinity was less when an equimolar mixture of NaCl and CaCl$_2$, or when a mixture of NaCl, Na$_2$SO$_4$, CaCl$_2$, and MgCl$_2$ (10:5:4:1) was used instead of NaCl alone (Nawaz *et al.*, 1986). Sulphate, rather than chloride, salinity results in higher concentrations of Na and lower concentrations of K, Ca and Mg as percent dry matter (Shimose and Sekiya, 1991). Treatment with seawater increased leaf Na$^+$, Ca^{2+}, Mg^{2+}, and Cl$^-$ but decreased K$^+$ in the experiment of Iyengar *et al.* (1978). Early literature and experiments on the effects of Ca^{2+} were reviewed by Azimov (1973). At present, the role of Ca^{2+} in plants under salinity stress is thought to be related mainly to membrane stabilization and intra-cellular signalling activity (Bressan *et al.*, 1998).

5.3.4   Nitrogen

High levels of NaCl salinity (nutrient solution with an osmotic pressure of -1.2 MPa) reduced ^{15}N (provided as ^{15}NH$_4$NO$_3$) uptake in cotton, whereas low or medium levels

(-0.4 and -0.8 MPa) did not have a significant effect on the ^{15}N absorption rate (Pessarakli and Tucker, 1985a). High salinity reduced incorporation of ^{15}N into proteins but increased the accumulation of amides, amino acids and total soluble-N in cotton shoots (Pessarakli and Tucker, 1985b). The synthesis of specific proteins induced by salinity was investigated by Momtaz *et al.* (1995).

Brugnoli and Björkman (1992) reported that total N decreased from 49.5 mg g^{-1} DW in controls to 32.4 mg g^{-1} DW at 55% seawater, whereas Silberbush and Ben-Asher (1987) found that salinity slightly increased the Michaelis-Menten coefficient for nitrate uptake, but otherwise had little effect on total N concentrations. In a variety of different experiments, and using a variety of methods for measuring N, it was found that N as a percentage of dry weight was usually higher in salt-treated cotton than in the controls (Esparon, Shafi, Armeanu, Memon, and Gorham, unpublished). This suggests that, although uptake per plant or unit root weight might be reduced, there is no real deficiency of N in salt-stressed cotton plants, up to at least 300 mol m^{-3} NaCl. Competition for uptake between chloride and nitrate has been reported in other crops (*e.g.* in tomato, *Lycopersicon esculentum* M.N., Kafkafi *et al.*, 1982). Increasing soil salinity decreased N, P, and K$^+$ uptake in pot experiments (Subbaiah *et al.*, 1995).

In vitro nitrate reductase activity showed similar reductions with increasing salinity of the assay medium for enzyme extracted from salt-sensitive bean and salt-tolerant cotton (Gouia *et al.*, 1994). However, *in vivo* activity was reduced in bean, but not in cotton, by 50 mol m^{-3} NaCl in the nutrient solution. This difference was attributed to better sub-cellular compartmentation of Na$^+$ and Cl$^-$ in cotton.

5.3.5   Phosphorus

Salinity (150 mol m^{-3} NaCl) reduced ^{32}P translocation from root to shoot and recirculation from cotyledons to young leaves of Acala SJ-2 (Martinez and Läuchli, 1991, 1994). Some of the effects of salinity on growth and ^{32}P accumulation and recirculation could be overcome by adding 10 mol m^{-3} CaCl$_2$ with the salt (Martinez and Läuchli, 1991). In older leaves total P was higher in salt-treated plants, but in young leaves there was less P than in unsalinized plants. Inhibition of ^{32}P uptake by salinity was more severe at low P concentrations than at 1 mol m^{-3} P in the nutrient solution. High NaCl inhibited ^{32}P uptake in the mature root zone, but enhanced uptake at the root tip (Martinez and Läuchli, 1994). Cultivar differences in the degree of inhibition of ^{32}P incorporation into organic phosphorus compounds by salt were reported by Nazirov *et al.* (1981), with cultivar 1306DV being much more affected than *G. hirsutum* ssp. *mexicanum*. Rathert (1983) reported that salt induced considerable increases in leaf phosphorylase activity in *G. barbadense*, although this was not observed in an earlier experiment (Rathert, 1982b).

5.3.6 Membrane Transport

It is accepted that a plasma membrane H$^+$-ATPase generates the potential required to drive in positive ions into the cell. The uptake then occurs by means of channels or more specific ion transporters. At least three types of K$^+$-uptake transporters has been described in plant root cells, showing different degrees of specificity (or discrimination capacity) for K$^+$ and Na$^+$ (Rodriguez-Navarro, 2000). In vacuolar membranes, another H$^+$-ATPase creates the potential required for the uptake of K$^+$ and/or Na$^+$ into the vacuole by the activity of transporters like a Na$^+$/H$^+$ antiport, which would regulate cytoplasmic Na$^+$ sequestering it into the vacuole. The proton-transporting ATPase of cotton root membrane vesicles was stimulated *in vitro* by K$^+$ in the experiments of Hassidim *et al.* (1986), but growth in 75 mol m^{-3} NaCl had no effect on the activity or other characteristics of this pump. This is in contrast to the halophytic chenopod *Atriplex nummularia*, where saline conditions were necessary for full activity (Braun *et al.*, 1986). Na$^+$/H$^+$ and K$^+$/H$^+$ antiporters were found in these two species, probably in both the plasmamembrane and the tonoplast (Hassidim *et al.*, 1990). Lin *et al.* (1997) found that cotton roots grown in 75 mol m^{-3} NaCl had increased plasmamembrane ATPase activity, but that vacuolar ATPase was not affected. The comparison with similar experiments using salt-tolerant lines in combination with studies with transporters of the plasmamembrane and tonoplast would provide a better understanding of the mechanisms leading to higher salinity tolerance. The manner in which different transport systems identified so far is integrated in the ion uptake process by plants cells is unknown (Rodriguez-Navarro, 2000) and at present work on ATPases and membrane transport provides no more clues to the mechanisms of salt tolerance and susceptibility in cotton than other physiological studies. ATPase activity in plasma membranes of three-day-old seedlings of cv. 108-F was stimulated by Mg^{2+} and K$^+$, but the activity was reduced by NaCl (Ivleva and Plekhanova, 1992).

5.4 Organic Solutes

In *G. barbadense* cultivars Dandara and Giza 45 carbohydrate accumulation was associated with high K/Na ratios (54/6) of chloride and, particularly, sulphate salts (Rathert, 1982b). Total carbohydrates increased slightly in roots and shoots of these two cultivars subjected to salt (20 mol m^{-3} KCl + 180 mol m^{-3} NaCl) stress, although changes in individual components (glucose, fructose, sucrose, and starch) were not consistent (Rathert, 1982a, b, 1983). Concentrations of sucrose, glucose, amino acids and proline increased in cotton, particularly when subjected to a gradual increase in salinity (Plaut and Federman, 1990). In this case proline concentrations increased from about 2 mol m^{-3} in the controls to 9 mol m^{-3} at 0.9 MPa. Lin *et al.* (1995) also obtained increases in proline concentrations in salt-stressed cotton seedlings. Kuznetsov *et al.* (1991) reported about 4 mmol g^{-1} DW in leaves of cotton cv. 149-F

leaves subjected to acute salt shock, but less in cvs. INEBR-85 and 133. In contrast, plants treated with heat shock followed by prolonged (40 days) drought stress accumulated up to 75 mol m^{-3} proline (Kuznetsov *et al.*, 1999). Ahmad and Abdullah (1980) reported increases in free proline concentrations in *G. hirsutum* and *G. arboreum* in response to seawater salinity, although the reported units (mg/g fresh weight) are improbable. Proline contents of cotyledons was higher in salt-resistant than salt-sensitive cotton cultivars treated with NaCl (Ivanina, 1991). A small increase in the low concentration of proline (about 1 mol m^{-3}) in Acala SJ-2 in saline hydroponic culture was observed (Gorham, unpublished). Shevyakova *et al.* (1998) reported an inverse correlation between the salt-tolerance and NaCl-induced proline accumulation in intact plants and cell cultures. Tolerance of intact plants to NaCl did not determine the salt-tolerance of their isolated cells, and the accumulation of proline in leaves was of no importance for adaptation or survival of the plants. No consistent changes in cell wall proline or hydroxyproline in response to salinity were found by Golan-Goldhirsh *et al.* (1990). Thus the concentrations of proline accumulated under salinity are very low in relation to whole plant osmotic adjustment, and could only have a significant physiological effect if proline was confined to the cytosol. However, salinity (200 mol m^{-3} NaCl + 10 mol m^{-3} CaCl$_2$) and water stress considerably increased the concentrations of glycine betaine in a cultivar of *Gossypium* genotypes, to around 100 mol m^{-3} in some cases (Gorham, 1996b). Osmotic considerations suggest that even in non-stressful conditions the large amount of glycine betaine present in cotton must be contained mainly within the vacuoles.

5.5 Distribution of Solutes Within Tissues

Limited accumulation of ions in young leaves was suggested as an adaptive response and correlated with the salt tolerance of cotton cultivars (Sokolova *et al.*, 1991). Slama (1991) concluded that stems did not protect leaves from Na$^+$ accumulation by extracting it from the xylem and acting as Na$^+$ traps. In a cultivar of cotton genotypes grown at 250 mol m^{-3} NaCl, Na$^+$ concentrations were highest in the stems, while K$^+$, Cl$^-$, and NO$_3^-$ were the most abundant ions in petioles. Leaves had higher concentrations of Mg^{2+}, Ca^{2+}, SO$_4^{2-}$, and malate (Gorham and Bridges, unpublished). Young leaves had higher concentrations of K$^+$ and glycine betaine, and lower concentrations of Na$^+$, Ca^{2+}, and Cl$^-$ than old leaves. This can largely be explained by the different ionic compositions of cytoplasm and vacuole, and the changing proportions of these compartments as tissues age. Profiles of leaf ion concentrations in *G. hirsutum* cv. Acala SJ2 and in *G. longicalyx* are shown in Fig. 13-2. Gradual replacement of K$^+$ by Na$^+$ in older leaves was also reported by Leidi and Saiz (1997).

The above observations also explain the differences in K$^+$/Na$^+$ ratio between leaf blades and petioles in cotton grown with saline irrigation water reported by Läuchli (1999). The K^+/Na^+ ratio did not change much in the leaf blades with time of exposure to salinity, but there was a decrease in the ratio in petioles as Na^+ accumulated and K^+ was recirculated to younger tissues, especially at lower salinities (400 to 3,000 ppm).

5.7   Antioxidants

In Chinese cotton cultivars Li *et al.* (1994, 1998) observed increased activities of peroxidase and superoxide dismutase in salt-stressed roots, but decreased activities in cotyledons. The difference in salt tolerance between a hybrid and its maternal parent was related to the increase in peroxidase activity but not to the accumulation of salt ions (Li *et al.*, 1998). However, an abrupt reduction in root antioxidant activity on exposure to salt was found by Kasumov *et al.* (1998). They also reported increased root respiration and uncoupling of respiratory oxidation and phosphorylation. Gossett *et al.* (1992, 1994a) and Lucas *et al.* (1993) found that a number of Acala cultivars (1517-88, 1517-91, 1517-SR2, and 1517-SR3) were more salt-tolerant than Deltapine 50 or Stoneville 825. The Acala lines had higher levels of enzymes such as catalase, peroxidase and glutathione reductase which are involved in protection against oxidative stress. Increased activity of these enzymes, and of superoxide dismutase and ascorbate peroxidase, were also found in callus tissue of Acala 1517-88 exposed to salt, but not in tissue of Deltapine 50 (Gossett *et al.*, 1994b, 1997, 1999). Ovule cultures from the more salt-tolerant cultivar Acala 1517-88 showed increased catalase and glutathione reductase activities when the salinity was increased from 75 to 150 mol m^{-3} NaCl, while those from the salt-sensitive Coker 312 did not (Rajguru *et al.*, 1996; Banks *et al.*, 1997). Salt treatment (100 mM NaCl) reduced ovule weight in all cultivars tested (Deltapine 50, MAR LBCBHGDPIS-1-91, Coker 312) except Acala 1517-88, and increases in peroxidase and superoxide dismutase activities were only recorded in Acala 1517-88 and MAR LBCBHGDPIS-1-91 (Rajguru *et al.*, 1999). A salt-tolerant cell line derived from a culture of Coker 312 callus showed greater activities of catalase, peroxidase, glutathione reductase, γ-glutamylcysteine synthetase, and glutathione-S-transferase than the original callus, and was more resistant to oxidative stress induced by paraquat (Gossett *et al.*, 1995, 1996; Lucas *et al.*, 1996). Increasing salinity resulted in more glutathione-S-transferase activity in callus cultures of Acala 1517-88 and other cultivars (Fowler *et al.*, 1996, 1997; Rajguru *et al.*, 1999). These results point to the possible involvement of antioxidation systems, and particularly the ascorbate-glutathione cycle, in resistance to salt toxicity. Construction of an *Agrobacterium*-mediated transformation vector containing sense and anti-sense glutathione reductase cDNA (Banks *et al.*, 1994) allowed the role of glutathione reductase in salt tolerance of cotton to be investigated (Banks *et al.*, 1999). Inhibitor studies showed that ascorbate oxidase and glutathione reductase activities increased because of *de novo* transcription of the genes (Banks *et al.*, 2000; Manchandia *et al.*, 1999).

5.8   Plant Growth Substances

The involvement of plant growth substances in the response of cotton to salinity has received little attention. Kefu *et al.* (1991) reported that transpiration was reduced and leaf, root, and xylem concentrations of abscisic acid (ABA) increased in cotton plants subjected to 75 or 150 mol m^{-3} NaCl. The ABA concentrations were higher after two days of salt treatment than after six days and decreased on transfer to non-saline conditions. ABA concentrations in seeds increased when they were soaked in salt solutions (Mert, 1993). Gadallah (1995) suggested that spraying cotton with proline or ABA reduced the effects of water stress, but did not examine responses to salt stress. ABA seems to be associated with the salt-induced increase in glutathione reductase activity in cotton (Gossett *et al.*, 2000).

Spraying salinized *G. barbadense* cultivars with GA_3, alone or in combination with boron, increased growth and monovalent cation contents, but decreased chloride content (Ibrahim, 1984). GA_3 alleviates the detrimental effects of salt on photosynthetic pigments in seedlings (Renu *et al.*, 1995). Heat shock (3 hours at 47°C) increased the tolerance of cotton to subsequent Na_2SO_4 salinity (Kuznetsov *et al.*, 1990, 1993), and the response was linked to changes in the production, at different times, of ethylene, proline and putrescine (Kuznetsov *et al.*, 1991).

Kinetin counteracts the inhibitory effects of salt on germination in cv. Carolina Queen (Bozcuk, 1990). Treatment (of imbibing seeds and/or as a foliar spray) with the cytokinin analog MCBuTTB improved the germination, growth, and yield of Paymaster 145 plants subjected to salinity (Stark, 1991). However, the effect was also observed in plants subjected to a mannitol-induced osmotic stress. MCBuTTB treatment resulted in ^{22}Na accumulation in the veins of leaves (Stark and Schmidt, 1991). Another cytokinin analogue, polystimuline K, alleviated the effects of salt on photosynthetic activity in two week old cotton plants (Ganieva *et al.*, 1998). Largely beneficial, but inconsistent, effects of chlorocholine chloride (CCC) treatment, as a pre-germination soak and foliar spray, were reported by Gabr and El-Ashkar (1977). Soaking seeds in 0.001% Na γ-phenylbutyrate increased growth, yield, seed weight, and oil content (Kariev, 1981). In pot and field trials in China, germination and cotton yields in saline soils were improved when seeds treated with vitamin B_6 (Wang *et al.*, 1991a). Various treatments with PHCA (polyhydroxycarboxylic acid) have also been reported to improve yields on saline soils (Munoz, 1994).

6. IRRIGATION

The use of saline irrigation water for cotton cultivation has been investigated by several groups (El-Saidi and Hegazy, 1980; Mantell *et al.*, 1985; Ayars and Schoneman, 1986; Ayars *et al.*, 1986, 1993, 1997, 2000; Rains *et al.*, 1987; Russo and Bakker, 1987; Bajwa et al, 1992; Knapp, 1992; Shennan *et al.*, 1995; Goyal *et al.*, 1999a,b). There is particular interest in cyclical use and blending of saline and non-saline water for irrgatioin (Bradford and Letey, 1992; Naresh *et al.*, 1993; Grattan *et al.*, 1994; Hamdy, 2000). Strategies using saline water alternated with fresh water during drought periods did not significantly affect seedcotton yield (Moreno *et al.*, 1998).

In the Negev region of Israel, higher yields of Acala SJ2 cotton were obtained with local well water with an EC of 3.2 dS m^{-1} than with water of EC 1.0, 5.4, or 7.3 dS m^{-1}, and good yields (>5,800 kg ha^{-1}) were obtained at all salinities (Mantell *et al.*, 1985; Freuke *et al.*, 1986;). A survey of 10 farms in the western Negev growing Acala cotton under trickle irrigation showed higher yields on saline than non-saline fields (Saden, 1992). Seedcotton yields could be increased by increasing plant density, mainly by using narrower rows (Francois, 1982; Keren and Shainberg, 1978; Keren *et al.*, 1983). Ayars *et al.* (1993) and Rains *et al.* (1987) concluded that long-term use of saline water could lead to problems of accumulation of salt, boron, or selenium in the soil. Changing the K:Na ratios of irrigation water from 1:9 to 1:4 did not improve yields of Egyptian cotton (Abd-Ella and Shalaby, 1993). In Pakistan furrow irrigation rather than flat bed flood irrigation improved salt movement in soil and was recommended for saline areas (Choudhry *et al.*, 1994), whereas in Tamil Nadu (India) Muthuchamy and Valliappan (1993) advocated the use of drip irrigation. Modelling, both of the cotton crop (Plant *et al.*, 1998) and of the soil (Singh and Singh, 1996; Smets *et al.*, 1997) is seen as a means of improving the efficiency of use of irrigation water.

6.1 High Water Tables

Since, with few exceptions (Khaddar and Ray, 1988; Ray and Khaddar, 1983, 1993; Bandyopadhyay and Sen, 1992), cotton cultivars are not tolerant of prolonged waterlogging, there is an interaction between salinity and the depth of the water table in poorly-drained soils (Hutmacher *et al.*, 1996). Hebbara *et al.* (1996) suggested thresholds of 5 to 6 dS m^{-1} soil salinity and 90 to 95 cm water table depth for cotton on medium-depth vertisols, and found the cultivar Sarvottam to be best out of 18 cultivars with shallow water tables. Yields of cotton cv. Menoufy increased with water table depth in the experiments of Moustafa *et al.* (1975). Pima cotton yields were increased with 2 or 3 irrigations after sowing on saline soil with a shallow water table in California (Munk and Wroble, 1995). Deficit irrigation was studied by Cohen *et al.* (1995), but reduced irriga-

tion did not stimulate root growth into deeper and wetter soil layers. A deeper water table increased seedcotton yield and fiber strength and length uniformity (Mohamed *et al.*, 1997).

6.2 Sprinkler Irrigation and Foliar Uptake of Salt

The effect of sprinkling with saline irrigation water on growth, and foliar uptake of salt, have been examined by a number of workers. Busch and Turner (1965, 1967) observed greater reduction in yield of cotton sprinkled with saline water during the day than when the sprinkling was done at night, or the water was applied in the furrows. On the other hand, Meiri *et al.* (1992) did not find any difference in dry matter production between drip and sprinkler irrigation at the same irrigation water salinity. Maas *et al.* (1982) found that sprinkling cotton (Deltapine cultivars) with 30 or 60 meq l^{-1} (ECe 3.4 and 6.5 dS m^{-1}) saline water (NaCl:CaCl$_2$ = 9:1) for 6 weeks (5 days per week) resulted in increased shoot fresh and dry weights compared to unsprinkled controls, despite some leaf tip and margin necrosis at the higher salt concentration. K$^+$ concentrations in the leaves decreased by 20% at 30 meq l^{-1}, while increases in Na$^+$ and Cl$^-$ concentrations were intermediate between those of tomato (greatest increase of 10 crop species tested) and sorghum (least increase). Cotton ranked with cauliflower, sugarbeet, and sunflower as being the most resistant of 20 crops to foliar salt injury (Maas, 1985). The amount of salt taken up through the leaves may be related to the length of time the salt stays in solution (Hofmann *et al.*, 1987), as was reported for other crops by Grattan *et al.* (1981). Memon and Gorham (unpublished) observed greater accumulation of Na$^+$ (and concomitant reduction in K$^+$) concentrations in leaves of cotton when the same concentration of salt was sprayed on the leaves than when it was applied to the soil.

7. OTHER ENVIRONMENTAL FACTORS

A number of environmental factors affect the response of cotton to salinity. Factors which increase transpiration - such as low atmospheric humidity and high temperature - increase sodium and chloride accumulation in the plant and decrease growth (Gale *et al.*, 1967; Hoffman *et al.*, 1971; Nieman and Poulsen, 1967). Leidi *et al.* (1991, 1992) found that cotton grew better with nitrate rather than ammonium as nitrogen source, but control (*i.e.,* in the absence of salt) carbon dioxide exchange and transpiration rates were higher with ammonium. In one study, stomatal conductance in control plants was much higher in ammonium-fed cotton, but lower than in nitrate-fed cotton when the plants were grown at 100 mol m^{-3} NaCl (Leidi *et al.*, 1991). In the later study (Leidi *et al.*, 1992) stomatal conductance was, with one exception, lower in ammonium-fed than nitrate-fed plants at all salinities. Lower K^+ uptake in ammonium-fed plants might be one of the reasons for higher sensitivity to salt in cotton (Leidi *et al.*, 1991). Cotton plants grown with NH_4^+ as N source had higher concentrations of Cl^- in leaves and stems than NO_3^--grown plants. Calcium concentrations in leaves, stems and roots were lower in NH_4^+-fed plants than in NO_3^--fed plants (Leidi *et al.*, 1991). Chlorophyll contents per m^2 leaf area were higher in ammonium-fed plants. McMichael *et al.* (1974) also found that growth was lower in ammonium-fed plants grown in the absence of salt. Total plant dry weight was highest with equimolar ammonium and nitrate, while seedcotton yield was highest at a 3:9 ratio. Increasing application of nitrogen fertilizer can reduce the apparent salt tolerance of cotton (*G. barbadense*) and corn (Khalil *et al.*, 1967), while salinity reduces the response to nitrogen fertilizer (Amer *et al.*, 1964; Gabr *et al.*, 1975).

A number of soil amendments have been reported to increase cotton yield on saline soils (Tiwari, 1994; Tiwari *et al.*, 1993, 1994a,b). These include addition of sand (Mathur *et al.*, 1983) and intercropping with *Sesbania* (Pan *et al.*, 1993). Mathur *et al.* (1983) reported that addition of gypsum had no effect on cotton yields.

8. NOVEL SOURCES OF SALT RESISTANCE FOR BREEDING

8.1 *GOSSYPIUM* SPECIES

Although intra-specific variation in *G. hirsutum* has long been exploited in the breeding of new cotton cultivars (Lee *et al.*, 1991; Liu *et al.*, 1998; Ashraf and Ahmad, 1999), there are also useful traits in other *Gossypium* species. Rana *et al.* (1980) reported greater survival of tetraploid cotton (*G. hirsutum*) than diploid cotton (*G. herbaceum*) in saline (ECe = 8 dS m^{-1}) and sodic (ESP 41, pH 9.4) soil. Seedling growth of *G. barbadense* cultivars was superior to that of lines of *G. hirsutum* and seedlings of both tetraploids were taller and had higher dry weights than seedlings of *G. arboreum*, although performance relative to unstressed controls was not reported (Abul-Naas and Omran, 1974). Similarly, Nawaz *et al.* (1986) found that *G. arboreum* (diploid) cultivars D-9 and Ravi were less tolerant than most tetraploid cultivars tested in hydroponics, soil culture or in the field. In contrast, the reduction in yield was less in D-9 than in the tetraploid cultivar M-100 in the experiments of Ahmad and Abdullah (1980). *G. herbaceum* cv. Sarvottam was less affected by salinity (percent reduction in yield etc.) than *G. arboreum* cv. G cot 15 or *G. hirsutum* cv. Laxmi (Uma and Patil, 1996).

Nazirov (1973) suggested that *G. davidsonii* and *G. hirsutum* ssp. *mexicanum* might be useful in breeding for

salt and drought tolerance, while Liu *et al.* (1993) identified two salt tolerant genotypes of *G. barbadense* and one of *G. herbaceum* in a survey of 4078 accessions of cotton. Of the D-genome diploids, *G. davidsonii*, *G. harknessii*, and *G. klotzshianum* displayed the greatest relative salt tolerance (relative to untreated controls), and *G. thurberi* and *G. trilobum* the least (Gorham and Bridges, unpublished). *G. longicalyx* (genome F) was also more salt tolerant than other species, including *G. hirsutum* Acala SJ2 and MNH 147. Although less vigourous than *G. hirsutum* in the absence of salt, some hexaploid interspecific hybrids between *G. hirsutum* and diploid wild species were relatively more salt-tolerant and had higher leaf K/Na ratios.

8.2 *In Vitro* Selection

Although *in vitro* cultures of cotton cells can tolerate quite high salinities, and selection can be made for salt-tolerant cell lines (Mukhamedkhanova *et al.*, 1991; Wang *et al.*, 1991b; Li *et al.*, 1992; Gossett *et al.*, 1996), the dangers of extrapolating cellular tolerance to the whole plant are well known (Dracup, 1991). Acala 1517-88 was more tolerant than Deltapine 50 both as whole plants and as callus (Millhollon *et al.*, 1993). In contrast, Kerimov *et al.* (1993) and Kuznetsov *et al.* (1991) found that INÉBR-85, a salt-tolerant cultivar of *G. hirsutum*, produced callus which was more sensitive to salt than that of cultivar 133, which was more sensitive as a whole plant. Zhang *et al.* (1993, 1995) reported the regeneration of salt-tolerant plants from callus of Coker 201 selected on medium supplemented with 10 g l^{-1} NaCl.

8.3 Mutation Breeding

Mutation breeding has produced a number of lines with increased salt tolerance, including AN402 and AN403 (Nazirov *et al.*, 1978, 1979). One of the most commercially successful mutant lines in Pakistan is NIAB 78 (Nawaz *et al.*, 1986; Khan *et al.*, 1995a,b,c; 1998a,b), It was produced by gamma irradiation of the F_1 hybrid between AC134 and a Deltapine cultivar (NIAB, 1987, 1992). NIAB 78 is reported to be more salt-tolerant than other cultivars (Nawaz *et al.*, 1986; Malik and Makhdum, 1987; Choudhry *et al.*, 1989; Ahmad *et al.*, 1991; Iqbal *et al.*, 1991), although a mutant of Stoneville-231 (NIAB-92) has higher yields (Iqbal *et al.*, 1994). The salt tolerance of NIAB 78 was associated with greater Na^+ retention in the roots of this cultivar than in more salt-sensitive cultivars.

9. SUMMARY

The effects of salinity on cotton depend on the exact circumstances in which salinity is encountered. Low salinities may result in higher or lower yield, but high salinity is invariably detrimental to cotton. Accumulation of Na^+ in young leaves may be low in field-grown cotton at low salinity where Ca^{2+} is readily available. The common cations (Na^+, K^+, Ca^{2+}, and Mg^{2+}) are to some extent interchangeable as vacuolar solutes, leading to conflicting views on the importance of K/Na ratios for salt tolerance in cotton.

Irrigation with saline water has been used with varying degrees of success for the cultivation of cotton, particularly on the sandy soils of the Negev.

New sources of variation for salt tolerance are being developed from wild *Gossypium* species, from *in vitro* selection, and from radiation breeding.

10   ACKNOWLEDGEMENTS

This document is, in part, an output from projects funded by the Department for International Development (DfID) Plant Sciences Research Programme, managed by the Centre for Arid Zone Studies, University of Wales, Bangor, U.K. for the benefit of developing countries. The views expressed are not necessarily those of DFID.

Chapter 14

PLANT RESPONSES TO MINERAL DEFICIENCIES AND TOXICITIES

Steven C. Hodges[1] and Greg Constable[2]

[1]NC State University, Raleigh, NC; and [2]CSIRO Cotton Research Unit, Narrabri, Australia

1. INTRODUCTION

To minimize crop yield losses from nutrient deficiencies, it is important to understand how mineral nutrient deficiencies affect metabolism, growth and development, and yield components. A better understanding of these interactions will lead to better diagnosis of deficiencies and improved nutrient management practices.

Mineral elements have numerous functions in plants including maintaining charge balance, electron carriers, structural components, enzyme activation, and providing osmoticum for turgor and growth. The effects of mineral nutrient deficiencies range from the dramatic – immediate cessation of root growth or massive disruption of membranes or cell walls – to the very subtle – such as small changes in the pH of the cytosol, reduced export of carbohydrates, or inability of an enzyme to align correctly with a reactant. Yet each of these can result in oxidative stress (photoinhibition and photooxidation), the ultimate destruction of chloroplasts, and the symptoms which we recognize as chlorosis and necrosis.

Deficiencies are affected by uptake, transport, utilization, and storage of nutrients within a wide range of tissues. Some mineral nutrients, such as N, P, K, and Mg, are readily remobilized from storage or metabolic pools, while others are tightly sequestered or excluded from transport in the high pH, highly concentrated, chemical environment of the phloem. To illustrate, about half the N in a cotton leaf at full expansion is lost from that leaf by senescence (Thompson et al., 1976). For mature cotton canopies, redistribution will most likely be to developing bolls. Using the proportion of a mineral removed at harvest to the total taken up as a relative index of mineral mobility, data from a large number of cotton experiments (Hodges, 1992) indicates removal values of 50% for N and P, 28% for Mg and K, 12% for S, and only 5% for Ca. Nutrient mobility and redistribution also affect the location where deficiency symptoms first become visible, and can aid in diagnosing a nutrient problem. For N, P, K, and Mg, deficiency symptoms appear on older leaves, whereas for Ca and most trace elements, the symptoms appear first on young leaves. Distance from the supply path also affects deficiency symptoms. Cells near a vein ending are more likely to be supplied with nutrients, while internal leaf cells some distance from a vascular element are more likely to become deficient since movement through or within cells may be restricted when concentrations are low or when there is a water deficit. Some tightly sequestered nutrients, such as copper, are remobilized from older leaves as they senesce. The mobility and redistribution of mineral nutrients can also influence fertilizer application strategies. For the relatively immobile trace elements, foliar applications may be effective at the time when symptoms appear, but repeated applications may be required.

Finally, we must realize that the demands of expanding tissues (sinks), e.g., shoot tips, fruit, and young leaves, have a high and continual demand for nutrients, whereas fully expanded tissue (sources) tend to balance imports and exports of nutrients. During vegetative growth, the plant partitions available supplies between the demands of expanding growing points, productive leaves, and stems in balance with demands of an expanding root. As squares and bolls begin to form at the base of the plant, these new sinks compete with expanding terminal growth and roots for resources, and greatly complicate the partitioning of nutrients (see Chapter 9).

Visual symptoms are the consequences of metabolic disturbances caused by inadequate levels of mineral nutrients to perform one or more of many essential roles. Although mineral stress may affect many different processes, the ultimate metabolic disturbance can result in a very similar expression of symptoms in the form of chlorosis, necrosis, or other abnormal alterations of the plant. Nonetheless, visible symptoms often provide the first clue of a mineral imbalance, and careful observation can help us to understand the nature of the problem. Visual symptoms reported here are compiled from Donald (1964), Hearn, (1981), Marcus-Wyner and Rains (1982), Grundon (1987), Hodges (1992), and Cassman (1993), and as well as from new observations. Where little is known about the unique responses to a deficiency or toxicity of a particular nutrient in the cotton plant, we have relied heavily on a number of excellent reviews, especially those of Longnecker (1994), Marshner (1995), Terry and Rao (1991), and chapters from the previous volume of this series. Interactions between mineral nutrients are particularly complex to understand and identify with the enormous range of chemical and physical properties of soils used for cotton production. The need for balanced mineral nutrition under such wide-ranging conditions has led Sumner (1977) and others to development of the DRIS system to determine ideal ratios for mineral nutrients in plant tissue for numerous crops. Unfortunately, this work is incomplete for cotton. Only interactions not previously reviewed (Hodges, 1992) are discussed below.

2. NITROGEN DEFICIENCY

2.1 Leaf Symptoms

Nitrogen is one of the most intensely studied of the mineral nutrients because of its numerous effects on plant growth and yield, and the almost universal growth and yield response to annual applications. Nitrogen deficiency affects fundamental processes and components such as water and solute uptake (*i.e.*, hydraulic conductivity), protein metabolism, photosynthesis, carbon partitioning, and enzyme and plant hormonal activity. These responses result in profound changes in growth rate, net photosynthate production, plant development, and yield.

Nitrogen deficiencies beginning early in the growing season may result in short, thin stalks, with shortened petioles, and small, pale green leaves which may hang vertically and fold inward toward the stem (Grundon, 1987). Stalks quickly develop a woody texture. In severe cases, lower stems and leaves may develop red pigmentation (Marcus-Wyner and Rains, 1982). Nitrogen is readily remobilized in the plant, so symptoms appear first, and are most intense, on older leaves. As the deficiency progresses, the oldest leaves will turn pale green, pale yellow, and then begin to develop brown necrotic areas, usually between the veins. Unlike K and Mg, the chlorosis is evenly distributed over the leaf, and necrotic spotting, if present, occurs only in the late stages of a severe deficiency. The youngest leaves will be small but remain green until the deficiency progresses up

the plant. Premature senescence and defoliation of the older leaves follows. Ishag *et al.* (1987) reported reddening of leaves in the middle of the canopy where N deficiency occurred during the boll-filling period on plants with a moderate to heavy boll load.

2.2 Metabolic Responses

Carbon partitioning is strongly affected by N deficiencies, with accumulation of soluble sugars and especially starch commonly observed, even before effects on leaf growth (Rufty *et al.*, 1988a). Recent work shows a strong positive correlation between CO_2 fixation and N content of fully matured, non-senescing leaves in cotton (Reddy *et al.*, 1996a) and a wide range of plant species as long as light is not limiting (see Longnecker, 1994). As N deficiency progresses, net photosynthesis declines (Rufty *et al.*, 1988a), and an increasing proportion of light energy is not used in photochemical reactions. Where light is not limiting, there is a corresponding increase in antioxidative defense mechanisms (Marschner, 1995), resulting in photooxidation of chloroplast pigments, and enhanced leaf senescence. Deficiencies in developing tissue lead to remobilization of reduced N (NO_3-N is not phloem mobile) from older tissue. This premature senescence and reduced leaf area duration (Longnecker, 1994; Marschner, 1995) causes visible leaf symptoms to appear most conspicuously in older leaves.

While N deficiency clearly has direct effects on photosynthesis (*e.g.,* Bondada *et al.*, 1996), the bulk of evidence suggests that decreased growth is the primary effect of N deficiency. Reduced growth under N-deficient conditions may result from decreased cell division (*i.e.*, fewer cells, MacAdam *et al.*, 1989) or limited cell expansion. Several studies document reduced cell numbers resulting from N deficiencies (Longnecker, 1994). In sugar beets, cell number is more strongly affected than cell size, while the opposite is true of epidermal cells in sunflowers (Radin and Parker, 1979).

Cell expansion requires loosening of the cell wall while maintaining protoplast turgor pressure. Radin and Parker (1979) concluded that metabolic functions of N in cell wall loosening were not affected by N deficiency, but found that cell walls of N-deficient cotton leaves were more rigid than those of control plants. This could lead to improved drought tolerance under N stress. Radin and Matthews (1988) found N stress increased the cell turgor and half-time for water exchange of cotton root cortex cells, along with an increased modulus of elasticity of the cell wall. These effects preceded decreases in leaf expansion rate. They concluded that biophysical properties of cell membranes, probably in the endodermis, limit the rate of water movement to the stele of N-stressed cotton roots. Radin and Mauney (1986) noted that N deficiency may alter the fatty acid composition of the root plasma membrane, which in turn may alter hydraulic conductivity. They concluded that reduced hydraulic conductivity and transpiration-generated water deficits inhibited daytime cell expansion resulting in lower leaf area growth rates under N-limited conditions.

In N-deficient plants, growth rate is affected before photosynthesis. Growth is typically inhibited prior to photosynthesis and to a greater extent. In slightly N-deficient, field-grown cotton Wullshleger and Oosterhuis (1990) found that the growth rate of new leaves decreased by 40% within five days of unfolding, and leaf area index decreased within 76 days of sowing, but net photosynthesis was not affected by the deficiency. Under conditions of low N there are notable decreases in stomatal and mesophyll conductance of CO_2 (Radin and Ackerson, 1981; Reddy *et al.*, 1996a), hydraulic conductivity (Radin and Parker, 1979; Radin and Boyer, 1982), leaf osmotic potential, turgor potential, transpiration (Reddy *et al.*, 1996a), leaf expansion, and leaf area. Reddy *et al.* (1996a) found a decrease in the content of sucrose and an increase in starch at low leaf N contents. Carbohydrate accumulation in roots is increased. These responses are very similar to those resulting from water stress (see Radin and Mauney, 1986). Reddy *et al.* (1996a) further determined that water use efficiency (mg CO_2 g H_2O^{-1}) was not affected by the N status of fully mature leaves.

The water potential of N-deficient leaves under mild water deficit drops sooner than adequately supplied leaves, and the conductance of CO_2 decreases as the water potential drops (Radin and Ackerson, 1981; Fernandez *et al.*, 1996a). However, Longstreth and Nobel (1979b) showed that the mesophyll conductance drops more than stomatal conductance in N deficient cotton leaves. These results are consistent with the interpretation that N deficiency decreases photosynthesis by negative feedback or by decreased content or activity of photosynthetic enzymes (Longnecker, 1994).

As reviewed by Radin and Mauney (1986), these similarities between N stress and water deficit allow growth to decrease through the effects of decreased hydraulic conductivity on growth very soon after reduced uptake of N occurs. As a result of reduced growth and sink demand, assimilate accumulates, with increased proportions directed toward the root, thereby enhancing root elongation. The trigger and mechanism for this switch in assimilate pathway remains unclear (see Longnecker, 1994). Assimilates accumulated in leaves are available for osmotic adjustment, and remobilization upon resupply of N. If N supply is permanently reduced, stored assimilate ensures early boll set and plant survival, although yields are reduced.

In addition to effects on growth and carbon assimilation, N deficiency results in decreased synthesis of amino acids, amides, galactolipids (structural) and other soluble N compounds, and decreases synthesis of chlorophyll and carotene (Marschner, 1995). Production of enzymes is also dramatically affected, particularly ribulose-1,5-biphosphate (RuBP) carboxylase (Reddy *et al.*, 1996a), which is a critical compound in the initial phases of CO_2 fixation. Up to 75% of total organic N in green leaf cells is located in chloroplasts, mainly as enzyme proteins. Machler *et al.*, (1988) concluded that photosynthesis is limited by the rate of RuBP carboxylase regeneration under N deficiency. In field studies, Guinn and Brummett (1989) also found that IAA of cotton fruiting branches was lower in N-deficient plants, and this may be a factor in decreased growth, flowering, and boll retention during cutout. Respiration tends to be decreased when N is deficient (Longnecker, 1994)

2.3   Growth Responses

In addition to effects on cell and leaf area expansion rate noted previously, N deficiency results in reduced leaf area index (Thompson *et al.*, 1976; Constable and Hearn, 1981; Jackson and Gerik, 1990; Bondada et al, 1996). The duration of leaf expansion is not affected (Radin and Boyer, 1982). Fernandez *et al.* (1996a), found N deficiency decreased whole plant cumulative leaf area by 40% as a result of reduced growth of main-stem and branch leaves, and decreased final area of individual main-stem and branch leaves. At the same time, N-deficient leaves exhibited delayed and decreased osmotic potential adjustment to water stress, and increased sensitivity of leaf growth inhibition in response to water deficits.

Shoot growth is more dramatically affected by N deficiency than root growth, thus the root to shoot ratio is seen to increase. Constable and Rawson (1982) concluded that greater amounts of ^{14}C applied to individual leaves were translocated to the roots under low N conditions. Fernandez *et al.* (1996a) found that decreased N in pot studies did not affect taproot growth, but lateral root growth was enhanced. The root to shoot ratio was increased under low N conditions, by both increased root growth and decreased shoot growth. N deficiency tends to decrease root branching, even though overall mass and length may be increased. High rates of N during early plant development can lead to a less extensively developed root system (see Marschner, 1995), while increasing shoot growth. The presence of strong fruit sinks during flowering can limit carbohydrate supply to developing roots, and reduce nutrient uptake. Cotton fruit forms are very strong sinks for N (as well as P, K, S, and Mg) and a notable decline in root system uptake occurs during flowering (Cappy, 1979). Uptake capacity of N by the root system can be specifically increased by a deficiency (Longnecker, 1994). Maintenance or growth of root mass preferentially over shoots should enhance survival through continued or increased uptake of N from the soil. In the presence of a localized supply of N, root branching and growth of N deficient plants is greatly enhanced, and uptake of other nutrients is strongly affected by this local root proliferation (Brouder and Cassman, 1994).

In addition to effects on leaf expansion rate and leaf area index, N deficiencies result in reduced vegetative branching (Wadleigh, 1944; Gardner and Tucker, 1967; Radin and Mauney, 1986; Bondada *et al.*, 1996), stem diameter (Gardner and Tucker, 1967), plant height (Gardner and Tucker, 1967; Oosterhuis *et al.*, 1983; Malik *et al.*, 1978; Bondada, *et al.*, 1996), main-stem node number (Thompson *et al.*, 1976: Bondada *et al.*, 1996), leaf number (Bondada *et al.*, 1996), fruiting branch development (Radin and Mauney, 1986), and total plant dry weight (Oosterhuis *et al.*, 1983; Wullschleger and Oosterhuis, 1990). Fernandez *et al.* (1996a) report that N deficiency decreased growth of leaves, petioles and branches, but not stems.

Nitrogen deficient plants initiate fewer fruiting sites, fruit for a shorter period, and produce less seed and lint weight than plants adequately supplied with N (Hearn, 1975; Thompson *et al.*, 1976; Malik *et al.*, 1978; Joham, 1986; Radin and Mauney, 1986; Bondada *et al.*, 1996).

Based primarily on data of Wadleigh (1944), N deficiency is reported to have little effect on node of first flower, time to flowering, fruiting index, (Joham, 1986; Radin and Mauney, 1986), and flowering interval (Radin and Mauney, 1986). Other studies (Malik, 1976; Thompson *et al.*, 1976) however, found delayed flowering, increased node of first flower, and increased flowering interval in N-deficient cotton. Bondada *et al.* (1996) found the node of the first fruiting branch was lower in plants receiving low N treatments, while Fernandez *et al.* (1996a) found partitioning of assimilate to squares (floral buds) was enhanced by N stress at the expense of leaves, petioles and branches, similar to the conclusions of Radin and Mauney (1986). Jackson and Gerik (1990) and Oosterhuis *et al.* (1983) found leaf N correlated with leaf area and boll number, but not boll weight or number of main-stem nodes. Shedding of squares and young bolls over a range of N concentrations was related to the ratio of actively growing bolls to the maximum boll carrying capacity. With decreasing N, leaf area per boll also decreased (Radin and Mauney, 1986; Bondada *et al.*, 1996). These results suggest that reductions in leaf area and number of fruiting sites determine the fraction shed rather than direct effects of N fertility. Deficient plants show extensive abortion of newest squares and many bolls may shed within 10 to 12 days after flowering (Cassman, 1993). Few mature bolls occur beyond the second lateral position of fruiting branches or at the top of the plant (Donald, 1964), especially in plants exhibiting mid-canopy N-deficiency symptoms (Ishag *et al.*, 1987).

Stewart (1986) and Ramey (1986) summarized the effects on low N status on yield components as decreased seed per boll, fiber length, lint weight, seed weight, and seed N content, along with increased percent oil content and lint to seed ratio (seed weight affected more than lint yield). Reduced storage of protein under N-limited conditions accounts for the reduced N content of seed (Leffler, 1986). Fiber maturity is little affected by low N conditions. The effects of N on fiber are probably an indirect effect of N deficiency on overall plant development rather than a specific role on N in fiber development.

Although N toxicity in cotton is not cited in the literature we surveyed, excess N applications may result in agronomically harmful effects on the plant and affect water quality. A number of studies (Bondada *et al.*,1996; Gerik *et al.*, 1998; Jackson and Gerik, 1990; Radin and Mauney,

1986;Thompson *et al.*, 1976) have indicated that higher N rates increase leaf area index, leaf size, vegetative branching, stem length, number of nodes, and internode length - essentially every component of vegetative growth. When excessive N application results in rank growth, light penetration to the lower canopy is reduced, and lower fruit forms may be shed or become more susceptible to loss from diseases such as the boll rot complex. While total boll numbers are generally increased, not all are harvestable, and a greater percentage of the harvestable bolls are set higher on the plant. Agronomic implications of excess N include increased insect and disease pressure, less effective coverage of plants with crop protection chemicals, delayed crop maturity, and more complicated defoliation and harvest management (Hearn, 1981; Hodges, 1992).

3. PHOSPHORUS DEFICIENCY

3.1 Leaf Symptoms

Symptoms of P deficiency are not strongly expressed. Growth rates may be affected if deficiency is severe, resulting in stunted, or spindly plants. The leaves may be lusterless, darker than normal, and hang at acute leaf angles from the stem. In severe cases, or in cool soils with marginal P levels, the foliage may exhibit reddish-purple tints (Donald, 1964). Because of high seed content of P and relatively low early demand, symptoms are rarely seen in early growth unless soils are cool, very acidic or P levels are extremely low. As deficiencies persist, older leaves begin to exhibit chlorosis, necrosis and leaf abscission (Marcus-Wyner and Rains, 1982). Phosphorus readily remobilizes from older tissue to new, actively growing tissue, so symptoms appear on older tissue first.

3.2 Metabolic Responses

Low P affects photosynthesis in a very complex manner. One of the most pronounced effects on cotton biochemical composition is the increase in non-phosphorylated sugars and starches, and the corresponding decrease in sugar phosphates (Ergle and Eaton, 1957). Assimilate accumulation may result from direct effects on photochemical capacity, enzyme driven reactions affecting carbon partitioning, or reduced assimilate transport (Terry and Rao, 1991),

Protein and chlorophyll per unit area are not much affected by P deficiency (Rao and Terry, 1989). Chlorophyll content may even increase. The purple or red coloration observed in some cases results from accumulation of anthocyanin (Marcus-Wyner and Rains, 1982). Since expansion of cells and leaves is retarded more than chloroplast and chlorophyll formation (Hecht-Buchholz, 1967), leaves of low-P plants will have dark green coloration. Furthermore, it appears that photosynthetic rate of low-P leaves is not limited by photochemical capacity, since P deficiency has little effect on photosynthetic electron transport and even less on photosynthetic quantum yield. In addition, effects on thylakoid function in the chloroplasts are mild and reversible (Brooks, 1986; Terry and Rao, 1991).

Based on significantly decreased rates of photosynthesis at a range of intercellular CO_2 partial pressures, Rao and Terry (1989) concluded that the major effect of low P on photosynthesis in sugar beet is on the enzymatic reactions occurring within the chloroplasts. Large decreases in sugar phosphates, including slow regeneration of RuBP appear to be the predominant effect of low P on photosynthesis. This switch in carbon partitioning pathways is induced when decreased levels of inorganic P cause increased activity of enzymes that synthesize starch and sucrose in preference to RuBP and other sugar phosphates (Terry and Rao, 1991; Marschner, 1995). The net result of this switch is to decrease levels of bound P and to increase the internal cycling of the limited inorganic P available. Under low P conditions, this permits substantial rates of photosynthesis and carbohydrate storage for later use while protecting the photosynthetic apparatus from photoinhibition. Thus, low-P plants do not immediately exhibit chlorosis. Terry and Rao (1991) suggest this may be a P conservation mechanism that allows plants to cope with periods of P deficiency without extensive damage to the plant.

Radin and Eidenbock (1984) found export of assimilate in P-deficient cotton was reduced much more so than photosynthesis, resulting in photosynthate accumulation. Movement of carbohydrates into the phloem may be reduced under low P conditions by decreases in ATP. While this may affect export, lack of growth and low demand for carbohydrate are likely more important (Terry and Rao, 1991).

3.3   Growth Responses

Growth rates are much more affected by P deficiency than rates of photosynthesis. The most striking visible effects of P deficiency are the reduction in leaf expansion rate and total leaf area (Freeden *et al.*, 1989). Decreases in leaf expansion are related strongly to extension of epidermal cells rather than decreased cell numbers, and are most likely a result of decreased hydraulic conductivity in the roots, similar to that observed with N deficiency (Radin and Eidenbock, 1984; Radin, 1990; Marshner, 1995). Expansion rates are severely limited during daylight, but very little at night. Thus, under low P conditions, water supply from the roots is decreased. During daylight hours, cells are unable to maintain sufficient turgor for expansion and growth because of transpiration losses. Small size and dark green color of leaf blades in P deficient plants are the result of impaired cell expansion and a correspondingly large number of cells per unit surface area (Hecht-Buchholz, 1967). In dicots, the number of leaves may also be reduced through a decrease in the number of nodes and reduced branching (Lynch *et al.*, 1991).

Root growth is much less affected by P deficiency than shoot growth, leading to an increase in the root:shoot dry weight ratio (Terry and Rao, 1991; Marschner, 1995). Deficiency induces dramatic increases in sucrose contents of roots as carbohydrates are preferentially partitioned into roots. Although root mass and length is increased, P deficiency decreased the number of newly emerging lateral roots (Skinner and Radin, 1994).

Reductions in growth under P deficiency extend to tissues such as lateral buds and fruiting branches and plant height. Delayed initiation of squares, reduced flowering, decreased boll set, early senescence, and delayed boll maturity of set bolls are apparent if P deficiency persists (Brown and Ware, 1958). According to Stewart (1986), effects of P deficiency are ultimately expressed in number of bolls set. For bolls retained by the plant, adequate P is generally available to mature the seed and lint. Sabino (1975) reported a slight decrease in fiber length under P limited conditions, which may be related to assimilate metabolism and membrane synthesis roles of P within the seed (fiber). Much of the P accumulated in the seed during the accumulation phase is retained as phytin rather than proteins or more complex metabolites (Stewart, 1986).

4.   POTASSIUM DEFICIENCY

4.1   Leaf Symptoms

Leaf symptoms of K deficiency in cotton are quite distinctive and were once termed cotton rust disorder before the true cause was known (Kerby and Adams, 1985). Symptoms usually occur first on the older, lower leaves since this element is readily remobilized within the plant. Symptoms typically begin as a yellowish-white mottling in the interveinal area and at leaf margins. Leaves become a light yellowish green with yellow specks appearing between leaf veins and the leaf margins. The specks become necrotic, causing the leaves to appear reddish-brown or dotted with brown specks at the leaf tip, the margin, and interveinal areas. As breakdown progresses the margins and leaf tip curl downward and shrivel, giving a ragged appearance. Finally the whole leaf becomes reddish-brown, dries and abscises prematurely. Symptoms proceed from the bottom to the top of the plant, eventually ending in death of the terminal in very severe cases. Strong light intensity accentuates these symptoms (Marschner, 1995). Plants low in K are more sensitive to drought stress (Kerby and Adams, 1985).

After flowering commences, K deficiency symptoms may develop on mature leaves of the upper canopy (Stromberg, 1960; Maples *et al.*, 1988; Cassman, 1993; Wright, 1999). We are not aware of a systematic comparison, but careful observation suggests there may be subtle differences in coloration and progression of historically described symptoms and these more recently recognized K deficiency symptoms. Maples *et al.* (1988) reported that the first sign of fruiting period deficiency is a deepening of the green color prior to appearance of chlorosis. Chlorotic symptoms begin as slight interveinal yellowing that rapidly moves to the leaf margins and changes to a orange or bronze coloration (Cassman, 1993). As the symptoms progress, the discoloration often covers the entire leaf surface except tissue adjacent to main veins. In published reports, this symptom is generally referred to as bronzing. In Australia, Wright (1999) indicated that leaves take on a distinctively red color as the symptoms progress. Leaves curl downward, become thickened and brittle. Necrotic patches with indistinct borders may then appear at the leaf margins. In severe cases, leaves prematurely defoliate. The marginal necrosis and bronze coloration distinguish K deficiency from Verticillium Wilt (*Verticillium dahliae* Kleb.), which produces interveinal lesions with a rich brown color and distinct borders (Cassman, 1993). Wright (1999) also reported that the underside of the leaf blade was not discolored.

4.2   Metabolic Response

Deficiency of K results in accumulation of soluble (hexose particularly) sugars and simple carbohydrates (Huber, 1985; Marshner, 1995; Oosterhuis, 1997; Bednarz and Oosterhuis, 1999). These simple carbohydrates build up in response to osmotic needs to maintain turgor, reduced capacity to transport assimilates, and reduced sink demand as growth slows.

Deficiency results in ATP accumulation in cotton (Bednarz and Oosterhuis, 1999) as ATPase activity and utilization decrease. Mengel (1985) considered the depolarization of membranes in response to ATPase pumping of H^+ one of the most important effects of K in plant metabolism. The effects of low K on this process in membranes of chloroplasts, roots and indeed, throughout the plant affect processes such as water uptake, water retention, meristem growth, and long distance transport of nutrients and assimilates in the xylem and phloem.

High concentrations of K in the cytoplasm are critical for maintenance of enzyme activity (see reviews by Leigh and Wyn Jones, 1986; Suelter, 1985). Potassium deficiency directly affects activation of some enzymes and indirectly affects many others through impaired regulation of cytosol pH, a critical factor for optimum enzyme activity (Leigh and Wyn Jones, 1986). Soluble N compounds also accumulate under low K conditions as a result of inhibition of protein synthesis. It is likely that K is involved in both the activation of nitrate reductase, and its synthesis (Marschner, 1995). Pettigrew and Meredith (1997) and Wright (1999) found decreased N utilization efficiency under low K conditions, perhaps because of impaired ammonium utilization. Inhibition of RuBP carboxylase in the chloroplasts particularly affects N assimilation and leads to accumulation of soluble N compounds. A number of other enzymes including hydroxylases and oxidases (*i.e.,* polyphenol oxidase) increase in activity as K becomes limiting, resulting in ac-

cumulation of superoxide radicals, and chlorosis. Chlorosis precedes the destruction of chloroplasts (Pissarek, 1973).

Effects of K deficiency on photosynthesis, respiration and the components of the photosynthetic apparatus have been reviewed by Huber (1985). Photosynthesis and translocation of assimilates are reduced where K is low (Bednarz and Oosterhuis, 1999, Wright, 1999), whereas rates of dark respiration are increased (Huber, 1985). Decreases in photosynthesis are brought about in two ways. First, reduced growth (discussed below) results in reduced total photosynthetic area. Secondly, diminished activity and capacity of the photosynthetic apparatus (not increased stomatal resistance) increases mesophyll resistance to CO_2 and ultimately reduces photosynthesis (Longstreth and Nobel, 1979, Wright, 1999). Huber (1985) concludes that effects of low K on reduced protein synthesis and developmental processes are likely causes. Potassium deficiency affects dark reactions much more than the photochemical reactions that generate ATP and NADP. Photorespiration increases, probably because of lower CO_2 depletion at the catalytic sites (Marschner, 1995).

Just as important as the biochemical (metabolic) roles of K, are the biophysical contributions to osmotic adjustments in the cytoplasm, cell vacuoles and other locations throughout the plant. In K-sufficient plants, turgor in the vacuole is maintained primarily by K salts, and even higher concentrations are required in the cytoplasm (Hsiao and Läuchli, 1986). A K deficiency can affect turgor and stomatal regulation in a number of ways. Since K is the primary charge balancing cation within the vacuole, inadequate K supply forces the plant to either maintain turgor by formation of organic solutes (energetically expensive, and in direct competition for assimilates) or lose turgor and wilt. Potassium also has a specific role in guard cells of stomata, which can only be replaced by organic solutes (Hsiao and Läuchli, 1986). Replacement by organic solutes, such as hexose sugars, can result in slower stomatal response to water stress, and incomplete closure and opening of stomata.

Transpiration is generally reduced by K deficiency at some stage (Huber, 1985, Hsiao and Läuchli, 1986; Wright, 1999). Sun *et al.* (1989) reported low K resulted in a poorly developed epidermal cuticle in cotton. In this instance they found an increase in transpiration related to deficiency-induced abnormalities in cuticle development. They also reported reduced hydraulic conductivity (of roots), and increased stomatal resistance. Cooper *et al.* (1967) reported that low K could also result in reduced numbers of stomates per leaf area in subterranean clover.

The many roles of K in transport are also negatively affected by deficiencies. Potassium is readily remobilized in the plant, and as much as half of the K supplied to expanding tissue is provided through phloem supply (Marschner, 1995). Ashley and Goodson (1972), found that assimilate transport was inhibited by low K in cotton. Transport of sugars may be directly affected by low K effects on phloem loading mechanisms, disruption of high pH regulation in phloem sieve tubes, or in reduced osmotic regulation within the sieve tubes which drive assimilate movement. Decreased activity of ATPase under low K conditions may also play a role (Huber, 1985).

Lignification of vascular bundles is impaired under low-K conditions, and may contribute to increased lodging potential. Cell walls tend to become thinner, and anatomy may be affected as well (Pissarek, 1973). In addition, xylem and phloem tissue formation were restricted by K deficiency in rape, but cortical tissue was minimally affected. This may increase lodging and reduce transport of water, minerals and assimilates. McCully (1994) analyzed developing xylem elements of cotton and other plants, and found these cells accumulated more K than any other type of cell in the root. Levels of K increased during development until cell death and maturation of the xylem element. As a major storage site of K in the plant, the pool of K in developing xylem elements should be considered in attempts to elucidate processes involving uptake, storage, remobilization, cycling, and exudation of K.

4.3   Growth Responses

Expansive growth is extremely sensitive to K deficiency (Hsiao and Läuchli, 1986; Longnecker, 1994). These effects appear directly related to K concentrations rather than an indirect result of limited assimilate supply (Mengel, 1985). In expanding leaves of bean plants deficient in K, turgor, cell size, and leaf areas were significantly lower than in leaves supplied with adequate K (Mengel and Arneke, 1982). Rosolem and Mikkelsen (1991), and Bednarz and Oosterhuis (1996) reported similar effects on dry matter accumulation in leaves of cotton. As summarized by Beringer and Nothdurft (1985), and Marschner (1995), cell extension is the consequence of accumulation of K^+ in the cell where it is required for increasing osmotic potential in the vacuoles for growth, and for stabilizing the pH in the cytoplasm which is required for cell wall extensibility and other metabolic functions. Deficiency during expansion of new tissue results in dramatic decreases in K content in conjunction with reduced leaf expansion rate, leaf area index, and meristem growth. The importance of K mediated turgor-induced growth in cotton has been shown for cotton fiber as well (Dhindsa et al., 1975). Leaf initiation is apparently unaffected by K deficiency.

Deficiencies of K reduced leaf area index, number of main-stem nodes, stem height, stem weight, and total plant dry weights (Cassman et al., 1990; Pettigrew and Meredith, 1997) because of lower total assimilate available for plant growth. Pissarek (1973) concluded that stem diameters or summer rape were reduced because of reduced activity of the cambium, and cells were poorly lignified.

In nutrient media studies, K deficiency increased the root/shoot dry weight ratio of cotton (Rosolem and Mikkelsen, 1991; Bednarz and Oosterhuis, 1999). In split-pot studies containing vermiculitic soils, Brouder and

Cassman (1994) found that cotton roots did not initiate compensatory root growth in response to localized supply of K, in contrast to localized supply of N and, to a lesser extent, P. A cultivar more susceptible to late-season K deficiency had less root branching and allocated less dry matter to roots relative to shoots. The presence of localized N increased root proliferation, increasing mean root area, and thus K uptake in soils with limited availability of K. Gulick et al. (1989) found that cotton roots were less able to exploit K in topsoil layers than barley. Differences were attributed to greater sensitivity of cotton to low soil water potential. Similarly, differences in root surface area between cultivars were strongly correlated with K uptake and differing sensitivity to late-season K deficiency (Brouder and Cassman, 1990). Wide fluctuations on soil matric potential in the upper 0.1 m limited root extension for both cultivars.

Developing fruit provides a strong sink for K (Leffler and Tubertini, 1976), with 65 to 70% of the K uptake found in reproductive tissues (Bassett et al., 1970, Mullins and Burmeister, 1990). In fields and cultivars subject to late-season K deficiency, fruit removal prevented expression of deficiencies (Cassman, unpublished data). In cotton given deficient levels of K, Rosolem and Mikkelsen (1991) found that sensitivity of tissue to K deficiency in the order leaves<bolls<roots<stems. In plants totally deprived of K after pinhead square, the sensitivity was in the order bolls<stems and petioles<leaves<roots (Bednarz and Oosterhuis, 1995). In the former case, deficiency effects were noted in other plant parts before leaves, whereas in the latter case deficiency effects were noted in other plant parts before bolls were affected.

Reduced leaf area duration and premature shedding of leaves results in decreased boll numbers, decreased bolls size, decreased boll mass, reduced lint yield, reduced seed yield, decreased oil percentage, and oil content per seed (Tharp et al., 1949; Kerby and Adams, 1985; Stewart, 1986). These small, immature bolls may fail to open (Kerby and Adams, 1985). In a recent comparison of eight cultivars, Pettigrew et al. (1996) found no cultivar differences in sensitivity to K deficiency, but found that low K decreased lint yield (9%), boll mass (7%), lint percentage (1%), and seed mass (4%) relative to treatments well supplied with K. Overall, there was no difference in the vegetative to reproductive dry matter ratio (harvest index) under K deficiency, indicating no K-related change in partitioning for these field grown plants. The differences noted are all affected by reduced assimilate availability and transport (Stewart, 1986).

Leffler (1986) showed that the burr and fiber are both strong sinks for K, but not the seed. Fiber demand for K increases linearly for five weeks following anthesis, suggesting K is required for both elongation and secondary wall deposition. Both processes require a high osmotic potential in the boll. Potassium malate forms the major osmoticum during fiber development (Dhindsa et al., 1975) and provides the turgor potential required for cell elongation. Effects of low K on fiber quality include reduced fiber length and uniformity (Bennett *et al.*, 1965; Sabino, 1975; Stewart, 1986). Effects on fiber diameter are not as consistent (Kerby and Adams, 1985), but several reports indicate a reduction in micronaire for K-stressed cotton (Sabino, 1975). In addition to decreased yield, fiber length and uniformity, Pettigrew and Meredith (1997) measured decreased 50% span length, micronaire, fiber maturity, and fiber perimeter.

Sun *et al.* (1989) reported a 276 kg ha⁻¹ increase upon K addition with resulting increase in weight of 1000 seeds from 53.2 to 6.7 to 69.5 g, seed viability from 51.5 to 63.7 to 71.0% and germination from 29.5 to 41 to 44.5%. K also increased the percentage of linolenic acid content of the oil.

Reduced disease resistance in cotton has been linked to declining plant K levels (Kerby and Adams, 1985; Hodges, 1992). In particular, resistance to Verticillium wilt, Fusarium wilt (*Fusarium oxysporum* f. sp. *vasinfectum*), and Alternaria leaf spot (*Alternaria macrospora*) (Hillocks and Chinodya, 1989) have been well documented. Based on the discussion above, reduced disease resistance at low K levels may result from lower enzyme activity, reduced lignification, accumulation of sugars and amino acids in leaves, sluggish stomatal response, decreased cuticle development, increased permeability of the (root) plasma membrane, and reduced rate of growth.

4.4 Comments On the Post-Flowering K Deficiency Syndrome

One of the more interesting questions in mineral nutrition of cotton today is the appearance of post-flowering K deficiency symptoms in the upper canopy (Cassman, 1993; Oosterhuis, 1995b). Similar symptoms have been noted in tomato, potato (Lingle and Lorentz, 1969; Hewitt, 1984) and summer rape (Pissarek, 1973). The central cause of this syndrome appears related to higher reproductive to vegetative ratios (greater than 1.0), and earlier fruit set in modern cultivars compared to older, lower yielding cultivars with similar total dry matter production (Meredith and Wells, 1989; Oosterhuis, 1997; Wright, 1999). The daily demand for K in these cultivars can be as high as 4.5 kg/ha/da (Wright, 1999). As a result, vegetative storage of K prior to boll set is reduced, and the fruit-setting period is compressed. Even though total demand may remain unchanged, the flux requirement after early fruit set may be much higher (Leffler and Tubertini, 1976; Bassett *et al.*, 1970; Mullins and Burmester, 1991; Wright, 1999), at a time when root activity is decreasing in response to competition for assimilate from reproductive sinks (Cappy, 1979), and especially from developing bolls in the middle part of the canopy (Bednardz and Oosterhuis, 1996). It seems clear that heavy boll set in a shorter time period predisposes cotton to K-deficiency-induced premature senescence, as noted by Wright (1999), however, the symptoms appear rather erratically and usually occur in a relatively small portion of the total crop areas. Unique environmental conditions during flowering such as K-fixing soils, low subsoil K levels, irrigation

induced rooting patterns, restricted rooting, soil compaction, and short term drought (Cassman, 1993; Oosterhuis, 1995b; Wright, 1999), may cause decreases in flux of K into the plant below prevailing levels seen during breeding and selection of these cultivars, and create a supply-demand imbalance. Cultivars selected under a wider range of these unique conditions would be better adapted, through more extensive root area generation (Cassman *et al.*, 1989a), for example, and less susceptible to post-flowering K deficiencies. The imposition of an often subtle flux-reducing condition on plants predisposed to K-deficiency by a high boll load would help to explain the extremely unpredictable appearance and disappearance of this syndrome in soils considered adequately supplied with K. But the lack, or delay, of K remobilization from the lower leaves during an induced deficiency remains to be satisfactorily explained.

As dominance for assimilate shifts to reproductive growth and root activity declines, uptake is either reduced or insufficient to meet K all demands, and remobilization must occur. In the oldest leaves we would expect substantial amounts of K, stored primarily in the vacuoles (balancing the charges of malate, nitrate and other anions, and maintaining leaf turgor) to be mobilized first, as in the historical symptom development. But in cases where K is withdrawn completely (Bednarz and Oosterhuis, 1996; Pissarek, 1973) or becomes limited (Rosolem and Mikkelsen, 1991) at a later stage, effects on K concentration and symptom development are clearly most evident in the upper leaves of the plant first, while lower leaves are slowly affected. Pissarek (1973) concluded that K is more readily mobilized from the most metabolically active leaves, and those with the highest K concentrations. These are typically the leaves nearer the top of the plant, where high K levels are required for expansive growth (Mengel and Arneke, 1982).

In non-expanding tissues, somewhat lower K levels are needed to maintain turgor. Remobilization of K may initially resupply K without loss of cell turgor. In more active leaves, this requirement may be partially met by Na (Bednarz and Oosterhuis, 1999) or, more commonly, production of organic solutes such as hexose sugars (Leigh and Wyn-Jones, 1986). Cutler and Rains (1978) found that osmotic adjustment in cotton during a water deficit was from organic solutes rather than K, even though K was not limiting. During similar periods of reduced expansion under K deficiency, a similar reaction seems likely. However, less metabolically active lower leaves would be less able to maintain turgor through production of organic solutes upon loss of K. A switch to this pathway would limit availability of assimilates to reproductive sinks, and is furthermore energetically unfavorable. If nitrate forms a significant component of the osmoticum in older leaves, cation exchange with Na, Ca, or Mg may be required to mobilize K. Interestingly, Wright (1999) reported leaves from plants experiencing premature senescence contained less K and P and more Mg, N, and S than non-symptomatic plants. Since nitrate is not phloem mobile, and activity of nitrate reductase is very low in older leaves, the rate of nitrate reduction may otherwise limit efflux of K from the vacuoles. Rapid remobilization of nutrients from older leaves clearly requires induced senescence. Yet these leaves have high levels of stored K and N, and are less likely to senescence than if they had been subject to season-long deficiency. Once remobilization begins, we must remember that developing bolls are strong sinks for nearby, photosynthetically active leaves. Assimilates from lower leaves may be used locally rather than transported long distances. Developing bolls in the middle of the canopy are the most dominant K sink, and may restrict K translocation to the upper canopy during this critical, high demand period (Bednardz and Oosterhuis, 1996). Indeed, this energetically favorable process is basic to increasing reproductive to vegetative ratios, and improving yields of modern cultivars.

On the other hand, as influx of K declines, K levels in expanding and recently mature leaves declines much more rapidly than in lower leaves (Rosolem and Mikkelsen, 1991; Bednarz and Oosterhuis, 1995). Cytosolic K levels must be maintained, even at the expense of growth (Leigh and Wyn-Jones, 1986). As K levels decline, growth is reduced, protein synthesis is reduced, carbohydrates accumulate, and transport of assimilates is reduced. Without resupply from lower leaves, this is a photosynthetic bomb waiting to explode. Marschner (1995) suggests that remobilization of highly phloem mobile nutrients like K can rapidly deplete vegetative shoots, and induce senescence in order to supply sink demands.

In summary, it appears the K deficiency syndrome or premature senescence is primarily a result of increased reproductive to vegetative ratios and early, more compact fruit set and maturation in modern cultivars. This reduces vegetative storage prior to boll set and increases the daily peak demand for K during early flowering. Environmental conditions that limit uptake rates at this time induce K deficiencies that are not remedied by remobilization from older leaves, with the result that symptoms appear in the mid- or upper-canopy. Remobilization from lower leaves to the upper leaves is likely limited by dominance of nearby sink demands and by cellular requirements for turgor maintenance.

5.   CALCIUM DEFICIENCY

5.1   Leaf Symptoms

Cotton seedlings grown in low Ca solutions are stunted, with thin stems and dark green leaves. Plants appear wilted, with stems bent over and leaves and petioles hanging limply down around the stem, even when adequate water is available (Donald, 1964). Symptoms are most pronounced in growing zones, and young tissues since Ca is not remobilized in the plant. Internodes at the top of the plant fail to elongate and give the plant a bushy or rosette appearance. Young buds in the terminal of the main shoot and the stem

immediately below it turn brown and die (Wiles, 1959). Grundon (1987) reported that lateral buds began to grow, but also turned brown and died. Similar brown lesions appeared on the base of the petioles, causing the leaves to die. In milder deficiencies, he observed that branching was severely reduced and few flowers and bolls were produced. There are no well-defined foliage symptoms for Ca deficiency in field-grown cotton.

5.2   Metabolic Responses

Cell division may be decreased by Ca deficiency in a number of crops (Longnecker, 1994). Effects of low Ca include abnormal mitosis, multinucleate cells, chromosome breakage, and incomplete separation of chromosomes without mitosis (Hewitt, 1984).

Roots cells deprived of external Ca supply cease growing within a few hours. Although Ca is involved with cell division, this response is primarily a result of reduced cell extension. The role of Ca in cell wall strengthening makes this response seem less than obvious. However, it appears that the roles of IAA in cell wall loosening and free cytosolic Ca concentrations in stimulating synthesis of cell wall precursors are the primary reason for reduced cell wall extension (Marshner, 1995; Longnecker, 1994).

Calcium deficiency has primary effects on destabilization of membrane and cell wall structures, and on enzyme activity (Marschner, 1995). Membranes are stabilized through Ca-bridges between phospholipids and proteins at membrane surfaces. Under conditions of low solution Ca, other ions such as H or Al (acid soils), Na (saline soils) can displace Ca, leading to membrane leakage and loss of ion uptake selectivity. Leakage of low molecular weight solutes can in turn displace plasma membrane bound Ca, leading to further degradation. As the deficiency becomes severe, general disintegration of membrane structures can occur (Marschner, 1995). A variety of Ca-deficiency symptoms result from leakage of water from cells, with some cells becoming desiccated and others becoming water-soaked. These are most likely a result membrane breakdown (Simon, 1978). Under conditions of low temperature or water-saturated soils, the membrane-protecting effect of Ca is particularly important for cotton (Christiansen *et al.*, 1970). Membrane leakage may cause secondary effects, such as increased respiration rates as respiratory substrate leaks from vacuoles into the cytoplasm of Ca-deficient tissues (Bangerth *et al.*, 1979).

Cell irregularities reported in Ca-deficient tissues include thickening of the cell walls of fiber and collenchyma tissues, premature vacuolation, enlarged or irregular cell shape, fragile cell walls, decreased lignification and changes in distribution of cell size and differentiation (Hewitt, 1984). Melanotic compounds may also accumulate in intracellular free space and in vascular tissue, where they can affect solute transport mechanisms (Bussler, 1963). Activity of polygalacturonase, which mediates the degradation of pectates, increases markedly under Ca deficiency. As a result, the strengthening effect of pectates in the middle lamella of cell walls is reduced, resulting in cell wall disintegration and plant tissue collapse. A number of other membrane-bound enzymes, including ATPases, are stimulated by Ca. Because of its effects on cell wall integrity, Ca deficiency can result in poor resistance to fungal and bacterial infections (Wiles, 1959).

5.3   Growth Responses

Plants deficient in Ca have poorly developed root systems. In the field, root tips of Ca-deficient cotton develop a brown coloration, root extension is inhibited, lateral branching is reduced, and taproots are small in diameter. Low subsoil Ca levels may result in poor water uptake, and symptoms of water stress resulting from poor root development and the inability to extract water from subsoils low in Ca (Howard and Adams, 1965). Both the secretion of mucilage and the gravitropic response mechanism of roots may be impaired by low solution levels of Ca (Bennet *et al.*, 1990).

Because of its very low phloem mobility and the low transpiration rate of expanding tissue (thus low provision via xylem), Ca deficiency can result in dramatic stunting of new tissue. A typical symptom of calcium deficiency is cell wall disintegration and collapse of affected tissue, such as petioles and upper parts of the stems (Bussler, 1963).

Lint yields of Ca-deficient cotton are very low (Grundon, 1987). Poor fruiting of surviving plants resulting in excessively large, vegetative plants have been reported from culture studies (Donald, 1964). Little is known about direct effects of Ca deficiency on flowering, however, pollen tube growth requires adequate Ca in the substrate, and direction of growth is chemiotrophically controlled by extracellular Ca gradient (see Marschner, 1995).

There are no reported studies on the effects of Ca deficiency on earliness, lint quality, or seed and oil quality of cotton.

6.   MAGNESIUM DEFICIENCY

6.1   Leaf Symptoms

In deficient plants, Mg is remobilized from older tissues to expanding tissues, and visual symptoms appear on older, fully expanded leaves. Symptoms progress up the plant as older leaves senesce and abscise prematurely. On older leaves, a pale green to yellow interveinal chlorosis develops. Frequently this is accompanied by purplish-red coloration, especially in young cotton growing in cool soils. The veins remain green and unaffected. Young leaves remain green and apparently healthy. If deficiency persists and becomes severe, small pale brown necrotic lesions develop within the interveinal areas in the lamina of the leaf.

Eventually necrotic regions join together, initially leaving the veins green and unaffected. As the deficiency progresses, the veins become affected, and the leaf dies and abscises from the plant (Grundon, 1987).

In the greenhouse, interveinal chlorosis occurs without the purple color (Marcus-Wyner and Rains, 1982), which may be related to lower ultra-violet radiation. However, Jayalalitha and Narayanan (1996) found cotton plants demonstrated purplish red and orange interveinal pigmentation in older leaves of cv. LK861 grown under hydroponic conditions.

6.2   Metabolic Responses

Very little attention has been given to the metabolic responses of cotton to Mg deficiency, and the comments in this section are summarized from the more detailed review of Marschner (1995). Since Mg is the central ion of the chlorophyll molecule, deficiency affects the size, structure and function of chloroplasts, including electron transfer in photosystem (PS) II. For most crops, growth reduction and visible deficiency symptoms occur once the proportion of chlorophyll-bound Mg to total leaf Mg exceeds 20 to 25%. While the rate of photosynthesis in Mg-deficient leaves is lower on both a unit leaf area and a unit chlorophyll basis, photosynthesis is less impaired than starch degradation in chloroplasts, sugar metabolism in cells, or phloem loading of sucrose. In spite of chloroplast pigment reductions, nonstructural carbohydrates (starch and sugars) accumulated in bean leaves under low Mg conditions (Fischer and Bussler, 1988). As summarized by Marschner (1995), impaired export of photosynthates from low Mg source leaves resulted in increased leaf dry weight, and distinct decreases in carbohydrate content in roots and in developing fruit. This results in a reduced root/shoot ratio in Mg-deficient plants (in contrast with N, P, and K). Impaired export also contributes to enhanced degradation of chlorophyll in source leaves.

Reduced export of photosynthates is most likely a result of reduced phloem loading of sucrose in Mg-deficient source leaves. Leaves deficient in Mg have inadequate Mg in the metabolic pool at the plasma membrane of sieve tubes to activate ATPase – a key enzyme in phloem loading. The resulting accumulation of photosynthates exerts feedback regulation on RuBP carboxylase/oxygenase synthesis in favor of the oxygen producing reactions. Thus formation of superoxide radicals and hydrogen peroxide is enhanced. In response, content of antioxidants, and the activity of superoxide radical and H_2O_2 scavenging enzymes increases. As a result, development of chlorosis and necrosis is strongly light sensitive.

The primary effect of Mg deficiency is, however, on protein synthesis. Low levels of Mg prevent aggregation of ribosome subunits, which is required for protein synthesis. As a result, non-protein N is increased, while protein N is decreased. The synthesis of RNA is particularly sensitive to low levels of Mg. This also explains why other plastid pigments are affected as well as chlorophyll.

6.3   Growth Responses

Jayalalitha and Narayanan (1996) noted plant height, leaf number, and leaf area of cotton were little affected by Mg deficiency, even though plant dry weight was decreased by 31%. In addition to decreased chlorophyll, they found increased leaf concentrations of S, K, and Cu and decreased levels of P and Mg.

Unlike N and P, Mg deficiency affects root growth more than shoot growth, and the shoot to root ratio is increased. If the root supply is permanently reduced, remobilization from leaves will eventually reduce leaf area duration and induce early senescence (Marschner, 1995).

Grundon (1987) reported that Mg-deficient cotton plants appeared stunted with short thin stems. Branching was severely reduced, and plants became single-stemmed (lacking vegetative branches). Fewer flowers and bolls were produced, and yields were reduced.

7.   SULFUR DEFICIENCY

7.1   Leaf Symptoms

Symptoms include short, slender-stemmed plants with small, pale green to yellow leaves similar to N deficiency. Although symptoms may occur in upper or lower parts of the canopy, younger leaves are the most sensitive indicator due to the low mobility of S in the plant. The veins are affected, but usually less so than surrounding tissue (Ergle and Eaton, 1951; Donald, 1964). In severe cases, pale brown necrotic lesions may develop in tissue adjacent to the margins of affected leaves, and the margins may become wavy or cupped upwards (Grundon, 1987). Leaves may be rigid and brittle (Ergle and Eaton, 1951).

7.2   Metabolic Responses

In sulfur deficient cotton plants, Ergle and Eaton (1951) documented decreases in total sugar, and protein N, and increases in NO_3-N, soluble organic N, and non-S containing amino acids. In reviewing the effects of S deficiency on plant metabolism, Marschner (1995) noted that the resulting shortage of cysteine and methionine not only inhibits protein synthesis, but also decreases chlorophyll content and leads to chlorosis. This should be expected since a high proportion of the leaf protein is located in the chloroplasts where chlorophyll molecules comprise a significant component of the light-harvesting complex. Marschner (1995) also indicated that reduced synthesis of S-containing compounds, especially glutathione, decreases the ability of the plant to detoxify oxidants and heavy metals. Sulfur is also an important structural component of sulfolipids, and is a structural constituent of all biological membranes. Sulfolipids

are especially important in the thylakoid membranes of the chloroplasts. In roots, sulfolipid level has been positively correlated with salt tolerance (Marschner, 1995).

Sulfur is a key component of several enzymes and co-enzymes, but the effects of deficiency and corresponding effects on plant growth are not well elucidated. Reduced S, primarily as glutathione, is transported in the phloem to the shoot apex, fruits and roots with demand for protein synthesis. In general, S is better distributed among older and newer leaves than N, and supply levels affect concentrations in both old and new leaves. Development of deficiency symptoms depends to some degree on the N supply. If N is deficient, symptoms may occur first in older leaves, while plants well supplied with N will exhibit symptoms in the newer, upper leaves. This demonstrates that the extent of remobilization and translocation from older leaves depends on the rate of nitrogen deficiency-induced senescence (Marschner, 1995).

7.3   Growth Responses

Deficiencies of S reduce leaf area as a result of both smaller cell size and fewer leaf cells. Root hydraulic conductivity, stomatal aperture, and net photosynthesis are all reduced with a few days of interruption of S supplies to the root (Marschner, 1995). Plant fresh weight is decreased. Shoot growth in S deficient plants is more depressed than root growth. Stems are shortened and thinner than cotton well supplied with S. If the deficiency is severe, the plants will have fewer vegetative and fruiting branches. Fruiting is about normal for the plant size, but boll size is smaller than normal (Ergle and Eaton, 1951; Donald, 1964).

8.   MANGANESE DEFICIENCY

8.1   Leaf Symptoms

Manganese is not readily remobilized, so symptoms are most evident in young leaves that are small and turn pale green. A faint yellow (yellowish gray to reddish-gray, Donald, 1964) interveinal chlorosis develops which becomes more distinct with time. Leaf veins remain green, and margins may cup downward or roll under the leaf. Small, brown necrotic lesions begin to appear within the chlorotic areas of young leaves as deficiency progresses. The tissue surrounding necrotic spots may become puckered and give the leaf a distorted appearance (Grundon, 1987). In some cases, leaves may appear wilted, even though soil water is adequate.

8.2   Metabolic Responses

Manganese deficiency severely reduces the supply of nonstructural carbohydrates. This is particularly evident in roots, and a key factor in reduced root growth under Mn-deficient conditions (Campbell and Nable, 1988).

While levels of nitrates, nitrites, amides, proteins and free amino acids are increased or unaffected by Mn deficiency, these are not a function of direct involvement of Mn in N assimilation (Campbell and Nable, 1988). Rather, they are secondary effects of limited supply of carbohydrates and reducing equivalents to the cytosol, and lower demand for reduced N. Protein synthesis is not specifically impaired in Mn-deficient tissues (Marschner, 1995).

Burnell (1986) reviewed the biochemical roles of Mn in plants. Under mild Mn deficiency, photosynthetic oxygen evolution decreases with a corresponding decrease in assimilates, high-energy phosphates, and reducing equivalents. Dry weight and chlorophyll content are little affected, and the effects are reversible upon resupply on Mn. Under more severe deficiency, chlorophyll content is decreased and the ultrastructure of the thylakoids is dramatically affected. These changes are difficult or impossible to restore. The effects on the thylakoid ultrastructure may result (primarily or secondarily) from release of free radicals and oxidative stress following impaired water oxidation (Campbell, and Nable, 1988), but are more likely caused by inhibition of the biosynthesis of lipids and carotenes (Marschner, 1995). Several experiments have shown Mn deficiency has little short-term effect on dark respiration of soybean and sugar beet. With severe and prolonged deficiency, a decline in respiration has been noted, most likely as a result of impaired photosynthesis and reduced supply of assimilates (Campbell and Nable, 1988).

Manganese deficiency lowers polyunsaturated fatty acids and glycolipid content of chloroplasts of soybeans, and decreased oil content and composition of seeds. Marschner (1995) attributes this reduction to the role of Mn in biosynthesis of fatty acids, carotenoids, and related compounds, and as noted above can strongly affect the thylakoids of the chloroplasts. Decreases in seed yield, oil content oleic acid content of soybeans may be secondary to supply of carbon skeletons for synthesis, or they may result from a direct role of Mn enzymes of the endoplasmic reticulum. Campbell and Nable (1988) concluded that the physiology of oil synthesis and maintenance of chloroplast membranes are major roles of Mn.

A number of reactions producing secondary metabolites in the shikimic acid pathway depend on Mn as a cofactor. The most important of these reactions include production of phenolics (such as the aromatic acids which form the backbone of the plant defense system, and the precursors of lignin) the final step in the synthesis of lignin, and regulation of IAA oxidase. Thus manganese deficiency may be an important factor in lower resistance of Mn-deficient plants to root infecting pathogens. High levels of Mn also appear to inhibit enzymes of some plant pathogens (Burnell, 1986).

Decreased levels of lignin in both shoots and roots of Mn-deficient wheat have been reported. Wilting observed in conjunction with Mn deficiency may be related to decreased structural support from poorly lignified schlerynchyma, or

from partially blocked xylem vessels (Campbell and Nable, 1988).

The plant Mn status exerts control on the level of IAA oxidase in cotton plants, and thus strongly affects the rate of IAA removal. Deficient plants (and those with toxic levels of Mn) contain high levels of IAA oxidase, with no detectable inhibitor activity (Taylor *et al.*, 1968). Inadequate or imbalanced levels of IAA could cause many of the symptoms of Mn deficiency or toxicity, including inhibition of growth, shortened internodes, leaf abscission, cessation of lateral root formation, decreased root extension and crinkle leaf (Campbell and Nable, 1988). Some 35 enzymes involve Mn as a cofactor or activator. Burnell (1986) gives an extensive review of the effects of Mn deficiency on these reactions.

8.3   Growth Responses

Rate of cell elongation is more affected by Mn deficiency than rate of cell division (Campbell and Nable, 1988), resulting in many, but non-vacuolated leaf cells dominated by cell wall (Abbot, 1967). Leaves prematurely senesce (Campbell and Nable, 1988). In Mn-deficient cotton, internodes are shortened and leaf abscission is increased (Morgan *et al.*, 1976).

Roots lacking an external supply of Mn show greatly reduced rates of growth. Inhibition of root growth in Mn-deficient plants may be caused by shortage of carbohydrates as well as direct requirement of Mn for growth (Campbell and Nable, 1988). Lateral root formation ceases completely in Mn-deficient tomatoes (Abbott, 1967). If part of the root system receives adequate Mn, the shoots receive an adequate supply, even though growth is less than optimal. The part of the root system without Mn will exhibit impaired synthesis of lignin and phenols (Nable and Loneragan, 1984), which may also decrease disease resistance. Manganese-deficient cotton has shown increased susceptibility to both Verticillium wilt and Rhizoctonia infections (Shao and Foy, 1982; Huber and Wilhelm, 1988).

Joham and Amin (1967) reported delayed flowering in Mn-deficient cotton. Lint yields of cotton were decreased under Mn-deficient conditions, although lint quality was virtually unaffected (Anderson and Boswell, 1968). Anter *et al.* (1976b) reported that fiber length was shorter in plants not supplied with foliar applications of Mn. Lower yields under Mn-deficient conditions most likely result from lower carbohydrate supply, but poor lignification and indehiscence in anthers has also resulted in male sterility in wheat and triticale (Campbell and Nable, 1988; Marschner, 1995). Reduced yields in soybean resulted from fewer fertile nodes (and pods) and lighter seeds (Gettier *et al.*, 1985). Seed yield and oil content of soybeans are reduced, and the linoleic acid increases while oleic acid content decreases (Campbell and Nable, 1988). Anderson and Worthington (1971) found no effect on oil or protein concentrations of cottonseed in plants affected by Mn-deficiency.

9 ZINC DEFICIENCY

9.1 Leaf Symptoms

In rapidly growing Zn-deficient cotton plants, the limited phloem mobility of Zn results in the appearance of "little leaf" symptoms in the upper part of the plant. Young plants typically have abnormally small, thickened and brittle leaves with interveinal chlorosis in the youngest leaves. Upward cupping or in-rolling of the leaf edges may also occur (Hinkle and Brown, 1968; Grundon, 1987). Small dots of necrotic tissue in the chlorotic area may expand to cover the entire leaf surface, producing a bronzing effect (Donald, 1964). Parallel elongation of the leaf tips may occur, resulting in a finger-like appearance (Hinkle and Brown, 1968; Marcus-Wyner and Rains, 1982). Some cultivars may accumulate anthocyanin in the petioles (Brown and Wilson, 1952). Although chlorosis and malformation of young leaves are reported as symptoms on severely deficient cotton, Grundon (1987) reported that symptoms first appear as brown necrotic lesions in the interveinal region of cotyledons and older true leaves without prior chlorosis. Veins are eventually affected as the lesions join together, and the leaves abscise. Young leaves of Zn-deficient cotton plants are very small, distorted, and may have holes or torn margins.

9.2 Metabolic Responses

In most cases, sugars and starch accumulate in Zn-deficient plants. Carbohydrate accumulation increases with increasing light, as new growth, particularly at shoot apices, is impaired (Marschner and Cakmak, 1989). Most evidence indicates that deficiency-induced changes in carbohydrate metabolism are not directly responsible for decreases in growth or visible plant responses.

Protein synthesis and protein content of Zn-deficient tissues is dramatically decreased whereas amino acids accumulate (Marschner, 1995). Contents of DNA and RNA in particular are decreased. Zinc is also a structural component of the ribosomes, the site of protein transcription and translation. In the absence of Zn, protein ribosome structure is compromised, and synthesis is inhibited. Deficiencies may particularly affect sites of protein synthesis, such as pollen tubes and the shoot meristems, where Zn concentrations may be five to ten times higher than in mature leaf blades (Marschner, 1995). Degradation of existing proteins may also be enhanced under low Zn conditions by increased enzyme activity. In particular, the increased activity of RNase, which enhances breakdown of RNA, is a very sensitive and early indicator of Zn deficiency (Marshner, 1995).

Plasma membrane permeability increases under Zn deficient conditions. In cotton roots affected by Zn deficiency, Cakmak and Marschner (1988a) reported increased leakage of low molecular weight solutes, decreased phospholipid content, and decreased degree of unsaturation of fatty acids in membrane lipids. Activity of superoxide generating oxidase enzymes in cotton roots was also increased, while activity of the Zn-containing superoxide dismutase (CuZn-SOD) was decreased (Cakmak and Marschner, 1988b). The resulting higher level of toxic radicals and related oxidants are major factors in peroxidation of membrane lipids and an increase in membrane permeability in cotton roots. The most obvious symptoms of zinc deficiency such as leaf chlorosis (from lipid peroxidation), necrosis, increased membrane permeability, and inhibited shoot elongation (IAA degradation) are most likely expressions of oxidative stress brought about by higher generation of superoxide radicals and a simultaneously impaired detoxification system in Zn-deficient plants (Marschner, 1995).

Cakmak and Marschner (1986) found that Zn deficiency in cotton enhanced P and uptake by roots and translocation to the shoots, but impaired retranslocation through the phloem. Retranslocation of ^{86}Rb and ^{35}Cl were not affected by Zn deficiency. Either induced P toxicity, or oxidative stress from impaired export of photosynthate from older leaves could cause chlorosis and necrosis in older leaves of Zn-deficient plants. Enhanced plasma membrane permeability for B (and Cl) could similarly lead to Zn-deficiency induced B toxicity (Marschner, 1995).

9.3 Growth Responses

Dramatic reductions in leaf size (termed little leaf) are seen in dicots subjected to Zn-deficiency. Little leaf effects are associated with compact a leaf arrangement that reduced intracellular spaces and appears to result from delayed differentiation (Hewitt, 1984). Internode shortening results in a bushy, or rosette, appearance at the terminal of affected plants (Hinkle and Brown, 1968; Grundon, 1987). Retarded stem elongation is strongly correlated with disturbances in the metabolism of IAA, and is most likely a result of enhanced oxidative degradation of IAA rather than inhibited synthesis (Cakmak *et al.*, 1989). Shoot growth is impaired more than roots, and root weights may even be increased initially (Zhang *et al.*, 1991). As deficiency progresses, root tips become enlarged and root hairs become crooked (Hewitt, 1984). In cotton, root exudates of low molecular weight solutes such as amino acids, sugars, phenolics and potassium are increased by low Zn availability (Cakmak and Marschner, 1988a).

If deficiency continues, growth may essentially cease for a period, accompanied by shedding of floral buds and flowers (Hinkle and Brown, 1968). Flower formation, fruit and seed formation are often abortive at several stages (Hewitt, 1984). Flowers often fall off before the bolls form, while bolls are small and sometimes blunt nosed. If the deficiency is mild, yields may be unaffected, but maturity will be delayed and fiber quality will be lower (Hinkle and Brown, 1968). Total dry matter production is less affected than yield, probably due to impaired pollen fertility in deficient plants (Marschner, 1995).

10. COPPER DEFICIENCY

10.1 Leaf Symptoms

Symptoms of Cu deficiency include severe stunting, and development of a dull yellow chlorosis on the older leaves. Veins and adjacent tissue remain dark green in color. In pot studies, plants may have a limp, wilted appearance, with leaves held almost vertical, and tips and margins cupped or pointing downward (Grundon, 1987; Rustamov, 1976; Brown and Jones, 1977). Observations of Cu deficiency symptoms in field-grown are very rare, but have been confirmed by plant analysis for soils of North Carolina with high organic matter content (Hodges, unpublished data).

10.2 Metabolic Responses

Marschner (1995) summarized the major metabolic responses resulting from Cu deficiency. Of primary importance, the activity of numerous Cu-containing enzymes decreases rapidly, with ensuing metabolic changes and inhibition of plant growth. As the amount of the Cu-containing protein plastocyanin decreases, the electron transport chain of PS I is strongly affected, and Cu-deficient plants assimilate less soluble carbohydrates than well-supplied plants. Levels of CO_2 fixation are lower than predicted based on PS I effects only, and appear to be a result of indirect effects of low Cu on chloroplasts affecting PS II. The reduction of CuZn-superoxide dismutase (CuZnSOD) is particularly important, since this enzyme helps to detoxify superoxide radicals. Under severe Cu deficiency, dramatic changes in the ultrastructure of the chloroplast occur, most likely because of inadequate detoxification of the superoxide radical, and resulting photooxidation of membranes within the chloroplasts. Impaired synthesis of carotenoids and quinones, along with disintegration of thylakoid membranes under low Cu conditions probably account for the greater than expected reduction in CO_2 assimilation.

Impaired lignification of cell walls is the most typical change induced by Cu deficiency in higher plants (Bussler, 1981). Activity of polyphenol oxidase and diamine oxidase is severely reduced under Cu-deficient conditions. As a result, phenolics rapidly accumulate and formation of melanotic compounds decreases (Marschner, 1995). These reactions give rise to the characteristic distortions of young leaves, the bending and twisting of stems and twigs, and the enhanced lodging seen in Cu-deficient plants (Vetter and Teichmann, 1968). Effects are especially prominent in schlerynchyma cells of stem tissue, and in severe cases, the xylem vessels may be inadequately lignified, resulting in drooping leaves or wilting. Disease resistance may also be decreased as poorly lignified tissues around wounds allow more rapid pathogen invasion. Impaired lignification, especially in the anther cell wall, may prevent rupture and release of pollen from the stamen, resulting in delayed or poor fruit set (Marschner, 1995). Increased levels of phenols and IAA (reduced activity of polyphenol oxidase and IAA oxidase) may also contribute to the delayed flowering and fruit maturation commonly reported for Cu-deficient plants (Reuter *et al.*, 1981; Hewitt, 1984).

10.3 Growth Responses

There is very little available information on the growth response to Cu deficiency by cotton. Generalizations from other crops indicate that yields are typically decreased due to decreased photosynthate production, reduced flowering, and impaired pollination. Reproductive growth is affected much more than vegetative tissue. Flower formation is typically delayed along with impaired seed and fruit formation. Micronaire of cotton plants receiving no Cu was higher than controls receiving Cu (Anter *et al.*, 1976a).

11. IRON DEFICIENCY

11.1 Visible Symptoms

Cotton is relatively iron efficient, and field observations of deficiency are uncommon. Iron is not readily remobilized in the plant, and symptoms are first seen as interveinal chlorosis on young, rapidly expanding leaves. As the deficiency continues, chlorosis becomes more pronounced, and each new leaf appears smaller and more bleached than the last. Veins will remain green in contrast to the chlorotic areas, which may become a yellowish white. Leaf margins may curl upward, but cupping is not seen. In the final stages the veins become chlorotic and collapse. Necrotic spots are infrequent or develop only in the late stages of symptom development (Hinkle and Brown, 1968; Hewitt, 1984). Symptoms may also appear as interveinal chlorosis in older leaves over time (Marcus-Wyner and Rains, 1982).

11.2 Metabolic Responses

Iron deficiency has major effects on the light harvesting and electron transport systems of photosynthesis (Terry and Rao, 1991), although several other metabolic functions are affected as well. Reduced chlorophyll content (chlorosis) results from direct effects of iron deficiency on synthesis (Pushnik and Miller, 1989). The primary effects of low Fe are especially prominent on the thylakoid membranes of the chloroplasts. Under low Fe conditions, chloroplast number per cell, and per leaf area are not affected. Chloroplasts apparently replicate normally, but contain fewer thylakoid membranes per chloroplast. Under Fe-deficient conditions, Nishio *et al.* (1985) found that galactolipids were reduced by 75%, thylakoid total protein was reduced by 60%, and total chlorophyll content was reduced by 90%. The loss of thylakoids under Fe deficiency reduces the amounts of electron carriers as well as chlorophyll, ferredoxin, and carotene. Xanthophyll content is affected much less than other pigments under Fe deficiency, so leaves are proportionally richer in xanthophyll. These reductions result in fewer photochemical units per leaf area, but electron transport mechanisms are reduced more than light absorption (Terry and Rao, 1991). These responses appear specific to the thylakoid membranes and are reversible upon restoration of Fe supply.

Iron deficiency also affects several enzymes. When photochemical capacity becomes limited by Fe deficiency, a decrease in photosynthesis is mediated through reduced regeneration of RuBP, as was noted earlier for N and P (Terry and Rao, 1991). In this case, the causal pathway involves a decrease in activity of Ru5P kinase. Other important enzymes affected include catalase, (involved in dismutation of peroxide and cell senescence), peroxidase (results in accumulation of phenolics in roots and disease resistance), nitrate reductase, and enzymes affecting ethylene synthesis (Marschner, 1995).

Iron-deficient plants shift from a normal mode of excess anion uptake to excess cation uptake. Under low-Fe conditions, dicots roots accumulate citric and malic acids, and phenolic compounds as they shift from excess anion uptake to excess cation uptake (Strategy I) (Marschner *et al.*, 1986). The resulting increase in reducing capacity and net excretion of protons can mobilizes Fe in the root zone and increase uptake.

Lime- or waterlogging-induced iron chlorosis has been reported or suspected in cotton from Greece (Vretta-Kouskoleka and Kallinis, 1968), India, Pakistan, Australia (Hodgson *et al.*, 1992) and California (Hinkle and Brown, 1968). This usually occurs in alkaline calcareous soils and is caused by high levels of bicarbonate that predispose expanding leaves to Fe deficiency (Mengel *et al.*, 1984; Dofing *et al.*, 1989). High P/Fe ratios (64 instead of 34 expected in well supplied plants) were associated with chlorotic symptoms in soybeans (review by Chen and Barak, 1982). Similar changes in P/Fe ratios have been observed in cotton following waterlogging-induced iron chlorosis (Constable and Rochester, unpublished). Further complications in mineral nutrient balance under waterlogging are high Mn/Fe ratios (Moraghan, 1979; Lucena *et al.*, 1990).

11.3 Growth Responses

In sugar beets, leaf structural attributes such as rate of growth, cells per leaf area, thickness, fresh weight per unit area, and average leaf cell volume are relatively unaffected by Fe-deficiency (Terry and Rao, 1991). Only under severe deficiency is cell division inhibited and leaf growth reduced. In such cases, the apical meristem is particularly sensitive, and cell division is arrested before cell expansion in peas (Longnecker, 1994). Hewitt (1984) reports that leaf

primordia production was halted in tomatoes, but the apical region remained apparently normal. Root elongation is inhibited by iron deficiency, and root diameter in the apical root zone is increased. Root hair production is stimulated (Marschner, 1995).

Specific studies of Fe deficiency effects on reproductive tissues of cotton are limited, but indicate that yield and fiber quality may be affected. Yields of cotton in high pH soils have been improved by supplement iron (Hinkle and Brown, 1968). Micronaire and fiber length were reduced in plants not supplied with foliar treatments of Fe (Anter *et al.*, 1976a,b).

12. BORON DEFICIENCY

12.1 Leaf Symptoms

Growth and development of young tissue throughout the plant is strongly affected since B is not readily remobilized in cotton. Rothwell *et al.* (1967) reported that the first visible sign of B deficiency in cotton was shortened flower corollas with ends of the petals folded inward. Corollas appeared water-soaked. The change from cream to pink color was irregular and took longer than the normal 24 hours. Petioles subsequently developed shortened and irregular thickened petioles on younger leaves, the most commonly reported symptom of B deficiency. They also noted chlorosis of floral bracts, but this is not commonly reported in other areas. Petioles may develop small ruptures, dark bands and necrotic pith (Donald, 1964; Hinkle and Brown, 1968; Rothwell *et al.*, 1967). Grundon (1987) reported that young plants became dark green, with young leaves hanging down, pointed toward the ground, and margins cupped so that the leaf appeared to fold around the upper stem. Excessive vegetative growth may occur where mildly deficient conditions result in poor square retention and shedding of young bolls (Donald, 1964; Rothwell *et al.*, 1967). Older leaves also may become thick and leathery. Under severe deficiency, upper internodes become shortened, and excessive branching produces a short bushy plant with many small, deformed leaves near the terminal. Young apical buds may be deformed or die, preventing further growth of the main-stem (Donald, 1964).

12.2 Metabolic Responses

The primary influence of B deficiency occurs at the primary cell walls and at plasma-membrane cell wall interface; changes here result in a cascade of secondary effects in metabolism, growth, and plant composition. Under conditions of low B, both cell division (DNA and RNA decrease) and cell elongation are reduced (Marshner, 1995). Birnbaum *et al.* (1974), working with cotton fibers cultured *in vitro*, found that B was required primarily for cell elongation and to prevent callusing of the epidermal cells. Marschner (1995) concluded that the first symptoms of B deficiency are modifications in the structure of primary cell walls rather than tissue differentiation. Under low B conditions, cell walls increase in diameter, and become roughen by irregular deposits of vesicular aggregations intermixed with the cell wall membrane. Higher concentrations of pectic and glucose are incorporated into callose precursors (B-1,3 glucan) which accumulate in the sieve tubes, impairing phloem transport.

Under low B conditions, there is a shift in metabolism toward the pentose phosphate cycle, which enhances phenol formation and accumulation while increasing the activity of polyphenol oxidase (Marschner, 1995). This leads to highly reactive intermediates in cell walls that are very effective in producing superoxide radicals potentially capable of damaging membranes by lipid peroxidation. Phenols can be effective root elongation inhibitors, but also enhance radial cell division, in a manner much like IAA (Marschner, 1995).

Cotton under B stress accumulates phenolics like caffeic acid that effectively inhibit IAA oxidase activity (Birnbaum *et al.*, 1977). Basipetal transport of IAA is inhibited, similar to conditions when Ca is deficient. This is most likely a result of impaired membrane integrity under low B conditions, which may be tied to phenol metabolism (Marschner, 1995).

Although B has been shown to increase uptake of sugars by leaves, and is frequently attributed a role to play in short and long distance transport of sugars (reviewed by Duggar, 1983), growing evidence suggests that the correlation between sucrose transport and B status may be correlative only. Observed effects are most likely secondary, such as the impaired export due to callose formation in the phloem or reduced sink activity in shoots or roots suffering from B deficiency (Longnecker, 1994). The effects of B deficiency on uptake of glucose and ions are related to its effects on reducing plasma membrane bound H$^+$ pumping ATPase activity, but it remains unclear if there is a direct link between B and the enzyme, or if the effects are perhaps mediated through improved integrity and functioning of the plasma membrane (Marschner, 1995).

12.3   Growth Responses

Grundon (1987) reported that meristem growth was inhibited by lack of B. Under severe deficiency, the meristem may die, giving rise to many lateral branches with short internodes. This crowding together at the shoot tip gives a bushy appearance, or rosette, which may occur even in the early stages of growth. In severe cases, the whole plant may die. But the most rapid response is inhibition of root elongation, giving roots a stubby or bushy appearance. Roots may become slimy and thickened with the tips necrotic (Johnson and Albert, 1967).

Reproductive tissue is also affected by B deficiencies. Rupturing at the base of squares, flowers, and early bolls

is common (Donald, 1964; Hinkle and Brown, 1968). Nectaries may exhibit excessive flow as membrane integrity is impaired. Many squares and young bolls desiccate and either remain on the stalk or fall to the ground. Incomplete fertilization of seeds results in bolls which are deformed, with a flat-sided or hook-bill appearance (Rothwell *et al.*, 1967). When cut with a knife, lint near the base will be discolored. Such bolls are reduced in size and tend to open only impartially. In severe cases, B deficiency may result in total failure to produce fruit (Donald, 1964; Hinkle and Brown, 1968).

Growth of pollen tubes occurs not by cell extension, but as new cell wall material is deposited at the growing point. Under conditions of low B in the stigma, pollen tube lengthening is impaired, apparently by synthesis of callose and phytoalexins in a defensive response similar to that found in response to microbial infection (Lewis, 1980). Boron may also increase pollen grain viability, and producing capacity of anthers. Indirectly, flowers may be made more attractive to pollinating insects through effects on amount and composition of sugars in the nectar (Marschner, 1995). Miley *et al.* (1969) reported that B applications to cotton enhanced the utilization of applied N fertilizer. This is probably a result of improved pollination, boll set, and fruit retention on cotton experiencing a season long or temporary B deficiency.

Elongation of cotton fibers required the presence of B (Birnbaum *et al.*, 1974). A decrease in fiber length and micronaire was reported for plants not receiving adequate B (Anter, 1976 a,b). Anderson and Worthington (1971) reported cottonseed oil and protein contents were unaffected, even though yields were increased by additions of B.

13.   MOLYBDENUM DEFICIENCY

13.1   Leaf Symptoms

If Mo deficiency is mild, symptoms may be similar to N deficiency, provided most N is taken up in the nitrate form (Marschner, 1995). In cotton, Mo deficiency symptoms begin as interveinal chlorosis followed by development of a greasy leaf surface and leaf thickening (Kallinis and Vretta-Kouskoleka, 1967). As the deficiency progresses, leaves become cupped, and develop necrotic spots and necrotic margins (Amin and Joham, 1960; Peterson and Purvis, 1961; Kallinis and Vretta-Kouskoleka, 1967). Plants appear stunted, with pale, withering leaves. Translocation of Mo within the plant occurs readily in most species, and symptoms appear first in lower or middle leaves (Hewitt, 1984).

13.2   Metabolic Responses

There is limited direct research on the role of Mo in cotton metabolism, plant growth or development. As a cofactor for nitrate reductase, the primary effect of Mo deficiency is a limited ability to convert NO_3–N to reduced forms. In the absence of Mo, nitrate reductase apparently has other catalytic properties that lead to metabolic disturbances similar to those produced by high levels of oxidative stress, such as peroxidation of membrane lipids (Marschner, 1995). Changes in plant biochemical composition (*i.e.,* accumulation of amino acids and organic acids), resulting from Mo deficiency are more complex than with many other elements, since the lack of available reduced N may be overshadowed by the effects of carbohydrate accumulation on reduced photosynthesis (Hewitt, 1984).

13.3   Growth Responses

Growth is rapidly reduced in Mo-deficient dicots, and size and leaf margins may become irregular because of local necrosis or insufficient differentiation of vascular bundles (Marschner, 1995). Molybdenum deficient wheat plants exhibited enhanced sensitivity to water-logging and low temperature (Vunkova-Radeva *et al.*, 1988). Delayed tasseling, reduced flower opening, decreased production capacity for pollen production, and poor pollen germination were reported for Mo-deficient maize (Agarwala *et al.*, 1979). In cotton, bolls developed but were abnormal in appearance, and did not mature or open normally (Kallinis and Vretta-Kouskoleka, 1967). McClung *et al.* (1961) and Mikkelsen *et al.* (1963) reported yield responses to Mo on Oxisols of the Brazilian Cerrado region, but there are no published reports of deficiency symptoms in field-grown cotton. Micronaire of untreated plants was lower than plants receiving foliar sprays containing Mo (Anter *et al.*, 1976a).

14.   TOXICITY

14.1   Aluminum Toxicity

14.1.1   Symptoms

Many plants show very stunted growth and delayed development when exposed to excess Al, but leaf symptoms of Al toxicity are often indistinct, and may resemble either those of P or Ca deficiency (Foy *et al.*, 1978). Plants may be stunted, have dark green leaves, and show delayed fruit set and maturity. Leaves and veins as well as stems may exhibit purple tints, with yellowing and eventual death of leaf tips as toxicity progresses. In other cases, young leaves will curl or roll, and growing points or petioles will collapse. Roots of Al injured plants are stubby and brittle. The root tips and lateral roots thicken and turn brown. The root system as a whole is coralloid in appearance, with many stubby lateral roots but lacking in fine branching (Fleming and Foy, 1968). Young seedlings are more susceptible to Al toxicity than older plants (Foy *et al.*, 1978). The extreme sensitivity

of cotton seedling roots to the presence of Al has been used to in a variety of studies ranging from the identification and chemical activity of biologically active forms of Al in solutions (Hue *et. al.,* 1986) to the role of landscape position on potential for Al toxicity (Beyrouty *et al.*, 2000).

14.1.2   Physiological Effects

Since the effects of excess Al occur primarily within the roots of cotton, toxicity can greatly reduce yield without producing identifiable symptoms in the shoots. Three primary effects of Al on the plant may result in toxicity symptoms: chemical reactions with other essential nutrients; disruption of membrane function; and interference with cell division. The first case is illustrated by the reduced ability of cotton plants to use P in the presence of free Al (Foy *et al.*, 1978), which probably results from formation of aluminum phosphate complexes at the soil-root interface or within the tissue (Naidoo *et al.*, 1978). The presence of free Al also interferes with uptake of Ca and Mg by roots (Marschner, 1995).

In the second case, changes in structure and function of the root cell plasmalemma have been reported in the presence of excess Al (Hecht-Buchholz and Foy, 1981). Aluminum can bind to membrane proteins and lipids (Foy *et al.*, 1978), and form cross-links between proteins and pectins within the cell wall, making it more rigid. Aluminum may also interfere with enzymes regulating the deposition of polysaccharides in the cell wall. Replacement of plasma membrane-bound Ca – or blocking of Ca channels – may also cause membrane disruption (Rengel, 1992). Root respiration is also reduced in the presence of toxic Al levels (Marschner, 1995).

At the cellular level, Al toxicity can increase Al concentrations within the nucleus (Foy *et al.*, 1978), where it forms strong complexes with nucleic acids. Through inhibition of DNA synthesis, most likely during replication, Al interferes with cell reproduction. Abnormal root growth has been attributed to disturbed mitotic processes.

Some plants are able to increase the pH of the rhizosphere to reduce solubility of Al. This may be accomplished by preferential uptake of NO_3-N over NH_4-N forms (Taylor and Foy, 1985). Production of chelating organic acids may also help protect plants from the effects of free Al (Mengel and Kirkby, 1987). Tolerant species have been shown to store Al in the roots rather than shoots (Foy *et al.* 1978).

14.2   Manganese Toxicity

14.2.1   Visible Symptoms

Toxic levels of Al and Mn tend to occur simultaneously in acid soils. Since Al toxicity primarily affects the roots, and Mn toxicity primarily affects the aerial portion of the plant, Mn has been given undue credit for reducing plant growth (Foy *et al.*, 1978). Typical symptoms of Mn toxicity on cotton include abnormally distorted or puckered leaves with irregular chlorotic mottling between veins and dark brown or purple or black necrotic spots along and between the veins (Foy *et al.*, 1995). As the deficiency continues, the leaves become thickened and brittle, with ragged leaf margins. Stems and leaf-stem junctions may have black streaking or sunken brown areas, and petiole tissues may collapse. In severe cases, plant roots turn brown, usually after tops have been severely injured (Adams and Wear, 1957). Symptoms must be quite severe before plant growth is limited solely by Mn toxicity. Foy *et al.* (1978) reported that leaf symptoms of Mn toxicity can be detected even at levels that produce little or no reduction in vegetative growth or yield.

14.2.2 Physiological Responses

Plants subject to Mn toxicity show early evidence of a decrease in net photosynthesis, perhaps as a result of interactions between Mn and enzymes involved in CO_2 fixation (Houtz *et al.*, 1988). As exposure continues, the rate of photosynthesis continues to decline along with a decrease in chlorophyll and total leaf protein. Inhibited DNA replication, errors in mitochondrial replication, and reduced protein synthesis all negatively affect photosynthesis and growth (Foy *et al.*, 1978). In cotton subjected to excess Mn, photosynthesis, respiration, ATP contents, and activity of catalase, ascorbic acid oxidase, and glutathione oxidase decreases while activity of IAA oxidase, peroxidase, and polyphenol oxidase increases (Morgan *et al.*, 1966; Sirkar and Amin, 1974; Morgan *et al.*, 1976). It is these latter reactions that have the most profound effect on leaf symptoms.

As levels of cytosolic Mn^{2+} increase, oxidization occurs, probably by a peroxidase system in the cell walls. This oxidation process accelerates other oxidative reactions that lead to alteration of membrane function, callose synthesis, and oxidative destruction of organic constituents, including IAA oxidase (Horst, 1988). The oxidation of Mn^{2+} by cell wall bound peroxidase systems appears to explain the accumulation of the oxides and deposition of callose in these less physiologically active regions. Callose formation appears to indicate disturbance of membrane function, but remains unclear if Mn stimulates callose formation directly in a manner similar to free Ca in the cytosol, or if this is a secondary effect of oxidative stress. These areas, which result in the brown necrotic spots common in older tissues, contain both Mn oxides and oxidized phenolics that contribute to the discoloration (Horoguchi, 1987). Oxidation of these phenolics produces ethylene, which may induce leaf abscission, and superoxide radicals, which may lead to lipid peroxidation, membrane destruction, and chlorosis. In cotton seedlings, Mn is readily transported into expanding growing zones where it stimulates IAA destruction. The resulting IAA deficiency not only results in poor transport of Ca from the root system to the growing zones (which appears related to downward movement of IAA), but may

limit cell expansion directly by reducing IAA-mediated cell wall extensibility (Horst, 1988). Bhatt *et al.*, (1976) reported that Mn-induced IAA deficiency stimulated cotton root growth at the expense of shoot growth. Manganese may also induce Fe deficiencies (Foy *et al.*, 1978). Management considerations and interactions with other nutrients are discussed in El-Jaoual and Cox (1998).

14.3 Zinc Toxicity

14.3.1 Visible Symptoms

In solution cultures containing high levels of Zn, Ohki (1975) observed that cotton plants developed an overall leaf chlorosis, but veins remained dark green. In a loamy sand with a Mehlich I extractable Zn level of 40 mg kg^{-1} and a pH of 5.7, field grown cotton was stunted, exhibited strong interveinal chlorosis, necrotic spotting of leaves, and reduced root elongation (Hodges, unpublished data).

14.3.2 Physiological Responses

The effects of Zn toxicity on cotton physiology have received little attention. In other plants, high concentrations of Zn can inhibit root elongation (Marschner, 1995), block xylem elements, and inhibit electron transport (Van Assche and Clijsters, 1986a). High Zn concentrations strongly reduce the Mn content of plants (Ruano *et al.*, 1988). Zinc replacement of Mn in thylakoid membranes inhibits PS II activity, and competition with Mg can also suppress RuBP carboxylase activity (Van Assche and Clijster, 1986b). Woolhouse (1983) indicated that Zn-toxicity might induce deficiencies of Mg or Fe. Tolerance to high Zn appears related to the ability of the plant to sequester excess Zn in non-metabolic pools within the vacuoles (Foy *et al.*, 1978).

14.4 Copper Toxicity

14.4.1 Visible Symptoms

The effects of excess Cu are localized in the roots. Leaf symptoms are usually not apparent until plants suddenly begin to wilt. In solution cultures, cotton exposed to toxic levels of Cu had wilted leaves and blackened roots (Sowell *et al.*, 1957).

14.4.2 Physiological Responses

Inhibition of root elongation and damage of the plasma membrane of root cells, as demonstrated by enhanced K efflux, are immediate responses to high Cu supply (Baker and Walker, 1989). In reviewing the effects of Cu toxicity, Marschner (1995) emphasized that root growth is inhibited before shoot growth, and may result from sharp increases in IAA following exposure to Cu. He also noted that root

Cu concentration rise in proportion to Cu supply (with as much as 60% of the total root Cu bound by the cell wall and the cell wall- plasma membrane interface), while restricted transport to shoots limits accumulation. Copper toxicity may also induce Fe deficiencies (Taylor and Foy, 1985). The toxic effects of Cu result primarily from increased permeability of the plasma membrane. This results from the high affinity of Cu for sulfhydryl groups of membrane constituents, and direct or indirect peroxidative degradation of lipids (Woolhouse, 1983; Vangronsveld and Clijsters 1994).

14.5   Boron Toxicity

14.5.1   Visible Symptoms

Cotton tolerates higher B levels in soils than many other crops (Eaton, 1944; Oertli and Roth, 1969), but application of excessive B or in planting in sites with very high soil B levels may result in B toxicity. Symptoms of boron toxicity appear very similar to salt damage, except the leaf reddening generally associated with salt damage does not develop. Leaf tips, then margins become chlorotic and yellow, and the yellowing may spread between the veins to the midrib. In severe cases of B toxicity, leaves will appear scorched, the whole leaf will exhibit symptoms, and senesce. Symptoms are most apparent on older leaves, where B accumulates. It is even possible to see toxicity on the old leaves, and deficiency on the new leaves according to Eaton (1944).

14.5.2   Physiological Responses

The low phloem mobility of B in cotton results in accumulation of B at toxic levels at the leaf margin (Eaton, 1944), the end of the transpiration stream. Accumulation is much greater in leaves than roots, and concentrations are significantly greater in areas of chlorosis and necrosis (Oertli and Roth, 1969). As reviewed by Nable *et al.* (1997), excess B decreased chlorophyll content, growth rates, leaf area , and CO_2 fixation prior to development of visible toxicity symptoms in squash plants. They noted that chlorosis may result from inhibition of ureide metabolism or complexation of ribonucleic acids. Leaf cupping likely results from inhibited expansion of cell walls through disturbances of cell wall cross-linkages (Loomis and Durst, 1992).

In general, tolerant crops will have lower B concentrations than non-tolerant crops supplied equal amounts of B in the substrate (Oertli and Roth, 1969). Nable *et al.* (1997) concluded that tolerance to B toxicity among and between species resulted primarily from differences in passive uptake of B, since there is no evidence of protein facilitated transport, or of internal tolerance mechanisms. They speculated that differences in passive uptake rates may result from plant-specific differences in allowing $B(OH)_3$ to cross the lipid bipolar layer.

14.6   Vanadium Toxicity

14.6.1   Plant Symptoms

Recent studies have shown that vanadium can be toxic to crops at levels less than 0.5 ppm. Kaplan *et al.* (1990) found that V toxicity symptoms for soybean included root darkening, reduced lateral root formation, and formation of short stubby roots with increased diameters. Chlorosis appeared in the upper canopy first and proceeded downward.

14.6.2   Physiological Responses

Accumulation of V is typically much greater in roots than in leaves. Vanadium replaces Mo in several compounds, and toxicity appears to result from interference with nitrate reductase activity and with plasma membrane ATPase (Bollard, 1983).

15.   SUMMARY

To improve our nutrient management strategies for cotton, it is important to understand the specific effects of nutrient disorders and how they affect yield components. In this chapter we have reviewed the effects of mineral nutrition disorders on visual symptoms, plant growth, and metabolic responses. Nutrient disorder symptoms in cotton are affected by uptake, transport, utilization, storage, and remobilization. Actual leaf symptoms, such as chlorosis and necrosis may be both direct, such as inhibition of chlorophyll formation by N, Mg, S, and Fe), or secondary reactions involving oxidative stress due to limited growth or inability to utilize photosynthate. Fortunately, a number of excellent studies on cotton responses to N, P, K deficiencies and Al toxicity are available, and important studies on Zn and B have improved our understanding of these important elements. Other than studies describing symptoms, yield responses and, occasionally, plant component responses, much less is known the metabolic and growth responses of cotton to Ca, Mg, S, Mn, Fe, Cu, and Mo, and we must extrapolate findings from other crops.

The importance of N, P and K in regulating cotton growth has been clearly established, as has their profound effects on plant components and yield. Studies of Zn-deficient cotton have greatly enhanced our understanding of oxidative stress and its effects on plants. Yet much work remains to be done. While recent work has helped us to better understand the role of mid-canopy sinks during flowering on K partitioning and development of upper-canopy K deficiency symptoms, further understanding of the metabolic processes and changes leading to altered K distribution in the whole plant during this period of high demand are needed to overcome this potential yield-limiting barrier. In light of recent advances in understanding the roles and functions of B in higher plants, and the importance of B in development of yield components, studies of B metabolism in this economically important plant deserve greater attention. With increasing pressure to apply sludge, animal manures, and other waste products to agricultural lands, further study of the toxic effects of Cu, Zn and non-nutrient trace metals on cotton is critical.

Chapter 15

AIR POLLUTION STRESS

Patrick J. Temple[1] and David A. Grantz[2]

[1]USDA Forest Service, PSW Research Station, Riverside, CA; and [2]Department of Botany and Plant Sciences, University of California, Riverside, CA and Kearney Agricultural Center, Parlier, CA

1.   INTRODUCTION

The atmosphere surrounding a plant is a complex mixture of gases and particles, some common and some present in only trace amounts. In addition to naturally-occurring gases such as N_2, O_2, CO_2, water vapor, and methane, the lower atmosphere also contains a wide array of natural and anthropogenic compounds whose presence in the air can strongly affect the growth of plants. Many of these potentially toxic compounds occur naturally in the lower troposphere, but when present at concentrations significantly in excess of background concentrations they are classified as air pollutants. Industrialization and urbanization over the past 200 years have greatly increased the concentrations of these toxic compounds in the atmosphere, increasing the potential for adverse effects of air pollution on the growth and productivity of crop plants and forests (Heck et al., 1988; Smith, 1990).

Primary air pollutants are those that are emitted directly into the atmosphere, often from industrial processes or combustion, and are usually highly local in origin and effects. Examples of primary gaseous air pollutants are sulfur dioxide (SO_2), from combustion of sulfurous coal or oil, and smelting and refining of metal ores; hydrogen fluoride (HF), emitted from aluminum refineries; ethylene (CH_2CH_2) emissions from polyethylene plants; or carbon dioxide (CO_2) and oxides of nitrogen (NOx) from a variety of combustion sources. Such pollutants can also arise from small local sources, such as accidental spills of liquid ammonia (NH_3) or chlorine compounds (Cl_2, HCl). Primary particulate air pollutants include fugitive dust from exposed soil and roadways, abrasion products such as automobile tire and brake particles, construction and demolition dusts, and even salt from marine spray and industrial cooling towers.

Secondary air pollutants are those produced in the atmosphere by reactions among natural or anthropogenic precursor compounds. The most significant secondary air pollutant for vegetation is tropospheric ozone (O_3) produced by photochemical reactions among O_2 and NO_x from natural sources and from autos and other combustion sources, and organic compounds emitted from auto exhaust, industrial processes, solvents, and vegetation. Ozone is produced in great quantities in the upper atmosphere through the photolysis of molecular oxygen, and in small amounts in the lower troposphere, by lightning or similar high-energy events. However, O_3 is sufficiently chemically reactive that natural background concentrations in the lower troposphere are low, and without the presence of excess NO_x and hydrocarbons, O_3 concentrations would be self-limiting (Finlayson-Pitts and Pitts, 1986). Other secondary air pollutants include peroxyacetyl nitrate (PAN), H_2O_2, and a variety of other oxygenated compounds. These are less abundant than ozone, but mole for mole may be more phytotoxic.

Polluted atmospheres also contain aerosols, particles, heavy metal vapors, acid precipitation, pesticides, herbicides, and numerous other potentially toxic materials. With a few exceptions, the discussion of air pollution effects in this chapter will be confined to the major gaseous air pollutants, particularly O_3 and SO_2. Descriptions of injury, lists of susceptible plant species, and color photographs of injury symptoms produced by exposure to these and to the minor air pollutants can be found in *Recognition of Air Pollution Injury to Vegetation: A Pictorial Atlas* (Flagler, 1998).

J.McD. Stewart et al. (eds.), *Physiology of Cotton*,
DOI 10.1007/978-90-481-3195-2_15, © Springer Science+Business Media B.V. 2010

2.   PLANT EXPOSURE TECHNIQUES

Our current state of knowledge of the effects of air pollutants on plants has been obtained from the integration of information derived across the spectrum of plant sciences applied to all levels of organization in the plant and its environment. The greatest progress in advancing the state of the science has come when techniques for the exposure of whole plants or plant parts have been applied at the appropriate level of organization within the plant. Information on the effects of acute (*i.e.,* high concentration, short-term) exposures on biochemical or physiological processes, such as photosynthesis, has primarily been obtained using leaf cuvettes. In cuvettes, whole leaves can be exposed to known concentrations of pollutants, and photosynthetic rates can be measured under defined environmental conditions (Legge *et al.,* 1979). Intermediate-term (days or weeks) exposures to study the effects of air pollutants on whole-plant processes, such as carbon partitioning or water relations can be carried out in growth chambers or in continuously-stirred tank reactor (CSTR) chambers, specifically designed for air pollution research (Rogers *et al.,* 1977; Heck *et al.,* 1978). Information on the effects of chronic exposures (several months or growing seasons) to low, medium, or high levels of pollutants to study effects on growth and yield of agronomic crops has been obtained primarily from open-top chamber studies, conducted in the field on plants growing directly in the ground or in large pots. One typical open-top chamber design (Heagle *et al.,* 1973), widely used in the National Crop Loss Assessment Network (Heck *et al.,* 1988), was 3 m in diameter, 3 m high, and was open at the top to allow natural precipitation and pollinators to enter (Fig. 15-1). Pollutant concentrations are controlled by filtering the air entering the chamber through activated charcoal to remove ambient pollutants, and then adding known amounts of pollutants such as O_3 or SO_2 to yield the desired concentrations and exposure regimes within the chambers.

All plant exposure systems come with a suite of advantages and disadvantages (Hogsett *et al.,* 1987; Manning and Krupa, 1992). For example, temperature and humidity in open-top chambers are often slightly higher than field conditions, and light intensity a shade lower. To date, the open-top chamber is the most widely used and accepted technique for the exposure of plants to gaseous air pollutants in the field. Most studies of the effects of air pollutants on the growth and yield of cotton have used open-top chambers, while physiological studies have often been conducted in CSTR and closed field exposure chambers (Musselman *et al.,* 1986).

3.   OZONE

3.1   Origin and Distribution.

Ozone is a colorless, odorless gas that is only moderately soluble in water. As mentioned earlier, excess tropospheric O_3 is formed by photochemical reactions among O_2, NO_x, and reactive hydrocarbons. Both NO_x and hydrocarbons are produced by the combustion of gasoline, so the production of excess O_3 is often associated with dense automobile traffic. Other factors that contribute to the formation of high concentrations of O_3 include the presence of an inversion layer or stagnant air mass, which traps the precursor and reaction product pollutants, and high temperature and light intensity, which increase the rate of the photochemical reactions (Finlayson-Pitts and Pitts, 1986). Although these conditions are normally associated with Los Angeles-type smog, high O_3 pollution occurs in major parts of the cotton-growing regions of the U.S., including the southern San Joaquin Valley, south-central Arizona, southeastern and central Texas, and the Cotton Belt from Alabama to North Carolina (Lefohn, 1992).

No specific information is available on O_3 concentrations in other cotton-growing regions of the world. However, given the close association among O_3 formation, high temperatures and light, and stagnant air masses (Finlayson-Pitts and Pitts, 1986), the potential is present for production of phytotoxic concentrations of O_3 in cotton-growing areas downwind of major urban areas around the world (Schenone, 1993).

Other photochemical oxidant air pollutants, particularly peroxyacetyl nitrate (PAN) are produced in polluted urban atmospheres if sufficiently high concentrations of precursor molecules are available. Cotton is resistant to the effects of PAN (Taylor and MacLean, 1970), and no known instances of PAN injury to cotton have been recorded in the field.

3.2   Entry of Ozone into the Leaf and Initial Toxicity

3.2.1   Transport

Ozone moves from the bulk atmosphere to the sites of action inside the sub-stomatal cavity by eddy transport in turbulent air and by molecular diffusion near the leaves.

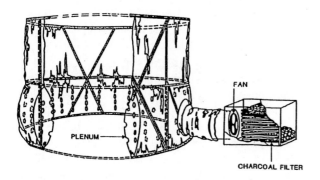

Figure 15-1. Diagram of an open-top chamber used to assess the effects of air pollutants on the growth and yield of field-grown crops. The open-top chambers used in the National Crop Loss Assessment Network were 3 m in diameter by 3 m high. (Diagram adapted from Heagle *et al.,* 1973)

Dry deposited O_3 is removed from the atmosphere through decomposition on soil and external plant surfaces, and on plant interior surfaces through oxidation of metabolites and enzymes and by degradation in the extracellular fluid. Ozone enters the plant through the stomata, and the rate of entry can be predicted from models of stomatal conductance and ambient O_3 concentrations (Grantz *et al.*, 1994; Musselman and Massman, 1999).

Cotton in the San Joaquin Valley of California was found to remove 32.5 mg m^{-2} of O_3 per day (Grantz *et al.*, 1994), based on aircraft measurements using the eddy covariance technique. This is a relatively high flux, attributed to the large stomatal conductance of cotton relative to other vegetation. For example, in the same study, orchards and vineyards removed about 25.8 mg m^{-2} per day (Grantz *et al.*, 1994). This removal by dry deposition to vegetation and other surfaces was sufficient to alter regional O_3 concentrations in large-scale model simulations. In general the stomatal pathway dominates O_3 removal from the atmosphere, and thus stomatal regulation is key to determination of total O_3 dose to a plant (Baldocchi *et al.*, 1987). The aircraft-based measurements of O_3 deposition to an extensive cotton field were closely related to directly measured stomatal conductance (Fig. 15-2). In general a correlation exists between radiation, stomatal opening, and atmospheric transport efficiency (Grantz *et al.*, 1997), which strengthens the relationship between stomatal conductance and O_3 deposition velocity (ratio of O_3 flux to ambient concentration). This may facilitate prediction of O_3 uptake by the cotton canopy.

3.2.2   Initial Toxicity

The site of action and the mode of initial toxicity of O_3 have been investigated for over four decades, yet these

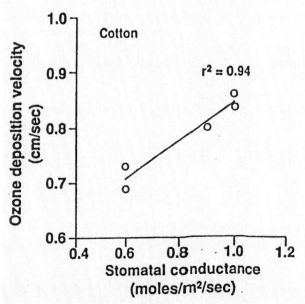

Figure 15-2. Relationship between ozone deposition velocity (total conductance for uptake) and stomatal conductance in upland cotton. (Data from Grantz *et al.*, 1994)

issues remain unresolved. Studies with animal cells demonstrate that little, if any, O_3 reaches the cytoplasm (Pryor, 1992), due to the reactivity of O_3 and its interactions with extracellular fluids. In plants, O_3 reacts with the cell wall and antioxidants and enzymes present therein. However, reaction products of O_3 with plant metabolites may be less reactive and thus penetrate to the cell membrane and possibly into the cytoplasm. That the oxidizing potential of O_3 is somehow communicated to the cytoplasm is made clear by the rapid changes observed in the chloroplasts. Chlorophyll fluorescence kinetics, mRNA for the primary carboxylase, RUBISCO, and ultrastructure of the thylakoid membranes, all change rapidly in response to O_3 exposure (Heath, 1988).

The plasma membrane, located just inside of the cell wall, is a logical site of initial phytotoxicity, either of the toxic reaction products or of O_3 itself (Heath, 1987). It is not clear how much O_3 actually survives the tortuous diffusion pathway through the wall space to the plasmalemma, given the reactivity of O_3 with aqueous solutions, and the presence of ascorbate, polypeptides, and other reactive compounds (Castillo and Greppin, 1988). Unsaturated lipids in the plasma membrane are highly reactive with O_3 in vitro, however, the significance of these reactions in vivo has been questioned. Similarly, hydrophilic domains of membrane-bound and surface proteins are reactive with O_3 (Heath, 1987) and could represent sites of primary toxicity.

Despite the uncertainties surrounding the mechanism of attack of O_3 on the plasma membrane, impacts on key membrane functions are observed very quickly following O_3 exposure. The increased permeability of the plasma membrane to monovalent cations was observed very early in studies of O_3 effects (Heath and Frederick, 1979; Evans and Ting, 1973). More recently (Castillo and Heath, 1990) the role of divalent cations, notably calcium, has been emphasized, as the importance of calcium as a messenger metabolite has become recognized in many plant processes (Leonard and Hepler, 1990). Exposure of isolated protoplasts to O_3 in solution (*e.g.*, guard cell protoplasts of *Vicia faba*; Torsethaugen *et al.*, 1999) significantly perturbed plasmalemma transport of monovalent cations. The inward rectifying K$^+$ channel that mediates guard cell swelling and stomatal opening was impaired, while the outward K$^+$ channel was unaffected. However, the divalent cation, Ca^{++}, may also be involved, if it serves as an intracellular second messenger of O_3 attack as it does for other abiotic stresses affecting stomatal conductance. Elevated cytosolic Ca^{++} concentration leads to an inhibition of the inward K$^+$ channel (*e.g.*, McAinsh *et al.*, 1996) similar to that observed in the O_3-treated guard cell protoplasts.

Indirect evidence for O3 interaction with the plasmalemma is the rapid impact in intact plants on phloem loading of recent photoassimilate in Pima cotton (Grantz and Farrar, 1999). This putative effect on sugar transport across the plasma membrane may be related to whole plant effects mediated by altered carbohydrate translocation (Cooley and Manning, 1987; Grantz and Yang, 1996). Oxidation of some yet unidentified membrane component, following initial oxidation of some compound in the cell wall space, is the likely initial phytotoxic effect of O_3 (Heath, 1987). Identification of these initial targets of O_3 action remains a high research priority.

3.3 Physiological Effects of Ozone

3.3.1 Visible Injury

The appearance of visual O_3 damage is not a good indicator of physiological or agronomic damage. Yield and visible symptoms are not strongly correlated in general (Runeckles and Chevone, 1992).

The visible injury responses of cotton foliage to O_3 are typical of most broad-leaved dicots. Rapid absorption of high concentrations of O_3 by cotton leaves produces an acute injury response characterized by degradation of cellular membranes, loss of membrane integrity, loss of turgor pressure, and cell death, particularly of palisades cells. The visible manifestation of this cellular injury is the appearance of small, irregularly-shaped chlorotic or necrotic lesions on the upper leaf surface, a symptom usually called chlorotic stipple or chlorotic flecking. As these lesions age, their color changes from yellow to various shades of reddish-purple to brown, as anthocyanin pigments accumulate in and around the lesion. Continuous exposure of cotton leaves to lower, but still toxic concentrations of O_3 produces a chronic injury condition. Ozone injury is usually most apparent on older leaves, as injury accumulates in response to continual exposure to O_3 and loss of chlorophyll, accelerated senescence, and premature leaf abscission on the older leaves that have already progressed toward senescence. The very youngest leaves do not exhibit much O_3 sensitivity, often exhibiting enhanced, possibly compensatory, rates of photosynthetic gas exchange. However, leaves approaching full expansion are more susceptible to O_3 injury than older leaves. In cotton, specifically, the period of maximum leaf susceptibility to O_3 follows maximum leaf expansion rates and the period of maximum cellular surface-to-volume ratio, and precedes cell wall lignification and secondary cell wall synthesis (Heath, 1975). For cotton, this period of maximum foliar susceptibility corresponds to a leaf age of 2 to 3 weeks (Ting and Dugger, 1968).

Symptoms of acute foliar injury on cotton may appear within 5 to 7 days following O_3 exposure of 0.25 ppm or higher for several hours (Taylor and Mersereau, 1963). Chronic O_3 injury symptoms, including chlorosis and accelerated foliar senescence and abscission, have been observed during several weeks or months of exposure to a daylight average of < 0.07 ppm in CA (Temple *et al.*, 1985) and < 0.05 ppm in the more humid areas of the southeast (Heggestad and Christiansen, 1982; Heagle *et al.*, 1986). Foliar O_3 injury symptoms on field-grown cotton have been observed in the San Joaquin Valley and in experimental cotton plantings in Riverside and Indio, CA, Phoenix, AZ (Taylor and

Mersereau, 1963; Brewer and Ferry, 1974; Temple *et al.*, 1988a), and Raleigh, NC (Heagle *et al.*, 1986).

3.3.2 Gas Exchange

Photosynthetic and respiratory gas exchanges are susceptible to O_3-inhibition. Plants susceptible to O_3, such as cotton, may experience a relatively rapid reduction in rates of net photosynthesis (Pn) when exposed to elevated concentrations of O_3 for short periods of time, even in the absence of visible foliar injury symptoms (Darrall, 1989; Runeckles and Chevone, 1992). Exposure to >0.20 ppm O_3 for >2 hr. induced a rapid, but reversible, reduction in Pn in susceptible plants (Hill and Littlefield, 1969; Dann and Pell, 1989). The direct effect of O_3 on Pn is presumed to occur through oxidative attack on the plasmalemma as discussed above, by O_3 or its cell wall space-derived free radical by-products, leading potentially to increased membrane permeability, loss of K^+ from cells and organelles, and loss of the pH gradient across the chloroplast membrane (Heath, 1987). As O_3 concentrations increase, or length of exposures increases, the ability of cells to repair this damage decreases, and the reduction in Pn could become irreversible. An additional, indirect, impact of chronic O_3 exposure on Pn has been suggested (Grantz and Farrar, 1999, 2000). End product inhibition and down regulation of photosynthetic enzymes may suppress Pn following O_3-inhibition of phloem loading and carbohydrate export from photosynthesizing leaves. In short term experiments (45 min exposure), carbohydrate export was inhibited considerably more than was Pn. In any case, the permanent reduction in Pn is associated with decreased RUBISCO concentrations, loss of leaf chlorophyll, increased chlorosis, accelerated leaf senescence, and foliar abscission (Pell *et al.*, 1992).

Reductions in Pn in response to increased exposure to O_3 have been documented for cotton and a number of other crop species in the field (Darrall, 1989; Runeckles and Chevone, 1992). Measurements of Pn on the youngest fully-expanded leaves of field-grown SJ-2 cotton were made on 2 September 1986, after three months of exposure to four levels of O_3 in open-top chambers in Riverside, CA (Temple *et al.*, 1988b). No visible O_3 injury was present on these leaves. Exposure to elevated O_3 significantly reduced Pn and the reduction in Pn was proportional to the O_3 dose (Fig. 15-3). However, while rates of carbon assimilation are clearly associated with plant growth and yield, direct correlation between leaf or whole canopy Pn and growth or yield is not straightforward (see Chapter 14). Measurements of whole-canopy photosynthesis of wheat exposed to a gradient of O_3 concentrations correlated well with grain yield reductions, but measurements of Pn on individual plant parts underestimated yield reductions (Amundson *et al.*, 1987). For field-grown soybeans (*Glycine max* Merr.), Pn of upper-canopy leaves was reduced proportionally more by O_3 in well-watered plants than was bean yield, but in drought-stressed plants yield was reduced more than Pn at low O_3 concentrations and slightly less at high O_3 concentrations

PHOTOSYNTHESIS

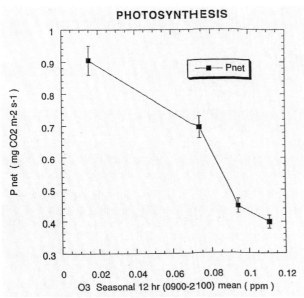

Figure 15-3. Reductions in net photosynthetic
rate (Pn) of the youngest fully-expanded leaf on the
main-stem of Acala (cv. 'SJ-2') cotton exposed to a range
of ozone concentrations in field exposure chambers. Each
data point is the mean of three plants from two replicate
chambers at each level of ozone, +/- one standard error.
Measurements were made in September, following three months
(chronic) exposure to ozone. (Data from Temple *et al.,* 1988b)

(Miller, 1988). In the SJ-2 cotton study mentioned above,
lint yield was highly correlated with reduced Pn as O_3 exposures increased in well-watered plants (R^2=0.94), but in
severely drought-stressed plants the correlation between Pn
and lint yield was not significant (Temple *et al.,* 1988b).
This interaction reflects both the dominant effect of water
deficit on yield, independent of O_3 exposure, and a reduced
effective dose (uptake) of O_3 in drought stressed plants because of stomatal closure. As moisture deficit is variable
across a field and unpredictable over time, these studies
suggest that measurements of Pn on field-grown cotton may
not be useful in predicting the effects of O_3 on yield.

Measurements of g_s in cotton leaves exposed to chronic
doses of O_3 have shown significant reductions in g_s both on
Upland cotton (SJ-2) growing in the field (Temple, 1986;
Temple *et al.,* 1988a) and in closed field exposure chambers (Grantz and McCool, 1992) and in Pima cotton (*G.
barbadense* L. cv. S-6) in CSTR experiments (Grantz and
Yang, 1996). However these instantaneous measurements
following long term exposure to O_3 do not distinguish direct effects of O_3 on stomatal guard cell function from indirect effects on g_s mediated by mesophyll metabolites or
elevated intercellular CO_2 concentrations resulting from
decreased Pn. A kinetics experiment, designed to determine
the direct effects of O_3 on stomatal function in cotton found
no effect of O_3 exposure on rates of stomatal opening or
closing in response to step changes in photon flux density
(Temple, 1986). Diurnal courses of g_s in O_3-exposed and
control leaves revealed similar rates of opening and closing,
and similar complete closure at night. The major difference

was that daily maximum g_s was significantly reduced in the
O_3-injured cotton plants. These results suggest that reduced
g_s may be an indirect effect of O_3-induced inhibition of Pn.

The role of stomatal conductance (g_s) in O_3-induced reduction in Pn has not been resolved. Experiments in which
g_s has been measured simultaneously with Pn during and
following plant exposure to O_3 have found in some instances concurrent reductions in both parameters [*e.g.* sunflower
(*Helianthus annuus* L.), Furukawa *et al.,* 1984], suggesting
increases in both stomatal and non-stomatal limitations to
Pn. Other studies have demonstrated a clear reduction in Pn
with no concomitant reduction in g_s [*e.g.* wheat (*Triticum
aestivum* L.), Lehnherr *et al.,* 1988]. This difference in O_3
response is not necessarily species-specific, as two out of
three lines of hybrid poplar (*Populus* spp.) exposed to O_3,
exhibited parallel reductions in g_s and Pn while in the third
Pn declined with no immediate effect on g_s (Furukawa *et al.,*
1984). In *Vicia faba* (Torsethaugen *et al.,* 1999) exposure to
0.10 ppm O_3 for 4 hr reduced g_s and the rate of stomatal
opening from darkness, without affecting Pn at saturating
intracellular CO_2 concentration. However, increasing the O_3
concentration to 0.18 ppm inhibited both Pn and g_s.

Relatively few studies have been conducted on the
effects of O_3 on plant respiration, and none on cotton. In
general, respiration in leaves increases in response to acute
or short-term (hours) O_3 exposures, particularly when leaf
injury occurs (Amthor, 1988; Dugger and Ting, 1970; Pell
and Brennan, 1973). This increase is not well understood
but is attributed to increased metabolic repair processes,
enzyme and antioxidant synthesis, and membrane reconstruction and energization (Chevrier *et al.,* 1990; Sutton
and Ting, 1977). Long-term exposure to O_3 may increase
or decrease (Miller, 1988; Runeckles and Chevone, 1992)
respiratory rates, both in leaves (Lehnherr *et al.,* 1988)
and roots (Hofstra *et al.,* 1981). Reductions may reflect
decreased availability of carbohydrate substrate in leaves
with impaired photosynthetic capacity and, in the case of
roots, reductions in phloem transport of carbohydrates from
leaves to roots (see below). Effects of O_3 on photorespiration (competitive fixation of O_2 rather than CO_2 by the photosynthetic carboxylase, RUBISCO) are largely unknown
(Runeckles and Chevone, 1992).

3.3.3   Leaf Area Responses: Senescence and Abscission

Exposure of cotton to mean daylight O_3 concentrations
of >0.05 ppm for several weeks can induce accelerated senescence and abscission of older leaves of well-watered
plants (Miller *et al.,* 1988; Temple *et al.,* 1988b). Leaf area
duration of 'McNair 235' grown in open-top chambers in
Raleigh, NC was reduced by 13 % at a seasonal O_3 concentration of 0.051 ppm and 28.5% at 0.073 ppm, compared
with charcoal-filtered air (CF) controls (Miller *et al.,* 1988).
Rates of foliar abscission for four cultivars of cotton grown
in Riverside, CA averaged 51% in CF open-top chambers,

60% in chambers receiving ambient O_3, and 75% in chambers receiving added O_3 (Temple, 1990b). Foliar abscission was significantly less in drought-stressed cotton exposed to O_3 in studies conducted in CA using cv. SJ-2 (Temple *et al.*, 1985; Temple *et al.*, 1988a). However, leaf abscission in response to O_3 in drought-stressed 'McNair 235' in Raleigh, NC did not differ from that of well-watered plants (Miller *et al.*, 1988). In the latter study, Miller *et al.* (1988) compared the relative loss of photosynthetic tissue due to foliar injury and leaf abscission with total lint and seed yield in response to O_3. They concluded that yield losses in cotton exposed to O_3 could not be explained solely on the basis of reductions in leaf area due to foliar abscission, but were also attributable to reductions in efficiency of net CO_2 assimilation in remaining leaves exposed to O_3.

Seedlings of Pima cotton (cv. S-6) exhibited a substantial decline in leaf area development and retention in response to O_3. Mechanical reduction of leaf area to simulate O_3-induced loss of photosynthetic tissues also showed that O_3 had other systemic effects, and plant responses were not mediated solely by loss of photosynthetic leaf area (Grantz and Yang, 1995). In these studies O_3 exposure and mechanical leaf area reduction produced similar reductions in whole plant biomass, but O_3 mediated an additional reduction in root system development on a leaf area basis that was not reproduced by an equivalent mechanical reduction in photosynthetic source tissue. The resulting O_3-specific effect on root hydraulic properties suggested a possible direct effect on carbohydrate translocation to the roots.

Plants can compensate for loss of photosynthetic tissues in a number of ways (Pell *et al.*, 1994). Increased g_s and Pn have been observed in the newly expanding leaves following O_3-induced foliar abscission (Beyers *et al.*, 1992; Greitner *et al.*, 1994). Plants exposed to O_3 also retain greater amounts of carbohydrates in leaves and stems, for use as substrate for repair and growth of new photosynthetic tissues (see below). For example, four cotton cultivars averaged 15.5 main-stem leaf nodes in open-top chambers equipped with charcoal filters (CF) to remove ambient O3, 16.0 in chambers supplied with ambient air, and 18.0 in O_3-added chambers (Temple, 1990b). The leaves produced on cotton plants exposed to O_3 are proportionally larger but thinner than those on control plants, as shown by the increase in specific leaf area on 'McNair 235' at elevated O_3 concentrations, relative to controls (Miller *et al.*, 1988). However, cotton cultivars can differ in their compensatory responses to O_3 injury. Increased branching in response to O_3 injury was observed in an indeterminate Acala cv. 'SJ-2' (Oshima *et al.*, 1979), but the determinate cvs. 'GC510' and 'SS2086' increased the number of main-stem leaves and showed no tendency to increase branching in response to O_3 (Temple, 1990b). Because boll yield in cotton is a function of number of sympodial branches (Oosterhuis and Urwiler, 1988), this difference in compensatory strategies among cotton cvs. may favor boll production in indeterminate cvs. growing in areas with high ambient O_3, though reduced carbohydrate availability may lead to abscission and incomplete development of these additional bolls.

3.3.4 Carbon Allocation and Biomass Partitioning

Plants can respond to environmental stressors that limit growth by altering patterns of carbon allocation and partitioning to favor maximum capture and utilization of resources (Chapin, 1991). In response to limited belowground resources, such as water or nutrient deficiencies, plants alter shoot-root ratios (SRR) to favor the growth of roots. Plants growing under limiting light conditions, or those that have lost leaf tissue through herbivory, retain greater amounts of carbohydrates in shoots, to support the growth of new photosynthetic tissues. In this respect, O_3 is similar to other above-ground stressors in that O_3-induced reductions in growth are proportionally more severe on roots than on shoots, so that SRR of O_3-injured plants are often higher than those of controls (Cooley and Manning, 1987; Reiling and Davison, 1992). The mechanism of this effect of O_3, and of allocation in general, remains very poorly characterized. A variety of experimental manipulations with CSTR-grown Pima (Grantz and Yang, 2000) ruled out two simple hypotheses. Source limitation caused by O_3 damage to Pn was simulated by progressive leaf pruning, but reduced total productivity without reproducing the enhanced SRR observed following O_3 exposure. Similarly, SRR was determined in plants of different sizes (various ages) chosen to correspond to the different sizes of plants of uniform age exposed to different O_3 concentrations. These studies showed that O_3-induced changes in SRR were not due to changes in rate of plant development during which changes in SRR are expected to occur.

The retention of newly-fixed carbon in the stems of plants exposed to O_3 was demonstrated by carbon isotope studies, in which beans (*Phaseolus vulgaris* L.) exported up to 57% less carbohydrate from leaves exposed to O_3 compared with controls (McLaughlin and McConathy, 1983). Primary bean leaves retained greater amounts of carbohydrates, thereby reducing the amount exported to roots, but upper-stem trifoliate leaves exported greater amounts of carbohydrates towards the shoot apex, increasing the amount of substrate available for growth (Okano *et al.*, 1984). The kinetics of export of recent photoassimilate were inhibited in Pima cotton by 45 min exposure to a range of high concentrations of O_3 (0 to 0.8 ppm; Grantz and Farrar, 1999, 2000). Compartmental analysis indicated that vacuolar storage and tonoplast function were not affected by O_3. In contrast, the labile carbohydrate, presumably the cytoplasmic pool mediated by plasmalemma transport, was substantially impacted. At the highest O_3 concentration the carbohydrate available for export to the root system was inhibited by about 80%, of which 20% was due to reduced Pn and about 60% due to impaired phloem transport.

Reduced root growth and increased SRR in cotton exposed to chronic O_3 have been demonstrated in a number of pot studies, both on Acala (Oshima *et al.*, 1979) and on Pima (Grantz and Yang, 1996). In a field study using 'SJ-2' cotton, root biomass of well-watered (WW) plants was reduced 20% more than stem biomass in the high-O_3 treatment, relative to CF controls. The SRR increased from 7.5 in CF controls to 11.1 in plants exposed to a seasonal mean O_3 level of 0.111 ppm. In severely drought-stressed (DS) plants, SRR increased from 3.8 in CF to 5.4 in the high-O_3 treatment (Temple *et al.*, 1988b). However, not all studies have reported significant effects of O_3 on SRR. In a field study using 'McNair 235', Miller *et al.* (1988) reported reduced root growth of plants exposed to elevated O_3, but no significant effect of O_3 on SRR. They attributed this difference to accelerated leaf abscission, which balanced losses in stem biomass with reductions in root growth so that SRR did not change across O_3 treatments.

Allocation of carbon to carbohydrate pools in plants exposed to O_3 has been investigated in a number of studies, but results appear to depend upon species, experimental design, and sampling procedures (Miller, 1988). Miller *et al.* (1989) reported that high O_3 reduced concentrations of reducing sugars, sucrose, and starch particularly in stems and roots of field grown well-watered (WW) 'McNair 235'. However, the reduction was dependent upon stage of growth of the plants, and was statistically significant only at mid-growing season. Starch levels in stems and roots showed the most consistent reductions throughout the growing season. Drought-stressed cotton plants generally had higher carbohydrate concentrations than WW, and no consistent effects of O_3 were observed in DS plants or in the combination O_3 x drought treatment.

In young source leaves of Pima cotton exposed to O_3 in CSTRs (Grantz and Yang, 2000), starch declined with increasing chronic exposure to O3, while soluble sugars increased. Over a range of acute exposures to O_3 (Grantz and Farrar, 2000) soluble sugars increased proportionally to the O_3 concentration except at the extreme value of 0.8 ppm a smaller increase was observed. These results are generally consistent with the model that exposure to O_3 reduces total carbohydrate pools in source leaves of cotton, and the available carbohydrates are partitioned primarily to repair and growth of new photosynthetic tissues. Reduced export of carbohydrate to distant sinks, including roots, seems to reflect a type of source limitation associated with inhibited phloem export from O_3-impacted source leaves.

Reduction in carbohydrate partitioning to roots and reduced root growth in plants exposed to O_3 can have significant effects on root physiology. Reduced nodulation in *Rhizobium*-infected legumes exposed to O_3 (Blum *et al.*, 1983), and reduced rates of endomycorrhizal infection in O_3-injured plants (McCool *et al.*, 1982) have been reported. Reductions in rates of infection and growth of root symbionts may reflect lower pools of available carbohydrate in roots of these O_3-injured plants (McCool, 1988).

3.3.5   Water Relations: Drought Stress, Root Hydraulics

Root Hydraulic Conductance. The inhibition of carbohydrate partitioning to roots that is induced by O_3 exposure undoubtedly has functional effects on root system performance. As 85% of the 20 species surveyed exhibited such reduction in root/shoot biomass ratio (Cooley and Manning, 1987), this effect seems to be of general importance. Potential effects of reduced root biomass are related to concomitant reductions in root surface area for absorption of water and mineral nutrients, and reduced exploration of the soil volume for acquisition of these soil resources. Other possible consequences of impaired root growth or function in plants exposed to O_3, such as reduced production of hormonal transducers such as cytokinins and abscisic acid (ABA), have not been investigated.

Ozone exposure may lead to an increase in hydraulic conductance per unit root biomass, as storage tissue in primary roots is reduced more than surface area of fibrous roots and root hairs [*e.g.*, red spruce (*Picea rubens* Sarg.), Lee *et al.*, 1990; Pima cotton, Grantz and Yang, 1996). Fibrous roots and root hairs are also substantially reduced by O_3 (Ogata and Maas, 1973). On a whole plant basis, however, the smaller root system and reduced conductive tissue lead to reduced hydraulic conductance following O_3 exposure.

A functional measure of root system performance is hydraulic conductance per unit transpiring leaf area (Yang and Tyree, 1993). This reflects the root system capacity to supply water (and by inference mineral nutrients and phytohormones) to the remaining leaf area. This parameter is reduced substantially by O_3 exposure in Pima cotton (Fig. 15-4; circles), even though leaf area expansion and retention are themselves reduced by O_3. While reduced root hy-

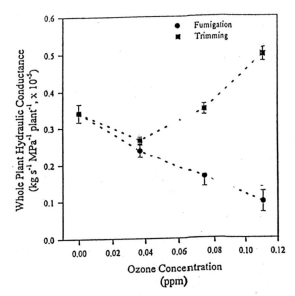

Figure 15-4. Effect of ozone on plant hydraulic conductance on a leaf area basis in Pima cotton grown in CSTRs. (Data from Grantz and Yang, 1995)

draulic conductance with reduced plant size are commonly observed on a per plant basis, the O_3 effect seems to be unique in inducing a similar relationship on a leaf area basis (Grantz and Yang, 1996). Following leaf pruning (Grantz and Yang, 1995, 2000) reduced photosynthetic capacity resulted in reduced whole plant biomass similar to that induced by a range of O_3 exposures. However, effects on root hydraulic conductance per leaf area were minimal (about 15%), and not similar to the large reductions (about 50%) induced by O_3 exposure.

Interactions of Ozone and Drought Stress. The common assumption that plants exposed to elevated O_3 may be more susceptible to drought because of reduced root growth has not been confirmed in the field. It is clear that drought stress (Temple *et al.*, 1985), like low nitrogen stress (Grantz and Yang, 1996), and elevated CO_2 (Heagle *et al.*, 1999), reduces the relative effect of O_3 exposure on yield and biomass production. As exposure to O_3 causes short-term stomatal closure in many cases, entry of O_3 is reduced and consequent phytotoxicity is minimized. However, the limitations posed by drought may sufficiently reduce plant performance to the extent that the relative benefit of increased resistance to O_3 is trivial in a practical sense.

Available evidence is conflicting regarding the effect of O_3 on transpiring leaf water potential, a potential marker for drought stress. Reduction in root hydraulic capacity might be expected to reduce water availability to the shoot, reducing midday water potential (increasing tissue water deficit and stress). On the other hand, many studies have shown that reduced root hydraulic conductance leads to nearly immediate and persistent reductions in stomatal conductance (*e.g.*, Fig. 15-5a), which reduces transpiration and maintains or improves shoot water status. In Pima cotton (Fig. 15-5b; Grantz and Yang, 1996), and Upland cotton (Temple, 1986, 1990a) leaf water potential increased or was stable with increasing exposure to O_3. Field-grown 'SJ-2' cotton exposed to a range of O_3 concentrations in open-top chambers maintained the same leaf water potentials in high-O_3 exposures as in control plants throughout the growing season (Temple, 1990a). It appeared that the lower canopy leaf area and lower rates of stomatal conductance and transpiration of the cotton plants in the high-O_3 treatment balanced the smaller root systems so that the plants were able to maintain adequate water supplies to the leaves (Temple *et al.*, 1988b; Temple, 1990a). This reflects integrated whole plant function. It is not clear whether this homeostasis of leaf water potential would be observed under non-experimental conditions of variable soil moisture supplies, particularly in environments subject to rapidly changing evaporative demand.

It is commonly considered that reductions of net carbon assimilation following exposure to O_3 lead directly to stomatal closure and to reduced carbon allocation to the roots. However, compensatory photosynthesis in newly emerging leaves may restore much of the whole plant carbon assimilatory capacity (Pell *et al.*, 1994). Long-term reduction of stomatal conductance may not be mediated by direct effects

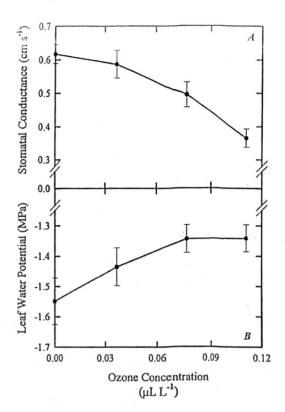

Figure 15-5. Effect of ozone on stomatal conductance (a) and transpiring leaf water potential (b) of Pima cotton grown in CSTRs. (Data from Grantz and Yang, 1996)

of O_3 on the components of leaf gas exchange. A recent modeling exercise has demonstrated that the O_3 effect on root hydraulic properties is sufficient to mediate observed reductions in stomatal conductance during chronic exposure to O_3 (Grantz *et al.*, 1997; Grantz and Yang, 2000). In this case, no direct effects on leaf gas exchange were incorporated into the model. Postulating a stomatal sensitivity to leaf epidermal water status and parameterizing this water potential in terms of leaf to air vapor pressure difference, soil water content, and root hydraulic conductance, resulted in model output that closely reflected observations of stomatal conductance. Scaling these observations to the canopy level, using a comprehensive soil-plant-atmosphere transport sub-model, reproduced O_3 fluxes observed in the field. While the mechanism of O_3 action on whole plants remains elusive, it is clear that a holistic, integrated view of the plant will be required.

3.4 Effects on Yield

The adverse effects of O_3 on yield of cotton have been under investigation for the past 30 years. Brewer and Ferry (1974) placed pairs of ventilated charcoal-filtered (CF) and non-filtered (NF) plastic-covered greenhouses over plots of field-grown 'SJ-1' at several locations in the southern San Joaquin Valley, CA in 1972 and 1973. Number of bolls and weight of lint plus seed were reduced 20% in NF plants compared with CF in areas with the highest ambient O_3 concentrations, and proportionally less in areas with lower ambient O_3. No effects of O_3 on fiber quality were observed in this study. However, in these experiments growth and yield of cotton plants inside the greenhouses differed from that of plants growing outside, prompting concerns that the reductions in yield may have been an artifact of the greenhouse enclosure. A two year study of the effects of O_3 on cotton yield was conducted on 'SJ-2' in Shafter, CA in 1981 and 1982, using open-top chambers to minimize differences between open and chambered plots (Temple *et al.*, 1985). This study was conducted as part of the National Crop Loss Assessment Network (NCLAN) program, designed to assess the economic impacts of air pollution on the major agronomic crops in the U.S. (Heck *et al.*, 1988). In the NCLAN experimental protocol the crops were exposed to a range of O_3 concentrations, from sub-ambient to levels 1.5 or 2.0 times greater than ambient. Ozone exposure-yield response data were then analyzed by regression analysis to produce exposure-response regression equations. Results from this NCLAN study were similar to those reported earlier: yields of well-watered (WW) plants in NF chambers were reduced 15 to 20%, relative to CF controls. Drought-stressed (DS) cotton had relatively little yield losses, except at O_3 levels higher than expected in the San Joaquin Valley. Yield losses to O_3 varied from year to year, being higher in a cool, humid summer, when plants were more susceptible to O_3, than in a year with a hot, dry growing season. Yield losses in cotton exposed to O_3 were attributable primarily to fewer numbers of bolls rather than reduction in weight per boll.

Ozone had no effect on fiber quality or in ratio of lint to seed weight in either year of the experiment, similar to results reported for 'SJ-2' in a greenhouse study (Oshima *et al.*, 1979). In Pima S-6 (Grantz and McCool, 1992) increasing ozone exposure led to a substantial decrease in yield and in fiber quality. Micronaire, length, and length uniformity were particularly impacted. An NCLAN study conducted on 'SJ-2' grown under three different irrigation regimes in Riverside, CA showed that ambient O_3 levels reduced lint yield of WW cotton 26.2% relative to CF plants. Ozone concentrations were higher in Riverside than in Shafter during the earlier study. Cotton grown with one-third less irrigation showed a 19.8% reduction in lint yield, and severely drought-stressed cotton had a 4.7% yield reduction compared with CF plants (Temple *et al.*, 1988a). As with previous studies with 'SJ-2', yield losses were due to fewer numbers of bolls per plant, and not reduction in weight per boll. Yield losses at ambient O_3 levels were also reported for two cultivars of cotton in NCLAN studies conducted in Raleigh, NC. Yields of the cv. 'Stoneville 213' declined 11% in NF relative to CF chambers (Heagle *et al.*, 1986). The cv. 'McNair 235' had a 15% reduction in number of harvested bolls and a 20% reduction in seedcotton weight in ambient O_3 chambers compared with CF plants. Drought-stressed plants showed no reductions in yield at comparable

O_3 exposures. Unlike results with 'SJ-2' in California, reductions in yield of both 'Stoneville 213' and 'McNair 235' were attributable both to fewer numbers of harvested bolls and to reductions in weight per boll. Minor changes in fiber quality were also observed at high O_3 concentrations in both cvs.

Relative seedcotton yield losses for the three cotton cultivars used in the five NCLAN studies are plotted in Fig. 15-6. Yield losses at a mean seasonal O_3 concentration of 0.06 ppm ranged from near zero to over 26%, relative to cotton grown at a background O_3 level of 0.025 ppm. The cv. 'SJ-2' had both the lowest loss, on DS plants in 1981, and the highest loss, on both WW and DS plants in 1982. This large difference in the yield response of a single cotton cultivar to similar concentrations of O_3 was attributable to weather conditions, which increased stomatal conductance and rendered the plants more susceptible to O_3 in 1982 than in 1981 (Temple *et al.*, 1985). The Deltapine cvs. 'Stoneville-213' and 'McNair-235' appeared to be intermediate in susceptibility, with somewhat greater losses in 'McNair'. Seasonal mean daylight O_3 concentrations between 0.05 and 0.06 ppm are common in many cotton-growing regions of the U.S. (Lefohn, 1992). This suggests that annual yield losses due to ambient O_3 for cotton cvs. similar in susceptibility to those used in the NCLAN studies may average between 10 to 15%. However, the variability in these data demonstrates the difficulty of providing accurate predictions of cotton yield losses, without reference to internal and external conditions that alter the physiological susceptibility of the plants to O_3 uptake and injury. Elevated concentrations

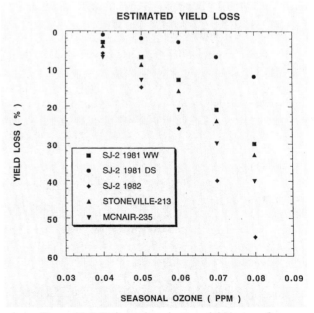

Figure 15-6. Estimated seedcotton yield losses of three cultivars of cotton from five NCLAN studies across a gradient of mean seasonal daylight ozone concentrations, relative to yields at a background ozone level of 0.025 ppm. WW = plants with optimum irrigation; DS = drought-stressed plants, with one-third less irrigation water. In 1982, WW and DS cotton did not differ in yields. (Data from Heck *et al.*, 1988)

of CO_2 in the atmosphere, for example, may reduce the potential impact of O_3 on growth and yield of cotton (Heagle *et al.*, 1999).

3.5   Relative Susceptibilities of Cotton Cultivars to Ozone

Variability in cotton cultivar responses to O_3 has been investigated in a small number of experiments. Growth and yield of eight cotton cultivars were compared in greenhouses supplied with ambient air or charcoal-filtered air in Beltsville, MD (Heggestad and Christiansen, 1982). The Acala type 'SJ-1' was the most resistant of the eight cultivars to O_3-induced yield losses, but the Delta type 'Stoneville-213' was almost as resistant as 'SJ-1'. The upland cultivar 'Paymaster-202' was the most susceptible, and had a 40% reduction in yield in the ambient-air greenhouse. Field trials conducted in the southern San Joaquin Valley have shown large differences in susceptibility to O_3 among CA cultivars (Brewer, 1979). The cv. 'SJ-2' had greater yield reductions in response to ambient O_3 than did the *Verticillium*-resistant 'SJ-5', and 'SJ-2' appeared to be slightly more resistant to O_3 than the earlier-released 'SJ-1'. Of the two Deltapine cvs. used in NCLAN experiments in NC, 'McNair-235' appeared to be more susceptible to O_3-induced yield losses than 'Stoneville-213', although environmental conditions during the two growing seasons may account for some of these differences (Heagle *et al.*, 1988). In a field study conducted in Riverside, CA, susceptibility to O_3-induced yield losses of four cotton cultivars was directly correlated with degree of determinance, so that the cvs. ranked in order of increased determinance and increased susceptibility to O_3: SJ-2 < C1 < GC510 < SS2086 (Temple, 1990b).

The physiological and biochemical mechanisms that determine cultivar susceptibility or resistance to O_3 are largely unknown. Because of the importance of stomata in regulating gas exchange it is reasonable to assume that differences in rates of stomatal conductance among cultivars could account for differences in relative susceptibility to O_3 or other air pollutants (Runeckles and Chevone, 1992). Indeed, greater susceptibility to O_3 among cultivars of common bean (*Phaseolus vulgaris* L.) (Knudson-Butler and Tibbets, 1979; Temple, 1991), tobacco (*Nicotiana* sp., Turner *et al.*, 1972), and petunia (*Petunia hybrida* Vilm., Thorne and Hanson, 1976) has been associated with higher rates of stomatal conductance in these cvs. However, no clear association between rates of stomatal conductance and susceptibility to O_3 injury or yield losses was observed in a field study of four cotton cvs. in Riverside, CA. Instead, differences in susceptibility to O_3 were attributed to differences in degree of earliness, because the more determinate cvs. flowered and set bolls earlier in the summer, when O_3 levels were highest and their short growing season did not allow recovery from O_3 injury, as did the indeterminate lines (Temple, 1990b). Pima cotton (*G. barbadense* L.) cv. S-6 appeared to be more susceptible to O_3 than the cv. SJ-1

(*G. hirsutum* L.), although the mechanisms for these differences have not yet been established (Grantz and McCool, 1992). *G. hirsutum* lines developed in the San Joaquin Valley of California were less susceptible than lines developed in the Mississippi Delta (Grantz, unpublished data). In general, lines developed in areas with low ambient O_3 concentrations, and introduced into production areas characterized by high ambient O_3 concentrations, such as the San Joaquin Valley may lack the constitutive O_3-resistance mechanisms present in cvs. selected for high yield potentials in areas with high ambient O_3 pollution.

Among the resistance predictors currently being studied intensively is ascorbic acid concentration in photosynthetically active leaves. In soybean lines exhibiting sensitivity (cv. Forrest) and resistance (cv. Essex) to O_3, the concentrations of total leaf ascorbate were greater in Essex. In addition the fraction of total ascorbate maintained in its reduced and thus protective form during exposure to O_3 was higher in Essex (Robinson and Britz, 2000). Recent simulations (Plochl *et al.*, 2000) suggest that interception and detoxification of O_3 in the cell wall apoplast by ascorbic acid, prior to its attack on the plasmalemma, could mediate plant resistance to O_3. In this case, a number of physiological or anatomical factors, including total leaf concentration of ascorbate, factors regulating diffusion, and capacity for cytosolic regeneration of ascorbic acid, might be related to the relative O_3 resistance of advanced cultivars of various species, including cotton.

3.6   Economic Losses

Translating crop yield reductions into economic losses is a complex problem in agricultural economics (Adams and Crocker, 1988). Reduction in total crop yield may increase crop unit value, so that the burden of economic losses due to O_3 air pollution falls disproportionally on consumers and not on producers of agricultural commodities. Indeed, models of economic losses from the impact of O_3 on major crops have estimated that consumer surplus losses comprise 50 to 100 % of total losses (Adams *et al.*, 1988). Regional differences in levels of O_3 pollution mean that the costs and benefits of improving air quality are not distributed evenly (Howitt and Goodman, 1988). Distortions in the free market economy introduced by government agricultural support programs also increase the difficulty of estimating the benefits to agriculture from cleaner air (McGartland, 1987). For cotton, benefits to the California economy from improvements in O_3 air quality have been estimated at a 12.4% increase in base benefits at a seasonal mean O_3 level of 0.06 ppm, to a 48% increase in benefits at a background O_3 level of 0.025 ppm (Howitt and Goodman, 1988). For the U.S. economy as a whole, the annual economic benefits accrued from a 25% reduction in O_3 concentrations in the major crop-growing regions of the country, including CA, the Mid-West, and the South-East were estimated at $1.7 x 10^9 (Adams *et al.*, 1986).

3.7   Agricultural Practices

Relatively little can be done to protect cotton from the adverse effects of O_3. Because O_3 flux is dependent upon the rate of stomatal conductance, it is theoretically possible to withhold irrigation water prior to a predicted O_3 episode, to close stomates and exclude O_3. This tactic seems unrealistic in common agricultural practice, and would not be available for non-irrigated cotton. Several agricultural chemicals have been developed that provide some protection from O_3 injury, including fungicides such as benomyl that provide ancillary protection from O_3, and various anti-oxidants such as citrate and ascorbate, that act as free-radical scavengers (Pryor *et al.*, 1982). The most widely studied "antioxidant" chemical is ethylene diurea [N-2-(2-oxo-1-imidazolidinyl)ethyl]-N'-phenylurea] or EDU. This chemical, supplied either as a soil drench or foliar spray, can provide many plant species almost complete protection from foliar O_3 injury (Carnahan *et al.*, 1978; Manning and Krupa, 1992). Apparently, EDU increases cellular activity of superoxide dismutase and catalase, thereby protecting foliage from O_3-induced accelerated senescence (Bennett *et al.*, 1984). However, because of its expense and the difficulty in adjusting dose (EDU may be toxic at high concentrations (Kostka-Rick and Manning, 1993)), and because the foliar O_3 protection afforded by EDU did not translate into significant increases in yield in many field trials (Heagle, 1989), EDU has not proven useful as a standard agricultural chemical.

Leaf wetness from dewfall was found to reduce ozone deposition to cotton (Grantz *et al.*, 1997), apparently by occlusion of adaxial (upper) stomatal pores. This suggests a possible management strategy in fields in which overhead sprinklers are available, and reliable air quality forecasts can be obtained. The blockage of upper stomata is a conjecture based, in part, on the opposite effect observed following dewfall in hypostomatous grape (Grantz *et al.*, 1995), in which leaf wetness increased ozone deposition to the canopy, apparently by increasing the reactivity of the upper surface of the leaves.

Breeding cotton lines for increased resistance to O_3 may have already occurred, inadvertently. Cultivars selected for optimum growth and yield in areas of relatively high ambient O_3, such as the San Joaquin Valley, have been selected under the environmental conditions prevalent in the area, including O_3. Cultivars originally developed in areas of low ambient O_3, such as Pima cotton, may suffer significant yield losses when introduced into high O_3 regions (Grantz and McCool, 1992). The increased susceptibility of highly determinate cotton lines to O_3-induced yield losses has already been discussed. In addition, breeding for one specific desirable characteristic, such as resistance to O_3, may not translate into increases in yield if the mechanism of resistance incurs metabolic costs, such as increased allocation of carbon into defensive antioxidant compounds, or reduc-

tions in assimilated carbon because of reduced stomatal conductance. Molecular engineering, involving transfer of genes for O_3-resistance to new cotton lines, is theoretically possible, and may become more economically viable if O_3 pollution increases in cotton-rowing regions of the world. The first stages in such a program would be to identify the biochemical and physiological mechanisms of susceptibility and resistance to O_3 in cotton, and then to identify and isolate the genes regulating those mechanisms. Given the complexities of the effects of O_3 on the different levels of organization within the plant (Darrall, 1989; Heath, 1988), this would appear to be a daunting task.

4.   SULFUR DIOXIDE

4.1   Origin and Distribution

Sulfur dioxide (SO_2) is an extremely acrid, irritating gas, emitted into the atmosphere from the combustion of sulfur-containing ores during roasting and smelting processes, and from the combustion of sulfur-contaminated fossil fuels, such as coal and natural gas. Natural emissions of SO_2 also occur during volcanic eruptions and from fumaroles and vents in tectonically-active regions around the world. With the exception of these natural emissions of SO_2, high concentrations of SO_2 are generally found around point sources, in the vicinity of ore-processing facilities or large coal-fired power plants. In the 19th and early 20th century unregulated emissions of SO_2 caused widespread damage to vegetation around these large industrial sources, most notably around Sudbury, Canada and Copper Basin, TN (Hursh, 1948).

The advent of environmental regulations and strict emission controls in North America and western Europe has substantially reduced the impact of SO_2 on vegetation, to the extent that SO_2 is no longer considered to be a significant air pollutant in most agricultural areas (Rosenbaum *et al.*, 1994). In some areas of the developed world sulfur deficiencies are again being observed in sensitive crops. In eastern Europe and in many developing countries, SO_2 continues to be a threat to agricultural crops and forest vegetation.

4.2   Mechanisms of Toxicity and
Symptomatology

Sulfur dioxide enters plant leaves through stomata, and because of its high degree of solubility in water, it immediately dissolves on the thin aqueous film surrounding cells in the sub-stomatal cavity (Mudd, 1975). The sulfite ion is readily oxidized to sulfate, and if SO_2 is absorbed slowly, it can act as a source of S nutrition for the plant (Thomas *et al.*, 1943; Olsen, 1957). If high concentrations of SO_2 (*e.g.*, 0.10 to 0.50 ppm for 8 hours) are absorbed rapidly by the plant, sulfite accumulates, and because sulfite is 30 times more toxic to plant cells than sulfate (Thomas *et al.*, 1943), it rapidly reaches toxic concentrations. Cotton leaves are highly susceptible to the effects of SO_2, and rapid absorption of the gas can produce plasmolysis of mesophyll cells directly underneath the lower epidermis, leading to a glazed or silvery appearance of the adaxial leaf surface (Barrett and Benedict, 1970). Continued exposure to toxic concentrations of SO_2 leads to interveinal tissue chlorosis and necrosis, producing bleached light tan to dark brown interveinal markings.

Foliar injury symptoms resembling SO_2 injury are occasionally observed on irrigated cotton. The symptoms usually appear shortly after irrigation, if the plants are subjected to periods of high temperatures and low humidity, producing rapid rates of transpiration. Results of chemical analyses of foliage show high concentrations of salts, particularly calcium sulfate (Barratt and Benedict, 1970). Because these SO_2-mimicking symptoms are often observed on plants growing on soils with high gypsum (hydrous calcium sulfate) content, rapid absorption of sulfate, translocation to leaves, and accumulation of sulfate salts at the termini of transpirational streams appears to be the mechanism of toxicity.

4.3   Effects on Yield

Brisley *et al.* (1959) conducted an extensive series of experiments on cotton (cv. 1517-C) determining the relationship between amount of leaf area injured by exposure to SO_2 and subsequent effects on yield. They concluded that exposure to concentrations of SO_2 that did not produce visible foliar injury had no effect on yield. Reductions in yield of cotton due to SO_2 injury resulted from fewer numbers of squares produced, because SO_2 injury had no effect on boll development or weight of seed cotton per boll. Exposures to SO_2 had no effect on incidence of plant diseases, particularly *Verticillium* wilt. The relationship between amount of leaf tissue injured by SO_2 (X) and percent of total seed cotton yield (Y) was: $Y = 99.26 - 0.68 X$. Fiber quality was unaffected by SO_2 injury to foliage (Davis *et al.*, 1965). A later NCLAN study found no effect of SO_2 on yield of 'Stoneville-213' at concentrations up to 0.35 ppm, and no interaction between SO_2 and O_3 on growth or yield of cotton (Heagle *et al.*, 1986). These data indicate that ambient levels of SO_2 in agricultural regions of the U.S. should have no effect on growth or yield of cotton. However, effects of SO_2 on cotton in developing countries are largely unknown.

5.   SUMMARY

A review of the effects of air pollution on cotton growth and yield has been presented. This included a general description of the sources of air pollutants, both primary air pollutants from industrial porcesses or combustion and secondary pollutants from natural or anthropogenic precursor compounds. Methods of exposing plants to air pollutants and for measurement of the effects on growth and yield were described. The bulk of the review focused on ozone: it's origin, entry into the plant, toxic effects, and a detailed account of the physiological effects of ozone including reductions in gas exchange, leaf area responses of senescence and abscission, altered carbon allocation and biomass partitioning, reduced hydraulic conductance, and the resulting detrimental effects on yield. A final mention was made of the origin, toxicity, and symptomology and effects on yield.

Chapter 16

RESPONSES OF COTTON TO CO_2 ENRICHMENT

Jack R. Mauney[1]
Research Consultant, Scottsdale, AZ

1. INTRODUCTION

The review of Krizek (1986) provides the background against which this review will proceed. The literature (Kimball,1983; Cure, 1986) suggested that cotton was one of the crops most strongly influenced by CO_2 enrichment. Based on greenhouse studies by Mauney *et al.* (1978) the yield response of cotton was placed at a doubling of yield with a doubling of CO_2 concentration.

With this background, cotton was considered a good candidate for exposures to CO_2 enrichment under field conditions. Because cotton is a woody perennial (Mauney, 1986a) it's responses to CO_2 enrichment were thought to be a good indicator of responses which might be seen in vegetative ecosystems such as forests and chaparral. With support from the U.S. Departments of Agriculture (USDA) and Energy (DOE) and the University of Arizona, field experiments were conducted from 1983 to 1991 exposing cotton to elevated levels of CO_2. In 1983 to 1987 the exposures were in clear plastic open-topped field enclosures through which flowed the CO_2-enriched air. From 1988 to 1991 the exposures were by means of a unique device called a Free-Air CO_2 Enrichment (FACE) device by which the crop was exposed to a controlled level of CO_2 enrichment without any surrounding enclosure. In addition, during these years there were attempts to determine if the strong yield response of cotton to CO_2 enrichment could be extended to grower-tended crops. Attempts were made to carry the CO_2 in irrigation water and in tubing placed within the cotton rows.

This review will report the findings of these studies and assess the future directions which cotton culture in CO_2-enriched atmospheres may take. CO_2-enrichment also allows an assessment of the physiological factors that limit production in field plantings. By increasing the photosynthetic capacity of the crop these experiments provide insight into the growth parameters of the species at high levels of productivity.

2. OPEN-TOPPED CHAMBER EXPERIMENTS

For five seasons, 1983 through 1987, experiments were conducted in Phoenix, AZ, using 3 m x 3 m clear plastic enclosures to expose cotton (cv. DPL70) to 650 ppm (v/v) CO_2. Controls were ambient air (350 ppm.) in similar plastic enclosures. In 1983, 1984, and 1985, an additional treatment at 550 ppm CO_2 was included in an attempt to establish a dose-response trendline for CO_2-enrichment. In 1985, 1986, and 1987 the 650 ppm treatment was also exposed to well-watered and water-stressed treatments. The fully watered plots received, on a three-day schedule, full replacement of evaporative demand as measured by a near-by pan evaporator while the stressed plots received 67% of that amount. In 1986 and 1987 nitrogen-stress was also included as a treatment (Kimball and Mauney, 1993; Kimball *et al.*, 1986, 1987).

Averaged over the five years of the open-topped chamber experiment, exposure to 650 ppm CO_2 induced a 65% increase in growth and yield of the crop (Kimball and Mauney, 1993). There was a linear response of the increase to the CO_2 concentration. Exposure to 550 ppm CO_2 induced approximately half the yield response as the higher dose.

Even though both water stress and nitrogen stress significantly limited growth and productivity of the crop in these chambers, they did not affect the relative response to CO_2-enrichment (Kimball and Mauney, 1993). This was unexpected since Mauney *et al.* (1978) had observed a strong interaction between nitrogen availability and response to CO_2 in a greenhouse, pot-culture setting, and the effects of CO_2 on stomatal closure was expected to reduce water demand in the enriched plots.

Limited space in the open-topped chambers prevented sufficient sampling for statistical testing of growth and yield data, but repeated sampling of net leaf photosynthesis in the chambers showed a statistically significant increase in CO_2 uptake of 45% (Kimball *et al.* 1987).

In 1985 Radin *et al.* (1987) and in 1986 Inoue *et al.* (1990) measured the stomatal conductance of CO_2 and water of cotton leaves from these chambers so as to calculated water use efficiency (WUE) of the photosynthetic mechanism under these conditions. Both observed an increase in net photosynthesis of 65 to 70% in the chambers exposed to near doubling to atmospheric CO_2 to 650 ppm.

Each of these groups observed a general decline in photosynthetic rate as the season progressed, with the decline in the rate in the enriched chambers being greater than that in the controls. Thus the photosynthetic enhancement of CO_2-enrichment was reduced to insignificance at the end of the season. Radin *et al.* (1987) attributed this to lower temperatures in September and October. By their calculation WUE was increased by both an increase in stomatal conductance of CO_2 and a decrease in conductance of water.

3.   SPAR EXPERIMENTS

K.R. Reddy and coworkers (Reddy *et al.*, 1994, 1996a, 1996b) used the Soil-Plant-Air-Research (SPAR) units at Starkville, MS, to study the relationship of air temperature and CO_2-enrichment. The SPAR units are clear plastic enclosures in which the plants can be exposed to various water, temperature, and CO_2 conditions while growing in a large soil bin under natural sunlight. They observed that CO_2-enrichment (to 720 ppm) increased the photosynthetic rate of cultivar DPL51 by 40%.

Under the conditions of these experiments they observed that growth was influenced to a greater extent by temperature than by CO_2 concentration and that at temperatures above 30°C the plants produced no harvestable bolls. Below 30°C the enriched plots had a yield increase of 40%, but enrichment did not alter the deleterious affects of the high temperatures. It is difficult to interpret the temperature values which were observed to be deleterious since this review reports temperatures up to 32°C and higher were recorded in experiments that were highly productive (Kimball *et al.*, 1992, 1993, 1994). Perhaps the leaf temperatures in the SPAR units were considerably higher than the air temperatures that were reported.

The important observations of these experiments was that above some maximum temperature the plants became unproductive and that the deleterious effects of that high temperature was not altered by CO_2 enrichment.

4.   FACE EXPERIMENTS

To transfer the greenhouse and open-topped chamber results to agricultural and natural ecosystems, field experiments were required in which there would be no 'chamber' effect. To do this a unique facility was constructed to produce a Free-Air CO_2 Enrichment (FACE) experiment. This facility is described in detail by Lewin *et al.* (1994). In essence it consisted of a torus plenum of pipe 25 m in diameter along the perimeter of which were place 32 vertical vent pipes. These vent pipes were 2 m in height, each fitted with five vent portals along its length. Each vent pipe had a control valve at its base. The CO_2 concentration within this circular area of crop was controlled by pumping CO_2-enriched air through the plenum and out through the vent portals. The CO_2 concentration was monitored at the center of the circular array once each second. Wind velocity and direction was monitored every 4 seconds.

The design concept of the FACE arrays was to inject CO_2-enriched air into the wind stream crossing the plots and let the wind distribute the enriched air throughout the plots. To accomplish this, the vents on the upwind half of the plenum were opened. Positioning of the opened or closed vents was updated every four seconds based on the wind velocity and direction. Studies by Evans and Hendry (1992) were convincing that the rapid excursions of CO_2 concentration created by this control system (Nagy *et al.*, 1994) did not alter the photosynthetic reaction of cotton leaves, compared to the average concentration, so long as the excursions were shorter than 60 seconds.

Examination of the response of the cotton crop to CO_2 concentration of 550 ppm in this experimental facility showed that the plots were surprisingly uniform in crop response within the center 20 m-diameter circle of the experimental space (Mauney *et al.*, 1994). This allowed an area of over 300 m^2 of crop for research sampling. This large area accommodated a large array of research objectives. Over the three-year span (1989 to 1991) of this experiment, over 30 research professionals made observations on the responses of the cotton crop to CO_2 enrichment (Hendry and Kimball, 1994).

In 1990 and 1991 each FACE array was divided into two irrigation treatments. The wet (W) plot was irrigated, by means of underground drip irrigation tubing, to fully replace evaporative demand of the crop as estimated by a nearby pan-evaporation device. The dry (D) half of each array received 67% of the evaporative demand.

4.1   Growth and Yield

The growth and yield response of the cotton crop (cv. DPL77) to CO_2 enrichment confirmed the observations previously made in open-topped chambers (Mauney *et al.*, 1992, 1994). Total crop biomass began diverging at about 30 days after planting (Day of Year, DOY, 165) and the difference was always statistically significant thereafter. In 1991, enrichment to 550 ppm increased total biomass by

37% in the W plots and 35% in the D plots (Fig. 16-1).

Leaf area index (LAI) and root-shoot ratio (R/S) showed varied responses dependant on the time-of-season and treatment interactions. LAI increased significantly early in the season in the FACE-W plots compared to all other treatments. But after 75 DAP the effect of moisture status dominated treatments. Both FACE-W and Control-W had LAI of 4, while both FACE-D and Control-D had maximum LAI of 2.5 (Mauney *et al.*, 1994; Fig. 16-2).

Due to the stimulation of LAI early by FACE, the R/S was greater in the Controls during the germination stage of growth (Fig. 16-3). During the juvenile stage of exponential growth (DOY 145 to 185) all plots had similar R/S (0.10 to 0.12). Thereafter, as growth of shoots was replaced by flowering and fruiting, the R/S increased more in the FACE plots (to 0.15) than in the Controls, due to the greater availability of carbohydrate to support roots in competition with the sink strength of bolls. These observations differ from those of Idso *et al.* (1988) who found that no change in R/S occurred as a result of CO_2-enrichments of cotton.

Prior *et al.* (1994) and Rogers *et al.* (1992) used soil probes to estimate root activity and confirmed greater root activity in the FACE plots during the boll maturation phase of the season. Runion *et al.* (1994) observed greater microbial activity in the soil surrounding the roots of the enriched plots, possibly due to the greater root length. Hendrix *et al.* (1994) found greater reserves of carbohydrate stored as starch in stems and roots beginning about DOY 190. This coincided with the change in R/S from higher in Controls to higher in FACE plots. Similarly, Huluka *et al.* (1994) observed that the C/N ratio in leaves and stems was greater in the enriched plots, both W and D. This increase in C/N ratio was attributed to greater C fixation and storage as starch (Hendrix, 1992) while supply of N via the transpiration stream remained the same in all treatments.

Typical of boll loading patterns from these experiments was that in 1991 (Fig. 16-4). This shows that the effect of the CO_2 enrichment was to prolong the period of maximum boll loading. The wet treatments experienced less severe "cut-out", that is, they continued loading bolls throughout the season at a slow rate.

Table 16-1 shows the seedcotton yield increases for the three years of the FACE study (Mauney *et al.*, 1992, 1994). The smaller yield increase in 1989 can be attributed to the return to growth and fruiting by the Control (C) plots late in the season while the FACE (F) plots, being more heavily loaded with bolls, remained on cut-out. This phenomenon was also observed in open-topped chambers (Kimball *et*

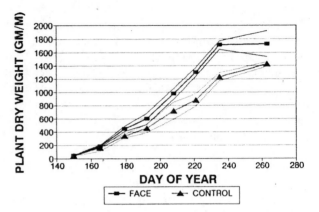

Figure 16-1. Dry weight (gm/m²) accumulation of the biomass of the cotton crop from the well-irrigated (Wet) Control and FACE plots grown in 1989. The crop was planted on DOY108 and emerged by DOY111. Plots were sampled by removing every third plant from 3 m of row of the plots. Each point is the average of four replications. Dashed lines represent confidence interval (p<.05). (From Mauney *et al.* 1992)

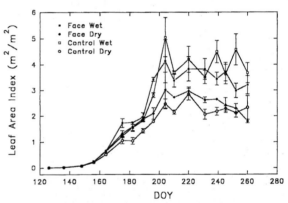

Figure 16-2. Leaf area index (LAI) of the cotton crop grown in Control or FACE conditions in the well-irrigated (Wet) and water-stressed (Dry) plots in 1991. The crop was planted on DOY106 and emerged by DOY116. Data are based on the biomass harvests of 1/3 of the plants from 3 m of row. Data points are the average of four replications. Vertical bars are the standard error of each data point. (From Mauney *et al.* 1994)

al., 1986, 1987). With an indeterminate plant such as cotton, which has a cyclical fruiting pattern, the length of the experiment may influence the relative effect of the treatments. Final yield response may not reflect physiological effects of the treatments, if the cycles got out of phase, and the Controls returned to fruiting while the FACE plots remained dormant.

The pattern of boll loading shown in Fig. 16-4 is the same as that observed by Mauney *et al.* (1978). The rate of boll increase is not the primary rationale for the increase of final yield. The greater numbers of bolls carried by the enriched crop came as the result of a delay in the time of cutout,

Table 16-1. Percentage increase in seedcotton and biomass yield, and water use efficiency (WUE) from FACE (F) plots compared to Controls (C) when irrigated with full replacement of evaporative demand (W) or with 67% of evaporative demand (D). WUE calculated using biomass data.

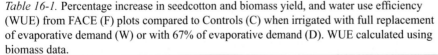

	FD/CD			FW/CW		
Year	Seedcotton	Biomass	WUE	Seedcotton	Biomass	WUE
1989	---			122	125	
1990	43	118	119	151	134	128
1991	142	135	137	143	137	139

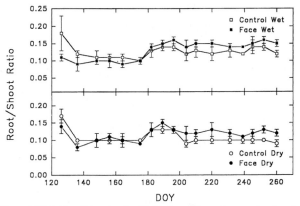

Figure 16-3. Root/shoot ratio (R/S) throughout the season for the cotton crop grown in Control or FACE conditions in the well-irrigated (Wet) or water-stressed (Dry) plots in 1991. The crop was planted on DOY106 and emerged by DOY116. Data are the ratio of the weights of dried roots which could be pulled from the soil divided by weights of stems, leaves, and bolls. Data are based on the biomass harvests of 1/3 of the plants from 3 m of row. Data points are the average of four replications. Vertical bars are the standard error for each data point. (From Mauney *et al.* 1994)

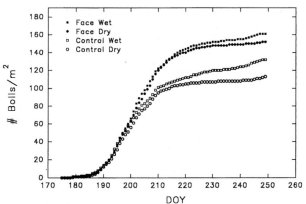

Figure 16-4. Cumulative boll load of harvestable bolls from cotton plants in the Control and FACE conditions in the well-irrigated (Wet) and water-stressed (Dry) plots. The crop was planted DOY106 and emerged by DOY116. Data are from flowers tagged each day that were retained to maturity in 1991. Data points are the average of four replications. (From Mauney *et al.* 1994)

the slowing and cessation of boll retention. Few bolls were added after DOY208 in the controls while the enriched plots continued to add bolls until DOY215. An explanation of this effect may be seen in the data for root/shoot ratio (Fig. 16-3). The R/S of the Controls (both CD and CW) began to decline at about DOY190 and maintained a significantly lower value than the enriched plots for the remainder of the season. Greater carbohydrate supply to the roots of the enriched plants may explain the delayed cutout and greater boll carrying capacity of the crop.

4.2 Water Use and Light Use Efficiency

Water use efficiency (WUE), as measured by kg of biomass accumulated per kg of water used, was increased 19%

and 28% (Table1) by F-D and F-W, respectively, compared to their Controls in 1990. In 1991 WUE was increased 37% and 39% by F-D and F-W. These increases were due to the increases in weight of the treated plots rather than reductions in water used. Dugas *et al.* (1994) using sap flow techniques, Hunsacker *et al.* (1994) using soil water depletion measurements, and Kimball *et al.* (1994) using energy-balance approaches all concluded that CO₂-enrichment did not change the evaporative demand of the crop. Measurements of leaf water potential by Bhattacharya *et al.* (1994) showed that in the Wet treatments enrichment had little effect on stomatal conductance or leaf water potential until late in the season (after DOY233) when the crop was mature. Since the water use was unchanged by FACE, the increased of WUE conformed closely to the biomass increase observed.

Factored into the interpretation of leaf area, transpiration, and water use by the crop was the data of Kimball *et al.* (1992, 1994) that recorded the average canopy temperature of the FACE and Control plots. Using infrared reflectance measurements they recorded canopy temperatures of 26° to 32°C during the growing season. The FACE plots had a canopy temperature that averaged 0.80°C higher than the Controls. These measurements were similar to those made by Idso *et al.* (1987a, b) in open-topped chambers. Thus, the stomatal closure induced by CO₂-enrichment was matched by higher leaf temperatures and greater leaf area, resulting in the same canopy transpiration and water use as the Controls.

Hileman *et al.* (1992, 1994) and Idso *et al.* (1994) measured the photosynthetic rate of individual leaves and of the canopy at various times during the three-year FACE experiment. Both concluded that leaf and canopy photosynthesis was increased by 20 to 30% due to CO₂ enrichment to 550 ppm. They observed no change in leaf or canopy transpiration rates. When they calculated WUE on the basis of CO₂ uptake compared to water vapor transpired, they found that the 20 to 40% increase in canopy WUE could be accounted for by the increase in photosynthetic CO₂ fixation.

Pinter *et al.* (1994a, b) calculated the light use efficiency (LUE) of the canopy in the FACE plots. This calculation is based upon the light energy absorbed compared to the dry weight accumulation of the crop. They found that the LUE calculated for the Control plots (1.51 g/MJ) was similar to that observed by Rosenthal and Gerik (1991), but that the FACE plots had 15 to 40% greater light absorption. They concluded that in both the W and D treatments the LUE was increased by an average of 25% across the three years of the measurements. There was no significant change in LUE due to the water deficit imposed by the D treatments.

5. ACCLIMATION TO ENRICHMENT

Mauney *et al.* (1978) observed that the photosynthetic rate of cotton leaves exposed to CO₂-enrichment decreased during the first three days of exposure as the starch content of the leaves increased. Both Mauney *et al.* (1978) and Sasek *et al.* (1985) observed that the leaf starch content was reduced and the photosynthetic rate response was restored when the plants were returned to ambient air for about three days.

The question of acclimation of plants to long term exposure to enrichment has been discussed by Arp (1991). He concluded that the evidence for acclimation can be attributed to experiments in which the plants have limited root volume. Drake and Leadley (1991) concluded that there is no theoretical reason for acclimation to occur.

Though the effect of CO_2 enrichment on the photosynthetic rate decreased as the season progressed from a 40% increase on DOY180 to 5% on DOY250, Radin *et al.* (1987) concluded that the change in temperature across that time span was likely the rationale for the reduced sensitivity and that cotton plants grown in the field show no evidence of photosynthetic acclimation (Radin, 1992).

6. CO_2 DELIVERY TO COTTON CROPS

The significant response of cotton to CO_2 enrichment suggests that use of CO_2 as a field fertilizer to enhance commercial production might be economically feasible if a suitable delivery technique were developed. Delivery of the CO_2 dissolved in irrigation water is an obvious possibility, but data from a 1946 experiment by Leonard and Pinckard (1946) suggested that a high concentration of CO_2 in the root zone of cotton plants was toxic. In that experiment, the authors observed that bubbling CO_2 through the nutrient solution into which cotton roots were submerged resulted in death of the plants.

When Mauney and Hendrix (1988) tested the effect of CO_2-saturated nutrient solutions applied to potted plants supported by potting mix they observed a very different outcome. The plant growth rate was stimulated and yield was increased by as much as 70%. Surprisingly, however, the enhancement could not be accounted for by any effect of the CO_2 on the photosynthetic uptake of the CO_2 itself. Instead, the greater growth and yield could be accounted for by the enhanced nutrient uptake by the cotton roots in the treated pots. Uptake of zinc and manganese were more active in the plants treated with CO_2-saturated nutrient solutions. With these deficiencies relieved the plants became more productive.

With the demonstration that CO_2-saturated water is not toxic to cotton roots in a soil medium, several attempts were made to determine if the yield enhancement could be economical (Kimball *et al.*, 1986, 1987). In those trials the CO_2, both as a gas and dissolved in the irrigation water, was delivered by means of drip irrigation tubing placed in each row of plants. Increases in yield of 10 to 20% were observed. However, when this technology was tried in commercial grower's fields (data not shown) the variation of yield response within fields and the difficulty of injecting the CO_2 into the water rendered the beneficial effects insufficient to cover costs.

7. FUTURE EFFECTS OF CO_2 ON COTTON CROPS

Reddy *et al.* (1996b) have concluded that the negative effects of increasing temperature which may occur due to the projected global warming caused (allegedly) by increasing atmospheric CO_2 would offset the enhancement of photosynthesis which will accompany this enrichment. If however, the planting and production season become adjusted to rising annual temperature, then the average temperature for the growing season for the crop may not change substantially. In the last century the CO_2 concentration in earth's atmosphere has increased by 25%. If we assume an equivalence of CO_2 enrichment and yield enhancement suggested by the FACE experiments where yields in some treatment were increased by 40% with a 50% enrichment of CO_2, then about 25% of the yield of modern cotton can be accounted for due to enrichment (about 200 lb of lint in 1998).

There is no reason to believe that the effect of enrichment will be any smaller as CO_2 increases another 50% in the next century.

8. SUMMARY

The effect of CO_2-enrichment is to relieve the photosynthetic limitations on carbohydrate production that always limit the productive potential of cotton. CO_2-enriched atmospheres elevate the mesophyll CO_2 concentration, thus enhancing photosynthesis. Soil applied CO_2 enhances nutrient uptake so that leaf nutrient levels are sufficient to enhance rates of photosynthesis.

The results from several methods of exposure to CO_2-enrichment are very similar. The crop responds in a linear fashion to enrichment and does not have a reduced demand for water or nutrients. Higher production efficiency of water use comes about due to the higher productivity rather than reduced demand.

The composite image which emerges from the many observations made in CO_2-enriched plots is of a cotton crop that has greater soluble carbohydrate and starch reserves in leaves and stems due to its greater net photosynthesis. The crop has greater leaf area early in the season and higher root/shoot ratio in mid-season. It functions at higher tissue temperature and consumes similar quantities of water to the ambient crop. These attributes lead to there being slightly longer time period before cessation of flowering at cutout. The additional time produces greater crop biomass and harvestable yield with no changes in harvest index.

These observations point to the central role that annual net carbohydrate production plays in limiting cotton production, the vital role of root/shoot ratio in supporting productivity, and the ability of cotton to control its leaf area to consume available water.

Chapter 17

INTER-PLANT COMPETITION: GROWTH RESPONSES TO PLANT DENSITY AND ROW SPACING

J.J. Heitholt[1] and G.F. Sassenrath-Cole[2]
Need affiliations for both authors

I.   INTRODUCTION

Optimizing cotton yield through manipulation of plant spacing has been the objective of many research efforts. Field research has shown that increasing plant population density (number of plants per unit ground area) consistently increases leaf area index (LAI) and light interception but its effects on yield have been inconsistent. Understanding how the arrangement of plants in the field affects cotton growth requires consideration of many interacting factors, such as genetics, physiology, and canopy structure.

Plant population density (hereinafter referred to as plant density) and row spacing are two management decisions that determine the spatial arrangement of plants within a field. Plant density is the number of plants per unit area and is inversely related to the distance between plants within a row. Row spacing is the distance between rows. Altering plant density and row spacing changes the radiative transport within the canopy. Changes in radiative transport alter the interception of photosynthetically active radiation (PAR), and the distribution and quality of light within the canopy. Changes in plant spacing have a distinct impact on physiology, morphology, canopy development, and boll and fiber growth although the specific physiological mechanisms are largely unknown. In order to summarize some of what is known about cotton and plant spacing, this chapter will discuss the effects of plant density and row spacing on: (1) PFD (photon flux density) quality, quantity, and distribution; (2) resource availability; and (3) growth, yield, and fiber quality.

2.   LIGHT QUALITY

2.1   Spectral Changes Due to Neighboring Plants

As plant density increases, PFD in the middle and lower sections of the canopy is greatly altered due to penumbral effects creating regions of shade and deeper shade as light passes through the canopy (Jones, 1992). Radiation is reflected from foliage and soil at lower wavelengths than incident, and contributes to the diffuse radiation within the canopy. The changes in incident and reflected radiation alter the ratio of red (600 to 700 nm) to far-red (700 to 800 nm) light within the canopy. With increased reflectance of far-red light from neighboring plants, an increase in the number and nearness of neighboring plants can decrease the red:far-red ratio of intercepted light (Sanchez *et al.*, 1993). Changes in the spectral composition of incident radiation directly impact carbohydrate metabolism (Casal *et al.*, 1995), but exert a more profound influence on crop performance through the photomorphogenetic responses of plant tissues.

2.2   Photomorphogenetic Responses to Plant Density

The changes in plant structure and canopy development with altered plant spacing observed in cotton canopies result, at least in part, from photomorphogenetic responses to the altered light quality within the canopy. The photomorphogenetic responses are mediated via the phytochrome system, which provides a mechanism for plants to sense and respond to the light environment of the canopy (Ballare

J.McD. Stewart et al. (eds.), *Physiology of Cotton*,
DOI 10.1007/978-90-481-3195-2_17, © Springer Science+Business Media B.V. 2010

et al., 1992). Documented photomorphogenetic responses to altered light quality include seed germination and cell elongation. The exact mechanisms by which plants translate light quality into physiological responses are not completely understood. Evidence indicates that specific photoreceptors are capable of translating photoexcitation of the chromophores into molecular signals triggering cellular responses (Aphalo and Ballare, 1995 and references therein). Recently, a novel phytochrome receptor has been identified (Carabelli *et al.*, 1996) that appears to mediate cell elongation by induction of a specific gene. The novel phytochrome is induced under red/far-red light conditions typically found at dawn/dusk and shade within a canopy.

Research on photomorphogenetic responses in dallisgrass *(Paspalum dilatatum* Poir.) and annual ryegrass *(Lolium multiflorum* Lam.) have shown that a lower red/far-red ratio (as would occur from increasing plant density) reduced tillering (Casal *et al.*, 1986). In several other species, greater plant density increased stem elongation, changed leaf shape and size, decreased specific leaf weight, and reduced chlorophyll per unit leaf area (Aphalo and Ballare, 1995 and references therein). In mustard *(Sinapis alba* L.) supplemental far-red light not only increased internode elongation but also increased internode concentrations of reducing sugars, starch, hemicellulose, and cellulose (Casal *et al.*, 1995). There is little data documenting specific changes in red to far-red light within cotton canopies due to altered plant spacing. However, decreasing the red:far-red ratio of incident light through the use of various colored mulches causes an increase in plant height (Kasperbauer, 1994), especially early in the season. For more details about spectral composition changes and the phytochrome response in other species, the reader is referred to other reviews (Ballare *et al.*, 1992; Sanchez *et al.*, 1993; Kendrick and Kronenberg, 1993; Aphalo and Ballare, 1995).

3.   GROWTH RESPONSES TO PLANT DENSITY

3.1   Leaf Area and Main-stem Development Responses to Plant Density

As in most crop species, cotton compensates for decreases in plant spacing by producing a greater number of leaves and fruits per plant (Constable, 1986; Morrow and Krieg, 1990) ostensibly through photomorphogenetic processes. Although previously published reviews have discussed cotton growth characteristics (Hearn and Constable, 1984; Mauney, 1986a) and its response to various management practices (Hearn and Fitt, 1992; Kerby *et al.*, 1996), only one review (Hearn, 1972) has specifically addressed the effects of plant density. In quantifying some of the effects, Constable (1986) showed that a high plant density (24 m^{-2}) had a slower rate of leaf appearance than a low plant density

(2 m^{-2}). The lower plant density also had larger leaves and more main-stem nodes than the high plant density. The low plant density had longer sympodial branches (~5.5 fruiting sites) than the high density (~1.7 fruiting sites). It is thought that this response is partly due to light quality changes and partly due to the amount of resources per plant.

Fowler and Ray (1977) showed that plant height at 149 days after planting decreased as plant density increased. In one year of a three-year study, increasing plant density (from 5 to 20 m^{-2}) resulted in taller plants early in the season but shorter plants later (Heitholt, unpublished). In the second year, plant height increased as density increased (from 2 to 15 m^{-2}), regardless of the stage of development. In that same study, the number of main-stem nodes increased as plant density decreased (Heitholt, 1995). Kerby (1990a) showed that a full-season Acala cultivar showed a negative relationship between plant height and density but there was no relationship for two early maturing "determinate" types. Because many findings appear to conflict, it appears that genotypic, environmental, and developmental factors confound the effects of plant density on plant height.

3.2   Reproductive Developmental Responses to Density

Because increasing plant density increases the number of main-stem nodes per unit area, researchers have proposed that high plant densities will increase the number of flowers per unit area. Guinn *et al.* (1981) showed that increasing plant density of 'Deltapine 61' from 5.2 to 9.4 plants m^{-2} of increased flower numbers by 5%, but decreased boll retention by 8% and yield by 6%. Heitholt (1995) showed that increasing plant density of DES 24-8ne okra and normal-leaf from 5 to 15 plants m^{-2} did not appreciably affect flower numbers, boll retention, or yield. Therefore, increasing plant density may, but does not necessarily, increase flower numbers.

Boll distribution is also an important consequence of the effects of plant density on yield. Increasing plant density from 3 to 15 plants m^{-2} increased the percentage of bolls found on first position fruiting sites from 40 to 80% (Kerby *et al.*, 1987). In a similar study, increasing plant density from 5 to 15 plants m^{-2} increased the percentage of bolls found on first-position fruiting sites from 48 to 71% (Kerby *et al.*, 1990a). The exact physiological causes for the altered boll distribution are not known. However, altered light environment in the "crowded" conditions experienced by plants grown at high density probably prevents sympodial branches from developing many distal fruiting sites as described earlier (Constable, 1986). The response is expected considering that a plant grown at high density generally develops a similar (Heitholt, 1994a) or a slightly lower (Constable, 1986) plant height and number of main-stem nodes by the end of the season than plants grown at lower density. However, plants grown at high densities produce fewer bolls per plant than plants grown at low densities (Guinn *et al.*, 1981).

3.3 Plant Density Effects on Yield

In general, the reported effects of plant density on yield have been small, inconsistent, or dependent upon environment (Buxton *et al.*, 1979; Guinn *et al.*, 1981; Kerby *et al.*, 1990a; Heitholt *et al.*, 1992). In one case where four low densities were used, four hybrid cottons had a 30% greater seedcotton yield at the two higher plant densities (2.8 and 5.5 plants m^{-2}) than at the two lower plant densities (1.4 or 1.8 plants m^{-2}) (Jadhao *et al.*, 1993). Using tillage and seeding rate to alter plant density in the Texas High Plains, Hicks *et al.* (1989) showed that yield of Paymaster 404 and Acala A246 increased as plant density increased from 2 to 8 plants m^{-2} but decreased above 8 plants m^{-2}. On a very fine sandy loam in Louisiana with Deltapine 41, Micinski *et al.* (1990) found that resulting plant densities explained yield differences found among planting dates.

In one of the most comprehensive summaries on cotton plant density and yield, Hearn (1972) demonstrated that optimal plant density increased as yield potential increased. This is consistent with the idea that with more resources a canopy can support more plants and is consistent with the skip-row results (presented later). However, this dogma does not hold for all situations. In contrast to Hearn (1972), the optimal plant density in a low yielding environment (800 kg ha^{-1}) for DES 24-8ne normal-leaf was 15 plants m^{-2} whereas the optimal density in a high yielding environment (1100 to 1400 kg ha^{-1}) was 5 plants m^{-2} (Heitholt, 1994a). These results can probably be reconciled by looking at LAI data. In Heitholt (1994a), the 15 plants m^{-2} was excessive when the resulting LAI was 5 or greater. The idea that optimal density is associated with resulting LAI was also advanced by Constable (1977). Kerby *et al.* (1996) also indicated that optimal plant density was greater (*i.e.*, 20 plants m^{-2} rather than 10 plants m^{-2}) under severely stressed conditions. Using an alternative approach to determine optimal plant density of Pima cotton, Kittock *et al.* (1986) measured plant height at plant densities ranging from 2 to 20 plants m^{-2}. The optimal plant density decreased by 1.1 plants m^{-2} for every 10-cm increase in plant height.

4. PLANT SPACING EFFECTS ON LIGHT QUANTITY

4.1 Sunlight Interception and Utilization

Differences in plant density and row spacing physically modify a canopy. Decreasing the distance between nearest neighbors alters the light environment within the canopy and the resources available to each plant. An ideal canopy would capture as much solar radiation as possible and opti-

mize utilization of that radiation for conversion to carbohydrate and production of harvestable product. The goal of the producer/researcher is to manipulate the inherent growth pattern of plants to optimize growth and development of harvested components. Directed manipulation of canopy architecture is a technique used extensively in many cropping systems to optimize radiation use and enhance yield. Fruit production systems regularly use direct manipulation of canopy structure (*e.g.*, through pruning), to optimize yield (Palmer, 1989). In soybean (*Glycine max* L. Merr.), Ikeda (1992) attempted to develop an optimal canopy structure by varying planting pattern so that land area per plant and plant density were maximized.

Because sunlight is essential to plant production, the canopy must optimize interception and utilization of radiation to achieve maximum growth. Some areas of the U.S. Cotton Belt (*e.g.*, Texas and Arizona) can plant earlier than other regions due to more favorable soil temperatures and therefore can optimize the interception of solar radiation. By planting earlier, canopy closure in these regions coincides with peak insolation periods (near 21 June) whereas in regions with later planting dates, canopy closure occurs later. By planting at higher plant densities or narrower row spacing, early-season light interception is increased and the crop can take advantage of the greater insolation levels during the early growing season. However, increased early-season light-capture does not necessarily result in increased yield.

Optimizing solar radiation interception has been the objective of many plant spacing studies. Radiation interception is dependent on the leaf area and can be defined as the absorption of solar radiation by the light-harvesting pigment-protein complexes in the leaf chloroplasts. Radiation that is not absorbed by the topmost layer of leaves passes through to the lower canopy layers. However, in passing through the upper canopy layer, the full sunlight intensity rarely reaches the lower canopy layers due to the penumbral effects of light movement through gaps in the canopy. For a thorough discussion of the movement of light through plant canopies, the reader is referred to Jones (1992) and Monteith and Unsworth (1990).

As LAI and light interception increase (with canopy development or increased plant density), the shape of the canopy foliage and the arrangement of leaves relative to one another determine solar radiation interception efficiency. In theory, a continuous layer of leaves at the top of a canopy, with LAI equal to 1.0, would allow complete radiation interception. Although this may appear to be an efficient canopy structure for the interception of radiation, this strategy does not result in an efficient use of sunlight. An LAI of 3 to 6, with evenly distributed leaves and more erectophile leaf angles, greatly increases the efficiency of sunlight utilization. Studies have demonstrated that canopy photosynthesis is enhanced when more PAR is allowed to reach the middle portion of the canopy as opposed to nearly complete radiation interception by the upper layer of leaves (Wells *et al.*, 1986; Aikman, 1989; Herbert, 1991).

Assuming 2000 μmol m^{-2} s^{-1} PFD is available and all leaf surfaces intercept equal PFD, theory suggests that optimal carbon uptake would occur at an LAI of 5 and a PFD of 400 μmol m^{-2} s^{-1}. This value is slightly greater than found in two Australian field studies where maximum crop growth rate and yield were found to occur at an LAI of 3.5 (Constable, 1977; Constable and Gleeson, 1977). In a Mississippi Delta field study, maximum crop growth rate and 95% light interception were reached at an LAI of 4.0 (Heitholt *et al.*, 1992). Plant densities of 2, 3, and 5 plants m^{-2} resulted in an LAI of 4.0 or less and a greater yield than plant densities of 10 and 15 plants m^{-2} that had an LAI of 5.0 or greater (Heitholt, 1994a). Although there is some disagreement on the optimal LAI, it is clear that plant spacings that allow a more even distribution of PAR throughout the canopy may optimize productivity.

Because leaf age affects the photosynthetic light response (Sassenrath-Cole *et al.*, 1996), the optimal LAI for maximal canopy carbon uptake may change during development. Peng and Krieg (1991) reported that leaf age, rather than PAR, was primarily responsible for reduced canopy carbon exchange rate (CER) late in the season. It has been suggested that extending the optimum CER phase of leaves by 10 days would lead to a significant increase in yield (Landivar *et al.*, 1983). Although a delay in the age-related decline in CER could indeed increase growth, two research groups have shown that other factors may prevent it from being realized. First, Wullschleger and Oosterhuis (1990a) showed that the CER decline of older leaves found lower in the canopy was limited to a significant extent by the intensity of PAR incident to the leaf surface (*i.e.,* CER and PAR were correlated) and not solely due to leaf aging *per se.* Second, we found that the loss of physiological activity due to leaf age accounted for less than one-half of the reduced CER that occurred in lower canopy leaves during the late season (Sassenrath-Cole and Heitholt, 1996). In that study, low PAR was the primary factor that reduced CER of lower canopy leaves. In many cotton canopies, the older, lower canopy leaves receive low PAR and therefore, contribute little carbon to the plant. Thus, distribution of radiation is an important consideration in designing an optimal canopy architecture. For a further discussion of photosynthesis within cotton canopies, the reader is referred to Chapter 7 (Temporal Dynamics of Leaves and Canopies).

In order to obtain a better PAR distribution, leaf distribution must be optimized. However, some disagreement exists as to whether leaf area is arranged more efficiently (less mutual shading) in low density or high density. Constable (1986) showed that canopy extinction coefficients at low density (2 m^{-2}) were greater than those from high density (24 m^{-2}). This suggested that leaf arrangement with low density exhibited less efficiency (or in the author's words greater "clumpyness") than at higher plant densities. In apparent contrast, Heitholt (1994a) showed that light interception per unit LAI was greater at lower density (5 m^{-2}) than at 20 m^{-2}. In this case, the lower plant density had a better leaf arrangement than the high density. The differences between these two studies may reflect the growth environment, suggesting that different strategies may be optimal under different growing conditions. For example, the high solar radiation environment found in the southwest U.S. results in a much denser canopy structure (Alarcon and Sassenrath-Cole, unpublished). Because improving the efficiency of leaf arrangement is important and there is not a consensus as to the best arrangement, future research directly testing the effects of plant density on leaf arrangement is needed.

4.2   Light Distribution Differences Caused by Genotype

Genetic variants with altered leaf and canopy structure can help reduce the radiation absorption by the upper canopy layers and allow greater PAR to reach lower canopy leaves. These simply-inherited leaf shapes range from the full-sized normal-leaf types to the narrow-lobed leaf types called okra-leaf. Leaf morphology variants, such as okra-leaf (discussed earlier by Stewart, chapter 2 and Wells, chapter 3) increase light and air movement within the canopy. There is a significant difference in PAR profile within canopies of normal vs. okra-leaf (Wells *et al.*, 1986). Although the leaves at the very top of the normal-leaf canopy receive more PAR, the PAR drops rapidly with decreasing height in the canopy. The decrease in PAR as a function of canopy depth is less severe in the okra-leaf canopy (*i.e.,* light penetrates further into the canopy) than in a normal-leaf canopy (Wells *et al.*, 1986).

At a given plant density, okra-leaf cottons usually develop a lower LAI than normal-leaf (Kerby *et al.*, 1980; Heitholt *et al.*, 1992). This raises the possibility that the optimal plant density for okra-leaf types may be greater than for normal-leaf types. Meredith (1985) compared the yield of okra-leaf, normal-leaf, their F$_1$, and their F$_2$ progeny using three seeding rates (9, 19, and 38 seeds m^{-2}) in four Mississippi Delta environments. Averaged across environments, yield of okra-leaf cotton types were unaffected by seeding rate but yield of the normal-leaf, F$_1$, and the F$_2$ progeny all declined as seeding rate increased above 9 seeds m^{-2}. In a later study, DES 24-8ne okra-leaf exhibited its greatest yield at 10 plants m^{-2} whereas yield of the normal-leaf isoline was greatest at 5 plants m^{-2} (Heitholt, 1994a). Kerby *et al.* (1990a,b) also showed that genotypes differed in optimal plant density. One short "determinate" type exhibited its greatest yield at 15 plants m^{-2} (and lowest yield at 5 plants m^{-2}) whereas yield of the more full-season type 'Acala SJC-1' (with a maximum LAI of 4.4 to 5.1) was unaffected by plant densities of 5, 10, and 15 plants m^{-2}. Based on these studies, it appears that increasing plant density above 10 plants m^{-2} for low leaf area, early-maturing types can sometimes increase yield whereas increasing plant density above 10 plants m^{-2} for normal-leaf, full-season types usually does not.

4.3   Growth Regulators

Directed modification of canopy architecture is possible through the use of plant growth regulators, such as mepiquat chloride (1,1-dimethyl-piperidinium chloride), a gibberellic acid (GA) synthesis inhibitor or conversely by GA-like substances (Oosterhuis and Zhao, 1995). Mepiquat chloride can inhibit vegetative growth (Kerby *et al.*, 1986; Cathey and Meredith, 1988; Fernandez *et al.* 1992; McConnell *et al.* 1992), reducing internode length and LAI. Gwathmey *et al.* (1995) showed that mepiquat chloride decreased the percentage of light intercepted in upper leaves and allowed greater light penetration to the middle portion of the canopy. This alteration in canopy light profile most probably arises due to a decrease in the penumbral effects with shortened internodes. Although mepiquat chloride can lower LAI and light interception (Heitholt *et al.*, 1996), it has also been found to increase yields in California (Kerby *et al.*, 1986) and in Mississippi when cotton was planted late (Cathey and Meredith, 1988; Heitholt *et al.*, 1996). The LAI in the Heitholt *et al.* (1996) Mississippi study was rather high (ranged from 4 to 5). Because mepiquat chloride is more likely to have a positive effect in canopies that would otherwise develop excess vegetative growth, this raises the possibility that mepiquat chloride would be more effective at high plant densities (Atwell, 1996; Atwell et al,. 1996). However, in four of five environments, mepiquat chloride did not increase yield at plant densities ranging from 3.7 to 13.6 plant m^{-2} (York, 1983). The exception occurred in a test that used 23 plants m^{-2} (a plant density often considered excessive) where mepiquat chloride increased yield by 51%.

4.4   Leaf Orientation

Factors altering leaf orientation may relate to the cotton plant's response to density. Light penetration within the canopy depends not only on LAI and leaf distribution (Constable, 1986), but also on the three-dimensional structure of the leaves and the heliotropic behavior (Lang, 1973; Ehleringer and Hammond, 1987; Sassenrath-Cole, 1995). Studies have demonstrated that changes in leaf angle in the upper canopy can redistribute PAR to lower canopy leaves and lead to increased canopy photosynthesis (Wells *et al.*, 1986; Aikman, 1989; Herbert, 1991; Sassenrath-Cole, 1995). Leaf angle in other species can also be altered by plant density (Aphalo and Ballare, 1995). Because upper leaves of a *G. hirsutum* canopy are diaheliotropic, tracking the sun particularly in the early morning and late evening, PAR striking the upper leaves of *G. hirsutum* during these times is greater than PAR striking leaves of *G. barbadense* that display no heliotropism (Ehleringer and Hammond, 1987). There is a significant difference in light penetration within mature canopies of diaheliotropic *G. hirsutum* from that in non-heliotropic *G. barbadense* (Sassenrath-Cole, 1995). Photon flux attenuation in the *G. barbadense* canopy declined gradually with decreasing height in the canopy,

whereas the lower third of the leaves in the *G. hirsutum* canopy received low PAR throughout the day (Sassenrath-Cole, 1995). Although the suntracking of *G. hirsutum* allowed upper canopy leaves to intercept more total radiation, the restriction of PAR to leaves lower in the canopy resulted in less efficient light distribution in canopies of *G. hirsutum* compared to that of *G. barbadense*. In agreement with this interpretation, leaf heliotropism was suggested to be beneficial to net canopy photosynthesis early in the season, but detrimental to photosynthesis due to a decreased distribution of PAR to lower canopy layers as LAI increased (Fukai and Loomis, 1976).

Although differences in heliotropic response contributed to differences in light environment within the two canopies, a more significant contribution was the degree of curvature or three-dimensional cupping (*i.e.,* a concave leaf shape) of the *G. barbadense* leaves (Sassenrath-Cole, 1995). This cupping of upper canopy *G. barbadense* reduced PFD incident to upper leaf surfaces but increased PFD penetration to lower canopy layers, more closely approximating an ideal canopy configuration (Kuroiwa, 1970). In addition to distributing more PAR to lower canopy layers, this strategy may be advantageous by reducing photoinhibition to upper canopy leaves by reducing the time and intensity of PAR exposure. Since upper leaves greatly reduce PAR received by older leaves lower in the canopy, PAR limitation, rather than age-induced physiological limitations, might be a more significant cause for reduced photosynthetic activity that occurs with aging (Sassenrath-Cole and Heitholt, 1996). The increase in PAR to lower canopy layers may significantly increase the potential carbon uptake of *G. barbadense*. Use of varieties with more erectophile orientation of the upper canopy foliage, achievable in cotton by greater three-dimensional "curling" of the leaves, would benefit canopy carbon uptake and potentially be better adapted to higher densities and narrower row-spacings. These changes in the vertical orientation of leaves have been successfully incorporated into corn (*Zea mays* L.) via traditional breeding methods.

4.5   Microclimate Changes as Affected by Plant Spacing

To understand differences in fruit development with plant spacing, we must consider the alteration of the microenvironment at the point of boll development. The infrared portion of the electromagnetic spectrum radiates heat to and from the earth's surface. This energy of both direct and diffuse PAR is attenuated (wavelength is increased but frequency and energy are decreased). This gives rise to differences in temperature profiles within cotton canopies as a function of canopy structure. Canopy characteristics that cause changes in radiative transfer, light penetration, distribution, and quality throughout the canopy will also change the heat balance within the canopy. Changes in radiative transport have been demonstrated in other cropping systems as a function of canopy modification (Baldocchi *et al.*, 1985). Alteration of radiative flux due to modification of the canopy architecture will drastically affect the microenvironment (temperature, humidity, and wind speed) within the crop canopy. These changes in temperature at the site of fruit development, while often subtle, can have profound influence on boll growth and fiber development when taken over the course of boll development (Hessler *et al.*, 1959; Gipson, 1986; Sassenrath-Cole and Hedin, 1996). Moreover, additional effects on air movement through the canopy alter the humidity that will affect the development of boll rot and other microorganisms. Preliminary evidence in cotton suggests that daytime mid-canopy temperatures are cooler in 51-cm rows than in 102-cm rows (Sassenrath-Cole, unpublished). Although extensive research has been conducted in other crop canopies examining differences in radiative transport as a function of canopy architecture (Baldocchi *et al.*, 1985), cotton has only recently received attention (Sassenrath-Cole, 1995).

In addition to the plant density and row spacing effects on mid-canopy boll temperatures, leaf morphology can also be important. Lower-canopy boll temperatures were found to be warmer during the day and cooler at night for okra-leaf plants compared to those measured in normal-leaf isolines (Sassenrath-Cole, unpublished). This difference in temperature in okra-leaf canopies would have significant impact on boll maturation over the course of the growing season and may account for the observed increased rate of boll maturation (*i.e.*, earliness) of okra-leaf types (Heitholt *et al.*, 1993; Heitholt *et al.*, 1996). Additionally, lint from okra-leaf plants exhibited lower strength than normal-leaf cotton (Heitholt *et al.*, 1993), which could result from the larger diurnal temperature fluctuations in the okra-leaf canopies. More data on the effect of plant density and leaf morphology on diurnal boll temperatures are needed to help explain differences in fiber quality.

5.   RESOURCE AVAILABILITY

When plant density is altered, the amount of resources (*e.g.*, water, nutrients, soil volume, etc.) available per plant is changed. At high density, leaf area per plant decreases and the volume of soil available for each plant decreases. The negative effects of reduced resources per plant that occur due to high density are likely to offset any gains realized from increases in LAI (Sanchez *et al.*, 1993). In this section, we cover the effects of plant density on physiological response to resource availability.

5.1   Soil Resources and Plant Density

Among the resources per plant that are potentially limiting at high density are mineral nutrients. In this discussion, we limit ourselves to how mineral nutrition might interact with plant density. For a more in-depth discussion of mineral nutrition, the reader is referred to chapter 14. Nitrogen fertility effects on cotton growth are well documented (Wullschleger and Oosterhuis, 1990b; McConnell *et al.*, 1993). Excess N can lead to unnecessary leaf area, delayed maturity, and lower fiber quality. When high soil N leads to excess vegetative growth (as described by Boman and Westerman, 1994), assimilates get distributed toward unneeded vegetative growing points and away from fruit. Although potassium deficiency can reduce plant biomass and limit yield (Cassman *et al.*, 1989a, 1990; Pettigrew *et al.*, 1996; Pettigrew and Meredith, 1997), there are no reported negative consequences from adding excess K. Deficient N or K results in an insufficient LAI (Wullschleger and Oosterhuis, 1990b; Bondada *et al.*, 1996; Pettigrew and Meredith, 1997) and a low number of potential fruiting sites. Increased density effects on root growth may reduce cotton's ability to obtain nutrients from the soil. In some desert shrubs, increased plant density reduced root proliferation and elongation rate (Mahall and Callaway, 1991, 1992).

Several studies have looked at the possible interaction between N and plant density but usually little interaction has been found. However, in one study, Yasseen *et al.* (1990) did report N x plant density interaction on plant height. At low N (60 kg N ha^{-1}), plant height was greater at 34 plants m^{-2} than at 17 plants m^{-2}. At 90 kg N ha^{-1}, no differences in plant height among densities were found. Rao and Weaver (1976) found no significant N x plant density interaction when three N levels (100, 134, and 168 kg N ha^{-1}) and three plant densities (2, 3, and 7 plant m^{-2}) were compared. Koli and Morrill (1976) found fiber properties were generally unaffected by N x plant density interaction.

In another study (Sadras, 1996a), plant density effects on compensatory growth (as a result of floral bud removal twice weekly until 80 days after sowing) were tested with 'Siokra S324' and 'CS7S'. Under low plant density and high nitrogen fertility, subsequent fruit growth rate of defruited plants was only 17% lower than the control. However, under high plant density and low N fertility (a stressed environment), the decrease in fruit growth rate was 50%. This decrease under stressed conditions was associated with an increased partitioning of dry matter to roots.

In order to better quantify a plant's effect on other plants, Sadras (1997b) calculated a factor for an individual plant referred to as "neighbor interference" which was equal to the leaf area of the adjacent plant divided by their distance from each other. A strong inverse linear relationship was found between "neighbor interference" and the percent of dry matter partitioned to fruit. This result supports previous observations by Heitholt (1994a) that showed high plant density and LAI was associated with reduced fruit numbers.

In two studies, low plant density (about 5 plants m^{-2}) combined with adequate N supply was compared to a high plant density (about 15 plants m^{-2}) with minimal N (Sadras 1996b, 1996c). In one study with Siokra V-15 (Sadras, 1996c), vegetative bud removal tended to increase yield of the high density treatment presumably by promoting greater root growth. The treatments did not affect yield in the low density treatment. Both low density treatments outyielded the high density treatments. One explanation for the results could be improved light quantity and quality striking floral meristems in the high density treatment. Another explanation could be that larger roots in the high density treatment allowed those plants to extract more water and nutrients than the untreated plants.

5.2 Water Availability and Plant Density

High plant densities may exhaust the available water earlier in the season than lower plant densities leaving insufficient water for the boll-filling period. Hearn (1972) demonstrated that the optimal plant density under irrigation was greater than under dryland conditions. Likewise, in the Texas High Plains, Staggenborg and Krieg (1993) found that the yield decline associated with high plant density was less severe when more water was available. In a separate study, Staggenborg and Krieg (1994) varied plant density and water supply and showed that the optimal density was one that supplied 30 to 35 kg H_2O per plant per season. The apparent interaction between water supply and plant density indicates that excessive plant density can cause soil water depletion that leads to lower yield.

6. PLANT DENSITY AND HARVEST VARIABLES

6.1 Effects of Plant Density on Harvest Maturity Date

Data from some studies have suggested that increasing plant density would increase earliness (Rao and Weaver, 1976) whereas other data have suggested the opposite (Kerby *et al.*, 1990a). Rao and Weaver (1976) showed that both the okra-leaf and normal-leaf isolines of 'Coker 201' matured earlier when grown at 6.5 plants m^{-2} than at 2.2 plants m^{-2}. Using three hand-picking dates, Kerby *et al.* (1990a) showed that a tall "indeterminate" cultivar ('Acala SJ-2') grown at 15 plants m^{-2} matured later (*i.e.,* 59% picked by first two harvests) than it did at densities of 5 and 10 plants m^{-2} (*i.e.,* 69% picked by first two harvests). Kerby *et al.* (1990a) also showed that the earliness of two "determinate" types was unaffected by plant density. Also using consecutive hand-pickings, in a three-year study (Heitholt, unpublished), the effects of plant density on earliness were examined. In 1991 and 1993, no effects were found. However, in 1992, DES 24-8 ne normal leaf at 2 and 3 plants m^{-2} reached 65% open bolls five days earlier than at 10 and 15 plants m^{-2}. Baker (1976) and Kostopoulos

and Chlichlias (1979) found that plant density did not affect earliness. Changes in earliness with altered plant density/ row spacing may depend on the differences in heat units experienced by the developing bolls due to an alteration of radiative transport within the modified canopy structures.

6.2 Effects of Plant Density on Fiber Quality

Developing bolls and subtending leaves on plants grown under high plant densities are more likely to be shaded than bolls and leaves on plants grown under lower densities. The lower leaves that subtend developing bolls have been suggested to be a primary assimilate source for those bolls (Constable, 1986). If inadequate radiation strikes these leaves, boll set or fiber quality of these bolls may be reduced (Pettigrew, 1995). Alternatively, if high plant density causes abscission of lower bolls, then subsequent bolls are more likely to be found in middle or upper portions of the canopy. The indeterminate growth habit of cotton results in bolls developing under very different environmental conditions throughout the growing season. Despite numerous studies that have been performed, most have found that plant density had little direct effect on fiber quality (Gannaway *et al.*, 1995; Minton and Supak, 1980; Kostopoulos and Chlichlias, 1979; Hawkins and Peacock, 1973). In two cases, increases in plant density were associated with decreases in micronaire and length (Fowler and Ray, 1977; Hearn, 1972).

7. ROW SPACING

7.1 Row Spacing and Light Distribution

The effects of row spacing on the ratio of percent light intercepted to LAI are more consistent than plant density effects. Row spacing is consistently and negatively related to light interception per plant (Heitholt *et al.*, 1992; Heitholt, 1994a). Given equal plant densities, narrow rows result in more equidistant plant spacing. This means more early season solar radiation is captured (Peng and Krieg, 1991; Heitholt *et al.*, 1992); leaves on narrow-row plants are less likely to shade each other (*i.e.,* a plant's nearest neighbor is further away and mutual shading is less likely). Although row spacing is consistently and negatively related to light interception per plant its effects on growth per plant and LAI range from no effect (Constable, 1977; Heitholt *et al.*, 1992) to increases in both (Peng and Krieg, 1991).

7.2 Row Spacing and Yield

The range of cotton row spacings that have been tested is large. Examples of row spacing comparisons include 18-cm vs. 102-cm (Constable, 1977), 33-cm vs. 67-cm vs. 102-

cm (Peng and Krieg, 1991), 51-cm vs. 102-cm (Heitholt *et al.*, 1992), and 76-cm vs. 102-cm (Heitholt *et al.*, 1996; Williford *et al.*, 1986). Unfortunately, many narrow-row studies have confounded plant density and row spacing (Andries *et al.*, 1969; Buxton *et al.*, 1979; Fowler and Ray, 1977; Minton, 1980) by using a greater plant density in the narrow row spacings. The confounding of plant density and row spacing in the studies listed above has made it difficult to separate the effects of the two factors in these studies. Confounding plant density and row spacing would be justified if the optimal density for narrow rows was shown to be greater than that for wide rows. This hypothesis is undoubtedly true for some environments. However, using one okra-leaf and one normal-leaf genotype, Heitholt (1994a) found that row spacing did not affect optimal plant density in 51-cm vs. 102-cm spacings or in 76-cm vs. 102-cm spacings.

Although narrow rows generally increase seasonal light interception by 10% for most crops, this increase does not necessarily provide production advantages for cotton (Constable, 1977; Heitholt *et al.*, 1996). The effects of row spacings, ranging from narrow rows (51-cm to 76-cm) to skip rows (defined later) on yield, are small in many environments (Constable, 1977; Heitholt *et al.*, 1992; Kostopoulos and Chlichlias, 1979; Williford *et al.*, 1986; Williford, 1992). The effect of row spacing is possibly dependent upon other management factors. In a three-factor study with mepiquat chloride (0 to 50 g ha^{-1}), irrigation vs. nonirrigated, and row spacing (76- vs. 96-cm) in the Missouri Bootheel, Tracy and Sappenfield (1992) reported that narrow rows outyielded wide rows if mepiquat chloride and irrigation were applied. However, row spacing did not affect yield if either mepiquat chloride or irrigation were withheld.

In addition to the conventional, solid planted row spacings mentioned above, a skip-row planting pattern (*e.g.*, a row pattern where each pair of 102-cm rows is separated by a 204-cm spacing), has been a commonly used variation of row spacing. The skip-row pattern allows a grower to plant fewer rows and to harvest fewer rows (harvesting being the most expensive input). Skip rows also allow use of soil water from the "unplanted" area (Quisenberry and McMichael, 1996). Like much of the research comparing solid-planted row spacings, skip-row comparisons to solid planting are often confounded by plant density (*i.e.*, skip rows use fewer plants per unit area). Although the yield of skip-row management per unit area is almost always lower than solid planted cotton, skip-rows may prove more economical than solid planting because the cost of producing a unit of lint is lower. The yield per plant in skip-row culture occurs because each row is essentially unbordered on one side. Therefore, lower leaves receive more light increasing the chances that yield per plant will increase. This claim is supported by a two-year study with three cotton cultivars that showed aluminum reflectors placed in the two row spaces (of solid planted cotton) bordering the yield row increased yield by 6% (Pettigrew, 1994). Quisenberry and McMichael (1996) also showed that the further the distance

between rows up to 204 cm also increased yield per plant. Yields of skip-row and solid cotton were reported to be similar when water was applied at a rate of 4 megaliters per hectare or less (Constable *et al.*, unpublished). However, at 6 and 8 megaliters of water per hectare, lint yields were about 500 kg ha^{-1} greater in the solid cotton. The skip-row culture appears to be useful when either economic or environmental resources are limited.

8.   SUMMARY

Plant density, row spacing, and genotype are three factors that alter the canopy PAR profile, light quality (*i.e.*, spectral composition), and resources available per plant. Plant density induced changes in light quality affect canopy development and yield. These changes in light quality can also affect fiber quality through photomorphogenetic effects on cell elongation and differentiation. Development mediated via the phytochrome system is also altered by plant density. High plant density can cause excessive PAR interception by the upper canopy and a reduction in the amount of light striking mid-canopy leaves. These changes subsequently reduce carbohydrate status of subtended bolls.

Commonly tested cotton plant densities range from 5 to 15 plants m^{-2} (Kerby *et al.*, 1990a,b) although data for densities on both sides of this range are available (Fowler and Ray, 1977; Hearn, 1972; Heitholt, 1994a; Hicks *et al.*, 1989; Sadras, 1996c). Increasing plant density results in greater LAI but less leaf area per plant and fewer bolls per plant. Plant densities that result in an LAI between 4 and 5 are usually associated with optimal yield. Increasing plant density also results in taller plants during early growth but at maturity the reverse may be true.

The effects of plant density on yield, maturation date, and fiber quality are often small. Optimal plant density was shown to be altered by soil water, nutrient availability, growth regulators, and genotype but not by row spacing. Assuming equal plant densities are used, narrow rows allow each plant to intercept more light. However, the effects of row spacing on biomass per plant are small compared to the effects of plant density.

Although this review indicates that many of cotton's agronomic responses to plant density have been thoroughly characterized, it is clear that the cotton plant's basic responses to plant density and row spacing, such as biophysical, canopy microclimate, and biotic changes remain largely uncharacterized and need to be researched.

9.   ACKNOWLEDGMENTS

The authors thank C.O. Gwathmey, K.E. Lege, W.T. Pettigrew, V.O. Sadras, and L.J. Zelinski for comments on early versions of the manuscript.

Chapter 18

COTTON HOST-MICROBE INTERACTIONS

A.A. Bell[1], C.R. Howell[1], and R.D. Stipanovic[1]
[1]USDA, ARS, Southern Plains Agriculture Research Center, College Station, TX

1. INTRODUCTION

Throughout its life the cotton plant is constantly confronted by microorganisms that may be either beneficial, commensal, or pathogenic. Some of these microbes are endophytes that live in all parts of the plant and are carried from one generation to the next in the seed. Other microbes, including many pathogens, are dormant in the soil until cotton or another suitable host stimulates their regrowth. Other bacteria and fungi in soil are active saprophytes that break down organic matter and also use exudates from the cotton root. Finally, there is a constant fall out of microbes from the air onto the plant surface. It is not uncommon to find microbial populations between one million and one trillion per gram of fresh tissue on either the root or the fiber soon after the boll opens. The cotton plant is exposed to enormous numbers of microbes that can influence its growth and usefulness by affecting nutrient availability; producing plant growth regulators; secreting antibiotics, phytotoxins, and mycotoxins; and invading plant tissues to cause disease.

The first section of this chapter discusses beneficial cotton-microbe interactions, especially as they relate to control of cotton pests. This is followed by a discussion of disease interactions in which specificity of interactions, influence of environment, mechanisms of parasitism and pathogenesis, responses to pathogens, and consequences of disease are considered. Commensal relationships are not discussed. "*Cotton and Microorganisms*" by Fischer and Domelsmith (1997) has dealt with the limited literature in this area.

2. BENEFICIAL COTTON-MICROBE INTERACTIONS

Some of the microbes associated with the cotton plant may act to prevent or alleviate damage caused by pests. This activity may take the form of improvement of overall plant health through increased nutrient uptake, occupation of niches commonly required by the pest in order to access the plant, or stimulation of defense systems in the host plant. Pests suppressed by beneficial microbes include pathogens, toxigenic microbes, and herbivorous mites.

2.1 Suppression of Pathogens by Mycorrhizal Fungi

A common example of prevention or remediation of disease caused by pathogens can be found in the activities of mycorrhizal fungi on the cotton root. Liu (1995) found that infection of cotton roots with the vesicular-arbuscular mycorrhizal fungi (VAMF) *Glomus mosseae* and *Glomus versiforme* reduced the incidence and severity of Verticillium wilt of cotton. VAMF inoculation reduced the numbers of germinable microsclerotia of the pathogen, *Verticillium dahliae,* in the mycorrhizosphere, and the percentage of arbuscle colonization in the roots was negatively correlated with disease grades. VAMF promoted seedling growth, advanced flowering, and increased the numbers of flowers and bolls and seedcotton yield. The efficacy of the VAMF infection, however, is dependent on the VAM fungus used. While *G. mosseae* and *G. versiforme* were effective in reducing the incidence and indices of Verticillium wilt, *Sclerocystis sinuosa* was inferior.

J.McD. Stewart et al. (eds.), *Physiology of Cotton*,
DOI 10.1007/978-90-481-3195-2_18, © Springer Science+Business Media B.V. 2010

Hu and Gui (1991) discovered similar variations in the efficacy of *Glomus* species. Inoculation of cotton seedlings with *G. mosseae* in sterile soil 4 weeks prior to transplanting into pots containing unsterilized soil suppressed Fusarium wilt when the pots were subsequently inoculated with *Fusarium oxysporum* f.sp. *vasinfectum* (*F.o.v.*). In contrast, wilt symptoms in plants infected with another VAM fungus, *G. intraradices*, were as severe as those in non-mycorrhizal plants. Infections of the roots by the VAM fungi were not affected by the presence of *F.o.v.*

The dynamics of cotton root colonization by VAM fungi and their effect on the activities of the plant parasitic nematode *Meloidogyne incognita* have also been studied. Saleh and Sikora (1984) found that stimulation of plant growth and suppression of the root knot nematode, *M. incognita*, was dependent on the inoculum concentration of *Glomus fasciculatum*. At concentrations of 30 to 480 chlamydospores per plant, no growth stimulation of cotton was observed, and the number of nematode eggs per plant was reduced only with the high spore numbers. A concentration of 750 chlamydospores per plant increased plant growth by 41% and reduced the egg and nematode numbers per gram of root by 59%. *G. mosseae* also reduced penetration (27 to 60%) and adult females of *Rotylenchulus reniformis* on cotton roots (Sikora and Sitaramaiah, 1980). *Glomus aggregatum* is even more effective than are either *G. fasciculatum* or *G. mosseae* in increasing fresh root weight and decreasing *M. incognita* reproduction on black henbane (Pandey *et al.*, 1999).

The activities of VAM fungi are often influenced by the microbial environment in which they are located. McAllister *et al.* (1994) found that infection of corn root systems by *G. mosseae* and the dry weight of VAMF colonized plants were reduced by prior inoculation with *Trichoderma koningii*, a known biocontrol agent. When plants were inoculated with *G. mosseae* 2 weeks before *T. koningii*, propagule numbers of the former on the root system were considerably reduced. Under these circumstances, colonization of the root by *G. mosseae* was not adversely affected by *T. koningii*, but its subsequent metabolic activities were. Therefore, the effect of a biocontrol agent on the VAMF root population should be considered before using the agent to control root diseases.

2.2 Suppression of Pathogens by Bacteria and Fungi

Bacteria found on the plant surface or within the plant may also be involved in the suppression of plant diseases. Pleban *et al.* (1995) isolated *Bacillus cereus, B. subtilis,* and *B. pumilus* from stringently sterilized seed. When these bacteria were injected into cotton plants, they reduced disease incidence caused by *Rhizoctonia solani* 51, 46, and 56%, respectively. When *B. cereus* was introduced into cotton plants, it could still be recovered from the root and stem 72 days after inoculation. Chen *et al.* (1995) isolated 170 bacterial strains from the internal tissues of cotton and screened them for biocontrol activity against Fusarium wilt of cotton caused by *F.o.v.* The bacteria were introduced into cotton stems 7 days after planting, and this was followed 10 days later by injection with the pathogen. Fifteen strains from 11 bacterial species reduced wilt symptoms. The bacteria survived in stems for up to 28 days.

A unique interaction in cotton between the endophytic bacterium *Agrobacterium* biovar I and the root pathogen *R. solani* has recently come to light (Bell, 2000a). The bacterial endophyte does not normally produce crown gall symptoms on cotton. However, when cotton plants inoculated with this bacterium were subsequently inoculated 4 to 14 days later with *R. solani*, a high frequency of crown gall tumors developed at the soil line. No tumors developed on plants inoculated with the bacterium alone or in concert with other fungi. These results indicate that *R. solani* somehow increased the transformation of cotton tissue by the bacterial endophyte.

Conversely, predisposition of the cotton plant to infection by *R. solani* is dependent on the strain of *Agrobacterium* that is present as an endophyte (Bell, 2000b). The presence of some strains allowed *R. solani* to kill 80% of the plants even 30 days after planting, whereas in the presence of other strains there was no plant death or damage.

The populations of endophytic bacteria found in the cotton plant may be influenced by the application of organic amendments to the soil. Hallmann *et al.* (1999) found that the addition of chitin to the soil resulted in long term nematode suppression, along with the appearance of the nematode biocontrol agent *Burkholderia cepacia* (Fravel, 1999) as the dominant endophyte in the plant.

Ectophytic bacteria have also been shown to suppress cotton diseases incited by various pathogens. Brannen and Backman (1994b) demonstrated that seed treatment of cotton with *Bacillus subtilis* strain GBO3 resulted in good colonization of the cotton root by the bacterium and a significant reduction in Fusarium wilt symptoms when the crop was planted in soil infested with *F.o.v.* and the root-knot nematode *Meloidogyne incognita*. This was confirmed by Zhang *et al.* (1996) who found that the incidence and severity of Fusarium wilt of cotton was suppressed by seed treatment with *B. subtilis* strains GBO3 and GBO7 and *Gliocladium virens* strains G-4 and G-6. All of these strains colonized the cotton root system and reduced colonization of the tap and secondary roots by *Fusarium* spp.

Isolate D1 of *Burkholderia cepacia* has given control of seedling damping-off in the field in Arizona (Zaki *et al.*, 1998). When applied as an in-furrow drench the bacterium was as effective as a mixture of three fungicides.

Several possible mechanisms involved in biocontrol by ectophytic bacteria have been proposed. *Enterobacter cloacae* suppresses the cotton seedling disease incited by *Pythium ultimum* (Nelson, 1987). Howell *et al.* (1988) showed that suppression was due to production of ammonia by the biocontrol agent. *E. cloacae* produced ammonia in the spermosphere by deaminating amino acids emanating from the germinating seed. Growth of *P. ultimum* is inhibited by low concentrations of ammonia, and production of ammonia in the spermosphere prevented infection of the seedling.

Pseudomonas fluorescens suppresses the cotton seedling disease incited by *R. solani*. Howell and Stipanovic (1979) demonstrated that suppression was due to the production of the antibiotic pyrrolnitrin by the biocontrol agent. This was confirmed by Hill *et al.* (1994) who showed that mutants of *P. fluorescens* deficient for pyrrolnitrin production were no longer effective as biocontrol agents, and that reintroduction of the genes coding for pyrrolnitrin synthesis restored biocontrol activity.

Fungi that colonize the cotton root system (Zhang *et al.*, 1996) also may exhibit biocontrol activity against seedling diseases of cotton. Strains G-6 and G-4 of *Trichoderma (Gliocladium) virens* control cotton diseases incited by *R. solani* (Howell, 1987) and *P. ultimum* (Howell, 1991), respectively. The mechanisms involved have not been fully determined, although mycoparasitism (Howell, 1987) and antibiosis (Howell and Stipanovic, 1995) do not appear to be important in the biocontrol of *R. solani*. Some of the beneficial effects of *Trichoderma* species might be due to their ability to solubilize phosphates and micronutrients (Altomare *et al.*, 1999) and produce extracellular chitinases (Baek *et al.*, 1999). Recent evidence indicates that resistance induction, in the form of increased phytoalexin synthesis in cotton roots, by *T. virens* is a primary factor in disease control (Howell *et al.*, 2000).

The phenomenon of cross protection, defined as host inoculation with low-virulent or avirulent strains of the pathogen to protect against virulent strains, has been employed in cotton in several instances. Schnathorst and Mathre (1966) protected cotton plants from a virulent strain of *Verticillium albo-atrum* by preinoculation with a mild strain of the fungus in greenhouse studies. This phenomenon was confirmed by Barrow (1969), who demonstrated that preinoculation of tolerant cotton varieties with a mild strain of *V. albo-atrum* in a field test protected the plants from a virulent strain that was present in the soil.

The same phenomenon occurs in Fusarium wilt of cotton. Hillocks (1986) showed that preinoculation of cotton plants with a nonpathogenic strain obtained from *F.o.v.* reduced the severity of symptoms induced in the host by a virulent strain of the pathogen. Because of the short time required for cross protection to become effective, protection was ascribed to the formation of vascular occlusions by the nonpathogen in the host vascular system, rather than to phytoalexin production. Phytoalexin production, however, may be an alternative explanation for the host resistance exhibited by cross protected plants. Bell and Presley (1969a) demonstrated that cross protection of cotton against Verticillium wilt by an avirulent strain was correlated to the production of terpenoid phytoalexins by the inoculated plant in response to infection by the mild strain.

Although many microbes are useful as biocontrol agents at low concentration, their use at higher concentration can have an adverse consequence on cotton seedling growth.

Howell and Stipanovic (1984) found that the biocontrol agent *Trichoderma* (*Gliocladium*) *virens* produced a phytotoxin, viridiol, in biocontrol preparations which, when used in high concentration, results in necrosis of the emerging seedling radicle. This problem can be alleviated by adding low (0.5 to 1 ppm) concentrations of the sterol-inhibiting fungicide triademenol to cultures and preparations of the biocontrol agent (Howell and Stipanovic, 1994), or by making mutants of the biocontrol agent that are deficient for production of the phytotoxin (Howell and Stipanovic, 1996).

2.3 Toxigenic Microbes

By far the most important toxigenic microbe associated with cotton culture is the fungus *Aspergillus flavus*. Propagules of the fungus are common contaminants in cotton field soils, and they are introduced into cotton bolls through portals made by insect damage. Growth of the fungus in cotton bolls results in contamination of the seed and fiber with the toxic metabolite aflatoxin B_1. If cottonseed or its products are subsequently fed to cattle, aflatoxin can be transferred to the milk. There have been several biological approaches to alleviating this problem. Cotty (1994) infested cotton field soils with an atoxigenic strain of *A. flavus* and demonstrated that it competitively excluded the toxigenic strain during cotton boll infection, thereby reducing aflatoxin contamination in the resulting crop. In another instance, Misaghi *et al.* (1995) isolated a strain of *Pseudomonas cepacia* from the surface of an immature cotton boll and demonstrated that simultaneous inoculation of cotton bolls with *P. cepacia* and *A. flavus* reduced infection by the latter 60 to 100% compared to inoculation with *A. flavus* alone. Suggested mechanisms for this effect are competitive exclusion, degradation of aflatoxin, or competition for nutrients required for aflatoxin production (Cotty and Bayman, 1993).

2.4 Herbivorous Mites

Interactions between microbes and herbivorous mites have also been shown in cotton. Karban *et al.* (1987) studied the effects of inoculation with the wilt pathogen *V. dahliae* or infestation with the spider mite *Tetranychus urticae* on cotton resistance to the other pest. They found that the fungal pathogen was less likely to cause leaf wilting or chlorosis in cotton plants if the seedlings had previously been exposed to spider mites. Likewise, spider mites grew less rapidly on seedlings that had been inoculated with *V. dahliae* than on uninoculated controls. These results show that highly unrelated organisms that share a host plant may interact strongly.

Karban *et al.* (1989) later demonstrated that the strength of resistance induced in cotton against spider mites or Verticillium wilt was related to plant density. Because induced resistance diminished as plant density increased, they concluded that overplanting may adversely affect induced resistance. However, thick stands alone increase resistance to Verticillium wilt (Bell, 1992), which may have reduced the potential for spider mites to increase resistance.

3.   DISEASE INTERACTIONS

Although most microbial relationships with cotton are beneficial or commensal, some lead to deleterious effects on the plant or its products, *i.e.,* disease. In this case the relationship is defined as pathogenic and the microbe as a pathogen. The most important diseases of cotton are those caused by soil-borne microbes that colonize the root or hypocotyl causing seedling diseases, root rots and wilts (Hillocks, 1992; Bell, 1999). Several nematodes also feed on cotton roots, causing damage to the plant. In addition, nematodes may aggravate diseases caused by other pathogens (Hillocks, 1992; Robinson, 1999).

The immature fiber is especially vulnerable to colonization by microbes. More than a hundred microorganisms are known to colonize fibers when introduced into immature bolls through wounds. A few microbes also invade the bolls without wounds. These infections may lead to boll and fiber rots (Cauquil, 1975), and to mycotoxin contamination of seed (Cotty, 1994). Once the boll opens, the fibers are rapidly colonized by other microorganisms, especially bacteria, that may contribute to endotoxin contamination of fibers (Fischer and Domelsmith, 1997). The microbes associated with fiber and seed also are responsible for heating and rapid loss of seed quality, especially the accumulation of free fatty acids, when seedcotton is stored moist in modules or carts.

Cottonseed is often infested with both pathogenic and nonpathogenic microbes. Pathogens such as *Fusarium* and *Agrobacterium* species enter the seed coat through xylem vessels, whereas boll rot pathogens, such as *Xanthomonas, Colletotrichum, Phoma, Fusarium*, and *Diplodia* species may directly penetrate the seed epidermis to invade the seed coat and sometimes the embryo. Other pathogens, such as *V. dahliae*, may occur as propagules that infest the lint of fuzzy seed. Infested seeds often serve as the primary source of inoculum for seedling infections and may be involved in moving pathogens from one geographical location to another.

In the following sections, specificity of disease interactions, influence of environment, mechanisms of pathogenesis, responses to pathogens and physiological consequences of disease will be discussed.

3.1   Specificity of Interactions

Most of the major pathogens attacking cotton have very broad host ranges and broad geographical distribution. Consequently, there is extensive variation in population structure within species. Many of these populations are genetically isolated from each other and show distinct differences in their interactions with cotton. Improvements

in genetic and molecular techniques have allowed many advances in the identification and understanding of the variation among isolates within pathogenic species. In some cases, these techniques have allowed the cloning of avirulence genes and the mapping of resistance genes that control the specificity of interactions between cotton cultivars and specific populations of the pathogen. Advances in understanding the specificity of interactions between cotton and the pathogens *Xanthomonas campestris, Verticillium dahliae, Fusarium oxysporum, Rhizoctonia solani, Aspergillus flavus, Meloidogyne incognita,* and *Agrobacterium* species will be discussed.

3.1.1   *Xanthomonas campestris*

The bacterial blight pathogen *X. campestris* attacks a wide range of plant species including many important agricultural crops. Individual isolates of the bacterium, however, vary considerably in their ability to attack different species and even cultivars within species. At present, an individual isolate is usually given a pathovar (pv.) name according to the system of Dye *et al.* (1980) to indicate the major plant species attacked by the isolate. Thus, the bacterial blight pathogen of cotton is named *X. campestris* pv. *malvacearum* (E. F. Smith) Dye (*X.c.m.*). All isolates in this group attack cotton and certain other malvaceous plants. An isolate is identified as being pathovar *malvacearum* if it causes a susceptible reaction on the cultivar 'Acala 44,' which is considered a universal suscept to the pathovar (Brinkerhoff, 1970; Hillocks, 1992).

Bacterial isolates within the pathovar *malvacearum* can be further divided into pathotypes called races based on the susceptibility of cultivars or breeding lines to the isolates. In 1968, Hunter *et al.* (1968) developed a set of 8 cotton lines that could be used as host differentials to identify 15 races of *X.c.m.* Later Hussain and Brinkerhoff (1978) added two more lines to the differentials to identify 18 races. Currently, 20 races of *X.c.m.* are known (Hillocks, 1992).

The specificity of interactions between host differentials and bacterial races depends on the presence of specific bacterial blight resistance genes (*B* genes) in each differential cotton line and of specific avirulence genes (*avr* genes) in the *X.c.m.* races. Genetic studies have shown that the host differentials contain the *B* genes listed in Table 1.

The B_6 gene from the NT 12/30 strain of *G. arboreum* also has been used in specificity studies. The *G. hirsutum* line S-295 is resistant to all 20 races of *X.c.m.,* and resistance is due to a single gene designated as B_{12} (Wallace and El-Zik, 1989). Wright *et al.* (1998) developed DNA markers diagnostic of seven bacterial blight genes and used these markers to show that five of the six *B* genes from *G. hirsutum* were located in the D genome. The specific locations of some of the genes were also determined. Gabriel *et al.* (1986) cloned 10 avirulence genes from races of *X.c.m.* and showed that five of these genes gave specific gene-for-gene interactions with B_2, B_3, B_6, B_N, and B_{In} genes in cotton lines.

Table Chapter 18-1. Cotton lines and *B* genes used to identify *X.c.m.* races.

Differential line or cultivar	*B* genes in line	Source of *B* genes
Acala 44	none	
Stoneville 2B-S9	minor polygenes	Stoneville 2B
Stoneville 20	B_7	Stoneville 2A
Mebane B-1	B_2	Mebane B-1
1-10 B	B_{ln}	*G. hirsutum*
20-3	B_N	Northern Star
101-102B	B_2,B_3	Uganda B31 (B_2) and Schoeder 1306 (B_3)
Gregg	unknown	Gregg 8
DPxP4	B_4	Multani Strain NT 12/30 of
Empire B4	B_4	*G. arboreum*

In extended studies, genes that control dispersal and pathogenicity, as well as host specificity, of *X.c.m.* have been cloned and characterized (De Feyter and Gabriel, 1991; Yang and Gabriel, 1992; De Feyter *et al.*, 1993; Gabriel *et al.*, 1994; Yang *et al.*, 1994). While most *avr* genes coded for proteins that triggered hypersensitive resistance (HR), the *avrb6* gene was essential for cotton-specific water-soaking ability and, thus, was also a virulence gene. De Feyter *et al.* (1998) obtained a genotype-specific necrosis in cotton lines containing B_4, B_7, and B_{ln} genes, when the $avrB_4$, $avrB_7$, and $avrB_{ln}$ genes, respectively, were introduced into cells of cotton leaves. The *avr* genes were joined to the cauliflower mosaic virus 35S promoter and incorporated into plasmid constructs that allowed them to be delivered by *Agrobacterium tumefaciens*, when it was used to inoculate cotton leaves. No necrosis from the *avr* genes occurred when plasmids directed expression of the *avr* protein within *A. tumefaciens* cells or within the apoplast of the plant tissue. The authors concluded that the specific host cell death coded by *B* genes results from the intracellular expression of the *X.c.m. avr* proteins.

Xanthomonas campestris pathovars also produce *hrp* (host resistance and pathogenicity) genes that are involved in host-microbe interactions. The gene *hrp X* has been cloned from *X.c.* pv. *campestris*, and the gene product has been shown to regulate other *hrp* genes involved in pathogenicity, non-host hypersensitivity and non-permissibility of certain defense responses (Oku *et al.*, 1998). The *hrp X* gene was conserved in 16 *X. campestris* pathovars indicating that it probably also occurs in *X.c.m.*

The proteinaceous defense elicitors coded by the *hrp* genes are called harpin proteins. Theoretically, the harpin protein should attach to a host membrane receptor to activate hypersensitive resistance (HR). A protein that may serve such a receptor function for the harpin – Pss protein from *Pseudomonas syringae* pv. *syringae* has been isolated and purified from cotton as well as tomato and sweet pepper (Lin *et al.*, 1997). The N-terminal amino acid sequence of the protein, referred to as an amphipathic protein (API), was conserved among the three crops. The API protein postponed harpin-Pss-mediated HR reactions in a dose dependent pattern. The authors conclude that the API protein may play a role in non-host responses caused by pathogens.

3.1.2 *Verticillium dahliae*

The fungal wilt pathogen *Verticillium dahliae* has a host range of several hundred species and occurs worldwide. Different isolates vary both in their host specificity and aggressiveness as pathogens. Isolates of *V. dahliae* from cotton have been organized into pathotypes (defoliating and non-defoliating) or races 1-3 that show similar patterns of virulence and pathogenicity to cotton species and cultivars (Bell, 1992). These groups are the result of vegetative incompatibility genes that have genetically isolated certain populations of *V. dahliae* from other populations. Several vegetative compatibility groups (VCGs) have been identified by genetic complementation tests of isolates of *V. dahliae* (Puhalla and Hummel, 1983; Bell, 1995; Daayf *et al.*, 1995). These studies show that *V. dahliae* isolates can be placed in one of three major VCGs, usually designated VCG-1, -2 and –4 (Bell, 1995; Daayf *et al.*, 1995) or in a fourth group that is self-incompatible presumably because the mycelium and conidia are diploid rather than haploid. Diploid isolates generally are obtained from crucifer crops (Subbarao *et al.*, 1995), but they also have been isolated from cotton in Russia (Puhalla and Hummel, 1983).

Each VCG can be divided into two or three subgroups: A, B, and A/B (Bell, 1995). Isolates in the VCG-1A subgroup are distinctly more virulent to 'Acala 4-42' or 'Acala 4-42-77' cotton cultivars than other isolates and have been called the defoliating pathotype or race 3. Isolates from VCG subgroups 1B, 2A, 2B, 4A, 4B, and 4A/B and diploid isolates are less virulent than those of VCG-1A and are collectively called the non-defoliating pathotype or races 1 and 2. Subgroups 2A and 4A/B are commonly isolated from cotton in many parts of the world, whereas 1A is prevalent in North America and 2B is prevalent in Europe and Asia. The VCG-1 group is probably indigenous to North America, although it has been found in recent years associated with severe Verticillium wilt of cotton in local areas in Spain (Bejarano-Alcazar *et al.*, 1996), Greece (Elena, 1999) and China (Xia *et al.*, 1998). Isolates of *V. dahliae* from cotton in Israel were assigned to two VCGs but did not include defoliating isolates (Bao *et al.*, 1998). The genetic bases of vegetative compatibility and host specificity in *V. dahliae* isolates have not been determined.

3.1.3 *Fusarium oxysporum*

The fungus *F. oxysporum* is often the most prevalent fungus found on the surface of cotton roots (Zhang, 1995; Zhang *et al.*, 1995). It also can be found readily in intact dead cotton roots, which remain after minimum tillage practices (Baird and Carling, 1998). Both pathogenic and saprophytic isolates of *F. oxysporum* occur on cotton roots. Only saprophytic isolates are present in some fields (Zhang *et al.*, 1995; Bell, unpublished), while the majority of isolates from other fields are pathogenic to cotton (Baird and Carling, 1998). The pathogenic isolates of *F. oxysporum* infect hundreds of plant species causing root rot and vascular wilt diseases. Different isolates vary in host-specific pathogenicity and, generally, are organized into formae speciales (f.sp.) according to the system of Snyder and Hansen (1940). Isolates that cause Fusarium wilt of cultivated cottons are assigned to f.sp. *vasinfectum* (*F.o.v.*) and most of these isolates also cause wilt of alfalfa (*Medicago sativa*), okra (*Abelmoschus elegans*), and lupine (*Lupinus luteus*). Moricca *et al.* (1998) have developed a polymerase chain reaction assay to specifically detect *F.o.v.* in cotton tissues. The two primers used in the assay are 5'-CCCCTGTGAACATACCTTACT-3' and 5'-ACCAGTAACGAGGGTTTTACT-3', which amplify a 400-bp DNA fragment (a spacer sequence between 18S, 5.8S, and 28S ribosomal DNAs) of all *F.o.v.* isolates tested. This detection system is reported to be very accurate and sensitive.

Isolates within *F.o.v.* show specific differences in pathogenicity. Armstrong and Armstrong (1980) used pathogenicity tests on six cotton cultivars (three *Gossypium* species), okra, alfalfa, and two cultivars of tobacco (*Nicotiana tobacum*) to distinguish six races of *F.o.v.* Variability within *F.o.v.* also has been studied using genetic complementation tests between *nit* (nitrate non-utilizing) mutants, restriction fragment length polymorphism (RFLP) (Fernandez *et al.*, 1994), random amplified polymorphic DNA (RAPD) analyses (Assigbetse *et al.*, 1994; Nirenberg, 1996), and fatty acid analyses (Hering *et al.*, 1999). Isolates belong to one of three groups that can be distinguished by pathogenicity to *Gossypium* species or by RAPDs (Shen, 1985; Assigbetse *et al.*, 1994; Nirenberg, 1996). Two of the groups include a single VCG and correspond to races 3 and 4 (Armstrong and Armstrong, 1980; Katan and Katan, 1988; Fernandez *et al.*, 1994). The third group, originally designated as races 1, 2, and 6 by Armstrong and Armstrong (1980) and more recently as race A by Assigbetse *et al.* (1994) and Fernandez *et al.* (1994), is heterogeneous, containing more than 20 VCGs (Bell and Decker, 1993; Fernandez *et al.*, 1994; Bell, 1995). All isolates in this group, which we prefer to call race 1, cause wilt of *G. hirsutum* 'Rowden' and 'Acala 44', *Abelmoschus elegans* 'Clemson Spineless', and *Medicago sativa* 'Grimm', which are not affected by races 3 and 4. Race 3 infects cultivars of *G. barbadense* and *G. arboreum*, and race 4 infects only cultivars of *G. arboreum*.

Races 3 and 4 have not been found in the USA or Australia, whereas race 1 occurs in all continents where cotton is grown. A new race distinct from races 1, 3, and 4 has been found in Australia, but it has not been given a number. This race attacks *G. hirsutum*, but is favored by dark grey heavy clay soils, even at pH levels as high as 8.5 (Wang *et al.*, 1999). In contrast, race 1 is favored by acidic sandy soils. The Australian isolates are not vegetatively compatible with *F.o.v.* isolates from any other country and appear to be unique (Kochman *et al.*, 1996; Bell, unpublished).

3.1.4 *Rhizoctonia solani*

The species named *R. solani* is used for the nonsporulating asexual form of several perfect (sexual) species that can be divided into groups that have binucleate or polynucleate hyphal cells. Two of the perfect species are further divided into distinct genetic populations called anastomoses groups (AG). Most pathogenic isolates from cotton belong to the polynucleate species *Thanatephorus cucumeris* and belong to AG4. AG11 also is virulent to cotton in the greenhouse (Carling *et al.*, 1994). The binucleate *Rhizoctonia* sp. CAG-5 has been identified as a cause of seedling disease of cotton in Georgia (Baird *et al.*, 1995). Methods for identification of *Rhizoctonia* species and AG groups are described in the book "*Identification of Rhizoctonia species*" (Snek *et al.*, 1991).

3.1.5 *Aspergillus flavus*

Studies of variability in cotton-microbe interactions of *A. flavus* have been conducted with the goal of using aflatoxin-deficient variants to displace the toxigenic strains (Cotty *et al.*, 1994). *A. flavus* strains generally are not specialized to particular hosts, either plant or animal (Leger *et al.*, 2000). However, most isolates from cotton fields can be placed in one of two distinct strains on the basis of sclerotial size, cultural characteristics, and virulence to cotton (Cotty, 1989). Strain L isolates produce large sclerotia (over 400 μm in diameter), and strain S isolates produce small sclerotia (less than 400 μm in diameter). Strain S isolates produce more aflatoxin than strain L in culture but not in cottonseed. This difference is due to the fact that strain L isolates are more aggressive than strain S isolates at deteriorating the locks of cotton bolls and growing within bolls. Geostatistics have been used to describe variability of strain S incidence over space and time (Orum *et al.*, 1999).

3.1.6 *Meloidogyne incognita*

Four host races of the root-knot nematode, *M. incognita*, have been identified based on the ability of the nematode isolate to parasitize *G. hirsutum* 'Deltapine 16' and *N. tobacum* 'NC 95' (Hartman and Sasser, 1985). Only host races 3 and 4 are able to parasitize cotton, apparently with similar aggressiveness (Starr and Veech, 1986). These two races are distributed worldwide with race 3 being the most common of the two on cotton. Race 4 can predominate in cotton, if cotton is grown in rotation with tobacco (Starr, 1998).

3.1.7 *Agrobacterium* species

The interactions between cotton and *Agrobacterium* species are of importance both for increasing the efficiency of transformations with foreign genes and in controlling diseases, such as bronze wilt, which may be caused by seed-borne endophytic strains of *Agrobacterium*. Both bacterial strain and plant genotype contribute significantly to the success of cotton genetic transformation. Velten *et al.* (1998) found that strain EHA105 of *A. tumefaciens* was much more efficient than strain LBA4404 at transforming seven different cotton cultivars. Four cultivars were more readily transformed than 'Coker 312' which has been used in most studies. Strain A281 was highly efficient at inducing tumors on cotyledons of Paymaster 1220 B/R, whereas strain B6 was only moderately efficient and strain C58 failed to induce tumors (Bell, unpublished). None of these three strains, however, were able to compete with the seed-borne endophytes from cotton in colonizing cotton roots.

Agrobacterium biovar 1 isolates obtained from cottonseed differ significantly from each other in three pathogenicity characteristics: 1) colonization of cotton roots, 2) severity of root necrosis and bronze wilt symptoms, and 3) predisposition of cotton hypocotyls to soreshin disease caused by *R. solani* (Bell, 2000b). *Agrobacterium* biovar 1 isolates from cottonseed also can be divided into several groups based on colony morphology, biochemical characteristics, fatty acid profiles, plasmid contents, and RAPD analyses using primers designed to study variability in *Rhizobium* (Cui *et al.*, 1997; Bell, 2000c, 2000d). A second group of *Agrobacterium* isolates, which are similar to biovar 2 except that they grow at 38°C, have also been obtained from cottonseed and are currently being evaluated for their effects on cotton (Bell, unpublished).

3.2   Influence of Environment

Cotton-pathogen interactions are often strongly influenced by the environment, particularly that around the roots. In this section, the effects of temperature, moisture, light, nutrition, agricultural chemicals, insects, nematodes, and other microorganisms on disease development will be considered. More detailed reviews are available on the effects of environmental stress on cotton pest interactions (Bell, 1982) and on the role of nutrition in diseases of cotton (Bell, 1989). Much of the research prior to 1990 also is reviewed in the book "*Cotton Diseases*" (Hillocks, 1992). Therefore, recent studies are emphasized here.

3.2.1   Temperature, Moisture, and Light

Three climatic factors (temperature, moisture, and light) are considered together because they normally interact in nature. That is, higher than normal temperatures often are associated with low moisture and maximum light exposure, whereas lower than normal temperatures are associated with increased moisture and low light exposure. In many cases all three components contribute to changes in cotton-microbe interactions, even though only one component may be the subject of a study.

Cotton originated in the tropics and has very poor adaptation to low temperatures. Growth and metabolism virtually stop when temperatures are lower than 15°C, and the plant becomes progressively subject to cold injury as

temperatures are lowered further. Because of cold sensitivity, planting of cottonseed is generally not recommended until soil temperatures reach 20°C. Even at this temperature, however, the active defense reactions of the plant are greatly diminished (Bell and Presley, 1969b). Several fungal pathogens that grow readily at 20°C take advantage of the diminished defenses and increased root exudation that occurs at low temperature. Seedling root rot and blight incited by *Pythium ultimum*, *Thielaviopsis basicola*, and *Phoma exigua* (*Ascochyta gossypina*) occur primarily under cool, moist soil conditions (Hillocks, 1992; Rothrock, 1992). Likewise, susceptibility to Verticillium wilt increases dramatically as temperature is lowered from 28°C to 22°C (Bell and Presley, 1969b). The same cultivar may be killed by *V. dahliae* at 22°C and be immune at 28°C. These changes are expected if the rate of active defense reactions relative to the rate of pathogen colonization is a primary determinant of cotton resistance to disease as proposed by Bell (1982).

Bacterial diseases caused by *X.c.m.* and *Agrobacterium* are favored by hot moist conditions. Day temperatures of 36°C or higher are usually most favorable for disease development. *Agrobacterium* biovar 1 concentrations in roots of some cultivars increase as much as 10-fold as soil temperature is increased from 30°C to 33°C or higher, and bronze wilt severity increases greatly as temperature is increased from 33°C to 36°C (Bell, 1999b). Resistance to bacterial blight conferred by single *B* genes generally is ineffective when day temperatures are 36°C or higher with cool 20°C night temperatures (Brinkerhoff, 1970). Contamination of cotton fiber by the saprophytic bacterium *Enterobacter agglomerans*, a primary source of endotoxins in mill dusts, also is favored by hot, moist conditions (Bell, 1997).

Aspergillus boll rots and Macrophomina root rot occur mostly under hot dry conditions. The pathogens in these cases are usually well adapted to grow at high temperatures with low water potential, but they are poor competitors with soil saprophytes in moist soils with moderate or low temperatures (Bell, 1982; Cotty, 1994).

High temperatures suppress many cotton pathogens. All of the pathogens favored by low temperatures are suppressed above 30°C (Hillocks, 1992). Even a few hours of temperatures above 42°C, such as occur in the Arizona desert, suppress development of Alternaria leaf spot of cotton (Cotty, 1987). The temperature-induced suppression is due to both reduced viability of spores and lysis of their germ tubes.

3.2.2   Soil Chemistry and Texture

Nutrient deficiencies or excesses, mineral toxicities, or soil texture may be the direct cause of abiotic diseases or affect the severity of infectious diseases. Potassium deficiencies and excess nitrogen are especially harmful in increasing the severity of infectious diseases. Potassium deficiencies apparently have an adverse effect on active defense reactions in cotton, while nitrogen stimulates growth and reproduction of pathogens. Studies dealing with nutrients have been reviewed (Bell, 1989, 1999a; Hodges, 1992; Cassman, 1993). Recent studies have been concerned with unusual cases of potassium and phosphorus deficiency that occur in aerial parts of the plant in spite of seemingly adequate supplies of these elements in the soil (Wright, 1999), the consequences of potassium deficiencies (Bednarz and Oosterhuis, 1999; Pettigrew, 1999b), and better methods of correcting potassium and phosphorus deficiencies (Bauer *et al.*, 1998; Howard *et al.*, 1998; Rosolem *et al.*, 1999).

Some of the unusual deficiencies of potassium and phosphorus may be the consequence of *Agrobacterium* infections of roots. The infections coupled with excess nitrogen and/or temperatures above 33°C lead to extensive necrosis of the root epidermis (Bell, 2000b, 2000c). This could impair the active uptake of potassium and phosphorus. Adding excess phosphorus as 0-46-0 granules has reversed much of the damage caused by *Agrobacterium* infections, even when soil analyses do not indicate a phosphorus deficiency. Likewise, high compared to moderate rates of potassium fertilization reduce the root damage caused by *Agrobacterium* inoculations (Bell, 2000d).

Soil texture and pH also have been shown to affect root damage caused by *Agrobacterium*. Disease severity in all cultivars is directly proportional to the clay percentage in soil, but the relative resistance of cultivars is not affected by clay percentage (Bell, 2000c, 2000d). The pH values that restrict availability of phosphorus in soils, especially pH 7.5 to 8.5, also aggravate root damage and bronze wilt symptoms.

Interference of phosphorus uptake by *Agrobacterium* infections is also indicated by the fact that plants showing bronze wilt symptoms have a greatly reduced capacity to actively "pump" xylem fluid when the shoot is excised. Applying phosphorus fertilizers 24 hours prior to excision partially restores the water pump. This indicates that ATP levels needed for active water uptake may be critically low in the root epidermis of bronze wilt-affected plants. Relationships between *Agrobacterium* infections of the cotton root surface and nutrient uptake need to be a focus of future studies.

3.2.3   Cultural Practice

In recent years, there has been a marked increase in minimum tillage practices in all crops. Diseases such as bacterial blight, Fusarium wilt, Verticillium wilt, Alternaria leaf spot, and Phoma blight would be expected to increase in severity with minimum tillage, because the pathogen in each case survives better in surface trash or undisturbed roots than in plowed under trash (Hillocks, 1992; Bell, 1999a). Baird and Carling (1998) studied the survival of pathogenic and saprophytic fungi on intact senescent cotton roots, such as would be left with a no-till practice. Several boll rot pathogens, seedling pathogens, and *F.o.v.* were obtained from the roots. *F. oxysporum* was routinely isolated from samples collected during the 2-year study, and 91%

of the *F. oxysporum* isolates were pathogenic (*i.e.*, *F.o.v.*). Thus seedling, boll rot, and wilt pathogens all overwintered on intact senescent roots.

Center pivot irrigation is another practice that may increase disease severity. This practice, compared to furrow irrigation, increases the frequency and severity of bacterial blight (Hillocks, 1992; Bell, 1997). The only reported cases of Sclerotinia blight on cotton have also occurred in association with this practice (Charchar, 1999). In each case, water splash and high humidity increase the percentage of successful infections by the pathogen. High moisture levels in soil also increase sclerotia formation by the Phymatotrichum root rot pathogen (Kenerley *et al.*, 1998).

Crop rotations often affect soil-borne disease severity by influencing the density of infectious propagules in soil (Hillocks, 1992; Bell, 1999a). Disease frequency and severity, in turn, are proportional to propagule density. Several studies have confirmed that incorporating green rye plants into soil reduces populations of *Meloidogyne* species and *Rotylenchulus reniformis* nematodes (Guertal *et al.*, 1998; Johnson *et al.*, 1998; McBride *et al.*, 1999). Although rye is an effective cover crop for suppressing nematodes, vetch and clover are not (Guertal *et al.*, 1998). Bacteria stimulated by decomposing green crops have been shown to lyse mycelium of *V. dahliae* and reduce cotton root colonization by the pathogen (Strunnikova *et al.*, 1997). Plow down of several green crops has effectively controlled Verticillium wilt (Bell, 1992).

3.2.4   Pests and Disease Complexes

Insects, nematodes, or disease complexes may affect the cotton-microbe interaction. Many studies have shown that nematodes can increase the severity of Fusarium wilt or seedling damping-off caused by *R. solani* (Starr, 1998). The root knot nematode can also increase the severity of *Thielaviopsis* root rot in cotton seedlings, especially at temperatures (24 to 28°C) that normally restrict the disease (Walker *et al.*, 1998). The nematode apparently allows a much higher frequency of invasion of the stele and pith by the fungal pathogen which, in turn, has a detrimental effect on the nematodes. Thrips infestations of cotton seedlings also may increase the severity of seedling diseases (Colyer and Vernon, 1991). In this case, the weakened plants apparently are not able to resist disease as effectively as uninfested plants.

Agrobacterium and *R. solani* show synergistic interactions in causing both crown gall and soreshin of cotton (Bell, 2000a, 2000b). The soreshin lesion acts as a selective medium for the *Agrobacterium* allowing it to develop very high populations (>10M/gm) in the lesion. The bacterium produces growth regulators and diamines that may suppress plant defense reactions. Disease complexes affecting cotton roots are common (Hillocks, 1992; Bell, 1999a) and need to be better understood.

3.2.5   Agricultural Chemicals

Pesticides and growth regulators used for cotton production can affect the cotton-microbe interaction. Many examples of increasing resistance to wilt diseases with growth regulators or herbicides have been reviewed (Bell, 1992a; Hillocks, 1992). More recent studies have been concerned with the mechanism of the induced resistance. Awadalla and El-Refai (1992) found that treatment of cottonseeds with prometryn and dalapon reduced Verticillium wilt of cotton, repressed mycelial growth of *V. dahliae*, and greatly increased terpenoid phytoalexin production in inoculated plants. The herbicide clomazone also causes accumulation of sesquiterpenoid phytoalexins when used to treat cotton seedlings (Duke *et al.*, 1991). Song and Zheng (1998) concluded that trifluralin increases resistance to Fusarium wilt of cotton by blocking ethylene synthesis. Spraying trifluralin-treated seedlings with ethephon increased both ethylene production and disease incidence, while spraying untreated seedlings with 2,4-dinitrophenol, an ethylene biosynthesis inhibitor, or silver thiosulfate, an ethylene action inhibitor, reduced Fusarium wilt incidence and severity.

The herbicide glyphosate is now applied directly to plants of many transgenic cotton cultivars that carry a bacterial gene resistant to the herbicide (Saroha *et al.*, 1998). When sprayed on transgenic cotton plants with three to six true leaves, glyphosate reduces boll retention of the first position of the first three fruiting branches, causing delayed maturity (Jones and Snipes, 1999). However, effects on diseases were not determined.

Studies on the effects of glyphosate on cotton-microbe interactions are needed for two reasons. First, glyphosate blocks a key step in the synthesis of phenylalanine, a key substrate for active defense reactions in cotton. Both lignins and condensed tannins which are important in cotton's defense against pathogens (Bell, 1986; Bell *et al.*, 1986; Bell *et al.*, 1992) are derived from phenylalanine and can make up as much as 20% of the dry weight of certain cotton organs or tissues. Therefore, the transgene needs to fully maintain the synthesis of tannins and lignins when cotton plants are sprayed with glyphosate and become infected. Second, the gene most used for glyphosate resistance, which codes for the enzyme 5-enolpyruvyl-3-phosphoshikimic acid synthase (EPSPS), was obtained from *Agrobacterium*. Thus, spraying with glyphosate might stimulate populations of endophytic *Agrobacterium* strains and bronze wilt severity by suppressing microbial antagonists that normally inhibit *Agrobacterium*.

3.2.6   Harvest Practices

Harvest practices are especially important in determining the interactions between cotton and microorganisms that produce endotoxins (Bell, 1997) or mycotoxins (Cotty, 1994). Endotoxins are produced by gram-negative bacteria that grow on fibers and trash in raw cotton. They are the major contributors to the human disease byssinosis which occurs in gin and mill workers. Mycotoxins in cotton are produced mostly by species of *Aspergillus* but also can be produced by *Fusarium* species. Mycotoxins can be produced either during maturation of infected bolls or during storage of mature raw cotton in carts or modules. Reducing green trash in cotton and keeping moisture levels low during storage are especially important to reduce growth of toxin-producing microbes and to prevent heating that may lead to loss of seed vigor (Curley *et al.*, 1990). Both *Aspergillus flavus* and *Enterobacter agglomerans* are able to grow at relatively high temperatures and low water potentials compared to other fungi and bacteria, respectively. This adaptation appears to be the reason for the growth advantage that they have in stored moist cotton.

3.3   Mechanisms of Parasitism and Pathogenesis

Pathogens that cause diseases of cotton can be divided into three classes based on cellular reactions in susceptible and resistant plants as shown in Table 2 (Bell, 1983). In all cases, resistance involves the rapid death of plant cells at the infection site. When a biotroph or transient biotroph incites the reaction, it often is referred to as a hypersensitive reaction (HR). The resistance reactions will be discussed in the next section. The differences among classes of pathogens are in their relationship with the susceptible host and hence in their mode of virulence (the ability to prevent or overcome resistance). The cotton-microbe relationships for each class of pathogen are discussed in the following sections.

3.3.1   Biotrophic Relationships

The biotrophic pathogen either actively suppresses or does not induce the HR response in the susceptible host. As a consequence, chemicals or enzymes associated with HR either are not formed or occur in very low concentrations, and genes required for HR are not fully activated in the susceptible host. Examples of biotrophic pathogens attacking cotton are rust and mildew fungi, root-knot and reniform nematodes, and viruses. In each case, the pathogen obtains nutrition from live host tissues and maintains the compatible conditions for several weeks. HR, however, occurs in resistant cultivars against the southwest cotton rust fungus, the tropical rust fungus, blue disease virus, anthocyanosis virus, reniform nematode and root-knot nematode (Bell *et al.*, 1986). The mechanisms of parasitism and pathogenesis have only been studied in the case of the nematodes.

Only races 3 and 4 of the root-knot nematode, *Meloidogyne incognita*, are able to establish a compatible (susceptible) feeding relationship with the cotton host (Veech, 1984; Starr, 1998). In the susceptible host, the nematode physically penetrates through the root cortex and establishes a feeding site, called the giant cell, in the peri-

Table Chapter 18-2. Classifications of pathogens based on cellular reactions in susceptible and resistant cotton plants.

Class of pathogen	Cellular reaction of plant	
	Susceptible	Resistant
Biotroph	Remains alive adjacent to pathogen	Rapid confined death adjacent to pathogen
Transient Biotroph	Delayed death adjacent to pathogen	Rapid confined death adjacent to pathogen
Necrotroph	Extensive death well beyond pathogen	Rapid confined death adjacent to pathogen

cycle. The animal becomes sedentary, so that maintenance of the giant cell is essential for feeding. The giant cell is metabolically hyperactive, contains numerous nuclei, and exhibits extensive protein synthesis with altered amino acid metabolism (Hedin and Creech, 1998). Active defense reactions are not apparent in or around the giant cell, and sterols do not influence susceptibility (Hedin *et al.*, 1995).

In resistant hosts, pericycle cells near the head of the sedentary nematode exhibit an HR response, and there is a marked reduction in numbers of egg masses, eggs/mass, and reproduction. Thus, reductions in eggs (or masses) per gram of root (Veech, 1982) or reproduction (Robinson *et al.*, 1997) are usually used as evidence of resistance.

Reniform nematodes are sedentary semi-endoparasites in that the anterior third of the body is embedded into the host root, while the posterior two-thirds remains exterior to the root (Starr, 1998). The nematode parasitizes the pericycle, endodermis, and phloem tissue of young roots. While invasion of roots by the root-knot nematode is usually restricted to root tips, the reniform nematode invades feeder roots at any point along their length. This nematode also induces host cells to differentiate into specialized nurse cells (the syncytium) at a permanent feeding site. The syncytium develops via dissolution of cell walls and coalescence of cytoplasm from several cells, whereas a giant cell develops from a single cell (Starr, 1998). Resistance to reniform nematode causes a reduction in nematode reproduction (Robinson *et al.*, 1997).

Carbon dioxide at critical concentrations in air attracts nematodes on artificial gels and in Baermann funnels (Robinson and Heald, 1991). Therefore, this gas may have a role in directing nematodes to plant roots.

3.3.2   Transient Biotrophic Relationships

Transient biotrophic pathogens also suppress HR in the susceptible host but only long enough to allow the pathogen to escape the resistance response, and progressively colonizes the host. Thus, biochemical reactions associated with HR occur more quickly and active defense products are at higher concentrations in resistant than susceptible hosts 12-72 hours after inoculation. In susceptible hosts, the slow defense response allows more extensive colonization by the pathogen and, eventually, a much greater number of cotton cells may show defense reactions. Consequently, by 1 to

2 weeks after inoculation, there may be more response chemicals in susceptible than in resistant plants. The fungal wilt pathogens *Fusarium* and *Verticillium* and the bacterial pathogen *Xanthomonas* are examples of important transient biotrophic pathogens of cotton.

A phylogenetic study of 13 species of *Verticillium*, including pathogens of insects, plants, mushrooms, nematodes and spiders, and saprobes, showed that the plant pathogens are a clan that is unique from the other species in producing diverse types of pectinase enzymes and broad-spectrum proteases that behave like trypsins in that they degrade Bz-AA-AA-Arg-NA substrates (Bz = benzoil; AA = various amino acids; Arg = arginine; NA = p-nitroanilide) (Bidochka *et al.*, 1999). Insect and mushroom pathogens do not produce pectinases, and they produce subtilisin-like proteases active against chymotrypsin substrates. Insect, mushroom, and nematode pathogens also are distinguishable from plant pathogens in their ability to produce chitinases. Pectinase and protease activity similar to that in plant pathogenic *V. dahliae* has also been demonstrated in the Fusarium wilt and bacterial blight pathogens (Hillocks, 1992). These enzyme groups may play important roles in the pathogenicity of the transient biotrophs.

The parasitic and pathogenic relationships of *V. dahliae* with its hosts have been reviewed by Bell (1992, 1993). Dormant microsclerotia in the soil germinate in response to root exudates (sugars and amino acids) and penetrate the root directly just back of the root tip. The fungus penetrates through the cortex, eventually entering the xylem vessel, where conidia are formed in the xylem stream. These conidia are carried to the xylem vessel end walls where they germinate and penetrate through perforation plates into the next vessel element or through vessel sidewalls into adjoining vessels. New conidia are again formed and the cycle is repeated until the entire vascular system is invaded. As terminal stem or leaf tissues die, the fungus invades cells surrounding the vessels and forms microsclerotia, which survive in plant trash or in soil to repeat the disease cycle. *Fusarium* has a similar infection pattern, except that it initially invades roots primarily through wounds or nematode feeding areas, and chlamydospores are formed in the moribund tissue. Fungal melanins play key roles in the survival of the dormant fungal propagules that survive apart from the host (Bell and Wheeler, 1986).

Various enzymes and toxins have been implicated in pathogenesis by *V. dahliae*. The pectinases may not be essential in the systemic spread of the fungi in xylem vessels (Puhalla and Howell, 1975); however, they probably are required for root penetration and growth in dead tissues (Bell, 1993). Phytotoxic protein-lipopolysaccharide complexes have been isolated from *V. dahliae* cultures (Meyer *et al.*,

1994). These complexes contain pectinase, cellulase and β-1,3-glucanase enzyme activities and readily cause wilting, necrosis, and inhibition of the H^+-ATPase activity of plasma membranes. They also elicit accumulation of pathogenesis-related proteins and defense reactions.

Heat-released soluble wall fragments and glycoproteins purified from cultures of *V. dahliae* also elicit defense responses (Dubery and Slater, 1997; Davis *et al.*, 1998). The triggering of defense reactions may be critical to development of wilt symptoms, because plugging of vessels with gels is part of the defense response and, if extensive, should lead to foliar wilting. Chlorosis and necrosis of leaves are results of complete plugging of some leaf veins, and crop losses are usually proportional to leaf damage including defoliation (Bell, 1992, 1993).

The fungus *F.o.v.* also produces various hydrolytic enzymes similar to those of *V. dahliae* (Puhalla and Bell, 1981). In addition, it produces the toxins fusaric acid, beauvericin, bikaverin and various naphthazarins (Bell *et al.*, 1996a, 1996b; Wheeler *et al.*, 2000). The specific role of these compounds in pathogenesis has not yet been determined, although some, such as fusaric acid, are extremely toxic to cotton plants (Puhalla and Bell, 1981).

The bacterium *X.c.m.* enters cotton leaves or stems through water-congested stomata or wounds, and it initiates growth on the surface of parenchyma cells (Hillocks, 1992). In susceptible cultivars, water-soaking of tissue occurs first, and tissues later become chlorotic and eventually necrotic (blighted). The gene $avrb_6$ has been cloned and shown to be essential for water-soaking and virulence (De Feyter and Gabriel, 1991; Gabriel *et al.*, 1994). The *opsX* gene, which affects lipopolysaccharide and extracellular polysaccharide synthesis, helps determine the host range of *X. campestris* (Kingsley *et al.*, 1993). Pathogenicity in *X.c.m.* involves the production of extracellular polysaccharide slime (Borkar and Verma, 1991; Pierce *et al.*, 1993) and pectinase enzymes (Papdiwal and Deshpande, 1983; Venere *et al.*, 1984). Bacterial populations in susceptible cotton tissues reach concentrations of ca 10^8 to 10^9 cfu/gm.

In resistant cultivars of cotton, the initial multiplication of bacteria is similar to that in susceptible cultivars, but then a rapid HR occurs giving rise to small clusters of dark brown cells. Bacterial populations are curtailed to about 10^6 to 10^7 /gm of fresh leaf. The role of *avr* genes in initiating this HR was discussed previously in the section on specificity of interactions.

Agrobacterium has only recently been discovered as a seedborne endophyte that parasitizes primarily cotton roots (Bell, 2000a-d). It is tentatively included with the transient necrotrophs until more is known about its interactions with cotton. The bacterium is often present in 90% or more of the seeds in any cottonseed lot. It occurs mostly in the seed coat but also may contaminate the nuclear membrane and/or embryo. The bacteria colonize the root surface and natural wounds created by the emergence of secondary roots. The bacteria gain access to the xylem vessels where they move throughout the plant and eventually enters vessels of the developing seed coat. The *Agrobacterium* in cotton, in may ways, behaves similarly to *Agrobacterium vitis* which causes crown gall and root rot of grape (Burr and Otten, 1999).

The mechanisms of *Agrobacterium* parasitism in cotton have not yet been ascertained. In other plants, the bacterium secretes cellulose to attach itself to host cells (Matthysse and McMahan, 1998; Solovova *et al.*, 1999), secretes a unique β-1,2-linked extracellular polysaccharide that acts as an immunosuppressant (York *et al.*, 1980), produces diamines that stimulate root formation (Bais *et al.*, 1999), produces IAA and cytokinin to regulate host cell division and differentiation (Barazani and Friedman, 1999, 2000), and rapidly fixes phosphorus from solution (Merzouki *et al.*, 1999a,b). *Agrobacterium* readily uses nitrate as a substrate and as an electron acceptor in place of cytochromes under anaerobic conditions, and it also can reduce nitrate, converting it to nitrogen gas (Merzouki *et al.*, 1999a,b). *Agrobacterium* also produces catalase as a virulence factor (Xu and Pan, 2000). Isolates from cottonseed appear to possess most of the same characteristics and, in addition, have exceptional adaptation to high temperatures (Bell, 1999b). Most strains show synergistic interactions with *R. solani*, both in causing crown gall and soreshin (Bell, 2000a). Pectic enzymes produced by the fungus might be facilitating transformation of cotton tissues (Alibert *et al.*, 1999). Some *Agrobacterium* strains, when infiltrated into leaves, also induce chlorosis in tobacco and necrosis in cotton, indicating production of a phytotoxin (Bell, unpublished). Populations in roots have been inhibited by certain *B* genes that affect *X.c.m.* (Bell, 2000c). Bacterial populations are usually similar among cultivars, but vary considerably with environmental variables such as N fertility and temperature (Bell, 2000c, 2000d).

3.3.3 Necrotrophic Relationships

The necrotrophic pathogens do not show a period of compatibility with the host but rather rapidly penetrate and overrun vulnerable tissues or kill tissues already approaching senescence. Only minor differences occur in the resistance of cultivars or species to the pathogen, but tissues of different ages show marked differences in resistance. The seed and young hypocotyl or radicle of all cultivars are susceptible to infection by pathogens such as *Rhizoctonia solani, Pythium ultimum, Thielaviopsis basicola*, and *Rhizopus arrhizus*, especially when environmental conditions favor the pathogen but are unfavorable to cotton (Hillocks, 1992; Bell, 1999a). Dormant propagules of these pathogens germinate in response to chemicals released by germinating seeds (Ruttledge and Nelson, 1997). Polygalacturonase enzymes are thought to be important in the killing and maceration of juvenile tissues.

Necrotrophs also attack stored seedcotton and cottonseed. The ability of *Aspergillus* species to invade seed appears to depend on their ability to avoid active defense reactions. These pathogens can attack seed at moisture levels as low as 12%, whereas active defense reactions do not occur until moisture levels reach 20% or higher (Halloin and Bell, 1979). The major protection of the seed comes from components that restrict rates of water imbibition, thus maintaining low seed moisture levels during alternate periods of wetting and drying in the field (Halloin, 1984).

Natural products of the cotton plant can influence the production of mycotoxins by necrotrophs. The alcohol, 3-methyl-1-butanol, decreased fungal growth, pigmentation, and sporulation but increased aflatoxin production of *A. flavus* and *A. parasiticus* in culture. In contrast, nonanol and the monoterpenes, camphene and limonene, suppressed aflatoxin production while having lesser effects on growth (Greene-McDowelle *et al.*, 1999). The importance of these compounds was not shown in the plant.

Cotton bolls show age-related changes in resistance to necrotrophic pathogens that cause boll rots, such as *Diplodia gossypina* and *Fusarium pallidoroseum*. These pathogens secrete numerous hydrolases that break down cell walls. In addition, *F. pallidoroseum* and *F. equiseti* produce the phytotoxins equisetin and epi-equisetin which inhibit germination and seedling growth (Wheeler *et al.*, 1999). Apparently, both enzymes and toxins are pathogenic factors in the necrotrophs that attack aging tissues. Several leaf spot pathogens, such as *Alternaria* species, are more severe on aging leaves and also produce hydrolytic enzymes and toxins (Hartman *et al.*, 1989). Therefore, most necrotrophs overpower vulnerable host tissues with a barrage of hydrolytic enzymes and toxins that kill tissues in advance of the invading pathogen.

The white fibers of cultivated cottons apparently have neither constitutive nor active defense systems and are attacked by a wide array of fungi and bacteria, if the pathogens gain access to the boll through injuries. Fibers of wild species usually deposit suberins and waxes within layers of the fiber cell to protect it from infections (Ryser and Holloway, 1985). These compounds, however, impart green or brown colors and cause the fibers to have other characteristics considered undesirable (Yatsu *et al.*, 1983).

3.4   Responses to Pathogens

Resistance is a relative term used to describe the ability of a plant to prevent, restrict, or retard the penetration and development of pathogens in host tissues. Some components of resistance are present in healthy plants and, therefore, are part of a constitutive defense. Other components are induced and formed following contact between the pathogen or its products and the host, and they are, therefore, parts of an active defense. Total defense is the sum of the constitutive and active components.

The biochemical events involved with active defense (or HR) are invariably invoked in "susceptible" as well as resistant plants in response to transient biotrophs or necrotrophs. From a biochemical perspective, susceptible plants also resist infection, but the speed or intensity of the response is inadequate, and the pathogen is able to progressively colonize the plant. The degree of resistance depends on the speed and intensity of the defense responses of the host relative to the speed of secondary colonization by the pathogen (Bell, 1980). Repeated sampling at intervals after inoculation is necessary to properly monitor biochemical mechanisms of defense.

In this section, defense mechanisms will be considered under different types of biochemical responses: oxidative bursts and PR proteins, mechanical barriers (cutin, lignin, suberin, wax), flavonoid antibiotics and enzyme denaturants, lytic enzymes, and phytoalexins. The final section will address recognition of pathogens.

3.4.1   Oxidative Bursts and PR Proteins

One of the first events in an active defense reaction is an "oxidative burst" that leads to the production of increased amounts of reactive oxygen species (ROS), such as superoxide anion, $O_2 \cdot^-$; hydrogen peroxide, H_2O_2; and hydroxyl radical, $OH \cdot$ (Apostol *et al.*, 1989; Alvarez *et al.*, 1998; Hutcheson, 1998). The "oxidative burst" is thought to be directly antimicrobial and to signal a variety of other cellular responses. At high concentrations, H_2O_2 is thought to trigger cell death at the site of infection, *i.e.*, HR, thus limiting the spread of infection. Low concentrations of H_2O_2 supposedly induce production of salicylic acid (SA) which moves systemically and induces a large group of defense genes collectively known as pathogenesis-related (PR) genes; products of the genes are PR proteins. Systemic induction of these defense responses leads to systemic acquired resistance (SAR). Salicylic acid also may bind to and inactivate catalase, preventing destruction of H_2O_2; thus, it maximizes H_2O_2-mediated defense responses.

An "oxidative burst" has been observed in cotton cotyledons infiltrated with an avirulent race of *X.c.m.* (Martinez *et al.*, 1998). Superoxide anion concentrations, NADH oxidation activity, H_2O_2 concentrations, and apoplastic peroxidase activity increased within three hours of inoculating the resistant cotton. The authors concluded that cotton cells undergoing the HR response to an avirulent race of *X.c.m.* produce superoxide anion through the activation of apoplastic peroxidase. The oxidative burst does not appear to develop an effective antibiotic environment against pathogens in cotton, because propagule numbers of *X.c.m.* and *V. dahliae* do not decline until later after inoculation, when toxic doses of phytoalexins accumulate (Bell, 1995a; Essenberg and Pierce, 1995; Gorski *et al.*, 1995). The role of H_2O_2 as a mediator in phytoalexin synthesis is suspect because highly purified elicitors from *V. dahliae* elicit phytoalexin synthesis but not an oxidative burst (Davis *et al.*, 1998). Only crude elicitor preparations elicited both responses.

Murray *et al.* (1997) cloned the glucose oxidase gene from the biocontrol fungus *Talaromyces flavus* and introduced it into cotton in an attempt to control fungal pathogens (Murray *et al.*, 1999). In the presence of glucose, this enzyme generates hydrogen peroxide which presumably would inhibit the fungal growth and confer resistance to the host. The transgenic lines showed some protection against *V. dahliae* but no protection against *F.o.v.* Unfortunately, high levels of expression of the glucose oxidase caused phytotoxicity, including reduced height, seed set, seedling germination, and lateral root formation. These severe limitations must be overcome before the glucose oxidase gene can be used effectively to control cotton diseases.

The possible roles of salicylic acid as a mediator in defense reactions of cotton have been studied with insects and nematodes. Bi *et al.* (1997) found that feeding by the cotton bollworm caused significant increases in salicylic acid and H_2O_2 in cotton. However, salicylic acid failed to inhibit the decomposition of H_2O_2 by catalase or ascorbate oxidase. Further, application of sprays of salicylic acid or methyl salicylate did not increase resistance to bollworms. Salicylic acid sprays did increase the concentration of terpenoid aldehydes (TA) in healthy and nematode-infected plants (Khoshkhoo *et al.*, 1993). However, the effects of these changes on the nematode were not studied, nor were induced TA distinguished from constitutive TA. Previous studies had shown that constitutive TA reside in the epidermis (Mace *et al.*, 1974), whereas nematode-elicited TA form in the pericycle around the head of the animal (Veech, 1979). Only the TA formed in the pericycle appear to have a role in resistance, and there is no evidence that they are affected by salicylic acid.

Both constitutive and pathogenesis-related (PR) proteins have been related to resistance to pathogens. Chung *et al.* (1997) purified and characterized basic proteins from cottonseed, which had antifungal activity against several seed and seedling pathogens. Most of the proteins were small cysteine-rich vicilin proteins, 9 to 16.3 kDa in size. Unlike antifungal proteins from seeds of other species, those from cotton were neither substrates nor inhibitors of signal transduction elements such as various protein kinases. Callahan *et al.* (1997) isolated a 14 kDa PR protein that was correlated with resistance to root-knot nematode. The protein occurred only in the nematode galls and only in a resistant cultivar.

Several PR proteins accumulated in cotton in response to infection by either vesicular-arbuscular mycorrhizal fungi (VAMF) or *V. dahliae* (Liu *et al.*, 1995; Hill *et al.*, 1999) or in response to a phytotoxic protein-lipopolysaccharide complex produced by *V. dahliae* (Meyer *et al.*, 1994). The PR proteins at certain concentrations were able to retard hyphal growth and kill conidia of *V. dahliae*. Genes associated with defense responses and PR proteins were identified from a cDNA library but no definite function was ascribed to them (Hill *et al.*, 1999). PR genes for which specific functions have been ascribed are discussed in the following sections.

3.4.2 Mechanical Barriers (Cutin, Suberin, Wax, and Lignin)

Cutin, suberin, and wax are polyester polymers that form a sheet over the surface of cells exposed to air. These sheets act as physical barriers against penetration by pathogens, prevent water retention on cell surfaces, and prevent leakage of water and nutrients from the cytoplasm to the cell surface. Lignin also is a polymer and is made up of cinnamyl alcohol moieties. The alcohols and lignin oligomers and polymers can bond to suberins, polysaccharides, and proteins in the cell wall, forming complexes that generally have greater physical strength and are more resistant to hydrolytic enzymes. The importance of these compounds in defense against pathogens has been reviewed (Bell, 1981; Bell and Mace, 1981; Bell *et al.*, 1986).

The structures of cutin, suberin, and wax in cotton have been studied in detail in green fibers of a mutant cotton strain (Ryser, 1992; Schmutz *et al.*, 1994a,b, 1996). The green color is associated with caffeoyl-fatty acid-glycerol esters, which occur in much lower concentrations in white fibers. The same esters occur in the epidermis of seed regardless of fiber color. Suberin and wax typically are deposited alternately with polysaccharides in concentric layers in secondary walls of fibers. The fatty acids in cotton suberin are predominantly the unique C22 acids, 22-hydroxydocosanoic and 1,22-docosanedioic. Cutin in cotton contains mostly C16 and C18 fatty acid derivatives and is very similar to cutin in other plants.

Halloin (1984) concluded that the major protection of seed in the boll against pathogens comes from components that restrict rates of water inbibition, thus maintaining low seed moisture levels during alternate periods of wetting and drying in the field. Suberin and associated wax probably are major components regulating water imbibition. In addition, the caffeoyl moiety may contribute radicle scavenging properties that would be useful against many necrotrophic pathogens.

The epicuticular wax content of leaves has been correlated with resistance to the cotton leaf curl virus. All resistant cotton lines had considerably higher wax content on the leaf surface than moderately-resistant or susceptible cultivars (Ashraf and Zafar, 1999; Ashraf *et al.*, 1999). The wax probably restricts the insect-vector of the virus, reducing the efficacy of transmission to new infection sites.

As much as 20% of the mature cotton stem and root may be lignin. In addition, most cells form complex appositions, including lignin, in response to attempted penetration of cells by fungi. The ratio of syringyl (3,5-dimethoxy-4-hydroxy cinnamyl) to guaiacyl (3-methoxy-4-hydroxy cinnamyl) units in cotton lignin varies with age, cotton species, and infection by microorganisms (Bell *et al.*, 1986). *G. barbadense* has a predominance of syringyl units, whereas *G. hirsutum* contains mostly guaiacyl units. Both internal condensation within the polymer and percentages of methoxyl groups increase with age and infection. About 75% of the bonds between cinnamyl alcohol units in the lignin polymer are alkyl-aryl ether bonds and 25% are carbon bonds.

The synthesis of lignin and accompanying polysaccharides has been studied as an active defense response in cotton against *V. dahliae* and *F.o.v.* Reinforcement of structural barriers with polysaccharides, including callose and cellulose, and with lignin typically occurs earlier in resistant cultivars than in susceptible cultivars (Shi *et al.*, 1991a, 1991b, 1992; Mueller and Morgham, 1993; Mueller *et al.*, 1994; Meyer *et al.*, 1994; Daayf *et al.*, 1997; Smit and Dubery, 1997). The induction of lignification correlated with, and was preceded by, a transient increase in levels of the following enzyme activities: phenyl alanine ammonia lyase (PAL), cinnamyl alcohol dehydrogenase and cell wall associated peroxidase (Smit and Dubery, 1997). The PAL enzyme from induced hypocotyls was purified and characterized (Dubery and Smit, 1994). The partial genes for PAL and for caffeic-O-methyl transferase, an enzyme used in synthesis of lignin, have been cloned and used to study the appearance of messenger RNA during the active defense reaction in cotton stele tissue (Joost, 1993; Cui *et al.*, 2000). PAL mRNA was constitutively expressed and mostly unchanged by infection. In contrast, levels of mRNA transcripts coding for caffeic-O-methyl transferase, were greatly elevated in *V. dahliae*-inoculated plants as compared to water-inoculated controls. Greater and earlier initial gene response was detected in the wilt resistant cultivar 'Seabrook Sea Island 12B2' than in the susceptible cultivar 'Rowden.' The highest concentrations of mRNA were present within 12 hours after inoculation in the resistant cultivar. Thus, lignification appears to be one of the early responses to wilt pathogens.

3.4.3   Flavonoid Antibiotics and Enzyme Denaturants

Previous reviews have dealt with the chemistry and biological significance of the flavonoids (Bell, 1986) and their polymers, the condensed proanthocyanidins (Bell *et al.*, 1992). Modern studies have not emphasized these compounds, probably because of the adverse effects of the condensed proanthocyanidins, also called condensed tannins, on the quality of cottonseed meal (Yu *et al.*, 1993) and because of the possible involvement of the tannins in the byssinosis disease of cotton textile workers (Rohrbach *et al.*, 1992; Specks *et al.*, 1995). Yet, it is important to remember that these compounds are an integral part of resistance to most cotton pests.

Cotton tannins contain mixtures of catechin, epicatechin, gallocatechin, and epigallocatechin moieties in the polymers. The free catechins also occur in cotton tissues, but in much lower concentrations than the tannins. The ratio of catechin to gallocatechin moieties in the polymer varies from 1:1 to 1:4 in different cultivars, species and tissues.

The tannins are potent protein denaturants and, consequently, act as enzyme inhibitors, antisporulants, and mild antibiotics (Bell *et al.*, 1986, 1992; Bell, 1995b). The tannins also are intermediates in the synthesis of dark brown melanoid pigments found in diseased tissues and seed coats. The melanins restrict water flow through seed coats (Halloin, 1982) and probably restrict water loss from wound sites. The tannins generally are located in vacuoles within cells of the endodermis and hypodermis and within scattered parenchyma cells of other tissues. Tannin concentrations greater than 20% of dry weight can occur in buds, young leaves, young bolls, and young bracts (Chan, 1985).

Tannins probably are synthesized by the endoplasmic reticulum, collecting in very small vacuoles that later coalesce to form the vacuoles found in mature tannin containing cells (Mueller and Beckman, 1976). Cells that contain tannins also have high levels of peroxidase and polyphenoloxidase activity, indicating that these enzymes may be involved in tannin synthesis and oxidation to melanins.

There is little evidence that tannins offer effective resistance to biotrophic pathogens (Bell *et al.*, 1986). However, they are involved in resistance to the transient biotrophs and, especially, to the necrotrophs. Infection with *V. dahliae* causes additional parenchyma cells to synthesize tannins, especially, in xylem ray cells of the stele (Bell and Stipanovic, 1978; Bell *et al.*, 1992; Bell, 1995a; Daayf *et al.*, 1997). Tannin levels in stele may increase from 0.5-1.0% to 5.0-10.0% dry weight following vascular invasion by the fungus. Tannin synthesis in the stele is activated more rapidly and occurs more intensely in resistant than susceptible cultivars inoculated with *V. dahliae* (Bell and Stipanovic, 1978; Bell, 1995a; Daayf *et al.*, 1997). Tannins have strong antibiotic and weak antisporulant activity against *V. dahliae* (Howell *et al.*, 1976; Bell *et al.*, 1992).

Leaves of cultivars resistant to *V. dahliae* often have higher concentrations of tannins than those of susceptible cultivars, and the tannins in the resistant cultivars have greater specific toxicity to *V. dahliae* (Bell, 1995a). Toxicity is largely a function of polymer length; intermediate sizes are the most toxic. Tannin synthesis also occurs in young, but not old, leaves in response to the fungus (Howell *et al.*, 1976).

Partial genes for phenylalanine ammonia-lyase (PAL), chalcone synthase (CHS), and dihydroflavanol reductase (DHF), which are required for tannin synthesis, have been cloned (Joost, 1993; Cui *et al.*, 2000). Accumulation of mRNA in stele tissue of resistant and susceptible cultivars inoculated with *V. dahliae* also was determined. Relatively high levels of mRNA from PAL and DHF genes occurred in healthy stele tissues and there was either no change or a decrease in mRNA in response to infection. In contrast, large increases in mRNA from CHS genes occurred in response to infection. The increase occurred more rapidly and was more intense in the resistant than in the susceptible cultivar from 10 to 72 hours after inoculation. Beyond 72 hours, CHS was down-regulated more rapidly in the resistant than in the susceptible cultivar. The down-regulation may minimize damage to the host from toxic flavonoid products in diseased tissues. The behavior of CHS genes was consistent with patterns of tannin accumulation in diseased tissue

(Bell, 1995a). Therefore, the CHS genes, which occur as a gene family, probably play a key role in regulating synthesis of condensed tannins.

Song and Zheng (1997) showed that tannin levels in cotton plants also increased in response to *F.o.v.* However, there were no differences between resistant and susceptible cultivars, or between plants treated with trifluralin to induce resistance and untreated plants. They concluded that tannin was not involved in resistance to the *F.o.v.*

A few observations indicate that flavonoids and tannins are involved in resistance to the bacterial blight pathogen, *X.c.m.* Dark brown pigments normally are formed in the necrotic cells associated with HR reactions to the bacterium in resistant cultivars (Essenberg and Pierce, 1995). In other systems, these pigments are the result of the synthesis and oxidation of condensed tannins and catechins (Bell *et al.*, 1992). A marked increase in extractable peroxidase activity occurs in resistant but not susceptible cultivars in response *X.c.m.*, and mixtures of peroxidase and catechin, a monomeric unit of tannin, were bactericidal (Venere, 1980). Even more toxicity might have been obtained if tannin oligomers had been used in place of catechin.

Essenberg *et al.* (2000) showed that the flavonoids, cyanidin-3-β-D-glucoside, and quercetin-3-β-D-glucoside (isoquercitrin), increase in epidermal cells next to those undergoing HR reactions and accumulating terpenoid phytoalexins. Isoquercitrin has an absorbance peak covering the photoactivating range for the phytoalexin 2,7-dihydroxycadalene and, thus, may protect the cells from the toxic effects of the phytoalexin that occur with light. The lysigenous pigment glands of cotton that contain high concentrations of toxic terpenoid aldehydes are enclosed by a layer of epithelial cells that also have high concentrations of flavonoids which may play a similar protective role (Bell and Stipanovic, 1977; Chan and Waiss, 1981).

The importance of flavonoids and tannins in resistance to necrotrophic pathogens have been reviewed previously (Bell, 1986; Bell *et al.*, 1986; Bell *et al.*, 1992). These compounds, especially in combination with peroxidase enzymes, help contain seedling pathogens, leaf spot pathogens, and boll rot pathogens. They both inactivate critical enzymes secreted by the necrotrophs and act as weak antibiotics. They have been shown to account for age-related changes in resistance to seedling and boll rot pathogens. The melanins formed from tannins are important for regulating water uptake by seeds and preventing deterioration of seed in the field and in storage.

3.4.4 Lytic Enzymes

The enzymes β-1,3-glucanase and chitinase lyse fragments from cell walls of fungi. The fragments, in turn, may act as signals for other defense reactions. The action of the enzymes, especially if coupled with appropriate proteases, may also be antibiotic. Both β-1,3-glucanase and chitinase have been implicated in resistance of cotton to Verticillium

wilt. The coordinated induction and accumulation of these enzymes were observed following elicitation in leaf discs with cell wall fragments from the fungus (Dubery and Slater, 1997). Increased enzyme activity was present within 12 hours after applying the elicitor. Three acidic isozymes of chitinase and one β-1,3-glucanase isozyme were found in the intercellular fluid.

Clones of the partial genes for β-1,3-glucanase and chitinase have been obtained and used to study the dynamics of mRNA accumulation in infected stele tissue (Joost, 1993; Cui *et al.*, 2000). Equivalent levels of β-1,3-glucanase mRNA were present in both fungal- and water-inoculated plants, and genes for this enzyme were expressed constitutively in both susceptible and resistant cultivars. Chitinase mRNA synthesis was strongly induced by inoculation with *V. dahliae*. Levels of mRNA continued to increase in susceptible 'Rowden' over a period of 96 hours, whereas they decreased rapidly in resistant 'SBSI' after 48 to 60 hours. Genes for both enzymes occurred as gene families with three or more copies in the genomes of both cultivars.

3.4.5 Phytoalexins

Cotton produces a diverse group of biologically active terpenoids that serve various roles in defense against pathogens and other pests. The plant tribe Gossypieae is divided from other malvaceous plants based on the presence of lysigenous glands which contain the terpenoid aldehyde gossypol in the seed embryos. In foliage, these glands contain complex mixtures of terpenoids which have been the subject of various reviews (Bell and Stipanovic, 1977; Bell *et al.*, 1987; Stipanovic *et al.*, 1999) and are discussed in this volume in Chapter 30, *Secondary Products*.

Infection of cotton tissues rapidly elicits synthesis of multiple terpenoids that have antibiotic activity against fungi and bacteria. Some of these compounds are intermediates in the pathway from (+)-δ-cadinene to gossypol. In this group, desoxyhemigossypol followed by hemigossypol are the most toxic antibiotics against fungi (Stipanovic *et al.*, 1991; Bell *et al.*, 1994). Other terpenoids formed from the (+)-δ-cadinene are members of a divergent pathway, leading to accumulation of lacinilenes. 2,7-Dihydroxycadalene is the most toxic terpenoid in this group to fungi (Mace *et al.*, 1987), and it is also the most antibiotic of all terpenoids against the bacterial pathogen *X.c.m.* (Abraham *et al.*, 1999). Various reviews have dealt with methods for studying cotton terpenoid phytoalexins (Bell *et al.*, 1993); the role of these compounds in resistance to wilt fungi (Stipanovic *et al.*, 1988; Bell, 1995b), bacterial blight (Essenberg and Pierce, 1995), and diseases in general (Bell and Stipanovic, 1978; Bell *et al.*, 1986); the toxicity and mode of action of desoxyhemigossypol (Stipanovic *et al.*, 1991); and structural variations of terpenoid phytoalexins and their effects on biological activity (Bell *et al.*, 1994; Stipanovic *et al.*, 1998).

Most studies of terpenoid phytoalexins have concentrated on diseases caused by transient biotrophic pathogens, such as *V. dahliae*, *F.o.v.*, or *X.c.m.* When inoculated with these pathogens, resistant cultivars generally synthesize the terpenoid antibiotics more rapidly and, sometimes, more intensely than susceptible cultivars. Also, avirulent or moderately virulent strains of the pathogen elicit earlier and more intense synthesis of the terpenoids than virulent or highly virulent strains (Bell, 1995b; Essenberg and Pierce, 1995). The necrotrophic pathogens generally elicit rapid, intense synthesis of terpenoids in tissues in good physiological condition, regardless of cultivar or pathogen strain. However, the intensity or magnitude of the response may be attenuated in juvenile or aging tissues (Bell *et al.*, 1986). Biotrophic pathogens generally do not elicit synthesis of terpenoid antibiotics except in resistant cultivars (Veech, 1979).

Terpenes are synthesized from isoprenoid units which, in turn, are derived from one of two pathways (Rohmer, 1999). Isoprenoid synthesis in one pathway is catalyzed by 3-hydroxy-3-methylglutaryl CoA reductase (HMGR). Joost *et al.* (1995) cloned a cotton HMGR DNA fragment and used it as a probe in Northern hybridizations to show that high levels of HMGR-mRNA are rapidly induced in the stele tissue of plants inoculated with *V. dahliae*. Induction was more rapid in a resistant cultivar than in a susceptible cultivar with the highest concentrations of mRNA being present as early as 10 hours after inoculation. Heat-killed conidia strongly activated HMGR in the resistant cultivar but not in the susceptible cultivar (Joost, 1993). In all cases, the patterns of HMGR-mRNA accumulation and down-regulation closely fit the patterns of terpenoid synthesis (Bell, 1995a). Two full-length genes (*hmg 1* and *hmg 2*) that encode for HMGR also have been cloned (Loguercio *et al.*, 1999), but their significance in active defense reactions is not known.

Eldon and Hillocks (1996) tested the importance of HMGR in active defense reactions by treating inoculated plants with compactin, a competitive inhibitor of HMGR. The treatment reduced production of the terpenoid phytoalexin, hemigossypol, by 48% and caused a break down in resistance to Verticillium wilt but not Fusarium wilt. Thus, HMGR activity is crucial for resistance to *V. dahliae*.

The first biosynthetic step committed to the formation of cyclic terpenoid antibiotics is the cyclization of linear farnesyl diphosphate (FPP) to form (+)-δ-cadinene (Benedict *et al.*, 1995, Davis and Essenberg, 1995). Five genes that encode for (+)-δ-cadinene synthase (CAD) have now been cloned, characterized, and placed into two gene families, *cad-1C* (including *cad1-C1*, *cad1-C2*, *cad1-C14*, and *cdn1*) and *cad1-A* (Chen *et al.*, 1995, 1996; Liu *et al.*, 1999; Meng *et al.*, 1999). The *cdn1* gene is from *G. hirsutum*, whereas the rest are from *G. arboreum*. The *cad1-C14* gene has been transferred into *Artemisia annua* L. via *Agrobacterium rhizogenes* and was expressed in hairy root cultures at the transcriptional level (Chen *et al.*, 1998). A full-length cDNA (*fps1*) encoding for farnesyl diphosphate synthase (FPS) also has been cloned from *G. arboreum* and the protein heterologously expressed in *E. coli* (Liu *et al.*, 1998, 1999).

Gene transcription and synthesis of FPS and CAD has been induced by *V. dahliae* and elicitors from it, and by infiltration of cotyledons or leaves with *X.c.m.* cells. In each case, the accumulation of the enzymes slightly precedes and parallels the accumulation of the sesquiterpenoid antibiotics (Davis *et al.*, 1996; Alchanati *et al.*, 1998; Bianchini *et al.*, 1999). The *fps1*-mRNA was also profoundly upregulated from 27 to 40 days post-anthesis in cotton seed embryos, which coincides with the period of gossypol accumulation (Liu *et al.*, 1998).

Three pathway variations of terpene biosynthesis in *Gossypium* species occur because of chemical modifications of desoxyhemigossypol (dHG). First, a methyl ether group can be added to the 3-hydroxyl group of dHG to form desoxyhemigossypol-3-methyl ether (dMHG). Subsequently, at least eight different terpenoid aldehydes are derived from dMHG. Second, a methoxyl group can be added to C-7 to form 7-methoxy-desoxyhemigossypol (MdHG) and, subsequently, the aldehyde raimondal (Stipanovic *et al.*, 1980). Synthesis of MdHG apparently is a two step process in which the oxygen group is first inserted and then a methyl group is transferred to it. Accordingly, it is controlled by two dominant genes (Kohel and Bell, 1999). Third, a hydroxyl group can be added to the C-5 position to form 5-hydroxy-desoxyhemigossypol and, subsequently, hemigossypolone (Bell *et al.*, 1987). This variation occurs in green tissues (*i.e.*, leaves, bracts, bolls, etc.) of cultivated *Gossypium* species, but is absent in most wild American diploid species that have the D genome. In the absence of modification, dHG is oxidized to hemigossypol and, subsequently, to gossypol.

A single variation occurs in the lacinilene branch of the terpenoid pathway. 2-Hydroxy-7-methoxycadalene (HMC) is present in *Gossypium* as well as 2,7-dihydroxycadalene (DHC) indicating that there may be a unique methyl transferase to form HMC from DHC. The genetics, biochemistry or biological significance of this variation has not been studied in detail.

The addition of the methyl ether groups to dHG or DHC and subsequent compounds reduces toxicity to most fungal and bacterial pathogens (Stipanovic *et al.*, 1988, 1991; Bell *et al.*, 1993, 1994; Essenberg and Pierce, 1995). Therefore, methyl ether synthesis appears to be undesirable for disease resistance. To test this hypothesis, Bell *et al.* (1994) introduced a recessive gene from *G. barbadense* or a dominant gene from *G. sturtianum* Willis into *G. hirsutum* to greatly increase the percentage of dMHG and derived terpenoids formed in foliage. In other cotton lines, they introduced two dominant genes from *G. raimondii* Ulbrich into *G. hirsutum* to give raimondal as the major terpenoid in foliage. Each of these genetic and biochemical changes significantly increased susceptibility to Verticillium wilt. Thus, phytoalexin potency is also a significant determinant of resistance in *Gossypium*.

Two approaches are being taken to improve phytoalexin potency. First, efforts are under way to disrupt the gene that encodes dHG-O-methyl transferase (dHG-OMT). This enzyme has been purified and characterized (Liu *et al.*, 1999). A gene encoding for dHG-OMT has recently been cloned and expressed in *E. coli* (Liu, Benedict, Stipanovic, Magill, and Bell, personal communication). Antisense or sense constructs of this gene may allow reduction in the percentage of methylated terpenoids.

In a second approach to improving potency of terpenoid phytoalexins, structural modifications of terpenoids in other species of the Malvaceae have been examined. Kenaf (*Hibiscus cannabinus* L.) is of special interest because it was reported to be highly resistant to Verticillium wilt in Russia (see references in Bell *et al.*, 1998). These observations were confirmed with USA isolates of *V. dahliae*, and structures and toxicity of some of the kenaf phytoalexins were determined (Bell *et al.*, 1998; Puckhaber *et al.*, 1998; Stipanovic *et al.*, 1998). o-Hibiscanone from kenaf is about ten times more toxic to *V. dahliae* than dHG, the most toxic terpenoid in cotton. o-Hibiscanone appears to be derived from the same precursors as *Gossypium* terpenoids. Thus, transfer of one or two genes from kenaf to cotton may allow synthesis of this highly toxic phytoalexin.

The toxicity of cotton phytoalexins is associated with their ability to readily undergo autoxidation , to form both free radicals and active oxygen species, such as hydrogen peroxide. Stipanovic *et al.* (1992) showed that dHG decomposed rapidly in solution to give hemigossypol. The oxygen inserted during the reaction comes from water rather than oxygen gas. The rate of decomposition was retarded by the reducing agents, ascorbic acid and reduced glutathione, and by catalase and the metal chelator diethylenetriamine pentacetic acid. The first two compounds also reduced toxicity of dHG to *V. dahliae* (Stipanovic *et al.*, 1991). Thus, oxidative products apparently are responsible for the antibiotic activity. The phytoalexin DHC readily undergoes oxidative degradation to lacinilene C, and this reaction can be enhanced by irradiation with 300 to 750 nm light. Photoactivated DHC inactivates virions of the cauliflower mosaic virus, presumably by cross-linking the nucleic acid binding domain of the coat protein to the viral DNA (Sun *et al.*, 1988). Photoactivated DHC also induced single-strand breaks in plasmid pBR322 DNA, destroyed catalytic activity of deoxyribonuclease I and malate dehydrogenase, and had enhanced activity against *X.c.m.* cell division (Sun *et al.*, 1989). Irradiation at 239 and 300nm was more effective than irradiation at 400, 500, or 600 nm for activating DHC, and scavengers of either reactive oxygen species or free radicals reduced biological activity. Thus, dHG and DHC appear to have similar modes of action. Differences in biological activity against microbial species may depend on how effective these compounds are in associating with or in intercalating membranes of specific microorganisms. Disruption of the plasmalemma is a primary effect of dHG in cells of *V. dahliae* (Mace *et al.*, 1992).

3.5   Recognition of Pathogens

Chemicals or treatments that effect active defense responses are referred to as elicitors, and may be either biotic or abiotic in origin. Abiotic elicitors of terpenoid and tannin synthesis in cotton include the following: chilling injury, ultraviolet irradiation, cupric ions, and pesticides (Bell *et al.*, 1986); the plant growth regulators salicylic acid, Pix, and Burst (Khoshkhoo *et al.*, 1993); and the herbicides prometryn, dalapon (Awadalla and El-Refai, 1992), and clomazone (Duke *et al.*, 1991). Biotic elicitors include live and dead fungal spores and bacterial cells (Bell and Stipanovic, 1978; Joost, 1993). Various heterogeneous polymers obtained from fungal and bacterial cell walls also act as elicitors. These include an extracellular preparation from *X.c.m.* (Bell and Stipanovic, 1978), a lipoprotein polysaccharide (LPS) from *V. dahliae* (Meyer *et al.*, 1994; Smit and Dubery, 1997), cell-free mycelial extracts (Zeringue, 1988, 1990), heat-released soluble cell-wall fragments (Dubery and Slater, 1997), and a purified glycoprotein (Davis *et al.*, 1998). Earlier and greater active defense reactions in resistant than in susceptible cultivars have been induced by heat-killed conidia of *V. dahliae* (Bell and Stipanovic, 1978; Joost, 1993), LPS from *V. dahliae* (Smit and Dubery, 1997), and extracellular preparations from *X.c.m.* (Bell and Stipanovic, 1978). Specific recognition, therefore, may involve some component of the outer cell wall of the pathogen.

Oligomers of chitosan and β-1,3-glucans are potent elicitors in many plant species. Chitin and the glucans are common constituents of fungal cell walls and are cleaved by chitinases and β-1,3-glucanases found in plants. Cotton contains multiple gene copies for both β-1,3-glucanase and chitinase, and mRNAs for these enzymes are either present or rapidly induced in diseased tissue (Joost, 1993; Cui *et al.*, 2000). A β-1,3-glucanase and three chitinase enzymes were found in cotton tissue treated with *V. dahliae* elicitor (Smit and Dubery, 1997). A β-N-acetyl-glucosaminidase is also present in cotton tissue (Bell *et al.*, 1986). Thus production of chitosan and glucan oligomers as elicitors is possible.

Avirulence genes that elicit hypersensitive reactions in cultivars carrying specific resistance genes (*B* genes) have been cloned from *X.c.m.* by Gabriel *et al.* (1986) and De Feyter and Gabriel (1991). These were discussed in the section on specificity of interactions.

Pectinase enzymes might serve as elicitors of HR in cotton. These enzymes cause browning and necrosis in cotton (Wang and Pinckard, 1972; Hooper *et al.*, 1975), and they release peroxidase from cotton cell walls (Strand and Mussell, 1975). Apoplastic peroxidase generates superoxide anions and H_2O_2 in cotton exhibiting HR response to *X.c.m.* (Martinez *et al.*, 1998), and H_2O_2 can elicit phytoalexin synthesis (Apostol *et al.*, 1989). Pectinase activity is suggested by disturbances of cell wall structure in cottons infected with *X.c.m.*; and these disturbances occur as early as two hours after inoculation of resistant, but not susceptible, cultivars (Cason *et al.*, 1978). Pectinase can detected be one to three days earlier in infected resistant tissue than in susceptible tissue (Venere and Brinkerhoff, 1974). Thus, pectinase is implicated as an elicitor, but more critical proof is needed.

The success of biotrophic relationships may reside with the ability of the pathogen to suppress active defense reactions. Certain pathogens produce antigens that cross-react with antibodies produced against cotton antigens (Charudattan and DeVay, 1972). These common antigens usually are not found in nonpathogenic organisms. A common antigen isolated from the Fusarium wilt pathogen of cotton was identified as a protein-carbohydrate complex similar to that found in the fungal cell walls (Charudattan and DeVay, 1981). Thus, virulent pathogens may synthesize outer cell walls that are similar to their host's. This might prevent breakdown of pathogen cell walls by the cotton enzymes to release oligomeric elicitors.

Apostol *et al.* (1987) found that citrate was a potent inhibitor of phytoalexin synthesis in cotton and soybean, if it was applied before elicitor treatments. Citrate also prevented the rapid changes in the plant cell membranes that are normally caused by elicitors. Other organic acids did not show similar activity in cotton.

3.6   Physiological Consequences of Disease

The end result of disease is reduced yield and quality of seed cotton. This results primarily from a disruption of water and mineral uptake, loss of effective photosynthetic area, and marked imbalances in plant hormones and ammonium ions. Because of these effects, the plant is no longer able to effectively convert photosynthate and minerals into seedcotton.

Verticillium wilt can be used as an example to show how disease influences the cotton plant. Symptom production is due to water deficits caused by plugging of xylem vessels, especially in leaves, and to premature senescence of leaves leading to defoliation and loss of photosynthetic area (DeVay, 1989). Damage from non-defoliating isolates of *V. dahliae* is due primarily to water deficits, whereas both factors contribute to damage caused by defoliating isolates. Defoliating isolates cause much larger increases in ethylene and abscisic acid concentrations in leaves than do non-defoliating isolates (Wiese and DeVay, 1970), and these changes apparently contribute to both increased plugging of vessels and defoliation (Misaghi *et al.*, 1978). The defoliating isolates also cause as much as a tenfold increase in ammonium ions in leaves compared to less than a 50% increase caused by non-defoliating isolates (Bell, 1991). Supplying exogenous ammonia caused symptoms typical of Verticillium wilt including defoliation. The severity of wilt symptoms shown by the plant is proportional to the amount of colonization by either pathovar in the vascular system. Thus, the speed or effectiveness of the active defense reactions ultimately determines the degree of plant damage.

There is increasing evidence that some cotton disease symptoms are due to the active defense responses. Browning, a classical symptom of plant disease, was once considered a passive process in which oxidases were allowed to mix with phenols released from vacuoles during breakdown of tissues by pathogens. Studies of Verticillium wilt, bacterial blight, and boll rots now clearly show that phenols (condensed tannins and flavanol glucosides) and even the oxidases involved in browning may be formed as part of the active defense (Bell *et al.*, 1986). Plugging of xylem vessels is primarily the consequence of active defense reactions initiated rapidly to contain the pathogen within the invaded vessel (Bell, 1995a). Even genetic lethal reactions that occur in interspecific *Gossypium* hybrids are associated with massive spontaneous active defense reactions (Mace and Bell, 1981). Recent studies indicate that the most effective resistance reactions are those in which active defense genes are rapidly down-regulated after containment of the pathogen (Joost, 1993; Joost *et al.*, 1995; Alchanati *et al.*, 1998; Cui *et al.*, 2000). This may be necessary to minimize damage from the defense response.

3.   SUMMARY

Microbes affect every aspect of the growth and usefulness of the cotton plant. Even before the boll is mature, the developing seeds are invaded by microorganisms that move in through the xylem vessels or penetrate the seed coat, usually following development of boll rots. Once the boll opens, there is rapid growth of bacteria on the drying fiber which is quickly followed by fungal colonization of the fiber under moist conditions. Microbes on the fiber may produce endotoxins that contribute to the brown lung (byssinosis) disease of mill and gin workers (Fischer and Domelsmith, 1997), and microbes in seed may produce mycotoxins that limit the usefulness of cotton seed as food and feed (Cotty *et al.*, 1994). Other microbes may reduce the quality and value of the cotton fiber (Cauquil, 1975). The seed-borne microbes also may attack the young seedling to establish early infections and eventually epidemics. Diseases such as bacterial blight, anthracnose, and Ascochyta blight were extremely important in cotton before acid-delinting and seed treatments became common practice to control early infection by seed-borne pathogens.

Other microbes are soil-borne and may establish either beneficial or detrimental relationships with the cotton plant. Bacteria such as *Pseudomonas fluorescence, Bacillus subtilis,* and *Burkholderia cepacia* have improved the growth of cotton by decreasing infection from soil-borne pathogens. Mycorrhizal fungi, such as *Glomus* species, *Trichoderma* species, *Talaromyces flavus, Fusarium* species, and *Rhizoctonia* species may improve the nutrition of the cotton plant by aiding the uptake of phosphorus, magnesium, iron, and zinc. These fungi also may limit infections by pathogens. Little is known about the biochemical or genetic interactions between beneficial microbes and cotton.

Many of the severe diseases of cotton also are caused by soil-borne pathogens that remain dormant until stimulated to attack an approaching root or germinating seed. Dormant propagules are formed again on the dying host tissue. Pathogens, such as *Pythium, Rhizoctonia, Thielaviopsis, F.o.v., V. dahliae, Phymatotrichum,* and nematodes are in this group. Very little is known about the genetics and biochemistry of pathogenesis in these microbes. New phytotoxins found in *Fusarium* species are just now being evaluated for roles in pathogenesis, and the roles of toxins in other cotton diseases remains unclear.

Agrobacterium was first discovered as a seed-borne endophytic parasite of cotton in 1997. Of greatest concern now is its possible role in bronze wilt. However, this bacteria, like other endophytes, probably has useful roles and may eventually be exploited as a beneficial organism.

In the past several years, we have learned a great deal about how active defense systems are developed in cotton in response to pathogenic microbes. Many of the key structural genes have been cloned, and it is now possible to genetically manipulate active defense responses. However, we still know very little about the biochemical regulation of defense systems and which genes are responsible for this regulation. Cotton-microbe interactions remain a frontier area in cotton physiology.

Chapter 19

ECOPHYSIOLOGY OF ARBUSCULAR MYCORRHIZAS IN COTTON

David B. Nehl[1] and Peter A. McGee[2]

[1]NSW Agriculture, Locked Bag 1000, Narrabri NSW 2390, Australia; and [2]School of Biological Sciences, A12, University of Sydney NSW 2006, Australia

1. WHEN IS A COTTON PLANT NOT A COTTON PLANT?

All the time! Healthy cotton plants growing in soil are always part of a mycorrhiza. A mycorrhiza is a symbiosis; a complex interaction that involves structural and physiological integration of a plant and one or more species of fungus. The symbiosis is based upon an exchange of nutrients and carbohydrates that, on balance, is mutually beneficial to both the host plant and the fungal symbiont(s). The host plant supplies carbohydrates to the mycorrhizal fungi (a substantial sink) and in return the fungi transfer mineral elements to the plant. At a macroscopic level mycorrhizal symbiosis may result in dramatic changes in the morphology of the plant (Fig. 19-1). At the ultrastructural level the roots and the fungi develop a specialised interface for exchange of nutrients. At the molecular level the metabolism and physiology of both symbionts are modified. In effect, a mycorrhiza is a dual (or multiple) organism whose component species differ functionally and physically from their isolated states.

Mycorrhizas are the result of co-evolution between plants and fungi and may have been a key factor that enabled plants to colonise the terrestrial environment (Smith and Read, 1997). To colonise land, plants needed a fine root system to explore the soil, particularly for acquisition of immobile elements such as P and Zn. To this end, plants appear to have exploited fungi as an extension of their root systems, as a form of controlled parasitism. Non-mycorrhizal plants appear to have evolved more recently (Trappe, 1987) and deploy alternative strategies to explore the soil for immobile elements. Hence, the growth benefit to plants from mycorrhizal fungi tends to decrease as root hairi-

Figure 19-1. Non-mycorrhizal cotton in fumigated soil (left) and mycorrhizal cotton in adjacent soil (right) at six weeks after sowing.

ness or fibrousness increases (Manjunath and Habte, 1991; Schweiger *et al.* 1995; Smith and Read, 1997).

The most common type of mycorrhiza, the arbuscular mycorrhiza (AM), occurs in about 60 to 70% of flowering plants including cotton and most other agriculturally important species (Hayman, 1987; Trappe, 1987). Arbuscular mycorrhizal (AM) fungi occur in all soils where plants grow; from deserts to alps, from Antarctica to the tropics. Despite the ubiquity of AM fungi and their profound influence on the ecology and physiology of plants, a basic understanding of arbuscular mycorrhizas has only been achieved during the last few decades. AM fungi are obligate biotrophs, only able to obtain their energy from living host cells, and this has clearly been an obstacle to their investigation. As a result, a good deal of research in the plant sciences lacks regard for the role of mycorrhizas.

J.McD. Stewart et al. (eds.), *Physiology of Cotton*,
DOI 10.1007/978-90-481-3195-2_19, © Springer Science+Business Media B.V. 2010

Cotton is intrinsically mycorrhizal and, therefore, it is appropriate that researchers and agronomists be aware of the implications. In this review, we endeavour to provide a brief overview of the ecology and physiology of arbuscular mycorrhizas in agricultural systems, with emphasis upon the available knowledge of cotton.

2. MORPHOLOGY OF ARBUSCULAR MYCORRHIZAS

The six genera of AM fungi, *Glomus, Sclerocystis, Acaulospora, Entrophospora, Gigaspora,* and *Scutellospora* are placed in the order Glomales (Zygomycetes). A unifying feature of the group is the formation of arbuscules (Fig. 19-2) in symbiosis with plants (Morton and Benny, 1990). Two genera, *Gigaspora* and *Scutellospora,* do not produce vesicles. Hence, the whole group is no longer referred to as vesicular arbuscular mycorrhizal fungi. The morphology and development of the asexual spores are important characters used to distinguish the species of AM fungi (Morton and Benny, 1990). From these data, approximately 150 fungi have been described. The genera most commonly associated with cotton are *Glomus, Gigaspora,* and *Acaulospora.*

Arbuscular mycorrhizal fungi are obligate symbionts and cannot grow in isolation of a host plant, either saprophytically on soil organic matter or on artificial media. When plant hosts are absent, AM fungi survive as spores, vesicles or hyphae in soil or moribund roots (McGee *et al.,* 1997).

Arbuscular mycorrhizas are initiated following colonisation of roots by the fungi. Infective hyphae branch as they approach the root and produce appressoria on the root surface (Friese and Allen, 1991; Fig. 19-2). Intracellular hyphae penetrate the root and then spread along the root within the cortex. Arbuscules are produced by lateral branches of hyphae that penetrate the wall and invaginate the plasma membrane of individual cortical cells. These hyphae branch dichotomously, forming a 'tree' of filaments that fill the cell. The arbuscule is a complex interface between the plant and the fungus. The plasma membrane surrounding the arbuscule is known as the periarbuscular membrane. The fungal membrane of the arbuscule, the periarbuscular membrane, and the apoplastic space between the two, play a specialized role in the transfer of nutrients (Smith and Read, 1997).

Some species of AM fungi produce vesicles in the root cortex and soil, particularly as colonies mature (Fig. 19-2d). Vesicles are bladder-like sacs that are usually terminal on intraradical hyphae. They contain lipids and numerous nuclei and are presumed to function as storage organs (Smith and Read, 1997).

As the intraradical colonies expand within the root, extraradical hyphae extend into the soil, developing a hierarchical network. Large 'runner' hyphae (Fig. 19-2) initiate secondary colonies elsewhere in the root system. The branches of the hyphal network differentiate into infective

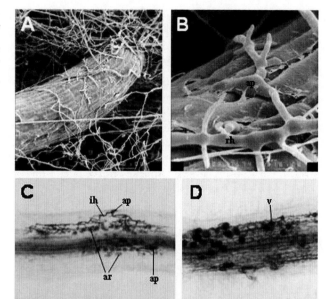

Figure 19-2. The morphology of arbuscular mycorrhizas: (a) Extensive hyphal network around a sorghum root in sand culture; (b) Runner hyphae and an appressorium on the sorghum root; (c) Young colonies in cotton, showing appressoria, intraradical hyphae and arbuscules close to the stele; (d) Older colonies in the same cotton root showing vesicles, extensive intraradical hyphae and a lack of arbuscules. (rh = runner hyphae, ap = appressorium, ih = intraradical hyphae, ar = arbuscule, v = vesicle).

hyphae, bearing asexual spores, and absorptive hyphae (Friese and Allen, 1991; Bago *et al.,* 1998).

3. PHYSIOLOGY OF ARBUSCULAR MYCORRHIZAS

3.1 Nutrient Uptake Versus Carbon Supply

Arbuscular mycorrhizal fungi transport a number of immobile nutrients from the soil to the host plant, including P, Zn, Cu, and NH_4^+ (Smith and Read, 1997). K and NO_3^- reach the roots by mass flow of water through the soil and their uptake by AM fungi has not been consistently observed. If the mobility of NO_3^- is slowed by conditions such as drought stress, then its uptake by AM fungi may become important (Tobar *et al.,* 1994). However, P is the most important element of exchange in AM symbiosis.

Phosphorus supply, not the capacity for P uptake by the roots, is the major limitation to P uptake by plants (Bolan, 1991). Diffusion of P and other 'immobile' elements is very slow in soil. A zone of depletion develops adjacent to roots (Fig. 19-3). By active uptake and transport of immobile elements to the plant, AM fungi increase the effective volume of soil exploited by the root system. The potential for plants to benefit from AM symbiosis depends upon the amount of

P in the soil solution (Bolan, 1991; Smith and Read, 1997) and hence the rate of replenishment of P, which is related to the P fixing capacity of the soil.

The uptake of P by plants will thus be determined, in part, by the length of hyphae, their spread from the plant, the longevity of hyphae exploring the soil, and the capacity of the hyphae to take up and transport P to the plant (Jakobsen, 1999). The cellular and molecular basis of P uptake has been reviewed (Smith and Read, 1997). AM fungi differ in their capacity to ramify through soil (Sanders *et al.,* 1977; Abbott and Robson, 1985). Estimates of the length of extraradical hyphae are sometimes in the order of tens of metres g^{-1} in undisturbed soil (Smith and Read, 1997), though probably much less in cultivated soils (*e.g.,* 0.5 m g^{-1} immediately following harvest of cotton in clay soil, McGee *et al.,* 1997). Estimates of hyphal P uptake and transport vary markedly. These differences appear to be related to differences in measuring hyphal length. Concentration of available P in soil, hyphal viability and species differences are also important (Jakobsen, 1999).

Although arbuscules are the major site for transfer of P and Zn to the host, transfer of photosynthate more likely occurs between cortical cells and intraradical hyphae (Smith and Read, 1997). Sucrose is transported to the roots in the phloem and then cleaved into glucose and fructose by extracellular invertase (Roitsch, 1999). Glucose and, to a lesser extent, fructose are transferred to intraradical hyphae and then rapidly transformed, initially into trehalose or glycogen and then to lipid that is either stored or exported to the extraradical mycelium (Pfeffer *et al.,* 1999; Bago, *et al.,* 2000). The rapid transformation of hexoses may serve to buffer host sugars and maintain a concentration gradient across the host fungus interface (Smith and Read, 1997; Pfeffer *et al.,* 1999). The cellular and molecular signals that affect rates of transfer and transport of sugars are unclear at this stage.

The carbon is used in fungal maintenance and growth or is stored in reproductive tissue. Estimates of the biomass of AM fungal spores in soil range from 10 to 100 kg ha^{-1}

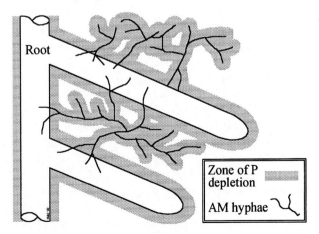

Figure 19-3. Diagrammatic representation of VAM fungal hyphae exploring the soil surrounding roots.

and even as high as 900 kg ha^{-1} (Sieverding *et al.,* 1989). Estimates of intraradical fungal biomass vary between 3 and 20% of root dry weight. Estimates of the biomass of extraradical fungus in the soil suggest that it may be as large as root biomass (Olsson and Johansen, 2000). Arbuscular mycorrhizal fungi are, therefore, a major sink for photosynthate and the symbiosis incurs a substantial cost to the plant; between 4 and 20% of net photosynthate (Smith and Read, 1997).

The cost of supporting the fungi is offset by the additional photosynthesis enabled by transfer of mineral nutrients to the plant, when those nutrients are in limiting supply. Cost benefit analysis can be applied to mycorrhizal symbiosis, although it is complicated by the responses of the plant and the fungi to mineral nutrition (Tinker *et al.,* 1994; Smith and Read, 1997). Estimation of mycorrhizal efficiency by cost benefit analysis is problematic, especially in the field (Tinker *et al.,* 1994). The growth response of plants to AM fungi, expressed as the proportion of growth of mycorrhizal plants enabled by the fungi, is the more frequent measure used and is termed mycorrhizal dependency (Plenchette *et al.,* 1983). Mycorrhizal dependency is a relative value and is rarely estimated in the field. Since mycorrhizal dependency is usually determined at one point in time, it does not account for temporal changes in environmental conditions or P demand by the plant.

If mycorrhizas are initiated by fungi from an established hyphal network then there is immediate potential for transfer of P and other elements to the host. When mycorrhizal colonies are initiated from individual propagules there is a period during which the net drain on photosynthate from plant to fungus outweighs the return in nutrients from fungus to plant. McGee *et al.* (1999) observed that the intraradical expansion of colonies of *Glomus mosseae* (Nicol. & Gerd.) Gerd. & Trappe in cotton roots occurred independently of propagule density in the soil, indicating that the propagules do not provide carbohydrate to the expanding colonies after their initiation. Hence, plant growth depression may occur during the early stages of mycorrhizal colonization (Bethlenfalvay *et al.,* 1982; Koide, 1985). Later the extra radical hyphal network explores enough soil to provide sufficient nutrients to the plant. The equilibrium of AM symbiosis is never constant but overall there is potential for mutual benefit.

3.2   Regulation of AM Symbiosis

The loss of carbon by the host is regulated by limiting colonization of roots. For a given host/fungus combination, the extent of AM colonization is determined by recognition stimuli, plant defence mechanisms and photosynthate dynamics. Signal exchange between the plant and the fungus occurs in the rhizosphere, at the point of attachment, and within the root (Koide and Schreiner, 1992; Smith and Read, 1997).

Before initiation of infection, extra radical hyphae respond to the quality and quantity of root exudates, differentiating hosts from non-hosts and the P status of the host. Appressoria form in response to chemical signals that are specific for host plants and for root epidermal cells (Smith and Read 1997; Douds and Nagahashi, 2000). The nature of these pre-infection signal molecules is unknown.

Within the root, plant defence responses are stimulated by AM fungi but are later depressed or at least operate at a low intensity compared to that induced by pathogens (Koide and Schreiner, 1992). Arbuscules are temporary structures that degenerate after a week or so, depending upon species interactions (Smith and Read, 1997). The mechanisms of degeneration are unclear but apparently involve host defence systems that enable resistance to pathogens. In AM the accumulation of defence-related products is restricted almost exclusively to the cortical cells containing arbuscules. Intracellular hyphae appear not to elicit defence responses (Smith and Read, 1997; Lambais, 2000). Pathogenesis-related proteins (PR proteins) are found in the apoplastic space of arbuscules but not in association with hyphae crossing cell walls. It is suggested that these PR proteins contribute to the control of arbuscule development and the spread of intraradical colonization (Gianinazzi-Pearson, 1996). Enhanced activity of phytoalexins, PR proteins, chitinases and acidic and basic ß-1, 3-glucanases, occurs in association with early events in AM development and diminishes strongly as colonization develops and is thought to be regulated by transcription, although the processes are not clear (Gianinazzi-Pearson, 1996; Lambais, 2000).

Salzer and Boller (2000) hypothesized that, initially, chitin elicitors from the fungus are only partially broken down by constitutive plant chitinases and are thought to bind to receptors that stimulate expression of PR proteins. Later, mycorrhiza-specific chitinases enhance the break down of elicitors and stop PR protein expression. Phytohormone-like molecules produced by the fungus may independently down-regulate defence responses as well (Lambais, 2000; Salzer and Boller, 2000).

The plant defence mechanisms activated by AM fungi may also affect carbohydrate dynamics. Invertase plays a crucial role in source-sink regulation of carbon flow and responds to a variety of stimuli and elicitors, including infection of roots by pathogens and host resistance responses (Roitsch, 1999). Intraradical spread of the fungi may thus be regulated by plant defence responses acting in concert with controls on carbon supply.

3.3 Other Physiological Effects

Although the reciprocal exchange of photosynthate for nutrients and its regulation is the dominant feature of AM symbiosis, the occurrence and function of many physiological responses are yet to be elucidated, especially at the molecular level. For example, expression of nitrate reductase can be down-regulated in AM plants while simultaneously fungal nitrate reductase is expressed in the roots, localized in arbuscules (Kaldorf *et al.,* 1998). Recent discoveries of novel proteins and metabolic pathways that are expressed in response to AM colonization illustrate further the complexity of the association (Balestrini, *et al.,* 1999; van Buuren, 1999; Walter *et al.,* 2000). Arbuscular mycorrhizal fungi do not transport water to the plant but they can affect the response of plants to water, through increases in stomatal conductance and photosynthesis (Smith and Read, 1997). Although increases in the hydraulic conductance of roots have been observed, the better drought tolerance of AM plants is largely attributable to improved P nutrition. Signal molecules released following colonization are thought to cause mycorrhizal roots to proliferate following initiation of AM (Yano *et al.,* 1998; Torrisi *et al.,* 1999) and to modify the interaction between pathogens and plants (Smith and Read, 1997).

The potential for AM to enhance nutrient uptake and yield is obviously of interest in crop production. However, the morphological and physiological processes of AM occur in the context of the soil ecosystem and are, therefore, subject to the effects of edaphic conditions and plant-microbe community dynamics. We now discuss the dynamics of AM in farming systems, including interactions with biotic and abiotic factors.

4. ECOLOGY OF ARBUSCULAR MYCORRHIZAS IN COTTON

4.1 Growth Response of Cotton

Cotton always develops AM in field soil (Rich and Bird, 1974; Jeffries *et al.,* 1988; Afek *et al.,* 1991; Nehl *et al.,* 1996). Positive effects of AM fungi on growth, nutrient uptake, and yield of cotton have been observed in inoculation and fumigation experiments (Fig. 19-1; Smith *et al.,* 1986; Afek *et al.,* 1991; Nehl *et al.,* 1994; Allen and Nehl, 1999).

Despite the shortcomings of measures, mycorrhizal dependency can be useful to compare relative growth responses. In general, the growth response of plants to AM fungi decreases as the supply of mineral nutrients, especially P, in the soil increases (Hayman, 1987; Smith and Read, 1997). Price *et al.* (1989) observed that the growth response of cotton to AM fungi at six weeks after sowing in loamy sand in pots decreased as P content (Mehlich No. 1 method) of the soil increased from 58 to 75 mg kg^{-1}. The corresponding mycorrhizal dependency values (calculated from their data) are 45 and 14% for cotton inoculated with *Glomus etunicatum* Becker and Gerd., and 52 and 5% with *G. margarita*. In contrast, mycorrhizal dependency of cotton grown in heavy clay soils in the field, where the native population of AM fungi was nil or low, declined with increasing P content (Colwell method) but was still relatively high in a soil with a P content of 88 mg kg⁻¹ (Table 19-1). These data support the notion that P fixing capacity, which is high in clay soils, determines the reliance of cotton on AM (Bolan, 1991; Smith and Read, 1997).

4.2   AM Colonization of Cotton

Although the rate of initiation of AM will vary according to the host and edaphic conditions, for a given soil and host, initiation is directly related to the quantity of fungal propagules in the soil (Pattinson and McGee, 1997; Nehl *et al.,* 1999; McGee *et al.,* 1999). Later, secondary colonies are initiated by the expanding extra radical mycelium. Colonization generally follows a pattern of logistic growth, with the proportion of colonized root increasing exponentially at first and then reaching a plateau (Rich and Bird, 1974; Pattinson and McGee, 1997).

Colonization spreads rapidly in roots of cotton (Torrisi *et al.,* 1999); the greater the inoculum, the faster the spread (McGee *et al.,* 1999). Furthermore, initiation of colonization is followed by striking proliferation of roots (Torrisi et al, 1999), probably caused by changes in the mitotic cell cycle at the root tip (Berta *et al.,* 1991). The extension of hyphae into the soil is also limited to the upper soil profile (Nehl *et al.,* 1999). Hence, few propagules occur in the soil at depth, even though AM fungi follow the root system down by secondary spread. Thus the effective root zone is limited to the upper profile, an aspect that is crucial for effective irrigation and fertilizer practices.

4.3   AM fungi

Under experimental conditions individual species of AM fungi can form mycorrhizas with almost any potential host (Smith and Read, 1997). As a striking example, *Glomus antarcticum* Cabello which occurs in an isolated ecosystem on the Danco Coast of Antarctica readily forms mycorrhizas with sorghum and alfalfa in pot culture (Cabello *et al.,* 1994). Despite the lack of absolute host specificity by AM fungi under experimental conditions, a degree of ecological host specificity is likely for most soils (Molina *et al.,* 1992). Since the response of plants to AM fungi varies according to the genotype of both the host and the fungus, the species composition of AM populations can potentially determine plant community structure and productivity (van der Heijden *et al.,* 1998).

In agriculture the plant community structure is consciously controlled, to varying degrees, whereas the microbial community structure is not. Continuous, high-input cropping has the capacity to alter the composition and density of populations of arbuscular mycorrhizal fungi in the soil (Douds *et al.,* 1993; Douds *et al.* 1995; Hendrix *et al.,* 1995). Frequent soil disturbance probably favours fungal species that sporulate prolifically (Kurle and Pfleger, 1994).

The selective effects of agriculture on arbuscular mycorrhizal fungi are important because the development and function of mycorrhizas can vary according to the species or clone of fungal symbiont (Smith and Read, 1997; Jakobsen, 1999). Crop plants can be more responsive to native populations of AM fungi than to introduced species (Dhillion, 1992). The ability of hyphae to absorb and transport P and Zn to the host varies among species of AM fungi (Bürkert and Robson, 1994; Jakobsen, 1999) and is related to the extent of the hyphal network and its interception of soil. For instance, uptake of Zn by cotton inoculated with *Glomus ambisporum* Smith & Schenck, was greater than with *Glomus intraradices* Schenck & Smith or *Gigaspora margarita* Becker & Hall (Smith and Roncadori, 1986).

Conversely, genotypic differences in the host may also determine the degree of response to AM fungi. Pugh *et al.* (1981) found that shoot growth increases caused by inoculation of five cotton cultivars with *G. margarita* ranged from 0 to 94%. As indicated earlier, these differences appear to be related to the root morphology.

4.4   The Rhizosphere Community

All plants normally grow in a complex, dynamic microbial community of which AM fungi are but one component. The presence of AM therefore, might be predicted to influence the composition and dynamics of a microbial community, either to the benefit or detriment of the host plant. Especially important are pathogens.

Arbuscular mycorrhizal fungi have the potential to attenuate plant disease, although this is not a ubiquitous phenomenon. In some circumstances, benefits to the host plant due to the suppression of pathogens by AM fungi may take precedence over P nutrition (Fitter and Garbaye, 1994; Pozo *et al.,* 1999). Protection from disease may be mediated through changes in host nutrition, physiology and cell structure, changes in the rhizosphere microflora (Linderman, 1994), or direct effects on the pathogen,

Table 19-1. Mycorrhizal dependency of cotton grown in the field at six weeks after sowing.

Site	Field 18ᶻ		Field 17ᶻ		Field 20ʸ	
Soil fumigation	0	-	+	-	-	-
Extended bare fallow (>6 yrs)	-	-	-	-	+	-
AM colonisation (%)	0 ± 0	58 ± 2.6	0	46 ± 1.4	5 ± 2.1	30 ± 2.4
Shoot dry matter (g plant⁻¹)	0.6 ± 0.1	7.0 ± 0.5	2.2 ± 0.4	9.9 ± 0.8	0.4 ± 0.04	1.0 ± 0.13
Shoot P (g kg⁻¹)	1.4 ± 0.3	3.9 ± 0.1	3.5 ± 0.4	4.1 ± 0.1	3.3 ± 0.2	4.4 ± 0.2
Mycorrhizal dependency (%)	92		78		57	
Soil P (mg kg⁻¹)	16		44		88	

ᶻ Data from Nehl *et al.* (1994).
ʸ Data from Allen and Nehl (1999).

perhaps via production of antimicrobial products (Smith and Read, 1997). The host defence responses induced by AM fungi probably do not affect pathogens directly because they are suppressed and localized near arbuscules. However, induction of systemic resistance by AM has been observed (Blee and Anderson, 2000). The mechanisms are unclear but may involve interaction between pathogens and chitosanases induced by AM (Blee and Anderson, 2000) or isoforms of ß-1,3-glucanases that are induced by the pathogen only when AM are present (Pozo *et al.*, 1999).

Inoculation of cotton with AM fungi has resulted in reductions in the severity of Verticillium wilt, Fusarium wilt and black root rot (Hu and Gui, 1991; Liu *et al.*, 1995; Schonbeck and Dehne, 1977). Liu *et al.* (1995) observed systemic production of PR proteins in cotton, associated with AM-enhanced resistance to *Verticillium dahliae* Kleb. In some cases AM may increase the severity of disease in cotton (Davis *et al.*, 1979). The influence of AM on pathogens probably depends upon the species of AM fungus. Hu and Gui (1991) observed that pre-inoculation with *G. mosseae* reduced the severity of Fusarium wilt in cotton, whereas pre-inoculation with *G. intraradices* did not. Thus the density and diversity of AM fungi may be critical in determining whether a pathogen causes significant disease. If the community of AM fungi is complex and at high densities, cotton seedlings will be immediately colonized by a variety of AM fungi each of which may interact with the various detrimental microbes to the overall benefit of the seedling.

Arbuscular mycorrhizal fungi also interact with other beneficial microbes. Synergism between arbuscular mycorrhizal fungi and rhizobia (Azcón *et al.*, 1991; Linderman, 1992), free-living nitrogen-fixing bacteria (Bagyaraj and Menge, 1978; Subba Rao, *et al.*, 1985; Negi *et al.*, 1990; Singh, 1992), plant growth-promoting rhizosphere bacteria (Meyer and Linderman, 1986), and phosphate solubilizing bacteria (Barea *et al.*, 1975) has been reported. Interactions between arbuscular mycorrhizal fungi and nutrient cycling bacteria are also possible but have received little investigation (Linderman, 1992).

Rhizosphere bacteria have the capacity to either inhibit or enhance the activity of AM fungi and, subsequently, the response of the host. Conversely, mycorrhizal colonization can affect root exudation and hence the composition and function of the rhizobacterial community, potentially altering the balance between deleterious and plant growth-promoting bacteria (Nehl *et al.*, 1997). Root border cells play an active role in rhizosphere community dynamics, through the selective effects of carbohydrates, extracellular enzymes, antibiotics, phytoalexins, and molecular signals (Hawes, 1990; Hawes *et al.*, 1999). Root border cell production appears to be correlated positively with the propensity for AM colonization in different plant hosts, although the mechanisms are not clear (Arriola *et al.*, 1997; Niemira *et al.*, 1996). It may be no coincidence that cotton produces large numbers of root border cells (Hawes, 1990; Nehl, unpublished) and is also highly mycorrhizal.

Cotton is affected by a disease complex that is associated with very heavy clay soils in which AM development, nutrient uptake, and growth are reduced (Nehl *et al.*, 1996). Bioassays during the period between successive crops indicated that the slow AM development was not due to a lack of fungal propagules but to physical and/or biological factors (Nehl *et al.*, 1998). This disease appears to involve deleterious rhizosphere bacteria that inhibit AM development (Nehl, unpublished).

4.5   Soil Structure

Hyphae of AM fungi grow into the soil from AM. The quantities of hyphae influence soil aggregation (Tisdall and Oades, 1979). A water stable and temperature stable glycoprotein (glomalin) produced by AM hyphae is abundant in soils and contributes to soil aggregate stability (Wright and Upadhyaya, 1998). Cultural practices that reduce the length of hyphae through changing the fungal population (see above) or hyphal viability by using fallows will also influence the aggregation of soil (Tisdall and Oades, 1979). Changes in fungal species diversity that reduce formation of hyphae and loss of hyphae during cultivation and fallows (Pattinson and McGee, 1997) may ultimately cause reduced aeration of soil following irrigation, soil compaction, which leads to reduction of root elongation (Nadian *et al.*, 1998), and increased soil erosion.

4.6   Cultural Practices

In cultivated soils the hyphal networks develop during the course of a crop but will inevitably be disturbed by cultivation, reducing the abundance of propagules, especially in light textured soils (Evans and Miller, 1990; Jasper *et al.*, 1991). There are reports of reduced AM colonization of cotton following cultivation of loamy soils (Zak *et al.*, 1998). In heavy textured soils the hyphal network may be less important as a source of inoculum. Cultivation of clay soils in eastern Australia did not reduce AM development or growth of cotton (Allen and Nehl, 1999). Moderate disturbance of these soils by sieving does not reduce the AM inoculum potential measured in bioassays (Pattinson and McGee, 1997; Nehl *et al.*, 1998), though severe disturbance may (McGee *et al.*, 1997).

The effects of soil disturbance on inoculum potential and subsequent AM development may also interact with environmental conditions or the species of host or fungal symbiont (McGonigle and Miller, 2000). Sufficient AM fungal propagules for normal AM development in cotton can survive dry storage of soil for at least 18 months, unless the soil is subjected to a combination of disturbance followed by wetting and drying cycles (Pattinson and McGee, 1997). These observations have implications for cropping sequences. Attrition of AM fungi in the absence of a host contributes to long fallow disorder. Long fallow disorder has been reported in cotton (Smith *et al.*, 1989; Brown *et al.,* 1990) but does not always occur (Hulugalle *et al.,* 1998; Allen and Nehl, 1999). For the AM fungi, rotation with non-mycorrhizal crops is equivalent to a bare fallow. Rotation with crops in the Brassicaceae is reported to have negative effects on AM development (Smith and Read, 1997), although this appears not to occur with cotton in rotation with canola in some soils (Allen and Nehl, 1999). Rainfall events and cultivation are likely to accelerate the declines in AM fungi during long bare fallows or rotations with non-mycorrhizal hosts (Pattinson and McGee, 1997).

The ultimate soil disturbance is topsoil removal. This can occur when cotton fields are being levelled to enable surface irrigation. Where more than 40 cm of soil is removed there is little or no AM development in cotton crops and yield losses occur (Allen and Nehl, 1999; Nehl *et al.,* 1999). This effect is usually transitory (one to two years) indicating that fungal population can re-establish from very small quantities of inoculum (McGee *et al.,* 1999).

5. CONCLUSIONS

The physiological basis of AM symbiosis is reasonably well characterised but much remains to be elucidated, particularly at the molecular and community levels. The regulation of AM symbiosis is intimately tied to the nutritional status, the carbon economy and the molecular defences of plants. Arbuscular mycorrhizal fungi, therefore, are a link in a structural and metabolic continuum between plants, micro-organisms and soil. It is the ecological specificity and functional compatibility of mycorrhizas under natural conditions that is most relevant to plant ecosystems (Molina *et al.,* 1992). Recognition of the function of AM will broaden the potential of research in the plant and soil sciences.

Since agricultural practices can affect the diversity of arbuscular mycorrhizal fungi, an understanding of the functional specificity of individual host-fungus combinations is important for the management of agricultural systems (Bethlenfalvay, 1992). Cotton management will be enhanced by considering the effects of field development, cultivation, fertilization, and cropping sequences on the quantity and quality of AM fungi and their role in the modification of host physiology. Arbuscular mycorrhizal symbiosis affects not only production of cotton but also the sustainability of cotton farming systems. A pertinent amendment of our initial rhetorical question would be: when are cotton farmers not cotton farmers? All the time; they are mycorrhiza farmers!

6. ACKNOWLEDGMENTS

Sincere thanks are extended to Dr Philip Wright for comments on the manuscript.

Chapter 20

MECHANISMS OF COTTON RESISTANCE TO ARTHROPOD HER-BIVORY

Víctor O. Sadras[1] and Gary W. Felton[2]
[1]CSIRO Plant Industry, Locked Bag 59, Narrabri, NSW 2390, Australia; Present address: South Australian Research and Development Institute, School of Agriculture, Food, and Wine, University of Adeliade, Adelaide, South Australia; and [2]University of Arkansas, Department of Entomology, Fayetteville, AR, U.S.A.; Present address: Department of Entomology, Pennsylvania State University, 501 Ag Sciences & Industries Building, University Park, PA, U.S.A.

1. INTRODUCTION

The cotton crop is host to a wide range of arthropod pests (Hargreaves, 1948; Room, 1979a). Key pests, *i.e.* those that are persistent, occur perennially, and usually reach economically damaging levels (Hearn and Fitt, 1992) have been characterized for most cotton-cropping systems worldwide. The identification and biology of key cotton pests, their economic importance, and methods for their control have been the focus of recent reviews by Frisbie *et al.* (1989), Hearn and Fitt (1992), Fitt (1994), Luttrel (1994), Luttrel *et al.* (1994), Ramalho (1994), Sugonyaev (1994), Vaissayre (1994), Matthews and Tunstall (1994), Hillocks (1995), and Pyke and Brown (1996).

Various terms have been used to describe the different kinds of plant responses to pests (Hooker, 1984). For the purposes of this chapter, we will consider that plant reactions to a given pest grade continuously from full resistance to the extreme sensitivity of those plants that are unprotected and unable to regrow after damage (Painter, 1951; Hooker, 1984; Belsky *et al.*, 1993). In this context, our aim is to analyse the mechanisms of cotton resistance to arthropod herbivory.

This chapter takes a "phytocentric" view of plant-herbivore relationships (Baldwin, 1993). This means that we shall concentrate on physiological and morphological plant and crop traits relevant to herbivory resistance; the effects of the cotton plant on its pests are only considered when necessary to characterise resistance mechanisms. Cotton resistance to arthropod herbivory has been the subject of reviews that emphasised general yield responses to pests (Brook, 1984;

Matthews, 1994a), chemical defenses (Bell, 1984b, 1986), breeding for resistance to arthropods (Thomson and Lee, 1980; Thomson, 1987; El-Zik and Taxton, 1989; Smith, 1992; Gannaway, 1994; Jenkins, 1994) and responses to reproductive damage (Sadras, 1995).

Understanding the physiological and morphological adaptations of cotton to arthropod pests is important for the improvement of pest management practices. For instance, a threshold for pest management of z insects per m^2 implies that (i) a fraction, f, of those insects will die due to various factors (*e.g.*, predation); and (ii) that the crop will be able to cope with the damage caused by the surviving $f \times z$ insects. Following with this example, an understanding of the plant and environmental factors that affect the responses of the crop to a given level of damage could assist in developing more precise thresholds for pest management. Thus, assuming that understanding the mechanisms of resistance to pests is important not only for breeding but also for pest and crop management, we have focussed on resistance traits broadly, irrespective of whether or not the traits are relevant for breeding purposes.

Ecological theories provide a valuable background to study the relationships between the cotton crop and its pests (*e.g.*, Gutierrez *et al.*, 1979a; Felton *et al.*, 1989; Sadras, 1996c). We have thus highlighted the ecological principles underlying these relationships. Within this framework, we have analysed the mechanisms of cotton resistance to arthropods including (i) **avoidance**, *i.e.*, escape in space and time, and chemical and morphological defenses; and (ii) **tolerance**, *i.e.*, recovery after damage. We have intentionally emphasised the analysis of tolerance mechanisms. This

J.McD. Stewart et al. (eds.), *Physiology of Cotton*,
DOI 10.1007/978-90-481-3195-2_20, © Springer Science+Business Media B.V. 2010

is because, in the past, most research on plant resistance to arthropods has focussed on avoidance traits. For instance, a whole chapter in the previous "*Cotton Physiology*" book dealt with the physiology of secondary products, but none dealt specifically with tolerance to herbivory. Our emphasis on tolerance is therefore not a matter of value, *viz.* we do not consider tolerance to be more or less important than avoidance.

Mechanisms of resistance to arthropods were analysed at various scales, from the molecular to the crop level. The relationship between avoidance and tolerance traits is discussed with emphasis on its implications for breeding varieties with enhanced resistance to pests. Multiple interactions between biotic and abiotic stresses are outlined and directions for further research on cotton responses to herbivory are indicated.

2.   ON PLANTS AND THEIR HERBIVORES

This section introduces two contrasting views of plant-herbivore relationships, briefly describes the general responses of crop yield to damage by herbivores, and outlines individual- and population-level mechanisms related to the adaptation of plants to herbivory.

2.1   Antagonism vs Mutualism

The sessile mode of life imposes obvious restrictions for plant survival, *e.g.*, plants cannot run or fly away when threatened by herbivores. Equally obvious is that plants survived because they developed traits for resistance to environmental stresses, including herbivory (Trewavas, 1981). Following this line of thought, relationships between plants and their natural enemies are usually regarded as the result of an antagonism between the ability of herbivores to attack their host and the ability of the host plant to resist such attack (Futuyama and May, 1991; Marquis and Alexander, 1992).

In contrast to this antagonistic view of the relationship between plants and herbivores, it has been proposed that, in some cases, herbivory may increase plant growth and fitness (Owen, 1980; Owen and Wiegert, 1976, 1981; Hilbert *et al.*, 1981; McNaughton 1983a; Paige and Whitham, 1987; Verkaar, 1988; Maschinski and Whitham, 1989; van der Meijden, 1990; Whitham *et al.*, 1991; Vail, 1992; Littler *et al.*, 1995). Owen's (1980) interpretation of the benefits that plants can obtain from their relationships with aphids is an example of this mutualistic view (section 3.2.6). Whether plants and herbivores have evolved a dominantly antagonistic relationship or some form of mutualistic relationship is a highly controversial issue that is out of the scope of this chapter. Readers interested in this debate may refer to Owen (1980), McNaughton (1983a, 1986), Crawley (1987),

Aarsen and Irwin (1991), Whitham *et al.* (1991), Bergelson and Crawley (1992), Belsky *et al.* (1993), Mathews (1994), and Vail (1994).

2.2   Crop Yield and Herbivores

General crop yield responses to herbivory have been reviewed by Jameson (1963), Bardner and Fletcher (1974), and Harris (1974). Plant growth and crop yield can be reduced, unaffected, or increased by herbivory. Harris (1974) emphasised the cases in which yield increases have been observed following insect damage. It is of course easier to find examples in which arthropod herbivory reduced rather than increased crop yield (Harris, 1974) but reports of no yield loss or moderate yield increases of cotton crops following pest or artificial damage are not uncommon (*e.g.,* Prokof'ev and Rasulov, 1975; Harp and Turner, 1976; Renou and Aspirot, 1984; Chen *et al.*, 1991; Brook *et al.*, 1992abc; Dyer *et al.*, 1993; see also Table 1 in Sadras, 1995). Physiological mechanisms underlying yield increases after damage in cotton have been discussed by Renou and Aspirot (1984), Gutierrez *et al.* (1979a), Brook *et al.* (1992b), and Sadras (1995). A detailed analysis of the main morphological and physiological responses of cotton plants and crops to different types of damage is presented in Section 3.2.

2.3   How Do Plants Cope with Herbivores?

While ecologists debate whether herbivory is beneficial to plants or not (Section 2.1), cotton growers usually regard pests as detrimental to yield (but see Dyer *et al.*, 1993). From an agronomic perspective, the conservative view of an antagonistic relationship between a crop and its herbivores is not surprising and, in many cases, justified as yield and economic losses due to pests may be quite large. This antagonistic view is, however, a primary obstacle for the implementation of integrated-pest management programmes.

Many schemes have been proposed to characterise the ways in which plants can avoid herbivory and recover after damage. These schemes place emphasis on different aspects of plant-herbivore interactions depending on the objectives of the analysis. For instance, the framework of Painter (1951) is particularly appropriate for entomological studies, the classification of Thomson (Thomson and Lee, 1980; Thomson, 1987) is useful for breeding purposes, while White (1993) emphasises the role of plants as a source of nitrogen for herbivores. A plant-centred approach, is more appropriate for the objectives of this chapter (see Introduction).

Plants may avoid damage via escape in time and space. They may also avoid damage through morphological and chemical defenses. General escape and defense strategies have been discussed in a number of studies (*e.g.,* Painter,

1951; Crawley, 1983; Chapin *et al.,* 1987; Karban and Myers, 1989; Zangerl and Bazzaz, 1992, Tuomi, 1992; Baldwin, 1993) and are further examined in section 3.1.

The strategy of tolerance depends on morphological and physiological traits that, rather than protect the plants from damage, allow them to regrow after damage has occurred (Belsky *et al.,* 1993). Interestingly, both plant- and animal-centred approaches define "tolerance" in similar terms (*cf.,* Belsky *et al.,* 1993 and Painter, 1951). Tolerance traits are important for recovery not only after damage caused by herbivores but also after damage caused by physical factors such as hail, wind, or fire (Belsky *et al.,* 1993). General tolerance traits have been reviewed recently by Belsky *et al.* (1993), Trumble *et al.* (1993) and Rosenthal and Kotanen (1994) and are discussed in section 3.2.

Since damage by pests in field crops is generally heterogeneous in space and time, population-level compensation needs to be considered. Population-level compensation, according to Crawley (1983), occurs when "herbivore attack on one individual allows another individual to grow more rapidly." Section 4 discusses population-level responses to herbivores in cotton.

Despite the commonly accepted view that cotton is highly susceptible to arthropod herbivory, the previous discussion highlights that (i) plants are not passive victims of herbivores, and (ii) in some cases herbivory may be neutral or even positive for plants.

3. RESISTANCE TO HERBIVORY: AT THE PLANT LEVEL

Analysis at the molecular level is required to understand some resistance traits (*e.g.,* chemical defenses). Other traits require to be analysed at the organ level (*e.g.,* okra leaf). Analyses at these levels of organisation are necessary and have been included in this section. They are, however, not sufficient to understand the role of these traits in plant resistance to herbivory, the level of the physiological unit at which they operate, the whole plant, needs to be considered. The discussion about using artificial diets as a method to investigate defensive compounds (section 3.1.2) highlights the risk of taking reductionist approaches in research of plant resistance to arthropod pests.

3.1 Avoidance

Avoidance mechanisms discussed in this section include escape, also referred to as phenological asynchrony and host evasion, and chemical and morphological defenses.

3.1.1 Escape

Earliness is a characteristic often sought in cotton cropping systems due to its implications for pest management.

Earliness, however, has also implications for other agronomic outcomes, including yield and quality. It is interesting to note, for instance, that earliness has been favoured during early stages of cotton domestication (Fryxell, 1978) and, more recently, by breeding and selection for high yield potential (Culp, 1994).

Genotype and crop management are the two keys for the manipulation of crop phenological development. Cotton genotypes selected for earliness may allow the plant to avoid pest damage (Smith, 1992). These genotypes are valuable for managing boll weevil, Heliothine (= budworm-bollworm complex; *Heliothis* spp.- *Helicoverpa* spp.) and pink bollworm populations in regions where cultivation of early maturing varieties is feasible (Walker and Niles, 1971; Gannaway, 1994). Fast-fruiting genotypes produce bolls that escape first-generation weevil damage (Walker and Niles, 1971). The resulting reduction in pesticide usage required for boll weevil control may also delay the build-up of Heliothines due to preservation of natural enemies (Smith, 1992). The utilisation of early maturing varieties has been successfully adopted in the U.S. and Brazil (Luttrell *et al.,* 1994; Luttrell, 1994; Ramalho, 1994).

In addition to early maturing varieties, time of maturity can be modulated (and thus pest avoidance can be achieved) through appropriate crop management practices. Watson *et al.* (1978), for instance, indicated that the impact of pink bollworm can be reduced by manipulating the time of last irrigation to terminate crops early. Chu *et al.* (1996) have assessed the results of mandatory short-season cotton management systems in the Imperial Valley of California. The aim of this program, established in 1989, is to reduce pink bollworm populations in the area by optimising the host-free period. The system comprises a number of components, including reference dates for (i) earliest sowing, (ii) defoliation, and (iii) stalk destruction and plow down. The authors of this study regard the short-season cotton system as very effective in reducing pink bollworm abundance and boll damage. These effects were partially confounded, however, with the effects of reduced cotton production in the Imperial Valley during the period of their assessment (Chu *et al.,* 1996).

Plant traits and cropping strategies that affect the fitness of pests have the potential for the selection of "resistance" in the target pest population. Insect resistance to insecticides is the most obvious example of this process (see Section 3.1.2 *Biotechnology and Plant Resistance* below). We are not aware of development of resistance to strategies based on earliness or phenological escape in cotton. Chu *et al.* (1996), however, pointed out that the short-season cotton system mentioned above implies a risk of selection for early diapausing pink bollworm larvae due to diminishing bollworm food sources. An interesting case of "resistance" to cropping strategies based on escape *via* crop rotations has been reported for corn (*Zea maize* L.) rootworm in the central U.S.A. (Karlen *et al.,* 1994). In a monoculture maize production system, rootworm reaches an economic threshold about 30% of the time, but in a 2-year maize/soybean (*Glycine max* L.) rotation, the threshold is reached less than 1% of the time. However, increased use of the 2-year maize/ soybean rotation has resulted in selection for rootworms with a 2-year (rather than the normal 1-year) diapause.

3.1.2   Defenses

The development of insect resistance to insecticides has greatly accelerated the emphasis on cotton resistance to herbivores during the last several decades (Smith, 1992). This research has culminated in the identification of scores of morphological and biochemical defenses (Bell, 1984b; 1986; Smith, 1992). It is not our purpose here to review the details of these traits; the reader is referred to several excellent reviews and historical perspectives (Hedin *et al.*, 1976; Thomson and Lee, 1980; Thomson, 1987; Benedict *et al.*, 1988; Jenkins *et al.*, 1991; Wilson, 1991; Gannaway, 1994; Smith, 1992; Summy and King, 1992). Morphological traits such as frego bract, nectariless, glabrousness, pilosity, okra-leaf, rugate bolls, reduced branching, stem-tip stiffness, red leaf coloration and reduced anther numbers are associated with plant resistance (Thomson and Lee, 1980; Thomson, 1987, 1994; Smith, 1992; Gannaway, 1994). Genotypes possessing these traits are often less preferred for feeding and/or oviposition and may be associated with greater arthropod mortality.

The cotton plant possesses a rich abundance of phenolic and terpenoid compounds that may reduce host suitability to arthropod pests. These phytochemicals may be directly toxic and/or interfere with the utilisation of essential nutrients. Phenolic compounds include condensed tannins, flavonoids (*e.g.*, quercetin glycosides, chrysanthemin; Hedin *et al.*, 1983; Hedin *et al.*, 1992), benzoic acids (*e.g.*, syringic acid; Benedict *et al.*, 1988), and cinnamic acids (*e.g.*, chlorogenic acid; Benedict *et al.*, 1988). Among the terpene aldehydes are gossypol, gossypolone, heliocides, and hemigossypol (Stipanovich *et al.*, 1988). Although extensive efforts on identifying sources of resistance to herbivory have been conducted in cotton, there is considerable germplasm, particularly with the Asiatic cottons, that remains to be tested for relevant traits (*e.g.*, Stanton *et al.*, 1994).

Despite these extensive research efforts several complications exist in employing these traits for enhancing plant resistance. First, conference of resistance to one pest may produce increased susceptibility to another. For example the frego bract trait confers resistance to the boll weevil, yet increases susceptibility to plant bugs. Likewise, gossypol may provide resistance to Heliothines, but increase susceptibility to thrips. Part of the contradiction between responses to quantitative traits may arise because ranges of these traits may exist that maximise resistance to a given pest species. This is the case, for instance, for aphids in relation to hairiness: they seem to prefer moderately hairy genotypes over glabrous or pilose surfaces (Jenkins, 1995). The multiple effects of a given resistance trait highlight the

need for breeding programs that focus (i) on broad spectrum resistance (*e.g.*, El-Zik and Thaxton, 1989; Calhoun *et al.*, 1994) and (ii) on positive balances whereby the enhancement of resistance associated with certain traits outweighs their detrimental effects. The successful inclusion of the okra-leaf trait in Australian varieties illustrates this point (Thomson, 1994).

Second, reduced quality and/or quantity in yields often accompany expression of these traits. For instance, high amounts of gossypol or tannins in cottonseed are undesirable due to their toxicity in feed and oil products (Yu *et al.*, 1993). Heavy pubescence responsible for enhanced resistance to some arthropods produces commercially unacceptable amounts of "plant trash" in mechanically harvested cotton (Smith, 1992). On the other hand, characters such as nectariless in certain genetic backgrounds do not negatively impact yields (Gannaway, 1994). The "penalties" potentially associated with resistance traits are further considered in section 3.3.

Third, artificial diet bioassays for assessing the toxicity of certain phytochemicals (*e.g.*, phenolics) may be ineffective for assessing their true role in resistance. For example, condensed tannins incorporated into the artificial diet of the bollworm/budworm strongly inhibit larval growth at concentrations above 0.2% dry weight (Reese *et al.*, 1982); however, larvae flourish on cotton tissues where tannin concentrations may exceed 10% dry weight. Furthermore, budworm growth and survival were not affected by several breeding lines selected for elevated tannin concentration (Smith *et al.*, 1992). Another similar discrepancy is with one of the major phenolic acids in cotton, chlorogenic acid. Amount of this phenolic increases significantly in cotton foliage following herbivory by the bollworm or budworm (Bi *et al.*, 1997; G. Felton, unpublished data). Because of this, and because chlorogenic acid is toxic in artificial diet to the budworm or bollworm (at concentrations exceeding 3 mmol g^{-1} wet weight; G. Felton, unpublished data), it is reasonable to suggest that it may have a defensive role in the plant. However, in tobacco (*Nicotiana tabacum* L.) plants which have been transformed to overexpress or underexpress chlorogenic acid, budworm growth is unaffected by this phenolic at concentrations ranging from 0.1 to 6.0 mmol g^{-1} wet weight (J. Bi, G. Felton, R. Dixon, C. Lamb, unpublished data). The discrepancy between plant and artificial diet assays may be due to many reasons including: (i) other phytoconstituents, missing in artificial diets, could interfere with the action of otherwise toxic compounds; (ii) changes in the concentration and/or distribution of putative toxic compounds among plant organs and with ontogeny (*e.g.*, Gubanov, 1966; Lane and Schuster, 1981); (iii) feeding pattern of arthropods in intact plants could allow them to avoid tissues with high concentration of toxic compounds (Parrott *et al.*, 1983); and (iv) behaviour of arthropods may affect the concentration of secondary compounds, *e.g.*, rolling over of leaves by caterpillars may reduce tannin concentration (Sagers, 1992). Whether artificial diets are truly suitable for assaying other cotton phytochemicals (*e.g.*, terpene aldehydes), will require similar tests using transgenic plants that have been specifically transformed for altered expression of the specified biosynthetic pathways. Otherwise, research efforts based solely upon evidence from artificial diets may be futile.

Induced Defenses. Induced defense is defined as situations where herbivory (or other agents) alters the physiology of the plant such that it becomes less suitable for pest feeding, growth, development, survival, and/or reproduction. From a complementary perspective, induced defense is viewed as a strategy that adjusts the defense level to the prevailing risk of herbivory in contrast to constitutive, invariant defense level (Åström and Lundeberg, 1994).

Induced defense in cotton has been established by the pioneering work of Karban and co-workers (Karban and Carey, 1984; Karban, 1987; 1988; Karban *et al.,* 1987; Karban and Meyers, 1989; Brody and Karban, 1989; 1992). They have demonstrated that induced defenses can markedly decrease the population growth of spider mites (Karban and Carey, 1984; Brody and Karban, 1989). Furthermore, defenses induced by one pest (*e.g.,* spider mites) may offer cross protection to multiple insect pests (*e.g., Spodoptera exigua;* Karban, 1988) and phytopathogens (Karban *et al.,* 1987).

Recent work on understanding the biochemical basis for induced defenses may aid in the development of varieties with "heightened" resistance (Hampton, 1990; Bi *et al.,* 1997). Brody and Karban (1992) have already shown that certain genotypes possess high levels of induced defenses, and thus could provide the genetic basis for further selection. Entomologists at the University of have been investigating the biochemical basis of resistance induced by bollworm or budworm feeding. The growth rate of larvae on previously fed-upon plants was significantly reduced compared to those feeding on unwounded plants. Bi *et al.* (1997) have found that herbivory on foliage or squares/bolls results in an extensive change in plant metabolism, as indicated by a decline in the nutritive content of cotton tissues accompanied by a shift to a more oxidative, antibiotic state. They have identified at least three proteins (*i.e.,* ascorbate oxidase, diamine oxidase, and peroxidase) that are strongly induced by herbivory. If these proteins are causally linked to defenses, then amplification of the genes encoding these proteins could be used to produce varieties with enhanced resistance. Overall, research on cotton proteins associated with arthropod or phytopathogen resistance has lagged behind most major crops (Liu *et al.,* 1995).

Furthermore, knowledge of the systemic signalling pathways for inducible defenses may offer an additional avenue for exploitation. For instance, in the tomato (*Lycopersicon esculentum* L.) plant, expression of the gene for the translocatable peptide, systemin, promotes the expression of defense genes for protease inhibitors and polyphenol oxidases resulting in enhanced insect resistance (Orozco-Cardenas *et al.,* 1993; Constabel *et al.,* 1995). In this case, expression of one gene may in fact lead to multicomponent resistance. Greater knowledge of the defense-signalling pathways in cotton is needed.

Another interesting approach for employing induced defenses involves the use of chemical elicitors. Several large agro-biotech firms are developing chemical elicitors (*e.g.* benzothiadiazole, 2,6-dichloroisonicotinic acid) for systemic acquired resistance to phytopathogens (Ward *et al.,* 1991; Gorlach *et al.,* 1996). A similar approach may be feasible for insect resistance. Preliminary data with cotton has shown that the application of minute concentrations of the signal compound, jasmonic acid, to cotton enhances resistance to the bollworm (G. Felton, unpublished data). This compound is essentially nontoxic and has been used for years as a component of many perfumes. Levels of the putative defense compound, gossypol, are increased by foliar applications of cytokinins (Hedin and McCarty, 1994b). Furthermore, defenses can be induced by application of spores from *Bacillus* spp. (Benedict *et al.,* 1988).

Price *et al.* (1980) suggested that enemies of herbivores may be considered "plant defenses." Following Price *et al.* (1980), the release of herbivory-induced volatile chemicals that attract parasitoids and predators to damaged plants (McCall *et al.,* 1994; Turlings and Tumlinson, 1992; Turlings *et al.,* 1995; Röse *et al.,* 1996) can also be considered, broadly, as a form of "induced defense". Thus, induced resistance may not only be directly targeted against the pest, but may also be indirect such that the impact of natural enemies on pest populations is increased (Turlings *et al.,* 1995). The chemical nature of compounds released by cotton plants in response to damage by insects and/or manual damage have been investigated in several laboratory studies (*e.g.* Röse *et al.,* 1996). There are, however, many questions remaining about the effectiveness of this form of indirect resistance, notwithstanding the question whether these same volatiles may attract herbivores. The defensive role of volatile chemicals is further discussed in section 4.1.

Biotechnology and Plant Resistance. Striking successes in enhancing plant resistance are possible with the advent of biotechnology. The development of transgenic cotton expressing the endotoxins from *Bacillus thuringiensis* (= Bt cotton) has been a remarkable story of years of basic research on microbial, plant, and insect biology, culminating in its commercial release (Perlak *et al.,* 1990; Benedict *et al.,* 1992, 1996; Wilson *et al.,* 1992, 1994; Carlton and Gawron-Burke, 1993; Fitt *et al.,* 1994; Cannon, 1995; Halcomb *et al.,* 1996). Nevertheless, the utilisation of Bt cotton is not without concerns, some emotional, but some scientifically valid (Fitt *et al.,* 1994; Raybould and Gray, 1994; Lefol *et al.,* 1995; Nap *et al.,* 1996). The potential for development of insect resistance to Bt cotton is a major, well recognised problem (Whalon and McGaughey, 1993; McGaughey, 1994; Tabashnik, 1994a). Strategies to manage or delay resistance to Bt have been delineated in several excellent reviews (Tabashnik, 1994b; Tabashnik *et al.,* 1991; Gould *et al.,* 1994; Forrester, 1994; Roush, 1994ab; Kennedy and Whalon, 1995) and will not be further addressed in this chapter. Other problems associated with the commercial use of Bt cotton that remain to be solved include changes in pest status associated with altered patterns of pesticide use, the declining Bt expression during the late stages of the crop cycle (Fitt *et al.*, 1994) and transient decline in Bt efficacy probably associated with environmental stresses (*e.g.,* Forrester and Pyke, 1997). Importantly, the introduction of *Bacillus thuringiensis* genes into cotton does not seem to have reduced the considerable capacity of the crop to tolerate insect damage (Sadras, 1998). Tolerance to damage in Bt cotton is obviously important as the crop remain vulnerable to non-lepidopteran insects and, when the efficacy of Bt toxins falls because of ontogenetic and/or environmental factors, to lepidopteran pests also.

Bt cotton is only the beginning. The continued identification of genes encoding insecticidal proteins will lead to new products as the cultivation of cotton continues through the 21st century. The fungal enzyme, cholesterol oxidase, has been shown to be a potent toxin against the boll weevil (Purcell *et al.*, 1993; Greenplate *et al.*, 1995) and to offer partial resistance to lepidopteran larvae (Purcell *et al.*, 1995). Whether cultivars expressing the cholesterol oxidase gene are ever commercially acceptable, of course, depends upon a host of factors such as health concerns regarding ingestion of cholesterol oxidation products in food or feed products, reduced crop yields, environmental impact, and non-target effects on natural enemies, pollinators, etc. This example, however, illustrates the enormous possibilities awaiting discovery and employment of new defense genes.

Expression of the gene for tryptophan decarboxylase from *Catharanthis roseus* into *Petunia hybrida* offers potential for resistance against the whitefly *Bemisia tabaci* and related pests (Thomas *et al.*, 1996). Expression of the gene in cotton is a goal of this research (J. Thomas, personal communication). However, one drawback may be that expression of this enzyme [in potato *Solanum tuberosum* L.)] results in suppression of phenolic biosynthesis (Yao *et al.*, 1995). The altered metabolism has the disastrous consequence of greatly increasing the susceptibility of potato to the pathogen *Phytophtora infestans* (Yao *et al.*, 1995). This illustrates an important concept; whenever plant metabolism is genetically redirected towards expression of novel gene products or towards overexpression of incumbent genes, there is a metabolic cost. This cost may translate to a reduction in primary metabolism resulting in reduced growth and yield and/or in altered secondary metabolism potentially causing increased susceptibility to other pest organisms. Trade-offs among various plant functions have been widely investigated and are beyond the scope of this chapter. For "trade-offs" or "costs" related to defenses readers are refereed to Gershenzon (1994), who reviewed chemical defenses in general, and Baldwin and colleagues, who emphasised inducible defenses (Baldwin and Ohnmeiss, 1994; Baldwin and Schmelz, 1994; Baldwin *et al.*, 1994; Ohnmeiss and Baldwin, 1994). Costs associated with improved resistance to herbivores in cotton are further discussed in section 3.3.

3.2   Tolerance

Plant recovery after damage depends on various physiological and morphological mechanisms that are at the core of this section. Nevertheless, due to the wide range of arthropod species that feed on cotton crops, general plant responses to damage are not straightforward. Differences in timing of attack, distribution of individuals on the plant, production of toxins, and feeding habit are some of the pest-specific characteristics that may generate variable plant responses. Despite this diversity, generalisations are necessary and possible, as shown by McNaughton (1983b), Boote *et al.* (1983), and Johnson (1987) in general and by Gutierrez *et al.* (1981), Sadras (1995), and Matthew (1994a) in cotton. Gutierrez *et al.* (1981) demonstrated that separating the effects of *Anthonomus grandis* and *Heliothis zea* was irrelevant to predicting final yield of cotton crops. Similarly, Sadras (1995) analysed cotton responses to reproductive damage and found that changes in the patterns of dry matter and nitrogen partitioning after loss of squares and bolls could account for most of the plant responses to insects, irrespective of the species involved. Considering the main organs attacked by cotton's major pests, Matthews (1994a) grouped quite diverse arthropod herbivores into eight categories.

The phytocentric focus of our chapter justifies the approach of dealing with groups of pests rather than with individual species. Boote *et al.* (1983) classified pest effects on plant growth into seven groups. The categories in this classification are not mutually exclusive; spider mites, for instance, are both photosynthetic rate reducers and leaf senescence accelerators (Sadras and Wilson, 1997a). Johnson (1987) grouped pests into two larger groups: those affecting radiation interception and those affecting radiation-use efficiency. Johnson's (1987) approach is particularly useful for models based on the resource capture concept (Rossing *et al.*, 1992; van Emden and Hadley, 1994).

The classifications of Boote *et al.* (1983) and Johnson (1987) provide (i) an interface to link pests and crop simulation models, and (ii) a scheme to analyse the main effects of pests on crop growth and yield. Importantly, both classifications are inappropriate to account for two major types of damage caused by common cotton pests: induction of shedding of reproductive organs, and reduction in lint quality. Two additional categories are proposed to account for these types of damage. Damage to seed is discussed together with lint quality (section 3.2.8), while effects of pests on oil and protein content of seeds (*e.g.* Roussel *et al.*, 1951; Wilson, 1993; Sadras and Wilson, 1996) are not considered explicitly.

For its greater detail, the classification of Boote *et al.* (1983) with our two additional categories has been used as a framework for this section; references to Johnson's (1987) classification will be made for comparative purposes when relevant.

3.2.1   Tissue Consumers

This section deals with tissue consumers that affect vegetative meristems and leaves. Pests that damage stems and roots are briefly considered in the group of turgor reducers (section 3.2.7), and pests that feed on square and young bolls in the group of abscission inducers (section 3.2.2). Pests that feed on sown-seed are included in the stand reducers group (section 4.2). Pests feeding on old bolls and developing seed are discussed in section 3.2.8.

Vegetative Meristems. Early in the season, thrips, mirids, and lepidopteran pests feed on cotton vegetative meristems causing, in many cases, the death of the growing apex. Yield response to vegetative bud damage in cotton ranges from considerable loss to moderate increase (Lane, 1959; Brook *et al.*, 1992b, Da Nóbrega *et al.* 1993; Evenson, 1969; Bishop *et al.*, 1977; Heilman *et al.*, 1981; Sadras, 1996a). A similar range of yield responses to the loss of the apical bud has been reported for many other species (Keep, 1969; Aarssen and Irwin, 1991). Interestingly, much of the debate about whether herbivores may have beneficial effects on plants (section 2.1) has been stimulated by studies dealing with apical damage (*e.g.* Paige and Whitham, 1987; van der Meijden, 1990; Aarssen and Irwin, 1991; Aarssen, 1995).

Dale and Coaker (1958) quantified the effects of feeding by *Lygus vosseleri* on the number and size of cells in cotton apices, separated mechanical and chemical (*i.e.* mediated by the saliva of the bug) effects of *Lygus* feeding, and assessed the insect pressure necessary to kill the meristem. The main recovery mechanism after loss of vegetative buds involves release of apical dominance (or "primigenic dominance", Bangerth, 1989) and activation of axillary buds. Aarsen (1995) presented a general discussion of the importance of apical dominance as a trait for recovery after vegetative bud loss, while ecologists at Lund University (Tuomi *et al.*, 1994, Nilsson *et al.*, 1996) developed a mathematical model to quantify plant responses to the loss of vegetative buds that could be modified for use in cotton studies. Using this model, they showed that selection will favour intermediate phenotypes having both dormant and active meristems in environments where risk of herbivory varies from year to year (Nilsson *et al.*, 1996). This agrees with preliminary experiments showing that intermediate degrees of apical dominance may maximise resistance to insects in cotton (Sadras and Fitt, 1997b).

Changes in the structure of the cotton plant induced by loss of vegetative buds have been described by Evenson (1969, his Table 3) and Heilman *et al.* (1981, their Fig. 4). Recovery of plants that are damaged during the reproductive stage, according to Lane (1959), relies more on greater elongation of existing fruiting limbs rather than on the growth of lateral branches. Thus, the great ability of cotton plants to recover after the loss of vegetative buds is the reflection of a substantial structural plasticity (*sensu* Rosenthal and Kotanen, 1994) that results from a large number of secondary meristems which are easily activated

after damage of the dominant apices (*cf.* Tuomi *et al.*, 1994). The concepts of branch autonomy and modular growth, that regard plant growth as the iteration of basic units with varying degrees of physiological integration, are also potentially useful to understand cotton structural changes after loss of vegetative buds (White, 1979; Franco, 1986; Hardwick, 1986; Spruegel *et al.*, 1991; Sachs *et al.*, 1993; Room *et al.*, 1994; Farnsworth and Niklas, 1995). Using these concepts, Room and colleagues developed a model of the cotton plant that can be used to investigate plant-herbivore interactions (Room *et al.*, 1994, 1996).

Some delay in maturity may occur in bud damaged cotton (*e.g.* Bishop *et al.*, 1977). If so, interactions between early (vegetative) and late (reproductive) damage can be important. For instance, Watts (1937) proposed that in "average and severe boll weevil years the delay in fruiting caused by thrips injury becomes of particular importance because much of the later fruit that otherwise would mature can be destroyed by the boll weevil."

In addition to the responses of individual plants to vegetative bud damage, population-level mechanisms may contribute to crop recovery (section 4).

Leaf. Lepidopterous larvae, Orthoptera, thrips, beetles, and other insects can reduce cotton leaf area at different crop stages (Gutierrez *et al.*, 1975; Harp and Turner, 1976; Bishop *et al.*, 1978; Forrester and Wilson, 1988; Quisenberry and Rummel, 1979; Rummel and Quisenberry, 1979; Russell *et al.*, 1993; Matthews, 1994a; Sadras and Wilson, 1998). Growth reduction due to this type of damage can be explained in terms of reduction in light interception.

The ability of the cotton crop to tolerate leaf area loss was well demonstrated by Lane (1959) who found almost no yield reduction after leaf losses up to 20% of controls. Depending on the stage of the crop, losses up to 75% of the total leaf area did not affect yield. Very extreme treatments that affect seedling survival, *i.e.* complete defoliation shortly after emergence, are required to severely reduce yield of cotton crops (Longer and Oosterhuis, 1995). Many studies confirmed Lane's early finding that cotton can indeed tolerate relatively severe defoliation with little yield reduction (Gutierrez *et al.*, 1975; Harp and Turner, 1976; Ferino *et al.*, 1982; Kerby and Keely, 1987; Russell *et al.*, 1993; Wilson *et al.*, 1994). Since changes in leaf area do not necessarily translate into changes in growth, relative leaf damage is probably not a useful measure of genetic resistance to thrips in cotton (Quisenberry and Rummel, 1979; Rummel and Quisenberry, 1979).

The mechanisms of cotton tolerance to leaf area loss remain speculative due to the fragmentary characterisation of the dynamics of plant recovery after damage (Kerby and Keely, 1987; Longer *et al.*, 1993; Gutierrez *et al.*, 1975; Russell *et al.*, 1993; Longer and Oosterhuis, 1995). In well-developed crops (leaf area index >3) moderate reductions in leaf area may have negligible effects on the amount of light intercepted by the canopy, and therefore growth and yield should not be affected. The spatial pattern of leaf damage, however, may have an important effect on the actual crop response, as discussed below. In cases of early damage and/ or severe defoliation that significantly reduces light interception with respect to undamaged crops, several mechanisms can contribute to attenuate the effects of defoliators. First, changes in partitioning, *viz.* increase in leaf area/leaf weight ratio and increase in leaf weight/shoot weight ratio may contribute to leaf area recovery. Second, new leaf addition can compensate, at least partially, for leaf loss due to insects and other agents such as hail (Lane, 1959; Bishop *et al.*, 1978). Third, if leaves are involved in apical dominance (Töpperwein, 1993; McIntyre, 1997), then enhanced branching following defoliation might also be a factor in the recovery of damaged crops. Importantly, enhanced branching after the release of apical dominance could contribute not only to the recovery of leaf area but also to a faster production of squares (Sadras and Fitt, 1997a). Fourth, there may be an increase in the photosynthetic rate of undamaged leaves in a damaged plant and/or in undamaged areas of damaged leaves, *i.e.* compensatory photosynthesis (Trumble *et al.*, 1993). Compensatory photosynthesis is further discussed in section 3.2.6.

For a given amount of leaf loss, the pattern of insect distribution within the plant and the pattern of feeding within the leaf both may affect the plant's photosynthetic response. The pattern of damage *among* leaves in a plant is important because leaf position influences the relative contribution of individual leaves to total plant photosynthesis (Constable and Oosterhuis, Chapter 7 this volume). Marquis (1988) showed in *Acer pennsylvanicum* that removal of 25% of the area from leaves subtending infrutescences reduced seed production of those infrutescences, while removal of area from leaves close, but not subtending, the infrutescence had no influence on seed production. To account for this kind of effect, the pattern of distribution of insects and of damage within the plant needs to be known (*e.g.*, Bishop *et al.*, 1978). The pattern of damage *within* the leaf, *e.g.* notches in leaves by weevils, perforations by borers, skeletonising by beetles, may also affect the photosynthetic rate of the remaining leaf tissue, probably because of variations in the proportion of wounded tissue (Morrison and Reekie, 1995). For a given insect, changes in the pattern of within-leaf damage with ontogeny could also be an important source of variation in whole leaf and plant responses (Zangerl and Bazzaz, 1992, their Fig. 16.4).

But leaves are more than sources of organic carbon. They are also the main site of nitrogen assimilation and a large reservoir of organic nitrogen in the cotton plant. Leaf properties (Bondada *et al.*, 1996) and leaf area may also affect the water economy of the crop. Leaf loss, according to McNaughton (1983a), may improve the water relations of remaining tissue and tissue newly synthetised due to a greater root/shoot ratio. For crops relying on stored soil water, early leaf loss may reduce transpiration and increase the proportion of soil available water during the fruit-growth period affecting therefore the crop's harvest index

(Passioura, 1977; Richards and Townley-Smith, 1987; Sadras and Connor, 1991). A full analysis of the effects of defoliators on cotton growth and yield should, therefore, take into account their effects on the carbon, nitrogen, and water economies of the crop.

Some ecologists concerned with plant-herbivore relationships have proposed that, in some cases, defoliation may increase plant growth and fitness (Verkaar, 1988). More conservatively, McNaughton (1983a) stated that "tissue destruction is rarely, if ever, translated monotonically into a proportional reduction of final yield". The previous discussion showed that McNaughton's general statement is also valid for cotton crops, and highlighted the importance of the largely unknown mechanisms underlying the capacity of cotton plants to recover after leaf loss.

3.2.2   Abscission Inducers

Key cotton pests feed preferentially on reproductive structures, which usually shed after damage (Hearn and Fitt, 1992). Plant and crop responses to reproductive damage have been recently reviewed (Sadras, 1995). Briefly, yield responses of crops that suffered reproductive damage, in comparison to protected controls, range from moderate gains to severe losses (Sadras, 1995, his Table 1). Loss of reproductive organs induces dramatic changes in the partitioning of plant resources, and in the structure and phenology of the crop. Damaged crops usually have, in relation to undamaged controls: (i) more carbon and nitrogen stores in vegetative organs, which are the primary result of increased vegetative growth (*i.e.* more dry matter in roots, stems, and leaves); greater concentrations of labile carbohydrates and organic nitrogen could also contribute to the greater stores in damaged plants; (ii) a potential to maintain greater radiation-use efficiency, due to changes in plant morphology that improve the distribution of light in the canopy (Sadras, 1996b), and (iii) a potential to intercept more light due to an extended period of leaf expansion (Brook *et al.*, 1992c; Sadras, 1996a).

Altogether, these changes suggest that damaged crops could have a larger yield potential than undamaged ones. The extent to which this potential is achieved depends on (a) the time available for recovery, and (b) the growing conditions. Different combinations of yield potential, as affected by reproductive damage, time for recovery, and growing conditions explain the wide range of yield responses of crops subjected to reproductive damage (Sadras, 1995, his Fig. 2; Sadras, 1996a).

Importantly, reproductive damage may affect not only the yield but also the earliness, harvestability and quality (see section 3.2.8) of cotton crops. The primary benefits sought in short-season cropping systems could therefore be lost by this type of damage. Interactions between plant responses to arthropod damage and growing conditions are further discussed in section 5.

3.2.3 Photosynthetic Rate Reducers

Photosynthesis of pest damaged cotton has received little attention in comparison to the many studies dealing with photosynthesis of healthy plants. Two recent studies investigated the effects of spider mites (*Tetranychus* spp.) on cotton photosynthesis at the cytological and leaf levels (Bondada *et al.*, 1995) and at the crop level (Sadras and Wilson, 1997a). Bondada *et al.* (1995) showed that spider mites disrupt leaf photosynthesis by reducing both stomatal and mesophyll conductance. At the crop level, these effects are reflected in reduced radiation-use efficiency, increased canopy temperature and reduced leaf nitrogen content due to mites (Sadras and Wilson, 1997a). Negligible responses of radiation-use efficiency to mites until a threshold of mite damage was achieved suggests some degree of compensatory photosynthesis that needs further evaluation (Sadras and Wilson, 1997a). Quantitative relationships between radiation-use efficiency and an index of mite damage have been developed that could be used to incorporate the effects of spider mites into cotton simulation models (Sadras and Wilson, 1997a). Further information on the general effects of arthropods herbivores on leaf gas exchange can be found in the review by Welter (1989).

3.2.4 Leaf Senescence Accelerators

Heavy infestations of *Tetranychus urticae* (Sadras and Wilson, 1997a) and *Bemisia tabaci* (Baluch, 1988) can accelerate leaf senescence. Faster leaf senescence implies reductions in leaf area duration (Watson, 1947), and potential reductions in light interception, growth, and yield (Monteith, 1977).

The effects of mites on crop growth are primarily related to reductions in photosynthesis but accelerated leaf senescence could further reduce growth by reducing light interception (Sadras and Wilson, 1997a). Mites have the potential to induce senescence. The mechanisms by which mites and other pests induce fast senescence and the existence of any tolerance mechanism are both unknown. We can speculate that new leaf addition and compensatory photosynthesis could partially compensate for this type of damage. However, given the high levels of infestation necessary for significant defoliation to occur, it might be too late for any compensation to be relevant, at least in the case of crops severely infested with mites.

Important changes in shoot nitrogen distribution can follow mite-induced leaf senescence. First, the process of nitrogen depletion typical of senescing leaves is accelerated in mite infested plants, with the consequent decline in leaf nitrogen concentration (Sadras and Wilson, 1997a, c). This reduction in leaf nitrogen concentration can be regarded as a negative feed-back on mite colonies, whose rapid decline in field crops coincided with leaf nitrogen concentrations dropping below 3.5-4.2% (Sadras and Wilson, 1997c). This agrees with independent laboratory tests showing that fecundity and developmental rate of *T. urtichae* are negatively affected when mites are fed cotton leaves with nitrogen concetration below *c.* 4% (Wilson, 1994). Reduced leaf nitrogen concentration, along with the red discolouration, may make mite infested plants less attractive to other herbivores. In parallel with the rapid reduction of leaf nitrogen, stems and fruits of mite infested cotton had greater nitrogen concentrations than uninfested controls (Sadras and Wilson, 1997c). This could may make mite-infested plants more suitable for a range of Lepidoptera, Coleoptera, and Hemiptera that feed preferentially on reproductive structures. A more detailed discussion of the many possible influences of biotic and abiotic factors on cotton responses to herbivores is presented in section 5.

3.2.5 Light Stealers

Homoptera (*e.g. Aphis gossypii* and *Bemisia tabaci*) produce honeydew and spider mites produce webs. Both honeydew with its fungal colonies and mite webs may reduce the light available for leaf photosynthesis. The importance of these barriers for light transmission has not been assessed in cotton and the mechanisms for tolerance, if any, are unknown. To separate the effects of honeydew from those related directly with the feeding activity of insects, experiments using artificial honeydew have been done with other species. The results of these experiments are ambiguous, probably reflecting the dependence of responses to honeydew on growing conditions and plant species. Hurej and van der Werf (1993) found no effects of artificial honeydew on sugar beet (*Beta vulgaris* L.) growth while Rossing and van de Wiel (1990) reported that wheat (*Triticuum aestivum* L.) dark respiration and mesophile resistance both increased in treated plants with respect to controls under hot-dry conditions but not under moderate temperature and humidity. It could also be that part of the differences between these experiments are related to variation in the chemical composition of honeydew, which depends on both the insect and plant species (Hendrix *et al.*, 1992). Measurements of light transmission by Wood *et al.* (1988) showed that sooty mould fungus grown on aphid honeydew could block up to 98% of incident PAR. Direct assessment of the effects of honeydew on cotton photosynthesis are necessary. Compensatory photosynthesis, we speculate, could be a relevant mechanism of tolerance to this type of damage (section 3.2.6).

3.2.6 Assimilate Sappers

Sucking pests absorb phloem or xylem contents from different organs. In the case of insects that feed preferentially on reproductive structures, the induction of shedding has probably more important effects on crop growth and yield than the actual loss of assimilates (section 3.2.2, cf. also Prokof'ev and Igamberdieva, 1971). This section concentrates on sucking pests that feed on vegetative organs.

Pollard (1973) described the damage caused by aphids at the tissue and cellular levels. At the plant and crop levels, sucking animals that feed on vegetative structures can be considered as additional "sinks" for carbohydrates and as such they are likely to affect source/sink relationships in the host plant (Crawley, 1983). The effect of sucking insects on cotton source/sink relationships has not been investigated. The notion that sucking insects are additional sinks to the plant implies that they have the potential to reduce vegetative and reproductive growth by diverting plant assimilates. Assuming (i) a crop growth rate of 15 g dry matter m^{-2} d^{-1} (Hearn and Constable, 1984), (ii) the energetic requirements for the synthesis of cotton dry matter given by Wall *et al.* (1994), and (iii) an intake of 50-400 mg sugar $aphid^{-1}$ day^{-1} (Hurej and van der Werf, 1993) we calculated that a reduction in growth rate of 10% would require at least 5000 aphids m^{-2}. Despite the many assumptions involved in this estimate, it suggests that a significant drain of carbohydrates in a well developed cotton crop could only occur at very high aphid densities and this is consistent with Smith (1992) who pointed out that "economic damage caused by the cotton aphid today is in dispute; yield losses are rarely definable". In addition, increased photosynthetic rate due to the sink activity of sucking pests might attenuate to some extent the losses due to sucking pests (Crawley, 1983). Evans (1993) summarized a number of studies showing that increased rate of carbon fixation may follow enhanced sink activity. The experiments of Quisenberry *et al.* (1994) suggest that photosynthesis of cotton leaves is responsive to source/sink ratios but direct evaluation is needed. Meyer (1993) and Meyer and Whitlow (1992) found in goldenrod (*Soldago* spp.) that plant growth and photosynthesis of leaves produced after insect removal were unaffected by a phloem-sap feeding aphid, but were severely reduced by a xylem-sap feeding spittlebug. Studies similar to the ones by Meyer and Whitlow (1992) are necessary in cotton.

In addition to sugars, sucking pests remove other plant nutrients, and they may also disrupt phloem-sap transport. Changes in C/N ratios in leaves infested with *Bemisia tabaci* (Baluch, 1988) indicate differential effects on the rate of synthesis, turnover and/or transport of carbohydrate and N-compounds in damaged cotton plants. Changes in C/N ratios, in turn, may have dramatic effects on important physiological processes from regulation of gene expression (Stock *et al.*, 1990) to phenological development of whole plants (Trewavas, 1985).

Importantly, indirect damage caused by succivorous insects, *i.e.* damage associated with (i) insect toxins, hormones and pathogens carried in the watery saliva (Bell, Chapter 18, this volume), and (ii) honeydew that potentially affects lint quality (section 3.2.8) can be comparatively more important than the damage caused by the actual drain of sugars and minerals. This is reflected in current recommendations for management of aphids in Australia: while 90% of infested plants is suggested as a threshold before boll opening, it drops to 10% during boll opening due to

the potential for severe downgrading of the lint value (Pyke and Brown, 1996).

In contrast with the previous discussion, in which effects of sucking insects has been implicitly considered deleterious to the plant, Owen (1980) has proposed some beneficial effect associated with the feeding activity of aphids (section 2.1). He suggested the following sequence to explain how plants may benefit from their association with aphids: (i) plants release "surplus" sugar by enlisting the "help" of aphids, (ii) free-living nitrogen-fixing bacteria develop beneath the aphid-infested plant, and (iii) more nitrogen is available for the infested plant. Owen's proposal is consistent with the hypotheses that (a) photosynthesis rarely limits growth (Went, 1974) and (b) that terrestrial plants have evolved a wide range strategies to dispose of excess carbohydrate (Thomas, 1994).

3.2.7   Turgor Reducers

In comparison with other crops, cotton suffers little from root-feeding animals (Matthews, 1994a). The cotton stem weevil (*Apion soleatum*), whose larvae feed on vascular tissues of main-stems and branches, can potentially reduce the growth and yield of cotton in the eastern producing regions of South Africa (Bennett, 1993). Vascular diseases that disrupt cotton water and nutrient economies are much more widespread and have the potential to cause severe yield reductions (*e.g. Rhizoctonia* spp., *Fusarium* spp., *Verticillium* spp.). For details of cotton responses to altered functioning of its root and vasular systems readers are referred to chapters in this book dealing with diseases (Bell *et al.*, Chapter 18) and several aspects of root growth (McMichael *et al.*, Chapter 6), mineral nutrition (Mullins and Burmester; Chapter 9; Hodges and Constable, Chapter 14), and water relations (Hake and Grimes, Chapter 23) in healthy plants exposed to stresses.

3.2.8   Lint Quality Reducers and Seed Consumers

This section deals with pests that typically affect cotton crops late in the season affecting, therefore, lint quality, seeds, and boll opening. Indirect effects of pests on lint quality are also considered.

Lint quality can be affected by pests that (i) affect plant growth and development, and/or (ii) stain or otherwise damage cotton fibres. Fibre quality depends on plant and environmental factors, as discussed by Turley and Chapman (Chapter 29), Haigler (Chapter 4), and Bradow and Bauer (Chapter 5) (this volume). In general, arthropods that reduce crop photosynthesis have the potential to affect lint quality, as illustrated by studies with spider mites (Roussel *et al.*, 1951; Canerday and Arant, 1964a, b; Leigh *et al.*, 1968; Duncombe, 1977; Wilson, 1993). Developmental delays, such as those caused by early-season vegetative damage (section 3.2.1) and loss of reproductive structures (section

3.2.2) could also affect lint quality due to larger proportions of bolls growing under less favourable environmental conditions. Reductions in fibre quality due to damage by *Heliothis* spp. that delays fruit growth (Wilson, 1981) illustrate this point.

Whiteflies and other sucking insects that excrete large amounts of honeydew are responsible for sticky cotton, characterised by little drops of honeydew - often crystallised - that are not eliminated during ginning. Saprophytic fungi that grow on honeydew further reduce lint quality. Details about sticky cotton can be found in Hector and Hodkinson (1989), Butler and Henneberry (1994), and Leclant and Deguine (1994). Lint quality can also be affected by cotton stainers *e.g. Dysdercus* spp. (Broodryk and Matthews, 1994) and by *Pectinophora* spp. (Ingram, 1994).

According to Matthews (1994a) the common stainer bugs and the cotton seed bugs, *Oxycarenus* spp., are able to feed on undamaged cotton seed, whilst scavengers of minor importance can often be found after primary damaged has occurred. Late seed damage is unlikely to affect yield seriously, but reductions in lint quality associated with these insects could be severe.

Whilst squares and young bolls usually shed after damaged (section 3.2.8), older bolls damaged by lepidopteran and other pests are normally retained in the plant. Damage to older bolls is rather localised and yield losses could be limited to the damaged locules which may remain closed at maturity. More often, however, the boll cavity is invaded by secondary fungi and the whole boll could be lost due to rotting (Matthews, 1994b).

Reduced seed number and/or viability is obviously important for plant fitness. We speculate, therefore, that cotton plants might have evolved some tolerance mechanisms for this type of damage. Irrespective of whether such recovery mechanisms exist, the limited time available for recovery (Sadras, 1995) makes them of restricted value from the agronomic viewpoint.

3.2.9   Summary

Crop yield reduction due to pest damage can be associated with reduction in growth, reduction in harvest index, or both. Reductions in growth, in turn, may be the result of less light interception and/or lower radiation-use efficiency. Many of the pests examined above affect growth by reducing radiation interception (tissue consumers, leaf senescence accelerators, light stealers, turgor reducers) or by reducing radiation-use efficiency (assimilate sappers, photosynthetic rate reducers, turgor reducers). In contrast, abscission inducers normally affect yield, despite increases in shoot growth in some cases, due to reductions in harvest index.

Mechanisms of tolerance related to damage that reduces growth include production of new leaves and compensatory photosynthesis. Increased harvest index can also be potentially important. Studies dealing with changes in dry matter partitioning after damage by different pests are scarce. Cotton crops severely damaged by spider mites had a significantly lower harvest index than undamaged controls but allometric analyses showed that reduction in harvest index was associated with small plant size rather than with true changes in partitioning (Sadras and Wilson, 1997b). Overall, the cotton crop seems to have a considerable ability to tolerate damage by herbivores but the actual mechanisms involved are largely unknown. Compensatory photosynthesis and changes in dry matter partitioning after damage can be indicated as two aspects of cotton physiological responses to pests that deserve closer attention.

3.3   Interactions Between Avoidance and Tolerance

Our aim in this section is not to review the research and achievements of breeding for improved cotton resistance to herbivores; this has been done recently by a number of authors (see Introduction). Instead, we will concentrate on traits and interactions between traits that, we believe, have been neglected.

Most research and breeding efforts on cotton resistance to pests have concentrated on avoidance traits. Thomson (1987) pointed out that besides these attributes, "it seems to be almost universally overlooked that increasing yield itself constitutes a form of breeding for host plant resistance". To highlight this point, Thomson (1987) compared two hypothetical cultivars, A and B, of different yielding ability but otherwise similar in their sensitivity to insects. The higher-yielding cultivar B can absorb nearly twice the insect damage as A can before it yields less than A. With the exception of Thomson's concept of "yield as a resistance factor", no consideration has been made of traits related to plant growth (or regrowth after damage) as factors in cotton resistance to pests. Only recently, studies have been designed to assess the importance of tolerance traits in cotton (Sadras and Fitt, 1997a, b). Preliminary experiments suggest that (i) considerable variability exists among *Gossypium* genotypes in their recovery capacity, and (ii) recovery capacity may be an important component of overall resistance to insect pests in the field.

The putative trade-off between allocation of resources to growth and defenses was investigated in many ways: theoretically (Bazzaz *et al.*, 1987; Tilman, 1990; Herms and Mattson, 1992), using simulation models (Basey and Jenkins, 1993; Yamamura and Tsuji, 1995) and in empirical studies including both intra- and inter-specific comparisons (*e.g.* van der Meijden *et al.*, 1988; Coley, 1988; Bryant *et al.*, 1989; Jing and Coley, 1990; McCanny *et al.*, 1990). Negative relationships, as predicted by theory, have been found in many studies whereby plants with high level of defenses had a limited capacity for growth (*e.g.*, Coley, 1988) or regrowth after damage (van der Meijden et al, 1988). Nevertheless, positive association (Bryant *et al.*, 1989) and no association (McCanny *et al.*, 1990) between growth and defenses have also been reported. Further references to relationships between defense and tolerance can be found in a recent review by Zangerl and Bazzaz (1992).

Cotton breeders are, of course, aware of the yield "penalties" that could be associated with enhanced plant resistance to pests (Thomson, 1987) and this was illustrated in the work by Wilson (1987). Cotton lines with enhanced resistance to insects yielded more than a "susceptible" control under high insect pressure but, in general yielded less than the control under low insect pressure. This suggests a "cost", in terms of yield, associated with enhanced resistance to herbivory. Moreover, for the range of resistance traits from 0 to 3, the number of resistance traits accounted for 74% of the variation in yield under low insect pressure. In contrast to the work by Wilson (1987), no association was found between chemical defenses and regrowth capacity in a collection of 25 *Gossypium* genotypes (Sadras and Fitt, 1997b). Furthermore, okra-leaf varieties combine avoidance characteristics with a very good recovery capacity derived, in part, from a high squaring rate (Thomson, 1994). Altogether, the information available for cotton suggests that (i) negative linkages may exist between tolerance and avoidance traits, but (ii) breakdown of the putative links seem feasible and genotypes could be bred that combine both kinds of traits.

Combination of tolerance and avoidance strategies could offer broader and ecologically more stable solutions to cotton's pest problems. On the one hand, the capacity of the crop for yield compensation can be severely limited in extreme environments, *i.e.* in sites with very high (Brook *et al.*, 1992a, b, c) or very low yield potential (Sadras, 1995, 1996c). Compensation is also of limited value for damage that occurs very late in the season (section 3.2.8). Thus, provided avoidance traits are effective over a wide range of conditions, cropping systems relying on both tolerance and avoidance could have advantages over those emphasising tolerance strategies. On the other hand, chemical defenses are, by definition, negative to the fitness of target pests. The negative effect of defenses on pests has the potential for the selection of resistant individuals that, eventually, could restrict the effectiveness of such defenses (see 3.1.2: *"Biotechnology and plant resistance"*). Since morphological defenses and phenological escape have the potential to affect the fitness of target pests, they could also trigger selective processes similar to those described for chemical defenses. To the best of our knowledge, development of resistance to morphological traits or escape strategies have not been reported. The case of the adaptation of rootworm populations to rotations the U.S.A. cornbelt discussed before, however, illustrates how pests could develop resistance to this kind of strategy (Section 3.1.1). Combination of tolerance and avoidance strategies would therefore be a more stable option than strategies relying solely on avoidance.

In summary, while avoidance traits are likely to remain a central component of plant resistance to herbivory, explicit consideration of recovery capacity could be a worthwhile aim for breeding programs. Importantly, a better understanding of the mechanisms of cotton resistance to herbivory are also essential for the development of improved pest management practices. In addition to ongoing research on avoidance traits, research is needed to: (i) fully assess the degree of intraspecific variability in recovery capacity in response to different intensities, types and times of damage, (ii) determine the inheritance of these traits, (iii) develop screening techniques suitable for breeding purposes, and (iv) investigate the link/s between tolerance and other traits relevant for resistance to herbivory and overall agronomic performance.

4. RESISTANCE TO HERBIVORY: AT THE POPULATION LEVEL

4.1   Avoidance

Injured plants may release volatile substances that play defensive roles. These include both elicitors of plant defenses and chemicals that attract natural enemies, as discussed before (section 3.1.2 *"Induced Defenses"*). In this section, we want to briefly emphasise that, due to the volatile nature of these "infochemicals", they could be regarded as defenses at the population level. This is because the benefits of such compounds are not restricted to the injured plant that has produced them, but could also be extended to undamaged neighbours.

The study of Bruin *et al.* (1992) showed that undamaged cotton plants can gain protection against mite herbivory by exposure to compounds released by mite-injured plants. They found that spider mites had oviposition rates on leaves previously exposed to volatiles from infested plants were 10% lower than on untreated controls (P = 0.008), and that (ii) predatory mites had a preference for volatiles related to uninfested plants or leaves that had been exposed to volatiles from mite-infested plants, compared to controls (P < 0.001).

The fact that volatile chemical signals operate at the population level is important in the consideration of the putative costs associated with their production. This was highlighted in a study of communication between the first (plants) and third ("beneficial") trophic levels by Godfray (1995), who pointed out that any signalling system in which there exists the possibility of a conflict of interest between signaller and receiver will require significant costs for evolutionary stability. Certainly more research is needed in this area.

In addition to plant traits involved in "attracting" predators and parasitoids, a number of morphological traits (*e.g.* domatia, leaf dispersion, petiole length, branching pattern) can influence (i) the ability of the plant to *retain* beneficials and (ii) the *accessibility* of herbivores on the plant to beneficials (Marquis and Whelan, 1996). Agrawal and Karban (1997) showed that the presence of leaf domatia may increase predator numbers, reduce populations of herbivorous mites, and enhance cotton yield. Thus, the ability of plants to attract beneficials should not be considered in isolation of such morphological traits that are likely to influence the overall effectiveness of herbivore control by beneficials.

4.2   Tolerance

Stand reductions can result from the activity of soil-dwelling arthropods but also from seedling diseases, or severe defoliation and meristem damage by early-season pests or hail. Interactions between these factors can be important, as illustrated by the work of Colyer *et al.* (1991) who showed that thrips infestations may increase both the severity of cotton seedling disease and stand reduction.

Tolerance to stand reduction depends on the capacity of the surviving plants to fill the gaps left by dead neighbours (*e.g.* Bardner and Fletcher, 1974). This response fits, in a broad sense, the definition of population-level compensation of Crawley (1983) (section 2.3). The detailed study of Hearn (1972) serves to illustrate the capacity of cotton for this type of compensation. For crops grown under extreme conditions of yield potential, a 10% reduction in maximum yield required (even) stand reductions of more than 70% with respect to the optimal density.

Matthews (1994b) indicated that, as a general rule, seedling pests are not important in Africa and pointed out that some compensatory growth often occurs if the plants are closely spaced (cf. also Pearson, 1958). When stand reduction occurs late in the season, as described by Tiben *et al.* (1990) for cotton crops attacked by the termite *Microtermres najdensis*, crop recovery is much more limited. Timing and heterogeneity of damage are therefore important factors influencing the capacity of cotton crops to compensate for stand reduction.

Significant compensation at the population level can be expected not only after death of plants but also when selective damage affects the competitive relationship between neighbouring plants. This form of compensation was investigated in cotton crops subjected to three treatments: (i) undisturbed controls, (ii) uniformly damaged, in which all plants were damaged, and (iii) non-uniformly damaged, in which every second plant was damaged (Sadras, 1996c). Damaged plants had their vegetative buds manually removed to simulate damage by early-season pests. Removal of vegetative buds did not reduce yield per unit ground area. In uniformly damaged crops, compensation was essentially the result of profuse branching after release of bud dormancy (section 3.2.1). In non-uniformly damaged crops, population level mechanisms acted that involved strong plant-plant interactions. Undamaged plants grown alongside damaged neighbours accumulated more shoot and tap root biomass and produced more seedcotton than undamaged plants in uniform crops. Changes in competitive relationships (Crawley, 1983) as well as early detection of and response to neighbour status (Aphalo and Ballaré, 1995) were likely involved in these responses (Sadras, 1996c; cf. also Watts, 1937). Similarly, Hurej and van der Werf (1993) reported that neighbouring plants of aphid-infested sugar beet plants were heavier than neighbouring plants of healthy controls.

Interactions between neighbours after non-uniform reproductive damage could also be expected. In cotton, as in other indeterminate plant species, fruit loss often counteracts the slowing down of vegetative growth that usually occurs during the stage of active reproductive growth (Section 3.2.2). Due to the relative increase in vegetative biomass, leaf area and plant height after fruit loss, plants that have suffered reproductive damage could be better able to intercept light and acquire soil resources than undamaged plants. If so, plants with damaged neighbours may grow less than their counterparts with smaller, undamaged neighbours (Sadras, 1997b). A study designed to test this hypothesis showed that : (i) as expected, damaged target plants had greater leaf area and more vegetative dry matter than undamaged targets; (ii) neighbour status did not affect vegetative growth; (iii) neighbour status had a substantial, asymmetric effect on the reproductive growth of target plants: while neighbour status did not affect the productivity of damaged targets, it had a significant effect on the production of mature fruit of undamaged targets; (iv) undamaged targets with damaged neighbours had 34% (low density) and 56% (high density) less open boll dry matter than their counterparts with undamaged neighbours; (v) the asymmetric response of target plants to neighbour status determined a reduction in the yield of non-uniformly damaged crops that was greater than expected from the additive effects of damage.

In summary, population-level compensation seems to be an important mechanism of tolerance to stand reduction and early-season bud damage. A substantial phenotypic plasticity (Bradshaw, 1965) and the modular organisation of the cotton plant enables damage to be repaired and differentiation to be adjusted to the availability of resources (Hardwick, 1986, see also section 3.2.1.). In contrast, non-uniform reproductive damage may have a two-fold effect in yield: yield may be reduced in the damaged plants and also in undamaged plants flanked by neighbours that are bigger, taller, and leafier after fruit loss. The impact of plant-plant interactions on yield losses due to insects in commercial crops will depend on the spatial distribution of insects and damage. Interactions between neighbouring plants that suffered different types, timings, and/or intensities of damage deserve further research.

5.   INTERACTIONS BETWEEN ARTHROPOD PESTS AND OTHER STRESSES

The previous section outlined the main effects of pests on the physiology and morphology of the cotton plant and described the main avoidance and tolerance mechanisms involved at the molecular, organ, plant, and crop levels. For analytical purposes, those mechanism were mostly considered in isolation of other factors. Plant-pest relationships in the field, however, are strongly influenced by other biotic (*e.g.* weeds, diseases, mycorrhiza, predators, and parasitoids) and abiotic (*e.g.* water and nutrient availability, temperature) factors. These influences are the subject of this section. Our aim is: (i) to illustrate, using selected examples, some of the many possible interactions between these factors, (ii) to discuss, briefly, approaches to investigate these interactions.

5.1   Some Examples

Baumgärtner *et al.* (1986) used a simulation model to investigate interactions between cotton and two pests, *Heliothis* spp and *Bemisia tabacci*. Their simulation experiments indicated that damage caused by *Heliothis* larvae to fruiting structures may increase whitefly numbers at the time of boll opening. This was related to model assumptions, justified by empirical evidence, that (i) reproductive damage extends the period of leaf growth, (ii) a leaf's nutritional value depends on its age, (iii) and host plant quality has an important effect on the population dynamics of whiteflies (Baumgärtner *et al.*, 1986). Other simulation studies on interactions between arthropods and cotton include Gutierrez *et al.* (1975, 1977, 1979ab), Wang *et al.* (1977), Room (1979b), Hearn and Room (1979), Hearn *et al.* (1981), Ives *et al.* (1984), Hearn and DaRoza (1985), Legaspi *et al.* (1989), and Baker *et al.* (1993).

In an interesting study, Simpson and Batra (1983) showed the interactions between air temperature, leaf-feeding insects, sucking insects, scavenger beetles, pink bollworm, and cotton boll rot caused by *Aspergillus flavus*. They showed that *A. flavus* develops better at moderate to high temperatures and that leaf-damaging insect deposits and honeydew promote the fungi's growth. Its transport to bolls is facilitated by scavenger beetles while actual boll penetration and infestation require wounds, such as those caused by pink bollworm. This study illustrates how yield losses due to, say sucking insects, will be influenced by the presence of other living organisms (bollworms, beetles, *A. flavus*) and environmental conditions (temperature) that could contribute to indirect yield losses due to boll rot. Another factor that adds complexity to these interactions is that gossypol can be induced by volatile constituents from leaves infected with *A. flavus* (Zeringue, 1987).

There are two main ways in which weeds and arthropod pests can interact with the crop. The first, and more widely investigated, is through the role of weeds as alternative hosts for, and sources of, pathogens, arthropod pests and beneficials. For instance, the importance of weeds as sources of spider mites has been demonstrated by Brandenburg and Kennedy (1981) for cropping systems in the U.S.A. and by Wilson (1995) for cotton crops in Australia. The second interaction between weeds and pests is related to the changes in cotton competitive ability after damage by herbivores (Sadras, 1997a). As discussed before (Section 3.2.1), early-season loss of vegetative buds transiently delays cotton growth and development and has the potential to reduce its competitive ability. Yield reductions due to weed interference and insect damage, therefore, could be greater than expected from the additive effects of weeds and damage acting separately. A preliminary experiment combining two levels of weed infestation (with and without weeds) and two levels of simulated insect damage (intact plants and plants with vegetative buds removed before squaring) showed a non-additive effect of weeds and damage that accounted for yield losses equivalent to 16% of the yield of control crops.

Interactions between aboveground herbivores and the mycorrhizal mutualists of plants have received little attention but the available data suggest: (i) that severe herbivory reduces root colonization by vesicular-arbuscular (VAM) and ectomycorrhyzal fungi, and that (ii) mycorrhizal fungi could deter herbivores, and interact with fungal endophytes influencing herbivores (Gehring and Whitham, 1994). For instance, larvae of *Helicoverpa zea* and *Spodoptera frugiperda* fed leaves from VAM-infected soybean grew more slowly, took longer to pupate, and had a greater mortality rate than larvae fed on non-mycorrhizal controls (Rabin and Pacovsky, 1985). This kind of interaction deserves to be investigated in cotton.

Hedin and McCarty (1991) showed that the concentration of secondary metabolites, such as gossypol, tannins, anthocyanin, and flavonoids in cotton leaves and squares, can be changed by exogenous plant growth regulators including commercially used cytokinins and mepiquat chloride. The direction, *viz.* decrease or increase in concentration, and magnitude of the change are largely unpredictable, as are most plant responses to exogenous applications of plant growth regulators. The report of Hedin and McCarty, however, is important in that analysis of crop responses to plant regulators in the field should take into account potential changes in plant-animal interactions mediated by changes in concentrations of plant secondary metabolites.

Interactions between insects, water availability and soil fertility have been widely investigated in cotton. In general, conditions that favour crop growth, including frequent irrigation and heavy fertilization, increase the abundance of both herbivorous and predacious arthropods (Joyce, 1958; McGarr, 1942; Mistric 1968; Leigh *et al.* 1970, 1974; Flint *et al.* 1994, 1995; Skinner and Cohen 1994). The overall response of the crop to different combinations of water supply, nutrient availability and pests is, however, difficult to predict due to the complexity of interactions involved. Interactions between the carbon, nitrogen, and water economies of cotton crops after reproductive damage have been outlined in Sadras (1995). Mistric (1968) with boll weevils and Sharma *et al.* (1989) with pink bollworms showed that heavily fertilised cotton: (i) attracted more insects, (ii) had more bolls and squares damaged by insects, but (iii) yielded more than poorly fertilised crops due to increased boll production. Joyce (1958) investigated the interactions between rainfall before sowing, soil and leaf nitrogen and development of jassids, thrips, and whiteflies in cotton crops in Sudan. Leaf nitrogen content and pre-sowing rains were negatively correlated, presumably due to leaching of soil nitrogen. This, together with a positive association between leaf nitrogen and rate of development of insects, resulted in a significant, negative association between rate of insect development and pre-sowing rainfall (Joyce, 1958). McGarr (1942) reported that nitrogen fertilisation increased aphid numbers when cotton was dusted with calcium arsenate but not in the absence of this treatment. Leggett (1992) found that the abundance of cotton insects, both pests and beneficials, was significantly affected by complex interactions between cultivar (Pima vs Upland), irrigation method (drip vs furrow) and sites in Arizona. Sadras *et al.* (1998) found that water deficit enhanced cotton resistance to spider mites in a comparison with well-watered crops. Other studies dealing with water availability-pest interactions include Ogborn and Proctor (1962), Kittock *et al.* (1983), Watson *et al.* (1978), and Ungar *et al.* (1992). Ungar *et al.* (1992) concluded after a five-year experiment that "irrigation scheduling and the control of pests that damage fruiting organs cannot be optimized independently". Leigh and colleagues conclusion after intensive studies (Leigh *et al.*, 1970, 1974) is that "where insect populations are not controlled, a highly complex relationship was found to exist between cotton lint production, vegetative growth, insect numbers, and water and nutritional management".

In addition to studies dealing with cotton, Waring and Cobb (1992) reviewed the general effects of nutrient (N, P, and K) and/or water stress on herbivore population dynamics and Kytö *et al.* (1996) reviewed the effects of soil fertilisation on phytophagous insects and mites on trees. More than 75% of the 450 studies reviewed by Waring and Cobb (1992) showed significant herbivore responses (positive, negative, or non-linear) to plant water and/or nutrient deficits. In general, they concluded, these stresses render plants poor resources for herbivores. However, quantitative and qualitative changes in plant defenses, changes in canopy temperature and effects on the populations of parasitoids and predators, are among the many factors that could be affected by nutrients and water stresses and could, in turn, affect the responses of herbivores to stressed plants, and *vice versa*. In fact, Kytö *et al.* (1996) proposed that enhanced nitrogen availability (i) usually benefits individual herbivores

by improving nutritional quality of the host plants, but (ii) it often has non-significant or negative effects on insects at the population level because it also affects higher trophic levels, *i.e.* parasitoids and predators. Changes in community structure, they suggest, override the effects of nutrients mediated by the improved quality of the host plant. These conclusions may not necessarily apply, however, to cropping systems in which the community of parasitoids and predators is significantly disturbed by the use of insecticides.

5.2 Approaches to Investigate Cotton Responses to Herbivory as Affected by Other Biotic and Abiotic Factors

The "limiting factor" concept is often used in agronomic studies (*e.g.* Paris, 1994) as well as in plant physiological research (Trewavas, 1986). This concept is obviously inappropriate to deal with the kind of interactions illustrated in the previous section. Parallel to the recognition of the restricted value of the "limiting factor" approach to analyse plant growth and development in most field situations (Körner, 1991; Gifford, 1992; Sinclair and Park, 1993; Sadras, 1995) the concept of "multiple-stresses" has been formally developed (Chapin *et al.*, 1987; Mooney *et al.* 1991).

It is tempting, in principle, to think of simulation models as a means to investigate the complex interactions involved in crop-pest relationships. Such models, we believe, could be useful as tools to assist in crop management, but they should be taken very cautiously in relation to their value as tools for understanding. Entomologists (Berryman, 1991; Liebhold, 1994) as well as plant scientists (Passioura, 1973, 1996; Sadras and Trápani, 1997) have discussed the main limitations of complex simulation models as tools for understanding. Furthermore, crop simulation models are often taken as a substitute for more appropriate frameworks, namely a suitable theory and, in some cases, models could be an obstacle rather than an aid for understanding biological processes (Sadras and Trápani, 1997). Simple rather than complex, fully testable models (Passioura, 1996) that are "transparent" because their simplicity allows the user to see how they work and what causes their outcomes (Berryman, 1991) are likely to be valuable tools for understanding the biological processes involved in crop-herbivore interactions. This is well illustrated by the models of Tuomi *et al.* (1994) and Nilsson *et al.* (1996) (section 3.2.1).

In summary, we propose that crop-herbivore relationships could be better understood by: (i) adopting a "multiple-stress" approach, (ii) incorporating current concepts of the biology of plant responses to stress, arthropod responses to stressed plants, and ecological theories on plant-herbivores relationships as a general framework (*e.g.* Jones and Coleman, 1991), (iii) developing simple, *ad hoc* models to suit specific research purposes, (iv) using complex crop-pest models with extreme caution, and (v) developing novel analytical methods. All these elements are required to overcome the problems summarised by Leigh *et al.* (1970) who, after intensive research concluded that "direct cause-and-effect relationships between lint production, plant growth, insect populations, and water and plant nutrition management were found difficult to identify."

6.   CONCLUSIONS AND DIRECTIONS FOR FURTHER RESEARCH

Despite the widespread view that cotton is highly susceptible to pests, we have shown that plants are not passive victims of herbivores and that, in some cases, cotton yield could be unaffected or even increased by mild, timely damage.

At the plant level, plant resistance to herbivores has two components: avoidance and tolerance. Avoidance strategies have been used widely in many breeding programs. Earliness in some regions of the U.S.A. and Brazil, okra-leaf genotypes in Australia, and transgenic cottons expressing *Bacillus thuringhensis* insecticidal proteins are some striking examples of the success achieved by breeders in improving cotton resistance to pests. Further biotechnological developments, including the continued identification of genes encoding insecticidal proteins, and manipulation of induced defenses, will certainly play a major role in breeding for improved cotton resistance to arthropod pests.

Not surprisingly, breeders have been more keen to include avoidance, rather than tolerance, traits in their programs of plant resistance. This is because: (i) we have a poorer understanding of tolerance traits than of avoidance traits, and (ii) avoidance traits are, in general, genetically simpler than tolerance traits. It is doubtful, however, whether a plant can be fully protected by its defenses against herbivores (Mc Naughton, 1983a). This, together with the ecological instability of chemical defenses, suggests that more attention should be paid to tolerance traits. Combination of tolerance and avoidance traits could offer broader and ecologically more stable solutions to cotton's pest problems.

At the population level, avoidance and tolerance mechanisms could also operate that can strongly affect the behaviour of the crop-herbivore-beneficial system, but they are poorly understood in comparison with mechanisms at lower levels of organisation. Widespread speculation, and some experimental evidence indicates that plants injured by herbivores may release chemicals that attract natural enemies of their herbivores (Godfray, 1995). These compounds, as well as volatile elicitors of plant chemical de-

fenses, need to be investigated in conjunction with morphological traits that affect both the ability of the plant to *retain* beneficials and the *accessibility* of herbivores to beneficials (Marquis and Whelan, 1996). Changes in competitive relationships between neighbouring plants brought about by damage that is uneven in space and/or time is an important determinant of the capacity of the crop to compensate for pest damage. Indirect evidence from plant density trials and recent studies with simulated damage support this view. Both positive and negative interactions between damaged and undamaged neighbours have been described, the direction of the response being dependent on the type of damage. Research is needed in which (i) the pattern of plant damage is assessed against the pattern of arthropod distribution and feeding in cotton fields, and (ii) crop yield is interpreted in terms of growth, development and yield of individual plants that have been exposed to different timings and/or intensities of damage.

A cautious, critical use of modelling tools, the consideration of ecological theories on plant-herbivore relationships, and novel conceptual frameworks (Jones and Coleman, 1991) are needed in the research of the relationships between cotton and its pests under varying environmental conditions and management practices. Better understanding of the cotton/pest system is important not only for breeding and selection of improved varieties but also for the development of more effective management practices.

Substantial improvement in cotton resistance to herbivores during the 21st century is likely with programmes fostering multi-disciplinary, basic and applied research. The multi-disciplinary team approach involving plant scientists and entomologists with a backgrounds ranging from molecular biology to agronomy, breeding, and ecology has proven to be a useful model for balancing the reductionist approach inherent to research in molecular biology, and for rapidly shortening the period of time between initial discovery and commercial application.

7.   ACKNOWLEDGMENTS

Our work on cotton physiology and entomology is supported by the Cotton Research and Development Corporation of Australia (VOS) and USDA-NRICGP (GWF). These sources of assistance are gratefully acknowledged. We thank our colleagues J. L. Bi who provided unpublished data, and G.P. Fitt, P.E. Reid, and L.J. Wilson who offered valued comments on the manuscript.

Chapter 21

EFFECTS OF ENVIRONMENT ON FIBER QUALITY

Judith M. Bradow[1] and Gayle H. Davidonis[1]
[1]USDA, ARS, Southern Regional Research Center, New Orleans, LA

1.   INTRODUCTION

"White as snow; strong as steel; fine as silk; long as wool, cheap as -- possible."
    Traditional cotton buyers' fiber quality specifications

The physiological responses of *Gossypium* species to the environment have been defined and described at the crop, whole-plant, or organ levels elsewhere in this book. Indeed, the profound and diverse effects of growth environment on cotton physiology are mentioned or implied in the title of every chapter in *Physiology of Cotton*. Bulk fiber yields have been used as the benchmark for treatment success, and environment-related yield components have been discussed. Clearly, the relationships between sub-optimal weather or management practices and reduced yields are much better understood than are the effects of growth environment on the 'quality' of the cotton fiber produced in response to the growth environment. Nevertheless, it is the *quality*, not the quantity, of the fiber ginned from the cotton seed that determines the end-use and economic value of the cotton crop and, consequently, the profits returned to both the producers and processors.

2.   WHAT IS FIBER QUALITY?

On a physiological basis, the fiber quality of any cotton genotype is a composite property determined by complex interactions among (1) the genetic potential of the genotype, (2) the environmental fluctuations experienced by the maternal plant from planting through harvest, and (3) the genetically controlled responses of the genotype to those environmental fluctuations. As do all metabolizing plant cells, a cotton fiber cell responds individually to fluctuations in the macro- and micro-environments so that the fibers on a single seed constitute a continuum of fiber lengths, shapes, cell-wall thicknesses, and maturities (Bradow *et al.*, 1996a, 1996b). Environmental variations within the plant canopy, among plants, and within and among fields assure that every bale of cotton contains a highly variable fiber population that encompasses broad ranges in fiber-quality properties. Thus, natural genetic and physiological variations in fiber cell shape, size, and maturity are modulated by fluctuations in the growth environment.

3.   WHY IS FIBER QUALITY IMPORTANT?

Successful processing of cotton fiber depends on highly variable fiber physical attributes which have been shown to affect finished-product quality and manufacturing efficiency (Bradow *et al.*, 1996a). If blending levels and spinning and dyeing processes are to be optimized for specific end-uses, production managers of textile mills require effective description and measurement of these highly variable fiber-quality properties (Moore, 1996). In the United States, the components of the cotton fiber-quality composite are those properties reported for every bale by the classing offices of the USDA, Agricultural Marketing Service (AMS). Fiber physical properties reported by the USDA, AMS classing offices are: micronaire, length, length uniformity index, strength, and trash measured by the High Volume Instrument (HVI), the classer's color and leaf grades, preparation (degree of roughness of ginned lint), and extraneous matter.

J.McD. Stewart et al. (eds.), *Physiology of Cotton*,
DOI 10.1007/978-90-481-3195-2_21, © Springer Science+Business Media B.V. 2010

The naturally wide variations in fiber quality and differences in cotton end-use requirements introduce significant variability into the value of the fiber. Therefore, a system of premiums and discounts has been established with respect to a specified 'base' quality. In general, cotton fiber value increases as the fiber increases in whiteness, length, strength, and micronaire. However, discounts are made for both 'low mike' (micronaire <3.5) and 'high mike' (micronaire >4.9). Traditionally, ideal fiber-quality specifications have been summarized thus: "as white as snow, as strong as steel, as fine as silk, as long as wool, and as cheap as hell." Current fiber-classing technology allows the quantitation of such qualitative fiber properties, the improvement of standards for end-product quality, and the beginnings of a fiber-quality 'language' and system of measurements that can be meaningful to producers and processors alike.

4. FIBER-QUALITY PROPERTIES UNDER GENETIC CONTROL

4.1 Genetic Control and Environmental Variability

Ongoing changes in textile processing, particularly the new, improved spinning technologies, have led to increased emphasis on breeding for *both* improved yield *and* fiber quality (Meredith and Bridge, 1972; Green and Culp, 1990; Meredith, 1990; Patil and Singh, 1995). Studies of gene action and heterosis have suggested that, within Upland cotton genotypes, there is little non-additive gene action in fiber length, strength, and fineness (Meredith and Bridge, 1972). Large interactions between combined annual environments and fiber strength have suggested that environmental variability can prevent full realization of genotype fiber-quality potential (Green and Culp, 1990.) However, early (pre-1980) statistical comparisons of the relative genetic and environmental influences on fiber strength suggest that fiber strength is conditioned by a few major genes only (May, 1999).

4.2 Genetic Potential and Environment

In reference to either fiber yield or fiber properties, genotype potential is the fiber quantity or quality level attained under *optimal* environmental conditions. The variability of fiber properties at the crop level can be used to ascertain genotype potential. Sorting bulk seedcotton samples into weight categories revealed that as seed weight increased fiber length and maturity increased while short fiber percentage decreased (Davidonis *et al.*, 1999). When seedcotton weight categories were compared by dye uptake, it was found that non-dyeing fiber was associated with low seedcotton weights (Kerby *et al.*, 1993). This genotype optimum changes in response to environmental fluctuations and modulations, including the inevitable seasonal shifts in

environmental factors such as temperature, day-length, and insolation. Such seasonal shifts in cotton metabolism and fiber properties have been seen in the higher growth rates of Upland and Pima bolls from July flowers, relative to the growth rates of bolls from August flowers on the same plants (Sassenrath-Cole and Hedin, 1996). The micronaire values and maturities of fibers from the July-flower bolls were also higher than those from the corresponding August-flower bolls (Bradow *et al.*, 1996b). Similar effects of environment on genotype potential have been quantified in fiber-quality plant maps of micronaire and maturity (Bradow *et al.*, 1996a).

In addition to modulations of genotype fiber properties at the crop and whole-plant levels, differences in fiber properties can be traced to variations in fiber properties on a single seed. Fiber-length array histograms from individual seeds have revealed that length variations occur in the micropylar, middle, and chalazal regions of seeds (Delanghe, 1986). Mean fiber lengths were shortest in the micropylar region of the seed in *G. hirsutum* L., *G. barbadense* L., and *G. arboreum* L. cultivars (Vincke *et al.*, 1985). The most mature fibers and those having the largest perimeters were also found in the micropylar region of the seed. The percentage of short fibers on a cotton seed after hand-ginning was extremely low; and it was concluded by Vincke and coauthors that, in baled cotton, short fibers with small perimeters did not originate in the micropylar region of the seed. Advanced Fiber Information System, Zellweger-Uster (AFIS) measurements of fiber from micropylar and chalazal regions of seeds revealed that the relative location of a seed within the boll was related to the magnitude of the differences in the properties of fibers from the micropylar and chalazal regions (Davidonis and Hinojosa, 1994). Motes (unfertilized ovules or aborted seeds) and seeds were examined 34 days post anthesis and showed micropylar and chalazal fiber property differences (Weis *et al.*, 1999).

There are also significant variations in other fiber properties that can be related to the seed position (apical, medial, or basal) within the boll (Porter, 1936; Iyengar, 1941). Degree of secondary wall thickening (quantified by AFIS as the fiber cell-wall maturity parameter, θ) is lowest in seeds at the apex (seed location 1) of the boll and highest in seeds at the pedicel or basal end (seed location 7) of the boll (Table 21-1). Fiber length and maturity also exhibit both seed and boll location effects. Porter (1936) examined fiber length in relation to seed position in the locule and found that seeds near the apical or basal end of the boll produced the shortest fibers. Fiber weight per unit length was greatest in seeds near the basal end of the boll (Iyengar, 1941). In Table 21-1, the least mature fiber occurred closest to the boll apex, whatever the plant fruiting node number. Fiber from plant nodes 9, 10, and 11 higher in the plant canopy was consistently longer and more mature than fiber from node 7 or lower on the plant. Thus, the different micro-environments within the boll and within the plant canopy had significant effects on the properties of fiber produced within the same macro-environment, *i.e.*, on the same plant in the same field in the same crop year.

Table 21-1. Effect of seed location within the locule on Upland 'DPL51' cotton fiber properties quantified by AFIS. Seed nearest the pedicel (basal) end of the boll is designated as location 7. Each value is an average of three bolls containing no motes (underweight seeds). All data are from first-position bolls. Data from nodes 9, 10, and 11 were pooled to obtain a statistically valid population. Cotyledonary node = 0. (Davidonis, unpublished).

Fiber property	Node number	Seed location						
		1	2	3	4	5	6	7
Length by weight, mm	7	24.6	25.1	25.6	25.4	25.1	24.6	24.9
Theta, θ		0.542	0.584	0.592	0.604	0.616	0.616	0.641
Immature fiber fraction (% with θ <0.250)		6.8	5.6	4.4	4.0	3.7	3.8	3.4
Length by weight, mm	9, 10,	26.2	26.7	26.7	26.7	26.9	26.9	26.4
Theta, θ	& 11	0.610	0.632	0.627	0.631	0.660	0.657	0.672
Immature fiber fraction (% with θ <0.250)		4.0	3.8	4.1	3.7	2.6	2.7	2.7

5. FIBER QUALITY, PLANT ARCHITECTURE AND SUBOPTIMAL GROWTH ENVIRONMENT

The effects of environment on cotton plant morphology and the correlations between plant architecture and yield are considered in other chapters in Parts II, III, and IV of this volume. In this chapter, linkages between canopy characteristics (both genotype and those induced by environmental factors) and fiber quality are considered on the basis that any modification of whole-plant morphology that significantly alters yield will also modify one or more fiber-quality properties in some way.

5.1 Canopy Architecture and Fiber Quality

Cotton canopy architecture, particularly plant height and branch formation, is modified by environmental factors such as temperature (Hanson *et al.*, 1956; Reddy *et al.*, 1990; Hodges *et al.*, 1993), growth-regulator application (Reddy *et al.*, 1990; Cadena and Cothren, 1996; Legé *et al.*, 1996), light intensity (Hanson *et al.*, 1956; Sassenrath-Cole, 1995), and herbivory (Terry, 1992; Rosenthal and Kotanen, 1994; Sadras, 1996c). Genotype canopy characteristics, such as solar tracking and leaf shape, and macro- and micro-environmental factors interact to modulate canopy light distribution which, in turn, alters photosynthetic activity within the canopy and the crop (Wells *et al.*, 1986; Reddy *et al.*, 1991; Sassenrath-Cole, 1995; Sassenrath-Cole and Heitholt, 1996). Thus, reduced photosynthetic rates and the modulation of other metabolic factors in association with lower light intensities resulted in lower micronaire, fiber strength, and yield (Pettigrew, 1996).

5.2 Boll Retention Patterns and Fiber Quality

Another obvious architectural linkage among environment and fiber yield and quality is seen in boll retention

patterns. Environmental conditions that induce boll drop alter fiber quality of the remaining bolls by modifying assimilate and metabolic resource partitioning within the reduced boll population. Assimilate partitioning, source/sink relationships and related topics are covered in Chapters 5, 14, and 17. The connection between boll retention and micronaire distribution patterns can be seen in Figures 21-1 and 21-2. Irrigation method was the macro-environment treatment in this study of PD3 grown in South Carolina in 1992 (Bradow *et al.*, 1997a; 1997b). The irrigation treatments were natural rainfall or water added through micro-irrigation tubing laid in the root zone under each row (in-row) or laid between alternate rows (alternate-row). Both the in-row and alternate-row irrigation treatments delivered a season total of 90 mm additional water in nine irrigation events.

In comparison to both the rainfed and alternate-row treatments, the in-row irrigation treatment skewed boll retention toward the lower nodes (Fig. 21-1). Both micro-irrigation methods increased boll retention on the upper branches and this trend was more evident in the alternate-row treatment. Overall, the rainfed plants retained 15% fewer bolls than did the plants in the micro-irrigation treatments, and irrigation method modulated the resulting boll retention patterns. Alternate-row irrigation resulted in greater boll retention at nodes 15 and above, and the increase in rainfed boll number at node 14 was associated with increased rainfall associated with a hurricane system passing to the south of the field in 1992.

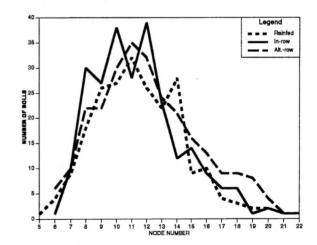

Figure 21-1. Boll retention patterns at harvest in rainfed, in-row, and alternate-row micro-irrigated PD3 cotton. Number of bolls = mean number of bolls at each node across branch positions from all plants in 1-m rows (with four replications; Bradow *et al.*, 1997b).

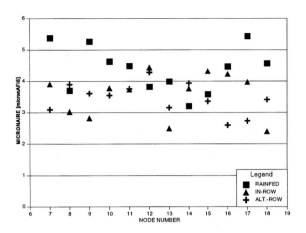

Figure 21-2. Node-by-node micronaire distributions from plant maps of PD3 Upland cotton irrigated by natural rainfall or in-row or alternate-row micro-irrigation (Bradow *et al.*, 1997a; 1997b).

The irrigation treatments did not significantly affect seed cotton yields or crop-average micronaire (Bradow *et al.*, 1997a; 1997b), but the macro-environment effects on the micronaire distribution patterns within the crop averages were apparent when micronaire was mapped according to node (Fig. 21-2). The rainfed micronaire distribution was bimodal with higher micronaire values occurring at the lower nodes within the main-crop (nodes 7 through 18) and a second high micronaire peak corresponding to the top of the main crop at nodes where only a single boll per plant persisted to harvest. Increased boll retention associated with in-row irrigation was correlated with marked decreases in micronaire. The low-micronaire bolls from nodes 13 and 14 were in peak fiber cell-wall deposition stage during a prolonged period of low insolation and increased rainfall associated with a hurricane in 1992.

Micronaire distributions in Figure 21-2 show the effects of both macro-environment and micro-environment on an economically important fiber property. Fluctuations in the environment increased fiber property variability and the frequency and proportion of fibers falling outside the fiber-quality range required by cotton processors, *i.e.*, 3.5 < micronaire > 4.9. Similar environment-related modulations of fiber maturity, cross-section, and length distributions have also been mapped within the whole-plant architectural framework (Bradow *et al.*, 1996a; Bradow *et al.*, 1997a; 1997b).

5.3   Seed Setting Efficiency and Fiber Quality

Marked variations in fiber properties were also found when within-boll architecture (or seed-setting efficiency) was considered as a subset of whole-plant architecture. The seed-setting efficiency, *i.e.*, the number of seeds produced compared to the number of ovules per ovary (Turner *et al.*,

1977; Davidonis *et al.*, 1996), is an indicator of the number of motes per boll where motes are developmentally arrested, non-viable seeds and their associated fiber (Verschaege, 1989; Davidonis *et al.*, 1996). Large numbers of long-fiber motes per boll reduced the degree of secondary wall thickening and, therefore, the relative maturity of fibers from the middle seeds in the same boll (Davidonis *et al.*, 1996).

6.   FIBER LENGTH

Due to the inherent variability of cotton fiber, there is no absolute value for fiber length within a genotype or within a testing sample (Behery, 1993). Even on a single seed, fiber lengths vary significantly with longer fibers occurring at the chalazal end of the seed and shorter fibers being found at the micropylar end. Coefficients of fiber-length variation, which also vary significantly from sample to sample, are on the order of 40% for cotton.

Variations in fiber length attributable to genotype and fiber location on the seed are, of course, modulated by micro- and macro-environmental factors (Bradow *et al.*, 1997a; 1997b). Environmental changes around the time of floral anthesis may limit fiber initiation or retard the onset of fiber elongation. Sub-optimal environmental conditions during the fiber elongation phase may decrease the rate of elongation per day or shorten the duration of the elongation period so that the genotype fiber length potential is not fully realized (Hearn, 1976). In addition, the causative environmental fluctuation need not occur during the affected growth stage. Thus, the results of the physiological responses may become evident at a later stage in fiber development.

6.1   Measurement of Fiber Length

Fiber lengths on individual seeds can be determined while the fibers are still attached to the seed (Gipson and Joham, 1969; Munro, 1987) or, after ginning, by hand-stapling or photo-electrical measurement (Munro, 1987; Behery, 1993). Traditionally, staple lengths have been measured and reported to the nearest thirty-second of an inch or to the nearest millimeter. The four Upland staple classes are: Short (<21mm), Medium (22 to 25 mm), Medium-Long (26 to 28 mm), and Long (29 to 34 mm). Pima staple is classed as Long (29 to 34 mm) and Extra-long (>34 mm).

The term staple length was used by cotton buyers and processors long before satisfactory methods for measuring fiber properties had been developed. Consequently, staple length has never been formally defined in terms of any statistically valid length distribution (Munro, 1987; Behery, 1993). Historically, fiber length was measured using the Baer diagram or Suter-Webb array methods. Both methods are based on sorting fibers, within a defined sample, according to length and/or weight. Banks of parallel combs segregate fibers into length arrays or length groupings at one-eighth inch intervals. In Suter-Webb testing, fibers in each length group are accurately weighed and the length-weight distribution is used in calculating various fiber length properties, including the mean length and upper quartile length by weight, *i.e.*, the fiber length exceeded by 25% of the fiber by weight in the test specimen.

Construction of Baer fiber-length diagrams must be done by hand and, consequently, is prohibitively labor- and time-intensive, particularly for classing-office use. Array construction with the Suter-Webb Duplex Cotton Fiber Sorter has been accepted as a standard test method for length and length distribution of cotton fibers (ASTM D 1440-90, 1994). The Suter-Webb array method physically sorts fibers of different lengths and serves as a benchmark to which other methods are compared. However, Test Method D 1440-90 is not commonly used for acceptance testing in commercial shipments. The Peyer Almeter AL-101, which reports fiber lengths by weight and by number (ASTM D 5332-92, 1994), is also used in the U.S., European, and Pacific Rim cotton industries (Bargeron, 1986; Behery, 1993).

Fiber length is directly related to yarn fineness, yarn strength, and spinning efficiency (Moore, 1996). Therefore, rapid, reproducible instrumental methods for fiber length measurement have been developed. Both length and length uniformity can be measured by the Fibrograph (ASTM D 1447-89, 1994). In Fibrograph testing, fibers are randomly caught on combs, and the beard formed by the captured fibers is scanned photoelectrically from base to tip (Behery, 1993). The amount of light passing through the beard is a measure of the number of fibers that extend various distances from the combs. Data are recorded as span length, *i.e.*, the distance spanned by a specific percentage of fibers in the test beard. Span lengths are usually reported as 2.5% and 50%, the 2.5% span length being the basis for machine settings at various stages during fiber processing. The uniformity ratio is the ratio between the two span lengths expressed as a percentage of the longer length. The Fibrograph provides a relatively rapid method for reproducibly measuring the length and length uniformity of fiber samples, and Fibrograph test data are used in research studies and in qualitative surveys such as checking commercial staple length classifications and assembling cotton bales into uniform lots.

Since 1980, USDA, AMS classing offices have relied increasingly on high volume instrumentation (HVI) for measuring fiber length and other fiber properties (Moore, 1996). The HVI length analyzer determines length parameters by photoelectrically scanning a test beard selected by a specimen loader and prepared by a comber/brusher attachment (Spinlab HVI, ASTM D 4605-86, 1994). (The Motion Control HVI, for which production ceased in 1995, pneumatically scanned the test beard [ASTM D 4604-86, 1994].) The fibers in the test beard are assumed to be uniform in cross-section so that the pressure drop across the beard is an estimate of the number of fibers in the airflow path. Scanning the pressure drop along the length of the beard provides a count of the fibers present at each point along the beard. These data are converted

to represent the percentage of the total number of fibers present at each length value and other length parameters such as mean length, upper-half mean length, and length uniformity (Behery, 1993). This test method for determining cotton fiber length is considered acceptable for testing commercial shipments when the testing services use the same reference standard cotton samples (Moore, 1996).

All of the fiber length measurement methods discussed above require from three to five grams of ginned fiber and were developed for classing relatively large bulk samples of cotton fiber. For analyses of small fiber samples, *e.g.*, the single-seed or single-locule samples collected in plant-mapping and boll-mapping studies, fiber property measurements from an electron-optical particle sizer, the Zellweger-Uster AFIS (Advance Fiber Information System), have been found acceptably sensitive, rapid, and reproducible. The AFIS Length & Diameter module (Bragg and Shofner, 1993) generates values for mean fiber length by weight, mean fiber length by number, fiber length histograms, upper quartile length, and short fiber contents by weight and by number (the percentage of fibers shorter than 12.7 mm), and also mean fiber diameters by number (Behery, 1993). AFIS is a count-based system and values are given on number basis while values given on a weight basis are calculated.

Although short fiber content (SFC) is not currently reported as part of USDA, AMS classing office information, SFC is increasingly recognized as a fiber property comparable in importance to fiber fineness, strength, and length (Deussen, 1992; Behery, 1993). The importance of SFC in determining fiber processing success, yarn properties, and fabric performance has led to serious consideration being given to the establishment of SFC standards similar to the micronaire premium and discount system. Although fiber length is primarily a genotype trait, SFC is dependent upon genotype, growing conditions, harvesting, ginning, *and* processing methods.

If the strong genetic component of fiber length is to be separated from the environmental components introduced by excessive temperatures and water and nutrient deficiencies, it is essential that cotton breeders, and physiologists understand the underlying concepts and limitations of the various methods used in fiber-length and SFC measurement. Genetic improvement of fiber length is fruitless if the genotype response to the growth environment prevents full realization of the enhanced genetic potential. The effects of separate environmental factors on fiber length and SFC at harvest are discussed in subsections that follow.

6.2 Fiber Length and Temperature

Maximum cotton fiber lengths were reached when night temperatures were around 19 to 20°C, depending on the genotype (Gipson and Joham, 1968; Gipson and Ray, 1970b). Early fiber elongation was highly temperature-dependent; late fiber elongation was temperature-independent (Gipson and Joham, 1969; Xie *et al.*, 1993). Fiber length (upper half mean length) was negatively correlated with the difference between maximum and minimum temperature (Hanson, 1956).

Field experiments on the Texas High Plains showed that a night temperature of 15°C caused a 4- to 5-day delay in lengthening of fibers, compared to a night temperature of 25°C (Gipson and Ray, 1968; 1969). Although the observed effects of cool night temperatures were not separated into delays in fiber initiation and early fiber elongation, field studies in India showed that fiber grown under 15°C conditions took three to five days longer to reach final fiber length than did control (24°C) fiber (Thaker *et al.*, 1989). Final fiber length was not significantly affected by temperature because reduced elongation rates were compensated by lengthening the elongation period by 8 to 24 days (Thaker *et al.*, 1989). Modifications of fiber length by growth temperatures have also been observed in planting-date studies in which later planting dates were associated with small increases in 2.5% and 50% span lengths (Aguillard *et al.*, 1980; Greef and Human, 1988). If the growing season is long enough and other inhibitory factors do not interfere with fiber development, early-season delays in fiber initiation and elongation can be counteracted by an extension of the elongation period (Bradow *et al.*, 1997b).

In addition to field studies, cotton ovule cultures have served as models for fiber growth and development (Meinert and Delmer, 1977; Haigler *et al.*, 1991; Xie *et al.*, 1993). Ovule cultures have been employed to differentiate the effects of cool temperatures on fiber initiation and early elongation. Ovules cultured under a 34°/15°C (12/12h) cycling temperature schedule showed delays in fiber initiation and early elongation. After fibers were 0.5 mm long, rates of elongation were similar in 34°C constant and 34/15°C cycling schedules (Xie *et al.*, 1993).

Variations in fiber length and the elongation period were also associated with relative heat-unit accumulations. Regression analyses showed that genotypes which produced longer fibers were more responsive to heat-unit accumulations than were genotypes that produced shorter fibers (Quisenberry and Kohel, 1975). However, genotype earliness was also a factor in the relationship between fiber length (or short fiber content by weight) and accumulated heat units (Bauer and Bradow, 1996; Bradow and Bauer, 1997b). Lower total cumulative heat units in 1992, compared to 1991, increased the short fiber content of the earliest genotype, 'DPL20' (Fig. 21-3). Higher total heat-unit accumulations in 1991 increased the short fiber content of the latest-maturing genotype, 'DPL5690'. Planting two weeks earlier than normal in the cooler spring of 1992 reduced the short fiber contents of 'DPL50' and 'DPL90'. The mean fiber length across genotypes and years was 21.6±0.4 mm, and the mean short fiber content by weight across genotypes and years was 13.8±2.7%.

Under well-watered conditions in closed-environment, sunlit growth chamber SPAR (Soil-Plant-Atmosphere-Research) units (Phene *et al.*, 1978) increased temperatures, decreased fiber length and short fiber percentage in ambient and elevated CO_2 environments (Reddy *et al.*, 1999).

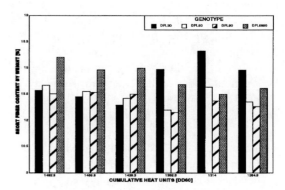

Figure 21-3. Relationships between Upland fiber short fiber contents by weight and total annual cumulative heat units (Degree Day 60) from a planting date study of 'DPL20', 'DPL50', 'DPL90', and 'DPL5690'. Planting and harvest dates were staggered so that the heat unit data on the x-axis above correspond, in descending order, to early planting date in 1991, normal planting date in 1991, late planting date in 1991, early planting date in 1992, normal planting date in 1992, and late planting in 1992 (Bauer and Bradow, 1996).

As temperature increased from 20° to 26°C fiber length frequency distributions were more uniform (Reddy *et al.*, 1999). Aware of the limitations imposed by growth chambers, Liakatas *et al.* (1998) varied day/night temperatures and found that the shortest fibers were observed under a 30/20°C regime and the longest in a 30/16°C regime. Length uniformity percentage was 42.5 for the 30/20°C regime and 54.3 for the 26/16.5°C regime (Liakatas *et al.*, 1998).

High temperatures promote the abscission of small bolls and abscission is more pronounced when the boll load is heaver (see Chapters 19 and 22) when cotton plants are grown in SPAR units boll retention decreased significantly at temperatures above 28°C (Reddy *et al.*, 1992c, 1995, 1999). As temperature increased the percentage of small, short-fiber motes increased (Reddy *et al.*, 1999). The low percentage of short fibers at higher temperatures may be related to increased assimilate availability to the fibers when there were fewer seeds per boll (see Chapter 14).

Boll shedding associated with high temperatures has also been attributed to pollen sterility (Powell, 1969; Fisher, 1973; 1975). Cotton pollen fertility was first linked with fiber length by Pressley (1937). However, subsequent research emphasis has been on the effects of pollen sterility on ovule abortion (Percy, 1986), particularly in interspecific crosses between *Gossypium hirsutum* and longer-fiber *G. barbadense* genotypes (Verschraege, 1989; Waller and Mamood, 1991) and in breeding programs directed at improving cotton heat tolerance and fertilization efficiency (Barrow, 1981; 1982; Rodriguez-Garay and Barrow, 1988; Gwyn and Stelly, 1989; Stelly *et al.*, 1990).

6.3   Fiber Length and Water

Cotton water relationships and irrigation have usually been studied with respect to yield (Hearn, 1976, 1995;

Ramey, 1986; Radin *et al.*, 1992). Grimes and Yamada (1982) concluded that fiber length was not affected unless the water deficit was great enough to lower the yield to 700 kg ha^{-1}. Fiber elongation was inhibited when the midday water potential was -2.5 to -2.8 mPa. The occurrence of moisture deficits during the early flowering period did not alter fiber length. However, when drought occurred later in the flowering period, fiber length was shorter (Marani and Amirav, 1971; Shimshi and Marani, 1971; Hearn, 1976).

Severe water deficits during the fiber elongation stage reduce fiber lengths (Hearn, 1995). Although water-deficit modulations of fiber length would appear to be linked simply and directly to the processes of cell expansion, the effects of water availability on duration and timing of flowering and boll set and upon fiber elongation result in complex physiological interactions between water deficits and fiber properties, including length. In the Coastal Plain of Texas, water deficits regularly occur during the mid- to late-flowering periods. When bolls containing zero to two small short-fiber motes were grown under rainfed and irrigated conditions in that region (Davidonis *et al.*, 1996), fiber lengths (both length by weight and length by number) were shorter in the mid-season population of rainfed bolls than in the mid- to late-season irrigated-bolls. The SFCs were the same for all irrigated bolls. In other studies, irrigation increased mean fiber length and upper-half mean length (Grimes *et al.*, 1969; Spooner *et al.*, 1958).

Drip irrigation and placement of the drip-irrigation tubing under or between the plant rows also modulated fiber length by weight (Bradow *et al.*, 1997a; 1997b). When the rainfed mean fiber length was 24.5±1.6 mm, the drip-irrigated fiber length by weight mean was 23.3±2.6 mm when the irrigation tubing was buried in the row under the plants and 23.5±2.6 mm when the tubing was buried between every other row. Fiber length distributions, according to fruiting site and within the locules, were also modified by irrigation method. The higher fiber length mean for the rainfed plants was related with the greater boll retention on nodes 13 and below (Fig. 21-1).

In India, moisture conservation practices [mulching] increased fiber length and yield (Singh and Bhan, 1993). However, under irrigated conditions, conservation tillage surface residues (Bauer *et al.*, 1995; Bauer and Busscher, 1996) did not affect any fiber property, including length.

6.4   Fiber Length and Light

Changes in the growth environment also alter canopy structure and the photon flux environment within the canopy. For example, loss of leaves and bolls resulting from unfavorable weather (wind, hail), disease, or herbivory and subsequent compensatory growth after loss of photosynthetic or reproductive organs can greatly affect both fiber yield and quality (Sadras, 1995). The light environment within the crop canopy is an important determinant of photosynthetic activity (Sassenrath-Cole, 1995) and, therefore, of the source-to-sink relationships that allocate photoassimilate within the canopy (Pettigrew, 1994; 1995). Eaton and Ergle (1954) observed that reduced light treatments increased fiber length. Shading during the first seven days after floral anthesis resulted in a 2% increase in the 2.5% span length of 'DES119', 'DPL5690', and 'Prema' genotypes (Pettigrew, 1995).

Shading (or prolonged periods of cloudy weather) and seasonal shifts in day-length also modulate temperature, which interactively modifies fiber properties, including length. Although commercial cotton genotypes are considered 'day neutral' with respect to both flowering and fruiting (Lee, 1984), incorporation of day-length data in Upland and Pima fiber quality models based on accumulated heat units increased the coefficients of determination for the length predictors from 30 to 54% for the Upland model and from 44 to 57% for the Pima model (Bradow *et al.*, 1997a; Johnson *et al.*, 1997). Kasperbauer (1994) also found that light wavelengths reflected from red and green mulches increased fiber length although plants grown over those mulches received lower reflected photosynthetic flux than did plants grown above white mulches. The longest fibers were harvested from plants that received the higher far-red/red ratios.

6.5   Fiber Length and Mineral Nutrition

Studies of mineral nutrition of cotton and the related soil chemistry usually emphasize increased yield and fruiting efficiency (Waddle, 1984; Joham, 1986; Radin and Mauney, 1986; Radin *et al.*, 1991; Bisson *et al.*, 1995). More recently, the effects of nutrient stress on boll shedding have also been examined (Jackson and Gerik, 1990; Heitholt, 1994b), and several mineral-nutrition studies have been extended to include fiber quality (Cassman *et al.*, 1990; Minton and Ebelhar, 1991: Bauer *et al.*, 1993; Matocha *et al.*, 1994; Bauer and Busscher, 1996; Pettigrew *et al.*, 1996). These studies investigated the effects of either potassium or nitrogen on fiber properties, including span length.

Reports of fiber property trends are often contradictory because of combined influences of genotype, climate, and soil conditions. Added potassium (112 kg K ha^{-1} yr^{-1}) does not affect the 2.5% span length of 'DES119'and 'STV825' when genotype was a significant factor in determining both 2.5% and 50% span lengths (Minton and Ebelhar, 1991). Genotype was not a significant factor in Acala fiber length, but an additional 480 kg K ha^{-1} yr^{-1} increased mean fiber lengths of the two Acala genotypes, 'SJ2' and 'GC510' when the K X genotype interaction was significant (Cassman *et al.*, 1990). Foliar applications of KNO_3 did not affect either yield or fiber length in Corpus Christi TX (Matocha *et al.*, 1994). Soil-applied KNO_3 did increase yields in two years out of three, but no potassium effects on fiber length were observed. In a Mississippi Delta study of eight genotypes (Pettigrew *et al.*, 1996), added potassium (112 kg K ha^{-1}) increased the length uniformity ratio and increased 50% span length, but not the 2.5% span length. The 2.5% span length was determined by genotype and the interaction, genotype X environment (crop year).

Added nitrogen and the nitrogen X genotype, nitrogen X potassium interactions had no effect on fiber span lengths or length uniformity (Pettigrew *et al.*, 1996). Environmental factors other than added nitrogen determined fiber span lengths in a South Carolina study of the effects of nitrogen and green manures on cotton fiber yield and quality (Bauer *et al.*, 1993). Nitrogen released from legume cover crops also had no effect on fiber span lengths (Bauer and Busscher, 1996).

6.6   Trends in Fiber Length

During the decade between 1985 and 1996, the U.S. Upland staple length trend has been toward greater fiber length (an increase of 1.8 mm per year) with two plateaus: an average length of 27.4 mm between 1985 and 1990 and an average length of 27.9 mm between 1991 and 1996 (Sasser and Shane, 1996). The apparent jump in staple length between the 1990 and 1991 crops has been attributed to the shift to 100% instrument testing of length in 1991. Annual fluctuations in fiber length were loosely correlated with variations in the growth environments. Weather extremes across the U.S. Cotton Belt in 1998 decreased average fiber length to 27.2 mm (Cotton Incorporated, 1999). In the twelve years during which length uniformity index data have been reported by USDA classing offices, length uniformity index has increased at a rate of approximately 0.1% per year but with the same kind of environment-related variability noted in the staple length trends during that period (Cotton Incorporated, 1999).

7.   FIBER STRENGTH

The inherent breaking strengths of the individual cotton fibers are considered the most important factor in determining the strength of the yarn spun from those fibers (Munro, 1987; Patil and Singh, 1995; Moore, 1996). Recent developments in high-speed yarn spinning technology, specifically open-end rotor spinning systems, have shifted the fiber-quality requirements of the textile industry toward higher strength fiber that can compensate for the decrease in yarn strength associated with open-end rotor spinning techniques (Patil and Singh, 1995). Compared to conventional ring spinning, open-end rotor-spun yarn production capacity is five times higher and, consequently, more economical. Rotor-spun yarn is more even than the ring-spun, but the rotor-spun yarn is 15 to 20% weaker than ring-spun yarn of the same thickness. Thus, fiber strength, together with fiber fineness, is given highest priority by mills using open-end rotor and friction spinning systems. Length and length uniformity, followed by fiber strength and fineness, remain the most important fiber properties in determining ring-spun yarn strength (Patil and Singh, 1995; Moore, 1996).

7.1   Estimating Fiber Strength

Historically, two instruments have been used to measure fiber tensile strength, the Pressley apparatus and the stelometer (Munro, 1987; ASTM D 1445-90, 1994). In both of these flat-bundle methods, a bundle of fiber is combed parallel and secured between two clamps. Force is applied to separate the clamps and gradually increased until the fiber bundle breaks. Fiber tensile strength is calculated from the ratio of the breaking load to bundle mass. Due to the natural inhomogeneity within a population of cotton fibers, bundle fiber selection, bundle construction and, therefore, bundle mass measurements, are subject to considerable experimental error (Taylor, 1994).

Fiber strength varies along the length of a fiber, as does fiber fineness (perimeter, diameter, or cross-section; Hsieh *et al.*, 1995). Further, the inherent variability within and among cotton fibers assures that two fiber bundles *of the same weight* will not contain the same number of fibers and that the clamps of the strength testing apparatus will not grasp the fibers in the bundle at precisely equivalent positions along the lengths. Thus, a normalizing length-weight factor has been included in bundle strength calculations. In the textile literature, fiber strength is reported as 'breaking tenacity' or grams of breaking load per tex where tex is fiber linear density in grams per kilometer (Munro, 1987; Taylor, 1994). Both Pressley and stelometer breaking tenacities are reported as 1/8 in. gauge tests, the 1/8 in. (or 3.2 mm) referring to the distance between the two Pressley clamps. Flat-bundle measurements of fiber strength are considered satisfactory for acceptance testing and for research studies of the influence of genotype, environment, and processing on fiber (bundle) strength and elongation. The relationships between fiber strength and elongation and processing success have also been examined using flat-bundle strength testing methods (Dever *et al.*, 1988). However, modern cotton fiber testing requires that procedures be rapid, reproducible, automated, and without significant operator bias (ASTM D 4604-86, 1994; ASTM D 4605-86, 1994; Taylor, 1994). Consequently, the HVI systems used for length measurements in USDA, AMS classing offices are also used to measure the breaking strength of the same fiber bundles (beards) formed in the length measuring process.

Originally, HVI strength tests were calibrated with the 1/8 in. gauge Pressley measurement, but the bundle-strengths of reference cottons are now established by stelometer tests, which also provide bundle elongation data. Fiber bundle elongation is measured directly from the displacement of the jaws during the breaking process, and fiber bundle strength and elongation data are usually reported together (ASTM D 4604-86, 1994). HVI bundle-strength measurements are reported in grams-force tex^{-1} and range from 30 and above (Very Strong) to 20 or below (Very Weak). In agronomic papers, fiber strengths are reported as kN m kg^{-1} where one Newton equals 9.81 kg-force (Meredith *et al.*, 1996a).

The HVI bundle strength and elongation test methods are satisfactory for acceptance testing and research studies when 3.0 to 3.3 g of blended fiber is available and the relative humidity of the testing room is controlled. A 1% increase in relative humidity and the accompanying increase in fiber moisture content will increase the strength value by 0.2 to 0.3 g tex^{-1}, depending on the fiber genotype and maturity. Further, classing office HVI measurements of fiber strength do not adequately describe the variations of fiber strength along the length of the individual fibers or within the test bundle. Thus, predictions of yarn strength based on HVI bundle-strength data can be inadequate and misleading (Taylor, 1994; Suh *et al.*, 1996). The problem of fiber-strength variance is being addressed by improved HVI calibration methods (Taylor, 1994) and by computer simulations of bundle-break tests where the simulations are based on large single-fiber strength data bases of more than 20,000 single fiber load-elongation curves obtained with MANTIS® (Suh *et al.*, 1996).

7.2 Fiber Strength, Environment, and Genotype

Growth environment and genotype responses to the growth environment all play a part in determining fiber strength and strength variability (Sasser and Shane, 1996). Early studies showed fiber strength to be significantly and positively correlated with maximum or mean growth temperature, maximum minus minimum growth temperature and potential insolation (Hanson *et al.*, 1956). Increased strength was correlated with a decrease in precipitation. Minimum temperature did not affect fiber strength. All environmental variables were interrelated, and a close general association between strength and environment was interpreted as an indication that fiber strength is more responsive to the growth environment than is either fiber length or fineness. (See section 8 - *Fiber Maturity* in this chapter.) Other investigators reported that fiber strength was correlated with genotype only (MacKenzie and Van Schaik, 1963; Greef and Human, 1988; Green and Culp, 1990).

Square removal did not affect either fiber elongation (Pettigrew *et al.*, 1992) or fiber strength (Terry, 1992; Pettigrew *et al.*, 1992). Shading, leaf-pruning, and partial fruit removal decreased fiber strength (Pettigrew, 1995). Early defoliation, at 20% open bolls, increased fiber strength (and length), but the yield loss due to earlier defoliation offset any potential improvement in fiber quality (Snipes and Baskin, 1994).

7.3 Fiber Strength, Mineral Nutrition, and Conservation Tillage

Acala fiber strength and elongation were positively correlated with the rate of added potassium (Cassman *et al.*, 1990). In 'SJ2' and 'GC510' fiber strength data, there were no significant genotype effects or interactions between genotype and potassium addition rates. However, the genotype main effect was significant for fiber elongation. Addition of potassium increased 'DES119' and 'STV825' fiber strength significantly and had a non-significant, but positive, effect on fiber elongation (Minton and Ebelhar, 1991). There were also strong genotype differences in the fiber strength and elongation of these two Upland genotypes. Added potassium and nitrogen did not affect fiber strength, but added potassium increased fiber elongation (Pettigrew *et al.*, 1996). Genotype differences in fiber strength were judged to be far more important than the level of nitrogen fertilization (MacKenzie and van Schaik, 1963). Supplemental boron had no effect on Upland fiber properties, including strength (Heitholt, 1994b).

Use of cover crops and tillage method had no effect on fiber strength, but significant differences in elongation were associated with winter cover type (rye) and/or tillage method (Bauer and Busscher, 1996). The influence of green manures on fiber strength tended to be small and inconsistent from year to year (Bauer *et al.*, 1993), but the authors reported that cotton planted in rye and fallow plots tended to reach cutout earlier and was ready for harvest before the other plots in the study. Linkages among maturation rate, planting date, and fiber strength were also reported when later planting resulted in increased fiber strength (Aguillard *et al.*, 1980; Greef and Human, 1988; Heitholt, 1993b). During 'Acala SJ2' fiber maturation, single-fiber breaking force and fiber linear densities increased markedly and in parallel at approximately 35 days post floral anthesis in the greenhouse (Hsieh *et al.*, 1995). No boll-position effects on single-fiber strength were observed above the fourth fruiting branch.

7.4 Trends in Fiber Strength

The 1996 Upland crop average fiber strength was 28.4 g tex^{-1}, and U.S. cotton fiber strength has increased 0.25 g tex^{-1} every year since 1980 when HVI strength data based became available and new calibration cottons were used (Cotton, Inc. 1999). The 1997 Upland crop average fiber strength was 28.4 g tex^{-1} and due to extreme weather conditions in 1998 the average fell to 28.0 g tex^{-1}. Most of the increases in fiber strength have been credited to the introduction of improved genotypes such as 'Prema' (strength >33 g tex^{-1}) in California (Patil and Singh, 1995). Differences in genotype grown, growth environment, and fiber fineness resulted in a range of fiber strength averages from 26.4 g tex^{-1} at the Corpus Christi, Texas, classing office to 32.6 g tex-1 at the Visalia, California office (Sasser and Shane, 1996). Since rotor spinning systems demand 1/8 in. fiber-bundle strengths of 28 to 30 g tex^{-1} or more, the demand for cotton genotypes that yield high strength fiber in a variety of growth environments remains strong.

8. FIBER MATURITY ([FIBER FINENESS, FIBER WALL THICKENING, AND MICRONAIRE)

Of the various fiber properties reported by USDA, AMS classing offices for use by the textile industry, fiber maturity is probably least well defined and understood. The term, fiber 'maturity', as used in cotton marketing and processing, is *not* an estimate of time elapsed between floral anthesis and fiber harvest (Lord and Heap, 1988). However, 'chronological' maturity can be a useful concept in studies that follow fiber development over time (Ramey, 1982; Bradow *et al.*, 1996b). On a physiological or physical basis, fiber maturity is generally accepted to mean degree (amount) of fiber cell-wall thickening relative to the diameter or fineness of the fiber (Perkins *et al.*, 1984: Munro, 1987).

8.1 Definitions and Related Estimates of Fiber Maturity

Classically, the mature fiber is one in which two times the cell wall thickness equals or exceeds the diameter of the fiber cell lumen, the space enclosed by the fiber cell walls (Ramey, 1982). However, this simple definition of fiber maturity is confounded by the cross-section of a cotton fiber never being a perfect circle and by fiber diameter being a genotype characteristic (Ramey, 1982; Lord and Heap, 1988; Matic-Leigh and Cauthen, 1994). Furthermore, fiber diameter varies along the length of the fiber, as does cell wall thickness. Thus, attempting to differentiate between naturally thin-walled or genetically fine fibers and truly immature fibers on a wall-thickness basis seriously complicates maturity comparisons among and within genotypes. For example, mean fiber diameters of Upland genotypes range from 21 to 29 μm, and diameters of genetically finer Pima fibers range from 17 to 20 μm (Ramey, 1982). On a locule-average basis and across fruiting sites within a single crop, PD3 Upland cotton fiber diameters (sample mean diameters by number determined by AFIS-L&D) ranged from 1.2 to 18.7 μm with a crop mean of 12.4±2.1 μm (Bradow and Bauer, unpublished). Within a single fiber sample examined by image-analysis, cell-wall thicknesses ranged from 3.4 μm to 4.9 μm when lumen diameters ranged from 2.4 μm to 5.2 μm (Matic-Leigh and Cauthen, 1994). Based on the (2 X cell-wall thickness > lumen diameter) formula for fiber 'maturity', 90% of the 40 fibers in that sample were mature, assuming there was no fiber-selection bias in the measurements.

8.2 Estimating Fiber Fineness

Fiber 'fineness' has long been recognized as an important factor in yarn strength and uniformity, properties which largely depend upon the average number of fibers in the yarn cross-section. Spinning finer fibers results in more uniform and stronger yarns (Ramey, 1982). However, direct determinations of 'biological' fineness in terms of fiber or lumen diameter and cell-wall thickness are precluded by the high expense in both time and labor, the non-circular cross-sections of dried cotton fiber, and the high degrees of variation in fiber fineness (Ramey, 1982; Munro, 1987). Advances in image analysis have improved determinations of fiber biological fineness and maturity (Matic-Leigh and Cauthen, 1994), but fiber image analyses remain too slow and sample-size limited for inclusion in the HVI classing process.

Initially, the textile industry adopted 'gravimetric' fiber fineness or linear density as an indicator of fiber spinning properties, which depend on fiber fineness and maturity combined (ASTM D 1769-77, 1977; Ramey, 1982). The gravimetric fineness testing method was discontinued in 1989, but the textile linear density unit of 'tex' persists. Tex is measured in grams per kilometer of fiber or yarn, and fiber fineness is usually expressed as millitex or micrograms per meter (Ramey, 1982; Munro, 1987). Direct measurements of fiber fineness, either biological or gravimetric, were subsequently replaced by indirect fineness measurements based on the resistance to air flow of a bundle of fibers.

The first indirect test method approved by ASTM for fiber maturity, linear density, and maturity index was the causticaire method in which the resistance of a plug of cotton to air flow was measured before and after a cell-wall swelling treatment with an 18% (4.5 M) solution of sodium hydroxide (ASTM D 2480-82, 1991). The ratio between the rate of air flow through an untreated and then treated fiber plug was taken as an indication of the degree of fiber wall development or maturity. The airflow reading for the treated sample was squared and corrected for maturity to serve as an indirect estimate of linear density. Causticaire method results were highly variable among laboratories, and the method was never recommended for acceptance testing before it was discontinued in 1992.

The arealometer was the first dual-compression airflow instrument developed for estimating both fiber fineness and fiber maturity from airflow rates through untreated raw cotton (ASTM D 1449-58, 1976; Lord and Heap, 1988). The arealometer measures the specific surface area of loose cotton fibers, *i.e.*, the external area of fibers per unit volume (approximately 200 mg each in four to five replicate samples). Empirical formulae were developed for calculating the approximate maturity ratio and the average perimeters, wall thicknesses, and weights per inch from the specific area data. The precision and accuracy of arealometer determinations are sensitive to sample preparation variations, repeated sample handling, and previous mechanical treatment of the fibers, *e.g.*, blending and opening conditions. The arealometer was never approved for acceptance testing, and the ASTM method was withdrawn in 1977 without replacement.

The variations in cotton fiber biological fineness and relative maturity described earlier cause the porous fiber plugs used in air-compression measurements of those properties to respond differently to compression and, consequently, to air flow (Lord and Heap, 1988). The IIC-Shirley Fineness/Maturity Test (Shirley FMT), a dual-compression instrument, was developed to compensate for this plug variation effect (ASTM D 3818-92, 1994). Air is drawn through a 4-g fiber plug at a rate of 4.0 L min[-1], and the initial pressure drop measured. The sample is then compressed to a higher density, and the pressure drop is measured at a flow rate of 1.0 L min[-1]. Fineness in millitex and a maturity index in percentage points are calculated according to empirically based formulae given in the ASTM method (ASTM D 3818-92, 1994; Ramey, 1982). The FMT is considered suitable for research, but not for acceptance testing, due to low precision and accuracy. Instead, micronaire has become the standard estimate of both fineness and maturity in the USDA, AMS classing offices.

8.3 Micronaire, an Indirect Estimate of Fiber Fineness and Maturity

Micronaire is the most commonly used instrumental cotton fiber property test (Lord and Heap, 1988; Moore, 1996). Micronaire is an indirect measure of the air permeability of a test specimen of known mass in a container of fixed dimensions. Initially, air permeability of the sample was thought to depend on the fiber linear density, and the empirically derived curvilinear micronaire scale was set in gravimetric fineness units of fiber weight per inch (Ramey, 1982; Lord and Heap, 1988). However, basic fluid-flow theory sets air permeability as inversely dependent on the square of the fiber surface area; and linear density units were subsequently dropped from the micronaire scale so that micronaire or 'mike' is now treated as a dimensionless fiber property.

Under standardized testing and calibration conditions, the micronaire test method incorporated in the HVI systems (ASTM D 4604-86, 1994; ASTM D 4605-86, 1994) is considered satisfactory for acceptance testing, if users of the test results consider micronaire readings as estimates of *both* fiber fineness and maturity. The micronaire test in the HVI system is relatively insensitive to sample preparation and *small* variations in relative humidity and temperature during testing, and standardized preconditioning is required at the USDA, AMS classing offices. For micronaire determinations in the HVI system, the minimum sample size is currently 10 g (ASTM D 4604-86, 1994; ASTM D 4605-86, 1994).

In the U.S., the 'acceptable' Upland micronaire range for which no price penalty is assessed is 3.5 to 4.9 with a premium range of 3.7 to 4.2. Empirical relationships between micronaire and cotton fiber processing properties have been developed, and bale micronaire readings are used by mills in bale selection and blending (Chewning, 1995; El Mogahzy and Gowayed, 1995a; 1995b).

The fineness factor of micronaire is considered more important in spinning, and fiber maturity is thought to have more effect on dyeability. However, the finer the fiber, the higher the number of reflective surfaces per unit area and, consequently, the higher the luster of the dyed fabric (Ramey, 1982). Immature fibers have thinner walls and are finer than mature fibers of the same variety. However, lower micronaire fibers stretch, tangle, and break more easily and do not impart the greater yarn strength and uniformity expected of finer fibers. The complex interactions among fiber fineness, fiber maturity, fiber spinning properties, and fiber dye-uptake characteristics are very difficult to interpret or predict and can result in confusion and frustration for breeders and physiologists who engage in research designed to improve fiber quality (Cooper *et al.*, 1996; Palmer *et al.*, 1996a, 1996b; Pellow *et al.*, 1996).

8.4 The Fiber Fineness/Maturity Complex

Various methodologies and instruments have been used to separate the causes and the effects of cotton fiber 'fineness' and 'maturity'. In addition to the previously discussed microscopic and image-analysis assays of fiber 'biological' fineness and estimates of fiber linear density, near-infrared transmission spectroscopy (NITS) has been used to describe a linear relationship between fiber fineness and the amount of light scattered (Montalvo *et al.*, 1989). The distribution of cotton fiber fineness as diameter by number can also be determined rapidly and reproducibly by the AFIS Length and Diameter (L&D) module (Bragg and Wessinger, 1993; Yankey and Jones, 1993).

The AFIS Fineness and Maturity (F&M) module uses scattered light to measure single-fiber cross-sectional areas (Bradow *et al.*, 1996a; Williams and Yankey, 1996). Algorithms have been developed for calculating the Fine Fiber Fraction (% of fibers for which the cross-sectional area by number is less than 60 μm^2), perimeter, and a micronaire analog, micronAFIS from fiber data collected by the AFIS-F&M. Newer AFIS instruments combine the L&D and F&M modules as the Length and Maturity (L&M) module that generates fineness data in millitex (Williams and Yankey, 1996). Near infra-red reflectance spectroscopy (NIRS) has also been used to examine fiber cross-sectional area, *e.g.*, fineness (Montalvo, 1991a; 1991b; 1991c).

8.5 Fiber Maturity and Dye Testing

Fiber fineness is most closely associated with spinning characteristics and the properties of the resulting yarn (Ramey, 1982). Fiber maturity affects the color of the fiber, both before and after dye application (Lord and Heap, 1988; Smith, 1991). Indeed, the anisotropic nature of the fibrillar cell walls of cotton fibers suggested the use of plane-polarized light microscopy for assessing cell wall developmental maturity (Lord and Heap, 1988). However, sorting fibers into maturity classes of thin-walled (violet-indigo), immature (blue), and thicker walled/more mature (yellow) is slow, strongly biased by differential color sensitivity of the classer, and insufficiently sensitive to the difference between mature fibers of small perimeter and immature fibers of large perimeter. Differential dye tests for assessing fiber maturity, including the Goldthwaite red-green dye test, have been found to be similarly biased and further confounded by differences in sample fiber fineness and affinity for the dyes used (Milnera, 1987; Lord and Heap, 1988). The Goldthwaite red-green dye test, in which redness is associated with maturity and an increasingly greenish coloration connotes decreasing fiber maturity, is still used (Pellow *et al.*, 1996). However, the results are qualitative and highly subjective since most dyed samples appear as a mat of mixed red and green fibers with the green coloration being strongly associated with boll suture lines in dyed intact, mature bolls. In dye uptake tests using a single dye, fibers appressed to boll sutures were also dye-resistant and, by inference, immature (Bradow *et al.*, 1996c).

8.6   Fiber Maturity and Circularity

As an estimate of fiber maturity, direct measurement of average cell wall thickness in traverse fiber sections is subject to numerous and serious biases, *e.g.*, insufficient sample size and non-circularity of cotton fibers (Lord and Heap, 1988; Matic-Leigh and Cauthen, 1994). Consequently, degree of thickening [θ] was defined as a measure of fiber maturity based on fiber cross-section and perimeter (Lord and Heap, 1988).

Degree of thickening is the cross-sectional area of the fiber wall divided by the area of a circle of the same perimeter. Thus, completely circular fibers of any perimeter have θ values equal to one. Mature, thick-walled fibers (56 DPA) collapsed into cross-sections shaped like kidney-beans with θ means approximating 0.576 for Upland genotypes and 0.546 for Pima (Bradow *et al.*, 1996b). Immature, thin-walled fibers (21 DPA) collapsed into flattened elliptical shapes with Upland θ means of 0.237 and Pima θ means of 0.221. Fruiting site and seed location within the locule also modulated fiber circularity and the degree of wall thickening (Table 21-1). In microscopic determinations of formalin-treated, air-dried *G. hirsutum* 'Gujaret 67' fibers, the 35 DPA circularity was 0.215 and the circularity at 63 DPA was 0.685 (Petkar *et al.*, 1986). In the same report, *G. barbadense* 'ERB4530' fiber circularity was 0.180 at 35 DPA and 0.567 at 56 DPA.

Degree of thickening can be directly quantified by image analysis (Matic-Leigh and Cauthen, 1994) or by AFIS (Bradow *et al.*, 1996a; Williams and Yankey, 1996). The AFIS-F&M also provides Immature Fiber Fraction (% of fibers with θ < 0.25; Bradow *et al.*, 1996a), and the AFIS-L&M reports Immature Fiber Content (defined as for Immature Fiber Fraction from the AFIS-F&M) and Immaturity ratio, which is the ratio of fibers with θ > 0.5 divided by the number of fibers with θ < 0.25 (Williams

and Yankey, 1996). In contrast to micronaire-based methods in which the fiber sample is held stationary in a porous plug when 'maturity' is measured at some arbitrary point on the long axis, the AFIS, in either configuration, estimates θ and cross-sectional area along the entire length of the fiber as up to 10,000 fibers per sample flow between the light source and the detector. The scattering of light in the near-infrared (NIR or near-infrared reflectance) is also used to quantify fiber maturity (Gordon, 1995; Thomassson *et al.*, 1995). A VIS/NIR diode-array HVI system is also in development (Buco *et al.*, 1995; 1996; Montalvo *et al.*, 1996).

8.7   Trends in Fiber Maturity (Micronaire)

In the U.S., the trend in Upland micronaire since 1985 has been a slight, but irregular, increase toward the 1995 crop average of 4.35 and the 1998 crop average of 4.47 (Cotton Inc., 1999). In 1995, the Upland cotton micronaire values in all western states were equal to or lower than the corresponding micronaire values from 1994 (Sasser and Shane, 1996). In every state east of Texas and Oklahoma, 1995 micronaire averages were lower than in 1994. This pattern was attributed to environmental influences on cotton micronaire. In 1998 micronaire averages from all classing offices were higher than the 1997 micronaire averages for those classing offices (Cotton Inc., 1999). Many areas of the U.S. Cotton Belt were plagued by hot, dry weather in 1998 (Wrona *et al.*, 1999).

8.8   Fiber Maturity and Environment

Whatever the method, direct or indirect, that is used for estimating fiber maturity, the fiber property being assayed remains the thickness of the cell wall. The primary cell wall and cuticle together (*ca.* 0.1 μm) make up approximately 2.4% of the total wall thickness (*ca.* 4.1 μm of a cotton fiber at harvest; Ramey, 1982; Ryser, 1985; Matic-Leigh and Cauthen, 1994). The remaining 98% of a fiber cell is the cellulosic secondary wall which is deposited during fiber maturation. Therefore, any environmental factor that affects photosynthetic carbon fixation and cellulose synthesis will also modulate cotton fiber wall thickening and, consequently, fiber maturation (Sassenrath-Cole and Hedin, 1996; Bradow *et al.*, 1996b; Murray, 1996; Murray and Brown, 1996; 1997). Please refer to Chapters 12, 14, and 19 for reviews of cotton carbon metabolism and assimilate partitioning.

8.9   Fiber Maturity and Temperature and Planting Date

The dilution, on a weight basis, of the chemically complex primary cell wall by secondary wall cellulose has been followed with x-ray fluorescence spectroscopy. This technique determines the decreases in relative weight ratios, over time, of calcium associated with the pectin-rich primary wall (Wartelle *et al.*, 1995; Bradow *et al.*, 1996a; 1996b; 1997b). Growth-environment differences between the two years of the studies cited significantly altered the maturation rates, quantified as rate of calcium weight dilution, of both Upland and Pima genotypes. The rates of secondary wall deposition in both Upland and Pima genotypes were closely correlated with growth temperature, *i.e.*, heat-unit accumulation (Johnson *et al.*, 1997; Bradow *et al.*, 1996b).

When temperature regimes were evaluated in SPAR units that differed by 2°C below or above the ambient temperature, it was found that as temperature increased theta, cross-sectional area and micronafis (micronaire) increased with increasing temperature (Reddy *et al.*, 1999). Under growth chamber conditions micronaire values were highest in a 30/20°C day/night temperature regime (Liakatas *et al.*, 1998). In temperature regimes in which the mean 24 h temperature was 22°C alteration in the maximum and minimum temperatures led to the conclusion that in a high maximum-low minimum regime (35-11°C), the high maximum effect overshadowed the effect of the low mean thus acting as a higher mean temperature (Liakatas *et al.*, 1998).

An early study on the effects of suboptimal temperatures on fiber development used 'micronaire fineness' to quantify the effects of heat-unit deficits (Hessler *et al.*, 1959). Temperature deficiencies (degree-hours per week below 21.1°C) in mid- or late-season reduced micronaire means so that late-season micronaire was in the penalty range, *i.e.*, below 3.5. Cell-wall thickness was not measured in this study, but cool night temperatures (15 to 28°C) modulated cellulose synthesis and secondary cell wall deposition (Haigler *et al.*, 1991; 1994; 1996).

Increases in micronaire over time were documented in maturing fibers by Hessler and coauthors (1959), and micronaire (micronAFIS) was also found to increase linearly with time for Upland and Pima genotypes (Bradow *et al.*, 1996a; 1996b). The rates of micronaire increase were correlated with heat-unit accumulation (Johnson *et al.*, 1997; Bradow *et al.*, 1997b). Rates of increase in fiber cross-sectional area were less linear than the corresponding micronaire rates, and rates of Upland and Pima fiber cell-wall thickening [quantified as θ by AFIS] were linear and without significant genotype effect (Bradow *et al.*, 1996b).

Environmental modulation of fiber maturity [micronaire] by temperature has most often been identified in planting and flowering date studies (Aguillard *et al.*, 1980; Greef and Human, 1988; Porter *et al.*, 1996; Bradow *et al.*, 1997b). Micronaire of four Upland genotypes decreased as the planting date advanced from early April to early June in Louisiana (Aguillard *et al.*, 1980). The effects of planting date on micronaire, FMT fiber maturity ratio, and fiber fineness (in millitex) were highly significant in a South African study (Greef and Human, 1988). Although genotype differences were detected among the three years

of the study, delayed planting generally resulted in lower micronaire. The effect of late planting was repeated in the FMT maturity ratio and fiber fineness data. Consistent with earlier reports (Bilbro and Ray, 1973; Cathey and Meredith, 1988), delaying planting until mid-June from an early-May planting norm decreased micronaire of Upland genotypes grown in coastal South Carolina (Porter *et al.*, 1996). Planting date significantly modified θ, Immature Fiber Fraction, cross-sectional area, and micronaire (micronAFIS) of four Upland genotypes, which were also grown in South Carolina (Bradow *et al.*, 1997b). In general, micronaire decreased with later planting, but early planting also reduced micronaire of 'DPL5690', a long-season genotype, in a year in which temperatures were suboptimal in the early part of the season. Harvest dates in this study were also staggered so that the length of the growing season was held constant within each year. Therefore, season-length should not have been an important factor in the relationships found between planting date and fiber maturity. However, micronaire was reduced by early defoliation in a Mississippi study (Snipes and Baskin, 1994).

8.10   Fiber Maturity and Source-Sink Manipulation

Variations in fiber maturity were linked with source-sink modulations related to flowering date (Bradow *et al.*, 1997b), fruiting site (Pettigrew, 1995; Davidonis *et al.*, 1996; Bradow *et al.*, 1997a; Murray and Brown, 1997), or seed position within the boll (Bradow *et al.*, 1996a; Davidonis *et al.*, 1996). However, manipulation of source-sink relationships by early-season square (floral bud) removal had no consistently significant effect on Upland cotton micronaire (Pettigrew *et al.*, 1992). Early-season square removal also did not affect fiber perimeter or wall thickness (measured by arealometer). Partial defruiting increased micronaire and had no consistent effect on Upland fiber perimeter in bolls from August flowers (Pettigrew, 1995). Based on an increase in micronaire detected under natural fruiting load, fibers in August-bloom bolls of the Upland genotype, 'DPL5415', matured more rapidly than did fibers from July-flower bolls of that genotype (Bradow *et al.*, 1996b; 1997b). Other investigators found that loss of flowers four weeks or more after flowering had commenced led to increased micronaire, but loss of flowers earlier in the season had no effect (Jones *et al.*, 1996). The effects of intra-boll source/sink dynamics on fiber maturity (θ, Immature Fiber Fraction and micronaire/micronAFIS) have also been quantified (Davidonis *et al.*, 1996).

8.11   Fiber Maturity and Water

Generous water availability can delay fiber maturation (cellulose deposition) by stimulating competition for assimilates between early-season bolls and vegetative growth (Hearn, 1995). Adequate water can also increase the maturity of fibers from mid-season flowers by supporting photosynthetic carbon fixation. In a year when rainfall was insufficient, initiating irrigation when the first bolls set were 20 days old increased micronaire, but irrigation initiation at first bloom had no effect (Spooner *et al.*, 1958). Irrigation and water-conservation effects on fiber fineness (millitex) were inconsistent between years, but both added water and mulching tended increase fiber fineness (Singh and Bhan, 1993). Aberrations in cell-wall synthesis correlated with drought stress have been detected and characterized by glycoconjugate analysis (Murray, 1996).

Adequate water supply in a growing season allowed maturation of more bolls at upper and outer fruiting positions, but the mote counts in those 'extra' bolls tended to be higher and the fibers within those bolls tended to be less mature (Hearn, 1995; Davidonis *et al.*, 1996). Rainfall during the blooming period and the associated reduction in insolation levels resulted in reduced fiber maturity (Bradow *et al.*, 1997b). Irrigation method also modified micronaire levels and distributions among fruiting sites. Early-season drought resulted in more mature fiber with higher micronaire from bolls in branch positions one and two on the lower fruiting branches of rainfed plants. However, reduced insolation and heavy rain reduced micronaire and increased Immature Fiber Fractions in bolls from flowers that opened during the prolonged rain incident. Soil water deficit and excess both may reduce micronaire if the water stress is severe or prolonged (Marani and Amirav, 1971; Ramey, 1986).

8.12   Fiber Maturity and Mineral Nutrition

Genotype differences, rather than added nitrogen, were responsible for micronaire treatment effects in an early study (MacKenzie and van Schaik, 1963). Green manures and added nitrogen had little consistent effect on fiber maturity, including micronaire (Bauer *et al.*, 1993; Bauer and Busscher, 1996). Nitrogen also did not affect fiber maturity index, micronaire, or perimeter of eight genotypes of differing relative earliness and regional adaptation. However, added potassium (112 kg ha^{-1}) significantly increased micronaire, fiber maturity index, and perimeter (Pettigrew *et al.*, 1996). That same level of added potassium did not affect micronaire in another study, but nematicide application did increase micronaire, probably through enhanced root growth (Minton and Ebelhar, 1991). Added potassium increased micronaire of two Acala genotypes, an effect the authors attributed to a potassium requirement for metabolic processes related to fiber secondary wall thickening (Cassman *et al.*, 1990). Genotype differences were noted in the relationship between micronaire and potassium availability. In a five-year study in which the fields were harvested twice, micronaire decreased with increasing nitrogen application rate (101 to 202 kg ha^{-1}; Ebelhar *et al.*, 1996). The decrease in micronaire was linear with increasing nitrogen for the first harvest only.

8.13   Fiber Maturity and Genetic Improvement

Micronaire or maturity data now appear in most cotton improvement reports (Green and Culp, 1990; Meredith, 1990; May and Green, 1994; Tang *et al.*, 1996). In a five-parent half diallel mating design, environment had no effect on HVI micronaire (Green and Culp, 1990). However, a significant genotype effect was found and associated with differences between parents and the F$_1$ generation and differences among the F$_1$ generation. The micronaire means for the parents were not significantly different although HVI micronaire means were significantly different for the F$_1$ generation as a group. HVI was judged to be insufficiently sensitive for detection of the small differences in fiber maturity resulting from the crosses. In another study, F$_2$ hybrids had finer fiber (lower micronaire) than the parents, but the improvements were deemed too small to be of value (Meredith, 1990). Unlike the effects of environment on the genetic components of other fiber properties, variance in micronaire due to the genotype X environment interaction can reach levels expected for genetic variance in length and strength (Meredith and Bridge, 1972; May and Green, 1994). Significant interactions were found between genetic additive variance and environmental variability for micronaire, strength, and span length in a study of 64 F$_2$ hybrids (Tang *et al.*, 1996).

The strong environmental components in micronaire and fiber maturity limit the use of these fiber properties as guides in studies of genotype differences in responses to growth environment. Based on micronaire, fiber maturity, cell-wall thickness, fiber perimeter, or fiber fineness data, row spacing had either no or minimal effect on okra-leaf or normal leaf genotypes (Heitholt, 1993b). Early planting reduced micronaire, wall-thickness, and fiber fineness of the okra-leaf genotype in one year of that study. In another study of leaf pubescence, nectaried *vs.* no nectaries, and leaf shape, interactions with environment were significant but of much smaller magnitude than the interactions among traits (Meredith *et al.*, 1996a).

Micronaire means for Bt transgenic lines were higher than the micronaire means of Coker 312 and MD51ne when those genotypes were grown in Arizona (Wilson *et al.*, 1994). In two years out of three, micronaire means of all genotypes, including the controls, exceeded 4.9. This apparent 'environmental' effect on micronaire may have been caused by a change in fiber testing methods in the one year for which micronaire were below the upper penalty limit. Genotype differences in bulk micronaire may be emphasized or minimized, depending on measurement method used (Meredith *et al.*, 1996b; Palmer *et al.*, 1996b; Pellow *et al.*, 1996).

9.   GRADE

In U.S. cotton classing, non-mandatory 'grade' standards were first established in 1909, but compulsory

Upland grade standards were not set until 1915 (Perkins *et al.*, 1984). Official Pima standards were first set in 1918. Grade is a composite assessment of three factors – *color*, *leaf*, and *preparation* (USDA, 1980; Munro; 1987; Moore, 1996). Color and trash (leaf residue) can be quantified instrumentally, but traditional, manual cotton grade classification is still provided by USDA, AMS, in addition to the instrumental HVI trash and color values. Thus, cotton grade reports are still made in terms of traditional color and leaf grades, *e.g.*, light spotted, tinged, strict low middling.

The color grade that is provided by USDA-AMS Cotton Classing Offices is determined by the classer on the basis of official AMS color grade standards (Edmisten, 1997b). Color refers to the 'whiteness' or 'yellowness' of the fiber. The numerical codes for American Upland color grades are found in Table 21-2. These are the Color Grade codes that appear in the USDA-AMS Cotton Division Universal Classification Data reports provided by the USDA-AMS Classing Offices for each bale. A special condition code of '96' is assigned to mixtures of Upland and Pima. Similarly '97' indicates "fiber damaged' color grade and '98' indicates 'water damaged' fiber.

9.1   Preparation

There is no instrumental measure of preparation, *i.e.*, the degree of roughness/smoothness of the ginned lint. Methods of harvesting, handling, and ginning cotton produce differences in roughness that are apparent in manual inspection, but no clear correlations have been found between degree of preparation and spinning success. The frequency of tangled knots or mats of fiber (neps) may be higher in 'high prep' ginned cotton, and the growth and processing environments can also modulate nep frequency (Perkins *et al.*, 1984). However, abnormal preparation occurs in less than 0.5% of the U.S. crop during harvesting and ginning (Moore, 1996).

9.2   Trash or Leaf Grade

Even under ideal field conditions, cotton lint becomes contaminated with leaf residues and other trash (Perkins

Table 21-2. Color Grade of American Upland Cotton (from Edmisten, 1997b).

	White	Light spotted	Spotted	Tinged	Yellow stained
Good Middling	11[z]	12	13	--	--
Strict Middling	21[z]	22	23[z]	24	25
Middling	31[z]	32	33[z]	34[z]	35
Strict Low Middling	41[z]	42	43[z]	44[z]	--
Low Middling	51[z]	52	53[z]	54[z]	--
Strict Good Ordinary	61[z]	62	63[z]	--	--
Good Ordinary	71[z]	--	--	--	--
Below Ordinary	81	82	83	84	85

[z] Physical Standards. All others are descriptive.

et al., 1984). Although most foreign matter is removed by the cleaning and drying processes during ginning, total trash extraction is impractical and can lower the quality of ginned fiber. In HVI cotton classing, trash in raw cotton is measured by a video scanner (Trash Meter), and the trash data are reported in terms of the total trash area and trash particle counts (ASTM D 4604, 1994; ASTM D 4605, 1994). These trash content data may be used for acceptance testing. In 1993, 'classer's grade' was split into color grade and leaf grade (Cotton Inc., 1999). Cotton fibers with the smallest amount of foreign matter, other factors being equal, have the highest value.

'Leaf' includes dried, broken plant foliage particles and can be divided into two general groups: large leaf and 'pin' or 'pepper' trash (Perkins *et al.*, 1984; Moore, 1996). The pepper trash is more expensive and difficult to remove and significantly lowers the value of the cotton to the manufacturer. Trash also includes stems, burs, bark, whole seeds, seed fragments, motes (undeveloped seeds), grass, sand, oil, and dust, all of which are found in ginned cotton. Growth environment obviously affects the amount of windborne contaminants trapped in the fibers, and environmental factors that affect pollination and seed development determine the frequency of undersized seeds and motes (Davidonis *et al.*, 1995; Davidonis *et al.*, 1996). Reductions in the frequencies of motes and small-leaf trash have also been correlated with semi-smooth and super-okra leaf traits (Novick *et al.*, 1991). Environment (year), harvest system, genotype, and second order interactions between those factors all had significant effects on leaf grade (Williford *et al.*, 1986). Delayed harvests resulted in lower grade fiber.

9.3   Fiber Color

Raw fiber stock color measurements are used in controlling the color of manufactured gray, bleached, or dyed yarns and fabrics (Nickerson and Newton, 1958). Of the three components of cotton grade, fiber color is most directly linked to growth environment. Color measurements are also related to overall fiber quality so that bright (reflective), creamy-white fibers are more mature and of higher quality than the dull, gray or yellowish fibers associated with field-weathering and generally low fiber quality (Perkins *et al.*, 1984). Although Upland cotton fiber is naturally white to creamy-white, pre-harvest exposure to weathering and microbial action can cause fiber to darken and to lose 'brightness' (Perkins *et al.*, 1984; Allen *et al.*, 1995). Premature termination of fiber maturation by frost or drought characteristically increases the saturation of the yellow fiber-color component. Other conditions, including insect damage and foreign matter contamination also modify fiber color (Moore, 1996).

The ultimate 'acceptance test' for fiber color, and also for finished yarns and fabrics is the human eye. Therefore, instrumental color measurements must be highly correlated with visual judgment. In the HVI classing system, color is quantified as the degree of reflectance (Rd) and yellowness (+b), two of the three tristimulus color scales of the Nickerson-Hunter colorimeter (Nickerson, 1950; Nickerson and Newton, 1958; ASTM D 2253-88, 1994; Thomasson and Taylor, 1995).

Munsell Color Space can be represented quantitatively as three mutually perpendicular unit vectors in which Rd (reflectance, ±L) is represented perpendicularly on the +white/-black Z-axis, and the chromaticity coordinates, ±a (+red/-green X-axis) and ±b (+yellow/-blue Y-axis) are represented in the horizontal plane. The USDA has established an official color grade diagram that relates Rd on the vertical axis and +b on the horizontal axis to the traditional color grades of cotton (Perkins *et al.*, 1984). The USDA Rd reflectance scale range is from +40 (darker) to +85 (lighter/brighter). The +b scale is from +4 to +18 with the higher +b indicating an increasing degree of yellow saturation. The third tristimulus Color Space scale, +a, indicates the degree of red saturation and is not reported in HVI color measurements.

Colorimeter measurements and the USDA color diagram have been empirically correlated with manual classer's color grades. Thus, a fiber sample with Rd = +70.7 and +b = 9.7 would fall in the light-spotted, strict low middling grade. HVI classing information also supplies a number code in which the first number refers to color, *i.e.*, white, light spotted, *etc.* and the second number refers to grade, *i.e.*, good middling, strict low middling, *etc.* The code for the fiber sample above would be 42-1 with the number after the hyphen describing more precisely the intersection of the Rd and +b vectors on the USDA color grade diagram. Samples of the of the USDA color chart can be found on page 456 of the Perkins *et al.* (1984) reference and on page 587 of the ASTM D 2253-88 (1994) method. Colorimeter data can also be used to quantify dye uptake success (Bradow *et al.*, 1996c).

Fiber maturity has been associated with dye variability in finished yarn and fabric (Smith, 1991; Bradow *et al.*, 1996c; Bradow and Bauer, 1997a; Bradow *et al.*, 1997b), but the color grades of raw fibers have seldom been linked to environmental factors or agronomic practices. In one year only of a three-year study, increased nitrogen fertilization and application of mepiquat chloride were associated with decreased Rd, which represented an undesirable graying of the raw fiber (Boman and Westerman, 1994). There was also an undesirable linear increase in +b (yellowing) with increasing nitrogen level, but mepiquat chloride did not affect fiber yellow saturation (Boman and Westerman, 1994; Ebelhar *et al.*, 1996). Environment (year), planting date, and genotype all significantly affected fiber Rd and +b in a South Carolina study (Porter *et al.*, 1996). Late planting (mid-June) had the most consistently negative effect on both Rd and +b. In undyed knit fabric, fiber reflectance (+L, brightness) was positively correlated with increasing cumulative heat units (Bradow and Bauer, 1997a). Undyed-fiber yellow chromaticity, +b, was negatively related to increasing heat-unit accumulation. Removal of trash from the lint increased reflectance (Rd) but did not affect +b (Thomasson, 1993; Nawar, 1995).

9.4   Trends in Cotton Grade

Since 1985, there has been a significant increase in the number of U.S. bales classed in the 'white' grade and a corresponding decrease in the number of bales in the light spotted grade (Sasser and Shane, 1996). Since 1988, approximately 78% of the U.S. bales have been in the white grade, but there continues to be considerable variation among classing offices (Cotton Inc., 1999). In 1997, light spotted bale frequencies were above 35% in Georgia while Alabama, Louisiana, Arkansas and Missouri had lighspotted bale frequencies below 35% (Cotton Inc., 1999).

Although not yet included in the USDA, AMS cotton fiber classing system, cotton stickiness is becoming an increasingly important problem (Perkins, 1991; Brushwood and Perkins, 1996). Two major causes of cotton stickiness are insect honeydew from whiteflies and aphids and abnormally high levels of natural plant sugars. Insect honeydew contamination is randomly deposited on the lint in heavy droplets and has a devastating, production-halting effect on fiber processing. The cost of clearing processing equipment halted by sticky cotton is so high that buyers have included "honeydew free" clauses in purchase contracts and have refused cotton from regions known to have insect-control problems. Rapid methods for instrumental detection of honeydew are under development for classing offices and mills (Frydrych *et al.*, 1995; Perkins and Brushwood, 1995; Ethridge and Hequet, 1999). Elevated levels of natural plant sugars have been associated with premature crop termination from frost or drought.

10.   RESEARCH NEEDS

Like all agricultural commodities, the value of cotton fiber responds to fluctuations in market supply-and- demand forces (Moore, 1996). Further, pressure toward specific improvements in cotton fiber quality, *e.g.*, the higher fiber strength needed for modern high-speed spinning, has intensified as a result of technological advances in textile production and increasingly stringent quality standards for finished cotton products. Changing fiber-quality requirements and increasing economic competition on the domestic and international levels has led to fiber quality becoming as important a factor in the value of cotton fiber as fiber yield (Ethridge, 1996; Hudson *et al.*, 1996). Indeed, it is the *quality*, not the quantity, of the fiber ginned from the cotton seed that decides the end use and economic value of a cotton crop and, consequently, the profits returned to the producer and processor.

Wide differences in cotton fiber quality and shifts in the demand for particular fiber properties based on end-use processing requirements have resulted in the annual creation of a price schedule of premiums and discounts for grade, staple length, micronaire, and strength (Deussen and Faerber, 1995; Ethridge, 1996). The price schedule was made possible by the development of rapid, quantitative methods for measuring those fiber properties considered most important in textile production (Chewning, 1995; Deussen and Faerber, 1995; Frye, 1995). Thus, with the arrival of fiber-quality quantitation by HVI, predictive models for ginning, bale-mix selection and fiber processing success were developed for textile mills (Chewning, 1995). Price analysis systems based on HVI fiber-quality data also became feasible (Deussen and Faerber, 1995; Ethridge, 1996; Hudson *et al.*, 1996). Quantitation, predictive modeling, and statistical analyses of previously qualitative fiber properties are now both practical and common in textile processing and marketing.

Field-production and breeding researchers, however, have failed to take full advantage of the fiber-quality quantitation methods developed for the textile industry. Most field and genetic improvement studies still focus upon *yield improvement* with little attention paid to fiber quality beyond obtaining bulk fiber length, strength, and micronaire averages for each treatment (*e.g.*, May and Green, 1994 Meredith *et al.*, 1996a, Porter *et al.*, 1996). Indeed, cotton crop simulation and mapping models of the effects of growth environment on cotton have been limited, almost entirely, to yield prediction and cultural-input management (*e.g.*, Boone *et al.*, 1995; Lemmon *et al.*, 1996; Wanjura *et al.*, 1996b; Chapter 38; Chapter 39). Some progress has been made in familiarizing breeders with AFIS data and their value in the late stages of cultivar development (Calhoun *et al.*, 1997).

Along the time line from cotton field to finished fabric, most field-production studies and the resulting quantitative fiber-quality databases terminate at the bale level. Fiber processing studies normally begin with selection of bales from the mill warehouse (Chewning, 1995). Although the experimental designs of field studies always include collection and analysis of environmental (weather) data, fiber processing studies begin to consider growth-environment factors *after* some significant processing defect cannot be attributed to post-harvest events. Very few integrated studies have attempted to follow fiber production and utilization from floral anthesis to finished yarn or fabric (*e.g.*, Bradow *et al.*, 1996c; Meredith *et al.*, 1996b; Palmer *et al.*, 1996a;

1996b; Pellow *et al.*, 1996). Physiological studies and textile processing models suggest that bulk fiber-property means at the bale, module, or crop level do not describe fiber quality with sufficient precision. Bulk fiber-property means do not adequately describe the variations in the fiber population response to environmental factors during the growing season. Such composite descriptors cannot accurately predict how highly variable fiber populations would perform during processing. Meaningful descriptions of the effects of environment on cotton fiber quality await high-resolution examinations of the variabilities, induced and natural, in fiber-quality means. Only then can the genetic and environmental sources of fiber-quality variability be quantified and modulated to produce the high quality cotton fiber demanded by the modern textile industry and, ultimately, the consumer. Only through increased understanding of the physiological responses to the environment that determine cotton fiber quality can real progress be made toward producing high yields of cotton fiber that is as white as snow, as strong as steel, as fine as silk, as long as wool, *and* as uniform as genotype response to the environment will allow.

11.   ACKNOWLEDGMENTS

The authors thank the following colleagues for their helpful comments during the preparation of this chapter: Dr. X. Cui; USDA, ARS, New Orleans, LA; Dr. R.M. Johnson, International Textile Center, Texas Tech, Lubbock, TX; Dr. O.L. May, USDA, ARS, Florence, SC; Dr. A.K. Murray, Glycozyme, Inc., Irvine, CA; Dr. K. Rajasekeran, USDA, ARS, New Orleans, LA; Dr. G.F. Sassenrath-Cole, USDA, ARS, Mississippi State, MS; Dr. Reiyao Zhu, International Textile Center, Lubbock, TX.

Trade names are necessary for reporting factually on available data. The USDA neither guarantees nor warrants the standard of the product or the service. The use of the name USDA implies no approval of the product or service to the exclusion of others that may be suitable.

Chapter 22

PHYSIOLOGICAL RESPONSES TO TILLAGE SYSTEMS, COVER CROPS, AND RESIDUE MANAGEMENT

C.O. Gwathmey[1], J.F. Bradley[2], A.Y. Chambers[3], D.D. Howard[1], and D.D. Tyler[1]
[1]*Dept. of Plant and Soil Science, Univ. of Tennessee, Jackson, TN;* [2]*Milan Experiment Station, Univ. of Tennessee, Milan, TN;* [3]*Dept. of Entomology and Plant Pathology, Univ. of Tennessee, Jackson, TN.*

1. INTRODUCTION

Tillage refers to mechanical manipulation of the soil, and a tillage system refers to a specific set of operations that manipulate the soil to produce a crop. Tillage operations modify the edaphic environment of cotton and may thus affect physiological determinants of yield. Winter annual cover crops and management of their residue may also affect the growth, development, and yields of subsequent cotton crops.

Different tillage and cover crop systems have evolved in various parts of the U.S. cotton belt and around the world in response to local soil and environmental conditions. These systems may have short- or long-term effects on the edaphic environment. Short-term effects mainly affect the same season's crop, whereas long-term effects may take several years of repeated tillage (or no-tillage) operations to become apparent. For instance, soil temperature, moisture content, or aeration may be rapidly modified by tillage. Soil organic matter, erosion, or bulk density may gradually change over several years of a continuous tillage system.

This chapter summarizes information available on cotton responses to tillage and residue management. The main tillage systems used for cotton are briefly reviewed, and their effects on stand establishment, vegetative growth, reproductive development, and yield are described. Some implications of tillage systems and cover crops on strategies of fertilizer management are discussed. Effects of residue from the main cover crops used with cotton are described, including stand establishment, allelopathic potential, nitrogen relations, reproductive development, and yield. A concluding summary suggests areas of research that may improve our understanding of observed responses from a physiological perspective.

2. COTTON TILLAGE SYSTEMS

In the U.S. cotton belt, the term "conventional tillage" has come to represent a set of intensive field operations that rapidly modify the soil environment, usually for purposes of weed control and seedbed preparation (Hutchinson, 1993b). In the southwestern U.S., conventional tillage is also a mandatory method of cultural control of pink bollworm [*Pectinophora gossypiella* (Saunders)]. Implements such as moldboard plows, chisels, and disc harrows are typically used to invert or mix the entire plow layer before planting (Dickey *et al.*, 1992). Intensive use of these implements typically leaves little or no plant residue from the previous crop on the soil surface to protect against erosion. Intensive tillage practices typically leave less than 15% residue cover after planting, or less than 500 lb/acre (560 kg ha[-1]) of small grain residue equivalent (Bradley, 1995). According to the Soil Science Society of America (1997), conventional tillage usually results in less than 30% cover of crop residues remaining on the soil surface.

Soil losses due to erosion have given rise to several systems that aim to conserve soil and water resources for crop production. Conservation tillage refers to any system that maintains at least 30% residue coverage of the soil surface after planting to reduce water or wind erosion (Hutchinson, 1993a; Bradley, 1995). Conservation tillage systems for cotton include no-tillage, ridge tillage, and mulch tillage systems, with many regional variations (Valco and McClelland, 1995).

In the mid-South cotton belt of the U.S., no-tillage is the predominant conservation tillage system used on highly erodible silt loam soils (Bradley, 1995). In this system, the crop is planted directly into untilled soil, and only the immediate seed zone is disturbed by the planter. Other surface residues are not displaced, and weed control is accom-

J.McD. Stewart et al. (eds.), *Physiology of Cotton*,
DOI 10.1007/978-90-481-3195-2_22, © Springer Science+Business Media B.V. 2010

plished mainly with herbicides. Some sandy clay loam soils in the southeast U.S. have limiting layers in the subsoil that restrict root growth and water movement (Busscher and Bauer, 1993; Bauer *et al.*, 1995). Root-limiting layers of this type may arise from previous tillage operations or other field traffic, but may also occur naturally. No-tillage may be combined with in-row subsoiling to disrupt these layers with minimal disturbance of surface residues. This system has been referred to as strip tillage (Bauer and Busscher, 1993). The Soil Science Society of America (1997) defines strip tillage as partial-width tillage operations performed in isolated bands separated by undisturbed bands of soil.

Ridge tillage systems are most common in furrow irrigated fields and some non-irrigated alluvial clay soils of the Mississippi Delta (Hutchinson, 1993b), and where wind erosion may occur (Valco and McClelland, 1995). With ridge tillage, the soil and residues are undisturbed until planting. The crop is planted in seedbeds prepared on top of raised beds or ridges with implements such as sweeps, coulters, or cultivator tines which disturb residue and remove weeds. The beds or ridges are rebuilt during cultivation. Beds rebuilt in the fall for spring planting are referred to as stale seedbeds.

Mulch tillage systems involve some soil and residue disturbance between harvest and planting, and are used mostly in Texas and other Southwest states where both soil and water conservation are essential to dryland cotton production (Valco and McClelland, 1995). Adequate residue cover (>30%) is obtained by rotation with high-residue crops, and by limited use of tillage implements. An effective method of mulch tillage uses chisels and sweeps to prepare seedbeds only in strips corresponding to the seed rows (Hutchinson, 1993a).

3. EARLY-SEASON RESPONSES TO TILLAGE

Seedbed preparation changes the edaphic environment for the seed and seedling hypocotyl. In much of the U.S. cotton belt, soils are too cool and moist for optimal seedling emergence in early spring. Conventional seedbed preparation is intended to improve the physical, chemical, and biotic environment in order to improve seedling emergence rates and establish optimal plant populations. Ridge tillage and planting on raised beds are intended to further improve the seed and seedling environment. The amount and depth of tillage affects soil density early in the growing season (Griffith *et al.*, 1992), which in turn affects soil temperature, moisture, and oxygen content. Temperature and moisture availability influence the rate and length of hypocotyl elongation (Wanjura, 1986).

Cotton seed germination and seedling growth are highly susceptible to chilling injury (Benedict, 1984; Christiansen and Rowland, 1986). Early growth phases most susceptible to chilling injury are initial seed hydration and after the radicle has elongated to 2 or 3 cm (Christiansen and Rowland, 1986). Temperatures below 15°C are deleterious to seed germination, while 20°C results in satisfactory radicle emergence (Wanjura, 1986). Chilling injures the radicle by growth rate reduction, membrane leakage, cortex sloughing, and increasing susceptibility to pathogens (Christiansen and Rowland, 1986). To avoid chilling injury under field conditions, it is recommended that soil temperatures at the 2-in (5-cm) depth exceed 20°C for three successive mornings before planting (Bradley, 1994). Combinations of cool spring weather, no tillage, and heavy cover crop residue can result in either suboptimal planting temperatures or planting delays.

Supply of available moisture to the seed depends on soil-seed contact, which may be controlled by choice of planting equipment for a given tillage system, seed placement, and closing of the seed furrow (Tompkins *et al.*, 1990). Cotton seed can be hydrated in 4 to 5 h at 30°C and 55% moisture, which is adequate for germination (Benedict, 1984). Excess moisture in the seed zone may be alleviated by conventional tillage of well drained soils, or by ridge tillage of poorly drained soils. Slower growth of seedlings has been observed under saturated soil conditions in pot experiments (Wanjura, 1986). Saturated soil conditions restrict oxygen diffusion to seeds and seedling radicles. Oxygen deficiency and soil compaction may interactively limit seedling root growth. In research cited by Christiansen and Rowland (1986), O_2 concentrations <10% limited growth at low soil bulk densities (<1.5 g cc^{-1}), whereas soil compaction was limiting at higher densities.

Abiotic stresses described above may increase susceptibility of seedlings to diseases such as *Rhizoctonia* and *Fusarium* spp., and these are also influenced by tillage practices (Colyer and Vernon, 1993). Seedling diseases injure the crop not only by reducing plant stands but also by stunting surviving plants. Reductions of stand and seedling vigor have been associated with incidence of seedling disease in reduced tillage. Chambers (1995a) found that plant stands were lower in cotton grown with no-tillage than with conventional tillage in 5 of 10 years of tests on silt loam soils in Tennessee. Stands were similar in the other years. In 1992, for instance, stands averaged 24,000 plants/acre (6 plants m^{-2}) in no-tillage, compared to 31,000 plants/acre (8 plants m^{-2}) in conventional tillage. Granular in-furrow fungicides reduced seedling disease, increased no-tillage plant stands by 30%, and improved seedling vigor. This long-term study showed that seedling disease severity was higher in no-tillage than conventional tillage, especially in earlier plantings and adverse weather. Fungicides applied in-furrow reduced disease incidence and increased stand establishment to levels comparable to conventional tillage.

Colyer and Vernon (1993) compared effects of three tillage systems on plant populations and seedling diseases on a Norwood very fine sandy loam in Louisiana. Conservation tillage involved paraplowing to 18 in (46 cm) in the fall with no spring tillage. Moldboard plowing in the fall was followed by smoothing of the seedbed in the spring. Conventional tillage involved discing and subsoiling to 14 in (36 cm) in the fall and hipping of beds in the spring. Conservation tillage produced lower plant populations than moldboard tillage in 2 years, and lower than conventional tillage in 3 years of this 4-year study. In the other year, dry conditions at planting minimized seedling disease incidence. Lower stands and higher incidence of seedling disease in conservation tillage were attributed in part to plant residue on the soil surface that may harbor pathogenic inoculum. Differences in soil temperature and moisture between tillage systems were also implicated.

Effects of early stand loss on yields depend largely on plant population at harvest. Plant populations greater than 28,000 to 32,000 per acre (7 to 8 plants m^{-2}) were found to be adequate for maximum yields in studies by Fowler and Ray (1977), Buxton *et al.* (1979), and Hicks *et al.* (1989). Lower yields in reduced tillage were associated with plant population densities below these thresholds in studies by Chambers (1995a) and Colyer and Vernon (1993). The practical implication of these findings is that plant stands adequate for optimum yields can be obtained in no-tillage by choice of planting date, and by use of an appropriate planter and in-furrow fungicides.

4.   FERTILIZER MANAGEMENT IN CONTRASTING TILLAGE SYSTEMS

Mineral nutrition requirements of cotton (discussed in Chapter 9) are similar in different tillage systems. However, fertilizer management strategies may differ somewhat between systems due to differences in nutrient availability.

Conventional tillage operations mix nutrients throughout the tilled portion of the soil, whereas conservation tillage provides little or no mixing of nutrients with other soil constituents (Mengel *et al.*, 1992). The lack of incorporation of fertilizer materials in continuous no-tillage may gradually lead to stratification of nutrients and organic matter near the soil surface (Mullen and Howard, 1992; Reeves *et al.*, 1993; Morrison and Chichester, 1994). A portion of the mineral nutrients taken up from lower soil horizons are deposited in crop residue on the soil surface, gradually magnifying the stratification (Mengel *et al.*, 1992). In no-tillage systems, lateral stratification can also occur when rows are planted close to the rows of the previous year (Mullen and Howard, 1992). These researchers speculated that as root systems decompose after harvest, nutrients return to the soil along the plant row. After several years, concentrations of phosphorus (P) and potassium (K) may be higher along the rows than between rows, which may bias soil test results.

Although less mobile elements such as P and K are likely to be stratified with reduced tillage, most research has shown that nutrient stratification does not adversely affect uptake or cotton response to surface-applied P or K fertilizers (Denton, 1993; Reeves *et al.*, 1993). One possible explanation is that root length density may be higher

in the surface layer (3 in or 7.6 cm) of soil due to higher moisture content, as has been demonstrated in no-tillage corn, *Zea mays* L. (Mengel *et al.*, 1992). Soil acidity may also stratify under continuous no-tillage, especially if high rates of acidifying fertilizer nitrogen are broadcast (Reeves *et al.*, 1993). Since surface-layer pH can decrease rapidly under these conditions, more frequent applications of lime may be required than with conventional tillage.

Tillage and residue can influence nitrogen (N) nutrition of cotton in several ways. Efficiency of N fertilization depends on the chemical form of N supplied and on its placement relative to organic residues. All N sources are subject to denitrification and leaching losses. Incorporation of urea ($[NH_2]_2CO$) by conventional tillage usually results in rapid ammonification and nitrification by soil microorganisms, making the N readily assimilated by roots. Ammonium nitrate (NH_4NO_3) also supplies N in directly assimilable forms. However, if urea is broadcast onto surface residues in no-tillage, losses by volatilization (to NH_3) may occur (Mengel *et al.*, 1992; Reeves *et al.*, 1993). This may result in N losses as large as 30% with heavy residue and warm, moist conditions followed by drying (Denton, 1993). Immobilization of N by microorganisms in organic residues can also temporarily reduce availability to the crop (Mengel *et al.*, 1992; Denton, 1993). Losses of surface-applied N by these processes, along with a high C:N ratio in residue of small grain cover crops, may lead to N deficiency in cotton grown with conservation tillage (Reeves *et al.*, 1993).

In situations where conservation tillage leaves the soil cooler and wetter (and sometimes less aerated) than conventional tillage in the spring, nutrient availability and uptake may be slowed. Since these conditions can also limit early-season root growth of cotton, considerable research has been conducted on starter fertilizers banded over or beside the seed row (Reeves *et al.*, 1993). Typical starter fertilizer rates for cotton are 15 to 30 lb N/acre (17 to 34 kg N ha^{-1}) and 15 to 50 lb P_2O_5 acre (7 to 24 kg P ha^{-1}). Responses to starter fertilizers have been variable, but they rarely decrease yields except when plant populations are greatly reduced. This may occur by ammonia toxicity if NH_4-bearing fertilizer is placed too close to the seed at planting (Reeves *et al.*, 1993). Starter fertilizer may be applied in a band 2 inches (5 cm) to the side and 2 inches below the dropped seed (2x2), in a surface band over the row at planting, or (at low rates) in the seed furrow.

Starter fertilizers more frequently increase yields in conservation tillage than in conventional tillage systems (Touchton *et al.*, 1986; Mengel *et al.*, 1992). In a 3-year study of conventional and no-tillage cotton in Alabama, Touchton *et al.* (1986) found that N and P starter fertilizers applied 2x2 increased early-season plant height on a sandy loam, but not on a silt loam soil. In Louisiana, Kovar *et al.* (1994) found that N-P starter produced greater early-season root length density with no tillage, but no differences were detected with conventional tillage of a Gigger silt loam. In neither the Alabama nor Louisiana study, however, were early-season growth responses to starters followed by increased cotton yields. On Memphis silt loam in Tennessee, Howard *et al.* (1995) found that broadcast-plus-band (2x2) applications of N and P increased lint yields 5%, on average, relative to broadcasting the same amounts of N and P with no tillage. Banding 15-15-0 (17 and 7.3 kg N and P ha⁻¹, respectively) produced the highest average yields, but response to P in the banded starter was greater in years of relatively high moisture stress. Burmester *et al.* (1993) also increased yields with a 15 lb N plus 50 lb P_2O_5/acre (17 and 24 kg N and P ha⁻¹, respectively) as a starter in a dry year and no tillage on a Decatur silt loam in Alabama, but not in a year with abundant rainfall or in conventional tillage.

5. TILLAGE EFFECTS ON YIELD AND EARLINESS

5.1 Conventional Tillage

Conventional tillage may have beneficial or detrimental effects on cotton yields. In either case, yield responses to conventional tillage probably result from effects on seeds or seedlings that occur shortly after the tillage operations. Beneficial effects may include warmer, better aerated seedbeds, and less weed interference relative to reduced tillage systems.

Improved seedling environment often results in higher plant populations with conventional tillage than with various reduced tillage systems, leading to higher yields (Brown *et al.*, 1985; Abaye *et al.*, 1995). In some cases, however, plant population and yield differences that were reported for early years of tillage studies disappeared in later years (*e.g.*, Buehring *et al.*, 1994). This may be due to rapid technological improvements during the early years of conservation tillage cotton.

Another short-term effect of conventional tillage is to reduce bulk density and soil strength in the tilled layer of some soils. These reductions have been associated with higher cotton yields (Burmester *et al*, 1993; Busscher and Bauer, 1993; Matocha, 1993). Tillage of a sandy soil which is low in organic matter tends to reduce mechanical impedance to root elongation, but may also result in soil compaction by subsequent wheel traffic (Bennie, 1991). Mechanically impeded roots are shorter, thicker, and more irregular in distribution. Reduction in root zone volume also tends to decrease aboveground vegetative growth (Carmi and Shalhevet, 1983), which can decrease the number of fruiting sites per unit ground area.

In a 5-year study on a Decatur silt loam in Alabama, conventional tillage cotton after wheat *(Triticum aestivum* L.) or cotton stubble residue was compared to no tillage (Burmester *et al.*, 1993). Yields with conventional tillage after stubble averaged about 11% higher than with no-tillage. After wheat, conventional tillage cotton averaged 5% higher yields than no-tillage. Yield differences were greater in dry years, and were associated with lower cone index (soil strength) in the top 12 in (30 cm) of the conventionally tilled plots. Burmester *et al.* (1993) suggested that less restricted root growth or increased water infiltration may have contributed to higher yields with conventional tillage in the drier years of this study. By contrast, in a comparison of no-tilled and tilled Lexington silt loam in Tennessee, Tyler *et al.* (1994) found that conventional tillage, equipment traffic, or both increased soil bulk density. However, by 60 days after emergence, differences in soybean *(Glycine max* [L.] Merr.) root distribution in the two tillage systems had disappeared, and no yield differences were detected.

Long-term degradation of the edaphic environment can occur by water erosion of cotton soils under conventional tillage, with detrimental effects on yield. In an 11-year study on a Providence silt loam in Mississippi, Mutchler *et al.* (1985) found that conventional tillage of cotton resulted in an average soil loss of 70 t/ha yr¹, compared to 19 t/ha yr¹ from conventional tillage soybean. These authors found that the root system of cotton was relatively ineffective at holding soil in place. Compared to no-tillage, continuous conventional tillage reduced average cotton yields by 17% in the last three years of this study. Yield reductions of this type are likely due to reductions of nutrient and moisture supplies to the plant, due to smaller soil volume accessible to roots or to lower nutrient content of the eroded soil.

5.2 Ridge Tillage

In ridge tillage, the crop is planted on raised beds or ridges that are rebuilt periodically (Hutchinson, 1993a). Tops of beds are prepared for planting with sweeps or other implements that disturb or remove plant residue. As with conventional tillage, ridge tillage effects often carry over from early in the season to affect yield. When ridge tillage was combined with a winter wheat cover crop in Missouri, early-season cotton growth was more vigorous and leaves intercepted 10% more light than in conventional (flat seedbed) tillage, although plant populations were equivalent (Stevens *et al.*, 1996). The growth response was associated with a reduction in wind speed at seedling canopy level in ridge-tillage cotton in wheat residue. Ridge-tillage plants produced more main-stem nodes, fruiting branches, and 15% to 29% more lint yield than conventional tillage cotton (Mobley and Albers, 1993). Ridge-tillage plants were also earlier maturing; they cut out about 2.5 days earlier than conventional tillage plants.

Ridge tillage resulted in higher plant populations, greater early-season plant height than conventional tillage with flat seedbeds on a Decatur silt loam in Alabama, but yields did not differ significantly (Reeves *et al.*, 1996). Cotton grown with ridge tillage tended to mature earlier than with conventional tillage, with 85% and 81 % open bolls before harvest, respectively.

5.3   Deep Tillage

Deep tillage can modify the subsoil environment with minimal disturbance of soil surface residues. In-row subsoiling of a sandy coastal plain soil disrupts a compacted, root-limiting layer above the B horizon (Bauer *et al,* 1995; Busscher and Bauer, 1995). Compacted subsoil layers may reduce cotton yields by two possible mechanisms: reduction in available soil volume which limits water or nutrient supply, or by reducing oxygen supply to the roots during periods of excess moisture (McConnell *et al.,* 1989; Unger and Kaspar, 1994). In a 3-year study on two silt loam soils with compacted subsoils in Arkansas, McConnell *et al.* (1989) found that tillage to 16 in (41 cm) reduced subsoil bulk densities. Cotton grown on subsoiled plots was taller, had more bolls, and higher yield than on non-subsoiled plots. Lint yield response ranged from 12% to 41%, with the greater response on the shallower soil. These authors suggested that yield reduction from tillage pans may be due to restriction of root access to moisture and/or nutrients, which may have limited plant height, fruiting sites, and harvestable bolls per plant. Growth and yield responses to subsoiling were investigated on Memphis and Collins silt loams in Tennessee in the 1970s (Tompkins *et al.,* 1979). Subsoiling to a depth of 16 to 20 in (41 to 51 cm) was done in the fall, and cotton planted the following spring in 40-inch rows directly over the subsoil slots. Plants were 7 to 10% taller and canopies were 11 to 13% wider in subsoiled plots than in conventionally tilled plots. However, subsoiling did not significantly affect yield on either soil in the three years of this study. More recently, Reeves *et al.* (1996) found that subsoil paratilling of a Decatur silt loam resulted in taller plants, but it also delayed maturity and tended to produce lower yields in the Tennessee Valley.

In a 6-year study of subsoiling frequency on sandy and silt loam soils in Mississippi, annual subsoiling increased lint yields by an average of 96 lb/acre (107 kg ha⁻¹) or 11% (Tupper *et al.,* 1989). Lint yields were associated with tap root length, which increased with subsoiling. Greater response in drier years was attributed to extraction of deep soil moisture reserves. On a sandy coastal plain soil, Bauer *at al.* (1995) found that subsoiling decreased subsoil strength, but that yields were unaffected in years when moisture was not limiting. Response to parabolic subsoiling was also less in years of abundant rainfall on a Tunica clay soil in Mississippi (Smith, 1992).

5.4   Mulch Tillage

Mulch tillage leaves at least 30% of surface residue undisturbed, often by use of chisels and sweeps to prepare seedbeds only in the row (Hutchinson, 1993a). A reduced tillage version of this system is used in dryland cotton to conserve soil moisture. In a long-term study on a Victoria clay soil in coastal Texas, reduced tillage was compared to conventional tillage cotton (Lawlor *et al.,* 1991, 1992).

Reduced tillage involved fall application of herbicides followed by row sweeps the following spring. Soil moisture content was higher with reduced tillage than with conventional tillage throughout a relatively dry season. Drought stress caused earlier cut-out and lower yields with conventional tillage, whereas reduced tillage cotton had more fruiting sites, more fruit, and higher yields. Lint yields were 32% higher with reduced tillage in a very dry year, and 11% higher in a year of intermittent drought stress.

5.5   No-Tillage

Among the three types of conservation tillage discussed here, no-tillage least disturbs the edaphic environment, and its long-term effects on cotton yields are most distinct from short-term differences. The first years of studies comparing conventional- and no-tillage frequently report lower yields from no-tillage cotton (Hoskinson *et al.,* 1982; Baker, 1987; Hart *et al.,* 1996). Some of the short-term effects are attributable to incorrect residue management, inadequate stands or poor weed control with no tillage (Hutchinson, 1993b; Touchton and Reeves, 1988). Poor stands of no-tillage cotton, especially after a winter legume cover crop, are often due to inadequate seed-soil contact, allelopathic residues, or biotic stresses imposed by diseases and insects.

In a 4-year irrigated study on a Gigger silt loam soil in Louisiana, yields were similar with no-tillage and conventional tillage (Hutchinson *et al.,* 1995). Maturity was not affected by tillage in three years out of four in this study, but was later with no-tillage in one year. Other long-term studies have shown that, with correct management, no-tillage yields are equivalent to (Bradley, 1994; Hoskinson and Gwathmey, 1996) or higher than (Triplett and Dabney, 1995; Triplett *et al.,* 1996) conventional tillage yields after several years. Hoskinson and Gwathmey (1996) found that cotton matured earlier and yielded more with no-tillage than with conventional tillage in recent years of a 13-year study on a Memphis silt loam in Tennessee (Table 22-1). In a 5-year study on loess silt loams in Mississippi, Triplett *et al.* (1996) found that yields with no-tillage were 19% lower in the first year, but 18 to 42% higher in later years than with conventional tillage (Table 22-2). Yields with ridge tillage remained consistently less than with conventional tillage (chisel followed by double discing, bedding up rows, and cultivation) in this study. No data were reported to indicate whether the edaphic environment deteriorated with time with conventional tillage, or if it improved with no-tillage. By the end of the study, however, more soil erosion was visible in conventionally tilled than no-tilled plots, which were situated in a field with slopes of 2 to 6% (G.B. Triplett Jr., 1996, personal communication). Loss of topsoil may have reduced soil volume accessible to roots and soil moisture reserves, which might account for the smaller shoot height with conventional tillage in the fifth year of this study. Triplett *et al.* (1996) indicated that an improved soil water regime was probably an important factor contributing to higher yields

Table 22-1. Lint yields and earliness of 'Deltapine 50' cotton grown with conventional tillage (CT) or no-tillage (NT) on a non-irrigated Memphis silt loam at the Milan (TN) Experiment Station, 1987 through 1993.

Year	Total lint yield			Percent first harvest[z]		
	CT	NT	P-value[y]	CT	NT	P-value
	-- (kg/ha) --			----- (%)----		
1987[x]	1104	1193	0.07	90	93	0.02
1988	773	859	0.05	86	89	0.06
1989	773	943	<0.01	77	87	0.01
1990	911	736	0.04	81	87	0.29
1991	1005	1112	0.20	78	90	<0.01
1992	1323	1337	>0.50	79	75	>0.50
1993	618	841	<0.01	89	92	0.11
Mean	931	1003	<0.01	83	87	<0.01

[z] Percent of total yield picked at the first of two mechanical harvests.

[y] Observed significance level of difference between tillage systems.

[x] 1987 = seventh consecutive year of tillage treatments.

Source: Hoskinson and Gwathmey (1996) and unpublished data of P.E. Hoskinson.

with no-tillage in the later years of their non-irrigated study. Bloodworth and Johnson (1995) reported higher yield and less drought stress in no-tillage cotton on a Grenada silt loam in Mississippi, and Stevens *et al.* (1992) attributed higher yield with no-tillage in one year to moisture conservation. Conversely, Matocha (1993) reported lower yields with no-tillage in wet years on a sandy clay loam in coastal Texas.

The available evidence suggests that physical, chemical, and biological soil conditions gradually improve with continuous no-tillage. Data from Indiana compiled by Griffith *et al.* (1992) shows that after several years in no-tillage corn, soil physical properties improved due to maintenance of soil organic matter, structural aggregation, and macropore space. Unger and Kaspar (1994) suggested that organic matter buildup with no-tillage may strengthen macropore formation, such that some untilled fields show no discernible compaction. Tyler *et al.* (1994) suggested that no-tillage aggregate stability was associated with improved load bearing capacity and less wheel compaction of a Lexington silt loam in Tennessee. Moreover, aggregate stability may

also contribute to erosion resistance. Mutchler *et at.* (1985) found that no-tillage reduced soil loss by 47% and increased cotton yields by 20% relative to long-term conventional tillage of a Providence silt loam in Mississippi. Stevens *et al.* (1992) found that no tillage reduced soil erosion by 70% on a Grenada silt loam with 2 to 3% slope in Mississippi.

In a 3-year study on a Gigger silt loam in Louisiana, no-tillage was compared to ridge tillage and conventional tillage systems for effects on cotton crop growth rate (CGR) and leaf area index (LAI) (Kennedy and Hutchinson, 1996). Pre-bloom CGR and LAI with no-tillage were equivalent or greater than with conventional tillage, and higher than with ridge tillage. Differences in CGR were related to LAI, not to net assimilation rate. The CGR effects generally carried over into reproductive growth with faster boll weight accumulation and larger first-position bolls. Triplett *et al.* (1996) found more fruiting sites and bolls per plant with no-tillage than with conventional tillage in the fifth year of their study.

Several studies in Tennessee have shown that no-tillage results in earlier cotton maturity than conventional tillage in most years (Hoskinson, 1988; Hoskinson and Gwathmey, 1996). Plant mapping studies suggested that the earliness of no-tillage cotton was associated with boll distribution. With no-tillage, 74% of bolls were produced at first-position sites versus 63% with conventional tillage (Hoskinson and Howard, 1992). Stevens *et al.* (1992) also found that no-tillage resulted in earlier maturity in two of three years, averaging 83% first harvest in no-tillage and 78% first harvest in conventional tillage. Triplett *et al.* (1996) found that no-tillage cotton reached 70% of final yield 6 to 10 days earlier than with conventional tillage in later years of their study (Table 22-2). Other investigators found no difference in earliness due to tillage (Buehring *et al.,* 1994), or that no-tillage resulted in later maturity than conventional tillage (Hutchinson and Sharpe, 1989). In the latter study, later maturity with no-tillage was attributed to suboptimal plant populations.

Aside from plant population effects, possible mechanisms by which no-tillage might affect earliness have not been elucidated. One hypothesis suggests that no-tillage cotton may cut out earlier due to inadequate nutrient uptake from the soil during reproductive development (D.O. Howard, unpublished data). An alternate hypothesis is based on evidence that higher soil strength restricts root growth in some no-tilled soils, resulting in more compact plants than in conventional tillage (Burmester *et al.,* 1993). Bulk density appears to increase more rapidly in sandy (Matocha, 1993) and ferruginous (Ike, 1986) soils with no tillage. Pot studies have shown that physical restriction of cotton root growth is accompanied by decreased vegetative growth, more compact plants and higher harvest index (Carmi and

Table 22-2. Yields and earliness of 'DES199' cotton growth with conventional tillage (CT), ridge tillage (RT), and no-tillage (NT) on a non-irrigated Grenada silt loam in Mississippi, 1988 through 1992.

Year	Seedcotton yield				Days to 70% final yield[z]			
	CT	RT	NT	LSD$_{0.05}$	CT	RT	NT	LSD$_{0.05}$
	---------------- (kg/ha) ----------------				---------------- (d)--------------------			
1988[y]	2074	1685	1671	234	156	159	170	4.0
1990	962	708	1315	247	130	126	123	4.4
1991	1728	1641	2460	267	139	136	133	3.3
1992	2967	2573	3514	265	157	157	147	3.5

[z] Days from planting to 70% of total yield estimated from weekly hand-harvest data.

[y] 1988 - first year of tillage treatments. No data from 1989 were presented for this experiment.

Source: Triplett *et al.* (1996).

Shalhevet, 1983). More compact plants tend to have fewer fruiting branches and thus cut out earlier in the season than taller plants. One possible reason for this response may be that the acropetal progression of flowering is more rapid in compact plants due to less competition by vegetative sinks for photoassimilates. Most studies have shown, however, that final plant height is not greatly affected by tillage system (Hart *et al.,* 1996; Hoskinson, 1989) or that no-tillage cotton may be slightly taller in some years (Hoskinson and Howard, 1992; Bloodworth and Johnson, 1995).

6. RESPONSES TO COVER CROPS AND RESIDUE MANAGEMENT

Winter annual cover crops and management of their residue may affect the growth, development, and yields of subsequent cotton crops. Commonly used winter cover crops include grasses such as wheat or rye *(Secale cereale* L.), or legumes such as crimson clover *(Trifolium incarnatum* L.) or hairy vetch *(Vicia villosa* Roth). These are grown for a number of reasons including erosion control on sloping land and to enhance soil quality by addition of organic matter. They may also alter soil moisture status, temperature, and fertility of the soil (Griffith *et al.,* 1992; Hutchinson, 1993b; Touchton and Reeves, 1988). Choice of cover crop species, and management of residues with herbicides or tillage implements, determine whether such changes to the edaphic environment are beneficial or detrimental to the subsequent cotton crop.

6.1 Stand Establishment

Effects of different tillage regimes and cover crops on cotton plant populations have been quite variable (Hutchinson and Sharpe, 1989; Bloodworth and Johnson, 1995). Both tillage and cover affect soil temperatures and soil water content, thus affecting seedling emergence, early growth, and disease incidence (Stevens *et al.,* 1992). Fields with greater than 20% residue cover may have lower soil temperatures than bare fields early in the growing season, due to reflection of solar radiation and delayed drying. However, cover crop residues have also been shown to reduce soil temperature fluctuations, resulting in warmer minimum and cooler maximum temperatures (Dabney, 1995). More than 30% residue cover after planting often increases soil moisture content during the season relative to clean cultivation, due to increased infiltration and/or decreased evaporation (Griffith *et al.,* 1992).

Most problems with stand establishment have been observed in no-tillage planting into residue of legume covers such as crimson clover (Rickerl *et al.,* 1984) or hairy vetch (Hutchinson and Sharpe, 1989; Stevens *et al.,* 1992). Fewer problems have been observed with grass covers such as wheat (Touchton and Reeves, 1988; Hutchinson, 1993b) or rye (Hoskinson *et al.,* 1982), perhaps due to rela-

tive ease of management. Early-season effects of residue on plant population density of cotton can strongly influence the number of squares per plant and fruit abscission later in the season (Stevens *et al.,* 1992). Effects on yield, however, may be buffered by the ability of cotton to compensate for variations in population between 28,000 and 56,000 plants/acre (7 and 14 plants m^{-2}) (Hicks *et al.,* 1989).

6.2 Allelopathic Potential of Cover Crops

Another possible influence of cover crop residue on seedling survival and growth is by allelopathy. In this context, allelopathy refers to the growth-inhibiting influence of chemicals released by decaying crop or weed residue (Hicks *et al.,* 1989; Bradow and Bauer, 1992). Allelopathy results from placing seed near decaying plant residue from which allelochemicals such as volatile ketones, alcohols, and aldehydes are released (Bradow and Connick, 1988). Laboratory experiments by these researchers demonstrated the inhibition of cotton seed germination and root elongation by organic volatiles released from decomposing winter legume covers such as hairy vetch and crimson clover. In the field, however, it is difficult to separate allelopathic effects from microbial and temperature effects, because residues also harbor pathogens and residue decay is temperature dependent (Griffith *et al.,* 1992).

Hicks *et al.* (1989) examined the effects of wheat straw on cotton germination, seedling growth and emergence. Cultivars differed in their tolerance of phytotoxins from wheat straw. In field studies, the cotton seed had to be in direct contact with residue for significant reductions in emergence to occur. They concluded that leaving residue on the surface would minimize any negative effects of wheat straw. White *et al.* (1989) investigated the effect of hairy vetch and crimson clover debris and aqueous extracts on emergence of cotton. Emergence decreased when debris was incorporated but was unaffected when legume debris was left on the surface.

6.3 Nitrogen Relations and Cover

Nitrogen availability to cotton can be affected by cover crops and can have a large effect on growth patterns. Considerable research across soil types indicates that more fertilizer N is needed for maximum yield following grass cover crops, but less or no N may be needed following a legume as compared to no cover (Brown *et al.,* 1985; Touchton and Reeves, 1988; Hutchinson *et al.,* 1995). The higher N requirement following grasses has usually been attributed to microbial immobilization of soil N during residue decomposition (Stevens *et al.,* 1992; Azevedo *et al.,* 1996). This is associated with a higher C:N ratio in residues of small grain crops than legumes (Reeves *et al.,* 1993).

Fertilizer N requirements following cover crops depend on cotton yield potential as well as interactions of the fertilizer with residue. Data reported by Brown *et al.* (1985)

suggested that about 34 kg ha^{-1} more N was required for maximum yield with no-tillage planting into rye than with conventional tillage, but yields in both of these systems were higher than in a no-till fallow system without cover crop residue. In this study, N fertilizer was banded next to the row about 25 days after planting, which may have reduced the amount of N immobilization by minimizing its contact with residue. Residues from leguminous crops may supply some or all of the N needed for the subsequent cotton crop (Hoskinson *et al.*, 1988; Varco, 1993; Hutchinson *et al.*, 1995). However, the timing of N release from legume residues may not coincide with peak N demand in cotton (Touchton and Reeves, 1988). Excessive N available in late season may account for the production of more fruiting sites per plant and later maturity of no-tillage cotton after vetch than after no cover (Stevens *et al.*, 1992). In a 4-year study on a Gigger silt loam in Louisiana, Hutchinson *et al.* (1995) found that optimum yields were obtained with N rates of 0 lb/acre after vetch, 70 lb/acre (78 kg ha^{-1}) after native cover, and 105 lb/acre (118 kg ha^{-1}) after wheat. Rates higher than these resulted in significant delays in maturity. Later maturity with excessive N may be due to changes in boll distribution within the canopy as described by Boquet *et al.* (1994b).

6.4 Vegetative Growth

Interactions of N, tillage and cover crop can affect plant height. In a 5-year study on a Sharkey clay in Louisiana, Boquet *et al.* (1994a) found that cotton was 10% taller after vetch than after native vegetation with conventional tillage, but not with no-tillage or ridge tillage. Except for one year, however, the taller cotton in conventionally tilled vetch plots did not produce higher yields than in no-tillage. Optimum N rates in this study were 90 lb/acre (101 kg ha^{-1}) without vetch, and 45 lb/acre (50 kg ha^{-1}) after a vetch cover. On a Decatur silt loam, Reeves *et al.* (1996) found greater internode lengths and plant height in no-tillage treatments planted in a killed rye cover crop with adequate N fertilization as compared to systems involving tillage. This may be due in part to etiolation of cotton seedlings induced by early shading by the rye cover residue. Burmester *et al.* (1993) found that no-tillage cotton was much taller after a wheat cover crop than after planting into cotton stubble, provided that 15 lb N and 50 lb P$_2$O$_5$ acre (17 kg N and 24 kg P ha^{-1}) had been applied as a starter fertilizer. Data are lacking, however, on differences in leaf area and root/shoot ratio that may accompany cotton plant height response to residue.

6.5 Reproductive Development

Effects of cover on plant height may be accompanied by changes in number and distribution of fruiting sites on the plant, and in fruit retention. In a 4-year study on an Ora fine sandy loam, Buehring *et al.* (1994) found more nodes/plant and harvestable bolls/plant on larger plants by planting with no-tillage into vetch than with conventional tillage without

vetch, when 80 lb N/acre (90 kg N ha^{-1}) was applied. Location of the first fruiting branch node and the percent bolls on first and second positions (traits associated with earliness of maturity) were not affected by tillage, cover, or N rate.

Stevens *et al.* (1992) found more fruiting sites per plant in cotton grown in conventionally tilled or no-tilled soil with no cover crop than in no-tillage cotton with wheat cover in the first year of their study. Cotton grown with no-tillage after vetch produced similar fruiting sites/plant as cotton seeded into tilled soil. However, in the next two years, no differences were observed in the total number of fruiting sites. Even with fewer fruiting sites in the first year with wheat cover, less fruit abscission occurred later in the season, and yields were not significantly different from conventional tillage. Fruit abscission patterns were not different in the second year of the experiment. Over three years, the number of fruiting sites on main-stem nodes five to eight, as a percentage of total, was significantly lower with no-tillage after wheat than with conventional tillage. This decrease corresponded to a 6- to 9-day delay in square initiation in no-tillage after wheat. Wheat delayed maturity more than vetch, but these effects were confounded to some extent by the amount of N applied to the wheat (110 lb/acre or 123 kg ha^{-1}) and vetch (50 lb/acre or 56 kg ha^{-1}) in this experiment. The authors speculated that earlier maturity with vetch may have been due to less shading of young cotton plants in vetch than in wheat residue.

Evidence supporting the hypothesis of delayed initiation of reproductive development in small grain residues was reported by Bauer *et al.* (1995). Their study involved winter fallow and rye covers that were either disced or killed with paraquat (1,1'-dimethyl-4,4'-bipyridinium chloride) before planting cotton. The cover crop did not influence the onset of flowering in disced plots. In the paraquat-treated plots, however, peak flower production was delayed several days in rye residue, relative to fallow. End-of-season plant mapping revealed that the delay was due in part to first sympodial branches located higher on cotton grown in rye mulch (data not shown).

6.6 Yield Response to Cover Crops

Yield responses to cover crops vary widely depending on cotton plant population, soil type, tillage, cover crop species, and residue management. Stevens *et al.* (1992) found similar yields in no-tillage cotton following wheat or vetch. Bloodworth and Johnson (1995) also found only minor yield differences after grass or legume cover crops over a three year period. By contrast, Hutchinson (1993a) reported lint yields, averaged across six years and three tillage systems, were 8% higher after hairy vetch and 11% higher after wheat than after native cover. The greatest yield response to wheat cover crop was with no-tillage or ridge tillage. He attributed the yield response to cover crops to improved soil moisture availability.

The three above-mentioned studies were on silt loam soils. On a Sharkey clay soil, Boquet *et al.* (1994a) obtained significantly higher yields after vetch than after native cover, using conventional or ridge tillage. On a Norfolk loamy sand, Bauer and Busscher (1996) found that rye cover produced higher cotton yields than crimson clover, hairy vetch, or fallow with conservation tillage, but not with conventional tillage. However, the cotton following legumes received no fertilizer N, whereas rye and fallow plots were sidedressed with 70 lb N/acre (78 kg N ha^{-1}) after planting. These investigators speculated that rye residues improved the soil water status during the cotton growing season more than the other covers. In this study, aboveground dry biomass of the rye averaged about 2200 lb/acre (2500 kg ha^{-1}), and did not affect cotton stand establishment. In a related study, however, rye biomass exceeded 5000 lb/acre (5600 kg ha^{-1}) in nondisked plots, which resulted in poor cotton stands (ca. 18,000 plants/acre or 4.5 plants m^{-2}) and lower yield than cotton after fallow or after rye that was disced under (Bauer *et al.*, 1995). These results illustrate the importance of residue management on performance of a subsequent cotton crop.

7.   SUMMARY

Different systems of tillage and residue management have evolved in various parts of the U.S. cotton belt and worldwide in response to local soil-related constraints to cotton production. The main categories of tillage include conventional, no-tillage, ridge tillage, deep tillage and mulch tillage systems. Conservation-tillage systems for cotton may involve use of winter cover crops for maximum soil and water conservation.

Tillage operations and residue management modify the edaphic environment and thus affect the types and severity of environmental stresses to which cotton may be exposed. Conventional and ridge tillage may alleviate early-season stresses by reducing seedling disease and weed interference, and by providing more favorable soil temperature and moisture regime. Thus, an optimal cotton plant population density may be established, and seedling vigor may be enhanced. Conservation-tillage systems that include effective cover crop management have long-term benefits to productivity by virtue of soil and water conservation, but may expose cotton to environmental stresses in early season that reduce plant stands and productivity. Optimal plant populations can be obtained with no-tillage by effective residue management, choice of planting date, appropriate planting equipment, and use of in-furrow fungicides.

Positive yield responses to conservation tillage have most often been demonstrated in environments where the volume of soil accessible to roots may limit crop growth

and development by severe soil erosion. Reduced available soil volume provides less soil water holding capacity and nutrient reserves for the plant, leading to drought stress or nutrient deficiency. Soil compaction may also result in compact shoot growth and earlier maturity. Excessively compact plants may limit yield potential, possibly by reducing the number of fruiting sites per unit ground area.

Deep tillage or subsoiling disrupts root-limiting layers from some coastal plain soils, thereby increasing soil volume accessible to roots. Mulch tillage may conserve soil moisture for the crop in the southwest U.S. cotton belt. No-tillage helps control soil erosion and maintain rooting volume, especially on loess soils. Winter cover crops are most effective in controlling soil erosion when used together with no tillage, and their residue acts as mulch to reduce evaporative water loss.

Mineral nutrition requirements of cotton are similar in different tillage systems, but these systems may affect nutrient availability and necessitate different fertilizer strategies. No-tillage may stratify some less mobile elements (P, K) and affect root distribution. Leguminous winter cover crops may supply some or all of the N needed by the subsequent cotton crop, but may impede stand establishment due to excessive mulch, and to possible allelopathic effects. Immobilization of N, or losses by volatilization, may lead to deficiency symptoms if not compensated by fertilizer application.

Some long-term studies have shown gradual improvement of cotton yields in no-tillage relative to conventional tillage. It remains to be seen whether these changes are due to gradual improvement in the edaphic environment in no-tillage, deterioration with conventional tillage, or both. This question pertains to the long-term sustainability of cotton cropping systems, so is worthy of a long-term research effort.

This chapter has attempted to draw a few physiological inferences from the many published reports on agronomic responses of cotton to tillage and residue management. It is evident that more research is needed on the physiological mechanisms of plant responses to changes in the edaphic environment due to tillage and cover crops, in order to explain the reasons for observed agronomic responses. For instance, there is an obvious need for basic root:shoot ratio and root length density distribution data for different tillage and cover crop systems. Water relations in the soil-plant-atmosphere continuum need to be described for these systems. Does tillage system or residue management affect water-use efficiency of cotton? More microclimate data are needed to address the variation in reported responses to residues; especially temperature, moisture, and light quality data. Variability in earliness response also needs to be resolved by more research on the mechanisms by which tillage and residue influence maturity.

Chapter 23

CROP WATER MANAGEMENT TO OPTIMIZE GROWTH AND YIELD

K.D. Hake[1] and D.W. Grimes[2]

[1]*Vice President Technology Development, Delta & Pine Land Company, Germantown, TN; and Agronomist, University of California, Kerney Agricultural Center, Parlier, CA*

1 OPTIMIZING PLANT-WATER RELATIONS AT DIFFERENT GROWTH STAGES

Due to the strong year-to-year variability in water supply and demand in most cotton growing regions, plant managers must be able to adapt their management (especially irrigation, fertilization, growth regulation, and pest management) to each year's climate and plant conditions. Only with a solid understanding of optimal plant-water relations at different growth stages and the ability to prioritize and assess risk can an optimal water-management program be developed.

The cotton plant undergoes a series of stages during its development from dormant seed to the production of mature bolls. While some of these stages are distinct and abrupt, others are gradual and overlap. Thus any consideration of the optimal plant-water relations for a specific time is complicated by the presence on the plant of organs and tissues in different stages of growth. For example, during long effective-bloom cycles (six weeks or more) a plant may contain both opening bolls and young bolls with economic value. However, the optimal plant-water relations for young bolls is substantially wetter than for opening bolls. Thus the optimal plant-water relations for the entire plant during early boll opening must incorporate the relative contributions of different fruit stages to economic value, where economic value includes such factors as yield, quality, risk, and additional input costs. Having said this, the most important factor to consider when designing optimal plant-water relations is the impact of water management decisions on the ability of the plant to initiate, retain, and mature harvestable bolls.

1.1 Germination and Early Shoot Growth

1.1.1 Water Availability

Not all soil water is equally available to the plant. Since water moves down an energy gradient the plant does not readily absorb salty water, or water held tightly by clay particles at the bottom of the root zone, which would have a low energy potential. In addition to the lowering of energy potential by salinity (ψ_s), soil texture (matric potential, ψ_m) and vertical depth (ψ_z), other limitations to water availability also important in cotton fields are: proximity to functional roots, soil gas exchange, soil temperature, and the presence of root pathogens and nematodes (Hillel, 1980a, 1980b; Jury *et al.*, 1991)

Soil texture alters water availability directly from the impact of soil particle size on matric potential (ψ_m), *i.e.* the force with which water is bound to the surface of soil particles. Small soil particles (clays) have substantially more surface area per mass than large soil particles (silts and sands), thus clayey soils have a substantially higher matric potential and reduced availability of soil water. The matric potential of clays is especially critical after the soil drains and some water has been removed by the plant roots, and is the reason that percent allowable depletions used in water budget irrigation scheduling are less in clay soils than in sandy soils. Matric potential is the dominant component of soil water potential measured in energy per mass (MPa), water in wet soil has a matric potential of -0.01 to -0.03 MPa and water in dry soil is less than -1.0 MPa. Salinity, solute potential (ψ_s), reduces soil water availability in much the same manner as matric potential, but salinity reduces availability even in

J.McD. Stewart et al. (eds.), *Physiology of Cotton*,
DOI 10.1007/978-90-481-3195-2_23, © Springer Science+Business Media B.V. 2010

wet soils. As plants remove relatively salt-free water from the soil, the salt concentration of the remaining soil water increases, thereby reducing the soil moisture availability.

1.1.2   Imbibition

Commercial cotton planting seed should be dry (less than 10% w/w water content) due to the sensitivity of seed vigor to storage under warm wet conditions. At 10% water content, the seed is very dry, approximately -200 MPa (Jordan, 1983), in comparison with either wet or dry soil at -1.5 to -0.01 MPa. Due to the large difference in energy potential, dry seed can pull water from soil that would not sustain life of even the most drought-hardy desert plant. However as seed hydrates its water potential rises and the soil water potential near the seed declines, thus the gradient for movement of water into the seed drops. For this reason, cotton planting seed needs to be placed in contact with moist soil, or else imbibition will not proceed to radicle emergence and hypocotyl elongation. Due to the slow movement of water in well-drained sandy soil, soil water availability must be very high for sufficient imbibition to permit seed germination in sandy soils.

Water imbibition is initiated through the cotton seed chalazal cap. After rehydration of the seed, metabolism increases with stored lipids and proteins providing the primary energy and food source during germination and early seedling development. Favorable conditions for rapid germination include a soil water content that is adequate to provide imbibition water, but not excessive so that aeration might be limiting. In climates where low temperature limits germination, excess soil water can further delay soil warming since the relatively high specific heat of water elevates the caloric energy required to warm wet soil. Approximately five times more energy is required to raise the temperature of water compared with soil minerals.

1.1.3   Radicle and Hypocotyl

The radicle is present in the seed and kept in a dessicated state of compression. With no subsequent cell division the radicle can elongate approximately 1 cm, a length that is often observed when chilling injury or damage during seed processing has occurred to the root tip (Hopper, 1999). Since approximately 48 hours are required from the time imbibition starts until radicle elongation, it is essential that the soil not dry to the seed depth for that period after planting. This is why producers in arid windy climates plant deeper than growers in humid or low wind environments–to minimize seed desiccation. If the root tip successfully elongates into moist soils the growth rate of the root tip is usually sufficient to maintain adequate water supply for hypocotyl elongation and emergence even under desiccating conditions. Once the cotyledons emerge through the soil surface and transpiration increases substantially, cotyledons can show transitory wilting during windy dry weather. If the roots are damaged by seedling disease then permanent wilting can lead to desiccation and plant death. Transitory wilting of the cotyledons is not a major cause for concern; however, permanent wilting associated with diseased roots usually results in seedling death, unless warm, low evaporation conditions (humid, light rainfall, cloudy, calm winds) allow plant recovery. Sprinkler irrigation is often used to decrease evaporative demand if the air temperature is warm, and thereby delay or prevent desiccation injury to the seedling. Due to the delay in the appearance of true leaves, during which time root growth continues, plants that develop a true leaf rarely desiccate and die if the soil moisture profile was full at planting.

1.1.4   Vegetative Growth

Although the plant growth rate is usually slow following emergence, new specialized cells are constantly being formed that will complete expansive and reproductive growth functions. Early growth and development are influenced by genetic, environmental, and cultural factors, but cell elongation is the most affected by water stress (Hsiao *et al.*, 1976). Turgor pressure (ψ_p) is essential for irreversible cell expansion and growth and a close correlation between the two has been shown. While the relationship between ψ_p and ψ_l (leaf water potential) is nonlinear for cotton (Gardner and Ehlig, 1965), expansive growth was linearly related to ψ_l when fairly long-term growth measurements (hours or days) were made. Expansive growth has been shown to be functionally dependent on ψ_l measured at predawn or midday (Cutler and Rains, 1977; Grimes and Yamada, 1982)

During the early vegetative growth period roots expand rapidly while leaf area expands slowly. This combination of increasing water supply relative to demand, results in rare occurrences where water supply is less than optimum during the pre-squaring period. Only where the soil moisture was lacking at the time of planting or roots are restricted due to soil compaction, salinity, or pathogens, is desiccation of pre-squaring cotton likely. The root pathogen *Thielaviopsis basicola*, causal agent of Cotton Black Root Rot, is one common cause of seedling desiccation. If the plant does not die, but rather resumes expansive growth before first square, the impact on final yield is often minimal. For this reason, seedling cotton is considered a stage that is, relative to later plant stages, more tolerant of drought.

Excess moisture, however, often limits seedling development. Frequent rain during cool springs in temperate climates further reduces soil temperature. Soil gas exchange, whereby oxygen for root growth and function is replenished from the atmosphere, is limited during periods of frequent rain or irrigation due to the low diffusions of oxygen through water. Under these conditions rooting is restricted to the oxygen-rich zone near the soil surface with subsequent vulnerability to drought during the mid-season growth.

Solar radiation interception is limited by an effective plant leaf surface in early season so it is important to avoid conditions that might limit leaf expansive growth during the squaring, prebloom period. High yield potentials require water-deficit stress avoidance during the prebloom construction of the vegetative framework. A comparable level of water stress during this growth stage leads to a greater reduction in yield than during any other stage of growth due to the restriction in fruiting sites and in the photosynthetic surface (both leaf area and activity) necessary later on to retain and mature bolls (Hake *et al.*, 1996; Krieg, 1988; Peng and Krieg, 1991). Even in dryland cotton, yield is most impacted by water supply prior to bloom (Miller *et al.* 1996). However, irrigating too early in the growing season can have adverse effects by lowering soil temperatures (Grimes *et al.*, 1978; Wanjura *et al.*, 1996). Early-season irrigation timing must balance the benefit of water stress alleviation with the resultant short-term, soil temperature reduction. For this reason irrigation regimes that either provide a full soil-moisture profile at planting or that only partially wet the soil surface during cool weather such as alternate furrow, Low Energy Precision Application (LEPA) and trickle/drop are preferred for pre-bloom cotton (Grimes *et al.*, 1978; Bordovsky and Lyle, 1999).

1.2 Root Growth and Extension

The plant's taproot with its associated lateral roots is capable of exploiting water and nutrients from a large soil volume. Rooting depth and root length density development over time follow a typical sigmoidal curve (Borg and Grimes, 1986) with exponential growth early if the root systems are not impeded by soil compaction and hardpans. If there are no restrictions, the taproot grows as much as 2.5 cm/day for several weeks after planting (Bassett *et al.*, 1970). Lateral roots are initiated about the time the cotyledons unfold. Soil type, moisture, aeration, and impeding zones determine how deep taproots penetrate. Although a few roots will grow as deep as 240 cm, normally about one-half of the total root length is confined to the top 60 cm of soil (Bassett *et al.*, 1970; Grimes *et al.*, 1975; Taylor and Klepper, 1978). Lateral roots may extend outward from the taproot to a horizontal length of 2 m (Taylor and Klepper, 1978). The total root length normally continues to increase with plant development until the onset of flowering, about eight to ten weeks after planting. As the season-end approaches, root length declines as older roots die.

Root death during the growing season involves both biotic and abiotic factors and can involve a major part of the root biomass. Adequate soil moisture is a major factor affecting root growth and extension (Grimes and El-Zik, 1982), but other factors may involve soil temperature, soil compaction, lack of oxygen in water-logged soils, nutrient availability, and pH. Root hairs on older roots decay and eventually the root develops a barky suberized surface that impede water uptake. Some uptake still occurs in older roots especially at the site where lateral roots emerge. During lateral root emergence the suberized bark is disrupted provid-

ing an entry site for soil moisture and associated nutrients.

1.2.1 Soil Strength and Moisture Limitations to Root Growth and Extension

Soil moisture limits root growth at either extremes (saturated or dry). When soils dry the plasticity of soil particles declines, thereby retarding the expansion of roots through drying soil (Jury et al, 1991). The growth of roots in soil can be related to the soil strength (or force required to push an object through the soil) (Grimes *et al.*, 1975). In addition to lower water content, soil strength is increased with an increase in bulk density and percent sand. Compacted sandy soils have high soil strength thereby retarding root elongation, even when wet to field capacity. Cotton production in these soils responds to tillage and avoidance of heavy compacting equipment that decreases bulk density (Carter, 1996).

1.2.2 Soil Oxygen Limitations to Root Growth and Extension

Cotton root growth is limited under conditions of reduced oxygen (anoxia). Since the oxygen content of the soil is a dynamic process with oxygen depletion by roots and microorganisms and replenishment through the soils mass from the atmosphere, predictions of oxygen content are difficult. Since the atmosphere is 20% oxygen, depleted levels usually occur only deep in the soil or where high oxygen consumption is combined with limited oxygen diffusion into and through the soil. Consumption is driven by temperature and microbial activity. Incorporation of decomposable organic material can result in oxygen depletion of warm soils if the flow of oxygen through the soil pores is impeded by water. The diffusion of oxygen through a water filled soil pore is dramatically less than through an equally sized pore devoid of water. In addition to reduced root growth when soil oxygen is low, water uptake, nutrient uptake, and conversions of ammonia to nitrate are reduced. Maintenance of macropores and avoidance of water logging are key management components to avoiding oxygen depletion. Since macropores serve to both channel water and oxygen rapidly into the soil, their maintenance under no-till cotton production is one of the reasons that no-till cotton is generally less damaged by either too much or too little rainfall.

1.3 Fruit Development and Retention

The indeterminate growth habit of cotton results in simultaneous initiation and development of squares (flower buds), flowers, and bolls over a lengthy period that varies with growing regions. Since each of these floral stages comprises different morphological and physiological processes, they have varying optimal water conditions and sensitivities to conditions outside of that optimal. Thus predicting the impact of water-deficit stress or soil-water excess at a point in time or describing optimum water conditions requires an appreciation of the distribution of the various age fruits and their potential contribution to yield (Hearn, 1995). First squares usually appear between 4 and 8 weeks after planting, depending on the cultivar and weather. A decline in squaring may be attributed to physiological factors related to an improper soil water supply, the environment, or pests. While some shedding of squares, flowers, and small bolls is normal, maintaining a desired water balance can avoid water stress induced shedding in irrigated regions. The initiation of squares and the flowering period duration depends on continued vegetative growth for new fruiting branches with new fruiting sites.

Abscission of fruiting forms in cotton is a complex phenomenon, but it is generally accepted that water deficits alter the normal hormonal balance of the abscission zone (Jordan, 1979; Guinn, 1979). Water-deficit stress increases the activity of cellulase and pectinase enzymes in the abscission zone. These enzymes weaken the cell wall and middle lamella between cells. Water-deficit stress may directly alter the hormones that regulate these enzymes (IAA, ABA, Ethylene) and indirectly through reductions in photosynthesis (Guinn, 1979, 1998).

Excess water, especially in fine textured soils, has been observed to promote fruit shed because of inadequate aeration of the roots (Longnecker and Erie, 1968). Anoxia may increase ethylene production by promoting synthesis in the roots of 1-aminocyclopropane-1-carboxylic acid (ACC, a precursor to ethylene). ACC is produced in oxygen starved roots and translocated to the leaves and fruit where it is converted to ethylene and may increase boll shedding. Rain or sprinkler irrigation during the morning of pollination can also cause low seed count bolls and small boll shed by rupturing pollen grains (Guinn, 1998). The economic impact of pollen rupture by sprinkler irrigation is negligible (<1%) under commercial irrigation frequencies of 4 days or longer due to the short duration of flower sensitivity to water in the morning and early afternoon (Pennington and Pringle, 1987). A third route whereby excess irrigation can decrease fruit retention is through the attractiveness of frequently irrigated cotton to plant bugs, *Lygus herperis* and *Lygus lineolaris*, (Leigh *et al.*, 1974)

Fruiting forms differ in their sensitivity to water-deficit stress that will trigger abscission (McMichael and Guinn, 1980; Guinn, 1998). As fruiting forms develop from the "pin head" stage to anthesis (flower), a time period of about 21 to 22 days, their sensitivity to water stress decreases (Guinn and Mauney as cited by McMichael, 1979). This was shown indirectly by Grimes *et al.* (1970) who observed a pronounced depression in flowering rate three weeks after alleviating a severe water stress interval. Young bolls, prior to reaching a size of about 2 cm in diameter, are sensitive to abscission due to water stress. McMichael *et al.* (1973) observed a marked increase in boll abscission for bolls smaller than 2 cm diameter as predawn leaf water potential (ψ_l), declined below -1.1 MPa in small container-grown plants and as midday

(ψ_l) declined below -1.9 MPa. Grimes and Yamada (1982) associated high cotton productivity to irrigation scheduling that did not allow midday ψ_l to decline below -1.9 MPa.

Early work in the San Joaquin valley in California demonstrated that peak flower was the most sensitive flowering stage to severe soil water deficit (Grimes *et al.*, 1970). At peak flower the plant attains the greatest number of young bolls - the fruiting form most sensitive to water-deficit induced abscission. In addition a less favorable carbohydrate supply coupled with a high assimilate demand puts the retention of small bolls at progressively greater risk of abscission during each week following peak flower. Irrigation strategies to minimize the fruit losses at this stage of extreme sensitivity to soil water deficit include shortening the irrigation cycle starting during the second or third week of flower and continuing until boll retention declines. From a practical standpoint this usually means applying less water each irrigation but more frequently since either water supply or soil infiltration limits the ability to increase the soil water content. Several mechanical systems facilitate frequent cycle irrigation: trickle/drip irrigation, solid-set sprinkers, and pivot irrigation with drops (LEPA, Low Energy Precision Irrigation). Although trickle/drip allows daily irrigation if needed, current commercial LEPA systems are limited to a minimum of 60 hour cycles – which may be sufficient for most fields.'

1.3.1 Optimum Soil Moisture for Square Initiation and Development

Square initiation at the primary (main-stem) position occurs under a wide range of plant conditions, including excess or deficit moisture. Moisture-deficits have less impact on the initiation and development of primary squares than on those formed at secondary and tertiary positions (Krieg *et al.*, 1993). Under prolonged, severe drought, only first-position fruit are initiated and mature. However, since the development of new main-stem nodes is required for the initiation of new fruiting branches and first-position squares, a moisture deficit that curtails new main-stem node production has a similar impact on first-position square initiation. In addition most production regimes depend on primary and secondary fruiting positions to develop an economically viable crop. For these reasons, during the initiation of economically important squares, leaf water potentials should be kept high (above -1.5 MPa; Hake *et al.*, 1996). Considering the lower daily water use prior to flowering when the leaf area is still expanding, if the soil profile is full prior to planting, severe water-deficit stress is unlikely until late squares are initiated.

1.3.2 Optimum Soil Moisture for Square and Boll Retention

Since soil moisture has a dramatic impact on both square and boll retention, this is one area of management where substantial improvements to yield can be made by careful attention to plant water needs and is the key reason that the highest yields worldwide are produced under full irrigation (*e.g.*, Arizona, Australia, California, Greece, Israel, Spain, Turkey, Xinjiang) (Hearn, 1995). Extremes of soil moisture, saturation, or desiccation, are always detrimental to square and boll retention and should be avoided during the fruit retention period. However the impact on retention of water-deficit stress varies widely depending on both the fruit stage considered and plant condition.

Fruiting forms (squares, flowers, and bolls) have relative degrees of competency to abscise. Young squares are moderately competent to abscise if the degree of plant stress is severe. Large squares are less competent to abscise, while flowers are incompetent of abscission. During the first two weeks post anthesis, bolls have their highest level of competency to abscise and with maturation they slowly loose competence to abscise. Based on research conducted with irrigated cotton, one could incorrectly conclude that under a certain level of water-deficit stress fruit retention will be nil. This has not been the common observation under dryland conditions on the High Plains of Texas. Even under severe water deficit (midday xylem pressure potentials of -2.7 MPa) that limits main-stem node production to 7 total nodes and plant height to 20 cm, 1 or 2 small first-position bolls are often retained and reach maturity per plant (K.D. Hake, unpublished data). The yield from these fields can be less than 100 kg/ha of fiber and the maximum nodes above white flower obtained is often only two. Another observation in these fields is the appearance of normal flowers fully turgid and leaves fully wilted. Clearly extrapolating plant response from irrigated research to dryland conditions is risky.

1.4 Boll Growth, Maturation, and Fiber Quality

Growth of bolls and fiber is much less sensitive to water stress than is expansive vegetative growth. Radulovich (1984) measured boll diameter and volume for field-grown plants as a function of mid-day ψ_l to find that boll size was not materially reduced by water stress until ψ_l dropped substantially below -2.2 MPa. Grimes and Yamada (1982), using procedures developed by Gipson and Ray (1969), measured the response of fiber elongation and weight increase to water stress. For field-grown plants neither elongation nor weight increase were affected until midday ψ_l declined to about -2.7 MPa, at which time a marked reduction for both parameters was observed. These responses indicate that bolls are a relatively strong photosynthate sink; indeed this has been demonstrated directly by Krieg and Sung (1979). Dry matter accumulation in reproductive plants parts will continue for a

period after expansive vegetative growth has stopped.

Although cotton fiber physical properties of length, strength, and fineness are generally considered to be under strong genetic control (Longenecker and Erie, 1968), both direct and indirect modifications of these properties occur from variations in water supply. Water stress sufficient to result in midday ψ_l 's of -2.4 MPa or less directly reduce fiber growth and development (Grimes, 1991). Fiber length is consistently reduced with severe water-deficit conditions. When water deficit reduces fiber length, micronaire is often elevated. However if the deficit conditions occur during the secondary wall thickening phase of boll development, micronaire has been observed to be greatly reduced. An insidious damage of water-deficit stress to fiber quality is the production of long fiber motes. These aborted seeds that produce elongated yet immature fiber increase yarn defects. Severe water-deficit stress increases the production of long fiber motes (Landivar *et al.*, 1997). Excess water may delay maturity and reduce fiber quality by causing bolls to develop under less favorable fall climatic conditions (Jackson and Tilt, 1968). Turner *et al.* (1979) found a decline in fiber mass from the second to sixth week of boll set where bolls were set in a six-week period. Night temperatures lower than 22°C during early boll development have been observed to reduce fiber growth rate (Gipson and Ray, 1969) and to be associated with reduced fiber length in commercial fields if low temperature coincides with early to peak flowering. To balance the beneficial effects of water-deficit stress on opening of mature bolls with the deleterious effect on immature boll development, late-season irrigation ideally permits a gradual decline in soil water availability. On soils with high water-holding capacity and surface irrigation this is accomplished with careful timing of the final irrigation. However with pressurized irrigation on low water holding soils a gradual reduction in both water volume and frequency is necessary to permit boll maturation without promoting excess late season vegetative growth and delay of boll opening. The opening of mature bolls is hastened by water-deficit stress that promotes leaf abscission, light penetration, air movement, and subsequent rapid drying of the fiber, seed, and boll wall. Table 1 shows the approximate impact of severe water-deficit stress on various stages of fruit development.

Table 23-1. Estimated impact of severe water-deficit stress (midday xylem pressure potential ≤ -2.5 MPa) on the economic value of various fruit stages.

Fruit stage	Impact on		
	Fruit retention	Fiber quality	Yield
Presquare initiation	minimal	minimal	minimal
Square initiation	moderate	minimal	loss, few bolls, small bolls
1st 30 days of a boll's development	severe	severe	loss, short staple, high micronaire
2nd 30 days of a boll's development	minimal	moderate	loss, immature fiber
Boll opening	none	minimal	hasten maturity

2.   PLANT-WATER INTERACTIONS

2.1   Nutrition

Plant nutrients reach absorption sites on roots by direct contact, diffusion, and mass flow. Since direct contact accounts for a very small part of absorbed nutrients, it is clear that nutrient movement to roots occur in the soil solution and is associated with an available water supply. For example: nitrogen side-dressed into relatively dry soil may experience only limited plant uptake before irrigation alleviates the soil-water deficit in arid or semi-arid regions (Grimes *et al.*, 1973; Zelinski, 1996).

2.1.1   Nutrient Uptake

The uptake of mineral nutrients by the cotton plant is almost exclusively via the root system from the soil solution. When the flow of soil solution into the roots is decreased from a root zone enriched in nutrients, uptake declines (Brouder and Cassman, 1990). This is a common occurrence under extremes of soil water availability (saturation or drought) but can also occur when evaporative demand is greatly reduced (high relative humidities, limited solar radiation) or when the xylem flow is blocked by vascular diseases.

The absorptive capacity of leaves to foliar-applied nutrients declines during their life due to changes in both the composition and thickness of the waxy cuticle layer (Oosterhuis *et al.*, 1991a). The absorptive capacity of roots has a similar age dependent relationship. Since soil water is absorbed primarily via root hairs and roots hairs persist for only a few days following formation in the root tip, the uptake of nutrients occurs in the zone of active or recent root elongation. This is one of the reasons the supply of nutrients from the soil to the plant is often limited during mid to late flowering - the growth of new roots slows at first flower. Management practices to over come the limitation to root uptake of nutrients during late flowering include: 1) storage of N and K in leaf tissue prior to flowering; and 2) split applications of fertilizer nutrients and foliar-applied nutrients.

Fertilizer nutrient activation is one of the roles of irrigation and rain. Water moves soluble nutrients into the root zone, promotes new root growth in the nutrient enriched soil and is necessary for the conversion of organic N to ammonia and nitrate. This time delay from water application to nutrient uptake can be as short as 1 to 2 days for mobile fertilizers such as ammonium nitrate to over a week for relatively immobile fertilizers such as anhydrous ammonia. Placement of fertilizer nutrients into moist root zone soil can result in uptake even without irrigation activation if the fertilized zone does not desiccate or suffer severe root injury.

The uptake of nitrogen is compounded by the various chemical transformations that nitrogen undergoes from a fertilizer or organic source to the most common uptake form nitrate (NO_3^-). Conversion of ammoniacal-N to nitrate-N from either fertilizer or organic source requires

soil oxygen, soil moisture, and temperature-dependent time. Under anaerobic conditions the conversion of organic-N to ammonia and then to nitrate is blocked. Under conditions of soil desiccation the conversion of organic-N to ammonia is also blocked, although this condition does occur in soils that are too dry to support new root growth. However the time delay for the release of nitrate from either inorganic or organic forms must be considered when application amounts and timing is determined. Due to the time delay in the release of nitrate from organic sources of N, their use for over half of the required annual N supply results in high levels of available soil N during late boll maturation and the resultant complications of poor defoliation and late-season vegetative growth (Weir *et al.* 1996).

Nutrients that are less soluble in the soil solution and thus less mobile in the soil require root interception for sufficient uptake. Thus the optimum placement of fertilizer P, K, Zn, Ca, and Mg is in a zone of sustained new root development. Under irrigated conditions this zone is relatively easy to identify but under rainfed conditions, depending on the rainfall pattern, this zone can move between the surface soil and the subsoil during the growing season. For example: following frequent light showers root expansion is limited to the soil surface where oxygen is replenished from the atmosphere. Nutrients placed deep in the soil will be largely unavailable until the water drains and oxygen can diffuse to deeper depths. On the contrary, an extended drought limits root expansion near the surface resulting in relatively more root proliferation deeper in the soil (if moisture is available in the subsoil). Under these conditions nutrients located near the soil surface will be largely unavailable until rain or irrigation sufficiently rewets the soil to allow new root growth. Management of these conditions is extremely difficult due to the unpredictability of weather patterns. However where high rainfed yields occur (in excess of 800 kg/ha of lint), it is prudent to maintain adequate fertility with regards to the less mobile and immobile nutrients (K, P, Zn, Ca, Mn) throughout the soil profile. With regards to the mobile nutrients (N, S, and B) greater flexibility in placement occurs because of their movement in water both into the soil profile and towards roots.

2.2   Salinity

Salinity in both soils and irrigation water are common in the semiarid and arid cotton production regions of the world with chloride, sulfate, and bicarbonate salts of Na, Ca, and Mg contributing in varying degrees to soil and water salinity. Cotton water-deficit stress, usually associated with a yield loss, may result from either insufficient soil water or an accumulation of salts in the soil since both osmotic and matric potential components together (and to a much lesser degree vertical potential) create the total potential against which plants must work to extract soil water. Although salts may be inherently present, especially on relatively unweathered soils, the accumulation of salts imported in irrigation water usually accounts for the majority of salinity problems in production agriculture. This is particularly common in irrigated regions where evaporation of water from the soil and plant transpiration far exceed rainfall amounts. Dissolved salts may affect crop growth by osmotic effects or specific-ion toxicity, but osmotic effects are the most common phenomenon associated with yield loss. Cotton is considered a salt-tolerant crop; a harvestable cotton crop can be produced on soil or with water that would preclude production of most vegetable and field crops due to salinity (Mass and Hoffman, 1977).

The potential crop yield loss from excess soil salinity is usually predicted with the equation:
$$Y = 100\% - B(EC_e - A)$$
where Y is % relative crop yield, EC_e is electrical conductivity of the soil water extract (decisiemens per meter, dS/m), A is the threshold of soil salinity for cotton tolerance (7.7 dS/m), and B is the rate of crop yield loss (5.2 %/dS/m) as EC_e exceeds A (Mass and Hoffman, 1977). This relationship shows that at an EC_e less than 7.7 dS/m yield is not limited by salinity and for every increase in soil salinity above 7.7 dS/m yield declines by 5.2%. Near the threshold of 7.7 dS/m, salinity decreases the availability of water to the plant. Soil solution containing salts has a lower energy potential, and thus flows less freely into the roots and the cotton plants reach higher levels of water-deficit stress. At higher levels of salinity, expansive growth is severely retarded and metabolic processes disrupted. In the range of most commercial cotton fields, the irrigator will observe plants that are droughty and more compact with smaller leaves, stems, and branches. Water management under saline conditions typically requires increases in frequency of irrigation, maintenance of a higher soil water content, and adjustments in planting pattern. Row widths of 70 cm (30 inches) or less have increased yields in saline fields due to the smaller plant size. Irrigation systems that can supply frequent precise amounts of water are often employed (*e.g.*, Trickle/Drip, LEPA and some sprinkler systems).

Some genotypic differences in salt tolerance and differences at various growth stages have been reported among Upland cotton types (Hayward and Wadleigh, 1949; Läuchli *et al.*, 1981). *Gossypium barbadense* cultivars may be somewhat more salt tolerant than Upland types (Hayward and Wadleigh, 1949).

Models relating crop productivity to water normally assume plant response to water and salt-induced stress to be the same except for specific ion effects (Soloman, 1985). This assumption appears reasonable and it has been used successfully (Meiri and Shalhevetk, 1973; Childs and Hanks, 1975). Howell *et al.* (1984) found that cotton plant water stress indices derived from canopy temperature, leaf diffusion resistance, and leaf water potential were sensitive measurements of salinity induced, water stress.

Ion toxicity may be associated with excess boron, chloride, and sodium, but boron toxicity is probably the most widespread. Cotton is considered to be semitolerant

to boron toxicity (Ayres, 1977). Where agriculture relies predominately on irrigation to meet the crop water needs, salinity in soils is increasing. This increase derives from the simple fact that salt into the field must equal the salt removed from the field plus the change in salinity in the field. Since irrigation water invariable contains salts that are difficult to fully remove from the field by leaching, the net result is a gradual increase is root zone salinity. This salt component of the soil influences both root growth and water extraction and thus must be considered in water management of irrigated fields. The impact of soil salinity on plant water relations is complex because of the various interaction between roots, soil matrix, specific elements and water.

Soil salinity and sodicity are two distinct characteristics of the soil with the former referring to total mass dissolved in the soil solution, while sodicity refers to the relative abundance of sodium in the soil solution compared with calcium and magnesium. Due to cotton's relatively high tolerance of sodium, sodicity alters cotton's plant water relations predominately by impacting the structure of the soil. Sodic soils usually are more susceptible to compaction and reduced internal drainage, both problems that limit root extension. The limited root volume of sodic soils and reduced drainage leads to premature drought and/or water logging. Higher irrigation volumes are often not the solution but rather judicious, more frequent irrigation combined with a long term program of soil amelioration by replacement of sodium with calcium and magnesium on the soil exchange surface.

During the germination and seedling growth stage the cotton plant is especially vulnerable to high salinity and sodicity. This often manifests itself as patchy stands in large areas of the field where cotton failed to emerge next to relatively productive areas in the field. Stand failure in saline soils is often caused by seedling disease injury to the slow growing seedlings. The young cotton plant's high level of susceptibility to saline and sodic soils can be minimized by several management practices: (1) Water with the lowest salt content available should be used for pre-irrigation or "irrigating up" a stand of cotton. Where winter rainfall occurs the reliance on rainwater for the preirrigation is recommended. Irrigation with saline or sodic water should be delayed until the plant is past the one-leaf stage, at which time susceptibility to salinity, sodicity, and seedling disease is minimal. (2) Where irrigation must be used to prewater or "irrigating up", water should be applied so that salts do not accumulate in the seed zone. Apply water and design planting bed configurations such that the salts wick away from the seed zone. Alternate row irrigation is one method to achieve this. Prewatering with high beds then removing the top of the bed is another method. In some areas, producers plant on a flattened bed and leave a 5-cm cap of soil above the seed; this soil cap is removed prior to growth of the hypocoty into the cap. (3) Increase the calcium content of the seed zone, by shallow incorporation of high purity gypsum (CaSO4) (Läuchli and Epstein, 1984). (4) Plant high vigor seed at elevated seeding rates to provide a sufficient final plant population for the reduced plant size and the seedling force necessary to break soil surface crusts. Limit the use of other salts need the seed such as starter fertilizers and granular insecticides.

2.3   Pathogens

Soil moisture may directly and indirectly affect soilborne plant pathogens, their antagonists, and the duration of the cotton plant's susceptibility to a disease. Soil temperature and aeration are soil environment variables controlled by soil moisture that influence pathogen activity. Relatively dry soil will minimize plant root activity and potentially lower pathogen-root infection rates. Seed rot and damping-off of cotton seedlings caused by *Pythium* spp. and *Rhizoctonia solani* can be reduced by maintaining a dry soil surface early in the growing season (Grimes and El-Zik, 1990). Timing of the first post-plant irrigation was observed to be closely tied to the severity of foliar Verticillium wilt symptoms (El-Zik, 1985; Huisman and Grimes, 1989). Huisman and Grimes (1989) found that delaying the first irrigation of the season had the greatest benefit on reducing Verticillium wilt severity. Percentage of defoliated plants at the end of the season had a linear, inverse relationship to the post-plant day of the first irrigation. Their results show that minimizing yield losses due to Verticillium wilt in cotton involves the balancing of two opposing trends. Delaying the first post-plant irrigation progressively decreased disease severity, but the associated water-deficit stress progressively lowered the yield potential. On soils with a high Verticillium wilt inoculum density, optimizing production requires a management balance between these opposing trends. In addition to seedling disease and Verticillium wilt the following cotton pathogens are increased in severity under frequent rainfall or sprinkler irrigation: Bacterial Blight caused by *Xanthomonas malvacearum*, Ramulosis caused by *Colletotrichum gossypii* var. *cephalosporioides* A.S. Costa, Leaf spots caused by *Alternaria macrospora*, *Stemphyllium* and *Ramularia* (CAB International, 1999)

Although most fungal pathogens of cotton are increased in severity when rainfall or irrigation is frequent viral pathogens vectored by aphids and whiteflies can be more severe under dry conditions that favor mobility and growth of the vector.

3.   MANAGING THE CROPS WATER SUPPLY

3.1   Water-Yield Production Functions

Water-yield functions provide a useful means of analyzing water-productivity relations provided they are based on data that utilize proper irrigation scheduling so that the least possible yield reduction results from a defined water deficit. Vaux and Pruitt (1983) pointed out that a variety of independent variables are used to indicate water input, namely,

evapotranspiration (ET), applied water, and soil moisture. Evapotranspiration has the greatest rigor and potential for transferability, however, applied water is the controlled variable and, economically, represents the cost consideration. Soil water status provides a link between ET and applied water.

For many crops and growing conditions the relationship between ET and yield is linear up to ET values that result in maximum productivity. This is especially true for crops where the above-ground biomass represents yield. Sammis (1981) reported a linear relation for cotton although longer-term studies (Grimes *et al.*, 1969; Grimes and El-Zik, 1982) suggest a slight curvature to the function. Considering the nature of cotton reproductive development and plant-water relations, a curvilinear function appears most appropriate for this crop.

While empirically-derived water-production functions are usually correct only for the site-specific conditions under which they are developed, Vaux and Pruitt (1983) pointed out that functions using relative yield (actual Y/maximum Y) and relative ET (actual ET/potential ET) offer some advantage for a more generalized function. Sammis (1981) compared his linear production function (YL vs. ET) for cotton to a similar function developed in California and concluded that such functions are not transferable. On the other hand, Grimes (1982) used a dimensionless (relative yield vs. relative ET) function developed in California to compare with cotton functions presented by Ayer and Hoyt (1981) to show reasonably good agreement. This result and that of Misra (1973) and Stewart and Hagan (1973) lend support to the concept for achieving a fair amount of transferability among geographic regions of contrasting soil and climatic conditions.

3.2   Production Economic Considerations

Since applied water (AW) is the cost input in a production economic solution, it is necessary to consider this as the independent variable in a yield function. Site-specific functions can be developed from empirical observations or developed from observed relationships between ET and AW as demonstrated by Grimes (1982). Limits of a "rational water use zone" can be developed from a yield-applied water function. Applied water to achieve maximum yield is the upper limit of this zone while AW required to reach a maximum average product (AP = Y_L/AW) is the basis for the lower limit. Heady (1952) presented the theoretical considerations justifying this lower limit and Grimes (1977) discussed the application to a cotton-AW response. Applied water to maximize profit will always fall within the limits of this AW input zone. Using traditional production economics, AW is used up to an amount that balances the cost of the last centimeter used with the value of the product resulting from this last increment of AW. Mathematically, this is accomplished by solving the equation:

$$dY_L/d(AW) = Pw/Py$$

for AW where Pw and Py are the respective water cost and

cotton lint price.

Cotton yield functions will also accommodate production inputs in addition to applied water. In arid and semiarid regions, water and nitrogen are frequent inputs that have the greatest influence on cotton productivity and these are usually combined into a single production function such as that presented by Grimes *et al.* (1969) and more recently by Zelinski (1996).

3.3 Managing Water Supply in Contrasting Regions

Optimal water management on a specific farm is related to many factors. In irrigated environments water managers must consider numerous aspects of the water supply (quality, timing, price, and reliability) in addition to multiple impacts on the field. Over years cotton growers develop experience with a crop growing on a field under a range of weather patterns that allows them to incorporate the many factors in the development of optimal water-management systems. However the most important factor is the availability of water, allowing general recommendations based on the relative abundance of water supply. Regardless of the water supply, one objective remains the same, to apply water as efficiently and controlled as possible.

3.3.1 Unlimited Water Availability

Where water supply is unlimited, the challenge to cotton irrigators is to not apply excess irrigation water. Due to the deleterious effect of excess irrigation on soil-borne diseases, spring and fall temperatures, and crop maturation, water should not be freely applied to the field, but rather applied in amounts and timing that maintain maximal dry weight accumulation into reproductive organs until the late stages of boll maturation and opening, when controlled water-deficit stress is desirable (Grimes and El-Zik, 1982). In temperate climates where low spring temperatures limit cotton growth, preirrigation is desired as a means to supply adequate moisture to the seedling plant while delaying the post plant irrigations that stimulate seedling disease, evaporative soil cooling, and weed germination. Once the preirrigation soil water has been partially depleted post plant irrigations are started with the timing and amount adjusted to the crop water use and soil water holding capacity. Avoidance of crop water stress is critical for maximal yield. Once cotton enters peak flowering the irrigation cycle can be shortened and volume adjusted appropriately to avoid sharp contrasts in water availability. Starting approximately two

weeks after the last effective flower, as determined either by crop stage or calendar date, irrigations should be reduced in amount to allow a gradual depletion of soil water while maintaining sufficient active leaf surface to mature the last set bolls on the plant. Although the actual irrigation timing and amounts used for diverse soils, climates, and water delivery systems will vary widely the basic objective outlined above has proven successful in many parts of the world where water availability is not limited (Hake *et al.*, 1996).

3.3.2 Limited Water Availability

The more common case with cotton production around the world is one of limited water availability. Cotton is grown in many water-limited regions due to its high economic value-per-megaliter (approximately US\$ 300 per megaliter) of applied water. Under limited water availability irrigators often rely on rainfall for a substantial part of the crop water supply. The element of uncertainty in rainfall events imposes an added level of complexity. Although comprehensive recommendations can not be given for the many diverse conditions, several principles apply across many regions. (1) The water supply must be used in a way that ensures sufficient water reserves to mature bolls. Yield and quality require boll maturation. Unless rainfall during the last two months of crop development is certain, water must be held in reserve to support boll maturation either off the field in dams or underground reservoirs or in the field beyond the utilization of early season root growth. (2) The decision to spread the limited water over a large area or small area is based on the rainfall pattern and the relative cost of inputs that accrue on an area basis versus those that accrue on a yield basis. Where input costs are high per area and rainfall is low, spreading the water on limited land so there is sufficient water supply for near maximal yield is a common recommendation. However, where rainfall is likely and purchased inputs per area are low, growers often plant substantial acres that can not be fully irrigated. (3) If more area is planted than can be fully irrigated, low density stands or wide/skip row patterns are often used that limit leaf area below that which is required for maximal photosynthesis. (4) Field management practices are employed that capture and preserve soil moisture. Depending on local customs, field conditions, and rainfall patterns, soils are managed to maximize infiltration while minimizing evaporation. Surface residue, where available substantially increases infiltration due to the avoidance of surface crusts and maintenance of macropores. Surface evaporation is often controlled in fine texture soils by lowering the unsaturated hydraulic conductivity with shallow surface tillage.

4. SUMMARY

This chapter provided a review of crop water management to optimize the growth and yield of cotton. This included a discussion of plant water relations at different growth stages and methods of optimizing growth, efficiency, and yield. The most important factor to consider when designing optimum plant-water relations is the impact of water management decisions on the ability of the plant to initiate, retain, and mature harvestable bolls. In addition, soil-plant water interactions were discussed including nutrient uptake, salinity, and pathogens. For management of crop water supply, water-yield functions provide a useful means of analyzing water use productivity relations. Applied water is the cost input in a production economic solution and it is necessary to consider this as the independent variable in a yield function for production economic considerations. Lastly, discussion was provided on the management of water supply in regions of contrasting water availability.

Chapter 24

INTERPRETATION OF PLANT MINERAL STATUS

Wayne E. Sabbe[1] and Steven C. Hodges[2]

[1]Univeristy of Arkansas, Fayetteville, AR (deceased), and [2]NC State University, Raleigh, NC;

1.   INTRODUCTION

The importance of mineral nutrition in cotton production is apparent by the number and frequency of reviews containing mineral nutrition and subsequent cotton fertilization as a topic (Hinkle and Brown, 1968; Sabbe and MacKenzie, 1973; Hearn, 1981: Joham, 1986; Sabbe and Zelinski, 1990). All have extensive literature citations and should be accessed for details on work published prior to about 1985. This chapter will focus on publications since the edition of *Cotton Physiology* (Mauney and Stewart, 1986).

The status of the cotton-fertilizer recommendations (Sabbe and Zelinski,1990) indicates that the majority of the cotton acreage receives nitrogen (N), with fewer acres receiving potassium (K) and phosphorus (P). Since N rates are directly related to yield, the greater potential production in the irrigated desert requires higher N rates than either irrigated humid or dryland areas. A correct fertilization program involves knowledge of the soil nutrient status, growth patterns of the cotton plant and the associated nutrient needs during the growing season. While general statements regarding the above can be made in terms of the long range properties of climate, soil series, and cultivars, a fertilization program must have the ability to vary according to, and during, the current growing season. Thus, management decisions on mineral status require accurate information based on long-term averages and analysis of short-term conditions including crop dry matter accumulations (Bassett *et al.*, 1970). Additionally, there is an interaction between cotton plant development (both vegetative and reproductive) and the concurrent mineral nutrition status of the plant (Table 24-1). Whereas the normal pattern for N, P, and K content (Mullins and Burmester, 1990) was to increase or remain constant in stems and fruits over the growing season,

the normal pattern for leaves was a late-season decrease. Concominant was an increase in whole-plant content of N, P, and K content. Sabbe and Zelinski (1990) concluded that the fertilization program that had the best chance of success included season-long N management under the specific environmental conditions, especially since cotton is often grown continuously on the same field. In the USA, the recommendation program for P and K fertilizers considers soil characteristics (clay content, fixation) and thus can be determined by soil test methodology. Deficiencies, and thus fertilizer recommendations, for the secondary and micronutrients are usually defined by either evaluation of the soil series or attention to specific cultural management practices. Mitchell

Table 24-1. Uptake and distribution of N, P, and K in plant parts (Bassett *et al.,* 1970).

	\multicolumn{4}{c}{Average number of days after planting}			
	60	90	120	165
	15 June	15 July	15 Aug	1 Oct
	\multicolumn{4}{c}{------------------(kg ha^{-1})----------------------}			
Nitrogen				
Stems	1.0	7.1	12.7	18.3
Leaves	9.3	33.3	45.5	37.9
Burrs and Seeds	—[z]	14.5	50.1	87.4
Total	10.3	54.9	109.3	143.6
Phosphorus				
Stems	0.2	1.3	2.2	3.1
Leaves	1.1	3.2	3.9	3.6
Burrs and Seeds	---	2.5	8.6	14.2
Total	1.3	7.0	14.7	20.9
Potassium				
Stems	1.6	13.7	28.5	27.5
Leaves	5.7	22.4	31.4	20.6
Burrs and Seeds	---	10.4	38.4	73.1
Total	7.3	46.5	98.3	121.2

[z] Component not yet formed.

J.McD. Stewart et al. (eds.), *Physiology of Cotton*,
DOI 10.1007/978-90-481-3195-2_24, © Springer Science+Business Media B.V. 2010

et al. (1995a) reported that 60 years of continuous cotton in Alabama showed that extremes in environmental factors could overwhelm annual fertilizer practices; thus a consistent year-to year fertilizer recommendation was advised.

2.   MINERAL NUTRIENT ROLES

Cultural-management practices recognized that cotton, an indeterminate oil-seed plant, presented a more complex nutrient-management scheme than a carbohydrate-seed plant. In particular was the distinct possibility that an incorrect nutrient balance would divert energy from reproductive to vegetative growth, thus resulting in negative benefits from the input. Awareness of the role of nutrients in the physiology of cotton allowed Radin and Mauney (1986) to suggest that improved N management may enhance drought resistance and earliness. Their approach was to utilize the positive effects of low early-season N rates via increased drought resistance and to increase yields by application of N at or after first flower.

The N source-soil relationship in cotton leaves involves at least three phases of utilization - rapid growth, photo-synethically active, and senescent (Thomspon *et al.*, 1976). The decline of N in the leaves during the latter stages was thought to occur because of leaf shedding and mobilization. However, research has shown that the leaves accumulated N from nutrient sources while at the same time N was being translocated from the leaf to the boll (Oosterhuis *et al.*, 1983; Rosolem and Mikkelsen, 1989). In fact the leaves were the major N source within the plant during this growth stage. The recommendation from Rosolem and Mikkelsen (1989) was to maintain an adequate soil-N level during boll and seed development even though most of their N was translocated from the leaves. A complementary conclusion stated that N fertilization during this stage of development had a low degree of efficiency, since retranslocation from the leaves was the primary N source. The prognosis for full development of bolls included an adequate supply of N throughout the growing season. The trend toward short-season cotton has focused on the efficiency of the early fruiting period. Joham (1986) emphasized that fruiting efficiency should be the goal of a nutrient program. Fruiting efficiency has two components; namely, relative fruitfulness (bolls weight per weight of leaves and stems) and fruiting index (*i.e.* harvest index) which measures the ability of the vegetative portion to support a fruit load. While fruiting efficiency is under genetic control, nutrients were separated into two groups - a nutrient group that affects fruiting efficiency and a nutrient group that has little or no effect on fruiting efficiency. The first nutrient group consists of P, K, Ca, Mg, B, and Zn; whereas, the second group consists of N, S, Mo, and Mn. A deficiency of a nutrient in the second group will decrease yield but restricts vegetative and fruiting equally, and thus by definition does not affect fruiting efficiency.

Two changes in cotton culture have affected nutrient management. In the mid-south, early maturation has been the focal point of recent fertilization trials (McConnell *et al.*, 1993). This change will be beneficial if neither yield nor quality is decreased (Mascagni *et al.*, 1993). The second change has been the movement of Pima cotton from southwestern USA into areas where Upland cottons have been the tradition (California). This movement resulted in lower lint yield for Pima under identical nutrient management programs. Unruh and Silvertooth (1996b) determined that the NPK requirement for each 100 kg lint for Upland cotton was 15 - 2.3 - 19 kg/ha; whereas, Pima cotton required a higher rate of 21 - 3.3 - 23 kg/ha. The higher requirement was attributed to Pima's lower harvest index (Unruh and Silvertooth, 1996a) and thus a higher K uptake was needed for Pima cotton to produce yields equal to Upland cotton.

3.   SAMPLING TECHNIQUES AND INTERPRETATION

The occurrence of deficiencies in cotton were categorized by Hearn (1981) as very common (N), common (P, K), occasional (Mg, S, Zn, B, Mn), rare (Ca, Fe, Cu) and unknown (Mo, C1, Na). These categories agreed with earlier observances of Hinkle and Brown (1968) who stated that most production fields have adequate levels of both secondary and micronutrients for optimum yields. The response to soil fertility and fertilizers varies from year-to-year because of plant response to other interactive environmental parameters which impact yields. Where the environmental parameters are more consistent among years, the response to fertilizer application becomes more predictable (Ahern, 1986) and a more reliable fertilizer rate can be recommended. Thus, the fertilizer practices for irrigated semi-desert areas can be formulated before the growing season; whereas, for the semi-humid / Upland areas, fertilizer practices may need tailoring by in-season monitoring systems. Two monitoring approaches have ben utilized; namely, (1) a determination of the plant nutrients (Constable *et al.*, 1991; Maples *et al.*, 1992), and (2) utilization of a crop-monitoring program to determine if normal growth and development occurred because the indeterminate growth habit of cotton creates a well-known flowering and fruiting sequence utilization of a crop (Bourland *et al.*, 1992; Oosterhuis, 1990)

4.   PLANT GROWTH AND DEVELOPMENT

The current use of flowering pattern to quantify the maturity of a cotton crop involves the first-position flower on fruiting branches and subsequent appearance of fruiting nodes (Bourland *et al.*, 1992). First flowers at about 60 days after planting (DAP) usually occur on main-stem nodes 5, 6, or 7. As the growing season progresses, the decline in main-stem nodes above the uppermost white flower (NAWF) correlates with a maturing of the plant such that NAWF = 5 generally indicates that further flowering will not contribute to yield (Bourland *et al.*, 1992). Thus, the effective fruiting period extends from 60 DAP to NAWF = 5 at about 80 days after maturing in the US mid-south. McConnell *et al.* (1995) concluded from N-rate studies at six locations that N fertilization had little effect on earliness of cotton even though the effective flowering period ranged from 4.5 days to 46.5 days among the locations. Perhaps a greater contribution of NAWF is as a signal for termination of practices such as insecticide application (Bernhardt *et al.*, 1986).

4.1 Petiole Analysis

Davis (1995) reviewed the history of petiole usage as an indicator of the cotton plant's nutrient status. The petiole-analysis program is primarily centered in the mid-south and southwestern cotton growing areas of the USA where foliar applications have been a successful practice for N fertilizer. In addition, Davis (1995) indicated that the daily variation and the transitory nature of petiole nitrates could result in altered recommendations. Therefore, it was necessary to obtain cotton petioles not only at the same time of day but also in the same relationship to the irrigation or rainfall schedule (Figs. 24-1 and 24-2). A consistent sampling time increased the probability of differentiating between K application rates, although the difference between no K and a K fertilizer rate was present regardless of sampling time. In Australia (Constable *et al.*, 1991), the interpretation of petiole nutrient data was improved by inclusion of the rate of nitrate decline at flowering. Also, they suggested that for plants very deficient in N, the petiole nitrate value at 750 degree days [(maximum + minimum temperature/2)-18°C] was a reliable indicator of N status. The correct time for sampling of K concentration in various plant parts was related to the application time of K fertilizer (Singh *et al.*, 1992). The optimal time could vary from peak flowering (70 days after planting) if fertilized at planting to 90-115 days after planting if fertilized at flowering. Additionally, the K concentration varied with both leaf and petiole location.

Lutrick *et al.* (1986) developed a model where the petiole nitrate values on a fine sandy loam in Florida were lower than comparative values in either Arkansas or Georgia. In New Mexico, Cihacek and Kerby (1991) concluded that, in late-season, low values of petiole nitrate could be disregarded since most of the cotton yield had already been determined. Phillips *et al.* (1987) reported that cotton petiole-nitrate values increased when an increase in yield occurred due to nitrogen application. However, at this Louisiana location, the petiole nitrate values varied when sampled at the same date among years which could be expected due to seasonal variation in temperature, rainfall, pests, and fruiting.

4.2 Leaf Analysis

Cotton leaves were preferentially chosen to estimate the mineral status of cotton for Upland and for more de-

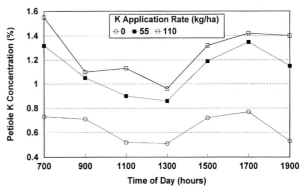

Figure 24-1. Effect of time of day and K application rate on petiole concentration (Davis, 1995).

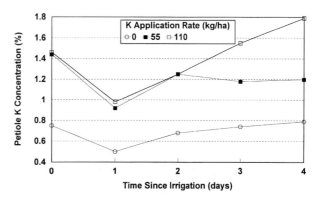

Figure 24-2. Effect of time since irrigation and K application rate on petiole K concentration (Davis, 1995).

terminate cultivars (Sabbe and Zelinski, 1990). The selection of leaves involved a more precise analytical procedure since current in-season remedies seemed improbable due to the short growing season, especially under rain-fed conditions. Additionally, the resultant analysis of the leaf material could include micronutrients plus the determination of total quantities of nutrients rather than only soluble nutrients as per usual petiole analysis.

Several reviews are available estimating the deficiency, sufficiency, and toxicity of nutrient concentration in cotton leaves (*e.g.*, Chapter 14, this volume). While most references utilize only the nutrient concentration for specific growth stages and locations on the plants (Sabbe and Zelinski, 1990; Jones *et al.*, 1991), Braud (1974) interjected leaf dry weight into the interpretation. Additionally, Hearn (1981) listed the nutrient uptake and removal for various lint yields. All of the above-mentioned citations have extensive references concerning growth stage, plant part, specific nutrients, cultivars, and cultural management practices on nutrient concentrations in cotton leaves.

Recent adaptations of the Minolta SPAD meter (Wood *et al.*, 1992) and the use of Near Infrared Spectroscopy (Hattey *et al.*, 1994) have the possibility of decreasing the time and labor necessary to determine N deficiency. Also, the ability to conduct field tests for plant-and soil-soluble N (Fig. 24-3) and K has been advanced by the use of portable ion-electrode instruments (Hodges and Baker, 1993). The advantage of in-field analysis coupled with an ability to distinguish among N rates during sampling periods complemented the technology.

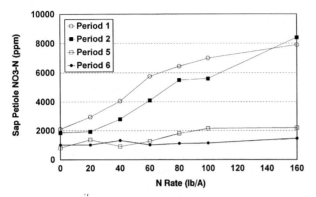

Figure 24-3. Effects of soil-N rate and sampling period on NO$_3$-N concentrations in fresh petiole sap extracts (Hodges and Baker, 1993). Periods refer to weeks after the first flower stage.

5.   NITROGEN FERTILIZATION

The indeterminate growth pattern of cotton coupled with a possible six-month uptake period, plus the transformations and possible losses of soil- and fertilizer-N during this time frame, plus the different yield potential of different years, has proven difficult for the adoption of a single N-application management practice. Each location appears to have a unique management system based on experience and long-term tradition (Mitchell *et al.*, 1995a). While the goal may be a singular pre-plant soil application, the incorporation of side-dressing and foliar application methods must remain as distinct annual possibilities. Therefore, utilization of diagnostic tools and criteria must be viewed as probable during any growing season (Baker *et al.*, 1992; Constable *et al.*, 1991). As management practices improve and as potential yields continue to increase, fertilizer N will increase (Maples *et al.*, 1987). Traditionally in the USA the Upland Texas areas, the humid southeast, and the irrigated desert have N-recommendations of less than 100, 100 to 150, and greater than 150 lb N/acre, respectively (Sabbe and Zelinski, 1990). The advantage of the multiple N application method is that a single large pre-plant application may result in excess N during that specific growing season. Unfortunately excess N rates usually delay maturity plus the resultant residual soil N affects subsequent fertilizer practices. The N fertilizer

recommendations for cotton differ depending on the preceding crop (Ahern, 1986; Constable *et al.*, 1992). When cotton was the preceding crop, the apparent recovery of fertilizer N rate was less than when the preceding crop was soybean, wheat, or fallow (Table 24-2). Thus, the fertilizer-N rate was increased to compensate for the decrease in both the percentage of apparent recovery and the available soil N. The reduced N uptake by cotton following cotton was overcome by applying N under the crop row two-to-four weeks prior to planting. Another effective application method was to sidedress the entire N rate after planting.

In Louisiana, Boquet (1995) reported that residual N occurred from N fertilizer rates as low as 25 lb N/acre. The residual amount was related to the N rate and the residual effects were present for at least two years. Therefore, residual N must be a component of a N fertilization program. In Arkansas, McConnell *et al.* (1989) concluded that among five irrigation methods the optimal N rate varied. However, for each irrigation method the optimal N rate was consistent over the five-year study. Excess N has a tendency to not only delay maturity but also increase residual soil N (Maples and Keogh, 1971; Boquet *et al.*, 1995) which should be quantified for the succeeding crop. Accordingly, the perfect N fertilizer rate would not increase nor delay maturity nor contribute to soil residual N (McConnell *et al.*, 1995). Thus, knowing the relative maturity of the crop at any one time would assist in predicting the future N needs of the crop with special emphasis on economic return of inputs (Bourland *et al.*, 1992).

The accumulation of N, P, and K for current cotton cultivars indicated that peak dry matter production and peak nutrient accumulation occurred during the same time interval (Mullins and Burmester, 1990). This time interval (*i.e.*, 63 to 98 days after planting) and the amounts of nutrients accumulated varied only slightly among the four cultivars tested. Redistribution of accumulated nutrients which occurred during the growing season with translocation from vegetative to fruiting sites being essentially similar among cultivars and the two locations differing in soil series. The average nutrient accumulation for the cultivar/locations were 19.9 kg N, 2.5 kg P, and 15.3 kg K for each 100 kg lint produced. However, the dry matter distribution varied between locations although the seedcotton yields were similar. This dissimilar dry matter distribution among locations confirmed earlier reports (Olson and Bledsoe, 1942; Christidis and Harrison, 1955).

Table 24-2. Example of estimation of cotton N requirements compared to other crops (Ahern, 1986).

	Previous crop			
	Cotton	Soybean	Wheat	Fallow
Target yield (kg ha^{-1})	1200	1480	1480	1480
Required uptake (kg N ha^{-1})	100	100	100	100
Available soil N (kg N ha^{-1})	45	57	72	81
Required fertilizer uptake (kg N ha^{-1})	55	43	22	19
Apparent recovery (%)	40	60	50	70
Fertilizer N required (kg N ha^{-1})	138	72	44	27

The amount of P, K, and S taken up by the cotton plant was not dependent on the N rate or when N was applied (Breitenbeck and Boquet, 1993); however, the N amount removed was increased as N rates increased (Table 24-3). The quantities of P, K, S, Ca, Mg, and micronutrients removed were not significantly increased as the N rate increased (0, 67, and 134 kg N/ha). Moore and Breitenbeck (1993) reported that the effects of N on uptake of other nutrients resulted in increased nutrient concentrations in the leaves (storage organ) with increasing N rates even though the seedcotton yields remained constant (Table 24-4). However, the concentrations of K, S, and B decreased, and the concentrations of Al, Cu, and P (data not shown) did not change over the range of N applications. Because of the major role of N in plant growth and the need for replenishment during the entire growing season, Gerik *et al.* (1993) suggested the inclusion of yield goals, soil tests, plant/tissue testing, and crop simulation models, if available, in the N-management program.

The introduction of reduced tillage fertilizer placement or crop rotation and associated N-fertilization programs in cotton-production systems have indicated that any advantages would be economic rather than agronomic (Ebelhar and Welch, 1989; Harmon *et al.*, 1989; Brown *et al.*, 1985). The extra acreage needed for rotations, the uncertainty of winter cover crops (cereals or legumes), plus the additional cultural-management operations have been a drawback to the few positive results from soil management inclusions (Pettiet, 1993a; Tupper *et al.*, 1993).

6. PHOSPHORUS AND POTASSIUM FERTILIZATION

The tap root system of cotton coupled with the relative immobility of soil P and K have re-focused P and K fertility management especially with the adaptation of shorter-statured, earlier-maturing cotton (Oosterhuis, 1994). The new fertilizer-management philosophy included the speculation that the great need for nutrients during an intense boll-filling period leads to a K deficiency because either the soil cannot release adequate K or the root system has a decreased foraging ability at this growth stage. Variations in soil P and K were associated with changes in relative fruitfulness (P) and fruiting index (P and K); however, the two nutrients may not affect earliness (Joham, 1986). In his report, the fruiting index was decreased from 0.92 to 0.25 by P deficiency and from 0.89 to 0.34 by K deficiency. Tewolde *et al.* (1994) determined that P deficiency in Pima cotton did not affect reproductive efficiency, thus suggesting that the desired earliness was better managed by N fertilizer than P fertilization.

Pettigrew *et al.* (1996) reported that cultivars of varied maturity were not affected differently by K deficiency under Mississippi cultural-management practices. All suffered yield reductions with approximately equal reductions in various growth parameters. This finding supported earlier research by Minton and Ebelhar (1991) that did not find a genotype K interaction for two cultivars. The ratio of P uptake between the upper and lower root zones varied during different growth stages (Nayakekorala and Taylor, 1990). However, this difference (between the upper and lower root zone) decreased as the plants matured. At the open-boll stage of growth the two zones had similar P fluxes. Hons *et al.* (1990) showed no effects on the emergence, germination percentage, survival, nor P and K nutrient concentrations of cotton seed from applications of fertilizer P and K. However, Mullins *et al.* (1994) reported that both in-row subsoiling and deep-applied K fertilizer increased cotton root growth at depths greater than 20 cm (Fig. 24-4); although, in general, deep placement was not superior to sur-

Table 24-3. Effect of N fertilization rate on the uptake of plant nutrients in aboveground plant biomass prior to defoliation (Breitenbeck and Boquet, 1993).

Nutrient	N applied	Ground litter	Lower plant	Upper plant	Total
	(kg/ha)	--------------(kg nutrient/ha)--------------			
N	0	17.6	46.1	27.0	90.6
	84	40.3	80.9	73.4	194.5
	168	60.4	70.0	109.3	239.7
P	0	3.5	11.0	8.3	22.7
	84	4.3	13.9	16.5	34.6
	168	6.3	10.5	17.2	33.9
K	0	15.5	51.7	32.9	100.1
	84	22.6	68.4	62.7	153.8
	168	29.0	56.2	62.2	147.4
S	0	10.4	10.2	9.7	30.4
	84	13.4	17.0	19.8	50.4
	168	17.6	12.1	20.7	50.4

Table 24-4. Effect of nitrogen rate on yield and nutrient concentrations in upper leaves of cotton (Moore and Breitenbeck, 1993).

N	Yield	N	K	Mg	Ca	S	Zn	Mn	Fe	B
-------- (kg/ha) -------		----------------------------------(g/kg)----------------------------------						----------- (mg/kg)--------------		
0	2100	18.9	13.9	5.9	33.4	10.0	17	49	62	73
28	2800	17.7	11.4	5.8	34.1	9.0	17	45	65	66
56	3600	20.7	11.7	6.3	35.7	10.3	18	49	79	63
84	3800	20.4	10.5	6.4	35.6	9.4	14	46	74	55
112	4000	25.1	9.5	7.5	35.6	8.6	16	52	81	55
140	4100	25.1	8.0	8.7	37.4	8.0	18	55	75	46
168	4000	27.4	7.9	10.0	3.8	7.1	16	57	105	41
LSD (0.05)	NA[z]	3.7	2.2	1.5	3.9	1.8	4	8	35	8

[z] NA = not available.

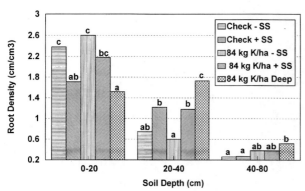

Figure 24-4. Cotton root length density at
three depths as affected by K rate and in-row subsoiling
(-SS, no subsoiling; +SS, subsoiling). Significance
at 0.20 level of probability (Mullins *et al.*, 1994).

face applications (Reeves and Mullins, 1995). In fact, deep application of K plus subsoiling produced a more severe leaf K deficiency than a surface application of K fertilizer.

In recent years the practice of foliar K application of K-deficient plants has produced inconsistent results (Oosterhuis, 1999b). The initial speculation indicated that at the peak of plant K required, K uptake from the soil is not sufficient to meet the plant's needs even on soils not considered K deficient (Oosterhuis, 1999b). However, the results from a three-year Beltwide study in 12 states that showed significant responses to foliar K 30% of the time were neither conclusive nor predictable (Oosterhuis *et al.*, 1994). These authors concluded that yield response to foliar-K fertilization was obtained where fertilizer-K was applied, but after two years of soil-K replenishment no response to foliar K was obtained. It may have taken two years to satisfy K fixation and to get K distributed adequately. Mitchell *et al.* (1995b) summarized a three-year study in Alabama by stating that responses from foliar K occurred less than 50% of the time and the positive response may range as low as 33%. Thus, they concluded that neither soil-test K level nor degree of deficiency were reliable indicators of expected response form foliar K. Nor did different K sources increase the predictability of yield response from foliar K (Miley and Oosterhuis, 1994; Mullins and Burmester, 1995).

Partial alleviation of a mid-season K deficiency occurred when a slow release K fertilizer was applied preplant (Davis, 1996); however, this alleviation appeared to be restricted to low K-supplying soils with high yield potential.

In the California San Joaquin Valley, a three-step process has been proposed upon which to base K fertilizer recommendation for subsequent cotton crops (Weir *et al.*, 1995): 1) preplant soil sample (5 to 15 inches) to be analyzed for extractable K; 2) if the extractable concentration was less than 95 ppm K then the soil should be evaluated for K fixation; and 3) a petiole sampled at both peak bloom and six weeks after peak bloom if the latter petiole sample contained less than 1.3% K and if low soil K levels existed, then a K fertilizer application was recommended for the next cotton.

The advent of both faster fruiting cotton cultivars and the intensive management needed for higher yield has led

to a reevaluation of K fertilizer practices. These concerns have resulted in research involving diagnosis of deficiency symptoms, fertilizer rates and placement, and consideration of the soil profile. Traditional K-deficiency symptoms have been complemented with symptoms that appeared on the upper portion of the plant (Maples *et al.*, 1988). The upper portion appeared to be the initial site for late-season deficiency symptoms with the subsequent progression downward. Since K deficiencies can manifest themselves through several avenues, Baker *et al.* (1992) concluded that both soil tests and plant tissue methods be used to evaluate K status. Utilization of multiple approaches will allow for the alleviation of a possible deficiency through soil application and, if necessary, by foliar application of K fertilizers. The favorable approach is to detect the possibility of a subsequent K deficiency either prior to planting (soil test) or early in the growing season (plant tissue analysis) so that a soil application has time to relieve the deficiency. Recognition of the K contribution of the subsoil and the mineralogy have led to inclusion of profile soil test for K fertilizer (Baker *et al.*, 1992; Cassman *et al.*, 1989b). If too late for soil-applied fertilizer, then foliar application may be feasible although the consistency of positive response has been elusive (Oosterhuis, 1992; Bednarz *et al.*, 1992). Most foliar application recommendations do not extend beyond the seventh week past first bloom, especially when the cotton is in cutout (Snyder *et al.*, 1995).

As with foliar applications, deep placement of K fertilizer has produced inconsistent yield improvements. Tupper (1992) advocated banding of K fertilizer at a 6- to 15-inch depth to correct deficiencies of increasing subsoil K on an Inceptisol soil in Mississippi. Subsoil soil samples were needed to not only detect K deficiency but also to ensure a pH of 5.7 or above. Over a period of four years the soil test K values in the 10- to 15-inch depth on this soil were increased 1.12 kg K per hectare for each 3.92 kg K_2O per hectare applied (Tupper *et al.*, 1992). Increase in cotton tap root length and cotton yield were produced by the combination of subsoiling and deep-banded K but not from deep banded and broadcast K, and generall when responses were measured they were small. Mullins *et al.* (1992b) reported that on ultisol soil in central Alabama subsoiling increased the water uptake and root growth from soil depths greater than 8 inches. The increase in rooting and water extraction resulted in a greater plant biomass from subsoiling and also from deep K placement (Table 24-5). However, higher yields occurred with surface application with in-row subsoiling rather than subsurface application of K plus subsoiling. Apparently the placement of K where the preponderance of the cotton root system occurred proved advantageous. In a companion study (Mullins *et al.*, 1992a), there was no yield advantage to deep-placed K fertilizer nor limestone compared to surface application except on a Norfolk sandy loam with a developed traffic pan. The authors concluded that two reasons subsoiling may not be advantageous were: 1) an acid subsoil, and 2) the volume of soil affected by

Table 24-5. Effect of subsoiling and deep placement of K fertilizer on seedcotton yields and the distribution of dry matter in the various plant parts (Mullins *et al.,* 1992b).

Treatment	Seedcotton yield		Dry matter distribution			
	1990	1991	Bolls	Stems	Leaves	Total
	---------- (kg/ha) --------		----------------------(grams/plant) ----------------------			
Control - no subsoiling	2397 a[z]	2900 a	82 b	31 b	18 b	131 b
Control - subsoiling	3063 b	2866 ab	119 ab	41 b	24 b	184 ab
101 kg K₂O - no subsoiling	2351 a	3448 ab	81 b	36 b	20 b	137 b
101 kg K₂O - subsoiling	3174 b	3687 b	138 a	85 a	41 a	264 a
101 kg K₂O - deep placement	3064 b	3284 ab	127 ab	48 b	26 b	201 ab
Significance[y]	0.6	0.6	0.05	0.05	0.05	0.05

[z] Numbers with the same letters within a column are not significant (P-0.05).

deep K placement was too small. Davis-Carter *et al.* (1992) indicated that cotton cultivars should not be overlooked as a parameter in non responsiveness to K fertilizer. The cultivars that are shorter seasoned and are faster fruiting may require a higher fertilizer rate in order to have an adequate supply during the intense fruiting period. Roberts (1992) summarized that the advantage of surface K fertilizer application was the number of options also pointed out the variation of the ability of cultivars to respond to K fertilizer.

7.   SUMMARY

Fertilizer recommendations for cotton should not be made without knowledge of certain plant growth parameters. While leaf analysis, petiole analysis, or soil-test values can be indicative of nutrient deficiencies, the resultant fertilizer recommendation (including method) must take into account cultivar characteristics, stage of growth, and ability of the plant to respond to the input. Advances in the technology to predict the plant's response by plant growth characteristics - such as the use of crop monitoring utilizing the number of nodes above the uppermost white flower (NAWF) - will avoid end-of-the-season maturity and harvesting problems. Since cotton is not rotated to as great an extent as other crops, responses over long term can be among the most valuable observations. Also, care must be exercised to include residual N in the fertilization program. The tools available for diagnostic purposes have their greatest value in predicting or detecting nutrient problems rather than aids in fertilization. Therefore, use of only measurements of plant nutrient status for fertilizer prediction should be exercised with care.

Chapter 25

FOLIAR FERTILIZATION OF COTTON

Derrick M. Oosterhuis[1] and Billy L. Weir[2]

[1]Department of Crop Soil and Environmental Sciences, University of Arkansas, Fayetteville, AR; and [2]Cooperative Extension, University of California, Merced, CA.

1. INTRODUCTION

Proper plant nutrition for optimal crop productivity in cotton requires that nutrient deficiencies be avoided. However, deficiencies often occur for a variety of reasons, most of which can be rectified by timely application of the deficient nutrient. In crop production, this usually entails a soil application, or foliar applications may be appropriate after canopy closure or when a specific nutrient is urgently required. Furthermore, foliar fertilization may lead to less concern about groundwater and surface water contamination, with nitrates in particular, and less scrutiny of the use of commercial fertilizers. This is particularly important because of current attention being focused on environmental protection. Farmers in the USA and elsewhere are using more commercial fertilizer than 20 years ago and major improvements have been made in how these fertilizers are managed. Foliar application of specific nutrients is a method used to improve the efficiency of fertilizer use and increase yields.

The increased use of foliar fertilizers in cotton production in the last decade is due in part to changes in production philosophy. The change to cotton cultivars which fruit in a shorter period of time and mature earlier (Wells and Meredith, 1984a) has placed greater emphasis on understanding plant uptake and utilization of nutrients. Current crop monitoring techniques (e.g., Danforth and O'Leary, 1998) also focus attention on plant development and make it easier to combine concomitant nutrient monitoring (McConnell et al., 1995) allowing remedial action on a more timely basis. Furthermore, cotton lends itself to foliar fertilization because of the large number of aerial applications that are already made for pest control.

There is a wealth of literature about foliar fertilization that was first used as long ago as 1844 to correct plant chlorosis with foliar sprays of iron (Gris, 1844). The practice has only caught on in cotton production in the last two decades, although there is still considerable speculation about the benefits and correct implementation of this practice. While there are many reports on research involving soil-applied fertilizer, there are relatively few definitive studies on the usefulness of foliar-fertilization. There have been a number of reviews on the subject (e.g., Neumann, 1988), but few specifically on cotton (Oosterhuis, 1995b). According to Neumann (1988) there are a number of different categories of chemicals that can be absorbed by plant leaves and affect plant development. These include atmospheric nutrients and pollutants, biological pesticides, growth regulator substances, film forming compounds, and fertilizer solution. However, only the latter category of fertilizer solutions are considered as true foliar fertilizers as these can supply a significant amount of a particular nutrient. This review will focus on available information about foliar fertilizers in cotton production; how they work and the factors that affect the efficiency of this practice. The leaf cuticle and its role in absorption of foliar-applied nutrients will be discussed. The article will also select the main nutrients that are used as foliar fertilizers in cotton and elaborate what is known about their use.

2. WHAT IS FOLIAR FERTILIZATION?

Foliar fertilization refers to the application of foliar sprays of one or more mineral nutrients to plants to supplement traditional soil applications of fertilizers. The emphasis should be on the foliar application supplementing soil avail-

J.McD. Stewart et al. (eds.), Physiology of Cotton,
DOI 10.1007/978-90-481-3195-2_25, © Springer Science+Business Media B.V. 2010

able nutrients and not on the foliar application serving as a main source of nutrition. A successful fertilizer plan starts with a soil based fertilizer program which necessitates a reliable soil analysis program (Westermann, 1990). Thereafter, any deficiencies in plant nutrients can be detected by tissue analysis (Sabbe and Hodges, 2009) and rectified on a timely basis with soil or foliar applications of the relevant fertilizer. There are a large number of fertilizer nutrients soluble in water that may be applied directly to the aerial portions of plants. The nutrients enter the leaf either by penetrating the cuticle or entering through the stomates before entering the plant cell where they can be used in metabolism.

Foliar fertilization can improve the efficiency and rapidity of utilization of a nutrient urgently required by the plant for maximum growth and yield (Oosterhuis, 1995b). When problems of soil fixation occur, foliar fertilization constitutes the most effective means of fertilizer placement. Foliar feeding provides for a more rapid material utilization and permits the correction of observed deficiencies in less time than can be accomplished by soil applications. However, the response is often only temporary, and thus multiple foliar applications may be needed. Foliar applications usually necessitate using smaller quantities of nutrient than a soil application. The most important use of foliar fertilization has been in the application of micronutrients where only small quantities of the nutrient is required. When larger amount of nutrients such as nitrogen, phosphorus, or potassium are required in foliar application, difficulty is often experienced in supplying adequate amounts without causing phytotoxicity. Additionally, an unduly large volume of solution or number of spraying operations may be required. Generally, it is necessary to use spray concentrations of only <1 to 2% to avoid injury to foliage (Havlin *et al.*, 1999). Nevertheless, foliar sprays may serve as excellent supplements to soil applications. The efficacy of foliar fertilizers is affected by environment, the condition of the crop, particularly plant water status, and the nature of the chemicals applied. This will be detailed later in this review.

Use of foliar fertilization to relieve physiological stress has great potential (Gray and Akin, 1984). This is because when a plant goes from the vegetative to reproductive stages, the amount of photosynthate produced by the leaves is preferentially translocated to the developing fruit and seed and the amount going to the roots is greatly reduced. The root still functions, but root extension and root growth declines and even ceases (Cappy, 1979) meaning that the root system operates at a lower efficiency and the uptake of nutrients is reduced. Therefore supplementing the nutrient supply to the developing fruit would be beneficial especially if any soil impedance to nutrient uptake exists.

3. MECHANISM OF FOLIAR FERTILIZATION

In order for a foliar applied fertilizer nutrient to be utilized by the plant for growth, the nutrient must first gain entry into the leaf prior to entering the cytoplasm of a cell within the leaf. To achieve this, the nutrient must effectively penetrate the outer leaf cuticle and the wall of the underlying epidermal cell. Of the different components of the pathway of foliar-applied nutrients, the cuticle is believed to offer the greatest resistance (Leece, 1976). Once penetration has occurred, nutrient absorption by leaves is probably not greatly different from absorption of the same nutrient by roots, the major difference being the environment in which each of these plant parts exists.

3.1 Entry of Nutrients into the Leaf

There are two possible channels for penetration of foliar-applied compounds into the leaf before they can produce a response. One is through the stomata and the other is through the external cuticle. It is generally accepted that most nutrient uptake occurs through the cuticle (Martin and Juniper, 1970), but solutes can also gain entry into the leaf indirectly through the stomata (Eichert *et al.*, 1998; Noggle and Fritz, 1983a). However, there is some controversy about the importance of stomatal penetration into the interior of the leaf. Prior to 1970 there was considerable debate (*e.g.*, Dybing and Currier, 1961; Middleton and Sanderson, 1965) about the importance of stomatal uptake of foliar-applied nutrients. This debate largely subsided since Schonherr and Bukovac (1972) showed theoretical proof that it was not possible for a water droplet to enter the stomata of leaves of higher plants due to the surface tension of water, the hydrophobicity of leaf surfaces, and the geometry of the stomate. Furthermore, ion uptake rates from foliar sprays are usually higher at night, when the stomata are closed, than during the day, when the stomata are open (Teubner *et al.*, 1957). Recently, however, Eichert et al (1998) provided new evidence for the uptake of large anions through stomata. It would appear that stomata might indeed represent a possible pathway through which a limited amount of the nutrient might gain entry into the leaf. Generally, however, it is assumed that all liquid uptake of water and dissolved substances occurs exclusively through the leaf cuticle provided there are no surfactants present. Surfactants in agrochemical sprays typically provide surface tensions of about 30 Mn m^{-1}, which would usually not be sufficient to enable stomata to be infiltrated. However, organosilicone surfactants can reduce aqueous surface tensions to about 20 Mn m^{-1} and allow nutrient entry via the stomates (Stevens *et al.*, 1992). Furthermore, stomatal penetration can occur only in the brief period after application while spray deposits remain liquid. Thereafter cuticle penetration remains the sole pathway of uptake (Stevens *et al.*, 1992). In cotton, it is unlikely that direct penetration of solutes from the leaf surface through open stomata into the leaf tissue plays an important role, because cotton has pronounced stomatal ledges that partly cover the mature stomate, as well as an *internal cuticular* layer which extends through the stomatal pore and partially covers the mesophyl cells in the sub-stomatal cavity closest to the stomate (Wullschleger and Oosterhuis, 1989a) (Plate 25-1). The cuticle is generally considered to be the rate-limiting step in the overall process of foliar penetration (Franke, 1967; Neumann, 1988).

3.2   Structure and Function of the Cuticular Layer

Aquatic plants inherently absorb nutrients through their leaves, but in terrestrial plants, such as cotton, the leaf surface severely restricts nutrient absorption. The outer surface of leaves consists of the cuticle located immediately above the epidermal cell layer. This biological barrier also enables plants to control the diffusion of molecules, a process essential to plant life (Kolattukudy, 1980), and constitutes the first line of defense against adverse environmental stresses, particularly water loss, and to some extent, protects against harmful chemical penetration such as occurs from atmospheric pollution, foliar fertilization, and defoliation.

The cotton cuticle constitutes a continuous waxy covering over the underlying epidermis (Plate 25-1A), interspersed with numerous stomates (Plate 25-1B) and glandular trichomes (Plate 25-1A) (Wullschleger and Oosterhuis, 1987b; Bondada and Oosterhuis, 2000). Superimposed on this is the epicuticular layer (Plate 1C) of predominantly lipid material (Oosterhuis *et al.*, 1991b). Cotton also has conspicuous waxy ledges extending part way over the stomatal pore and an internal cuticle extending through the stomatal pore and covering some of the substomatal mesophyll cells (Plate 25-1D) (Wullschleger and Oosterhuis, 1989a). These

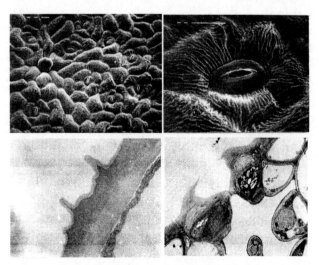

Plate 25-1. The cotton cuticle. (A) Adaxial sympodial leaf surface scanning electron micrograph (x700), (B) stoma with epicuticular wax ridges (x3000), (C) cross section of a stoma showing epicuticular ledges and internal cuticle (x10,000), and (D) cross section of an adaxial cotton stomate showing the pronounced stomatal ledges that partly cover the mature stomate and the cuticular layer extending through the guard cell into the substomatal cavity (X1000). From Oosterhuis *et al.*, 1991a (A), Wullschelger and Oosterhuis, 1989

latter two anatomical features are probably associated with the zerophytic habitat from which cotton originates. The characteristics of the cotton leaf epidermal surface including the stomata, the shape and number of epidermal cells, the cuticle and trichomes play a pivotal role in evaporative losses and gas exchange (Bondada and Oosterhuis, 2000).

The cuticle of different plants varies considerably in waxiness and is usually thicker in plants growing in dry habitats. In cotton the leaf cuticle is about 0.46 µm thick (Plate 25-1C; Oosterhuis *et al.*, 1991a). In general, there is an inverse relationship between cuticle thickness and uptake of chemicals (Hull, 1964). The cuticle is composed of a framework of insoluble polymeric cutin with soluble waxes embedded in the cutin matrix and extruding from its surface. The cuticle is subdivided into two layers (Fig. 25-1); a thin outer layer, the *cuticle proper*, built of cutin only, and the larger *cuticle layer* below consisting of cutin and wall materials (Fahn, 1990). The cuticle layer consists of cutin, deposited between the cellulose microfibrils of the outermost layers of the epidermal cell wall, and also pectin, hemicellulose, and wax. Pectin, hemicellulose, and cellulose dilute the cutin as the cuticle merges with the outer wall of the epidermal cell (van Overbeek, 1956). The *pectin layer* cements the cuticle layer and the cell wall. Cutin is an insoluble heterogenous high molecular weight lipid polymer composed of various characteristic long chain substituted aliphatic acids. In general, cutin is a biopolyester composed largely of various combinations of members in two groups of fatty acids, a group having 16 carbon atoms and a group having 18 carbons (Kolattukudy, 1980; Holloway, 1982) most having two or more hydroxyl groups. The polymeric nature of the cutin arises from the ester bonds joining the various fatty acids. Small amounts of phenolic compounds present in cutin are thought to bind by ester linkages the fatty acids to the pectins of the epidermal cell walls (Salisbury and Ross, 1992). This layer of wax and cutin, formed from condensation of C_{18} hydroxy fatty acids has a semihydrophilic property that affects plant water relations as well as chemical retention and absorption. The waxes affect the deposition and retention of spray chemicals, whilst the composition and thickness of the cuticle as a whole are likely to influence the penetration of agricultural chemicals into the plant (Silva Fernandes *et al.*, 1964). Oosterhuis *et al.* (1991a) showed that increased epicuticular wax on cotton leaves from water-deficit stress decreased the uptake of foliar-applied defoliants by 34%.

Wax is also deposited on the surface of the cuticle as epicuticular wax (Holloway, 1982), arising from precursors in the underlying epidermal cells (Bondada and Oosterhuis, 1994a; Fig. 25-1). The wax precursors are synthesized by the living protoplasts of the epidermal cells (Kolattukudy, 1970) and the wax and/or precursors probably move to the surface through ubiquitous natural cuticular pores and/or transcuticular channels (DeBarry, 1871; Miller, 1985). These epidicuticular waxes are a mixture of diverse long chain hydrocarbons, which, upon crystallization, form a hy-

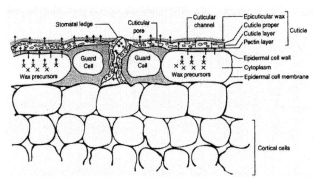

Figure 25-1. Schematic representation of surface wax formation in the external wall of an epidermal cell showing the precursors arising in the underlying epidermal cells (Bondada and Oosterhuis, 1994).

drophobic boundary layer at the surface of aerial portions of higher plants (Fisher and Bayer, 1992). The epiculticular wax morphology of the cotton leaf has been described by Bondada and Oosterhuis (2000) as fine, slender striations that run randomly on the periclinal walls of both adaxial and abaxial surfaces of the epidermal cells (Plate 1A). These authors also described the wax striations on the adaxial surface as circumscribing the stomata (Plate 1B) and as four "wings" of striations emanating from abaxial stomata. Epicuticular wax in the form of striations also occurs on the leaves of other plants such as apples (Leece, 1978). The magnitude of leaf wax has been shown to vary with leaf age as measured by gravimetric analysis (Bondada *et al.*, 1997). Bondada and Oosterhuis (2000) showed that leaf striations were hardly discernable on the periclinal walls of the epidermal cells of 10-day-old leaves. However, the wax deposition which increased with leaf age, was easily visible in 20-day-old leaves, densely distributed in 40-day-old leaves, and in 60-day-old leaves the entire periclinal walls of epidermal cells were eclipsed by wax striations.

Genotypic variation in epicuticular wax has been documented in rice (O'Toole and Cruz, 1983), red clover (Moseley, 1983), sorghum (Jordan *et al.*, 1983), wheat (Clarke *et al.*, 1994) and other crops. Furthermore, environmental factors such as radiation, temperature, humidity, and water stress have been shown to affect the amount of surface wax in plants. Leaf glaucousness (waxiness) is a characteristic that has been cited as a plant adaptation to drought (Johnson *et al.*, 1983; Shantz, 1927). The wax content of cotton leaves increases under drought (Weete *et al.*, 1978; Oosterhuis *et al.*, 1991). Increased epicuticular wax production has been suggested as a selection criterion for breeding drought resistance, *e.g.* in sorghum (Jordan *et al.*, 1984) and alfalfa (Galeano *et al.*, 1986). However, for this, a genetic analysis of epicuticular wax production is necessary (Jefferson *et al.*, 1988) which has not been performed for cotton.

There have only been a few reports on the nature of the wax composition of the cotton leaf cuticle. Hanny and Gueldner (1976) reported that the surface lipids of a glabrous *Gossypium hirsutum* L. strain Bayou SM1 consisted mainly of C_{27}-C_{38} *n*-alkanes (49.9%) with *n*-nonacosane as the major

constituent (28.7%), *n*-primary alcohols (5.5%) with *n*-octacosanol predominating, and nineteen sterols and triterpenoids (44.6%). Oosterhuis *et al.* (1991a) documented the epicutucular wax constituents of leaves of *G. hirsutum* L. cultivar Stoneville 506 and also showed that the composition changed in water-stressed leaves to a greater concentration of higher molecular weight waxes which increased the hydrophobicity of the leaf surface (Table 25-1). Bondada and Oosterhuis (1996) confirmed that water stress increased the number and levels of long-chain, high molecular weight alkanes in the leaf, whereas the water-stressed boll wax contained only the long-chain alkanes. They documented that the main constituents of the long-chain alkanes were *n*-octacosane (2.7%), *n*-nonacosane (1.8%), *n*-triacontane (2.1%), dotriacontane (4.6%), and *n*-tetratriacosane (24.5%). These authors also characterized the wax composition of the boll wall and the associated bracts, reporting that the boll had considerably more total epicuticular wax (+ 66%) than the leaf, with the bract being more similar to the leaf. Information is needed on genotypic and species differences and the effect of environmental stresses on the wax composition of cotton leaves.

The major function of the hydrophobic surface cuticle is to protect the leaf from excessive water loss by transpiration to the drier atmosphere (Nobel, 1991). Mineral elements entering the leaves via the xylem from the roots are released into the apoplasm of leaf tissue before uptake by individual leaf cells (Pitman *et al.*, 1974). Therefore, the other main function of the hydrophobic surface layer of leaf surfaces is protection against excessive leaching of inorganic and organic solutes from the leaves by rain. The relative importance of these two functions depends on climatic conditions, *i.e.*, arid versus humid regions.

3.3 Movement of Nutrients Through the Cuticle

The structure of plant surfaces and the their penetration by various solutes have been extensively reviewed (Swietlick and Faust, 1884; Cutler *et al.*, 1982; Juniper and Jeffree, 1983; Schonherr and Riederer, 1989). However,

Table 25-1. Epicuticular wax composition from the adaxial cuticles of well-watered and water-stressed cotton leaves. (From Oosterhuis *et al.*, 1991a).

Epicuticular constituent	Molecular formula	Epicuticular wax composition	
		Well-watered	Water-stressed
Tricosane	$C_{23}H_{48}$	+[z]	-
n-Tetracosane	$C_{24}H_{50}$	+	-
Pentacosane	$C_{25}H_{52}$	+	-
Hexacosane	$C_{26}H_{54}$	+	+
Octocosane	$C_{28}H_{58}$	+	tr
n-Nonacosane	$C_{29}H_{60}$	tr	0
Decasane	$C_{30}H_{62}$	tr	++
Octocosanol	$C_{28}H_{58}O$	+	++
Fucosterol	$C_{29}H_{48}O$	+	++

[z] - wax absent, + wax present, ++ increased quantity, tr trace present.

these reviews have not focused on or discussed cotton specifically. In general, the pathways and controlling mechanism for absorption of foliar-applied nutrients are poorly defined (Hunt and Baker, 1982). The fate of chemicals once they are deposited on leaf surfaces depends primarily on the molecular configuration of the chemical, which in turn determines its chemical and physical properties, as well as on the nature of the leaf surface (Ashton and Crafts, 1981). Environmental conditions strongly influence the amount of a compound that enters a plant, but penetration rate is generally determined largely by physical properties of the cuticle (Hunt and Baker, 1982). Plant characteristics and the site of application also have important implications on absorption.

There are two pathways by which exogenous chemicals may traverse the distance from the leaf surface into the symplast; a lipoidal route and an aqueous pathway (Crafts, 1956). Compounds that penetrate the cuticle in the lipoidal-soluble form do so principally in the non-polar, undissociated form, whereas compounds that enter via the aqueous route move in slowly, and their penetration is greatly benefitted by a saturated atmosphere (Ashton and Crafts, 1981). Absorption is partially by passive diffusion of molecules through the largely lipoidal cuticle and partially by a dynamic process of uptake dependent upon metabolic activity of the plant (Sargent, 1965). Passive diffusion is believed responsible for most of the penetration of exogenous chemicals through the cuticle and underlying membranes (Neumann, 1988). The movement probably follows Fick's first law whereby the rate of diffusion across a membrane is proportional to the concentration gradient across it (Neumann, 1988), although current thinking is that the process is much more complex than this law entails. In accordance with Fick's law, the higher the concentration of solute which can be applied to a leaf surface without causing damage and the longer time it remains in an active state on the leaf surface, *i.e.*, as a solution, the greater the likely rate and amount of penetration (Neumann, 1988). Diffusion of a nutrient occurs mainly because of a gradient in concentration from the external leaf surface to the free space in the cell wall and in the cytoplasm within the cell. A distinct gradient occurs from low to high charge density, from the hydrophobic external surface toward the hydrophilic internal cell walls (Yamada *et al.*, 1964). Ion penetration across the cuticle is therefore favored along this gradient, an important factor for both uptake from foliar sprays and losses by leaching. Uptake of nutrient elements via the cuticle depends on whether the element it is in inorganic form or combined in an organic form, its ionic concentration, and on the existing environmental conditions which influence how long the nutrient remains in solution on the leaf. The cuticular layer can also function as a weak cation exchanger attributable to the negative charge of the pectic material and nonesterified cutin polymers.

It was proposed earlier that the movement of solutes across the cuticular layer took place in channels of a nonplasmic nature called *ectodesmata* or *teikodes* (Franke, 1967, 1971; Schonherr and Bukovac, 1970) in which the fibrillar structure is more loose than elsewhere in the wall. These bundles of interfibrillar spaces may be filled with coarse reticulum of cellulose fibrils extending from the plasmalemma to the cuticle (Esau, 1977) and serve as polar pathways in foliar absorption and excretion (Franke, 1967; Merkens *et al.*, 1972). However, experimental evidence for the existence of these structures is lacking. It is now more commonly believed that plant cuticles are traversed by numerous hydrophilic pathways that are permeable to water and small solute molecules such as mineral elements and carbohydrates (Marschner, 1995). The majority of these pores in the cuticle have a diameter of <1 nm and, with a density of about 10^{10} pores cm^{-2} (Schonherr, 1976), and are readily accessible to low molecular weight solutes such as urea (radius 0.44 nm) but not to larger molecules such as synthetic chelates (Marschner, 1995). These pores are lined with negative charges increasing in density from the outside to the inside of the cuticle, which facilitates the movement of cations along this gradient (Tyree *et al.,* 1990). Uptake of cations, therefore, is faster than anions and is also particularly rapid for small, uncharged molecules such as urea. The density of cuticular pores is highest in the cell wall system between the guard cells and subsidiary cells and serves as pathways for cuticular or peristomatal transpiration (Maier-Maercker, 1979). This further explains the commonly observed positive correlation between the distribution of stomata and the intensity of nutrient uptake from foliar sprays (Franke, 1967; Levy and Horesh, 1984).

Following penetration of the cuticle, the uptake of solutes into the cell interior depends primarily on the electrochemical concentration gradient from outside into the cell, but also on the plasma membrane permeability coefficient of the molecule and the degree of cell-mediated active uptake (Neumann, 1988). There are various fates of foliar-applied fertilizer nutrients including volatilization and loss to the atmosphere, crystallization and retention on the outer plant surface, loss in dew, or dripping from the leaf. There may also be absorption and retention inside the treated leaf including in the cuticle in the lipoidal layer, and then movement into the aqueous phase of the apoplast and diffusion into the inner leaf structure, penetration of the cuticle and then movement with the transpiration water into the mesophyll and into the symplast, and translocation out of the leaf via the petiole (Ashton and Crafts, 1981; Boyton, 1954). Foliar-applied nutrients are transported through the phloem and take the pathway of photosynthetic assimilates (Marshall and Wardlaw, 1973). In a review of transport of foliar-applied nutrients, Kannan (1986) concluded that solutes absorbed by cells within the leaf can take either the apoplastic or symplastic pathways to reach the vascular tissues for outward translocation, and are transported from the pallisade and mesophyll cells to the vascular tissues in a passive way. The most important consideration for efficient and profitable foliar fertilization is that this practice will only be affective if the applied nutrient ultimately reaches the target site for its use, *i.e.*, the growing points in a vegetative cotton plant and the developing fruit in a more mature reproductive plant.

4. RATE AND TIMING OF FOLIAR FERTILIZERS

The timing of foliar sprays, particularly in regard to growth stage, can be critical in relation to the optimum efficacy of the foliar treatment, and more attention should be given to it (Alexander, 1986). This is because the seasonal pattern of uptake of nutrients varies with growth rate and growth stage (Bassett *et al.*, 1970) but generally follows a sigmoidal pattern with sharp increases occurring as the boll load develops (Mullins and Burmester, 1991). The developing boll load has a high requirement for nutrients, N, P, and K in particular, and this demand is not always completely met by the soil especially when adverse conditions occur, and as root growth declines. Rate and timing of individual foliar fertilizers will be covered below.

5. NUTRIENTS APPLIED BY FOLIAR FERTILIZATION IN COTTON

Although a large number of nutrients are foliar-applied in horticulture (*e.g.,* Swietlick and Faust, 1984; Weinbaum, 1988), only a few are currently used in cotton production. Furthermore, the use of foliar fertilizers in cotton varies greatly around the world and even within the U.S. Cotton Belt. Nitrogen and boron are regularly applied as foliar fertilizers in the U.S. Cotton Belt to supplement soil nutrients (Hake and Kerby, 1988). More recently, potassium has been used as a foliar spray to correct late-season deficiencies in the U.S.A. and Greece (Oosterhuis, 1995b). Foliar applications of zinc to cotton are used widely in Australia and occasionally in the U.S. There has been some isolated use of sulfur applied to cotton foliage on a small scale. However, nitrogen is the most widely and commonly used foliar fertilizer in cotton production.

5.1 Foliar Fertilization with Nitrogen

5.1.1 Cotton Nitrogen Requirements

The nitrogen (N) requirement for cotton is about 15 kg N/100 kg lint (Miley and Oosterhuis, 1989) although this varies greatly from 10 to 29 kg N/100 kg lint (Gerik *et al.,* 1998). Peak daily uptake rates range from 1.5 to 4.6 kg N/ha under irrigation and 0.6 to 5.7 kg/ha under rainfed conditions (Mullins and Burmester, 2001). A recent compilation of recommendations for N fertilization in the U.S. Cotton belt suggested that 56 kg of N from all sources are required per 218 kg bale of cotton fiber produced (Nicholls, 1996). Plant N usually comes from mineralization of the soil organic matter and from soil-applied fertilizer. However, absorption from the soil is decreased by many factors, including drought, leaching losses, soil compaction, and poor root growth. When one or more of these occur, deficiency symptoms can appear (Hodges and Constable, 2007) and yields will be reduced. Unless extended drought is the problem, foliar applications of N offer an alternate method of supplying N to the plant, especially after the period for successful soil application has passed (Gerik *et al.,* 1998). The seasonal pattern of N distribution in the cotton plant shows that N requirement varies with growth rate and growth stage (Thompson *et al.*, 1976; Bassett *et al.,* 1970; Oosterhuis *et al.,* 1983; Mullins and Burmester, 1991). At maturity, the fiber and seed removed in harvest contain almost half the total N (42 %) accumulated in the shoot during the growing season (Oosterhuis *et al.,* 1983). Thus, cotton N requirements are highest during boll development when N available to the cotton plant typically diminishes (Gerik *et al.,* 1998) with decreased root activity (Cappy, 1979).

The timing and method of N fertilization in the U.S. Cotton Belt differs greatly among regions. Nitrogen fertilizer was typically applied in a split application with about half the total amount applied before planting and the remainder applied before flowering (Maples *et al.*, 1990). However, less than 10% of the U.S. cotton acreage presently receives N at planting and less than 5% of the acreage receives N as a foliar treatment (Gerik *et al.,* 1998). The practice of foliar feeding with N varies widely from single applications to multiple applications, and is usually applied in combination with other foliar applications of agrochemicals. A recent e-mail survey indicated that the use of foliar fertilization with N varied widely from as much as 50% in certain counties in California, 20% in Arkansas, 15% in Georgia, 10% in Arizona, Louisiana, and North Carolina, and 5% or less in Alabama and Tennessee (Oosterhuis, unpublished). Urea is often mixed with insecticides to improve uptake, in which case foliar fertilization with urea may be considered to be applied on many more acres, *e.g.,* over half the acres in Arkansas, although this practice only supplies a small amount of nitrogen (W. Robertson, personal communication). Foliar-applied N is used mainly on irrigated cotton. In west Texas, foliar fertilization with N only occurs on dryland cotton when adequate rainfall occurs (D. Krieg, pers. commun.). The practice of foliar fertilization with N is not used much in other cotton growing countries, except for Greece where about 15% of the cotton area receives foliar-applied N (S. Galanapoloulou, pers. commun.). In Australia, foliar N fertilization has proved beneficial when applied before waterlogging (Hodges and Constable, 2001). In some regions, such as Arizona and Greece, N is often applied with the irrigation water, *i.e.,* fertigation, which for overhead irrigation is a form of foliar fertilizer.

5.1.2 Foliar-Applied Nitrogen

Foliar application of N to cotton has been widely used in the U.S. Cotton Belt (Anderson and Walmsley, 1984; Hake and Kerby, 1988; Miley and Maples, 1988) although some scepticism has arisen as to the benefit of this procedure

(Keisling *et al.,* 1995; Heitholt, 1994c, Parker *et al.,* 1993) with some negative results reported (Bednarz *et al.,* 1997). It has been suggested that foliar-applied N may serve as a N supplement to cotton plants to alleviate N deficiency caused by low soil N availability, to meet the high demand for N by the rapidly developing bolls, and to avoid the possible hazards of rank growth from excessive soil N (Hake and Kerby, 1988; Zhu and Oosterhuis, 1992). The effects of foliar-N applications on cotton yield, however, have generally been inconsistent (Anderson and Walmsley, 1984; Smith *et al.,* 1987), although there have been reports of yield increases from foliar feeding depending on the plant and soil N status (McConnell *et al.,* 1998; Oosterhuis and Bondada, 2001). McConnell *et al.* (1998) reported that yield increases from foliar-applied N tended to be greater in irrigated production than under dry land conditions, and that increases occurred most frequently when soil-applied N rates were low (*i.e.,* 0 to 67 kg N/ha). Oosterhuis and Bondada (2001) demonstrated that the response to foliar-applied N was strongly influenced by the size of the boll load as well as the soil N level. Furthermore, there is a substantial decrease in the amount of N absorbed with increased individual leaf age (Bondada *et al.,* 1997) and with progression of the season (Bondada *et al.,* 1994). These results show that N-deficient cotton can benefit from foliar-applied N. However, indiscriminate application of N without due consideration of soil N availability and plant N status can be wasteful.

Chokey and Jain (1977) suggested that the recommended growth stages in cotton for foliar-applied N were at the pinhead and first flower stages, and at peak boll development. Whereas, Maples *et al.* (1977) suggested that cotton was more responsive to foliar fertilization with N during the boll development period. The developing boll load has a high requirement for N, which is apparently not completely met by the subtending leaves (*i.e.,* from soil uptake), thereby indicating a requirement for additional N from foliar applications (Zhu and Oosterhuis, 1992; Maples *et al.,* 1977).

5.1.3   Foliar Nitrogen Fertilizers

Urea is the most popular form of N for foliar application, and it has also been widely used as a foliar fertilizer in various crops. Yamada (1962) reported that the greater effectiveness of N in urea when applied to foliage resided in its non-polar organic properties. Other sources of N that have been used for foliar fertilization in cotton include ammonium nitrate and calcium nitrate. Low biuret forms of N, such as Tricert (consisting of 33% triazone N plus urea) have shown some merit although the costs may not always warrant their use. Recently, there has been some interest in slow release foliar-N fertilizers, but the results have been disappointing (Oosterhuis *et al.,* 2000).

5.1.4   Foliar Uptake and Movement of N Within the Cotton Plant.

The stable isotope ^{15}N has been employed to measure rates of absorption and translocation of foliar-applied N, because it permits direct determination of the uptake and movement of foliar-applied N (Hauck, 1982). In a series of field studies in Arkansas, Oosterhuis *et al.* (1989) clearly demonstrated about 60% absorption of foliar-applied ^{15}N-labeled urea in cotton at a rate of 11.2 kg N/ha. Foliar-applied N was rapidly absorbed by the leaf to which it was applied (30% within one hour) and translocated into the closest boll within 6 to 48 hours after application (Fig. 25-2). The N moved progressively into adjoining bolls for the next few days with no translocation to other leaves. These results are similar to those of Boynton *et al.* (1953) with apples, in which 50% of the foliarly-applied urea was absorbed within 8 hours and 88% was absorbed within 4 days, and Klein and Weinbaum (1984) with olive leaves (*Olea europaea* L.) where 60 to 70% of applied urea-N was absorbed within 24 hours after application. Similarly, Cain (1956) reported that in cacao approximately 80% of N was absorbed in two hours, whereas coffee averaged 50% and banana only 20% in the same length of time. In tobacco about 33% of the N was absorbed by the leaf within 24 hours (Volk and McAuliffe, 1954).

Foliar absorption of N in cotton is complete in about 48 hours (Zhu, 1989) although this would depend on plant and atmospheric conditions at the time of application. Cain (1956) found that N absorption was virtually complete in less than 24 hours in coffee, cacao, and banana. In cotton, the developing bolls were the dominant sinks for foliar-applied ^{15}N (Zhu, 1989) with about 60% of the N being absorbed by the leaf and very little remaining in the vegetative plant parts (Oosterhuis *et al.,* 1989). Similarly in soybean, the seed were the dominant sinks for foliar-applied ^{15}N, where approximately 94% of the recovered ^{15}N was ultimately found (Vasilas *et al.,* 1980) while a small percentage, less than 1.6% of the applied N, was translo-

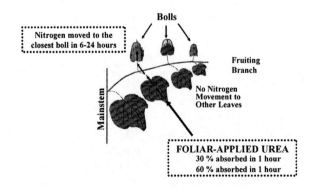

Figure 25-2. The uptake of foliar-applied ^{15}N labeled urea by a leaf on a cotton branch and subsequent movement to the closest developing boll (Redrawn from Miley and Oosterhuis, 1989).

cated to the root. Bondada *et al.* (1997) showed that foliar-applied ^{15}N was compartmentalized within the boll mainly to lint and seed, and much less was partitioned to the capsule wall, implicating the lint and seed as the major sinks.

5.1.5 Suggestions for Optimum Foliar Fertilization with Nitrogen

The cotton crop is most responsive to foliar-applied N during the boll development period and applications made earlier, particularly on seedling cotton are ineffectual. However, applications made after peak boll development may be less effective because of decreased absorption of N as the canopy leaf age increases and leaf waxiness increases. The response to foliar N is strongly influenced by the soil N status as well as by the size of the boll load. Foliar fertilization with N is more effective on irrigated, well-watered cotton. Urea is the most widely used source of N for foliar fertilization, and optimum rates of foliar N are about 8.5 kg N/ha because phytotoxicity can occur above 11 kg N/ha. The use of an adjuvant with the foliar spray is apparently not necessary.

5.2 Foliar Fertilization with Potassium

Speculation has surrounded the explanation for the widespread appearance of K deficiency in recent years across the U.S. Cotton Belt. The occurrence of a complex of K-deficiency symptoms in cotton was first recognized in California during the early 1960s (Ashworth *et al.*, 1982; Brown *et al.*, 1973). These deficiencies have manifested themselves during the latter half of the season in a range of soils and cotton cultivars. Cotton appears to be more sensitive to low K availability than most other major field crops, and often shows signs of K deficiency on soils not considered K deficient (Cassman *et al.*, 1989b; Cope, 1981). However, the explanation for these deficiencies is unclear and a considerable amount of research and speculation has surrounded this phenomenon. It has been proposed that K deficiencies are related to soils with K availability problems (Cassman *et al.*, 1990), the relative inefficiency of cotton at absorbing K from the soil compared to most other crop species (Cassman *et al.*, 1989b), the incidence of Verticillium wilt (Wakeman *et al.*, 1993), the decrease in root activity after the start of flowering (Cappy, 1979), or because modern cotton cultivars may have less K in storage prior to boll development (Bednarz and Oosterhuis, 1995). In addition, in recent years more cotton has been planted on poorer soils low in available K (Kerby and Adams, 1985). It has also been speculated that the deficiency symptoms may be associated with the use of higher-yielding and faster-fruiting cotton cultivars creating a greater demand than the plant root system is capable of supplying and the increased use of nitrogen in cotton management (Oosterhuis *et al.*, 1990). Many K deficiencies can be corrected through preplant soil applications, or partially corrected using midseason sidedress applications of K. Foliar applications of

K may offer the opportunity of correcting these deficiencies more quickly and efficiently, especially late in the season when soil application of K may not be effective.

There have been a number of reviews of K nutrition in cotton (*e.g.*, Hearn, 1981; Kerby and Adams, 1985) and more recently on the foliar fertilization with K (Oosterhuis, 1995b). Potassium plays a particularly important role in cotton fiber development because it is a major solute in the single-celled fibers necessary in providing the turgor pressure for fiber elongation (Dhindsa *et al.*, 1975). A shortage of K will, therefore, result in poorer fiber quality and lowered yields (Cassman *et al.*, 1990). If K is in limited supply during active fiber growth, there will be a reduction in the turgor pressure of the fiber resulting in less cell elongation and shorter fibers at maturity. As K is associated with the transport of sugars, it is likely implicated with secondary wall deposition in fibers and, therefore, related to fiber strength and micronaire. Xi *et al.* (1989) reported poor cuticle development in cotton plants grown without sufficient K, which may have resulted in increased water loss by nonstomatal transpiration. Potassium has been reported to reduce the incidence of Verticillium wilt (Hafez *et al.*, 1975) although the physiological reasons for this are not clear.

5.2.1 Cotton Potassium Requirements

Potassium is required in large quantities by cotton, *i.e.*, about 13 kg K/100 kg lint (Mullins and Burmester, 2007) or about 2 to 5 kg K ha[-1] day[-1] (Halevy, 1976; Bassett *et al.*, 1970; Mullins and Burmester, 1991). An average mature cotton crop is estimated to contain between 110 and 250 kg/ha (Hodges, 1992) with about 50% of the K in the boll (Rimon, 1989) and 24% in the seed and lint (Mullins and Burmester, 2007). However, only about 20 kg of K are needed to produce one 218 kg bale of cotton fiber, with about 2.5 to 6 kg being removed mainly by the seeds (Hodges, 1992; Rimon, 1989). At maturity, the capsule wall of the boll accounts for over 60%, the seed about 27% and the fiber about 10% of all the K accumulated by the boll (Leffler, 1986). Plant uptake of K during the season follows a pattern similar to dry weight accumulation (Bassett *et al.*, 1970), except that dry matter continues to increase until maturity, whereas maximum K accumulation is reached in about 110 days after which there is a decrease (Halevy *et al.*, 1987). Potassium is absorbed more rapidly than dry matter is produced, as evidenced by the higher concentration of K in young plants (Bassett *et al.*, 1970). Also, K can be taken up in luxury amounts (Kafkafi, 1990) and this could possibly confuse tissue diagnostic recommendations (Oosterhuis, 1995b). The need for K rises dramatically when the boll load begins to develop (Halevy, 1976) because the bolls are the major sinks for this element (Leffler and Tubertini, 1976). However, most fertilizer programs utilize a single preplant application of K, surface applied or shallowly incorporated into the topsoil, even though previous research in California (Gulick *et al.*, 1989) showed that cotton root systems fail to adequately exploit available K in the topsoil. There has been some interest in deep placement of K (Tupper *et al.*, 1988) although yield responses from this method of K placement have been inconsistent (T. Keisling, personal commun.).

5.2.2   Foliar Fertilization with Potassium

While there are many reports on research involving soil applied K (*e.g.*, Kerby and Adams, 1985), there are no definitive studies available on the usefulness of foliar-applied K. Earlier research in southern Africa (Oosterhuis, 1976) indicated that foliar-applications of K could significantly increase seedcotton yield. Halevy and Markovitz (1988) in Israel reported increased lint yield and average boll weight from foliar sprays containing N, P, K, and S in locations where the soil fertility was low. In the U.S., there have been a number of recent reports showing improved petiole K, lint quality and yield from foliar-applied K (Oosterhuis *et al.*, 1990; Weir *et al.*, 1992; Snyder *et al.*, 1998). However, there have also been a number of reports indicating no responses to foliar applied K (Matocha *et al.*, 1994; Harris *et al.*, 2000) or inconsistent responses (Mullins and Burmester, 1994; Oosterhuis *et al.*, 1994).

Earlier studies in Zimbabwe showed significant yield responses to foliar-applied K (Oosterhuis, 1976), and subsequent research in Arkansas (Oosterhuis *et al.*, 1990) and California (Weir *et al.*, 1992) further developed and supported this practice. Oosterhuis *et al.* (1991c) reported yield increases of 73 kg lint/ha or 6.3% and Weir *et al.* (1992) reported yield increases of 63 to 167 kg lint/ha (6.3 to 16%) compared to the recommended soil-applied K treatment. However, a subsequent three-year, twelve location study in the U.S. Cotton Belt evaluating the effect of foliar-applied KNO_3 compared to soil-applied KCl on cotton yield and fiber quality showed inconsistent responses to foliar-applied K with significant responses only 40% of the time (Oosterhuis *et al.*, 1994). Generally, responses to foliar-applied K have been poor when soil K status is adequate, although in some cases responses to foliar fertilizer on cotton growing in soils with a higher K status have been recorded (Weir and Roberts, 1994). Oosterhuis (1995b) suggested that yield increases from foliar-applied K can be expected on soils with a relatively low soil K status of less than 125 ppm K. Yield increases from foliar-applied K were associated with increased boll weight.

Potassium is intimately involved in fiber development and fiber quality (Dhindsa *et al.*, 1975) and a shortage of K will result in poorer fiber quality (Cassman *et al.*, 1990). Foliar application of KNO_3 was shown to increase fiber quality (Oosterhuis *et al.*, 1990) with the main effect being primarily on fiber length, uniformity, and strength (Pettigrew, 1999a) and much less on micronaire. Oosterhuis *et al.* (1991c) showed that the K concentration and the K content of the fibers were increased by foliar-K application. With the national emphasis on lint quality (Sasser, 1991) and the introduction of high volume instru-

mentation classification, the positive effect of foliar-applied K on lint quality may be of paramount importance.

Adverse environmental conditions often experienced during seedling development have focused some interest on the possibility of foliar fertilization of seedlings to enhance their growth during this critical stage. However, research in Alabama (Edmisten *et al.*, 1993) and Arkansas (Holman *et al.*, 1992a) showed that cotton yield was not influenced by foliar applications of fertilizers to seedling cotton. It has been speculated that foliar sprays of K could have a positive effect on droughted cotton seedlings because of the important role that K plays in the water relations of plants. Holman *et al.* (1992b) reported that foliar-applied K did not improve the growth or drought tolerance of stressed cotton seedlings. However, subsequent research has suggested that applying foliar fertilizers, including K, after relief of the drought stress to stimulate recovery and enhance growth, may be beneficial (E.M. Holman, unpublished).

5.2.3   Optimum Rate and Timing of Applying Foliar Potassium

Most reported studies on foliar fertilization with K have used rates of 4.48 kg K/ha (11.2 kg KNO_3/ha) (Oosterhuis, 1995b; Weir *et al.*, 1992). No visible injury of cotton leaves has been observed at 11.2 kg KNO_3/ha and at rates of up to 22.4 kg KNO_3/ha (Oosterhuis *et al.*, 1990) in application volumes of 94 L water/ha. However, solubility in cold water may be a problem at high rates near 10 kg KNO_3/ha. The optimum timing of foliar application of K to cotton is during the boll development period starting soon after flowering and continuing at periodic intervals of 7 to 10 day intervals for about 6 weeks (Oosterhuis, 1995b; Weir *et al.*, 1992). Weir and Roberts (1993) showed that the optimum stage of response to foliar application of KNO_3 was three weeks after first flower (Fig. 25-3). Applying KNO_3 according to the K use uptake curve of the developing boll load (*i.e.*, applying at weekly intervals for eight weeks starting at first flower with and increasing slowly to peak flowering , and then decreasing again resulted in the largest increase in yield compared to four two-weekly applications of 11.2 kg KNO_3/ha (Oosterhuis, 1995b).

5.2.4   Sources of Potassium for Foliar Fertilization

A three-year field comparison of the major K source fertilizers was conducted in Arkansas (Miley and Oosterhuis, 1994) on a silt loam soil with a moderate K status of 200 to 237 kg K/ha in the surface 15 cm and 197 to 176 kg K/ha in soil 16 to 30 cm deep. Salts of K used included nitrate, sulfate, thiosulfate, chloride, and carbonate, applied at a rate equivalent to 11.2 kg KNO_3/hectare in 93 liters of solution per hectare. For the control and each treatment containing a source other than KNO_3, 1.5 kg N/hectare as urea was applied to equal the nitrogen rate supplied by the KNO_3 treat-

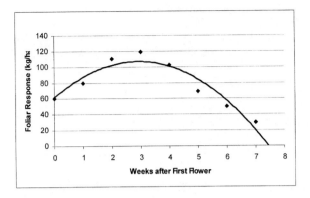

Figure 25-3. Typical response to foliar-applied potassium at various times after the start of flowering (from Weir and Roberts, 1993) ($y = 67.4 + 29.86x - 5.26x2$, R=0.979, p=0.03).

ment. Results showed a trend for KNO_3 to increase yield the most, followed closely by potassium thiosulfate and potassium sulfate. Potassium chloride had no effect on yield, and potassium carbonate significantly decreased yield (Miley and Oosterhuis, 1994). The detrimental effects of potassium carbonate on yield and the lack of effect on yield of potassium chloride were related to physiological effects on leaf photosynthesis and cell membrane integrity (Oosterhuis, unpublished). Only very minor and non significant visual symptoms of foliar burn were observed following foliar application of any of the K fertilizers. Symptoms consisted of a few small spots on the leaf particularly with potassium carbonate. Chang and Oosterhuis (1995b) confirmed these findings and showed clear differences between K sources foliar applied in affect on leaf burn, leaf expansion, K absorption, and yield. However, Mullins and Burmester (1994) found no differences among four sources, nitrate, chloride, sulfate, and thiosulfate, on a sandy loam testing medium for K.

There has been some interest in using KCl because it is a cheaper source of K. Pettiet (1993b) reported that KCl dissolved more easily than KNO_3 and was more easily absorbed by the leaf; however, the effects on yield were not recorded. There have been six field tests across the Cotton Belt in recent years comparing KNO_3 and KCl and in all but one test, KCl had either no effect on yield or decreased yield (W.R. Thompson, personal communication). A salt index has been used as a measure of the effect of a fertilizer on the osmotic potential of the soil solution (Rader *et al.*, 1943) and may give some insight into the possible effect that a K fertilizer could have on leaf tissue. The salt index is defined as the ratio of the increase in osmotic potential produced by a fertilizer material as compared to that produced by an equal weight of sodium nitrate based on the relative value of 100. The salt index for KNO_3 is 73.6 compared to 116.2 for KCl, further indicating that KCl in large concentrations could possibly have a negative influence on leaf tissue. In support of this it has been shown that foliar-applied KCl adversely affected membrane integrity of leaf discs compared to the untreated control and the foliar KNO_3 treatment (Oosterhuis, unpublished).

5.2.5 Foliar Uptake and Movement of K in the Cotton Plant

Understanding the and translocation of foliar-applied K in the cotton plant is important in order to be able to predict how rapidly, and in what amounts, the foliar-applied K is taken up by the leaf, and how quickly it moves to the developing boll. Using $^{42}KNO_3$ applied to the midrib of cotton by micropipette, Kafkafi (1992) in Israel showed that foliar-applied K moved into the leaf and to the boll within 20 hours. However, no information was provided on the quantity taken up by the leaf or the time intervals for translocation to the boll. Preliminary studies in Arkansas in 1990, using Rubidium to monitor K movement into the leaf, indicated that K first entered the leaf within 6 hours and then in greater quantities between 6 and 48 hours after application and was translocated to the developing bolls with little delay during the same period (Oosterhuis and Hurren, unpublished). Further evidence that foliar-applied K is translocated to the boll was provided by Oosterhuis *et al.* (1991c) who demonstrated in field studies that foliar-applied KNO_3 increased fiber dry weight, as well as the K concentration and K content of the fibers compared to the untreated check (Fig. 25-4). More detailed information on the time course of K uptake by the leaf and translocation to the developing boll, as well as factors that effect this process, is needed. The capsule wall contained the highest amount of K in the boll, between 32 and 60% (Bassett *et al.*, 1970; Kerby and Adams, 1985) and probably acts in a storage capacity.

5.2.6 Suggestions for Optimum Foliar Fertilization with Potassium

Research results and practical experience in commercial fields have led to some recommended practices for foliar fertilization of cotton with K (Snyder *et al.*, 1991) with the following general principles. The optimum stage of response to foliar application of KNO_3 was three weeks after first flower. The petiole threshold level of K will decrease

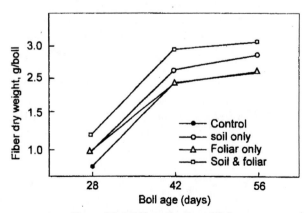

Figure 25-4. Effect of soil- and foliar-applied KNO_3 on fiber dry weight during boll development. (From Oosterhuis *et al.*,1991a).

from about 5.0% at first flower to about 2.0% near open boll. When called for, three to four foliar applications of K should be made during the first five weeks of boll development at 7 to 10 day intervals starting at the commencement of flowering. A minimum rate of approximately 4.5 kg/ha of K should be used at each application. The recommended source of K for foliar fertilization is KNO_3, although K_2SO_4 or $K_2S_2O_4$ appear to work almost as well. Attention should be given to possible solubility problems in cold water. The use of an adjuvant with the foliar spray will increase leaf K uptake but may not necessary result in increased yields, although it may permit the use of a lower rate of K per application. Further insight into the practical applications of foliar fertilization of cotton with K are give by Roberts *et al.* (1993).

5.3   FOLIAR APPLICATION OF BORON IN COTTON

Boron is an essential micronutrient in higher plants, required for cell wall synthesis, integrity of plasma membranes and pollen tube growth (Hodges and Constable, 2007). The important role of B in pollen germination and pollen tube growth for successful fruit set of higher plants (Gupta, 1993) suggests that B deficit during flowering and fruiting may significantly reduce boll retention, resulting in lower yields. Zhao and Oosterhuis (2000) showed that B deficiency in cotton decreased leaf photosynthesis and carbohydrate transport from leaves to developing fruit, and depressed plant growth and dry matter resulting in increased fruit abscission. Boron deficiency is common in highly leached, acidic sandy soils in cotton growing regions in the world (Shorrocks, 1997). The critical level in dicotyledonous species (*e.g.*, clover) is 20 to 70 mg B kg^{-1} dry weight (Marschner, 1995) whereas in cotton, a range of critical values has been reported. Sun and Xu (1986) suggested that the critical soil B concentration affecting yield was 0.55 mg/kg, and Jiang *et al.* (1986) reported that the critical B value for cotton was 0.4 mg/kg or 0.5 mg/kg for heavy soils. El-Gharably and Bussler (1985) documented that the critical B levels in roots, young leaves and stems of cotton plants were 103, 61, and 78 mg/kg, respectively. Recently, Zhao and Oosterhuis (2000) reported that the critical leaf B value for diagnostic purposes was 30 mg/kg in the upper fully-expanded main-stem leaf located four nodes from the terminal. Subsequent research has indicated a critical concentration of 15 mg/kg in the third main-stem leaf at the seedling and squaring stages and 20 mg/kg at flowering (Zhao and Oosterhuis, unpublished). Xie *et al.* (1992) investigated the absorption, translocation and distribution of B in cotton and reported that uptake of B from the soil by roots was more rapid than absorption and translocation of B applied as a foliar spray.

In the past thirty years, there have been many reports about growth and yield responses of cotton to soil or foliar application of B. Foliar applications have included boric acid, sodium borates, borax, and sodium tetraborate. The

narrow concentration range between B deficiency and toxicity (Hodges and Constable, 2007) requires special care in the application of B fertilizers. Rosolem and Costa (1999) reported that although B deficiency reduced total dry matter production, plant height, and number of reproductive structures under the greenhouse conditions, foliar application of B did not correct the deficiency because the mobility of B in cotton was limited. Reports of cotton yield response to foliar application of B have been inconsistent. Smithson (1972) showed that under B deficiency, five foliar applications of B, each of 1.4 kg borax/ha at weekly intervals from 10 weeks after sowing, increased seedcotton yields 36%. Howard *et al.* (1998) and Miley *et al.* (1969) also reported that foliar sprays of B significantly increased cotton lint yield in the Mid-South. In Brazil, a field experiment showed that six foliar sprays, of 0.075 or 0.15 kg B/ha, at early growth stages increased yield and fiber length, whereas foliar applications during flowering were effective only at the higher rate (Carvalho *et al.*, 1996). In China, several studies have shown that foliar B application during square formation, as well as at early and peak flowering, could improve seedcotton yield (Dong, 1995; Jiang *et al.*, 1986) and staple length (Sun and Xu, 1986). In contrast, other studies have shown positive (Anderson and Ohki, 1972; Oosterhuis and Venter, 1976) and negative effects on cotton yield from foliar sprays of B (Heitholt, 1994c). These contrasting results may be associated with soil texture, soil pH, soil fertility, and soil B level because all these factors influence B uptake by plants and crop yield response to supplemental B application (Gupta, 1993).

5.4   FOLIAR FERTILIZATION WITH SULFUR

Cotton requires about 4 kg sulfur (S) per 218 kg bale of cotton fiber (Mascagni *et al.*, 1990). Historically, S was added to the soil as part of the standard fertilizer makeup, but modern high analysis fertilizers contain little or no S and tighter control of industrial emissions have reduced the S content of the atmosphere. Gypsum contains 19% sulfate-S and is commonly used to alleviate probable S deficiency problems. Under conditions of S deficiency, inhibition of protein synythesis is correlated with an accumulation of soluble organic and nitrate (Marshner, 1995). Furthermore, the extent of S redistribution from older leaves depends on the rate of N deficiency-induced leaf senescence, although these relationships have not been documented in cotton. A plant S content of <0.25 % indicates S deficiency in cotton (Miley *et al.*, 1990). These authors also reported that sulfate-S, the plant available form, is soluble and in soils with sandy subsoils, may be leached out of the root zone resulting in a temporary S deficiency. However, there have generally been little or no yield responses to soil applications of S in the U.S. (Mascagni *et al.*, 1990). Some S deficiencies in cotton were noted in Arkansas on sandy subsoils and yield responses were obtained from foliar applications of S as

$MgSO_4$ (Miley *et al.*, 1990). Nevertheless, foliar sprays of S have only been considered as an emergency measure after peak fruiting if a S deficiency is noted. Furthermore, irrigation water can often supply some of the plants S requirement.

5.5 FOLIAR FERTILIZATION WITH ZINC

Zinc deficiency occurs widely among plants grown in highly weathered acid soils and in calcareous soils of high pH and is often linked with iron deficiency. Generally, in leaves the critical deficiency levels are below 15-20 µg Zn g^{-1} dry weight and cotton is reported to be particularly sensitive to Zn deficiency compared to some other crops such as wheat, oat, or pea (Marschner, 1995). The use of foliar-applied Zn varies widely across the U.S. Cotton Belt; from virtually none used in the Eastern U.S. and Mid-South to up to half a million hectares per year in West Texas. In Arizona, Zn is a common constituent when foliar fertilization in cotton is used (Silvertooth, pers. commun.). In Australia, Zn has been widely used as a foliar fertilizer because of the lack of Zn availability in their alkaline soils (Constable, pers. commun.).

5.6 FOLIAR FERTILIZATION WITH "MIXTURES" OF FERTILIZER NUTRIENTS

Generally, there have not been consistent or significant responses to foliar applications of mixed fertilizer "cocktails" to cotton. For example, Bednarz *et al.* (1997) reported no yield benefits from foliar sprays of a range of foliar fertilizers including N, Ca, Zn, Fe, B, and Mn. Silvertooth *et al.* (1998) found no growth, tissue nutrient concentration, or yield responses of cotton to foliar applications of mixed fertilizers containing B, Mn, and Zn or Mg, Mn, Zn, Cu, Fe, Mo, and S. This practice has been used with more success with horticulture plants.

5.7 FACTORS AFFECTING THE EFFICACY OF FOLIAR-APPLIED FERTILIZERS

Plant uptake and response to foliar-applied fertilizers will depend on a number of external (environmental) and internal (plant) factors as well as on various practical considerations. External factors include nutrient concentration, valency, molecular size and charge, temperature, and humidity, whereas internal factors encompass metabolic activity, plant water status, and cuticular thickness and composition. Many of these factors affect the efficiency of foliar-applied nutrients because they influence the length of time that the nutrient solution stays in aqueous form (Boynton, 1954; Kannan, 1986). Practical phenomena including phytotoxicity, compatibility with other chemicals, location of the spray in the canopy, leaf age, and diurnal time of application also need to be taken into consideration in deciding whether or not to foliar fertilize. Although the practice of foliar fertilization holds much potential in cotton production, there are many circumstances where the practice will not work or where the efficiency will be severely reduced.

5.7.1 Environmental Effects on Nutrient Absorption and Volatilization

The time a foliar-applied chemical remains in an aqueous form on the leaf surface will determine the absorption of nutrients concerned. Wittwer *et al.* (1963) suggested that the adherence of an aqueous solution to the cuticle resulted in the dehydration and swelling of the leaf cuticle, whereby the wax platelets interspersed throughout the cuticle were spread further apart facilitating nutrient penetration. The period during which a foliar-applied chemical remains in an aqueous state on the leaf surface will depend on the chemical's concentration, molecular size, charge, and also the environmental conditions. The prevailing environmental conditions, particularly temperature and humidity, but also air movement, will effect the period during which the effective concentration of the solute can be maintained on the leaf surface and, therefore, the absorption by the leaf. Low humidity, high temperature, and air movement will decrease the rate and extent of nutrient absorption by the leaf (Volk and McAuliffe, 1954; Zhu, 1989). Furthermore, environmental conditions such as radiation, temperature and humidity (Baker, 1982), and also drought (Clark and Levitt, 1956) affect the amount of surface wax on plant leaves (addressed below). Long photoperiods and cool nights have been associated with increased wax production in tobacco (Wilkinson and Kasperbauer, 1980). The thickness of the rice cuticle was affected by shade (Takeoka *et al.*, 1983) although this has not been demonstrated for cotton where nutrient absorption by leaves lower in the canopy may be decreased because of increases in leaf wax in older leaves (Bondada *et al.*, 1997). Volatilization of fertilizer solutions from the leaf surface will concentrate the chemical on the leaf and often cause crystallization. However, dew can enhance the uptake of residue from the foliar fertilizers remaining on the leaf after excessive evaporation (Zhu, 1989). The rates of uptake by intact leaves are much lower than the corresponding rates for roots for a given concentration of mineral nutrient since the extremely small pores in the cuticle severely restrict diffusion from the external leaf surface to the leaf apoplasm (Marschner, 1995). Glaucousness is considered a drought adaptation in plants such as wheat (*Triticum turgiolum* L.) (Johnson *et al.*, 1983) and sorghum (*Sorghum bicolor* L. Moench) (Jordan *et al.*, 1983) because plants with higher wax production, attributable to increased reflection of radiation and lower tissue temperature (Blum 1975; Richards *et al.*, 1986) and because of increased cuticular resistance to water loss (Jordan *et al.*, 1984; Wullschleger and Oosterhuis, 1987b).

5.7.2   Metabolic Activity

The nutrient requirements of a crop will depend on the developmental phase of the crop, the physiological activity of the leaves, the size of the developing boll load, and of course on plant water status. Generally, tissues and organs that are in a more rapid phase of development will provide a stronger sink for available resources such as nutrients. It has been well documented that the nutrient requirements of the cotton crop follows a general pattern with a low requirement during the seedling stage, increasing during squaring, peaking during early boll development and then dropping off during boll maturation (Mullins and Burmester, 1991; Bassett *et al.*, 1970). Therefore, foliar-applied nutrients will be more effective during these periods of high demand, with a great chance of effectiveness, especially if any impedance of root activity occurs. The developing boll load is a major sink for nutrients (Leffler, 1989) and the size of the boll load has a major influence on plant response to foliar-applied N (Oosterhuis and Bondada, 2001).

5.7.3   Plant Water Status, Drought, and Cuticular Wax

Leaf water status plays an important role in the absorption of foliar-applied nutrients. It has been proposed that water status of the leaf affects the physical structure of leaf cuticle and consequently affects the absorption of the foliar-applied nutrients (Wittwer *et al.*, 1963; Boynton, 1954; Kannan, 1986). These researchers suggested that when the leaf was dry, the structure of the cuticle constricted and impeded the penetration of foliar-applied nutrients. Increases in wax content with drought have been reported for cotton (Weete *et al.*, 1978; Oosterhuis *et al.*, 1991a; Bondada and Oosterhuis, 1996), soybean (Clark and Levitt, 1956), wheat (Fisher and Wood, 1979; Johnson *et al.*, 1983), oat (Bengston *et al.*, 1978), and sorghum (Jordan *et al.*, 1984). Increasing wax content of the cuticle helps plants resist water loss (Marschner, 1995), but also reduces the penetration of agrochemicals such as herbicides (Sherrick *et al.*, 1986), defoliants (Oosterhuis *et al.*, 1991a), and foliar fertilizers (Zhu, 1989). Oosterhuis *et al.* (1991a) showed that water deficit increased cotton leaf cuticular thickness, weight of epicuticular wax, and chemical composition, and thereby reduced the penetration of foliar-applied chemicals. Details of the cotton cuticle and its impact on evaporative water loss have been documented (Wullschleger and Oosterhuis, 1987b). Weete *et al.* (1978) reported that wax synthesis in cotton leaves was inhibited by severe water stress (-2.4 MPa), and that after rehydration, previously stressed leaves in cotton produced more wax than leaves prior to stressing. The thickness of the leaf cuticle of the water-stressed leaves was approximately 0.59 µm compared to 0.46 µm for the well-watered control (Oosterhuis *et al.*, 1991a). Zhu (1989) showed that ^{15}N absorption from foliar fertilization of field-grown cotton with urea was reduced by water stress by 45%,

34%, and 23% after 1, 6, and 24 h post urea application, respectively. Furthermore, the translocation of foliar-applied ^{15}N to the developing bolls was also reduced by water stress.

The qualitative composition of the cuticle wax may be of greater significance than cuticular thickness as suggested by Norris (1974) working with 2,4-D penetration. In cotton, qualitative differences occur in leaf epicuticular wax composition between water deficit-stressed and non-stressed treatments (Oosterhuis *et al.*, 1991a; Bondada *et al.*, 1996) (Table 1). Water stress increased the number and levels of long-chain, high-molecular weight alkanes in the leaf and bract wax (Bondada *et al.*, 1996). The leaf-wax extract of well-watered leaves contained the shorter-chain n-alkanes, tricosane, n-tetracosane, pentacosane, hexacosane, and octocosane, whereas, the water-stressed plants had more high molecularweight long-chain alkanes, fucosterol, n-nonacosane, n-octacosane, n-tiracontane, dotriacontane, and n-tetratriacontane (Bondada *et al.*, 1996; Oosterhuis *et al.*, 1991a). This trend towards longer-chain waxes would result in a greater hydrophobicity of the cuticle (Leon and Bukovac, 1978), which would contribute to the reduced penetration of the leaf by defoliants. Hanny and Gueldner (1976) provided the only other report on wax composition of glabrous cotton but did not relate this to water stress or chemical penetration.

5.7.4   Foliar Burn

Phytoxicicity from foliar fertilization can occur when excessively high spray concentrations are used. However, accept for urea in certain circumstances, this is not usually a problem at the recommended rates of foliar fertilizer application. Foliar application of KNO_3 at double the normal recommended rates, up to 22.4 kg KNO_3/ha, did not cause any appreciable foliar burn to field-grown cotton (Oosterhuis *et al.*, 1990). Rapid drying of urea solution on the leaf surface increases the relative concentration and is a major cause for leaf injury in foliar fertilization. Urea may undergo rapid hydrolysis by urease producing ammonia as one of the end products (Hinsvark *et al.*, 1953) and excessive accumulation of ammonia in leaf tissue can result in leaf injury (Noggle and Fritz, 1983a). The ammonia ion was reported to uncouple photophosphorylation and inhibit photosynthesis in spinach (Krogmann *et al.*, 1959) and cause a pH imbalance in higher plants resulting in an alteration of cellular metabolites such as organic acids (Raven and Smith, 1976). Zhu (1989) described the progressive development of foliar burn in cotton as increasing with time after urea application (11.2 kg N/ha in 92 L water to flowering cotton at midday) from a slight darkening of the affected adaxial and abaxial epidermis, to brown necrotic spots and dead tissue. Using transmission electron microscopy Zhu (1989) revealed that ultrastructural alterations in the cotton leaf included cytological distortion consisting of cytoplasmic vacuolation, disruption of cellular membranes, plasmolysis, initial degradation of the grana and cytoplasm, and accumulation of electron dense osmiophilic phenolic compounds from vacuoles released into the cytoplasm. The cytoplasmic degradation was probably from hydrolytic enzymes released from vacuoles upon the breakdown of the tonoplast. Final cellular changes from urea burn included collapse of stomates, epidermal cells, palisade and spongy mesophyll, distortion of vascular bundles, and compression of the palisade and spongy mesophyll cells. Leaf burn from urea was associated with water loss probably resulting from the disruption of cell membrane function causing the loss of cell turgidity and the collapse of urea burned leaf tissue. Even mild foliar burn from urea will probably reduce the photosynthetic rate of leaves because it causes degradation of grana and eventually complete breakdown of the chloroplast structure. To avoid foliar burn, lower concentrations of the nutrient should be used, and foliar applications of urea should not be made at mid-day, especially to water-stressed plants.

Researchers have long been interested in using urease inhibitors to prevent N losses from urea fertilizer applications, although in general this practice has only been successful in decreasing soil N losses, but not losses from foliar-applied urea. Urease inhibitors have been successfully used in a number of crops to improve recovery of N from soil-applied urea, but there has been limited success with urease inhibitors in foliar fertilizers. For example, Rawluk et al (2000) reported that the urease inhibitor [N-(n-butyl) thiophosphoric triamide (NBPT] improved the recovery of N by wheat plants from soil-applied urea but not from foliar applications. The urease inhibitor NBPT is probably the most effective compound currently available for retarding hydrolysis of urea fertilizer in the soil (Bremner, 1995). In a review, Bremner (1995) reported that leaf burn in soybean, following foliar-fertilization with urea, was increased rather than decreased because of the accumulation of toxic amounts of urea in the leaves rather than from the formation of toxic amounts of ammonia from urea hydrolysis by urease. Research is needed to evaluate the benefits of urease inhibitors with foliar urea applications in cotton for improved absorption efficiency.

5.7.5 Diurnal Time of Foliar Application

The absorption of foliar-applied nutrients by plants varies with the time of the day in which they are applied. This is related to the environmental conditions at the time of application (Zhu, 1989) and the effect on the length of time the nutrient solution stays in aqueous form on the leaf surface (Boynton, 1954; Kannan, 1986). Volk and McAuliffe (1954) working with tobacco and Teubner *et al.* (1953) with beans and tomato reported that the effects of foliar applications of nutrients made at different times of the day were inconsistent, and that absorption was most rapid at night when the leaf surface was wet. Zhu (1989) showed clear diurnal differences in foliar-applied ^{15}N absorption by well-watered cotton leaves (Fig. 25-5), with uptake being highest early morning (0600 h) and late evening (1930 h), and significantly reduced at midday. He showed that the percentage ^{15}N re-

covery in the leaf was 40%, 34%, and 38% for the morning, midday, and evening treatments, respectively. Water deficit decreased ^{15}N-uptake from foliar applications, particularly at midday when uptake was reduced by about 80%. This was associated with lower relative humidity at midday and crystallization of the urea on the leaf surface. Zhu (1989) documented that crystals appeared on the leaf one hour after foliar application in the well-watered treatment, and 30 minutes after application in the water-stressed leaf. It has been suggested that the thin vapor films formed by leaf transpiration aid in the absorption of foliar-applied nutrients (Boynton, 1954; Kannan, 1986, Wittwer *et al.*, 1963). Zhu (1989) reported that the highest absorption of foliar-applied ^{15}N coincided with the least negative leaf water potential. Obviously, relative humidity as well as leaf water status, plays a central role in nutrient absorption from foliar sprays.

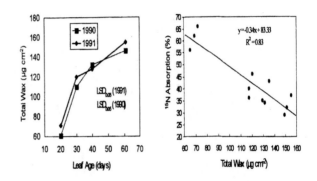

Figure 25-5. The uptake of foliar-applied ^{15}N-labeled urea as affected by water-deficit stress and diurnal timing of application. Values within the same treatment with the same letter are not significantly different (P>0.05). Redrawn from Bondada *et al.* (1987).

5.7.6 Leaf age and Nutrient Absorption

Zhu (1989) reported that absorption of foliar-applied ^{15}N was more rapid in young cotton leaves than in old leaves. A similar observation was made by Cain (1956) for coffee, cacao, and banana. However, Boynton (1957) and Cook and Boynton (1952) reported no significant differences in the absorption of foliar-applied urea by apple leaves of different ages. Miller (1982) subsequently found that the cuticle of apple fruit was relatively hydrophilic during the first 2 weeks after anthesis, after which it became hydrophobic and less capable of absorbing water. He suggested that in terms of foliar-applied nutrients, the permeability of the cuticle to water and solutes was a function of its maturity. Bondada et al (1997) showed that cotton leaf cuticle thickness increased with leaf age from a maximum of 80% in 20-day-old leaves to 20% in 60 day-old leaves, with a concomitant decrease in absorption of foliar-applied ^{15}N-urea (Fig. 25-6). Increasing average leaf age in the cotton canopy (Wullschleger and Oosterhuis, 1992) has been used to explain the reports of

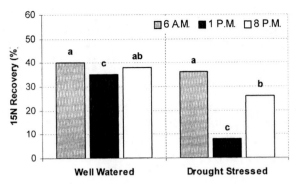

Figure 25-6. Effect of leaf age and leaf wax content on uptake of foliar-applied nitrogen. A. Change in total leaf epicuticular wax content with increase in leaf age of field-grown cotton. B. Relationship between leaf ^{15}N absorption and total wax content during leaf ontogeny for field-grown cotton. From Zhu (1989).

decreased effectiveness of foliar-applied urea three weeks after flowering (Keisling *et al.*, 1995) because increasing leaf age would also mean increased average leaf cuticle thickness and decreased absorption of foliar-applied ^{15}N (Bondada *et al.*, 1994). Bondada and Oosterhuis (1996) also reported that there was a progressive development of the cuticle in leaves, bracts, and bolls, which supports the observations of Zhu (1989) that foliar absorption decreased in older bracts and bolls as well as in older leaves.

5.7.7 Absorption of Foliar-Applied Nutrients by Different Plant Organs

All aerial parts of the cotton plant are capable of absorbing foliar-applied urea including leaves, bracts and bolls (Zhu, 1989). The amount of foliar-applied N absorbed directly by cotton bolls, like leaves, decreases with age, as the demand for N by the developing fruit declines after reaching a maximum at about 20 to 30 days after anthesis (Zhu and Oosterhuis, 1992; Pinkhasov and Tkachenko, 1981). Bondada and Oosterhuis (1996) showed that boll stomates became occluded with wax about two weeks after fertilization, and that the physical structure of the boll cuticle was modified with age. Miller (1982) suggested that, the permeability of the cuticle of apple fruits to water and solutes was a function of maturity. Obviously, the leaves are the major sites of foliar absorption, but all aerial parts of the cotton plant can absorb nutrients directly. However, the location within the canopy where the foliar-N spray is directed from tractor or backpack applications can affect absorption and yield. For example, there was a significant yield increase when ^{15}N-urea was applied to the top of the cotton canopy compared to the bottom, five weeks after flowering which was associated with the larger N requirement of the bolls developing in the upper part of

the canopy late in the season (Oosterhuis, unpublished).

5.7.8 Genotypic Differences in Response to Foliar-Applied Fertilizers

Genotypic differences in the penetration of nutrients through the cuticle and epidermal cell wall have been reported for wheat and rye (Romheld and El-Fouly, 1999). For cotton, there have only been a few studies investigating genotypic responses to foliar fertilization in cotton. With K fertilization, most of the research on genotypic responses to K fertilization has been conducted using soil-applied K (Cassman *et al.*, 1990) with only one reference to foliar-applied K. Janes *et al.* (1993) found no significant differences among six *Gossypium hirsutum* L. cultivars in response to foliar-applied KNO_3 which may have been expected because of a lack of difference in total nutrient uptake (Mullins and Burmester, 1991) and limited differences in cuticle thickness (Meek and Oosterhuis, unpublished) between modern cotton cultivars.

5.7.9 Compatibility of Foliar Fertilizers with Other Chemical Sprays

Fertilizer materials are often added to foliar applications of pesticides, but questions have arisen about the compatibility and subsequent efficacy of these nutrients with insecticides. Long *et al.* (1991) reported decreased efficacy and a 35 to 75% reduction in the amount of pyrethroid insecticide recovered when tank mixed with urea (23% N) solution caused by separation of the pyrethroid out of suspension in the urea solution mixture and not due to chemical degradation of the pyrethroid (Graves, pers. commun.). It has been suggested that similar problems may occur with tank mixes of KNO_3 and insecticides. Baker *et al.* (1994) showed that the pyrethroid insecticide Cymbush 3EC (Zeneca Agricultural Chemicals Co., Wilmington, DE) mixed with KNO_3 remained well dispersed with mild agitation and did not pose a physical compatibility problem. These authors report some problems with urea-insecticide mixtures which could be lessened by first mixing the insecticide with water to provide a stable emulsion dispersion. They further suggested that because of the unique behavior of specific emulsified insecticides, each insecticide/fertilizer combination should be evaluated separately for compatibility to obtain a high level of assurance of the mixture stability before use. Tank mixes of N and K fertilizers is another production practice that has raised some questions about compatibility. Oosterhuis and McConnell (unpublished) demonstrated that mixtures of 11.2 kg N/ha as urea and 11.2 kg KNO_3/ha had no detrimental effect on yield.

5.7.10 pH of the Foliar Spray Solution

Chang and Oosterhuis (1995) showed that lowering the pH of foliar K solutions to between 4 and 6 significantly increased the absorption of K, its subsequent accumulation in the boll, and the seedcotton yield. This was confirmed by Howard *et al.* (2000) who reported that buffering K or B foliar fertilizer solutions to a pH of 4 increased cotton lint yields. Chang and Oosterhuis (1995) demonstrated that applying KOH, K_2CO_3, $K_2S_2O_3$, $KHCO_3$, and CH_3COOK as foliar fertilizers to cotton at their standard alkaline pH values caused significant leaf burn, whereas foliar applications of K_2SO_4, KNO_3, and KCl caused either none or minimal leaf burn. Lowering their pH to 7 significantly reduced the phytotoxicity of highly alkaline solutions such as KOH and K_2CO_3. The pH of foliar fertilizer solutions has an important role in altering phytotoxic effects as well as on absorption and translocation of K to the bolls. These findings may help to explain the inconsistent response to foliar fertilization with K. Additional research is needed to determine the optimum spray solution pH for leaf absorption and to implement these findings into production practices.

5.7.11 Use of Adjuvants with Foliar-Applied Potassium

Surfactants can markedly increase the permeability of plant cuticles for both water and solutes (Schonherr and Bauer, 1992). Adjuvants have generally not been used with foliar fertilizers in cotton. However, Heitholt (1994d) and Howard *et al.* (1993) reported increased cotton leaf and petiole K concentrations after spraying with KNO_3 and a surfactant, although the final yield of lint was not increased. Howard *et al.* (1998) and Oosterhuis (1999a) have since reported increased lint yields from foliar application of KNO_3 plus a surfactant. Howard *et al.* (1993) speculated that the use of adjuvants with foliar sprays of K may improve the efficacy of the foliar-applied K fertilizer, and thereby provide the potential for decreasing the quantity of K applied per application. The advent of organosilicate surfactants has greatly enhanced the performance of agrochemicals, compared to conventional surfactants, by virtue of their wetting properties attributable to their unique structure, ability to lower the surface tension to very low values, fast kinetics of adsorption at interfaces, high affinity for hydrophobic surfaces, and the favorable orientation and structure of the absorbed molecules (Goddard and Padmanabhan, 1992). It has been shown that organosilicate adjuvants can even enable stomatal penetration by spray liquids due to their ability to produce exceptionally low surface tensions of about 20 mN m^{-1} compared to about 30 mN m^{-1} for conventional non-organosilicone surfactants (Buick *et al.*, 1992; Neumann and Prinz, 1974; Stevens *et al.*, 1992). The organosilicate surfactant Penetrator Plus significantly increased uptake of foliar-applied KNO_3 in cotton (Oosterhuis, 1999a). It seems likely that in the future increased use will be made of surfactants with foliar fertilization in cotton production.

6. TISSUE DIAGNOSES FOR FOLIAR FERTILIZATION

Analyses of soil and plant samples offer a means of determining the nutrient status of a crop. Soil sampling and analytical methods of assessing nutrient availability in the soil have been reviewed by Sabbe and Zelinski (1990). Knowledge of the soil being used is important because the mineralogy, organic matter, and level of K depletion for a specific soil can significantly affect the fate and availability of applied fertilizer K (Roberts, 1992). In cotton, tissue tests have become a valuable diagnostic tool for assessing crop nutrient status, for determining fertilizer recommendations during the growing season, and for detecting potential K deficiency (Baker *et al.*, 1992). Foliar fertilization is usually only appropriate when a nutrient deficiency is observed or detected by tissue analysis. The petiole is generally considered more indicative of cotton plant N and K status than the leaf blade. For K, this is partly because of the more rapid decline in K concentration in the petiole, compared to the leaf, during the boll development period (Hsu *et al.*, 1978). Critical or threshold levels for the major nutrients have been studied and documented (Hodges and Constable, 2007) although some uncertainty exists about sufficiency levels in the leaf or petiole for certain elements, *i.e.*, boron (Zhao and Oosterhuis, 2003), as these values may be appreciably altered by the environment, plant genetics, and sampling procedure.

The patterns of nutrient uptake and use by the cotton plant have been well documented (Bassett *et al.*, 1970; Mullins and Burmester, 1991) and soil and plant sampling techniques have been established to check if crop nutrient status is adequate for the particular soil, environment and yield level. However, there is always an element of uncertainty because of changes in the environment and plant demand, the ability of the soil to meet this demand, declines in root growth during boll development, increases in nematode populations, and other production problems. Therefore, mid-season tissue diagnoses and foliar fertilization provides a mechanism of providing timely inputs of needed nutrients. Accurate soil analysis coupled with mid-season plant tissue analysis are needed to formulate and manage a suitable fertilizer program.

7. ADVANTAGES AND DISADVANTAGES OF FOLIAR FERTILIZATION WITH POTASSIUM

The advantages of using foliar feeding with K include low cost, a quick plant response (increased tissue K concentration and fewer new deficiency symptoms), use of only a small quantity of the nutrient, quick grower response to plant conditions, compensation for the lack of soil fixation of K, independence of root uptake problems, increased yields and improved fiber quality. On the other hand, the disadvantages are that only a limited amount of nutrient can be applied in the case of severe deficiencies, and the cost of multiple applications can be prohibitive unless incorporated with other foliar applications such as pesticides. Other disadvantages when using high concentrations of K include the possibility of foliar burn, compatibility problems with certain pesticides, and low solubility of certain K salts, especially in cold water. Another restraint is the lack of a full understanding of this technology, specifically the optimum rate and timing, tissue threshold levels to predict the need for foliar-applied K, the physiological mechanism of absorption, and the effect of plant condition and environmental factors on absorption.

8.   METHODS OF FOLIAR FERTILIZATION

Generally, foliar fertilizer application in the U.S. Cotton Belt is accomplished by aerial application and sometimes by tractor mounted spray equipment. Foliar fertilization can be accomplished by a variety of means of overhead sprinklers and by application through equipment used to spray pesticides. In many cotton-producing countries backpack sprayers are used. Fertigation is widely practiced especially with drip irrigation, but is not considered a method of foliar fertilization. Slow-release fertilizers (*e.g.*, Tricert and CoRoN) are being touted by some in the U.S. Cotton Belt, but there is doubt about their effectiveness (Oosterhuis *et al.*,). Of more importance, perhaps, is the use of low biuret N fertilizers for foliar feeding. For reasons of economics and practicality, foliar fertilizers are frequently incorporated with other spray applications, particularly with insecticides. In these cases consideration of compatibility to ensure chemical efficacy is needed. In all cases, consideration is needed for maximizing plant absorption and growth response. This depends on the method of delivery, and includes considerations of plant water status, time of day, weather conditions, chemical concentration,

and physical considerations of droplet size should be carefully controlled, since all of these affect crop response.

9   SUMMARY

Foliar fertilization can be used to improve the efficiency and rapidity of utilization of a nutrient urgently required by the plant for maximum growth and yield. However, foliar fertilization should only serve as a supplement to traditional soil applied fertilizer for a sufficient supply of nutrients to the developing cotton crop for optimum yields and fiber quality. In general, foliar applications should be made either early morning or late evening for maximum efficiency, and no foliar applications should be made to water-stressed plants. It would appear that the use and success of foliar fertilization with nutrients other than N, B, and K, have been isolated and depend on the specific site related soil conditions. Foliar applications have the advantage of allowing producers to add the necessary K, when tissue analysis indicate a pending shortage, and thereby correct the deficiency and prevent yield loss. In general there is a lot of conflicting information about the benefits of foliar fertilization, but the scientific evidence to date and the widespread practical use of this phenomenon indicate that it is a viable and useful practice for improved cotton production. However, indiscriminate application of foliar-applied nutrients without due consideration of soil N availability and plant N status can be wasteful.

Information is still needed on the quantification of environmental effects, humidity and temperature in particular, on the absorption by cotton of foliar-applied fertilizers. Research needs to determine the amounts of nutrients that actually reach their targets (*e.g.*, the bolls) and how much of the total budget of these organs this constitutes. Studies on the use of adjuvants with foliar fertilizes need to be continued with all the major nutrients for improved efficiency and more economical returns. Lastly, details of the actual movement and pathways of the major nutrients foliarly applied in cotton is lacking. An improved understanding of these fundamental aspects of foliar additions of nutrients in cotton would help to more precisely predict when to use foliar fertilization for maximum benefit and enhance the success and reliability of this practice.

Chapter 26

USE OF GROWTH REGULATORS IN COTTON PRODUCTION

J. Tom Cothren[1] and D.M. Oosterhuis[2]

[1]*Soil and Crop Sciences Dept., Texas A&M University, College Station, TX 77843 and* [2]*Crop, Soil & Environmental Sciences Dept., University of Arkansas, Fayetteville, AR 72701*

1. INTRODUCTION

Cotton (*Gossypium hirsutum* L.), a perennial woody shrub with an indeterminate growth habit, evolved in tropical, relatively dry areas of the world. Through adaptive changes, accomplished through breeding and selection, cotton is now widely grown under both semi-arid and humid conditions. However, despite these adaptive changes, cotton continues to exhibit many attributes of its tropical origin. The crop grows best under warm temperatures and high light intensity, is somewhat drought tolerant, and often continues or resumes growth late in the season. Because of these growth habits, alterations in growth and development of the crop are often agronomically desirable. These alterations may be accomplished through the use of plant growth regulators (PGRs). PGRs are classified as organic compounds that alter the growth and development of plants. PGRs are biologically active at very low concentrations, and elicit responses similar to those observed from plant hormones. Unlike plant hormones which are produced by the plant, PGRs may be either produced naturally by the plant or synthetically by a chemist. However, responses to PGRs are complicated by the interaction of environment and cultural practices. For a PGR to be widely accepted, it must perform consistently in a given production scheme.

Since most plant growth and development processes are regulated by natural plant hormones, these processes may be manipulated either by 1) altering the plant hormone level or 2) by changing the capacity of the plant to respond to its natural hormones. In recent years, synthetic

Names of products are included for the benefit of the reader and do not imply endorsement or preferential treatment by the authors or their universities.

plant growth regulators have been investigated for their ability to alter cotton growth and development in an attempt to improve production. PGRs are management tools in the producer's arsenal that can be used to ensure efficient production. These compounds are diverse in both their chemistry and use. The objective of this review is to document the use of PGRs in cotton production. A brief synopsis of the roles of the endogenous, or naturally-occurring plant hormones, precedes the discussion on PGRs.

2. PLANT HORMONES

A plant hormone is an organic compound that is synthesized in one part of a plant and translocated to another part, where it causes a physiological response. Hormones serve as chemical messengers in plants to aid in coordination of developmental and environmental responses. A very low concentration of hormone is required to cause a response due to a signal transduction pathway that produces a cascade effect. For example, a simple hormone molecule may lead to the activation of an enzyme that produces hundreds of molecules of a second messenger. Hormone binding causes the levels of one or more second messengers to be elevated, and results in the activation or deactivation of enzymes, especially protein kinases and phosphatases. Figure 26-1 illustrates the present understanding of the signal transduction network in plant cells.

Until recently, five major classes of plant hormones were commonly recognized. Auxins, gibberellins, and cytokinins are usually classified as growth-promoting hormones, and abscisic acid and ethylene are usually classified as growth-inhibiting hormones. Today brassinosteroids, jasmonate, and salicylic acid are also recognized as plant

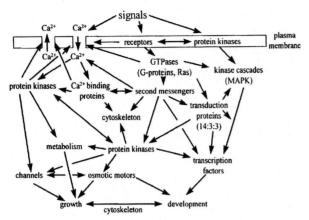

Figure 26-1. Signal transduction network in plant cells. The figure indicates that signals can alter membrane potential, occupy and activate plasma membrane receptors, or modify plasma membrane protein kinases. For the sake of clarity, some important details, including soluble receptors, second-messenger synthesis, transduction proteins, and kinase cascades, have been omitted. Nevertheless, the figure indicates the extent of interactions that collectively constitute the signal transduction network (Trewavas and Malhó, 1997).

hormones. Many other naturally occurring growth-promoting substances have been isolated including turgorins and strigol. These growth-promoting hormones may become classes of hormones in the future (Gross and Parthier, 1994). The biologically active form and classical role of the currently known plant hormones are presented in Table 26-1.

Although classical roles have been assigned to each hormone class, the activity of one hormone is usually not isolated from the other hormones. Plant growth regulation is accomplished through the interaction of several hormones. For example, brassinosteroids, a newer class of hormones, interact with auxin to induce ethylene formation (Sakurai and Fujioka, 1993). The complexity and multiplicity of the known regulatory activities of plant hormones and their roles in plant growth and development are illustrated in Table 26-2. A developmental process is ultimately controlled by the ratio of the promoter to inhibitor hormones. Note the regulatory activity of the classical five hormones on processes such as growth rate, flower initiation, abscission, and senescence. The lack of a hormone in a particular process may not be absolute; instead it may indicate that the hormone has not been positively identified with the process at the present time.

2.1   Auxins

Auxin is the only plant growth hormone that is transported polarly, and it travels from the apex to the base (basipetally). The shoot apex serves as the primary source of auxin for the whole plant. The formation of an auxin gradient from the shoot to the root affects various developmental processes including stem elongation, apical dominance, wound healing, vascular differentiation, floral bud formation, fruit development, and leaf senescence. While low levels of aux-

in promote root elongation, higher concentrations of auxin may inhibit root growth. At the cellular level, auxin increases the concentration of the nucleotides DNA and RNA, affects protein and enzyme synthesis; increases proton exchange, membrane charge, and potassium uptake (Marre, 1977), and rapidly changes gene activity (Guilfoyle, 1986; Key, 1989).

Auxins stimulate cell elongation and cause cell wall loosening, a term describing the more rapidly extensible or plastic nature of walls from cells treated with auxins (Salisbury and Ross, 1992). At least three mechanisms have been considered in the last 30 years to explain wall loosening. The most popular of these mechanisms, the acid-growth hypothesis (Ray, 1987), proposes that auxins cause receptor cells in stem sections to secrete H^+ into their surrounding primary walls. The accumulating H^+ ions reduce the pH resulting in wall loosening and accelerated growth. The low pH presumably allows activation of certain cell wall-degrading enzymes that are inactive at a higher pH. These cell wall-degrading enzymes purportedly break bonds in the wall polysaccharides, and allow the cell walls to stretch more easily.

2.2   Gibberellins

Gibberellins are a large group of related compounds with over 136 identified in various fungi and plants (MacMillan, 2001). All gibberellins are comprised of the ent-gibberellane structure, and the majority of the gibberellins are precursors or inactivation products formed during the formation of the biologically active forms GA_1 and GA_3. Gibberellins exhibit many physiological effects, suggesting that they have more than one primary site of action. They stimulate cell division in the shoot apex (Liu and Loy, 1976), promote cell growth by inducing various hydrolases (Noggle and Fritz, 1983b), often increase cell wall plasticity (Taylor and Cosgrove, 1989), and increase leaf size of a number of different plants (Taiz and Zeiger, 1998). Additionally, gibberellins are well known for promotion of internode growth, seed germination, fruit set, fruit growth, and seed germination. Gibberellins also are linked to changes in juvenility and flower sexuality.

2.3   Cytokinins

Cytokinin is a generic name for substances that typically stimulate cell division (cytokinesis). Chemically, they are related to adenine, a purine base found in both DNA and RNA. Cytokinins are most abundant in the young, rapidly dividing cells of the shoot and root apical meristems. Besides cell division, cytokinins participate in seed development, chloroplast maturation, cell enlargement, embryo development, senescence, and apical dominance (Hare *et al.*, 1997). Cytokinins may regulate nucleic acid metabolism and protein synthesis. In particular, cytokinins stimulate the synthesis of specific chloroplast proteins that are encoded by nuclear genes and synthesized by cytoplasmic ribosomes (Binns, 1994, Taiz and Zeiger, 1998).

Table 26-1. Major classes of currently recognized plant hormones and their classical roles in plant development.

Class	Biologically active form	Classical role(s)	Structure
Growth promoters			
Auxins	Indole Acetic Acid (IAA)	Cell elongation Acid growth	
Gibberellins	Gibberellic Acid 1 (GA_1)	Stem elongation Release dormancy	
Cytokinins	Zeatin	Cell division	
Growth inhibitors			
Ethylene	Ethylene	Senescence Fruit ripening	$CH_2{=}CH_2$
Abscisic acid	Abscisic Acid (ABA)	Stomatal regulation Dormancy	
Recent discoveries			
Jasmonates	Jasmonic Acid (JA)	Herbivore defense	
Proteinase inhibitors			
Salicylic Acid	Salicylic Acid (SA)	Pathogen defense Thermogenecity	
Systemic Acquired Resistance			
Brassinosteroids	Brassinolide	Light regulation of genes	

Table 26-2. Growth-regulating compounds and their elicited responses in cotton. (From Cathey with modifications, 1983 update).

Compound	Response	Reference
I. Germination and Seedling Development		
Oleic acid D-α-tocopherol	Reduced herbicide damage caused by trifluralin	Christiansen and Hilton, 1974
Indole-3-acetic acid (IAA) Kinetin	Overcome trifluralin damage	Hassawy and Hamilton, 1971
Gibberellic acid	Induced more rapid emergence and taller seedlings	Ergle, 1958
	Reduced chilling injury in germinating cotton	Cole and Wheeler, 1974
Adenosine-3'-5'-cyclic (AMP) Monophosphate	Reduced chilling injury in germinating cotton	Cole and Wheeler, 1974
Antitranspirants	Prevented chilling injury to seedling cotton	Christiansen and Ashworth, 1978
Mepiquat chloride	Reduced wind-blown damage to cotton	Gausman et al., 1981
(1,1-dimethylpiperidinium chloride)	Increased drought resistance of seedling cotton	Xu and Taylor, 1992
	Seed treatment reduced chlorosis and necrosis of cotton treated with fluometuron	Corbin and Frans, 1991
II. Vegetative Development		
Chlormequat (CCC0 [(2-chloroethyl)-trimethyl ammonium chloride]	Reduced plant height and yield	Thomas, 1964
	Reduced plant height but not yield	DeSilva, 1971; Thomas, 1972
	Promoted drought resistance	Singh, 1975
	Increased fruit loss	Kittock et al., 1974; Thomas, 1975
	Seed treatment reduced chlorosis and necrosis of cotton treated with fluometuron	Corbin and Frans, 1991
	Decreased growth rate	Marani, 1973
	Increased yields	Singh, 1970
N-dimethyl-N-B-chloroethyl hydrazonium chloride (CMH)	Decreased growth rate	Marani, 1973
2,3,5 triiobdobenzoic acid (TIBA)	Reduced plant size and dry weight; reduced seedcotton yield	Thomas, 1967
Mepiquat chloride	Darker green leaves	Gausman et al., 1978; Gausman et al., 1980a; Walter et al., 1980
	Height reduction under luxuriant growth conditions	Heilman, 1981
	Reduction in canopy width	Walter et al., 1980
	Reduced internode length	
	Reduced bolls/plant; individual boll weight increased	Feaster et al., 1980
	Smaller bolls; increased flower production	Briggs, 1981
	Erratic yield response	Gausman et al., 1980a
	Thicker leaves with reduced surface area	Gausman et al., 1980b
	Stimulation of CO_2 uptake	Cappy and Cothren, 1980
	Rate of incrase in root density significantly less	Hodges et al., 1991
	Increased gross canopy photosynthesis	Fernandez et al., 1992;
	Lack of effect on carbon use efficiency in well-watered and water-deficit plants, helped maintain leaf turgor potential in water-stressed plants	Fernandez and Cothren, 1990
	Increase in fine root hairs	Fernandez et al., 1991
	More favorable water status, higher leaf water potential	Stuart et al., 1980, 1981, 1984
	Reduced boll rot	Snow et al., 1981

continued

Table 26-2. Continued

Compound	Response	Reference
Mepiquat chloride (continued)	Mitigates symptom expression of Verticillium wilt	Erwin et al., 1979a, b
Indolebutyric acid (IBA)	Increased NADH oxidase activity in treated roots, indicating increase in root respiration	Urwiler and Oosterhuis, 1986
III. Reproductive Development		
Chlordimeform [N-(chloro-O-tolyl)-N, N-dimethylformamide]	Increased initiation and retention of fruit	Phillips et al., 1977
	No difference in fruiting response under insect free environment	Cathey and Bailey, 1987
Acephate (O-S-dimethyl acetylphosphoramidothioate)	Yield increase	Bauer and Cothren, 1990
	Cytokinin-like activity	Cathey, 1979 (unpublished)
	Increased flowering rate, boll production and size, and yield; no effect on previously mentioned parameters	Cathey et al., 1981
Guthion (O-O-dimethyl S-(4-oxo-1, 2, 3,-benzo triazin - 3 (4H-yl) methyl phosphorodithioate)	Produced more flowers and more bolls; longer maturation period	Hacskaylo and Scales, 1959
Gibberellic acid (GA$_3$)	Significant increase in boll set; erratic yield results	Walhood, 1958
	Taller plants; slightly less yield	Lane, 1958
	Flower bud and boll abscission reduced; yields not affected	Subbiah and Mariakulandia, 1972
Napthalene acetic acid (NAA)	Increase in boll numbers; 8 to 10% increase in yield at first harvest	Negi and Singh, 1956
	Decreased boll shedding; increased yields	Murty et al., 1976
	Increased fiber length	Patel, 1992
2,3,5 triiobdobenzoic acid (TIBA)	Slight increase in boll set and slightly earlier termination; no significant yield effect	Thomas, 1967
	Yield increases; increased boll size and number of bolls per plant; lowered position of first fruiting branch	Freytag and Coleman, 1973
Cytozyme Crop$^+$ (bacterial enzyme plus nutrients)	Significant decrease in transpiration and volatile nitrogen loss; trends, but non-significant increases in yield	Cothren and Cotterman, 1980
Chlormequat	Low concentrations - coarser fiber; higher concentrations - increased length and fineness; decreased strength maturity; and yield	Bhatt et al., 1992
	Significant yield increase	Rao et al., 1980
Indoleacetic acid (IAA)	Improved length and fineness of fiber	Bhatt et al., 1972
Naphthalene acetic acid (NAA)	Low concentrations increased fiber fineness; high concentration decreased fiber fineness	Bhatt et al., 1972
Gibberellic acid (GA$_3$)	Increased fiber length; no effect on fiber length	Bhatt and Ramanujam, 1971; Sitaram and Abraham, 1973
Mepiquat chloride	Increased boll weight, seed and lint index, lint yield and earliness	Sawan and Sakr, 1990
	Increased levels of petiole nitrates compared to controls	Stedman et al., 1982
Paclobutrazol ([2RS-3RS]-1-[4-chlorophenyl]-4-4-dimethyl-2-1,4-triazol-yl-pentan-3-ol)	Reduced plant height, number of main-stem nodes, and dry weight of stems and leaves	Ben-Porath et al., 1988
Catechin	Applied at flowering to seeds soaked in chlormequat chloride or succinic acid increased drought tolerance, increased root volume, higher seedcotton yield	Shanmugham, 1992
Triacontanol	Increased lint yield, increased fiber length	Patel, 1992
Polyhydroxycarboxylic acid (PHCA)	Yield increase	Munoz, 1994

2.4 Abscisic Acid

Abscisic acid (ABA) was isolated from cotton fruits in the early 1960s. Addicott and his colleagues first identified and chemically characterized ABA while studying compounds responsible for abscission of cotton fruit (Ohkuma *et al.*, 1963). ABA is synthesized in almost all cells that contain plastids. ABA, which is classified as a growth inhibitor, inhibits auxin-induced cell growth by preventing cell loosening (Zeevaart and Creelman, 1988). ABA also inhibits growth by interfering with nucleic acid synthesis, reducing the rate of cell enlargement, and reducing the rate of cell division. ABA is a promoter of bud and seed dormancy, and mutants that are deficient in ABA are viviparous (Pilet and Barlow, 1987). In addition, ABA is considered the plant's signal for water stress. The root synthesizes more ABA and translocates it to the shoot under water stress conditions. ABA levels of the leaf can increase 50-fold during water stress. The increased concentration of ABA in turn induces the closure of the guard cells of the stomata because high concentrations of ABA cause potassium and other ions to leave the guard cell. After the ions leave the guard cell, the guard cell loses turgidity and the stomata close (MacRobbie, 1997). In addition to closing stomata, ABA increases hydraulic conductivity of the root and increases the root: shoot ratio at low water potentials. (Taiz and Zeiger, 1998).

2.5 Ethylene

Ethylene, a gaseous molecule formed in most plant organs, is associated with inhibition of stem and root growth. Ethylene also promotes leaf bending, or epinasty, abscission, and stem swelling. The biosynthesis of ethylene is triggered by various developmental processes, auxins, environmental stresses, and wounding (Sticher *et al.*, 1997; Taiz and Zeiger, 1998). In cotton production, we are especially interested in ethylene's effects on fruit ripening and dehiscence. Ethylene production increases dramatically prior to abscission and the climacteric rise in ethylene production is the trigger for leaf abscission (Morgan *et al.*, 1992).

2.6 Brassinosteroids

Brassinosteroids include over 60 steroidal compounds (Davies, 1995). Brassinosteroids are classified as growth-promoting substances, and they accelerate cell division and cell elongation (Adam and Marquardt, 1986; Clouse and Sasse, 1998). The brassinosteriods are also involved in light-regulated development, and brassinosteroid-induced cell growth is light dependent (Li *et al.*, 1996).

2.7 Jasmonates

Jasmonic acid and its methyl ester, substrates of the biosynthesis of jasmonates, are considered powerful se-

nescence-promoting substances (Ueda *et al.*, 1991; Gross and Parthier, 1994). Jasmonates accelerate senescence by reductions in chlorophyll content and degradation of chloroplast proteins, especially rubisco (ribulose 1,5-bisphosphate carboxylase/oxygenase) (Creelman and Mullet, 1995; Beltrano *et al.*, 1998). They are also known to promote tuber formation, fruit ripening, pigment formation, and tendril coiling while inhibiting growth and seed germination (Gross and Parthier, 1994; Davies, 1995). Recent work has focused on the role of jasmonate in promoting abscission. Ueda *et al.* (1996) report that jasmonic acid and its methyl ester affect sugar metabolism in the abscission zone in bean petioles, in which the increase in cellulase activity involves the degradation of cell wall polysaccharides. In addition, jasmonates are involved in the plant's defense against water stress, wounding, insect attack, and pathogen attack (Creelman and Mullet, 1995; Baron and Zambryski, 1995; Creelman and Mullet, 1997). Jasmonates also induce ethylene formation (van Loon *et al.*, 1998).

2.8 Salicylic Acid

Salicylic acid, which is chemically related to aspirin, belongs to a diverse group of plant phenolics (Raskin, 1995). Like jasmonates, salicylic acid may be involved in the resistance to pathogens because it induces the production of pathogenesis-related proteins. However, the salicylic acid defense pathway is independent of the jasmonate defense pathway (van Loon *et al.*, 1998). Pathogenesis-related proteins are protein compounds with antimicrobial and antifungal activities; eleven pathogenesis related protein families have been characterized (Sticher *et al.*, 1997). Research with transgenic tobacco plants lacking the ability to produce salicylic acid, showed that these plants were unable to induce a resistance mechanism, called systemic acquired resistance, to certain plant diseases (Delaney *et al.*, 1994; Ryals *et al.*, 1996; Baker *et al.*, 1997; Hammerschmidt and Becker, 1997). Salicylic acid also enhances flower longevity, inhibits ethylene biosynthesis, inhibits seed germination, blocks the wound response, and reverses the effects of ABA (Gross and Parthier, 1994; Davies, 1995).

Salicylic acid serves as a trigger for increasing the activity of alternative respiration (Kapulnik *et al.*, 1992). Alternative respiration refers to a minor respiratory pathway in plants that is not sensitive to cyanide, unlike conventional respiration. Alternative respiration represents approximately 27 to 30% of the electron flow through the electron transport chain (Lennon *et al.*, 1997; Ordentlich *et al.*, 1991). Because alternative respiration does not produce much energy in the form of ATP, most of the energy produced in alternative respiration is released as heat. The heating of plant tissue caused by an increase in alternative respiration is coined thermogenecity. Thermogenecity plays a key role in increasing the temperature of Araceae (Arum family) inflorescences by as much as 25°C. Increasing the temperature of the influorescence volatilizes amine compounds, and the odor given off from the amines attracts insect pollinators (Taiz and Zeiger, 1998). The alternative oxidase associated with alternative respiration has been implicated in various biochemical processes including a role in lowering mitochondrial reactive oxygen production in tobacco cells (Ribas-Carbo *et al.*, 2000) to alleviate stress, as an avenue for improving tolerance to chilling injury, and as a means for improving resistance to various pests.

3. PLANT GROWTH REGULATORS

Numerous plant growth regulators (PGRs) are commercially available for cotton production. Table 26-1 is a comprehensive, but by no means exhaustive, list of PGRs and the diverse responses that they elicit. In cotton, PGRs are commonly organized into groups based on the developmental stage at which they elicit a response: germination and seedling development, vegetative development, reproductive development, and harvest aids.

The use of exogenously applied compounds such as PGRs to regulate cotton growth and development has previously been reviewed by Walhood and Addicott, 1968; Namken and Gausman, 1978; Cathey, 1983, 1986; Guinn, 1984b. 1986; Cathey and Thomas, 1986. Information in Table 26-2 was adapted from Cathey (1983) with modifications to illustrate the broad array of uses and times of application for PGRs.

3.1 Areas of Research Emphasis

Like many other production programs, PGR use begins at planting and continues through harvest. Timeliness of application, rates, and interaction with cultural inputs significantly impact the crop response and the potential for success with PGR use. Due to the observed benefits of using PGRs on cotton, they have been the focus of many research projects. Table 3 lists some of the areas of major emphasis for PGR research in cotton. Interpretation of PGR studies is difficult because experimental inputs are confounded by differences in cultural inputs and environmental conditions. Cotton responds differently when production variables are changed, and adding a PGR as another variable further complicates the expression of the crop's genetic potential. Conclusive evidence for increased boll set, retention, and weight is especially difficult due to the inherent variability of individual plants. Currently, PGR use in cotton has been documented in germination and seedling development, root enhancement, root growth, canopy architecture, photosynthate partitioning, nutrient uptake, and harvesting. Before discussing specific effects of PGRs on cotton, we will briefly identify and explain the potential areas for PGR use.

3.1.1 Germination and Seedling Development

Cotton is sensitive to chilling, and is adversely affected by low temperature at the early stages of growth.

Because cotton is sensitive to chilling injury during germination, compounds that improve seed germination and seedling vigor could contribute significantly to increased yield. Chilling temperatures during initial hydration of cotton seed can be extremely damaging. Chilling (<10°C) for as little as four hours at the onset of hydration can kill a seed or cause high incidence of aborted root tips (nub root) (Christiansen, 1967). Once the radicle has elongated two to three centermeters, chilling causes cortex sloughing, slowing of early growth, and long-term growth reduction and flowering delay (Christiansen and Thomas, 1969).

Broadening the base of tolerance to chilling injury with PGRs could enhance the yield potential. Although total yield is not significantly influenced by chilling, the crop value is significantly reduced because of lower fiber quality (Christiansen and Thomas, 1969) and delayed maturity. Wanjura *et al.* (1969) correlated speed of cotton emergence and productivity and reported that a more rapidly emerging crop was more productive than when emergence was delayed. Thus, rapid plant emergence and crop vigor are important factors in predicting crop yield.

3.1.2 Early Flower Production and Increased Fruit Retention

Early flower production and increased fruit retention has become increasingly important with the widespread adoption of faster-fruiting cotton cultivars. With a narrower production window, less time exists for recuperation from fruit loss due to the environment, cultural practices, or pests. Since a large percentage (66 to 75%) of the yield is produced on first-position fruiting sites (Jenkins *et al.*, 1990a), retention and maturation of these bolls is critical. Use of PGRs may increase boll retention at the first fruiting sites, enhance and accelerate crop maturity, promote an earlier harvest, and improve lint quality, and potentially alter membrane properties associated with enhanced tolerance to deviation in temperature.

Enzymes influence all physiological processes, and enzyme activity is highly dependent upon temperature. One way to describe the difference in enzyme activity involves the K_m, or Michaelis-Menton, constant. The K_m value is equal to the concentration of substrate for an enzyme when it is operating at half its maximum rate; a large increase in the K_m indicates a loss of catalytic efficiency (Kidambi *et al.*, 1990). One way in which K_m values are used is for determining the thermal kinetic window, or TKW (Burke *et al.*, 1988). The TKW describes the temperature range where the K_m value lies within 200% of the minimum K_m value for that enzyme. Enzymes function optimally at temperatures within the TKW. Temperatures outside the TKW correspond to dramatic increases in K_m and decreases in enzyme catalysis. For cotton, the TKW lies between 23.5 and 32°C. Burke *et al.* (1988) estimates that cotton grows within its TKW roughly 30% of the growing season. The amount of time which cotton experiences temperatures within the TWK is positively correlated with dry matter production (Burke *et al.*, 1990). Positive correlation between the TKW and the rate of synthesis for the light-harvesting complex for Photosystem II (Burke and Oliver, 1993) and recovery from fluorescence (Ferguson and Burke, 1991) have been identified. Cotton root growth also occurs more rapidly when growth temperatures are within the TKW (Burke and Upchurch, 1995). Furthermore, cotton leaves increase transpiration, or water movement through the plant, to cool the canopy and maintain foliage temperature near 26 to 27°C with sufficient soil moisture (Burke *et al.*, 1989). The TKW has also been used to determine irrigation timing (Wanjura *et al.*, 1988; 1989).

3.1.3   Canopy: Improved Canopy Development, Control of Vegetative Growth, Improved Partitioning

Considerable effort has been expended on ways to improve the efficiency of the canopy for light interception and the conversion of light energy to photosynthates. Evidence has been presented to show that photosynthesis limits the yield of cotton, even under optimum environmental conditions (Guinn *et al.*, 1976). Because of its perennial growth habit, balancing vegetative and reproductive growth in cotton can be difficult, especially under high nitrogen fertility and irrigation regimes. Height-to-node ratios (Bourland and Watson, 1990; Bourland *et al.*, 1994) and position of the uppermost white flower defined by node counts (Bernhardt *et al.*, 1986; Bourland *et al.*, 1992) provide means for monitoring transient growth and developmental changes of cotton to assist in the formulation of timely management decisions. PGRs have been successfully used to reduce height, to produce plants with more desirable height-to-node ratios (Hickey, 1994), to reduce excessive leaf area production (Walter *et al.*, 1980; Stuart *et al.*, 1984), and to alter partitioning of assimilates (Fernandez *et al.*, 1991).

Increased cotton yields have primarily resulted from improved partitioning of dry matter from vegetative to reproductive structures. A comparison of five obsolete cultivars to twenty more-recently developed cultivars showed that modern cultivars invest a greater proportion of dry matter to reproductive rather than vegetative structures. Consequently, the modern cultivars possess more lint than the older cultivars (Meredith and Wells, 1989). Although this study suggested that yield increases through the use of conventional breeding methods were likely to be achieved by continued partitioning of dry matter from vegetative to reproductive structures, the limits of this potential were questioned. The point was made that surely at some point, further reductions in leaves and/or stems would cease to improve yields. They suggested that yield increases would be obtained only through some other source of variation, such as photosynthesis.

A further avenue for improving yield investigated by breeders involves leaf morphology. Meredith (1984) compared the influence of three mutant leaf types, (super okra, okra, and sub okra) to a normal leaf type for lint yield. The

mutant leaf types, often referred to as "open-canopy" cottons, permits greater light penetration and air movement into the lower parts of the canopy than normal-leaf cotton. Of the three open-canopy cottons, only sub okra yielded significantly more lint (4.8%) than its corresponding normal population. Okra and super okra types may eliminate too much leaf area, thereby reducing photosynthetic capability, even though they increase light penetration. The challenge then becomes to achieve the best balance between light interception and photosynthetic production.

As the season progresses, excessive canopy production (leaf area) can lead to shading of the lower canopy. Guinn (1974) indicated that levels of photosynthetically active radiation (PAR) reaching the lower part of the plant canopy in high density populations of cotton might severely limit photosynthesis. As developing bolls obtain most of their photosynthesis from subtending leaves, bracts, and leaves one node removed (Ashley, 1972; Wullschleger and Oosterhuis, 1990d), reduced PAR levels in the lower part of the canopy could influence boll retention at the bottom of the canopy. Increased boll retention and development can be accomplished by altering the shape of the plant. This may be accomplished by restructuring the plant architecture to improve light interception in the lower canopy and to produce rapid leaf area expansion early in the season to reduce the amount of solar radiation reaching the soil.

3.1.4   Root Growth and Nutrient Uptake

Increased yields and faster fruiting rates cannot be realized unless the plant has the ability to supply the nutrients necessary to meet the demands for increased flowering and boll retention. Alteration of root:shoot ratios (*i.e.* higher root:shoot) could potentially benefit the plant by providing the larger root mass required to meet the needs of the aboveground biomass. PGRs cause increased carbohydrate partitioning to the root system. The increase in carbon allocated to the root is illustrated by the higher root:shoot ratios (Walter *et al.*, 1980; Fernandez *et al.*, 1991; Zhao and Oosterhuis, 1997) and higher mass of root tissue (Atkins *et al.*, 1992; Oosterhuis and Zhao, 1994) expressed by cotton plants treated with several PGRs. PGRs also alter the branching pattern of the root system; more lateral roots (Oosterhuis and Zhao, 1994; Oosterhuis, 1995a) and more fine roots (Fernandez *et al.*, 1991) have been observed in plants treated with PGRs. Also, increasing the physiological activity of the root for nutrient uptake would be beneficial. One PGR, PGR-IV, has been documented to increase the activity of the dehydrogenase enzyme in the root, and the increased dehydrogenase activity may be correlated with additional capacity for nutrient uptake by the plant (Clark *et al.*, 1992). Increased nutrient uptake has been observed in cotton treated with PGRs (Zhang *et al.*, 1990; Atkins *et al.*, 1992; Guo *et al.*, 1994; Oosterhuis and Zhao, 1994; Weir *et al.*, 1994). Synchronization of root activity with fruit production is critical. Increased root activity during the later stages of boll filling is crucial for supplying needed minerals and water to the developing fruit. However, prolonged root activity can lead to serious problems with late-season vegetative growth near to or following defoliation/desiccation. Prolonged root activity complicates leaf removal at harvest and introduces the potential for regrowth.

3.1.5 Harvest

Producers strive for the proper balance between nutrients, especially nitrogen, water, and a good fruit load to make harvest easier. Provided the proper management decisions are made, the crop usually depletes water and nitrogen sources by the end of the season. Depleted nitrogen and water supplies at the end of the season allow for easier and more effective defoliation/desiccation. Compounds that are used for harvest aids are categorized as either hormonal or herbicidal. Cotton cultivars that are machine picked are typically treated with hormonal-type harvest aids while stripper varieties are treated with herbicidal products. Use of harvest aids allows for more timely removal of the crop from the field to minimize losses from weathering and weather-related problems. A harvest aid also increases efficiency of harvest and reduces the amount of trash associated with the harvest process. Both of these factors potentially increase net returns to the producer.

3.2 PGR Modification of Cotton Growth and Development

For a cultural practice to be widely adopted, its use must produce consistent responses. The use of PGRs, like most other cultural inputs, is influenced by the environment and other production inputs. Changes in parameters, such as cultivar selection, row spacing, plant density, fertility, planting date, temperature, and seed treatments have the potential to alter PGR response.

3.2.1 Cultivar Response

Numerous studies on cultivar response to mepiquat chloride have been conducted, including those of Briggs, 1981; York, 1983a, Bader and Niles, 1986; Niles and Bader, 1986; Landivar *et al.*, 1992; and Boquet and Coco, 1993. A three-year study conducted in eight environments with 14 cultivars indicated that cultivar selection should not be a consideration in deciding whether to apply mepiquat chloride (York, 1982). Similar conclusions were made by Cathey and Meredith (1983, 1988) in Mississippi. A study of the morphological and phenological variables of a short- and a full-season cotton cultivar to mepiquat chloride suggested full-season types are more flexible in their response to mepiquat chloride for maturity modification, despite having similar changes in morphological characters following treatment with the compound (Bader and Niles,

1986). In this study, mepiquat chloride exhibited a trend for increased yield in the full-season cultivar, but reduced yield in the short-season cultivar (Niles and Bader, 1986).

One of the more recent advancements in cotton production involves the commercial release of transgenic cultivars. Genetically-altered cultivars give producers an alternative management strategy to conventional pest and weed control; however, the effect of genetic alteration on the physiology of the crop is not documented. Thus, concern arose over the application rate for PGRs, especially mepiquat chloride. Some researchers and producers even claimed that more mepiquat chloride may be required to elicit the same response as in conventional cultivars (Wrona *et al.*, 1997; Jones *et al.*, 1996). Viator *et al.* (1999) stated that planting herbicide-resistant cultivars, such as BXN™ and Roundup Ready™ cotton, should not affect implementation of PGR strategies. Furthermore, the concentration of mepiquat chloride based on plant biomass required to suppress vegetative growth did not differ between the insect-resistant cultivar DPL 33B® (*Bt* cotton) and its conventional parent DPL 5415®. Although the transgenic cultivars were numerically taller, no differences in the growth rate were detected between the transgenic and conventional cultivars (Underbrink *et al.*, 1998; 1999). In fact, the optimum mepiquat chloride concentration of 8 to 12 ppm for conventional cotton (Landivar *et al.*, 1995) also served as the optimum mepiquat chloride concentration for the transgenic cultivar (Underbrink *et al.*, 1999).

3.2.2 Row Spacing

The effect of row spacing and mepiquat chloride treatment on earliness of eight cultivars was inconsistent (Boquet and Coco, 1993). Earliness of the cultivar DPL 20 was unaffected by row spacing without mepiquat chloride, but with mepiquat chloride, maturity was earlier at the 30-inch row spacing compared to the 40-inch row spacing. On the other hand, Stoneville LA 887 matured earlier in 40-inch rows than in 30-inch rows when treated with mepiquat chloride.

More recent studies are re-investigating the use of ultra-narrow row spacings to improve light interception and to enhance yield potential. Post-emergence weed control was a major limitation to ultra-narrow row cotton (UNRC) in the past, but the availability of transgenic cultivars, such as BXN® and Roundup Ready® cotton, has alleviated this problem (Snipes, 1996; Gerik *et al.*, 1998). PGRs, especially mepiquat chloride, have reduced the problems associated with harvesting and rank growth by suppressing vegetative growth in UNRC (Atwell, 1996). Thus, UNRC production is becoming a more attractive alternative for producers. The UNRC production scheme allows for rapid canopy closure. Jost *et al.* (1998) estimated that cotton planted in 7.5-in. rows approached 50% canopy closure by match-head square compared to 10% canopy closure for the conventional spacing. Accelerated canopy closure and rapid early-season leaf area development in UNRC reduce weed competition, increase light interception, and decrease soil water evaporation (Krieg, 1996; Heitholt *et al.*, 1992). Decreasing plant spacing reduces plant height, boll size, number of bolls per plant, and number of nodes (Fowler and Ray, 1977). Decreasing the number of bolls per plant may lead to an earlier harvest for UNRC because the bolls are set earlier in the season (Buxton *et al.*, 1979; Jost and Cothren, 1999). Thus, the potential for canopy manipulation with PGRs in an UNR system makes this production practice more attractive.

3.2.3   Plant Density and Fertility

Other inputs for obtaining optimal lint yields include higher plant populations and nitrogen (N) fertility. Although high N rates and high plant populations should increase cotton yields, research has shown that an optimal N ratio and plant population exist, and that yield may be decreased if these optimum levels are exceeded (Bridge *et al.*, 1973; Smith *et al.*, 1979). Since both higher populations and higher N rates result in excessive vegetative growth, it seemed logical to use mepiquat chloride with these production inputs to suppress excessive growth and possibly increase yields. However, a three-year study at eight locations showed that mepiquat chloride produced no effect on the optimum N rate or plant population (York, 1983b). These results suggest that the benefits of mepiquat chloride may be evident only when environmental conditions favor excessive vegetative growth and delayed maturity.

Robertson and Cothren (1991) reported that lint yields were influenced by the interaction of mepiquat chloride, row spacing, and N rate. Gordon *et al.* (1986) found that mepiquat chloride increased yield, but yield increase was similar for all N rates. However, McConnell *et al.* (1992) and Boman and Westerman (1994) failed to show a significant positive interaction between N rates and mepiquat treatment. When above optimum N rates were used in Texas, the significant interaction for mepiquat chloride and N application rate was at the highest nitrogen rate (180 kg ha^{-1}) and the unfertilized control (0 kg ha^{-1} N) (Han *et al.*, 1990). Petiole nitrate-N concentration of mepiquat chloride-treated cotton was higher than untreated cotton at three sampling dates and suggests an improved nutrient fertility status of the crop (Maples, 1981) from mepiquat chloride treatment.

3.2.4   Planting Date

Mepiquat chloride has been found to mitigate the adverse effects of increased vegetative growth caused by delayed planting (Cathey and Meredith, 1983, 1988). For three planting dates (mid-April, early May, and mid-May), mepiquat chloride reduced lint yield by 4.5% in the early planted plots, but increased yield by 5.4 and 12.7% in the optimum and late plantings, respectively. Boll weight and seed index were also found to increase with mepiquat chloride treatment at all plantings, but flower production was increased only in the late planting. Lint percentage was reduced in all mepiquat chloride-treated plots from the three dates of planting.

3.2.5   Temperature

Temperature stress (both high and low) can adversely affect the physiology and subsequent productivity of cotton. Chilling of cotton plants caused reduced growth even when plants were exposed later to favorable temperatures. The reduction in growth was directly proportional to the duration of chilling at 10°C (Christiansen, 1963, 1964). Mepiquat chloride reportedly increases heat and cold tolerance of cotton (Huang *et al.*, 1981; Huang and Gausman, 1982 a, b). Electrolyte leakage, an indicator of membrane integrity, showed that nontreated leaf discs leaked twice as much electrolytes than did mepiquat chloride-treated discs (Huang and Gausman, 1982a). Ultrastructural observations revealed that in mepiquat chloride-treated leaves, the plasma membrane was altered by protein aggregation, but the plastidal envelope and the thylakoidal system remained intact. These three membrane systems in the nontreated leaf showed extensive degeneration. At elevated temperatures (55°C) cotton plants previously treated with mepiquat chloride showed increased heat resistance compared to the untreated control (Huang and Gausman, 1982b; Reddy *et al.*, 1990). The mepiquat chloride-treated leaves had larger starch grains in their chloroplasts than control leaves, which suggests a difference in photosynthetic activity (Reddy *et al.*, 1996). Urwiler (1981) suggested that mepiquat chloride held potential for use as a cryoprotectant.

3.2.6   Seed Treatments

Although mepiquat chloride is not labeled as a seed treatment, it is capable of inducing physiological changes in the seed (Urwiler, 1981; Albers and Cothren, 1981, 1983; Zhang *et al.*, 1990). Germination tests at a suboptimal temperature regime (15°C) indicated membrane permeability was influenced (the membranes were less leaky) by mepiquat chloride treatment (Albers and Cothren, 1981). After cottonseed were germinated for 96 hours in germination paper at 15°C, mepiquat chloride-treated seed showed a significant increase in germination rate. Morphological development of cotton from treated seeds was similar to a response following foliar application of mepiquat chloride (*i.e.*, enhanced root/shoot ratio). Emergence tests under simulated crusting conditions with high and low quality seed were conducted for six cotton cultivars grown at two locations in Arkansas (Albers and Cothren, 1983). In this study, mepiquat chloride-treated seedlings displayed enhanced emergence when compared to untreated seedlings under controlled environments. In addition, chilling injury was lessened by mepiquat chloride seed treatment. However, reduced chilling injury was detected only in the more sensitive cultivars from the "unhardened" location. Zhang *et al.* (1990) evaluated the effect of mepiquat chloride on early plant growth of cotton when seeds were treated with 0, 0.02, and 2.0 g a.i. mepiquat chloride. All mepiquat chloride treatments significantly decreased the number of nodes, leaves, and squares, as well as dry weight of leaves, stems, and roots, as compared to control plants at 28 days after emergence. Plant height and total leaf area of mepiquat chloride seed treatments were also significantly reduced compared to controls.

Studies with other PGRs applied as a seed treatment or in-furrow have generally been inconclusive. For example, use of the multiple entity PGR-lV applied in-furrow was shown to enhance root and shoot growth of developing cotton seedlings up to six weeks after planting (Oosterhuis and Zhao, 1994;), although Oosterhuis and Steger (1999) subsequently found no positive effects of PGR-IV or Early Harvest on emergence, seedling development, or yield.

3.2.7   Alteration of Nutrient Uptake

Alteration in the distribution of nutrient uptake by mepiquat chloride-treated cotton plants has been reported (Cothren *et al.*, 1977; Nester, 1978; Heilman, 1981, 1985; Zhang *et al.*, 1990; Han *et al.*, 1991), although the changes reported for specific ions have not always been consistent. Increases of calcium, magnesium, potassium, and phosphorus were reported from mepiquat chloride foliar treatments in cotton plant tissues (Cothren *et al.*, 1977). Nester (1978) also reported that calcium, magnesium, and phosphorus increased in leaves of plants treated foliarly with mepiquat chloride, and potassium and phosphorus increased in the roots. Seed and foliar treatments of mepiquat chloride applied singly or in combination generally increased levels of calcium, potassium, and magnesium (Cothren *et al.*, 1983). Heilman (1981) reported that nitrogen and phosphorus concentration in leaves was unaffected by mepiquat chloride treatments; however, the percent of calcium and magnesium was significantly increased (Heilman, 1985). Results of cotton seed treated with varying rates (0, 0.2, 1.0, and 2.0 g a.i.) of mepiquat chloride showed that, in general, the highest rate of mepiquat chloride resulted in greater concentrations of calcium, phosphorus, and nitrogen in plant leaves and stems and also higher concentrations of magnesium, phosphorus, and nitrogen in roots (Zhang *et al.*, 1990). There have been a number of reports showing increased nutrient uptake in cotton treated with PGRs other than mepiquat chloride (Zhang *et al.*, 1990; Atkins, 1992; Guo *et al.*1994; Oosterhuis and Zhao, 1994; Weir *et al.*, 1994) but the results have been inconsistent and PGRs are not generally sold for this attribute.

3.2.8   Yield Response

Mepiquat chloride has been shown to increase yields (Erwin *et al.*, 1979a), to increase yield in some tests while decreasing yields in others (Armstrong, 1982; York, 1983a, 1983b), to have little effect on yield (Heilman, 1981; Stuart *et al.*, 1984) or to reduce yield (Thomas, 1975; Crawford, 1981).

Results from 35 experiments conducted over a five-year period in the San Joaquin Valley of California indicated that yield responses occurred only when control plant heights exceeded 1.10 m at maturity or when the length of the growing season was short (Kerby, 1985). Plant mapping from eleven experiments conducted from 1981 to 1984 in the San Joaquin Valley showed that mepiquat chloride treatment produced 3.1% fewer fruiting positions than untreated plants (Kerby *et al.*, 1986). However, mepiquat chloride stimulated early boll load to a peak at nodes 9 and 10, while the boll load declined continuously after node 10. This decrease in late season boll load was apparently due to increased abortion of fruiting forms rather than to a limited initiation of fruiting positions.

3.3   Insecticides as Growth Regulators

Ample information suggests that some insecticides have physiological effects on the growth and development of cotton (Brown *et al.*, 1961; Brown *et al.*, 1962; Roark *et al.*, 1963; Lincoln and Dean, 1976; Phillips *et al.*, 1977; Campbell *et al.*, 1979; Weaver and Bhardwaj, 1985; Bauer and Cothren, 1990). However, testing of insecticides for plant growth regulation is often difficult because it is hard to distinguish between plant responses to the chemical *per se* and to control of insects. In early reports on the growth regulating properties of insecticides, some hydrocarbons were shown to increase fruit set and early maturity, and some organophosphates tended to induce lateness (Brown *et al.*, 1961; Brown *et al.*, 1962; Roark *et al.*, 1963). Early season applications of the insecticide chlordimeform [N-4-chloro-O-tolyl-N,N-dimethylformamidine] increased cotton lint yields above those expected from the pesticidal properties of the chemical (Lincoln and Dean, 1976; Phillips *et al.*, 1977). Campbell et al (1979) reported a 25% increase in yield for chlordimeform-treated cotton and described the effect as a combination of insecticidal control and physiological yield enhancement. Other researchers failed to show influences on growth and development of cotton treated with chlordimeform, and thus concluded that the product did not increase yields beyond yields associated with actual pest control (Cathey and Bailey, 1987; Guthrie, 1987; Durant, 1989; Youngman *et al.*, 1990).

Some responses of cotton to chlordimeform, including modification of petiole ion levels (Weaver and Bhardwaj, 1985) and increased reproductive growth (Lloyd and Krieg, 1987), prompted speculation that the chemical may interact with endogenous plant hormones (Bauer and Cothren, 1990). Cytokinin-like activity has been reported for several agricultural chemicals, including the photosynthesis-inhibiting triazine, uracil, and phenylurea herbicides, which retard senescence in intact maize (*Zea mays* L.) leaves at subtoxic dosages (Hiranpradit and Foy, 1973); the cotton defoliant thidiazuron (N-phenyl-N-1, 2, 3-thidiazol-5-yl urea), which promotes growth of sieva bean (*Phaseolus lunatus* L.) callus culture (Mok *et al.*, 1982); and the fungicide triadimefon [1-(4-chlorophenoxy)-3, 3-dimethyl-1-(1H-1, 2, 4-triazol-1-yl)-2-butanone], which exhibits cytokinin-like activity in detached barley (*Hordeum vulgare* L.) leaves (Forster *et al.*, 1980). In Bauer and Cothren's (1990) study, chlordimeform was compared with the naturally occurring cytokinin zeatin in two hormone bioassays - [radish (*Raphanus sativa* L.)] cotyledon expansion and cytokinin-depleted soybean (*Glycine max* L.) callus growth. Chlordimeform induced radish cotyledon expansion when the cotyledons were incubated under continuous light, but did not induce growth when the cotyledons were incubated in the dark. In addition, high concentrations of chlordimeform inhibited chlorophyll synthesis in the radical expansion test. Although the data suggested that chlordimeform had plant physiological activity, the biological activity was not identical to that of zeatin.

3.4   Cytokinin and Cytokinin-like PGRs

Several cytokinin and cytokinin-like compounds (Burst, Cytozyme, Cytokin, Stimulate, Early Harvest, and Triggrr) have been tested for PGR activity in cotton. Specific modes of action have not been elucidated, but these compounds theoretically promote fruit set and retention, and increase the ability of the plant to fill existing fruit, especially in the middle of the plant canopy and on the first fruiting position (Mayeux, 1992; Mayeux and Kautz, 1992). A two-year study in Arkansas with Cytozyme Crop⁺, a mixture of complexes containing various bacterially-active cytokinins, auxins, and amino-acid chelated minerals, showed a numerical trend toward increased yields, but these changes were not significantly different from the control (Cothren and Cotterman, 1980). Laboratory tests performed on Cytozyme Crop⁺ in controlled environmental chambers revealed appreciable decreases in foliar nitrogen loss and increases in carbon dioxide fixation (Cotterman and Cothren, 1979). Namken (1984) applied single and multiple foliar applications of a cytokinin product to cotton at first one-third-grown square and at first bloom; treated cotton produced a significantly higher lint cotton yield than the untreated cotton. Foliar application of this cytokinin product was reported to promote bud initiation and development leading to increased plant fruitfulness and increased efficiency of the plant to develop and fill the fruit (Mayeux, 1985; Mayeux *et al.*, 1986). Significant increases in boll number, especially in first-position fruiting sites (Mayeux *et al.*, 1986; Mayeux *et al.*, 1987) were also reported for Burst. However, Urwiler *et al.* (1987) reported no significant effect of Burst on cotton yields. A comprehensive review of results with a similar product, Foliar TRIGGR, showed increases in cotton yields that averaged 44 pounds of lint/A in 83 research trials conducted in nine states over a nine-year period (Parker and Salk, 1990). Additional studies by Bednarz and van Iersel (1998) with Stimulate and Early Harvest, products containing cytokinin, failed to show significant influence on photosynthesis or any associated parameters.

3.5   Ethylene and Fruiting Modification

In cotton, ethephon, which is cleaved to yield ethylene in the plant, has been successfully used and is widely accepted as a harvest aid to accelerate boll dehiscence prior to harvesting (Cathey *et al.*, 1982; Gwathmey and Hayes, 1996; Stewart *et al.*, 1998; Jones *et al.*, 1998; Lemon *et al.*, 1999). The use of ethylene forming compounds in harvesting and crop termination will be discussed later in this chapter. Ethephon may also be used to raise the node level of the first flower in Pima cotton to a higher position, thus potentially increasing the efficiency mechanical harvesting (Pinkas, 1972). Other studies have been directed toward early-season applications of ethephon in cotton to induce early fruiting bud, or square, loss to obtain a better understanding of the growth and development of the crop relative to fruiting (Pettigrew *et al.*, 1992).

The loss of early squares has been suggested as a way to deprive overwintering insects of a food source and to delay the development of certain insect infestations (Bariola *et al.*, 1988; Henneberry *et al.*, 1988; King *et al.*, 1990; King *et al.*, 1992; Namken and King, 1991). An additional reason for examining early fruit removal is that cotton may compensate for the loss of early squares by increasing the subsequent rate of square initiation (Kletter and Wallach, 1982; Kennedy *et al.*, 1986; Ungar *et al.*, 1987; Sheng *et al.*, 1988; Kennedy *et al.*, 1991; Henneberry *et al.*, 1992; Pettigrew *et al.*, 1992; Moss and Bednarz, 1999). Sheng *et al.* (1988) and Kennedy *et al.* (1991) suggest that the increased rate of square initiation permits cotton to overcompensate for early season square abscission induced by ethephon application, and sometimes overcompensation of early square loss may result in greater yields. However, ethephon-treated plants do not always exhibit increased yields, and yield results are quite variable. Yields are affected by the year, genotype, and degree of fruit removed (Kennedy *et al.* 1986; Ungar *et al.*, 1987; Sheng *et al.*, 1988; Pettigrew *et al.*, 1992). Bariola *et al.* (1989) and King *et al.* (1990) found that yield was not affected by early fruit removal. Yield decreases or no yield change, depending on the type of early fruit removal, were reported by Kennedy and Jones (1990). It has been found that ethephon application lowers net assimilation rate and carbon exchange rate (CER) (Pettigrew *et al.*, 1993). However, no meaningful increase in photosynthesis was observed during the period when the ethephon treated plots were compensating for the early square loss. Thus, Pettigrew *et al.* (1992) concluded that early square removal should not be used in current production practices. Although reproductive development was delayed by early square removal and cotton compensated for early square removal in one year, it did not overcompensate for early reproductive loss to produce higher yields. Also, there were no overall improvements in fiber quality traits to justify the occasional loss in yield from the early square removal.

3.6   Multiple Entity PGRs

PGR-IV is a multiple entity plant growth regulator that contains indolebutyric acid (IBA), an auxin, and gibberellic acid (GA) in a nutrient solution blend. The purported mode of action of PGR-IV involves an alteration of plant hormone balance that affects the plant's growth. Various effects of PGR-IV on plant growth have been reported including increases in shoot dry matter, nutrient uptake, and square retention (Oosterhuis and Zhao, 1993; Oosterhuis, 1995a). Earlier unsubstantiated reports that suggested PGR-IV positively impacted root development (Atkins, 1992; Clark et al, 1992) were strengthened by growth room studies in which in-furrow applications of PGR-IV dramatically increased root length (47%), root dry weight (29%), the number of lateral roots (75%), and nutrient uptake one week after planting (Oosterhuis and Zhao, 1994). Enhanced uptake of potassium, manganese, iron, and copper at three and five weeks after planting was noted for PGR-IV treated plants, although the results were variable (Oosterhuis and Zhao, 1994). Oosterhuis and Zhao (1993) reported increased photosynthesis following PGR-IV application and suggested that increased photosynthesis was related to improved partitioning of assimilates to the developing bolls. Zhao and Oosterhuis (1998a) also observed a decrease in fruit abscission with PGR-IV for cotton plants grown in shaded conditions; again, improved carbohydrate assimilate translocation caused by PGR-IV was implicated as the source of increased fruit retention. Whole plant assimilation studies at 20°C showed increases of 29, 37, and 24% in gross carbon uptake, respiration, and net carbon uptake, respectively, for PGR-IV treated plants (Cadena *et al.*, 1994). The increases in carbon uptake and respiration suggest better growth of PGR-IV-treated plants at lower temperatures. However, carbon use efficiency was reduced because a greater proportion of the carbon fixed was diverted toward respiration; therefore, respiratory maintenance costs limited plant growth. Zhao and Oosterhuis (1994, 1995) showed that an application of PGR-IV decreased the detrimental effects of flooding on cotton growth and improved photosynthesis and dry matter production in flooded cotton plants. Improvements in photosynthesis of PGR-IV treated plants compared to untreated plants were also observed under water-deficit stress. Increases in photosynthesis caused by PGR-IV are more evident for cotton that is grown under dryland and low fertility conditions (Cadena and Cothren, 1995). Increased boll retention and other favorable growth responses have resulted in lint yield increases for PGR-IV-treated plants (Urwiler and Stutte, 1988; Cothren and Oosterhuis, 1993; Oosteruis and Zhao, 1993; Livingston and Parker, 1994; Cothren *et al.*, 1996; Hickey, 1996). Lege *et al.* (1996) also detected a significant increase in the seed weight and boll size of PGR-IV-treated plants. Nepomuceno *et al.*(1997a) showed that the duration of activity of an early-flowering application of PGR-IV or mepiquat chloride was about two weeks, information needed for timing subsequent applications if they are needed.

A combination of mepiquat chloride and *Bacillus cereus* was reported by Parvin and Atkins (1997) to increase the number of main-stem nodes, total boll number, and earliness with an associated yield increase compared to mepiquat chloride alone. Zhao and Oosterhuis (1998a) reported that the combination of mepiquat chloride and *Bacillus cereus,* subsequently named MepPlus and later changed to Pix Plus, significantly decreased plant height similar to mepiquat chloride, but had no effect on the number of main-stem nodes. Wells and Edmisten (1998) did not find any significant yield differences between plants treated with mepiquat chloride compared to mepiquat chloride with various *Bacillus cereus* combinations. Zhao and Oosterhuis (1998a) showed an increase in leaf stomatal conductance and net photosynthetic rate of field-grown cotton treated with mepiquat chloride or a mepiquat chloride-*Bacillus cereus* combination (MepPlus). However, these authors also showed that Pix Plus improved dry matter partitioning to the fruit compared to mepiquat chloride alone (Zhao and Oosterhuis, 2000). Wells and Edmisten (1998) did not find any significant differences in canopy photosynthesis between cotton treated with mepiquat chloride-*Bacillus cereus* combinations compared to mepiquat chloride alone, but they reported higher reproductive-to-vegetative dry weight ratios in various mepiquat chloride-*Bacillus cereus* combinations compared to a mepiquat chloride treatment.

3.7   Pest Resistance and PGRs

Plant growth regulators may be effective in reducing pest populations because they alter the morphological and biochemical characteristics of cotton (Graham *et al.*, 1987). As discussed earlier, PGRs have been used both early and late in the growing season to remove vegetative and reproductive parts, and removal of plant parts deprives the insects of a food supply and a habitat. In addition, the application of PGRs may also increase the biosynthesis of compounds, including secondary plant metabolites that are detrimental to pests. Zummo *et al.* (1984) reported that mepiquat chloride applied at rates used to control excessive vegetative growth increased resistance to bollworm [*Heliothis zea* (Boddie)] damage in cotton. The increased resistance was attributed, in part, to increased tannin and terpenoid production after the mepiquat chloride treatment. Mulrooney *et al.* (1985), on the other hand, concluded from larval growth studies in the laboratory that mepiquat chloride does not enhance cotton's resistance to second stage tobacco budworm (*Heliothis virescens* F.). Instead, mepiquat chloride may actually increase larval growth and decrease cotton's natural resistance in an ideal growing season. Growth rates of second and third stage tobacco budworm larvae increased slightly when grown on leaves treated with mepiquat chloride at either the recommended or twice the recommended rate. Jenkins *et al.* (1987) and Graham *et al.* (1987) did not believe the changes in allelochemicals from mepiquat chloride treatments were sufficient to increase the natural resistance to tobacco budworm. In their studies, gossypol was significantly increased in squares and leaves in two of three years, but was offset by a decrease in anthocyanin and flavonoid concentration. Mepiquat chloride treatment alone or with a commercial cytokinin preparation significantly increased gossypol and one or more of the other allelochemicals in cotton (Hedin and McCarty, 1991). Mepiquat chloride-treated plants expressed a 10 to 13% increase in gossypol content (McCarty and Hedin, 1994). Subsequent studies with natural bioregulators (kinetin, kinetin riboside, indole-3-acetic acid, and gibberellic acid) did not appear effective for increasing yield or allelochemicals (Hedin and McCarty, 1994a).

3.8   Crop Termination/Harvest Aids

Plant growth regulators are used at the end of the growing season for different purposes, including chemical termination. Chemical termination refers to the technique of applying certain PGRs to terminate plant fruiting, remove late-season green bolls, and reduce the number of diapausing pink bollworm [*Pectinophora gossypiella* (Saunders)] and bollworm/budworm (*Heliothis* spp.) larvae that overwinter (Bariola *et al.*, 1976; Thomas *et al.*, 1979; Bariola *et al.*, 1990). Chemical termination combined with early irrigation cutoff has produced acceptable yields and reduced bollworm populations when applications were properly timed (Bariola *et al.*, 1981). Ethephon and thidiazuron are two chemicals that are effective in terminating plant fruiting (Hopkins and Moore, 1980; Bariola *et al.*, 1986; Jones *et al.*, 1990). Eleven varieties of Upland cotton grown in a short season were evaluated for their response to ethephon and thidiazuron applied as chemical terminators (Bariola and Chu, 1988). Yield was unaffected by chemical termination treatments, and all treatments significantly reduced the number of green bolls remaining at harvest. Leaf shed was significantly greater in treatments with thidiazuron or thidiazuron plus ethephon than in the untreated check or those treated with ethephon alone. Kittock *et al.* (1973) reported that ethephon caused an 87 to 96% decrease in diapausing pink bollworm larvae [*Pectinophora gossypiella*] (Saunders) by reducing the number of green bolls remaining after harvest by more than 90%. The rates used in this study, however, were excessive, and declines in yield and quality were too severe to warrant further consideration. Aside from reducing overwintering worm populations, harvest aids used as chemical terminators also reduce the food supply for overwintering boll weevils (*Anthonomus grandis* Boheman). In the Rolling Plains of Texas, untreated plants contained 18,000 to 20,000 squares and small green bolls per acre following regrowth after termination caused by a freeze. Approximately 12,000 bolls per acre in the untreated plots were punctured by weevils. Plants treated with the harvest aid Ginstar contained significantly lower regrowth squares and green bolls than the control, and the food supply for overwintering boll weevils was depleted (Clark *et al.*, 1996). Montandon *et*

al. (1994) also observed a significant reduction in the number of late-season squares damaged by boll weevils.

A second use of PGRs at the end of the season involves boll opening and preconditioning the plant for defoliation. The indeterminate growth habit of cotton often forces producers to harvest more than once or to postpone harvest for several weeks. Ethephon, 2-chloroethyl phosphonic acid, breaks down to ethylene inside plant cells. Thus, application of ethephon causes the concentration of ethylene to increase inside bolls, ethylene stimulates cellulase and other hydrolase enzymes which weaken and dissolve cell walls (Abeles, 1969). A buildup of internal pressure causes carpels to split apart. The carpels immediately start to dry and fold backwards, allowing bolls to open naturally. Ethephon causes immature bolls to open, and this allows for a greater percentage of the crop to be harvested during the first harvest or in a once-over harvest (Cothren, 1980; Weir and Gaggero, 1982; Sawan *et al.*, 1984; Smith *et al.*, 1986; Scott, 1990). Dunster *et al.* (1980) reported that rates of ethephon from 1.12 to 2.24 kg a.i. ha^{-1} applied when 20 to 60% of the mature bolls were opened, consistently caused the unopened bolls to dehisce. Effects of ethephon on fiber quality and yield are inconsistent. Ethephon effected a greater percent first harvest, but resulted in reduced micronaire, boll weight, and seed weight in the bolls that were unopened at time of treatment (Cathey and Luckett, 1980). In Louisiana, Crawford (1980) reported that ethephon did not affect seed cotton yields, but ethephon did reduce micronaire in approximately the last 10% of the total yield when a significant acceleration of boll opening occurred. In Arkansas, ethephon applied to cotton with 12 to 25% opened bolls did not reduce seed cotton yields relative to application to cotton with 48 to 72% opened bolls (Smith *et al.*, 1986); no consistent detrimental effects on fiber quality were detected for fiber collected from the first harvest or from a once-over harvest. Williford (1992) reported that ethephon used to accelerate boll opening significantly reduced yield and grade if applied at the 40% or 60% open stage, but no detrimental effect on yield or grade was observed if ethephon was applied at the 80% open stage.

Research reports have indicated that PGR-IV can be used as a mid-to-late season application to accelerate boll growth for earlier boll opening, with increased yield (Hickey and Landivar, 1997; Oosterhuis *et al.*, 1997). These authors suggested the use of PGR-IV as remedial application applied mid- to late-flowering to help compensate for excessive shedding of fruit earlier in the season through accelerated growth, enhanced boll retention, earlier maturity.

Vegetative regrowth frequently presents problems at the end of the growing season, especially following chemical defoliation. Terminal and axillary buds of actively growing plants are often activated following defoliation. These new juvenile leaves are less responsive to defoliation treatments and can cause problems in picking efficiency and green staining of lint. Glyphosate (N-phosphonomethyl glycine) was shown to suppress regrowth development for as long as seven weeks when used alone or in combination with a de-foliant (Cathey and Barry, 1977). Effectiveness of defoliant chemicals was also enhanced with glyphosate, but deleterious effects on seeds from immature bolls were found at the rates used in this study. The use of a sub-lethal rate (0.50 pt/A, 114 g ai/A) of glyphosate was recently shown to suppress regrowth for as long as 55 days (or more) after application at 10% open boll without a significant effect on lint yield or fiber quality (Landivar *et al.*, 1994). The cotton defoliant thidiazuron shows excellent regrowth suppression (Taylor, 1981).

4.   SUMMARY

A review of the recent literature indicates that using plant growth regulators in cotton production remains a viable option for effectively modifying plant growth and development. Success with growth retardants, yield enhancers, and crop terminating compounds makes managing the crop an easier task. Suppressing excessive vegetative growth allows for better control of insects and assists with harvest. Identification of compounds that enhance photosynthetic activity and greater partitioning of carbon to developing fruit remains as one of the primary focuses of growth regulator research. However, compounds that increase root activity, ion uptake, and water use efficiency are equally important. Consistency of PGR performance is complicated because these compounds interact with the heritable characteristics of the crop and the environment.

Considerable interest exists for the use of plant growth regulators (PGRs) in cotton production. Plant growth regulators are organic compounds, that affect physiological processes of plants when applied in small concentrations. These compounds represent diverse chemistries and modes of action, and provide numerous possibilities for altering crop growth and development. Their time of use extends from early season when they are applied either in-furrow or as seed treatments during planting to late season in preparation for harvest. Overall benefits from plant growth regulator use in cotton include yield enhancement, improved fiber quality, and greater ease of harvest. More specific benefits include alteration of carbon partitioning, greater root:shoot ratios, enhanced photosynthesis, altered nutrient uptake, improved water status, and altered crop canopy. Plant response to growth regulators reflects the interaction of heritable characteristics, cultural inputs, and the environment. Because of this complex interaction, crop response to PGRs is not always predictable. Techniques have been developed to monitor the growth and development of the crop, with specific emphasis on the fruiting characteristics. One such technique, plant mapping, provides detailed information on fruiting rates and potential, fruit retention, and distribution of fruit set relative to PGR treatment. Since over 80% of the yield is produced on first-position fruiting sites, retention and maturation of first-position bolls is critical. Increased boll retention at the early fruiting sites hastens crop maturity, allows quicker harvesting, and improves lint quality. In summary, strategies for using PGRs in cotton production include numerous options for beneficially modifying crop response in an effort to improve yield and manageability of the crop.

Chapter 27

PHYSIOLOGICAL RATIONALES IN PLANT MONITORING AND MAPPING

Thomas A. Kerby[1], Fred M. Bourland[2], and Kater D. Hake[1]
[1]Delta and Pine Land Company, Scott, Mississippi, USA; [2]University of Arkansas, Keiser, Arkansas, USA

1.   INTRODUCTION

Cotton is an indeterminate crop that produces both vegetative and reproductive growth at the same time. Yield, earliness, and quality factors can be influenced by the balance between the two sinks. As a result, growth, development, and yield are responsive to changes in environment, and management adjustments must be designed to optimize the environment. Cotton management expertise has historically been an art acquired through years of experience. Skilled managers anticipate how the crop will respond to changing environments. Cotton monitoring and mapping provide a means to quantify growth and development and increase the precision of management decisions. It offers a tool to take cotton management from an art to a science.

Plant monitoring and plant mapping are frequently used interchangeably. However, plant mapping specifically refers to the recording and evaluating of plant structure and the distribution and retention of fruit on plants at a specific time, whereas plant monitoring refers to the evaluation of growth and development of the plant (Oosterhuis et al., 1996). Plant height, number of nodes, percent retention of bolls in specific positions or groups of positions are commonly used plant mapping parameters. Examples of plant monitoring parameters include various vigor indices, nodes above white flower, and nodes above cracked boll. The terms are sometimes ambiguous because some techniques include both plant mapping and plant monitoring information, and plant mapping parameters can be used to monitor plants when data are evaluated over time. Also, when used in a general sense, the term "monitoring" (as opposed to "plant monitoring") may refer to collection of data from any of a number of diagnosis techniques, including plant monitoring, plant mapping, pest species/density, tissue testing, soil moisture, etc.

No single formula exists to optimize management. Optimization must take into account year, location, and management constraints. While general guidelines can be developed that describe normal development, optimal development must be calibrated for the specific conditions of a field in a given year. Optimal plant map parameters will not be the same for the short-season management areas of Missouri and Tennessee compared to full-season areas of south Alabama and Georgia. Even within the same growing area, management adjustments should be made based upon timeliness of the crop planting and favorableness of early-season weather for growth (i.e., changes in the effective length of the growing season).

Plant development follows an orderly pattern (Oosterhuis, 1990). Production of early nodes is primarily under the constraint of temperature. The onset of fruit development is determined by the node of the first sympodia, which, in most cases, occurs between the fourth and seventh node. Following the initiation of sympodia, new squares develop at successive first sympodial positions up the plant at about twice the rate they do between successive squares on the same sympodia. Once a fruit load is present, carbon supply and assimilate competition become the primary constraints.

In commercial production, plant mapping and plant monitoring data are only valuable if they help determine management decisions that will improve yield and/or profitability. Data taken during the season can be used to determine whether developmental rates and fruit retention are within acceptable ranges (Bourland et al., 1997). Along with other information, the cause(s) of deviant patterns can usually be ascertained and corrective action can be suggested. Obviously, data from plant mapping done just prior to harvest cannot help with management decisions within that growing season. The value of these data is to document

J.McD. Stewart et al. (eds.), Physiology of Cotton,
DOI 10.1007/978-90-481-3195-2_27, © Springer Science+Business Media B.V. 2010

growth of a particular field so the effects of production strategies or management inputs can be quantified, which provides the opportunity to learn from the experience.

2. HISTORY OF COTTON PLANT MONITORING AND MAPPING

Generally, development of plant mapping techniques predated the development of plant monitoring in cotton. Cotton plant mapping essentially began with the work of McClelland (1916), Martin *et al.* (1923), and Bailey and Trought (1926) that provided summaries of plastochrons for both sympodial and monopodial branches. Although they may not have considered their work as "plant mapping", their plant diagrams required the collection of information from individual fruiting sites. Their reports provided the framework for the development of modern plant mapping techniques.

Two major restraints to the development of plant mapping were (1) the inability to handle massive sets of data, and (2) the absence of defined variables which could be statistically analyzed. Typically, nodes and fruit from the entire plant (whole-plant mapping) were included on a two-dimensional stick diagram, with each plant mapped on a single page of paper. If mapping were done during the growing season, age (*e.g.*, square, open flower, young boll) of the fruit was coded. To help summarize these whole-plant data, Munro and Farbrother (1969) developed the "composite plant diagram" technique. By including data from numerous plants on a single stick diagram, effects of treatments could be visualized. However, they did not establish variables that could be easily analyzed. Kerby (Kerby and Buxton 1976, 1978, 1981) used a modified system of Munro and Farbrother (1969) to collect developmental sequence data and square and boll retention data by position as influenced by leaf type and plant density. Each plant required a separate sheet of paper to record development data. Labor requirements for collection and reduction of data limited the use of plant monitoring and mapping.

The innovation of the computer has been the catalyst for the advance of cotton plant mapping. Computer programs, such as DIAGRAMMER and BOLOCATE (Kennedy *et al.*, 1990) and PMAP (Landivar, 1993) were developed to facilitate handling data from whole-plant maps. Bourland and Watson (1990) developed COTMAP, a modified whole-plant mapping program which uses a simple code to map fruit in prime fruiting positions (closest two position from the main axis) are mapped, then number of fruit in other positions are accumulated by category. Up to eight plant structure variables and nine yield-related variables can be analyzed using the COTMAP data. Plant and Kerby (1995) used the same code as devised for COTMAP in their CPM (Cotton Plant Mapper) program. Data from CPM (primarily for end-of-season plant mapping) and CCM (California Cotton Manager, primarily for in-season plant mapping) can be entered into the CottonPro software to attain full

analytical and archival use (Plant and Bernheim, 1996). With its graphics and variable options, the CottonPro programs are better suited than COTMAP as a management tool in production fields. Each of these computer programs has greatly increased the ability to summarize mapping data from multiple plants with respect to time and nodal position.

Cotton plant monitoring techniques usually address either early-season or late-season development of the plant. Several vigor indices (*e.g.*, height-to-node ratio, average length of top five internodes, and growth rate) have been developed to evaluate the rate of vegetative growth in early-season (Silvertooth *et al.*, 1996; Kerby *et al.* 1997). Typically, these indices focus on internode expansion by examining changes in plant height with respect to number of main-stem nodes. Recently, University of Arkansas researchers developed the SQUAREMAP technique which can be used to collect first-position square retention data in addition to plant vigor parameters (Slaymaker *et al.*, 1995). SQUAREMAP data are analyzed by the SQUAREMAN computer program, which generates several vigor indices, compares nodal development to a target development curve, and evaluates square sheds.

The most widely used and accepted parameter for plant monitoring after flowering is nodes-above-white-flower (NAWF). NAWF measures the maturity of plants by the progress of first-position white flowers to the plant terminal. Waddle (1974) was perhaps the first to use this concept. Among several mid-summer measurements of earliness, he noted that "number of nodes above the last white flower" was the "quickest and easiest count to make." Furthermore, he indicated that it was a good earliness indicator, and that it had a degree of genetic control. Unfortunately, the NAWF concept received little attention until the late 1980's with simultaneous work in Arkansas and California. The NAWF parameter became the nucleus of the BOLLMAN expert system which is primarily used to assist with end-of-season decisions including insecticide termination, timing-of-defoliation, and sequencing fields for harvest (Zhang *et al.*, 1994). Together, BOLLMAN and SQUAREMAN form the COTMAN system for monitoring plants from first square to cutout (Cochran *et al.*, 1996). Another late-season plant monitoring technique is nodes-above-cracked-boll (NACB), which like NAWF, can be used to time the application of defoliants (Kerby *et al.*, 1992b).

Using plant mapping and plant monitoring data for calibration, cotton crop simulation models have been developed. GOSSYM/COMAX was designed as a complete physiologically based simulation model that generates expected fruiting profiles based upon soil, weather, and plant information (McKinion *et al.*, 1989). In contrast to plant mapping and plant monitoring techniques, which measure plant responses that have occurred, simulation models attempt to predict how the plant will respond. Hopefully, plant mapping and plant monitoring programs can be more closely linked with simulation programs to provide simultaneous reading and projecting of plant responses.

Using data from plant mapping and plant monitoring techniques, many researchers have contributed greatly to our knowledge of cotton plant growth and development. In 1989, BASF Corporation provided a grant to The Cotton Foundation that provided funding for The Cotton Physiology Program as a function of the Technical Services Department of the National Cotton Council of America. This has provided a mechanism for interested parties from across the Cotton Belt to share techniques and concepts, coordinate data collection and standardize terms.

3.   PLANT MONITORING TERMS AND CONCEPTS

3.1   Plant Height

Plant height is usually measured as the distance from the cotyledons to the growing point (terminal mass) on the main-stem. In many instances the terminal leaf is above (taller than) the terminal mass, but this is not the correct reference for plant height. Average height and number of nodes should be determined from at least 20 plants (*e.g.*,5 plants from 4 locations in the field). Sample size should increase for large fields, or in fields with a high degree of variation. If areas of a field are managed differently, height measurements should be made according to the area of a field that receives unique management.

In fields where plant stresses are at a minimum, growth in height between early square to early flower averaged 2.5 cm for each 20.6 DD_{60} (Kerby *et al.*, 1987). Maximum growth per day occurred at first flower with an average of 2.7 cm per day (Kerby and Hake, 1996). Data from Tharp (1960) indicated a value of 2.8 cm per day.

Marani and Ephrath (1985) investigated light penetration into cotton canopies. They reported a close association between plant height development and leaf area index (LAI) development ($R^2 = 0.94$). LAI is a ratio of leaf area per unit land area (m^2 leaf surface/m^2 land area). Radiation penetration through the cotton canopy was also found to be highly associated with LAI ($R^2 = 0.83$) and plant height ($R^2 = 0.90$). Since height is easier and more accurately measured than LAI, they concluded that it is the better predictor of radiation penetration. Development of plant height with its associated leaf area development is an important factor affecting canopy photosynthesis (Baker and Meyer, 1966). Net assimilation rate (NAR) was 97% of the maximum rate, for an average plant stand of 8.2 plants m^{-2} grown under 1.0 m row spacings, with a LAI of 2.77 and plant height of 83 cm (Kerby *et al.*, 1987).

3.2   Number of Nodes on the Main-stem

Cotyledonary nodes occur opposite each other on the main-stem and are not counted as a node (counted as node zero). A new node is counted when a new leaf at the main-stem terminal either has unfolded or reached 2.5 cm in diameter. During formation of the first several nodes, terminal leaves are small and may not reach this size. Until the plant has six nodes, a node is counted if the leaf has unfolded.

Main-stem nodes are detectable on the cotton plant soon after emergence and continue to appear throughout the period of vegetative growth (Mauney, 1986a). During the early season (prior to anthesis), vegetative growth consisting of root and shoot development and leaf growth provides the dominant carbohydrate sink (Oosterhuis, 1990). Hesketh *et al.* (1972) found that the rate of nodal development in cotton broadly supports the concept of temperature-based physiological time. Early-season node development is highly linearly related ($R^2 = 0.95$) to DD_{60} accumulation (Kerby *et al.*, 1997). From emergence until about the 15th node, a new node is produced on average every 43 DD_{60}. This relationship between number of main-stem nodes and accumulated degree days, prior to assimilate competition by bolls and under irrigated conditions, is stable across a wide range of field conditions. After anthesis, assimilate competition by developing bolls influences both node growth rates and the final number of nodes (Kerby *et al.*, 1987; Kerby *et al.*, 1997).

Node development between square initiation to early flowering in relation to days after planting are reported in Figure 27-1. Data for the San Joaquin Valley (SJV) of California were regressed from mean data presented by Kerby *et al.*, 1987 (R^2 not possible). The SJV requires more days to develop the first six nodes due to the large diurnal fluctuation in temperatures with low DD_{60} accumulation per day during the early portion of the season (Kerby *et al.*, 1987). Node number as related to days after planting for picker cotton in the mid-south and southeast (actually Texas to Virginia) are represented by the mean of all varieties at a location regressed over years and locations (n=425). Number of days per node between nodes 6 and 16 averaged about 3.5 for both sets of data (Fig. 27-1). Assimilate competition by bolls begins to slow rate of node development near the time of first flower (Kerby *et al.*, 1987; Kerby *et al.*, 1997).

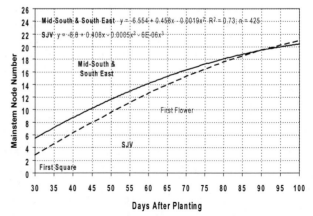

Figure 27-1. Node development as related to days after planting. SJV data from Kerby *et al.*, 1987. Data for the Mid-South and Southeast was collected from over 313 locations between 1994 to 2000.

Mauney (1986) reviewed the literature regarding the number of days between successive first-position flowers up the main-stem (vertical flowering interval or VFI) and between successive flowers on the same sympodium (horizontal flowering interval or HFI). He reported VFI was between 2.2 and 4.0 days and HFI was between 5.8 and 8.5 days. Plants in Arizona required an average of 2.7 days VFI at early flower compared to 3.5 days three weeks later (Kerby and Buxton, 1978). After peak flower, HFI increased from a minimum of 7.4 days to 8.1 days. Within the COTMAN system, the target development curves assumes a VFI of 2.7 days between first square and first flower (Danforth *et al.*, 1995). This assumption is based on the sequential development of new fruiting branches as defined by Tharp (1960). Under controlled environmental conditions, VFI was 5 days at an average temperature of 20°C, but only 2 days at 35°C (Reddy *et al.*, 1997c). HFI averaged 9 days at 20°C and 5 days at 35°C. Differences in VFI and HVI are undoubtedly related to differences in DD_{60} accumulation per day for the typical planting dates of the region. Node development represents the physiological clock during vegetative development.

3.3   Height-to-Node Ratio

Medical doctors use weight and height of a child as related to age to help determine if development is normal. A similar concept has been developed for evaluating normal development in cotton (Kerby *et al.*, 1997). Number of main-stem nodes developed is indicative of plant age. To determine if the plant is the proper "size" (plant height) for its age (number of nodes), height is divided by number of nodes. This is the height-to-node ratio (HNR). Typical HNR by growth stage has been reported for California, Arizona, and Texas (Silvertooth *et al.*, 1996).

HNR is sensitive to a number of factors including soil temperature early in the season. HNR for nodes developed prior to squaring will reflect spring temperatures more than management decisions. Above normal temperatures increase the value of HNR while cool growing conditions will decrease the value. Prior to seven nodes, a low HNR value may not limit yield potential because the important leaves that will supply the majority of assimilates to bolls are not yet developed (Kerby *et al.*, 1992a). Changes in HNR following early squaring become more important because the health and size of these leaves is related to assimilate supply to these important bolls.

Some studies suggest the rate of nodal development in cotton is less affected by environmental stress than is the rate of growth in height. Kerby and Keeley (1987) conducted leaf removal experiments involving 20 treatment combinations of removal of one or both cotyledons, removal of zero, one, or two first true leaves, and removal at 0, 5, 10, 15, or 20 days after emergence. Plants were harvested approximately 50 days after emergence and plant height, number of main-stem nodes, leaf area, leaf weight, and to-

tal dry weight were measured. Eight removal combinations significantly affected plant height, whereas only 2 removal combinations, namely removal of both cotyledons at 0 or 5 days after emergence, significantly affected number of main-stem nodes. These observations were confirmed in the studies published by Longer and Oosterhuis (1999).

Since number of nodes is closely associated with physiological age (temperature) and height is influenced by stress factors sooner and to a greater extent than number of main-stem nodes, HNR should provide an early season quantification of stress level. The effect of water stress, use of mepiquat chloride, and cultivar on HNR was reported by Kerby *et al.* (1998).

Different HNR for cultivars grown side-by-side indicate that the cultivars do not respond identically to an environment. It is also a common observation that the HNR for a cultivar depends on the environment where it is grown. Kerby *et al.* (1997) presented data where two cultivars with an 18% difference in HNR had a differential response to management. However, the appropriate management could be predicted by knowing the HNR, independently of cultivar. HNR values for a specific field in a given year should dictate management, not average HNR for the cultivar. A short early-fruiting cultivar will have a low HNR and requires management practices that do not limit growth. Conversely, a vigorous full-season cultivar, under the same growing conditions, will have a high HNR and may require management that controls vegetative growth. Thus, if managers make decisions based on HNR, optimum management would be indicated by the value itself. HNR properly recommends cultural practices to fit the cultivar and growing conditions of a specific field and a separate index for different cultivars should not be necessary. Optimum HNR should vary by region and year depending upon the length of time to grow the crop and expected weather factors that will influence HNR.

For 104 fields mapped between 1982 and 1991 in the San Joaquin Valley, the highest yielding 20 fields averaged 2017 kg lint ha^{-1}, produced 22.4 nodes, and had a final HNR of 4.45 cm (Kerby and Hake, 1996). For 90 fields mapped in Australia during the 1992/1993 and 1993/1994 growing season, the highest yielding 25 fields averaged 1887 kg lint ha^{-1}, produced 22.8 nodes, and had a final HNR of 4.32 (unpublished data, Kerby, 1994).

The HNR value represents the average value of height divided by all of the nodes developed to date. As such, this integrates the entire season up to the time of a measurement. This limits the usefulness of HNR because the manager needs to know how the crop is changing. It is possible to get an impression of change in growth by observing how HNR, taken at multiple sample dates, compares to regionally developed normal values. Only the terminal 4 to 5 nodes elongate. When subsequent observations parallel the regional HNR curve through time, this indicates the growth rate of new nodes is equal to the rate of the reference. If the observed slope is greater (steeper) than the reference curve, current growth rate is greater than the reference. This could suggest management decisions designed to slow the rate of vegetative development.

3.4　　Growth Rate

If plant height and node development data are taken every 7 to 10 days, a field growth rate (GR) can be established. The change in height between two sample dates divided by the change in node number for the two sample dates provides a measure of current growth rate. For example, assume plants averaged 50 cm at 13 nodes, but 10 days later they were 74 cm with 16 nodes. Height changed 24 cm while nodes changed by 3, therefore growth rate is 24 divided by 3, or 8 cm per new node. Data were collected in this manner for 13 tests over a four year period in California (Kerby and Hake, 1996). Figure 27-2 represents the average change in height per node for data collected on a weekly basis. The average growth rate of 8.0 in the example represents the "y" value. The value 8.0 is plotted against the "x" value or mid-point node number (the average of the two sample dates or 14.5).

The growth rates in Figure 27-2 represent a crop that is growing with minimum environmental stresses and has an accumulating boll load to help limit growth. Yield averaged 1478 kg lint ha[-1] in these tests. This growth rate may not necessarily be the ideal growth rate , but it represents the potential for an indeterminate cultivar (Acala SJ-2) to grow in an environment that is conducive to growth. In fact, for most conditions, the growth rates in Figure 27-2 would be greater than desired. The optimal value would vary by the length of the growing season, fruit retention, plant density, row spacing, row configuration (skip row or solid plant), nitrogen supply late in the season, and probability for excess late-season moisture.

Growth rates calculated in this manner require a minimum of two sample dates. This places limitations on field use of the technique. Additionally, there is variation associated with measurements on both sample dates.

If the time interval between sample dates is small, great variation can be noted in calculated growth rates. A single time measure of growth rate would be desirable.

Landivar *et al.* (1996) evaluated the elongation of nodes from the node with the uppermost fully unrolled leaf to nodes with the first-position flower one week post anthesis. They noted internode elongation was complete in 12 to 15 days depending upon temperature. Reddy *et al.* (1997c), under controlled environments, indicated it required 12 days for the terminal node to be fully elongated at a temperature that averaged 32°C and 16 days at 20°C. Kerby (unpublished data, 1993) measured average internode distance between each node for nodes 10 to 5 from the plant apex. The uppermost node was defined as one having a leaf with minimum size of 2.5 cm, and was counted as node 1. To determine the relative contributions of elongation rates between specific internodes to the overall growth rate, internode distance was measured for the same plants at node 10 and at node 12. The average internode distance between nodes 10, 9, 8, 7, 6, and 5 at 12 main-stem nodes was respectively 193, 150, 125, 106, and 100% of the internode distance measured when these same plants had 10 main-stem nodes. The distance between the fourth and fifth node below the terminal at main-stem node 10, counting the terminal node as 1, (distance between main-stem node 6 and 7) was 94% of the distance between the same nodes measured later when plants had 12 main-stem nodes. There was no change in internode length between the two dates for nodes that were five or six nodes below the terminal at 10 main-stem nodes. Thus, elongation (increase in plant height) takes place in the terminal node and the four internodes below the terminal. Assuming 3 days per node and normal temperatures, internodes would elongate for 12 days. Landivar *et al.* (1996) and Reddy *et al.* (1997c) measured elongation of internodes at the time of flowering. Their results confirm internode elongation occurs in the terminal node and three to four nodes below the main-stem terminal node. This reference is referred to as the maximum internode distance (MID) or internode distance between the fourth and fifth node below the terminal node. which has a leaf that has expanded to at least to 2.5 cm diameter and the terminal node is counted as 1.

Height of the terminal five nodes (referred to as ALT 5 by Landivar *et al.*, 1996) and MID provide a growth rate measure with a single measure in time. Determination of average internode distance for all nodes taken at the end of the season provides a history of the crop growth rate through time.

Growth rate provides a quantitative measure of assimilate supply to sustain vegetative growth as well as presence of vegetative growth stress factors. Growth rates considered along with fruit retention, crop stage of growth, remaining season, and potential for sustained growth (soil moisture and nitrogen) provides useful information in making management decisions. Reference values are needed for all cotton growing regions. The implications of growth rate in management decisions such as use of mepiquat chloride will be discussed later in the chapter.

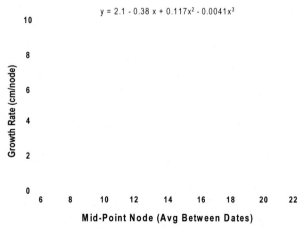

$$y = 2.1 - 0.38 x + 0.117x^2 - 0.0041x^3$$

Figure 27-2. Expected growth rate of Acala SJ-2 at various stages of development; average of 13 locations between 1982 and 1989 under favorable conditions. Taken from Figure 24.12 of Kerby and Hake, 1996.

3.5 Node of the First Sympodium

Determination of the first sympodium is most easily and accurately determined early in the squaring season. Monopodial branches produce squares that are approximately 10 days younger than the first sympodial square (Kerby *et al.*, 1987). With a moderate to high plant density, vegetative branches are small and barren, and often abscise. The first sympodial branch is normally small and less developed than subsequent sympodia. Later in the season it can be difficult to tell the difference between a small monopodial branch and a small sympodial branch that only produced one fruiting position and that position aborted.

Node of the first sympodium should be determined for a minimum of five plants selected from four areas of the field, and recorded as nodes above the cotyledons (counted as zero) to the node with a sympodial branch. An easy way to tell the difference between a monopodial and sympodial branch is that sympodia always have a leaf and fruiting position at the first node (Mauney, 1986; Oosterhuis and Jernstedt, 1999). These are approximately 180 degrees opposite each other at the same distance from the main-stem. A new node on a sympodium initiates from an axillary bud and thus produces a typical zig-zag pattern. Monopodial branches have a pattern of development similar to the main-stem. They seldom have anything more than a leaf at the first node on a monopodial branch and, if the branch is well developed, sympodia can be initiated at subsequent nodes as is the case for the main-stem. However, these sympodia (off from monopodial branches) are weak and generally produce only one fruiting position each.

Kerby *et al.* (1990b) demonstrated that the capacity for a crop to sustain vegetative growth is related to fruit growth rates and the leaf area expansion rate between node 10 and 18. Node of the first sympodium, when combined with early square retention, will determine the leaf area produced at the time when vegetative and reproductive sink begin competition for assimilates. A more determinate growth pattern can be expected when the first sympodium is developed at a low node and fruit retention is normal causing reproductive assimilate competition to occur at an earlier stage of growth (and at lower LAI).

The number of nodes produced before the first sympodial branch is formed depends on the cultivar and the environment. Plant map data have been collected for many commercial cultivars at 140 locations over the period 1994 to 1996 (Kerby, unpublished data, 1996). Node of the first sympodium averaged 5.7 for 'Sure-Grow 125' (lowest) and 6.4 for 'Deltapine 90' (highest). While these differences due to cultivar can be important, environment can have an equal or greater influence.

Factors affecting formation of early floral buds were summarized by Mauney (1986). Differentiation of the first floral bud begins 10 to 14 days after emergence when plants have two to three leaves. At controlled temperatures, node of the first sympodium averaged 6 at 24°C compared to 10 at 31°C. Kerby (unpublished data, 1996) noted high values for the node of the first sympodium under late planted cotton in the sandy soils of southern Georgia in 1996. Soil and air temperatures were much higher than normal. Bourland and White (1986) also reported a higher node of the first fruiting branch for late-planted cotton at Mississippi State in 1983 and 1984. This suggests that when soil and average air temperatures are above 24°C during the two to three leaf stage, node of the first sympodium will be higher than normal.

Soil temperatures and average daily air temperatures are normally well below 24°C during the two to three leaf stage. Kerby and Hake (1996) demonstrated when temperatures are low during the two to three leaf stage (15°C), node of the first fruiting branch is higher than when temperatures were warm (21°C). Apparently, there is an optimum temperature range that reduces the node of the first sympodium. Lower or higher temperatures raise the node of the first sympodium. Quantitative controlled environment data are lacking to describe the optimum range as well as the rate at which node of the first sympodium increases in response to either lower or higher than optimal temperatures.

3.6 Retention in the Top Five Sympodial Positions

Early in the season, assimilate supply is generally not a limiting factor in square retention. Constable and Rawson (1980b) demonstrated that squares produce nearly 50% of their assimilate requirements from photosynthesis of their bracts. A recent study by Zhao and Oosterhuis (1999) showed bracts produced 56% of the squares assimilate. Kerby *et al.* (1987) demonstrated that only 2% of the net assimilation rate (NAR) at first flower is allocated to fruiting structures. Small square shed in the terminal five nodes can mostly be attributed to pest damage until the late flowering period when assimilate demand by bolls exceeds NAR (Wullschleger and Oosterhuis, 1990d; Kerby *et al.*, 1987).

To determine average retention in the top five sympodial positions, select five plants from each of four areas of a field. Record the number of surviving first sympodial position fruit in the terminal five nodes and divide this by the number of first sympodial positions in the top five nodes. Less than 90% survival of these young squares early in the season suggests some level of pest damage. Additional information should be taken to determine if losses are enough to initiate control measures and to determine the cause for the loss.

Retention in the top five first sympodial positions provides an early evaluation of square production and survival. Previous studies indicate Lygus bugs (*Lygus hesperus* Knight) prefer to feed on young squares (Leigh *et al.*, 1988). A square of the same age on any sympodial or monopodial position is equally attractive to Lygus for feeding. However, not all squares on the plant have equal value in contributing to yield (Kerby *et al.*, 1987; Jenkins *et al.*, 1990a; Constable, 1991). Fruit at first sympodial positions on nodes 7 to 16

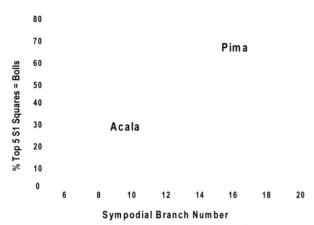

Figure 27-3. Percentage of the top five first sympodial positions (S1) that produce harvestable bolls, as related to sympodial branch number. Taken from Figure 24.13 of Kerby and Hake, 1996.

are the most important to yield and fiber quality. Square monitoring should focus on these important positions.

Dynamic thresholds for Lygus control based upon retention of the top five first sympodial positions have been established for California (Kerby and Hake, 1996). California thresholds were established by averaging first sympodial position retention for the top 20 yielding fields from 124 map data sets from 1981 through 1991 (Fig. 27-3). This represents the probability that the positions will produce harvestable bolls. A working threshold was established by adding 10% to the value in Figure 27-3. This 10% represents the average decline in retention of first sympodial positions (S1) from early square until harvest for California conditions. The top 5 retention value listed by sympodial branch number is the average retention of that sympodial branch and the previous four. As the season progresses, the value of the top five first sympodial squares diminishes. The threshold values for protection of squares should be based upon boll load, length of season remaining, and the ability of the plant to set and mature additional bolls.

Australia uses a threshold of 60% first sympodial position retention prior to anthesis (personal communication, Forrester 1996). This threshold represents the minimal value that did not delay maturity, and provided the greatest opportunity to avoid early-season beneficial pest disruption.

Danforth *et al.* (1995), using SQUAREMAP, demonstrated a 1.45 day delay in cutout for each one percent decline in average first sympodial position retention at first flower. Consultants in the U.S. use a wide range of top five first sympodial position thresholds. It is expected that square retention thresholds should vary according to the possibility for boll loss to pests, length of season a region has to produce a crop, late-season harvest conditions, growth potential of the region (summer rainfall patterns), and pest disruption issues. Additional data in the different cotton growing regions would improve the value of top five first sympodial position retention in management decisions.

3.7 Retention in the Bottom Five First Sympodial Positions

Until the plant has more than 5 sympodial branches, retention of the bottom five is the same as retention for the top five first sympodial positions. This zone represents the oldest fruiting forms and provides an early indication of the reproductive sink strength. When the plant is nearing the first flower stage (9 to 10 sympodia), squares in the bottom five first sympodial positions are larger and less attractive to Lygus feeding (Leigh *et al.*, 1988), but they may be very attractive to worm feeding damage. During the first two weeks of flowering when assimilate supply should not limit boll retention, about 90% of the first sympodial positions that reach anthesis can set a boll. Large squares at this early stage of development generally do not shed unless pests are involved.

Retention percentage on the bottom five sympodia at early flower is indicative of the early boll load. This information should be used to help make adjustments near the time of early flowering in irrigation scheduling, fertilizer requirements, and the value of mepiquat chloride. Once bolls reach 10 days of age, it is rare for them to shed (Guinn, 1982a). There is no reason to continue to collect data for this plant zone later in the season as the number of retained bolls should not vary. The plant has already made significant assimilate investment in the bolls (Wullschleger and Oosterhuis, 1990d), and other plant monitoring data provides more insight into the plant's response to its environment.

3.8 Number of Squaring Nodes

At the time of first flower, only 2% of measured above ground dry matter increase was associated with fruit growth (Kerby *et al.*, 1987). When plant stresses are not present, node development rate averaged one node per 43 DD_{60} between first square and first flower (Kerby *et al.*, 1997). In the San Joaquin Valley of California, with wide diurnal temperature fluctuations, 43 DD_{60} per node translated into an average of 3.5 days per new node (Fig. 27-1). During the early weeks of flowering, first-position flowers appeared at an average rate of 2.7 days per node in Arizona (Kerby and Buxton, 1978).

If plant stresses do not limit assimilate production prior to anthesis, temperature should be the limiting factor in the rate of node production. Node development under normal field conditions from early square to early flower was closely associated with DD_{60} accumulation (Kerby *et al.*, 1997).

The number of sympodia developed with the passage of time (days or DD_{60}) can serve as an early indication of normal (or abnormal) development. California used node development per DD_{60} to determine if growth proceeded in a normal way (Kerby *et al.*, 1997). Arkansas developed a term called "number of squaring nodes" (Oosterhuis *et al.*, 1996) that is as helpful as nodes per DD_{60}, but much easier to use. Squaring nodes prior to first flower are the number of sympodia and are equal to nodes above white flower (NAWF) after first flower. The Arkansas data (Oosterhuis *et*

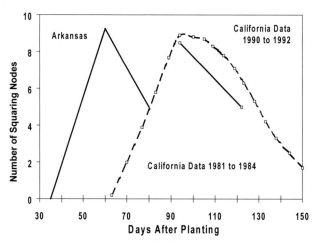

Figure 27-4. Average number of squaring nodes by days after planting for Arkansas (Oosterhuis *et al.*, 1996) and California (Kerby *et al.*, 1987; Kerby *et al.*, 1996).

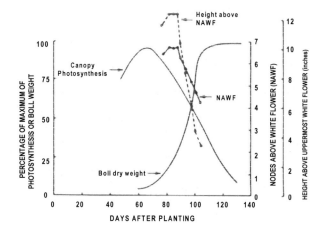

Figure 27-5. Change in canopy photosynthesis and boll dry weight as related to height above the uppermost white flower and number of NAWF with time after planting. Taken from Figure 1 of Bourland *et al.*, 1992.

al., 1996) indicated first square occurs on average at 35 days after planting, new nodes develop at 2.7 days per node, first flower occurs on average at 60 days after planting, and there are 9.25 squaring nodes at first flower. Data for 313 combined locations from 1994 to 2000 from Texas to Virginia indicate an average first square 30 days after planting, new nodes appear at an average rate of 3.5 days per node, and first flower occurs at approximately 63 days after planting (Fig. 27-1). First square and first flower requires more days in California due to fewer DD_{60} per day early in the season.

If fields develop normally, 8 to 10 squaring nodes would be expected at first flower (Benson *et al.*, 1995; Danforth *et al.*, 1995; Kerby *et al.*, 1987, 1996). Kerby *et al.* (1996) presented regression data for 35 tests between 1990 and 1992 to demonstrate that the intercept for NAWF (average number of squaring nodes on the day of first flower) averaged 8.5. Figure 27-4 is a graphic representation of squaring nodes through the season as presented in COTMAN for Arkansas (Oosterhuis *et al.*, 1996) and California for the data of Kerby *et al.* (1987, 1996). Days from planting until squaring averaged 12 more days in California. From first square to first flower required an average of 28 days in California compared to 25 for Arkansas. From first flower to NAWF = 5.0 required 20 days in Arkansas compared to 36 days for California data collected between 1981 and 1984 (Kerby *et al.*, 1987) and 29 days for California data collected between 1990 and 1992 (Kerby *et al.*, 1996). The two California data sets differ because of increased early fruit retention and use of mepiquat chloride in the latter data set.

3.9   Nodes Above White Flower (NAWF)

Bolls have a higher priority for assimilates than vegetative growth (Constable and Rawson, 1980b; Kerby *et al.*, 1987, 1996; Wullschleger and Oosterhuis, 1990d). Once bolls are present, they compete with vegetative growth for assimilates. As allocation to bolls increases with additional boll set, competition increases. If boll set

is high, ultimately boll growth exceeds the canopy NAR (Wullschleger and Oosterhuis, 1990d; Kerby *et al.*, 1997).

Previously produced first sympodial positions reach anthesis at a constant DD_{60} rate. Assimilate deficits for vegetative growth (new node formation) results in first-position flowers moving up the plant faster than new nodes are produced. Thus, changes in NAWF following first flower accurately reflects assimilate supply to vegetative growth. This source-to-sink balance is clearly illustrated in Figure 27-5, where the association between NAWF, boll dry weight accumulation, and canopy photosynthesis can be observed (Bourland *et al.*, 1992). If number of squaring nodes at first flower are lower than 8.0, assimilate supply has limited rate of new node production and some early season stress has been experienced (Benson *et al.*, 1995; Danforth *et al.*, 1995; Kerby *et al.*, 1987, 1996).

The concept of using NAWF to determine the last effective flower was first reported by Bernhardt *et al.* (1986). Research to use NAWF to indicate when physiological cutout occurred was taking place in California and Arkansas at approximately the same time. Neither group was aware of the activity of the other until preliminary results began to be published. Bourland *et al.* (1992) concluded that the number of effective sympodia was determined when NAWF = 5.0. Kerby *et al.* (1996) collected final plant map data from 104 field studies between 1982 and 1991 and described cutout as the sympodial number where 95% of all harvestable bolls had been set (Kerby and Hake, 1996). In 71% of the fields, the last effective sympodia was when NAWF = 5.0. At NAWF = 4.0, 91% of the fields had reached cutout. Pima (*Gossypium barbadense*) cotton had the last effective boll set when NAWF = 3.5. A three year, 4 location study (GA, AR, LA, and VA) confirmed the use of NAWF = 5.0 for cutout, but a lower NAWF (*e.g.*, 4.0) for fields treated with mepiquat chloride or fields with low N (Abaye *et al.*, 2000). Physiological cutout represents the time when assimilate demand by the reproductive sink is equal to the plant assimilate production and additional boll set in minimal.

3.10   Number of Effective Sympodia

Based upon final plant map data, the node and position on the sympodia can be determined for all harvestable bolls. The effective fruiting period is defined as the time required to set 95% of all harvestable bolls. A boll is counted as harvestable from final map data taken near the time of defoliation if the boll is well developed, near maturity, and presumed to open in time for mechanical harvest. In COTMAN, the number of effective sympodia is represented by the number of sympodia between the first sympodium and the sympodium where a first position has a white flower and there are five nodes above this position (NAWF = 5).

Properly managed cotton should cutout. Acala cotton in the San Joaquin Valley averaged 10.7 sympodia for the 95% zone in 104 trials over a 10 year period with average yield of 1575 kg ha^{-1} (Kerby and Hake ,1996). The average 95% zone occurred at main-stem-node 16.3. The highest yielding 20 fields averaged 2017 kg ha^{-1} with a 95% zone at main-stem-node 16.4 compared to 16.0 for the lowest yielding 20 fields which had an average yield of 1100 kg ha^{-1}. This represents an effective flowering period of 37 and 31 days (node of first sympodium was 6.0 for high and 6.9 for low) for the high and low yielding fields. There were only modest differences in plant height, number of nodes, or HNR. Fields with high yield not only had more sympodia (11.4 for high compared to 10.1 for low) for the 95% zone, but this zone retained 62.1% of all first positions in the 95% zone compared to 46.9 for the low yielding fields.

Table 27-1. Contribution of nodes to yield and the approximate date first sympodial positions reach anthesis and an open boll based on long term historical data. Mississippi data was developed from Jenkins *et al.*, 1990a and California data developed from Kerby and Hake, 1996.

Node	% Sympodial yield		Cumulative % yield		~SJV-CA date 1st S1	
	MS	CA	MS	CA	Anthesis	Open boll
21		0.3	100.0	99.8	24 Aug	6 Nov
20	0.2	0.5	100.0	99.5	19 Aug	29 Oct
19	0.5	0.8	99.8	99.1	14 Aug	22 Oct
18	1.2	1.6	99.3	98.2	10 Aug	15 Oct
17	2.3	2.4	98.1	96.7	6 Aug	9 Oct
16	3.5	4.0	95.9	94.3	2 Aug	4 Oct
15	5.5	5.3	92.3	90.3	29 Jul	29 Sep
14	7.7	6.4	86.8	85.0	26 Jul	24 Sep
13	9.1	7.9	79.1	78.6	23 Jul	19 Sep
12	10.0	8.9	70.0	70.7	19 Jul	15 Sep
11	10.6	9.9	60.0	61.8	16 Jul	11 Sep
10	11.2	10.8	49.4	51.9	13 Jul	7 Sep
9	11.0	11.5	38.3	41.1	10 Jul	4 Sep
8	10.6	11.7	27.3	29.6	7 Jul	1 Sep
7	9.1	10.7	16.7	17.9	4 Jul	29 Aug
6	5.5	7.2	7.6	7.2	1 Jul	26 Aug
5	2.0		2.0	0.0		

Jenkins *et al.* (1990a) reported yield by branch position and node number for eight cultivars grown at Starkville, MS. They counted the cotyledonary node as one which most now count as zero. When one is subtracted from their node counts, averaged over cultivars and years, the node representing 95% of the yield could be calculated. Table 27-1 compares the position yield contribution for Mississippi and California (Jenkins *et al.*, 1990a; Kerby and Hake, 1996). Mississippi had 95% of yield accumulated by node 15.8 compared to 16.3 for California. Based on final plant map data from 191 tests, conducted from 1993 to 1996 from Texas to North Carolina and averaged over cultivars, the 95% zone averaged 17.3 (unpublished data, Kerby, 1997). Constable (1991) reported yield contribution by node for Australia. While it was not possible to calculate the 95% zone from their figure, visual inspection of the figure suggests 95% of the yield was accumulated near main-stem node 17. The number of nodes for the 95% zone averaged 17.6 for 318 Australian production fields grown over a four year period (unpublished data, Kerby 1994).

These data from different areas and different years indicate 95% of the yield comes from sympodia on the first 16 to 18 main-stem nodes. While there may be variation in the number of nodes for the 95% zone, the value is indicative of the time required to produce the crop. Small values would be associated with an early crop, and large values with a crop that required more time. The data of Kerby and Hake (1996) clearly indicate high yields can result from bolls produced on 17 main-stem-nodes or less. Since physiological cutout occurs when NAWF is approximately 5.0, plants with high yield potential would generally not require more than 21 to 22 main-stem-nodes.

3.11   First Position Retention in the Effective Fruiting Zone

The effective fruiting zone, for purposes of this discussion, is the 95% zone. Percent retention in this zone is the average retention of all first sympodial positions in the 95% zone.

First sympodial fruiting forms have the greatest leaf area to support a boll (Horrocks *et al.*, 1978) and a larger supply of assimilates (Wullschleger and Oosterhuis, 1990d). In addition, they are the oldest fruiting forms on sympodia and are more competitive for assimilates than other positions on the sympodia (Kerby and Buxton, 1981). They have higher retention percentages and contribute more to yield than other positions on the sympodia (Kerby *et al.*, 1987; Constable, 1991; Jenkins *et al.*, 1990a). Additionally, first sympodial boll size and fiber quality is superior to positions further out on the sympodia (Kerby *et al.*, 1987; Jenkins *et al.*, 1990b; Constable, 1991; Kerby and Ruppenicker, 1989).

3.12   Location of Harvestable Bolls

Bourland and Watson (1990) devised a convenient method to chart bolls on the important sympodial positions. Only four possible outcomes occur with respect to presence of a harvestable boll at the first two positions on a sympodial branch. Codes 0 = no boll at position one or two; 1 = a boll at position one but not two; 2 = a boll at position two but not one; and 3 = a boll at both position one and two. If additional harvestable bolls are present beyond the second position, in CottonPro (Plant and Bernheim, 1996) these can be recorded by adding a "+" to code 0, 1, 2, or 3 for each boll beyond the second position. There are several software programs that track these events including COTMAP (Bourland and Watson, 1990), BOLOCATE (Kennedy *et al.*, 1990), PMAP (Landivar *et al.*, 1994), and CottonPro (Plant and Bernheim, 1996).

Low plant densities produce a smaller proportion of their crop at first sympodial positions. The percentage of total bolls on first sympodial positions increased in a linear way between 42% at a plant density of 4.0 plants m^{-2} to 86% at 15 plants m^{-2} (Kerby *et al.*, 1987). As plant density increases one plant m^{-2}, percent of total yield accounted for by sympodial position one increases by an average 3.9%. Percent of the total crop in the first sympodial position, according to the average plant density for other studies are in close agreement with these regression relationships (Baker, 1976; Fowler and Ray, 1977; Galanopoulou-Sendouka *et al.*, 1980; Jenkins *et al.*, 1990a; Constable 1991).

3.13 Nodes above Cracked Boll (NACB)

Bolls on a plant have an age relationship with each other according to node and branch position. Boll set ceases on monopodial branches and secondary positions on sympodial branches earlier than first sympodial positions. Hence, the youngest bolls will be at the top of the plant and will normally be at first sympodial positions. A coordinated research project between California, Texas, Oklahoma, and Mississippi evaluated boll and fiber development of all positions above a first-position cracked boll (Kerby *et al.*, 1992b; Supak *et al.*, 1993). To determine NACB, only plants that had a boll at the first sympodial positionand were in the process of opening (cracking) were selected. The main-stem-node with a first-position cracked boll is counted as "0". Count nodes above this node to the uppermost node with a harvestable boll. Do not count nodes above the cracked boll to the terminal, only to the node with the last harvestable boll. If the uppermost harvestable boll is a second position boll, add "2" to the NACB value. This is because a second position boll is the same age as a first-position boll two fruiting branches higher on the plant. Sample size should reflect the size and variability of a field. A minimum of 5 plants should be selected from four areas of the field. The number of plants sampled from an area should represent the percentage of the total field represented by the growth of that particular area.

Fields with NACB of 4.0 or less can be defoliated without yield or fiber quality loss (Kerby *et al.*, 1992b; Supak *et al.*, 1993). Micronaire proved to be the most sensitive fiber parameter to premature defoliation. Boll weight loss and micronaire were affected in a similar manner by premature defoliation. Growers could initiate defoliation on a few fields to begin the harvest season at NACB = 5.0 with only minor changes in yield and micronaire. Defoliation when NACB = 6.0 or more is discouraged as yield loss and micronaire reduction could be substantial.

For bolls at NACB = 8.0, we noted a 15% increase in dry weight of these bolls following defoliation (Kerby *et al.*, 1992b; Supak *et al.*, 1993). Thus, there is some dry weight transfer into immature bolls following defoliation. NACB = 4.0 assumes defoliation with continued boll development that is equivalent to 100 DD$_{60}$'s. Therefore if fields are to be desiccated compared to defoliation, NACB should be 2.0 to have no effect on the last harvestable bolls. When NACB = 2.0 dry weight transfer is complete, and the boll is in the process of drying down.

4. IN SEASON MANAGEMENT DECISIONS

Plant monitoring during the season provides data that should influence management decisions. It is used to: a) determine if growth is normal; b) determine if crop progress is adequate for the calendar date; c) determine if fruit retention is adequate; d) measure the balance between vegetative and fruit growth; and e) determine when to stop pest control, irrigations, and apply harvest aids.

4.1 Early-Season Growth and Development

Early-season growth and development are measured by height and number of nodes. Early node development is closely associated with accumulation of thermal units while internode length is affected by a wide range of stress factors (*e.g.*, cool temperature, aphids, thrips, salinity, disease, nematodes, fertility, soil water, etc.). Internode length is more sensitive to environmental factors than is node development. Leaf removal studies and controlled environment studies both suggest carbon supply is not generally limiting until assimilates are partitioned to fruit (Kerby *et al.*, 1987; Baker *et al.*, 1972.). During the early-season, prior to anthesis, vegetative growth is the dominant carbohydrate sink (Kerby *et al.*, 1987; Oosterhuis and Jernstedt, 1999).

For cotton grown in temperate areas, HNR and GR prior to squaring is generally more reflective of temperature than potential stress factors. Warm temperatures increase HNR and GR while cool temperatures limit internode growth (Kerby and Hake, 1996). Until assimilate competition occurs following anthesis, HNR and GR provide measures of source strength and detects differences in determinacy of cultivars as well as differences between environments. A comparison of cultivars grown at 110 field test locations across the U.S. cotton belt over a two-year period, indicated the contribution to variation in HNR was 4.2 times greater for location than cultivar (Kerby *et al.*, 1996). Management decisions should be modified by the environment as well as growth of a cultivar based on progress of the season.

Low HNR values indicate stresses have limited the development of the plant, and if not corrected by early flowering when assimilate competition by bolls intensifies, new leaf area growth will not be adequate to sustain a large boll load (Kerby *et al.*, 1990b). High HNR near flowering indicates high leaf area renewal which will sustain high net assimilation rates. With high HNR, high early fruit retention is necessary to provide the assimilate competition with vegetative growth. HNR and early fruit retention must both be considered when considering adjustments in water supply, fertilizer applications, use of plant growth regulators, and pest control thresholds.

HNR has a weakness in that it is a cumulative measurement. GR (change in height per change in nodes) between two sample dates can help. However, it requires a minimum of two dates to determine the value and there is variation associated with measurements from both dates. Single time measurements that consider the HNR of terminal nodes address this limitation. Two methods discussed earlier in this chapter consider the height of the terminal five sympodia (ALT 5) and internode distance between the 4th and 5th node below the main-stem terminal node.

Edmisten (1995) provided guidelines for mepiquat chloride application that consider HNR and internode length for the period between two weeks after first square to two weeks after first flower. The value of these single time measures of growth rate will increase as additional data becomes available that describe normal values for a region, and as treatment effects (PGR's, water, nutrients, and boll load) are quantified regarding their effect on GR.

4.2   Balance Between Source and Sink

Peak NAR does not occur until near the time of first flower due to limited leaf area (Kerby *et al.*, 1987). Between node 6 and 17, the percentage of maximum NAR increased by 2.6% per day (Fig. 27-6). Peak NAR occurred at node 18, but was within 5% of the peak value between node 16 to 20. NAR began to decline two weeks after first flower, which corresponded to a decrease in new LAI from 0.64 to 0.27 per 100 DD_{60} during the first three weeks of flowering (Kerby *et al.*, 1987). Five weeks after first flower, plants were in cutout and no new LAI developed. NAR declined by 1.6% per day for 8 weeks beginning 12 days after first flower (Fig. 27-6). Data presented in Figure 27-6 were based on NAR per 100 DD_{60} to remove seasonal temperature differences as a factor.

Leaf age has been shown to be a primary determinate of leaf carbon assimilation capacity (Constable and Rawson, 1980b; Krieg, 1988; Landivar *et al.*, 1988; Wullschleger and Oosterhuis, 1990; Sassenrath-Cole and Heitholt, 1996). Both Constable and Rawson (1980b) and Krieg (1988) demonstrated peak net photosynthesis occurred when a leaf was

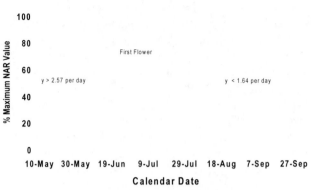

Figure 27-6. Percentage of maximum net assimilation rate through the growing season. Developed from Kerby *et al.*, 1987.

20 days old (from the time it was 2.5 cm in diameter). As age increased to 60 days, photosynthetic activity decreased 1.70% per day (Constable and Rawson, 1980b) and 1.76% per day (Krieg, 1988). This single leaf decline in net photosynthesis due to leaf aging is very close to the decline in NAR noted in Figure 27-6. Other data indicate that assimilate supply by the canopy declines rapidly at a time when demand by bolls is the greatest (Wullschleger and Oosterhuis, 1990d; Peng and Krieg, 1991). These studies emphasize that cotton source and reproductive sink are not well synchronized.

The use of environmental stress to control vegetative growth has been discouraged (Kerby, 1991). When growth control is achieved by imposing plant stresses, carbohydrate requirements for boll growth must be supplied by a smaller and older leaf area. When assimilate demand by bolls equals the total of assimilates available for growth, cutout will occur. For short-season determinate cultivars and long-season indeterminate cultivars, each grown under variable stress levels and variable fruit retention values, capacity to sustain assimilate supply to bolls was associated with leaf area growth rates between week 2 to 5 of flowering (Kerby *et al.*, 1990b). In this study, conditions which reduced LAI growth (water stress or small determinate cultivars) brought about an early cutout. Yield potential of early determinate cultivars was reduced more than that for older more indeterminate cultivars when environmental stresses were present.

Therefore, the cultural practice of imposing early-season stress to delay vegetative growth results in a high reproductive carbohydrate demand taking place when plants have reduced photosynthesis. While reduced early-season vegetative growth reduces productivity, excessive early season vegetative growth, coupled with insufficient fruit retention can result in excessive growth rates during flowering, and reduced yield unless the season length is long.

Cotton management requires that an appropriate vegetative growth rate be matched to the different lengths of growing seasons available for boll development. When the season length is short, fruit retention should be high and growth rates should be balanced. When the season length is long, growth rates should be higher to sustain vegetative growth during early boll set. In addition, for long seasons, plants can compensate for some early fruit loss (Kerby and Buxton, 1981; Ungar *et al.*, 1987; Constable, 1991). These studies indicate compensation normally occurs at higher sympodia instead of additional positions on the same sympodia.

4.3   Interpretation of Crop Growth

HNR and GR provide a measure of vegetative tendency that can be useful in management decisions prior to first flower (Kerby *et al.*, 1997). Following flowering, HNR and GR are affected by competition for assimilates as well as other plant stresses that may be present. Kerby and Goodell (1990) and Bourland *et al.* (1992 and 1997) indicate that NAWF provides a useful measure of source-sink relationships after anthesis.

Benson *et al.* (1995) reported NAWF data during the season for determinate and indeterminate cultivars grown with or without irrigation over a three year period. Earliness of the crop could be predicted by NAWF at 60 days after planting (high value = later maturity), or by days to NAWF = 5.0. Differences between cultivars was difficult to detect under conditions that resulted in late maturity (seasonal cutout, NAWF = 5.0 occurring after the latest possible cutout date). Full-season cultivars progressed towards cutout at the same rate as short-season cultivars, but took longer to reach NAWF = 5.0 because NAWF at 60 days after planting was higher than for short-season cultivars. Irrigation was shown to delay days to NAWF = 5.0 by as much as 11 to 13 days. For environments experiencing physiological cutout (NAWF = 5.0 occurring before the latest possible cutout date), indeterminate cultivars required 6 more days to reach NAWF = 5.0 and had initial NAWF values that averaged 1.1 more than determinate cultivars.

Early bloom NAWF for Deltapine cultivars were evaluated at 112 locations during 1994, 1995, and 1996. Differences between early- and full-season cultivars varied as much as 0.9 NAWF at early flower (unpublished data, T.A. Kerby, 1997). These tests support the conclusion of Benson *et al.* (1995) that differences in genetic earliness of cultivars can be detected by early flower NAWF. Genetic differences in NAWF at early bloom can be overwhelmed by environmental differences between test locations (Benson *et al.*, 1995; Johnson and Bourland, 1996; Kerby *et al.*, 1995). Early flower NAWF is a good indicator of the level of stress accumulated prior to anthesis. Once squares are initiated they progress toward anthesis based upon heat accumulation with little sensitivity to assimilate supply (Kerby and Hake, 1996; Kerby *et al.*, 1997). If assimilate supply is limited during early development, NAWF at first flower will be low. Normal NAWF at first flower should be between 8.0 and 10.0 (Benson *et al.*, 1995; Kerby *et al.*, 1996).

NAWF declines during the flowering period. Rate of change is dependent on assimilate competition by bolls as well as condition of the field to support sustained vegetative growth. For environments with physiological cutout, NAWF declines by approximately 1.0 each week (Bourland *et al.*, 1992; Benson *et al.*, 1995; Kerby *et al.*, 1996). In fields having seasonal cutout due to early fruit loss and/or exceptionally strong vegetative growth, NAWF may be high at early flower and have a slow rate of decline. Growth control management is especially important with these delayed fruiting patterns to avoid the consequences of late maturity. Plotting squaring nodes as the crop moves towards bloom, then plotting NAWF (same as squaring nodes) following first bloom provides an early warning regarding the need for growth control (Bourland *et al.*, 1997). Managers can compare squaring nodes to a calendar date and project approximate maturity of the crop. If maturity is projected to be later than desired, growth control measures can be applied. If cutout is projected to be earlier than desired, management to stimulate growth could increase yield potential.

4.4   Crop Termination Decisions

Since the decline in NAWF is related to assimilate supply and demand by bolls (Bourland *et al.*, 1992), it seems reasonable to expect there should be a NAWF value that indicates cutout (Kerby, 1994; Kerby and Hake, 1996; Oosterhuis *et al.*, 1999). Bernhardt *et al.* (1986) demonstrated that when the uppermost white flower (UWF) was within 4, plants were considered in cutout. A value of UWF of 4.0 is equivalent to NAWF of 5.0. Cutout is the time when the source is equal to the reproductive demand. The visible field evidence is a cessation in new node development, and sheds of many squares and small bolls.

Subsequent work (Bourland *et al.*, 1992; Benson *et al.*, 1995; Lammers *et al.*, 1996; Kerby and Goodell, 1990; Kerby and Hake, 1996) indicates on average fields are in cutout when NAWF = 5.0. Kerby and Hake (1996) expressed NAWF for cutout as the difference between number of nodes at cutout and the number of nodes for the 95% zone (zone where 95% of the harvestable bolls were set). Based on 104 field tests between 1982 to 1991, Acala cotton on average reached cutout when NAWF = 5.7. However, there was variation between fields in NAWF at cutout. On average 71% of fields were in cutout by NAWF = 5.0 and 91% were in cutout by NAWF = 4.0. Pima did not reach cutout on average until NAWF = 3.5. Results from a Regional study (GA, VA, LA, and AR) reported by Oosterhuis *et al.* (1998) demonstrated NAWF for cutout is less when mepiquat chloride is used or when nitrogen availability is low.

Plant map data were collected from 210 Australian cotton fields. Average cutout was at NAWF = 4.5 (Kerby, 1994). In 1990/1991, with adequate water, average cutout occurred at NAWF = 5.8. Subsequent years had less water available, and crops could not be managed for high yield. In 1991/1992, water was somewhat limited and the last effective boll set was NAWF = 4.9. The 1992/1993 year had severe water limitations and the average crop cutout was NAWF = 3.7.

Benson *et al.* (1995) indicated irrigation could extend the time to NAWF = 5.0 by as much as 11 to 13 days. Lammers *et al.* (1996) indicated indeterminate cultivars set their last effective boll at NAWF values that were up to 2.0 more than determinate cultivars. Their work also indicated high nitrogen increased NAWF for cutout while use of mepiquat chloride decreased the NAWF for cutout. Unpublished data (T.A. Kerby, 1994 to 1996) indicated that mepiquat chloride use can decrease NAWF by 0.5 to 1.0 depending on rate, and slightly reduces the NAWF value for cutout. Presumably, cutout at a lower NAWF value is associated with mepiquat chloride decreasing the number of nodes produced more than it shifts fruit set patterns (Kerby, 1985; Kerby *et al.*, 1986). Calculating plant map data from the data presented by Kerby *et al.* (1986) indicated use of mepiquat chloride at 1.2 L ha^{-1} decreased the node for the 95% zone from 17.6 to 16.5.

For the Australian data set (Kerby, 1994), 58% of the variation in NAWF for cutout could be explained by five variables: number of fruiting branches for the 95% zone; final plant height; first flower value of NAWF; rate of NAWF decline; and lint yield. NAWF for cutout tended to be high when vegetative growth was strong before and after first flower (approximately from main-stem node 10 to 18) especially when accompanied by high fruit retention (high yield). Low NAWF for cutout was accompanied by conditions that resulted in modest vegetative growth and low fruit retention (low yield). While NAWF = 5.0 is a good estimate for cutout, differences will vary by field conditions. In addition, there may be regional differences. For example, stripper cultivars on the High Plains of Texas have cutout at NAWF = 3.0 to 4.0 (K. Hake, personal communication).

Two plant monitoring approaches are used to aid defoliation timing decisions. The COTMAN expert management system suggests defoliation when NAWF = 5.0 plus 850 DD$_{60}$ (Danforth and O'Leary, 1998). Another method uses nodes above a first sympodial branch cracked boll (NACB) to the last harvestable boll (Kerby *et al.* 1992b; Supak *et al.*, 1993; Kerby and Hake, 1996). When the value is within 4.0, it is safe to defoliate. The method also provides information to predict yield loss and fiber quality effects of premature defoliation. Figure 27-7 demonstrates the relationship between nodes above a first sympodial cracked boll and per-

cent yield loss for an early maturing crop with 17 nodes for the 95% zone. Micronaire was the most sensitive fiber parameter to premature defoliation. Micronaire declined from a maximum value of 4.20 to 4.08 for defoliation at NACB = 8.0 with a response curve that was identical to Figure 27-7.

Defoliation has typically been scheduled when 60 to 65% of the crop is open. For normally fruited crops with 16 to 17 nodes for the 95% zone, NACB = 4.0 will be close to 65% open. However, using an example of a field with high fruit retention and a 95% zone of only 13, the crop was only 50% open at NACB = 4.0. Short, early, well fruited fields may be physiologically ready for defoliation well before 65% open boll. The other extreme would be a crop characterized by prolonged and/or delayed fruiting and maturity. An actual field that represents this case had low fruit retention where the 95% zone was not reached until main-stem node 21. In this case the field did not reach NACB = 4 until 82% of the bolls were open. Fields with low early boll retention and sustained vegetative growth during the flowering period would be defoliated prematurely when using a 65% open boll guideline. Defoliation schedules using either NAWF = 5.0 plus 850 DD$_{60}$, or NACB = 4.0 considers the fruiting pattern and growth habit of a field.

Use of NACB to schedule defoliation will help managers defoliate as soon as the crop is ready. The action of cotton defoliants is most favorable under warm temperatures, which generally occurs earlier in the fall. From a producer perspective, there may be other economic advantages to defoliating as soon as the crop is ready. Early in the season there are more hours per day with favorable picking conditions. The best grades are generally produced from on time defoliation because it reduces the weather exposure of open bolls. The best economic decision may be to accept a small decrease in yield to get expensive pickers operating at a time of the year when picking days are long, grades are at their peak, and harvest losses can be kept to a minimum.

5.   LESSONS LEARNED FROM END OF SEASON MAPPING

Plant growth and yield is the accumulative expression of its environment for the year. End-of-season monitoring documents the crop history and assists in interpretations of what went well and what may have detracted from yield. They are very helpful as a teaching tool to understand how the field responded to the environment that particular year and what influence management decisions had on the final outcome. This can be determined from final plant maps.

Final maps will not improve the profitability or aid management in the same year. They do provide a measure of crop growth and development through the season and help teach managers how the crop responds to weather and management variables. Their value is in comparing results from a particular field to a reference. Reference values exist for HNR, GR, squaring nodes, NAWF, and branch contribution

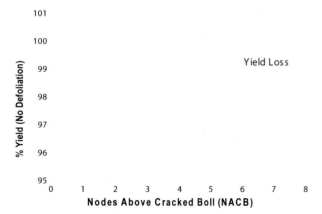

Figure 27-7. Yield loss associated with timing of crop defoliation using NACB. Taken from Figure 24.17 of Kerby and Hake, 1996.

to yield by sympodial position (Kerby *et al.*, 1987; Kerby and Goodell, 1990; Jenkins *et al.*, 1990a, 1990b; Constable, 1991; Bourland *et al.*, 1992; Benson *et al.*, 1995; Danforth *et al.*, 1995; Edmisten, 1995; Johnson and Bourland, 1996; Kerby and Hake, 1996; Lammers *et al.*, 1966; Landivar *et al.*, 1996; Silvertooth *et al.*, 1996; Kerby *et al.*, 1997).

Fruit retention can be compared against pests pressure and control decisions. If node development or position of first sympodial flowers is documented for a field through the season, estimated dates of anthesis can be assigned to all positions on the plant. Departures from normal can be associated with a weather or management event.

Figure 27-8 is a graphic representation of an early maturing fruiting pattern using the data of Kerby *et al.* (1987). This reference value is very similar to the values reported by Jenkins *et al.* (1990a, 1990b) and by Constable (1991). The line with small dashes represents a field with a typical fruit retention pattern except for nodes 10 to 13. Retention was decreased in these intervals. The decline could have been associated with pest pressure, drought, a period of cloudy weather, or other possible factors. The manager would probably be able to associate it with an unfavorable event. Final retention profiles provide a view of departures from normal, approximate time of the departure from normal fruit set, and the consequences of the departure in maturity and yield potential.

The prolonged, delayed fruiting curve illustrated in Figure 27-8 represents a field with a combination of low fruit retention and sustained vegetative growth. Yield potential would depend on length of the season and harvest conditions. Management practices to review would include possible combinations of fruit loss to pests, high vegetative growth rates as a result of excessive moisture, nitrogen, and failure to make a timely mepiquat chloride application at a sufficient rate to moderate growth. Bourland *et al.* (1997) provides additional examples with interpretations and applications of various growth patterns.

6. SUMMARY

Cotton is an indeterminate, tropical, perennial plant that is grown commercially as a temperate annual. Cotton yield is influenced by the balance between vegetative and reproductive sinks. The optimum balance between source and sink is not constant but varies by length of the growing season and many environmental and management factors. The level of crop inputs may define many management options. For example, without supplemental irrigation, management that limits source while encouraging a high sink will result in early cutout and decreased yield potential un-

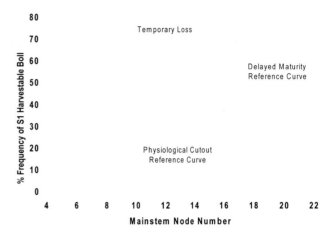

Figure 27-8. End of season first sympodial (S1) fruit retention percentages by node number for different fruiting patterns. Physiological cutout curve was developed from Kerby *et al.*, 1987.

less the season length is a major yield factor. Conversely, with a good and consistent supply of irrigation water, the size of source can be managed by water, fertility, and mepiquat chloride to optimally fit an expected boll load.

HNR and growth rate (including internode length between the 4th to 5th node below the terminal) are reliable measures of source size at early flowering. They provide useful indicators so that an optimal quantity of vegetative growth can be produced for the specific growing conditions for a given field/year combination. Once plants reach anthesis, NAWF and internode length provide data that indicate the source to sink balance. If the source to sink balance is not on target with the length of the growing season left to finish the crop, management factors can be adjusted (irrigation, nitrogen, mepiquat chloride, and fruit retention thresholds) that can help improve the balance. NAWF and NACB provide tools to indicate the timing of cutout and to schedule defoliation. At the end of the season, a crop history can be obtained with a final plant map. This provides the contribution of all fruiting positions to yield. Average internode length can be compared against reference growth curves to determine degree and timing of growth departure from ideal values. Associations between growth departures and the environment that caused the departure are useful in understanding more about how cotton responds to variable environments.

In recent years plant monitoring tools have been refined and are being increasingly used in managing fields. In addition, as researchers compare a treatment effect, plant monitoring data are being collected and reported. This provides the necessary information to quantify the growing conditions of a test location, as well as provide the necessary data to develop reference values for HNR, growth rate, internode length, and NAWF for the many different areas where cotton is grown.

Chapter 28

PHYSIOLOGICAL SIMULATION OF COTTON GROWTH AND YIELD

Juan A. Landivar[1], K. Raja Reddy[2], and Harry F. Hodges[2]

[1]*Texas Agricultural Experiment Station, The Texas A&M University Ssytem Research and Extension Center, Corpus Christi, TX, USA; and* [2]*Department of Plant and Soil Sciences, Mississippi State University, Mississippi State, MS, USA*

I.   INTRODUCTION

Scientists early in the twentieth century sought ways to describe and predict plant growth. Gregory (1917) and Blackman (1919) developed methodology called "growth analysis" to describe net assimilation rate, and compared dry matter accumulation to compound interest. By the middle of the century, leaf area index and light interception were recognized as important parameters for estimating photosynthesis in crop stands and were thus related to canopy dry matter growth. During the 1960s and 1970s, leaf and canopy photosynthesis were described using commercially available gx analyzers, radiation sensors, and other devices for measuring environmental conditions that facilitated studies remarkably. There were extensive debates regarding leaf angles, radiation attenuation, carbon accumulation, and maintenance and growth respiration. Several laboratories became interested in relating information on environmental conditions to photosynthesis, plant growth, and harvestable yield. The development of simulation models of the various processes had become feasible. The objectives and validation methods varied widely among crop modelers.

The extent of detail incorporated into a model depends largely on the available data, the conceptual understanding the modeler has of growth processes, and the intended end user of the model. Most models simulate both the gradual growth of the whole crop and the individual organs (leaves, stems, branches, roots, and fruit). A crop model should not just predict a final condition such as yield or biomass, but it should contain appropriate functions for the underlying contributing processes. Ideally, the mathematical relationships of the responses of various growth and growth-related processes accurately depict what occurs in real plants over a wide range of environmental conditions. If growth of the or-

gans is properly simulated, then canopy development, light interception and photosynthesis are likely to also be well simulated. If, on the other hand, the model is designed to simulate photosynthesis and growth on an empirical basis, its accuracy under a wide range of conditions will be reduced.

Unfortunately, complex models are expensive to develop and may require information that is not readily available. This has led several scientists and groups to develop simpler, usually site-specific, models that simulate development under some "typical" conditions. These simpler models, however, cannot encapsulate current understanding of growth interactions. Such models allow the user to compare an actual crop with the simulation under "typical" conditions. Atypical conditions can cause a field-grown crop to develop faster or slower than the simulated crop. Weather may have been cooler or warmer, or the crop may have experienced stress conditions that slow growth. Evaluation of the reasons for the differences can be instructive and growth of an actual crop may be correctable with a management tactic. Such models may provide a basis for scheduling various management practices such as irrigation, nitrogen (N) application, plant growth regulators, crop termination chemical application, insect control measures, and others. However, it should always be recognized that each model can simulate a crop only as well as the information on which it is based, and model use should be tempered with careful examination of the actual crop and environmental conditions.

Our objectives are to provide a brief history of cotton model development and the types of models available for a specific task; to provide the necessary mathematical relationships that describe physiological, growth, and developmental processes; and to show ways to integrate all these processes into a coherent system. As biological systems are often non-linear and subject to many sources of uncertainty, models that use robust mathematical methods to describe those processes will provide insight into growth interactions and be useful in both research and production. Hopefully, these realistic models can be applied across environments with a minimum of the tuning necessary for the less complicated, site-specific models.

Early crop model development was stimulated by Stapleton (Stapleton *et al.*, 1973), an agricultural engineer interested in applying computer technology to better utilize cotton production machinery. He invited several cotton production scientists to a workshop to itemize the steps and processes required for cotton production. This group became interested in predicting the time requirement and availability of various processes. From that beginning, SIMCOT, the first cotton simulation model, evolved (Duncan *et al.*, 1967; Duncan, 1972). SIMCOT calculated daily photosynthesis and simulated the distribution of the photosynthate produced among the plant parts. This model included logic for triggering fruit abscission and delays in leaf initiation due to carbohydrate shortage. McKinion *et al.* (1975) reported an incremental upgrade, designated SIMCOT II, by adding a nitrogen balance subroutine, a plant map that kept track of the age, weight, and nutritional status of each organ, and changing the basis for estimating potential growth. Considerable work on testing and improving crop models was conducted during this era (Brown *et al.*, 1985). Lambert and Baker (1984) reported the addition of RHIZOS, a simulator of root growth and soil processes, to SIMCOT II. The RHIZOS was designed to serve as a general rhizosphere model that estimated soil water potential, the metabolic sink strength in roots, and nitrogen uptake. They called this more detailed and combined model GOSSYM, an acronym from the words *Gossypium* and simulation. It attempted to incorporate many of the physical properties unique to individual soils into cotton crop simulation. These improvements represented a major advance over other models because they allowed quantitative characterization of many soil properties to be included that more realistically represented the uniqueness of field situations (Baker *et al.*, 1983). With the advent of personal computers at that time, Baker and colleagues began testing the model on private farms with on-farm computers. Producers recognized the value of information available from the simulation, but complained of the time required for data input and interpreting the reports. An expert system, called COMAX, was developed that facilitated calling weather stations and downloading information, constructing graphical information instead of the tabular format, and automating repeated model runs with "what if" questions concerning different management practices or weather scenarios (Lemmon, 1986). A complete discussion of GOSSYM/COMAX was published by Hodges *et al.* (1998, see also Chapter 5). Producers across the Cotton Belt tested GOSSYM/COMAX for several years. Many growers claimed GOSSYM/COMAX improved their knowledge of cotton growth and aided management decisions. Other growers using GOSSYM/COMAX were dissatisfied be-

cause the model was complex, sometimes inaccurate, and lacked technical support (Ladewig and Thomas, 1992).)

Parallel to the development of SIMCOT and GOSSYM, other scientists sought easier and/or better ways to simulate crop production. Hearn (1994) described OZCOT as a model that linked a temperature driven model of fruiting dynamics to the soil water balance of RItchie (1972). Others also attempted similar models of cotton growth (Wallach, 1978; Brown *et al.*, 1985). Several scientists attempted to computerize the optimization of production decisions by simplifying GOSSYM, and this resulted in more site-specific simulators. These products are combinations of models and expert systems that recommend practices suitable for specific regions [*e.g.*, CALGOS (Marani *et al.*, 1992a, b, c), and ICEMM (Landivar, 1991)]. In addition, several cotton models were developed over the years for special purposes. For example, Mutsaers (1984) proposed a detailed morphogenetic model to serve as a guide for plant breeders who wanted to simulate developmental responses modified by genetic means. Wall *et al.* (1994) developed "COTCO2: a cotton growth simulation model for global change" in an attempt to simulate changes in cotton growth caused by changes in atmospheric CO_2 concentrations.

We have learned several principles during this period of crop model development. One, it is exceedingly difficult to adequately represent crop growth, development, and yield in a wide array of environments with mathematical equations. We know many processes well, but other processes are poorly understood and available data on the quantitative relationships are insufficient. Because growth processes are interdependent, errors in one part of a model may be easily misinterpreted and associated with other processes, causing considerable time and effort to correct a problem. Two, many people are needed to provide a complex model to farmers who are relatively unfamiliar with the use of computers. Users are generally unable to separate software errors from data input errors, from errors caused by factors not modeled but definitely a production concern, or from errors in conceptual design of the model. This causes considerable frustration among both users and developers. To the user, an error simply prevents reasonable results. The developer may find a "software" error to be software-hardware incompatibility. Errors caused by factors not modeled may appear as water or nutrient deficiencies in the field, but are really caused by poor rooting due to herbicidal injury. Any of these errors may be misinterpreted and assumed to be the result of a serious conceptual model design problem. The solution may require long hours of study to identify the real culprit.

We are convinced that a solution lies somewhere between the difficult-to-use, complex model that simulates the crop over a wide range of environmental and management conditions and the easier-to-use but less reliable model. It is obvious that the data input must be automated and kept quick and simple. The model output information must be meaningful and easy to interpret. Ideally, the model itself should be based on sufficient detail that a complete range of conditions to which the crop may be exposed will be imbedded in the model equations. If unexpectedly high or low temperatures or other extreme environmental variables occur, the model should be sensitive to those conditions and respond appropriately. The major physiological processes should be incorporated into the model's growth and development routines. If herbicide injury is an important production problem, its effects on water and nutrient uptake should be modeled.

The problems with complex models may be partially overcome by delivering the information provided by the model through simpler, site-specific growth and development curves and data management systems such as CALEX (Plant, 1989), PMAP (Landivar, 1993; Landivar and Benedict, 1996), and COTMAN (Zhang *et al.*, 1994). As producers develop crop yield maps, they will become more concerned about the reasons crops yielded poorly in certain areas. A crop model that is appropriately designed with responses to nutrient deficiency symptoms and growth responses will likely provide a useful diagnostic tool. Similarly, models that have appropriate water relations subroutines will likely provide useful information to assist in the diagnosis of production problems.

Since the publication of Baker *et al.* (1983), numerous experiments have ben conducted to extend the information on which GOSSYM is based to simulate modern cotton cultivars more accurately (Reddy *et al.*, 1993b, 1997b, c). The varieties used in recent times are quite different from those on which the original GOSSYM was based. Therefore, the growth coefficients are different. The temperature range has been extended significantly so the impact of temperature extremes is much different from that originally modeled. As we learn more about the growth of leaves and internodes, it becomes feasible to model their growth more realistically. Also, new algorithms were developed and incorporated for plant growth regulators and crop termination chemicals as their use became a major production practice across the U.S. Cotton Belt (Reddy, 1995; Reddy *et al.*, 1995a).

It is not the purpose of this chapter to compare cotton models, but rather it is to provide quantitative functions for cotton responses to environmental variables. Growth and development of plants in the natural environment are the result of the interactions of three major factors: the genetic potential of individual plants, the external environment, and their interaction. Potential growth and developmental rates for a particular species or a genotype are defined as the maximum rates achievable at a given temperature. Controlled-environment facilities are necessary sources of process-rate data, because they allow one to vary one environmental factor while maintaining other factors in non-limiting conditions. Data obtained in this manner are less ambiguous and allow understanding of the responses to environmental variables and nutrient status. After potential rates of development or growth for a particular genotype have been established, the actual rates may be delayed or reduced by environmental and nutritional (including carbon) stresses. If major changes in genetic potential occur among

cotton cultivars, then the model needs to be adjusted to reflect those changes. Typically, differences that are unique to the cultivar can be simulated with minor calibration adjustments to potential rates of the given species. Information required for such adjustments has been expensive to obtain in the past, but less expensive techniques may be feasible.

Process-level simulation models are difficult to build from field-collected data, because many factors interact to affect growth and development processes. Many environmental and biological factors are often significant covariates, which makes accurate assessment of causes and effects difficult. In this chapter, we have elected to use data collected from controlled-environment chambers to develop functional relationships between plant processes and state variables. Unique sunlit plant growth chambers were designed by Phene *et al.* (1978) which utilized approximately 95% of the radiation from the sun while controlling other aspects of the environment. Numerous experiments were conducted in these naturally-lit plant growth chambers in which temperature, atmospheric carbon dioxide concentration $[CO_2]$, water, and nutrients were varied independently while other factors were maintained at near optimal conditions (Reddy *et al.*, 1992b, d; 1993a; b; 1995b; 1997b, c; 1998). allowed determination of the relationships between the varying factors and crop response. These experiments were usually conducted in environments in which only the variable tested was limiting. Plants were typically grown in well-watered sand and were adequately fertilized with all essential mineral nutrients. Plants were also grown in higher than ambient CO_2 to enhance photosynthesis and avoid carbon deficiencies and the plants were not subjected to disease or insect infestations. Potential CO_2-exchange rates may be estimated from relationships developed in this manner. The potential rates can be then corrected daily or hourly by stress factors known to occur in the natural environment (*e.g.*, nutrient or water limitations). Unique stress factors can be developed by measuring crop responses to known levels of water or nutrient status conditions. Thus, potential CO_2-exchange rates may be estimated from the temperatures the plants experienced, and actual rates can be simulated by reducing the potential rates with daily or hourly "stress factors."

2.   DEVELOPMENTAL PROCESSES

Crop growth and development are driven by canopy temperature but are modulated by water, radiation, and nutritional supplies (Gepts, 1987; Hodges, 1991; Kiniry *et al.*, 1991). Plant phenology is the study of the time between developmental events, or the duration of processes. Such events include the intervals between mainstem or branch leaves on a plant, and between plant emergence and formation of a flower bud, flower, or mature fruit. Duration of a process might include the period between unfolding of a leaf, or the appearance of an internode, and the time the leaf or internode reaches maximum size. Phenological information is essential for the analysis and understanding of developmental processes as well as growth and source:sink interactions in plants.

Hesketh and his co-workers conducted a series of experiments in the late 1960s and early 1970s on many phenological events of cotton cultivars (Hesketh and Low, 1968; Moraghan *et al.*, 1968; Low *et al.*, 1969; Hesketh *et al.*, 1972). Subsequently, Reddy *et al.*, (1993b, 1997 a, b) studied the phenological rate functions for modern Upland and Pima varieties and added duration of expansion or elongation functions for leaves and internodes that were not previously available. The data of Reddy *et al.* (1993b, 1995b, 1997a, b) are in the form of daily developmental increments. This is the form needed by modelers for building process-level dynamic simulation models. Also, their data were collected in a wide range of temperatures and in both ambient CO_2 and twice ambient CO_2 levels to extend applicability to model development. Genetic improvements in recent years resulted in cultivars that are earlier in maturity than cultivars grown 25 to 30 years ago (Moraghan *et al.*, 1968; Hesketh *et al.*, 1972; Wells and Meredith, 1984a, b, c). Models developed from such a database are useful both in present-day crop production environments and in the future environments hypothesized to be warmer and to have high CO_2. Crop simulation models that were developed to study global climate (Wall *et al.*, 1994) or used to study the possible impacts of climate change on crop production (Adams *et al.*, 1990; Curry *et al.*, 1990; Fisher *et al.*, 1995) did not use data collected in potential growing conditions. Those studies should be considered preliminary, since the crop models were based on a narrow temperature range and the phenological information was collected at present-day or slightly lower atmospheric CO_2.

2.1 Reproductive Initiation

In cotton, induction of reproduction occurs within a few weeks of germination. The appearance of first square (flower bud) marks the start of a period in which the plant produces both vegetative and reproductive growth simultaneously. Commercially grown cotton cultivars are very sensitive to temperature, but not sensitive to photoperiod, so square and floral development in these cultivars are easily predicted. The appearance of a morphologically distinct square, 3 mm in length, coincides with the unfolding of the subtending leaf. The number of days from emergence to the appearance of first square may be expressed both directly as a duration and as the inverse of the duration (Table 28-1, Eq. 1) at a range of temperatures. Expressing developmental rates as the reciprocal of duration allows one to predict square formation by adding daily developmental rates determined by daily temperatures until the sum equals one. The base temperature, below which no progress occurs toward flower bud formation, is about 15°C. The maximum rate of progress occurs at about 30°C, with rates of progress declining at temperatures above 30°C.

Table 28-1. Response of leaf and fruit growth variables to temperature and leaf N concentration. Normal (350 μl l⁻¹) and enriched (700 μl l⁻¹) CO_2 environments did not differ, therefore the values were averaged.

Equation number	Physiological function	Dependent or y variable	Independent or x variable	Slope or x coefficient	Quadratic coefficient (x^2)	Intercept	R^2	Reference	Range x values
1	Time from emergence to first flower bud	1/days	Temp (°C)	1.14×10^{-2}	-1.95×10^{-4}	-0.13	0.98	Reddy *et al.*, 1993b	19-34°C
2	Time from square to open flower	1/days	Temp (°C)	9.17×10^{-3}	-1.43×10^{-4}	-0.11	0.94	Reddy *et al.*, 1993b	18-34°C
3	Flower to open boll	1/days	Temp (°C)	9.95×10^{-4}	–	-0.006	0.92	Reddy *et al.*, 1993b	16-32°C
4	Main-stem phyllochron	1/days	Temp (°C)	5.7×10^{-3}	-6.76×10^{-4}	-0.67	0.94	Reddy *et al.*, 1993a	18-37°C
5	Sympodial phyllochron	1/days	Temp (°C)	3.39×10^{-2}	-5.2×10^{-4}	-0.36	0.84	Reddy *et al.*, 1993a	17-36°C
6	Node/leaf development rate	1/days	Leaf N Conc.	0.856	-0.16	-0.805	0.86	Reddy *et al.*, 1997b	1.5-2.5 g m²
7	Leaf unrolling to full size leaf	1/days	Temp (°C)	1.07×10^{-2}	-1.7×10^{-4}	-9.3×10^{-2}	0.95	Reddy *et al.*, 1993a	17-35°C
8	Internode initiation to full length internode	1/days	Temp (°C)	7.38×10^{-3}	-1.05×10^{-4}	-4.3×10^{-2}	0.96	Reddy *et al.*, 1993a	17-35°C
9	Maximum relative leaf elongation rate	1/days	Temp (°C)	2.04×10^{-2}	–	-3.39×10^{-2}	0.95	Reddy *et al.*, 1993a	17-36°C
10	Maximum relative internode elongation rate	1/days	Temp (°C)	1.66×10^{-2}	–	1.4×10^{-3}	0.97	Reddy *et al.*, 1993a	17-36°C

The effect of temperature on development fate has often been described in terms of growing degree days, a procedure that relates developmental rate to temperature above a species- or cultivar-specific base temperature below which developmental rate is zero (Hodges, 1991; Ritchie and NeSmith, 1991). The growing degree days concept has been used for simulating days to first square and several other developmental events in cotton (Jackson, 1991; Hearn, 1994). This procedure has been only moderately successful because of variability and lack of linearity in responses of different processes to temperature. It may be satisfactory if temperature thresholds are correctly chosen, and if temperatures do not exceed the linear phase of the response curve. If temperatures are outside the range of linearity, growing degree days alone do not accurately predict development (Table 28-1, Eq. 1). Developmental response curves are nearly linear to about 27°C, above which the rate of progress toward square appearance is slower. Setting the upper threshold above 27°C results in accumulating degree days too rapidly on hot days, whereas phenological progress is in fact slower. The errors in the estimation of thermal time are cumulative and vary depending on the temperature thresholds used. Reddy *et al.* (1993b) and Cogn'ee (1988) found that a quadratic function best described the relationship of developmental rate to temperature in cotton; however, such relationships may be complicated by acclimation and the effect of short-term extreme events on phenology.

Above-optimum temperatures delay progress toward square initiation and extend the vegetative period, even in well-watered conditions. In the natural environment, high temperatures are often associated with water deficits that partially close stomates and cause higher leaf temperatures. Therefore, progressively more adverse environments lead to extremely high canopy temperatures and delayed developmental processes.

2.2   Square Maturation Period

The equation for the reciprocal of days from square formation to open flower as a function of temperature is presented in Table 28-1 (Eq. 2). Rate of progress toward flowering is sensitive to temperature as rate of square formation throughout the temperature range except at the above-optimum temperature for growth, 27°C. Similar temperature response functions for the rate of flower development were published by Hesketh and Low (1968) for the cotton cultivars used two or three decades ago. At 27°C, the plants of Hesketh and Low (1968) required 20 days from square to bloom, while in the work of Hesketh *et al.* (1972), it took about 26 days. The modern cultivars studied by Reddy *et al.* (1993b) required about 24 days. After flowering begins, bolls become major photosynthetic and mineral-nutrient sinks in cotton. As fruit load increases, the rate of vegetative growth and development decreases because carbon becomes limiting, young fruit abscission increases, and vegetative growth eventually stops completely. This process

is often described as "cutout" (Guinn, 1986a; Landivar, 1991; Sadras, 1995) and will be discussed in more detail in a later section. If insects or environmental conditions cause significant boll abscission, vegetative growth may resume. Vegetative growth will also resume if weather is suitable, when bolls mature and are no longer energy sinks.

2.3   Boll Maturation Period

Unlike square initiation and square maturation periods, daily progress from flower to mature fruit (cracked boll) is linear from 16 to 32°C (Table 28-1, Eq. 3). Unlike the square to flower response, development rate from flower to open boll was greater at 32°C than at 28°C. Similar results were found by Hesketh and Low (1968).

2.4   Leaf Appearance Rates

The rates of addition of main-stem and fruiting branches determine the number of leaves produced and canopy development, and therefore they help to determine interception of photosynthetically active radiation. The rate of leaf appearance is defined as the time from one main-stem leaf unfolding to the next most apical unfolding leaf. Reddy *et al.* (1997c) defined a leaf to be unfolded when three main veins were visible. Defined in this way, leaf appearance can be used as a discrete event. Others used the Haun scale of leaf emergence over short periods, but that method estimates leaf emergence rate by comparing the size of the emerging leaf relative to that of the preceding leaf

The rates of leaf appearance on the main-stem and fruiting branches are functions of temperature (Reddy *et al.*, 1997c). We accumulated daily developmental rates until a summed value of one was reached, and that was used to predict a newly unfolded leaf on the main-stem or fruiting branches. Developmental rates were not linear. Leaf unfolding rates of both main-stem and fruiting branch leaves increased with increasing temperature. At 30°C, only 2.2 days were required to produce a new leaf on the main-stem, while at 20°C, 5 days were needed (Table 28-1, Eq. 4). Fruiting branches, on the other hand, required 6 days at 30°C and 9.5 days at 20°C to produce a leaf (Table 28-1, Eq. 5). The rate of leaf formation on fruiting branches is considerably slower than on the main-stem because the branch primordium develops a flower bud and an axillary meristem from which the next leaf, internode, and flower is produced (Mauney, 1984). Thus, the ratio of main-stem and fruiting branch leaf unfolding interval is not constant at different temperatures as assumed by others (Hearn, 1969a; Mutsaers, 1983a). Similarly, Hesketh *et al.* (1972) reported that 2.4 days were required to produce a leaf on the main-stem at 27°C, while modern Upland cultivars required 2.7 days (Reddy *et al.*, 1997c).

Fruiting or sympodial branch leaf-unfolding interval at 27°C was 7 days for cultivars used by Hesketh *et al.* (1972), while the modern cultivars reported by Reddy *et al.*,

(1997c) required only 5.8 days. Unlike many other phenological events, Upland and Pima cotton cultivars did not differ in rates of new leaf development (Reddy *et al.*, 1993b).

Development rates of leaves prior to main-stem node five were slower than main-stem leaves produced at higher positions. This is apparently not caused by limited assimilates since CO_2 enrichment did not speed their development. Leaf developmental rates at positions above node 16 were also slower than between nodes five and 15. This was probably the result of competition for assimilates by developing bolls. Leaf-unfolding intervals, generally referred to in the literature as phyllochrons, were not different from the square appearance intervals (Hesketh *et al.*, 1972; Reddy *et al.*, 1993a). Squares normally appear when the leaf at a given node unfolds exposing the main veins. Defined in this way, the leaf response rate functions can be used to predict square intervals. Other researchers have distinguished between plastochrons, the time between two successive leaf primordial initiations that can be observed with a dissecting microscope, and phyllochrons. Phyllochrons are more easily verifiable in the field than plastochrons.

2.5 Water and Nitrogen Effects on Development

The rate of cotton node or leaf development increases as leaf N increases (Table 28-1, Eq. 6; Reddy *et al.*, 1997b). Plants grown at twice ambient atmospheric CO_2 did not differ in response to N; therefore, the data from both CO_2 concentrations were combined to generate a quadratic relationship. One could speculate that, since N deficits decrease photosynthesis, the observed delays in forming new leaves was the result of carbon deficiency. However, the delays were not different in plants grown in 700 μl l^{-1} CO_2 from those grown in 350 μl l^{-1}.

The maximum rate of node development was achieved at about 2.5 g•N•m^{-2}, and development was projected to stop below 1.25 g•N•m^{-2}, although leaf N never reached such a low concentration. The rate of development from square to flower was not affected by N nutrition. Several others have reported no differences between N treatments for time to first flower, or time between flower and open boll (Waldeigh, 1944; Tewolde *et al.*, 1993; Gerik *et al.*, 1989). This suggests that once the organs are formed, the rate of development is governed by canopy temperature, and that canopy nutrient status rarely gets so low that developmental rates are altered.

The number of bolls and squares produced may be limited by N effects on node development, because both vegetative and fruiting branch growth are limited in a nutrient-deficit condition. Root media N concentration up to 2 mM caused more fruiting sites to be produced, but at higher concentrations additional increases were very small. Similar results were observed by several others concerning the effect of N deficits on production of fruiting sites (Jackson and Gerik, 1990; Gerik *et al.*, 1989).

Potential development rates were calculated as functions of canopy temperatures under optimum water and nutrient conditions (Table 28-1). Maximum development was assumed to occur at maximum leaf-N, thus, development rates at leaf-N below the maximum were expressed as a fraction of the maximum rate. These newly scaled values ranged from zero, where no development occurred, to one, where potential growth or development occurred at maximum leaf-N. The relationships calculated in this way can be used as factors to estimate the decrease in potential development rates in nitrogen-deficient environments.

Factors limiting growth responses to multiple variables are not easily separable, nor can one always be confident of the most appropriate way to model stress results. Acock (1990) discussed several processes that may control partitioning of photosynthates and other plant growth resources into different plant parts. For an extended discussion of how plants respond to two or more simultaneously varying factors see Sinclair (1992). Compensation occurs when an apparent, single-most-limiting nutrient, *e.g.* nitrogen, is supplied with some other nutrient also limiting growth, *e.g.* carbon. Bloom *et al.* (1985) proposed that an increase in simultaneously-limiting factors controlling plant growth increases growth in proportion to the extent the deficit of either nutrient is overcome.

2.6 Expansion Duration of Organs

The reciprocal of duration of leaf expansion and internode elongation is a measure of the rate at which these processes are completed. Duration of leaf expansion or internode elongation is defined as the time from leaf unfolding until final size is obtained. Internodes typically take less time to elongate at all temperatures than leaves (Table 28-1, Eqs. 7 and 8). Leaf petiole elongation occurs simultaneously with lamina expansion. Duration of leaf expansion at a particular temperature is nearly independent of leaf position on the main-stem (Reddy *et al.*, 1993b).

3. GROWTH PROCESSES

3.1 Leaf Area Expansion and Internode Elongation Rates

To mechanistically simulate plant height and leaf area development throughout the season, it is essential to simulate potential leaf and internode growth rates. The mechanism of internode elongation is similar in both dicots and monocots, although development is acropetal in dicots and basipetal in monocots because of the positions of intercalary meristem (Evans, 1965; Kaurman *et al.*, 1965; Sachs, 1965; Morrison *et al.*, 1994). Growth is defined as an increase in mass, area, or length. Since much of the plant is not growing, one needs to simulate the responses of only the growing organs. Such detail is necessary because each individual organ has a sigmoidal growth pattern, and expansion is influenced by environmental conditions only during that growth period. Thus, to simulate plant height, one should model the potential responses of elongating internodes, including the duration of the elongation process, to the conditions that prevail during elongation of each internode. For the following discussion, we assume that plants grown in the natural environment are to be simulated, and the primary role of solar radiation is to drive photosynthesis and transpiration.

Reddy et al (1997c) calculated growth rates by plotting relative leaf expansion rate (RLER) and relative internode elongation rate (RIER) as functions of days after leaf unfolding. These functions were calculated from daily measurements of leaf area and subtending internode lengths for each leaf and internode on plants grown under a range of temperature and CO_2 environments. The RLER and RIER decreased with age. The linearly-extrapolated intercepts provided estimates of the maximum RLER ($cm^2 \cdot cm^{-2} \cdot d^{-1}$) or RIER ($cm \cdot cm^{-1} \cdot d^{-1}$) on day one. The slopes of the RLER or RIER with age for each leaf ($cm^2 \cdot cm^{-2} \cdot d^{-1}$) or internode ($cm \cdot cm^{-1} \cdot d^{-1}$) were also calculated. The maximum RLER or RIER and slopes are functions of temperature (Table 28-1, Eqs. 9 and 10). The intercepts and slopes for leaves and internodes changed progressively and inversely with temperature. The maximum RLER (day 1) was 23% higher than the maximum RIER at all temperatures, while the slope, or the rate of growth reduction with age, was 5% lower for the leaves than the internodes. The effect of temperature on final leaf area or final internode length was the net result of temperature effects on both duration and rates of expansion (Table 28-1, Eqs. 7 to 10).

3.2 Leaf Area and Internode Length at Leaf Unfolding

At unfolding, the initial leaf area of each leaf was progressively greater up the main-stem until the first square (nodes 7 and higher) was formed then decreased at higher positions (Table 28-2, Eqs. 11 and 12; Reddy *et al.*, 1997b). Similarly, the length of internodes at leaf unfolding incrased up the main-stem until the first flower was formed after which it decreased (Table 28-2, Eqs. 13 and 14). Mature leaf areas followed a similar pattern on the main-stem node number increased up to node 7, then subsequent mature main-stem leaves were progressively smaller with higher node positions. The first square was produced when the fruiting branch formed on node six. A possible explanation is that processes involved in determining initial leaf areas and leaf area expansion compete with similar processes in branches, roots,

Table 28-2. Response of leaf and internode length to selected variables.

Equation number	Physiological function	Dependent or y variable	Independent or x variable	Slope or x coefficient	Quadratic coefficient (x^2)	Intercept	R^2	Reference	Range x values
11	Leaf area at unfolding nodes 1-6	cm^2	Main-stem node no.	1.81	–	6.06	0.91	Reddy *et al.*, 1997b	1-6
12	Leaf area at unfolding nodes 7 and above	cm^2	Main-stem node no.	-0.523	–	18.38	0.95	Reddy *et al.*, 1997b	7-26
13	Internode length at unfolding nodes 1 to 14	cm	Main-stem node no.	-5.605×10^{-2}	–	5.738×10^{-2}	0.93	Reddy *et al.*, 1997ab	1-14
14	Internode length at unfolding nodes 15 and above	cm	Main-stem node no.	-4.1×10^{-2}	–-	1.36	0.91	Reddy *et al.*, 1997a	15-26
15	Leaf area at unfolding (main-stem node 11)	cm^2	Temp (°C)	2.19	-3.81×10^{-2}	-18.6	0.62	Reddy *et al.*, 1997b	NA
16	Internode length at leaf unfolding (main-stem node 11)	cm	Temp (°C)	0.1077	-2.031×10^{-3}	-6.853×10^{-2}	0.11	Reddy *et al.*, 1997b	NA
17	Initial leaf size at 350 ppm CO_2	cm^{-2}	Leaf N conc. $g\ m^{-2}$	13.4476	-2.8023	-6.5774	0.81	Reddy *et al.*, 1997b	1.5-2.5 g m^2
18	Initial leaf size at 700 ppm CO_2	cm^{-2}	Leaf N conc. $g\ m^{-2}$	12.149	-2.4555	-4.8477	0.69	Reddy *et al.*, 1997b	1.5-2.5 g m^2
19	Relative leaf expansion rate	$cm^{-2}\ cm^{-2}\ day^{-1}$	Leaf N conc. $g\ m^{-2}$	0.5063	9.8×10^{-2}	-0.5374	0.84	Reddy *et al.*, 1997b	1.5-2.5 g m^2
20	Stem elongation rate at 350 ppm CO_2	$cm\ day^{-1}$	Leaf N conc. $g\ m^{-2}$	11.633	-2.389	-10.967	0.91	Reddy *et al.*, 1997b	1.5-2.5 g m^2
21	Stem elongation rate at 700 ppm CO_2	$cm\ day^{-1}$	Leaf N conc. $g\ m^{-2}$	11.2	-2.282	-10.238	0.90	Reddy *et al.*, 1997b	1.5-2.5 g m^2

and reproductive structures for available photosynthates. Squares formed when the leaves at nodes five to seven unfolded; then, perhaps, fruiting branches and other reproductive structures initiate more rapidly with time and compete for the same resources. However, increasing the supply of photosynthates by increasing the level of CO_2 to 700 ppm in the growth chamber did not affect the pattern of leaf and internode initiation or development (Reddy *et al.*, 1997c). Increased rates of photosynthesis often lead to increased growth rates and initiation of new organs. This enhanced growth due to increased CO_2 was offset by assimilate demand. Consequently, a constant supply:demand ratio and constant level of competition between organs kept leaf sizes similar. Similar patterns in mature leaf characteristics were observed in field (Constable and Rawson, 1980b; Constable, 1986) and growth chamber studies (Mutsaers, 1983a). Mutsaers (1983a) found a positive relationship between leaf sizes and cell number. If one accepts hypothesis of a limited-nutrient mechanism for the control of organ size, then these results suggest that leaf size expansion is more sensitive to available nutrients than internode length expansion.

Initial internode length, on the other hand, increased linearly as node number increased, until the plants started producing bolls, and then each succeeding internode was shorter than the previous internode. Bolls were first produced when the leaves at nodes 15 to 17 unfolded. Mature internode lengths also followed a pattern on the main-stem similar to internode lengths at leaf unfolding ($R^2 = 0.78$). Potential internode lengths and leaf areas likely were determined by the time of leaf unfolding. Similar results were observed for mature internode lengths in growth chamber studies by Mutsaers (1984).

Leaf area and internode lengths at leaf unfolding for main-stem nodes 10, 11, and 12 increased (Table 28-2, Eqs. 15 and 16). as temperature increased to about 27°C and declined at higher temperatures (Reddy *et al.*, 1997c)

3.3   Nitrogen and Water Deficit on Leaf and Stem Expansion

Nitrogen deficit effects on the processes underlying leaf area expansion rates are only occasionally documented in mathematical terms in the literature. Leaf growth parameters such as initial leaf sizes, relative leaf expansion rates, expansion duration, and leaf appearance rates, are essential information for modeling canopy development in field situations where leaf N content varies. Leaf area at leaf unfolding was greater in plants with more leaf N (Reddy *et al.*, 1997b), and plants grown in high CO_2 had significantly larger leaves at unfolding than plants grown at ambient CO_2 when plotted as a function of leaf N concentration (Table 28-2, Eqs. 17 and 18). Therefore, the parameters that affect leaf growth will also cause variation in initial leaf sizes (Terry, 1970; Dale, 1972; Robson and Deacon, 1978; Tolley-Henry and Raper, 1986; Gerik *et al.*, 1993). The RLER increased as the leaf N concentration increased to the highest concen-

tration (Table 28-2, Eq. 19). Similar increases in leaf area with increased leaf N were reported by others (Radin and Sell; 1975; Oosterhuis *et al.*, 1983; Jackson and Gerik, 1990; Gerik *et al.*, 1989; Fernandez *et al.*, 1993). The RLER was not different between the 350 and 700 μL l^{-1} CO_2 treatments, so the data were combined to generate the rate parameters.

Stem elongation rates also increased as leaf N increased (Table 28-2, Eqs. 20 and 21), and they followed a quadratic trend similar to that of leaf expansion rates (Reddy *et al.*, 1997b). Plants grown in high-CO_2 environments had consistently higher stem elongation rates at all N levels compared to plants grown in ambient-CO_2 environments (higher intercept). The shapes of the curves in both CO_2 levels, however, were similar (*i.e.*, similar x and x^2 coefficients). These data suggest that stem elongation was also carbohydrate limited, consistent with the sensitivity of stem growth to boll load.

3.4   Effects of Water Deficits on Growth Processes

Plants growing in the natural environment often do not reach their full genetic potential for yield. Environmental stresses have been estimated to reduce crop yields in the United States by about 71% from maximum achievable yields (Boyer, 1982). Similar problems are encountered worldwide since much of the world's cotton production is grown in arid or semiarid climates. With irrigation, attempts are made to optimize moisture conditions by controlling amounts and timing of water application (Chapter 23). Within the United States, for example, yields in Arizona are approximately double those of the nation as a whole (USDA, 1989). Such high yields are the result of growing crops in a high radiation environment and correcting the most limiting environmental constraint, water deficits.

When a plant experiences a shortage of water, its water content decreases and tissue water potentials (Ψ_w) become more negative. Lower tissue Ψ_w causes reduced tissue expansion, lower photosynthetic rates and lower stomatal conductance. Also, Ψ_w causes proportionally more dry matter to be partitioned to roots. Plants that are exposed to relatively long cycles between irrigations are less able to extract water from the soil than frequently irrigated plants (Radin *et al.*, 1989). This is particularly true during the fruiting period when intraplant competition is strong between fruiting structures and roots for limited supplies of carbohydrates. Roots of water-stressed plants have increased hydraulic resistance compared to roots of non-water-stressed plants. In addition, water stress causes greater intraplant competition for resources limiting root growth and delayed recovery after watering. Such delayed recovery extends the negative effects of drought on crop productivity.

Leaf Ψ_w potential is commonly accepted as an index of plant water status, and is relatively easy to measure. It is an algebraic expression of leaf cell turgor, osmotic concentration, and matrix potential. It is a function of both availability of water in the root profile of the soil and prevailing atmospheric demand. Extended exposure to drought conditions slows growth and increases cell osmotic pressure delaying some physiological effects of the drought. Leaf water potential is also influenced by the hydraulic conductivity of the system which may be influenced by the age of the plants and their previous environmental conditions. Under extremely dry conditions, the pressure differential may cause air to be pulled into xylem vessels. Such embolisms interrupt the water flow and increase stem resistance. Recovery from such a drought may be slow (Tyree and Sperry, 1989).

3.5 Modeling the Effects of Water Status on Tissue Expansion

Water deficits have very large effects on tissue expansion in both stems and leaves. Rates of increase in plant height versus midday leaf water potential are illustrated by Equation 22 (Table 28-3). Rates of stem elongation and leaf expansion declined in response to declines in leaf Ψ_w. Leaf growth rate was maximum (m^2 d^{-1}) at about -1.2 MPa and decreased to zero at -2.4 MPa (Table 28-3, Eq. 23).

3.6 Specific Leaf Weight

Leaf expansion characteristics and specific leaf weight (mass per unit area) determine the contribution of leaves to total sink strength. During expansion, leaf mass increases as new cell walls and cell constituents are added. After expansion ceases, cell walls thicken, and starch, sugars, and other cell constituents continue to accumulate. These additions change specific leaf weight and provide nutrient reserves that may be utilized for metabolism when other nutrients are not limiting. Leaf density, or specific leaf weight, is a function of temperature and other growing conditions (Table 28-3, Eqs. 24 and 25). Specific leaf area (the inverse of specific leaf weight) was maximum when leaf N concentration was low (Reddy *et al.*, 1997b) and decreased as leaf N increased (Table 28-3, Eq. 26).

3.7 Internode Mass Accumulation Rate

Internode mass under optimum water and nutrient conditions increased from 1 to 1.2 g during its 110 days of growth (Reddy *et al.*, 1997b). Unlike elongation duration and rate, internode mass accumulation rate (g d^{-1}) continuously increased throughout the growing season. The accumulation rate followed two distinct patterns, one during the expansion phase (0 to 20 days, Table 28-3, Eq. 27) and the other during the rest of the growth period (21 days and above, Table 28-3, Eq. 28). The mass accumulation rate from 21 days onwards was 338% more than the rate during the expansion phase, 0 to 20 days.

Table 28-3. Response of several cotton growth variables to leaf N concentration.

Equation number	Physiological function	Dependent or y variable	Independent or x variable	Slope or x coefficient	Quadratic coefficient (x^2)	Intercept	R^2	Reference	Range x values
22	Stem elongation rate	cm day^{-1} water potential (MPa)	Midday leaf	-0.6491	-0.9737	4.2904	0.77	Marani *et al.*, 1995	-2.4 to -1.2 MPa
23	Leaf expansion rate	m^2 day^{-1} water potential (MPa)	Midday leaf	34.0	6.38	44.89	0.73	Reddy *et al.*, 1997b	-2.4 to -1.2 MPa
24	Specific leaf weight at 350 ppm CO_2	g dm^2	Temp (°C)	-0.101	1.593 x 10^{-3}	2.05	0.95	Reddy *et al.*, 1997b	17-36°C
25	Specific leaf weight at 700 ppm CO_2	g dm^2	Temp (°C)	-0.224	3.692 x 10^{-3}	3.884	0.94	Reddy *et al.*, 1997b	17-36°C
26	Specific leaf area	cm^2 g^{-1} (g N m^{-2})	Leaf N conc.	1.438	-0.264	-0.946	0.80	Reddy *et al.*, 1997b	1.5-2.5 g m^{-2}
27	Internode mass accumulation rate (0 to 20 d)	g cm^{-1} day^{-1}	Age (days)	2.895 x 10^{-3}	—	5.032 x 10^{-3}	0.46	Reddy *et al.*, 1997b	0-20 d
28	Internode mass accumulation rate (21 d and later)	g cm^{-1} day^{-1}	Age (days)	1.178 x 10^{-2}	—	-1.828 x 10^{-2}	0.91	Reddy *et al.*, 1997b	21 d and more
29	Square growth rate	g^{-1}	Temp (°C)	1.25	-1.95 x 10^{-2}	10.2	0.98	Reddy *et al.*, 1997b	17-36°C

3.8 Square and Boll Growth

Reddy *et al.*, (1997b) derived equations for square and boll growth from data obtained at ambient and twice ambient CO_2. Growth response of developing squares to temperature was curvilinear; growth rate increased up to 22°C, but did not change at higher temperatures (Table 28-3, Eq. 29). Square growth rate was low and less sensitive to high temperature than boll growth. Boll growth was much more rapid, had a distinct optimum temperature at about 25°C, and declined sharply at higher temperature (Table 28-3, Eq. 30). Mature boll size, which is a product of boll growth rate and boll maturation period, showed an inverse relationship with temperature. Bolls were larger at low temperatures, but took more time to fill that mass (Reddy *et al.*, 1999). Boll mass increased with age of the bolls, but no difference in weight per boll occurred due to different CO_2 environments. Plants grown in elevated CO_2, however, produced more fruit mass per plant, and the differences in total fruiting structure mass were the result of more bolls and squares produced, rather than to more mass per boll (Reddy *et al.*, 1995c).

With increasing temperature, the duration of boll filling became shorter and the rate of filling slowed above 28°C (Table 28-1, Eq. 3). Faster crop developmental rates resulted in less time available for processes or events to occur. Above-optimal temperature caused less time to be available for a fruit to develop and reduced the rate of growth per day., Thus, cotton bolls grown at high temperatures were smaller than those grown at optimal temperatures. This result was also true in rice (Baker *et al.*, 1992) and wheat (Wardlaw and Wrigley, 1994).

4. PHOTOSYNTHESIS

A primary need of all plant models is to estimate light capture and photosynthesis. Modeling light interception has taken several forms and been discussed by several authors. Baker *et al.* (1978a) used a ratio of plant height to row spacing to estimate light interception. They found LAI was a poor predictor of canopy light interception. Internode lengths and leaf canopy density varied depending on whether the crop was under water-deficit conditions or in moist, well-watered conditions. Canopies produced in dry conditions had shorter internodes and captured a lower percent of the incoming radiation. Therefore, they reasoned that the ratio of plant height to the distance between rows is a better index of ground cover and a better indicator of light interception in cotton. However in a mature, senescing canopy this relationship does not remain true. They also empirically adjusted the light capture relationship from field data so that during maturation, when the LAI became less than 3.1, light interception decreased linearly with LAI. Others (Acock, 1991; Boote and Loomis 1991 and references cited therein) described light interception in row crops and also in closed canopies. They computed the shadow projected by the rows of the canopy as a function of canopy height, canopy width and time of day, day of the year, latitude, and row azimuth. The leaf area was stipulated to intercept light only in the shadow projected zone. Estimating light interception is important only prior to the establishment of a complete canopy. After canopy cover is accomplished, essentially all the photosynthetically active radiation is intercepted and photosynthesis is determined by environmental conditions and the biology of the crop.

One must also decide whether to model photosynthesis at the leaf level or canopy level. The rationale for using a canopy model, which is relatively empirical in nature, includes several considerations for the alternatives. The basic physiology of photosynthesis is controlled at the leaf level; however, there is considerable difficulty in scaling up to a canopy level, in part because of the problem of defining leaf environment. The quantity of radiation that strikes a leaf varies with position in the canopy and its orientation (angle) to the sun. Leaf angle is modified by the heliotropic nature of cotton leaves and also by leaf water potential. In addition, the age and nutritional status of individual leaves are critical factors for photosynthesis. However, validation of a leaf element model is difficult because data are not readily available. For these reasons, Baker *et al.* (1983) chose to model photosynthesis on a canopy basis. Their approach was to simulate an appropriate increment of dry matter produced each day of the growing season. That increment was a function of the plant age, size, physiological status, and environmental conditions, and to distribute it to the growing points in the plant.

Responses of canopy gross photosynthesis (P_g) to radiation are shown in Fig. 28-1 (also Table 28-4, Eq. 31). Gross photosynthesis is equal to net photosynthesis plus respiration. Respiration is a function of biomass and temperature. Canopy photosynthesis was measured under a range of temperatures, vapor pressure deficits, and fertility conditions, but with adequate soil moisture. These individual factors significantly influenced photosynthesis, but none changed the estimate of photosynthesis by 5% over the conditions tested. Therefore, Baker *et al.* (1983) used the relationship from daily total radiation to simulate daily potential P_g in GOSSYM, a cotton crop model. They did not adjust P_g for N status because N was not limiting in any of their tests. Reddy *et al.* (1991, 1995b) grew cotton in naturally-lit, controlled-temperature chambers at ambient CO_2. They measured CO_2 fixation in canopies that intercepted 95 to 98% of the solar radiation at different temperatures for 49 days during the fruiting period. No differences in photosynthesis level due to temperature occurred during the early portion of the growth period, but differences appeared and increased as the growth period increased. This result supports the observations of Baker *et al.* (1983) that photosynthesis in cotton is not very sensitive to temperature. The small apparent differences were probably caused by differences in growth rates among temperatures resulting in younger leaves at the tops of some canopies. The reported effects of temperature on photosynthesis of individual leaves vary.

Table 28-4. Response of photosynthesis to solar radiation and leaf N concentration.

Equation number	Physiological function	Dependent or y variable	Independent or x variable	Slope or x coefficient	Quadratic coefficient (x^2)	Intercept	R^2	Reference	Range x values
30	Boll growth rate	g day⁻¹	Temp (°C)	2.79×10^{-2}	$-5.8 \times 10{-4}$	-0.233	0.98	Reddy *et al.*, 1997b	17-32°C
31	Canopy photosynthesis	g CO_2 m⁻² day⁻¹	Solar radiation (ly)	0.1379	5.414×10^{-5}	0.24	–	Baker *et al.*, 1983	0-700 ly
32	Leaf photosynthesis at 350 ppm CO_2	mg CO_2 m⁻² s⁻¹	Leaf nitrogen (g N m⁻²)	2.0116	-0.3172	-1.664	0.88	NA	1.5-2.2 g N m⁻²
33	Leaf photosynthesis at 700 ppm CO_2	mg CO_2 m⁻² s⁻¹	Leaf nitrogen (g N m⁻²)	2.196	-0.3092	-1.72	0.90	NA	1.3-2.2 g N m⁻²
34	Leaf photosynthesis at 1600 µmol m⁻² s-1 and at 350 ppm CO_2	mg CO_2 m⁻² s⁻¹	Leaf water potential (MPa)	1.6755	–	8.732	0.67	Reddy *et al.*, 1997c	1.1-3.6 MPa
35	Leaf photosynthesis at 700 ppm CO_2	mg CO_2 m⁻² s⁻¹	Leaf water potential (MPa)	1.2033	–	6.058	0.64	Reddy *et al.*, 1997b	1.4-3.4 MPa

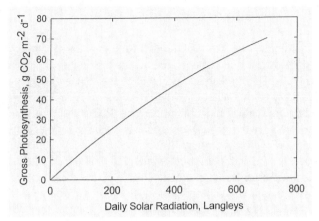

Figure 28-1. Daily canopy photosynthesis as a function of solar radiation (Baker *et al.*, 1983).

Recent data not incorporated into the latest version of GOSSYM, shows that N deficiency leads to carbon deficiency, because N deficiency reduces photosynthesis and leaf size. Photosynthetic rates are uniquely related to leaf N concentration in several species (Reddy *et al.*, 1997b, c). Photosynthesis increased three-fold as leaf N increased from 15 to 23 g kg⁻¹ (Table 28-4, Eq. 32). A similar response was observed at twice ambient CO_2 (Table 28-4, Eq. 33).

Canopy photosynthesis declined linearly as midday leaf water potentials decreased (Table 28-4, Eqs. 34 and 35). Maximum photosynthesis occurred when midday leaf water potential was -1.25 MPa or above. Fifty percent of the maximum photosynthesis rate occurred when leaf water potential was -2.5 MPa. The lowest photosynthesis rate in the water-stressed plants was about 25% of maximum, which occurred when the fourth leaf from the top was at -3.5 MPa.

Light-response curves generally indicate that light utilization efficiency (quantum efficiency) is greatest when intensity is limited. At light-saturation (P_{max}) further increase in photosynthetic photon flux density (PPFD) fails to increase photosynthesis. Quantum efficiency and P_{max} are affected by temperature, CO_2, and other environmental factors that may also influence these through effects on leaf thickness and N concentration.

Environmental history of the crop affects canopy light utilization efficiency and canopy conductance. For example, leaves produced in low PPFD exhibit lower light saturated photosynthetic rate than leaves developed in a high PPFD environment. A "feed-back" mechanism may slwo photosynthesis when plants are stressed. We, and many others, have found photosynthesis to be low under conditions in which growth is retarded due to low temperature, water, or nutrients. However, direct evidence that this results from feedback inhibition of photosynthesis in cotton *per se* is lacking.

5.   RESPIRATION

Cotton plants can accumulate only very limited quantities of sugars. Sugar compounds generated by photosynthesis are quickly converted to starch or translocated to growing organs anywhere in the plant. Approximately 50% of the sugar each growing organ receives is incorporated into the cell walls as cellulose or other structural compounds. Much of the remaining 50% of the sugar is used to synthesize proteins and lipids. Respiration is sensitive and positively correlated to temperature. Therefore, plants growing in a high temperature environment require more sugars for respiration than similar plants in lower temperatures. Tissue of water-stressed plants may be warmer than tissue of well-watered plants due to stomatal closure. This indirect effect of the water deficit causes a greater need for sugars because of the higher temperatures. A lower sugar supply also results because closed stomates reduce photosynthesis. Leaf N deficiency also results in a shortage of sugars required for respiratory-supported growth. Plants stressed because of drought, nutrient deficits, or high temperature may produce insufficient sugars to support all their structures. This may result in the death of tissues that were previously actively growing.

6. CUTOUT AND CROP TERMINATION

Cotton is a perennial plant with an indeterminate growth habit that has been adapted and managed as an annual crop. As long as the main-stem continues to grow, the number of fruiting branches increases, and each fruiting branch can continue to grow and initiate leaves and fruiting structures. Thus, the potential capacity of the nutrient sink formed by these organs increases in direct proportion to their number. In contrast, the carbohydrate supply does not increase proportionally to the number of leaves, but reaches a limit set by light interception. The rate of plant development depends on the environmental conditions such as temperature, light, and nutrients (Table 28-1, Eqs. 1 thru 8). Therefore, the sink potential eventually exceeds the carbohydrate supply, causing reproductive parts to compete for carbohydrates. This phenomenon has been recognized as the cause of cessation of vegetative growth, commonly referred to as cutout (Guinn, 1984c, 1985).

Guinn and his coworkers reported a series of detailed studies on the causes of cutout. Hormonal effects, such as ethylene evolution (Guinn, 1976) and levels of abscisic acid in growing fruit (Guinn, 1982b, 1986a, b) showed that nutritional stress induces boll shedding through increased ethylene production. However, these results did not explain other aspects of cutout such as the observed decrease in growth and flowering. Guinn (1984c, 1985) subsequently tested the hypothesis that cutout is caused by an imbalance between demand for and supply of photosynthate. Results of the experiments reviewed here validated his hypothesis. We conclude that vegetative growth, flowering, and boll retention decrease when the demand for photosynthate increase and exceed the supply. Additional factors related to cutout were reviewed by Sadras (1995).

Several other workers have also argued that the disparity between supply and demand for carbohydrates was a mechanism for the induction of cutout (Hearn, 1976; Mauney et al., 1978; Baker et al., 1983; Baker and Acock, 1986; Sadras, 1995). The most comprehensive study is the work of Baker and his co-workers in their cotton growth and yield modeling efforts in cotton simulation model, GOSSYM (Baker et al., 1983; see also Chapter 8).

The basic strategy used in GOSSYM to simulate the causes and consequence of cutout is accomplished by calculation of a nutritional stress parameter. This term is defined as the product of the supply:demand ratio for carbohydrates and a similar term for N (Fig. 28-2). The demand for carbohydrates is calculated from the sum of the growth potential of each organ on the plant (Fig. 28-3). The growth potential of organs is controlled by temperature, age of the organ, and turgor (Table 28-1). The calculations to determine demand are large, but are not complex. Baker et al., 1983) estimated carbohydrate supply from photosynthesis which is a function of the light intercepted by the canopy, leaf turgor, N, and other nutrient concentrations (Table 28-4, Eqs. 31 thru 35). The N supply is an estimate of the N taken up on any day based on transpiration plus any that may have been taken up but not used on previous days (Fig. 28-4). The N "reserve" is assumed to be stored in the leaf. These ratios, which are calculated daily, are used in GOSSYM as carbohydrate (CH_2O) and N-stress indices. When CH_2O stress and N stress are equal to one, the various organs are assumed to not be limited by carbon or nitrogen. As the ratio of supply to demand for carbon or nitrogen becomes less than unity, growth is limited by the respective nutritional deficits. The magnitude of these stress terms are used in GOSSYM to simulate the reduced rate of new organ initiation. The magnitude of these stresses are also used to adjust actual organ growth rate from the potential rates estimated to occur without these nutritional stresses and to simulate abscission of squares and young green bolls (Fig. 28-2).

This cotton modeling strategy allowed Baker and his coworkers to successfully simulate the delays in plant development induced by metabolic stresses encountered in field-grown plants. Baker et al. (1973) analyzed the relationship of photosynthetic efficiency and yield, using the supply: demand ratio to reduce potential growth of the plant organs, to abort fruit, and to induce delays in morphogenesis. These delays were shown to cause sigmoidal development patterns found in field-grown crops.

Thus, the experimental work of Guinn and his coworkers and the modeling efforts of Baker and his colleagues provide complementary support for the causes of cutout. These studies show that when the supply for carbohydrates is less than demand, fruit abscission increases and vegetative growth and flowering ceases. Meanwhile, existing leaves age and eventually senesce. This further reduces the supply of carbohydrates and accelerates the onset of cutout.

Modem cotton cultivars mature earlier and partition more photosynthate to fruiting structures, and there-

Nutritional Stress

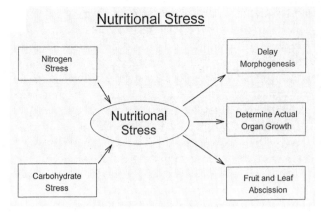

Figure 28-2. Relationships among nutritional stress (carbon and nitrogen) and plant growth, development or organ abscission (from Baker *et al.*, 1983).

Carbohydrate Stress

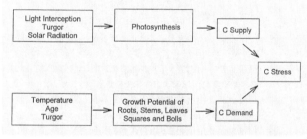

Figure 28-3. Factors that control available carbon. When the C-supply:demand ratio is less than one, plant growth is carbon limited (from Baker *et al.*, 1983).

Nitrogen Stress

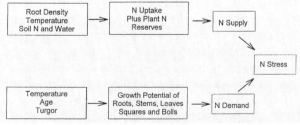

Figure 28-4. Factors that control nitrogen supply and nitrogen requirements of plants. When the supply:demand ratio is less than one, plant growth is nitrogen limited (from Baker *et al.*, 1983).

fore exhibit a greater synchrony with assimilatory activity (Wells and Meredith, 1984a, b, c). Hearn (1969b) suggested that existing reproductive growth was favored over would-be new vegetative growth. As this occurs, vertical and horizontal flowering intervals are delayed by carbohydrate stress (Baker and Acock, 1986). As fruit mature and less carbohydrate is consumed by developing fruiting structures, stem, and leaf growth resume.

7.   ASSEMBLING THE DATA FOR MODEL DEVELOPMENT

A physiologically-based cotton model can be assembled using the equations described. The simulated crop follows the above-described cutout scenarios and is sensitive to various conditions that occur in the field. The equations are based on data collected from controlled-environment studies in which only the variable being tested was limiting. Therefore, the equations provide a basis to model crop growth and development rates. The quantitative information only briefly introduces how one might model water- and nutrient-deficit conditions.

The step-by-step model development and applications for cotton crop are as follows:

- The time required to produce the first square from emergence can be calculated by summing the daily developmental rates from Eq. 1 for Upland cultivars until the summed value equals one.
- Once the squares are formed, their potential developmental rates to open flower can be calculated from Eq. 2.
- The boll maturation period, or the time from flower to open boll, can be calculated from Eq. 3 by summing daily developmental rates, similar to the procedure used in the calculation of time to first square.
- The daily potential rates of leaf unfolding on the main-stem and on fruiting branches can be estimated from Eqs. 4 and 5, respectively. Vegetative branch-leaf unfolding intervals follow the main-stem axis. Nutrient-deficit effects may be added to account for morphogenetic delays due to inadequate carbon or nitrogen. Equation 6 is used to simulate N effects on leaf development.
- Once leaves and internodes are initiated, their potential growth can be simulated with three rate functions: growth duration, maximum relative leaf expansion rate (RLER) or maximum relative internode elongation rate (RIER), and rates of growth reduction with age (Eqs. 7 to 10).
- Variable internode lengths or leaf areas on the main-stem and on branches can be simulated by using initial values, size at the time of leaf unfolding, from Eqs. 11 to 16 assuming ontogenetic patterns seen in the mature internodes and leaves are set at or before leaf unfolding.
- Potential growth of plant height or whole-plant leaf area can be predicted by integrating the growth rates of successive internodes on the main-stem or all the leaves on both the main-stem and branches capable of growth.

- Nitrogen-deficit effects on leaf and stem growth can be added using Eqs. 17 to 21. These growth processes can be decremented by water-deficit effects on stem elongation and leaf area development, based on Eqs. 22 and 23.
- Potential leaf weight per unit area increase can be calculated from Eqs. 24 and 25. The effect of nutrient status on specific leaf weight can be calculated with Eq. 26.
- Potential internode growth rates can be calculated from Eqs. 27 and 28, and total main-stem weights can be calculated by summing growth rates over all internodes. A similar approach can be employed for fruiting branch internodes.
- Potential boll and square growth rates can be calculated from Eqs. 29 to 30.
- Potential canopy photosynthesis can be calculated as a function of solar radiation from Eq. 31. The effects of water and nutrient deficit could be estimated from Eqs. 32 to 35.

8. SUMMARY AND FUTURE RESEARCH DIRECTIONS

In this chapter we provided the quantitative growth and developmental responses of cotton plants grown un-der optimal conditions as well as under N and water limiting conditions. These responses were obtained from plants grown where environmental factors affecting growth and development, other than the factors being studied, were not limiting. Models developed with such response rates have been robust in predicting appropriate crop responses in field environments where many factors change simultaneously and affect growth (Hodges *et al.*, 1998; McKinion *et al.*, 1989; Reddy *et al.*, 1995c). Thus, models built with sufficient mechanistic detail can provide simulated information that cannot be readily observed or measured. The model's deficiencies include a need for mechanistic approaches for estimating leaf and fruit abscission, heat stress effects, photosynthesis, yield, and yield components, fiber properties, energy balance, water relations, and root growth.

9. ACKNOWLEDGEMENTS

Part of the research was funded by the USDOE National Institute for Global Environmental Change through the South Central Regional Center at Tulane University (DOE cooperative agreement no. ED-RCO3-09ER 61010).

Chapter 29

ONTOGENY OF COTTON SEEDS: GAMETOGENESIS, EMBRYOGENESIS, GERMINATION, AND SEEDLING GROWTH

Rickie B. Turley[1] and Kent D. Chapman[2]
[1]Cotton Physiology and Genetics Research Unit, USDA-ARS, P.O. Box 345, Stoneville, MS 38776; and [2]Department of Biological Sciences, Division of Biochemistry and Molecular Biology, University of North Texas, Denton, TX.

1. INTRODUCTION

Production of a viable seed is of paramount importance to the survival of a plant species. Many plants, including cotton, amass large reserves of storage protein and oil in their seeds (oilseeds) to use during dormancy, germination, and seedling growth. These reserves are mobilized after imbibition when environmental conditions are favorable for growth. Due to their rich supply of protein and oil, *e.g.*, approximately 80% of the dry weight of cottonseed kernels consists of protein and oil reserves (Doman *et al.*, 1982), oilseeds have also been used as a human food source. The nutritional value of each oilseed variety or species differs; therefore, minor changes in either protein or oil composition could result in major nutritional improvements and increased economic value of the crop. Improving seed quality is not restricted to increasing its nutritional value; it can also be accomplished by increasing seed vigor, viability, cold tolerance, or resistance to disease. A better understanding of the cellular processes involved in the ontogeny of cottonseeds will lead to new approaches in seed improvement. Discussion of developmental events from gametogenesis to seedling growth will be presented and will focus on important processes involved in producing a viable cottonseed and plant.

2. MEGASPOROGENESIS AND MICROSPOROGENESIS

Development of ovules, or megasporogenesis, is perceived as the transitional passage between the sporophytic (2n) and gametophytic (n) generations in plants (Angenent and Colombo, 1996). In cotton, ovules originate from rounded masses of meristematic cells along the placental edge of the carpel walls (Balls, 1905; Gore, 1932). Each of these masses enlarge forming an anatropous, bitegmic structure which contains a single megasporocyte–the progenitor of the embryo sac. The megasporocyte undergoes two meiotic divisions generating four linearly aligned spore cells; the three micropylar spore cells degenerate and appear crushed by the enlarging chalazal or germinal spore (Haig, 1990). The cell ultimately destined to become the germinal spore is apparent after the first meiotic division as a preferential accumulation of organelles occurs in this cell (Schulz and Jensen, 1971). This is accompanied by a unilateral loss or lack of callose, β-1,3 glucan, in the cell wall of the germinal spore (Rodkiewicz, 1970; Reiser and Fischer, 1993). Loss of callose possibly facilitates a greater inflow of nutrients from the nucellus. A diagram displaying the temporal development of the reproductive structures of cotton is shown in Figure 29-1. Megasporogenesis proceeds with a *Polygonum*-type development pattern, common in 70 to 80% of flowering plants (Gore, 1932; Haig, 1990; Reiser and Fischer, 1993). A slight variation from the pattern occurs in the mature egg sac of cotton; the antipodals, postulated to function in nutrient absorption (Diboll, 1968), disintegrate between 3 and 0 days before anthesis (DBA; Gore, 1932; Jensen, 1963). It is not known how the loss of antipodals is compensated in cotton embryos.

Development of the male sporophytic generation–microsporogenesis (production of pollen), begins earlier than megasporogenesis as shown in Figure 29-1. Androceium (Greek for "male house") primordia can be identified as early as 36 DBA in the flower bud (Mauney, 1984). Delineation of the stamen primordia occurs independently from either sepal or petal primordia. Neither mutations (*Arabidopsis*) which influence development of petals or

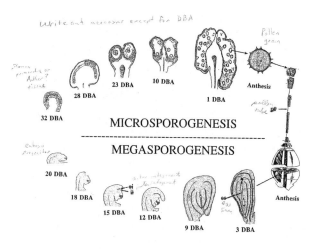

Figure 29-1. Temporal development of megasporogenesis and microsporogenesis in cotton. The ages of gametoenesis were extrapolated from Gore (1932) for megasporogenesis and Mauney (1984) for microsporogenesis, and the Figure is modified from Joshi *et al.* (1967). Abbreviations: ii, inner integuments; oi, outer integuments; es, egg sac; pt, pollen tube: and DBA, days before anthesis.

sepasl (Coen, 1991; Coen and Meyerowitz, 1991), or the surgial removal of their primordia (Hicks and Sussex, 1971) affect the initiation or differentiation of stamen primordia. By 30 DBA, the stamen primordia have formed buds on the androecium. Shortly thereafter, this primordia differentiates into the anther and filament compartments. Meiosis of the pollen mother cells begins by approximately 23 to 22 DBA (Mauney, 1984). Also at 22 DBA, mean relative humidity was positively correlated with anther number in mature flowers (Meyer, 1966) indicating that some prestamenal structures degenerate due to arid environmental conditions (stamen primordia were formed 8 days earlier at 30 DBA). The effects of environmental conditions on production of stamen and viable pollen were reviewed previously (Stewart, 1986). The reduction division of microspores occurs upon normal telosynaptic lines (Balls, 1905; Denham, 1924). The mitotic divisions of the pollen begin approximately 10 DBA and continue until anthesis (Balls, 1905). Cotton pollen grains are large and unusually rich in carbohydrates, lipids, and proteins (Fisher *et al.*, 1968).

As outlined above, the cellular development of mega- and microspores in cotton is well documented at the microscopic level. However, very little information exists in cotton as to the biochemical and physiological processes which are involved in gametophyte development. Many of the gene/gene products participating in these processes are being elucidated with mutants of gametogenesis from *Arabidopsis* and other species. Important gametogenesis gene discoveries include: *Bell1 (bel1)* and *SHORT INTEGUMENTS1 (sin1)* in *Arabidopsis* which develop abnormal integuments (Robinson-Beers *et al.*, 1992); *FBP11* from petunia (*Petunia violacea* Lindl.) which is expressed in the ovule primordia during early development then later

found exclusively in the endothelium surrounding the egg sac (Angenent *et al.*, 1995; Colombo *et al.*, 1995); and three homeotic genes *APETAL3, PISTILLATA,* and *AGAMOUS* in *Arabidopsis* or their analogues *DEFICIENS, GLOBOSA,* and *PLENA* in snapdragon (*Antirrhinum majus* L.) which control stamen development (Coen and Carpenter, 1993; Goldberg *et al.*, 1993; Okamura *et al.*, 1993). Any loss or mutation to these genes eliminates the production of normal ovules or stamen. In many (cases, of these mutants), the sporophytic tissue is replaced with a modified or different floral organ type (Goldberg *et al.*, 1993; Angenent *et al.*, 1995). The roles of numerous other genes are being examined for their function in gametogenesis (Coen and Carpenter, 1993; Goldberg *et al.*, 1993; Okamura *et al.*, 1993; Okada and Shimura, 1994; Chasen, 1995).

Endrizzi *et al.* (1984) produced a listing of the known variants (mutants) of cotton; a few of these loci are associated with ovule and stamen development, including: g - no ovary; stg - sterile female; st_1 - club stigma-style; st_2 - mini stigma-style; st_3 - invert stigma-style; $ms_{1-3,5,6,8,9}$ - recessive, and $MS_{4,7,10,11}$ - dominant male sterile. External ovules, *EO*, have also been identified by Meyer and Buffet (1962). External ovules form at the tip of the staminal tube on structures resembling carpel walls. With the discovery of the genes from *Arabidopsis* and other plants, rapid characterization of variant cotton lines is possible. Use of heterologous genes will allow for the screening of cotton homologs and may illuminate how environmental factors or additives (chemicals) affect seed production and viability.

3. FERTILIZATION

A few hours before pollination, the cytoplasm of stigmatic trichomes degenerate as the stigma become receptive to pollen. Pollen grains are dispersed onto the stigma, where they germinate, sending growing tubes down the surface of the trichomes. Pollen tubes enter the stigmatic tissue through the apoplast, then grow through stigma and the transmitting tissue of the style (Jensen and Fisher, 1969). The paths of pollen tube growth appear to be established by preexisting routes of the least mechanical resistance determined by (morphology/anatomy) of cells in the stigma and style (Jensen and Fisher, 1968, 1970). Pollen tube growth through the style does not induce degeneration of the transmitting tissue until the tube has grown past these cells. Degeneration may result from accelerated production of ethylene by the stigma and style. Lipe and Morgan (1973) reported that greater than 50% of the ethylene produced by 1 day postanthesis (DPA) flowers was produced by the stigma, style, and stamen. In petunia, ethylene production associated with pollination had no effect on pollen tube growth in the style (Hoekstra and Roekel, 1988). Also, as the pollen tube grows through the style, it is common for one synergid to degenerate; this occurs 6 to 8 hours after pollination

(Jensen and Fisher, 1968). Synergid degeneration also occurred in unfertilized ovules which were removed from the plant and cultured (Jensen *et al.*, 1977). This degeneration is observed when ovules are fixed with glutaraldehyde (Jensen and Fisher, 1968), but not when prepared for microscopic analyses by freeze substitution (Fisher and Jensen, 1969). The pollen tube grows through the micropylar end of the ovule and into the degenerated synergid (Jensen, 1965a). It is not understood how the pollen tube is guided toward the micropylar opening, but chemotropism appears to be involved (Mascarenhas, 1993). No universal chemotrophic compound has been identified in plants; however, a few compounds have been proposed *i.e.*, calcium, (Malhó and Trewavas, 1996), glucose, or possibly a water soluble protein of 14 kDa (Reger *et al.*, 1992). After the pollen tube enters the synergid, it ceases growth and forms a small pore for the discharge of the sperm cytoplasm and nuclei. Only the two sperm nuclei will leave the synergid, entering the egg and central cell directly (Jensen and Fisher, 1968). A diagram of pollination at anthesis is shown in Figure 29-1. Fertilization is completed approximately 20 to 30 hours after pollination.

4   EMBRYOGENESIS

Embryogenesis begins at the successful double fertilization of the egg and central cell with two sperm nuclei. Fertilization of the egg results in the creation of the zygote, whereas that of the central cell forms the nonembryonic endosperm tissue. These fertilization events begin a series of cellular changes essential for survival of the zygote. Very little is known about the gene networks that control (or regulate) embryo form or which orchestrate the sequential steps in embryonic development of plants (Goldberg *et al.*, 1994). These networks establish polarity of the embryo and induce the development of morphological patterns and tissue differentiation. Embryonic development sequentially determines the organization of the plant's structure and prepares the embryo for both dormancy and germination (West and Harada, 1993).

The chronology of the major events occurring during embryogenesis in cotton is divided into three overlapping stages of development, *i.e.*, morphogenesis, maturation, and desiccation (West and Harada, 1993). Embryo morphogenesis begins at the creation of the zygote (1 to 2 DPA) and lasts approximately until 25 DPA, as the full embryo length is obtained (Reeves and Beasley, 1935). Embryo maturation follows between 20 and 45 DPA and is characterized by the deposition of protein and oil reserves in embryonic cotyledons. The final stage, embryo desiccation, commences about 35 DPA and is characterized by preparation of the embryo to enter into a desiccated, quiescent state. The timing of these stages is dependent on the varieties used and environmental conditions during development.

4.1   Embryo Morphogenesis

The zygote is formed by the fertilization of the egg, which thereafter begins a period of cellular shrinkage as the zygote is reduced to about half its original size. Two to three days after fertilization (3 to 5 DPA) the zygote undergoes its first cell division (Jensen, 1963, 1968). This division is usually horizontal and results in the formation of a small terminal and a large basal cell. The basal cell develops into the suspensor, which is very small in cotton, consisting of 2 to 4 cells (Gore, 1932, Pollock and Jensen, 1964). The suspensor likely promotes growth of the zygote by transporting nutrients or synthesizing growth regulators (Yeung and Meinke, 1993). The terminal cell becomes the embryo through a series of random cell divisions. Consequently, no absolute cell lines can be established from specific embryo cells. The synergids disintegrate by 6 to 7 DPA, before this, they probably act as nurse cells to the egg (Jensen, 1965a). Passage of the zygote through the globular, heart, and torpedo stages are well documented (Reeves and Beasley, 1935; Mauney, 1961; Pollock and Jensen, 1964). Variations in the reported temporal development of cotton embryogenesis likely result from the use of different varieties and environmental conditions (Mauney, 1961; Forman and Jensen, 1965). For example, the formation of the heart stage was reported to be 6 to 9 DPA (Reeves and Beasley, 1935), 7 to11 DPA (Mauney, 1961), 12 to 14 DPA (Mauney *et al.*, 1967), and 15 to 17 DPA (Pollock and Jensen, 1964). Therefore, research on the temporal development of specific stages of embryogenesis needs to specify the genotypes and environmental conditions used.

Total embryo size remains below the original size of the egg until the zygote reaches approximately 75 cells (Jensen, 1963). As the embryo enters the heart stage, 12 to 15 DPA, the first signs of provascular tissue can be seen in the cotyledon (Reeves and Beasley, 1935). Succinate dehydrogenase (SDH), a respiratory enzyme, was distributed uniformly throughout the globular stage. The activity of SDH gradually shifted to areas of rapid growth in the cotyledon and basal hypocotyl regions during the heart and torpedo stages of embryo development (Forman and Jensen, 1965). The first lysigenous glands developed by 15 DPA with the first evidence of gossypol by 18 DPA (Reeves and Beasley, 1935). By 18 DPA, all embryonic organs and provascular tissue were being rapidly developed (Reeves and Beasley, 1935). Also by 18 DPA, semi-quantitative measurements indicated an increased accumulation of oil in the cotyledons (Reeves and Beasley, 1935). Turley and Trelease (1990) reported activity, protein, and mRNA levels of three glyoxysomal enzymes, *i.e.*, malate synthase (MS), isocitrate lyase (ICL), and catalase (Cat) in 18 DPA embryos. Embryos continue to expand until approximately 25 DPA (Reeves and Beasley, 1935; Mauney, 1961).

The second fertilization event consists of the union of sperm with the two polar nuclei in the central cell. Fusion of the two polar nuclei usually occurs before fertilization

(Jensen, 1965b). Unlike the zygote, the endosperm nucleus divides within hours after fertilization (Jensen, 1963, 1965b) forming a coencytic tissue (Schulz and Jensen, 1977). Free nuclear division continues until a mass of multinucleate endosperm surrounds the zygote and synergids (Schulz and Jensen, 1977). During the rapid growth of the endosperm there is a decline in starch and lipid reserves. Endosperm becomes cellular by about 10 DPA (Schulz and Jensen, 1977), whereas, in the culturing of unfertilized ovules this occurs by 3 DPA (Jensen *et al.*, 1977). Mauney *et al.* (1967) found that the major organic acid in 12 to 14 DPA endospenn was malate (7 to 10 mg/ml). The dry weight of the endosperm (endosperm, inner integuments, and nucellus) was reported to increase until 20 to 24 DPA, then gradually decrease and stabilize by 35 DPA (Stewart, 1986).

4.2 Embryo Maturation

The embryo maturation phase (20 to 45 DPA) overlaps with morphogenesis of the embryo (0 to 25 DPA) and is characterized by the parenchyma cells of the developing cotyledons undergoing a rapid and intense accumulation of storage lipids and proteins (Figure 29-2). The lipids, which are almost exclusively triacylglycerols, are the major carbon reserve for germinating and growing seedlings (Trelease *et al.*, 1986). Mature storage proteins of cotton embryos are predominantly of two subunit sizes (48 and 52 kDa) which appear closely related with each other and with many other seed storage proteins (Dure and Chlan, 1981, 1985; Galau *et al.*, 1983; Galau, 1986). Cotton storage proteins serve as a source of nitrogenous compounds for seed gennination and seedling growth (Huang *et al.*, 1983). In the maturing cottonseed, the temporal accumulation of storage lipids coincides with the accumulation of storage proteins (see Figure 29-2), which is not necessarily the case with other oilseeds, *e.g.*, *Brassica* (Murphy *et al.*, 1993; Bewley and Black 1994).

4.2.1 Triacylglycerol (TAG) Biosynthesis and Lipid Body Ontogeny

The synthesis of fatty acids *de novo* and their assembly into seed triacylglycerols have been the subject of much attention in recent years. This is likely a reflection of the recognized nutritional and economic importance of plant oils (Ohlrogge, 1994). The triacylglcerols of cottonseed are comprised mostly of palmitic (16:0) and linoleic (18:2) acids, with modest amounts of oleic (18:1) acid (Table 1; Jones and King, 1996).

Compartmentation of fatty acid and oil biosynthesis are outlined in Figure 29-3. The first step in fatty acid biosynthesis is the carboxylation of acetylCoA to form malonyl-CoA, which is catalyzed by the enzyme, acetylCoA carboxylase (ACCase; Ohlrogge *et al.*, 1993). In dicots, there appears to be two isoforms of the enzyme (Alban *et al.*, 1994). One isozyme is localized in plastids and is a multisubunit complex of approximately 700 kDa. The other isozyme

OIL BIOSYNTHESIS IN DEVELOPING COTTON SEEDS

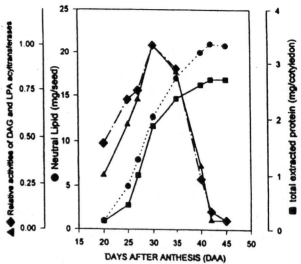

Figure 29-2. Time course of accumulation of storage oil (circles), protein (squares; Chapman, unpublished results) reserves in cotyledons of developing cottonseeds. Triacylglycerol biosynthetic capacity judged by specific activity of *lyso*-phosphatidic acid (LPA) acyltransferase (diamonds) and diacylglycerol (DAG) acyltransferase (triangles) is highest in cotton embryos between 25 and 35 days after anthesis (DAA; Chapman, unpublished results).

Table 29-1. Fatty acid composition of total lipids, storage (nonpolar) lipids, and membrane (polar) lipids isolated from dry cottonseeds (cv. Stoneville 7A glandless, 1993 harvest).

	Fatty acid (no. carbons:no. double bonds)				
	14:0	16:0	18:0	18:1	18:2
	--------------(mol %)---------------				
Total seed lipids	0.8	21.6	2.1	15.4	58.5
Nonpolar (mostly triacylglycerols)	0.7	20.2	2.3	14.7	61.4
Polar (mostly phospholipids)	0.8	23.5	9.7	13.1	53.1

Ratio of saturated to unsaturated fatty acids (membrane lipids much different from storage lipids).
Nonpolar 1:3.30
Polar 1:1.99
Total 1:3.12

is likely cytosolic with a single multifunctional polypeptide, similar to the mammalian liver enzyme. The plastid ACCase of dicots is not yet fully characterized; however, this large complex is composed of subunits encoded in both the nuclear and plastid genomes (Ohlrogge and Browse, 1995). The plastid isozyme supplies malonylCoA to the fatty acid synthase complex whereas the cytosolic form is believed to be involved in elongation of C18 fatty acids outside the plastid. In cottonseed TAGs, this latter activity likely is not important, since very little C20 or longer fatty acids are found in cottonseed lipids. Based on quantitative analyses of metabolite pools *in vivo*, the activity of the plas-

Compartmentation of Oil Biosynthesis in Seeds

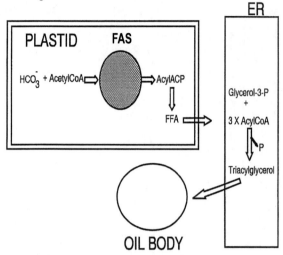

Figure 29-3. Compartmentation of fatty acid and oil biosynthesis in cottonseeds. Abbreviations: FAS, fatty acid synthase; FFA, free fatty acid; and ER, endoplasmic reticulum.

tid ACCase is considered to be a key regulatory step in the synthesis of fatty acids (Post-Beittenmiller *et al.,* 1992).

Fatty acids are assembled in two-carbon increments from malonylCoA through a series of acylation, condensation, and reduction reactions (Ohlrogge *et al.,* 1993). The fatty acid chain is elongated while it is covalently attached to an acyl carrier protein (ACP) platform. The enzymes required to complete one cycle of elongation are components of a type II fully dissociable fatty acid synthase (FAS) complex located in the stroma of the plastid (Ohlrogge and Browse, 1995). Fatty acid chain length is increased until terminated by a soluble thioesterase, or until transferred to a glycerol-3-phosphate (G-3-P) by acyl-ACP *sn*-glycerol-3-phosphate acyltransferase (Somerville and Browse, 1991). Fatty acids transferred to G-3-P are destined to remain in the plastid membranes and this route of synthesis is referred to as the "prokaryotic" pathway (Roughan and Slack, 1982; Somerville and Browse, 1991). Fatty acids released from acyl-ACP by a soluble thioesterase are exported to the endoplasmic reticulum for incorporation into membrane lipids or TAGs, and this route of synthesis is referred to as the "eukaryotic" pathway. The first double bond in C18 unsaturated fatty acids is introduced by a soluble desaturase (Δ-9 stearoylACP desaturase) before the fatty acid (oleic acid, 18:1) is released from the ACP. A cDNA sequence reported to encode the cotton homolog of the plastidial Δ-9 desaturase has been isolated (Reddy, 1995). The mechanism(s) for transport of fatty acids out of the plastid and to the endoplasmit reticulum (ER) are unknown.

Compartmentation of storage lipid biosynthesis is generally known for oilseeds (Murphy *et al.,* 1993; Bewley and Black, 1994). It is difficult to separate the synthesis of storage lipids from that of membrane lipids because the

biosynthetic pathways are interrelated, utilizing the same enzymatic machinery (Somerville and Browse, 1991). Furthermore, phosphatidylcholine (PC) is the major membrane lipid, and also an intermediate in the synthesis of polyunsaturated fatty acids of storage TAGs (Roughan and Slack, 1982; Heinz, 1993). The eukaryotic pathway continues in the ER as two fatty acyl moieties are esterified in a two step process to the *sn-l* and -2 position of G-3-P to form phosphatidic acid (Frentzen, 1993). Glycerol-3-phosphate acyltransferase and lysophosphatidic acid acyltransferase are two distinctly different membrane-bound enzymes which catalyze the transfer of the fatty acid from acylCoA (Frentzen, 1993). In some species, these acyltransferases display a clear substrate preference, and hence are responsible for the positional distribution of fatty acids in TAGs (Stymne and Stobart, 1993). Phosphatidic acid is either activated to form CDP-diacylglycerol for phosphotidylinositol (PI) and phosphatidylserine (PS) biosynthesis, or it can be de phosphorylated to produce a diacylglycerol (DAG) molecule (Moore, 1982; Kinney, 1993). The *sn-3* hydroxyl of this DAG can be acylated with a third fatty acid to produce a TAG, catalyzed by diacylglycerol acyltransferase (DAGAT; Frentzen, 1993). DAGAT is the only enzyme unique to the TAG biosynthetic pathway. Alternatively, DAG can be converted to a membrane phospholipid (PC or PE) by the transfer of a phosphoryl choline or a phosphorylethanolamine head group to the *sn-3* position via cholinephosphotransferase (CPT) or ethanolaminephosphotransferase (Moore, 1982; Kinney, 1993). In species with polyunsaturated fatty acids in their TAGs, the acyl moieties of PC are the substrates for the introduction of additional double bonds. Embryo specific isoforms of the Δ-12 desaturase have been cloned from developing soybean (Heppard *et al.,* 1996) and cottonseeds (Liu *et al.,* 1997). The polyunsaturated fatty acids are removed from PC and returned to the acylCoA pool where they can be incorporated into TAGs via DAGAT (Heinz, 1993). There is evidence that CPT may function in reverse to provide additional DAG (from PC) for TAG biosynthesis (Slack *et al.,* 1983).

The TAGs synthesized in ER are compartmentalized into oil bodies. Structural features of seed oil bodies are similar and account for the stable deposition of nonpolar lipids in the aqueous compartment of the cytoplasm. The current model proposed by Huang (1992) builds on the concept originally put forward by Yatsu and Jacks (1972). This model is based on a thermodynamically-stable "half-unit" membrane, a monolayer of phospholipid, surrounding a matrix of TAG. Embedded in this membrane is a ubiquitous, highly-conserved protein termed oleosin (Huang *et al.,* 1993). Oleosins are small molecular weight proteins which are believed to cover the surface of the lipid body and prevent their coalescence. About one-third of the protein extends into the matrix of the lipid body in a "hairpin-like" configuration. Recently, both cDNA and genomic clones have been isolated for oleosin from maturing cottonseeds (Hughes *et al.,* 1993), and its seed specific expression confirmed.

The precise mechanism by which oil bodies are formed is the matter of some debate (Napier *et al.,* 1996). Two alternative mechanisms have been proposed. In both models, TAGs are synthesized in the ER and oil-rich bodies vesiculate from the ER surface, retaining a phospholipid monolayer. In one model (Huang *et al.,* 1993) oleosins are inserted into the ER membrane and help to sequester the neutral lipid bodies for vesiculation. Lipid bodies "pinch" off into the cytoplasm at their mature size. Alternatively, oleosins are inserted into nascent lipid bodies which coalesce in the cytoplasm after vesiculation from the ER (Hills *et al.,* 1993). It is worth mentioning that some of the contradictory evidence regarding lipid body ontogeny may be due to differences in mechanisms among plant species, as appears to be the case with the biogenesis of other organelles (protein bodies and glyoxysomes, see below).

Utilization of newly synthesized fatty acids for membrane or storage lipid biogenesis depends largely upon cellular/physiological conditions. For example, in developing cottonseeds, a large portion of fatty acids is exported to the ER for storage. However, in cells of non oil-storing tissues, fatty acids are predominantly required for membrane lipid assembly, with different subcellular membranes exerting temporally distinct demands for lipids (Thompson, 1995). The control mechanisms by which plants partition fatty acids into storage or various cellular membrane lipids remain largely unknown.

In terms of storage lipids, manipulation of fatty acid composition in seeds has become commonplace for several crops (Ohlrogge, 1994). The commercial production of high lauric acid rapeseed (*brassica rapa* L.) indicates the maleability of seed storage lipid composition (DelVecchio, 1996). Thus far, little or no adverse agronomic effects have been noted, even with drastic changes in seed oils. *Brassica,* with predominantly C18 unsaturated fatty acids, was changed to primarily C12 saturated fatty acids with little change in oil yield or crop performance (DelVecchio, 1996). Thus far, field trials indicate that satisfactory yields for transgenic soybean plants with altered fatty acid composition (Kinney, 1997). Manipulation of fatty acid composition could improve the nutritional and economical value of cottonseed.

4.2.2   Storage Protein Biosynthesis and Protein Body Ontogeny

In dicot seeds, the two major classes of storage proteins are globulins and albumins, which differ in solubility properties (Shewry *et al.,* 1995). Globulins are soluble only in dilute salt solutions, whereas albumins are soluble in water. Both globulins and albumins are synthesized and compartmentalized in storage protein vacuoles during cottonseed maturation (Dure and Chlan, 1981).

Cottonseed storage proteins have received considerable attention over the last 30 years. These proteins are not only valuable for their economic importance (meal) but also because they are expressed at high levels in a seed-specific manner (Galau *et al.,* 1983; Dure and Chlan, 1985; Chlan *et al.,* 1986); thus, genes encoding seed storage proteins are likely to provide important clues as to the regulation of gene expression during embryogenesis. In previous reviews on this topic (Galau, 1986), cDNA clones encoding the major cottonseed storage proteins, the globulins, had been isolated and characterized. Further, a scheme for the synthesis and processing of the major seed storage proteins during embryo maturation and seed germination was proposed (Dure and Chlan, 1985). Now, genomic clones have been described for cottonseed vicilin (Chlan *et al.,* 1987) and legumin genes (Galau *et al.,* 1991). Upland cotton contains several closely related vicilin genes in each of its A and D genomes (Chlan *et al.,* 1987; Galau *et al.,* 1988), while only single copies of two distantly related legumin genes are present (Galau *et al.,* 1991). Products of these genes are specifically and highly expressed during seed maturation. Regulatory segments believed to be important in controlling the transcription of these genes include a seven nucleotide sequence (A/GTTTTTA/G) identified in vicilin genes (two copies) and legumin genes (nine copies). Also a 28-nucleotide legumin segment was identified in the 3' end of the cotton LegA-D gene, analogous to an element present in many *Fabaceae* legumin genes (Shirsat *et al.,* 1989).

Galau and coworkers have isolated and characterized cDNA and genomic clones encoding 2S cottonseed albumins (Galau *et al.,* 1992). The cotton 2S albumins share similarities characteristic of other known 2S albumins, including identity at leucine and cysteine residues (Shewry *et al.,* 1995). Of particular interest are the cottonseed albumins, which unlike albumins of many other species, are enriched in methionine (10% of the total amino acid content). A single copy of the gene is present in both the A and D genomes; however, only cDNA sequences corresponding to the Mat5-D were recovered, suggesting that only one alloallele is actively transcribed (Galau *et al.,* 1992).

Considerable efforts have focused on the genetic modification of seed storage proteins. Generally, two strategies have been employed, both with success. One strategy has been to introduce heterologous genes encoding proteins with desired amino acid profiles into transgenic plants (*e.g.,* expression of the "high-methionine" 2S albumin from Brazil nut into soybeans; Couglan *et al.,* 1997). The other strategy has been to alter the coding sequence of endogenous storage protein genes to elevate the percentage of desired amino acids (*e.g.,* increased methionine content of phaseolin, Dyer *et al.,* 1995). In both cases, it is important to consider factors which determine the proper folding, assembly, and packaging of these modified storage proteins. Cotton may benefit from such efforts to manipulate amino acid profiles in the seed meal; however, concerns over levels of the secondary metabolite, gossypol, must first be addressed before changes in nutritional properties of the meal will be of real value (Jones and King, 1996).

Protein body formation in cotton likely proceeds in a manner similar to other non-leguminous dicots, *i.e.,* sunflower, pumpkin, *Brassicas,* and castor bean, wherein globulins and albumins are inserted into the ER cotranslationally and transported to the vacuoles via the Golgi apparatus (Shewry *et al.,* 1995; Robinson and Hinz, 1996). Glycosylation and processing is predicted to occur similar to globulins and albumins of other species; however, little direct experimental evidence exists in maturing cottonseeds. Discrete packaging and organization in protein body vacuoles is not well understood, but likely plays an important role in the biogenesis of this organelle. Mechanisms of protein body formation in seeds differs among species and tissues (Shewry *et al.,* 1995). In fact, packaging of storage proteins into different types of protein bodies can occur within the same cell, as was documented in developing endosperm of oat (Lending *et al.,* 1989).

The precise organization and mechanism of protein body formation in maturing cottonseeds remains largely unknown. Molecular chaperones of the ER are believed to play an important role in storage protein assembly and packaging in other species such as maize (Zhang and Boston, 1992; Li *et al.,* 1993). It is not yet known whether chaperone proteins are involved in the ontogeny of cottonseed protein bodies; however, cottonseed calnexin, a membrane-bound ER chaperone, was recently isolated and determined *in vitro* to bind target proteins in an ATP-dependent manner. Future research on the formation/assembly of cottonseed protein bodies will be important in designing stra.tegies to modify the storage protein composition for improved nutritional value.

4.3   Embryo Desiccation

Storage protein synthesis persists until 45 to 47 DPA, well into the stage of embryo desiccation. Desiccation begins around 35 DPA as maximum seed weight is realized (Reeves and Beasley, 1935). After 35 DPA, the fresh weight of the seeds decreases until it stabilizes around 40 DPA. Seed dry weights, however, continued to increase, until about 40 DPA (Reeves and Beasley, 1935). Desiccation of cotton embryos is correlated with the rise of abscisic acid (ABA) in cottonseeds (Davis and Addicott, 1972). Trace levels of ABA were reported at 30 DPA, whereas, the highest level was found in cottonseeds by 40 DPA (2.5 µg/fruit). ABA levels dropped to approximately 1.25 µg/fruit by 50 DPA.

Desiccation of a plant usually leads to serious physiological stresses; seeds, however, are programmed to enter into a desiccated, quiescent state. Seeds, as do whole plants, induce numerous cellular adjustments which protect the cell from desiccation (Oliver, 1996). The most serious problem results from the vulnerability of membranes (Tanford, 1978). As water is lost from the cell, the membranes become more fluid and deformable. Therefore, in order to protect the membranes, dielectric constraints on membranes have to be maintained by molecules other than water. Caffrey *et al.* (1988) reported that sucrose

could act as a replacement for water in seeds, stabilizing the structural integrity of membranes. Sucrose is a widespread constituent of dry seeds, however, it is not always the ideal replacement for water. As the seed dries, sucrose has the tendency to crystallize (Amuti and Pollard, 1977). Crystallization of sucrose can be prevented by the addition of raffinose, another storage sugar commonly associated with mature seeds (Caffrey *et al.,* 1988). In cottonseeds, raffinose is the common storage sugar (Doman *et al.,* 1982; see also *Chapter 10*) and can only be found in advanced stages of desiccation (Shiroya, 1963). Stachyose, the third most predominant storage saccharide in cottonseeds (Doman *et al.,* 1982; Shiroya, 1963), is formed by the transgalactosidation of raffinose and may also play a role in prevention of sucrose crystallization during desiccation.

Proteins have also been implicated in conferring desiccation tolerance to cottonseeds. Large macromolecules, such as proteins, would have a major advantage over the much smaller storage sugars – the ability to span and protect greater areas. In the first edition of Cotton Physiology, Galau (1986) reviewed an emerging group of polypeptides called *L*ate *e*mbryogenesis *a*bundant (*Lea*) proteins. These were originally referred to as subset 5 and characterized by transcript and protein accumulation during late embryogenesis (Dure *et al.,* 1983; Galau, 1986). In all, 18 *Lea* proteins have been classified (Dure *et al.,* 1981; Galau *et al.,* 1986) with six these inducible by exogenous ABA (Hughes and Galau, 1989). The kinetics of transcript/protein accumulation are of two classes, *Lea* and *LeaA* (Hughes and Galau, 1989). The *Lea* class is expressed at low levels (some as early as 22 DPA) during embryo maturation, but transcript levels rapidly increase in the embryos after 45 to 47 DPA (Hughes and Galau, 1989). The period of 45 to 47 DPA correlates with abscission of the funiculus (referred to as postabscission by Hughes and Galau) which terminates nutrition and water transport to the seed from the mother plant (Benedict *et al.,* 1976). The *LeaA* class have an ABA-associated expression as well as postabscission expression (Hughes and Galau, 1989).

The hypothesis remains that many of these *Lea* or *LeaA* proteins are functionally involved in eliciting desiccation tolerance in the seed (Galau *et al.,* 1986; Hughes and Galau, 1987). Three findings suggest this function: first, postabscission expression *Lea* transcript and protein (Hughes and Galau, 1989); second, *Lea* transcript and protein levels drop dramatically after the inception of imbibition (Hughes and Galau, 1987; Hughes and Galau, 1989); and third, expression of these *Lea* transcripts is not required for precocious germination (Dure *et al.,* 1981; Hughes and Galau, 1991). Additionally, homologs of the *Lea* and *LeaA* proteins are found to be induced during salt stress, chilling injury, and water stress; all conditions when cells encounter severe osmotic stress or periods of desiccation (Dure, 1993; Close, 1996; Oliver 1996).

Putative functions of the *Lea* and *LeaA* proteins have been deduced by secondary and tertiary structure of the proteins and from the theoretical needs of desiccated cells. Interestingly, none of the *Lea* or *LeaA* proteins appear capable of forming globular entities; several families appear to exist as random coils or as amphiphilic α helices (Dure, 1993). The cotton *Lea* genes can be categorized into six multigene families, only four of which have cataloged homologs from other species (Dure, 1993; Dure 1994). These four families include D-19, D-113, D-11, and D-7. The D-19 family is highly hydrophilic with high glycine content and which structurally forms a random coil (70%) through much of its length (Dure, 1993). The Em wheat protein, a D-19 homolog, putatively functions in desiccated tissue by binding water and thereby resisting total desiccation (McCubbin *et al.*, 1985). The D-l13 family is characterized by high glycine, alanine, and threonine content, which contribute to the amino terminal α helix structure (Baker *et al.*, 1988; Dure, 1993). This protein is presently being tested for a function in preserving membrane structures in drying cells (Dure, 1993). The D-11 family, also called dehydrins, is a very diverse group of proteins which are generally characterized by a highly conserved 14 to 15 amino acid sequence (Dure, 1993; Close, 1996). The conserved region is designated as the K sequence and is highly enriched in lysine residues (example K sequence – EKKGIMDKIKEKLPG) with 1 to 11 copies present in dehydrin proteins (Close, 1996). Two other motifs are common but not universal: first the S segment which consists of a tract of 6 to 9 serines; and second, the Y segment, which is related to the binding sites of chaperons and consists of the residue sequence (V/T)DEYGNP. The D-7 family is characterized by a tandem repeating 11-mer which likely functions to sequester salts during desiccation (Dure, 1993, 1994), protecting the cell from salt precipitation or crystallization.

Biogenesis of cellular organelles also occurs during embryo maturation and desiccation. Peroxisomes, which are involved in the metabolism of fatty acids, are reported to persist through maturation and desiccation into growing seedlings (Kunce *et al.*, 1984). In maturing cottonseeds, the peroxisomes were originally characterized as incomplete glyoxysomes, because activity of the glyoxylate cycle enzyme ICL was found to be low (Trelease *et al.*, 1986). However, accurate measurements ICL activity in maturing seeds has been difficult due to apparent inhibitors in embryo homogenates (Theimer, 1976; Frevert *et al.*, 1980; Fusseder and Theimer, 1984). Turley and Trelease (1990) established that transcripts and protein of both glyoxylate cycle enzymes, MS and ICL, existed in embryos from 17 DPA throughout cottonseed maturation and desiccation. These levels were however, relatively low when compared to transcript and protein levels from 2 to 3 day-old seedlings. Doman and Trelease (1985) used Protein A-gold immunocytochemical staining to detect ICL protein in 50 DPA cottonseed glyoxysomes. These data are consistent with other oilseed species indicating the coexistence of MS and ICL in immature seeds. Allen *et al.* (1988) detected ICL

mRNA in maturing sunflower seeds using immunostaining techniques (MS mRNA was not examined). However, they could not detect ICL protein at these same stages of seed development on western blots using antiserum against cottonseed ICL. Comai *et al.* (1989) identified transcripts of ICL and MS in maturing *Brassica napus* L. seeds.

Taylorson and Hendricks (1977) reviewed a few of the cellular changes which occur during seed drying. These included ER disappearance or fragmentation; absence of Golgi bodies; lipid bodies form a lining on the plasmalemma, and mitochondrial cristae become less abundant. These responses are only partially seen in mature cottonseeds. The electron microscopic analyses of both immuno- and cytochemical staining of 50 DPA mature cottonseeds demonstrated that peroxisomes and lipid bodies could be easily identified (Kunce *et at.*, 1984; Doman and Trelease, 1985). Other organelles were more difficult to visualize due to the small size of the published micrographs.

5.   GERMINATION

In a reversal of storage product deposition during embryogenesis, germination begins a period of mobilization of oil, protein, and carbohydrate reserves. This reversal begins at rehydration (imbibition) of the seed. Imbibition, therefore, is a key event which signals the deviation from quiescence into a state of high metabolic activity and growth. One of the first actions of the imbibing seed is to repair damage of the cellular machinery incurred during desiccation or inflow of water into the cells. Cellular damage to the seed is usually measured by leakage of ions (Oliver, 1996). Reconstruction or repair of protein synthesis is required during early stages of imbibition. In imbibing wheat embryos, Obendorf and Marcus (1974) found that the concentration of ATP increased 5-fold by 30 min and 10-fold by 60 min., thereafter remaining at a constant level through germination. The requirement for ATP during the early hours of imbibition is consistent with reinitiating protein synthesis and biogenesis of protein structures (Spiegal and Marcus, 1975).

Early energy requirements are likely fulfilled from the storage carbohydrates raffinose, sucrose, and stachyose (Kuo *et al.*, 1988). These three sugars exist at a combined weight of 6 mg/cotyledon pair in dry cottonseeds. The concentrations of raffinose and stachyose drop 10-fold in the first 24 h seedling growth, whereas, concentrations of sucrose increas for 24 h before stabilizing (Doman *et al.*, 1982). A similar drop of raffinose was observed when using *G. herbaceum* seeds (Shiroya, 1963).

An unusual membrane phospholipid, N-acylphosphatidylethanolamine (NAPE) has been identified in cotyledons of germinated cottonseeds (Chapman and Moore, 1993a). NAPE is a derivative of the more common membrane phospholipid, phosphatidylethanolamine (PE), with a third fatty acid moiety attached to the ethanolamine head group via an amide bond. NAPE accumulates in animal cells under conditions which favor tissue damage (*e.g.*, ischemia), and biophysical studies with aqueous dispersions indicate that NAPE has membrane bilayer stabilizing properties (Domingo *et al.*, 1994). Due to its unusual enzyme activity, it was proposed that NAPE synthase may scavenge free fatty acids in membranes under certain physiological conditions to help stabilize membrane bilayers and maintain cellular compartmentation (McAndrew *et al.*, 1995; Chapman, 1996). In a comprehensive examination of NAPE biosynthesis in cotton cotyledons at various stages of development, including embryogenesis, germination, post-germinative growth (with or without light), and senescence, the highest NAPE synthase activity was associated with germinating seeds (Chapman and Sprinkle, 1996). Also, cool growth temperatures appeared to induce NAPE synthase activity in cotyledons of germinated seeds. Both desiccation and temperature stresses are known to involve the release of free fatty acids in membranes, and NAPE biosynthesis in cotton cotyledons may be an endogenous protective mechanism to help minimize the cellular damage associated with these stresses.

Cottonseed NAPE was identified and characterized structurally by a combination of biochemical and biophysical approaches and the *de novo* pathway leading to its synthesis was elucidated (Chapman and Moore, 1993a). A microsomal enzyme, designated NAPE synthase, catalyzes the direct acylation of PE with unesterified free fatty acids (Chapman and Moore, 1993b; Chapman *et al.*, 1995). NAPE synthase was photoaffinity labeled with a photoreactive fatty acid analogue and a 64 kDa protein was correlated with the unusual enzyme activity (McAndrew *et al.*, 1995). Cottonseed microsomal NAPE synthase recently was purified to homogeneity and confirmed to comprise a 64 kDa protein (Cai *et al.*, 1995). Purified NAPE synthase catalyzed the ATP- and CoA-independent formation of NAPE from PE and palmitic (or linoleic) acid (Chapman and McAndrew, 1997).

6.   POSTGERMINATIVE GROWTH

A period of seedling growth follows germination and is characterized by the rapid synthesis of gluconeogenic precursors from stored lipids. Mobilization of lipids becomes measurable approximately 16 h after the initiation of imbibition (Doman *et al.*, 1982). A concomitant appearance of glyoxylate cycle and fatty acid β-oxidation enzymes also occurs. A detailed discussion of the glyoxylate cycle in cotton was outlined previously (Trelease *et al.*, 1986), with a schematic of the glyoxylate cycle and the gluconeogenic pathways shown in Figure 29-4. Metabolites which are produced through these pathways are responsible for supplying energy and building blocks for seedling growth.

6.1   Glyoxysome Biogenesis

It is well documented that the activities of glyoxysomal enzymes increase dramatically, peaking a few days after seed imbibition (Choinski and Trelease, 1978; Kunce *et al.*, 1984; Turley and Trelease, 1990). This increase in enzyme activity is accompanied by a concomitant increase in glyoxysomal size. Kunce *et al.* (1984) performed morphometric analyses of cotton glyoxysomes during maturation, germination, and seedling growth. Glyoxysomes remained nearly spherical throughout maturation and imbibition, then dramatically increased in size, forming pleiomorphic organelles by 48 h of seedling growth. No substantial increase in glyoxysomal number or in cell size was found; neither was there any evidence of glyoxysomal degradation. This increase in size is not unique to cotton; glyoxysomes from watermelon also enlarge during postgerminative growth. However, unlike cotton, watermelon glyoxysomes appear to be formed in close association or derived from ER (Wanner *et al.*, 1982). No electron microscopic evidence for ER appendages on enlarging glyoxysomes was found in cottonseeds.

Several models have been proposed to explain the biogenesis of glyoxysomes in oilseeds (Bewley and Black, 1994). The experimental evidence accumulated for cottonseed glyoxysome biogenesis overwhelmingly favors the hypothesis proposed by Trelease (1984). Glyoxysomal enzymes were shown to be incorporated into organelles posttranslationally without proteolytic processing or glycosylation (Trelease *et al.*, 1987; Turley and Trelease, 1987; Kunce *et al.*, 1988). No cleavable signal sequence could be identified from sequences reported for MS (Turley *et al.*, 1990a), ICL (Turley *et al.*, 1990b) or catalase (Ni *et al.*, 1990; Ni and Trelease, 1991a). Chapman *et al.* (1989) reconciled the controversial localization of malate synthase and catalase in ER as a cosedimentation of aggregated enzymes with ER vesicles not an ontogenetic relationship between ER and glyoxysomal enzymes. This was shown both biochemically and immunocytochemically. A cDNA clone encoding a glyoxysomal membrane protein, identified as ascorbate peroxidase, revealed no ER targeting sequences and immunolocalization of this protein demonstrated an exclusive glyoxysomal membrane compartmentation

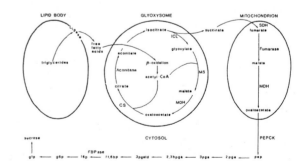

Figure 29-4. Glucogenesis from storage lipid during oilseed germination. Abbreviations: ICL, isocitrate lyase; MS, malate synthase; MDH, malate dehydrogenase; CS, citrate synthase; and SDH, succinate dehydrogenase. Figure is from Trelease *et al* (1986).

(Bunkelman and Trelease, 1996). Perhaps most convincing, the transcription and translation of glyoxysomal proteins in transgenic tobacco (*Nicotiana tabacum* L.) cells (Banjoko and Trelease, 1995) or plants (Olsen *et al.,* 1993) *in vivo*, resulted in the accumulation of proteins in peroxisomes and not in compartments of the secretory pathway.

Other problems arise with the models of glyoxysomal biogenesis which attempt to incorporate ER function in glyoxysomal membrane synthesis. In cottonseeds, glyoxysomal membrane lipids were not synthesized directly from the ER. Radiolabeling experiments in cotyledons indicated that membrane phospholipids were synthesized in the ER and transported to mitochondria, but not to enlarging glyoxysomes (Chapman and Trelease, 1990; 1991a, 1991b). The glyoxysomes also did not have the capacity to synthesize their own membrane lipid. Instead, because of their unusual enrichment in nonpolar lipids, glyoxysome membranes were proposed to be derived from nonpolar and polar lipids of lipid bodies during mobilization of stored lipid (Chapman and Trelease, 1991b). In support, radiolabeled triolein and phosphatidylcholine were transferred to cottonseed glyoxysome membranes from lipid bodies *in vitro*. The precise mechanisms for incorporation of proteins and lipids into enlarging cottonseed glyoxysomes remain largely unknown and warrant further investigation.

During postgerminative growth of cottonseeds, the lipid and protein reserves are depleted. This depletion is accompanied by a drop in the enzymatic activities and transcript levels of MS and ICL, as plants prepare to become photosynthesizing units. The enzyme Cat is generally not considered a glyoxylate cycle enzyme, however, its activity also drops (Ni and Trelease, 1991b). Cat is a marker enzyme for peroxisomes and is responsible for eliminating hydrogen peroxide which is produced in the cell during metabolism *e.g.*, hydrogen peroxide is a product of β-oxidation and photorespiration (both peroxisomal functions). Upon further investigation of the decrease of Cat activity, it was found that multiple isoforms of Cat exist in cotton cotyledons (Kunce and Trelease, 1986). These isoforms are formed by the combination of two different genes referred to as SU1 and SU2 (Ni and Trelease, 1991b). The SU1 form is expressed during the maturation of embryos and early seedling growth (Kunce and Trelease, 1986; Ni and Trelease, 1991b). After one day of seedling growth, the transcript levels of SU1 decrease; the transcripts of SU2 (leaf-type isoform) begin increasing by three days of seedling growth (Ni and Trelease, 1991b). A combination of these two isoforms interact to form five isoforms of cottonseed catalase (Ni *et al.,* 1990). Future work with Cat could add insights into the understanding the transition of seed-type glyoxysomes to the leaf-type peroxisomes.

7. PRACTICAL IMPLICATIONS

In summary, cottonseed ontogeny can affect both producers and consumers of cotton products. Prooucers are generally more interested in production of fiber–which accounts for approximately 80% of the value of the cotton crop. Genetic improvements have therefore focused on producing lines which produce more fiber. This has led to selecting high lint yielding lines which has been correlated with smaller seed size. A reevaluation of cottonseed and their products could lead to enhanced economic value of the cotton crop. Improving seed quality would enhance the profits associated with cotton production. An example is cottonseed oil which is recognized for its flavor enhancing properties and stability factors in the cooking/frying and food processing industries. However, with health concerns over the consumption of saturated fatty acids, criticism has been raised to the use of cottonseed oil in many of its traditional dietary applications. Certainly, cottonseeds and their products could benefit from biotechnology and genetic efforts to decrease saturated fatty acid composition. Another concern has been the concentrations of gossypol in seeds. A decrease in this harmful compound would also facilitate and increase the use of cottonseed products. Therefore, an understanding of the important events in embryogenesis, germination, and seedling growth will likely lead to improved cotton genotypes.

8. ACKNOWLEDGMENTS

The authors would like to thank Drs. Richard N.Trelease, Glen Galau, Judith M. Bradow, and Kevin C. Vaughn for reviewing this manuscript.

CHAPTER 30

SECONDARY PRODUCTS

Robert D. Stipanovic[1], Howard J. Williams[2], and Alois A. Bell[1]
[1]*United States Department of Agriculture, Agricultural Research Service, Southern Plains Agricultural Research Center, College Station, Texas 77845; and* [2]*Texas A & M University, Departments of Chemistry and Entomology, College Station, Texas 77843*

1.   INTRODUCTION

In spite of their name, ""secondary" products are essential for plant survival. They are required for basic cell functions as well as communicating the plant's presence to the surrounding environment and providing defense against pests (*i.e.*, diseases, nematodes, insects, and plant competitors). Significant information has been gathered since Bell (1986) reviewed secondary products in *Cotton Physiology* (Mauney and Stewart, 1986). This new information deals with analytical techniques used to identify and quantify secondary compounds, biological activity, factors controlling synthesis, discovery of new compounds, and mechanisms involved in biosynthesis. Particularly relevant to future research is the identification of the genes and gene products that produce these compounds and regulate their synthesis.

2.   VOLATILES

Cotton plants produce a variety of volatile chemicals. Some are responsible for the "green" odor associated with cotton leaves, while others produce the diverse odors that are characteristic of the whole plant. Some volatiles can be used as genetic markers in breeding programs. Due to their obvious importance in cotton ecological interactions, volatile chemicals have been extensively studied. The field might be exhausted except for advances in modern analytical techniques that have allowed a continuous expansion of knowledge as chemical determinations are made on smaller and smaller amounts of material.

2.1   Extraction Techniques

Early analytical methods to identify volatiles in cotton required the isolation of large amounts of chemical due to rudimentary techniques available at the time. Steam distillation was often the method of choice for these preparations. A secondary method also in use was collection from refrigerated coils in the vicinity of growing cotton followed by solvent extraction (Minyard *et al.*, 1965, 1966, 1967, 1968; Hedin *et al.*, 1971). Steam distillation tends to be a relatively harsh procedure, causing pyrolysis, acid-base degradations, and Diels-Alder, dehydration, or other thermal reactions. In cold trapping of plant volatiles, a very large amount of water is collected which contains only minute quantities of the volatile organic compounds of interest.

As amounts of chemicals required for analysis decreased, solvent extraction or solid phase adsorption replaced the methods described above. Low boiling solvents such as hexane, ether, methylene chloride, and ethyl acetate, or mixtures of these with a given polarity have been used, usually to extract fresh green plant material (Elzen *et al.*, 1985; Bell *et al.*, 1987, 1988). Air drying or freeze drying of the plant material prior to extraction is avoided because these procedures may remove volatile compounds of interest. The polarity of the solvent or mixture should be high enough to extract alcohols and other compounds of interest, but not so polar as to extract interfering compounds such as chlorophylls, gossypol and its related terpenoids, tannins, and other high molecular weight compounds; these compounds can destroy expensive gas chromatography (GC) columns. Preparative thin layer (TLC) or column chromatography utilizing silica gel, florisil, polyamide, reverse phase, or mixtures of these gels can be used to reduce the amounts of components deleterious to the GC.

J.McD. Stewart et al. (eds.), *Physiology of Cotton*,
DOI 10.1007/978-90-481-3195-2_30, © Springer Science+Business Media B.V. 2010

Direct analysis of the resulting extracts produces best results in quantification, but if concentration of the extract is required, the practice of removing the solvent using a stream of nitrogen gas should be avoided. Sample concentration with a short packed or spinning band column results in an extract more nearly representative of the original.

Solid phase extraction of volatiles involves passing purified air through a chamber containing plant material and then through a collection tube packed with porous polymer beads (usually Porapak Q® or Tenax®). These beads remove organic chemicals from the air stream but allow water vapor and other polar compounds to pass through. The volatile organics can then be removed from the collection tube either by solvent extraction (Chang *et al.*, 1988) or by heat desorption, usually directly onto a GC column for analysis (Zeringue and McCormick, 1989). Solid phase extraction is especially useful in determining changes in volatile evolution occurring over time (*e.g.*, morning *vs.* afternoon, day *vs.* night) or due to plant stress of various kinds (insect damage, water stress, temperature) (Loughrin *et al.*, 1994), and with some effort can be used in the field as well as the laboratory. Both solvent and solid phase extraction can be used to determine differences of volatiles in old versus young leaves, different parts of the plant, etc.

2.1.1 Method Comparison

Fresh green plant extraction produces very different quantitative and even qualitative patterns of volatile compounds than solid phase extraction. Solid phase extracts of volatiles from air contain higher quantities of the more volatile green leaf chemicals and monoterpenes and relatively smaller amounts of the less volatile sesquiterpenes. If the solid phase extraction tubes are placed a long distance from the plants, the collection of less volatile components is further reduced, and components of low volatility are often absent altogether. However, if the purpose of a study is to determine the effect of volatile chemicals at a distance on an organism, solid phase extraction can be the method of choice. The sample in this case more closely resembles the chemical blend perceived by an organism located some distance from the plant. For other studies, such as those concerning natural product biosynthesis, chemical taxonomy, close range attraction and effects of ingestion of volatile chemicals on herbivores, the more complete picture is given by solvent extraction.

2.2 Volatile Identification Procedures

Identification of the volatile compounds found in cotton essential oils has depended largely on gas chromatography, coupled when necessary with infrared (IR), mass (MS), and nuclear magnetic resonance (NMR) spectroscopy for determination of new structures or confirmation of identifications tentatively made by GC analysis. Samples have been purified for analysis mainly by preparative GC. Capillary GC is a major advance. Green leaf chemicals and monoterpenes are best determined on relatively polar columns usually related to OV-17 or the carbowax series, while sesquiterpenes are analyzed on non-polar GC stationary phases. Unknown compounds are identified by comparisons of GC retention time and mass spectra of unknown compounds with those of standards. Identifications based on comparisons of mass spectra of unknown compounds with those in an MS spectral library without the use of standards are less definitive. Many terpenoids produce similar spectra and spectra obtained on a given MS instrument are not totally comparable to those obtained elsewhere. Rare compounds and compounds not previously reported, which are of greatest interest, will not be found in the spectral library. NMR spectroscopy and especially ^{13}C NMR is the most reliable method for identifying unknown compounds by comparison with published data. However, ^{13}C NMR suffers from poor sensitivity, and tends to be under-reported by workers in the field. Most identifications of new compounds now rely mainly on ^{1}H and ^{13}C NMR data.

2.2.1 Green Leaf Compounds

The "green leaf" compounds are the most volatile of the chemicals found in extracts of cotton. These compounds are derived from fatty acid metabolism and include small alcohols, aldehydes, and related compounds. The major components of the cotton volatile mixture tend to have 6 to 8 carbons and are often unsaturated. GC-MS analyses shows some may be methyl branched, indicating interaction with another biosynthetic pathway or possibly methyl migration during formation. Representative examples are given below.

The number of individual "green leaf" compounds tentatively identified in cotton extracts from various sources exceeds one hundred. The total green leaf fraction and many of its individual components have a sharp, plant aroma like that of new mown grass. Plant damage by insects and fungal infections may increase the production of these compounds even in undamaged tissue (Röse *et al.*, 1996; Zeringue, 1996).

2.2.2 Monoterpenes

Terpenoids are compounds derived from mevalonic acid (Torssell, 1983) or 1-deoxy-D-xylulose-5-phosphate (Eisenreich *et al.*, 1997; Lois *et al.*, 1998; Sprenger *et al.*, 1997; Takahashi *et al.*, 1998) which are enzymatically converted to isopropenyl diphosphate (IDP). Isopropenyl diphosphate can be isomerized to dimethylallyl diphosphate (DMADP), the second "isoprene" building block, which then reacts with IDP to form head-to-tail isoprene polymers. Terpenoids with two isoprene units (C-10 compounds) are designated as monoterpenes. Geranyl diphosphate can form acyclic monoterpenes or can undergo cyclization, oxidation, and other processes, producing many volatiles of several structural and chemical types. Limonene, α- and β-pinene, β-myrcene, and trans-β-ocimene are found in most mod-

ern cotton varieties, while δ- and γ-terpinene occur in some primitive strains of *G. hirsutum* (Bell *et al.*, 1995, 1996).

Monoterpenoid fragrance is varied, ranging from the piney aromas of the pinenes, to the citrus odor of limonene, and the sharp unpleasant fragrances of myrcene and β-ocimene. In addition to their role as fragranc-

es, β-ocimene and myrcene are precursors of the heliocides that are discussed elsewhere in this chapter.

2.2.3 Sesquiterpenes

Combination of three isoprene units by the process discussed above leads to the production of sesquiterpenes (C-15 compounds). Through cyclization, oxidation, and conjugation, the sesquiterpenes of cotton are quite diverse. Most cotton cultivars produce α-copaene, β-caryophyllene, α-humulene, δ-cadinene, γ-bisabolene, and their oxygenated products (Scheme 1). Newly identified compounds α- and

Scheme I

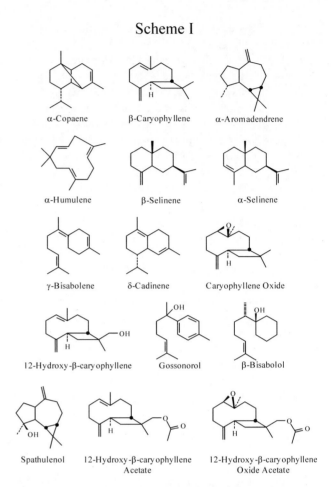

β-selinene (Williams *et al.*, 1995) and 12-hydroxy-β-caryophyllene and its acetate and oxide (Williams *et al.*, 1996) are useful taxonomic markers for geographical varieties and *Gossypium* species. Aromadenedrene and its derivative spathulenol belong to another structural type showing the wide variety of cyclization products possible in cotton. Among the sesquiterpene basic types reported from cotton sources are acyclic copaene, caryophyllene, humulene, cadinene, eudesmene, and aromadendrene (Williams *et al.*, 1995). Hydroxylation of terpenoids to form β-bisabolol and 12-hydroxy-β-caryophyllene probably involves P-450 type enzymes, but some aromatization and epoxidation may result from air oxidation because compounds of these types tend to occur only in old or dead leaf tissue, and in some cases these compounds accumulate in extracts left at room temperature in light. Stipanovic *et al.* (1996) prepared labeled δ-cadinene for studies on the biochemical pathway leading to gossypol type compounds discussed later in this chapter.

2.2.4 Biological Activity

Individual chemicals from the volatile blend are active in a variety of biological assays. Green leaf compounds attract insects and affect growth and aflatoxin production of *Aspergillus flavus* (Zeringue and McCormick 1989; Zeringue, 1990, 1996). Green leaf and terpenoid components attract parasitoids which protect the plant by killing herbivorous species (Elzen *et al.*, 1984; Vinson *et al.*, 1987; Williams *et al.*, 1988; Vinson and Williams, 1991; Turlings and Tumlinson, 1991; Turlings *et al.*, 1991; Dicke *et al.*, 1993; Röse *et al.*, 1996). Certain terpenoids attract insect herbivores and affect feeding behavior (Bell, 1986; Dickens, 1986). Terpenoids also may affect rate of insect development, with low concentrations stimulating growth and feeding while high concentrations act as antifeedants (Gunasena *et al.*, 1988; Williams *et al.*, 1987; Vinson *et al.*, 1987). Toxicity of gossypol is enhanced by some of these compounds. Volatiles from cotton plants also can increase attraction toward pheromone baits of both boll weevil and boll worm. In glanded plants, the volatile terpenes and the gossypol type compounds are found in high concentration in subepidermal glands (Elzen *et al.*, 1985; Loughrin *et al.*, 1994). These glands are avoided by feeding herbivorous insects, especially in their early instars (Parrott *et al.*, 1983, 1989; Johnson, 1984; Montandon *et al.*, 1987; Hedin *et al.*, 1991). In addition to green leaf compounds, insect feeding damage can cause production of acyclic and homo-terpenoids that are absent or present in only minute quantities in undamaged plants and are not found in the subepidermal glands (Loughrin *et al.*, 1994; McCall *et al.*, 1994; Donath and Boland, 1995; Röse *et al.*, 1996; Paré and Tumlinson, 1998). These volatiles can be collected from areas not subject to herbivore damage, indicating a systemic effect on the plant. Labeling studies with ^{13}C, in which some labeling is observed on green leaf compounds and exceptionally higher levels are found in terpenoids, indicate that the compounds are synthesized following damage (Paré and Tumlinson, 1998). Donath and Boland (1995) have shown that 4,8-dimethylnona-1,3,7-triene is formed by lysis of nerolidol in *G. herbaceum*. This reaction can occur in response to certain lytic enzymes of the feeding herbivore (*e.g.*, β-glucosidase).

3. HIGHER MOLECULAR WEIGHT TERPENOIDS

3.1 Analytical Techniques

3.1.1 High Performance Liquid Chromatography (HPLC)

The recognition that many structurally related terpenoids are produced in glands (Bell *et al.*, 1987) or in response to disease infection (Bell *et al.*, 1993), led to the development of HPLC methods to identify and quantitate these compounds. Lee *et al.* (1986), Bell *et al.* (1993), and Zhang *et al.* (1993) published procedures for the analysis of *Verticillium dahliae*-induced terpenoids in stele infected tissue. An HPLC analysis of foliar gland terpenoids has also been published and compared to the aniline method (Stipanovic *et al.*, 1988). This HPLC method was also used for analyzing gossypol in seed (Stipanovic *et al.*,

1988; Khoshkhoo *et al.*, 1994). The HPLC method provided a more accurate measure of total terpenoids, but the aniline method, because of its speed, favorable economics, and high correlation with the data obtained by the HPLC method, remains a useful procedure. Schmidt (1989) also published an HPLC method that is applicable to terpenoids in seed. Another publication (Schmidt and Wells, 1990) describes an HPLC method for analyzing gossypol in seed, that is especially applicable to glandless cottonseed and other seeds from the family Malvaceae (Schmidt and Wells, 1990). With this method Schmidt and Wells (1990) detected gossypol in velvetleaf seed (6 ppm), okra (70 ppm), prickly sida (60 ppm), venice mallow (8 ppm), glandless Stoneville 213 cottonseed (50 ppm), and glanded Stoneville 213 cottonseed (2500 ppm). Based on the same method Abu-Tarboush and Ahmed (1996) reported 680 ppm of free gossypol in karkade (*Hibiscus sabdariffa*) seed flour. However, in a detailed HPLC study, Jaroszewski *et al.* (1992) were unable to detect gossypol [with a detection limit of 10 ppm (0.001%)] in any seed, aerial parts, or roots of the Malvaceae listed above, except Gossypieae. Thus, the presence of gossypol in species outside the tribe Gossypieae is uncertain. An HPLC method was published for analysis of gossypol in plasma (Wang, 1987).

Gossypol exists as two enantiomers due to hindered rotation around the binaphthyl bond. Because these enantiomers differ in biological activity (discussed later in this chapter), considerable effort has been directed toward developing methods to resolve them for both analytical quantitative studies and large scale preparative separation. Zheng *et al.* (1985) reacted (R)- or (S)-methylphenethylamine with (±)-gossypol to produce two diastereomers in a 1:1 ratio [*i.e.*, RR and RS if the (R)-methylphenethylamine is used]. The mixture could be resolved, although with apparent low resolution, on a silica column. Acid hydrolysis provided (+)- and (-)-gossypol with less than 5% of their antipodes. Matlin *et al.* (1985) utilized (R)-(+)-phenylethylamine and Sampath and Balaram (1986) utilized L-phenylalaninol to form diastereoisomeric Schiff bases and separated these on reverse phase HPLC columns. Matlin *et al.* (1987) utilized the Schiff base of (+)-phenylalanine methyl ester to obtain multi-gram quantities of each enantiomer. This Schiff base has also been used to determine enantiomeric ratios in various species of cotton (Cass *et al.*, 1991). Matlin *et al.* (1990) investigated the use of fluorescent amines to form Schiff bases in order to improve sensitivity. Kim *et al.* (1996) used

(S)-(+)-Gossypol

(+)-2-amino-1-propanol as an adduct to form a Schiff base. This method, with a detection limit of 2 ng, was used to analyze gossypol enantiomers in tissues from lambs fed cottonseed. Cass *et al.* (1999) developed a direct resolution of gossypol enantiomers by HPLC on a chiral stationary phase.

Of the various cotton species examined, only *G. barbadense* contains an excess of the (-)-enantiomer [30% more than the (+)-enantiomer] in one cultivar, *G. barbadense brazilians* (Tussac) (Cass *et al.,* 1991; Hron *et al.,* 1999). Xiang and Yang (1993) reported similar observations in 21 wild species of cotton.

3.1.2 Gas Chromatography (GC), Mass (MS), Ultraviolet (UV), and Nuclear Magnetic Resonance (NMR) Spectroscopy

Mass spectral analysis of gossypol and some of its derivatives using positive EI and CI, and negative CI have been reported (Phillips *et al.,* 1990; Phillips and Hedin, 1990). A method using second-derivative UV has been used to analyze for gossypol in seed (Botsoglou, 1991). This method does not require reaction with a chromogenic reagent such as aniline, since the second-derivative transformation of the absorption band at 300 nm permits direct quantitation of gossypol in seed extracts. The complete ^{1}H- and ^{13}C-NMR spectrum of δ-cadinene (Davis *et al.,* 1996) and for some desoxyhemigossypol derivatives (Alam *et al.,* 1994) have been published.

3.2 Synthesis

3.2.1 Gossypol and Analogs

Carbon-14 labeled gossypol has been prepared by feeding radiolabelled acetate to cold shocked cotton seedling roots (Rojas *et al.,* 1989). A chemical method for preparing tritiated gossypol has also been published (Stipanovic *et al.,* 1987). The diverse biological activity of gossypol has prompted a number of synthetic studies including: a) the gossypol analog (**1**) (Ognyanov *et al.,* 1989); b) substituted naphthols such as (**2**) where R is OH or OCH$_3$, R$_1$ is H, Br, COOH, COOCH$_3$, CH$_2$OH or CHO, R$_2$ is CO$_2$Et, CHO, CH$_2$OH or CH$_3$ and R$_3$ is H, NO$_3$, CO$_2$CH$_3$, CHO, or CH$_2$OH and dimers of 2 (Dallacker *et al.,* 1989); c) peri-acylated gossylic nitriles (**3**) where R=C$_2$H$_5$, C$_3$H$_7$ and C$_4$H$_9$ (Royer *et al.,* 1986), and d) dehydroxygossypol (**4**) and dehydroxy-

gossylic acid (**5**) (Royer *et al.,* 1995). The anti-malarial activity of the divalerate of gossypol nitrile was comparable to that of gossypol. The absence of the *peri*-hydroxyl groups in (**4**) and (**5**) did not significantly affect anti-HIV activity, but it did reduce cytotoxicity compared to gossypol.

Fish *et al.* (1995) reported that the Schiff bas-

3

4) R=CHO
5) R=COOH

es of gossypol can be photo-epimerized. This provides an efficient method to recycle the biologically less active (+)-enantiomer of gossypol after resolution.

(S)-(+)-gossypol with an enantiomeric purity of 99.2:0.8 has been prepared in 9% overall yield from a readily available starting material (Meyers and Willemsen, 1998). The authors note a slight modification of this procedure will provide the (R)-(-)-isomer with equal agility.

3.2.2 Other Sesquiterpenes

Gant and Meyers (1993) reported a new synthesis of the cotton phytoalexin, 2-hydroxy-7-methoxycadalene (9).

3.3 Identification and Occurrence

3.3.1 New Sesquiterpenoids

Davila-Huerta *et al.* (1995) reported the isolation of two new sesquiterpenoids, *cis*- and *trans*-7-hydroxycalamenen-2-one, (6) and (7), respectively. These compounds and *trans*-7-hydroxycalamenene (8) were isolated from cotton leaves following inoculation with the bacterial pathogen *Xanthomonas campestris* pv. *malvacearum*. The authors suggest these may be the biosynthetic precursors to the phytoalexins 2,7-dihydroxycadalene (9) and lacinilene C (10).

Raimondalone (11) was isolated from the leaves of a cotton plant derived from an interspecific hybrid (*G. hirsutum* x *G. raimondii* Ulbr.) (Stipanovic *et al.,* 1994). It is found uniquely in progeny from such a cross. *G. raimondii* produces raimondal (12) but not hemigossypolone (13), whereas *G. hirsutum* produces 13 but not 12 (Altman *et al.,* 1990, 1991). Analysis of F$_1$, F$_2$ and backcross progeny from a cross of a CAMD-E line homozygous for raimondal (12) production and the genetic marker line TM-1 (*G. hirsutum*), showed segregation for raimondal was consistent with this character being controlled by two dominant genes with epistasis [*i.e.* hydroxylation at C-2 and methylation (Bell *et al.,*

1

2

1994; Kohel and Bell, 1999)]. Raimondalone apparently is formed by genes from both *G. raimondii* (C-2 hydroxylation

and methylation) and *G. hirsutum* (C-4 hydroxylation). Zhang *et al.* (1998) reported the identification of two new sesquiterpenoid glycosides (14) and (15) from cotton oil cake.

3.3.2 Foliar Sesquiterpenoid Aldehydes

HPLC analysis of leaves from 58 accessions of 18 *Gossypium* species showed relatively few differences in terpenoid aldehydes (Altman *et al.*, 1990). Exceptions included raimondal (Stipanovic *et al.*, 1980) which was found uniquely in *G. raimondii*, and the absence of all terpenoid aldehydes (TA) except gossypol in leaves of some species. Fourteen genotypes of *G. hirsutum*, with terpenoid contents that varied from normal to very high, were grown at five diverse locations in Texas over 2 years (Altman *et al.*, 1989). HPLC was used to analyze squares and terminal leaves for gossypol, p-hemigossypolone (HGQ), and heliocides H_1, H_2, H_3 and H_4. Location means showed differences between the high and low values ranging from less than twofold for flower bud H_2 to as high as eightfold for leaf HGQ. Genetic X environment interactions gave smaller variance components than genetic variance components in all instances. Broad-sense heritability averages were 0.46 and 0.94 for leaf TAs on a plant-basis and on an entry mean-basis, respectively, and 0.61 and 0.93 for flower bud TAs on a plant-basis and on an entry mean-basis, respectively. With a regression coefficient for a stable genotype defined as 1.00, high terpenoid regression coefficients varied from 0.06 to 2.11, while commercial cultivars had values from 0.11 to 1.05. The authors concluded that selection for higher terpenoid levels is an attainable goal.

3.3.3 Seed Gossypol

Transgenic cottonseed have been analyzed for gossypol and cyclopropenoid fatty acid. The content of these compounds in glyphosate-tolerant cottonseed (Nida *et*

al., 1996) and in insect-protected cottonseed (Berberich *et al.*, 1996) were not different than those of the parental or other commercial cotton varieties. Seeds of all known Australian *Gossypium* species were surveyed for gossypol content (Brubaker *et al.*, 1996). No detectable level of gossypol was found in seeds from *G. sturtianum* Willis and *G. robinsonii* Muellea and the three species in section *hibiscoidea*. *G. stocksii* Masters and *G. somalense* Hutchinson have pigment glands in their seeds, but have very low gossypol concentrations (Xiang and Yang, 1993).

3.3.4 Animal Toxicity and Biological Activity of Gossypol

The presence of gossypol in seed lowers the economic value of both the protein and the oil derived from it. Although a source of high protein, cottonseed meal cannot be fed in large amounts to non-ruminant animals such as swine, poultry, and fish because of the presence of gossypol, which is toxic. Cottonseed can be fatal to dairy cattle when fed in high amounts (2.7 to 4.5 kg/cow/day) (Smalley and Bicknell, 1982). The toxicology of gossypol and its effect on reproduction of livestock has been reviewed (Randel *et al.*, 1992).

The potential value of glandless cottonseed has long been recognized. However, attempts to grow commercial glandless cotton, which did not produce gossypol or the related terpenoids in the seed or foliage, respectively, were a disaster. Glandless cottons suffered severe losses not only from insects, which are normally recognized as pests of cotton, but also from non-traditional insect pests (Lukefahr *et al.*, 1966). The glandless cottons lacked the natural defense compounds that are present in the foliar glands. These compounds include the heliocides H_1, H_2, H_3, and H_4 and hemigossypolone in addition to small amounts of gossypol (for structures of these compounds see Scheme II). Studies by Hedin *et al.* (1992), showed that heliocides H_1 and H_2 and hemigossypolone were more closely associated with resistance than gossypol and other gland components that they measured. However, gossypol acts as an antifeedant to first instar of *Heliothis virescens* (Parrott *et al.*, 1983, 1989; Johnson, 1984; Montandon *et al.*, 1987), and is toxic to later instars (Montandon *et al.*, 1987; Hedin *et al.*, 1988). The early failure of glandless cottons to resist insects led to evaluations of alternative approaches to achieve a gossypol free (or benign gossypol) cottonseed, without disrupting the plants natural defenses.

The seminal finding by Waller *et al.* (1983) that the (+)-enantiomer of gossypol was inactive as a male antifertility agent (confirmed by Wang *et al.*, 1987) was particularly important. This finding and the ability to isolate the individual enantiomers resulted in studies on the biological activity of both enantiomers.

Zheng *et al.* (1985) reported that although they differ in their antifertility action, both enantiomers inhibited lactate dehydrogenase-X, the proposed specific target for antifertility action. Yao *et al.* (1987) confirmed these findings.

The enantiomers also vary in other biological activities. In a broad range of tests the (-)-enantiomer generally was more active against the amoebic parasite, *Entamoeba histolytic*, than either the racemate or the (+)-enantiomer (Gonzalez-Garza *et al.*, 1993). Polsky *et al.* (1989) showed that gossypol was active against the HIV virus. However, again it was the (-)-enantiomer that was the active constituent (Lin *et al.*, 1989). Similarly the (-)-enantiomer had better antitumor activity (Liang *et al.*, 1995) and was more active against melanomas than the (+)- enantiomer (Blackstaffe *et al.*, 1997).

Differences exist in the disposition and metabolism of the enantiomers in mammals. In studies on rats, Chen *et al.* (1987) found the half-lives of free (+)- and (-)-gossypol after IV injection were 7.80 hr and 3.96 hr, respectively. The half-lives of the (+)- and (-) enantiomers in the gastrointestinal tract were 18.4 and 13.5 hr, respectively. In further antifertility studies, Yu (1987) found a single intratesticular injection of 200 μg of (-)-gossypol caused a 70% decrease in sperm count and atrophy of the testes in rats. Neither a significant decrease in sperm count nor atrophy of the testes was observed with a similar injection of the (+)-enantiomer. Kim *et al.* (1996) found that the ratio of (+)- and (-)-gossypol differed in various tissue when cottonseed was fed to lambs.

The toxicity of (+)-gossypol vs. (-)-gossypol is under investigation. In one study, the weights of broiler chicks fed a diet incorporating 5% of cracked cottonseed meal with 80% (+)-gossypol were not statistically different than those fed a control diet; whereas those fed a diet with 62% (-)-gossypol were significantly smaller than the controls (C. Bailey, personal communications). If (+)-gossypol is significantly less toxic than its antipode to livestock and poultry, a new approach to eliminate gossypol toxicity in cottonseed may be available.

The process that determines the ratio of (+)- and (-)-gossypol in the plant is not clear. Veech *et al.* (1976) showed peroxidase converts hemigossypol to gossypol. Since gossypol does not racemize easily, and optically pure (+)-gossypol has been found in *Thespesia populnea* (Jaroszewski *et al.*, 1992), the process of dimerization appears to be under enzymatic control, but the gene(s) controlling this process is not known. The identification of the gene controlling the dimerization of hemigossypol may be a desirable research goal. The cloned gene might be used in cotton to direct synthesis to the (+)-enantiomer as occurs in *T. populnea*.

The antifertility activity of various gossypol derivatives and analogs have been compared in rats (Wang *et al.*, 1990). Gossypol-6,6-dimethyl ether, which occurs naturally in cottonseed, especially *G. barbadense* (Stipanovic *et al.*, 1975), was among the compounds tested. In a rat antifertility bioassay it was approximately one-third as effective as gossypol when injected, and had no effect when administered orally. The inactivity of this compound when administered orally was subsequently confirmed (Xue *et al.*, 1992). In general, the methyl ether derivatives of the cotton terpenoids are less toxic than the conformed. Hemigossypol and desoxyhemigossypol are more toxic

than their methyl ether derivatives to the wilt pathogens *Verticillium dahliae* (Mace *et al.*, 1985) and *Fusarium oxysporum* f. sp. *vasinfectum* (Zhang *et al.*, 1993). Likewise, heliocides H_1 and H_2, and hemigossypolone are more toxic than their methyl ethers to the tobacco budworm *Heliothis virescens* (Stipanovic *et al.*, 1977). Thus, the methyl-transferase gene may offer a second strategy to convert gossypol to a benign substance in cottonseed. Progress in this area, and a discussion of a third strategy (*i.e.,* prevention of gossypol formation in the seed using an anti-sense construct) are presented in this chapter under "Biosynthesis."

The mode of action of gossypol and related compounds as antifertility agents remains uncertain. Both enantiomers of gossypol are potent inhibitors of lactate dehydrogenase-X in sperm (Zheng *et al.*, 1985; Yao *et al.*, 1987). Gossypol causes peroxidation of sperm membrane lipids (Kaur *et al.*, 1995), and the (-)-enantiomer may be better suited to cause such peroxidation. This would explain its shorter half-life. The toxicity of desoxyhemigossypol to *V. dahliae* has been correlated with its ability to form both free radicals and hydrogen peroxide (Stipanovic *et al.*, 1992; Mace and Stipanovic, 1995). These reactive species may oxidize critical components causing membrane disruption and inhibition of membrane-bound enzymes.

3.4   INDUCTION

3.4.1   By Bioregulators

Environmental interactions may have influenced the mixed results observed when bioregulators were used to alter the production of allelochemicals (Hedin and McCarty, 1991). During a study that extended from 1986 to 1991, urea and the commercial bioregulators Burst (a mixture of cytokinins including zeatin), Foliar Trigger (contains cytokinin), Maxon (contains cytokinin), FPG-5 Foliar (contains micronutrients, cytokinin, IAA auxin and gibberellic acid), and PG-IV (contains micronutrients, indolebutyric acid and gibberellic acid) caused increases of terpenoid aldehydes in flower buds (Hedin and McCarty, 1994). However, a statistically significant increase occurred only in some years and the increase was observed with all of the treatments including urea. None of the treatments in any year statically changed the concentrations of condensed tannins or flavonoids. Also yield, lint, boll size or seed index did not increase in yield, lint, boll size, or seed index in any year. Mepiquat chloride (PIX) caused increases in gossypol in squares in only two of nine years (McCarty and Hedin, 1994). The authors note that although yields decreased in most years as a result PIX treatment, earlier maturation of PIX treated plant may more than offset any loss in yield.

Bioregulators elicited changes in terpenoid content of cotton roots, especially those infected with the root-knot nematode (Khoshkhoo *et al.*, 1993). Roots of two susceptible (glanded and glandless) and two resistant (glanded and glandless) varieties were analyzed. Gossypol was the major terpenoid found, and concentrations in the healthy root were approximately the same in each variety. Treatment with PIX or salicylic acid increased the concentration of gossypol in the roots of all inoculated plants by at least two-fold, and in some cases the increases were over five-fold. The bioregulator Burst also increased gossypol, especially in the susceptible plants.

Feeding by *Heliothis zea* larvae enhanced levels of salicylic acid over three-fold compared to the control (Bi *et al.,* 1997). However, exogenous application of salicylic acid or methyl salicylate failed to increase resistance to *H. zea.* The concentrations of terpenoid aldehydes were not reported.

Gossypol production also increased in roots when cottonseed was treated with the herbicides prometryn or dalapon before planting (Awadalla and El-Refai, 1992). Gossypol levels were further increased when the plants were inoculated with *V. dahliae.* Duke *et al.* (1991) found that etiolated cotyledons from cotton seedlings infused with 0.5 mM clomazone {2-[(2-chlorophenyl)methyl]-4,4-dimethyl-3-isoxazolidinone} accumulated gossypol in greater amount (7340 ± 170 µg/cotyledon) than control cotyledons (4240 ± 540 µg/cotyledon).

3.4.2 By Stress

2,7-Dihydroxycadalene, 2-hydroxy-7-methoxy-cadalene and their optically active oxidation products lacinilene C and lacinilene C 7-methyl ether have been isolated from cotton leaves and cotyledons after inoculation with *Xanthomonas campestris* pv *malvacearum* (*Xcm*) (Essenberg *et al.,* 1990). These same investigators (Davila-Huerta *et al.,* 1995) reported 7-hydroxycalamenene is formed in bacterial inoculated cotton along with the phytoalexin 2,7-dihydroxycadalene and its 7-methyl ether. The authors suggested 7-hydroxycalamenene may be the biosynthetic precursor of 2,7-dihydroxycadalene. Beckmann and Heitefuss (1998) reported *Xcm* elicits the synthesis of dHG and HG in cotyledons; this was confirmed by Abraham *et al.* (1999). Induction of terpenoid synthesis in cotton tissues by microbial infections has been discussed in a series of in reviews (Bell *et al.,* 1986, 1993; Essenberg *et al.,* 1992; Bell, 1995a). This topic is also reviewed in the chapter on host-parasite relationships in this book. The induction of terpenoid synthesis by pathogens has been especially beneficial for biochemical and genetic studies. The biocontrol agent *Trichoderma virens* induces the biosynthesis of gossypol, desoxyhemigossypol, hemigossypol, desoxyhemigossypol-6-methyl ether, and hemigossypol-6-methyl ether (for structures of these compounds see Scheme II) in the root (Howell *et al.,* 2000). The induction of these compounds may play a critical role in the biocontrol efficacy of *T. virens.* This subject is covered in detail in the chapter entitled Cotton Host-Microbe Interactions (Chapter 18).

3.5 Synthesis

3.5.1 Gossypol

Early work based on incorporation studies with mevalonate-2-^{14}C, neryl-2-^{14}C-diphosphate and E,Z-farnesyl-2-^{14}C-diphosphate indicated that Z,Z-farnesyl diphosphate (16) was the biosynthetic precursor of gossypol (Heinstein *et al.,* 1970). However, Essenberg *et al.* (1985) using (1,2-^{13}C$_2$) acetate found that 2,7-dihydroxycadalene (9) is formed in *Xanthomonas*-infected cotton by a folding pattern consistent with Z,E-farnesyl diphosphate (ZE-FDP) (17) or an intermediate equivalent to nerolidol diphosphate (NDP) (18). This finding led to a reevaluation of (16) as the precursor to gossypol. Studies in which ^{14}C-labelled mevalonate (Masciadri *et al.,* 1985) or (1,2-^{13}C$_2$) acetate (Stipanovic *et al.,* 1986) was fed to intact cotton showed that gossypol is formed by the same intermediate as 2,7-dihydroxycadalene (9) and not from (16). Akhila and Rani (1993) confirmed these findings in their study of gossypol biosynthesis in *Thespesia populnea.* Thus, the biosynthesis of the cotton phytoalexins and gossypol agree with the

early steps established for cadinene sesquiterpenoids in other species. Davis and Essenberg (1995), working with *Xanthomonas campestris* pv. *malvacearum*-infected cotton cotyledons, showed that (+)-δ-cadinene (19) is the product of cyclization from E,E-FDP (20). They also showed that δ-cadinene synthase (δ-CS) is the first enzyme unique to the synthesis of the terpenoid aldehydes including gossypol.

Utilizing a *G. arboreum* cell suspension culture, Chen *et al.* (1995) isolated two cDNA clones that contain open reading frames coding for proteins of 554 amino acids with M$_r$ 64,096 and 64,118. The encoded protein from the XpC1 cDNA was expressed in *Escherichia coli* and purified. The protein synthesized (+)-δ-cadinene from FDP (20). Purification of this enzyme has also been reported by Davis *et al.* (1996). A very similar (+)-δ-cadinene synthase gene has been identified and cloned from *G. arboreum* L. cv. Nanking (Chen *et al.,* 1996). On the basis of sequence similarities, these three genes were grouped into two subfamilies, the *cad*1-C that includes C1 and C14, and the *cad*1-A gene. Davis *et al.* (1998) isolated a cDNA from *G. hirsutum* (*cdn*1) that was over 95% identical to the *G. arboreum cad*1-C1 and *cad*1-C14. Meng *et al.* (1999) isolated a new member of the (+)-δ-cadinene synthase family from *G. arboreum.* The cDNA clone encodes a protein that is 80% identical to the CAD1-C1 and C14 clones of (+)-δ-cadinene synthase from *G. arboreum.* Heterologous expression of this cDNA produced a 64kD protein that catalyzed the cyclization of farnesyl diphosphate to (+)-δ-cadinene. The identification of the genes coding for (+)-δ-cadinene synthase provides an opportunity to specifically delete gossypol production from seed using an antisense construct coupled to a seed specific promoter. Attempts to accomplish this task are under active investigations in several laboratories. However, the task may be difficult since at least six copies of the δ-cadinene synthase gene appear to occur in cultivated Upland cotton (Cui *et al.,* 1996).

3.5.2   Heliocides

Heliocides H_1 and H_4, and H_2 and H_3 are the products of a Direls-Alder reaction involving hemigossypolone and either β-ocimene or myrcene, respectively (Scheme II). The heliocides have been prepared in the laboratory by dissolving hemigossypolone and either β-ocimene or myrcene in a nonpolar organic solvent and allowing them to stand for a few days (Stipanovic *et al.,* 1977; 1978). The glands appear to offer a similar reaction "vessel" in that hemigossypolone (Gray *et al.,* 1976) and both β-ocimene and myrcene (Elzen *et al.,* 1985) are in the glands dissolved in an oily matrix. In the laboratory preparation, the ratio between heliocides H_3 and H_2, and between H_4 and H_1 are about 0.66 and 0.50, respectively. However, in a study of fourteen cotton varieties at five different geographical locations the H_4/H_1 ratio varied from 0.14 (± 0.01) to 0.33 (± 0.02) in flower buds and from 0.09 (± 0.01) to 0.24 (± 0.04) in leaves. The H_3/H_2 ratio varied from 0.32 (± 0.01) to 0.40 (± 0.01) in flower buds and from 0.34 (± 0.01) to 0.47 (± 0.01) in leaves (Stipanovic *et al.,* 1988). Some external factor appears to be influencing these product ratios. Stipanovic (1992) suggested that an enzyme may be involved. Evidence to support this proposal has been provided by the work of Oikawa *et al.* (1992) with *Chaetomium subaffine*, and the

Scheme II†

† HG = hemigossypol; dHG = desoxyhemigossypol; dMHG = desoxyhemigossypol-6-methyl ether;
MHG = hemigossypol-6-methyl ether; HGQ = hemigossypolone; MHGQ = hemigossypolone-6-methyl ether;
HH_1, HH_2, HH_3, HH_4 = heliocides H_1, H_2, H_3, H_4; HB_1 = heliocide H_1-6-methyl ether; HB_4 = Heliocide H_4-6-methyl ether.

work by Oikawa *et al.* (1995) reporting the isolation of a so-called Diels-Alderase enzyme from *Alternaria solani.*

3.5.3 O-Methylation of Terpenoids

Addition of a methyl group (Scheme II) to the phenolic group at the C-6 position dramatically reduces the biological activity of the terpenoids in cotton. Thus, methylation causes the heliocides to be less toxic to *Heliothis virescens*, desoxyhemigossypol to be less toxic to *Verticillium dahliae* and *Fusarium*, and gossypol-6,6'-dimethyl ether to be a less effective spermatocide. Work by Alchanati *et al.* (1994) identified the enzyme responsible for methylation of desoxyhemigossypol and showed that desoxyhemigossypol is the unique substrate for this enzyme (*i.e.*, hemigossypol did not act as substrate). The enzyme has been isolated, purified and characterized (Liu *et al.*, 1999). The native enzyme has a mass of 81.4kD and dissociates into two subunits of 41.2 kD on sodium dodecyl sulfate-polyacrylamide gel electrophoresis. This enzyme controls the production of several compounds in cotton (right side of Scheme II). Genetic manipulation of this gene could be beneficial. For example, inhibiting the production of methylated phytoalexins and heliocides could increase resistance to diseases and insects, respectively. Conversely, increasing its expression in seed could convert gossypol to gossypol-6,6'-dimethyl ether, which may be less toxic and provide a more nutritious and valuable cottonseed.

3.5.4 Polyphenols

Condensed tannins occur in most parts of the cotton plant. On hydrolysis they yield cyanidin and often delphinidin. The tannins co-occur with the flavan-3-ols (+)-catechin and (+)- gallocatechin. The chemistry and biological significance of these compounds in cotton have been reviewed (Bell *et al.*, 1992; Ismailov *et al.*, 1994). The occurrence of these compounds has been correlated with disease resistance and extracts of tannins from cotton are toxic to *V. dahliae* (Bell *et al.*, 1992). However, some studies suggest tannins play only a minor role in disease resistance (Song and Zheng, 1997). Studies on the relationship between flavanol concentrations and insect resistance have provided mixed results. Confounding factors include: high variations normally obtained during analysis, high protein content of tissue that can offset the anti-nutritive value of the tannins (Yokoyama and Mackey, 1987; Yokoyama *et al.*, 1987), and the presence or absence of other antifeedant/toxicants (*i.e.*, the terpenoid aldehydes) in various tissues. The observation that insect larvae, especially neonates, feed selectively on tissue with low tannin content supports the contention that tannins are involved in insect resistance. Furthermore, cotton tannins reduce growth of young larvae of *H. virescens* (Parrott *et al.*, 1987). However, elevated levels of tannins did not prevent damage to fruiting structures even though insect growth was inhibited (Schuster *et al.*, 1990;

Zummo *et al.*, 1983). Larvae of *H. virescens* larvae fed flower buds from *G. arboreum* plants with diversely colored flower petals, and therefore higher levels of certain flavonoids, grew equally well as those fed buds from commercial *G. hirsutum* lines (Hedin *et al.*, 1992). However, gossypetin 8-O-rhamnoside and gossypetin 8-O-glucoside present in *G. arboreum* flower petals were toxic to *H. virescens* with ED_{50}'s of 0.007% and 0.024%, respectively.

A comparison between polyphenols in callus tissue and plants has been reported (Karimdzhanov *et al.*, 1997). The cotton flavonoids quercetin-3-rutinoside and kaempferol induced the accumulation of the melanin metabolite 2,5-didroxy-1,4-naphthoquinone when added to *V. dahliae* cultures (Navrezova *et al.*, 1986). Quercitin-3-glucoside co-accumulates with the phytoalexin 2,7-dihydroxycadalene in cotton tissue infected with *X. campestral* pv. *malvacearum*. The flavanol stabilizes this phytoalexin against oxidation by UV light (Essenberg *et al.*, 2000).

3.5.5 Sterols

The total sterol content of roots of two near-isogenic lines was compared after inoculation with *Meloidogyne incognita* (Hedin *et al.*, 1995). The control plants (uninoculated) of the susceptible cultivar Stoneville 213 had higher total sterol content (1.62 mg/g dried tissue) than the control resistant line 81-249 (1.35 mg/g). Eight days after inoculation the total sterol content had decline in the susceptible line (1.07 mg/g) while that in the resistant line had increased slightly (1.69 mg/g). The major sterols were stigmasterol and sitosterol that together accounted for over 80% of the sterols in both control and treated plants of both lines. Campesterol was the next most prevalent (9-13%) with minor amounts of cholesterol, lanost-8-en-3-ol, isofucosterol, stigmata-4,22-dien-3-one, stigmastan-4-en-3-one and a trace of stigmastanol. Because the amounts and distribution of the sterols in the susceptible and resistant lines were similar, the authors concluded that the sterols did not appear to be a factor in resistance to this nematode. Sitosterol was the major sterol in leaves of the Russian variety 108-F (Rashkes *et al.*, 1994). α-Tocopherol, undecaprenol, and dedicaprenol were also reported as major constituents. Rashkes *et al.* (1997) studied the levels of sterols and α-tocopherol during maturation in leaf blades and petioles in several cotton lines.

3.5.6 Lipids

With a highly sensitive and accurate method for analyzing cyclopropane fatty acids as phenacyl derivatives by HPLC, Wood (1986) found cyclopropane and cyclopropene fatty acids were located almost completely, if not completely, in the axis of the embryo. They were essentially absent in the cotyledons.

The relative concentrations of oil and fatty acids in germinating cottonseed changes over time. Six days after imbibing water, two varieties of *G. hirsutum* (Hybrid-4 and Laxmi) and 1 variety of *G. barbadense* (Suvin) showed a decrease in oil content, but the cyclopropane fatty acids decreased over 90% during this time (Pandey and Subrahmonyam, 1986). Banerji and Dixit (1995) reported changes in fatty acid content of mature and immature cottonseed oil.

Rikin *et al.* (1993) observed a relationship between a circadian rhythm for resistance to extreme temperatures and changes in fatty acid composition in cotton seedlings. Increased levels of polyunsaturated fatty acids coincided both with low-temperature-induced chilling resistance (5°C) and with the chilling resistance that develops rhythmically.

3.5.7   Miscellaneous Compounds

The turnover of ascorbic acid has been studied in developing cotton bolls (Bhatt and Renganayagi, 1986). The levels of free ascorbic acid and its enzymatic utilization were higher in the early stages of boll growth and declined progressively with age. A study of polar lipids in leaves of glanded and glandless cotton genotypes found phosphatidylethanolamine in the glanded genotypes but absent in glanded genotypes (Reine and Stegink, 1988). Eleven other polar lipids occured in both genotypes.

Glycinebetaine [$(CH_3)_3N^+CH_2COO^-$] increases to concentrations in excess of 100 mM in response to water deficit or salinity stress in *Gossypium* (Gorham, 1996b; see also Chapter 16), with highest concentrations in young tissue. In drought-stressed plants it amounted to 30 mmol/kg dry weight which represents 8 to 10% of the total nitrogen. Zuniga and Corcuera (1987) suggested that increased gly-

cinebetaine increases the susceptibility of water-stressed plants (*i.e.,* barley) to aphids (*Schizaphis graminum*). Shakhidoyatov *et al.* (1997) recently reviewed the chemical composition of cotton leaves including the hydrocarbons, amino acids, hydroxy acids, alcohols, triterpenes, phenols, carotenoids, sugars, and sterols. They identified over 75 different secondary compounds. John and Keller (1996) transfered *phb*B and *phb*C genes into cotton by particle bombardment. The fiber of the resulting cotton plants contained poly-(3R)-hydroxybutanoate (PHB), which gave the fiber improved insulating properties.

4.   CONCLUSION

The cotton industry has seen many advances in knowledge of plant secondary products since Cotton Physiology was published in 1986. Advances in molecular biology are providing new tools for the agricultural scientists to utilize secondary products. As agrochemical research moves into the 21st century, these tools provide the potential to make unprecedented advances fashioning value-added products, enhancing pest control and increasing productivity and yield. The launch pad for many of these advances will continue to be our fundamental chemical and biochemical knowledge-base of plant secondary products – what they are, how they work, how they are synthesized in nature, and how they attract pest species, augment pest resistance and affect biological control activity.

Chapter 31

NEUTRAL NONSTRUCTURAL CARBOHYDRATES

Donald L. Hendrix[1]
[1]*Western Cotton Research Laboratory, USDA-ARS, Phoenix, Arizona*

1. INTRODUCTION

The photosynthetic creation of carbohydrates and their subsequent metabolism are at the heart of growth and productivity of the cotton plant. Carbohydrate metabolism in cotton has received considerable attention since the pioneering work of Phillis and Mason (1933) and Ergle (1936), and it has played a central role in various models of cotton growth and yield (Baker *et al.*, 1972; Harley *et al.*, 1992; Wall *et al.*, 1994). This chapter deals with some of the cellular and tissue-level processes involved in nonstructural carbohydrate metabolism. In addition, problems associated with the analysis of soluble carbohydrates in tissues are discussed, and the metabolism and excretion by phloem-feeding insects of plant sugars which cause cotton fiber stickiness are examined briefly.

2. NONSTRUCTURAL LEAF CARBOHYDRATES

Nonstructural carbohydrates are those saccharides which are not metabolically inert, structural polymers, such as cellulose. The major nonstructural carbohydrates in cotton leaf blades are glucose, fructose, sucrose and starch. Free (*i.e.*, not a part of other molecules) fructose is usually present in cotton leaf tissues in much lower concentrations than other soluble sugars (Hendrix and Peelen, 1987; Miller *et al.*, 1989). Leaf starch and sucrose constitute the largest pools of nonstructural carbohydrate, and the sizes of these pools exhibit a more pronounced diurnal cycling than do leaf monosaccharide pools (Hendrix and Huber, 1986). During a photoperiod, cotton leaves typically accumulate starch as a reserve, unlike many other crop spe-

cies which accumulate carbohydrate reserves in the form of sucrose (Goldschmidt and Huber, 1992). When cotton leaves are illuminated, starch is rapidly formed in their chloroplasts after a brief period during which the cytoplasmic sucrose pool is filled (Hendrix and Grange, 1991). If cotton plants receive sufficient illumination for several successive light periods, the greatest proportion of the nonstructural carbohydrate in their leaves at the end of a photoperiod is usually in the form of starch. Leaf starch created during the day is largely broken down during subsequent dark periods. Chang (1982) showed that this degradation initially was by α-amylase, followed by a period during which degradation was catalyzed by α-amylase and starch phosphorylase acting together. During dark periods, sucrose from starch dissolution constitutes the major portion of the carbohydrates exported from the leaf.

The rate of carbohydrate export from cotton leaves during the day depends upon the sucrose concentration in the leaf, but the rate during dark periods typically depends upon the amount of starch accumulated during the preceding light period. If sufficient starch has accumulated by the end of a photoperiod, carbohydrate export will continue during the dark at the same rate as during the preceding light period. However, if sufficient starch does not accumulate, the rate of carbohydrate export from leaves during the subsequent dark period will be strongly decreased (Hendrix and Grange, 1991). Export of leaf carbohydrate during dark periods is particularly important because high demand processes such as boll growth, stem growth, and leaf expansion occur primarily at night (Radin, 1983).

The sizes of soluble sugar and starch pools (*i.e.*, concentrations) in cotton leaves are strongly influenced by a number of environmental variables. For example, monosaccharide content (amount per leaf area) is depressed by increasing atmospheric ozone concentrations, but sucrose content is relatively unaffected by this pollutant (Miller *et al.*, 1989). Shading decreases leaf starch and sucrose much more than leaf hexose content (Zhao and Oosterhuis, 1998b). Cool air temperatures (20°C) can cause an increase in both structural and nonstructural carbohydrates in leaves (Thompson *et al.*, 1975; Warner and Burke, 1993), apparently by inhibiting carbohydrate export. Soil type and fertilization (Fig. 31-1) can alter the soluble carbohydrates in leaves and other parts of the cotton plant (Ergle, 1936; Ergle *et al.*, 1938; Radin *et al.*, 1978; Ackerson, 1985). Carbohydrate translocation, lint quality, and incorporation of glucose into cotton fibers are all reported to be especially sensitive to tissue boron content (Eaton, 1955; Anderson and Boswell, 1968; Birnbaum *et al.*, 1974; Dugger and Palmer, 1980). However, Heitholt (1994b) reported only very slight effects of boron fertilization on field-grown cotton. Soil potassium deficiency increases leaf sugars (Bednarz *et al.*, 1994).

Phosphorus can have a strong influence upon leaf starch (Barrett and Gifford, 1995). Ackerson (1981) and Ackerson and Hebert (1981) found that when growth chamber-grown plants were water stressed (Ψ_w = -2 MPa) by withholding

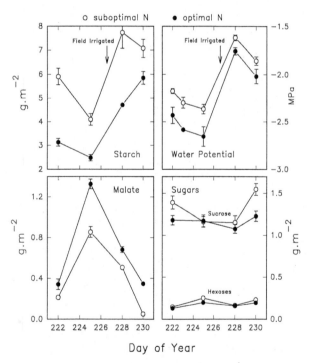

Figure 31-1. Cotton leaf carbohydrate, malate, and water potential, pre and post irrigation, in a flood-irrigated field in Phoenix, Ariz. Symbols represent means of five replicates ± SEM of leaves from a field with optimal N fertilization (●) and a field with suboptimal (no N added to the soil) N fertilization (○). Nonstructural carbohydrates were determined as in Hendrix (1993). Hexose refers to fructose plus glucose. Malate was determined by a microplate assay employing malate dehydrogenase. Leaf water potential was measured by a pressure chamber technique (Jordan and Ritchie, 1971). Unpublished data of D. L. Hendrix and G. Guinn.

nutrient solutions, the soluble sugar and starch contents of their leaves was two to four times greater than leaves from well-watered plants. However, these experiments were conducted by withholding nutrient solution, so the leaves being compared differed from those of control plants not only in their water content but also in their content of P and other nutrients. Later, Ackerson (1985) concluded that water stress caused a significant increase in starch in leaves containing low P, in contrast to decreased starch in leaves from water-stressed plants irrigated with 5 mM P. He also found that during water stress, high leaf phosphorus was associated with increased leaf monosaccharide content.

Nitrogen fertilization also strongly influences cotton leaf starch, which is more sensitive to N fertilization and leaf water stress than are the other soluble sugars (Fig. 31-1). Starch content decreases as leaves become water stressed and increases following irrigation. Hendrix and Guinn (Fig. 31-1; unpublished data) found that less starch accumulated in leaves of field-grown cotton plants given optimal N than in leaves of mildly N-deficient plants. Similar results were reported by Reddy *et al.* (1996b), who reported a decrease in starch with increasing leaf N-content in N-stressed Pima plants grown in sand-filled pots. Unlike starch, Hendrix and Guinn (Fig. 31-1) found that malate, a metabolite linked to osmotic adjustment in cotton leaves (Cutler and Rains, 1978) and elongation of fibers (Basra and Malik, 1983), accumulated in leaves under water stress and declined when plants were irrigated. At any given leaf water potential less malate was present in nitrogen-stressed leaves than in well-fertilized leaves, the reverse of the behavior exhibited by leaf starch.

The concentrations of sucrose and monosaccharides in field-grown cotton leaves are altered by relatively severe (Ψ_w < -3 MPa) drought (Eaton and Ergle, 1948; Souza *et al.*, 1983; Abdullah, 1985; Chang and Wetmore, 1986). Moderate water stress (Ψ_w 2 to 3 MPa) has little effect on these sugars (Miller *et al.*, 1989) but does increase leaf starch (Fig. 31-1). Eaton and Ergle (1948) found that at very low water potentials, cotton leaves accumulated soluble carbohydrate. They found large reductions in starch and increases in soluble hexoses and sucrose, a result later confirmed by Cutler and Rains (1978) and Timpa *et al.* (1986). Eaton and Ergle (1948) concluded that severe drought inhibited the conversion of soluble carbohydrates to starch even more than it inhibited photosynthesis.

Soluble sugar and starch contents depend upon leaf age and upon the stage of plant development (Chang, 1980; Constable and Rawson, 1980b; Souza and da Silva, 1987; Hendrix *et al.*, 1994; Wullschleger and Oosterhuis, 1990a). Kimball *et al.* (1987; Fig. 31-2) found that once bolls appeared on field-grown plants, the dawn and dusk starch content of leaves declined with increasing boll set. During this period of decline, the amount of starch degraded each night (*i.e.*, the difference between the evening leaf starch content and that in the same leaf next morn-

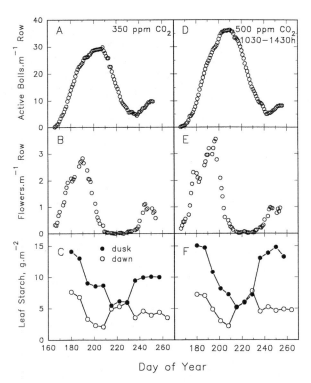

Figure 31-2. Plot of the number of bolls (A,D) actively receiving photosynthate from the plant (*i.e.*, no more than 40 days old) and the number of flowers which opened on each day of this period (B,E) per meter of row in a field located in Phoenix, Ariz. The diurnal variation in leaf starch content of the fifth main-stem leaves from the top of the plant (C,F). Plants were watered by drip-irrigation to eliminate water stress (cf. Fig. 1). Most recently fully-expanded leaves were sampled at dusk and at dawn at weekly intervals. At each date, leaves were sampled by removing 6 punches at dusk (●) and 6 punches at dawn (○) from the same leaf (Hendrix, 1993). The set of punches taken at the dawn sampling were adjacent to those taken the previous evening.

Panels labeled D, E, and F represent data from plants fumigated (beginning day 168) with CO_2 from 1030 to 1430 MST to increase the atmospheric CO_2 content surrounding the leaves to approximately 500 ppm. Each data point represents the mean of samples taken from four different plants. Starch was analyzed as described in Hendrix (1993). Data from Kimball *et al.* (1987).

ing) remained relatively constant until the boll load[1] on the plants reached a maximum (Fig. 31-2). After plant boll load peaked, diurnal cycling of leaf starch synthesis and degradation ceased. The cycling resumed the last week of August (day 235) when flowering resumed and active boll load again increased. Interestingly, this collapse in leaf starch cycling did not occur in adjacent plots of Pima cultivars which did not undergo a pronounced cutout (*i.e.*, cessation of flowering during periods of heavy boll load).

---

[1] Boll load refers to the relative number of bolls drawing upon photoassimilate provided by the plant leaves via the phloem. The flow of photoassimilate entering developing bolls essentially ceases approximately 40 days from anthesis (Benedict *et al., 1976*).

The cotton canopy consists of leaves of widely differing age (Wullschleger and Oosterhuis, 1992), leaf orientations and photosynthetic responsiveness (Baker *et al.*, 1978a), and leaf carbohydrate export varies with each of these parameters. The amount of light intercepted by a leaf helps to determine not only its photosynthetic activity but also its nonstructural carbohydrate concentration or composition (Eaton and Rigler, 1945; Eaton, 1955). In a typical field of closely-spaced cotton plants, leaves in the top of the canopy receive a much higher irradiance than those further down the plant where most maturing bolls are located. The carbohydrate used in the development of a particular boll comes from multiple leaf sources. While it is generally agreed that the leaf subtending a boll can potentially contribute a significant proportion of the carbohydrate for that boll's development (Benedict and Kohel, 1975), Constable and Rawson (1980) concluded from plant carbon budgets that a major portion of the carbohydrates utilized by developing bolls must come from leaves other than the subtending sympodial leaf.

Elevated atmospheric CO_2 increases leaf carbohydrate production. Compared to leaves of other species, cotton leaves are capable of accumulating very high levels of starch; this starch accumulation is greatly stimulated by CO_2 concentrations above ambient (Fig. 31-2; Mauney *et al.*, 1979; Wong, 1990; Hendrix, 1992; Barrett and Gifford, 1995). Elevated CO_2 increases leaf starch more than other leaf carbohydrates. The starch content of leaves exposed for a few days to 1,000 ppm CO_2 can exceed 50% of leaf dry weight (J. R. Mauney, personal communication). The accumulation of high levels of leaf starch under CO_2 enrichment has been associated with decreased photosynthetic capacity (Mauney *et al.*, 1976; Harley *et al.*, 1992). However, accumulated leaf starch *per se* does not appear to be responsible for this photosynthetic inhibition (Goldschmidt and Huber, 1992). The degree of stimulation of starch accumulation by elevated CO_2 depends upon the stage of plant development (Fig. 31-2; Hendrix *et al.*, 1994). Kimball *et al.* (1987) found that after cutout in field-grown cotton, CO_2 enrichment increased leaf starch (Figs. 31-2 and 31-3). They also found that elevated CO_2 increased the number of flowers and boll load per plant. In spite of these changes, the seasonal patterns and diurnal cycling of leaf starch in fields with increased atmospheric CO_2 (500 ppm) remained remarkably similar to that in control fields (350 ppm CO_2).

Much of the carbohydrate released in the remobilization of starch is rapidly exported rather than used within the leaf. Chang (1980) found that starch degradation in cotton leaves during dark periods did not lead to an increase in soluble sugar content unless the petioles of the leaves were girdled to block phloem transport. He also found that the extent of starch breakdown during the night depended upon leaf age. In young leaves, 70 to 80% of the starch was broken down at night, but in older leaves only 30% of was remobilized.

Several enzymes are especially important for the metabolism of soluble carbohydrates. Sucrose phosphate synthase (SPS), the enzyme which synthesizes sucrose 6-

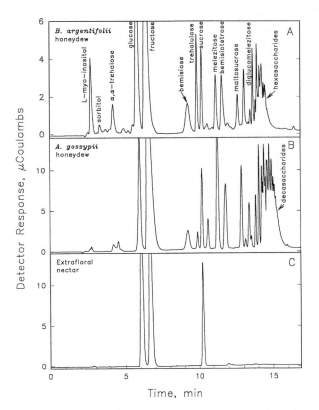

Figure 31-3. HPLC chromatograms of (A) honeydew from the silverleaf whitefly feeding upon cotton leaves,
(B) honeydew from the cotton aphid feeding upon cotton leaves and (C) sugars secreted by cotton nectaries. For HPLC elution and detection details, refer to Hendrix and Wei (1994).

phosphate, the immediate precursor to sucrose, may have a regulatory role in the creation of sucrose and in the export of carbon from plant leaves (Huber, 1983). SPS activity is related to the buildup of leaf starch. In a number of crop species, Huber (1981) found a highly significant inverse relationship between leaf starch and extractable SPS activity. Sucrose synthase (SS), an enzyme which usually degrades sucrose (Giaquinta, 1978), was also quite active in mature cotton leaves (Hendrix and Huber, 1986). Unlike most crops, the ratio of SS:SPS in mature cotton leaves is relatively high and the activities of both enzymes have strong diurnal fluctuations (Hendrix and Huber, 1986; Salerno, *et al.*, 1989; Tognetti *et al.*, 1989).

Another important control point in the synthesis of sucrose, which has not been investigated in cotton, is the conversion of fructose 1,6-bisphosphate to fructose 6-phosphate (Stitt *et al.*, 1983). This metabolic step is unusual in that the forward reaction is catalyzed by a different enzyme than the reverse reaction. Both enzymes, phosphofructokinase and fructosebisphosphatase, are strongly regulated by fructose 2,6-bisphosphate, the former being stimulated and the latter inhibited by this metabolite. The net result of this double enzyme control system is a tight regulation of sucrose synthesis from photoassimilate.

Other enzymes implicated as important control points in leaf soluble carbohydrate metabolism include α- and β-

amylase and invertase. These enzymes probably control mobilization of starch reserves (Chang, 1982; Souza *et al.*, 1990). Alpha-amylase randomly digests the 1,4 bonds between glucose units in starch but not the 1,6 branch points. Beta-amylase starts at one end of a starch molecule and cleaves two glucose units at a time, releasing maltose which is rapidly converted to glucose by maltase (α-glucosidase). Invertase is a fructohydrolase, which hydrolyzes the bond linking fructose to oligosaccharides. In plants, invertase mainly hydrolyzes sucrose into glucose and fructose. Souza *et al.* (1990) found that pruning main-stem leaves, sympodial leaves, or both, increased leaf β-amylase activity but decreased the diameter of cotton stems, the starch content of the roots, and the ratio of root to shoot dry weights. Souza *et al.* (1983) also demonstrated that removing flower buds from field-grown plants stimulated leaf invertase activity. From these experiments they concluded that β-amylase and invertase activity in leaves could be used to screen cotton cultivars for better carbohydrate utilization.

3.   CARBOHYDRATES IN OTHER ORGANS

The cotton plant must translocate large amounts of carbohydrate from leaves to maturing bolls, primarily as sucrose (Phillis and Mason, 1933; Tarczynski *et al.*, 1992). This is converted to UDP-glucose within the boll before it is incorporated into fiber cellulose (Carpita and Delmer, 1981; Nolte *et al.*, 1995). The fiber in a boll can increase in dry weight by as much as 15% per day (Schubert *et al.*, 1986). Nearly all of this increase is due to carbohydrate since mature cotton fibers are 95% cellulose (Timpa and Triplett, 1993). The accumulation of sugars in bolls is strongly influenced by night temperature (Conner *et al.*, 1972). Both sucrose import and the incorporation of glucose into cotton fibers is stimulated by boron (Dugger and Palmer, 1980).

Even though most (ca. 80%) of the dry weight of cotton seeds is lipid and protein, seeds also accumulate significant amounts of sugars during maturation. The galactosides raffinose and stachyose constitute the most abundant sugars in cotton mature seeds (Shiroya, 1963; Doman *et al.*, 1982; Kuo *et al.*, 1988; Hendrix, 1990), but are absent in developing seeds (Hendrix, 1990). Raffinose and stachyose are synthesized by the addition of galactose to sucrose or raffinose by the enzymes galactinol:sucrose 6-galactosyl transferase and galactinol:raffinose 6-galactosyl transferase, respectively (Kandler and Hopf, 1982). In maturing seeds these galactosides are formed after assimilate movement to bolls has ceased (Benedict *et al.*, 1976; Hendrix, 1990). The sucrose utilized in the synthesis of raffinose and stachyose must, therefore, originate from sucrose synthesized within the seed, not imported from the phloem. The raffinose content of mature cotton seeds is greater than that of stachyose (Shiroya, 1963; Hendrix, 1990), the reverse of that in the seeds of most legumes and melons (Kuo *et al.*, 1988). In addition to stachyose, several other sucrosyl-containing tetrasaccharides accumulate in cotton seeds (Kato *et al.*, 1979; Muller and Jacks, 1983). Details of their formation and possible utilization during germination remain unknown (for further information see Chapter 29, *Ontogeny of Cottonseed*).

The pattern of carbohydrate accumulation in developing cotton seeds suggests the following scenario. Sucrose that enters seeds the via the phloem is converted by SS to UDP-glucose, the preferred substrate for the synthesis of cellulose in cotton fibers (Carpita and Delmer, 1981; Hendrix, 1990; Amor *et al.*, 1995; Nolte *et al.*, 1995; Ruan *et al.*, 1997). A significant amount of the carbon from sucrose entering developing seeds is deposited in seedcoat tissues as starch (Hendrix, 1990). This starch is broken down during the maturation of this tissue, and its carbon appears later in the developing embryo as sucrose and starch. Unlike other seeds, the activity of SS in developing cotton seeds does not seem to correlate with starch synthesis, but it does correlate with fiber formation (Ruan *et al.*, 1997). Much of the embryo's starch disappears during the final stages of seed maturation, when raffinose and stachyose appear in the embryo. Galactose-containing saccharides in mature seeds are very rapidly converted to monosaccharides and sucrose during germination (Shiroya, 1963; Doman *et al.*, 1982).

Plant growth requires an expanding root system to supply nutrients and water. The rate of root growth decreases after flowering due to competition for assimilate between roots and developing bolls (Cappy, 1979). Radin *et al.* (1978) found that cotton roots compete poorly with other tissues for assimilate sugars. They concluded that the allocation of photoassimilate to growing roots was one of the most important factors limiting root system growth. They found that the shoot:root ratio was inversely related to the soluble sugar content of the roots. They also found that the sugar content of roots was significantly higher in plants fertilized with nitrate compared to urea.

Hendrix *et al.* (1994) found that increasing the atmospheric CO_2 concentration to 500 ppm increased soluble carbohydrates in cotton leaves, stems and roots of field-grown cotton. However, the increases in stems and taproots were much more pronounced than in the leaves. The soluble carbohydrates of roots and stems underwent periodic fluctuations of large amplitude during a growing season and the amplitude of the fluctuations were magnified by exposure to increased atmospheric CO_2. The nonstructural carbohydrate content of both tissues reached a minimum during maximal boll load, suggesting that developing bolls draw upon carbohydrate stored in the plant stem and taproot.

4. CARBOHYDRATES SECRETED BY COTTON EXTRAFLORAL NECTARIES

Extrafloral nectaries occur on the abaxial side of the main vein near the petiole of cotton leaves and at the base of bolls during the first few weeks of their development (Eleftheriou and Hall, 1983a,b). These structures secrete a highly concentrated solution of glucose, fructose, and sucrose which serves as food for a number of insects (Stapel *et al.*, 1997 and references therein). Floral nectar is secreted only on the day of anthesis, but extrafloral nectaries secrete nectar for several days (Tyler, 1908).

Only three sugars, glucose, fructose, and sucrose, were detected when the extrafloral nectar from greenhouse-grown upland cotton was analyzed by HPLC (Fig. 31-3C). Correcting for the relative response of the HPLC detector to each of these sugars (Larew and Johnson, 1988), the maximum calculated concentration of sugars in this exudate, for samples obtained over a several month period, was 4.2 moles per liter. This corresponds to an extraordinary water potential for such excretions of approximately -10 MPa. The ratios among the individual sugars in these samples were highly consistent. On the basis of moles of sugar per volume of fluid, the most abundant sugar in the secretions analyzed for Figure 31-3C was fructose [47.2 ± 0.7% of the total]. Glucose (31.1 ± 0.3%) and sucrose (21.7 ± 0.8%) were consistently less concentrated than fructose. Similar results were found by Butler *et al.* (1972) who used gas-liquid chromatography to analyze nectary secretions from field-grown upland cotton. They found that glucose and fructose were nearly equally abundant in nectar from upland cotton. In Pima secretions, however, they found that the concentration of glucose significantly exceeded that of fructose. They noted that the volume of nectary secretions collected from field-grown leaves decreased to a minimum around noon, suggesting a positive relationship between secretion rate and leaf water potential (*cf.* Jordan and Ritchie, 1971). They also found a decrease in the sugar content of nectar from both cotton species from samples collected during the flowering period (June through August). For any time during this period, the total concentration of sugars in these secretions was at a maximum at noon, when it reached as high as 700 mg·mL^{-1}.

Assuming the carbohydrates in nectary secretions come only from phloem sap, the origin of monosaccharides in these secretions is unclear as monosaccharides are found only at very low concentrations in phloem (Ziegler, 1975). Reducing sugars such as glucose and fructose are toxic to living cells above ca. 10 mM. The only sugar in cotton phloem, sucrose, is nonreducing (Tarczynski *et al.,* 1992). Since sucrose typically occurs in phloem sap at concentrations of 0.3 to 0.4 M (Peel, 1975), considerable concentration must occur in the nectaries. Findlay (1982) proposed that this concentrating occurred by evaporation of secreted material. The process which creates the extrafloral nectary secretion from phloem sap in cotton must, therefore, involve the conversion of sucrose into monosaccharides by enzymes in the nectary cells. This hypothesis agrees with that of Baker *et al.* (1978b) for extrafloral nectaries of castor bean (*Ricinus communis*), another plant which translocates sucrose as its only phloem sugar. Like cotton, castor bean extrafloral nectary secretions are a mixture of glucose, fructose, and sucrose of approximately equal proportions by weight.

Eleftheriou and Hall (1983a,b) carefully examined the path of sugars from the phloem to nectar excreted by extrafloral nectaries in upland cotton. From anatomical observations, they concluded that during secretion sugars moved from the phloem through nectary cells (a symplastic route). They concluded that this nectar was eliminated from the nectary cells via vesicles which fused with the cell membrane, releasing their contents outside the cell. From the high concentration of reducing sugars in nectar, it seems logical to assume that the hydrolysis of sucrose in nectar formation does not take place within the cytoplasm of nectary cells.

5.   HONEYDEW SUGARS

The sugars excreted by phloem-feeding insects living on cotton leaves (Fig. 31-3A and B) are distinctly different from sugars found in the plant. Also, honeydews (concentrated sugar secretions by insects following the ingestion of phloem sap) from different species of phloem-feeding insects have distinctive sugar compositions, and the sugar profile depends both upon the species of insect and the plant upon which the insect feeds (Hendrix *et al.*, 1992). In the United States, two common honeydew-producing cotton insects are the silverleaf whitefly (*Bemisia argentifolii* Bellows and Perring) and the cotton aphid (*Aphis gossypii* Glover). Honeydew falling from these insects feeding upon leaves above open bolls often accumulates on cotton fiber, causing severe problems when it is processed by cotton gins and textile mills. The presence of honeydew is thus a major detriment to lint quality and marketability. The honeydews from *A. gossypii* and *B. argentifolii* consists of at least 20 different sugars, some with very unusual structures (Fig. 31-3A and B; Hendrix and Wei, 1992, 1994; Wei *et al.*, 1996; Wei *et al.*, 1997). Most of the oligosaccharides in silverleaf whitefly honeydew contain α-D-glucose moieties; a few of them contain a single D-fructose, as well (Wei *et al.*, 1997). The predominant sugar found in the honeydew when these insects feed on cotton (Fig. 31-3A) is the unusual disaccharide trehalulose [α-D-glucose-(1↔1)-D-fructose]. This sucrose isomer constitutes approximately 40% of the total sugars in this excreta (Wei *et al.* 1996). Trehalulose is not easily detected by some commonly used assays for reducing sugars (Hendrix and Wei, 1992; Hendrix *et al.*, 1996), which can cause silverleaf whitefly honeydew contamination of cotton fiber to be underestimated.

Honeydew excreted by the cotton aphid feeding upon cotton (Fig. 31-3B) contains only traces of trehalulose, but it does contain substantial quantities of melezitose (Hendrix *et al.*, 1992). Melezitose [α-D-glucose-(1→3)-β-D-fructose-(2←1)-α-D-glucose], a trisaccharide characteristic of many aphid honeydews (Hendrix and Wei, 1992), is a relatively minor component of immature silverleaf whitefly honeydew and is not secreted by silverleaf whitefly adults (Davidson *et al.*, 1994). Other significant differences exist between the honeydew secreted by the silverleaf whitefly and the cotton aphid. For instance, silverleaf whitefly honeydew contains oligosaccharides as large as hexasaccharides but aphid honeydews typically contain oligosaccharides as large as decasaccharides (Fisher *et al.*, 1984; Wei *et al.*, 1997; Fig. 31-3A,B).

Since neither trehalulose nor melezitose occur in plant tissues (Bacon and Dickinson, 1957; Cookson *et al.*, 1987), their presence on cotton fiber can be taken as an indicator of insect honeydew contamination. Other factors besides honeydew could cause cotton fibers stickiness. Sugar on cotton fiber from nectary secretions could cause fiber stickiness but these sugars can be distinguished from honeydew since they consist entirely of glucose, fructose and sucrose (Fig. 31-3A and B vs. Fig. 31-3C). Immature fibers also contain elevated levels of these three saccharides, but the sugars in such fibers would be evenly distributed along the fibers rather than in localized spots as in nectary or honeydew contamination.

6.   CARBOHYDRATE MEASUREMENT

Many methods have been utilized to determine soluble sugars and starch in (and on) cotton tissues, including gas chromatography (Guinn and Hunter, 1968; Butler, *et al.*, 1972), colorimetric and enzymatic techniques (Chang, 1980; Hendrix and Peelen, 1987; Miller *et al.,* 1989; Salem and Cothren, 1990; Hendrix, 1993; Timpa *et al.*, 1995), high-performance liquid chromatography (HPLC) (Doman *et al.*, 1982; Lee *et al.*, 1983; Timpa *et al.*, 1986; Kuo *et al.*, 1988; Tarczynski *et al.*, 1992) and infrared reflectance spectroscopy (Hattey *et al.*, 1994). Gas chromatographic methods are extremely sensitive but great care and skill are needed to analyze sugars by these techniques. Before their analysis by gas chromatography, carbohydrates must be carefully converted to volatile derivatives. Derivatives becomes increasingly difficult to prepare as the size of the oligosaccharide increases. Therefore, very large carbohydrates, such as the larger fragments resulting from the digestion of starch, can not be analyzed by gas chromatography. Nearly all reactions used to produce volatile carbohydrate adducts yield two isomers from each sugar, giving rise to two peaks for every sugar analyzed. This significantly increases the complexity of the resulting chromatographs. In addition, this type of analysis is destructive (*i.e.*, it is not possible to recover the analyzed sample for additional tests).

Recent advances in HPLC detection, especially the amperometric detection of carbohydrates (Rocklin and Pohl, 1983; Hicks, 1988), have led to much wider use of HPLC for carbohydrate analysis. HPLC techniques are quite sensitive and are not destructive; samples can be recovered after analysis and reanalyzed by other methods. In addition, derivatization of carbohydrates prior to HPLC analysis is usually not necessary and carbohydrates as large as 10,000 D can be analyzed by HPLC (Chatterton *et al.*, 1993; Henshall, 1996). However, the expense of HPLC equipment, the technical knowledge required for its use and maintenance, and the time necessary for sample preparation and analysis, limits the use of HPLC for carbohydrate determinations.

Colorimetric methods to quantify carbohydrates have been popular for many years. These can be coupled to enzyme reactions to provide a very selective, and often very sensitive, method of analyzing individual carbohydrates. These methods can easily be adapted for use in analysis of cotton tissues (Hendrix and Peelen, 1987; Salem and Cothren, 1990; Hendrix, 1993; Timpa *et al.*, 1995). They can be employed in large scale field experiments since they are very rapid and simple to perform. They are also relatively inexpensive and easily adaptable to microscale. However, several potential problems must be kept in mind when employing such assays. First, these methods depend upon proteins whose enzymatic actions can be strongly inhibited by various chemicals which occur in plant tissues at high levels (Blunden and Wilson, 1985; Hendrix and Peelen, 1987). Such inhibition can be difficult to detect without careful testing. Secondly, due to their extreme sensitivity (a glucose 6-phosphate dehydrogenase-based assay is capable of detecting 10 ng of glucose) only a minute amount of tissue is needed for these assays. Care is needed when employing such assays to insure a representative sample is analyzed. Starch, for instance, is not uniformly distributed across field-grown cotton leaves (Fig. 31-4), therefore, a number of locations must be sampled across each leaf for accurate analyses. Finally, unlike gas chromatography or HPLC, enzyme-linked assays provide data only for the substrate of the particular enzyme reactions employed.

Unlike many plant tissues, the cotton leaf contains an exceptionally active invertase and an active sucrose synthase which are not inactivated by freezing (Hendrix and Peelen, 1987). Therefore, sucrose is very quickly degraded to glucose and fructose when frozen leaf tissue is thawed or when lyophilized leaves are rehydrated. Up to half of the sucrose of cotton leaves can be converted to monosaccharides within an hour after thawing. Improper drying, either by heating or by lyophilization, can also lead to erroneous sugar determinations. For very precise sugar measurements, a step which stops enzyme activity should be included shortly after tissue isolation such as immersion in liquid nitrogen or aqueous alcohol. The time of day (Hendrix and Huber, 1986), time of season (Fig. 31-2) and plant water status and fertilization (Fig. 31-1) are important parameters to be noted when sampling for nonstructural carbohydrates. In sampling cotton leaves for leaf starch, it is also important to note the illumination history of the plant (Hendrix and Grange, 1991).

The detection of honeydew sugars on cotton fiber requires specialized techniques. Brushwood and Perkins (1993) described a modification of Benedict's reducing-sugar test and a pH indicator spray test which allowed an estimation of honeydew on contaminated fibers. Others have suggested using a clinical variation of Benedict's

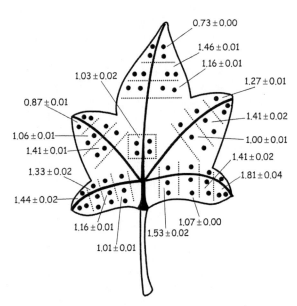

Figure 31-4. Variation in starch content across a single fifth main-stem leaf (from the top of the plant) growing on a cotton plant in a flood-irrigated field in Phoenix, Ariz., with suboptimal N fertilization (*i.e.*, no N added to the field). This figure represents data from a single, representative, leaf sampled at 1100 h from the most recently-expanded leaf of a mature cotton plant during the month of July. Data for a single leaf are presented to emphasize the variation in starch per leaf area across each leaf and between leaves due to variation in shading and leaf orientation. Dotted lines on the leaf diagram refer to areas where 3 or 4 punches removed (ca. 0.42 cm² each) and pooled for a starch sample (digest). These samples were immediately extracted with ice-cold aqueous 80% ethanol and analyzed for starch according to the method of Hendrix (1993). Numbers represent the mean ± SE starch content (g m⁻²) of the results of analysis of 4 aliquots of each starch digest (sample). SE shown only to indicate precision of starch analysis.

reducing sugar test to detect stickiness (Topping and Broughton, 1997). However, the ferricyanide reducing-sugar test appears to give the best results for detection of silverleaf whitefly honeydew sugars on cotton fiber and for correlation of sugar content with fiber stickiness (Brushwood and Perkins, 1993). Unlike Benedict's test, the ferricyanide test readily detects trehalulose.

Methods employing infrared spectra have been proposed to detect honeydew sugar and therefore the relative stickiness of honeydew-contaminated cotton fiber (Perkins, 1993; Anthony *et al.*, 1995). These methods are rapid and can be used on relatively small samples with little sample preparation. However, calibration of machines, so that their output is reliably proportional to the stickiness of the lint for a wide variety of samples, remains a problem.

The most reliable method for predicting the stickiness of honeydew-contaminated fiber in the textile mill is the minicard test (Brushwood and Perkins, 1993; Perkins, 1993). This is basically a miniature carding machine which functions very much like that in textile mills. However, due to the cost of the carding machine and the time and skill necessary for its use, in recent years the thermodetector has replaced the minicard for the detection of honeydew stickiness. In the thermodetector, a thin web of cotton contaminated with honeydew is heated between two sheets of aluminum foil. When the web is removed, the charred honeydew spots on the foil are recorded. The correlation between the minicard and the thermodetector is fairly high (Brushwood and Perkins, 1993). However, because both of these methods require considerable time, several machines have been introduced recently to automate stickiness sampling (Hequet and Frydrych, 1997; Mor, 1997).

7.   SUMMARY

The formation and utilization of nonstructural carbohydrates in cotton constitutes much of the essential metabolism underlying cotton growth and development. Photosynthesis provides sucrose for export via the phloem to growing points throughout the plant. If leaves contain sufficient sucrose, they accumulate starch, much of which is degraded nocturnally to supply sucrose for export and growth. These processes can be altered by various environmental parameters, including temperature, atmospheric CO_2, or fertilization. Carbohydrate is translocated from leaves to developing bolls as sucrose. The enzyme sucrose synthase is important for the conversion of sucrose to UDP-glucose for fiber growth in developing seeds. A number of analytical methods are available for the quantification of nonstructural carbohydrate in cotton tissues. Enzyme-coupled colorimetric assays are particularly useful for this purpose, provided one is aware of their limitations and the precautions for their use. Honeydew excreted by cotton-feeding insects contains a complex mixture of both plant and insect-derived carbohydrates which can cause 'sticky fiber,' a problem for gins and textile mills.

CHAPTER 32

BIOCHEMISTRY OF THE FIBER

Carolyn Zieher[1]
[1] *Tucson, Ariz.*

1.   INTRODUCTION

The use of cotton as a textile fiber dates as far back as 2000 BC and remains today the most widely used textile fiber. Its predominance comes from its soft, unique spinning characteristics; the mature fiber when it dries twists in such a way that enables fine strong threads to be spun. With the advancements in spinning technologies, the textile industry is placing an increasing emphasis on fiber quality for manufacturing. Premium prices are given for fiber with increased length, fineness, uniformity and strength, properties that determine quality. Improvements in fiber quality with the selection of longer staple varieties and introduction of new crop management techniques have met with limited success. Our poor understanding of the basic mechanistic processes that determine the physical properties of the fiber that, in turn, define fiber quality, continues to limit advancements in this area. Understanding the underlying biochemical and molecular processes that control fiber differentiation and growth promises to facilitate efforts towards fiber quality improvement. Recent advances in the transformation and genetic engineering of commercial cultivars of cotton have established that directed modification of gene expression is possible (McCabe and Martinell, 1993). Increasing efforts are being directed at identifying suitable target genes whose modifications might lead to the creation of cotton plants with improved fiber properties or at adding foreign genes whose expression in fibers would provide novel traits (John and Stewart, 1992). Development of effective strategies for fiber quality improvement and novel modification of the fiber requires knowledge of the rate-limiting biochemical events associated with fiber differentiation and growth. This knowledge may be particularly important when adding novel traits, as the overexpression of foreign genes could significantly alter fiber metabolism at the expense of fiber quality that may or may not, depending on the fiber's use, affect its commercial potential.

This chapter focuses on recent advances and their importance to our understanding of the biochemistry of the cotton fiber and its regulation. Recent findings from other experimental systems will be included when deemed appropriate because of their potential relecance to the fiber cell. For more detailed discussion on earlier work, the reader is referred to reviews by Basra and Malik, (1984), Ryser (1985), and Seagull (1990b) and to chapters by Berlin, (1986), De Langhe (1986), Kosmidous-Dimitropoulou (1986), and Stewart (1986) in *Cotton Physiology Vol. I.*

2.   KINETICS OF FIBER GROWTH

Cotton fibers are single-celled trichomes that develop from specific epidermal cells on the outer integument of the ovule (see also Chapter 4). Fiber development can be divided into four distinct stages: initiation, elongation, secondary wall deposition, and maturation. The epidermal cells that differentiate into fiber cells protrude above the ovule surface the day of anthesis marking initiation. The fiber initials enter immediately into the elongation phase of fiber growth. During the first 24 to 36 h post-anthesis, the fibers expand isodiametrically, thereafter, they switch to anisotropic expansion (Stewart, 1975). Elongation continues for about 15 to 30 days reaching a final length of between 20 and 35 mm that is determined by the rate and duration of elongation (Natithani *et al.*, 1982) and influenced by genetic (Basra and Malik, 1984) and environmental (water, temperature, K^+; Ramey, 1986; Cassman *et al.*, 1990) factors. Fiber extension can have a significant impact on the quality of the resulting yarn; the rate and duration of elongation can ultimately determine length, fineness, uniformity, and to some extent fiber strength (Wilkins, 1992).

Between 20 and 25 days post-anthesis (DPA) wall synthesis and the cellulose content of the wall increases, indicative of secondary wall deposition (Meinert and Delmer, 1977). Fiber elongation will continue for some time after the initiation of secondary wall deposition; the extent of this overlap dependent on the cultivar (Naithani *et al.*, 1982; Schubert *et al.*, 1973) and species (Schubert *et al.*, 1976) of cotton. By 50 to 60 DPA, fiber maturation is evidenced by the splitting of the capsule wall, and the desiccation of the fiber that result in the collapse of the fiber cell into a flattened twisted ribbon. The degree of secondary wall thickening laid down prior to harvest largely determines the tensile strength of the fiver that is decided upon by genetics (Basra and Malik, 1984) but also influenced by the environment (*e.g.*, low temperature; Roberts *et al.*, 1992).

3.   HORMONES AND FIBER DEVELOPMENT

3.1   Hormone Requirements and Their Metabolism During Fiber Development

Studies using ovules cultured *in vitro* established that auxins and gibberellins are required for maximum fiber production (Beasley and Ting, 1973, 1974). Both hormones are required in the culture medium the first few days post anthesis (Beasley *et al.*, 1974; Kosmidou-Dimitropoulou, 1986) but auxin is also required later in the culture period (Kosmidou-Dimitropoulou, 1986). These hormone requirements are dependent on the fertilization of the ovule. Fertilized ovules have sufficient endogenous IAA but require additional GA while unfertilized ovules require both hormones for maximum fiber growth (Beasley and Ting, 1973; 1974). Fertilization thus apparently triggers the synthesis of IAA while GA is predominately synthesized elsewhere, possibly the subtending leaf (Peeters *et al.*, 1991) and transported to the developing ovule. Epidermal cells destined to become fiber initials develop the full capacity to respond to auxin and GA in culture 2 days before anthesis (DBA) and maintain this capacity up to the time of anthesis (*i.e.*, 2 days) but thereafter fail to produce fibers (Graves and Stewart, 1988a). Fiber initiations and anthesis are couple with differentiated fiber cells *in vivo*, remaining latent until a hormonal stimulus associated with anthesis is perceived. Addition of abscisic acid (ABA) with the other two hormones inhibits fiber growth *in vitro* (Beasley and Ting, 1973, 1974; Dhindsa *et al.*, 1976) but only when added the first few days postanthesis (Dhindsa *et al.*, 1976).

The hormone requirements for fiber production *in vitro* are supported by detection of these hormones *in vivo*. Auxin

extracted from fibers (Naithani *et al.*, 1982) and measured by the straight growth coleoptile biossay was elevated at the initiation of fiber elongation (5 to 10 DPA). Nayyer *et al.* (1989) showed higher levels of IAA between 20 and 30 DPA. Elevated levels of gibberellin-like substances were, likewise, detected in lint plus seed extracts between 10 and 15 DPA (Rogers, 1981) and GA₃ (Nayyar et.al.,) 1989) peaked in elongating fibers at 15DPA (Nayyar *et al.*, 1989). However, no relationship between endogenous hormone concentration and biological response is frequently observed (see Davies, 1995 for review). This poor correlation may be attributed to the rate of consumption/degradation of the hormone being a more important determinant than hormone concentrations (Naithani *et al.*, 1982); the influence of other endogenous hormones; the sequestration of the hormone in a cell compartment separate from the hormone recptor site; and /or a change in hormone sensitivity (*i.e.* change in receptor concentration and/or in receptor affinity for the hormone; Davies, 1995) of the fiber.

IAA levels in the fiber during development are in part a balance between synthesis and degradation and as a result, the relative activities of these two pathways may contribute to the control of fiber growth. A comparison of developmental activity profiles for enzymes involved in IAA synthesis and catabolism indicates a shift from predominately IAA synthesis during elongation to degradation during secondary wall deposition. Indoyl-3-actealdehyde, an immediate precursor to IAA and derived from tryptophan via the indole-3-acetaldehyde, an immediate precursor to IAA and derived from tryptophan via the indole 3-pyruvic acid and tryptamine pathways (Bandurski *et al.*, 1995), is oxidized to IAA indoyl-3-acetaldehyde dehydrogenase (Thaker et. Al., 1986a). This enzyme increases significantly (25 fold) in activity during fiber elongation and decreases a similar amount during secondary wall deposition. IAA oxidative catabolism is an irreversible chemical modification of the indole nucleus that results in the loss of auxin activity (Bandurski, *et al.*, 1995). IAA oxidase activity increases just prior to cessation of fiber elongation (Jasdanwala *et al.*, 1977; Rama Rao *et al.*, 1982a) as indoyl-3-acetaldehyde dehydrogenase activity declines (Thaker, *et al.*, 1986a). 0-diphenol oxidase activity, likewise, increases during this time period (Naithani *et al.*, 1981). Diphenols, such as 0-diphenols, inhibit IAA oxidation (Bandurski *et al.*, 1955).

The response of the fiber call to any hormone is dependent on the concentration of other hormones present (Beasley and Ting, 1973, 1974; Dhindsa, 1978; Nayyar *et al.*, 1989). For example, GA₃, and IAA added to ovule culture medium resulted in greater fiber production than when the hormones were added individually (Dhindsa, 1978). An increase in IAA or GA₃ concentration overcame inhibition of fiber growth induced by ABA (Beasley and Ting, 1973, 1974). In fact, Nayyar *et al.*, (1989) argues that the ratio of endogenous levels of ABA to GA and IAA may be more important in determining the extent of fiber growth than the levels of IAA and GA. ABA levels in fibers of

Gossypium arboreum L., a short staple cotton (final fiber length ca. 17 mm), were several fold higher than IAA and GA_3 during the period of rapid fiber elongation (10 and 15 DPA). Fiber growth increased when the endogenous ratio of ABA to IAA and GA_3 was decreased by culturing ovules in the presence of fluridone, an ABA biosynthesis inhibitor. The balance between the concentration of growth promoting and inhibiting hormones could therefore play a regulatory role in determining fiber length and contribute to the staple length differences between *Gossypium* spp.

3.2 Hormonal Influence on Transcription

The nucleolus is a discrete sub-nuclear area that is the site of rRNA synthesis and pre-ribosomal assembly. During differentiation of *G. hirsutum* L. fibers, the nucleolus (Nu) increases in size from less than 1 μm in diamerer at 4 DBA (Berlin, 1986) to a maximum size of between 7 and 11 μm by ca. 3 to 6 DPA (De Langhe *et al.*, 1978), that occurs sometimes early but usually later during this time period (Kosmidou- Dimitropoulou, 1986). Maximum nucleolar size is maintained for several days followed by a rapid decrease due to a decline of Nu-material that eventually slows such that the nucleolus reaches a minimum size at the end of fiber growth (De Langhe *et al.*, 1978; Kosmidou-Dimitropoulou, 1986; Peeters *et al.*, 1987). Similar observations were made in fibers of other *Gossypium* spp., although the timing and extent of nucleolar enlargement and vacuolation varied between species (Peeters *et al.*, 1987). Nucleolar size reflects the activity of nucleolar genes (Shaw and Jordon, 1995) that, in turn, presumably influences the subsequent size of the fiber. A close relationship between nucleoli size at early stages of fiber growth and final fiber dimensions for different cultivars is reported (De Langhe *et al.*, 1978; Berlin, 1986; Peeters *et al.*, 1987, Peeters *et al.*, 1988; Peeters *et al.*, 1991), although this relationship did not hold when different *Gossypium* spp. were compared (Peeters *et al.*, 1987). Examination of the frequency distribution of nucleolar volume within a fiber population *in situ* indicated that large and small nucleoli develop simultaneously and were associated with large and small fibers, respectively (Peeters *et al.*, 1988).

The morphology and size of the nucleolus in the fiber initial are strongly influenced by hormones; GA and auxin differentially affect nucleolar activity. GA_3 stimulated the synthesis of Nu material and strongly inhibited Nu vacuolation (De Langhe *et al.*, 1978; Kosmidou- Dimitropoulou, 1986). Nucleolar size was reduced when an endogenous source of GA (*i.e.* subtending leaf) was removed (Peeters *et al.*, 1991). Auxin induced an earlier increase in nucleolar size and earlier Nu vacuolation but the final size was less than observed with GA_3 (De Langhe *et al.*, 1978; Kosmidou-Dimitropoulou, 1986). Based on these morphological differences, De Langhe *et al.* (1978) suggested that GA regulates ribosome synthesis while auxin controls ribosome trans-

port and use, although the molecular details are lacking to substantiate these observations. In other tissues, ribosomal RNAs, RNA polymerase I, and ribosomal proteins, components necessary for ribosomal assembly (Shaw and Jordon, 1995), were identified as auxin responsive genes (Garbers and Simmons, 1994). The mechanism of auxin action is unclear.

Comparison of the timing of increase transcription and of hormonal requirements for fiber differentiation provides further indirect evidence of hormonal influence on transcription. Transcriptional activity is most prevalent the first 6 DPA (Berlin, 1986) at the same time IAA and GA are absolutely required for maximum fiber production *in vitro* (Beasley *et al.*, 1974; Kosmidou-Dimitropoulou, 1986). IAA and GA can stimulate RNA synthesis (Brock and Kaufman, 1991) and presumably are involved in upregulating genes required for fiber differentiation and growth, although these genes have not be identified. Nevertheless, the evidence collectively suggests that the potentiation for fiber growth is determined during these first few days post anthesis.

3.3 Hormone Effect on Fiber Extension

The enlargement of the fiber cell consists of an increase in cell volume and the extension of the existing wall. This process is described by the following equation:

$$dV / dt = m (P-Y)$$

where the change in cell volume (dV / dt) is the product of the extensibility of the wall (m) an the effective growth turgor (P-Y) with P representing the turgor pressure of the cell and Y the wall yield threshold (the critical pressure that turgor must exceed for wall extension). A hormone, irrespective if its mechanism of action is gene activation or activation of enzyme activity, can only increase the rate of cell enlargement by increasing m, by increasing P, or by decreasing Y or by any combination of the three and vice versa if the rate is inhibited (Cleland, 1986, 1995).

The requirement for auxin in the medium for maximum fiber length (Kosmidou- Dimitropoulou, 1986) strongly indicates that auxin is responsible for stimulating fiber extension. According to acid growth theory (see Rayle and Cleland, 1992 for review), auxin acts on susceptible cells by stimulating the excretion of protons via plasma membrane (PM) H^+-ATPase into the apoplast that decrease apoplastic pH. This decrease in pH activates cell wall loosening processes that disrupt the load bearing network of the cell wall and thereby causes an increase in cell wall extensibility. Based on the use of exogenous acids in other plant systems, this acid induced growth occurs only transiently (1 to 4 h) and thus, for elongation to continue for extended period, auxin must also mediate other accessory processes that may include osmoregulation, cell wall synthesis, and maintenance of cell wall extensibility (Rayle and Cleland, 1992). Evidence of these pheonomena during fiber elongation include the accumulation of K^+ and malate for the maintenance of turgor (Dhindsa *et al.*, 1975; Basra and Malik,

1983), the constant deposition of cellulose (1 ng/ mm; Meinert and Delmer, 1997), and the ability of the cell wall for 30 days (Naithani *et al.*, 1982). Presently, the signals that coordinate these cellular events and how tightly these events are coupled to wall acidification are not well-defined.

One mechanism of GA action suggested during fiber initiation is to stimulate malate synthesis, although the evidence is not strong. Dhindsa (1978b) showed that the activity of the malate synthesizing enzymes, PEP carboxylase (PEPC) and malate dehydrogenase (MDH), and $H^{14}CO_3$ incorporation into organic acids increased upon addition of unfertilized ovules to a culture medium containing IAA and GA_3 and attributed this stimulation to GA_3. However, in another study a similar stimulation in PEPC activity was not observed when extraction conditions were optimized for maximum recovery of PEPC activity (Ramsey, 1996). Furthermore, in the study of Dhindsa (1978b) a stimulation caused by increased ovule growth was distinguished from that caused by increased fiber growth. In fact, in another study where he used inhibitors of hormones, Dhindsa (1978a) concluded that GA mainly promoted ovule growth while IAA was responsible for fiber growth.

The inhibition of fiber growth by ABA is also suggested to be due, in part, to a reduction in malate synthesis. The presence of ABA in ovule culture medium inhibited PEPC and MDH activity (Dhindsa, 1978b) and decreased malate levels (Dhindsa *et al.*, 1976) in fibers coincident with ABA inhibition of fiber growth (Dhindsa *et al.*, 1976). Similar findings were observed by Barsa *et al.* (1993) in 15 DPA fibers after incubating seed clusters in ABA for several hours, although the ABA was applied at a time when ABA had no effect on fiber growth *in vitro* (Dhindsa *et al.*, 1976). However, these studies did not establish whether this inhibition of malate synthesis is the cause (due to decrease in P) or the result (due to decrease in m) of reduced growth rates. Since cell walls are considered the major controlling factor in cell extension (Cosgrove, 1993a), a primary action of ABA inhibition of auxin induced cell elongation in maize coleoptiles is attributed to reduced cell wall extensibility rather than a decrease in cell turgor (Kutschera and Schopfer, 1986).

3.3.1   Carbohydrate Metabolism

Carbohydrates transported from a source leaf to the developing boll and imported into fiber can have a significant impact of fiber differentiation and growth. Fiber growth is dependent on the oxidation of carbohydrates for the generation of ATP that drives cellular metabolism and biosynthesis of organic compounds (*e.g.*, isoprenoids, amino acids, polysaccharides, fatty acids) that serve as the building blocks for the macromolecules (*e.g.*, hormones, proteins, cellulose, lipids) of the fiber cell. Sugar molecules can also serve as key regulatory molecules that control metabolism, gene expression, and development (Koch, 1996).

3.4   Import of Carbohydrates into the Fiber Cell

Sucrose is transported to the developing boll via the phloem and can either enter the fiber cell wall free space (apoplast) after passage through the carpel wall or move into the developing seeds through the funiculi. (Buchala and Meier, 1985). In the later case, the sucrose is transported from the neighboring seed cells to the fiber by either a symplastic (through plasmodesmata) or apoplastic (from cell wall free space) pathway. As the fiber cell matures, it becomes constricted in the zone in contact with other epidermal cells and develops an extended base (foot) below this zone (Fryxell, 1963; Ryser, 1985). Numerous pit-like structures occur in the base that are believed to facilitate transport of nutrients into the fibers (Ryser, 1992). Electron micrographs of the basal part of the fiber cell during secondary wall deposition reveals that pits found in the periclinal (between epidermal, and mesophyll cells) and in the anticlinal (between epidermal cells) walls contain plasmodesmata. The plasmodesmata frequency in these respective walls of the white lint cultivar of *G. hirsutum* were approximately similar to ordinary epidermal cells estimated as 22 and 27 $\mu m^{-2.}$ The assimilate flux through the plasmodesmata at the fiber base is estimated to be 10^{-3} pg $\mu m^{-2} sec^{-1}$ (Ryser, 1992) which is the same order of magnirtude for symplastic flux observed in other cells (Sauter and Kloth, 1986). Suberization of the fiber base of *G. arboretum* and the green lint mutant of *G. hirsutum* during secondary wall deposition excludes movement of assimilate via the apoplastic pathway (Ryser, 1992). For the white linted cultivar of *G. hirsutum*, where the suberin deposits at the fiber base are more variable (only abot 50% of fibers suberized), the mechanism of assimilate transport may be more complex, involving both the apoplastic and symplastic pathways.

Sucrose in the apoplast is transported across the plasma membrane as hexoses after hydrolysis by an extracellular invertase or as sucrose. Buchala (1987) showed by light microscopy and histochemical staining that an acid invertase was localized at the cell surface of *G. arboretum* fibers. By removing sucrose, this invertase may contribute to phloem unloading and carbohydrate portioning by steepening the sucrose concentration gradient and enhancing flow to a given sink (Roitsch and Tanner, 1996). Extracellular cleavage of sucrose, however, is not a prerequisite for sucrose uptake. Feeding asymmetrically labeled sucrose or labeled 1- flurosucrose analogue not hydrolyzed by invertase, revealed that sucrose could be taken up without inversion (Buchula, 1987). Transport of carbohydrate into the fiber cell can therefore occur by two different mechanisms, possibly involving proton-sucrose and proton-glucose symports (Bush, 1989).

3.5   Sucrose Breakdown

The sucrose and hexoses imported into the fiber cell may enter metabolism in the cytosol or transported into the vacuole for temporary storage. Sucrose in the vacuole could be hydrolyzed to glucose and fructose by an acid invertase. Jacquet *et al.* (1982) showed that of the methanol soluble neutral sugars, glucose and fructose and their phosphorylated derivatives rather than sucrose accumulated in *G. hirsutum* or *G. arboretum* fibers during development. Carpita and Delmer (1981) estimated that at the initiation of secondary wall deposition 89% of total reducing sugars was in a storage pool, presumably the vacuole. Invertase activity was detected in a soluble fraction from *G. arboreum* fibers possibly localized to the vacuole (Black *et al.*, 1995) based on its acidic pH optimum (Buchala, 1987). Sugar accumulation in the vacuole can provide a carbohydrate reserve when external sources become limiting and contribute to the osmoregulation of the fiber cell during elongation.

Fiber cells contain two types of enzymes in the cytosol that are capable of cleaving sucrose. An alkaline invertase was detected in *G. hirsutum* fibers that was active at pH 8.0 (Wafler and Meier, 1994) distinguishable from its vacuolar and cell wall counterparts which show very little activity above pH 6.0 (Buchala, 1987). This alkaline invertase, however, was not detected in *G. arboreum* fibers (Buchala, 1987). The other enzyme detected in fibers that cleaves sucrose is sucrose synthase (Amor *et al.*, 1995; Notle *et al.*, 1995; see also Chapter 10). Although this enzyme catalyzes a freely reversible reaction, the high levels of this enzyme and the steady state measurements of its substrates and products in non-photosynthetic tissue indicate that it functions primarily in the direction of sucrose degradation and UDP-glucose synthesis (Xu *et al.*, 1990; Geigenberger *et al.*, 1993). The free glucose and fructose produced by these reactions are phosphorylated to hexose-6-phosphate by hexose kinases and funneled into intermediary metabolism via glycolysis and the pentose phosphate pathways, used for cell wall biosynthesis, or hydrolyzed to free hexoses by phosphatases. The UDP-glucose produced by sucrose synthase may be either converted to hexose phosphates by a UDP-glucose pyrophosphorylase (UGPase) and enter the glycolytic and pentose phosphate pathways or used for the synthesis of cellulose and other cell wall polysaccharides and for oligosaccharide synthesis. UDP-Glucose is also used in the synthesis of β-glucosides (Ohana *et al.*, 1992).

Although sucrose synthase is generally thought to be a cytosolic enzyme (Kruger, 1990), Amor *et al.* (1995) showed in fibers that over 50% of total sucrose synthase was tightly associated with the plasma membrane, possibly complexed with cellulose synthase and/or callose synthase. This complex enables carbon from sucrose (via UDP-glucose) to be directly channeled to the growing glucan polymer. The possible advantages of such a complex were recently reviews by Delmer and Amor (1995). The alternative source of UDP-glucose for cellulose biosyntheisis is UGPase. The reaction catalyzed by this enzyme is readily reversible with the direction and flux rate determined *in vivo* largely by PPi concentration (Kleczkowski, 1994). Although this enzyme is quite active during secondary wall formation (Wafler and

Meier, 1994) the prevailing evidence suggests that it is an unlikely source of UDP-glucose for cellulose synthesis. In contrast to sucrose synthase, UGPase occurs almost exclusively in the cytosol (Kleczkowski, 1994). Furthermore, in sink tissues undergoing active sucrose consumption, PPi-dependent phosphofructokinase (PFK) is thought to maintain PPi levels by operating primarily in the direction of PPi synthesis (Rea and Poole, 1993) that would shift UGPase catalyzed reaction in the direction of UDP-glucose degradation (Kleczkowski, 1994). Radiolabeling studies further showed that UDP-glucose was a poor substrate for cellulose biosynthesis in disrupted fiber cells (Amor *et al.*, 1995).

3.6 Glycolysis and the Oxidative Pentose Phosphate Pathway

Glycolysis and the pentose phosphate pathway operate in cotton fibers and the extent of their operation varies with the respiratory demands of the fiber cell (Basra and Malik, 1984). The oxidation of [1-^{14}C] and [6-^{14}C] glucose and measurements of enzyme activity and intermediates of these pathways indicate that these pathways are most active during fiber extension (Basra and Malik, 1984; Wafler and Meier, 1994). The increased flux through these two pathways during this period reflects a high requirement for energy and reducing power needed to sustain cell growth. When the growth rate of the fiber slows, enzyme activity, metabolites and turnover rates of ^{14}C-glucose decline due to a shift in metabolic priorities (Basra and Malik, 1984; Wafler and Meier, 1994). During secondary wall deposition, a greater proportion of the imported carbon is channeled towards the synthesis of cellulose and β-1,3 glucans (Mutsaers, 1976; Carpita and Delmer, 1981) than used for respiratory activity (Carpita and Delmer, 1981).

The fiber cell, being non-photosynthetic, requires the oxidative pentose phosphate pathway for the production of NADPH for reductive biosynthesis and precursors for other metabolic pathways. Specifically, ribose-5-phosphate will be required for nucleic acid biosynthesis and erythose-4-phosphate for synthesis of aromatic amino acids. Flux through this pathway will be largely controlled by NADPH/NADP$^+$ ratios with both glucose-6-phosphate (Glu-6-P) dehydrogenase and 6-phosphogluconate (GP) dehydrogenase sensitive to NADPH inhibition (Miernyk, 1989).

In plant cells, glycolytic and oxidative pentose phosphate pathway can occur independently in two subcellular compartments, the cytosol and the plastids (Dennis and Greyson, 1987; Plaxton, 1996). The extent to which these pathways are completely duplicated in plastids of the fiber cell is not known but plastids from other non-photosynthetic tissue depending on the tissue can possess enzymes of part of all of these pathways (Dennis and Miernyk, 1982; Plaxton, 1996). Since very little carbohydrate is stored as starch in the fiber cell (Carpita and Delmer, 1981; Ryser, 1985), the carbon precursors for plastid glycolysis must come from the cytosol. The prime function of glycolysis and oxidative pentose phosphate pathway in non-photosynthetic plastids is to generate carbon precursors, reductant, and ATP required for anabolic pathways such as synthesis of fatty acids and aromatic amino acids (Emes and Tobin, 1993; Hermann, 1995; Plaxton, 1996). During fiber elongation, increased fatty acid synthesis will be required to support membrane biogenesis and formation of waxes found in the cuticle. Based on geometric calculations of the elongated fiber, the surface area of the plasma membrane and tonoplast is estimated to increase over 2000 fold by 3 DPA (Berlin, 1986). The shikimate pathway interfaces with carbohydrate metabolism in plastids through the action of 3-deoxy-D-arabino-heptulosonate-7-phosphate (DAHP) synthase which utilizes PEP produced by glycolysis and erythrose-4-phosphate produced by oxidative pentose phosphate pathway (Hermann, 1995). Shikimate pathway synthesizes aromatic amino acids that serve not only as building blocks for protein synthesis but are also precursors of phenolics. Morphological and radiolabeling studies with [³H]-phenylalanine indicate that phenolic biosynthesis is active in fiber cells on 1 DPA (Berlin, 1986). Phenolic biosynthesis is suggested to play a key regulatory role during differentiation of fiber primordial, for instance, inhibiting IAA oxidation.

The glycolytic pathway in the cytosol of fibers as in all plant cells consists of a series of parallel reactions rather than a strict linear pathway. Parallel enzymatic reactions, as summarized in Figure 32-1 can occur at sucrose as previously discussed, glyceraldehyde 3-phosphate, fructose-6-phosphate (Fru-6-P), and phosphoenolpyruvate (PEP) (Plaxton, 1996). The phosphorylation of Fru-6-P to fructose-1, 6-bisphosphate (Fru-1, 6-P_2) can be achieved by two distinct enzymes, phosphofructokinase (PFK) and PFP with PFP strictly a cytosolic enzyme. Both enzymes are most active during fiber elongation with PFP at peak activity several fold more active than PFK (Wafler and Meier, 1994). High levels of PFP activity are commonly observed in sucrose importing tissues that require sucrose metabolism (Xu *et al.*, 1990). PFK catalyzes an irreversible reaction and is highly regulated sensitive to allosteric inhibition by PEP and this inhibition is relieved by the activator Pi (Plaxton, 1996). The concentration ratio of Pi:PEP is therefore thought to play an important role in the control of carbon flow through glycolysis *in vivo*. Unlike, PFK, PFP catalyzes a freely reversible reaction utilizing PPi as the energy source rather than ATP and contributes to carbon flow by providing an alternative route to the highly regulated PFK. During active sucrose consumption, PFP is thought to operate in the direction of PPi synthesis but may work in the opposite direction when ATP becomes limiting (Rea and Poole, 1993; Plaxton, 1996) either due to environmental stress (Plaxton, 1996) or increase metabolic demands. The freely reversible reaction catalyzed by PFP may, in addition, serve to equilibrate the triose and hexose phosphate pools (Dennis and Greyson, 1987). Both triose and hexose phosphate pools can be funneled into the pentose phosphate pathway that provides carbon skeletons for other biosynthetic pathways; the hexose phosphates can in addition be used for cell wall biosynthesis. PFP is activated by fru-2,6-P_2 whose concentration increase with increased rates of biosynthesis (Dennis and Greyson, 1987).

Two alternative metabolic routes can be employed for converting PEP to pyruvate. The most direct route is a reaction catalyzed by pyruvate kinase (PK) but pyruvate can also be generated by coupling PEPC, aubiquitously plant cytosloic enzyme, with cytosolic MDH and mitochondrial NAD-dependent malic enzyme (NAD ME). This route can provide an important supplement to the mitochondrial pyruvate pool because of the slow transport of pyruvate through the inner mitochondrial membrane (Lance and Ruskin, 1984) or when PK activity becomes ADP limited (Plaxton, 1996). Although these enzymes, with the exception of NAD⁺ malic enzyme have been measured in fibers (Basra and Malik, 1983; 1984; Corcoran and

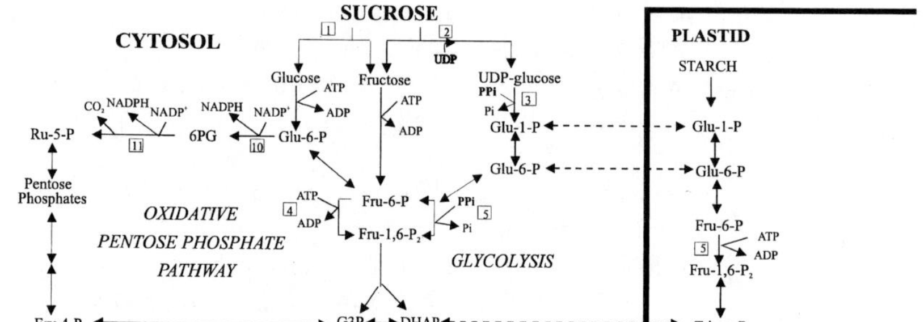

Figure 32-1. Summary of the major biochemical pathways involved in sucrose metabolism in the fiber cell. The number reactions are catalyzed by the following enzymes: 1. invertase; 2. sucrose synthase; 3. UGPase; 4. PFK; 5. PFP; 6. PK; 7. PEPC; 8. MDH; 9. NAD+-ME; 10. Glu-6-P dehydrogenase; 11. 6PG dehydrogenase; and 12. DAHP synthase. The abbreviations used are as in the test or as follows: Glu-1-P, glucose-1-phosphate; Ru-5-P, ribulose-5-phosphate; G3p, glyceraldehydes-3-phosphate; DHAP, dihydroxyacetone phosphate; 3-PGA, 3-phosphoglycerate; OAA, oxaloacetate; Ery-4-P, erythrose-4-phosphate.

Zeiher, 1995; Wafler and Meier, 1994), very little is known about what controls carbon flux between these two metabolic routes. In germinating *Ricinus communis* L. cotyledons, the PK and PEPC are regulated independently with PK showing a weaker response to metabolites involved in amino acid and energy metabolism than PEPC and the two enzymes are inversely affected by pH with PK have a more acidic pH optimum (Podesta and Plaxton, 1994).

Primary control of glycolysis occurs at the terminal reaction utilizing PEP with secondary regulation exerted at the level of Fru-6-P utilization (Plaxton, 1996). For example, stimulation of cytosolic glycolysis as a result of increased H^+ efflux in *Chenpodium rubrum* L. cell suspension cultures was initially accompanied by decreases in PEP and 3-PGA indicating that PK and/or PEPC were activated (Hatzfeld and Stitt, 1991). A similar increase in H^+ efflux is expected during auxin stimulated fiber elongation. Consistent with this observation is the potent allosteric inhibition of many plant PFKs by PEP (Plaxton, 1996). Stimulation of PK or PEPC activity relieves PEP inhibition of PFK and allows glycolysis from hexose-P to proceed. A conceivable benefit of regulation at the terminal reactions is that it allows considerable freedom to adjust the triose phosphate and hexose phosphate pools required for biosynthesis without drastically affecting respiration (Hatzfeld *et al.*, 1990).

3.6.1 Dark Metabolism of Fixed Carbon Dioxide

Cotton fibers develop within the confines of the capsule wall protected agasint environmental perturbations. Respiratory losses of CO_2 from cotton fruit during ontongeny are substantial and vary with fruit age (Wullschleger and Oosterhuis, 1990c). Stomata of the capsule wall function early in development but decline in older fruit presumably because they become occluded by waxes that decrease its permeability to water and CO_2. As a result of this decrease permeability, there is a significant elevation of internal CO_2 concentration within the fruit (Ryser, 1985; Wullschleger and Oosterhuis, 1990c). This internal CO_2 can be refixed by the fiber seed complex through dark CO_2 fixation (Dhindsa *et al.*, 1975; Basra and Malik, 1983).

The enzyme primarily responsible for dark CO_2 fixation in the fiber is PEPC. The action of PEPC coupled with cytosolic MDH produces malate that can play a versatile role in fiber metabolism. Malate dianions contributes to the osmoregulation of the fiber cell by providing a charge balance to K^+ imported from the apoplast (Dhindsa *et al.*, 1975; Basra and Malik, 1983). The accumulation of these osmotically active solutes control the influx of water and thus, the turgor pressure that drives fiber elongation. Malate imported into the mitochondria serves an anapleortic role by replenishing tricarboxylic acid cycle (TCA) intermediates depleted during amino acid biosynthesis (Lance and Rustin, 1984) and can be used, as previously indicated, for the synthesis of pyruvate required for respiration. The synthesis and degradation

of malate are components of a biochemical pH stat mechanism that contribute to the control of cytoplasmic pH. Malate synthesis releases H^+ to the cytoplasm and its degradation produces OH^- (Davies, 1973). Due to the sequential action of PEPC, MDH, and $NADP^+$ malic enzyme ($NADP^+$ ME), NADPH required for reductive biosynthesis can be generated via a transhydrogenase reaction (Basra and Malik, 1983).

3.7 PEP Carboxylase

The initial dedicated step in malate synthesis is the β-carboxylation of PEP to form oxaloacetate, the immediate precursor of malate, and Pi; an essentially irreverdible reaction under physiological conditions catalyzed by PEPC (Latzko and Kelly, 1983). This enzyme is most active during fiber elongation and declines during secondary wall deposition (Basra and Malik, 1983; Zeiher *et al.*, unpublished observations). PEPC subunit composition is quite heterogeneous during fiber development with the polypeptide banding pattern changing both temporally and spatially (Corcoran and Zieher, 1995). Five polypeptide bands ranging in molecular masses between 122 and 106 kD cross-reacted with maize leaf antiserum. The 106kD, 111kD and 113 kD polypeptides were present during elongation and secondary wall deposition whereas the 119 and 122 kD polypepetides were most prevalent during secondary wall deposition. Comparison of these banding patterns in fibers and delinted ovules 10 DPZ revealed that the 106 kD and 113 kD polypeptides were enriched in the fiber. This banding pattern was not due to polymorphisms associated with the allotetraploid *Gossypium* spp. As similar banding patterns were observed in ovules of allotetraploid and diploid species (Corcoran and Zieher, unpublished observations) as well as tissues of diploids of other taxonomic origins (Matsuoka and Hata, 1987). Extraction under denaturing conditions also established that this heterogeneity was not due to proteolysis during extraction (Corcoran and Zieher, unpublished observations). Although the origin of these multiple polypeptide bands is unclear, several possibilities could explain the differences in the electrophoretic mobility of these immunostained bands. This heterogeneity may be the result of PEPCD being post-translational modified (Schultz *et al.*, 1993; Chollett *et al.*, 1996); being a heterotetramer composed of subunits of different molecular masses (Law and Plaxton, 1995) and/or representing different PEPC isoforms (*i.e.*, each isoform a homotetramer composed of subunits with the same molecular mass; Cushman *et al.*, 1990; Lepiniec *et al.*, 1994).

The catalytic properties of the fiber PEPC are similar to those observed for other plant PEPCs (Denecke *et al.*, 1993; Podesta and Plaxton, 1994). The enzyme has pH optimum around pH 8 and shows a marked decline in activity below pH 7.5 (Corcoran and Zieher, unpublished observations). The K_m (PEP) is pH dependant; at pH 8, the K_m (PEP) is 0.10 mM and increases to a K_m (PEP) of 1.3 mM at pH 7.3 (Menke and Zeiher, unpublished observations). As with other PEPCs (Schuller *et al.*, 1990; Law and Plaxton, 1995) the fiber enzyme is sensitive to malate inhibition in a pH dependant manner. At pH 7.3 and 2.5 mM PEP, the fiber PEPC has and I_{50} of 0.43 mM whereas at pH 8 only a 23% inhibition was detected at 10 mM malate (Menke and Zeiher, unpublished observation). These catalytic properties are in agreement with the proposed role for PEPC in a biochemical pH stat mechanism. During auxin stimulated fiber growth, H^+ excretion is presumably stimulated, leading to a rise in cytoplasmic pH. This more alkaline pH favors the synthesis and accumulation of malate that results in the release of H^+ into cytosol lowering cytosolic pH. PEPC is considered the pH regulated step in the synthesis of malate (Davies, 1973). The increased affinity for PEP and the lowered sensitivity to malate inhibition favors the stimulation of this enzyme with alkalization of cytoplasmic pH. The pH regulation of the enzyme may be further affected by the phosphorylation state of the enzyme. Echevarria *et al.* (1994) showed that phosphorylation of the recombinant C_4 PEPC isoform amplified the regulatory influences of pH on enzyme activity.

Presently, there is no evidence that the fiber enzyme is reversibly phosphorylated, although there is now convincing evidence that reversible phosphorylation of N-terminal domain of plant PEPC is widespread, including the C_4, CAM, C_3, and non-photosynthetic isoenzymes (Chollet *et al.*, 1996). Phosphorylation of PEPC fine tunes the activity and allosteric properties of the enzyme. The phosopho-enzyme is less sensitive to L-malate inhibition and more active when assayed at near physiological conditions (sub-saturating PEP and neutral pH) (Jiao and Chollet, 1988; Schuller and Werner, 1993; Echevarria *et al.*, 1994; Duff and Chollet, 1995). The classical test for *in vivo* phosphorylation of PEPC is a decrease in malate sensitivity (Chollet *et al.*, 1996); yet, there is no consistent evidence available that the malate sensitivity of PEPC changes during fiber development when the enzyme is assayed at suboptimal conditions (Menke and Zeiher, unpublished observations). However, these assays are difficult with crude fiber extracts because of the low specific activity and may therefore, not be reliable enough to detect differences in malate inhibition; a problem recently encountered by Duff and Chollet (1995) with their work the wheat leaf PEPC. In addition, this enzyme can be altered by limited proteolysis at its N-terminus during extraction that decreases its sensitivity to malate but does not have a major influence on its electrophoretic mobility (Chollet *et al.*, 1996), confounding the interpretation of the results. Given that this appears to be a common mechanism for regulating PEPC, it is highly likely that the fiber PEPC will be phosphorylated; but confirmation awaits a partial purification to enrich the enzyme, and the assay and extraction conditions must be better defined.

3.8   Changes in K^+ and Malate During Fiber Elongation.

Elongation of the fiber for extended time periods (*i.e.* days) requires the uptake and production of osmotically active solutes for the maintenance of turgor. It is well established that K^+ imported for the apoplast and electrochemically balanced in part by malate dianions synthesized intracellularly are the major solutes that produce the necessary turgor to maintain fiber elongation (Dhindsa *et al.*, 1975; Basra and Malik, 1983; see Fig. 32-2). Potassium and malate fluctuate in parallel in relation to growth rate, with peak levels achieved when growth rate is at a maximum. This coordinate increase in K^+ and malate indicates a strong interrelation between ion transport and carbon metabolism, although the signals responsible for synchronizing theses two events are not well defined. The activation of the plasma membrane (PM) H^+-ATPase by auxin can cause an extracellular acidification and hyperpolarization of the PM (Schroeder and Hedrich, 1989). Hyperpolarization to potentials more negative than around -80 mV can result in the opening of inward rectifying K+ channels that provide a pathway for K^+ uptake and of high abundance in cell types that undergo pronounced volume changes (Hedrich and Dietrich, 1996), as occurs in a fiber cell (can increase as much as 3000 fold relative to its diameter; Basra and Malik, 1984). It is well established in plants that PM H^+- ATPase can polarize the membrane to values more negative than -100 mV providing the necessary driving force for K^+ uptake (Schroeder and Hedrick, 1989).

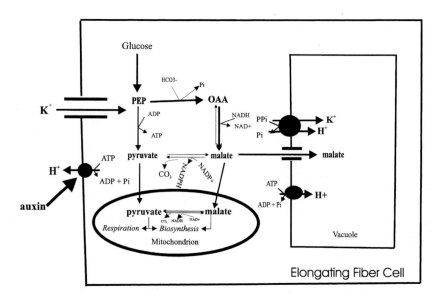

Figure 32-2. The accumulation of K+ and malate during turgor-driven fiber elongation. The more intense arrows indicate the preferred direction of carbon and ion flux during this phase of growth.

The influx of K$^+$ driven by auxin stimulated H$^+$ - ATPase activity may activate PEPC. PEPC activity isolated from *Vicia faba* L. guard cell protoplasts was stimulated by K$^+$ (Schnabl and Kottmeier 1984). Auxin stimulated H+ extrusion may in addition increase cytoplasmic pH and activate PEPC, a response possibly amplified by a change in phosphorylation state of the enzyme (Echevarria *et al.*, 1994).

3.9 Ion Transport across the Vacuolar Membrane

In a elongating fiber cell, approximately 905 of the volume is occupied by large central vacuole (V; Carpita and Delmer, 1981; Berlin, 1986). This multifunctional compartment serves not only as a storage and hydrolytic compartment but plays a key role during osmoregulation and in cytoplasmic homeostasis. During fiber elongation, the storage of inorganic and organic ions and carbohydrates creates a positive osmotic pressure that drives the uptake of water and generates the necessary turgor pressure required for cell extension. The transport of solutes across the vacuolar membrane (tonoplast) is mediated by two electrogenic pumps, a proton-translocating inorganic pyrophosphatase and a H$^+$ ATPase (Barkla and Pantoja, 1996). These two pumps derive energy from the hydrolysis of the high energy metabolites, PPi and ATP and generates a proton motive force that drives the secondary transport of solutes into the vacuole. Histochemical localization of ATPase activity in the outer integument of the cotton ovule showed that the only significant ATPase activity was in the tonoplast of elongating fiber cells (Joshi *et al.*, 1988) and suggest that at least one of these proton pumps may be important to turgor regulation during fiber elongation. Further evidence of a pivotal role for V-ATPase in cell elongation came with the isolation of V-ATPAse expression mutants in transgenic carrots (Gorgarten *et al.*, 1992). When the expression of tonoplast V-ATPase catalytic subunit was inhibited with antisense mRNa, the carrot plants exhibited a dwarf morphology that resulted from reduced cell expansion.

3.9.1 V-ATPase

V-ATPase is a large multimeric complex of 450 to 750 kD comprised of 7 to 10 different subunits (Barkla and Pantoja, 1996). This complex can be subdivided into two sectors, the hydrophilic catalytic complex (V$_1$) located on the cytosolic face on tonoplast membrane and hydrophobic integral membrane complex (V$_0$). The principal subunits of the V$_1$ complex are the A subunit that contains catalytic nucleotide binding site and the B regulatory subunit, each present as 3 copies per functioning enzyme. The hydrophobic complex is comprised of 616 kD proteolipid subunits that spans the tonoplast membrane and functions as the proton pore.

cDNa clones encoding the three major subunits of V-ATPase have been isolated and characterized from developing cotton ovules (Wilkings, 1993; Wan and Wilikns, 1994; Hasenfratz *et al.*, 1995). The presence of small mutigene families for the A,B, and proteolipid subunits were identified by isolation of cDNA clones representing unique members of the gene family (Wilkins, 1992; Wan and Wilkins, 1994; Hasenfratz *et al.*, 1995). Hybridization of clone specific probes to unique restriction fragments on DNA blots in A and D diploid as well as allotetraploid species established that the cDNA clones for at least the A and proteolipid subunits represented distinct isoforms rather than alloalleles of the same isoform (Wilkins, 1992; Hasenfratz *et al.*, 1995). RNA analysis using clone specific probes for genes coding for 2 16 kD proteolipid isoforms showed that the two transcripts differentially accumulate in different tissues of cotton and increase dramatically in tissues undergoing rapid expansion, particularly in anthers, ovules and petals. The CVA16.4 proteolipid transcript was most prevalent of the two proteolipid messages in 10 DPA ovules when fibers were undergoing rapid elongations (Hasenfratz *et al.*, 1995). The transcript levels for the catalytic subunit were also prevalent in developing ovules during rapid elongation declining to basal levels at 25 DPA coincident with the cessation of elongation (Wilkins, 1992). The induction of V-ATPase subunit gene expression in tissues undergoing rapid expansion is attributed in part to the increased biochemical machinery needed to support turgor driven cell expansion (Hasenfratz *et al.*, 1995). Increased V-ATPase activity is expected to privide the required energy to drive the uptake of osmotically active solutes through secondary transport.

3.9.2 Malate and K Transport In and Out of the Vacuole

While it is well accepted that malate and K$^+$ are the principle osmoregulatory solutes in fibers (Dhindsa *et al.*, 1978; Basra and Malik, 1983) essentially nothing is known about the transport mechanisms that control their movement in and out of the vacuole. Significant progress in the identification of these transport mechanisms in other plant systems has been made in recent years. Because many of these mechanisms appear to be ubiquitously present in plant cells, they will briefly be mentioned and their regulation discussed, although it should be emphasized that there will probably be differences in the transport capacity and regulation of these common transport mechanisms to meet the physiological needs of the cell in which they are found.

A voltage activated, inward rectifying malate channel has been identified in vacuoles from a diversity of plant species (see Barkla and Pantoja, 1993 for review). The channel present in vacuoles from sugar beet cell suspensions is 6-10 times more selective for mal^2- than K$^+$ (Panjota *et al.*, 1992). Two mechanisms have been described for transporting K$^+$ into the vacuole. The H$^+$ -translocating pyrophosphatase, ubiquitously present in plant cells, catalyzes the coordinate translocation of both H$^+$ and K$^+$ from the cytoplasm into the vacuolar lumen and its postulated that this enzyme may contribute significantly to the turgor regulation of a cell (see Rea and Poole, 1993 for review). Unlike the V-ATPase, this enzyme is modulated by cytosloic $[^{Ca2+}]_{free}$, subject to pronounced inhibition by micromolar concentrations. The other possible mechanism is a K^+/H^+ antiporter identified in tonoplast membranes isolated from cotton roots (Hassidim *et al.*, 1990).

Associated with a reduction in fiber elongation rate is a decrease in cell turgor (Dhindsa *et al.*, 1975) and a decline in K^+ and malate concentration (Dhindsa *et al.*, 1975; Basra and Malik, 1983). A fast-activating vacuolar (FV) channel has been described that could release K^+ from the vacuole (see Barkla and Pantoja, 1996 for review). This ion channel is voltage independent and is highly selective for cations over anions. In guard cells this channel plays an important role in mediating the release of K^+ from the vacuole into the cytoplasm with the reduction in turgor during stomatal closure and is activated by micromolar concentrations of Ca^{2+} (Ward and Schroeder, 1994). The mechanism of malate efflux from the vacuole, on the other hand, is not well-defines. Several possible mechanisms have been proposed but it is unclear whether any of these mechanisms have widespread occurrence amongst different cell types. These mechanisms include a 4, 4'-diisothiocyano-3,3'-stilbenedisulfonic acid sensitive protein carrier identified in tonoplast vesicles isolated form *Catharanthus roesus* cells (Bouyssou *et al.*, 1990) and passive diffusion of H^2mal form from vacuoles of the CAM plant *Kalanchoe daigremontiana* Raym-Hamet. and Perr. (Luttge and Smith, 1984). A slow-acting vacuolar channel has also been suggested as a mechanism for releasing malate from guard cells during stomatal closure (Hedrich *et al.*, 1988) and from vacuoles of the CAM plant *Graptopetalum paraguayense Walther* (Iwasaki *et al.*, 1992). Although this channel appears to be present in all plant cells (Hedrich *et al.*, 1988), the significance of this channel is unclear since it mainly opens at non-physical positive tonoplast potentials (Barkla and Panjota, 1996).

3.10 Fate of K^+ and Malate During Secondary Wall Deposition

Potassium released from the vacuole could subsequently be transported into the extracelluar space by an outward rectifying K^+ channel (Tester, 1990) and translocated to burs and seeds (Leffler

and Tubertini, 1976). This channel is present in the plasma membrane of many different cell types from diversity of plant species and opens upon depolarization of the PM (see review Tester, 1990). In guard cells where there is a large efflux of K^+ during stomatal closure, membrane depolarization may be the result of inhibition of the H^+-ATPase pump, anion efflux, and/ or Ca^{2+} entry (Assmann, 1993).

Malate released from the vacuole in fiber cells can be decarboxylated by a $NADP^+$-ME to pyruvate with the reduction of $NADP^+$, transported into the mitochondria or exported to the apoplast (summarized in Fig. 32-3). In the mitochondria, malate can replenish tricarboxylic acid cycle intermediates depleted during biosynthesis or used for the generation of mitochondrial pyruvate via NAD+ ME. Basra and Malik (1983) reported that decarboxylation of malate is most prevalent during secondary wall deposition while malate synthesis via a PEPC-MDH couple most active during fiber elongation increasing 10-fold between 15 and 20 DPA. Reexamination of the developmental profiles of $NADP^+$ ME, however, failed to show this large increase in activity concomitant with secondary wall deposition (Corchoran and Zeiher, unpublished observations). The reason for this discrepancy is unclear. In the later study, the enzyme showed a high affinity for $NADP^+$ ($Km = 6\ \mu M$) and a requirement for a divalent cation; properties that distinguish NADP malic enzyme from mitochondrial NAD-ME and NADP-MDH, respectively (Edwards and Andreo, 1992). Compared to PEPC which has a more alkaline pH optimum NADP-ME is quite active at acidic pHs (pH optimum 8.8-6.9; Corcoran

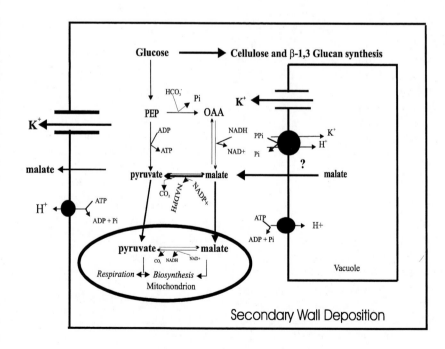

Figure 32-3. The release of K+ and malate catabolism during secondary wall deposition. The more intense arrows indicate the preferred direction of carbon and ion flux during this phase of growth.

and Zeiher, unpublished observations). NADP- ME isolated from glumes of developing wheat grains is inhibited by NADPH that can be over come by increasing concentrations of malate (Dhillon *et al.*, 1985). The cytoplasmic environment and possibly the phosphorylation state of PEPC may therefore contribute to the temporal separation of malate synthesis and catabolism between the two phases of fiber growth. Changes in cytosolic pH, PEP and malate concentration, and NADPH/NADP$^+$ ratio may determine whether malate synthesis or catabolism predominates.

During secondary wall deposition, the majority of the carbon passing through the glucose metabolic pool is used for synthesis of cellulose and β-1, 3-glucans rather than oxidized to pyruvate via glycloysis (Mutaser, 1976; Carpita and Delmer, 1981). The associated decline in PEP concentration would reduce the substrate available for malate synthesis. Increased malate catabolism during this phase of fiber growth could provide a source of pyruvate for respiration or for biosynthetic processes utilizing C-3 acids (*e.g.* amino acid biosynthesis via transamination reactions). In addition, the NADPH generated by NADP$^+$ ME could be either oxidized by the mitochondria through an external mitochondrial NADPH dehydrogenase resulting in the production of ATP or utilized for biosynthetic reaction (Edwards and Andreo, 1992). Malate during this phase of fiber growth could therefore serve as a carbon and energy source when oxidation of glucose becomes limiting.

3.10.1   Cell Walls and Cell Wall Metabolism

Fiber cells can be distinguished from non-fiber forming epidermal cells by their ability to elongate and differentiate. The coordinate expansion and differentiation of individual fiber cells is achieved by altering the structure of the developing wall, the mechanical determinant of the size and shape of the cell. Two kinds of cell wall, the primary and secondary, are formed during the differentiation and growth of the fiber cell. They can be distinguished by their composition and when they are formed.

The primary wall surrounds the actively expanding cell and has a composition similar to other dicots that include cellulose, xyloglucans, pectins and proteins (Meinert and Delmer, 1977; Hayashi and Delmer, 1988). The primary wall, based on current models of wall structure, is comprised of three interwoven polymeric domains. The first domain is composed of the cellulose-xyloglucan framework that is embedded into the second domain of a gelled matrix of pectic polysaccharides ionically linked by calcium bridges. The third domain is comprised of structural proteins that are covalently cross-linked with one another and possibly to other elements of the wall matrix (Talbot and Ray, 1992; Carpita and Gibeaut, 1993).

The secondary wall is deposited as the fiber cell matures and by definition, is the wall deposited after elongation has ceased. This wall is deposited between the primary wall and the plasma membrane and is compose

almost entirely of cellulose (94% at maturity) with only minor amounts of other components (β-1, 3-glucans) present (Meinert and Delmer, 1977). The secondary wall is a polylamellate structure with helical arrays of parallel cellulose microfibrils exhibiting various orientations that can be subdivided into the S1 and S2 layers (De langhe, 1986). The first layer (S1), often referred to as the winding layer, is deposited between the primary cell wall and the oriented cellulose fibrils of the S2 layers. The S1 layer can be distinguished form the S2 layers in that it has a wider fibril angle and the S1 fibril helix orientation is opposite to that of the subsequent S2 layers. The fibril angle of each of the S2 layers that follow the S1 layer decline, so that the fibrils in the layer closed to the plasma membrane are deposited essentially parallel to the longitudinal axis of the fiber cell.

3.11   Dynamics of the Cell Wall During Fiber Development

The cell wall surrounding the fiber protoplasm is a dynamic structure as evidenced by its continually changing composition and degree of polymerization of wall polymers during fibers development (Meinert and Delmer, 1977; Huwyler *et al.*, 1979; Timpa and Triplett, 1993). The cellulose molecule present in both the primary and secondary wall is a linear chain of D-glucopyranose residues linked by β-(1,4) glycosidic linkage that are associated laterally to form microfibrils. The amount of cellulose deposited during primary wall formation is constant at about 1 ng/mm but increases to 130 ng/mm during deposition of the secondary cell wall (Meinert and Delmer, 1977). Differences in the degree of polymerization are observed for cellulose between the primary and secondary cell wall. The cellulose present in the primary wall is relative heterogeneous and has a low degree of polymerization (DP) of between 2000 and 6000 (Marx-Figini and Schulz, 1966). This low DP cellulose is synthesized during the elongation phase but ceases shortly after initiation of the secondary wall deposition. A more homogenous, high DP (14,000) is observed for cellulose in the secondary cell wall (Marx-Figini and Schulz, 1966; Timpan and Triplett, 1993). Timpa and Triplett (1993), using less caustic isolation techniques, identified polymers with a DP similar to secondary wall cellulose low present at the earliest stages of fiber elongation indicating that high DP is not restricted to secondary wall synthesis. However, this work did not distinguish whether this high DP was cellulose or other high molecular weight wall polymers. These high DP polymers appear to decline during the later stages of elongation and then undergo a massive increase with the initiation of secondary wall deposition, indicating a possible turnover of these high molecular weight species.

Xyloglucans are present in the primary walls of fiber cells and have a structure typical of known xyloglucans (Buchala and Meier, 1985). The main chain is composed of (1, 4)-β-D linked glucpyranosyl residues and side chains containing xylosyl- galactoxyosyl-, and fucoxylgalactosylxylosyl units. The average molecular weight of xyloglucan isolated from fibers is 80,000 (Hayashi and Delmer, 1988). Xyloglucans in the primary wall discontinually coat the cellulose microfibrils and determine the spacing between microfibrils (McCann and Roberts, 1991). The association of xyloglucans with microfibrils is proposed to prevent fasciation of the microfibrils into larger bundles and to help the microfibrils slide during cell expansion because of the mutual affinity of xyloglucans (Hayashi, 1991). In addition, xyloglucan oligosaccharins can be formed by partial hydrolysis of the preformed polysaccharide (McDougal, and Fry, 1991) that can serve as signal molecules that control cell growth (Fry wt al., 1993). The deposition of xyloglucans is limited to the elongation phase of fiber growth (Hayashi and Delmer, 1988) which agrees well with its proposed role of contributing to the extensibility of the primary wall (Hayashi, 1991).

The modification of xyloglucans can change with fiber development. Using monoclonal antibodies that recognize the terminal fucose reside of xyloglucans, Wilkins and Tiwari, (1994) showed by immunofluorescence microscopy that fucosylated xyloglucans were readily detected on the inner surface of the cell wall in ovule epidermal cells prior to anthesis but not after the formation of fiber initials, although they were detected in the secretory system of the fiber initial en route to the cell wall. The authors suggested that the fructose epitope is absent because it is wither masked as it becomes cross-linked in the cell wall or the terminal fucose residue is cleaved in the extracelluar matrix. The presence of a fucose residue in the xyloglucan molecule may be important in cell-cell adhesion (McCann and Roberts, 1994). Conformational analysis of xyloglucans suggests that fucosylated trisaccharide side chain may wrap around the glucan backbone to facilitate binding to the cellulose microfibrils (Levy et al., 1991). This may be particularly important during cell-cell adhesion where fucose residues of extralong xyloglucans may facilitate the cross-linkage of cellulose microfibrils between neighboring cells and thereby, increase the tensile strength of the wall (McCann and Roberts, 1994). The cell-cell adhesion may be predominant between ovule epidermal cells prior to anthesis but becomes restricted at anthesis with fiber initiation. The removal of fucose may facilitate wall loosening by weakening the binding of xyloglucans to the cellulose microfibrils (Hoson, 1993), allowing the differentiated fiber primordial to expand.

Pectin fraction, consisting of glucose and uronic acids, and proteins are also prevalent during the elongation phase of fiber development (Meinhert and Delmer, 1977; Huwyler et al., 1979); both declining abruptly between 16 and 18 DPA due to increased turnover (Meinert and Delmer, 1977). The pectin domain may perform several functions in the primary wall. It determines the wall's porosity and provides a charged surface that modulates pH and ionic environment of the wall (McCann and Roberts, 1991; Carpita and Gibeaut, 1993). The orientation of the pectin matrix may also be involved in determining the direction of fiber expansion. Immunofluorescene studies using monoclonal anti-

body specific for methyl eterified pectins of the extracelluar matrix revealed that these pectins undergo a change during fiber initiation (Wilkins and Tiwari, 1994). At anthesis, these pectins switch from amorphous material to a highly organized matrix that is fibrillar in nature and appears as concentric rings forming a helical continuum from the base to the apex of the fiber initial. Between anthesis and 3 DPA, the methyl esterified pectins are orientated in a transverse orientation relative to the elongation axis of the expanding fiber but reorientate to a more oblique orientation by 5 DPA; coincident with the switch form isotropic to anisotropic growth. The identity of the proteins associated with the fiber cell wall during elongation is limited, but presumably they include structural proteins and enzymes that contribute to the formation and modification of the cell wall during fiber differentiation and growth. The enzymes identified to date include acid invertase (Buchala, 1987), malate dehydrogenase (Wafler and Meier, 1994); peroxidase (Rama Rao et al., 1982a; Ryser, 1985; Wafler and Meier, 19940; acid phosphatase (Wafler and Meier, 1994); esterase (Thaker et al., 1986a and b); and β-(1,3)-glucanase (Bucheli et al., 1985).

The onset of secondary wall deposition is marked by an increase in non-cellulose glucose; the majority of which is β-1,3 glucan or callose (Maltby et al., 1979; Jacquest et al., 1982; Rowland et al., 1984). β-1, 3-glucan deposition is a part of the normal development of the fiber rather than induced by a stress as occurs in other plant systems (Delmer, 1987). The percentage of total cell wall represented by β-1, 3-glucan during secondary wall deposition varies between 7 and 10.5% at the onset of secondary wall deposition to <1% at the cessation of growth (Maltby et al., 1979; Jacquet et al., 1982; Rowland et al., 1984). This decline is due to dilution as a result of the massive increase in cellulose content rather than turnover of β-1, 3-glucan (Rowland et al., 1984). Localization studies indicate callose is present as an inner layer between the plasma membrane and the cellulose microfibrils and is observed throughout secondary wall thickening (Waterkyn, 1981). Although several roles are proposed for this polymer that include temporary glucose storage polymer (Pillonel et al., 1980), a role in the plasticity of the wall (Maltby et al., 1979, Waterkyn 1981), or a permeability barrier (Seagull, 1990b), its function remains unknown.

3.12   Extension of the Primary Wall

3.12.1   Diffuse versus Tip Growth

Elongation of fiber cells requires that the load-bearing network in the walls be rearranged to reduce wall stress and cell turgor pressure thereby allowing the cell to take up water, extend the cell wall, and increase in volume. This increase in volume may occur by either diffuse or intercalary growth such that irreversible expansion occurs in all parts of the wall parallel to the elongation axis or is due to more localized growth where irreversible extension occurs in patches of the wall as is observed with tip growth. Radiolabeling cell walls *in vivo* and subsequent radioautography showed that fibers cells deposit cellulose throughout the length of the cell (O'Kelly, 1953; Ryser, 1977; Delmer *et al.*, 1992). Ultrastructural studies revealed that the orientation of organelles (*i.e.* secretory vesicles versus mitochondria) within an elongating fiber cell are not distributed into distinct zone (Ryser, 1979; Tiwari and Wilkins, 1995) as occurs in tip growth (Steer and Steer, 1989). In addition, the cortical microtubules are oriented perpendicular to the growing axis of the cell and are found evenly distributed throughout the cell, including the extreme apical region (Tiwari and Wilikns, 1995) where microtubules are rarely observed in tip-growing cells (Tiwari and Polito, 1988). The sensitivity of fiber growth to microtubule depolymerizing agents but not to actin-disrupting agents further established that rapidly elongating fibers undergo diffuse growth (Tiwari and Wilkins, 1995).

This growth pattern, however, may change during the overlap period when elongation and secondary wall deposition occur concurrently. During this overlap period up to 90% of the final weight of the fiber wall is deposited before cessation of fiber elongation is observed (Schubert *et al.*, 1976). Since by definition the secondary wall is deposited after elongation has ceased the mechanism which allows this significant overlap is poorly understood. Asynchronous development within a fiber population on a given ovule is largely discounted as a possible explanation (De Langhe, 1986). Alternatively, deposition of the primary and secondary wall could occur at distinct localized region on the surface of the fiber cell, allowing the two processes to occur simultaneously. For instance, the fiber during this overlap period may continue to elongate at the tip while the secondary wall is deposited behind the tip region, as suggested by Meinert and Delmer (1977). Another possibility is that the orientation of the cytoskeleton which, in turn, determines the pattern of cellulose microfibril deposition may control this overlap period (Delmer *et al.*, 1992). The cellulose microfibrils may continue to be deposited in a random or transverse orientation allowing cell extension to continue (Meinert and Delmer, 1977). Some evidence of this later possibility is observed for fibers grown *in vitro* where microfibril deposition was shown to be in a transverse orientation through 19 DPA (Seagull, 1992b) that coincided with this overlap period (12-16 DPA; Meinert and Delmer, 1977).

3.12.2 Mechanisms Controlling Primary Wall Extension

Turgor-driven fiber extension requires loosening of the load-bearing network of the wall that allows the cellulose microfibrils to separate and permits the incorporation of new wall material. Xyloglucans serve as molecular tethers between cellulose microfibrils and are considered the principal tension-bearing molecules in the longitudinal axis of an enlonging cell (Carpita and Gibeaut, 1993). Disruption

of xyloglucan polymer or its interaction with microfibrils would allow slippage of the microfibrils and render it more susceptible to turgor-driven growth. The biochemical mechanism responsible for this wall loosening in fiber cells is poorly understood. Endo-β-(1-4)-glucanases by their ability to irreversibly cleave the xyloglucan polymer is suggested as one mechanism to facilitate cell wall extension. Although their activity in other plant systems was correlated with cell growth (Hayashi, 1991), the activity of (1-4)-β-glucanases in fiber homogenates were negligible throughout fiber development when carboxymethylcellulose was used as substrate (Bucheli *et al.*, 1985). However, it is possible that the enzyme requires another specific substrate. Matsumoto *et al.*, (1997) isolated a xyloglucan specific endo-1, 4-β-glucanase from an apoplast fraction of auxin-treated pea stems that could not hydrolyze carboxymethylcellulose. Nevertheless, the involvement of glucanases in wall loosening *per se* has recently been disputed since extension of isolated walls was not correlated with the hydrolysis of matrix polysaccharides (Cosgrove and Durachko, 1994). McQueen-Mason *et al.* (1992) recently isolated a 29 and 30 kD protein, termed expansins, from cucumber walls that induce wall extension of heat inactivated cell walls isolated from dicot stems. These proteins induce wall extension in the same pH range observed for endogenous wall extension activity (both requiring acidic pHs) (McQueen-Mason *et al.*, 1992), consistent with acid growth theory (Rayle and Cleland, 1992), and act by disrupting hydrogen bonding, possibly between xyloglucans and cellulose microfibrils (McQueen-Mason and Cosgrove, 1995).

With the completion of fiber elongation, the primary wall becomes locked into shape by cell wall tightening processes that likely involve the formation of cross-linkages between cell wall polymers (Fry, 1986). These cross-linkages can form from the oxidation of phenolic residues of cell wall polysaccharides and structural proteins by peroxidase. An association between and increase in ionically bound wall peroxidase activity and the timing of cessation of fiber elongation is observed in fibers (Rama Rao *et al.*, 1982a; Thaker *et al.*, 1986a). Peroxidase activity, shown by histochemical staining, was localized to the primary but not the secondary cell wall (Ryser, 1985). In ovule cultures, peroxidase synthesized *de novo* (Mellon and Triplett, 1989) was secreted into the culture medium (Ryser, 1985; Mellon and Triplett, 1989). Further support for the involvement of cross-linkages contributing to wall rigidity comes from examining the development profiles of nonspecific wall-located esterases that cleave ester link phenols. These esterases maintain higher levels of activity during fiber elongation than secondary wall thickening (Thaker *et al.*, 1986b). Cleavage of the ester-linked phenols during elongation could restrict formation of cross-linkages in the primary wall allowing extension to continue.

3.13   Cellulose Biosynthesis

The secondary wall of cotton fiber is almost exclusively cellulose and thus, cotton fibers are considered an excellent experimental system for studying celluose biosynthesis. However, characterization of the process of cellulose synthesis in fibers has been hindered by the inability to detect cellulose synthesis *in vitro*. Cotton fiber membranes isolated at a developmental stage lacking substantial xyloglucan synthesis activity (Delmer, 1987) or using a fiber system that is perturbed (*e.g.*, vacuum infiltration, change in temperature, or disturbing the plasma membrane by plasmolysis or ionic detergents; (Buchala and Meier, 1985) typically make callose over cellulose from UDP-glucose or glucose. Recently, with the refinement of membrane isolation and enzyme reaction conditions, progress towards the synthesis of cellulose *in vitro* has been made with fiber membranes. Brown and his colleagues using a variety of physical and biochemical techniques demonstrated that β-(1,4) glucan could be synthesized *in vitro* (Okuda *et al.*, 1993; Li and Brown, 1993; Kudlicak *et al.*, 1995). Using acetic/nitric acid insolubility as an indicator of cellulose, Li and Brown (1993) showed that cellulose synthase had an essential requirement for Mg^{2+} ($K_a = 100$ μM). Both cellobiose ($K_a = 3.26$ mM) and cyclic-3': 5'-GMP ($K_a = 100$ μM) were required in the reaction mixture to stimulate activity as well as low concentration of Ca^{2+} ($K_a = 90$ μM). These reaction conditions, however, have been criticized because they resemble closely reaction conditions commonly used for callose synthesis (Delmer *et al.*, 1993a). Two of the effectors (Ca^{2+} and cellobiose) identified, although effective activators of callose synthase, are not effectors of the bacterial celluose synthase. The acetic/nitric acid insoluble product produced by digitonin-solubilized membranes, however, yielded an x-ray diffraction pattern characteristic of cellulose II (Okuda *et al.*, 1993), the allomorph of cellulose typically observed in *in vitro* preparations (Kudlicka *et al.*, 1995). The final yield of the insoluble product was 4% (Okuda *et al.*, 1993). Modification of the membrane isolation procedure resulted in the synthesis of cellulose I, the allomorph of cellulose observed *n vivo*, and improved the final yields of cellulose production to 32.1% (Kudlicka *et al.*, 1995). The morphology of the cellulose I produced consisted of extended fibrils with dimensions of 1.2 to 1.8 nm and was distinguishable from the disorganized aggregates identified as cellulose II (Okuda *et al.*, 1993; Kudlicka *et al.*, 1995).

The *in vitro* cellulose synthesis rates observed with isolated membranes, however, were much less than *in vivo* rates (Delmer *et al.*, 1993b). Those *in vitro* studies were conducted using UDP- glucose as substrate (Okuda *et al.*, 1993; Li and Brown, 1993; Kudlicka *et al.*, 1995). Evidence of PM sucrose synthase (Amor *et al.*, 1995) and the possibility that it might be tightly associated with cellulose synthase suggested that sucrose might be a better substrate. Measurement of cellulose synthase activity using radiolabeled sucrose as substrate and digitonin permeabilized detached fibers demonstrated that cellulose was synthesized from sucrose at rates that approached *in vivo* rates whereas callose was synthesized from either UDP-glucose or sucrose (Amor *et al.*, 1995). Thus, cellulose synthase appears to preferentially accept UDP-Glc channeled through sucrose degradation, whereas callose synthase can accept it in either form. Addition of Ca^{2+} and cellobiose stimulated callose synthesis; whereas chelation of Ca^{2+} enhanced cellulose synthesis. These findings are contrary to the results of Kudlicka *et al.*, (1995), who showed that Ca^{2+} and cellobiose stimulated the synthesis of both polymers from UDP-glucose but are consistent with what is thought to occur *in vivo*, where an increase in intracellular Ca^{2+} is believed to stimulated callose over cellulose synthesis (Delmer and Amor, 1995). The reaction rates observed using radiolabeled sucrose and permeabilized cotton fibers were, however, quite variable, possibly indicating a higher level of organization was needed for celluose synthesis, disrupted when the fibers were detached and permeabilized (Amor *et al.*, 1995). This liability may also account for the poor *in vitro* yields reported by Kudlicka *et al.* (1995).

3.13.1   Identification of Catalytic and Non-catalytic Subunits of Cellulose Synthase

Because of the inability to measure cellulose synthase *in vitro*, the purification and characterization of this enzyme has been slow. However, using either a biochemical or molecular approach significant progress has been made towards the identification of the catalytic subunit. Li *et al.* (1993) identified a polypeptide of 37 kD by photoaffinity labeling with azido-UDP-glucose and differential product entrapment. This protein showed Mg^{2+} dependent interaction with the substrate and was suggested as a possible candidate for the catalytic subunit. The same size polypedtide enriched by product entrapment was observed in the solubilized fraction that produced the cellulose I allomorph *in vitro* (Kudlicka *et al.*, 1995). Pear *et al.* (1996) used a molecular approach to identify the catalytic subunit. They randomly sequenced selected cDNAs derived from a library prepared from mRNA harvested from 21 DPA fibers undergoing secondary wall synthesis. This developmental stage was chosen because the secondary wall is almost pure cellulose, and no other wall polysaccharides (*e.g.* xyloglucans, β-(1, 3)-D-glucans) are synthesized this late in fiber development (Meinert and Delmer, 1977). Transcripts for cellulose synthase were expected to be prevalent at this time. Using this approach two distinct cDNA clones were identified and shown based on sequence homology with the bacterial *cel A* genes to code for true homologs of the bacterial cel A protein (Pear *et al.*, 1996). In fibers, these genes are temporally expressed with high levels observed during active secondary wall synthesis, and the level at maximum expression is approximately 500 times that observed in other tissues, constituting 1 to 2% of fiber MRNA.

Hydropathy plots indicate the cel A proteins have 8 transmembrane helices, two at the N-terminus and six at the C-terminus, that could anchor the protein in the membrane (Pear *et al.*, 1996). The central region of these proteins are more hydrophilic and predicted to reside in the cytoplasm and contain the site(s) of catalysis. Four conserved amino acids subdomains (U-1, U-2, U-3, and U-4) were found in the hydrophilic region that contain the conserved Asp residues (in U-1, U-2, and U-3) and the motif QXXRW (in U-4) proposed by hydrophobic cluster analysis of β-glycosyl transferases (including cellulose synthase) to be involved in UDP-glucose binding and catalysis (Saxena *et al.*, 1995). A recombinant polypeptide containing the four amino acid subdomains binds UDP-glucose in Mg^{2+} dependent manner with only, at most, weak binding observed in the presence of Ca^{2+} (Pear *et al.*, 1996). A recombinant protein in which the U-1 subdomain was deleted did not bind UDP-glucose.

Based on deduced primary amino acid sequence, the cel A protein appears to be different from the 37 kD isolated by photoaffinity labeling. The opening reading frame of *cel A1* gene codes for a larger intergral membrane protein of 109.6kD (Pear *et al.*, 1996) whereas the 37 kD protein can be solubilized from membranes with 0.05% digitonin, a detergent concentration that releases peripherally associated membrane proteins (Kudlicka *et al.*, 1995). These differences indicate that either the cel A protein and the 37 kD are distinct or that the 37 kD protein is a proteolytic product of the cel A protein that possibly contains the hydrophilic region with the substrate binding site.

Cellulose synthase complex in the bacterium *Acetobacter xylinum*, a prokaryotic model system for studying cellulose biosynthesis, is coded by an operon of four genes with only one of these genes coding for the catalytic subunit (Wong *et al.*, 1990; Saxena *et al.*, 1994). Characterization of these genes in combination with either biochemical or insertional mutagenesis studies indicates the *A. xylinum* cellulose synthase complex also contains non-catalytic subunits that include the regulatory subunit that binds c-di- CMP (Wong *et al.*, 1990; Mayer *et al.*, 1994). The number of non-catayltic subunits associated with the fiber complex is difficult to estimate because the protein has not been purified or homologs to the bacterial genes isolated. Amor *et al.* (1991) identified two polypeptides with molecular masses of 83 and 43 kD in fiber membranes that bind c-di-GMP with high affinity and specificity and were antigentically related to the bacterial c-di-GMP binding protein. These proteins increased markedly during the transition between primary and secondary wall synthesis (Amor *et al.*, 1991) coincident with the increase in rates of cellulose synthesis (Meinert and Delmer, 1977).

3.14   Callose Biosynthesis

The biochemical mechanism for synthesizing callose and cellulose, although originally thought to involve the same protein (Delmer, 1987), now appears to be different. The reaction requirements for callose synthesis are distinguishable from cellulose synthesis in the Mg^{2+} (Hayashi *et al.*, 1987; Okuda *et al.*, 1993) and cyclic-3':5'-GMP (Okuda *et al.*, 1993) are not required for maximal catalytic activity. Mg^{2+} in the reaction causes a conformational change or aggregation of the enzyme and a greater production of alkali insoluble glucan (Hayashi *et al.*, 1987). Callose synthase requires β-glucosides and Ca^{2+} for maximal activity. The effectors act by raising the V_{max} of the enzyme and lowering the apparent K_m for UDP-Glucose (> 1 mM to 0.2 to 0.3 mM; Hayashi *et al.*, 1987). The reported K_a values for cellobiose ranges between 0.13 mM and 1.16 mM (Hayaski *et al.*, 1987; Li and Brown, 1993) and for Ca^{2+} is 0.25 mM (Li and Brown, 1993). Mg^{2+} enchances the enzyme's affinity for Ca^{2+} but not for the β-glucoside (Hayashi *et al.*, 1987). B-furfurly-β-glucoside (FG) is the endogenous β-glucoside responsible for activating callose synthase in fibers (Ohana *et al.*, 1992). This compound is present at a concentration of 10 µg/g fr wt (Ohana *et al.*, 1992) when callose deposition is maximum (Maltby *et al.*, 1979) and does not affect cellulose synthesis in intact plant cells or bacterial cellulose synthase activity *in vitro* (Ohana *et al.*, 1992). FG is sequestered in the vacuole and can be released by elevation of cytoplastmic Ca^{2+} or a decrease in cytoplasmic pH that results in an increase in callose deposition *in vivo* (Ohana *et al.*, 1993).

3.14.1   Identification of Catalytic and Non-Catalytic Subunits of Callose Synthase

A likely candidate for the catalytic subunit for callose synthase has been isolated for cellulose synthase. The UDP-glucose-binding polypeptide of callose synthase was identified as a 52 kD protein by photoaffinity labeling with either α-[^{32}P] UDP-glucose (Delmer *et al.*, 1991) or [β-^{32}P] 5'-azido-UDP-glucose (Li *et al.*, 1993). The labeled polypeptide co-migrated with callose synthase activity on glycerol gradients (Delmer *et al.*, 1991) and binding of the labeled probe required Ca^{2+} (Delmer *et al.*, 1991; Li *et al.*, 1993) not Mg^{2+} as observed for cellusoe synathase (Li *et al.*, 1993; Pear *et al.*, 1996). This polypeptide was also the most abundant protein in a product entrapment fraction that favored β-(1,3) glucan synthesis *in vitro* (Li *et al.*, 1993).

Several additional polypeptides have been identified that may interact with the catalytic subunit. Andrawis *et al.* (1993) showed that an annexin purified from fibers binds to an influences callose synthase activity. This protein is suggested to function as a Ca^{2+} channel that may allow Ca^{2+} entry into the fiber cell and cause localized activation of the callose synthase complex (Delmer and Amor, 1995). In another study, using monoclonal antibody capable of immobilizing callose synthase activity, 65 kD polypeptide was identified that reacts with the antibody and associates with the callose synthase in a cation-specific manner (Delmer *et al.*, 1993b). Indirect immunofluorescence localization showed that this polypeptide co-localized with sites of high callose deposition but not with sites of high cellulose deposition *in vivo*. Although the function of this protein is unknown, recent cloning of the gene coding for this polypeptide indicates that it shares strong homology with the Ca^{2+} binding protein, calnexin (Delmer and Amor, 1995).

3.15   Cell Wall Lipids

3.15.1   Chemical. Composition and Ultrastructure of Cutin and Suberin

The outer surface of the fiber cell is covered with a hydrophobic cuticle. The principle function of this cuticle is to provide the fiber cell with a permanent biological barrier that minimizes the diffusion of water and other molecules and protects the fiber from bacterial and fungal pathogens (Kolattukudy, 1980; Post-Beittenmiller, 1996). The waxy cuticle consists of several stratified layers of lamellae laid down principally during the early stages of fiber development (first few days post anthesis), as revealed by freeze fracturing; but as the fiber elongates, the cuticle stretches and thins such that more than one layer is only occasionally observed at later stages of development (Willison and Brown, 1977). This hydrophobic surface of a mature fiber is about 10 to 20 nm in thickness (Yatsu et al., 1983; Ryser and Holloway, 1985).

Cutin, the principal polymer of the cuticle, is composed of long chain fatty acids linked together by ester bonds to form a three-dimensional network. The major fatty acid monomers identified by GLC- MS after LiA1D4 depolymerization are the isomeric 9, 16-and 10, 16-hexadeconoic acids (Yatsu et al., 1983), the cutin component commonly observed in rapidly expanding cells (Kolattukudy, 1980). Embedded within the cutin polymers are soluble waxes that can be separated from the aliphatic components of the cutin polymer by brief immersion in an organic solvent such as chloroform or hexane (Post-Beittenmiller, 1996). The major classes of wax in the cuticle are the free 1-alkanols (c18-C36) and free alkanoic acids (C14-C38), a wax composition typical of intracuticular waxes but distinct from epicuticular waxes that are secreted onto the cell surface (Ryser and Holloway, 1985).

It was recognized as early as 1941 that green fibers of a dominant color mutant (Lg) of cotton had a much higher wax content (14 to 17% of dry weight) than white cotton fibers (0.4 to 07% of dry weight; Conrad, 1941). Subsequent analysis established that the secondary walls of the green fivers were suberized (Ryser et al., 1983; Yatsu et al., 1983; Ryser and Holloway, 1985). It now appears that suberization is a general feature of the epidermal cells of the seed coats in the genus *Gossypium*, but only the fibers of wild species are suberized (Ryser and Holloway, 1985). Of the lines of fiber color mutants in *G. hirsutum* examined, suberin was only associated with fibers containing the gene Lg.

Suberin is deposited in several concentric layers in the secondary cell walls of green fiber, with each layer separated by cellulose (Ryser et al., 1983). Only in the pits found regularly at the base of the cotton fibers is a compact layer of suberin observed; here; cellulose synthesis is inhibited but suberin synthesis is not (Ryser, 1992). Each layer of suberin is lamellated with a periodicity of about 4.2 nm (Ryser et al., 1983; Schmutz et al., 1994a). The electron translucent part of the lamellae is 3 nm thick whereas the electron opaque part is more variable being on average about 1 nm in thickness (Schmutz et al., 1994a). The number of concentric layers deposited correspond to the number of days of secondary wall deposition (Ryser et al., 1983). This periodic deposition appears to be controlled by the plant's endogenous clock as culturing ovules at constant temperature and in the dark did not affect this periodicity.

The chemical composition of suberin of the green lint fiber and of the outer epidermis of seed coats of both the green and white varieties of *G. hirsutum* are similar (Ryser and Holloway, 1985). The main aliphatic monomer of suberin is ω-hydroxyalkanoic representing 79% and 76% of the total monomers in fibers and seed coats, respectively. The two differed in the relative proportion of 1-alkanols and α,ω-alkanedioic acids; the fibers contained less 1-alkanols and more α,ω-alkanedioic acid than seed coats. For both the green lint fiber and seed coats the predominant homologue of the aliphatic monomers was C_{22}. This is markedly different from the periderm of the same plants which is comprised mainly of C_{16} and C_{18} compounds.

Glycerol is a monomer in green fiber suberin. Purified cell walls isolated from mature green fibers contain 1.0% (of fiber dry weight) glycerol; 0.9% extracted from the wax fraction and the remainder 0.1% extracted from cell wall residues (Schmutz et al., 1993). White cotton fibers, by comparison, contain 0.03% glycerol; 0.01% with the cell wall and the remainder 0.02% with the cytoplasm. Since white cotton fibers do not contain suberin (Ryser and Holloway et al., 1985), the glycerol in the cell wall residues suggests that glycerol may also be a monomer in cutin. Further support for glycerol as a component of suberin comes from inhibitor studies. Fibers exposed to 2-aminoindan-2-phosphonic acid (AIP), an inhibitor of phenylalanine ammonia lyse, during secondary wall deposition contained normal amounts of bound glycerol in the wax fraction, but glycerol content in the cell wall residue was inhibited 95% (Schmutz et al., 1993). The molar ratio of polymer bound fatty acids to glycerol in green fibers is 2:1 (Schmutz et al., 1993).

The main phenolic compound in the suberin polymer of green fibers identified by GC-MS is caffeate (Schmutz et al., 1994a). Caffeate in the suberin polymer can be distinguished from that in the wax in that at least one of the phenolic hydroxyls of caffeate is bound whereas in the wax they are both free. Culturing ovules of the green lint mutant in AIP inhibited the deposition of suberin into the cell walls (Schmutz et al., 1993). Electron micrographs of the inhibitor treated fibers showed that the electron opaque regions in suberin layers were replaced with empty spaces and the coherence between the secondary wall layers was lost (Schmutz et al., 1993). The authors suggest that the cinnamic acid derivatives in the suberin polymer, most likely caffeate, either plays a structural role in the suberin polymer or is involved in cross-linking suberin to cellulose.

The waxes embedded in the suberin polymer resolvable by chromatography were similar in composition to the waxes found associated with cutin from white cotton fibers (Ryser and Holloway, 1985). The major difference between the two cultivars is in the free alkanoic acid fraction, with the green cultivar containing a much greater concentration of the C_{22} homologue. Phenols are present in the wax associated with suberin of green fibers. A chloroform methanol extract of mature green fiber cell walls was separated by reversed phase thin layer chromatography into several yellow compounds and one colorless fluorescent compound; compounds not present in white cotton fibers (Schmutz *et al.*, 1993; Schmutz *et al.*, 1994a). The colorless compound was purified and identified using ultraviolet spectroscopy, FAB-MS and GC-MS as a molecule containing caffeate, 22-hydroxydocosanoic acid and glycerol with caffeate esterified to the fatty acid and the fatty acid esterified to glycerol (Schmutz *et al.*, 1994a). The yellow compounds were composed mainly of hydroxyl and dicarboxylic fatty acids and unidentified cinnamic acid derivatives (Schmutz *et al.*, 1994a). Their UV and visible spectra were similar to the caffeic acid ester, and their biosynthesis was inhibited with AIP (Schmutz *et al.*, 1993).

Schmutz *et al.* (1993) proposed, based on suberin's lamellated structure and the presence of glycerol and fatty acids, that they hydrophobic part of the long chain fatty acids defined the thickness of the electron translucent lamellae and the hydrophyllic ends of the fatty acids, glycerol, and cinnamic derivatives determined the size of the electron opaque region. To test this hypothesis, Schmutz *et al.*, (1996) exposed cultured ovules to EPTC to reduce the fatty acid chain length and alter the thickness of the translucent region. EPTC is a specific inhibitor of the ER associated fatty acid elongases that produce very-long chain fatty acids from C_{16} to C_{18} fatty acid precursors derived from plastids. In the presence of 100 μM EPTC, the synthesis of very long chain alkanols, alkanoic acids, ω-hydroxyalkanoic acids and α,ω-alkanedioic acids in suberin of green fibers was inhibited. With the exception of α,ω-alkanedioic acids, these long chain monomers were replaced with homologus monomers of shorter chain length. Both α,ω-alkanedioic acids and glycerol were reduced to similar low amounts in suberin of EPTC treated fibers. Based on this observation, the authors suggested that glycerol and α,ω-alkanedioic acid are probably secreted together into the cell wall space as part of an suberin oligomer and the glycerol content of this oligomer is dependent on the availability of α,ω-alkanedioic acid.

The ultrastructure of the suberin layer was significantly altered by shortening the fatty acids and altering monomer composition of the suberin in EPTC treated fibers (Schmutz *et al.*, 1996). The number of lamellae per suberin layer was significantly reduced, containing on average one or two lamellar periods. A broad distribution of thickness of suberin lamelle was observed in the presence of EPTC, with the size distribution showing a periodicity of 2.2. nm. This periodicity corresponds approximately to the length of esterfied C_{16} ω-hydroxyalkanoic acids and

supports the hypothesis that fatty acid chain length determines the thickness of the electron translucent suberin lamelle. Schmutz *et al.* (1996) reconciled the variation in thickness in the EPTC-treated fibers to the diminution of α,ω-alkanedioic acid and glycerol and some cross-linking maintained by mono-. di-, and oli-gomers of esterified ω-hydroxyalkanoic acids. The size of the electron translucent lamellae was always about 0.3 nm greater than the length of the extended fatty acid chain (Schmutz *et al.*, 1996). This was sufficient space for the carboxylic group of hydroxycinnamic acids and glycerol and provided further evidence that the electron-translucent and the electron-opaque lamellae were composed of aliphatic and aromatic suberin constituents, respectively, analogous to biological membranes. Schmutz *et al.* (1996) suggested that the function of glycerol may be to cross-link the two types of lamellae.

3.15.2   Role of Lipid Transfer Proteins in Cuticle Formation

Lipid transfer proteins (LTP) are small, abundant basic proteins that are thought to get involved in the deposition of extracellular lipophyllic material including cutin (Post-Beitenmiller, 1996; Kader, 1996). A cDNA clone was isolated from a cotton fiber cDNA library by differential screening and identified as coding for a LTP based on deduced amino acid sequence homology with other plant LTPs (Ma *et al.*, 1995). A characteristic of all plant LTPs (Kader, 1996), including the fiber LTP (Ma *et al.*, 1995), is they contain a transmembrane signal peptide at the N-terminus which allows it to move vectorally into the lumen of the endoplasmic reticulum and enter the secretory pathway where it could ultimately be deposited on the outer surface of the cell. Immunocytochemical studies using polyclonal antibody against a fusion protein corresponding to the LTP1 gene of *Arabidopsis thaliana* L. showed that LTP was localized to the cell wall (Thoma *et al.*, 1993). Maximum expression of LTP mRNA in cotton fibers occurs at 15 DPA (Ma *et al.*, 1995) when the fiber is elongating at a rapid rate (Schubert *et al.*, 1973; Naithani *et al.*, 1982). Associated with fiber extension is the synthesis of cutin polymers and their deposition on the outer surface of the fiber cell and thus, LTP could be involved in transport cutin monomers across the plasma membrane to the active site of cutin polyermization. Cutin synthesis is typically restricted to peripheral cell layers (*e.g.* epidermal cells) (Kader, 1996). Consistent with this observation was the significantly higher levels of LTP mRNA observed in elongating fibers compared to other organs (leaves, roots, flowers; Ma *et al.*, 1995).

4.   SUMMARY

The development of the cotton fiber from differentiation of fiber primordial to maturation is controlled by the interplay of a complex series of biochemical and molecular events that occur either coordinately or sequentially. Some key biochemical and cytological events that occur at distinct stages in fiber development are summarized in Figure 32-4. In this diagram the transition period between fiber elongation and secondary wall deposition is designated as a separate stage (stage III) because of distinct biochemical changes that occur during this phase of growth.

It is clear from *in vitro* studies and measurement of endogenous hormone levels that auxin, GA and ABA are signals are responsible for controlling the differentiation and growth of the fiber, although other unidentified signals may also be involved. The mechanisms controlling the concentration of these hormones at the site of perception, the identity of the signal receptors and their linkage to second messenger pathways that transmit theses signals still need to be defined. Both Ca^{2+} and pH have been implicated as second messengers. Changes in cytoplasmic Ca^{2+} or pH was shown to be involved in controlling callose synthase activity by releasing β-furfuryl-β-glucoside, an effector of callose synthase, from the vacuole and Ca^{2+} itself is and effector of callose synthase. Ca^{2+}, in addition, may control the movement of K^+ to and from the vacuole, preventing a futile cycling of K^+ (Rea and Poole, 1993). A change in cytoplasmic pH was also implicated in coordinating wall loosening via extracelluar acidification and osmoregulation. Given the likelihood the PEPC is phosphorylated,

the isolation and characterization of the protein kinase could be a good starting point for further defining the signal transduction pathway coordinating these two events.

Cotton fibers undergo cell elongation for extended time periods (*i.e.* 30 days) making it a useful biological system for studying the biochemical processes coordinating wall extension with the synthesis and deposition of new wall polymers. Yet, essentially nothing is known about these processes and their mechanism of regulation. Given that fiber extension contributes to several determinants (length, uniformity, fineness, and strength) of fiber quality, this particular area is in need of attention. Progress in our understanding of extension of walls of other dicots (Cosgrove 1993b) should facilitate these efforts.

Significant progress towards *in vitro* synthesis of cellulose came with the refinement of membrane isolation procedures and the identification of a PM sucrose synthase. With the likely identification of catalytic and non-catalytic subunits of cellulose synthase, improvements in the *in vitro* synthesis of cellulose should continue and will ultimately enable the regulatory mechanisms responsible for controlling polymerization and crystallization of cellulose to be defined. Finally, cotton fibers by providing a unique biological system for studying the biochemical processes associated with cell differentiation and growth should ultimately yield information of biological and agricultural significance that can be extended beyond improvement in fiber quality.

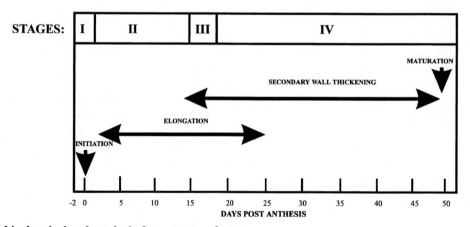

Key biochemical and cytological events at each stage:

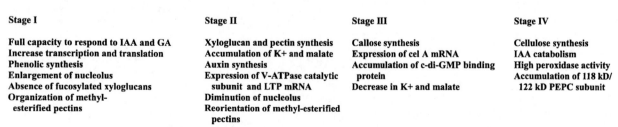

Stage I	Stage II	Stage III	Stage IV
Full capacity to respond to IAA and GA	Xyloglucan and pectin synthesis	Callose synthesis	Cellulose synthesis
Increase transcription and translation	Accumulation of K+ and malate	Expression of cel A mRNA	IAA catabolism
Phenolic synthesis	Auxin synthesis	Accumulation of c-di-GMP binding	High peroxidase activity
Enlargement of nucleolus	Expression of V-ATPase catalytic	protein	Accumulation of 118 kD/
Absence of fucosylated xyloglucans	subunit and LTP mRNA	Decrease in K+ and malate	122 kD PEPC subunit
Organization of methyl-	Diminution of nucleolus		
esterified pectins	Reorientation of methyl-esterified		
	pectins		

Figure 32-4. Summary of biochemical and cytological events at different stages of fiber development.

CHAPTER 33

COTTON REGENERATION

Norma Trolinder
South Plains Biotechnologies, Inc., Lubbock, TX 79404

The perception of the value of regeneration of any crop plant via somatic embryogenesis has evolved over the last one and a half decades. In the early eighties, somaclonal variation produced during the embryogenesis process was touted as being of significant value to plant breeding. And indeed, various positive agronomic attributes have been identified, stress tolerance, herbicide tolerance, earlier fruiting habits, disease resistance, etc. Protoplast fusion, followed by regeneration, as a means of making wide crosses was also touted as a benefit, as was the possibility of artificial seeds (Ammirato, 1987). During this period, Fraley *et al.* (1983) and Horsch *et al.* (1984) demonstrated that *Agrobacterium* could transfer foreign genes to plant cells and whole plants regenerated that express the foreign gene. With gene transfer demonstrated, the attention turned to somatic embryogenesis as a vehicle for making transgenic plants. However, somaclonal variation proved to be undesirable when a true-to-type transgenic plant was the ultimate target. Obtaining a true-to-type tranformant remains problematic in many species when somatic embryogenesis is the vehicle for transformation. Additionally, somatic embryogenesis has proven to be highly genotype dependent in most of our most important agronomic crop species. Cotton is among those (Trolinder and Xhixian, 1989).

Cotton has been grown as an important fiber crop for more than 3000 years in some parts of the world and was one of the first crops cultivated in America (Beasley, 1940). Cotton fiber constitutes approximately 50% of the world's textiles today. In the United States, the textile industry accounts for almost 14% of the Gross National Product. Economic inputs to cotton production in terms of labor, fertilizers, equipment, and chemicals is greater per acre than any other crop. Breeding efforts to improve the agronomic characteristics have been intensive for generations.

Cotton is in the genus *Gossypium*. There are approximately 30 diploid species (2n=2x=26) of cotton and 4 tetraploid (2n=4x=52). Two diploid and two tetraploid species are commercially grown. The diploid species contain six genomes, A, B, C, D, E, and F. The origins of A, B, E, and F are Africa and Asia, whereas C originated in Australia and the D genome in America. The chromosomes of the American D genome are smaller than their Asian and African counterparts. Three tetraploid species are found in America and were derived from the A and D genomes as amplidiploids, having 26 large chromosomes and 26 smaller ones (Fryxell, 1980; Beasley, 1942; Phillips, 1962, 1966, 1976; Wilson *et al.*, 1968).

Four species have commercially desirable fibers, two diploids (*G arboreum* and *G. herbaceum*) and two tetraploids (*G. hirsutum* and *G. barbadense*). *G. hirsutum* constitutes the bulk of commercially grown cotton throughout the world today. While cultivated cottons are the only ones with substantial fiber content, wild species such as *G. harknessii* and *G. thurberi* have contributed to the improvement of cultivated cottons. It was the search for improving survival of progeny from such wide crosses that began the efforts in tissue culture of cotton.

Regeneration of cotton was first attempted as embryo rescue from wide crosses. The early work of Lofland (1950), Weaver (1955), and Beasley (1940) resulted in plants when embryos were excised at 27 days post anthesis or older but failed to produce plants from younger immature embryos. This was problematic because in wide crosses, embyro abortion occurs far earlier. Lofland (1950) reported that 15 to 20 day embyros only callused and finally died. Joshi and Johri (1972) and Joshi and Pundir (1966) reported that cotton ovules often formed callus. In none of these early studies callus was not desirable and no effort was made to maintain the growth of calli.

J.McD. Stewart et al. (eds.), *Physiology of Cotton*,
DOI 10.1007/978-90-481-3195-2_33, © Springer Science+Business Media B.V. 2010

In the early sixties, Mauney (1961, Mauney *et al.*, 1967) was successful in recovering plants from immature embryos 12 to 14 days post-anthesis instead of callus. Mauney's work offered some of the first information that would later be relevant to optimization of cotton tissue culture. He reported the optimum temperature for embryo growth to be 90°F. This temperature is now known to be near optimum for growth of any cotton plant or tissue. Growth was more normal in the light versus the dark, and an increase in all salt concentrations over that in early tissue culture media was shown to be important to growth rate. In these studies, sucrose proved superior over glucose. The use of coconut milk and casein hydrolysate was shown to be beneficial but not necessary. All plant-growth substances inhibited normal growth of cotton embryos at very low concentrations and promoted browning and callus formation. The most important aspect of embryo culture proved to be osmotic pressure. An osmotic pressure of 10 atm. (added as 0.7% NaCl) was necessary to maintain viability of these very small embryos. Sequential lowering of the osmoticum was necessary to obtain mature embryos. Recovery of plants from these small embryos was extremely variable from experiment to experiment.

Although the purpose of studies by Mauney, Joshi, Lofland, Weaver, and Beasley was to rescue the progeny of wide *Gossypium* crosses, these studies formed the basis for cotton tissue culture. In 1972, Schenk and Hildebrandt reported the successful initiation of callus from cotton (Auburn 56) hypocotyl on agar containing sucrose, 2,4-D, and kinetin. The calli were successfully subcultured to medium containing sucrose, NAA, and kinetin. Several reports of callus induction and maintenance were rapidly reported thereafter, beginning with the work of Davis *et al.* (1974). Davis *et al.* (1974) reported callus formation from leaf discs of Deltapine 16 using sucrose, 2,4-D, and kinetin in the medium. In all these reports substantial browning of the tissue and very slow growth was problematic.

In 1975, Sandstedt obtained callus from Deltapine 16 and Pima S-4 stem segments using 1AA and kinetin and glucose instead of sucrose. Glucose reportedly reduced tissue browning. This was a first step in the successful establishment of cotton callus by several laboratories (Rani and Bhojwani, 1976; Katterman *et al.*, 1977; Smith *et al.*, 1977; Price and Smith, 1979; Price *et al.*, 1977). Smith *et al.* (1977) began a series of reports in 1977 on the culture of several *Gossypium* species. It was these studies that established the basic tenet of cotton tissue culture and difficulties with callus induction, browning of the explant, and subculture were overcome. However, attempts to regenerate plants proved futile with the exception of two plants. The breakthrough came in 1979, when Price and Smith reported the establishment of embryogenic cell suspensions of a wild species, *Gossypium klotzschianum* Anderss. A series of publications followed about embryogenesis in *G. klotzschianum* that defined this system but whole plants were not recovered (Finer and Smith, 1984; Finer *et al.*, 1987). This work subsequently influenced the successful regeneration

of cotton by several laboratories. Still, it was not until 1983 that Davidonis and Hamilton first reported regeneration of whole plants from cotyledon explants of *G. hirsutum*, cv. Coker 310. Their plants were recovered after a lengthy time in culture and the path to regeneration was unclear. Within a short time frame, Rangan *et al.* (1984), Trolinder and Goodin (1985), Mitten (1985), and Shoemaker *et al.* (1986) reported successful regeneration of cotton. In all of these reports, with the exception of Rangan *et al.* (1984), the Coker varieties appeared to be superior. Although this common thread was shown, at the time, the notion of genotype dependent regeneration was not fully developed. A series of papers by Trolinder and Goodin (1987, 1988a and b) defined the culture parameters for somatic embryogenesis in the Coker varieties and suggested the importance of genotype. Trolinder and Xhixian (1989) subsequently clearly showed that cotton somatic embryogenesis is genotype dependent. Although the culture environment affected the quality of embryogenesis, the deciding factor was genotype. Sister lines from a cross between Deltapine 15 and Coker 100W were apparently inadvertently selected for high regeneration capability, resulting in lines closely derived from the sister lines being regenerative. Coker lines proving highly regenerable were 201, 310, 312, 315, and 5110. Rangan et al (1984) regenerated the California Acala lines SJ1, SJ4, SJ5, and SJ2. Trolinder (unpublished) also regenerated these varieties, adding another Acala type, Germain 510, in addition to several varieties from China. Further screening of most of the American cultivars for the embryogenesis response resulted in the identification of only a few embryogenic varieties beyond these (Trolinder, unpublished) and these produced embryos at a low frequency.

Genetic analysis of cotton somatic embryogenesis by Gawel and Robacker (1990) substantiated that the trait was heritable but a model could not be developed from the data. However, Scheneschal (1994), working with elite regenerable Coker lines selected for their high regenerability (Trolinder and Xhixian, 1989) and non-regenerable lines, defined a working genetic model for the trait. The data for a genetic model in which the expression of either one of two blocker genes inhibits the expression of the embryogenesis trait. Because the trait was heritable at such low frequencies, crossing the trait into more desirable germplasm seemed unfeasible. It did suggest that individual plants not carrying the blocker genes might be selected from within a cultivar population at very low frequency. Although selection for embryogenic types within a large population seems possible, this requires a massive screening effort and considerable time until elite plants are identified and enough progeny tested and bulked to be useful (Trolinder, unpublished). A second possibility is that a window during development might exist when the blocker genes are inactive. Firoozabady and Deboer (1993) indicated that hypocotyl sections from precociously germinated seeds produced embryogenic calli in three non-Coker varieties whereas explants taken from mature seeds did not. This window during development has not been demonstrated for other varieties.

In the early days of cotton regeneration, excitement about the future of somaclonal variation for enhancement of plant breeding led several laboratories to explore the possibility of selecting for improved agronomic traits. Somaclonal variation was documented in cotton regenerants obtained by Rangan, Mitten, and Trolinder (Stelly *et al.*, 1989; Li *et al.*, 1989; Liu *et al.*, 1985). Plants were recovered by Rangan with increased resistance to the pathogens, increased yield, and improved fiber properties (personal communication). Blair *et al.* (1988) isolated photoautrophic cell lines for use in photosynthesis studies. Trolinder and Shang isolated cell lines resistant to high (1991) or low temperature (unpublished data). Plants regenerated from the high temperature was infertile but those resistant to low temperature were fertile. Cell lines with resistance to low temperature were stable over time. Proteins unique to the selected cell lines during exposure to low temperature were identified by 2-D gel electrophoresis. Differential display techniques revealed unique mRNA species as well. Progeny were identified with a heritable resistance to low temperature (B. Davis, personal communication). Although somaclonal variation proved useful in selecting herbicide resistance, pathogen resistance, temperature tolerance, and other improved traits, it was detrimental to maintaining variety purity during the transfer of a single desired foreign gene. Length of time in culture appeared to be a major factor to increased variation and optimization of regeneration to decrease time in culture became a priority. The first cotton regenerants obtained by Davidonis and Hamilton (1983) were in culture for over two years. Optimization has resulted in plants routinely regenerated in six months if not undergoing transformation and from six to eight months with transformation. The procedure has become rather simple as long as the elite Coker varieties are used.

One centimeter hypocotyl sections are isolated from about 7-day old seedlings and placed onto Murashige and Skoog's basal medium supplemented with Gamborg's B5 vitamins, 30 g/l glucose, 180 g/l myoinositol, eitehr 0.1 mg/l 2,4-D + 0.1 (T1) or 0.1 mg/l 2,4-D + 0.5 mg/l kinetin (T2), or 2 mg/l NAA + 0.1 mg/l kinetin (2 nk) at pH 5.8, solidified with 2 g/l gelrite and 0.75 g/l MgCl2. Either of the growth regulator regimes work quite well. Callus initiates a bit more rapidly with T1 medium but less variability with T2. The Acala varieties respond better to the NAA regime.

Callus formation occurs within a matter of weeks. Calli are transferred to the same medium but without growth regulators or gelling agent and shaken on a rotary shaker at 120 rpm under low light. Embryogenic suspensions form readily within two to three weeks. Cell suspensions are sieved and the 30 mesh fraction resuspended at 1 ml settled cells per 10 ml medium and one ml/plate pipetted to embryo development medium. Embryo development medium is Murashige and Skoog's medium supplemented with B5 viatmins, 30 g/l glucose, myoinositol and 1.9 g/l KNO_3 and solidified with 2.5 g/l gelrite. Embryos developed within two to four weeks. Fully developed embyros are removed and placed onto maturation medium containing MSNH.

Protoplasts have been recognized by many as ideal for direct gene transfer and for wide species hybridization. Attempts were also made to develop protoplast systems for cotton during the time period that somatic embryogenesis in cotton was being defined. In early work, micro colonies or callus were formed but no plants were regenerated (Bhojwani *et al.*, 1977; Finer and Smith, 1982; Firoozabady and DeBoer, 1986; Saka *et al.*, 1987) until the work of Jianming *et al.* (1991) and Peeters *et al.* (1994). Jianming *et al.* (1991) and Peeters *et al.* (1994) successfully regenerated plants from the protoplasts isolated from embryogenic cell suspension of cultures of Coker 312. The use of cotton protoplasts for gene and/or organelle transfer or somatic hybridization also depends upon the genotype dependent nature of somatic embryogenesis in cotton and is subject to somaclonal variation with extended culture periods.

The knowledge that only certain cultivars are amenable to regeneration through somatic ambryogenesis, induced efforts in several laboratories to develop other methods that would allow transformation of any cultivar of cotton. McCabe and Martinell (1993) demonstrated that transgenic plants could be produced by particle bombardment of a newly germinated cotton embryo axis, albeit at very low frequency. This required that the DNA be incorporated into germ line cells in the embryonic shoot tip meristem. Shoot tip culture had long been advocated as a means of propagating elite cotton cultivars. In 1985, Yuefang and Yuanling (1985) reported the first successful shot tip culture from field-grown cotton. These researchers advocated an approach to batch production of cotton from shoot tips. However, their data suggested that several morphogenetic events took place at the base of the shoot tip, including adventitious buds, lateral bud break, and the formation of embroids. The cultivar used in these studies was an Acala, SJ1, known to be embryogenic so their description of multiple shoot formation is clouded by the question of whether it was adventitious bud break, lateral buds, or embryos. No histological data was presented. Lee (1987) reported adventitious shoot formation from the stump of field-grown cotton that had been cut below the cotyledonary nodes. A callus formed and subsequently shoots. Once again, the cultivar was one known to be embryogenic, Coker 201, so the origin of the shoots is unclear.

The search for embryogenic cultivars in Trolinder's laboratory led indirectly to another approach to transformation. As a part of Trolinder's (unpublished) elite selection, the 0.5 cm shoot tip of *in vitro* germinated seedlings from many cultivars were labeled and placed in rooting medium for shoot development and the corresponding labeled hypocotyl used in testing. The shoot tips developed into shoots and rooted quite readily at near 100% efficiency. Shoot tips dissected in half or in quarters also readily developed shoots and rooted. The advantage of dissection is the direct exposure of a greater number of germ line cells. Transformation of germ line cells in the meristem requires that the DNA or *Agrobacterium* penetrate six cell layers to reach the 1.2 germ line layer. This is reflected in the low frequency of germ line transformants obtained by McCabe and Martinelli (1993) when whole embryo axis were bombarded. With dissection, the two dormant nodes at the base of each cotyledon are directly exposed as well as the apical meristem. Therefore, ten germ line areas are exposed to the transformation vector in place of one indirectly exposed as in McCabe and Martinelli (1993). The efficiency of this approach for transformation remains to be determined. In 1991, Gould *et al.* showed that a 3 mm section of the shoot tip could be readily regenerated as well. Further, this group demonstrated

that *Agrobacterium* will infect plant meristematic tissue. The efficiency of this system also remains to be determined.

Although successful transformation via particle gun bombardment and *Agrobacterium* infection of teh apical meristem has been demonstrated, the low frequency of germ line transformants and the excessive labor and greenhouse space needed makes these methods unattractive for many laboraotories. Transformation of cotton via somatic embryogenesis remains the most highly efficient system available today. Thus the search for the window of opportunity and/or embryogenic individuals within cotton varieties continues.

CHAPTER 34

CURRENT STATUS OF COTTON MOLECULAR BIOLOGY

Lynnette M.A. Dirk and Thea A. Wilkins[1]
[1]*Department of Agronomy and Range Science, University of California - Davis, One Shields Avenue, Davis, CA 95616-8515 USA*

1. INTRODUCTION

A plethora of cotton genes are being isolated at escalating rates in response to the increased availability of heterologous gene probes and improved screening techniques. Because the primary focus has been on the isolation of cDNA clones and structural genes, most reports are limited to publication of the gene sequence and/or preliminary expression work. With only a few exceptions, cotton research generally lags behind other systems in terms of determining the functional role of genes beyond speculation based on homology to other genes. In cotton, functional studies are often hampered because it is a polyploid species and is far from being an ideal experimental system, especially from a geneticist's perspective. Yet, considerable strides have been made in exploring the molecular genetics of cotton from the standpoint of evolution, as well as the regulation of metabolic and biosynthetic pathways related to the growth and development of cotton in response to developmental, physiological, and environmental stimuli. In conjunction with molecular mapping, the door of opportunity is now open to novel molecular strategies for the genetic improvement of cotton. Because of the diversity of cotton genes isolated to date, this chapter is structured according to the presumed functional role of a gene product as it relates to a particular developmental and/or physiological process. The chapter is divided into sections emphazing the molecular biology of: I. Reproductive Tissues, II. Organelles, III. Physiological Processes, and IV. Genes, Genome Organization and Evolutionary Studies. In recognition of biotechnology efforts in cotton, discussions related to experiments utilizing transgenic plants have been incorporated into the relevant sections.

2. REPRODUCTIVE TISSUES

2.1 Anthers

The development of the next generation of plants is dependent, in part, on the availability of fertile pollen. Environmental factors, such as humidity and temperature, interact with genetic factors in determining pollen fertility (Stewart, 1986). Development of cotton pollen proceeds in a manner similar to other dicots, and yet, to date only a single gene specifically expressed during anther development is isolated and characterized from cotton (John and Petersen, 1994). The deduced amino acid sequences of both a cDNA (*CPA-G9*; *Gossypium hirsutum* L.) and a genomic (*G9*; *G. barbadense* L.) clone were significantly homologous to known polygalacturonases (PG). The PG activity, which results in the hydrolysis of pectin polymers, probably facilitates pollen tube growth permitting elongation by partial disassembly of the pectic portion of the pollen tube cell wall itself or of cell walls of the styler cells through which the pollen tube grows (John and Petersen, 1994). *G9* transcripts began to accumulate in the developing anthers 12 days before anthesis (DBA), reaching maximum amounts on the day of anthesis. Tobacco (*Nicotiana tabacum* L.) plants transformed with *G9* promoter/β-glucuronidase (GUS)-reporter gene constructs confirmed the anther-specificity and temporal regulation of cotton *G9* expression (John and Petersen, 1994). Thus, expression patterns of the reporter gene driven by the *G9* promoter are similar in the heterologous system, which indicates that this *G9* promoter may be universally anther-specific and capable of driving expression of transgenes in an anther-specific manner.

J.McD. Stewart et al. (eds.), *Physiology of Cotton*,
DOI 10.1007/978-90-481-3195-2_34, © Springer Science+Business Media B.V. 2010

2.2   Embryos

2.2.1   Late Embryogenesis-Abundant Proteins

Late embryogenesis-abundant proteins (*Leas*), which are strongly hydrophilic proteins with a predicted amphiphilic helical structure, accumulate to concentrations approaching 30% of non-storage proteins during the late stages of cotton embryogenesis (Bewley and Black, 1994; Dure *et al.*, 1983, 1984). *Lea* transcripts are induced by abscisic acid (ABA) in seeds (Galau *et al.*, 1986), as are highly homologous proteins, the so-called dehydrins and ABA regulated proteins (RABs), which accrue in vegetative tissues under water-deficit conditions (Bewley and Black, 1994). Eighteen cotton *Lea* cotton cDNAs (Galau *et al.*, 1988) are categorized into two sets, termed 5A and 5B (six members of the 5A family and 12 members in the 5B family; Bewley and Black, 1994), based on the differential accumulation of 5A and 5B transcripts (Galau, 1986; Baker and Dure, 1987; Galau *et al.*, 1987; Baker *et al.*, 1988). The *Lea* 5A transcripts increase transiently twice during early maturation (25 to 27 and 32 to 34 days post anthesis [DPA]) of the embryo and reach a maximum at three days (40 to 50 DPA) before the commencement of seed desiccation. The *Lea* 5B transcripts, which are produced only during late maturation (44 DPA), peak either three days prior to or during seed desiccation. Changes in *Lea* gene expression in response to developmental and/or environmental signals, as well as the predicted protein structure of *Leas*, are consistent with *Leas* functioning to protect certain cellular components against desiccation damage. The amphiphilic helical structure predicted for *Leas* is congruous with the hypothesized interaction of *Leas* with cytoplasmic membranes, which require protectants and stabilizers to retain integrity during desiccation. Cellular ionic strength of the cytoplasm increases in any desiccating cell and the charged amino acids of *Leas*/dehydrins/RABs would be available for forming salt bridges to avoid ion crystallization which would disrupt cytoplasmic constituents (Baker *et al.*, 1988). However, the functional role of *Leas* in any plant remains in question because, for instance, seeds which do not survive desiccation still have increased levels of *Leas* and *Lea* transcripts (Bewley and Black, 1994).

2.2.2   Storage Proteins

Two classes of storage protein, the albumins and globulins, are characterized in developing cotton embryos and represent the major nitrogen reserve of seeds (Galau *et al.*, 1991, 1992b; Chlan *et al.*, 1986, 1987). Albumins (2S) are synthesized as preproteins which are sequentially processed to yield a 11.9 kD methionine-rich protein (Galau *et al.*, 1992a). The two alloallelic albumin genes (*Mat5-A* and *Mat5-D*) are expressed such that the mRNAs comprise 2% of total RNA in a mature embryo. Cleavage sites are conserved based on alignment with other albumin sequences (Galau *et al.*, 1992a).

Similar to other plant globulins, newly synthesized cotton globulin preproteins require post-translational processing to form the mature protein. The two predominant sets of storage proteins in cotton, with mature molecular weights of 48 and 51 kD (Dure and Chlan, 1981) are globulins. Comparisons of the sizes and glycosylation features of four representative globulin clones allows the genes to be grouped into two families of globulins, termed α and β, most closely related to vicilins and legumins, respectively. Each family is further divided into A and B subfamilies (Chlan *et al.*, 1986). Expression of the cotton α globulin gene family accounts for 15% of mature seed mRNA. These α-globulins are synthesized as 69 kD preproteins which are processed to yield 46.5 and 52 kD mature proteins. This 5.5 kD difference between α-globulins distinguishes the subfamilies (B subfamily encodes the larger of the two proteins) and is determined by a single glycosylation site (Chlan *et al.*, 1987). The family of β-globulins are synthesized as 60 kD preproteins and are subsequently cleaved into three mature proteins between 10 to 25 kD (Chlan *et al.*, 1986). The expression of β-globulin genes from both A and B subfamilies (genes *Legumin A* and *Legumin B*) constitute 15% and 5%, respectively, of mature seed mRNA. Genes encoding representatives of the two β-globulin subfamilies are more divergent than genes encoding the two α-globulin subfamilies (Chlan, 1995; Galau *et al.*, 1991).

2.2.3   Oleosin

During biogenesis of oil bodies in developing embryos, low molecular mass proteins (15 to 26 kD) called oleosins become embedded in the single lipid bilayer surrounding the oil bodies (Bewley and Black, 1994). Oleosins, presumably, function during embryo development to prevent oil bodies from coalescing. Oleosins may also function as the binding sites of lipases which mobilize the triacylglycerols in the oil bodies at the onset of germination. These provide the primary energy and carbon sources after germination is initiated and before the seedlings become fully autotrophic (Bewley and Black, 1994).

Initially identified based on 55% deduced amino acid identity to carrot oleosin, two cotton oleosin genes (*MatP6* and *MatP7*) and their corresponding cDNAs (D129 and D103) encode for 18.1 and 16.4 kD proteins, respectively, which are 77% identical at the amino acid level of the gene sequence (Hughes *et al.*, 1993). A 26-nucleotide motif similar to a *legumin A* element, may indicate gene regulation by either maturation or stress signals, although this remains to be tested (Hughes *et al.*, 1993).

2.3   Fibers

A summary of fiber development is beyond the scope of this chapter (Wilkins and Jernstedt, 1999), but the genes to be discussed will be put in context of the appropriate developmental stage - differentiation, expansion, secondary wall synthesis, and maturation, whenever possible. In many instances, accumulation of an mRNA transcript is assumed to correspond to the developmental stage at which the translational product would be functioning. Consequently, such accumulation is used to delineate possible functions based on the observed subcellular changes associated with the specific stage. However, careful analysis of the expression of the one gene, *H6*, and its product, for example, showed that post-transcriptional and post-translational regulation occurs (John and Keller, 1995). Thus, expression patterns based on mRNA may not always be the most telling clue of a protein's functional role.

Fiber-enriched expression of five cDNAs, isolated by differential screening of a 15 DPA fiber cDNA library, span development from 0 through 28 DPA (exact days tested are unknown; John and Crow, 1992). Patterns of mRNA accumulation, and in some cases protein accumulation, are reported for four of these fiber-enriched cDNAs: E6, H6, Fb-B6, and Fbl-2A (John and Crow, 1992; John and Keller, 1995; John, 1995, 1996; Rinehart *et al.*, 1996). Differential screening (unknown stage of the cDNA library) was also used to isolate a clone encoding a putative lipid transfer protein (LTP) (Ma *et al.*, 1995). Using mRNA differential display, two partial clones, GHACP2 and Gh-1, were identified which showed greater expression in fibers transfer tissues (Dang *et al.*, 1996; Song *et al.*, 1996). Two other types of clones, *Rac* homologs (Delmer *et al.*, 1995) and putative cellulose synthase clones (Pear *et al.*, 1996) have been specifically isolated to determine a probable function in secondary cell wall synthesis during fiber development. Because the plant type and fibers used for RNA extraction and expression analysis vary from report to report, the genes which are deemed to contribute to any particular stage of fiber development may become obscured by this lack of consistency. Many other genes are expressed during fiber development but only the fiber-enriched genes are presently discussed.

Of the genes showing fiber-enriched expression, the gene encoding an acyl carrier protein (ACP) (Song and Allen, 1997) is expressed as early as 2 DPA, although expression of other genes has not been tested this early in fiber development. The acidic protein, ACP, contains a prosthetic group which links acyl chains as thioesters, such that the acyl chains are carried through fatty acid synthesis (Ohlrogge, 1987). Steady-state amounts of ACP mRNA are greatest at 6 to 8 DPA, and decline to barely detectable amounts by 20 DPA (Song and Allen, 1997). This expression pattern correlates well with periods of the highest fiber-expansion rates, adding credence to the hypothesis that ACP functions in membrane synthesis for both the tonoplast and plasma membrane (Song and Allen, 1997).

Expression patterns of the genes represented by the LTP and Fbl-2A clones show some similarity in that expression is first detected (or tested) at 5 DPA, although the maximum accumulations of LTP transcripts peak at 15 DPA and Fbl-2A at 25 to 35 DPA (Ma *et al.*, 1995; Rinehart *et al.*, 1996). The deduced amino acid sequence for LTP is comparable to other known LTPs and shows the characteristic signal sequence of secretory proteins. This implies that the protein is exported into the cell wall to transport cutin monomers across the plasma membrane for synthesis of cutin during rapid expansion (Ma *et al.*, 1995). The deduced amino acid sequence of Fbl-2A has four 55 amino acid repeats, which by analogy to a similar motif in cold-acclimation-related proteins and *Leas*, may protect the fiber's essential cellular functions during desiccation (Rinehart *et al.*, 1996).

The other five cDNAs were used as probes in northern analysis of fiber RNA during development, and accumulation of the mRNAs appears to span mid- to late-expansion and early secondary cell wall synthesis. The *H6* transcripts accumulate somewhat early, starting at 10 DPA and peaking at 22 DPA (John and Keller, 1995). The *H6* deduced amino acid sequence is proline-rich and shares structural features to arabinogalactan proteins, which led the authors to propose a function in matrix assembly of the secondary cell wall (John and Keller, 1995). Analysis of expression of Fb-B6 occurred during primary and secondary cell wall synthesis (John, 1995), and thus, may also begin to accumulate during the mid-expansion periods. The *E6*-gene was maximally expressed during late expansion and early secondary cell wall synthesis stages (15 to 22 DPA), however, immunodetected protein amounts were highest between 5 and 15 DPA (John and Crow, 1992). Based on the lack of a significant change in phenotype from expression of antisense constructs in transgenic cotton, which reduced mRNA amounts to as low as 2% of normal, E6 is unlikely to be an essential factor in fiber development (John, 1996).

The Rac homolog, *Rac13*, is preferentially expressed in fibers, with the greatest accumulation occurring at 17 DPA (Delmer *et al.*, 1995). The most probable role proposed for this particular GTP-binding protein, because of the role of the mammalian counterparts in cytoskeletal organization, is in the signal transduction for the spatial switch of the cellulosic microfilaments from transverse to a sharp helical orientation with respect to the axis of growth. The timing of expression corresponds to such a spatial switch which has, in turn, been related to the orientation of both the microtubule and actin networks of the cell.

Beginning in late expansion (17 DPA), the expression of two genes (*celA1* and *celA2*), which putatively encode the catalytic subunit of cellulose synthase, increases from low, constitutive expression before 17 DPA at high levels throughout secondary cell wall synthesis (Pear *et al.*, 1996). Evidence which lends some credence to the proposed function of this gene is the Mg^{2+}-dependent binding of UDP-glucose to a renatured fusion protein of the expected subdomain (Pear *et al.*, 1996). Additional evidence is that the database clones which are consistently identified as having the greatest identity to the *celA1* and *celA2* genes are bacterial genes encoding the cellulose synthase catalytic subunit.

On a more biotechnological note, the promoters of the E6, Fb-B6, and Fbl-2A genes, were used successfully to express transgenes in cotton fibers (John and Crow, 1992; John, 1995; John and Keller, 1996; Rinehart *et al.*, 1996).

The promoters of other genes, Gh-1, GHACP1, and LTP, await description as to their effectiveness in directing expression in fibers during a specified stage of development (Dang *et al.*, 1996; Song *et al.*, 1996; Ma *et al.*, 1997).

3.   ORGANELLES

3.1   Vacuoles

The vacuole of a plant cell performs multiple functions, including storing ions and metabolites, maintaining turgor, hydrolyzing proteins, and maintaining cellular homeostasis (Salisbury and Ross, 1992). The V-type H$^+$-ATPase (EC 3.6.1.34), one of two vacuolar proton pumps, generates a proton-chemical gradient across the tonoplast to effect vacuolar acidification. Such a gradient also facilitates active secondary transport of both ions and metabolites in and out of the vacuole. During rapid expansion of fibers, these functions and the regulation of the major V-ATPase subunits at the molecular level have been studied using cDNAs encoding subunits for 69 Kd catalytic A (CVA69.75 - *vat69A* and CVA69.24 - *vat69B*; catalytic; Wilkins, 1993; Wilkins *et al.*, 1994), 60 kDa regulatory B (CVA55.29 and pAT3P; partial cDNA pAT33R; regulatory; Wan and Wilkins, 1994), and 16 kD proteolipid c (CVA16.2 and CVA16.4; Hasenfratz *et al.*, 1995). In 10 DPA ovules, transcripts accumulate to greater amounts for subunit c than for subunits A and B (Hasenfratz *et al.*, 1995). Although differentially expressed, coordinate regulation of the three subunits is still deemed possible, considering the stoichiometric requirement of two proteolipid (c) subunits for every catalytic (A) or regulatory (B) subunit for holoenzyme assembly (Hasenfratz *et al.*, 1995). Northern blot analysis using total RNA from different tissues showed greater accumulation of subunit c transcripts in tissues undergoing rapid expansion such as roots, petals, pollen, and fibers (Hasenfratz *et al.*, 1995). The subunit c transcripts corresponding to CVA16.4 consistently accumulated to greater amounts compared to transcripts corresponding to CVA16.2 in 10 DPA ovules (Hasenfratz *et al.*, 1995).

The other vacuolar proton pump, vacuolar H$^+$-pyrophosphatase, as well as plasma membrane H$^+$-ATPase, are differentially regulated at both the transcriptional and post-transcriptional levels, based on comparison of the amounts of message, protein, and proton pump activities present during fiber development (Smart *et al.*, 1996). Therefore, the functional regulation of vacuolar proton pumping occurs at multiple levels, as expected for such a critical component of the growth process of plants.

3.2   Nucleus

MYBs are a class of transcriptional factors which function in the regulation of gene expression in diverse plant tissues (Martin and Paz-Ares, 1997). A group of six cDNA clones encoding novel MYB proteins was identified by PCR-based screening of a cDNA library of a -3 DPA cotton ovule. Based on expression patterns, the six MYB cDNAs correspond to different gene family members that belong to one of two types, I or II, representing ubiquitously expressed genes (*GhmybA*, *GhmybD*, and *GhmybG*), and genes (*GhmybJ*, *GhmybN*, and *GhmybO*) exhibiting differentially-regulated tissue expression, respectively. Significantly, *GhmybA* transcripts, although from a type I gene, show subtle increases which are coincident with key stages of fiber development (Wilkins and Jernstedt, 1999) at -9 DPA (possibly related to differentiation), -1 DPA (stage I of expansion), 5 DPA (stage II of expansion), and 15 DPA (stage III of expansion). Failure to detect type II *GhmybJ* transcripts whatsoever in developing fibers indicates that *GhmybJ* is the sole gene not expressed in developing fibers. However, temporal regulation of type II *GhmybN* gene expression is noted in fibers as a two-order magnitude decrease between 3 and 5 DPA, the stage at which fibers begin their greatest rate of expansion (stage II of expansion). A slight upregulation of type II *GhmybO* expression is detected during differentiation, rapid polar expansion (stage II of expansion), and again during secondary cell wall synthesis. Thereby, the first inroad has been forged into unravelling the regulation of gene expression during cotton fiber growth by a small collection of unique cotton regulatory proteins, although the genes targeted by these MYBs remain unknown.

3.3   Glyoxysomes

As a class of peroxisomes abundant in oil seed storage tissues, the glyoxysome functions to oxidize fatty acids and synthesize succinate via the glyoxylate cycle (Bewley and Black, 1994). Saturated fatty acids are degraded by β-oxidation and the acetyl-CoA as a product is converted sequentially into citrate and isocitrate. Two enzymes of the glyoxylate cycle are required to form malate; isocitrate lyase (EC 4.1.3.1), which cleaves the isocitrate to succinate and glyoxylate, followed by malate synthase (EC 4.1.3.2) to condense another acetyl-CoA with the glyoxylate to form malate. The post-translational import into glyoxysomes, of isocitrate lyase and malate synthase from the cytoplasm after synthesis from nuclear-encoded genes does not require proteolytic processing, but is likely to depend only on sorting signals (Bewley and Black, 1994) or a "piggy-back" mechanism (Lee *et al.*, 1997). The complete cDNA clones for both glyoxysome-specific enzymes, isocitrate lyase (Icl) and malate synthase (Ms), are isolated from cotton cotyledons and contain the putative, C-terminal-tripeptide peroxisomal sorting signals, ARM and SKL, respectively (Turley *et al.*, 1990a, 1990b).

Few integral membrane proteins of glyoxysomes have been functionally characterized. The deduced amino acid sequence of a full-length cDNA clone (PMP31), isolated by immunoscreening using an antibody against a glyoxysomal-membrane protein, carries a potential C-terminal membrane-spanning region and is significantly homologous to cytoplasmic ascorbate peroxidases (Bunkelmann and Trelease, 1995; 1996). The membrane-bound ascorbate peroxidase is speculated to scavenge the hydrogen peroxide that escapes from catalase within the soluble matrix, as well as regenerate NAD⁺ used in β-oxidation of fatty acids for continual operation of the glyoxylate cycle (Bunkelmann and Trelease, 1996).

3.4   Peroxisomes

As cotton cotyledons emerge from the soil and become photosynthetically competent, the glyoxysomes, which function in the degradation of oil reserves, are converted into peroxisomes by the import of newly translated enzymes (Ni and Trelease, 1991a). The peroxisome is required during photosynthesis to recycle carbon during photorespiration (Salisbury and Ross, 1992). Catalase (EC 1.11.1.6), which breaks down hydrogen peroxide formed during the recycling process, undergoes a switch in isoforms from SU1 to mostly SU2 during the functional changeover of the microbody (Ni *et al.*, 1990a).

Two cDNA clones, *Su1* and *Su2*, encode two of five peroxisomal isoforms of cotton catalase (Ni *et al.*, 1990b; Ni and Trelease, 1991a). The three other isoforms present in cotton (Ni *et al.*, 1990a) are postulated to be heterotetramers comprised of the two subunits encoded by *Su1* and *Su2* (Ni and Trelease, 1991a). Although peak steady-state mRNA amounts corresponding to each clone parallels accumulation of the protein, run-on transcriptional assays determined that both *Su* transcripts were essentially synthesized at the same rate during the five days following the start of imbibition. Thus, regulation of the relative abundance of the five catalase isoforms is thought to occur post-transcriptionally, although light is purportedly necessary for the accumulation of *Su2* transcripts (Ni and Trelease, 1991b). These results are consistent with the corresponding increase in protein accumulation observed as cotyledons become photosynthetically competent, even though the evidence for such a proposition cannot be readily evaluated. This is due to the fact that northern blot analysis did not include data for dark-grown cotyledons for all time points sampled, leaving the possibility of inherent, temporal-regulating factors being responsible for the differential accumulation of *Su2* mRNA.

3.5   Chloroplasts

Sequences of genes encoding the small (*rbc*S; Sagliocco *et al.*, 1991) and large subunits (*rbc*L; Giannasi *et al.*, 1992; Gulov *et al.*, 1990; Chase *et al.*, 1993) of ribulose-1,5-bisphosphate carboxylase/oxygenase (EC 4.1.1.39) (rubisco) have been reported for cotton. Based on southern blot analysis, the cotton rubisco small subunit is encoded, as usual in higher plants, by a large multigene family (Sagliocco *et al.*, 1991). Similarly, the existence of multiple cotton *rbc*L sequences also suggests the existence of a multigene family for the large subunit, leading to the possibility of utilizing

*rbc*L sequences in phylogenetic studies (Chase *et al.*, 1993; Giannasi *et al.*, 1992; Nickrent and Soltis, 1995; also see Section 5.1 Gene Families). The predictable expression pattern for the cotton rubisco small subunit is reported only in terms of variability among tissues. There are no *rbc*S transcripts detected in non-photosynthetic tissues (roots, young embryos, or mature seeds), only low amounts in the stem, and substantial amounts in the major photosynthetic organs, such as cotyledons of germinated seeds and leaves (Sagliocco *et al.*, 1991). The light-regulated motifs, I- and G-box, are completely conserved in the promoter region of cotton genes compared with other *rbc*S genes (Sagliocco *et al.*, 1991).

The cotton genes, *Lhcb1*1*, *cab-151*, and *Lhcb3*1*, encode each of three types (Type I, II, and III) of chlorophyll a/b-binding protein (CAB), respectively, which form part of the light-harvesting complex (specifically LHCII) in chloroplasts, and as, expected, are differentially expressed (Sagliocco *et al.*, 1992; Anderson *et al.*, 1993). According to northern blot analysis, the greatest expression of all three gene family representatives occurs in photosynthetically competent leaves and cotyledons, with limited expression in tissues, such as bracts, stems, and the pericarp, which exhibit a transient photosynthetic capacity (Sagliocco *et al.*, 1992; Anderson *et al.*, 1993). Although no comprehensive expression study of all three genes has been conducted, transcript amounts are ten-fold greater for type III compared to that for type I (Anderson *et al.*, 1993). Even so, the type I promoter in a *Lhcb1*1* promoter-GUS construct is only five-fold less active, as determined by relative GUS activity in transgenic cotton cotyledons, than a GUS-reporter construct containing a double CaMV promoter and a tobacco mosaic virus omega translational enhancer (Anderson *et al.*, 1993). A characteristic G-box element sequence, present in other *cab* and *rbc*S genes and likely functioning in light-induced transcription (Gilmartin *et al.*, 1990), is only present in type II genes (Sagliocco *et al.*, 1992). Detailed experimental analysis of *cis* elements for light regulation of photosynthetic genes, rubisco, and CAB, has not been done in cotton but could proceed based on current available information.

Photosystem II is composed of three complexes, the light-harvesting antennae, a reaction center core, and an oxygen-evolving complex, consisting of proteins (both nuclear and chloroplastic gene-encoded), chlorophyll *a*, β-carotene, membrane lipids, ions, plastiquinone, and pheophytin (Salisbury and Ross, 1992). The interaction of light with the three complexes involves the oxidation of water during carbon dioxide fixation (Salisbury and Ross, 1992). The 32 kD thylakoid membrane protein, Q_B or D1 protein, is one of two reaction center proteins of photosystem II in the chloroplast (Ulmasov *et al.*, 1990). Mutagenesis, by either spontaneous mutation or biotechnological means, in which a single amino acid of Q_B protein is altered, has been used in other plant species for studying herbicide (specifically atrazine) resistance (Goloubinoff *et al.*, 1984). A chloroplastic operon of *psb*A and *trn*H, which encodes the Q_B protein and tRNA[His] (GUG), respectively, has been sequenced from cotton chloroplast DNA fragments, although no complete clone for the Q_B protein could be reconstructed from the fragments (Ulmasov *et al.*, 1990). This failure is taken as evidence that this protein is toxic to bacterial cells.

Many small MW proteins, ranging from 3.6-10 kD, are also associated with photosystem II (Kapazoglou *et al.*, 1995). An intronless cotton gene, *psbT* encodes a 11 kD protein. Based on *in vitro*-translated protein and isolated pea chloroplasts, the intricate processing steps required for import into the chloroplast was investigated, and the final protein was reported to be the smallest polypeptide of photosystem II identified at that time (Kapazoglou *et al.*, 1995). Upon transport into the stroma, the precursor is cleaved into a 7.5 kD intermediate. The precise sequence of further events is unclear, but involves cleavage to a 3.8 kD intermediate, before final cleavage to the mature protein of three kD and transport into the lumen of the thylakoid (Kapazoglou *et al.*, 1995). Thus, a number of cotton genes and partial cDNAs functioning in photosystem II are available for further cellular and molecular study of this important photosynthetic complex.

During embryo development, the proplastid, which will form oil bodies in the mature embryo, is the likely target for the stearoyl-acyl-carrier-protein desaturase (EC 1.14.99.6) polypeptide carrying an N-terminal 33-amino acid transit peptide (cDNA, pGh9-1; Liu *et al.*, 1996). The *cis*-double bond introduced between carbons nine and ten of C18 fatty acids by the desaturase produces monounsaturated oleic acid, and thus, regulates the amount of unsaturated lipid in membranes and seed oils (Harwood, 1988; Bewley and Black, 1994). Another recently cloned cDNA, ACBP, encoding an acyl-CoA-binding protein, which is involved in regulating triacylglycerol and phospholipid synthesis by shuttling long-chain acyl-CoAs between intracellular compartments, has no known signal peptide sequence, which is consistent with its role in transport, and its presumed cytoplasmic location (Reddy *et al.*, 1996).

3.6   Cell Walls

The dynamic nature of the cell wall enables it to assume a regulatory role in the expansion of all plant cells. *GrmM4* encodes a peroxidase (EC 1.11.1.7) with a putative 23 amino acid signal sequence and is purported to function in lignin polymerization in the cotton primary cell wall (Ritter *et al.*, 1993; Salisbury and Ross, 1992, pp.323). Studied initially during embryo development, the mRNA increases to ca. 1% of the total mRNA in a mature embryo, decreases after boll dehiscence, and increases again during the first two days of imbibition (Hughes and Galau, 1989). *GrmM4* is also expressed in roots and induced in leaves of salt-stressed plants (Ritter *et al.*, 1993).

Proteins presumably functioning in the enzymatic alteration of the primary cell wall architecture have been characterized for relative abundance of their respective transcripts during cotton fiber development (Shimizu *et al.*, 1997). Endo-1,4-β-glucanase (EC 3.2.1.4) and expansin transcripts, as detected by southern blot analysis of reverse transcribed-polymerase chain reaction (RT-PCR) products generated with gene-specific primers, are greatest during cell elongation and decline during the transition to secondary cell wall synthesis. The pattern of transcript accumulation of endo-1,3-β-glucanase (EC 3.2.1.6) has the opposite trend (Shimizu *et al.*, 1997). No evidence is provided (Shimizu *et al.*, 1997), however, to add credence to the supposition that the endo-1,3-glucanase is present to transform the 1,3-β-glucan intermediate deposited for cellulose synthesis (Amor *et al.*, 1995). Xyloglucan endotransglycolase (XET) (EC 2.4.1.207) transcripts are in essentially constitutive amounts during both primary and secondary cell wall synthesis (Shimizu *et al.*, 1997). The peak accumulation of xyloglucans in the fiber primary cell wall occurs at 15 DPA, although the structure of the xyloglucan fraction has not been determined (Shimizu *et al.*, 1997), and so it is unknown whether XET protein amount and/or activity regulates the xyloglucan structure. Although a membrane-bound form of sucrose synthase (EC 2.4.1.13) is purported to function in cell wall cellulose synthesis (Amor *et al.* 1995), the amount of transcript detected with gene-specific primers in RT-PCR experiments does not increase solely at stages of fiber development during which cellulose synthesis would be greatest (Shimizu *et al.*, 1997). Using a partial clone, *Ss3*, of sucrose synthase for southern and northern blot analysis and a polyclonal antibody for western analysis, accumulating data indicates two genes for sucrose synthase, both of which are expressed in etiolated shoots, but only one of which is expressed as the sole isozyme in developing fibers (Ruan *et al.*, 1996).

Although 1,3-β-glucanase (cDNA, *pGLU18.1*) transcripts remain relatively constant in leaves treated with ethyphon or salicylic acid, such treatments induces the accumulation of mRNA for the enzyme chitinase (EC 3.2.1.14) that hydrolyzes β-1,4 linkages of N-acetyl-D-glucosamine polymers (chitin; Hudspeth *et al.*, 1996a). Isolation of discrete cDNA and genomic clones, *Chi1;1*, *Chi2;1*, and *Chi2;2*, (Levorson and Chlan, 1996; 1997) and southern blot analysis (Hudspeth *et al.*, 1996a) establishes that chitinase genes are organized in the genome as a small gene family. Presumably, the physiological role that chitinase plays in plant defense against phytopathogenic fungal species involves the hydrolysis of the fungal cell wall polysaccharide to limit pathogen growth, although this still requires stringent investigation. The activity of chitinase increases in cotton leaves in response to ethylene (Boller *et al.*, 1983) and an ethylene-responsive chitinase gene has been successfully cloned from cotton (Levorson and Chlan, 1997). Elicitors, such as salicylic acid and ethylene, increase pathogen resistance (van Loon, 1977; White, 1979) and chitinase activity (Derckel *et al.*, 1996) in plants.

Annexins are a class of proteins that interact with membranes and bind Ca^{2+} in the presence of phospholipids. Two cDNA clones, *AnnGh1* and *AnnGh2*, encoding annexins are 70% similar at the nucleotide level, and this permits gene-specific primers to be used in RT-PCR reactions to determine the expression patterns of the different genes represented by the cDNAs (Potikha and Delmer, 1997). From western blot analysis, high amounts of annexin are detected in most cotton tissues. During fiber development, expression of the *AnnGh1* gene is eight- to ten-fold greater than that of the *AnnGh2* gene at each stage examined, although both are expressed in greater abundance during expansion and decrease gradually during the onset of secondary cell wall synthesis (Potikha and Delmer, 1997). Evidence from the purified protein, against which the antibodies used for expression screening were made, suggests these annexins bind to and influence the activity and or location of callose synthase (Andrawis *et al.*, 1993). This may be the primary function in developing fibers, but other roles in growth and development of plant tissues are proposed (Battey and Blackbourn, 1993; Calvert *et al.*, 1996; Gidrol *et al.*, 1996; McClung *et al.*, 1994).

4. PHYSIOLOGICAL PROCESSES

4.1 Biotic Stresses (see Chapters 26 and 27)

4.1.1 Resistance Genes

The so-called R (resistance) genes confer a characteristic type of resistance against individual strains of a specific pathogen a genetic resource for breeding resistant cultivars. Given that resistance to a specific pathogen is usually inherited as a single gene trait, the gene-for-gene model (Flor, 1946) describes the interaction of a resistance gene with a specific "target" gene of the pathogen (avirulence (*avr*) gene) to produce the characteristic type of resistance. Strains of the causal agent of bacterial blight disease, *Xanthomonas campestris* pv. *malvacearum* (Brinkerhoff, 1970), carrying the avirulence gene, *avrB4*, are effectively and equally resisted by cotton plants carrying either of two unlinked resistance gene loci, B1 and B4 (De Feyter *et al.*, 1993). The cotton locus, B1, in fact, may confer gene-for-gene resistance to isogenic bacterial strains carrying any one of the following avirulence genes, *avrB4*, *avrb6*, or *avrB102* (De Feyter *et al.*, 1993) and represents an avenue of research requiring attention.

4.1.2 *Bacillus thuringiensis* Delta Endotoxin

After years of breeding attempts, the approach taken to develop cotton resistance against Lepidopteran species has been the introduction of a transgene for the delta endotoxin from *Bacillus thuringiensis*, the so-called Bt toxin (Umbeck *et al.*, 1987). Numerous greenhouse and field studies (Benedict *et al.*, 1996; Fitt *et al.*, 1994; Flint *et al.*, 1995; Jenkins *et al.*, 1993; Rummel *et al.*, 1994; Wilson *et al.*, 1992, 1994; and others) have been conducted with

Monsanto transgenic cotton plants containing truncated forms of the gene, *cryIA(b)* and *cryIA(c)* from *B. thuringiensis* var. *kurstaki* and strains, HK-1 and HD-73, respectively. The first commercial planting in 1996 became infested with bollworms and such damage initially threatened all further cotton biotechnological efforts (Macilwain, 1996). The need for pesticide application in addition to the technology fee, and the increased seed price reduces the economic attractiveness of Bt cotton to growers in areas where damage due to the worm complex is not extensive. Evaluations of the gains depend on the viewpoint taken. For growers, the yield benefits accrued from planting Bt cotton is dependent on the insect load and water conditions in the field. For the research scientists, the results of large-scale Bt cotton production allows evaluation of the performance/efficacy of the promoters in the cotton field. Although the concern of insects developing resistance to the Bt toxin have been well-studied prior to commercial release, and the strategies for managing the crop to delay the establishment of resistance are in place (Deaton, 1996; others), the overall benefits of the effort remain equivocal.

4.1.3 Phytoalexin Synthesis

The formation of terpenoid aldehydes (TA) within stem xylem and surrounding parenchyma cells is part of the response of cotton plants to pathogenic fungi (Mace *et al.*, 1985, 1989). The isoprenoid compound, desoxyhemigossypol, is derived from mevalonic acid, which is formed *in vivo* by the activity of 3-hydroxy-3-methylglutaryl CoA reductase (HMGR; EC 1.1.1.34). Using PCR fragments from conserved regions of the gene, Joost *et al.* (1995) localized the HMGR message within stem stele tissues induced in a *Verticillium* wilt-resistance cotton variety ten hours after inoculation with *Verticillium dahliae* conidia, either live or heat-killed. An increase in the same message in a susceptible variety was not detected until the next sampling at 24 hours. Because steady-state mRNA amounts for HMGR were ultimately greater in the susceptible variety, the difference in tolerance to the fungal spores was attributed to the lag of the susceptible plants' response to accumulate an active enzyme for increasing the mevalonic acid pool required to synthesize the deterrents (Joost *et al.*, 1995).

Southern analysis with the cloned HMGR PCR fragment reveals seven to nine distinct members of a HMGR gene family (Joost *et al.*, 1995), but sequence analysis and southern blot analysis together suggest there are at least two small gene families (Loguercio and Wilkins, unpublished). Based on gene-specific probes in ribonuclease protection assays, one gene family is upregulated in roots, while the other family has increased expression during embryo development. A sequenced pseudogene appears to be a chimera of the two families (Loguercio and Wilkins, unpublished). No correlation between TA content or synthesis and HMGR gene expression is observed in wild type, glandless cotton mutants exhibiting reduced TA, or a cotton cultivar with a high TA content (Loguercio and Wilkins, unpublished).

The linking of isoprenoid units involves synthases that produce phosphorylated intermediates, which are then cyclized in the ultimate formation of the sesquiterpene phytoalexin. Although the clone, *fps1*, for farnesyl pyrophosphate synthase (FPS) (EC 2.5.1.10) has been isolated from *G. arboreum* L., confirmation of FPS's involvement and relevance in the linkage of isoprenoid units awaits publication of the gene expression and protein accumulation data (Liu *et al.*, 1997).

Cadinene synthase (EC 4.6.1.11), an enzyme which catalyzes the cyclization of farnesyl diphosphate to (+)-delta-cadinene, exists in two different isozymes, CAD1-A and CAD1-C, in cotton (Chen *et al.*, 1996). Although the isozymes catalyze the same reaction, *in vitro* synthesized CAD1-A has a higher pH optima (8.7 vs. 7 to 7.5) and a lower optimal Mg^{2+} concentration (2.5 mM vs. 15 mM) than those of CAD1-C (Chen *et al.*, 1996). Gene-specific primers for RT-PCR reactions to detect the CAD messages, indicate that mRNAs in cell suspensions treated with *Verticillium dahliae* elicitor are induced eight-fold after six hours for CAD1-A and ten-fold after eight to ten hours for CAD1-C (Chen *et al.*, 1995, 1996). The only supposition Chen *et al.* (1996) presented for the need of multiple CAD isoforms involved in phytoalexin production responding to the same elicitor was tissue specificity, however, this could not be addressed by their cell culture approach.

4.2 Abiotic Stresses (see Chapters 19, 20, 21, 22, and 28)

4.2.1 Water / Anaerobic Stresses

Under conditions of low oxygen, cotton grows poorly. The lack of oxygen leads to ethanolic fermentation to allow glycolysis to continue for ATP production and thus, alcohol dehydrogenase (EC 1.1.1.1) activity (and mRNA transcripts) increases under such a stress to regenerate ethanol and NAD^+ and to maintain cytoplasmic pH (Millar and Dennis, 1996). Of the three isozymes of alcohol dehydrogenase in cotton, ADH2 is the isozyme which increases after the initiation of anaerobic stress (Millar *et al.*, 1994). Motifs present in the *Adh2* promoter are similar to those which increase transcription in other species under similar low oxygen conditions (Millar and Dennis, 1996). The cDNA clone for this isozyme has been introduced into transgenic cotton to alter ADH expression with the goal of creating a plant more tolerant of anaerobic conditions (Millar and Dennis, 1996), but the results of these attempts are still forthcoming.

4.2.2 Metal Stresses

Metallothioneins are low-molecular weight proteins that detoxify metals in humans and other animals (Salisbury and Ross, 1992). The expression of two cotton genes, *MTl-A* and

MTl-B, encoding metallothionein-like proteins is, as expected, greatest in the roots with little expression in any other tissue (Hudspeth *et al.*, 1996b). Along with this expression pattern, three conserved Cys-X-Cys motifs are present in both the N- and C-termini that classify the small gene family as type 1 metallothionein-like. A 640-bp promoter fragment from *MTl-A* directs high expression of the GUS-reporter gene in transformed cotton root tips. The expression pattern in the rapidly dividing, metabolically active region infers that the protein functions in protecting against toxicity following heavy metal uptake by roots (Hudspeth *et al.*, 1996b).

4.2.3 Broadleaf Herbicide Stresses

Accidental drift from the application of the herbicide, 2,4-dichlorophenoxyacetic acid (2,4-D), from neighboring cereal crops presents a serious problem to cotton crops in Australia (Lyon *et al.*, 1993) and the United States (Bayley *et al.*, 1992). Introduction of a bacterial gene encoding an enzyme, 2,4-D monooxygenase, capable of degrading 2,4-D, into cotton offers a biotechnological approach to detoxify the artificial auxin. Tolerance in two transgenic cotton cultivars to the herbicide increases at least 90 times (Lyon *et al.*, 1993) and exceeds 3-fold the recommended application rate of 2,4-D application for wheat, corn, sorghum, and pastures (Bayley *et al.*, 1992). This level of tolerance is deemed sufficient to protect the cotton crop against drift from normal field applications in neighbouring cereal crops, but remains insufficient against higher doses used for problem weeds (Lyon *et al.*, 1993).

5. GENES, GENOME ORGANIZATION, AND EVOLUTIONARY STUDIES

5.1 Gene Families

5.1.1 *Lea* and Legumin

Like most gene families in allotetraploid cotton, both legumin storage protein (A and B) and *Lea* (18 unique) cDNAs are found in two copies in *G. hirsutum* (Galau *et al.*, 1988). However, the assignment of a particular gene to either the A or D subgenomes by restriction length polymorphisms (RFLPs) is more direct than usual for these genes because each of the diploid species studied has only a single RFLP, and each of these RFLPs are found in the tetraploid.

5.1.2 Branched-Chain Amino Acid Synthetic Enzyme

The enzyme, acetohydroxyacid synthase (AHAS) (EC 4.1.3.18), functions in branched-chain amino acid (valine, isoleucine, and leucine) synthesis. The importance of AHAS in metabolism is demonstrated by plant susceptibility to AHAS specfic herbicides and by development of herbicide-resistance with a single amino acid substitution in endogenous and transgenic forms of the enzyme (Haughn *et al.*, 1988; Lee *et al.*, 1988). Because of the biotechnological potential, how the gene encoding this enzyme is organized, inherited and expressed has been studied in cotton. Six genes encoding AHAS demonstrates the complexity of this gene family in cotton relative to other plants (Grula *et al.*, 1995). Two of the AHAS genes are designated as housekeeping forms, based on constitutive expression in leaves, pericarp, seeds, and cotton callus, and on alignment with a sequence from a constitutively expressed, housekeeping form from *Brassica napus* L. (Ouellet *et al.*, 1992). The four other cotton AHAS genes are arranged as tandem pairs with the genes of a pair separated by two to three kb. The downstream gene of each pair is constitutively expressed at low amounts, and the upstream gene of each pair is expressed only in the anthers. The duplication of these genes probably predates the divergence of the A and D genomes of cotton because the two genes within a pair are less similar (85%) to each other than the upstream or downstream genes of each pair are to each other (more than 95% for both upstream and downstream gene comparisons). Southern blot analysis suggests that each of the A and D subgenomes contributes one gene of the housekeeping form and one of the tandem pair of genes in tetraploid cotton (*G. hirsutum*).

5.1.3   Vacuolar H+-ATPase

A superfamily comprising of two multigene families for the catalytic subunit (subunit A) of the vacuolar H+-ATPase has been identified based on the comparison of sequence and restriction enzyme digestion patterns of PCR products amplified from A, D, and AD genomes (Wilkins *et al.*, 1994). Each of the families has at least four to six members derived from both A and D subgenomes. Thus, the organization of the two gene families is not related to polyploidization. Members from the A subgenome were noted to encode a more diverse group of subunits. Nucleotide substitution estimates indicated that the two families are at least as old as the genus *Gossypium* itself. Preliminary analysis using southern blots also shows a multigene family organization for both the regulatory subunit, B, and the proteolipid subunit, c (Hasenfratz *et al.*, 1995; Wan and Wilkins, 1994).

5.1.4   Ribosomal Proteins and Sequences

Eukaryotic ribosomes consist of three to four rRNAs and between 70 to 80 proteins. The construction of ribosomes begins with synthesis of the proteins in the cytoplasm and subsequent transport into the nucleus to form ribonucleoprotein particles with the rRNA. Further processing occurs to form ribosomes, once this complex is transported out into the cytoplasm (Perry, 1976). Genes encoding four types of ribosomal proteins (L41, S16, S4e, and RL44) have been isolated from *G. hirsutum* (Turley *et al.*, 1994a, 1994b, 1995; Hood *et al.*, 1996). To corroborate the initial import of the protein into the nucleus, a bipartite nuclear targeting sequence has been identified in three ribosomal protein genes, with the exception of the S16 ribosomal protein gene, which has two bp missing from the spacer region of the nuclear targeting sequence with an unknown effect on nuclear import (Turley *et al.*, 1994a). A restriction fragment length polymorphism (RFLP) is detected in cotton for each of the ribosomal protein clones. Not surprisingly for cotton, an allotetraploid, each of the protein types is encoded by individual members of a multigene family.

Fluorescent *in situ* hybridization (FISH) has been used to detect loci of both the 5S and the 18-28S rDNA loci in *G. hirsutum* (Crane *et al.*, 1993; Hanson *et al.*, 1996) and close relatives of its putative diploid ancestors (A-genome members: *G. herbaceum* L. and *G. arboreum* L; and D-genome members: *G. raimondii* Mueller and *G. thurberi* Todaso) (Hanson *et al.*, 1996). The initial report (Crane *et al.*, 1993) mapped four 18S-28S loci, and two 5S loci, however, with the improved sensitivity of the method, the number of 18S-28S rDNA loci in allotetraploid cotton was increased to 11 (Hanson *et al.*, 1996). These were categorized as major (three), intermediate (one), and minor (seven) loci. Dual-label FISH allows mapping of these 18S-28S ribosome loci together with the 5S tandem repeat loci (Hanson *et al.*, 1996).

The presumed inheritance from progenitor diploid species includes three 18S-28S loci from the A-subgenome and four loci from the D-subgenome, leaving four loci unaccounted for in the progenitor relatives (Hanson *et al.*, 1996). Each of the two 5S loci identified in *G. hirsutum* is on the same chromosome as the major 18S-28S loci, as is the single 5S locus of the A progenitor relative, *G. herbaceum* (Hanson *et al.*, 1996). In contrast, the lack of additivity of loci is exemplified by the single 5S loci in *G. raimondii*, the D progenitor relative, mapping to the chromosome with an intermediate 18S-28S locus (Hanson *et al.*, 1996). The lack of additivity between the diploid ancestors and the tetraploid species suggests that rDNA loci in cotton are more complex and dynamic than initially considered.

Analysis of different species of tetraploid cotton in terms of types and copy number of repetitive DNA elements (Zhao *et al.*, 1995) provides evidence supporting the hypothesed, relatively-recent origin of tetraploid cotton and promotes the use of isolated, repetitive DNA clones for both species and genome relationship studies. Thus, the evolutionary studies of speciation of cotton (Cronn *et al.*, 1996; Hanson *et al.*, 1996; Martsinkovskaya *et al.*, 1996; Wendel *et al.*, 1995a, 1995b), as well as angiosperms (Nickrent and Soltis, 1995; Soltis *et al.*, 1997) and eukaryotes (Qi *et al.*, 1988), benefit greatly from the characterization of rDNA loci, as well as the copia-like retrotransposable elements (VanderWiel *et al.*, 1993).

Additionally, such evolutionary questions of speciation of cotton have been studied using restriction length polymorphisms (RFLPs) of nuclear and/or chloroplast DNA and random amplified polymorphic DNA (RAPD) markers (Brubaker *et al.*, 1993; Brubaker and Wendel, 1994; DeJoode and Wendel, 1992; Multani and Lyon, 1995; Wang *et al.*, 1995; Wendel and Albert, 1992; Wendel and Percival, 1990; Wendel *et al.*, 1991). Prior to, and even after, the evidence for the utility of such molecular characterization, such questions of speciation in cotton are addressed by morphological and allozyme analyses (*e.g.*, Stanton *et al.*, 1994; Wendel *et al.*, 1989, 1992, 1994; Wendel and Percy, 1990) and remain important as corroborative evidence.

5.1.5   Repetitive DNA Analysis

The genomes of most higher plants largely consists of repetitive DNA (Thompson and Murray, 1981), including microsatellites, *copia*-like retrotransposons, and rDNA. Microsatellites are tandemly repeated DNA less than 100 bp with repeat unit lengths of six bp or less (Thompson and Murray, 1981). *Copia*-like retrotransposons are one of three classes of mobile genetic elements that replicate by reverse transcription and accumulate to high copy numbers in plant genomes (Flavell *et al.*, 1992; Voytas *et al.*, 1992). The rDNA sequences in plants are highly repeated such that the tandemly repeated arrays of the 5S gene including the nontranscribed spacer occurs in thousands of copies in the genome, depending on the species (Sastri *et al.*, 1992).

The repetitive DNA of the genomes of two allotetraploid species of cotton, *G. hirsutum* L. (Baker *et al.*, 1995) and *G. barbadense* L. (Zhao *et al.*, 1995), when characterized by two different methods and yield consistent estimates with each other (35.6%, Baker *et al.*, 1995; and 32.2%, Zhao *et al.*, 1995) and an earlier estimate using the kinetics of DNA reassociation (39.5%, Walbot and Dure, 1976).

In addition to these estimates, conclusions about genome organization are inferred from these data. Baker *et al.* (1995) suggests that the cotton genome is organized with closer grouped repetitive DNA and larger stretches of single-copy DNA (Baker *et al.*, 1995) than previously predicted (Walbot and Dure, 1976). The categorization of 313 clones of nuclear repetitive sequences into 103 families determines that interspersed repetitive elements are the most abundant class of repetitive DNA, and this agrees with observations in most eukaryotic genomes (Zhao *et al.*, 1995). The presence of a few tandem repeat families with high copy number ($>10^4$) is unique to cotton, compared to other higher plant genomes. Methylation is more apparent in interspersed repeats than tandem repeats, possibly reflecting a difference in base composition of DNA or chromsomal location between the two types of repeats. Isolated repetitive DNA clones provide tools for physical mapping using yeast artificial chromosomes (YACs) (Zhao *et al.*, 1995).

5.2   Mapping

An RFLP map has been generated for *G. hirsutum* X *G. barbadense* and provides evidence that two polyploidiza-

tion events occurred to create the current n = 26 genomes, one about 1.1 to 1.9 million years ago to account for the n = 13 to n = 26 event and an earlier polyploidization about 25 million years ago to create the n = 13 genome (Reinisch *et al.*, 1994). In addition to the feasibility of map-based cloning, the cotton map, with 705 RFLP loci assembled into 41 linkage groups, presents an opportunity to study both chromosome evolution and speciation; not only are over half of the linkage groups assigned to a particular chromosome, but most RFLPs are also associated to a subgenome. A suggestion that the larger A subgenome has recombinationally inert repetitive DNA was forwarded (Reinisch *et al.*, 1994).

Considerable effort has been expended on RFLP and RAPD analysis and identification of markers for breeding purposes (Brubaker *et al.*, 1993; Brubaker and Wendel, 1994; DeJoode and Wendel, 1992; Multani and Lyon, 1995; Shappley *et al.*, 1996 and references therein; Wang *et al.*, 1993, 1995; Wendel and Albert, 1992; Wendel and Percival, 1990; Wendel *et al.*, 1991). More classical mapping techniques, using genetic crosses for instance, are still useful for these mapping efforts (to mention only a few: Endrizzi and Ray, 1992; Saha and Stelly, 1994; Samora *et al.*, 1994).

6.   SUMMARY

A multitude of genes have been isolated from cotton which greatly advanced biochemical, physiological, and genetic studies of cotton growth. For the most part, the function of the isolated genes has been assessed solely on the basis of homology with database genes. Only a few in-depth studies address physiological relevance, primarily because the initial gene isolation from cotton is necessary to determine the impact of different levels of regulation on cotton physiological processes. Given that this resource is now firmly launched, a flourish, not only of continued growth of the "cotton gene bank", but also of detailed investigations into function and levels of regulation are anticipated. Although limited numbers of cotton researchers will proceed with such studies, the scarcity of molecular work associated with the cytoskeleton, signal transduction pathways, and hormone perception and action is astounding given the significance of these components to growth and production of cotton. Additionally, many meaningful questions remain for cotton geneticists about the placement on the genetic or physical maps of most of the isolated genes, many of which are members of small gene families belonging to A and D subgenomes. Results from ongoing mapping studies utilizing newer technologies such as YACs and BACs (yeast and bacterial artificial chromosomes, respectively) are also expected to spearhead molecular plant breeding efforts. Thus, the isolated cotton cDNAs/genes represent an excellent resource, not only for continued basic plant science investigations, but also for future biotechnological improvement of cotton, both of which hold great promise for many fruitful years.

7. ACKNOWLEDGMENTS

Any omission of published data through June of 1997 is strictly an oversight on the part of the authors. The authors wish to thank Dr. Bruce Downie for his critical comments of the manuscript and gratefully acknowledge the grant support of the U.S. Department of Energy (De-FG03-92ER20067) and Cotton Incorporated (92-815) to T.A.W.

CHAPTER 35

GENETIC ENGINEERING APPLICATIONS IN CROP IMPROVEMENT

M.E. John[1] and J.McD. Stewart[2]
[1]Monsanto, Madison Campus, Madison, Wis.; and [2]Department of Crop, Soil, and Environmental Sciences, University of Arkansas, Fayetteville, Ark.

1.   INTRODUCTION

Genetic engineering of crop plants is one of, or perhaps, the most influential technologies in agriculture for plant improvement of the 21st century. The food and fiber requirements of approximately 10 billion people need to be met with existing resources that are currently limited. Biotechnology, and specifically genetic engineering, can greatly expand those limits. The potential of agricultural biotechnology and its ability to complement and enhance traditional plant breeding are just beginning to be understood.

Plant genetic engineering refers to the manipulation of genetic materials of plants to effect gene expression. Thus, today we are able to isolate and modify specific pieces of DNA from any source and introduce it into plants. While there are technical limitations in terms of size of DNA to be introduced, location of integration, plant species amenable to transformation, and detection of phenotype/genotype alterations, by and large these limitations can be expected to be overcome within the next several years. The process of genetic engineering of plants involves three major steps: gene manipulation, transformation, and regeneration. The gene manipulation is now a routine step where genes are isolated, cloned, and modified using general molecular biology techniques (Maniatis, 1989; Ausubel *et al.,* 1987). The next step, introduction of DNA into plants, transformation, is much less well developed. We will discuss transformation as it pertains to cotton. Regeneration of cotton is discussed in Chapter 33.

2.   COTTON TRANSFORMATION

Cotton transformation is a process by which an identifiable DNA segment is inserted into a cotton chromosome and which may or may not be transcriptionally active. In most cases the input DNA is transmitted to sexual offspring. Thus, transformation enables us to alter gene expression to enhance specific agronomic or fiber traits. Certain marker genes are used in transformation, thus, enabling one to distinguish between transformed and non-transformed plants.

2.1   Reporter and Selectable Genes (Markers)

The products of the genes currently in use are normally stable in plant cells, have no endogenous counterparts, have well defined assays available, and do not interfere with the normal development or yield of the plant. The most commonly used selectable marker genes in transgenic cotton are neomycin phosphotransferase II *(npt II)* that allows antibiotic selection of transformed tissues on kanamycin and C-glucuronidase (*UIDA* or *GUS*) that allows histochemical detection of transformed tissue. The npt II gene was isolated from transposon 5 of *Escherichia coli* K12 and encodes aminoglycoside 3-phosphotransferase II (APH) activity. Kanamycin, a bactericidal agent impairs the protein synthesis of plant mitochondria and chloroplasts through interaction with 30S ribosomal subunits. Thus, plant cells not expressing the npt II gene undergo chlorosis, while the enzyme in transformed tissue phosphorylates the kanamycin and prevents its interaction with the 30S ribosomal subunit, thereby preventing toxicity to cells (Dickie *et al.,* 1978).

The *GUS* reporter gene from *E. coli* encodes a glucuronidase that catalyzes the hydrolysis of various C-glucuronides. Indigogenic substrate X-gluc (5-bromo-4-chloro-3-indolyl b-D-glucuronic acid) upon hydrolysis forms 5-bromo-4-chloro-indoxyl and 5-bromo-4-chloro-y-indoxyl. These two soluble forms dimerize in the presence of oxygen to form blue-colored, insoluble compounds. Thus, the histochemical detection of GUS activity can be easily accomplished by incubating plant tissue with X-gluc (Jefferson *et al.,* 1987).

Selection of transformed tissue on a medium containing a herbicide such as glyphosate, phosphinothricin, or 2,4-dichlorophenoxyacetic acid are other possibilities, although at the present time the protocols for them are not well developed. There are two major methodologies that are well developed for cotton transformation, *Agrobacterium*-mediated transformation and particle bombardment.

2.2   *Agrobacterium tumefaciens*

The causative agent of many plant tumors, *i.e.,* crown gall disease, is the soil-borne bacterium *Agrobacterium tumefaciens.* The bacterium infects a wound site in a plant, and transfers a segment (T-DNA) of a plasmid (Ti) into a plant cell. The ability of *A. tumefaciens* to transfer DNA into plant cells can be exploited to transform plants with useful genes (Barton *et al.,* 1983; Fraley *et al.,* 1986) because any DNA segment is transferred so long as it is flanked on either side by the consequence recognized by the *A. tumefaciens* endonuclease/transferase. A gene conferring antibiotic or herbicide resistance is also incorporated into the T-DNA so that transformed cells can be selected from among the many non-transformed cells. Plant tissue is infected with the *A. tumefaciens* carrying the engineered T-DNA. Infected tissue is then treated with a selection agent to kill non-transformed cells, and the remaining transformed cells are regenerated into whole plants. The progeny of the transformed plant inherit the transgene. One of the major drawbacks of *Agrobacterium*-mediated transformation of cotton is that only a limited number of cultivars can be efficiently regenerated into fertile plants through tissue culture. Although *Agrobacterium* infects most varieties of cotton, the next step (regenerating cells from the callus) has proven to be recalcitrant in the majority of them. Only Coker varieties have been consistently regenerated (Trolinder and Goodin, 1987; Firoozabady *et al.,* 1987, Umbeck *et al.,* 1987), although Acala varieties are reported to be amenable to regeneration and transformation (Rajasekaran *et al.,* 1996).

2.3   **Particle Bombardment**

Rapid commercialization of genetically engineered cotton requires a reliable method of introducing genes directly into elite varieties of germplasm, circumventing the need for tissue culture, regeneration, and time-consuming plant breeding. Particle bombardment (gene gun), a method for physically transferring DNA through cell walls, meets the above objective. Klein *et al.* (1987) demonstrated that tungsten particles coated with DNA can be accelerated through *Allium cepa* (onion) cell walls, using a modified gun. The epidermal cells expressed the introduced genes. This method was refined by McCabe and colleagues (Agracetus Inc.,

Madison, Wis.) to introduce genes into a number of recalcitrant crops (McCabe *et al.,* 1988; Christou *et al.,* 1992; McCabe and Martinell, 1993). Gold particles (1 to 3 mm in diameter) were coated with DNA and accelerated toward plant tissues with steam generated by an electric discharge through a water droplet. The extent of particle penetration to specific cell layers (L1, L2, and L3) and damage to the tissue could be controlled, since the electric discharge gun allowed fine tuning of the discharge strength. The microparticle-carrying DNA is lodged in the cytoplasm by the propulsion. The DNA adhering to the microparticles is then taken up by the nucleus and if the DNA codes for a gene(s), the gene is transcribed by the cell within hours. Several thousand copies of each plasmid (gene) are resident on each microparticle. It appears that many of these copies contribute to transient expression. However, only a few copies are eventually integrated into the plant genome (Cooley *et al.,* 1995). Particle bombardment in the transformation of elite cotton cultivars has been very successful (McCabe and Martinell, 1993; John, 1995b; John, 1996). In short, the meristems of embryonic axes were bombarded with microparticles, containing plasmid DNA including a marker gene, *GUS.* After bombardment, the seed axes were allowed to develop for 2 to 3 weeks, and each leaf was tested for GUS activity by histochemical staining. Plants expressing GUS were mapped to identify nodes or axillary buds subtending the transformed leaves and pruned to induce the growth of the appropriate buds. The selective pruning resulted in plants that expressed *GUS* in each leaf (McCabe and Martinell, 1993).

A disadvantage of particle barbardment as a transformation method is the frequency with which tandem repeats of the inserted DNA occur. Multiple copies of a gene within the genome often results in gene silencing with the mistaken result that the percentage of transformation appears to be less than it actually is. In subsequent generations, if the tandem repeats have the capacity to segregate, expression may appear.

This phenomenon also occurs when multiple transformation events occur within the genome that are not tandem inserts. When multiple non-tandem transformation events occur it is possible to recover each single event through segregation in subsequent generations.

The particle acceleration mode of gene transfer in cotton results in two types of transformants, epidermal and germline. Various plant tissues arise from three physiological cell layers (L1, L2, and L3) of the meristem (Sussex, 1989). Selective pruning and forcing of the axillary buds cause a few transformed cells from L1, L2, or L3 layer to gain ascendancy and populate an entire tissue layer which normally arises from that cell layer (McCabe and Martinell, 1993). If the transformed tissue layer is responsible for germline, the transgenes will be passed on to the progeny. Those plants are referred to as germline transformants. Germline transformants are detected by histochemical staining for *GUS* in the vascular tissues and in pollen. In epidermal transformants, only the epidermal cell layers (Ll) contain the transgene and therefore plants possessing only its epidermal layer transformed do not pass on the transgene to their progenies.

The epidermal transformants are useful for gene screens. For example, a herbicide- or an insect-resistant gene can be inserted into cotton and epidermal transformants can be test-ed for corresponding resistance in leaf biomass. Similarly, the effects of fiber modification genes can be tested in the fibers of epidermal transformants, as fibers are of epidermal origin (John, 1996). The following cotton cultivars have been transformed by particle bombardment: DP 50, DP 90, C312, Suregrow 125, Suregrow 1001, Pima S6, Pima S7, Sea Island, Acala, and El Dorado. The heritability of the input genes was confirmed by Southern analysis of several generations. Each generation showed that the marker gene migrating as high molecular weight DNA when undigested genomic DNA was analyzed by gel electrophoresis sug-gesting integration of genes into cotton genome. The analy-sis of progenies from various generations indicated normal Mendelian inheritance (McCabe and Martinell, 1993; John, unpublished results). The transformation frequencies for cotton by particle bombardment is low. Only 0.1 to 0.2% of bombarded meristems are recovered as transgenic plants. However, the many positive attributes of this technology, summarized in Table 35-1, make it a very useful method of obtaining transformed cotton. Further technological re-finements, which would increase transformation frequency would broaden its applications to routine cotton transfor-mation in academic laboratories. Development of an an-tibiotic or herbicide selection regimen after particle bom-bardment may be useful in this regard (Chlan *et al.*, 1995). Particle bombardment of embryogenic tissue of Coker and Acala may also be useful in the generation of transgenic plants (Finer and Mullen, 1990; Rajasekaran *et al.*, 1996).

3.   THE SCOPE OF BIOTECHNOLOGY IN COTTON

Genetic engineering is likely to affect every facet of cot-ton crop cultivation, fiber, and seed processing. The fields that are under current considerations of genetic engineers fall under three overlapping categories: I) resistance to bio-logical/chemical/environmental factors; II) quality (fiber, oil, and protein meal); and III) yield. Published reports in-dicate ongoing investigations into the first two categories.

Table 35-1. Characteristics of cotton transformation methodologies.

Characteristics	Agro-bacterium	Protoplast gene transfer	Particle bombardment
Tissue culture/ regeneration/ somoclonal variation	Yes	Yes	No
Antibiotic selection	Yes	Yes	No
Transgenic plant quality	Variable	Variable	Consistent
Leaf for testing	2-6 months	2-6 months	3-4 months
Fiber for testing	14-16 months	14-16 months	6 months
R1 Homozygote	14-18 months	14-18 months	11-12 months
Multigene transfer	Limited	Not limited	Not limited
Efficiency	High	Limited data	Low
Commercialization	7-10 years	No data	5-6 years*

z * = based on estimation (no data).

3.1   Resistance to Biological, Chemical, and Environmental Factors

Broad spectrum properties including resistance to in-sect pests, viral and bacterial agents, herbicides, and tem-perature and water stress management are being inves-tigated. Improvements of these traits will result in better crop management, increased yield, reduced cultivation costs, and reduced impact of chemicals on the environment.

3.1.1   Insect Resistance

Phytophagous insects devour approximately 20 to 30% of total production by world farmers despite the $10 billion spent annually for crop protection including chemical insec-ticides (Oerke, 1994). Table 35-2 shows major cotton pests and the corresponding plant part upon which each group of pests inflicts damage. Thus, in addition to the traditional chemical insecticides, new technologies for insect manage-ment are warranted. One such new development is the trans-fer of insecticide principles from bacteria, plants, and ani-mals to crops through genetic engineering. Commercial suc-cess of this technology was demonstrated in cotton and other crops through the introduction of the Bt bacterial toxin gene.

Many bacterial species produce protein toxins that are effective against insects. For example *Bacillus thuringien-sis* Berliner produces several insecticidal metabolites. A protein toxin, C-exotoxin is found to be toxic to both insects and mice (Keieg, 1971). A protein secreted by the vegeta-tive cells, C-exotoxin (thuringensin) is active against both invertebrates as well as vertebrates (Sebesta *et al.*, 1981). The best known toxin from *Bacillus thuringiensis* is the C-endotoxin (Bt) that is produced as a protoxin. Based on their amino acid sequences and target pests, they are classi-fied into six basic groups (Table 35-3). The Bt protoxin (130 kDa) is insoluble under acidic conditions but is solubilized in the alkaline milieu of the insect gut. Proteases present in the gut cleave the protoxin to form active toxins (65 to 75 kDa). The lipids of the midgut epithelium membranes inter-act with hydrophobic and amphiphilic regions of the toxin resulting in the formation of channels through the mem-branes. Sodium and potassium ions pass through these channels creating an osmotic gradient and allowing water influx which ruptures the cells (Himeno *et al.*, 1985; Dronieski and Ellar, 1988; Gill *et al.*, 1992; English and Slatin, 1992). Deprived of nutri-tion, the insect larvae starve to death. The in-teraction of the Bt toxin and the membranes of the gut cells show unusual specificity. Table 35-3 summarizes the efficacy of vari-ous Bt toxin classes towards different pests. The Bt has been used as a safe and effective microbial insecticide for several decades.

The cloning and sequencing of one of the Bt genes enabled development of

Table 35-2. Common cotton pests (insects and nematodes).

Boll weevil	Adult	Boll, squares, boll, terminals
Cotton bollworm (corn earworm)	Larvae	Squares, young bolls
Tobacco budworm	Larvae	Squares, young bolls
Plant bugs (Tarnished, Clouded, Rapid, Cotton fleahopper)	Nymphs, adults	Stems, tender terminals, squares
Thrips (Western flower, Onion)	Larvae, adults	Seedling leaves & terminals
Aphids (Cotton, Greenpeach)		Leaves, bracts
Armyworms (Beet, Fall)	Larvae	Leaves, stems, squares, bolls
Pink bollworm	Larvae	Bolls
Mites (Two-spotted, strawberry)	Adults	Leaves
Loopers	Larvae	Leaves
Stinkbugs	Nymphs, adults	Young bolls
European corn borer	Larvae	Stems, bolls
Cutworms	Larvae	Seedling stalk (hypocotyl)
Leafcutter	Larvae	Leaves
Nematodes (Root-knot, Reniform, Lance)	Adults, juveniles	Roots

Table 35-3. Potential genes conferring pest resistance in cotton.

Toxin gene	Target pest	Reference
Bt vegetative insecticide protein (Vip3A)	Black cutworm, Fall armyworm, Beet armyworm, Tobacco budworm, Corn earworm	Estruch *et al.,* 1996
Bt Cry I	Lepidoptera	Feitelson *et al.,* 1992
Bt Cry II	Lepidoptera, Diptera	Feitelson *et al.,* 1992
Bt Cry III	Coleoptera	Feitelson *et al.,* 1992
Bt Cry IV	Diptera	Feitelson *et al.,* 1992
Bt Cry V, VI	Nematodes	Feitelson *et al.,* 1992
Polyphenol oxidase		Felton *et al.,* 1992
Proteinase inhibitors		Kanost & Jiang, 1996
C-amylase inhibitor		Shah *et al.,* 1995
Chitinase, (chitosanase)		Shah *et al.,* 1995
Lectins		Chrispeel & Raikhel, 1991
Neurotoxins		Becerril *et al.,* 1996
Cholesterol oxidase	Boll weevil	Purcell *et al.,* 1993

transgenic plants expressing Bt (Schnepf and Whiteley, 1981; Vaeck *et al.,* 1987; Fischhoff *et al.,* 1987). A number of gene modifications designed to optimize the nucleotide sequence for plant expression resulted in enhanced Bt synthesis in plants. These include changing the G/C ratio, adaptation of crop-specific codon usage, modification of mRNA secondary structures and truncations of the protein coding sequences. Ensuring post-transcriptional events such as proper protein folding and targeting to specific cellular organelle /compartment may enhance transgene expression (Perlak *et al.,* 1991; Koziel *et al.,* 1993; Barton and Miller, 1993; Koziel *et al.,* 1997). Transgenic cotton containing such a modified Bt gene (Bollgard™) was marketed by Monsanto in 1996. Approximately 1.6 million acres of Bt cotton were planted the first year it was available (Cotton Physiology Newsletter, 1996).

Generally the Bt cotton performed as expected in controlling tobacco budworm and pink bollworm. However, in a number of instances crop damage was observed due to cotton bollworm. Part of the problem may have been an unusually large bollworm population from previous year's corn crop (Damaske, 1997). A number of new Bollgard varieties along with herbicide-resistant varieties (Roundup Ready and Buctril) of cotton were grown in 1997 (Cotton Grower, 1997).

A number of other genes have been tested for their insecticidal efficacy. Proteins that interfere with the digestion and absorption of nutrition of insects have been identified and include polyphenol oxidase, proteinase inhibitors, lectins, and C-amylase inhibitors (Felton *et al.,* 1992; Hilder *et al.,* 1987; Huesing *et al.,* 1991). Their utility in insect resistance is somewhat limited by the requirement of high doses for efficacy and the requirement for chronic exposure. A member of the acyl sterol oxidases, cholesterol oxidase was shown to be active against boll weevil larvae (Purcell *et al.,* 1993; Corbin *et al.,* 1994; Greenplate *et al.,* 1995). The cholesterol oxidase oxidizes cholesterol from the midgut epithelium membranes to produce ketosteroids and hydrogen peroxide leading to membrane attenuation and cytolysis (Corbin *et al.,* 1995). Generally this mode of action is not specific to a particular group of insects, therefore has a broader range of affected insects, including beneficials, than the Bt toxins. Genes encoding neurotoxic peptides from scorpion venom have been isolated (Bougis *et al.,* 1989; Becerril *et al.,* 1996). Their use in transgenic plants poses questions regarding public acceptance as well as safety.

Proteinase inhibitors are a class of defensive proteins that inhibit enzymes in the predator's digestive system that are responsible for breaking proteins into amino acids. This slows feeding and further development of the insect is delayed or arrested. In China the cowpea trypsin inhibitor (CpTI) gene has been transformed into cotton to increase the effectiveness of the Bt gene (Zhao *et al.,* 1997).

Lectins are produced by many plants and possess a broad range of antimicrobial and insecticidal properties (Cavlieri *et al.,* 1995). The lectins bind to complex carbohydrates; each lectin tends to be specific to the carbohydrate to which it will bind. The insect toxicity of lectins relates to this ability to bind to carbohydrates. The midgut membranes of insects possess complex carbohydrates and when a lectin binds to these, absorption of nutrients by the insects is impaired, thereby inhibiting their growth. Rajgura *et al.* (1998) observed that Coker 312 plants grown from regenerated from tissue transformed with lectin genes, when fed to neonate larvae of *Heliothis virescence*, inhibited larvae growth.

3.1.2   Nematode Resistance

The best known genes that confer resistance to nematodes are Bt genes of the Cry5 or Cry6 type (Feitelson *et al.*, 1992). The mode of action of these toxins appears to be much the same as that of the Cry 1 group of Bt toxins in Lepidoptera (Marroqauim *et al.*, 2000), that is there are specific receptors for a portion of the toxin in the gut wall of the host. Although these genes hold potential for control of pathogenic nematodes, no commercial product is yet available in which resistance to pathogenic nematodes is based on genetic engineering. In cotton it is likely that natural sources of nematode resistance (Creech *et al.*, 1995; Yik and Birchfield, 1984) will be used rather than genetic engineering.

3.1.3   Microbial Pathogen Resistance

Microbial diseases cause large economic losses each year despite the extensive use of fungicides and antibiotics to reduce crop damage. Plant genetic engineering has the potential to reduce losses due to pathogens with efficacious genes to confer resistance. Because plants are stationary they must have endogenous mechanisms to defend against the many microbies that would otherwise attack the plants. Some transgenic strategies for plant disease resistance enhancement include expression of plant defense response pathway components (Cao *et al.*, 1998). A number of hydrolytic enzymes known as pathogenesis related (PR) proteins also have antmicrobial activity. Examples of these include chitinase (Punja, 2001), chitosanase (El Quakfaoui *et al.*, 1995; Stewart, unpublished), and C-glycosidase (Nakamura *et al.*, 1999). The mechanisms of action of the hydrolytic enzymes are thought to be related to their activity on the cell wall of invading fungi. However, the role of the hydrolytic enzymes may also be to enhance pathogenesis-related signal transduction. Theoretically the endogenous pathways can be exploited by genetically modifying plants to over-express an extra-cellular enzyme capable of degrading the fungal cell walls to produce small sugar compounds that stimulate a plant defense response that could ward off the invading fungi more rapidly as occurs in wildtype plants. Chitosan is one of the key structural elements found in many fungal cell walls and has been shown to be an effective elicitor of plant defense pathways when applied to tomatoes, tobacco and other plants. Exploitation of these pathways may be possible by genetically modifying plants to over-express an extra-cellular enzyme such as chitosanase capable of degrading the chitosan found in fungal cell walls and stimulating a plant defense response

that could ward off the invading fungi. Stewart and coworkers (unpublished) cloned a bacterial chitosanase, modified it for plant expression, added extracellular transport signal sequences and expressed it in *Nicotiana tabacum* 'Xanthi.' Genomic integration, transcription, and expression of active protein were confirmed with Southern blot, northern blot, and activity assays, respectively. Isolated polymeric fungal cell wall extracts were applied to T1-generation transgenic or wild type plants through cut stems. Pathogenesis-related responses were assessed via transcriptional induction of three phenylalanine ammonia lyase (pal) genes at 0, 2, or 4 hours after treatment application with quantitative real-time PCR. The relative expression values of two pal genes (AB008199, X78269) increased an average of 0.28 and 0.43 log2 expression units per hour after treatment application in the transgenic plants, whereas the expression changes in wild type plants were not different from zero after 1 hour (Hendrix and Stewart, unpublished results.)

A second group of antimicrobial genes code for small peptides that interact with the microbial membrane and result in sepsis that causes death of the cell microbe cell. A number of genes have been identified whose products have anti-microbial activity which are generally classified into a groups such as the magainins, cecropins, defensins, protegrins, and tachyplesins (Hancock and Lehrer,1998; Osusky *et al.*, 2000). The antimicrobial peptides of plants and animals are typically cationic (positively charged) amphipatic molecules composed of 12 to 45 amino acid residues. Structurally they are C-helical peptides [cecropins (Van Hofsten *et al.*, 1985) and magainins (Zasloff, 1987)], or C-sheet peptides [defensins, protegrins, and tachyplesins (Hancock and Lehrer, 1998; Osusky *et al.*, 2000)]. They generally act in concert to form pores in the membranes (Ludtke *et al.*, 1996). Although work on these compounds is progressing, no cotton containing any of these genes is commercially available.

3.1.4   Environmental Stress Tolerance

Since tolerance to environmental stress is thought to be multigenic, it is unlikely that a single gene will confer much resistance. However, this has not stopped several research groups from attempting to identify such genes. Superoxide dismutase (SOD) is one such gene (Gupta *et al.*, 1993) that has been isolated and transformed into cotton (Allen, 1995). The mode of action of SOD is to remove any free radical oxygen species that may are generated by environmental stress so that functioning membranes (*e.g.*, chloroplast membranes) are not damaged and continue to function during the stress.

Several other genes have been tested in model systems and have been found to enhance resistance to some abiotic stress. Usually these genes are involved in the production of osmoprotectants. Examples of such genes include those involved in sunthesis of glycinebetaine, such as choline oxidase to enhance resistance to cold in (Alia *et al.*, 1998a) heat (Alia *et al.*, 1998b) or salt stress in *Arabidopsis* (Hayashi *et al.*, 1997); trehalose-6-phosphate synthase for overproduction of trehalose in tobacco to enhance drought tolerance (Holmstrom *et al.,* 1996; Romero *et al.,* 1998); and mannitol-1-phosphate dehydrogenase for overproduction of polyols to enhance salinity tolerance in tobacco (Tarczynski *et al.,* 1993) and *Arabidopsis* (Thomas *et al.,* 1995). Certain transcription factors known to be regulated by abiotic stress, particularly low temperature, have been genetically engineered with constitutive promoters to regulate the endogenous stress tolerance mechanisms. This includes the Cold-induced Binding Factors (CBF) of Arabidopsis (Jaglo-Ottosen *et al.,* 1998) among others. While increased tolerance to the abiotic stress being investigated was observed in certain test plant species in others there was no effect. Thus, there is no guarantee that the engineered gene would be effective in cotton. Probably the positive effects occur when the tolerance mechanism being tested is a mechanism the plant already has and the engineered plants have enhanced expression of the genes involved in the mechanism.

3.1.5 Herbicide Tolerance

Although cotton growers in the U.S. spend more than $200 million per year on herbicides, an estimated $600 million worth of crops are lost due to ineffective weed control. New strategies based on genetically engineered cotton to confer resistance to herbicides that are effective against broadleaf weeds have been developed. Examples are 2,4-dichlorophenoxyacetic acid (2,4-D); bromoxynil and glyphosate resistant cotton. 2,4-D is used primarily to control dicot weeds in grass crops, pastures, forests, and lawns. 2,4-D resistant cotton was developed to reduce damage by drift-levels of application to nearby monocotyledonous crops that are naturally resistant to this herbicide (*e.g.,* rice, sorghum). Genetically engineered broadleaf herbicide protection for 2,4-D in cotton was obtained by incorporating an *Alcaligenes eutrophus* gene encoding 2,4-D monooxygenase (tfdA). Transformants, containing tfdA exhibited 50- to 100-fold more tolerance to 2,4-D compared to nontransformed controls (Bayley *et al.,* 1992). Bromoxynil is a post emergence benzonitrile herbicide that controls cotton weeds such as morningglory, cocklebur, and velvetleaf. It is an inhibitor of electron transport in the photosystem II of chloroplasts. A gene encoding bromoxynil nitrilase from *Klebsiella ozaenaen* was cloned and introduced into tobacco and cotton (Stalker *et al.,* 1996). Nitrilase acts on nitrils present in bromoxynil, thereby inactivating the herbicide. The bromoxynil resistant cotton received regulatory approval and was marketed in the 1998 production season.

Glyphosate is a slow-acting, non-selective, post-emergence herbicide. It is translocated from the site of uptake to metabolic sinks, such as meristems, leaf buds, and storage organs. It inhibits 5-enolpyruvylshikimate-3-phosphate synthase (EPSPS) of the shikimate pathway involved in the biosynthesis of the aromatic amino acids phenylalanine, tryptophan, and tyrosine. Resistance to glyphosate in plants was achieved by overproducing EPSP

synthase (Shah *et al.,* 1986). Additionally, a mutated bacterial EPSPS that is insensitive to glyphosate was also shown to be resistant to the herbicide. The glyphosate resistant cotton has probably been one of the most rapidly adopted technologies that has ever occurred in agriculture.

Several groups of herbicides target the inhibition of amino-acid biosynthesis in plants. Triazolopyrimidine sulfonanilides, sulfonylureas, and imidazolinones inhibit branched-chain amino acid synthesis. These herbicides act on acetolactate synthetase (ALS or AHAS) which is the first enzyme in the biosynthetic pathway of branched-chain aminoacids, valine, leucine, and isoleucine. Its inhibition results in toxic accumulation of an AHAS substrate, C-ketobutyrate and termination of protein synthesis (Ray, 1984). The ALS gene family in cotton has been characterized (Grula *et al.,* 1995), and a mutant form of the cotton AHAS gene was shown to confer resistance to both imidazolinone and sulfonylreas in Coker and Acala varieties of cotton (Rajasekaran *et al.,* 1996). A mutant tobacco ALS (csr) also has been introduced to cotton to generate sulfonylurea-resistance (Saari and Mauvais, 1996). One of the problems associated with the commercial use of transgenic plants, especially those of herbicide resistant varieties, is the potential for transgene inactivation in the field. Studies in transgenic tobacco carrying *A. thaliana* (acetohydroxyacid synthase: AHAS) csr-1-1 gene showed that the activities of the transgene, as well as the endogenous AHAS genes, surA and surB were suppressed (Brandle *et al.,* 1995). Also, resistance to a herbicide that depends upon a single nucleotide mutation generally has a short life span because of the strong in-field selection pressure among targeted weeds for the mutation.

Glufosinate is a non-selective herbicide that is toxicologically and environmentally benign. *Streptomyces* spp. produce bialaphos, a natural analog of glufosinate, an antibiotic consisting of two alanine residues linked to phosphinothricin (PPT), a glutamic acid analog that inhibits glutamine synthase. Glutamine synthetase is critical for the detoxification of ammonia in plants (Miflin and Lea, 1977). Another possible mechanism of bialaphos action in plants is glyoxylate toxicity. Glyoxylate accumulates in plants treated with glufosinate due to the reduced levels of amino acids required for the photorespiratory glyoxylate transamination (Devine *et al.,* 1993). A gene conferring resistance to bialaphos (bar) was identified from *S. hygroscopicus.* It encodes a phosphinothricin acetyltransferase (PAT) that acetylates the free NH_2 group of PPT, thereby preventing autotoxicity. A second gene (pat) from *S. viridochromogenes* with significant sequence homology has also been cloned and introduced into tobacco and carrots (Strauch, 1988; Droge *et al.,* 1992). The bar gene was introduced into DP50, Coker 312, El Dorado, and Pima S6 by particle bombardment and transgenic lines were recovered (Keller *et al.,* unpublished). Herbicide (BastaR, Hoechst AG, Frankfurt) tolerance up to 15,000 ppm was demonstrated in greenhouse trials. The basta resistance was inherited as a Mendelian fashion in subsequent generations.

3.2 Quality

About 66% of world cotton fiber is utilized by the apparel industry. The remaining portion is used for home furnishing, floor covering, and medical and industrial applications (Industry study, 1992). Thus, properties impacting textile applications is a major focus of technological improvements. Some traits, such as strength, length, uniformity, and yield are amenable to improvement through plant breeding; whereas, improvements in thermal absorbency, dye binding, or fire retardancy may be brought about only through genetic engineering. The area of fiber improvements through genetic engineering is in its infancy. The first report of fiber modification appeared only as recently as 1996. Thus, the following is a brief review of technology that has been developed for fiber modifications along with a few examples of application. First, a brief description of the fiber traits important for textile applications seems appropriate (See Chapters 4 and 21 for detailed discussion). This is followed by descriptions of specific examples of genetic engineering targeted towards modification of fiber traits.

3.2.1 Fiber Properties

Cotton fiber is a differentiated single epidermal cell of the outer integument of the ovule. Strength, length, micronaire, uniformity, and maturity of cotton fiber are some of the most important physical properties necessary for textile applications. Fiber strength is a factor contributing to yarn and fabric strength (Table 35-4). Strength is determined (70.9%) by the genetics of the plant, but is also influenced by environmental factors (Meredith, 1986). In addition to the obvious advantage of superior product performance, increased fiber strength also contributes to increased spinning efficiency. Fiber strength is also important for manufacturing wrinkle-free apparel, where chemical treatment results in the reduction of fabric strength (Sasser, 1992). Superior fiber strength is an objective for both biotechnology and conventional breeding programs. The fiber strength has shown steady increase of 0.25 grams-force/tex per year since 1980 as reported by U.S. Department of Agriculture (Sasser, 1992). However at the present time, there are no reports of breakthroughs in increasing fiber strength through genetic engineering.

Fiber length, another important quality parameter, is measured by High Volume Instrument (HVI) or Stelometer in terms of "mean" or "upper half mean" (UHM) lengths. The "mean" is the average length of all the fibers, and UHM is the average of the length of the longer half of the fiber distribution.

The micronaire value of fibers denotes the fiber fineness (linear density) and maturity (wall thickness). Micronaire is measured by compressing a plug of fibers to a fixed volume and subjecting it to air flow. The resistance to the air flow is read on a scale calibrated in micronaire units. The environment has a greater influence on the micronaire of fiber, than on strength (Meredith, 1986). Cotton maturity, the total cell wall thickness related to the diameter or width of the fiber is important in dye uptake.

3.2.2 Opportunities for Fiber Modification through Genetic Engineering

In textile applications, it is essential to preserve the strength, length, micronaire, and surface properties of fiber while enhancing other attributes such as dye binding, wrinkle and shrinkage resistance. However, in certain product applications, for example, super absorbent fibers, it may not be necessary to retain all of the original fiber properties.

A number of different genetic strategies are obvious for making modifications of the physical and chemical properties of cotton fiber. Identification of genes responsible for strength, length, etc. may lead to superior fibers. Unfortunately, the genes responsible for these and other fiber traits are unknown at this time. Thus, characterizations of genes expressed in fiber and their correlation to any of the fiber properties is a first step towards genetic engineering (John and Stewart, 1992; Triplett, 1992; John, 1996). A second plausible strategy is to select known genes from other sources (animals, plants, bacteria, or fungi) that may confer a new trait to fiber cotton (John, 1994a & 1994b). For example presence of a new enzyme in fiber may catalyze formation of a new polymer (John and Keller, 1996).

3.2.3 Examples of Fiber Modifications

Genetic engineering for fiber modification requires genes that confer a desired trait to the fibers, promoters and other regulatory elements to enable or control express the genes in the developing fibers, and a methodology to introduce the genes and promoters into a cotton plant. The transformation methods useful for cotton were discussed earlier. A brief over-view of promoter elements that are known to be useful in fiber expression follows.

3.2.3.1 Promoters for Expression of Genes in Fibers

Promoters are DNA sequences that determine the time, tissue-specificity, and level of gene expression. Genetic engineering of cotton for agronomic or product traits requires different promoters. For example, for herbicide or insect resistance a constitutive promoter is required. Whereas, fiber modification may require only fiber-specific expression. The most widely used constitutive promoter in plant genet-

Table 35-4. Cultivar differences in fiber properties (Stelometer).

Cultivar	Strength	Length	Micronaire
Coker 312			
DP50	18.3±0.83	1.16±0.043	4.1±0.40
DP90	21.83±0.93	1.15±0.040	3.85±0.30
SG125			
Pima S6	26.9±1.3	1.22±0.04	3.6±0.24
Sea Island	27.8±1.8	1.5±0.08	2.9±0.13

ic engineering is the cauliflower mosaic virus 35S promoter (Ow *et al.,* 1986). This promoter directs a high level of gene expression in many plant tissues. A transcriptional enhancer situated upstream of the TATA box (RNA polymerase binding site essential for gene transcription) can be duplicated to increase the gene expression further (Kay *et al.,* 1987).

Targeting the expression of foreign genes to the tissue of interest has a number of advantages. For example, it avoids detrimental effects expression of the transgene may have on normal growth and morphology of the plant. A large numbers of plant promoters that exhibit predominant expression in certain tissues have been isolated. They are referred to as tissue-specific, although in many instances low levels of expression can be detected in other tissues. Examples include promoters that direct expression of a gene only in the anthers (Paul *et al.,* 1992), pollen (Twell *et al.,* 1991; John and Peterson, 1994), or seeds (Roberts *et al.,* 1989).

Genes expressed (as mRNA) predominantly in cotton fiber are identified through differential and subtractive cDNA library screens (John and Crow, 1992). The corresponding genes for fiber-specific mRNAs were then isolated from genomic libraries. DNA fragments containing putative promoters can be identified from nucleotide sequence comparison between cDNA and genomic clones. Table 35-5 lists some of the promoters that are shown to be useful regulating gene expression in fibers.

3.2.3.2 Modification of Thermal Properties

The synthetic fiber industry developed a value-added product, the fiber with two components. Bicomponent fibers contain a core polymer surrounded by a sheath polymer. The properties of the two polymers are combined and, in some instances are a manufacturing necessity (Bach *et al.,* 1990). This technical innovation can be duplicated in cotton fiber. If the cotton plant can be genetically engineered to produce a second useful polymer in the fiber lumen, such a fiber would have the properties of cellulose along with that of the new polymer.

This hypothesis was tested by genetically engineering cotton to produce an aliphatic polyester compound, polyhydroxybutyrate (PHB) in the fiber (John and Keller, 1996). PHB is a polyhydroxyalkanoate (PHA) which is a natural biodegradable thermoplastic with chemical and physical properties similar to polypropylene (Steinbuchel, 1991). PHAs are produced by many genera of bacteria as inclusion bodies to serve as reserve carbon sources and as electron sinks. The formation of PHB in bacteria involves three enzymes, C-ketothiolase, NADPH-dependent acetoacetyl-CoA reductase, and PHA synthase. Two molecules of acetyl-CoA are joined by C-ketothioase to form acetoacetyl-CoA. Acetoacetyl CoA is reduced by acetoacetyl-CoA reductase to R-(-)- 3-hydroxybutyryl-CoA. This activated monomer is then polymerized by PHA synthase to form PHB. It has been shown that PHB can be synthesized in transgenic plants (Poirier *et al.,* 1992). C-ketothiolase, is involved in the synthesis of mevalonate, and is ubiquitous in plants. Therefore, the production of PHB in plants requires only two additional enzymatic activities, acetoacetyl Co-A reductase and PHA synthase. These genes were re-engineered for expression in fiber by linking them to fiber-specific promoters and introduced into cotton by particle bombardment. Transformants were selected based on the expression of the marker gene, GUS. Transgenic fibers were examined by a number of analytical methods to determine the presence of PHB. These studies are summarized in Table 35-6.

As expected the presence of PHB in fibers resulted in measurable changes in its physical and chemical properties. For example, the rate of heat uptake and cooling was slower and heat uptake capacity was higher for transgenic fibers than controls. These results suggest that the transgenic fibers containing PHB have better insulating properties, and therefore may be suitable for winter wear and other textile applications where superior insulating properties are advantageous (John and Keller, 1996). These studies point out the potential for fiber modifications through genetic engineering.

Table 35-5. Promoters Useful in Transgene Expression in Fiber.

Promoter	Tissue-specificity	Transgenic system	Reference
Cotton E6	Fiber	Carrot extensin, phaB, iaaM, iaaH in cotton	John &Crow, 1992, John & Keller, 1996
Cotton FbL2A	Fiber	phaB & phaC in cotton	Reinehart *et al.,* 1996
Cauliflower 35s	All tissues	phaC in cotton	John & Keller, 1996
Cotton B8 root	Fiber, stem,	phaB in cotton	John, 1996. Patent #5521078 & unpublished data
Cotton Gh-1	Fiber*	GUS: tobacco and cotton	Dang *et al.,* 1995 & personal communication
Cotton H6	Fiber	phaB in cotton	John, 1996. Patent#5521078 & unpublished data
Arabidopsis cab	All tissues	GUS in cotton	John unpublished data
Cotton 4-4	Fiber	Indigo gene /tryptophanase	McBride *et al.,*1996. Patent application #W09640924
Cotton Rac13	Fiber		McBride *et al.,* 1996 Patent application #W09640924
Cotton LTP	Fiber		McBride *et al.,*1996 Patent application #W09640924
Kapok E6	Fiber		John, 1996; Patent #5521078 & unpublished data

Table 35-6. Evidence for the synthesis of new biopolymer in transgenic fibers.

Epifluorescence microscopy	Fluorescent granules in fiber cytoplasm
Transmission electron microscopy	Electron lucent granules in fiber cytoplasm (0.1 to 0.2 um)
HPLC	Chloroform extract contains Crotonic acid
GC	Peak corresponding to ethylester of C-hydroxyacid
MS	Mass fragmentation pattern of ethylester derivative of trans-genic fiber extract is identical to reference compound, ethylester hydroxybutyrate
Gel permeation chromatography & HPLC	Sixty-eight % of PHB granules are 0.6x106 Da or more
TGA, DSC	Heat absorption/dissipation properties different than control fibers

3.2.3.3   Manipulating Hormone Levels in Developing Fibers

Various plant growth regulators such as giberellins, auxins, cytokinins, abscisic acid, and ethylene influence fiber development (reviewed in Kosmidou-Dimitropoulou, 1986). Indoleacetic acid (auxins: IAA) is one of the several phytohormones that collectively regulate growth and differentiation of plant cells. Shoot gravitropism, maintenance of apical dominance, and differentiation of vascular tissue were shown to be dependent on IAA (Tamas, 1987; Aloni, 1988). Higher levels of auxins in relation to giberellins was postulated to favor fiber elongation and smaller nucleoli (Beasley and Ting, 1973; Popova *et al.,* 1979; Kosmidou,1976). Fibers from different cultivars differ in their strength, length, and micronaire. In general, the *G. bar-badense* varieties have superior fiber properties compared to the upland cultivars (*G. hirsutum*) as shown in Table 35-4.

The levels of free IAA in developing fibers of DP50, Pima S6, and Sea Island measured by solvent partitioning and high-performance liquid chromatography, followed by gas chromatography-mass spectrometry using selected-ion monitoring (Li *et al.,* 1992). Isotope dilution was used to correct for incomplete recovery. The *G. barbadense* varieties appears to have higher levels of free IAA during fiber development (John, 1997). The effect of higher levels of free IAA on fiber properties of DP 50 was tested by incorporating Agrobacterium genes, iaaM and iaaH, necessary for auxin synthesis into DP50. The iaaM gene encodes a tryptophan monooxygenase that converts tryptophan to in-dole-3-acetamide (IAM). This is the first step in the two-step IAA biosynthesis pathway. IAM is further hydrolyzed to IAA by indoleacetamide hydrolase (iaaH). A number of investigators have shown that plants transformed with iaaM and iaaH genes expressed high levels of IAA (Klee *et al.,* 1984; Sitbon *et al.,* 1991). In order to avoid any growth abnormalities by the constitutive production of auxins, the iaaH and iaaM genes were linked to fiber-specific E6 promoter and introduced into DP50 by particle bombardment (John, 1994). Several transformants were identified by GUS screening. All of the transformants showed normal morphology and growth. This was expected as the E6 promoter is active only in fiber cells. The engineered fibers contained 2- to 8-fold higher levels of IAA. However, the three fiber properties measured, length, strength, and micronaire showed no difference in comparison to DP50 fibers (not shown).

Naithani *et al.* (1982) measured the auxin levels in developing fibers from three Indian cotton cultivars. No positive relationship was found between auxin and rate of elongation. Similarly, Nayyar *et al.* (1989) found no relationship between IAA levels and extent of fiber elongation in *G. arboreum*. These results are in agreement with our transgenic cotton results that showed no discernible effect of increased free IAA content in DP50.

A second set of experiments were conducted by introducing another hormone gene, isopentenyl transferase (ipt) into cotton. The *A. tumefaciens* ipt gene is involved in the biosynthesis of cytokinin and catalyzes the condensation of AMP and isopentenyl pyrophosphate to form isopentenyl AMP. AMP is the precursor for all other cytokinins (Akiyoshi *et al.,* 1984). Over-expression of ipt in transgenic plants produced frequent incited shoots, suppression of root formation, increased growth of auxiliary buds and short intense green plants (Smigocki *et al.,* 1993; Klee *et. al.,* 1987; Smigocki, 1991). The ipt gene was linked to the 35S constitutive promoter, and introduced into DP50 plants. Transformants were selected based on GUS expression. The transformants exhibited a range of phenotypes that included bushy, short plants with large numbers of auxiliary branches. The leaves were dark green in color. Some of the transformants were severely stunted and did not produce flowers (not shown). We did not measure cytokinin levels in these plants, but only made observations regarding their phenotypes. However, plants that produced flowers and boll exhibited no changes in fiber traits (unpublished data).

4.   COLORED COTTON

Dyeability for color is a critical attribute of cotton fiber and is critical for the textile industry. However, the dyeing process also increases cost and has negative environmental implications due to toxic waste. Thus, naturally colored cotton is attractive to the industry as well as to consumers. Naturally colored (mocha, brown, red, and green) cotton varieties are being grown by various private companies and growers both in this country as well as abroad. More importantly, genetic engineering of cotton to produce colored fibers has received much attention. Genes responsible for production of melanins as well as indigo have been inserted into cotton (McBride *et al.,* 1996). Melanins are dark brown pigments found in microorganisms, plants, and animals that arise from the action of tyrosinase on tyrosine. A second gene (ORF 438) is also involved in melanin formation. These genes were cloned from *Streptomyces antibioticus* (Bernan *et al.,* 1985; della-Cioppa *et al.,* 1990), and expression of the genes in cotton resulted in some color formation (McBride *et al.,* 1996), although the color intensity was not sufficient for commercial use.

Indigo is generated from indole by naphthalene dioxygenase or by xylene oxygenase from toluene to form indole dihydrodiol which then was converted to indoxyl. Air oxidation of indoxyl results in the formation of indigo (Ensley *et al.,* 1983; Keil *et al.,* 1987; Suen *et al.,* 1993). Tryptophanase efficiently converted tryptophan to indole (Deeley and Yanofsky, 1981). Hart *et al.* (1992) and Murdock *et al.* (1993), through metabolic pathway engineering involving 9 genes, showed that recombinant *E. coli* could synthesize indigo from glucose. Two genes, tryptophanase (tna) and indole oxygenase (pig), were introduced into cotton using fiber-specific promoters. A faint blue color was observed in transgenic fibers expressing tna and pig (McBride and Stalker, 1996). These results imply the potential of producing colored fibers through genetic engineering.

Specifics of other research programs on fiber modifications in corporate laboratories are not known. But announcements from industry, government, and public organizations clearly suggest that research towards improving fiber traits is a priority goal in cotton (John, 1994b; Clune, 1994; Kiser,

1995). In this regard, the genetic tools necessary to engineer cotton plants with better fiber traits are now available.

5.   OIL AND PROTEIN MEAL

Genetic engineering with the objective of modifying cotton oil is subject only to the economic value gained by genetically engineering the oil. The genes for modifying plant oils have been isolated (Huang *et al.,* 2001). Currently there is no economic incentive to modify cotton oil. This is especially true since the oil is consumed by humans, thus the cost of meeting regulatory requirements are probably too high to justify the expense.

The same might be said for modification of the protein cake from cotton seeds. Currently a very high percentage of cottonseed is fed to cattle because of its high caloric content. Obviously much of the seed that is fed to cattle is transgenic but the transgenes and the proteins they encode are deemed to be innocuous to the cattle or to humans consuming the milk or meat of the animals which consume the cottonseeds. To genetically engineer the cotton seed for enhanced protein quality would require direct human consumption of the protein. The economics and regulatory costs of such an endeavor are prohibitive.

BIBLIOGRAPHY

Aarssen, L.W. 1995. Hypotheses for the evolution of apical dominance in plants: Implications for the interpretation of overcompensation. Oikos 74:149-156.

Aarssen, L.W. and D.L. Irwin. 1991. What selection: herbivory or competition? Oikos 60:261-262.

Abaye, O., J. Bryant, D. Oosterhuis, C. Bednarz, and M. Holman. 2000. Characterization of the fruiting growth curve: A Regional Report. p. 678. *In:* P. Dugger and D. Richter (eds.). Proc. Beltwide Cotton Conf., National Cotton Council of America, Memphis, Tenn.

Abaye, A.O., J.C. Maitland, and W.B. Wilkinson. 1995. Influence of cover crops and tillage practices on yield and quality of cotton. p. 1349 *In:* Proc. Beltwide Cotton Conf., National Cotton Council of American, Memphis, Tenn.

Abbott, A.J. 1967. Physiological effects of micronutrient deficiencies in isolated roots of *Lycopersicon esculentum*. New Phytol. 66:419-437.

Abbott, L.K. and A.D. Robson. 1985. Formation of external hyphae in soil by four species of vesicular-arbuscular mycorrhizal fungi. New Phytol. 99:245-255.

Abd-Ella, M.K.A. and E.E. Shalaby. 1993. Cotton response to salinity and different potassium/sodium ratio in irrigation water. J. Agron. Crop Sci. 170:25-31.

Abeles, F.B. 1969. Abscission: Role of cellulase. Plant Physiol. 44:447-452.

Abeles, F.B., P.W. Morgan, and M.E. Saltveit. 1992. Ethylene in Plant Biology, 2nd ed., New York.

Abraham, K.J., M.L. Pierce, and M. Essenberg. 1999. The phytoalexins desoxyhemigossypol and hemigossypol are elicited by *Xanthomonas* in *Gossypium* cotyledons. Phytochemistry 52: 829-836.

Abu-Tarboush, H.M. and S.A.B. Ahmed. 1996. Studies on karkade (*Hibiscus sabdariffa*): Protease inhibitors, phytate, in vitro protein digestibility and gossypol content. Food Chern. 56:15-19.

Abul-Naas, A.A. and M.S. Omran. 1974. Salt tolerance of seventeen cotton cultivars during germination and early seedling development. Z. Ack. Pflanzenbau 140:229-236.

Abdullah, M.M. 1985. Effect of drought conditions at different stages of growth on phenol and carbohydrate content of cotton leaves. Z. Acker. Pflanzenbau. 155:246-252.

Abdullah, Z. and R. Ahmad. 1986. Salinity induced changes in the reproductive physiology of cotton plants. *In:* R. Ahmad and A. San Pietro (eds.). Prospects for Biosaline Research. Dept. Botany, Univ. of Karachi, Pakistan. pp. 125-137.

Acevedo, E., E. Fereres, T.C. Hsiao, and D.W. Henderson. 1979. Diurnal growth trends, water potential, and osmotic adjustment in maize and sorghum leaves in the field. Plant Physiol. 64:476-480.

Acevedo, E. and T.C. Hsiao. 1974. Plant responses to water deficit, water use efficiency and drought resistance. Agri. Meteorol. 14:59-84.

Ackerson, R.C. 1981. Osmoregulation in cotton in response to water stress: II. Leaf carbohydrate status in relation to osmotic adjustment. Plant Physiol. 67:489-493.

Ackerson, R.C. 1984. Regulation of soybean embryogenesis by abscisic acid. J. Exp. Bot. 35:403-413.

Ackerson, R.C. 1985. Osmoregulation in cotton in response to water stress. III. Effects of phosphorus fertility. Plant Physiol. 77:309-312.

Ackerson, R.C. and R.R. Herbert. 1981. Osmoregulation in cotton in response to water stress: I. Alterations in photosynthesis, leaf conductance, translocation, and ultrastructure. Plant Physiol. 67:484-488.

Ackerson, R.C., D.R. Krieg, C.L. Haring, and N. Chang. 1977a. Effects of plant water status on stomatal activity, photosynthesis, and nitrate reductase activity of field grown cotton. Crop Sci. 17:81-84.

Ackerson, R.C., D.R. Kreig, and R.E. Zartman. 1977. Water relations of field-grown cotton and sorghum: Temporal and diurnal changes in leaf water, osmotic, and turgor potential. Crop Sci. 17:76-80.

Acock, B. 1990. Effects of carbon dioxide on photosynthesis, plant growth, and other processes. pp. 45-60. *In:* B.A. Kimball (ed.). Impact of Carbon Dioxide, Trace Gases, and Climate Change on Global Agriculture. ASA Special Publication no. 53, Madison, WI.

Acock, B. 1991. Modeling canopy photosynthetic responses to carbon dioxide, light interception, temperature, and leaf traits. pp. 41-55. *In:* K.J. Boote and R.L. Loomis (eds.). Modeling Canopy Photosynthesis - From Biochemistry to Canopy. CSSA Special Publication No. 19, Madison, WI.

Acock, B. and M.C. Acock. 1989. Calculating air leakage rates in controlled-environment chambers containing plants. Agron. J. 81:619-623.

Adam, G. and V. Marquardt. 1986. Brassinosteroids. Phytochem. 25:1787-1799.

Adams, F. and Z.F. Lund. 1966. Effect of chemical activity of soil solution aluminum on cotton root penetration of acid subsoils. Soil Sci. 101 193-1 98.

Adams, F. and J.I. Wear. 1957. Manganese toxicity and soil acidity in relation to crinkle leaf of cotton. Soil Sci. Soc. Am. Proc. 21:305-308.

Adams, R.M. and T.D. Crocker. 1988. Model requirements for economic evaluation of pollution impacts upon agriculture. *In:* W.W. Heck, O.C. Taylor, and D.T. Tingey (eds.). Assessment of Crop Loss from Air Pollutants. Elsevier Appl. Sci., London. pp. 463-471.

Adams, R.M., J.D. Glyer, and B.A. McCarl. 1988. The NCLAN economic assessment: approach, findings and implications. *In:* W.W. Heck, O.C. Taylor, and D.T. Tingey (eds.). Assessment of Crop Loss from Air Pollutants. Elsevier Appl. Sci., London. pp. 473-504.

Adams, R.M., S.A. Hamilton, and B.A. McCarl. 1986. The benefits of pollution control: The case of ozone and U.S. agriculture. Am. J. Agric. Econ. 68:886-893.

Adams, R.M., C. Rosenzweig, R.M. Peart, J.T. Ritchie, B.A. McCarl, J.D. Glyer, R.B. Curry, J.M. Jones, K.J. Boote, and L.H. Allen, Jr. 1990. Global climate change and U.S. agriculture. Nature 345:219-224.

Afek, U., J.A. Menge, and E.L.V. Johnson. 1991. Interaction among mycorrhizae, soil solarization, metalaxyl, and plants in the field. Plant disease. 75:665-671.

Afzal, M. 1964. Sodium chloride as fertilizer. W. Pakistan Agric. Res. 2:111-112.

Agrawal, A.A. and R. Karban. 1997. Domatia mediate plant-arthropod mutualism. Nature 387:562-563.

Agarwala, S.C., C. Chatterjee, P.N. Sharma, C.P. Sharma, and N. Nautiyal. 1979. Pollen development in maize plants subjected to molybdenum deficiency. Can. J. Bot. 57:1946-1950.

Aguillard, W., D.J. Boquet, and P.E. Schilling. 1980. Effects of planting dates and cultivars on cotton yield, lint percentage, and fiber quality. Louisiana State University Agric. Expt. Sta. Bulletin No. 727. 23 pp.

Ahern, A.B. 1986. Effect of preceding crop on the nitrogen requirement of irrigated cotton (*Gossypium hirsutum* L.) on a Vertisol. Field Crops Res. 13:159-175.

Ahmad, M. and M.I. Makhdum. 1992. Effects of salinity-sodicity on different phases of cotton plant, its fibre quality and oil contents - A review. Agricultural Reviews 13:107-118.

Ahmad, N., S. Ahmad, and M.A. Iqbal. 1990. Genotypic cotton response to salinity. J. of Agric. Res. Lahore 28:39-45.

Ahmad, N. and G.R. Chaudhry. 1988. Irrigated Agriculture of Pakistan. Shahzad Nazir, Lahore 810 pp.

Ahmad, R. and Z. Abdullah. 1980. Biomass production of food and fiber crops using highly saline water under desert conditions. *In:* A. San Peitro (ed.). Biosaline Research. Plenum Press, New York. pp. 149-163.

Ahmad, M., A. Rauf, and M.I. Makhdum. 1991. Growth performance of cotton under saline-sodic field conditions. J. Drainage and Reclamation 3:43-47.

Ahmed, F.M. 1994. Effect of saline water irrigation at different stages of growth on cotton plant. Assiut J. of Agric. Sci. 25:63-74.

Aiken, R.M. and A.J.M. Smucker. 1996. Root system regulation of whole plant growth. Annu. Rev. Phytopathol. 34:325-326.

Aikman, D.P. 1989. Potential increase in photosynthetic efficiency from the redistribution of solar radiation in a crop. J. Exp. Bot. 40:855-864.

Akhila, A. and K. Rani. 1993. Biosynthesis of gossypol in *Thespesia populnea.* Phytochemistry 33:335-340.

Aladin, N.V., I.S. Plotnikov, and W.T.W. Potts. 1995. The Aral Sea desiccation and possible ways of rehabilitating and conserving its Northern part. Environmetrics 6:17-29.

Aladin, N.V. and W.T.W. Potts. 1992. Changes in the Aral Sea ecosystems during the period 1960-1990. Hydrobiologia 237:67-79.

Alam, T.M., M. Rosay, L. Deck, and R. Royer. 1994. Complete assignment of the ¹H and ¹³C NMR spectra of desoxyhemigossypol derivatives. Magn. Reson. Chem. 32:561-563.

Albers, D.W. and J.T. Cothren. 1981. Electrical conductivity, ion leakage, and subsequent germination studies on cotton seed treated with mepiquat chloride. p. 23. *In:* W.C. Wilson (ed.). Proc. Eighth Ann. Plant Growth Regul. Soc. Conf., Plant Growth Regulator Soc. Am., Lake Alfred, FL.

Albers, D.W. and J.T. Cothren. 1983. Influence of seed source and mepiquat chloride on cotton seed emergence. pp. 38-39. *In:* J.M. Brown (ed.). Proc. Beltwide Cotton Prod. Res. Conf., National Cotton Council of America, Memphis, Tenn.

Alchanati, I., C.R. Benedict, and R.D. Stipanovic. 1994. The enzymatic conversion of desoxyhemigossypol to desoxy methyl hemigossypol in cotton stems: dHG-O-methyltransferase. pp. 35-39. *In:* 1994 Proceedings Biochemistry of Cotton Workshop, Galveston, Texas. Cotton Incorporated, Raleigh, N.C.

Alchanati, I., J.A.A. Patel, J. Liu, C.R. Benedict, R.D. Stipanovic, A.A. Bell, Y. Cui, and C.W. Magill. 1998. The enzymatic cyclization of nerolidyl diphosphate by δ-cadinene synthase from cotton stele tissue infected with *Verticillium dahliae*. Phytochemistry 47:961-967.

Alexander, A. 1986. Optimum timing of foliar nutrient sprays. pp. 44-60. *In:* A. Alexander (ed.). Foliar Fertilization. Martinus Nijhoff Publishers, Dordrecht.

Ali, A., N. Ahmad, K.H. Gill, M.I.Makhdum, and S.M. Shah. 1992. The effect of exchangeable sodium percentage on growth performance of cotton plant. Sarhad J. Agric. 8:369-372.

Alia, H., T.H. Hayashi, H. Chen, and N. Murata. 1998a. Transformation with a gene for choline oxidase enhances the cold tolerance of *Arabidopsis* during germination and early growth. Plant, Cell and Environment 21:232-239.

Alia, H., A. Sakamoto, and N. Murata. 1998b. Enhancement of the tolerance of *Arabidopsis* to high temperatures by genetic engineering the synthesis of glycinebetaine. Plant Journal 16:155-161.

Alibert, B., O. Lucas, V. Le Gall, J. Kallerhoff, and G. Alibert. 1999. Pectolytic enzyme treatment of sunflower explants prior to wounding and cocultivation with *Agrobacterium tumefaciens* enhances efficiency of transient *beta*-glucuronidase expression. Physiologia Plantarum 106:232-237.

Alimov, K.H. and S. Ibragimov. 1976. Trace elements in different cotton cultivars. Field Crop Abstracts: 29(9):7404.

Allen, R.D. 1995. Dissection of oxidative stress tolerance using transgenic plants. Plant Physiol 107:1049-1054.

Allen, S.J., P.D. Auer, and M.T. Pailthorpe. 1995. Microbial damage to cotton. Textile Res. J. 65:379-385.

Allen, S.J. and D.B. Nehl. 1999. Final Report to CRC for Sustainable Cotton Production: Maximising mycorrhizal infection in cotton. NSW Agriculture, Narrabri, NSW, Australia.

Altman, D.W., R.D. Stipanovic, and A.A. Bell. 1990. Terpenoids in foliar pigment glands of A, D, and AD genome cottons: Introgression potential for pest resistance. J. Hered. 81:447-454.

Altman, D.W., R.D. Stipanovic, and A.A. Bell. 1991. Terpenoids of asiatic and western hemisphere diploid and tetraploid cottons. pp. 534-537. *In:* Proc. Beltwide Cotton Prod. Res. Conf., National Cotton Council of America, Memphis, Tenn.

Altman, D.W., R.D. Stipanovic, and J.H. Benedict. 1989. Terpenoid aldehydes in upland cottons. II. Genotype-environment interactions. Crop Sci. 29:1451-1456.

Altomare, C., W.A. Norvell, T. Bjorkman, and G.E. Harman. 1999. Solubilization of phosphates and micronutrients by the plant-growth-promoting and biocontrol fungus *Trichoderma harzianum* Rifae 1295-22. Applied and Environmental Microbiology 65:2926-2933.

Alvarez, M., R. Pennell, P. Meijer, A. Ishikawa, R. Dixon, and C. Lamb. 1998. Reactive oxygen intermediates mediate a systemic signal network in the establishment of plant immunity. Cell 92:773-784.

Amer, F., M.M. El-Gabaly, and A.B. Monem. 1964. Cotton response to fertilization on two soils differing in salinity. Agron. J. 56:208-211.

American Society for Testing and Materials. 1994. Standard test method for breaking strength and elongation of cotton fibers (Flat bundle method). pp. 392-397. *In:* Annual Book of ASTM Standards, 07.01: ASTM Standard D 1445-90.

American Society for Testing and Materials. 1994. Standard test method for color of raw cotton using the Nickerson-Hunter cotton colorimeter. pp. 584-588. *In:* Annual Book of ASTM Standards, 07.01: ASTM Standard D 2253-88.

American Society for Testing and Materials. 1994. Standard test method for fiber Length and Length Distribution of Cotton Fibers. pp. 753-756. *In:* Annual Book of ASTM Standards, 07.02: ASTM Standard D 5332-92.

American Society for Testing and Materials. 1994. Standard test method for length and length distribution of cotton fibers (array method). pp. 377-382. *In:*

Annual Book of ASTM Standards, 07.01: ASTM Standard D 1440-90.

American Society for Testing and Materials. 1994. Standard test method for length and length uniformity of cotton fiber by Fibrograph measurement. pp. 400-401. *In:* Annual Book of ASTM Standards, 07.01: ASTM Standard D 1447-89.

American Society for Testing and Materials. 1977. Standard test method for linear density of cotton fibers (Array sample). pp. 320-323. *In:* Annual Book of ASTM Standards, ASTM Standard D 1769-77.

American Society for Testing and Materials. 1994. Standard test method for linear density and maturity index of cotton fibers (IIC-Shirley Fineness/Maturity Tester). pp. 143-146. *In:* Annual Book of ASTM Standards, 07.02: ASTM Standard D 3818-92.

American Society for Testing and Materials. 1991. Standard test method for maturity index and linear density of cotton fibers by the causticaire method. pp. 677-681. *In:* Annual Book of ASTM Standards, ASTM Standard D 2480-82.

American Society for Testing and Materials. 1994. Standard test methods for measurement of cotton fibers by high volume instruments (HVI). Annual Book of ASTM Standards, 07.02: ASTM Standard D 4604-86, pp. 475-485; ASTM Standard D 4605-86.

American Society for Testing and Materials. 1976. Standard test method for specific area and immaturity ratio of cotton fibers (Arealometer method). pp. 278-284. *In:* Annual Book of ASTM Standards, ASTM Standard D 1449-58.

Amin, J.V. and H.E. Joham. 1960. Growth of cotton as influenced by low substrate molybdenum. Soil Sci. 89:101-107.

Ammirato, P.V. 1987. Organizational events during somatic embryogenesis. pp. 57-81. *In:* C.E. Green, D.A. Somers, W.P. Hackett, and D.D. Biesboer (eds.). *Plant Tissue and Cell Culture.* Alan R. Liss, Inc.

Amor, Y., C.H. Haigler, S. Johnson, M. Wainscott, and D.P. Delmer. 1995. A membrane-associated form of sucrose synthase and its potential role in synthesis of cellulose and callose. Proc. Natl. Acad. Sci. USA 92:9353-9357.

Amor, Y, R. Mayer, M. Benziman, and D. Delmer. 1991. Evidence for a cyclic diguanylic acid-dependent cellulose synthase in plants. Plant Cell 3:989-995.

Amthor, J.S. 1988. Growth and maintenance respiration in leaves of bean (*Phaseolus vulgaris* L.) exposed to ozone in open-top chambers in the field. New Phytol. 110:319-325.

Amundson, R.G., R.J. Kohut, A.W. Schoettle, R.M. Raba, and P.B. Reich. 1987. Correlative reductions in whole-plant photosynthesis and yield of winter wheat caused by ozone. Phytopathology 77:75-79.

Anderson, D., J. Pellow, J. Palmer, J. Grula, H.B. Cooper, and K. Rajasedaran. 1997. Field evaluation of cotton transformed for tolerance to Imidazolinone herbicides. Proc. Beltwide Cotton Conf. Vol. 1:412-414.

Anderson, D.B. and T. Kerr. 1938. Growth and structure of cotton fiber. Industr. Engng. Chern. 30:48-55.

Anderson, D.J. and M.R. Walmsley. 1984. Effects of eight different foliar treatments on yield and quality of an unfertilized short season variety in the Texas coastal bend. pp. 128-130. *In:* Proc. Beltwide Cotton Prod. Res. Conf., National Cotton Council, Memphis, Tenn.

Anderson, D.M., R.L. Hudspeth, S.L. Hobbs, and J.W. Grula. 1993. Chlorophyll *a/b*-binding protein gene expression in cotton. Plant Physiol. 102:1047-1048.

Anderson, O.E and F.C. Boswell. 1968. Boron and manganese effects on cotton yield, lint quality, and earliness of harvest. Agron. J. 60:488-493.

Anderson, O.E. and K. Ohki. 1972. Cotton growth response and B distribution from foliar application of B. Agron J. 64:665-667.

Anderson, O.E. and R.E. Worthington. 1971. Boron and manganese effects on protein, oil content, and fatty acid composition of cotton seed. Agron. J. 63:566-569.

Anderson, R.N. and W.L. Koukkari. 1978. Response of velvetleaf (*Abutilon theoprasti*) to bentazon as affected by leaf orientation. Weed Sci. 26:393-395.

Anderson, R.N. and W.L. Koukkari. 1979. Rhythmic movements of some common weeds. Weed Sci. 27:401-415.

Anderson, W.K. 1971. Responses of five cotton varieties to two field soil temperature regimes at emergence. Cotton Growers Rev. 48:42-50.

Andrawis, A., M. Solomon, and D.P. Delmer. 1993. Cotton fiber annexins: a potential role in the regulation of callose synthase. Plant Journal 3:763-772.

Andreu, L., F. Moreno, N.J. Jarvis, and G. Vachaud. 1994. Application of the model MACRO to water movement and salt leaching in drained and irrigated marsh soils, Miasmas, Spain. Agric. Water Manage. 25:71-88.

Andries, J.A., J.E. Jones, L.W. Sloane, and J.G. Marshall. 1969. Effects of okra leaf shape on boll rot, yield, and other important characteristics of upland cotton, *Gossypium hirsutum* L. Crop Sci. 9:705-710.

Angenent, G.C. and L. Colombo. 1997. Molecular control of ovule development. Trends in Plant Sci. 1:228-232.

Anter, F., M.A. Rasheed, A. Abd. El-Salam, and A.I. Metwally. 1976a. Effect of foliar application of certain micronutrients on fiber qualities of cotton. I. Application of copper, zinc, molybdenum, and boron. Ann. Agric. Sc. (Moshtohor) 6:303-311. (cited by Stewart, 1986).

Anter, F., M.A. Rasheed, A. Abd. El-Salam, and A.I. Metwally. 1976b. Effect of foliar application of certain micronutrients on fiber qualities of cotton. II. Iron and manganese. Ann. Agric. Sc. (Moshtohor) 6:311-319. (cited by Stewart, 1986).

Anthony, W.S., R.K. Byler, H. Perkins, and J. Askew. 1995. A new method to rapidly assess the stickiness of cotton. Appl. Eng. Agr. 3:415-419.

Aphalo, P.J. and C.L. Ballare. 1995. On the importance of information-acquiring systems in plant-plant interactions. Functional Ecology 9:5-14.

Apostol, I., P. Heinstein, and P. Low. 1989. Rapid stimulation of an oxidative burst during elicitation of cultured plant cells: Role in defense and signal transduction. Plant Physiology 90:109-116.

Apostol, I., P.S. Low, P. Heinstein, R.D. Stipanovic, and D.W. Altman. 1987. Inhibition of elicitor-induced phytoalexin formation in cotton and soybean cells by citrate. Plant Physiology 84:1276-1280.

Appel, R.D., J.-C. Sanchez, A. Bairoch, O. Golaz, M. Miu, J.R. Vargas, and D.F. Hochstrasser. 1993. Swiss-2-D PAGE: a database of two-dimensional electrophoresis images. Electrophoresis 14:1232-1238.

Armstrong, G.M. and Albert, W.B. 1931. A study of the cotton plant with special reference to its nitrogen content. J. Agric. Res. 42:689-703.

Armstrong, G.M. and J.K. Armstrong. 1980. Race 6 of the cotton-wilt *Fusarium* from Paraguay. Plant Disease 64:596.

Armstrong, J.R., D. Glat, B.B. Taylor, and H. Buckwalter. 1982. Heat units as a method for timing Pix applications. p. 57. *In:* J.M. Brown (ed.). Proc. Beltwide Cotton Prod. Res. Conf., National Cotton Council of America, Memphis, Tenn.

Arndt, C.H. 1945. Temperature-growth relations of the roots and hypocotyls of cotton seedlings. Plant Physiol. 20:200-219.

Arndt, W. 1965. The nature of the mechanical impedance to seedlings by soil surface seals. Aust. J. Soil Res. 3:45-54.

Arp, W.J. 1991. Effects of source-sink relations on photosynthetic acclimation to elevated CO_2. Plant Cell and Environ. 14:869-875.

Arriola, L., B.A. Niemira, and G.R. Safir. 1997. Border cells and arbuscular mycorrhizae in four Amaranthaceae species. Phytopathology 87:1240-1242.

Arthur, J.C. 1990. Cotton. pp. 118-141. *In:* J.I. Kroschwitz (ed.). Polymers: Fibers and Textiles, A Compendium. John Wiley and Sons, NY.

Ashley, D.A. 1972. ^{14}C-labelled photosynthate translocation and utilization in cotton plants. Crop Sci. 12:69-74.

Ashley, D.A. and R.D. Goodson. 1972. Effect of time and plant K status on ^{14}C-labeled photosynthate movement in cotton. Crop Sci. 12:686-690.

Ashour, N.I. and A.M. Abd El-Hamid. 1970. Relative salt tolerance of Egyptian cotton varieties during germination and early seedlings development. Plant and Soil 33:493-495.

Ashraf, M. and S. Ahmad. 1999. Exploitation of intra-specific genetic variation for improvement salt (NaCl) tolerance in upland cotton (*Gossypium hirsutum* L.). Hereditas 131:253-256.

Ashraf, M. and S. Ahmad. 2000. Influence of sodium chloride on ion accumulation, yield components and fibre characteristics in salt-tolerant and salt-sensitive lines of cotton (*Gossypium hirsutum* L.). Field Crops Res. 66:115-127.

Ashraf, M. and Z.U. Zafar. 1999. Some physiological characteristics in resistant and susceptible cotton cultivars infected with cotton leaf curl virus. Biologia Plantarum 42:615-620.

Ashraf, M., Z.U. Zafar, T. McNeilly, and C.J. Veltkamp. 1999. Some morpho-anatomical characteristics of cotton (*Gossypium hirsutum* L.) in relation to resistance to cotton leaf curl virus (CLCuV). Journal of Applied Botany-Angewandte Botanik 73:76-82.

Ashton, F.M. and A.S. Crafts. 1981. Absorption and translocation of herbicides. pp. 20-39. *In:* Mode of Action of Herbicides. Second Edition. John Wiley and Sons.

Ashworth, L.J., A.D. George, and O.D. McCutcheon. 1982. Disease-induced potassium deficiency and Verticillium wilt in cotton. California Agriculture 36(9-10):18-20.

Assigbetse, K.B., D. Fernandez, M.P. Dubois, and J.P. Geiger. 1994. Differentiation of *Fusarium oxysporum* f.sp. *vasinfectum* races on cotton by random amplified polymorphic DNA (RAPD) analysis. Phytopathology 84:622-626.

Assman, S.M. 1993. Signal transduction in guard cells. Ann. Rev. Cell Biol. 9:345-375.

Association of Official Seed Analysts. 1988. Rules for testing seeds. J. Seed Technol. 12:1-122, Revised in 1994.

Åström, M. and P. Lundberg. 1994. Plant defense and stochastic risk of herbivory. Evol. Ecol. 8:288-298.

Atalla, R.H. and D.L. VanderHart. 1984. Native cellulose: a composite of two distinct crystalline forms. Science 223:283-285.

Atkins, R.R. 1992. Performance of PGR-IV in cotton. p. 1061. *In:* D.J. Herber and D.A. Richter (eds.). Proc. Beltwide Cotton Conf., National Cotton Council of America, Memphis, Tenn.

Atwell, S.D. 1996. Influence of ultra narrow row on cotton growth and development. Proc. Beltwide Cotton Conf., National Cotton Council, Memphis, Tenn. pp. 1187-1188.

Atwell, S.D., R. Perkins, B. Guice, W. Stewart, J. Harden, and T. Odeneal. 1996. Essential steps to ultra narrow row cotton production. Proc. Beltwide Cotton Conf., National Cotton Council of America, Memphis, Tenn. pp. 1210-1212.

Awadalla, O.A. and I.M. El-Refai. 1992. Herbicide-induced resistance of cotton to Verticillium wilt disease and activation of host cells to produce the phytoalexin gossypol. Canadian Journal of Botany 70:1440-1444.

Ayars, J.E., M.E. Grismer, and J.C. Guitjens. 1997. Water quality as design criterion in drainage water management systems. J. Irrig. Drain. Eng. 123:154-158.

Ayars, J.E., R.B. Hutmacher, R.A. Schoneman, S.S. Vail, and T. Pflaum. 1993. Long-term use of saline water for irrigation. Irrig. Sci. 14:27-34.

Ayars, J.E., R.B. Hutmacher, R.A. Schoneman, S.S. Vail, and D. Felleke. 1986. Drip irrigation of cotton with saline drainage water. Trans. A.S.A.E. 29:1668-1673.

Ayars, J.E., R.M. Mead, R. Soppe, D.A. Clark, R.A. Schoneman, C,R. Camp, E.J. Sadler, and R.E. Yoder. 2000. Weighing lysimeters for shallow ground water management studies. *In:* Evapotranspiration and Irrigation Scheduling. San Antonio, Texas, USA, November 3-6 1996. pp. 825-830.

Ayars, J.E. and R.A. Schoneman. 1986. Use of saline water from a shallow water table by cotton. Trans. A.S.A.E. 29:1674-1678.

Ayer, H.W. and P.G. Hoyt. 1981. Crop-water production functions: Economic implications for Arizona. Arizona Agric. Exp. Stn. Tech. Bull. 242.

Ayers, R.S. 1977. Quality of water for irrigation. J. Irrig. Drain. Div. Am. Soc. Civ. Eng.l03(IR2):l35-l54.

Azcón, R., R. Rubio, and J.M. Barea. 1991. Selective interactions between different species of mycorrhizal fungi and *Rhizobium meliloti* strains, and their effects on growth, N_2-fixation (^{15}N) and nutrition of *Medicago sativa* L. New Phytol. 117:399-404.

Azevedo, D.M., J.A. Landivar, R.m. Vieira, and D.W. Moseley. 1996. The effect of cover crop and crop rotation on cotton: Soil-plant relationship. pp. 1405-1410. *In:* Proc. Beltwide Cotton Conf., National Cotton Council of America, Memphis, Tenn.

Azimov, R.A. 1973. Physiological role of calcium in salt tolerance of cotton. Fan, Tashkent, 204 pp. (Russian).

Baath, E. and D.S. Hayman. 1984. No effect of VA mycorrhizae on red core disease of strawberry. Trans. British Mycol. Soc. 82:534-536.

Babu, V.R., S.N. Prasad, A.M. Babu, and D.S.K. Rao. 1987. Evaluation of cotton genotypes for tolerance to saline water irrigations. Indian J. Agron. 32:229-231.

Bacon, J.S.D. and B. Dickinson. 1957. The origin of melezitose: a biochemical relationship between the lime tree (*Tilia* spp.) and an aphis (*Eucallipterus tiliae* L.). Biochem. J. 66:289-299.

Bader, R.F. and G.A. Niles. 1986. Response of short- and full-season cotton cultivars to mepiquat chloride. I. Morphological and phenological variables. Proceedings Beltwide Cotton Conf., National Cotton Council, Memphis., pp. 513-517.

Baek, J.M., C.R. Howell, and C.M. Kenerley. 1999. The role of an extracellular chitinase from *Trichoderma virens* Gv29-8 in the biocontrol of *Rhizoctonia solani*. Current Genetics 35:41-50.

Bagnall, D., J. Wolfe, and R.W. King. 1983. Chill-induced wilting and hydraulic recovery in mung bean plants. Plant, Cell & Environ. 6:457-464.

Bago, B., C. Azcón-Aguilar, and Y. Piché. 1998. Architecture and developmental dynamics of the external mycelium of the arbuscular mycorrhizal fungus *Glomus intraradices* grown under monoxenic conditions. Mycologia 90:52-62.

Bago, B., Y. Sharchar-Hill, and P.E. Pfeffer. 2000. Dissecting carbon pathways in arbuscular mycorrhizas with NMR spectroscopy. p. 111-126. *In:* G.K. Podila, and D.D. Douds (eds.). Current Advances in Mycorrhizae Research. APS Press, St Paul, MN.

Bagyaraj, D.J. and J.A. Menge. 1978. Interaction between a VA mycorrhiza and *Azotobacter* and their effects on rhizosphere microflora and plant growth. New Phytol. 80:567-573.

Bailey, J.E. and S.R. Koenning. 1997. Disease management in cotton. *In:* 1997 North Carolina Cotton Production Guide, Center for IPM, North Carolina State University, Raleigh N.C. Available http// ipmwww.ncsu.edu/Production_Guides/cotton/ chptr9.html

Bailey, M.A. and T. Trought. 1926. The development of the Egyptian cotton plant. Tech. Sci. Serv. Bull. No. 60, Min. Agric. Egypt.

Baird, R.E., T.B. Brenneman, and D.K. Bell. 1995. First report of *Rhizoctonia* sp. CAG-5 on cotton in Georgia. Plant Disease 79:320.

Baird, R. and D. Carling. 1998. Survival of parasitic and saprophytic fungi on intact senescent cotton roots. J. Cotton Science 2:27-34.

Bais, H.P., J. George, and G.A. Ravishankar. 1999. Influence of polyamines on growth of hairy root cultures of witloof chicory (*Cichorium intybus* L. cv. Lucknow local) and formation of coumarins. J. Plant Growth Regulation 18:33-37.

Bajwa, M.S., O.P. Choudhary, and A.S. Josan. 1992. Effect of continuous irrigation with sodic and saline-sodic waters on soil properties and crop yields under cotton-wheat rotation in northwestern India. Agric. Water Manage. 22:345-356.

Baker, A.J.M. and P.L. Walker. 1989. Physiological responses of plants to heavy metals and the quantification of tolerance and toxicity. Chem. Speciation Bioavail. 1:7-17.

Baker, B., P. Zambryski, B. Staskawicz, and S.P. Dinesh-Kumar. 1997. Signaling in plant-microbe interactions. Science 276:726-733.

Baker, D.N. 1965. Effects of certain environmental factors on net assimilation in cotton. Crop Sci. 5:53-56.

Baker, D.N. 1966. Ann. Rep. Southern Branch USDA-ARS in cooperation with Miss. Agr. Exp. Sta. 60 pp.

Baker, D.N. and B. Acock. 1986. A conceptual model of stress effects. pp. 245-258. *In:* J.R. Mauney and J.M. Stewart (eds.). Cotton Physiology. The Cotton Foundation, Memphis, Tenn.

Baker, D. N., R.R. Bruce, and J.M. McKinion. 1973. An analysis of the relation between photosynthetic efficiency and yield of cotton. Proc. 1972 Cot. Prod. Res. Conf. pp. 110-114.

Baker, D.N., J.L. Hall, and J.R. Thorpe. 1978b. A study of the extrafloral nectaries of *Ricinus communis.* New Phytol. 81:129-137.

Baker, D.N. and J.D. Hesketh. 1969. Respiration and the carbon balance in cotton (*Gossypium hirsutum* L). pp. 60-64 *In:*1969 Proc. Beltwide Cotton Conferences. National Cotton Council, Memphis.

Baker, D.N., J.D. Hesketh, and W.G. Duncan. 1972. The simulation of growth and yield in cotton. I. Gross photosynthesis, respiration and growth. Crop Sci. 12:431-435.

Baker, D.N., J.D. Hesketh, and R.E.C. Weaver. 1978a. Crop architecture in relation to yield. p. 110-136. *In:* U.S. Gupta (ed.). Crop Physiology. Oxford & IBH Publishing Co. New Delhi, Bombay, Calcutta.

Baker, D.N., J.R. Lambert, and J.M. McKinion. 1983. GOSSYM: A simulator of cotton crop growth and yield. South Carolina Agric. Stn. Bull. 1089.

Baker, D.N. and R.E. Meyer. 1966. Influence of stand geometry on light interception and net photosynthesis in cotton. Crop Sci. 6:15-19.

Baker, D.N., V.R. Reddy, J.M. McKinion, and F.D. Whisler. 1993. An analysis of the impact of lygus on cotton. Comput. Electron. Agric. 8:147-161

Baker, E.A. 1982. Chemistry and morphology of plant epicuticular waxes. pp. 139-165. *In:* D.J. Cutler, K.L. Alvin and C.E. Price (eds.). The Plant Cuticle. Academic Press, London.

Baker, J., C. Steele, and L. Dure, III. 1988. Sequence and characterization of 6 *Lea* proteins and their genes from cotton. Plant Mol. Biol. 11:277-291.

Baker, J.C. and L. Dure, III. 1987. DNA and amino acid sequences of some ABA responsive genes expressed in late embryogenesis in cotton. pp. 51-62. *In:* J.E. Fox (ed.). *Molecular Biology of Plant Growth Control.* Alan R. Liss, New York, N.Y., USA

Baker, J.T. and L.H. Allen, Jr. 1993. Effects of CO_2 and temperature on rice: A summary of five growing seasons. J. Agric. Meterol. (Toyko) 48(5):575-582.

Baker, J.T., L.H. Allen Jr., and K.J. Boote. 1992. Temperature effects on rice at elevated CO_2 concentrations. J. Expt. Bot. 43:959-964.

Baker, R.J., J.L. Longmire, and R.A. Van Den Bussche. 1995. Organization of repetitive elements in the upland cotton genome (*Gossypium hirsutum*). J. Hered. 86:178-185.

Baker, S.H. 1976. Response of cotton to row patterns and plant populations. Agron. J. 68:85-88.

Baker, S.H. 1987. Effects of tillage practices on cotton double cropped with wheat. Agron. J. 79: 513-516.

Baker, W.H., J.S. McConnell, R.L. Maples, and J.J. Varvil. 1992. Soil and plant methods for diagnosing K deficiency in cotton. Proc. Beltwide Cotton Conf., Nashville, Tenn. pp. 67-70.

Baker, W.H., J.S. McConnell, J. Varvil, R. Bagwell, and N.P. Tugwell. 1994. The physical stability of insecticide mixed with foliar urea nitrogen and potassium nitrate fertilizers. pp. 56-61. *In:* D.M. Oosterhuis (ed.). Proc. 1993 Cotton Research Meeting and Summaries of Research in Progress. Univ. Arkansas Agric. Exp. Stn. Special Report 162.

Baldocci, D.B., B.B. Hicks, and P. Camara. 1987. A canopy stomatal resistance model for gaseous deposition to vegetated surfaces. Atmos. Environ. 21:91-101.

Baldocchi, D.D., S.B. Verma, N.J. Rosenberg, B.L. Blad, and J.E. Specht. 1985. Microclimate-plant architectural interactions: Influence of leaf width on the mass and energy exchange of a soybean canopy. Agric. Forest Meteorol. 35:1-20.

Baluch, A.A. 1988. A review on the management of cotton whitefly. Pakistan Cottons 32:214-233.

Baldwin, I.T. 1993. Chemical changes rapidly induced by folivory. pp. 1-23 *In:* Insect-Plant Interactions. CRC Press, Boca Raton, Florida.

Baldwin, I.T., M.J. Karb, and T.E. Ohnmeiss. 1994. Allocation of N-15 from nitrate to nicotine: Production and turnover of a damage-induced mobile defense. Ecology 75: 1703-1713.

Baldwin, I.T. and T.E. Ohnmeiss. 1994. Coordination of photosynthetic and alkaloidal responses to damage in uninducible and inducible *Nicotiana sylvestris.* Ecology 75: 1003-1014.

Baldwin, I.T. and E.A. Schmelz. 1994. Constraints on an induced defense - the role of leaf area. Oecologia 97: 424-430.

Balestrini, R., S. Perotto, E. Gasverde, P. Dahiya, L. Guldmann, N.J. Brewin, and P. Bonfante. 1999. Transcription of a gene encoding a lectinlike glycoprotein is induced in root cells harbouring arbuscular mycorrhizal fungi in *Pisum sativum.* Mol. Plant-Microbe Interact. 12:785-791.

Ball, R.A. 1992. Root dynamics, shoot growth and solute accumulation in cotton (*Gossypium hirsutum* L.) during water deficit stress and recovery. M.S. Thesis, University of Arkansas, Fayetteville.

Ball, R.A., D.M. Oosterhuis, and R.W. Brown. 1990. Measurement of root and leaf osmotic potential using the vapor pressure osmometer. American Soc. of Agron. Agron. Abst. p. 118.

Ball, R.A., D.M. Oosterhuis, and A. Mauromoustakos. 1994. Growth dynamics of the cotton plant during water-deficit stress Agron. J. 86:788-795.

Ballaré, C.L., R.A. Sanchez, A.L. Scopel, J.J. Casal, and C.M. Ghersa. 1987. Early detection of neighbour plants by phytochrome perception of spectral changes in reflected sunlight. Plant Cell Environ. 10:551-557.

Ballaré, C.L., A.L. Scopel, and R.A. Sanchez. 1989. Photomodulation of axis extension in sparse canopies: Role of stem in the perception of light-quality signals of stand density. Plant Physiol. 89:1324-1330.

Ballare, C.L., A.L. Scopel, R.A. Sanchez, and S.R. Radosevich. 1992. Photomorphogenic processes in the agricultural environment. Phytochem. Photobiol. 56:777-788.

Balls, W.L. 1905. Sexuality of cotton. *In:* Yearbook of the Khedival Agricultural Society, Cairo, Egypt.

Balls, W.L. 1919a. The cotton plant in Egypt. MacMillan and Co., London. 202 pp.

Balls, W.L. 1919b. The existence of daily growth-rings in the cell wall of cotton hairs. Proc. R. Soc. Lond. B. Biol. Sci. 90:542-555.

Bandurski, R.S., J.D. Cohen, J.P. Slovin, and D.m. Reineke. 1995. Auxin biosynthesis and metabolism. pp. 39-65. *In:* P.J. Davies (ed.). Plant Hormones. Kluwer Academic Publishers, Netherlands.

Bandyopadhyay, B.K. and H.S. Sen. 1992. Effect of excess soil water conditions for a short period on growth and nutrition of crops on coastal saline soil. J. Indian Soc. Soil Sci. 40:823-827.

Banerji, R. and B.S. Dixit. 1995. Variation in fatty acid composition of mature and immature seed oils from *Gossypium arboreum* Linn. 1. Oil Technol. Assoc. India (Bombay) 27:169-170.

Bangerth, F. 1979. Calcium-related physiological disorders of plants. Ann. Rev. Phytopathol. 17:97-122.

Bangerth, F. 1989. Dominance among fruits/sinks and the search for a correlative signal. Physiol. Plant. 76:608-614.

Banks, S.W., F.E. Gordon, S.N. Rajguru, T.J. Minova, D.R. Gossett, M.C. Lucas, P. Dugger, and D. Richter. 1999. Cotton callus tissue transformed with glutathione reductase cDNA derived from *Pisum sativum* L. exhibit increased glutathione reductase activity. Proc. Beltwide Cotton Conf., Orlando, Florida, USA, 3-7 January, 1999: Vol. 1, 542-546.

Banks, S.W., D.R. Gossett, M.C. Lucas, and E.P. Millhollon. 1994. Construction of a transformation vector for cotton containing glutathione reductase cDNA. Proc. Beltwide Cotton Conf., San Diego 1994, 1318-1319.

Banks, S.W., D.R. Gossett, A. Manchandia, B. Bellaire, M.C. Lucas, E.P. Millhollon, P. Dugger, and D. Richter. 2000. The influence of alpha-amanitin on the induction of antioxidant enzymes during salt stress. 1998 Proc. Beltwide Cotton Conf., San Diego, California, USA, 5-9 January 1998 1393-1395;

Banks, S.W., S.N. Rajguru, D.R. Gossett, and E.P. Millhollon. 1997. Antioxidant response to salt stress during fiber development. 1997 Proc. Beltwide Cotton Conf., New Orleans, LA, USA, January 6 10, 1997:Vol. 2 1422-1426.

Bao, J.R., J. Katan, E. Shabi, and T. Katan. 1998. Vegetative-compatibility groups in *Verticillium dahliae* from Israel. European J. Plant Pathology 104:263-269.

Barakat, M.A., S.I. Fakhry, and M.A. Khalil, M.A. 1971. Relative salt tolerance in five varieties of Egyptian cotton. Agric. Res. Rev. Cairo 49:191-200.

Barazani, O. and J. Friedman. 1999. Is IAA the major root growth factor secreted from plant-growth-mediating bacteria? J. Chemical Ecology 25:2397-2406.

Barazani, O. and J. Friedman. 2000. Effect of exogenously applied L-tryptophan on allelochemical activity of plant-growth-promoting rhizobacteria (PGPR). J. Chemical Ecology 26:343-349.

Barber, S.A. 1984. Soil nutrient bioavailability: A mechanistic approach. Wiley, New York.

Bardner, R. and K.E. Fletcher. 1974. Insect infestations and their effects on the growth and yield of field crops: A review. Bull. Ent. Res. 64:141-160.

Barea, J.M., R. Azcón, and D.S. Hayman. 1975. Possible synergistic interactions between Endogone and phosphate -solubilizing bacteria in low-phosphate soils. p. 409-417. *In:* F.E. Sanders, B. Mosse, and P.B. Tinker (eds.). Endomycorrhizas. Academic Press, London.

Bargeron, J.D., III. 1986. Preliminary investigation of the length measurement of cotton fibers with the Peyer Texlab system: Comparability and repeatability. Textile Res. J. 56:121-123.

Bariola, L.A. and C.C. Chu. 1988. Response of cotton varieties grown in a short season to ethephon and thidiazuron in the arid Southwest. pp. 302-303. *In:* J.M. Brown (ed.). Proc. Beltwide Cotton Prod. Res. Conf., National Cotton Council of America, Memphis, Tenn.

Bariola, L.A., C.C. Chu, and T.J. Henneberry. 1990. Timing applications of plant growth regulators and last irrigation for pink bollworm (Lepidoptera: Gelechiidae) control. J. Econ. Entomol. 83:1074-1079.

Bariola, L.A., T.H. Henneberry, C.C. Chu, T. Meng, Jr., and B. Deeter. 1989. Effects of early-season ethephon applications on initiation of pink bollworm infestation and yield. pp. 231-233. *In:* J.M. Brown (ed.). Proc. Beltwide Cotton Prod. Res. Conf., National Cotton Council of America, Memphis, Tenn.

Bariola, L.A., T.J. Henneberry, J.L. McMeans, and C.M. Brown. 1988. Effect of early-season applications of ethephon on cotton fruiting and pink bollworm, *Pectinophora gossypiella* (Saunders), populations. Southwest. Entomol. 13:153-157.

Bariola, L.A., T.J. Henneberry, and T. Meng, Jr. 1986. Plant growth regulators for pink bollworm and boll weevil control. pp. 235-238. *In:* T.C. Nelson (ed.). Proc. Beltwide Cotton Prod. Res. Conf., National Cotton Council of America, Memphis, Tenn.

Bariola, L.A., D.L. Kittock, H.F. Arle, P.V. Vail, and T.J. Henneberry. 1976. Controlling pink bollworms: Effects of chemical termination of cotton fruiting on populations of diapausing larvae. J. Econ. Entomol. 69:633-636.

Bariola, L.A., D.L. Kittock, and T.J. Henneberry. 1981. Chemical termination and irrigation cutoff to reduce overwintering populations of pink bollworms. J. Econ. Entomol. 74:106-109.

Barkla, B.J. and O. Pantoja. 1996. Physiology of ion transport across the tonoplast of higher plants. Ann. Rev. Plant Physiol. Plant Mol Biol. 47:159-184.

Baron, C. and P.C. Zambryski. 1995. The plant response in pathogenesis, symbiosis, and wounding: Variations on a common theme? Annu. Rev. Genetics 29:107-129.

Barratt, D.H.P., P.N. Whitford, S.K. Cook, G. Butcher, and T.L. Wang. 1989. Analysis of seed development in *Pisum sativa* L. VII. Does abscisic acid prevent precocious germination and control storage protein synthesis? J. Exp. Bot. 40:1009-1014.

Barrett, D.J. and R.M. Gifford. 1995. Acclimation of photosynthesis and growth by cotton to elevated CO_2: Interactions with severe phosphate deficiency and restricted rooting volumes. Aust. J. Plant Physiol. 22:955-963.

Barrett, T.W. and H.M. Benedict. 1970. Sulfur dioxide. *In:* J.S. Jacobson and A.C. Hill (eds.). Recognition of Air Pollution Injury to Vegetation: a Pictorial Atlas. Air Pollut. Control. Assoc., Pittsburgh, PA. pp. C1-C17.

Barrow, J.R. 1969. Cross protection against field infestation of Verticillium wilt. pp.32-33. *In:* Proc. Cotton Production Research Conf., National Cotton Council, Memphis, Tenn.

Barrow, J.R. 1981. A new concept in assessing cotton pollen germinability. Crop Sci. 21:441-443.

Barrow, J.R. 1982. Comparisons among pollen viability measurement methods in cotton. Crop Sci. 23:734-736.

Barrs, H.D. 1971. Cyclic variation in stomatal aperture, transpiration, and leaf water potential under constant environmental conditions. Annu. Rev. Plant Physiol. 22:223-236.

Basey, J.M. and S.H. Jenkins. 1993. Production of chemical defenses in relation to plant growth rate. Oikos 68:323-328.

Baskin, C.C., N.W. Hopper, G.R. Tupper, and O.R. Kunze. 1986. Techniques to evaluate planting seed quality. pp. 519-534. *In:* J.R. Mauney and J.McD. Stewart (eds.). Cotton Physiology. The Cotton Foundation, Memphis, Tenn.

Basra, A.S., R. Dhillon-Grewal, Sarlack, R.S. and C.P. Malik. 1993. Influence of ABA and calcium on dark fixation of carbon dioxide in cotton fibers. Phyton 33:1-6.

Basra, A.S. and C.P. Malik. 1983. Dark metabolism of CO_2 during fibre elongation of two cottons differing in fibre length. J. Exp. Bot. 34:1-9.

Basra, A.S. and C.P. Malik. 1984. Development of the cotton fiber. Int. Rev. Cytol. 89:65-113.

Basra, A.S., R.S. Sarlach, H. Nayyar, and C.P. Malik. 1990. Sucrose hydrolysis in relation to development of cotton *(Gossypium* spp.) fibres. Indian J. Exp. Biol. 28:985-988.

Bassett, D.M.; W.D. Anderson, and C.H.E. Werkhoven. 1970. Dry matter production and nutrient uptake in irrigated cotton (*Gossypium hirsutum*). Agron. J. 62:299-303.

Bassett, D.M., J.R. Stockton, and W.L. Dickens. 1970. Root growth of cotton as measured by P^{32} uptake. Agron. J. 60:200-203.

Battey, N.H. and H.D. Blackbourn. 1993. The control of exocytosis in plant cells. New Phytol. 125:307-338.

Bauer, P.M. and J.M. Bradow. 1993. Cover Crops. *In:* M.R. McClelland, T.D. Valco, and RE. Frans (eds.).Conservation Tillage Systems for Cotton. University of Arkansas Agric. Expt. Sta. Spec. Rpt. 169:18-22. Fayetteville, Ark.

Bauer, P.J. and J.M. Bradow. 1996. Cotton genotype response to early-season cold temperatures. Crop Sci. 36:1602-1606.

Bauer, P.J. and W.J. Busscher. 1993. Effect of winter cover on soil mosture content in conventional and strip tillage cotton. p. 8-10. *In:* Proc. South. Conserv. Tillage Conf. Sust. Agric., Monroe, La, 13-17 June 1993. Louisiana State Univ. Agric. Center, Baton Rouge, La.

Bauer, P.J. and W.J. Busscher. 1996. Winter cover and tillage influences on coastal cotton production. J. Prod. Agric. 9:50-54.

Bauer, P.J., W.J. Busscher, and J.M. Bradow. 1995. Cotton response to reduced tillage and cover crops in the southeastern coastal plain. pp. 100-102. *In:* Proc. South. Cons. Tillage Conf. Sust. Agric., Miss. State Univ., Mississippi State, Miss.

Bauer, P.J., J.J. Camberato, and S.H. Roach. 1993. Cotton yield and fiber quality responses to green manures and nitrogen. Agron. J. 85:1019-1023.

Bauer, P.J. and J.T. Cothren. 1990. Growth-promoting activity of chlordimeform. Agron. J. 82:73-75.

Bauer, P.J., J.R. Frederick, W.J. Busscher, and J.M. Bradow. 1995. Effect of tillage and surface residues on cotton fiber properties. pp. 14-17. *In:* M.R. McClelland, T.D. Valco, and R.E. Frans (eds.). Conservation-Tillage Systems for Cotton. Univ. Arkansas Agric. Exper. Sta. Special Report No. 169.

Bauer, P.J., O.L. May, and J.J. Camberato. 1998. Planting date and potassium fertility effects on cotton yield and fiber properties. J. Prod. Agric. 11:415-420.

Baumgärtner, J, V. Delucchi, R. Arx, D. Rubli, and R. Von-Arx. 1986. Whitefly (*Bemisia tabaci* Genn., Stern.: Aleyrodidae) infestation patterns as influenced by cotton, weather and Heliothis: hypotheses testing by using simulation models. Agric. Ecosyst. Environ. 17:49-59.

Bauske, E.M., K.M. Harper, P.M. Brannen, and P.A. Backman. 1994. Effects of residual acidity from delinting and fungicide seed treatments on colonization of biological control agents. pp. 240-243. *In:* Proc. Beltwide Cotton Conf., National Cotton Council of America, Memphis, Tenn.

Bayley, C., N. Trolinder, C. Ray, M. Morgan, J.E. Quisenberry, and D.W. Ow. 1992. Engineering 2,4-D resistance into cotton. Theor. Appl. Genet. 83:645-649.

Bazzaz, F.A., N.R. Chiarello, P.D. Coley, and L.F. Pitelka. 1987. Allocating resources to reproduction and defense. BioSci. 37: 58-67.

Beasley, C.A. 1979. Cellulose content in fibers of cottons which differ in their lint lengths and extent of fuzz. Physiol. Plant. 45:77-82.

Beasley, C.A. and J.P. Ting. 1973. The effects of plant-growth substances on *in vitro* fiber development from fertilized cotton ovules. Amer. J. Bot. 60:130-139.

Beasley, C.A. and J.P. Ting. 1974. Effects of plant-growth substances on *in vitro* fiber development from unfertilized cotton ovules. Amer. J. Bot. 61:188-194.

Beasley, C.A., I.P. Ting, A.E. Linkins, and E.H. Birnbaum. 1974. Cotton ovule culture: A review of progress and a preview of potential. pp. 169-192. *In:* Tissue Culture and Plant Science. Academic Press, New York.

Beasley, J.O. 1940. The origin of American tetraploid *Gossypium* species. Amer. Nat. 74:285-286.

Beasley, J.O. 1942. Meiotic chromosome behavior in species, species hybrids, haploids and induced polyploids of *Gossypium*. Genetics 27:25-54.

Beckmann, B. and R. Heitefuss. 1998. Formation of gossypol and related substances in glandless cotton after inoculation with *Xanthomonas campestris* pv. *malvacearum.* J. Phytopathol. 146:459-463.

Bednarz, C.W., M.G. Hickey, and N.W. Hopper. 1992. Effects of foliar fertilization on Texas southern high plains cotton. Proc. Beltwide Cotton Conf., Nashville, Tenn. pp. 1154-1157.

Bednarz, C.W., N.W. Hopper, and M.G. Hickey. 1997. Effects of foliar fertilization of Texas high plains cotton: Leaf phosphorus, potassium, zinc, iron, manganese, boron, calcium, and yield distribution. Marcel Dekker.

Bednarz, C.W. and D.M. Oosterhuis. 1995. Plant potassium partitioning during progression of deficiency symptoms in cotton (*Gossypium hirsutum*). Better Crops. Potash and Phosphate Institute. Atlanta GA 79:12-14.

Bednarz, C.W. and D.M. Oosterhuis. 1996. Partitioning of potassium in the cotton plant during the development of a potassium deficiency. J. Plant Nutr. 19:1629-1638.

Bednarz, C.W. and D.M. Oosterhuis. 1999. Physiological changes associated with potassium deficiency in cotton. J. Plant Nutr. 22:303-313.

Bednarz, C.W., D.M. Oosterhuis, and D.L. Hendrix. 1994. Physiological aspects of potassium nutrition in cotton. p 1330. *In:* Proc. Beltwide Cotton Proc. Res. Conf. National Cotton Council of America, Memphis, Tenn.

Bednarz, C.W. and M.W. van Iersel. 1998. Semi-continuous carbon dioxide exchange rates in cotton treated with commercially available plant growth regulators. J. Cotton Sci. 2:136-142.

Behery, H.M. 1993. Short fiber content and uniformity index in cotton. ICAC Review Article on Cotton Production Research No. 4. CAB International, Wallingford UK.

Bejarano-Alcazar, J., M.A. Blanco-Lopez, J.M. Melero-Vara, and R.M. Jimenez-Diaz. 1996. Etiology, importance, and distribution of Verticillium wilt of cotton in southern Spain. Plant Disease 80:1233-1238.

Bell, A.A. 1980. The time sequence of defense. pp. 53-73. *In:* J.G. Horsfall and E.B. Cowling (eds.). Plant Disease: An Advanced Treatise, Vol. V, How Plants Defend Themselves. Academic Press, New York.

Bell, A.A. 1981. Biochemical mechanisms of disease resistance. Annual Review of Plant Physiology 32:21-81.

Bell, A.A. 1982. Plant pest interaction with environmental stress and breeding for pest resistance: Plant diseases. pp. 335-363. *In:* M.N. Christiansen and C.F. Lewis (eds.). Breeding Plants for Marginal Environments. John Wiley & Sons, Inc., New York.

Bell, A.A. 1983. Physiological responses of plant cells to infection. pp. 47-69. *In*: R. Moore (ed.). Vegetative Compatibility Responses in Plants. Baylor University Press, Waco, TX.

Bell, A.A. 1984a. Cotton protection practices in the USA and world. Pp. 288-309 *In*: R.J. Kohel and C.F. Lewis (eds.). *Cotton*. Amer. Soc. Agronomy, Madison.

Bell, A.A. 1984b. Morphology, chemistry and genetics of Gossypium adaptations to pests. *In*: Phytochemical adaptations to stress, B.N. Timmerman, C. Steelink and F.A. Loews (eds.), pp. 197-230. Plenum New York.

Bell, A.A. 1986. Physiology of secondary products. pp. 597-621. *In*: J.R. Mauney and J.McD. Stewart (eds.). Cotton Physiology. The Cotton Foundation, Memphis, Tenn.

Bell, A.A. 1989. Role of nutrition in diseases of cotton. pp. 167-204. *In*: A.W. Engelhard (ed.). Soilborne Plant Pathogens: Management of Diseases with Macro- and Micro-elements. APS Press, St. Paul, MN.

Bell, A.A. 1991. Accumulation of ammonium ions in *Verticillium*-infected cotton and its relation to strain virulence, cultivar resistance, and symptoms. p. 186. *In*: Proc. Beltwide Cotton Conferences, National Cotton Council of America, Memphis, Tenn.

Bell, A.A. 1992. Chapter 3, Verticillium wilt. pp. 87-126. *In*: R.J. Hillocks (ed.). Cotton Diseases. CAB International, Wallingford, UK.

Bell, A.A. 1993. Biology and ecology of *Verticillium dahliae*. pp. 147-210. *In*: S.D. Lyda and C.M. Kenerley (eds.). Biology of Sclerotial-Forming Fungi. Texas A and M University Press, College Station, TX.

Bell, A.A. 1995a. Mechanisms of diseases resistance in *Gossypium* species and variation in *Verticillium dahliae*. pp. 225-235. *In*: G.A. Constable and N.W. Forrester (eds.). Challenging the Future: Proceedings of the World Cotton Research Conferences – 1. CSIRO, Melbourne, Australia.

Bell, A.A. 1995b. Genetic variation among *Fusarium oxysporum* isolates from the Regional Wilt Nursery at Shorter, Alabama. pp. 216-217. *In*: Proc. Beltwide Cotton Conferences, Volume 1, National Cotton Council of America, Memphis, Tenn.

Bell, A.A. 1997. Effects of cultural practices on bacteria in and on cotton. pp. 52-72. *In*: J.J. Fischer and L. Domelsmith (eds.). Cotton and Microorganisms. U.S. Department of Agriculture, ARS-138.

Bell, A.A. 1999a. Chapter 3.4, Diseases of cotton. pp. 553-593. *In*: C.W. Smith and J.T. Cothren (eds.). Cotton: Origin, History, Technology, and Production. John Wiley & Sons, Inc., New York, NY.

Bell, A.A. 1999b. Agrobacterium bronzing and wilt: Cultivar reactions and effects of temperature. pp. 117-120.

In: Proc. Beltwide Cotton Conferences, National Cotton Council of America, Memphis, Tenn.

Bell, A.A. 2000a. Synergism between *Rhizoctonia solani* and *Agrobacterium tumefaciens* in causing crown gall of cotton. pp. 175-177. *In*: Proc. Beltwide Cotton Conferences, National Cotton Council of America, Memphis, Tenn.

Bell, A.A. 2000b. Role of *Agrobacterium* in bronze wilt of cotton. pp. 154-160. *In*: Proc. Beltwide Cotton Conferences, National Cotton Council of America, Memphis, Tenn.

Bell, A.A. 2000c. Variability and heritability of bronze wilt resistance in cotton cultivars. pp. 138-144. *In*: Proc. Beltwide Cotton Conferences, National Cotton Council of America, Memphis, Tenn.

Bell, A.A. 2000d. Agrobacterium bronzing and wilt: Epidemiology and control. pp. 927-930. *In*: U. Kechagia (ed.). New Frontiers in Cotton Research, Proceedings of the World Cotton Research Conferences – 2. September 6-12, 1998. Athens, Greece.

Bell, A.A. and K. Decker. 1993. Relationships among vegetative compatibility, race designation, and virulence in isolates of *Fusarium oxysporum* f.sp. *vasinfectum* from cotton in the USA. p. 221. *In*: Proc. Beltwide Cotton Conferences, Volume 1, National Cotton Council of America, Memphis, Tenn.

Bell, A.A., K.M. El-Zik, and P.M. Thaxton. 1992. Chemistry, biological significance, and genetic control of proanthocyanidins in cotton (*Gossypium* spp.). pp. 571-595. *In*: R.W. Hemingway and P.E. Laks (eds.). Plant Polyphenols: Synthesis, Properties, Significance. Plenum Press, New York.

Bell, A.A., J. Liu, R.D. Stipanovic, and H. Orta. 1996a. Toxin production of *Fusarium oxysporum* f.sp. *vasinfectum*, pp. 270-271. *In*: Proc. Beltwide Cotton Conferences, Volume 1, National Cotton Council of America, Memphis, Tenn.

Bell, A.A., J. Liu, R.D. Stipanovic, and H. Orta. 1996b. Production of polyketide toxins by *Fusarium oxysporum* f.sp. *vasinfectum*. Phytopathology 86:S19.

Bell, A.A. and M.E. Mace. 1981. Biochemistry and physiology of resistance. pp. 431-486. *In*: M.E. Mace, A.A. Bell, and C.H. Beckman (eds.). Fungal Wilt Diseases of Plants. Academic Press, New York.

Bell, A.A., M.E. Mace, and R.D. Stipanovic. 1986. The biochemistry of cotton (*Gossypium*) resistance to pathogens. pp. 36-54. *In*: M.A. Green and P.A. Hedin (eds.). Natural Resistance of Plants to Pests: Roles of Allelochemicals. ACS Symposium Series 296, American Chemical Society, Washington, D.C.

Bell, A.A., A.E. Percival, Jr., and H.J. Williams. 1996. Volatile terpene profiles of A, D, and AD genome cottons: Implications for origin of AD species. pp.

1181-1182. *In:* P. Dugger and D. Richter (eds.). Proc. Beltwide Cotton Conferences, Volume 2. National Cotton Council of America, Memphis, Tenn.

Bell, A.A. and J.T. Presley. 1969a. Heat-inhibited or heat-killed conidia of *Verticillium albo-atrum* induce disease resistance and phytoalexin synthesis in cotton. Phytopathology 59:1147-1151.

Bell, A.A. and J.T. Presley. 1969b. Temperature effects upon resistance and phytoalexin synthesis in cotton inoculated with *Verticillium albo-atrum*. Phytopathology 59:1141-1146.

Bell, A.A. and R.D. Stipanovic. 1977. The chemical composition, biological activity, and genetics of pigment glands in cotton. pp. 244-258. *In:* Proc. Beltwide Cotton Production Research Conf., National Cotton Council, Memphis, Tenn.

Bell, A.A. and R.D. Stipanovic. 1978. Biochemistry of disease and pest resistance in cotton. Mycopathologia 65:91-106.

Bell, A.A., R.D. Stipanovic, G.W. Elzen, and H.J. Williams, Jr. 1987. Structural and genetic variation of natural pesticides in pigment glands of cotton (*Gossypium*). pp. 477-490. *In:* G.R. Waller (ed.). Allelochemicals: Role in Agriculture and Forestry. ACS Symposium Series 330, American Chemical Society, Washington, D.C.

Bell, A.A., R.D. Stipanovic, and M.E. Mace. 1993. Cotton phytoalexins: A review. pp. 197-201. *In:* Proc. Beltwide Cotton Production Research Conf., Volume 1, National Cotton Council of America, Memphis, Tenn.

Bell, A.A., R.D. Stipanovic, M.E. Mace, and R.J. Kohel. 1994. Genetic manipulation of terpenoid phytoalexins in *Gossypium*: Effects on disease resistance. pp. 231-249. *In:* B.E. Ellis, G.W. Kuroki, and H.A. Stafford (eds.). Genetic Engineering of Plant Secondary Metabolism, Recent Advances in Phytochemistry. Volume 28. Plenum Press, New York.

Bell, A.A., R.D. Stipanovic, A.E. Percival, and H.J. Williams. 1988. Qualitative variations in the volatile terpenes produced by selected Texas race stocks of *Gossypium hirsutum*. pp. 553-554. *In:* Proc. Beltwide Cotton Prod. Res. Conf., National Cotton Council of America, Memphis, Tenn.

Bell, A.A., R.D. Stipanovic, and H.J. Williams. 1995. Variation and genetic control of volatile terpenes in Upland cotton (*Gossypium hirsutum* L.). p. 8. *In:* Proc. 1995 Meeting of the Phytochemical Society of North America.

Bell, A.A., R.D. Stipanovic, J. Zhang, M.E. Mace, and J.H. Reibenspies. 1998. Identification and synthesis of trinorcadalene phytoalexins formed by *Hibiscus cannabinus*. Phytochemistry 49: 431-440.

Bell, A.A. and M.H. Wheeler. 1986. Biosynthesis and functions of fungal melanins. Annual Review of Phytopathology 24:411-451.

Belsky, A.J., W.P. Carson, C.L. Jensen, and G.A. Fox. 1993. Overcompensation by plants: herbivore optimization or red herring? Evolutionary Ecology 7:109-121.

Beltrano, J., M.G. Ronco, E.R. Montaldi, and A. Carbone. 1998. Senescence of flag leaves and ears of wheat hastened by methyl jasmonate. J. Plant Growth Regul. 17:53-57.

Ben-Porath, A., I. Levin, and M. Meron. 1988. Effects of paclobutrazol growth regulator on drip-irrigated narrow-row cotton. p. 60. *In:* J.M. Brown (ed.). Proc. Beltwide Cotton Prod. Res. Conf., National Cotton Council of America, Memphis, Tenn.

Benedict, C.R. 1984. Physiology. pp. 151-200. *In:* R.J. Kohel and C.F. Lewis (eds.). Cotton. Amer. Soc. Agronomy, Madison WI.

Benedict, C.R., I. Alchanati, P.J. Harvey, J. Liu, R.D. Stipanovic, and A.A. Bell. 1995. The enzymatic formation of δ-cadinene from farnesyl diphosphate in extracts of cotton. Phytochemistry 39:327-331.

Benedict, C.R. and R.J. Kohel. 1975. Export of ^{14}C-assimilates in cotton leaves. Crop Sci. 15:367-372.

Benedict, C.R., R.J. Kohel, and G.M. Jividen. 1992. Cellulose polymers: crystallinity and cotton fiber strength. pp. 227-246. *In:* C.R. Benedict (ed.). Proc. Cotton Fiber Cellulose: Structure, Function, and Utilization Conference. National Cotton Council, Memphis Tenn.

Benedict, C.R., R.J. Kohel, and A.M. Schubert. 1976. Transport of ^{14}C-assimilates to cottonseed: integrity of funiculus during seed filling stage. Crop Sci. 16:23-27.

Benedict, C.R., S. Pan, S.L. Madden, R.J. Kohel, and G.M. Jividen. 1994. A cellulose cotton fiber mutant: effect on fiber strength. pp. 115-120. *In:* G.M. Jividen and C.R. Benedict (eds.). Proc. Biochemistry of Cotton Workshop. Cotton Incorporated, Raleigh NC.

Benedict, J.H., D.W. Altman, P.F. Umbeck, and D.R. Ring. 1992. Behavior, growth, survival, and plant injury by *Heliothis virescens* (F.) (Lepidoptera: Noctuidae) on transgenic Bt cottons. J. Econ. Entomol. 85:589-593.

Benedict, J.H., D.F. Clower, W.F. Kitten, and M.F. Schuster. 1988. Host plant resistance in cotton to Heliothis spp. pp. 1-12 *In:* G.A. Herzog, S. Ramaswamy, G. Lentz, J.L. Goodenough, and J.J. Hamm (eds.). Theory and Practice of Heliothis Population Management. Southern Cooperatives Series Bull. 337.

Benedict, J.H., E.S. Sachs, D.W. Altman, W.R. Deaton, R.J. Kohel, D.R. Ring, and S.A. Berberich. 1996. Field performance of cotton expressing transgenic CryIA insecticidal proteins for resistance to *Heliothis virescens* and *Helicoverpa zea* (Lepidoptera: Noctuidae). J. Econ. Entomol. 89: 230-238.

Bengston, C.S., S. Larsson, and C. Liljenberg. 1978. Effect of water stress on cuticular transpiration rate and amount and composition of epicuticular wax in seedlings of six oat varieties. Physiol. Plant. 44:319-324.

Bennett, M.D., J.B. Smith, and J.S. Heslop-Harrison. 1982. Nuclear DNA amounts in angiosperms. Proc. R. Soc. Lond. B 216:179-199.

Bennett, O.L., D.A. Ashley, and B.D. Doss. 1964. Methods of reducing soil crusting to increase cotton seedling emergence. Agron. J. 56:162-165.

Bennett, O.L., R.D. Rouse, D.A. Ashley, and B.D. Doss. 1965. Yield, fiber quality, and potassium content of irrigated cotton plants as affected by rates of potassium. Agron. J. 57:296-299.

Bennet, R.J., C.M. Breen, and V.H. Bandu. 1990. A role for Ca 2+ in the cellular differentiation of root cap cells: a re-examination of root growth control mechanisms. Environ. Exp. Bot. 30:515-523.

Bennett, A.L. 1993. Biology of *Apion soleatum* Wagner (Coleoptera: Apionidae) relative to cotton production in South Africa. African Entomology 1:35-47.

Bennett, J.H., E.H. Lee, and H.E. Heggestad. 1984. Biochemical aspects of plant tolerance to ozone and oxyradicals: superoxide dismutase. *In:* M.J. Koziol and F.R. Whatley (eds.). Gaseous Pollutants and Plant Metabolism. Butterworths Sci. Publ., London. pp. 413-424.

Bennie, A.T.P. 1996. Growth and mechanical impedance. pp. 453-470. *In:* Y. Waisel, A. Eshel, and U. Kafkafi (eds.). Plant Roots: The Hidden Half. Marcel Dekker.

Benson, N.R., E.D. Vories, and F.M. Bourland. 1995. Variation in growth patterns among cotton cultivars using nodes-above-white-flower. p. 524-530. *In:* D.A. Richter and J. Armour (eds.). Proc. Beltwide Cotton Conf., National Cotton Council of America, Memphis, Tenn.

Berberich, S.A., J.E. Ream, T.L. Jackson, R. Wood, R. Stipanovic, P. Harvey, S. Patzer, and R.L. Fuchs. 1996. The composition of insect-protected cottonseed is equivalent to that of conventional cottonseed. J. Agri. Food Chem. 44:365-371.

Bergelson, J., and Crawley, M.J. 1992 Herbivory and *Ipomopsis aggregata*: the disadvantages of being eaten. Am. Nat. 139:870- 882.

Beringer, H. and F. Nothdurft. 1985. Effects of potassium on plant and cellular structure. *In:* R.D. Munson (ed.). Potassium in agriculture. Amer. Soc. Agron. Madison, WI. pp. 351-368.

Berlin, J.D. 1986. The outer epidermis of the cotton seed. pp. 375-414. *In*: J.R. Mauney and J.M. Stewart

(eds.). Cotton Physiology. The Cotton Foundation, Memphis, Tenn.

Bernhardt, J.L., J.R. Phillips, and N.P. Tugwell. 1986. Position of the uppermost white bloom defined by node counts as an indicator for termination of insecticide treatments in cotton. J. Econ. Entomol. 79:1430-1438.

Berryman, A.A. 1991. Population theory: an essential ingredient in pest prediction, management, and policy making. Am. Entomol. 37:138-142.

Berta, G., A.M. Tagliasacchi, A. Fusconi, D. Gerlero, A. Trotta, and S. Scannerini. 1991. The mitotic cycle in root apical meristems of *Allium porrum* L. is controlled by the endomycorrhizal fungus *Glomus* sp. strain E3. Protoplasma 161:12-16.

Bethlenfalvay, G.J. 1992. Mycorrhizae and crop productivity. p. 1-27. *In:* G.J. Bethlenfalvay and R.G. Linderman (eds.). Mycorrhizae in Sustainable Agriculture. American Society of Agronomy, Madison, Wisconsin.

Bethlenfalvay, G.J., M.S. Brown, and R.S. Pacovsky. 1982. Parasitic and mutualistic associations between a mycorrhizal fungus and soybean: Development of the host plant. Phytopathology 72:889-893.

Bethlenfalvay, G.J. and R.G. Linderman. 1992. Mycorrhizae in Sustainable Agriculture. ASA Special Publication Number 54, Madison, WI.

Bewley, J.D. and M. Black. 1978. Physiology and Biochemistry of Seeds I: Development, Germination, and Growth. pp.106-131. Springer-Verlag, New York.

Bewley, J.D. and M. Black. 1994. Seeds: Physiology of Development and Germination. Second Ed. Plenum Press, New York, N.Y., USA.

Beyers, J.L., G.H. Reichers, and P.J. Temple. 1992. Effects of long-term ozone and drought on the photosynthetic capacity of ponderosa pine (*Pinus ponderosa* Laws.). New Phytol. 122:81-90.

Beyrouty, C.A., J.K. Keino, E.E. Gbur, and M.G. Hanson. 2000. Phytotoxic concentrations of subsoil aluminum as influenced by soils and landscape position. Soil Science 165:135-143.

Bhatt, J.G. and E. Appukuttan. 1971. Nutrient uptake in cotton in relation to plant architecture. Plant Soil. 35:381-388.

Bhatt, J.G., C.V. Raman, T.G. Sankaranarayanan, and S.K. Iver. 1972. Changes in lint character of cotton varieties by growth regulators. Cotton Grow. Rev. 49:160-165.

Bhatt, J.G. and T. Ramanujam. 1971. Some responses of a short-branch cotton variety to gibberellin. Cotton Grow. Rev. 48:136-139.

Bhatt, J.G. and R. Renganayagi. 1986. Ascorbic acid turnover, carbohydrates and mineral nutrients in the cotton plant *(Gossypium hirsutum* L.). Turrialba 41 :316-320.

Bhatt, K.C., P.P. Vaaishnav, Y.D. Singh, and J.J. Chinoy. 1996. Reversal of gibberillic acid-induced inhibition of root growth by manganese. Bichem. Physiol. Pflanz. 170:453-455.

Bhattacharya, N.C., J.W. Radin, B.A. Kimball, J.R. Mauney, G.R. Hendry, J. Nagy, K.F. Lewin, and D.C. Ponce. 1994. Leaf water relations of cotton in a free-air CO_2-enriched environment. 1994 Agric. For. Met. 70:171-182.

Bhojwani, S.S., J.B. Powers, and E.C. Cocking. 1977. Isolation, culture and division of cotton callus protoplasts. Plant Sci. Letters 8:85-89.

Bi, J.L., J.B. Murphy, and G.W. Felton. 1997. Does salicylic acid act as a signal in cotton for induced resistance to *Helicoverpa zea*? J. Chemical Ecology 23:1805-1818.

Bianchini, G.M., R.D. Stipanovic, and A.A. Bell. 1999. Induction of δ-cadinene synthase and sesquiterpenoid phytoalexins in cotton by *Verticillium dahliae*. J. Agric. and Food Chemistry 47:4403-4406.

Bidochka, M.J., R.J. St. Leger, A. Stuart, and K. Gowanlock. 1999. Nuclear rDNA phylogeny in the fungal genus *Verticillium* and its relationship to insect and plant virulence, extracellular proteases and carbohydrases. Microbiology-UK 145:955-963.

Bielorai, H. and P.A.M. Hopmans. 1975. Recovery of leaf water potential, transpiration, and photosynthesis of cotton during irrigation cycles. Agron. J. 67:629-632.

Bilbro, J.D. and L.L. Ray. 1973. Effect of planting date on the yield and fiber properties of three cotton cultivars. Agron. J. 65:606-609.

Binns, A. 1994. Cytokinin accumulation and action: Biochemical, genetic, and molecular approaches. Annu. Rev. Plant Physiol. Plant Mol. Biol. 45:173-196.

Bird, L.S. 1986. Seed quality and stand establishment. pp. 543-550. *In:* J.R Mauney and J.McD. Stewart (eds.). Cotton Physiology. The Cotton Foundation, Memphis, Tenn.

Bird, L.S. 1997. Total biology of cotton planting seed quality relative to obtaining productive plant populations. pp. 438-442. *In:* Proc. Beltwide Cotton Conf., National Cotton Council of America, Memphis, Tenn.

Birnbaum, E.H., C.A. Beasley, and W.M. Dugger. 1974. Boron deficiency in underfertilized cotton (*Gossypium hirsutum*) ovules grown in vitro. Plant Physiol. 54:931-935.

Birnbaum, E.H., W.M. Dugger, and C.A. Beasley. 1977. Interaction of boron with components of nucleic acid metabolism in cotton ovules cultured in vitro. Plant Physiol. 59:1034-1038.

Bishop, A.L., P.R.B. Blood, R.E. Day, and J.P. Evenson. 1978. The distribution of cotton looper (*Anomis flava* Fabr.) larvae and larval damage on cotton and its relationship to the photosynthetic potential of cotton leaves at the attack sites. Aust. J. Agric. Res. 29: 319-325.

Bishop, A.L., R.E. Day, P.R.B. Blood, and J.P. Evenson. 1977. Effect of damaging main stem terminals at various stages of flowering, on yield of cotton in south-east Queensland. Aust. J. Exp. Agric. Anim. Husb. 17: 1032-1035.

Bisson, P., M. Crétenet, and E. Jallas. 1995. Nitrogen, phosphorus, and potassium availability in the soil -- physiology of the assimilation and use of these nutrients by the plant. pp. 115-124. *In:* G.A. Constable and N.W. Forrester (eds.). Challenging the Future. Proc. World Cotton Conference I, CSIRO, Australia.

Black, C.C., T. Loboda, J. Chen, and S.S. Sung. 1995. Can sucrose cleavage enzyme serve as markers for sink strength and is sucrose a signal molecule during plant sink development? pp. 49-64. *In:* H.G. Pontis, G.L. Salerno, and E.J. Echeverria (eds.). Sucrose Metabolism, Biochemistry, Physiology, and Molecular Biology. American Society of Plant Physiologists, Rockville, Md.

Black, C.C., L. Mustardy, S.S. Sung, P.P. Kormanik, D.P. Xu, and N. Paz. 1987. Regulation and roles for alternate pathways of hexose metabolism in plants. Physiol. Plant. 69:387-394.

Blackman, V.H. 1919. The compound interest law and plant growth. Ann. Bot. 33:353-360.

Blackstaffe, L., M.D. Shelley, and R.G. Fish. 1997. Cytotoxicity of gossypol enantiomers and its quinone metabolite gossypolone in melanoma cell lines. Melanoma Research 7(5):364-372.

Blair, L., C.J. Chastain, and J.M. Widholm. 1988. Initiation and characterization of a cotton (*Gossypium hirsutum* L.) photoautoautrophic cell suspension culture. Plant Cell Reports 7:266-269.

Bland, W.L. 1993. Cotton and soybean root system growth in three soil temperature regimes. Agron. J. 85:906-911.

Bland, W.L. and W.A. Dugas. 1989. Cotton root growth and soil water extraction. Soil Sci. Soc. Am. J. 53:1850-1855.

Blanton, R.L. and C.H. Haigler. 1996. Cellulose biogenesis. pp. 57-76. *In:* M. Smallwood, J.P. Knox, and D.J. Bowles (eds.). Membranes: Specialized Functions in Plants. BIOS Scientific Pub., Oxford UK.

Blanton, R.L. and D.H. Northcote. 1990. A 1,4-β-D-glucan synthase system from *Dictyostelium discoideum*. Planta 180:324-332.

Blee, K.A. and A.J. Anderson. 2000. Defense responses in plants to arbuscular mycorrhizal fungi. p. 27-44. *In:* G.K. Podila and D.D. Douds (eds.). Current Advances in Mycorrhizae Research. APS Press, St Paul, MN.

Bloodworth, L.H. and J.R. Johnson. 1995. Cover crops and tillage effects on cotton. J. Prod. Agric. 8:107-112.

Bloodworth, M.E. 1960. Effect of soil temperature on water use by plants. Trans. Seventh Int'l Cong. Soil Sci. 1:153-163.

Bloom, A.J., F.S. Chaplin III, and H.A. Mooney. 1985. Resource limitation in plants - an economic analogy. Annu. Re. Eco. Sys. 16:363-392.

Blum, A. 1975. Effect of the Bm gene on epicuticular wax deposition and the spectral characteristics of sorghum leaves. SABRAO J. 7:45-52.

Blum, U., E. Mrozek, and E. Johnson. 1983. Investigation of ozone (O_3) effects on ^{14}C distribution in ladino clover. Environ. Exptl. Bot. 23:369-378.

Blunden, C.A. and M.F. Wilson. 1985. A specific method for the determination of soluble sugars in plant extracts using enzymatic analysis and its application to the sugar content of developing pear fruit buds. Anal. Biochem. 151:403-408.

Boffey, S.A., J.R. Ellis, G. Sellden, and R.M. Leech. 1979. Chloroplast division and DNA synthesis in light-grown wheat leaves. Plant Physiol. 64:502-505.

Bohm, W. 1979. Methods of Studying Root Systems. Ecol. Stud. 33.Springer-Verlag Publisher. New York.

Bolan, N.S. 1991. A critical review of the role of mycorrhizal fungi in the uptake of phosphorus by plants. Plant Soil 134:189-207.

Bolger, T.P., D.P. Upchurch, and B.L. McMichael. 1992. Temperature effects on cotton root hydraulic conductance. Env. Expl. Bot. 32:49-54.

Bollard, E.G. 1983. Inorganic plant nutrition. pp. 695-755. *In:* A. Läuchli and R.L. Bieleski (eds.). Encyclopedia of plant physiology, New Series. Vol. 15B. Springer-Verlag, Berlin and NewYork.

Boller, T., A. Gehri, F. Mauch, and U. Vögeli. 1983. Chitinase in bean leaves: Induction by ethylene, purification, properties, and possible function. Planta 157:22-31.

Bolwell, G.P. 1993. Dynamic aspects of the plant extracellular matrix. Int. Rev. Cytol. 146:261-324.

Boman, R.K. and R.L. Westerman. 1994. Nitrogen and mepiquat chloride effects on the production of nonrank, irrigated, short-season cotton. J. Prod. Agric. 7:70-75.

Bondada, B.R. and D.M. Oosterhuis. 1996. Effect of water stress on the epicuticular wax composition and ultrastructure of cotton leaf, bract, and boll. J. Environ. Exp. Bot. 36:61-69.

Bondada, B.R. and D.M. Oosterhuis. 2000. Comparative epidermal ultrastructure of cotton leaf, bract, and capsule wall. Annals Bot. 86:1143-1152.

Bondada, B.R., D.M. Oosterhuis, J.B. Murphy, and K.S. Kim. 1996. Effect of water stress on the epicuticular wax composition and ultrastructure of cotton

(*Gossypium hirsutum* L.) leaf, bract, and boll. Env. Exp. Bot. 36: 61-69.

Bondada, B.R., D.M. Oosterhuis, and R.J. Norman. 1997. Cotton leaf age, epiculticular wax, and nitrogen-15 absorption. Crop Sci. 37:807-811.

Bondada, B.R., D.M. Oosterhuis, R.J. Norman, and W.H. Baker. 1996. Canopy photosynthesis, growth, yield, and boll ^{15}N accumulation under nitrogen stress in cotton. Crop Sci. 36:127-133.

Bondada, B.R., D.M. Oosterhuis, N.P. Tugwell, and K.S. Kim. 1995. Physiological and cytological studies of two spotted spider mite, *Tetranychus urticae* K., injury in cotton. Southwest. Entomol. 20:171-180.

Bondada, B.R., D.M. Oosterhuis, S.D. Wullschleger, K.-S. Kim, and W.M. Harris. 1994. Anatomical considerations related to photosynthesis in cotton (*Gossypium hirsutum* L.) leaves, bracts and the capsule wall. J. Exp. Bot. 45:111-118.

Boone, M.Y.L., J.M. McKinion, and J.W. Willers. 1995. Baker's Plant Mapping Technique -- An alternative procedure for GOSSYM. pp. 443-444. *In:* Proc. Beltwide Cotton Conf., National Cotton Council of America, Memphis, Tenn.

Boote, K.J., J.W. Jones, J.W. Mishoe, and R.D. Berger. 1983. Coupling pests to crop growth simulators to predict yield reductions. Phytopatology 73:1581-1587.

Boote, K.S. and R.L. Looms. 1991. The prediction of canopy assimilation. pp. 109-140. *In:* K.J. Boote and R.L. Loomis (eds.). Modeling Crop Photosynthesis - From Biochemistry to Canopy. CSSA Special Publication No. 19, Madison, WI.

Booth, J.E. 1968. Principles of Textile Testing. 3rd ed. Butterworths, London. 583.

Boquet, D.J., G.A. Breitenbeck, and A.B. Coco. 1994a. Cotton yield and growth on clay soil under different levels of tillage, nitrogen and cover crop. pp. 1513-1515. *In:* Proc. Beltwide Cotton Conf., National Cotton Council of America, Memphis, Tenn.

Boquet, D.J., G.A. Breitenbeck, and A.B. Coco. 1995. Residual N effects on cotton following long-time applications of different N rates. Proc. Beltwide Cotton Conf., San Diego, CA. pp. 1362-1364.

Boquet, D.J. and A.B. Coco. 1993. Cotton yield and growth interactions among cultivars, row spacings, and soil types under two levels of Pix. pp. 1370-1372. *In:* D.J. Herber and D.A. Richter (eds.). Proc. Beltwide Cotton Conf., National Cotton Council of America, Memphis, Tenn.

Boquet, D.J., E.B. Moser, and G.A. Breitenbeck. 1994b. Boll weight and within-plant yield distribution in field-grown cotton given different levels of nitrogen. Agron. J. 86:20-26.

Bordovsky, J.P. and W.M. Lyle. 1999. Evaluation of irrigation interval on high plains cotton production with

LEPA systems. *In:* Proc. Beltwide Cotton Prod. Res. Conf. National Cotton Council, Memphis, Tenn. pp. 372-375.

Borg, H. and D.W. Grimes. 1986. Depth development of roots with time: An empirical description. Trans. ASAE. 29:194-197.

Borkar, S.G. and J.P. Verma. 1991. Inhibition of susceptible water soaking and/or hypersensitive reaction in cotton by exopolysaccharide of avirulent *Xanthomonas campestris* pv. *malvacearum*. Folia Microbiologia 36:173-176.

Botsoglou, N. 1991. Determination of "free" gossypol in cottonseed and cottonseed meals by second-derivative ultraviolet spectrophotometry. J. Agri. Food Chem. 39:478-482.

Boulanger, J. and D. Pinheiro. 1971. Evolution de la production cotonnière au nord-est du Brésil. Cot. et Fib. Trop. 26:319-353.

Bourland, F.M., D.M. Oosterhuis, and N.P. Tugwell. 1992. Concept for monitoring the growth and development of cotton plants using main-stem node counts. J. Prod. Agric. 5:532-538.

Bourland, F.M., N.P. Tugwell, D.M. Oosterhuis, and M.J. Cochran. 1994. Cotton plant monitoring: the Arkansas system (an overview). pp. 1280-1281. *In:* D.J. Herber and D.A. Richter (eds.). Proc. Beltwide Cotton Conf., National Cotton Council of America, Memphis, Tenn.

Bourland, F.M., D.M. Oosterhuis, N.P. Tugwell, M.J. Cochran, and D.M. Danforth. 1997. Interpretation of crop growth patterns generated by COTMAN. University of Arkansas Agricultural Experiment Station Special Report 181.

Bourland, F.M. and C.E. Watson, Jr. 1990. COTMAP, a technique for evaluating structure and yield of cotton plants. Crop Sci. 30:224-226.

Bourland, F.M. and B.W. White. 1986. Variation in early fruiting and boll retention in cotton. p. 117-120. *In:* J.M. Brown (ed.). Proc., Beltwide Cotton Production Research Conferences, National Cotton Council of America, Memphis, Tenn.

Bourne, H.R., D.A. Sanders, and F. McCormick. 1991. The GTPase superfamily: conserved structure and molecular mechanism. Nature 349:117-127.

Bouyssou, H., H. Canut, and G. Marigo. 1990. A reversible carrier mediates the transport of malate at the tonoplast of *Catharanthus roseus* cells. FEBS Lett. 275:73-76.

Bouzaidi, A. and S.E. Amami. 1980. Irrigation of two cotton varieties in field trials Physiologie Vegetale 18:35-44.

Box, J.E. and G.W. Langdale. 1984. The effects of in-row subsoil tillage on corn yields in the southeastern Coastal Plains of the United States. Soil and Tillage Res. 4:67-78.

Boyer, J.S. 1964. Effects of osmotic water stress on metabolic rates of cotton plants with open stomata. Plant Physiol. 40:229-233.

Boyer, J.S. 1970. Leaf enlargement and metabolic rates in corn, soybean, and sunflower at various leaf water potentials. Plant Physiol. 46:233-23 5.

Boyer, J.S. 1982. Plant productivity and environment. Science 218:443-448.

Boylston, E.K., D.P. Thibodeaux, and J.P. Evans. 1993. Applying microscopy to the development of a reference method for cotton fiber maturity. Text. Res. J. 63:80-87.

Boynton, D. 1954. Nutrition by foliar application. Ann. Rev. Plant Physiol. 5:31-54.

Boynton, D., D. Margolis, and C.R. Gross. 1953. Exploratory studies on nitrogen metabolism by McIntosh apple leaves sprayed with urea. Proc. Amer. Soc. Hort. Sci. 62:135-146.

Bozcuk, S. 1981. Effects of kinetin and salinity on germination of tomato, barley and cotton seeds. Ann. Bot. 48:81-84.

Bozcuk, S. 1990. Interaction between salt and kinetin on seed germination of some crop plants. Doga, Turk Botanik Dergisi 14:139-149.

Bradford, K.J., D.M. May, B.J. Hoyle, S. Skibinski, S.J. Scott, and K.B. Tyler. 1988. Seed and soil treatments to improve emergence of muskmelon from cold or crusted soils. Crop Sci. 28:1001-1005.

Bradford, S. and J. Letey. 1992. Cyclic and blending strategies for using nonsaline and saline waters for irrigation. Irrig. Sci. 13:123-128.

Bradley, J.F. 1994. Success with no-till cotton. Nat. Cons. Tillage Digest 1:34-36.

Bradley, J.F. 1995. Overview of conservation tillage on cotton production in the mid-South. pp. 200-203. *In:* Proc. Beltwide Cotton Conf., National Cotton Council of America, Memphis, Tenn.

Bradow, J.M. 1990a. Chilling sensitivity of photosynthetic oil-seedlings. I. Cotton and sunflower. J. Exp. Bot. 41:1585-1593.

Bradow, J.M. 1990b. Chilling sensitivity of photosynthetic oil-seedlings. II. Cucurbitaceae. J. Exp. Bot. 41:1595-1600.

Bradow, J.M. 1991. Cotton cultivar responses to suboptimal postemergent temperatures. Crop Sci. 31:1595-1599.

Bradow, J.M. 1993. Inhibitions of cotton seedling growth by volatile ketones emitted by cotton cover crops. J. Chem. Ecol. 19:1085-1108.

Bradow, J.M. and P.J. Buaer. 1992. Inhibition of cotton seedling growth by soil containing LISA cover crop residues. p. 1175 *In:* Proc. Beltwide Cotton Conf., National Cotton Council of America, Memphis, Tenn.

Bradow, J.M. and P.J. Bauer. 1997. Fiber-quality variations related to cotton planting date and temperature. pp. 1491-1493. *In:* Proc. Beltwide Cotton Conf.,

National Cotton Council of America, Memphis, Tenn.

Bradow, J.M., P.J. Bauer, O. Hinojosa, and G.F. Sassenrath-Cole. 1997a. Quantitation of cotton fibre-quality variations arising from boll and plant growth environments. Eur. J. Agron. 61:191-204.

Bradow, J.M., P.J. Bauer, G.F. Sassenrath-Cole, and R.M. Johnson. 1997b. Modulations of fiber properties by growth environment that persist as variations of fiber and yarn quality. pp. 1351-1360. *In:* Proc. Beltwide Cotton Conf., National Cotton Council of America, Memphis, Tenn.

Bradow, J.M. and W.J. Connick, Jr. 1988. Inhibition of cotton seedling root growth by rhiosphere volatiles. pp. 90-91. *In:* Proc. Beltwide Cotton Conf., National Cotton Council of America, Memphis, Tenn.

Bradow, J.M., G.H. Davidonis, O. Hinojosa, L.H. Wartelle, K.J. Pratt, K. Pusateri, P.J. Bauer, B. Fisher, G.F. Sassenrath-Cole, P.H. Dastoor, J.A. Landivar, D. Locke, and D. Moseley. 1996c. Environmentally induced variations in cotton fiber maturity and related yarn and dyed knit defects. pp. 1279-1284. *In:* Proc. Beltwide Cotton Conf., National Cotton Council of America, Memphis, Tenn.

Bradow, J.M., O. Hinojosa, L.H. Wartelle, and G. Davidonis. 1996a. Applications of AFIS fineness and maturity module and X-ray fluorescence spectroscopy in fiber maturity evaluation. Textile Res. J. 66:545-554.

Bradow, J.M., G.F. Sassenrath-Cole, O. Hinojosa, and L.H. Wartelle. 1996b. Cotton fiber physical and physiological maturity variation in response to genotype and environment. pp. 1251-1254. *In:* Proc. Beltwide Cotton Conf., National Cotton Council of America, Memphis, Tenn.

Bradshaw, A.D. 1965. Evolutioinary signifficance of phenotipic plasticity in plants. Adv. Genet. 13:115-155.

Brady, N.C. and R.R. Weil. 1996. The nature and properties of soils. 11th edition. Prentice Hall, Inc., New Jersey.

Bragg, C.K. and F.M. Shofner. 1993. A rapid, direct measurement of short fiber content. Textile Res. J. 63:171-176.

Bragg, C.K. and J.D. Wessinger. 1993. Cotton fiber diameter distribution as measured by AFIS. pp. 1125-1128. *In:* Proc. Beltwide Cotton Conf., National Cotton Council of America, Memphis, Tenn.

Brandenburg, R.L. and G.G. Kennedy. 1981. Overwintering of the pathogen *Entomophtora floridana* and its host, the twospotted spider mite. J. Econ. Entomol. 74:428-431.

Brannen, P.M. and P.A. Backman. 1994a. Decrease in *Fusarium oxysporum* f. sp. *vasinfectum* incidence through use of *Bacillus subtilis* seed inoculants. pp. 244-245. *In:* Proc. Beltwide Cotton Conf., National Cotton Council of America, Memphis, Tenn.

Brannen, P.M. and P.A. Backman. 1994b. Suppression of Fusarium wilt of cotton with *Bacillus subtilis* hopper box formulations. pp. 83-85. *In:* M.H. Ryder, P.M. Stephens, and G.D. Bowen (eds.). Improving Plant Productivity with Rhizosphere Bacteria, Third International Workshop On Plant Growth-Promoting Rhizobacteria. CSIRO, East Melbourne, Victoria, Australia.

Braud, M. 1974. Le controle da nutrition minerale du cotonnier par analyses foliares. Coton Fibes Trop. 29:215-225.

Braun, Y., M. Hassidim, H.R. Lerner, and L. Reinhold. 1986. Studies on H$^+$--translocating ATPases in plants of varying resistance to salinity. I. Salinity during growth modulates the proton pump in the halophyte *Atriplex nummularia*. Plant Physiol. 81:1050-1056.

Breitenbeck, G.A. and D.J. Boquet. 1993. Effects of N fertilization on nutrient uptake by cotton. Proc. Beltwide Cotton Conf., New Orleans, LA pp. 1298-1300.

Bremner, J.M. 1995. Recent research on problems in the use of urea as a nitrogen fertilizer. Fertilizer Research 42:321-329.

Bressan R.A., P.M. Hasegawa, and J.M. Pardo. 1998. Plants use calcium to resolve salt stress. Trends in Plant Sci. 3:411-412.

Brewer, R.F. 1979. The effects of present and potential air pollution on San Joaquin Valley cotton. Final Report, ARB Agreement A7-119-39, CA Air Resources Board, Sacramento, CA.

Brewer, R.F. and G. Ferry. 1974. Effects of air pollution on cotton in the San Joaquin Valley. Calif. Agric., June 1974.

Bridge, R.R. and W.R. Meredith, Jr. 1983. Comparative performance of obsolete and current cotton cultivars. Crop Sci. 23:949-952.

Bridge, R.R., W.R. Meredith, Jr., and J.F. Chism. 1971. Comparative performance of obsolete varieties and current varieties of upland cotton. Crop Sci. 11:29-32.

Bridge, R.R., W.R. Meredith, Jr., and J.F. Chism. 1973. Influence of planting methods and plant population on cotton (*Gossypium hirsutum* L.). Agron. J. 65:104-109.

Briggs, R.E. 1981. Varietal response to PIX-treated cotton in Arizona. p. 47. *In:* J.M. Brown (ed.). Proc. Beltwide Cotton Prod. Res. Conf., National Cotton Council of America, Memphis, Tenn.

Briggs, R.E., L.L. Patterson, and G.D. Massey. 1967. Within- and between-row spacing of cotton. Arizona Annual Report. pp. 6-7. University of Arizona Agric. Ext. Serv.

Brinkerhoff, L.A. 1970. Variation in *Xanthomonas malvacearum* and its relation to control. Ann. Rev. Phytopathology 8:85-110.

Brisley, H.R., C.R. Davis, and J.A. Booth. 1959. Sulphur dioxide fumigation of cotton with special reference to its effects on yield. Agron. J. 51:77-80.

Brock, T.G. and P.B. Kaufman. 1991. Growth regulators: An account of hormones and growth regulation. pp. 277-340. *In:* Plant Physiology a treatise, Vol. X: Growth and Development. Academic Press, New York.

Brody, A.K. and R. Karban. 1989. Demographic analysis of induced resistance against spider mites (Acari: Tetranychidae) in cotton. J. Econ. Entomol. 82:462-465.

Brody, A.K. and R. Karban. 1992. Lack of a tradeoff between constitutive and induced defenses among varieties of cotton. Oikos 65:301-306.

Broodryk, S.W. and G.W. Matthews. 1994. *Dysdercus* (Hemiptera: Pyrrhocoridae) and other heteroptera. pp. 267-284. *In:* G.A. Matthews and J.P. Tunstall (eds.). Insects Pests of Cotton. CAB International, Wallingford.

Brook, K.D. 1984. Review of the effects of damage on the cotton plant. Australian Cotton Growers Res. Conf, Toowoomba, pp. 228-235.

Brook, K.D., A.B. Hearn, and C.F. Kelly. 1992a. Response of cotton to damage by insect pests in Australia: Pest management trials. J. Econ. Entomol. 85:1356-1367.

Brook, K.D., A.B. Hearn, and C.F. Kelly. 1992b. Response of cotton, *Gossypium hirsutum* L., to damage by insect pests in Australia: Manual simulation of damage. J. Econ. Entomol. 85:1368-1377.

Brook, K.D., A.B. Hearn, and C.F. Kelly. 1992c. Response of cotton to damage by insect pests in Australia: Compensation for early season fruit damage. J. Econ. Entomol. 85:1378-1386.

Brooks, A. 1986. Effects of phosphorus nutrition on ribulose-1,5-biphosphate carboxylase activation, photosynthetic quantum yield and amounts of some Calvin-cycle metabolites in spinach leaves. Aust. J. Plant Physiology 13:221-237.

Brouder, S.M. and K.G. Cassman. 1990. Root development of two cotton cultivars in relation to potassium uptake and plant growth in a vermiculitic soil. Field Crops Res. 23:187-203.

Brouder, S.M. and K.G. Cassman. 1994. Cotton root and shoot response to localized supply of nitrate, phosphate, and potassium: Split-pot studies with nutrient solution and vermiculitic soil. Plant Soil 161:179-193.

Brouwer, R. and Hoagland, A. 1964. Responses of bean plants to root temperatures II. Anatomical aspects. Meded. Inst. Biol. Scheik. Onderz. Land. Gewass. 236:23-31

Brown, A.L., J. Quick, and G.J. DeBoer. 1973. Diagnosing potassium deficiency by soil analysis. California Agriculture 27:13-14.

Brown, H.B. and J.O. Ware. 1958. Cotton. 3rd ed. McGraw-Hill, New York.

Brown, J.C. and W.E. Jones. 1977. Fitting plants nutritionally to soils II. Cotton. Agronomy J. 69:405-409.

Brown, J.F., S.J. Allen, and G.A. Constable. 1990. Mycorrhizas and plant nutrition: long fallow disorder and cotton. p. 67-72. *In:* Proc. 5th Australian Cotton Conference. Australian Cotton Growers Research Association, Broadbeach, QLD, Australia.

Brown, K.J. 1968. Translocation of carbohydrate in cotton: movement to the fruiting bodies. Ann. Bot. 32:703-713.

Brown, K.J. 1971. Plant density and yield of cotton. Cotton Grow. Rev. 48:255-266.

Brown, K.J. 1973a. Factors affecting translocation of carbohydrates to fruiting bodies of cotton. Cotton Grow. Rev. 50:32-42.

Brown, K.J. 1973b. Effect of selective defoliation on development of cotton bolls. Cotton Grow. Rev. 50:106-114.

Brown, L.C., G.W. Cathey, and C. Lincoln. 1962. Growth and development of cotton as affected by toxaphene-DDT, methyl parathion, and calcium arsenate. J. Econ. Entomol. 55:298-301.

Brown, L.C., C. Lincoln, R.E. Frans, and B.A. Waddle. 1961. Some effects of toxaphene-DDT and calcium arsenate on growth and development of cotton. J. Econ. Entomol. 54:309-311.

Brown, L.C. and C.C. Wilson. 1952. Some effects of zinc on several species of *Gossypium* L. Plant Physiol. 27:812-817.

Brown, L.G., J.W. Jones, J.D. Hesketh, J.D. Hartsog, F.D. Whisler, and F.S. Harris. 1985. COTCROP: Computer simulation of growth and yield. Mississippi Agricultural and Forestry Experiment Station, Information Bulletin 69, Mississippi State, MS.

Brown, L.R. 1991. The Aral Sea. Disaster area and interdisciplinary solution. Interdisciplinary Science Reviews 16:345-350.

Brown, M.S. and M.Y. Menzel. 1950. New trispecies hybrids in cotton. J. Hered. 41:291-295.

Brown, M.S. and M.Y. Menzel. 1952a. The cytology and crossing behavior of *Gossypium_gossypioides*. Bull. Torrey Bot. Club 79:110-125.

Brown, M.S. and M.Y. Menzel. 1952b. Additional evidence on the crossing behavior of *Gossypium gossypioides*. Bull. Torrey Bot. Club 79:285-292.

Brown, R.M.J., I.M. Saxena, and K. Kudlicka. 1996. Cellulose biosynthesis. Trends in Plant Science 1:149-156.

Brown, S.M., T. Whitwell, J.T. Touchton, and C.H. Burmester. 1985. Conservation tillage systems for cotton production. Soil Sci. Soc. Am. J. 49:1256-1260.

Browning, V.D., H.M. Taylor, M. Huck, and B. Klepper. 1975. Water relations of cotton: A rhizotron study. Auburn Univ. Exp. Sta. Bul. 467 pp 70.

Brubaker, C.L. 1996. Occurrence of terpenoid aldehydes and lysigenous cavities in the 'Glandless' seeds of Australian *Gossypium* species. Aust. J. Bot. 44:601-612.

Brubaker, C.L., C.G. Benson, C. Miller, and D.N. Leach. 1996. Occurrence of terpenoid aldehydes and lysigenous cavities in the "glandless" seeds of Australian *Gossypium* species. Aust. J. Bot. 44:601-612.

Brubaker, C.L., J.A. Koontz, and J.F. Wendel. 1993. Bidirectional cytoplasmic and nuclear introgression in the New World cottons, *Gossypium barbadense* and *G. hirsutum* (Malvaceae). Amer. J. Bot. 80:1203-1208.

Brubaker, C.L., A.H. Paterson, and J.F. Wendel. 1999. Comparative genetic mapping of allotetraploid cotton and its diploid progenitors. Genome 42:184-203.

Brubaker, C.L. and J.F. Wendel. 1993. On the specific status of *Gossypium lanceolatum* Todaro. Genet. Res. Crop Evol. 40:165-170.

Brubaker, C.L. and J.F. Wendel. 1994. Reevaluating the origin of domesticated cotton (*Gossypium hirsutum*: Malvaceae) using nuclear restriction fragment length polymorphisms (RFLPs). Amer. J. Bot. 81:1309-1326.

Bruce, R.R. and M.J.M. Römkens. 1965. Fruiting and growth characteristics of cotton in relation to soil moisture tension. Agron. J. 57:135-140.

Brugnoli, E. and O. Björkman. 1992. Growth of cotton under continuous salinity stress - influence on allocation pattern, stomatal and nonstomatal components of photosynthesis and dissipation of excess light energy. Planta, 187:335-347.

Brugnoli, E. M. and Lauteri. 1991. Effects of salinity on stomatal conductance, photosynthetic capacity, and carbon isotope discrimination of salt-tolerant (*Gossypium hirsutum* L.) and salt-sensitive (*Phaseolus vulgaris* L.) C3 non-halophytes. Plant Physiol. 95:628-635.

Bruin, J., A.T. Groot, M.W. Sabelis, and M. Dicke. 1992. Mite herbivory causes better protection in downwind uninfested plants. Proc. 8th Int. Symp. Insect-Plant Relationships, Dodrdrecht, Kluwer Acad. Publ., S.B.J. Menken, J.H. Visser, and P. Harrewijn (eds.). pp. 357-358.

Brushwood, D.E. and H.H. Perkins, Jr. 1993. Cotton stickiness potential as determined by the minicard, thermodetector and chemical methods. pp. 1132-1135 *In:* Proc. Beltwide Cotton Prod. Res. Conf., National Cotton Council of America, Memphis, Tenn.

Brushwood, D.E. and H.H. Perkins, Jr. 1996. Cotton plant sugars and insect honeydew characterized by high performance liquid chromatography. pp. 1310.1313. *In:* Proc. Beltwide Cotton Conf., National Cotton Council of America, Memphis, Tenn.

Bryant, J.P., P.J. Kuropat, S.M. Cooper, K. Frisby, and N. Owen-Smith. 1989. Resource availability hypothesis of plant antiherbivore defense tested in a South African savanna ecosystem. Nature 340:227-229.

Buchala, A.J. 1987. Acid β-fructofuranoside fructohydrolase (invertase) in developing cotton (*Gossypium arboreum* L.) fibres and its relationship to β-glucan synthesis from sucrose fed to the fibre apoplast. J. Plant. Physiol. 127:219-230.

Buchala, A.J. and H. Meier. 1985. Biosynthesis of β-glucans in growing cotton (*Gossypium arboreum* L. and *Gossypium hirsutum* L.) fibers. pp. 220-241. *In:* C.T. Brett and J.R. Hillman (eds.). Biochemistry of Plant Cell Walls. Cambridge Univ. Press, Cambridge UK.

Bucheli, P., M. Durr, A.J. Buchala, and H. Meier. 1985. β-Glucanases in developing cotton (*Gossypium hirsutum* L.) fibres. Planta 166:530-536.

Buco, S.M., J.G. Montalvo, Jr., S.E. Faught, R. Grimball, E. Stark, and K. Luchter. 1996. Determination of maturity/fineness by FMT and diode array HVI. Part 2. Data analysis and results. pp. 1279-1281. *In:* Proc. Beltwide Cotton Conf., National Cotton Council of America, Memphis, Tenn.

Buco, S.M., J.G. Montalvo, Jr., S.E. Faught, E. Stark, and K. Luchter. 1995. Determination of wall thickness and perimeter by FMT and diode array HVI. Part 1. Data analysis and results. pp. 1289-1292. *In:* Proc. Beltwide Cotton Conf., National Cotton Council of America, Memphis, Tenn.

Buehring, N.W., W.F. Jones, and G.A. Jones. 1994. Cotton response to conventional tillage and no-till cover crop system. pp. 1508-1510. *In:* Proc. Beltwide Cotton Conf., National Cotton Council of America, Memphis, Tenn.

Buick, R.D., B. Robson, and R.J. Field. 1992. A mechanistic model to describe organosilicon surfactant promotion of triclopyr uptake. Pesticide Science 36:127-133.

Bunkelmann, J.R. and R.N. Trelease. 1995. Molecular cloning and characterization of ascorbate peroxidase localized to the glyoxysome membrane of cotton cotyledons. Plant Physiol. Suppl. 108:67.

Bunkelmann, J.R. and R.N. Trelease. 1996. Ascorbate peroxidase. Plant Physiol. 110:589-598.

Bunning, E. 1973. The physiological clock, revised 3rd Edn. Springer, Berlin.

Burke, J.J. 1990. Variation among species in the temperature dependence of the reappearance of variable fluorescence following illumination. Plant Physiol. 93:652-656.

Burke, J.J. 1994. Enzyme adaptation to temperature. *In:* N. Smirnoff (ed.). Environment and Plant Metabolism. BIOS Scientific Publishers Ltd.

Burke, J.J., J.L. Hatfield, R.R. Klein, and J.E. Mullet. 1985. Accumulation of heat shock proteins in field-grown cotton. Plant Physiol. 78:394-398.

Burke, J.J., J.L. Hatfield, and J.R. Mahan. 1989. Quantifying thermal stress in cotton through the thermal kinetic window. pp. 44-45. *In:* J.M. Brown and D.A. Richter (eds.). Proc. Beltwide Cotton Prod. Conf., National Cotton Council of America, Memphis, Tenn.

Burke, J.J., J.L. Hatfield, and D.F. Wanjura. 1990. A thermal stress index for cotton. Agron. J, 82:526-530.

Burke, J.J., J.R. Mahan, and J.L. Hatfield. 1988. Crop-specific thermal kinetic windows in relation to wheat and cotton biomass production. Agron. J. 80:553-556.

Burke, J.J. and M.J. Oliver. 1993. Optimal thermal environments for plant metabolic processes (*Cucumis sativus* L.): Light-harvesting chlorophyll a/b pigment-protein complex of Photosystem II and seedling establishment in cucumber. Plant Physiol. 102:295-302.

Burke, J.J. and D.R. Upchurch. 1995. Cotton rooting patterns in relation to soil temperatures and the thermal kinetic window. Agron J. 87:1210-1216.

Bürkert, B. and A.D. Robson. 1994. ^{65}Zn uptake in subterranean clover (*Trifolium subterraneum* L.) by three vesicular-arbuscular mycorrhizal fungi in a root-free sandy soil. Soil Biol. Biochem. 26:1117-1124.

Burmester, C.H., M.G. Patterson, and D.W. Reeves. 1993. No-till cotton growth characteristics and yield in Alabama. pp. 30-36. *In:* Proc. South. Conserv. Tillage Conf. Sust. Agric., Monroe, LA, 13-17 June 1993. Louisiana State Univ. Agric. Center, Baton Rouge, LA.

Burmester, C.H., M.G. Patterson, and D.W. Reeves. 1995. Challenges of no-till cotton production on silty clay soils in Alabama. *In:* M.R. McClelland, T.D. Valco, and R.E. Frans (eds.). Conservation-Tillage Systems for Cotton. University of Arkansas Agric. Expt. Sta. Spec. Rpt. 169:5-7, Fayetteville, Ark.

Burnell, J.N. 1986. The biochemistry of manganese in plants. *In:* R.D. Grahham, R.J. Hannam, and N.C. Uren (eds.). Manganese in soils and plants. Kluwer Academic Publishers, Boston. pp. 125-137.

Burnside, C.A. and R.H. Bohning. 1957. The effect of prolonged shading on the light saturation curves of apparent photosynthesis in sun plants. Plant Physiol. 32:61-63.

Burr, T.J. and L. Otten. 1999. Crown gall of grape: Biology and disease management. Ann. Rev. Phytopathology 37:53-80.

Busch, C.D. and F. Turner, Jr. 1965. Sprinkling cotton with saline water. Prog. Agric. Ariz. 17:27-28.

Busch, C.D. and F. Turner, Jr. 1967. Sprinkler irrigation with high salt-content water. Trans. A.S.A.E. 10:494-496.

Buser, C. and P. Matile. 1977. Malic acid in vacuoles isolated from *Bryophyllum* leaf cells. Z. Pflanzenphysiol. 82:462-466.

Bush, D.R. 1993. Proton-coupled sugar and amino acid transporters in plants. Annu. Rev. Plant Physiol. Plant Mol. Biol. 44:513-542.

Busscher, W.J. and P.J. Bauer. 1993. Cone index and yield in conventional- and conservation-tilled cotton. pp. 1367-1369. *In:* Proc. Beltwide Cotton Conf., National Cotton Council of America, Memphis, Tenn.

Busscher, W.J. and P.J. Bauer. 1995. Soil strength of conventional- and conservation-tillage cotton growth with a cover crop. *In:* T.D. Valco and M.R. McClelland (eds.). Conservation-tillage Systems for Cotton. University of Arkansas Agricultural Experiment Station Special Report 169:18-20. Fayetteville, Ark.

Bussler, W. 1963. Die entwicklung von calcium-mangelsymptomen. Z. Pflanzernernaer., Dueng. Bodenkd. 100:53-58.

Bussler, W. 1981. Physiological functions and utilization of copper. *In:* J.F. Loneragan, A.D. Robson, and R.D. Graham (eds.). Copper in Soils and Plants. Academic Press, London. pp. 213-234.

Butler, G.D. Jr. and T.J. Henneberry. 1984. *Bemisia* and *Trialeurodes* (Hemiptera: Aleyrodidae). pp. 325-352. *In*: G.A. Matthews and J.P. Tunstall (eds.). Insect Pests of Cotton. CAB International, Wallingford.

Butler, G.D., G.M. Loper, S.E. McGregor, J.L. Webster, and H. Margolis. 1972. Amounts and kinds of sugars in the nectars of cotton (*Gossypium* spp.) and the time of their secretion. Agron. J. 64:364-368.

Butter, N.S., B.K. Vir, K. Gurdeep, T.H. Singh, and R.K. Raheja. 1992. Biochemical basis of resistance to whitefly *Bemisia tabaci* Genn. (Aleyrodidae: Hemiptera) in cotton. Trop. Agric. 69:119-122.

Buxton, D.R., R.E. Briggs, L.L. Patterson, and S.D. Watkins. 1977. Canopy characteristics of narrow-row cotton as influenced by plant density. Agron. J. 69:929-933.

Buxton, D.R., P.J. Melick, L.L. Patterson, and E.J. Pegelow, Jr. 1977. Relationships among cottonseed vigor and emergence. Agron. J. 69:677-681.

Buxton, D.R., L.L. Patterson, and R.E. Briggs. 1979. Fruiting pattern in narrow-row cotton. Crop Sci. 19:17-22.

Buxton, D.R., P.J. Sprenger, and E.J. Pegelow, Jr. 1976. Periods of chilling sensitivity in germinating Pima cottonseed. Crop Sci. 16:471-474.

CAB International. 1999. Crop Protection Compendium. CAB International, Wallingford, UK.

Cabello, M., L. Gaspar, and R. Pollero. 1994. *Glomus antarcticum* sp. nov., a vesicular-arbuscular mycorrhizal fungus from Antarctica. Mycotaxon 51:123-128.

Cadena, J. and J.T. Cothren. 1995. Yield response of cotton to nitrogen, irrigation, and PGR-IV regimes. pp. 1142-1150. *In* D.A. Richter and J. Armour (eds.). Proc. Beltwide Cotton Prod. Res. Conf., National Cotton Council of America, Memphis, Tenn.

Cadena, J. and J.T. Cothren. 1996. Carbon balance of PGR-IV treated cotton plants grown under two irrigation regimes. pp. 1225-1232. *In:* Proc. Beltwide Cotton Conf., National Cotton Council of America, Memphis, Tenn.

Cadena, J., J.T. Cothren, and C.J. Fernandez. 1994. Carbon balance of PGR-IV-treated cotton plants at two temperature regimes. pp. 1309-1313. *In:* D.J. Herber and D.A. Richter (eds.). Proc. Beltwide Cotton Conf., National Cotton Council of America, Memphis, Tenn.

Cain, D. 1956. Absorption and metabolism of urea by leaves of coffee, cacao and tobacco. Proc. Amer. Soc. Hort. Sci. 67:279-286.

Cakmak, I. and H. Marschner. 1986. Mechanism of phosphorus-induced zinc deficiency in cotton. I. Zinc deficiency-enhanced uptake rate of phosphorus. Physiol. Plant. 68:483-490.

Cakmak, I. and H. Marschner. 1988a. Increase in membrane permeability and exudation of roots of zinc deficient plants. J. Plant Physiol. 132: 356-361.

Cakmak, I. and H. Marschner. 1988b. Enhanced superoxide radical production in roots of zinc-deficient plants. J. Exp. Bot. 39:1449-1460.

Cakmak, I., H. Marschner, and F. Bangerth. 1989. Effect of zinc nutritional status on growth, protein metabolism, and levels of indole-3-acetic acid and other phytohormones in bean (*Phaseolis vulgaris*). J. Exp. Bot. 40:405-412.

Calahan, J.S. and H.E. Joham. 1974. Sodium and calcium interactions in the salt tolerance of cotton. Proc. Beltwide Cotton Prod. Res. Conf., Memphis, 38-39.

Calhoun, D.S., J.D. Bargeron, and W.S. Anthony. 1997. An introduction to AFIS for cotton breeders. pp. 418-424. *In:* Proc. Beltwide Cotton Conf., National Cotton Council of America, Memphis, Tenn.

Calhoun, D.S., J.E. Jones, W.D. Caldwell, E. Burris, B.R. Leonard, S.H. Moore, and W. Aguillard. 1994. Registration of La. 850082FN and La. 850075FHG, two cotton germplasm lines resistant to multiple insect pests. Crop Sci. 34:316-317.

Callahan, F.E., J.N. Jenkins, R.G. Creech, and G.W. Lawrence. 1997. Changes in cotton root proteins correlated with resistance to root knot nematode development. J. Cotton Science 1:38-47.

Calvert, C.M., S.J. Gant, and D.J. Bowles. 1996. Tomato annexins p34 and p35 bind to F-actin and display nucleotide phosphodiesterase activity inhibited by phospholipid binding. Plant Cell 8:333-342.

Camp, C.R., P.J. Bauer, and P.G. Hunt. 1997. Subsurface drip irrigation lateral spacing and management for cotton in the southeastern Coastal Plant. Trans ASAE 40:993-999.

Camp, C.R. and Z.F. Lund. 1964. Effect of soil compaction on cotton roots. American Society of Agronomy, Crops and Soils. Nov: pp.8-9.

Campbell, L.C. and R.O. Nable. 1988. Physiological functions of manganese in plants. *In:* R.D. Grahham, R.J. Hannam, and N.C. Uren (eds.). Manganese in soils and plants. Kluwer Academic Publishers, Boston. pp. 139-154.

Campbell, W.R., C.J. Counselman, H.W. Ray, and L.I. Terry. 1979. Evaluation of chlordimeform (Galecron) for *Heliothis virescens* control in cotton. pp. 122-125. *In:* J.M. Brown (ed.). Proc. Beltwide Cotton Prod. Res. Conf., National Cotton Council of America, Memphis, Tenn.

Canerday, T.D. and F.S. Arant. 1964a. The effect of late season infestations of the strawberry spider mite, *Tetranychus atlanticus*, on cotton production. J. Econ.Entomol. 57:931-933.

Canerday, T.D. and F.S. Arant. 1964b. The effect spider mite populations on yield and quality of cotton. J. Econ.Entomol. 57:553-556.

Cannon, R.J.C. 1995. *Bacillus thuringiensis* in pest control. Plant Microbial Biotechnol. Res. Series. 4:190-200.

Cannon, W.A. 1925. Physiological features of roots with special reference to the relation of roots to the aeration of the soil. Carnegie Institution, Washington D.C. Pub. No. 368.

Cao, H., X. Li, and X.N. Dong. 1998. Generation of broad-spectrum disease resistance by overexpression of an essential regulatory gene in systemic acquired resistance. Proc. National Academy of Sciences USA 95:6531-6536.

Cappy, J.J. 1979. The rooting patterns of soybean and cotton throughout the growing season. Ph.D. Dissertation. Univ. of Arkansas, Fayetteville.

Cappy, J.J. and J.T. Cothren. 1980. Root growth and development of Pix-treated cotton plants. p. 22. *In:* J.M. Brown (ed.). Proc. Beltwide Cotton Prod. Res. Conf., National Cotton Council of America, Memphis, Tenn.

Carabelli, M., G. Morelli, G. Whitelam, and I. Ruberti. 1996. Twilight-zone and canopy shade induction of the Athb-2 homeobox gene in green plants. Proc. Natl. Acad. Sci. 93:3530-3535.

Carling, D.E., C.S. Rothrock, G.C. MacNish, M.W. Sweetingham, K.A. Brainard, and S.W. Winters. 1994. Characterization of anastomosis group 11

(AG-11) of *Rhizoctonia solani*. Phytopathology 84:1387-1393.

Carlton, B.C. and C. Gawron-Burke. 1993. Genetic improvement of *Bacillus thuringiensis* for bioinsecticide development. pp. 43-61 *In:* L. Kim (ed.). Advanced Engineered Pesticide. Marcell Dekker Inc., New York.

Carmi, A. 1986. Effects of root zone volume and plant density on the vegetative and reproductive development of cotton. Field Crops Res. 13:25-32.

Carmi, A., Z. Plaut, and A. Grava. 1992. Establishment of shallow and restricted root systems in cotton and its impact on plant response to irrigation. Irrig. Sci. 13:87-91.

Carmi, A., Z. Plaut, B. Heuer, and A. Grava. 1992. Establishment of shallow and restricted root systems in cotton and its impact on plant response to irrigation. Irrig. Sci. 12:87-91.

Carmi, A., Z. Plaut, and M. Sinai. 1993. Cotton root growth as affected by changes in soil water distribution and their impact on plant tolerance to drought. Irrig. Sci. 13:177-182.

Carmi, A. and J. Shalhevet. 1983. Root effects on cotton growth and yield. Crop Sci. 23:875-878.

Carnahan, J.E., E.L. Jenner, and E.K.W. Wat. 1978. Prevention of ozone injury to plants by a new protectant chemical. Phytopathology 68:1225-1229.

Carpita, N. and D. Delmer. 1981. Concentration and metabolic turnover of UDP-glucose in developing cotton fibers. J. Biol. Chem. 256:308-315.

Carpita, N.C. and D.M. Gibeaut. 1993. Structural models of primary cell walls in flowering plants: consistency of molecular structure with the physical properties of the walls during growth. Plant J. 3:1-30.

Carter, L.M. 1996. Tillage. *In:* S.J. Hake, T.A. Kerby, and K.D. Hake (eds.). Cotton Production Manual. Univ. of Calif. Oakland. pp. 175-186.

Carter, O.G.1980. The nutrition of Crops. *In:* J.E. Pratley (ed.), Principles of field crop production, Sydney Univ. Press, Sydney. pp. 221-249.

Carvalho, L.H., N.M. Silva, S.M. Brasil, J.I. Kondo, and E.J. Chiavegato. 1996. Application of boron to cotton by sidedressing and foliar spray. Revista Brasileira de Ciencia do Solo. 20:265-269.

Casal, J.J., R.A. Sanchez, and V.A. Deregibus. 1986. The effect of plant density on tillering: the involvement of R/FR ratio and the proportion of radiation intercepted per plant. Environ. Exper. Bot. 26:365-371.

Casal, J.J., R.A. Sanchez, A.R. Paganelli-Blau, and M. Izaguirre. 1995. Phytochrome effects on stem carbon gain in light-grown mustard seedlings are not simply the result of stem extension-growth response. Physiol. Plant. 94:187-196.

Cason, E.T., Jr., P.E. Richardson, M.K. Essenberg, L.A. Brinkerhoff, W.M. Johnson, and R.J. Venere.

1978. Ultrastructural cell wall alterations in immune cotton leaves inoculated with *Xanthomonas malvacearum*. Phytopathology 68:1015-1021.

Cass, Q.B., A.L. Bassi, and S.A. Matlin. 1999. First direct resolution of gossypol enantiomers on a chiral high-performance liquid chromatography phase. Chirality 11:46-49.

Cass, Q.B., E. Tiritan, S.A. Matlin, and E. Freire. 1991. Gossypol enantiomer ratios in cotton seeds. Phytochemistry 30:2655-2657.

Cassman, K.G. 1993. Cotton. *In:* W.F. Bennett (ed.). Nutrient deficiencies and toxicities in crop plants. American Phytopathological Society, APS Press, St. Paul, MN. pp. 111-119.

Cassman, K.G.; T.A. Kerby, B.A. Roberts, D.C. Bryant, and S.M. Brouder. 1989a. Differential response of two cotton cultivars to fertilizer and soil potassium. Agron. J. 81:870-876.

Cassman, K.G., T.A. Kerby, B.A. Roberts, D.C. Bryant, and S.L. Higashi. 1990. Potassium nutrition effects on lint yield and fiber quality of Acala cotton. Crop Science 30:672-677.

Cassman, K.G.; B.A. Roberts, T.A. Kerby, D.C. Bryant., and S.L. Higashi. 1989b. Soil potassium balance and cumulative cotton response to annual potassium additions on a vermiculitic soil. Soil Sci. Soc. Am. J. 53:805-812.

Castillo, F.J. and H. Greppin. 1988. Extracellular ascorbic acid and enzyme activities related to ascorbic acid metabolism in *Sedum album* L. leaves after ozone exposure. Environ. Exp. Bot. 28:231-238.

Castillo, F.J. and R.L. Heath. 1990. Calcium transport in membrane vesicles from pinto bean leaves and its alteration after ozone exposure. Plant Physiol. 94:788-795.

Cathey, G.W. 1979. Unpublished data.

Cathey, G.W. 1983. Cotton. pp. 233-252. *In:* L.G. Nickell (ed.). Plant Growth Regulating Chemicals. Vol. I. CRC Press, Inc., Boca Raton, FL.

Cathey, G.W. 1986. Physiology of defoliation in cotton production. pp. 143-153. *In:* J.R. Mauney and J.M. Stewart (eds.). Cotton Physiology. Vol. One. The Cotton Foundation Reference Book Series. The Cotton Foundation, Memphis, Tenn.

Cathey, G.W. and H.R. Barry. 1977. Evaluation of glyphosate as a harvest-aid chemical on cotton. Agron. J. 69:11-14.

Cathey, G.W. and J.C. Bailey. 1987. Evaluation of chlordimeform for cotton yield enhancement. J. Econ. Entomol. 80:670-674.

Cathey, G.W. and K. Luckett. 1980. Some effects of growth regulator chemicals on cotton earliness, yield and quality. p. 45. *In:* J.M. Brown (ed.). Proc. Beltwide Cotton Prod. Res. Conf., National Cotton Council of America, Memphis, Tenn.

Cathey, G.W., K.E. Luckett, and S.T. Rayburn. 1982. Accelerated cotton boll dehiscence with growth regulator and desiccant chemicals. Field Crops Res. 5:113-120.

Cathey, G.W. and W.R. Meredith, Jr. 1983. Effects of planting date on response of five cotton cultivars to mepiquat chloride. pp. 251-252. *In:* A.R. Cook (ed.). Proc. Tenth Meet. Plant Growth Regul. Soc. Am. Conf., Plant Growth Regul. Soc. Am, Lake Alfred, FL.

Cathey, G.W. and W.R. Meredith, Jr. 1988. Cotton response to planting date and mepiquat chloride. Agron. J. 80:463-466.

Cathey, G.W., B.W. Ross, and A.J. Harvey. 1981. Absorption, translocation and accumulation of ^{14}C-acephate in cotton plants. p. 165. *In:* W.C. Wilson (ed.). Proc. Plant Growth Regul. Working Group Conf., Plant Growth Regulator Working Group, Longmont, CO.

Cathey, G.W. and T.O. Thomas. 1986. Use of plant growth regulators for crop modification. pp. 137-142. *In:* J.R. Mauney and J.M. Stewart (eds.). Cotton Physiology. Vol. One. The Cotton Foundation Reference Book Series. The Cotton Foundation, Memphis, Tenn.

Catsky, J., I. Ticha, and J. Solarova. 1976. Photosynthetic characteristics during ontogenesis of leaves. 1. Carbon dioxide exchange and conductances for carbon dioxide transfer. Photosynthetica 10:394-402.

Catsky, J. and I. Ticha. 1982. Photosynthetic characteristics during ontogenesis of leaves. 6. Intercellular conductance and its components. Photosynthetica 16:253-284.

Cauquil, J. 1975. Cotton Boll Rot: Laying Out a Trial of a Method of Control. Amerina, New Delhi. 143 pp.

Cawley, N.M. 1999. Growth and yield response of cotton to ultra-narrow row spacing in North Carolina. M.S. thesis. North Carolina State Univ., Raleigh, NC.

Chambers, A.Y. 1990. Effects of no-till planting on severity of cotton seedling disease. p. 30. *In:* Proc. Beltwide Cotton Conf., National Cotton Council of America, Memphis, Tenn.

Chambers, A.Y. 1995a. Comparative effects of no-till and conventional tillage on severity of three major cotton diseases in Tennessee. *In:* T.D. Valco and M.R. McClelland (eds.). Conservation-tillage Systems for Cotton. University of Arkansas Agricultural Experiment Station Special Report 169:96-99. Fayetteville, Ark.

Chambers, A.Y. 1995b. Conventional tillage vs. conservation tillage and their effects on disease severity. pp. 203-204. Proc. Beltwide Cotton Conf., National Cotton Council of America, Memphis, Tenn.

Chan, B.G. 1985. From L-phenylalanine to condensed tannins: A brief review of tannin chemistry and biochemistry and a report on its distribution in cotton plant parts. pp. 49-52. *In:* W.P. Wakelyn and R.R. Jacobs (eds.). Proc. Ninth Cotton Dust Conference, National Cotton Council of America, Memphis, Tenn.

Chan, B.G. and A.C. Waiss, Jr. 1981. Evidence for acetogenic and shikimic pathways in cotton glands. pp. 49-51. *In:* Proc. Beltwide Cotton Production Research Conf., National Cotton Council of America, Memphis, Tenn.

Chang, A.M. and D.M. Oosterhuis. 1995. Efficacy of foliar application to cotton of potassium compounds at different pH levels. pp. 1364-1366. *In:* Proc. Beltwide Cotton Conferences. National Cotton Council, Memphis, Tenn.

Chang, C.W. 1980. Starch depletion and sugars in developing cotton leaves. Plant Physiol. 65:844-847.

Chang, C.W. 1982. Enzymatic degradation of starch in cotton leaves. Phytochem. 21:1263-1269.

Chang, C.W. and H.E. Dregne. 1955. Effects of exchangeable Na on soil properties and on growth and cation content of alfalfa and cotton. Soil Sci. Soc. Am. Proc. 19:29-35.

Chang, C.W. and J.A. Wetmore. 1986. Effects of water stress on starch and its metabolizing enzyme activities in cotton leaves. Starch 38:311-313.

Chang, J.F., J.H. Benedict, T.L. Payne, and B.J. Camp. 1988. Volatile monoterpenes collected from the air surrounding flower buds of seven cotton genotypes. Crop Sci. 28:685-688.

Chang, K. and J.K.M. Roberts. 1989. Observation of cytoplasmic and vacuolar malate in maize root tips by ^{13}C-NMR spectroscopy. Biochim. Biophys. Acta. 109:29-34.

Chang-lie, J. and S. Sonobe. 1993. Identification and preliminary characterization of a 65 kDa higher-plant microtubule-associated protein. J. Cell Sci. 105:891-901.

Chanzy, H., K. Imada, and R. Vuong. 1978. Electron diffraction from the primary wall of cotton fibers. Protoplasma 94:299-306.

Chapin, F.S. 1991. Effects of multiple environmental stresses on nutrient availability and use. *In:* H.A. Mooney, W.E. Winner, and E.J. Pell (eds.). Response of Plants to Multiple Stresses. Academic Press, San Diego. pp. 67-88.

Chapin, F.S. III, A.J. Bloom, C.B. Field, and R.H. Waring. 1987. Plant responses to multiple environmental factors. BioSci. 37:49-57.

Charchar, M.J.D., J.R.N. dos Anjos, and E. Ossipi. 1999. Occurrence of a new disease of irrigated cotton, in Brazil, caused by *Sclerotinia sclerotiorum*. Pesquisa Agropecuaria Brasileira 34:1101-1106.

Charudattan, R. and J.E. DeVay. 1972. Common antigens among varieties of *Gossypium hirsutum* and isolates of *Fusarium* and *Verticillium* species. Phytopathology 62:230-234.

Charudattan, R. and J.E. DeVay. 1981. Purification and partial characterization of an antigen from *Fusarium oxysporum* f.sp. *vasinfectum* that cross-reacts with antiserum to cotton (*Gossypium hirsutum*) root antigens. Physiological Plant Pathology 18:289-295.

Chase, M.W., D.E. Soltis, R.G. Olmstead, D. Morgan, D.H. Les, B.D. Mishler, M.R. Duvall, R.A. Price, H.G. Hills, Y.-L. Qiu, K.A. Kron, J.H. Rettig, E. Conti, J.D. Palmer, J.R. Manhart, K.J. Sytsma, H.J. Michaels, W.J. Kress, K.G. Karol, W.D. Clark, M. Hedrén, B.S. Gaut, R.K. Jansen, K.-J. Kim, C.F. Wimpee, J.F. Smith, G.R. Furnier, S.H. Strauss, Q.-Y. Xiang, G.M. Plunkett, P.S. Soltis, S.M. Swensen, S.E. Williams, P.A. Gadek, C.J. Quinn, L.E. Eguiarte, E. Golenberg, G.H. Learn Jr., S.W. Graham, S.C.H. Barrett, S. Dayanandan, and V.A. Albert. 1993. Phylogenetics of seed plants: An analysis of nucleotide sequences from the plastid gene *rbc*L. Ann. Missouri Bot. Gard. 80:528-580.

Chatterton, N.J., P.A. Harrison, W.R. Thornley, and J. H. Bennett. 1993. Separation and quantification of fructan (insulin) oligomers by anion exchange chromatography. pp. 93-99. *In* A. Fuchs (ed.). Inulin and Inulin-Containing Crops. Elsevier Science.

Chaudhry, F.I., M.N.A. Malik, M.I. Makhdum, and S.I. Hussain. 1989. Salt tolerance of nine cotton cultivars at germination stage. The Pakistan Cottons 33:72-75.

Chen, C., E.M. Bauske, G. Musson, R. Rodriguez-Kabana, and J.W. Kloepper. 1995. Biological control of Fusarium wilt on cotton by use of endophytic bacteria. Biological Control 5:83-91.

Chen, D.-H., Y.-L. Meng, H.-C. Ye, G.-F. Ji, and X.-Y. Chen. 1998. Culture of transgenic *Artemisia annua* hairy root with cotton cadinene synthase gene. Acta Botanica Sinica 40:711-714.

Chen, D., R. Yu, D.M. Chen, and R.P. Yu. 1996. Studies on relative salt tolerance of crops. II. Salt tolerance of some main crop species. Acta Pedologica Sinica 33:121-128.

Chen, P., Z.Q. Zhang, W. Xu, K. Wang, G. Zhu, L. Lu, and H. Liu. 1991 Effects of chemical control of the cotton aphid during the early season on cotton plants, natural enemies and yield. Journal of Applied Entomology 111:211-215.

Chen, Q.Q., H. Chen, and H.P. Lei. 1987. Comparative study on the metabolism of optical gossypol in rats. J. Ethnopharmacol. 20:31-37.

Chen, T. and K.H. Kreeb. 1989. Combined effects of drought and salt stress on growth, hydration and pigment composition in *Gossypium hirsutum* L. *In:* K.H. Kreeb, H. Richter, and T.M. Hinckley (eds.). Structural and Functional Responses to Environmental Stresses: Water Shortage. SPB Academic Publishing, The Hague. pp. 165-177.

Chen, X.-Y., Y. Chen, P. Heinstein, and V.J. Davisson. 1995. Cloning, expression, and characterization of (+)-δ-cadinene synthase: A catalyst for cotton phytoalexin biosynthesis. Archives of Biochemistry and Biophysics 324:255-266.

Chen, X.-Y., M. Wang, Y. Chen, V.J. Davisson, and P. Heinstein. 1996. Cloning and heterologous expression of a second (+)-δ-cadinene synthase from *Gossypium arboreum*. J. Natural Products 59:944-951.

Chen, Y. and P. Barak. 1982. Iron nutrition of plants in calcareous soils. Adv. Agron. 35:217-240.

Cheng, W., D.C. Colemen, and J.E. Box, Jr. 1991. Measuring root turnover using the minirhizotron technique. Agric. Ecosys. Environ. 34:261-267.

Chevrier, N., Y.S. Chung, and F. Sarhan. 1990. Oxidative damages and repair in *Euglena gracilis* exposed to ozone: II. Membrane permeability and uptake of metabolites. Plant Cell Physiol. 31:987-992.

Chewning, C.H. 1992. Cotton fiber management using high volume instrument testing and Cotton Incorporated's Engineered Fiber Selection System. pp. 29-42. *In:* C.R. Benedict (ed.). Proc. Cotton Fiber Cellulose: Structure, Function, and Utilization Conference. National Cotton Council, Memphis Tenn.

Chewning, C.H. 1995. Cotton fiber management using Cotton Incorporated's Engineered Fiber Selection System and High Volume Instrument testing. pp. 109-115. *In:* Proc. Beltwide Cotton Conf., National Cotton Council of America, Memphis, Tenn.

Chien, J.C. and I.M. Sussex. 1996. Differential regulation of trichome formation on the adaxial and abaxial leaf surfaces by gibberellins and photoperiod in *Arabidopsis thaliana* (L.) Heynh. Plant Physiol. 111:1321-1328.

Childs, S.W. and R.J. Hanks. 1975. Model of soil salinity effects on crop growth. Soil Sci. Soc. Am. J. 39:617-622.

Ching, P.C. and S.A. Barber. 1979. Evaluation of temperature effects on K uptake by corn. Agron. J. 71:1040-1044.

Ching, T.M. 1972. Metabolism of germination seeds. pp. 103-218. *In:* T.T. Kozlowski (ed.). Seed Biology, Volume II, Academic Press, New York.

Chlan, C.A. 1995. Structural similarities between the legumins of *Gossypium hirsutum*: Sequence of the *legumin B* Gene 1 (Accession No. U43727) (PGR95-141). Plant Physiol. 110:714.

Chlan, C.A., K. Borroto, J.A. Kamalay, and L. Dure, III. 1987. Developmental biochemistry of cottonseed embryogenesis and germination. XIX. Sequences and genomic organization of the α globulin (vicilin) genes of cottonseed. Plant Mol. Biol. 9:533-546.

Chlan, C.A., J.B. Pyle, A.B. Legocki, and L. Dure, III. 1986. Developmental biochemistry of cottonseed embryogenesis and germination. XVIII. cDNA and amino acid sequences of members of the storage protein families. Plant Mol. Biol. 7:475-489.

Choinski, J.S., Jr., R.N. Trelease, and D.C. Doman. 1981. Control of enzyme activities in cotton cotyledons during maturation and germination: III. *In-vitro* embryo development in the presence of abscisic acid. Planta 152:428-435.

Chokey, S. and N.K. Jain. 1977. Critical growth stages of American rainfed cotton for foliar feeding of nitrogen. Indian Journal of Agronomy 22(2):87-89.

Chollet, R., J. Vidal, and M.H. O'Leary. 1996. Phosph*oenol*pyruvate carboxylase: a ubiquitous, highly regulated enzyme in plants. Annu. Rev. Plant Physiol. Moll. Biol. 47:273-298.

Choudhry, M.R., M.A. Gill, M.S. Arshad, and M. Asif. 1994. Surface water application techniques for cotton crop to alleviate waterlogging and salinity. Sarhad J. Agric. 10:461-467.

Chourey, P.S., Y.C. Chen, and M.E. Miller. 1991. Early cell degeneration in developing endosperm is unique to the shrunken mutation in maize. Maydica 36:141-146.

Chowdhury, K.A. and G.M. Buth. 1971. Cotton seeds from the Neolithic of Egyptian Nubia and the origin of Old World cotton. Biol. J. Linnean Soc. 3:303-312.

Chrispeels, M.J. and C. Maurel. 1994. Aquaporins: the molecular basis of facilitated water movement through living plant cells? Plant Physiology 105:9-13.

Christiansen, M.N. 1963 Influence of chilling upon seedling development of cotton. Plant Physiology 38:520-522.

Christiansen, M.N. 1964. Influence of chilling upon subsequent growth and morphology of cotton seedlings. Crop Sci. 4:584-586.

Christiansen, M.N. 1967. Periods of sensitivity to chilling in germinating cotton. Plant Physiology 42:431-433.

Christiansen, M.N. 1968. Induction and prevention of chilling injury to radicle tips of imbibing cottonseed. Plant Physiology 43:743-746.

Christiansen, M.N. and E.N. Ashworth. 1978. Prevention of chilling injury to seedling cotton with antitranspirants. Crop Sci. 18:907-908.

Christiansen, M.N., H.R. Carns, and D.J. Slyter. 1970. Stimulation of soluble loss from radicles of *Gossypium hirsutum* L. by chilling, anaerobiosis, and low pH. Plant Physiol. 46:53-56.

Christiansen, M.N. and J.L. Hilton. 1974. Prevention of trifluralin effect on cotton with soil supplied lipids. Crop Sci. 14:489-492.

Christiansen, M.N. and R.P. Moore. 1959. Seed coat structural differences that influence water uptake and seed quality in hard seed cotton. Agron. J. 51:582-584.

Christiansen, M.N. and R.A. Rowland. 1986. Germination and stand establishment. pp. 535-541. *In:* Mauney, J.R., and J. McD. Stewart (eds.), Cotton Physiology. The Cotton Foundation, Memphis, Tenn.

Christiansen, M.N. and R.O. Thomas. 1969. Season-long effects of chilling treatments applied to germinating cottonseed. Crop Science 9:672-673.

Christidis, B.G. and G.J. Harrison. 1955. Cotton growing problems. McGraw-Hill Book Co., Inc., New York.

Chu, C., T.J. Henneberry, R.C. Weddle, E.T. Natwick, J.J. Carson, C. Valenzuela, S.L. Birsdall, and R.T. Staten. 1996. Redustion of pink bollworm (Lepidoptera: Gelechiidae) populations in the Imperial Valley, California, following mandatory short-season cotton management systems. J. Econ. Entomol. 89: 175-182

Chu, Y.N., C.G. Coble, and W.R. Jordan. 1991. Hypocotyl elongation and swelling of cotton as affected by soil temperature, moisture, and physical impedance. Crop Sci. 31:410-415.

Chung, R. P.-T., G.M. Neumann, and G.M. Polya. 1997. Purification and characterization of basic proteins with *in vitro* antifungal activity from seed of cotton, *Gossypium hirsutum*. Plant Science 127:1-16.

Cihacek, L.J. and T.A. Kerby. 1991. Effects of residual soil nitrogen and urea on yield and petiole nitrate of cotton. J. Proc. Agric. 4:193-197.

Clark, G.B. and S.J. Roux. 1995. Annexins of plant cells. Plant Physiol. 109:1133-1139.

Clark, J.A. and J. Levitt. 1956. The basis of drought resistance in the soybean plant. Physiol. Plant. 9:598-606.

Clark, L.E., J.E. Slosser, E.P. Boring, T.W. Ruchs, and R.R. Minzenmayer. 1996. Evaluation of harvest-aid chemicals for early fall termination of cotton as a boll weevil management strategy. p. 977-982. *In:* P. Dugger and D.A. Richter (eds.). Proc Beltwide Cotton Prod Conf., National Cotton Council of America, Memphis, Tenn.

Clark, T.H., R.K. Ball, C.A. Stutte, and C. Guo. 1992. Root activity in cotton as affected by stress and bioregulants. p. 1028. *In:* D.J. Herber and D.A. Richter (eds.). Proc. Beltwide Cotton Conf., National Cotton Council of America, Memphis, Tenn.

Clarke, J.M., T.N. McCaig, and R.M. DePauw. 1994. Inheritance of glaucousness and epicuticular wax in durum wheat. Crop Sci. 34:327-330.

Cleland, R.E. 1986. The role of hormones in wall loosening and plant growth. Aust. J. Plant Physiol. 13:93-103.

Cleland, R.E. 1995. Auxina and cell elongation. pp. 214-227. *In:* P.J. Davies (ed.). Plant Hormones, Kluwer Academic Publishers, Netherlands.

Clouse, S.D. and J.M. Sasse. 1998. Brassinosteroids: Essential regulators of plant growth and development. Ann. Rev. Plant Physiol. Plant Mol. Biol. 49:427-451.

Clowes, F.A.L. and H.E. Stewart. 1967. Recovery from dormancy in roots. New Physiol. 66:115-123.

Cochran, M., F. Bourland, N.P. Tugwell, and D.M. Oosterhuis. 1996. COTMAN overview, update and future. University of Arkansas Agricultural Experiment Station Special Report 178:23-24.

Coen, E.S. 1991. The role of homeotic genes in flower development and evolution. Annu. Rev. Plant Physiol. Plant Mol. Biol. 42:241-279.

Coen, E.S. and E.M. Meyerowitz. 1991. The war of the whorls: Genetic interactions controlling flower development. Nature 353:31-37.

Cogn'ee, M. 1988. Temperature and development in Mediterranean cotton cultivation. Cot. Fib. Trop. XLIII:85-100.

Cohen, Y., Z. Plaut, A. Meiri, and A. Hadas. 1995. Deficit irrigation of cotton for increasing groundwater use in clay soils. Agron. J. 87:808-814.

Cole, D.F. and J.E. Wheeler. 1974. Effect of pregermination treatments on germination and growth of cottonseed at suboptimal temperatures. Crop Sci. 14:451-454.

Coley, P.D. 1988. Effects of plant growth rate and leaf lifetime on the amount and type of anti-herbivore defense. Oecologia 74:531-536.

Collins, G.H. and J.D. Warner. 1927. Root development of cotton on Cecil sandy loam during 1926. Agron. J. 19:839-842.

Colyer, P.D., S. Micinsky, and P.R. Vernon. 1991. Effects of thrips on the development of cotton seedling diseases. Plant Disease 75:380-382.

Colyer, P.D. and P.R. Vernon. 1991. Effect of thrips infestation on the development of cotton seedling diseases. Plant Disease 75:380-382.

Colyer, P.D. and P.R. Vernon. 1993. Effect of tillage on cotton plant populations and seedling diseases. J. Prod. Agric. 6:108-111.

Conner, J.W., D.R. Kreig, and J.R. Gipson. 1972. Accumulation of simple sugars in developing cotton bolls as influenced by night temperatures. Crop Sci. 12:752-754.

Conrad, C.M. 1941. The high wax content of green lint cotton. Science 94:113.

Constabel, C.P., D.R. Bergey, and C.A. Ryan. 1995. Systemin activates synthesis of wound-inducible tomato leaf polyphenol oxidase via the octadecanoid defense signaling pathway. Proc. Natl. Acad. Sci. USA 92:407-411.

Constable, G.A. 1977. Narrow row cotton in the Namoi Valley. I. Growth, yield and quality of four cultivars. Aust. J. Exp. Agric. Animal Husbandry 17:135-142.

Constable, G. A. 1986. Growth and light receipt by mainstem cotton leaves in relation to plant density in the field. Agric. For. Meteorol. 37:279-292.

Constable, G. A. 1991. Mapping the production and survival of fruit on field-grown cotton. Agron. J. 83:374-378.

Constable, G.A. 1994. Predicting yield responses of cotton to growth regulators. *In:* G.A. Constable and N.W. Forrester (eds.) Challenging the future. CSIRO, Australia. pp. 3-5.

Constable, G.A. and A.C. Gleeson. 1977. Growth and distribution of dry matter in cotton. Aust. J. Agric. Res. 28:249-256.

Constable, G.A. and A.B. Hearn, 1981. Irrigation for crops in a sub-humid environment. VI. Effect of irrigation and nitrogen fertilizer on growth, yield and quality of cotton. Irrig. Sci. 3:17-28.

Constable, G. A. and H.M. Rawson. 1980a. Effect of leaf position, expansion and age on photosynthesis, transpiration and water use efficiency of cotton. Aust. J. Plant. Physiol. 7:89-100.

Constable, G. A. and H.M. Rawson. 1980b. Carbon production and utilisation in cotton: inferences from a carbon budget. Aust. J. Plant. Physiol. 7:539-553.

Constable, G.A. and H.M. Rawson. 1980c. Photosynthesis, respiration and transpiration of cotton fruit. Photosynthetica 14:557-563.

Constable, G.A. and H.M. Rawson. 1982. Distribution of ^{14}C label from cotton leaves: Consequences of changed water and nitrogen status. Austral. J. Plant Physiol. 9:735-747.

Constable, G.A., I.J. Rochester, J.H. Betts, and D.F. Herridge. 1991. Prediction of nitrogen fertilizer requirement in cotton using petiole and sap nitrate. Commun. Soil Sci. Plant Anal. 22:1315-1324.

Constable, G.A., I.J. Rochester, and J.B. Cook. 1988. Zinc, copper, iron, manganese and boron uptake by cotton on cracking clay soils of high pH. Aust. J. Exp. Agric. 28:351-356.

Constable, G.A., I.J. Rochester, and I.G. Daniels. 1992. Cotton yield and nitrogen requirement is modified by crop rotation and tillage method. Soil and Tillage Res. 23:41-59.

Cook, C.G. and K.M. El-Zik. 1992. Cotton seedling and first-bloom characteristics: relationships with drought-influenced boll abscission and lint yield. Crop Sci. 32:1464-1467.

Cook, J.A. and D. Boynton. 1952. Some factors affecting the absorption of urea by McIntosh apple leaves. Proc. Amer. Soc. Hort. Sci. 59:82-90.

Cookson, D., P.S.J. Cheetham, and E.B. Rathbone. 1987. Preparative high-performance liquid chromatographic purification and structural determination of 1-*O*-α-D-glucopyranosyl-D-fructose (trehalulose). J. Chromatog. 402:265-272.

Cooley, D.R. and W.J. Manning. 1987. The impact of ozone on assimilate partitioning in plants: A review. Environ. Pollut. 47:95-113.

Cooper, G., D. Delmer, and C. Nitsche. 1986. Photo affinity analog of herbicide inhibiting cellulose biosynthesis: Synthesis of [³H]-2,6-dichlorophenyazide. J. Labeled Cmpds Radiopharm. 24:759-761.

Cooper, H.B. 1992. Cotton for high fiber strength. pp. 303-314. *In:* C.R. Benedict (ed.). Proc. Cotton Fiber Cellulose: Structure, Function, and Utilization Conference. National Cotton Council, Memphis Tenn.

Cooper, H.B., Jr., J. Pellow, J. Palmer, K. McRae, and D. Anderson. 1996. Evaluation of the processing efficiency of two Acala cottons using bale lots. pp. 1687-1689. *In:* Proc. Beltwide Cotton Conf., National Cotton Council of America, Memphis, Tenn.

Cooper, R.B., R.E. Blaser, and R.H. Brown. 1967. Potassium nutrition effects on net photosynthesis and morphology of alfalfa. Soil Sci. 31:231-235.

Cope, J.T., Jr. 1981. Effects of 50 years of fertilization with phosphorus and potassium on soil test levels and yields at locations. Soil Science Society of America Journal 45:342-347.

Corbin, B.R., Jr. and R.E. Frans. 1991. Protecting cotton (*Gossypium hirsutum* L.) from fluometuron injury with seed protectants. Weed Sci. 39:408-411.

Corcoran, C.J. and C. Zeiher. 1995. Regulation of PEP carboxylase during cotton fiber elongation. Plant Physiol. Suppl. 108:120(610).

Corcoran, C.J., G.R. Zepeda, and C.A. Zeiher. 1993. Identification of PEP carboxylase isoforms differentially expressed during cotton fiber developement. Plant Physiol. Suppl. 102:123 (694).

Cornish, K. 1988. Why does photosynthesis decline with leaf age? pp. 50-55. Proc., Beltwide Cotton Production Research Conferences. National Cotton Council of America; Memphis.

Cornish, K. and J.A.D. Zeevart. 1985. Abscisic acid accumulation by roots of *Xanthium strumarium* L. and *Lycopersicon esculentum* Mill. in relation to water stress. Plant Physiol. 79:653-658.

Cosgrove, D.J. 1993a. Wall extensibility: its nature, measurement and relationship to plant cell growth. New Phytol. 124:1-23.

Cosgrove, D.J. 1993b. How do plant cell walls extend. Plant Physiol. 102:1-6.

Cosgrove, D.J. 1997. Relaxation in a high stress environment: the molecular bases of extensible cell walls and cell enlargement. The Plant Cell 9:1031-1041.

Cosgrove, D.J. and D.M. Durachko. 1994. Autolysis and extension of isolated walls from growing cucumber hypocotyls. J. Exp. Bot. 45:1711-1719.

Cosgrove, D.J. and D.P. Knievel, eds. 1987. Physiology of cell expansion during plant growth. Amer. Soc. Plant Phys., Rockville MD. 318.

Cothren, J.T. 1980. Boll opening responses of cotton to Ethrel and GA 776141. p. 83. *In:* E.S. Sullivan (ed.). Proc. Seventh Ann. Meet. Plant Growth Regul. Working Group, Plant Growth Regulator Working Group, Longmont, CO.

Cothren, J.T. 1994. Use of growth regulators in cotton production. *In:* G.A. Constable and N.W. Forrester (eds.) Challenging the future. CSIRO, Australia. pp. 6-24.

Cothren, J.T., D.W. Albers, M.J. Urwiler, and D.S. Guthrie. 1983. Comparative growth analysis of mepiquat chloride-treated cotton cultured under controlled environment. pp. 253-261. *In:* W.C. Wilson (ed.). Proc. Tenth Ann. Meet. Plant Growth Regul. Soc. Am., Plant Growth Regulator Soc. Am., Lake Alfred, FL.

Cothren, J.T. and C.D. Cotterman. 1980. Evaluation of Cytozyme Crop⁺ as a foliar application to enhance cotton yields. Ark. Farm. Res. 29:2.

Cothren, J.T., J.A. Landivar, and D.M. Oosterhuis. 1996. Mid-flowering application of PGR-IV to enhance cotton maturity and yield. p. 1149. *In:* P. Dugger and D.A. Richter (eds.). Proc. Beltwide Cotton Prod. Res. Conf., National Cotton Council of America, Memphis, Tenn.

Cothren, J.T., P.R. Nester, and C.A. Stutte. 1977. Some physiological responses to 1,1-dimethyl-piperidinium chloride. p. 204. *In:* E.S. Sullivan (ed.). Proc. Fourth Ann. Meet. Plant Growth Regul. Working Group, Plant Growth Regulator Working Group, Longmont, CO.

Cothren, J.T. and D.M. Oosterhuis. 1993. Physiological impact of plant growth regulators. pp. 128-132. *In:* D.J. Herber and D.A. Richter (eds.). Proc. Beltwide Cotton Conf., National Cotton Council of America, Memphis, Tenn.

Cothren J.T. and D.M. Oosterhuis. 2009. Use of growth regulators in cotton production. pp. 291-305. *In:* J.M.Stewart, D.M. Oosterhuis, J.J. Heitholt, and J.R. Mauney (eds.). Physiology of Cotton. National Cotton Council of America, Memphis,Tenn. pp. Springer, London.

Cotterman, C.D. and J.T. Cothren. 1979. Field and laboratory studies using Cytozyme Crop + in cotton. pp. 80-85. *In:* E.S. Sullivan (ed.). Proc. Sixth Ann. Meet. Plant Growth Regul. Working Group, Plant Growth Regulator Working Group, Longmont, CO.

Cotton Incorporated. 1999. Quality Summary of 1996 through 1998 USA Upland Crops. Cotton

Incorporated home page. Available http://www.cottoninc.com

Cotty, P.J. 1987. Temperature-induced suppression of Alternaria leaf spot of cotton in Arizona. Plant Disease 71:1138-1140.

Cotty, P.J. 1989. Virulence and cultural characteristics of two *Aspergillus flavus* strains pathogenic on cotton. Phytopathology 79:808-814.

Cotty. P.J. 1994. Influence of field application of an atoxigenic strain of *Aspergillus flavus* on the populations of *A. flavus* infecting cotton bolls and on aflatoxin content of cottonseed. Phytopathology 84:1270-1277.

Cotty, P.J. and P. Bayman. 1993. Competitive exclusion of a toxigenic strain of *Aspergillus flavus* by an atoxigenic strain. Phytopathology 83:1283-1287.

Cotty, P.J., P. Bayman, D.S. Egel, and K.S. Elias. 1994. Agriculture, aflatoxins, and *Aspergillus*. pp. 1-27. *In*: K.A. Powell, A. Renwick, and J.F. Peberdy (eds.). The Genus Aspergillus: From Taxonomy and Genetics to Industrial Application. Plenum Press, New York.

Crafts, A.S. 1956. Weed control: Applied botany. Am. J. Bot. 43:548-556.

Cramer, G.R., A. Läuchli, and E. Epstein. 1986. Effects of NaCl and CaCl$_2$ on ion activities in complex nutrient solutions and root growth of cotton. Plant Physiol. 81:792-797.

Cramer, G.R., A. Läuchli, and V.S. Polito. 1985. Displacement of Ca^{2+} by Na$^+$ from the plasmalemma of root cells. A primary response to salt stress? Plant Physiol. 79:207-211.

Cramer, G.R., J. Lynch, A. Läuchli, and E. Epstein. 1987. Influx of Na$^+$, K$^+$, and Ca^{2+} into roots of salt-stressed cotton seedlings. Effects of supplemental Ca^{2+}. Plant Physiol. 83:510-516.

Cramer, M.D., O.A.M. Lewis, and S.H. Lips. 1993. Inorganic carbon fixation and metabolism in maize roots as affected by nitrate and ammonium nutrition. Physiologia Plantarum. 89:632-639.

Crane, C.F., H.J. Price, D.M. Stelly, D.G. Czeschin Jr., and T.D. McKnight. 1993. Identification of a homeologous chromsome pair by *in situ* DNA hybridization to ribosomal RNA loci in meiotic chromosomes of cotton (*Gossypium hirsutum*). Genome 36:1015-1022.

Craven, L.A., J. McD. Stewart, A.H.D. Brown, and J.P. Grace. 1994. The Australian wild species of *Gossypium*. Pp. 278-281 *In*: G.A. Constable and N.W. Forrester (eds.). Challenging the Future. Proc. World Cotton Research Conference-1. CSIRO, Melbourne, Australia.

Crawford, S.H. 1980. Acceleration of cotton boll opening with GAF-7767141. pp. 35-36. *In*: J.M. Brown (ed.). Proc. Beltwide Cotton Prod. Res. Conf., National Cotton Council of America, Memphis, Tenn.

Crawford, S.H. 1981. Effects of mepiquat chloride on cotton in Northeast Louisiana. pp. 45-46. *In*: J.M. Brown (ed.). Proc. Beltwide Cotton Prod. Res. Conf., National Cotton Council of America, Memphis, Tenn.

Crawley, M.J. 1983. Herbivory. The Dynamics of Animal-plant interactions. Studies in Ecology, Vol. 10. Blackwell Scientific Publications, London, 437 pp.

Crawley, M.J. 1987. Benevolent herbivores?Trends Ecol. Evol. 2: 167-168.

Creech, R.G., J.N. Jenkins, B. Tang, G.W. Lawrence, and J.C. McCarty. 1995. Cotton resistance to root-knot nematode. I: Penetration and reproduction. Crop Sci. 35:365-368.

Creelman, R.A. and J.E. Mullet. 1995. Jasmonic acid distribution in plants: Regulation during development and response to biotic and abiotic stress. Proc. Natl. Acad. Sci. USA 92:4114-4119.

Creelman, R.A. and J.E. Mullet. 1997. Biosynthesis and action of jasmonates in plants. Annu. Rev. Plant Physiol. Plant Mol. Biol. 48:355-381.

Cronn, R.C., X. Zhao, A.H. Paterson, and J.F. Wendel. 1996. Polymorphism and concerted evolution in a tandemly repeated gene family: 5S ribosomal DNA in diploid and allopolyploid cottons. J. Mol. Evol. 42:685-705.

Crowther, F. 1934. Studies in growth analysis of the cotton plant under irrigation in the Sudan. I. The effects of different combinations of nitrogen applications and water supply. Ann. Bot. 48:877-913.

Cui, Y., A.A. Bell, O. Joost, and C. Magill. 2000. Expression of potential defense response genes in cotton. Physiological and Molecular Plant Pathology 56:25-31.

Cui, Y., A.A. Bell, and C. Magill. 1996. Differential induction of cotton defense pathways by *Verticillium*: cloning defense response genes. Phytopathology (supplement) 86:545.

Cui, Y., H. Orta, A.A. Bell, C. Magill, and C. Gonzalez. 1997. Characterization of plasmids in the *Agrobacterium* causing root rot and wilt of cotton. Phytopathology 87:S21.

Culp, T.W. 1992. Simultaneous improvement of lint yield and fiber quality in upland cotton. pp. 247-288. *In*: C.R. Benedict (ed.). Proc. Cotton Fiber Cellulose: Structure, Function, and Utilization Conference. National Cotton Council, Memphis Tenn.

Culp, T.W. 1994. Genetic contributions to yield in cotton. pp. 321-360 *In*: G.A. Slafer (ed.). Genetic Improvement of Field Crops. CRC Press, New York.

Culpepper, A.S. and A.C. York. 2000. Weed management in ultra-narrow row cotton (*Gossypium hirsutum* L.). Weed Tech. 14:19-29.

Cummings, B.G. and E. Wagner. 1968. Rhythmic processes in plants. Ann. Rev. Plant Physiol. 19:381-416.

Cure, J.D. and B. Acock. 1986. Crop responses to carbon-dioxide doubling: A literature survey. Agric. For. Met. 38:127-145.

Curley, R., B. Roberts, T. Kerby, C. Brooks, and J. Knutson. 1990. Effect of moisture on moduled seed cotton. pp. 683-686. *In*: Proc. Beltwide Cotton Production Research Conf., National Cotton Council of America, Memphis, Tenn.

Curry, R.B., R.M. Peart, J.W. Jones, K.J. Boote, and L.H. Allen, Jr. 1990. Response of crop yield to predicted change in climate and atmospheric CO_2 using simulation. Trans. ASAE 33:1383-1390.

Curtis, P.S. and A. Läuchli. 1986. The role of leaf area development and photosynthetic capacity in determining growth of kenaf under moderate salt stress. Aust. J. Plant Physiol. 18:553-565.

Cushman, J.C., C.B. Michalowski, and H.J. Bohnert. 1990. Developmental control of crassulacean acid metabolism inducibility by salt stress in the common ice plant. Plant Physiol. 94:1137-1142.

Cutler, D.F., K.L. Alvin, and C.E. Price. 1982. The Plant Cuticle. Linnean Soc. Symp. Academic Press, London. 461 pp.

Cutler, J.M. and D.W. Rains. 1977. Effects of irrigation history on responses of cotton to subsequent water stress. Crop Sci. 17:329-335.

Cutler, J. M. and S.W. Rains. 1978. Effects of water stress and hardening on the internal water relations and osmotic constituents of cotton leaves. Physiol. Plant. 42:261-268.

Cyr, R.J. and B.A. Palevitz. 1989. Microtubule-binding proteins from carrot. 1. Initial characterization and microtubule bundling. Planta 177:245-260.

Da Nóbrega, L.B., N.E.D. Beltrao, D.J. Vieira, M.D. Diniz, and D.M.P. De Azevedo. 1993. Effect of plant spacing and of apical bud removal period on herbaceous cotton. Pesquisa Agr. Brasil. 28:1379-1384.

Da Silva, M.J., J.G. De Souza, M.B. Neto, and J.V. Da Silva. 1992. Selection on 3 cotton cultivars for tolerance to germination under saline conditions. Pesquisa Agropecuaria Brasileira 27:655-659.

Daayf, F., M. Nicole, B. Boher, A. Pando, and J.P. Geiger. 1997. Early vascular defense reactions of cotton roots infected with a defoliating mutant strain of *Verticillium dahliae*. European J. Plant Pathology 103:125-136.

Daayf, F., M. Nicole, and J.P. Geiger. 1995. Differentiation of *Verticillium dahliae* populations on the basis of vegetative compatibility and pathogenicity on cotton. European J. Plant Pathology 101:69-79.

Dabney, S.M. 1995. Cover crops in reduced tillage systems. p. 58-60. *In:* T.D. Valco and M.R. McClelland (eds.). Conservation-Tillage Systems for Cotton. University of Arkansas Agricultural Experiment Station Special Report 169. Fayetteville, AR.

Dale, J.E. 1972. Growth and photosynthesis in the first leaf of barley. I. The effect of time of application of nitrogen. Ann. Bot. 36:967-979.

Dale, J.E. and T.H. Coaker. 1958. Some effects of feeding by *Lygus vosseleri* Poop. (Heteroptera, Miridae) on the stem apex of the cotton plant. Ann. appl. Biol. 46:423-429.

Dallacker, V.F., P. Leuther, and K. Wabinski-Westrop. 1989. Gossypol und hemigossypol, II. Darstellung von 8-Hydroxy-4-isopropyl-naphtho[2,3-d][1 ,3]diox-ol-derivaten. Chemiker-Zeitung 113:97-102.

Damp, J.E. and D.M. Pearsall. 1994. Early cotton from coastal Ecuador. Econ. Bot. 48:163-165.

Danforth, D.M., M.J. Cochran, N.P. Tugwell, F.M. Bourland, D.M. Oosterhuis, and E.D. Justice. 1995. Statistical relationship between SQUARMAP and earliness. p. 522-524. *In:* D.A. Richter and J. Armour (eds.). Proc. Beltwide Cotton Conf., National Cotton Council of America, Memphis, Tenn.

Danforth, D.M. and P. O'Leary. 1998. COTMAN crop management system. Version 5.0. Publ. Cotton Incorporated and Univ. Arkansas, Gary, NC.

Dang, P.M., J.L. Heinen, and R.D. Allen. 1996. Expression of a "cotton fiber specific" gene, *Gh-1*, in transgenic tobacco and cotton. Plant Physiol. Suppl. 111:55.

Daniels, M.J., F. Chaumont, T.E. Mirkov, and M.J. Chrispeels. 1996. Characterization of a new vacuolar membrane aquaporin sensitive to mercury at a unique site. Plant Cell 8:587-599.

Dann, M.S. and E.J. Pell. 1989. Decline of activity and quantity of ribulose bisphosphate carboxylase/oxygenase and net photosynthesis in ozone-treated potato foliage. Plant Physiol. 91:427-432.

Darrall, N.M. 1989. The effect of air pollutants on physiological processes in plants. Plant Cell Environ. 12:1-30.

Darwin, C.A. and F. Darwin. 1897. The power of movement in plants. Appleton, New York, N.Y.

Davidonis, G. 1993a. A comparison of cotton ovule and cotton suspension cultures: Response to gibberellic acid and 2-chloroethylphosphonic acid. J. Plant Physiol. 141:505-507.

Davidonis, G. 1993b. Cotton fiber growth and development *in vitro*: effects of tunicamycin and monensin. Plant Science 88:229-236.

Davidonis, G.H. and R.H. Hamilton. 1983. Plant regeneration from callus tissue of *Gossypium hirsutum* L. Plant Sci. Letters 32:89-93.

Davidonis, G. and O. Hinojosa. 1994. Influence of seed location on cotton fiber development *in planta* and *in vitro*. Plant Science 203:107-113.

Davidonis, G.H., A. Johnson, J. Landivar, and O. Hinojosa. 1996. Influence of low-weight seeds and motes on the fiber properties of other cotton seeds. Field Crops Res. 48:141-153.

Davidonis, G.H., A. Johnson, J.A. Landivar, and K.B. Hood. 1999. The cotton fiber property variability continuum from motes through seeds. Textile Res. J. 69:754-759.

Davidonis, G., J. Webb, S. May, and O. Hinojosa. 1995. Monitoring fiber quality during the ginning process using AFIS. pp. 1287-1289. *In:* Proc. Beltwide Cotton Conf., National Cotton Council of America, Memphis, Tenn.

Davidson, E.W., B.J. Segura, T. Steele, and D.L. Hendrix. 1994. Microorganisms influence the composition of honeydew produced by the silverleaf whitefly, *Bemisia argentifolii.* J. Insect Physiol. 40:1069-1076.

Davies, D.D. 1973. Control of and by pH. Symp. Soc. Exp. Biol. 27:513-529.

Davies, P.J. 1995. *Plant Hormones: Physiology, Biochemistry, and Molecular Biology*. 2ⁿᵈ edition. Kluwer Academics Publishers, Boston, MA.

Davies, P.J. 1995. The plant hormone concept: Concentration, sensitivity, and transport. pp. 13-38. *In:* P.J. Davies (ed.). Plant Hormones. Kluwer Academic Publishers, Netherlands.

Davies, W.J., U. Schurr, G. Taylor, and J. Zhang. 1987. Hormones as chemical signals involved in root to shoot communication of effects of change in the soil environment. pp. 201-2016. *In:* G.V. Hoad, M.B. Jackson, .L. Lenton, and R. Atkin (eds.). Hormone Action in Plant Development-A Critical Appraisal. Butterworths, London.

Davila-Huerta, G., H. Hamada, G.D. Davis, R.D. Stipanovic, C.M. Adams, and M. Essenberg. 1995. Cadinane-type sesquiterpenes induced in *Gossypium* cotyledons by bacterial inoculation. Phytochemistry 39(3):531-536.

Davis, C.R., G.W. Morgan, and D.R. Howell. 1965. Sulphur dioxide fumigation of cotton and its effect on fiber quality. Agron. J. 57:250-251.

Davis, D.A., P.S. Low, and P. Heinstein. 1998. Purification of a glycoprotein elicitor of phytoalexin formation from *Verticillium dahliae*. Physiological and Molecular Plant Pathology 52:259-273.

Davis, D.G., K.E. Dusabek, and R.A. Hoerauf. 1974. *In vitro* culture of callus tissues and cell suspensions from okra (*Hibiscus esculentus* L.) and cotton (*Gossypium hirsutum* L.). In Vitro (9) 6:395-398.

Davis, E.M., Y.-S. Chen, M. Essenberg, and M.L. Pierce. 1998. cDNA sequence ofa (+)-deltacadinene syn-

thase gene induced in *Gossypium hirsutum* L. by bacterial infection. Plant Physiol. 116:1191.

Davis, E.M., J. Tsuji, G.D. Davis, M.L. Pierce, and M. Essenberg. 1996. Purification of (+)-δ-cadinene synthase, a sesquiterpene cyclase from bacteria-inoculated cotton foliar tissue. Phytochemistry 41:1047-1055.

Davis, G.D. and M. Essenberg. 1995. (+)-δ-cadinene is a product of sesquiterpene cyclase activity in cotton. Phytochemistry 39:553-567.

Davis, J.G. 1995. Impact of time of day and time since irrigation on cotton leaf blade and petiole nutrient concentration. Commun. Soil Sci. Plant Anal. 26:2351-2360.

Davis, J.G. 1996. Provision of mid-season potassium requirements in cotton with slow release potassium applied preplant. J. Plt. Nutr. 19:1-14.

Davis, L.A. and F.T. Addicott. 1972. Abscisic acid: Correlations with abscission and with development in the cotton fruit. Plant Physiol. 49:644-648.

Davis-Carter, J.G., S.H. Baker, and C. Hodges. 1992. Potassium fertilization of irrigated cotton on sandy soils. Proc. Beltwide Cotton Conf., Nashville, Tenn. pp. 1147-1150.

Davis, R.M. 1980. Influence of *Glomus fasciculatus* on *Thielaviopsis basicola* root rot of citrus. Plant Dis. 64:839-840.

Davis, R.M. and J.A. Menge. 1981. *Phytophthora parasitica* inoculation and intensity of vesicular-arbuscular mycorrhizae in citrus. New Phytol. 87:705-715.

Davis, R.M., J.A. Menge, and D.C. Erwin. 1979. Influence of *Glomus fasciculatus* and soil phosphorus on verticillium wilt of cotton. Phytopathology 69:453-456.

Davis, R.M., J.J. Nunez, K.D.Marshall, D.S.Munk, R.N. Vargas, W.L. Weir, and S.D. Wright. 1994. Chemical control of cotton seedling disease in California. pp. 252-253. *In:* Proc. Beltwide Cotton Conf., National Cotton Council of America, Memphis, Tenn.

Davis, W.J. and J. Zhang. 1991. Root signals and the regulation of growth and development of plants in drying soils. Annu. Rev. Plant Physiol. and Plant Mol. Biol. 423:76.

Deaton, R.W. 1996. Controlling cotton pests. Science 273:1641.

DeBarry, A. 1871. Ueber die Washsuberzuge bei Pflanzen; ein Studdium mit dem Oberflachenmikroskop. Botanisches Archiv 19:461-473.

DeFeyter, R. and D.W. Gabriel. 1991. At least six avirulence genes are clustered on a 90-kilobase plasmid in *Xanthomonas campestris* pv. *malvacearum*. Molecular Plant-Microbe Interactions 4:423-432.

DeFeyter, R., H. McFadden, and L. Dennis. 1998. Five avirulence genes from *Xanthomonas campestris* pv. *malvacearum* cause genotype-specific cell death when expressed transiently in cotton. Molecular Plant-Microbe Interactions 11:698-701.

DeFeyter, R., Y. Yang, and D.W. Gabriel. 1993. Gene-for-gene interactions between cotton R genes and *Xanthomonas campestris* pv. *malvacearum avr* genes. Molecular Plant-Microbe Interactions 6:225-237.

Dehne, H.W. 1 982. Interactions between vesicular-arbuscular mycorrhizal and plant pathogens. Phytopath. 72:1115-1119.

DeJong, T.M. and J. Goudriaan. 1989. Modeling peach fruit growth and carbohydrate requirements: re-evaluation of the double-sigmoid growth pattern. J. Amer. Soc. Hort. Sci. 114:800-804.

DeJoode, D.R. 1992. *Molecular insights into speciation in the genus Gossypium L. (Malvaceae)*. M.S. Thesis, Iowa State University, Ames.

DeJoode, D.R. and J.F. Wendel. 1992. Genetic diversity and origin of the Hawaiian Islands cotton, *Gossypium tomentosum*. Amer. J. Bot. 79:1311-1319.

Delaney, T.P., S. Uknes, B. Vernooij, L. Friedrich, K. Weymann, D. Negrotto, T. Gaffney, Manuela gut-Rella, H. Kessman, E. Ward, and J. Ryals. 1994. A central role of salicylic acid in plant disease resistance. Science 266:1247-1249.

DeLanghe, E.A.L. 1986. Lint development. pp. 325-350. *In:* J.R. Mauney and J.M. Stewart (eds.). Cotton Physiology. The Cotton Foundation, Memphis, Tenn.

DeLanghe, E., S.K. Kosmidou-Dimitropoulou, and L. Waterkyn. 1978. Effect of hormones on nucleolar growth and vacuolation in elongating cotton fibers. Planta 140:269-278.

Delcourt, H.R. and W.F. Harris. 1980. Carbon budget for the southeastern U.S. biota: analysis of historical change in trend from source to sink. Science. 210:321-323.

Deleuze, C. and F. Houllier. 1995. Prediction of stem profile of *Picea abies* using a process-based tree growth model. Tree Physiol. 15:113-120.

Delmer, D.P. 1987. Cellulose biosynthesis. Ann. Rev. Plant Physiol. 38:259-290.

Delmer, D.P. 1990. Role of the plasma membrane in cellulose synthesis. pp. 256-268. *In:* C. Larsson and I.M. Moller (eds.). The Plant Plasma Membrane: Structure, Function, and Molecular Biology. Springer Verlag, Berlin.

Delmer, D.P. 1994. The potential role of membrane-associated sucrose synthase in cellulose synthesis and of a small G-protein in cytoskeletal organization in the developing cotton fiber. pp. 105-108. *In:* G. Jividen and C. Benedict (eds.). Proc. Biochemistry of Cotton Workshop. Cotton Incorporated, Raleigh, NC.

Delmer, D.P. 1999. Cellulose biosynthesis in developing cotton fibers. pp. 85-112. *In:* A.S. Basra (ed.). Cotton Fibers: Developmental Biology, Quality Improvement, and Textile Processing. Food Products Press, Binghamton, NY.

Delmer, D.P. and Y. Amor. 1995. Cellulose biosynthesis. Plant Cell 7:987-1000.

Delmer, D.P., Y. Amor, A. Andrawis, T. Potikha, M. Solomon, and L. Gonen. 1992. Synthesis of the cotton fiber secondary cell wall and its relation to fiber quality: A discussion of what we know and do not know. pp. 153-162. *In:* C.R. Benedict (ed.). Proc. Cotton Fiber Cellulose: Structure, Function, and Utilization Conference. National Cotton Council, Memphis Tenn.

Delmer, D.P., P. Ohana, L. Gonen, and M. Benziman. 1993a. In vitro synthesis of cellulose in plants still a long way to go. Plant Physiol. 103:307-308.

Delmer, D.P., J.R. Pear, A. Andrawis, and D. Stalker. 1995. Genes encoding small GTP-binding proteins analogous to mammalian rac are preferentially expressed in developing cotton fibers. Mol. Gen. Genet. 248:43-51.

Delmer, D.P., S.M. Read, and G. Cooper. 1987. Identification of receptor protein in cotton fibers for the herbicide 2,6dichlorobenzonitrile. Plant Physiol. 84:415-420.

Delmer, D.P., M. Solomon, and S.M. Read. 1991. Direct photolabeling with [^{32}P]UDP-glucose for identification of a subunit of cotton fiber callose synthase. Plant Physiol. 95:556-563.

Delmer, D.P., M. Volokita, M. Solomon, U. Fritz, W. Delphendahl, and W. Herth. 1993b. A monoclonal antibody recognizes a 65 kDa higher plant membrane polypeptide which undergoes cation dependent association with callose synthase in vitro and co-localizes with sites of high callose deposition in vivo. Protoplasma 176:33-42.

Delouche, J.C. 1986. Harvest and post-harvest factors affecting the quality of cotton planting seed and seed quality evaluation. pp. 483-518. *In:* J.R Mauney and J.McD. Stewart (eds.). Cotton Physiology. The Cotton Foundation, Memphis, Tenn.

Delouche, J.C., C. Guevara, and B.C. Keith. 1995. Development, release, and characteristics of the hard seed condition in cotton. pp. 1121-1125. *In:* Proc. Beltwide Cotton Conf., National Cotton Council of America, Memphis, Tenn.

Denecke, M., M. Schultz, C. Fischer, and H. Schnabl. 1993. Partial purification and characterization of stomatal phosphoenolpyruvate carboxylase from *Vicia faba*. Physiol. Plant. 87:96-102.

Denham, J.H. 1924. The cytology of the cotton plant. I. Microspore formation in Sea Island cotton. II. Chromosome numbers of old and new world cottons. Annal. Bot. 38:407-438.

Dennis, D.T. and J.A. Miernyk. 1982. Compartmentation of nonphotosynthetic carbohydrate metabolism. Ann. Rev. Plant Physiol. 33:27-50.

Dennis, D.T. and m.F. Greyson. 1987. Fructose-6-phosphate metabolism in plants. Physiol. Plant 69:295-404.

Dennis, R.E. and R.E. Briggs. 1969. Growth and development of the cotton plant in Arizona. Univ. of Arizona Agric. Exp. Stn. Bull. A-64.

Denton, P. 1993. Fertilization practices in conservation tillage. pp. 113-116. *In:* Proc. Beltwide Cotton Conf., National Cotton Council of America, Memphis, Tenn.

Derckel, J.-P., L. Legendre, J.-C. Audran, B. Haye, and B. Lambert. 1996. Chitinases of the grapevine (*Vitis vinifera* L.): Five isoforms induced in leaves by salicylic acid are constitutively expressed in other tissues. Plant Sci.119:31-37.

DeSilva, W.H. 1971. Some effects of the growth retardant chemical CCC on cotton in Uganda. Cotton Grow. Rev. 48:131-135.

Deussen, H.1992. Improved cotton fiber properties–The textile industry's key to success in global competition. pp. 43-63. *In:* C.R. Benedict (ed.). Proc. Cotton Fiber Cellulose: Structure, Function, and Utilization Conference. National Cotton Council, Memphis Tenn.

Deussen, H. and C. Faerber. 1995. A cotton valuation model. pp. 431-433. *In:* Proc. Beltwide Cotton Conf., National Cotton Council of America, Memphis, Tenn.

DeVay, J.E. 1989. Physiological and biochemical mechanisms in host resistance and susceptibility to wilt pathogens. pp. 197-217. *In*: E.C. Tjamos and C.H. Beckman (eds.). Vascular Wilt Diseases of Plants. NATO ASI Series H: Cell Biology, Volume 28, Springer-Verlag, New York.

Dever, J.K., J.R. Gannaway, and R.V. Baker. 1988. Influence of cotton fiber strength and fineness on fiber damage during lint cleaning. Textile Res. J. 58:433-438.

deWit, C.T. 1965. Photosynthesis of leaf canopies. Agr. Res. Rep. 663. Inst. for Biol. and Chem. Research on Field Crops and Herbage. Wageningen.

Dhillion, S.S. 1992. Host-endophyte specificity of vesicular-arbuscular mycorrhizal colonization of *Oryza sativa* L. at the pre-transplant stage in low or high phosphorus soil. Soil Biol. Biochem. 24:405-411.

Dhillon, S., S.K. Suneja, S.K. Sawhney, and R. Singh. 1985. Properties of NADP-malic enzyme from glumes of developing wheat grains. Phytochem. 24:1657-1662.

Dhindsa, R.S. 1978a. Hormonal regulation of enzymes of non-autotrophic CO_2 fixation in unfertilized cotton ovules. Z. Pflanzenphysiol. 89:355-365.

Dhindsa, R.S. 1978b. Hormonal regulation of enzymes of nonautotropic CO_2 fixation in unfertilized cotton ovules. Z. Pflanzenphysiol. 89:355-362.

Dhindsa, R.S., C.A. Beasley, and I.P. Ting. 1975. Osmoregulation in cotton fiber accumulation of potassium and malate during growth. Plant Physiol. 56:394-398.

Dhindsa, R.S., C.A. Beasley, and I.P. Ting. 1976. Effects of abscisic acid on in vitro growth of cotton fiber. Planta 130:197-201.

Diboll, A.G. 1968. Fine structural development of the megagametophyte of *Zea mays* following fertilization. Am. J. Bot. 55:787-806.

Dicke, M., P. van Baarlen, R. Wessels, and H. Dijkman. 1993. Herbivore induces systemic production of plant volatiles that attract predators of the herbivore: Extraction of endogenous elicitor. J. Chem. Ecol. 19:581

Dickens, J.C. 1986. Orientation of boll weevil, *Anthronomus grandis* Boh. (Coleoptera: Curculionidae), to pheromone and volatile host compound in the laboratory. J. Chem. Ecol. 12:91-98.

Dickey, E.C., J.C. Siemens, P.J. Jasa, V.L. Hofman, and D.P. Shelton. 1992. Tillage system definitions. pp. 5-7. *In:* Conservation Tillage Systems and Management. MidWest Plan Serv. Pub. MWPS-45. Iowa State Univ. Ames, IA.

Dickinson, C.D., T. Altabella, and M.J. Chrispeels. 1991. Slow-growth phenotype of transgenic tomato expressing apoplastic invertase. Plant Physiol. 95:420-425.

Dixon, D.C., R.W. Seagull, and B.A. Triplett. 1994. Changes in the accumulation of a- and b-tubulin isotypes during cotton fiber development. Plant Phys. 105:1347-1353.

Dofing, S.M., E.J. Penas, and J.W. Maranville. 1984. Effect of bicarbonate on iron reduction by soybean roots. J. Plant Nutr. 12:797-802.

Dolan, L., K. Janmaat, V. Willemsen, P. Linstead, S. Poethig, K. Roberts, and B. Scheres. 1993. Cellular organization of the *Arabidopsis thaliana* root. Development 119:71-84.

Doman, D.C., J.C. Walker, R.N. Trelease, and B.D. Moore. 1982. Metabolism of carbohydrate and lipid reserves in germinated cotton seeds. Planta 155:502-510.

Domsch, K.H., W. Gams, and T.H. Anderson. 1980. Compendium of soil fungi. Vol. 1. Academic Press, London, U.K.

Donald, L. 1964. Nutrient deficiencies in cotton. pp. 59-98. *In:* H.B. Sprague (ed.). Hunger signs in crops, 3rd ed. McKay. New York.

Donath, J. and W. Boland. 1995. Biosynthesis of acyclic homoterpenes: enzyme selectivity and absolute configuration of the nerolidol precursor. Phytochem. 39:785-790.

Dong, J.F. 1995. The yield increasing ability of spraying cotton with boron. Henan Nongye Kexue. 3:6 (in Chinese).

Doorenbos, J. 1976. Agrometeorolgical-field stations. Irrigation and Drainage Paper no. 27, FAO, United Nations, Rome, Italy. 94 pp.

Douds, D.D., L. Galvez, R.R. Janke, and P. Wagoner. 1995. Effect of tillage and farming system upon populations and distribution of vesicular-arbuscular mycorrhizal fungi. Agric. Ecosyst. Environ. 52:111-118.

Douds, D.D., R.R. Janke, and S.E. Peters. 1993. VAM fungus spore populations and colonization of roots of maize and soybean under conventional and low-input sustainable agriculture. Agric. Ecosyst. Environ. 43:325-335.

Douds, D.D. and G. Nagahashi. 2000. Signaling and recognition events prior to colonization of roots by arbuscular mycorrhizal fungi. p. 11-18. *In:* G.K. Podila and D.D. Douds (eds.). Current Advances in Mycorrhizae Research. APS Press, St Paul, MN.

Downward, J. 1992. Rac and rho in tune. Nature 359:273-274.

Dracup, M. 1991. Increasing salt tolerance of plants through cell culture requires greater understanding of tolerance mechanisms. Aust. J. Plant Physiol. 18:1-15.

Drake, B.G. and P.W. Leadley. 1991. Canopy photosynthesis of crops and native plant communities exposed to long term elevated CO_2. Plant, Cell Environ. 14:853-860.

Drew, M.C. and L.H. Stolzy. 1991. Growth under oxygen stress. pp.331-350. *In:* Y. Waisel, A. Eshel, and U. Kafkafi (eds.). Plant Roots: The Hidden Half. Marcel Dekker, New York.

Dubery, I.A. and V. Slater. 1997. Induced defense responses in cotton leaf discs by elicitors from *Verticillium dahliae*. Phytochemistry 44:1429-1434.

Dubery, I.A. and F. Smit. 1994. Phenylalanine ammonia-lyase from cotton (*Gossypium hirsutum*) hypocotyls: Properties of the enzyme induced by a *Verticillium dahliae* phytotoxin. Biochimica et Biophysica Acta 1207:24-30.

Duckett, C.M. and C.W. Lloyd. 1994. Gibberellic acid-induced microtubule reorientation in dwarf peas is accompanied by rapid modification of an α-tubulin isotype. Plant J. 5:363-372.

Duff, S.M. and R. Chollet. 1995. In vivo regulation of wheat-leaf phosphoenolpyruvate carboxylase by reversible phosphorylation. Plant Physiol. 107:775-782.

Dugas, W.A., M.L. Heuer, D. Hunsaker, B.A. Kimball, K.F. Lewin, J. Nagy, and M. Johnson. 1994. Sap flow measurements of transpiration from cotton grown under ambient and enriched CO_2 concentrations. Agric For. Met. 70:231-246.

Duggar, W.M. 1983. Boron in plant metabolism. *In:* A. Läuchli and R.L. Bieleski (eds.). Encyclopedia of Plant Physiology, New Series. Vol.15B. Springer-Verlag, Berlin. pp. 626-650.

Dugger, W.M. and R.L. Palmer. 1980. Effect of boron on the incorporation of glucose from UDP-glucose into cotton fibers grown *in vitro*. Plant Physiol. 65:266-272.

Dugger, W.M., Jr. and I.P. Ting. 1970. Air pollution oxidants-their effects on metabolic processes in plants. Ann. Rev. Plant Physiol. 21:215-234.

Duke, S.O., R.N. Paul, J.M. Becerril, and J.H. Schmidt. 1991. Clomazone causes accumulation of sesquiterpenoids in cotton (*Gossypium hirsutum* L.). Weed Science 39:339-346.

Dukhovny, V.A. 2000. The regional water strategy as mechanism and set of measures for sustainable water management of the Aral Sea Basin. Water and Land Resources Development and Management for Sustainable Use Denpasas, Bali, Indonesia, 19-26 July, 1998.

Duncan. W.G. 1972. SIMCOT: A simulator of cotton growth and yield. pp. 115-118. Proc. Workshop on Tree Growth Dynamics and Modeling. Duke university, Oct. 11-12, 1971.

Duncan, W.G., R.S. Loomis, W.A. Williams, and R. Hanau. 1967. A model for simulating photosynthesis in plant communities. Hilgardia, J. Agric. Sci. 38:181-205.

Duncombe, W.G. 1977. Cotton losses caused by spider mites (Acarina: Tetranychidae). Rodhesia Agric. J. 74:141-146.

Dunster, K.W., R.L. Dunlap, and F.J. Gonzales. 1980. Influence of Ethrel plant regulator on boll opening and defoliation of Western cotton. pp. 317-321. *In:* J.M. Brown (ed.). Proc. Beltwide Cotton Prod. Res. Conf., National Cotton Council of America, Memphis, Tenn.

Durant, J.A. 1989. Yield response of cotton cultivars to early-season applications of chloridimeform and aldicarb. J. Econ. Entomol. 82:626-632.

Dure, L., III. 1975. Seed formation. Annu. Rev. Plant Physiol. 26:259-278.

Dure, L., III. 1993. Structural motifs in *Lea* proteins of higher plants. *In:* T.J. Kohel and E.A. Bray (eds.). Response of Plant to Cellular Dehydration during Environmental Stress. Amer. Soc. Plant Physiol., Rockville, Md.

Dure, L., III. 1994. Structure/function studies of *Lea* protein. *In:* G. Coruzzi and P. Puigdomenech (eds.). NATO ASI Series Vol. H81, Plant Molecular Biology. Springer Verlag, Berlin

Dure, L., III and C.A. Chlan. 1981. Developmental biochemistry of cottonseed embryogenesis and germination. XII. Purification and properties of the principal storage proteins. Plant Physiol. 68:180-186.

Dure, L., III, C.A. Chlan, J.C. Baker, and G.A. Galau. 1984. Gene sets active in cottonseed embryogenesis. pp. 591-599. *In:* E.H. Davidson and R.A. Firtel (eds.). *Molecular Biology of Development.* Alan R. Liss, New York, N.Y., USA.

Dure, L., III, G.A. Galau, C.A. Chlan, and J. Pyle. 1983. Developmentally regulated gene sets in cotton embryogenesis. pp. 331-342. *In:* R.B. Goldberg (ed.). *Plant Molecular Biology.* Alan R. Liss, New York, N.Y., USA.

Dure, L., III, S.C. Greenway, and G.A. Galau. 1981. Developmental biochemistry of cottonseed embryogenesis and germination: Changing messenger ribonucleic acid populations as shown by *in vitro* and *in vivo* protein synthesis. Biochemistry 20:4162-4168.

Dybing, C.D. and H.B. Currier. 1961. Foliar penetration by chemicals. Plant Physiol. 36:169-174.

Dye, D.W., J.F. Bradbury, M. Goto, A.C. Hayward, R.A. Lelliot, and M.N. Schroth. 1980. International standards for naming pathovars of phytopathogenic bacteria and a list of pathovar names and pathotype strains. Rev. Plant Pathology 59:153-168.

Dyer, D. and D.A. Brown. 1980. *In situ* root observation using fiber optic/video and fluorescence. American Society of Agronomy. Agronomy Abstracts. P. 80.

Dyer, M.I., C.L. Turner, and T.R. Seastedt. 1993. Herbivory and its consequences. Ecological Applications 3:10-16.

Eaton, F.M. 1944. Deficiency, toxicity, and accumulation of boron in plants. Agric. Res. 69:237-277.

Eaton, F.M. 1955. Physiology of the cotton plant. Ann. Rev. Plant Physiol. 6:299-328.

Eaton, F.M. and J.E. Bernardin. 1964. Mass-flow an salt accumulation by plants on water versus soil cultures. Soil Sci. 97:411-416.

Eaton, F.M. and D.R. Ergle. 1948. Carbohydrate accumulation in the cotton plant at low moisture levels. Plant Physiol. 23:169-187.

Eaton, F.M. and D.R. Ergle. 1953. Relationship of seasonal trends in carbohydrate and nitrogen levels and effects of girdling and spraying with sucrose and urea to the nutritional interpretation of boll shedding in cotton. Plant Physiol. 28:503-520.

Eaton, F.M. and D.R. Ergle. 1954. Effects of shade and partial defoliation on carbohydrate levels and the growth, fruiting, and fiber properties of cotton plants. Plant Physiol. 29:39-49.

Eaton, F.M. and N.R. Rigler. 1945. Effect of light intensity, nitrogen supply, and fruiting on carbohydrate uti-

lization by the cotton plant. Plant Physiol. 20:380-411.

Ebelhar, M.W. and R.A. Welch. 1989. Cotton production in rotation systems with corn and soybean. Proc. Beltwide Cotton Conf., Nashville, Tenn. pp. 509-512.

Ebelhar, M.W., R.A. Welch, and W.R. Meredith, Jr. 1996. Nitrogen rates and mepiquat chloride effects on cotton lint yield and quality. pp. 1373-1378. *In:* Proc. Beltwide Cotton Conf., National Cotton Council of America, Memphis, Tenn.

Echevarria, C., V. Pacquit, N. Bakrim, L. Osuna, B. Delgado, M. Arrio-Dupont, and J. Vidal. 1994. The effect of pH on the covalent and metabolic control of C_4 phosphoenolpyruvate carboxylase from sorghum leaf. Arch. Biochem. Biophys. 315:425-430.

Edmisten, K.L. 1994. The use of plant monitoring techniques as an aid in determining mepiquat chloride rates in rain-fed cotton. *In:* G.A. Constable and N.W. Forrester (eds.) Challenging the future. CSIRO, Australia. pp. 25-28.

Edmisten, K.L. 1995. Making the most of the management moment - Early square to peak bloom: Southeast, sand and rain. p. 55-56. *In:* D.A. Richter and J. Armour (eds.). Proc. Beltwide Cotton Conf., National Cotton Council of America, Memphis, Tenn.

Edmisten, K. 1997a. The cotton plant. *In:* 1997 North Carolina Cotton Production Guide, Center for IPM, North Carolina State University, Raleigh, N.C. Available http://ipmwww.ncsu.edu/Production_ Guides/cotton/chptr2.html

Edmisten, K. 1997b. Cotton classification. *In:* 1997 North Carolina Cotton Production Guide. Available http://ipmwww.ncsu.edu/Production_Guides/ Cotton/chptr18.html

Edminsten, K. 1997c. Planting decisions. *In:* 1997 North Carolina Cotton Production Guide, Center for IPM, North Carolina State University, Raleigh, N.C. Available http://ipmwww.ncsu.edu/Production_ Guides/cotton/chptr4.html

Edmisten, K.L., C.W. Wood, C.H. Burmester, and C.C. Mitchell. 1993. Foliar fertilization of seedling cotton. pp. 1304-1306. *In:* D.L. Herbert and D.A. Richter (eds.). Proc. Beltwide Cotton Conferences. National Cotton Council of America, Memphis, Tenn.

Edwards, G.A., J.E. Endrizzi, and R. Stein. 1974. Genome DNA content and chromosome organization in *Gossypium*. Chromosoma 47:309-326.

Edwards, G.E. and C.S. Andreo. 1992. NADP-malic enzyme from plants. Phytochem. 31:1845-1857.

Edwards, G.A. and M.A. Mirza. 1979. Genomes of the Australian wild species of cotton. II. The designation of a new G genome for *Gossypium bickii*. Can. J. Genet. Cytol. 21:367-372.

Ehleringer, J.R. and S.D. Hammond. 1987. Solar tracking and photosynthesis in cotton leaves. Agric. For. Meterol. 39:25-35.

Ehlig, C.F. and R.D. LeMert. 1973. Effects of fruit load, temperature and relative humidity on boll retention of cotton. Crop Sci. 13:168-171.

Eichert, T., H.E. Goldbach, and J. Burkhardt. 1998. Evidence for the uptake of large anions through stomatal pores. Bot. Acta. 111:461-466.

Einspahr, K.J. and G.A. Thompson, Jr. 1990. Transmembrane signaling via phosphatidylinositol4,5-bisphosphate hydrolysis in plants. Plant Physiol. 93:361-366.

Eisenberg, A.J. and J.P. Mascarenhaus. 1985. Abscisic acid and the regulation of synthesis of specific seed proteins and their messenger RNAs during culture of soybean embryos. Planta 166:505-514.

Eisenreich, W., S. Sagner, M.H. Zenk, and A. Bacher. 1997. Monoterpenoid essential oils are not of mevalonoid origin. Tetra. Lett. 38:3889-3892

El Mogahzy, Y.E. and Y. Gowayed. 1995a. Theory and practice of cotton fiber selection. Part I: Fiber selection techniques and bale picking algorithms. Textile Res. J. 65:32-40.

El Mogahzy, Y.E. and Y. Gowayed. 1995b. Theory and practice of cotton fiber selection. Part II: Sources of cotton mix variability and critical factors affecting it. Textile Res. J. 65:75-84.

El Quakfaoui, S., C. Potvin, R. Brzezinski, and A. Asselin. 1995. A *Streptomyces* chitosanase is active in transgenic tobacco. Plant Cell Rep. 15: 222-226.

El-Fouly, M.M. and H.A. Moustafa. 1969. Growth, yield and nitrogen content of cotton plants as affected by gibberellic acid. Z. Planzenernahr. Budenk. 123:106-113.

El-Gharably, G.A. and W. Bussler. 1985. Critical levels of boron in cotton plants. Zeitschrift fur Pflanzenernahrung und Bodenkunde. 148: 681-688.

El-Gharib, E.A. and W. Kadry. 1983. Effect of potassium on tolerance of cotton plants to salinity of irrigation water. Ann. Agric. Sci. Moshtohor 20:27-34.

El-Jaoual, T. and D.A. Cox. 1998. Manganese toxicity in plants. J. Plant Nutr. 21:353-386.

El-Saidi, M.T. and W.A. Hegazy. 1980. Effect of using saline water for irrigation at different growth stages on yield and some physiological processes of cotton plant. Agric. Res. Rev. Cairo, 58:337-355.

El-Sharkawy, M. and J.D. Hesketh. 1965. Photosynthesis among species in relation to characteristics of leaf anatomy and CO_2 diffusion resistances. Crop Sci. 5:517-521.

El-Sharkawy, M.A., J.D. Hesketh, and H. Muramoto. 1965. Leaf photosynthetic rates and other growth characteristics among 26 species of *Gossypium*. Crop Sci. 5:173-175.

El-Sharkawy, H.M., F.M. Salama, and A.A. Mazen. 1986. Chlorophyll response to salinity, sodicity, and heat stress in cotton, rama, and millet. Photosynthetica 20:204-211.

El-Zahab, A.A.A. 1971a. Salt tolerance of eight Egyptian cotton varieties Part 1. At germination stage. Z. Acker- und Pflanzenbau 133:299-307.

El-Zahab, A.A.A. 1971b. Salt tolerance of eight Egyptian cotton varieties. Part II. At the seedling stage. Z. Acker- und Pflanzenbau 133:308-314.

El-Zik, K.M. 1985. Integrated control of Verticillium wilt of cotton. Plant Dis. 69:1025-1032.

El-Zik, K.M., C.R. Howell, P.M. Thaxton, and A.D. Brashears. 1993. Influence of strain, carrier, and seed sticker on the capacity of the biocontrol agent *Gliocladium virens* to affect cotton seedling disease, stand, yield, and fiber quality. pp. 187-191. *In:* Proc. Beltwide Cotton Conf., National Cotton Council of America, Memphis, Tenn.

El-Zik, K.M. and P.M. Thaxton. 1989. Genetic improvement for resistance to pests and stresses in cotton. pp. 191-224. *In:* R.E. Frisbie, K.M. El-Zik, and L.T. Wilson (eds.). Integrated Pest Management Systems and Cotton Production. John Wiley & Sons, New York.

El-Zik, K.M. and P.M. Thaxton. 1992. Simultaneous improvement of yield, fiber quality traits, and resistance to pests of MAR cottons. pp. 315-332. *In:* C.R. Benedict (ed.). Proc. Cotton Fiber Cellulose: Structure, Function, and Utilization Conference. National Cotton Council, Memphis Tenn.

El-Zik, K.M. and P.M. Thaxton. 1994. Genetic improvement and factors affecting cotton fiber quality traits. pp. 51-58. *In:* G. Jividen and C. Benedict (eds.). Proc. Biochemistry of Cotton Workshop. Cotton Incorporated, Raleigh, NC.

Eldon, S. and R.J. Hillocks. 1996. The effect of reduced phytoalexin production on the resistance of Upland cotton (*Gossypium hirsutum*) to Verticillium and Fusarium wilts. Annals of Applied Biology 129:217-225.

Eleftheriou, E.P. and J.L. Hall. 1983a. The extrafloral nectaries of cotton. I. Fine structure of the secretory papillae. J. Exp. Bot. 34:103-119.

Eleftheriou, E.P. and J.L. Hall. 1983b. The extrafloral nectaries of cotton. II. Cytochemical localization of ATPase activity and Ca^{2+}-binding sites, and selective osmium impregnation. J. Exp. Bot. 34:1066-1079.

Elena, K. 1999. Genetic relationships among *Verticillium dahliae* isolates from cotton in Greece based on vegetative compatibility. European J. Plant Pathology 105:609-616.

Elmore, C.D. 1973. Contributions of the capsule wall and bracts to the developing cotton fruit. Crop Sci. 13:751-752.

Elmore, C.D., J.D. Hesketh, and H. Muramoto. 1967. A survey of rates of leaf growth, leaf ageing and leaf photosynthesis rates among and within species. J. Ariz. Acad. Sci. 4:215-219.

Elzen, G.W., H.J. Williams, A.A. Bell, R.D. Stipanovic, and S.B. Vinson. 1985. Quantification of volatile terpenes of glanded and glandless *Gossypium hirsutum* L. cultivars and lines by gas chromatography. J. Agric. Food Chem. 33:1079-1082.

Elzen, G.W., H.J. Williams, and S.B. Vinson. 1984. Isolation and identification of cotton synomones mediating searching behavior by parasitoid *Campoletis sonorensis.* J. Chem. Ecol. 10:1251-1254.

Emes, M.J. and A.K. Tobin. 1993. Control of metabolism and development in higher plant plastids. Int. Rev. of Cyt. 145:149-216.

Emons, A.M.C. 1991. Role of particle rosettes and terminal globules in cellulose synthesis. pp. 71-98. *In:* C.H. Haigler and P.J. Weimer (eds.). Biosynthesis and Biodegradation of Cellulose. Marcel Dekker, New York.

Endrizzi, J.E. and D.T. Ray. 1992. Mapping of the cl_1, R_1, yg_1, and Dw loci in the long arm of chromosome 16 of cotton. J. Hered. 83:1-5.

Endrizzi, J.E., E.L. Turcotte, and R.J. Kohel. 1985. Genetics, cytology, and evolution of *Gossypium.* Adv. Genet. 23:271-375.

Ergle, D.R. 1936. Carbohydrate content of cotton plants at different growth periods and the influence of fertilizer. Agron. J. 28:775-786.

Ergle, D.R. 1958. Compositional factors associated with the growth response of young cotton plants to gibberellic acid. Plant Physiol. 33:344-346.

Ergle, D.R. and F.M. Eaton. 1951. Sulfur nutrition of cotton. Plant Physiol. 26:639-654.

Ergle, D.R. and F.M. Eaton. 1957. Aspects of phosphorus metabolism in the cotton plant. Plant Physiol. 32:106-113.

Ergle, D.R., L.E. Hessler, and J.E. Adams. 1938. Carbohydrates of the cotton plant under different seasonal conditions and fertilizer treatment. J. Am. Soc. Agron. 30:951-959.

Erwin, D.C., S.D. Tsai, and R.A. Kahn. 1979a. Growth retardants mitigate Verticillium wilt growth and increase yield of cotton. California Agric. 33:8-10.

Erwin, D.C., S.D. Tsai, and R.A. Kahn. 1979b. Growth retardants mitigate Verticillium wilt and influence yield of cotton. Phytopathology 69:283-287.

Esau, K. 1977. Anatomy of seed plants. 2nd Edition. John Wiley and Sons. 550 pp.

Essenberg, M., G.D. Davis, M. Pierce, H. Hamada, and G. Davila-Huerta. 1992. Biosynthesis of sesquiterpenoid phytoalexins in cotton foliar tissue. Environ. Sci. Res. 44:297-304.

Essenberg, M., P.B. Grover, Jr., and E.C. Cover. 1990. Accumulation of antibacterial sesquiterpenoids in bacterially inoculated *Gossypium* leaves and cotyledons. Phytochem. 29:3107-3113.

Essenberg, M., J.A. Hall, and A.A. Bell. 2000. Protection of cotton foliar tissue from its own photoactivated phytoalexins by epidermal flavonol glucosides. pp. 609-612. *In:* Proc. Beltwide Cotton Conf., National Cotton Council of America, Memphis, Tenn.

Essenberg, M. and M.L. Pierce. 1995. Sesquiterpenoid phytoalexins synthesized in cotton leaves and cotyledons during the hypersensitive response to *Xanthomonas campestris* pv. *malvacearum.* pp. 183-198. *In:* M. Daniel and P.R. Purkayastha (eds.). Handbook of Phytoalexin Metabolism and Action. Marcel Decker, New York.

Essenberg, M., A. Stoessl, and J.B. Stothers. 1985. The biosynthesis of 2,7-dihydroxycadalene in infected cotton cotyledons: the folding pattern of the farnesol precursor and possible implications for gossypol biosynthesis. J. Chem. Soc., Chem. Commun. pp. 556-557.

Ethridge, D. 1996. Valuing HVI quality differences in U.S. Cotton. pp. 78-83. *In:* Proc. Beltwide Cotton Conf., National Cotton Council of America, Memphis, Tenn.

Ethridge, M.D. and E.F. Hequet. 1999. Prospects for rapid measurements of stickiness in cotton. pp. 56-60. *In:* Proc. Beltwide Cotton Conf., National Cotton Council of America, Memphis, Tenn.

Evans, D.G. and M.H. Miller. 1990. The role of the external mycelial network the effects of soil disturbance upon vesicular-arbuscular mycorrhizal colonization of young maize. New Phytol. 114:65-71.

Evans, L.S. and G.R. Hendry. 1992. Responses of cotton foliage to short-term fluctuations in CO_2 partial pressures. Crit. Rev. Plant Sci. 11:203-212.

Evans, L.S. and I.P. Ting. 1973. Ozone-induced membrane permebility changes. Amer. Jour. Bot. 60:155-162.

Evans, L.T. 1993. Crop evolution, adaptation and yield. Cambridge Univ Press, Cambridge, 500 pp.

Evans, P.S. 1965. Intercalary growth in aerial shoots of *Elocaris actua* R. Br., Prodr. Ann. Bot. (N.S.) 29:205-217

Evenson, J.P. 1969. Effects of floral and terminal bud removal on the yield and structure of the cotton plant in the Ord Valley, North Western Australia. Cotton Grow. Rev. 46:37-44.

Fahn, A. 1990. Plant Anatomy. 4th ed. Pergammon Press, New York. 588.

Farage, P.K., S.P. Long, E.G. Lechner, and N.R. Baker. 1991. The sequence of change within the photosynthetic apparatus of wheat following short-term exposure to ozone. Plant Physiol. 95:529-535.

Farnsworth, K.D. and K.J. Niklas. 1995. Theories of optimization, form and function in branching architecture. Functional Ecol. 9:355-363.

Fasheun, A. and M.D. Dennett. 1981. Interception of radiation and growth efficiency in field beans (*Vicia faba* L). Agric. Meteorol. 26:221-229.

Feaster, C.V., R.E. Briggs and E.L. Turcotte. 1980. Pima cultivar responses to Pix. p. 81. *In:* J.M. Brown (ed.). Proc. Beltwide Cotton Prod. Res. Conf., National Cotton Council of America, Memphis, Tenn.

Felton, G.W., R.M. Broadway, and S.S. Duffey. 1989. Inactivation of protease inhibitor activity by plant-derived quinones: complications for host-plant resistance against noctuid herbivores. J. Insect Pysiol. 35:981-990.

Feng, G., D.S. Bai, M.Q, Yang, X.L. Li, and F.S. Zhang. 1999. Effects of salinity on VA mycorrhiza formation and of inoculation with VAM fungi on salt tolerance of plants. Chinese J. App. Ecol. 10:79-82.

Ferguson, D.L. and J.J. Burke. 1991. Influence of water and temperature stress on the temperature dependence of the reappearance of variable fluorescence following illumination. Plant Physiol. 97:188-192.

Ferguson, D.L., R.B. Turley, and R.H. Kloth. 1997. Characterization of a δ-TIP cDNA clone and determination of related A and D subfamilies in *Gossypium* species. Plant Mol. Biol. 34: 111-118.

Ferguson, D.L., R.B. Turley, B.A. Triplett, and W.R.J. Meredith. 1996. Comparison of protein profiles during cotton (*Gossypium hirsutum* L.) fiber cell development with partial sequences of two proteins. J. Agric. Food Chem. 44:4022-4027.

Ferguson, J.M. 1991. Cotton seed quality and stand establishment. pp. 26-30. *In:* North Carolina Agric. Extension Service Bull. 417, revised, North Carolina State University, Raleigh, N.C.

Ferino, M.P., F.B. Calora, and E.D. Magallona. 1982. Population dynamics and economic threshold level of the cotton semi-looper, *Anomis flava flava* (Fabr.) (Noctuidae, Lepidoptera). Philipp. Ent. 5:401-446.

Fernandez, C.J. and J.T. Cothren. 1990. Water use efficiency and biomass partitioning of mepiquat chloride-treated cotton plants. Agron. Absts. 82:121.

Fernandez, C.J., J.T. Cothren, and K.J. McInnes. 1991. Partitioning of biomass in well-watered and water-stressed cotton plants treated with mepiquat chloride. Crop Sci. 31:1224-1228.

Fernandez, C.J., J.T. Cothren, and K.J. McInnes. 1992. Carbon and water economies of well-watered and water-deficient cotton plants treated with mepiquat chloride. Crop Sci. 32:175-180.

Fernandez, C.J., J.T. Cothren, and K.J. McInnes. 1993. Whole-plant photosynthetic rates of cotton under nitrogen stress. pp. 1256-1258. 1993 Proc.

Beltwide Cotton Confs. National Cotton Council of America, Memphis, Tenn.

Fernandez, C.J., J.T. Cothren, and K.J. McInnes. 1996a. Partitioning of biomass in water-and nitrogen-stressed cotton during pre-bloom stage. J. Plant Nutrit. 19:595-617.

Fernandez, C.J., K.J. McInnes, and J.T. Cothren. 1996b. Water status and leaf area production in water- and nitrogen-stressed cotton. Crop Sci. 36:1224-1233.

Fernandez, D., K. Assigbetse, M. DuBois, and J.P. Geiger. 1994. Molecular characterization of races and vegetative compatibility groups in *Fusarium oxysporum* f.sp. *vasinfectum*. Applied and Environmental Microbiology 60:4039-4046.

Ferreira, L.G.R. and M.A.A. Reboucas. 1992. Influence of hydration dehydration on cotton seeds on overcoming the effects of salinity on germination. Pesquisa Agropecuaria Brasileira 27:609-615.

Fevre, M. and M. Rougier. 1981. β-1-3- and β-1-4-glucan synthesis by membrane fractions from the fungus *Saprolegnia.* Planta 151:232-241.

Findlay, N. 1982. Secretion of nectar. pp. 667-683. *In:* F.A. Loewus, and W. Tanner (eds.). Encyclopedia of Plant Physiology, New Series, Vol. 13A. Springer-Verlag, New York.

Finer, J.J. 1988. Plant regeneration from somatic embryogenic suspension cultures of cotton (*Gossypium hirsutum* L.). Plant Cell Reports 7:399-402.

Finer, J.J., A.A. Reilley, and R.H. Smith. 1987. Establishment of embryogenic suspension cultures of a wild relative of cotton (*Gossypium klotzchianum* Anderss). In Vitro Cellular and Developmental Biology 21 (10):717-722.

Finer, J.J. and R.H. Smith. 1982. Isolation and culture of protoplasts from cotton (*Gossypium klotzchianum* Ancerss) callus cultures. Plant Sci. Letters 26:147-151.

Finer, J.J. and R.H. Smith. 1984. Initiation of callus and somatic embryos from explants of mature cotton (*Gossypium klotzchianum* Anderss). Plant Cell Reports 3:41-43.

Finlayson-Pitts, B.J. and J.N. Pitts, Jr. 1986. Atmospheric Chemistry: Fundamentals and Experimental Techniques. John Wiley Sons, N.Y.

Firoozabady, E. and D.L. Deboer. 1986. Isolation, culture, and cell division in cotyledon protoplasts of cotton (*Gossypium hirsutum* and *G. barbadense*). Plant Cell Reports 5:127-131.

Firoozabady, E. and D.L. Deboer. 1993. Plant regeneration via somatic embryogenesis in many cultivars of cotton (*Gossypium hirsutum* L.). In Vitro Cell. Dev. Biol. 23:166-173.

Fischer, E.S. and W. Bussler. 1988. Effects of magnesium deficiency on carbohydrates in *Phaseolus vulgaris*. Z. Pflanzenernahr. Bodenk. 151:295-298.

Fischer, J.J. and L. Domelsmith (eds.). 1997. Cotton and Microorganisms. U.S. Department of Agriculture, Agricultural Research Service, ARS-138. 164 pp.

Fischer R.A. and J.T. Wood. 1979. Drought resistance in spring wheat cultivars. III. Yield associations with morpho-physiological traits. Aust. J. Agric. Res. 30:1001-1020

Fiscus, E.L. 1983. Water transport and balance within the plant: resistance water flow in roots. *In:* H.M. Taylor, W.R. Jordan, and T.R. Sinclair (eds.). Limitations to efficient water use in crop production. American Society of Agronomy, Madison, WI.

Fish, R.G., P.W. Groundwater, and J.J.G. Morgan. 1995. The photo-epimerisation of gossypol Schiffs bases. Tetrahedron Asymmetry 6:873-876.

Fisher, D.A. and D.E. Bayer. 1992. Thin sections of plant cuticles, demonstrating channels and wax platelets. Canadian J. Bot. 50:1509-1511.

Fisher, D.B., W.A. Jensen, and M.E. Ashton. 1968. Histochemical studies of pollen: Storage pockets in the endoplasmic reticulum (ER). Histochemie 13:169-182.

Fisher, D.B., J.P. Wright, and T.W. Mittler. 1984. Osmoregulation by the aphid *Myzus persicae*: A physiological role for honeydew oligosaccharides. J. Insect Physiol. 30:387-393.

Fisher, G., K. Frohberg, M.L. Parry, and C. Rosenzweig. 1995. Climate change and world food supply, demand, and trade. *In:* C. Rosenzweig, J.T. Ritchie, J.W. Jones, G.Y. Tsuji, and P. Hildebrand (eds.). Climate Change and Agriculture: Analysis of Potential International Impacts. ASA Special Publication No. 59. Madison, WI.

Fisher, W.D. 1973. Association of temperature and boll set. pp. 72-73. *In:* Proc. Beltwide Cotton Conf., National Cotton Council of America, Memphis, Tenn.

Fisher, W.D. 1975. Heat induced sterility in upland cotton. p. 85. *In:* Proc. Beltwide Cotton Conf., National Cotton Council of America, Memphis, Tenn.

Fitt, G.P. 1994. Cotton pest management: Part 3. An Australian Perspective. Annu. Rev. Entomol. 39:543-562.

Fitt, G.P., C.L. Mares, and D.J. Llewellyn. 1994. Field evaluation and potential ecological impact of transgenic cottons (*Gossypium hirsutum*) in Australia. Biocontrol Sci. Technol. 4: 535-548.

Fitter, A.H. and J. Garbaye. 1994. Interactions between mycorrhizal fungi and other soil organisms. p. 123-132. *In:* A.D. Robson, L.K. Abbott, and N. Malajczuk (eds.). Management of Mycorrhizas in Agriculture, Horticulture and Forestry. Kluwer Academic Publishers, Dordrecht, The Netherlands.

Fitter, A.H. and R.K.M. Hay. 1987. Environmental Physiology of Plants. 2nd ed. Academic Press, NY. 423 pp.

Flagler, R.B. (ed.). 1998. Recognition of Air Pollution Injury to Vegetation: A Pictorial Atlas, 2nd Ed. Air&Waste Management Assoc., Pittsburgh, PA.

Flavell, A.J., E. Dunbar, R. Anderson, S.R. Pearce, R. Hartley, and A. Kumar, A. 1992. *Ty1-copia* group retrotransposons are ubiquitous and heterogeneous in higher plants. Nucl. Acids Res. 20:3639-3644.

Fleming, A.L. and C.D. Foy. 1968. Root structure reflects differential aluminum tolerance in wheat varieties. Agron. J. 60:172-176.

Fletcher, D.C., J.C. Silvertooth, E.R. Norton, B.L. Unruh,, and E.A. Lewis.1994. Evaluation of a feedback vs. scheduled approach to PIX application. Proceedings Beltwide Cotton Conf., National Cotton Council, Memphis., pp. 1259-1262.

Flint, E.A. 1950. The structure and development of the cotton fiber. Biol. Rev. 25:414-434.

Flint, H.M., T.J. Henneberry, F.D. Wilson, E. Holguin, N. Parks, and R.E. Buehler. 1995. The effects of transgenic cotton, *Gossypium hirsutum* L., containing *Bacillus thuringiensis* toxin genes for the control of the pink bollworm, *Pectinophora gossypiella* (Saunders) and other arthropods. Southwest. Entom. 20:281-292.

Flint, H.M., J.W. Radin, N.J. Parks, and L.L. Reaves. 1995. The effects of drip or furrow irrigation of cotton on *Bemisia argentifolii* (homoptera: Aleyrodidae). J. Agric. Entomol. 12: 25-32.

Flint, H.M., F.D. Wilson, D. Hendrix, J. Leggett, S. Naranjo, T.J. Henneberry, and J.W. Radin. 1994. The effect of plant water stress on beneficial and pest insects including the pink bollworm and the sweetpotato whitefly in two short-season cultivars of cotton. Southwest. Entomol. 19:11-21.

Flor, H.H. 1946. Genetics of pathogenicity in *Melampsora lini*. J. Agric. Res. 73:335-357.

Flowers, T.J., M.A. Hajibagheri, and A.R. Yeo. 1991. Ion accumulation in the cell walls of rice plants growing under saline conditions:evidence for the Oertli hypothesis. Plant, Cell Environ. 14:319-325.

Forrester, N.W. 1994. Resistance management options for conventional *Bacillus thuringiensis* and transgenic plants in Australian summer field crops. Biocontrol Sci. Technol. 4: 549-553.

Forrester, N.W. and B. Pyke. 1997. The Bt report. The Australian Cottongrower 17: 23.

Forrester, N.W. and A.G.L. Wilson. 1988. Insect pests of cotton. Department of Agriculture New South Wales, Agdex 151/620, 18 pp.

Forster, H., H. Buchenauer, and F. Grossman. 1980. Side effects of the systemic fungicides triadimefon and triadimenol on barley plants. II. Cytokinin-like effects. Z. Pflanzenkrankl. Pflanzenschutz 84:640-653.

Fowler, J.L. 1979. Laboratory and field response of precon-ditioned upland cottonseed to minimal germina-tion temperatures. Agron J. 71:223-228.

Fowler, J.L. and L.L. Ray. 1977. Response of two cotton genotypes to five equidistant spacing patterns. Agron. J. 69:733-738.

Fowler, T., C. Lucas, and D. Gossett. 1996. Glutathione-S-transferase activity in salt adapted cotton callus. Plant Physiol. 111:No. 2SS, 256.

Fowler, T., C. Lucas, and D. Gossett. 1997. Glutathione S-transferase isozymes in control and salt-adapted cotton callus. 1997 Proceedings Beltwide Cotton Conferences, New Orleans, LA, USA, January 6 10, 1997:Vol. 2 1377-1379;

Fowler, J.L. and L.L. Ray. 1977. Response of two cotton genotypes to five equidistant spacing patterns. Agron. J. 69:733-738.

Foy, C.D., R.L. Chaney, and M.C. White. 1978 The physi-ology of metal toxicity in plants. Ann. Rev. Plant Physiol. 29:511-567.

Foy, C.D., R.R. Weil, and C.A. Coradetti. 1995. Differential manganese tolerance of cotton genotypes in nutri-ent solution. J. Plant Nutr. 18:685-706.

Fraley, R.T., S.G. Rogers, R.B. Horsch, P.R. Sanders, J.S. Flick, S.P. Adams, M.L. Bittner, L.A. Bran, C.L. Fink, J.S. Fry, G.R. Galluppi, S.B. Goldberg, N.L. Hoffman, and S.C. Woo. 1983. Expression of bac-terial genes in plant cells. Proc. Natl. Acad. Sci. USA 80:4803-4807.

Franco, M. 1986. The influence of neighbours on the growth of modular organisms with an example from trees. Phil. Trans. R. Soc. Lond. B313:209-225.

Francois, L.E. 1982. Narrow row cotton (*Gossypium hirsu-tum* L.) under saline conditions. Irrig. Sci. 3:149-156.

Franke, W. 1967. Mechanisms of foliar penetration of solu-tions. Ann. Rev. Plant. Physiol. 18:281-300.

Franke, W. 1971. The entry of residues into plants via ec-todesmata (ectocythodes). pp. 81-115. *In:* F.A. Gunther and J.D. Gunther (eds.). Residue Reviews. Vol. 38. Springer-Verlag, New York.

Frankenberger, W.T. and M. Arshad. 1995. Phytohormones in soils: microbial production and function. Marcel Dekker, New York.

Fraps, G.S. 1919. The chemical composition of the cotton plant. Texas Agric. Exp. Stn. Bull. 247.

Fravel, D.R. 1999. Commercial biocontrol products for use against soilborne crop diseases. USDA, ARS, Biocontrol of Plant Diseases Laboratory. Commercial Biocontrol Product List (on Line): http://www.barc.usda.gov/psi/bpdl/bpdlprod/bio-prod.html.

Freeden, A.L.; I.M. Rao, and N. Terry. 1989. Influence of phosphorus nutrition on growth and carbon par-titioning in *Glycine max* (L.) Merr. Planta 181: 399-405.

Freuke, H., A. Mantell, and A. Meiri. 1986. Drip Irrigation of cotton with saline-sodic water. Hassadeh 66:664-667.

Freytag, A.H. and E.A. Coleman. 1973. Effect of multiple applications of 2,3,5-triiodobenzoic acid (TIBA) on yield of stormproof and nonstormproof cotton. Agron. J. 65:610-612.

Friedhli, H., H. Lotscher, H. Oeschger, U. Siegenthaler, and B. Stauffer. 1986. Ice core record of the $^{13}C/^{12}C$ ratio of atmospheric CO_2 in the past two centuries. Nature. 324:237-238.

Friese, C.F. and M.F. Allen. 1991. The spread of VA mycor-rhizal fungal hyphae in soil: Inoculum types and external hyphal architecture. Mycologia 83:409-418.

Frisbie, R.E., K.M. El-Zik, and L.T. Wilson. 1989. Integrated pest management systems and cotton production. John Wiley & Sons, New York, 437 pp.

Fry, S.C. 1986. Cross-linking of matrix polymers in the growing cell walls of angiosperms. Annu. Rev. Plant Physiol. 37:165-186.

Fry, S.C., S. Aldington, P.R. Hetherington, and J. Aitken. 1993. Oligosaccharides as signals and substrates in the plant cell wall. Plant Physiol. 103:1-5.

Frydrych, R., E. Hequet, and C. Brunissen. 1995. The high speed stickiness detector: Relation with the spin-ning process. pp. 1185-1189. *In:* Proc. Beltwide Cotton Conf., National Cotton Council of America, Memphis, Tenn.

Frye, E. 1995. Why HVI sells cotton. pp. 115-116. *In:* Proc. Beltwide Cotton Conf., National Cotton Council of America, Memphis, Tenn.

Fryxell, P.A. 1963. Morphology of the base of seed hairs of *Gossypium.* I. Gross Morphology. Bot. Gaz. 124:196-199.

Fryxell, P.A. 1965a. Stages in the evolution of *Gossypium.* Adv. Frontiers Pl. Sci. 10:31-56.

Fryxell, P.A. 1965b. A further description of *Gossypium tri-lobum.* Madroño 18:113-118.

Fryxell, P.A. 1967. *Gossypium trilobum*: an addendum. Madroño 19: 117-123.

Fryxell, P.A. 1968. A redefinition of the tribe Gossypieae. Bot. Gaz. 129:296-308.

Fryxell, P.A. 1971. Phenetic analysis and the phylogeny of the diploid species of *Gossypium* L. (Malvaceae). Evolution 25:554-562.

Fryxell, P.A. 1976. Germpool utilization: *Gossypium*, a case history. USDA, ARS-S-137.

Fryxell, P.A. 1978. *Gossypium turneri* (Malvaceae), a new species from Sonora, Mexico. Madroño 25:155-159.

Fryxell, P.A. 1979. The natural history of the cotton tribe. Texas A&M University Press, College Station, Texas.

Fryxell, P.A. 1980. the natural history of the cotton tribe. Texas A&M Press, College Station, Texas.

Fryxell, P.A. 1992. A revised taxonomic interpretation of *Gossypium* L. (Malvaceae). Rheedea 2:108-165.

Fryxell, P.A., L.A. Craven, and J.McD. Stewart. 1992. A revision of *Gossypium* sect. *Grandicalyx* (Malvaceae), including the description of six new species. Syst. Bot. 17:91-114.

Fryxell, P.A. and S.D. Koch. 1987. New or noteworthy species of flowering plants from the Sierra Madre del Sur of Guerrero and Michoacan, Mexico. Aliso 11:539-561.

Fuchs, M. 1990. Infrared measurement of canopy temperature and detection of water stress. Theor. Appl. Climatol. 42: 253-261.

Fukai, S. and R.S. Loomis. 1976. Light display and light environments in row-planted cotton communities. Agric. Meteorol. 17:353-379.

Funderberger, E.R. 1988. Effects of starter fertilizer on cotton yields in Mississippi. pp. 496-497. *In:* J.M. Brown (ed.). Beltwide Cotton Prod. Res. Conference., New Orleans, LA. 3-8 Jan., 1988. National Council of America. Memphis, Tenn.

Furukawa, A., M. Katase, and T. Ushijima. 1984. Inhibition of photosynthesis of poplar species and sunflower by O_3. Res. Report Natl. Inst. Environ. Studies, Japan 65:77-86.

Futuyama, D.J. and R.M. May. 1991. The coevolution of plant-insect and host-parasite relationships. pp. 139-166 *In:* R.J. Berry, T.J. Crawford, and G.M. Hewitt (eds.). Genes in Ecology. The 33rd symposium of the British Ecological Soc., Univ. East of Anglia, Blackwell Scientific Publications, Oxford.

Fye, R.E., V.R. Reddy, and D.N. Baker. 1984. The validation of GOSSYM: Part I. Arizona conditions. Agr. Sys. 14:85-105.

Gabr, A.I. and S.A. El-Ashkar. 1977. The effect of different combinations of soil salinity and CCC on dry matter accumulation and yield of cotton plants. Biol. Plant. 19:391-393.

Gabr, A.I., M. El-Kadi, M.T. El-Saidi, and M.A. Kortam. 1975. The combined effect of soil salinity and nitrogen levels on dry matter accumulation and yield of cotton plants. Z. Acker- und Pflanzenbau 141:151-159.

Gabriel, D.W., A. Burges, and G.R. Lazo. 1986. Gene-for-gene interactions of five cloned avirulence genes from *Xanthomonas campestris* pv. *malvacearum* with specific resistance genes in cotton. Proc. National Academy of Science, USA 83:6415-6419.

Gabriel, D.W., M.T. Kingsley, Y. Yang, J. Chen, and P. Roberts. 1994. Host-specific virulence genes of *Xanthomonas*. pp. 141-158. *In*: C.I. Kado and J.H. Crosa (eds.). Molecular Mechanisms of Bacterial Virulence. Kluwer Academic Publishers, The Netherlands.

Gadallah, M.A.A. 1995. Effect of water stress, abscisic acid, and proline on cotton plants. J. Arid Environ. 30:315-325.

Galanopoulou-Sendouka, S., A.G. Sficas, N.A Fotiadis, A.A. Gagianas, and P.A. Gerakis. 1980. Effect of population density, planting date, and genotype on plant growth and development of cotton. Agron. J. 72:347-353.

Galau, G.A. 1986. Differential gene activity in cotton embryogenesis. Chapter 28. pp. 425-439. *In:* J.R. Mauney and J. McD. Stewart (eds.). *Cotton Physiology Number One,* The Cotton Foundation, Publisher, Memphis, Tenn., USA.

Galau, G.A., H.W. Bass, and D.W. Hughes. 1988. Restriction fragment length polymorphisms in diploid and allotetraploid *Gossypium*: Assigning the late embryogenesis-abundant (*Lea*) alloalleles in *G. hirsutum*. Mol. Gen. Genet. 211:305-314.

Galau, G.A., N. Bijaisoradat, and D.W. Hughes. 1987. Accumulation kinetics of cotton late embryogenesis-abundant mRNAs and storage protein mRNAs: coordinate regulation during embryogenesis and the role ofabscisic acid. Dev. Biol. 123:198-212.

Galau, G.A., C.A. Chlan, and L. Dure, III. 1983. Developmental biochemistry of cottonseed embryogenesis and germination. XVI. Analysis of the principle cotton storage protein gene family with cloned cDNA probes. Plant Mol. Biol. 2:189-198.

Galau, G.A., D.W. Hughes, and L. Dure, III. 1986. Abscisic acid induction of cloned cotton late embryogenesis-abundant *(Lea)* mRNAs. Plant Mol. Biol. 7:155-170.

Galau, G.A., H.Y.-C. Wang, and D.W. Hughes. 1991. Sequence of the *Gossypium hirsutum* D-genome alloallele of *Legumin A* and its mRNA. Plant Physiol. 97:1268-1270.

Galau, G.A., H.Y.-C. Wang, and D.W. Hughes. 1992a. Cotton *Mat5-A* (C164) gene and *Mat5-D* cDNAs encoding methionine-rich 2S albumin storage proteins. Plant Physiol. 99:779-782.

Galau, G.A., H.Y.-C. Wang, and D.W. Hughes. 1992b. Cotton *Lea4* (D19) and *LeaA2* (D132) group 1 *Lea* genes encoding water stress-related proteins containing a 20-amino acid motif. Plant Physiol. 99:783-788.

Gale, J., H.C. Kohl, and R.M. Hagan. 1967. Changes in the water balance and photosynthesis of onion, bean, and cotton plants under saline conditions. Physiol. Plant. 20:408-420.

Galeano, R., M.D. Rumbaugh, D.A. Johnson, and J.L. Bushnell. 1986. Variation in epiculticular wax content of alfalfa cultivars and clones. Crop Sci. 26:703-706.

Galway, M., J. Masucci, A. Lloyd, V. Walbot, R. Davis, and J. Schiefelbein. 1994. The *TTG* gene is required to specify epidermal cell fate and cell patterning in *Arabidopsis* root. Dev. Biol. 166:740-756.

Ganieva, R.A., S.R. Allahverdiyev, N.B. Guseinova, N.I. Kavakli, and S, Nafisi. 1998. Effect of salt stress and synthetic hormone Polystimuline K on the photosynthetic activity of cotton (*Gossypium hirsutum*). Turkish J. Bot. 22:217-221.

Gannaway, J.R. 1994. Breeding for insect resistance. *In*: Insects pests of cotton, G.A. Matthews and J.P. Tunstall (eds.), pp. 431-453. CAB International, Wallingford.

Gannaway, J.R. and J.K. Dever. 1992. Development of high quality cottons adapted to stripper-harvested production areas. pp. 333-340. *In:* C.R. Benedict (ed.). Proc. Cotton Fiber Cellulose: Structure, Function, and Utilization Conference. National Cotton Council, Memphis Tenn.

Gannaway, J.R., K. Hake, and R.K. Harrington. 1995. Influence of plant population upon yield and fiber quality. pp. 551-556. *In:* D.A. Richter and J. Armour (eds.). Proc. Beltwide Cotton Conf., National Cotton Council of America, Memphis, Tenn.

Gant, T.G. and A.I. Meyers. 1993. Oxazoline-mediated synthesis of the *Gossypium* sesquiterpene lacinilene C-7 methyl ether and a structurally related HIV-1 reverse-transcriptase inhibitor. Tetra. Lett. 34:3707-3710.

Garber, RH. 1994. Biological and environmental effects on development of soilborne cotton seedling diseases. pp. 225-226. *In:* Proc. Beltwide Cotton Conf., National Cotton Council of America, Memphis, Tenn.

Garber, R.H., J.E. DeVay, C.R.Howell, R.J. Wakeman, and S.A. Wright. 1995. Field performance of chemical and biological cotton seed treatments and the basis for their effectiveness or failure. pp. 229-232. *In:* Proc. Beltwide Cotton Conf., National Cotton Council of America, Memphis, Tenn.

Garbers, C. and C. Simmons. 1994. Approaches to understanding auxin action. Trends Cell Biol. 4:245-250.

Gardner, W.R. and C.F. Ehlig. 1965. Physical aspects of the internal water relations of plant leaves. Plant Physiol. 40:705-710.

Gardner, B.R., D.C. Nielsen, and C.C. Shock. 1992. Infrared thermometry and the crop water stress index. II. Sampling procedures and interpretation. J. Prod. Agric. 5:466-473.

Gardner, B.R. and T.C. Tucker. 1967. Nitrogen effects on cotton. I. Vegetative and fruiting characteristics. Soil Sci. Soc. Amer. Proc. 31:780-785.

Garrels, J.I., B. Futcher, R. Kobayashi, G.I. Latter, B. Schwender, T. Volpe, J.R. Warner, and C.S. McLaughlin. 1994. Protein identification for a *Saccharomyces cerevisiae* protein database. Electrophoresis 15:1466-1486.

Gausman, H.W., P.S. Baur, Jr., M.P. Porterfield, and R. Cardenas. 1972. Effects of salt treatments of cotton plants (*Gossypium hirsutum* L.) on leaf mesophyll cell microstructure. Agron. J. 64:133-136.

Gausman, H.W., J.E. Quisenberry, J.J. Burke, and C.W. Wendt. 1984. Leaf spectral measurements to screen cotton strains for characters affected by stress. Field Crops Res. 9:373-381.

Gawel, N.J. and C.D. Robacker. 1990. Genetic control of somatic embryogenesis in cotton petiole callus cultures. Euphytica 49:249-254.

Geever, R.F., F.R.H. Katterman, and J.E. Endrizzi. 1989. DNA hybridization analyses of a *Gossypium* allotetraploid and two closely related diploid species. Theor. Appl. Genet. 77:553-559.

Gehring, C.A. and T.G. Whitham. 1994. Interactions between aboveground herbivores and the mycorrhizal mutualists of plants. Trends. Eco. Evol. 9: 251-255.

Geigenberger, P., S. Langerberger, I. Wilke, D. Heineke, H. Heldt, and M. Stitt. 1993. Sucrose is metabolized by sucrose synthase and glycolysis within the phloem complex of *Ricinus communis* L. seedlings. Planta 190:446-453.

Gemtos, T.A. and Th. Lellis. 1997. Effects of soil compaction, water and organic matter contents on emergence and initial plant growth of cotton and sugar beet. J. Agric. Engng. Res. 66:121-134.

Gepts, P. 1987. Characterizing plant phenology: growth and development scales. pp. 3-24. *In:* K. Wisol and J.D. Hesketh (eds.). Plant Growth Modeling for Resource Management. II. Quantifying Plant Processes. CRC Press, Inc., Boca Raton, FL.

Gerard, C.J. and E. Hinojosa. 1973. Cell wall properties of cotton roots as influenced by calcium and salinity. Agron. J. 65:556-560.

Gerard, C.J., P. Sexton, and G. Shaw. 1982. Physical factors influencing soil strength and root growth. Agron. J. 74:875-879.

Gerik, T.J., R.G. Lemon, K.L. Favor, T.A. Hoelewyn, and M. Jungman. 1998. Performance of ultra-narrow row cotton in Central Texas. pp. 66-67. *In:* P. Dugger and D.A. Richter (eds.). Proc. Beltwide Cotton Conf., National Cotton Council of America, Memphis, Tenn.

Gerik, T.J., R.G. Lemon, K.L. Faver, T.A. Hoelwyn, and M. Jungman. 1998. Performance of ultra-narrow row cotton in central Texas. pp. 1406-1409. *In:* P. Dugger and D. Richter (eds.) Proc. Beltwide Cotton Conf., National Cotton Council of America, Memphis, Tenn.

Gerik, T.J., J.E. Morrison, and F.W. Chichester. 1987. Effects of controlled-traffic on soil physical properties and crop rooting. Agron. J. 79:434-438.

Gerik, T.J., D.M. Oosterhuis, and W.H. Baker. 1993. Cotton Monitoring: Crop Nutrition and Fertilizer Management. Proc. Beltwide Cotton Conf., New Orleans, LA. pp. 1184-1190.

Gerik, T.J, D.M. Oosterhuis, and H.A. Tolbert. 1998. Managing cotton nitrogen supply. Adv. Agron. 64:115-147.

Gerik, T.J., W.D. Rosenthal, C.O. Stockle, and B.S. Jackson. 1989. Analysis of cotton fruiting. pp. 64-69. *In:* D.J. Herber and D.A. Ritcher (eds.). Proc. Beltwide Cotton Conf., National Cotton Council, Memphis, Tenn.

Gershenzon, J. 1994. The cost of plant chemical defense against herbivory - A biochemical perspective. pp. 105-173 *In:* E.A. Bernays (ed.). Insect-Plant Interactions, Vol 5.

Gerstel, D.U. 1953. Chromosomal translocations in interspecific hybrids of the genus *Gossypium*. Evol. 7:234-244.

Gettier, S.W., D.C. Martens, and T.B. Brumback, Jr. 1985. Timing of foliar manganese application for correction of manganese deficiency in soybean. Agron. J. 77:627-630.

Ghassemi, F., A.J. Jakeman, and H.A. Nix. 1995. Salinization of Land and Water Resources. Human Causes, Extent, Management and Case Studies. CAB International, Wallingford. ISBN 0 85198 906 3. 526 pp.

Gianinazzi-Pearson, V. 1996. Plant cell responses to arbuscular mycorrhizal fungi: getting to the roots of the symbiosis. Plant Cell 8:1871-1883.

Giannasi, D.E., G. Zurawski, G. Learn, and M.T. Clegg. 1992. Evolutionary relationships of the Caryophyllidae based on comparative *rbc*L sequences. Syst. Bot. 17:1-15.

Giaquinta, R.T. 1978. Source and sink leaf metabolism in relation to phloem translocation. Carbon partitioning and enzymology. Plant Physiol. 61:380-385.

Giddings, T.H. and L.A. Staehelin. 1991. Microtubule-mediated control of microfibril deposition: A re-examination of the hypothesis. pp. 85-91. *In:* C.W. Lloyd (ed.). The Cytoskeletal Basis of Plant Growth and Form. Academic Press, New York.

Gidrol, X., P.A. Sabelli, Y.S. Fern, and A.K. Kush. 1996. Annexin-like protein from *Arabidopsis thaliana* rescues *ΔoxyR* mutant of *Escherichia coli* from H_2O_2 stress. Proc. Natl. Acad. Sci. USA 93:11268-11273.

Gifford, R.M. 1992. Interaction of carbon dioxide with growth-limiting environmental factors in vegetation productivity: Implications for the global carbon cycle. Advances in Bioclimatology 1:25-58.

Gijsman, J., J. Floris, M.V. Van Noordwijk, and G. Brouwer. 1991. An inflatable minirhizotron system for root observations with improved soil/tube contact. Plant Soil. 134:261-269.

Gilmartin, P.M., L. Sarokin, J. Memlink, and N.-H. Chua. 1990. Molecular light switches for plant genes. Plant Cell 2:369-378.

Gipson, J.R. 1986. Temperature effects on growth, development, and fiber properties. *In:* J.R. Mauney and J. McD. Stewart (eds.), Cotton Physiology, The Cotton Foundation, Memphis. pp.47-56.

Gipson, J.R. and H.E. Joham. 1968. Influence of night temperature on growth and development of cotton (*Gossypium hirsutum* L.). I. Fruiting and development. Agron. J. 60:292-295.

Gipson, J.R. and H.E. Joham. 1969. Influence of night temperature on growth and development of cotton (*Gossypium hirsutum* L.). III. Fiber elongation. Crop Sci. 9:127-129.

Gipson, J.R. and L.L. Ray. 1968. Fiber elongation rates in different varieties of cotton. pp. 212-217. *In:* Proc. Beltwide Cotton Conf., National Cotton Council of America, Memphis, Tenn.

Gipson, J.R. and L.L. Ray. 1969a. Fiber elongation rates in five varieties of cotton (*Gossypium hirsutum* L.) as influenced by night temperature. Crop Sci. 9:339-341.

Gipson, J.R. and L.L. Ray. 1969b. Influence of night temperature on boll development and fiber properties of five varieties of cotton. pp. 117-118. *In:* Proc. Beltwide Cotton Conf., National Cotton Council of America, Memphis, Tenn.

Gipson, J.R. and L.L. Ray. 1970. Temperature variety interrelationships in cotton. I. Boll and fiber development. Cotton Grow. Rev. 47:257-271.

Girma, F.S. and D.R. Krieg. 1985. Osmotic adjustment - cause and effect relationships in cotton leaves. pp. 43-44. Proc. 1985 Cotton Prod. Res. Confs. New Orleans, LA. National Cotton Council of America, Memphis, Tenn.

Glantz, M.H., A.Z. Rubinstein, and I. Zonn. 1993. Tragedy in the Aral Sea basin - Looking back to plan ahead. Global Environmental Change - Human and Policy Dimensions 3:174-198.

Glinski, J. and J. Lipiec. 1990. Soil Physical Conditions and Plant Roots. CRC Press, Inc., Boca Raton, FL.

Goddard, E.D. and K.P.A. Padmanabhan. 1992. A mechanistic study of the wetting, spreading, and solution properties of organosilicone surfactants. pp. 373-383 *In:* C.L. Foy (ed.). Adjuvants for Agrichemicals. CRC Press, Boca Raton, FL.

Godfray, H.C.J. 1995. Communication between the first and third trophic levels: An analysis using biological signalling theory. Oikos 72:367-374.

Godoy, A.S. and G.A. Palomo. 1999. Genetic analysis of earliness in upland cotton (*Gossypium hirsutum* L.): I. Morphological and phenological variables. Euphyt. 105:155-160.

Gogarten, J.P., J. Fichmann, Y. Braun, L. Morgan, P. Styles, S.L. Taiz, K. DeLapp, and L. Taiz. 1992. The use of antisense mRNA to inhibit the tonoplast H$^+$ ATPase in carrot. Plant Cell 4:851-864.

Golan-Goldhirsh, A., B. Hankamer, and S.H. Lips. 1990. Hydroxyproline and proline content of cell walls of sunflower, peanut, and cotton grown under salt stress. Plant Sci. 69:27-32.

Goldacre, P.L., A.W. Galston, and R.L. Weintraub. 1953. The effect of substituted phenols on the activity of IAA oxidase of peas. Arch. Biochem. Biophys. 43:358-373.

Goldschmidt, E.E. and S.C. Huber. 1992. Regulation of photosynthesis by end-product accumulation in leaves of plants storing starch, sucrose and hexose sugars. Plant Physiol. 99:1443-1448.

Goloubinoff, P., M. Edelman, and R.B. Hallick. 1984. Chloroplast-coded atrazine resistance in *Solanum nigrum*: *psbA* loci from susceptible and resistant biotypes are isogenic except for a single codon change. Nucl. Acids Res. 12:9489-9496.

Golovina, N.N., D. Ye Minskiy, Ye.I. Pankova, and D.A. Solov'-yev. 1992. Automated air photo interpretation in the mapping of soil salinization in cotton-growing zones. Mapping Sci. Remote Sensing 29:262-268.

Gonzales, R.E., C.J. deMooy, and J. Olsen. 1979. Effect of soil temperature and moisture on cotton germination under arid and semi-arid conditions. pp. 209-224. *In:* Improv. Irrig. Water Manage. Farms. Annu. Tech Rept., Colorado State University, Fort Collins, Co.

Gonzalez-Garza, M.T., S.A. Matlin, B.D. Mata-Cardenas, and S. Said-Fernandez. 1993. Differential effects of the (+)- and (-)- gossypol enantiomers upon *Entamoeba histolytica* axenic cultures. J. Pharm. Pharmacol. 45:144-145.

Goodwin, T.W. and E.I. Mercer. 1983. Introduction to Plant Biochemistry. pp. 230-33. Pergamon Press, Oxford, UK.

Gordon, S.G. 1995. Applications of cotton maturity testing by NIR and other methods. pp. 1231-1235. *In:* Proc. Beltwide Cotton Conf., National Cotton Council of America, Memphis, Tenn.

Gordon, W.B., D.H. Richerl, and J.R. Touchton. 1986. Nitrogen fertilizer and PIX interactions with continuous cotton and cotton rotated with soybeans. pp. 399-400. *In:* T.C. Nelson (ed.). Proc. Beltwide Cotton Prod. Res. Conf., National Cotton Council of America, Memphis, Tenn.

Gore, U.R. 1932. Development of the femal gametophyte and embryo in cotton. Amer. J. Bot. 19:795-807.

Gorham, J. 1996a. Mechanisms of salt tolerance of halophytes. *In:* R. Choukr-Allah, C.V. Malcolm, and A. Hamdy (eds.). Halophytes and Biosaline Agriculture. Marcel Dekker, New York. pp. 31-53.

Gorham, J. 1996b. Glycine betaine is a major nitrogen-containing solute in the Malvaceae. Phytochemistry, 43:367-369.

Gorham, J. and J. Bridges. 1995. Effects of calcium on growth and leaf ion concentrations of *Gossypium hirsutum* grown in saline hydroponic culture. Plant and Soil 176:219-227.

Gorham. J., A. Lauchli, and E.O. Leidi. 2009. Plant responses to salinity. pp. 130-142. *In:* J.M.Stewart, D.M. Oosterhuis, J.J. Heitholt, and J.R. Mauney (eds.). Physiology of Cotton. National Cotton Council of America, Memphis,Tenn. pp. Springer, London.

Gorlach, J., S. Volrath, G. Knauf-Beiter, G. Hengy, U. Beckhove, K.-H. Kogel, M. Oostendorp, T. Staub, E. Ward, H. Kessmann, and J. Ryals. 1996. Benzothiadiazole, a novel class of inducers of systemic acquired resistance, activates gene expression and disease resistance in wheat. Plant Cell 8:629-643.

Gorski, P.M., T.E. Vickstrom, M.L. Pierce, and M. Essenberg. 1995. A 13C-pulse-labeling study of phytoalexin biosynthesis in hypersensitively responding cotton cotyledons. Physiological and Molecular Plant Pathology 47:339-355.

Gossett, D.R., S.W. Banks, E.P. Millhollon, and M.C. Lucas. 1996. Antioxidant response to NaCl stress in a control and an NaCl-tolerant cotton cell-line grown in the presence of paraquat, buthionine sulfoxime, and exogenous glutathione. Plant Physiol. 112:803-809.

Gossett, D.R., S.W. Banks, E.P. Millhollon, M.C. Lucas, and T. Czeerneicki. 1995. Antioxidant responses in cotton callus grown in the presence of glutathione inhibitors and exogenous glutathione. Proc. Beltwide Cotton Conferences, San Antonio. 1995, 2:1094-1096.

Gossett, D.R., B. Bellaire, S.W. Banks, M.C. Lucas, A. Manchandia, and E.P. Millhollon. 1997. Induction of antioxidant enzyme activity in cotton. 1997 Proc. Beltwide Cotton Conf., New Orleans, LA, USA, January 6-10, 1997:Volume 2 1374-1377.

Gossett, D.R., B. Bellaire, S.W. Banks, M.C. Lucas, A. Manchandia, E.P. Millhollon, P. Dugger, and D. Richter. 1999. Specific ion effects on the induction of antioxidant enzymes in cotton callus tissue. Proc. Beltwide Cotton Conf., Orlando, Florida, USA, 3-7 January, 1999:Volume 1 540-542.

Gossett, D.R., B. Bellaire, S.W. Banks, M.C. Lucas, A. Manchandia, E.P. Millhollon, P. Dugger, and D. Richter. 2000. The influence of abscisic acid on the induction of antioxidant enzymes during salt stress. Proc. Beltwide Cotton Conf., San Diego, California, USA, 5-9 January 1998 1396-1399.

Gossett, D.R., M.C. Lucas, E.P. Millhollon, W.D. Caldwell, and A. Barclay. 1992. Antioxidant status in salt stressed cotton. Proc. Beltwide Cotton Physiol. Conf., Memphis 1992, vol 3., pp. 1036-1039.

Gossett, D.R., E.P. Millhollon, and M.C. Lucas. 1994a. Antioxidant response to NaCl stress in salt-tolerant and salt-sensitive cultivars of cotton. Crop Sci. 34:706-714.

Gossett, D.R., E.P. Millhollon, M.C. Lucas, S.W. Banks, and M.-M. Marney. 1994b. The effects of NaCl on antioxidant enzyme activities in callus tissue of salt-tolerant and salt-sensitive cotton cultivars (*Gossypium hirsutum* L.). Plant Cell Reports 13:498-503.

Gouia, H., M.H. Ghorbal, and B. Touraine. 1994. Effects of NaCl on flows of N and mineral ions and on NO_3^- reduction rate within whole plants of salt-sensitive bean and salt-tolerant cotton. Plant Physiol. 105:1409-1418.

Gould, J., S. Banister, O. Hasegawa, M. Fahima, and R.H. Smith. 1994. Regeneration of *Gossypium hirsutum* and *G. barbadense* from shoot apex tissues for transformation. Plant Cell Reports 10:12-16.

Gould, F., P.A. Follet, B. Nault, and G.G. Kennedy. 1994. Resistance management strategies for transgenic potato plants. pp. 255-277 *In:* G. Zehnder, R.K. Janssen, M.L. Powelson, and K.V. Raman (eds.). Potato Pest Management: a Global Perspective, APS, St. Paul, MN.

Goyal, S.S., S.K. Sharma, D.W. Rains, and A. Läuchli. 1999a. Long-term reuse of drainage waters of varying salinities for crop irrigation in a cotton-safflower rotation system in the San Joaquin Valley of California - A nine-year study: I. Cotton (*Gossypium hirsutum* L.). J. Crop Production 2:181-213.

Goyal, S.S., S.K. Sharma, D.W. Rains, and A. Läuchli. 1999b. Long-term reuse of drainage waters of varying salinities for crop irrigation in a cotton-safflower rotation system in the San Joaquin Valley of California - A nine-year study: II. Safflower (*Carthamus tinctorius* L.). J. Crop Production 2:181-213.

Graham, C.T., Jr., J.M. Jenkins, J.C. McCarty, Jr., and W.L. Parrott. 1987. Effects of mepiquat chloride on natural plant resistance to tobacco budworm in cotton. Crop Sci. 27:360-361.

Graham, J.H. and D.S. Egel. 1988. *Phytopthora* root development on mycorrhizal and phosphorous-fertilized nonmycorrhizal sweet orange seedlings. Plant Dis. 72:611-614.

Grajal, A. 1999. Biodiversity and the nation state: Regulating access to genetic resources limits biodiversity research in developing countries. Conserv. Biol. 13:6-10.

Grantz, D.A. and J.F. Farrar. 1999. Acute exposure to ozone inhibits rapid carbon translocation from source leaves of Pima cotton. Jour. Exp. Bot. 50:1253-1262.

Grantz, D.A. and J.R. Farrar. 2000. Ozone inhibits phloem loading from a transport pool: Compartmental efflux analysis in Pima cotton. Aust. Jour. Plant Physiol. 27:859-868.

Grantz, D.A. and P.H. McCool. 1992. Effect of ozone on Pima and Acala cottons in the San Joaquin Valley. *In:* Proc. 1992 Beltwide Cotton Conferences, Vol. 3. Natl. Cotton Council Amer., Memphis, Tenn. pp. 1082-1084.

Grantz, D.A. and S. Yang. 1995. Ozone effects on cotton below ground: Water relations. *In:* Proc. 1995 Beltwide Cotton Conferences, Vol. 2. Natl. Cotton Council Amer., Memphis, Tenn. pp. 1130-1132.

Grantz, D.A. and S. Yang. 1996. Effect of O_3 on hydraulic architecture in Pima cotton. Plant Physiol. 112:1649-1657.

Grantz, D.A. and S. Yang. 2000. Ozone impacts on allometry and root hydraulic conductance are not mediated by source limitation nor developmental age. Jour. Exp. Bot. 51:919-927.

Grantz, D.A., J.I. MacPherson, W.J. Massman, and J. Pederson. 1994. Study demonstrates ozone uptake by SJV crops. Calif. Agric. July-August 1994. pp. 9-12.

Grantz, D.A., X.J. Zhang, W.J. Massmann, A. Delany, and J.R. Pederson. 1997. Ozone deposition to a cotton (*Gossypium hirsutum* L.) field: Stomatal and surface wetness effects during the California Ozone Deposition Experiment. Agric. For. Meteorol. 85:19-31

Grantz, D.A., X.J. Zhang, W.J. Massman, G. den Hartog, H.H. Neumann, and J.R. Pederson. 1995. Effects of stomatal conductance and surface wetness on ozone deposition in field-grown grape. Atmos. Env. 29:3189-3198.

Grattan, S.R. and C.M. Grieve. 1992. Mineral element acquisition and growth response of plants grown in saline environments. Agriculture, Ecosystems and Environ. 38:275-300.

Grattan, S.R., E.V. Maas, and G. Ogata. 1981. Foliar uptake and injury from saline aerosol. J. Environ. Qual. 10:406-409.

Grattan, S.R., C. Shennan, D. May, B. Roberts, M. Borin, and M. Sattin. 1994. Utilizing saline drainage water to supplement irrigation water requirements of tomato in a rotation with cotton. Proc. 3rd Congress European Soc. Agron., Padova University, Abano Padova, Italy, 18 22 September 1994 802-803;

Graves, D.A. and J.M. Stewart. 1988a. Chronology of the differentiation of cotton *(Gossypium hirsutum* L.) fiber cells. Planta 175:254-258.

Graves, D.A. and J.M. Stewart. 1988b. Analysis of the protein constituency of developing cotton fibers. J. Exp. Bot. 39:59-69.

Gray, J.R., T.J. Mabry, A.A. Bell, and R.D. Stipanovic. 1976. *para-Hemigossypolone:* a sesquiterpenoid aldehyde quinone from *Gossypium hirsutum.* J.C.S. Chem. Comm. pp. 109-110.

Gray, R.C. and G.W. Akin. 1984. Foliar fertilization. pp. 579-584. *In:* R.D. Hauck (ed.). Nitrogen in Crop Production. Publ. Amer. Soc. Agron., Madison, Wisconsin.

Greef, A.I. and J.J. Human. 1988. The effect of date of planting on the fibre properties of four cotton cultivars grown under irrigation. S. Afr. J. Plant Soil 5:167-172.

Green, C.C. and T.W. Culp. 1990. Simultaneous improvements of yield, fiber quality, and yarn strength in Upland cotton. Crop Sci. 30:66-69.

Greene-McDowelle, D.M., B. Ingber, M.S. Wright, H.J. Zeringue, D. Bhatnagar, and T.E. Cleveland. 1999. The effects of selected cotton-leaf volatiles on growth, development and aflatoxin production of *Aspergillus parasiticus.* Toxicon 37:883-893.

Greenplate, J.T., N.B. Duck, J.C. Pershing, and J.P. Purcell. 1995. Cholesterol oxidase: an oostatic and larvicidal agent against the cotton boll weevil, *Anthonomus grandis.* Entomol. Exp. Appl. 74:253-258.

Gregory, F.G. 1917. Physiological conditions in cucumber houses. Third Annu. Report. Experiment and Research Station, Chestnut, England. pp. 19-28.

Greitner, C.S., E.J. Pell, and W.E. Winner. 1994. Analysis of aspen foliage exposed to multiple stresses - ozone, nitrogen deficiency, and drought. New Phytol. 127:579-589.

Griffith, D.R., J.F. Moncrief, D.J. Eckert, J.B. Swan, and D.D Breitbach. 1992. Crop response to tillage systems. pp. 25-33. *In* Conservation Tillage Systems and Management. MidWest Plan Serv. Pub. MWPS-45. Iowa State Univ., Ames, IA.

Grimes, D.W. 1977. Physiological response of cotton to water and its impact on economical production. *In:* Proc. West. Cotton Prod. Conf. pp. 22-25.

Grimes, D.W. 1982. Water requirements and use patterns of the cotton plant. *In:* Proc. West. Cotton Prod. Conf. pp. 27-30.

Grimes, D.W. 1991. Water management for quality cotton. *In:* Proc. Beltwide Cotton Prod. Res. Conf. National Cotton Council, Memphis, Tenn. pp. 52-54.

Grimes, D.W., W.L. Dickens, H.Yamada, and R.J. Miller. 1973. A model for estimating desired levels of nitrate-N concentration in cotton petioles. Agron. J. 65:37-41.

Grimes, D.W., W.L. Dickens, and H. Yamada. 1978. Early-season water management of cotton. Agron. J. 70:1009-1012.

Grimes, D.W. and K.M. El-Zik. 1982. Water management for cotton. Univ. California Div. Agric. Sci. Bull. 1904.

Grimes, D.W. and K.M. El-Zik. 1990. Cotton. *In:* B.A. Stewart and D.R. Nielsen (ed.). Irrigation of agricultural crops. Agronomy 30. pp. 741-773.

Grimes, D.W., R.J. Miller, and L. Dickens. 1970. Water stress during flowering of cotton. Calif. Agric. 24(3):4-6.

Grimes, D.W., R.J. Miller, and P.L. Wiley. 1975. Cotton and corn root development in two field soils of different strength characteristics. Agron. J. 67:519-523.

Grimes, D.W. and H. Yamada. 1982. Relation of cotton growth and yield to minimum leaf water potential. Crop Sci. 22:134-139.

Grimes, D.W., H. Yamada, and W.L. Dickens. 1969. Functions for cotton (*Gossypium hirsutum* L.) production from irrigation and nitrogen fertilization variables: I. Yield and evapotranspiration. Agron. J. 61:769-773.

Gris, E. 1844. Nouvelles experiences sur l'action des composis ferrugineux solubles, appliques a la vegetation et specialment au traitement de la chlorose at a la debilite des plant. Compt. Rend. (Paris) 19:1118-1119.

Gross, O. and B. Partheir. 1994. Novel natural substances acting in plant growth regulation. J. Plant Growth Regul. 13:93-114.

Grula, J.W., R.L. Hudspeth, S.L. Hobbs, and D.M. Anderson. 1995. Organization, inheritance, and expression of acetohydroxyacid synthase genes in the cotton allotetraploid *Gossypium hirsutum.* Plant Mol. Biol. 28:837-846.

Grundon, N.J. 1987. Hungry crops: A guide to mineral deficiencies in field crops. Queensland Dept. Primary Industries. Brisbane, Australia.

Gubanov, G.Y. 1966. Physiology of the opening of cotton bolls. Soviet Plant Physiol. 13: 756-761.

Guertal, E.A., E.J. Sikord, A.K. Hagan, and R. Rodriguez-Kabana. 1998. Effect of winter cover crops on populations of southern root-knot and reniform nematodes. Agriculture Ecosystems & Environment 70:1-6.

Guilfoyle, T. 1986. Auxin regulated gene expression in higher plants. *CRC Critical Reviews in Plant Sciences.* 4:247-277.

Guinn, G. 1971 Chilling injury in cotton seedlings: Changes in permiability of cotyledons. Crop Sci. 11:101-102.

Guinn, G. 1974. Abscission of cotton floral buds and bolls as influenced by factors affecting photosynthesis. Crop Sci. 14:291-293.

Guinn, G. 1976. Nutritional stress and ethylene evolution by young cotton bolls. Crop Sci. 16:89-91.

Guinn, G. 1979. Hormonal relations in flowering, fruiting, and cutout. *In:* Proc. Beltwide Cotton Prod. Res. Conf. National Cotton Council, Memphis, Tenn. pp. 265-276.

Guinn, G. 1982a. Causes of square and boll shedding in cotton. USDA ARS Tech. Bull. 1672.

Guinn, G. 1982b. Fruit age and changes in abscisic acid content, ethylene production, and abscission rate of cotton fruits. Plant Physiol. 69:349-352.

Guinn, G. 1984a. Abscisic acid and cutout in cotton. Plant Physiol. 77:16-20.

Guinn, G. 1984b. Potential for improving production efficiency with growth regulants. pp. 67-71. *In:* J.M. Brown (ed.). Proc. Beltwide Cotton Prod. Res. Conf., National Cotton Council of America, Memphis, Tenn.

Guinn, G. 1985. Fruiting of cotton. III. Nutritional stress and cutout. Crop Sci. 25:981-985.

Guinn, G. 1986a. Abscisic acid and cutout in cotton. Plant Physiol. 77:16-20.

Guinn, G. 1986b. Hormonal relations during reproduction. pp. 113-136 *In:* J.R. Mauney and J.McD. Stewart. Cotton Physiology. Cotton Foundation, Memphis, Tenn.

Guinn, G. 1998. Causes of square and boll shedding. *In:* Proc. Beltwide Cotton Prod. Res. Conf. National Cotton Council, Memphis, Tenn. pp. 1355-1364.

Guinn, G. and D.L. Brummett. 1989. Fruiting of cotton. IV. Nitrogen, abscisic acid, indole-3-acetic acid, and cutout. Field Crops Res. 22:257-266.

Guinn, G. and M.P. Eidenbock. 1982. Catechin and condensed tannin contents of leaves and bolls of cotton in relation to irrigation and boll load *Lygus hesperus, Heliothis* spp., insect resistance. Crop Sci. 22: 614-616.

Guinn, G., J.D. Hesketh, K.E. Fry, J.R. Mauney, and J.W. Radin. 1976. Evidence that photosynthesis limits yield of cotton. pp. 100-105. *In:* J.M. Brown (ed.). Proc. Beltwide Cotton Prod. Res. Conf., National Cotton Council of America, Memphis, Tenn.

Guinn, G. and R.E. Hunter. 1968. Root temperature and carbohydrate status of young cotton plants. Crop Sci. 8:67-70.

Guinn, G. and J.R. Mauney. 1984. Fruiting of cotton. II. Effects of plant moisture status and active boll load on boll retention. Agron. J. 76:94-98.

Guinn, G., J.R. Mauney, and K.E. Fry. 1981. Irrigation scheduling and plant population effects on growth, bloom rates, boll abscission, and yield of cotton. Agron. J. 73:529-534.

Gulati, A.N. and A.J. Turner. 1928. A note on the early history of cotton. Ind. Cent. Cotton Committee, Tech. Lab. Bull. No. 17.

Gulick, S.H.; K.G. Cassman, and S.R. Grattan. 1989. Exploitation of soil potassium in layered profiles by root systems of cotton and barley. Soil Sci. Soc. Am. J. 53:146-153.

Gulov, M.K., T.N. Ulmasov, K.A. Aliev, V.M. Andrianov, and E.S. Piruzian. 1990. Nucleotide sequence of the chloroplast *rbc*L gene from cotton *Gossypium hirsutum*. Nucl. Acids Res. 18:185.

Gunasena, G.H., S.B. Vinson, H.J. Williams, and R.D. Stipanovic. 1988. Effects of caryophyllene, caryophyllene oxide, and their interaction with gossypol on the growth and development of *Heliothis virescens* (F.) (Lepidoptera: Noctuidae). J. Econ. Entomol. 81:93-97.

Guo, C., D.M. Oosterhuis, and D. Zhao. 1994. Enhancing mineral nutrient uptake with plant growth regulators. pp. 244-251. Proc. 21st Annual Meeting Plant Growth Regulator Soc. of Am.

Gupta, U.C. 1993. Factors affecting boron uptake by plant. pp. 87-104 *In:* G.C. Gupta (ed.). Boron and Its Role in Crop Production. CRC Press, Inc.

Guthrie, D.S. 1987. Fruiting profile and petiole nitrate levels following chlordimeform applications. pp. 73-74. *In:* T.C. Nelson (ed.). Proc. Beltwide Cotton Prod. Res. Conf., National Cotton Council of America, Memphis, Tenn.

Guthrie, D.S. 1991. Cotton response to starter fertilizer placement and planting dates. Agron. J. 83:836-839.

Gutierrez, A.P., R. Daxl, G. Leon Quant, and L.A. Falcon, L.A. 1981. Estimating economic thresholds for bollworm, *Heliothis zea* Boddie, and boll weevil, *Anthonomus grandis* Boh, damage in Nicaraguan cotton (*Gossypium hirsutum* L). Environ. Entomol. 10:872-879.

Gutierrez, A.P., L.A. Falcon, W. Loew, P.A. Leipzig, and R. van der Bosch. 1975. An analysis of cotton production in California: A model for Acala cotton and the effects of defoliators on yield. Environ. Entomol. 4:125-136.

Gutierrez, A.P., T.F. Leigh, Y. Wang, and R. Cave. 1977. An analysis of cotton production in California: *Lygus hesperus* (Heteroptera: Miridae) injury-an evaluation. Can Ent 109: 1375-1386.

Gutierrez, A.P., Y. Wang, and R. Daxl. 1979a. The interaction of cotton and boll weevil (Coleoptera:curculionidae) - a study of coadaptation. Can. Ent. 111:357-366.

Gutierrez, A.P., Y. Wang, and U. Regev. 1979b. An optimization model for *Lygus hesperus* (Heteroptera: Miridae) damage in cotton: The economic threshold revisited. Can Ent 111: 41-54.

Gwathmey, C.O. and R.M. Hayes. 1996. Ethephon effects on boll opening and earliness of early- and late-planted 'Deltapine 50'. p. 1161. *In:* P. Dugger and D.A. Richter (eds.). Proc. Beltwide Cotton Prod. Res. Conf., National Cotton Council of America, Memphis, Tenn.

Gwathmey, C.O., C.E. Michaud, R.D. Cossar, and S.H. Crowe. 1999. Development and cutout curves for ultra-narrow and wide-row cotton in Tennessee. *In:* P. Dugger and D. Richter (eds.). Proc. Beltwide Cotton Conf. 3-7 Jan., 1998, Orlando, FL. Natl.

Cotton Counc. of Am., Memphis, Tenn. pp. 630-632.

Gwathmey, C.O., O.M. Wassel, and C.E. and Michaud. 1995. Pix effects on canopy light interception by contrasting cotton varieties. p. 1153. *In:* D.A. Richter and J. Armour (eds.). Proc. Beltwide Cotton Conf., National Cotton Council of America, Memphis, Tenn.

Gwyn, J.J. and D.M. Stelly. 1989. Method to evaluate pollen viability of upland cotton: Tests with chromosome translocations. Crop Sci. 29:1165-1169.

Hafez, A.A.R.; P.R. Stout, and J.E. DeVay. 1975. Potassium uptake by cotton in relation to verticillium wilt. Agron. J. 67:359-361.

Haig, D. 1990. New perspectives on the angiosperm female gametophyte. Bot. Rev. 56:236-274.

Haigler, C.H. 1985. The functions and biogenesis of native cellulose. pp. 30-83. *In:* T.P. Nevell and S.H. Zeronian (eds.). Cellulose Chemistry and its Applications. Ellis Horwood, Chichester England.

Haigler, C.H. 1991. Relationship between polymerization and crystallization in microfibril biogenesis. pp. 99-124. *In:* C.H. Haigler and P.J. Weimer (eds.). Biosynthesis and Biodegradation of Cellulose. Marcel Dekker, Inc., New York.

Haigler, C. 1992. The crystallinity of cotton cellulose in relation to cotton improvement. pp. 211-226. *In:* C.R. Benedict (ed.). Proc. Cotton Fiber Cellulose: Structure, Function, and Utilization Conference. National Cotton Council, Memphis Tenn.

Haigler, C.H., A.S. Holaday, L.K. Martin, and J.G. Taylor. 1996. Mechanisms of cool temperature inhibition of cotton fiber cellulose synthesis. pp. 1174-1175. *In:* Proc. Beltwide Cotton Conf., National Cotton Council of America, Memphis, Tenn.

Haigler, C.H., N.R. Rao, E.M. Roberts, J.-Y. Huang, D.R. Upchurch, and N.L. Trolinder. 1991. Cultured ovules as models for cotton fiber development under low temperatures. Plant Physiol. 95:88-96.

Haigler, C.H., J.G. Taylor, and L.K. Martin. 1994. Temperature dependence of fiber cellulose biosynthesis: Impact on fiber maturity and strength. pp. 95-100. *In:* G. Jividen and C. Benedict (eds.). Proc. Biochemistry of Cotton Workshop. Cotton Incorporated, Raleigh NC.

Hacskaylo, J. and A.L. Scales. 1959. Some effects of guthion alone and in combination with DDT and of a dieldrin-DDT mixture on growth and fruiting of the cotton plant. J. Econ. Ent. 52:396-398.

Hake, K and T. Kerby. 1988. Nitrogen fertlization. pp. 1-20. Cotton Fertilization Guide. University of California, Bakersfield, CA. No. KC-880.

Hake, S.J., D.W. Grimes, K.D. Hake, T.A. Kerby, D.J. Munier, and L.J. Zelinski. 1996. Irrigation scheduling. *In:* S.J. Hake, T.A. Kerby, and K.D. Hake (eds.). Cotton Production Manual. Univ. of Calif. Oakland. pp. 228-247.

Halcomb, J.L., J.H. Benedict, and D.R. Ring. 1996. Survival and growth of bollworm and tobacco budworm on nontransgenic and transgenic cotton expressing a CryIA insecticidal protein (Lepidoptera: Noctuidae). Environ. Entomol. 25:250-

Halevy, J. 1976. Growth rate and nutrient uptake of two cotton cultivars grown under irrigation. Agron. J. 68:701-705.

Halevy, J.; A. Marani, and T. Markovitz. 1987. Growth and N P K uptake of high-yielding cotton grown at different nitrogen levels in a permanent-plot experiment. Plant Soil. 103:39-44.

Halevy, J. and T. Markovitz. 1988. Foliar NPKS application to cotton from full flowering to boll filling. Fertilizer Research 15:247-252.

Hall, A.J., D.M. Whitfield, and D.J. Connor. 1990. Contribution of pre-anthesis assimilates to grain-filling in irrigated and water-stressed sunflower crops. II. Estimates from a carbon budget. Field Crops Res. 24:273-294.

Hall, J.C. 1996. Are cycling gene-products as internal zeitgebers no longer the zeitgeist of chronobiology? Neuron 17:799-802.

Hall, W.C. 1952. Evidence on the auxin-ethylene balance hypothesis of foliar abscission. Botan. Gaz. 113:310-322.

Hallmann, J., R. Rodriguez-Kabana, and J.W. Kloepper. 1999. Chitin-mediated changes in bacterial communities of the soil, rhizosphere and within roots of cotton in relation to nematode control. Soil Biology and Biochemistry 31:551-560.

Halloin, J.M. 1976. Inhibition of cottonseed germination with abscisic acid and its reversal. Plant Physiol. 57:454-455.

Halloin, J.M. 1982. Localization and changes in catechin and tannins during development and ripening of cottonseed. New Phytologist 90:651-657.

Halloin, J.M. 1984. The influence of seed coat permeability to water on weathering, stand establishment, and yield of cotton. p. 28. *In:* Proc. Cotton Production Research Conf., National Cotton Council of America, Memphis, Tenn.

Halloin, J.M. and A.A. Bell. 1979. Production of nonglandular terpenoid aldehydes within diseased seeds and cotyledons of *Gossypium hirsutum* L. J. Agric. and Food Chemistry 27:1407-1409.

Hamdy, A. 2000. Cotton growth and salt distribution in soils under alternate application of irrigation water of different quality. Proc. 16th ICID European regional conference technological and socio-economical impacts on agricultural water management. 1992, 85-91.

Hammerschmidt, R. and J.S. Becker. 1997. Acquired resistance to disease in plants. Hort. Rev. 18:247-289.

Hamner, K.C. 1960. Photoperiodism and circadian rhythms. Cold Spring Harbor Symp. Quant. Biol. 25:269-277.

Hampton, R.E. 1990. Phenolic metabolism in cotton : Physiological and ecological aspects related to fruit abscission. Thesis (Ph. D.).University of Arkansas, Fayetteville, 164 pp.

Hampton, R.E., D.M. Oosterhuis, J.M. Stewart, and K.S. Kim. 1987. Anatomical differences in cotton related to drought tolerance. Ark. Farm Res. 36:4.

Han, T., J.T. Cothren, and F.M. Hons. 1990. Effect of nitrogen fertilizer applications of Pix treatments on cotton growth and development. pp. 654-655. *In:* J.M. Brown and D.A. Richter (eds.). Proc. Beltwide Prod. Res. Conf., National Cotton Council of America, Memphis, Tenn.

Han, T., J.T. Cothren, and F.M. Hons. 1991. Cotton yield and nutrient uptake as affected by Pix and rate and time of nitrogen application. p. 1028. *In:* D.J. Herber and D.A. Richter (eds.). Proc. Beltwide Cotton Conf., National Cotton Council of America, Memphis, Tenn.

Hancock, R.E.W. and R. Lehrer. 1998. Cationic peptides: A new source of antibiotics. Trends in Biotechnology. 16:82-88.

Hanny, B.W. and R.C. Guelder. 1976. An investigation of the surface lipids of the glabrous cotton (*Gossypium hirsutum* L.) strain, Bayou SM1. J. Agric. Food Chem. 24:401-403.

Hansen, J., C. Black-Schaefer, W. Shafer, and L. Larson. 1996. Effects of Ryzup (GA$_3$) on the growth and development of early season cotton. Proc. Beltwide Cotton Conf. 9-12 Jan, 1996. Nashville, Tenn. Natl. Cotton Counc. of Am., Memphis, Tenn. pp. 38-40.

Hanson, R.E., M.N. Islam-Faridi, E.A. Percival, C.F. Crane, Y. Ji, T.D. McKnight, D.M. Stelly, and H.J. Price. 1996. Distribution of 5S and 18S-28S rDNA loci in a tetraploid cotton (*Gossypium hirsutum* L.) and its putative diploid ancestors. Chromosoma 105:55-61.

Hanson, R.G., E.C. Ewing, and E.C. Ewing, Jr. 1956. Effect of environmental factors on fiber properties and yield of Deltapine cottons. Agron. J. 48:576-581.

Hardwick, R.C. 1986. Physiological consequences of modular growth in plants. Phil. Trans. R. Soc. Lond. B 313:161-173.

Hare, P.D., W.A. Cress, and J. van Staden. 1997. The involvement of cytokinins in plant responses to environmental stress. Plant Growth Regul. 23:79-103.

Hargreaves, H. 1948. List of recorded cotton insects of the world. Commonweakth Institute of Entomology, London.

Harland, S.C. 1939. Genetical studies in the genus *Gossypium* and their relationship to evolutionary and taxonomic problems. Proc. 7th Genet. Congr. Edinburgh, pp. 138-143.

Harley, P.C., R.B. Thomas, J.F. Reynolds, and B.R. Strain. 1992. Modeling photosynthesis of cotton grown in elevated carbon dioxide. Plant Cell Environ. 15:271-282.

Harman, W.L., G.J. Michaels, and A.F. Wise. 1989. Conservation tillage system for profitable cotton production in the central Texas high plains. Agron J. 81:615-618.

Harp, S.J. and V.V. Turner. 1976. Effects of thrips on cotton development in the Texas Blacklands. The Southern Entomologist 140:40-45.

Harris, G., C. Bednarz, and G. Gascho. 2000. Using split applications and foliar potassium to increase cotton yields. 1999 Cotton Research and Extension Report. University of Georgia, Tifton, GA. UGA/CPES Research-Extension report #4, pp. 125-127.

Harris, P. 1974. A possible explanation of plant yield increases following insect damage. Agro Ecosystems 1:219-225.

Hart, W.E., F.D. Tompkins, M.S. Palmer, and J.F. Bradley. 1996. Influence of crop cultivation on cotton physiology. pp. 1643-1649. *In:* Proc. Beltwide Cotton Conf., National Cotton Council of America, Memphis, Tenn.

Hartman, K.M. and J.N. Sasser. 1985. Identification of *Meloidogyne* species on the basis of differential host test and perineal pattern morphology. pp. 69-77. *In:* K.R. Barker (ed.). An Advanced Treatise on *Meloidogyne*, Volume 2, Methodology, North Carolina State University Graphics, Raleigh, NC.

Hartman, P.E., C.K. Suzuki, and M.E. Stack. 1989. Photodynamic production of superoxide *in vitro* by altertoxins in the presence of reducing agents. Applied and Environmental Microbiology 55:7-14.

Hartmann, H. and D. Kester. 1984. Plant propagation: Principles and Practices. Prentice Hall, Engelwood, CA.

Harwood, J.L. 1988. Fatty acid metabolism. Ann. Rev. Plant Physiol. Plant Mol. Biol. 39:101-138.

Hasenfratz, M., C. Tsou, and T.A. Wilkins. 1995. Expression of two related vacuolar H$^+$-ATPase 16-kilodalton proteolipid genes is differentially regulated in a tissue-specific manner. Plant Physiol. 108:1396-1404.

Hassawy, G.S. and K.C. Hamilton. 1971 Effects of IAA, kinetin, and trifluralin on cotton seedlings. Weed Sci. 19:265-268.

Hassidim, M., Y. Braun, H.R. Lerner, and L. Reinhold, L. 1986. Studies on H$^+$-translocating ATPases in plants of varying resistance to salinity. II. K$^+$ strongly promotes development of membrane potential in vesicles from cotton roots. Plant Physiol. 81:1057-1061.

Hassidim, M., Y. Braun, H.R. Lerner, and L. Reinhold. 1990. Na$^+$/H$^+$ and K$^+$/H$^+$ antiport in root membrane vesicles isolated from the halophyte *Atriplex* and the glycophyte cotton. Plant Physiol. 94:1795-1801.

Hastings, J.W. and H.G. Schweiger. 1976. The molecular basis of circadian rhythms. Dahlem Konferenzen, Berlin, 1976.

Hatfield, J.L., J.E. Quisenberry, and R.E. Dilbeck. 1987. Use of canopy temperatures to identify water conservation in cotton germplasm. Crop Sci. 27:269-273.

Hattey, J.A., W.E. Sabbe, G.D. Balter, and A.B. Blakeney. 1994. Nitrogen and starch analysis of cotton leaves using Near Infrared Reflectorial Spectroscopy (NIRS). Comm. Soil. Sci. Plant Anal. 25:1855-1863.

Hatzfeld, W.D., J. Dancer, and M. Stitt. 1990. Fructose-2,6-bisphosphate, metabolites, and 'coarse' control of pyrophosphate: fructose-6-phosphate phosphotransferase during triose phosphate cycling in heterotropic cell-suspension cultures of *Chenopodium rubrum*. Planta 180:205-211.

Hatzfeld, W.D. and M. Stitt. 1991. Regulation of glycolysis in heterotrophic cell suspension cultures of *Chenopodium rubrum* in response to proton fluxes at the plasmalemma. Physiol. Plant. 81:103-110.

Hau, B., J. Lançon, and D. Dessauw. 1997. Les cotonniers. Pp. 241-165. *In:* A. Charrier, M. Jacquot, S. Hamon, and D. Nicolas (eds.). L'amélioration des plantes tropicales. Montpellier, France. Collection Repères, CIRADORSTOM.

Hauck, R.D. 1982. Nitrogen: Isotope-ratio analysis. pp. 735-779. *In:* A.L. Pace, P.M. Miller, and D.R. Keeney (eds.). Methods of Soil Analysis, Second Edition, Amer. Soc. Agron., Madison, Wisconsin.

Haughn, G.W., J. Smith, B. Mazur, and C. Somerville. 1988. Transformation with a mutant *Arabidopsis* acetolactate synthase gene renders tobacco resistant to sulfonylurea herbicides. Mol. Gen. Genet. 211:266-271.

Havlin, J.L., J.D. Beaton, S.L. Tisdale, and W.I. Nelson. 1999. Soil Fertility and Fertilizers. An Introduction to Nutrient Management. Prentice Hall, Upper Saddle River, New Jersey.

Hawes, M.C. 1990. Living plant cells released from the root cap: A regulator of microbial populations in the rhizosphere. Plant Soil 129:19-27.

Hawes, M.C., L.A. Brigham, F. Wen, H.H. Woo, and Y. Zhu. 1999. Function of root border cells in plant health: Pioneers in the rhizosphere. Annu. Rev. Phytopathology 36:311-327.

Hawkins, B.S. and H.A. Peacock. 1973. Influence of row width and population density on yield and fiber characteristics of cotton. Agron. J. 65:47-51.

Hawkins, R.S. 1930. Development of cotton fibers in the Pima and Acala varieties. J. Agric. Res. 40:1017-1029.

Hayashi, T. 1991. Biochemistry of xyloglucans in regulating cell elongation and expansion. pp. 131-144. *In:* C.W. Lloyd (ed.). The Cytoskeletal Basis of Plant Growth and Form. Academic Press, New York.

Hayashi, H., H. Alias, L Mustardy, P. Deshnium, M. Ida, and N. Murata. 1997. Transformation of *Arabidopsis thaliana* with the *codA* gene for choline oxidase. Accumulation of glycinebetaine and enhanced tolerance to salt and cold stress. Plant Journal 12:133-142.

Hayashi, T. and D.P. Delmer. 1988. Xyloglucan in the cell walls of cotton fiber. Carbohydr. Res. 181:273-277.

Hayashi, T., S.M. Read, J. Bussell, M. Thelen, F.C. Lin, R.M. Brown, Jr., and D.P. Delmer. 1987. UDP-glucose: (1-3)-β-glucan synthases from mung bean and cotton. Differential effects of Ca^{2+} and Mg^{2+} on enzyme properties and on macromolecular structure of the glucan product. Plant Physiol. 83:1054-1062.

Hayman, D.S. 1987. VA mycorrhizas in field crop systems. p. 171-192. *In:* G.R. Safir (ed.). Ecophysiology of VA Mycorrhizal Plants. CRC Press, Boca Raton, FL.

Hays, S.M. 1996. Foreign genes for domestic cotton - collection of Asiatic cotton germplasm. Feb. Agric. Res.

Hayward, H.E. 1938. The structure of economic plants. The MacMillan Co., New York.

Hayward, HE. and C.H. Wadleigh. 1949. Plant growth on saline and alkali soils. Adv. Agron. 1:1-38.

Heady, E.O. 1952. Economics of agricultural production and resource use. Prentice-Hall, Englewood Cliffs, NJ.

Heagle, A.S. 1989. Ozone and crop yield. Ann. Rev. Phytopathol. 27:397-423.

Heagle, A.S., D.E. Body, and W.W. Heck. 1973. An open-top field chamber to assess the impact of air pollution on plants. J. Environ. Qual. 2:365-368.

Heagle, A.S. and W.W. Heck. 1980. Field methods to assess crop losses due to oxidant air pollutants. *In:* P.S. Teng and S.V. Drupa (eds.). Crop Loss Assessment. Proc. E. C. Stakman Commemorative Symp. Univ. Minnesota Agric. Exp. Stn. Misc. Publ. #7.

Heagle, A.S., W.W. Heck, V.M. Lesser, J.O. Rawlings, and F.L. Mowry. 1986. Injury and yield response of cotton to chronic doses of ozone and sulfur dioxide. J. Environ. Qual. 15:375-382.

Heagle, A.S., J.E. Miller, F.L. Booker, and W.A. Pursley. 1999. Ozone stress, carbon dioxide enrichment, and nitrogen fertility interactions in cotton. Crop Sci. 39:731-741.

Heagle, A.S., J.E. Miller, W.W. Heck, and R.P. Patterson. 1988. Injury and yield response of cotton to chronic doses of ozone and soil moisture deficit. J. Environ. Qual. 17:627-635.

Hearle, J.W.S. 1985. Mechanical properties of cellulosic textile fibers. pp. 480-504. *In:* T.P. Nevell and S.H. Zeronian (eds.). Cellulose Chemistry and Its Applications. Ellis Horwood, Ltd., Chichester, UK.

Hearn, A.B. 1969a. The growth and performance of cotton in a desert environment. I. Morphological development. J. Agric. Sci., Camb. 73:65-74.

Hearn, A.B. 1969b. The growth and performance of cotton in a desert environment. II. Dry matter production. J. Agric. Sci., Camb. 73:75-86.

Hearn, A.B.1969c. The growth and performance of cotton in a desert environment. III. Crop performance. J. Agric. Sci., Camb. 73:87-97.

Hearn, A.B. 1972.Cotton spacing experiments in Uganda. J. Agric. Sci. 78:13-25.

Hearn, A.B. 1975. The response of cotton to water and nitrogen in a tropical environment. I. Frequency of watering and method of application of nitrogen. J. Agric. Sci., Camb.73:407-417.

Hearn, A.B. 1976. Crop physiology. *In:* M.H. Arnold (ed.). Agriculture Research for Development. Cambridge University Press, London, UK.

Hearn, A.B. 1976. Response of cotton to nitrogen and water in a tropical environment. III. Fibre quality. J. Agric. Sci. 86:257-269.

Hearn, A.B.1979. Water relationships in cotton. Outlook Agric. 10:159-166.

Hearn, A.B. 1981. Cotton Nutrition. Field Crop Abstracts. 34:11-34.

Hearn, A.B. 1994. OZCOT: a simulation model for cotton crop management. Agric. Sys. 44:257-259.

Hearn, A.B. 1995. The principles of cotton water relations and their application in management. *In:* G.A. Constable and N.W. Forrester (eds.). Proc. World Cotton Research Conference (1st:1994 Brisbane, Queensland). CSIRO, Melbourne, Australia. pp. 66-90.

Hearn, A.B. and G.A. Constable. 1984. Cotton. pp. 495-527. *In:* P.R. Goldsworthy and N.M. Fisher (eds.). The Physiology of Tropical Field Crops, John Wiley and Sons, New York.

Hearn, A.B. and G.D. da Roza. 1985. A simple model for crop management application for cotton (*Gossypium hirsutum* L). Field Crops Res. 12:49-69.

Hearn, A.B. and G.P. Fitt. 1992. Cotton cropping systems. pp. 84-142. *In:* C.J. Pearson (ed.). Ecosystems of the World. 18. Field Crop Ecosystems. Elsevier, Amsterdam.

Hearn, A.B., P.M. Ives, P.M. Room, N.J. Thomson, and L.T. Wilson. 1981. Computer-based cotton pest management in Australia. Field Crops Res. 4: 321-332.

Hearn, A.B. and P.M. Room. 1979. Analysis of crop development for cotton pest management. Prot. Ecol. 1:265-277.

Heath, R.L. 1975. Ozone. *In:* J.B. Mudd and T.T. Kozlowski (eds.). Responses of Plants to Air Pollution. Academic Press, N.Y. pp. 23-55.

Heath, R.L. 1987. The biochemistry of ozone attack on the plasma membrane of plant cells. Rec. Adv. Phytochem. 21:29-54.

Heath, R.L. 1988. Biochemical mechanisms of pollutant stress. *In:* W.W. Heck, O.C. Taylor, and D.T. Tingey (eds.). Assessment of Crop Loss from Air Pollutants. Elsevier Appl. Sci., London. pp. 259-286.

Heath, R.L. and P.E. Frederick. 1979. Ozone alteration of membrane permeability in Chlorella. Plant Physiol. 64:455-459.

Hebbara, M., S.G. Patil, M.V. Manjunatha, and R.K. Gupta. 1996. Performance of cotton (*Gossypium* species) genotypes under different salinity and water-table conditions. Ind. J. Agric. Sci. 66:446-454.

Hebert, J.J. 1992. An X-ray study of crystallinity in developing cotton fibers. pp. 193-198. *In:* C.R. Benedict (ed.). Proc. Cotton Fiber Cellulose: Structure, Function, and Utilization Conference. National Cotton Council, Memphis Tenn.

Hebert, J.J. 1993. Strength of the primary wall of cotton fibers. Text. Res. J. 63:695.

Hecht-Buchholz, C. 1967. Uber die Dunkelfarbung des Blattgruns bei Phosphormangel. A. Pflanzenernahr. Bodenk. 118:12-22.

Heck, W.W., O.C. Taylor, R.M. Adams, G. Bingham, E.M. Preston, and L.H. Weinstein. 1982. Assessment of crop loss from ozone. J. Air Pollut. Control Assoc. 32:353-362.

Heck, W.W., R.B. Philbeck, and J.A. Denning. 1978. A continuous stirred tank reactor (CSTR) system for exposing plants to gaseous air pollutants. USDA, ARS, Publ. No. ARS-5-181, Washington DC.

Heck, W.W., O.C. Taylor, D.T. Tingey (eds.). 1988. Assessment of Crop Loss from Air Pollutants. Elsevier Appl. Sci., London.

Hector, D.J. and I.D. Hodkinson. 1989. Stickiness in cotton. CAB International, Wallingford, 43 pp.

Hedin, P.A., F.E. Callahan, D.A. Dollar, and R.G. Creech. 1995. Total sterols in root-knot nematode *Meloidogyne incognita* infected cotton *Gossypium hirsutum* (L.) plant roots. Comparative Biochemistry and Physiology 111B:447-452.

Hedin, P.A. and R.G. Creech. 1998. Altered amino acid metabolism in root-knot nematode inoculated cotton plants. J. Agric. and Food Chemistry 46:4413-4415.

Hedin, P.A., J.N. Jenkins, D.H. Collum, W.H. White, W.L. Parrott, and M.W. MacGown. 1983. Cyanidin 3-b-glucoside, a newly recognized basis for resistance in cotton to the tobacco budworm *Heliothis virescens* (Fab.) (lepidoptera: Noctuidae). Experientia 39:799-801.

Hedin, P.A., J.N. Jenkins, and W.L. Parrott. 1992. Evaluation of flavonoids in *Gossypium arboreum* (L.) cottons as potential source of resistance to tobacco budworm. J. Chem. Ecol. 18:105-114.

Hedin, P.A. and J.C. McCarty, Jr. 1991. Effects of kinetin formulations of allelochemicals and agronomic traits of cotton. J. Agric. and Food Chem. 39:549-553.

Hedin, P.A. and J.C. McCarty, Jr. 1994a. Effects of natural bioregulators on cotton production. pp. 1345-1347. *In:* D.J. Herber and D.A. Richter (eds.). Proc. Beltwide Cotton Conf., National Cotton Council of America, Memphis, Tenn.

Hedin, P.A. and J.C. McCarty, Jr. 1994b. Effects of several commercial plant growth regulator formulations on yield and allelochemicals of cotton (*Gossypium hirsutum* L.). J. Agric. Food Chem. 42:1355-1357.

Hedin, P.A., W.L. Parrott, and J.N. Jenkins. 1991. Effects of cotton plant allelochemicals and nutrients on behavior and development of tobacco budworm. J. Chem. Ecol. 17:1107-1119.

Hedin, P.A, W.L. Parrott, and J.W. Jenkins. 1992. Relationships of glands, cotton terpenoid square aldehydes and other allelochemicals to larval growth of *Heliothis virescens* (Lepidoptera: Noctuidae). J. Econ. Entomol. 85:359-364.

Hedin, P.A, W.L. Parrott, J.W. Jenkins, J.E. Mulrooney, and J.J. Menn. 1988. Elucidating mechanisms of tabacco budworm resistance to allelochemicals by dietary tests with insecticide synergists. Pest. Biochem. Physiol. 32:55-61.

Hedin, P.A., A.C. Thompson, and R.C. Gueldner. 1976. Cotton plant and insect constituents that control boll weevil behavior and development. Recent Adv. Phytochem. 10:271-350.

Hedin, P.A., A.C. Thompson, R.C. Gueldner, and J.P. Minyard. 1971. Constituents of the cotton bud. Phytochemistry 10:3316- 3318.

Hedrich, R., H. Barbier-Byrgoo, H. Felle, U.I. Flugge, U. Luttge, F.J.M. Maathuis, S. Marx, H.B.A. Prins, K. Raschke, H. Schnabl, J.I. Schroeder, I. Struve, L. Taiz, and P. Ziegler. 1988. General mechanisms for solute transport across the tonoplast of plant vacuoles: A patch-clamp survey of ion channels and proton pumps. Bot. Acta 101:7-13.

Hedrich, R. and P. Dietrich. 1996. Plant K$^+$ channels: Similarity and diversity. Bot. Acta 109:94-101.

Heggestad, H.E. and M.N. Christiansen. 1982. Effects of air pollution on cotton. *In:* J.S. Jacobson and A.A. Miller (eds.). Symposium on the effects of air pollution on farm commodities. Izaak Walton League Amer., Arlington, VA. pp. 9-32.

Heilman, M.D. 1981. Interactions of nitrogen with PIX on the growth and yield of cotton. pp. 47. *In:* J.M. Brown (ed.). Proc. Beltwide Cotton Prod. Res. Conf., National Cotton Council of America, Memphis, Tenn.

Heilman, M.D. 1985. Effect of mepiquat chloride and nitrogen levels on yield, growth characteristics, and elemental composition of cotton. J. Plant Growth Regul. 4:41-47.

Heilman, M.D., L.N. Nakmen, and R.H. Dilday. 1981. Tobacco budworm: Effect of early-season terminal damage on cotton lint yield and earliness. J. Econ. Entomol 74:732-735.

Heinstein, P.F., D.L. Herman, S.B. Tove, and F.H. Smith. 1970. Biosynthesis of gossypol. J. Biol. Chem. 245:4658-4665.

Heitholt, J.J. 1993a. Cotton boll retention and its relation to lint yield. Crop Sci. 33:486-490.

Heitholt, J.J. 1993b. Growth, boll opening rate, and fiber properties of narrow-row cotton. Agron. J. 85:590-594.

Heitholt, J.J. 1994a. Canopy characteristics associated with deficient and excessive cotton plant population densities. Crop Sci. 34:1291-1297.

Heitholt, J.J. 1994b. Supplemental boron, boll retention, ovary carbohydrates, and lint yield in modern cotton genotypes. Agron. J. 86:492-497.

Heitholt, J.J. 1994c. Effect of foliar urea- and triazone-nitrogen, with and without boron, on cotton. J. Plant. Nutr. 17:57-70.

Heitholt, J.J. 1994d. Comparison of adjuvant effects on cotton leaf potassium concentration and lint yield. J. Plant Nutr. 17:221-233.

Heitholt, J.J. 1995. Cotton flowering and boll retention in different planting configurations and leaf shapes. Agron. J. 87:994-998.

Heitholt, J.J. and W.R. Meredith Jr. 1998. Yield, flowering, and leaf area index of okra-leaf and normal-leaf cotton isolines. Crop Sci. 38:643-648.

Heitholt, J.J.; W.R. Meredith, Jr., and J.R. Williford. 1996. Comparison of cotton genotypes varying in canopy characteristics in 76-cm and 102-cm rows. Crop Sci. 36:955-960.

Heitholt, J.J., W.T. Pettigrew, and W.R. Meredith, Jr. 1992. Light interception and lint yield of narrow-row cotton. Crop Sci. 32:728-733.

Heitholt, J.J., W.T. Pettigrew, and W.R. Meredith, Jr. 1993. Growth, boll opening rate, and fiber properties of narrow-row cotton. Agron. J. 85:590-594.

Hendrix, D.L. 1990. Carbohydrates and carbohydrate enzymes in developing cotton ovules. Physiol. Plant. 78:85-92.

Hendrix, D.L. 1992. Influence of elevated CO_2 on leaf starch of field-grown cotton. Crit. Rev. Plant Sci. 11:223-226

Hendrix, D.L. 1993. Rapid extraction and analysis of non-structural carbohydrates in plant tissues. Crop Sci. 33:1306-1311.

Hendrix, D.L. and R.I. Grange. 1991. Carbon partitioning and export from mature cotton leaves. Plant Physiol. 95:228-233.

Hendrix, D.L. and S.C. Huber. 1986. Diurnal fluctuations in cotton leaf carbon export, carbohydrate content and sucrose synthesizing enzymes. Plant Physiol. 81:584-586.

Hendrix, D.L., J.R. Mauney, B.A. Kimball, K.F. Lewin, J. Nagy, and G.R. Hendry. 1994. Influences of elevated CO_2 and mild water stress on non-structural carbohydrates in field-grown cotton tissues. Agric. For. Met. 70:153-162.

Hendrix, D.L. and K.K. Peelen. 1987. Artifacts in the analysis of plant tissues for soluble carbohydrates. Crop Sci. 27:710-715.

Hendrix, D.L. and J.W. Radin. 1984. Seed development in cotton: feasibility of a hormonal role for abscisic acid in controlling vivipary. J. Plant Physiol. 117:211-221.

Hendrix, D.L., J.W. Radin, and R.A. Nieman. 1987. Intracellular pH of cotton embryos and seed coats during fruit development determined by ^{31}P nuclear magnetic resonance spectroscopy. Plant Physiol. 85:588-591.

Hendrix, D.L., T.L. Steele, and H.H. Perkins, Jr. 1996. *Bemisia* honeydew and sticky cotton. pp. 189-199. *In:* D. Gerling and R.T. Mayer (eds.). *Bemisia* 1995: Taxonomy, Biology, Damage and Management. Intercept, Andover, Hants.

Hendrix, D.L. and Y-A. Wei. 1992. Detection and elimination of honeydew excreted by the sweetpotato whitefly feeding upon cotton. pp. 671-673. *In:* Proc. Beltwide Cotton Res. Conf., National Cotton Council of America, Memphis, Tenn.

Hendrix, D.L. and Y.-A. Wei. 1994. Bemisiose: An unusual trisaccharide in *Bemisia* honeydew. Carbohydrate Res. 253:329-334.

Hendrix, D.L., Yuan-an Wei, and J.E. Leggett. 1992. Homopteran honeydew sugar composition is determined by both the insect and plant species. Comp. Biochem. Physiol. 101B: 23-27.

Hendrix, J.W., B.Z. Guo, and Z.-Q. An. 1995. Divergence of mycorrhizal fungal communities in crop production systems. Plant Soil 170:131-140.

Hendry, G.R. and B.A. Kimball. 1994. The FACE program. Agric. For. Met. 70:3-14.

Henneberry, T.J., L.A. Bariola, C.C. Chu, R. Meng, Jr., B. Deeter, and L.F. Jech. 1992. Early-season ethephon applications: Effect of cotton fruiting and initiation of pink bollworm infestations and cotton yields. Southwest. Entomol. 17:135-147.

Henneberry, T.J., T. Meng, W.D. Hutchison, L.A. Bariola, and B. Deeter. 1988. Effects of ethephon on boll weevil (Coleoptera: Curculionidae) population development, cotton fruiting, and boll opening. J. Econ. Entomol. 81:628-633.

Henshall, A. 1996. Analysis of starch and other complex carbohydrates by liquid chromatography. Cereal Foods World 41:419-424.

Henson, C.A., S.H. Duke, and W.L. Koukkari. 1986. Rhythmic oscillations in starch concentrations and activities of amylolytic enzymes and invertase in *Medicago sativa* nodules. Plant Cell Physiol. 27:233-242.

Hequet, E. and R. Frydrych. 1997. The use of high speed stickiness detector on a large range of cotton coming from different countries. pp. 1649-1653. *In:* Beltwide Cotton Prod. Res. Conf., National Cotton Council of America, Memphis, Tenn.

Herbert, T.J. 1991. Variation in interception of the direct solar beam by top canopy layers. Ecol. 72:17-22.

Hering, O., H.I. Nirenberg, S. Kohn, and G. Deml. 1999. Characterization of isolates of *Fusarium oxysporum* Schlect f.sp. *vasinfectum* (Atk.) Synd. & Hans., races 1-6, by cellular fatty acid analysis. J. Phytopathology 147:509-514.

Hermann, K.M. 1995. The shikimate pathway: Early steps in the biosynthesis of aromatic compounds. Plant Cell 7:907-919.

Herms, D.A. and W.J. Mattson. 1992. The dilemma of plants: to grow or defend. Quart. Rev. Biol. 67:283-335.

Herth, W. 1985. Plasma-membrane rosettes involved in localized wall thickening during xylem vessel formation of *Lepidium sativum* L. Planta 164:12-21.

Herth, W. 1989. Inhibitor effects on putative cellulose synthetase complexes of vascular plants. pp. 795-810. *In:* C. Schuerch (ed.). Cellulose and Wood: Chemistry and Technology. John Wiley and Sons, New York.

Hesketh, J.D. 1968. Effect of light and temperature during plant growth on subsequent leaf CO_2 assimilation rates under standard conditions. Aust. J. Biol. Sci. 21:235-241.

Hesketh, J.D., D.N. Baker, and W.G. Duncan. 1972. Simulation of growth and yield in cotton: II. Environmental control of morphogenesis. Crop Sci. 12:436-439.

Hesketh, J.D., H.C. Lane, R.S. Alberte, and S. Fox. 1975. Earliness factors in cotton: new comparisons among genotypes. Cotton Grow. Rev. 52:126-135.

Hesketh, J.D. and A. Low. 1968. The effect of temperature on components of yield and fiber quality of cotton varieties of diverse origin. Cotton Growing Rev. 45:243-257.

Hess, D. and D. Bayer. 1974. The effect of trifuralin on the ultrastructure of dividing cells of the root meristem of cotton (*Gossypium hursutum* L. 'Acala 4-42') J. Cell Sci. 15:429-441.

Hessler, L.E., H.C. Lane, and A.W. Young. 1959. Cotton fiber development studies at suboptimum temperatures. Agron. J. 51:125-128.

Hewitt, E.J. 1984. The effects of mineral deficiencies and excesses on growth and composition. *In:* C. Bould, E.J. Hewitt, and P Needham (eds.). Diagnosis of mineral disorders in plants. Vol I. Principles. Chemical Publishing, NY. pp. 54-110.

Hickey, J.A. 1994. A practical system of plant growth regulation in cotton production. pp. 1269-1271. *In:* D.J. Herber and D.A. Richter (eds.). Proc. Beltwide Cotton Conf., National Cotton Council of America, Memphis, Tenn.

Hickey, J.A. 1996. Farm verification of active bloom applications of PGR-IV to enhance yield and maturity. p. 1150. *In:* P. Dugger and D.A. Richter (eds.). Proc. Beltwide Cotton Prod. Res. Conf., National Cotton Council of America, Memphis, Tenn.

Hickey, J.A. and J.A. Landivar. 1997. The application of PGR-lV on commercial cotton crops during the boll filling period. pp. 1400. *In:* P. Dugger and D.A. Richter (eds.). Proc. Beltwide Cotton Prod. Res. Conf., National Cotton Council of America, Memphis, Tenn.

Hicks, G.S. and L.M. Sussex. 1971. Organ regeneration in sterile culture after median bisection of the flower primordia or *Nicotiana tabacum*. Bot. Gaz. 132:350-363.

Hicks, K.B. 1988. High-performance liquid chromatography of carbohydrates. Adv. Carb. Chem. Biochem. 46:17-72.

Hicks, S.K., C.W. Wendt, J.R. Gannaway, and R.B. Baker. 1989. Allelopathic effects of wheat straw on cotton germination, emergence, and yield. Crop Sci. 29:1057-1061.

Hilbert, D.W., D.M. Swift, J.K. Detling, and M.I. Dyer. 1981. Relative growth rates and the grazing optimization hypothesis. Oecologia 51:14-18.

Hileman, D.R., N.C. Bhattacharya, P.P. Ghosh, P.K. Biswas, K.F. Lewin, and G.R. Hendry. 1992. Responses of photosynthesis and stomatal conductance to elevated carbon-dioxide in field-grown cotton. Crit. Rev. Plant Sci. 11:227-232.

Hileman, D.R., G. Huluka, P.K. Kenjige, N. Sinha, N.C. Bhattacharya, P.K. Biswas, K.F. Lewin, J. Nagy, and G.R. Hendry. 1994. Canopy photosynthesis and transpiration of field-grown cotton exposed to free-air CO_2 enrichment (FACE) and differential irrigation. Agric. For. Met. 70:189-208.

Hill, A.C. and N. Littlefield. 1969. Ozone: Effect on apparent photosynthesis, rate of transpiration and stomatal closure in plants. Environ. Sci. Technol. 3:52-56.

Hill, D.S., J.I. Stein, N.R. Torkewitz, A.M. Morse, C.R. Howell, J.P. Pachlatko, J.O. Becker, and J.M. Ligon. 1994. Cloning of genes involved in the synthesis of pyrrolnitrin from *Pseudomonas*

fluorescens and role of pyrrolnitrin synthesis in biological control of plant disease. Applied & Environmental Microbiology 60:78-85.

Hill, M.K., K.J. Lyon, and B.R. Lyon. 1999. Identification of disease response genes expressed in *Gossypium hirsutum* upon infection with the wilt pathogen *Verticillium dahliae*. Plant Molecular Biology 40:289-296.

Hillel, D. 1980. Fundamentals of Soil Physics. Academic Press, New York.

Hillel, D. 1980. Applications of Soil Physics. Academic Press, New York.

Hillis, D.M., B.K. Mable, and C. Moritz. 1996. Applications of molecular systematics. Pp. 515-543 *In:* D.M. Hills, C. Moritz, and B.K. Mable (eds.). *Molecular Sytematics*, Sinauer, Sunderland, MA.

Hillman, W.S. 1976. Biological rhythms and physiological timing. Annu. Rev. Plant Physiol. 27:159-179.

Hillocks, R.J. 1986. Cross protection between strains of *Fusarium oxysporum* f.sp. *vasinfectum* and its effect on vascular resistance mechanisms. J. Phytopathology 117:216-225.

Hillocks, R.J. (ed.). 1992. Cotton Diseases. CAB International. Wallingford, UK. 415 pp.

Hillocks, R.J. 1995. Integrated pest management of insect pests, diseases and weeds of cotton in Africa. Integrated Pest Management Reviews 1:31-47.

Hillocks, R.J. and R. Chinoyda. 1989. Relationship between Alternaria leaf spot and potassium deficiency causing premature defoliation of cotton. Plant Pathol. 38:502-508.

Hinkle, D.A. and A.L. Brown. 1968. Secondary and micronutrients. *In:* F.C. Elliot, M. Hoover, and W.K. Porter (eds.). Advances in Production and Utilization of Quality Cotton: Principles and Practices. Iowa State University Press. Ames, Iowa. pp. 282-320.

Hinsvark, O.N., S.H. Wittwer, and H.B. Tukey. 1953. The metabolism of foliar-applied urea. 1. Relative rates of $C^{14}O_2$ production by certain vegetable plants treated with labeled urea. Plant Physiology 28:70-76.

Hirai, A., F. Horii, and R. Kitamaru. 1990. CP/MAS 13C NMR study of never-dried cotton fibers. J. Polym. Sci: Part C: Polymer Letters 28:357-361.

Hiranpradit, H. and C.L. Foy. 1973. Retardation of leaf senescence in maize by subtoxic levels of bromacil, fluometuron, and atrazine. Bot. Gaz. 134:26-31.

Ho, L.C., R.G. Hurd, L.J. Ludwig, A.F. Shaw, J.H.M. Thornley, and A.C. Withers. 1984. Changes in photosynthesis, carbon budget and mineral content during the growth of the first leaf of cucumber. Ann. Bot. 54:87-101.

Hocking, P.J., D.C. Reicosky, and W.S. Meyer. 1985. Nitrogen status of cotton subjected to two short term periods of waterlogging of varying severity using a sloping plot water-table facility. Plant and Soil 87:375-391.

Hodges, H.F., V.R. Reddy, and K.R. Reddy. 1991. Mepiquat chloride and temperature effects on photosynthesis and respiration of fruiting cotton. Crop Sci. 31:1302-1308.

Hodges, H.F., K.R. Reddy, J.M. McKinion, and V.R. Reddy. 1993. Temperature effects on cotton. Miss. Agri. and For. Exp. Sta. Bull. 990. February.

Hodges, H.F., F.D. Whisler, S.m. Bridges, K.R. Reddy, and J.M. McKinnion. 1998. Simulation in crop management - GOSSYM/COMAX. *In:* R.M. Peart and R.B. Curry (eds.). Agricultural Systems Modeling and Simulation. Marcel Dekker, Inc., New York, NY.

Hodges, S.C 1991. Nutrient uptake by cotton: A review. 1991 Proc. Beltwide Cotton Prod. Res. Conf. pp. 938-940. National Cotton Council, Memphis, Tenn.

Hodges, S.C. 1992. Nutrient Deficiency Disorders. *In:* R. Hillocks (ed.). Cotton Diseases. CAB International, Wallinford UK. pp. 355-403.

Hodges, S.C. and G. Constable. 2009. Plant responses to mineral deficiencies and toxicities. pp. 143-162. *In:* J.M.Stewart, D.M. Oosterhuis, J.J. Heitholt, and J.R. Mauney (eds.). Physiology of Cotton. National Cotton Council of America, Memphis,Tenn. pp. Springer, London.

Hodges, S.C. and S. Baker. 1993. Correlation of plant sap extracts of nitrate-N and -K with dried petiole extracts. Proc. Beltwide Cotton Conferences, New Orleans, LA. pp. 1335-1337.

Hodges, T. 1991. Temperature and water stress effects on phenology. p. 7-13. *In:* T. Hodges (ed.). Predicting Plant Phenology. CRC Press., Inc., Boca Raton, FL.

Hodgson, A.S. and K.Y. Chan. 1982. The effect of short-term waterlogging during furrow irrigation on cotton in a cracking gray clay. Aust. J. Agric. Res. 33:109-116.

Hodgson, A.S., J.S. Holland, and E.F. Rogers. 1992. Iron deficiency depresses growth of furrow irrigated soybean and pigeon pea on vertisols of Northern NSW. Aust. J. Agric. Res. 43: 635-644.

Hoffman, G.J. and C.J. Phene. 1971. Effect of constant salinity levels on water-use efficiency of bean and cotton. Trans. Amer. Soc. Agric. Eng. 14:1103-1106.

Hoffman, G.J. and S.I. Rawlins. 1970. Infertility of cotton flowers at both high and low relative humidities. Crop Sci. 10:721-723.

Hoffman, G.J., S.L. Rawlins, M.J. Garber, and E.M. Cullen. 1971. Water relations and growth of cotton as influenced by salinity. Agron. J. 63:822-826.

Hofmann, W.C., M.M. Karpiscak, and P.G. Bartels. 1987. Response of cotton, alfalfa, and cantaloupe to foliar-deposited salt in an arid environment. J. of Environ. Qual. 16:267-272.

Hofstra, G., A. Ali, R.T. Wukasch, and R.A. Fletcher. 1981. The rapid inhibition of root respiration after exposure of bean (Phaseolus vulgaris L.) plants to ozone. Atmos. Environ. 15: 483-487.

Hogsett, W.E., D. Olszyk, D.P. Ormrod, G.E. Taylor, Jr., and D.T. Tingey. 1987. Air pollution exposure systems and environmental protocols: Vol. 1: a review and evaluation of performance. U.S. EPA Publ. 600/3-87/037a. Environ. Protect. Agency, Corvllis, OR.

Holloway, P.J. 1982. The chemical composition of plant cuticles. *In:* D.F. Cutler, K.L. Alvin, and C.E. Price (eds.). The Plant Cuticle. Academic Press, London. 461 pp.

Holman, E.M. 1993. Foliar fertilization of environmentally stressed cotton seedlings. MS. Thesis. University of Arkansas. Fayetteville, AR.

Holman, E.M. and D.M. Oosterhuis. 1992. Effect of foliar-applied nitrogen on the growth and drought tolerance of cotton seedlings. Proc. Beltwide Cotton Prod. Cof., National Cotton Council, Memphis, Tenn. p. 1085.

Holman, E.M., D.M. Oosterhuis, and R.G. Hurren. 1992a. Effect of foliar-applied N and K on vegetative cotton. Arkansas Farm Research 41(1):4-5.

Holman, E.M., D.M. Oosterhuis, and R.G. Hurren. 1992b. Effect of foliar fertilizer on droughted cotton seedlings. *In:* W.E. Sabbe (ed.). Arkansas Soil Fertility Studies 1990. Univ. of Arkansas Agr. Exp. Stn., Research Series 411:141-143.

Holmstrom, K.O., E. Mantyla, B.Welin, A. Mandal, E.T.Palas, O.E. Tunnela, and J. Londesborough. 1996. Drought tolerance in tobacco. Nature 379:683-684.

Holtz, B.A., A.R. Weinhold, and B.A. Roberts. 1994. Populations of *Thielaviopsis basicola* in San Joaquin Valley field soils and the relationship between inoculum density and disease severity of cotton seedlings. pp. 247-249. *In:* Proc. Beltwide Cotton Conf., National Cotton Council of America, Memphis, Tenn.

Holubec, V. 1990. African *Gossypium* as a genetic resource for cotton breeding. Proc. 12th plenary meeting of AETFAT, Mitt. Inst. Allg. Bot. Hamburg 23a:73-79.

Hons, F.M., N.W. Hopper, and T.V. Hicks. 1990. Applied phosphorus and potassium effects on the emergence, yield, and planting seed quality of cotton. J. Prod. Agric. 3:337-340.

Hons, F.M. and B.L. McMichael. 1986. Planting pattern effects on yield, water use and root growth in cotton. Field Crops Res. 13:147-158.

Hood, G., R.B. Turley, and J. Steen. 1996. Ribosomal protein RL44 is encoded by two subfamilies in upland cotton *(Gossypium hirsutum* L). Biochem. Biophys. Res. Comm. 226:32-36.

Hooker, A.L. 1984. The pathological and entomological framework of plant breeding. pp. 177-208 *In:* J.P. Gustafson (ed.). Gene manipulation in plant improvement. 16th Stadler Genetics Symposium, Plenum, New York.

Hooper, D.G., R.J. Venere, L.A. Brinkerhoff, and R.K. Gholson. 1975. Necrosis induction in cotton. Phytopathology 65:206-213.

Hopkins, A.R. and R.F. Moore. 1980. Thidiazuron: effect of application on boll weevil and bollworm population densities, leaf abscission, and growth of the cotton plant. J. Econ. Entomol. 73:768-770.

Hopper, N.W. 1999. The Cotton Seed. *In:* C.W. Smith and J.T. Cothren (eds.). Cotton. John Wiley & Sons. New York.

Horiguchi, T. 1987. Mechanism of manganese toxicity and tolerance of plants II. Deposition of oxidized manganese in plant tissues. Soil Sci. Plant Nutr. 33:595-606.

Horrocks, R.D., T.A. Kerby, and D.R. Buxton. 1978. Carbon source for developing bolls in normal and super-okra cotton. New Phytol. 80:335-340.

Horsch, R.B., R.T. Fraley, S.G. Rogers, P.R. Sanders, A. Lloyd, and N. Hoffman. 1984. Inheritance of functional foreign genes in plants. Science 223:496-498.

Horst, W.J. 1988. The physiology of manganese toxicity. pp. 175-188. *In:* R.D. Grahham, R.J. Hannam, and N.C. Uren (eds.). Manganese in Soil and Plants. Kluwer Academic Publishers, Boston.

Hoskinson, P.E. 1988. No-till cotton cultivar tests. Tenn. Farm & Home Sci. 145: 3-7.

Hoskinson, P.E. 1989. Effects of planting dates on cotton cultivars no-tilled into wheat residue. p. 122. *In:* Proc. Beltwide Cotton Prod. Res. Conf., National Cotton Council of America, Memphis, Tenn.

Hoskinson, P.E. and C.O. Gwathmey. 1996. Fifteen years of testing cotton response to tillage systems. Tenn. Agri Sci. 179:18-20.

Hoskinson, P.E. and D.O. Howard. 1992. Influence of tillage on fruiting patterns of Deltapine 50 cotton. p. 603. *In:* Proc. Beltwide Cotton Conf., National Cotton Council of America, Memphis, Tenn.

Hoskinson, P.E., D.D. Tyler, and R.M. Hayes. 1982. Influence of tillage and cover crops on lint yield and maturity. p. 147. *In:* Proc. Beltwide Cotton Prod. Res. Conf., National Cotton Council of America, Memphis, Tenn.

Hoskinson, P.E.; D.D. Tyler, and R.M. Hayes. 1988. Long-term no-till cotton yields as affected by nitrogen. p. 504. *In:* Proc. Beltwide Cotton Prod. Res. Conf., National Cotton Council of America, Memphis, Tenn.

Hoson, T. 1993. Regulation of polysaccharide breakdown during auxin-induced cell wall loosening. J. Plant Res. 106:369-381.

Hoson, T. and Y. Masuda. 1992. Relationship between polysaccharide synthesis and cell wall loosening in auxininduced elongation of rice coleoptile segments. Plant Sci. 83:149-154.

Houtz, R.L., R.O. Nable, and G.M. Cheniae. 1988. Evidence for effects on the in vivo activity of ribulose-biphosphate carboxlyase/oxylase during development of Mn toxicity in tobacco. Plant Physiol. 94:300.

Howard, D.D., and F. Adams. 1965. Calcium requirement for penetration of subsoils by primary cotton roots. Soil Sci. Soc. Am. Proc. 29:558-562.

Howard, D.D., M.E. Essengton, C.O. Gwathmey, and W.M. Percell. 2000. Buffering of foliar potassium and boron solutions for no-tillage cotton production. J. Cotton Sci. 4:237-244.

Howard, D.D., C.O. Gwathmey, and C.E. Sams. 1998. Foliar feeding of cotton: Evaluating potassium sources, potassium solution buffering, and boron. Agronomy Journal 90:740-766.

Howard, D.D., P.E. Hoskinson, and P.W. Brawley. 1993. Evaluation of surfactants in foliar feeding cotton with potassium nitrate. Better Crops Summer 1993. pp. 22-25.

Howard, D.D., P.E. Hoskinson, and C.O. Gwathmey. 1995. Effect of starter nutrient combinations and nitrogen rate on no-till cotton. pp. 67-69. *In:* T.D. Valco and M.R. McClelland (eds.). Conservation-Tillage Systems for Cotton. University of Arkansas Agricultural Experiment Station Special Report 169. Fayetteville, AR.

Howard, D.D., D.M. Oosterhuis, A. Steger, E.M. Holman, and C.W. Bednarz. 1998. Programmed soil fertilizer release to meet crop N and K requirements. Proc. Beltwide Cotton Conf. San Diego, CA. National Cotton Council, Memphis, Tenn. Vol. 1. pp. 672-673.

Howell, C.R. 1987. Relevance of mycoparasitism in the biological control of *Rhizoctonia solani* by *Gliocladium virens.* Phytopathology 77:992-994.

Howell, C.R. 1991. Biological control of Pythium damping-off of cotton with seed-coating preparations of *Gliocladium virens.* Phytopathology 81:738-741.

Howell, C.R. 1994. Fungal antagonists of cotton seedling disease pathogens. pp. 238-239. *In:* Proc. Beltwide Cotton Conf., National Cotton Council of America, Memphis, Tenn.

Howell, C.R., R.C. Beier, and R.D. Stipanovic. 1988. Production of ammonia by *Enterobacter cloacae* and its possible role in the biological control of Pythium preemergence damping-off by the bacterium. Phytopathology 78:1075-1078.

Howell, C.R., A.A. Bell, and R.D. Stipanovic. 1976. Effect of aging on flavonoid content and resistance of cotton leaves to Verticillium wilt. Physiological Plant Pathology 8:181-188.

Howell, C.R., L.E. Hanson, R.D. Stipanovic, and L.S. Puckhaber. 2000. Induction of terpenoid synthesis in cotton roots and control of *Rhizoctonia solani* by seed treatment with *Trichoderma virens*. Phytopathology 90:248-252.

Howell, C.R. and R.D. Stipanovic. 1979. Control of *Rhizoctonia solani* on cotton seedlings with *Pseudomonas fluorescens* and with an antibiotic produced by the bacterium. Phytopathology 69:480-482.

Howell, C.R. and R.D. Stipanovic. 1984. Phytotoxicity to crop plants and herbicidal effects on weeds of viridiol produced by *Gliocladium virens*. Phytopathology 74:1346-1349.

Howell, C.R. and R.D. Stipanovic. 1994. Effect of sterol biosynthesis inhibitors on phytotoxin (viridiol) production by *Gliocladium virens* in culture. Phytopathology 84:969-972.

Howell, C.R. and R.D. Stipanovic. 1995. Mechanisms in the biocontrol of *Rhizoctonia solani*-induced cotton seedling disease by *Gliocladium virens*: Antibiosis. Phytopathology 85:469-472.

Howell, C.R. and R.D. Stipanovic. 1996. Mechanisms in cotton soreshin biocontrol by *Trichoderma virens*: Viridiol production. p. 271. *In*: Proc. Beltwide Cotton Conf., Volume 1, National Cotton Council of America, Memphis, Tenn.

Howell, T.A., J.L. Hatfield, J.D. Rhoades, and M. Meron. 1984. Response of cotton water-stress indicators to soil salinity. Irrig. Sci. 5:25-36.

Howitt, R.G. and C. Goodman. 1988. Economic impacts of regional ozone standards on agricultural crops. Environ. Pollut. 53:387-395.

Hron, R.J., H.L. Kim, M.C. Calhoun, and G.S. Fisher. 1999. Determination of (+)-, (-)-, and total gossypol in cottonseed by high-performance liquid chromatography. J. Am. Oil Chem. Soc. 76:1351-1355.

Hsiao, T.C. 1973. Plant responses to water stress. Annu. Rev. Plant Physiology. 24:519-570.

Hsiao, T.C., E. Acevedo, E. Fereres, and D.W. Henderson. 1976. Stress metabolism: Water stress, growth, and osmotic adjustments. Phil Trans. R. Soc. (London) 273:479-500.

Hsiao, T.C. and Läuchli, A. 1986. Role of potassium in plant-water relations. *In*: B. Tinker and A. Läuchli. Advances in plant nutrition 2:281-312. Paeger Publishers, NY.

Hsiao, T.C., E.C. Oliveira, and R. Radulovich. 1982. Physiology and productivity of cotton under water stress. p. 60. Proc. 1982 Cotton Prod. Res. Confs. Las Vegas, NV. National Cotton Council of America, Memphis, Tenn.

Hsieh, Y.-L. 1994. Single fiber strength in developing cotton. pp. 109-114. *In*: G. Jividen and C. Benedict (eds.). Proc. Biochemistry of Cotton Workshop. Cotton Incorporated, Raleigh NC.

Hsieh, Y.-L., E. Honik, and M.M. Hartzell. 1995. A developmental study of single fiber strength: greenhouse grown SJ-2 Acala Cotton. Text. Res. J. 65:101-112.

Hsieh, Y.-L., X.-P. Hu, and A. Nguyen. 1997. Strength and crystalline structure of developing Acala cotton. Text. Res. J. 67:529-536.

Hsu, C.L. and J.M. Stewart. 1976. Callus induction by (2-chloroethyl) phosphonic acid on cultured cotton ovules. Physiol. Plant. 36:150-153.

Hsu, H.H., J.D. Lancaster, and W.F. Jones. 1978. Potassium concentration of leaf blades and petioles as affected by potassium fertilization and stage of maturity of cotton. Commun. In Soil Sci and Plant Anal. 9:265-277.

Hu, X.-P. and Y.-L. Hsieh. 1996. Crystalline structure of developing cotton fibers. J. Polym. Sci.: Part B: Polymer Physics 34:1451-1459.

Hu, Z.J. and X.D. Gui. 1991. Pretransplant inoculation with VA mycorrhizal fungi and Fusarium blight of cotton. Soil Biology and Biochemistry 23:201-203.

Huang, S.Y. and H.W. Gausman. 1982a. Ultrastructural observations on increase of cold tolerance in cotton plant (*Gossypium hirsutum* L.) by mepiquat chloride. J. Rio Grande Valley Hort. Soc. 35:35-41.

Huang, S.Y. and H.W. Gausman. 1982b. Ultrastructural observations on temperature resistance increase in cotton plants (*Gossypium hirsutum* L.) by mepiquat chloride. pp. 106-111. *In*: W.C. Wilson (ed.). Proc. Eight Ann. Meet. Plant Growth Regulator Soc. Am. Conf., Plant Growth Regulator Soc. Am., Lake Alfred, FL.

Huang, S.Y., H.W. Gausman, R.F. Rittig, D.E. Escobar, and R.R. Rodriguez. 1981. Increase of cold tolerance in cotton plant (*Gossypium hirsutum* L.) by mepiquat chloride. pp. 202-209. *In*: W.C. Wilson (ed.). Proc. Eight Ann. Meet. Plant Growth Regulator Soc. Am. Conf., Plant Growth Regulator Soc. Am., Lake Alfred, FL.

Huang, Y.S., P. Mukerji, T. Das, and D.S. Knutzon. 2001. Transgenic production of long-chain polyunsaturated fatty acids. Pp. 243-248. *In*: T. Hamazaki and H. Okuyama (eds.). Fatty Acids and Lipids – New Findings. World Rev. Nutr. Diet. Basel, Karger. Vol. 88.

Huber, S.C. 1981. Interspecific variation in activity and regulation of leaf sucrose phosphate synthase. Z. Pflanzenphysiol. 102:443-450.

Huber, S.C. 1983. Role of sucrose-phosphate synthase in partitioning of carbon in leaves. Plant Physiol. 71:818-821.

Huber, S.C. 1985. Role of potassium in photosynthesis and respiration. *In:* R.D. Munson (ed.). Potassium in agriculture. Amer. Soc. Agron. Madison, WI. pp. 369-396.

Huber, S.R. 1999. Nutrient management, nutrient flux and root density for cotton production on a Limestone Valley soil under varying irrigation regimes. M.S. Thesis, Auburn Univ., Auburn, AL.

Huber, D.M. and N.S. Wilhelm. 1988. The role of manganese in resistance to plant diseases. *In:* R.D. Grahham, R.J. Hannam, and N.C. Uren (eds.). Manganese in soils and plants. Kluwer Academic Publishers, Boston. pp. 155-173.

Huck, M.G. 1970. Variation in taproot elongation rate as influenced by composition of the soil air. Agron. J. 62:815-818.

Huck, M.G. 1983. An overview of root growth, distribution and function for modelers of soil- plant systems. p. 88-101 *In:* D.G. DeCoursey (ed.). Proc. Natural Resources Modeling Symp. Pingree Park, CO. Oct. 16-21.

Hudson, D., D. Ethridge, and C. Chen. 1996. Producer prices received and mill prices paid for quality in Southwest cottons: Similarities and differences. pp. 437-440. *In:* Proc. Beltwide Cotton Conf., National Cotton Council of America, Memphis, Tenn.

Hudspeth, R.L., S.L. Hobbs, D.M. Anderson, and J.W. Grula. 1996a. Characterization and expression of chitinase and 1,3-β-glucanase genes in cotton. Plant Mol. Biol. 31:911-916.

Hudspeth, R.L., S.L. Hobbs, D.M. Anderson, K. Rajasekaran, and J.W. Grula. 1996b. Characterization and expression of metallothionein-like genes in cotton. Plant Mol. Biol 31:701-705.

Hue, N.V., G.R. Craddock, and F. Adams. 1986. Effect of organic acids on aluminum toxicity in subsoils. Soil Sci. Soc.Am. J. 50:28-34.

Hughes, D.W. and G.A. Galau. 1989. Temporally modular gene expression during cotyledon development. Genes Dev. 3:358-369.

Hughes, D.W. and G.A. Galau. 1991. Developmental and environmental induction of *Lea* and *LeaA* mRNAs and the postabscission program during embryo culture. Plant Cell 3:605-618.

Hughes, D.W., H.Y.-C. Wang, and G.A. Galau. 1993. Cotton (*Gossypium hirsutum*) *MatP6* and *MatP7* oleosin genes. Plant Physiol. 101:697-698.

Huisman, O.C. and D.W. Grimes. 1989. Cultural practices: The effect of plant density and irrigation regimes on Verticillium wilt of cotton. *In:* E.C. Tgamos, and C.H. Beckman (eds.). Vascular Wilt Diseases of Plants. Springer-Verlag, New York. pp. 537-542.

Hull, H.M. 1964. Leaf structure as related to penetration of organic substances. pp. 47-93. *In:* J. Hacskaylo

(ed.). Symp. 7th Annual Southern Section, Amer. Soc. Plant Physiol., Emory Univ., Atlanta, GA.

Hulugalle, N.R., P.C. Entwistle, J.L. Cooper, S.J. Allen, and D.B. Nehl. 1998. Effect of long-fallow on soil quality and cotton lint yield in an irrigated, self-mulching, grey Vertosol in the central-west of New South Wales. Austr. J. Soil Res. 36:621-639.

Huluka, G., D.R. Hileman, P.K. Biswas, K.F. Lewin, J. Nagy, and G.R. Hendry. 1994. Effects of elevated CO_2 and mild water stress on mineral concentration of cotton. Agric. For. Met. 70:141-152.

Hunsaker, D.J., G.R. Hendry, B.A. Kimball, K.F. Lewin, J.R. Mauney, and J. Nagy. 1994. Cotton evapotranspiration under field conditions with CO_2 enrichment and variable soil moisture regimes. Agric. For. Met. 70:247-258.

Hunt, G.M. and E.A. Baker. 1982. Developmental and environmental variations in plant epicuticular waxes: Some effects on the penetration of naphthylacetic acid. pp. 279-301. *In:* D.F. Cutler, K.L. Alvin, and C.E. Price (eds.). The Plant Cuticle. Academic Press, London.

Hunter, R.E., L.A. Brinkerhoff, and L.S. Bird. 1968. Development of a set of upland cotton lines for differentiating races of *Xanthomonas malvacearum*. Phytopathology 58:830-832.

Hurej, M. and W. van der Werf. 1993. The influence of black bean aphid, *Aphis fabae* Scop., and its honeydew on leaf growth and dry matter production of sugar beet. Ann. Appl. Biol. 122:201-214.

Hursh, C.R. 1948. Local climate in the Copper Basin of Tennessee as modified by the removal of vegetation. U.S. Dep. Agric. Circ. 774:1-38.

Hush, J.M. and R.L. Overall. 1992. Re-orientation of cortical F-actin is not necessary for wound-induced microtubule re-orientation and cell polarity establishment. Protoplasma 169:97-106.

Hussain, T. and L.A. Brinkerhoff. 1978. Race 18 of the cotton bacterial blight pathogen, *Xanthomonas malvacearum*, identified in Pakistan in 1977. Plant Disease Reporter 62:1085-1087.

Hutcheson, S. 1998. Current concepts of active defense in plants. Ann. Rev. Phytopathology 36:59-90.

Hutchinson, J.B. 1943. A note on *Gossypium brevilanatum* Hochr. Trop. Agric. 20:4.

Hutchinson, J.B. 1951. Intra-specific differentiation in *Gossypium hirsutum*. Heredity 5:161-193.

Hutchinson, J.B. 1954. New evidence on the origin of the Old World Cottons. Heredity 8:225-241.

Hutchinson, J.B. 1962. The history and relationships of the world's cottons. Endeavour 21:5-15.

Hutchinson, J.B. and H.L. Manning. 1945. The Sea Island cottons. Empire Journal of Experimental Agriculture. 13:80-92.

Hutchinson, J.B., R.A. Silow, and S.G. Stephens. 1947. *The Evolution of Gossypium*. Oxford Univ. Press, London.

Hutchinson, R.L. 1993a. Conservation tillage overview and terminology. pp. 108-110. *In:* Proc. Beltwide Cotton Conf., National Cotton Council of America, Memphis, Tenn.

Hutchinson, R.L. 1993b. Overview of conservation tillage. pp. 1-9. *In:* T.D. Valco and M.R. McClelland (eds.). Conservation-Tillage Systems for Cotton. University of Arkansas Agricultural Experiment Station Special Report 169. Fayetteville, AR.

Hutmacher, R.B., J.E. Ayars, S.S. Vail, A.D. Bravo, D. Dettinger, and R.A. Schoneman. 1996. Uptake of shallow groundwater by cotton: Growth stage, groundwater salinity effects in column lysimeters. Agric. Water Manage. 31:205-223.

Hutchinson, J.B. 1951. Intra-specific differentiation in *Gossypium hirsutum.* Heredity 5:161-193.

Hutchinson, R.L., G.A. Breitenbeck, R.A. Brown, and W.J. Thomas. 1995. Winter cover crop effects on nitrogen fertilizer requirements of no-till and conventional-tillage cotton. pp. 73-76. *In:* T.D. Valco and M.R. McClelland (eds.). Conservation-Tillage Systems for Cotton. University of Arkansas Agricultural Experiment Station Special Report 169. Fayetteville, AR.

Hutchinson, R.L. and T.R. Sharpe. 1989. A comparison of tillage systems and cover crops for cotton production on a loessial soil in northeast Louisiana. pp. 517-519. *In:* Proc. Beltwide Cotton Prod. Res. Conf., National Cotton Council of America, Memphis, Tenn.

Huwyler, H.R., G. Franz, and H. Meier. 1979. Changes in the composition of cotton fibre cell walls during development. Planta 146:635-642.

Ibrahim, A.A. 1984. Effect of GA3 and boron on growth, yield and accumulation of Na, K, and Cl in cotton grown under saline conditions. Ann. Agric. Sci. Moshtohor 21:519-531.

Idso, S.B., B.A. Kimball, and J.R. Mauney. 1987a. Atmospheric carbon dioxide enrichment effects on cotton midday foliage temperature: Implications for plant water use and crop yield. Agron. J. 79:667-672.

Idso, S.B., B.A. Kimball, and J.R. Mauney. 1987b. Effects of atmospheric CO_2 enrichment on plant growth: The interactive role of air temperature. Agric. Ecosys. Environ. 20:1-10.

Idso, S.B., B.A. Kimball, and J.R. Mauney. 1988. Effects of atmospheric CO_2 enrichment on root: shoot ratios in carrot, radish, cotton, and soybean. Agric. Ecosys. Environ. 21:293-299.

Idso, S.B., B.A. Kimball, G.W. Wall, R.L. Garcia, R. LaMorte, P.J. Pinter, Jr., J.R. Mauney, G.R. Hendry, K.F. Lewin, and J. Nagy. 1994. Effects of free-air CO_2 enrichment on the light response curve of net photosynthesis in cotton leaves. Agric. For. Met. 70:183-188.

Ihle, J.N. and L.S. Dure, III. 1972. The developmental biochemistry of cottonseed embryogenesis and germination. III. Regulation of the biosynthesis of enzymes utilized in germination. J. Biol. Chem. 247:5048-5055.

Ike, I.F. 1986. Soil and crop responses to different tillage practices in a ferruginous soil in the Nigerian savanna. Soil & Tillage Res. 6:261-272.

Ikeda, T. 1992. Soybean planting patterns in relation to yield and yield components. Agron. J. 84:923-926.

Inamdar, R.S., S.B. Singh, and T.D. Pande. 1925. The growth of the cotton plant in India. I. The relative growth-rates during successive periods of growth and the relation between growth-rate and respiratory index throughout the life cycle. Ann. Bot. 39:281-311.

Ingham, R.E. 1988. Interactions between nematodes and VA mycorrhizae. Agric. Ecosyst. Environ. 24:169-182.

Ingram, W.R. 1994. *Pectinophora* (Lepidoptera: Gelechiidae). pp. 107-149 *In:* G.A. Matthews and J.P. Tunstall (eds.). Insects Pests of Cotton. CAB International, Wallingford.

Inoue, Y., B.A. Kimball, J.R. Mauney, R.D. Jackson, P.J. Pinter, Jr., and R.J. Reginato. 1990. Stomatal behavior and relationship between photosynthesis and transpiration in field-grown cotton as affected by CO_2 enrichment. Jap. Jour. Crop Sci. 59:510-517.

Iqbal, R.M.S., M.B. Chaudhry, M. Aslam, and A.A. Bandesha. 1991. Economic and agricultural impact of mutation breeding in cotton in Paksitan. A review. *In:* P.H. Kitto (ed.). Plant Mutation for Crop Improvement. IAEA, Vienna. pp. 187-201.

Iqbal, R.M.S., M.B. Chaudhry, M. Aslam, and A.A. Bandesha. 1994. Development of a high yielding cotton mutant, NIAB-92 through the use of induced mutations. Pakistan J. Bot., 26:99-104.

Iraki, N.M., R.A. Bressan, P.M. Hasegawa, and N.C. Carpita. 1989. Alteration of the physical and chemical structure of the primary cell wall of growth-limited plant cells adapted to osmotic stress. Plant Physiol. 91:39-47.

Ishag, H.M. A.T. Ayoub, and M.B. Said. 1987. Cotton leaf reddening in the irrigated Gezira. Exp. Agric. 23:207-212.

Ismailov, A.I., A.K. Karimdzhanov, S.Y. Islambekov, and Z.B. Rakhimkhanov. 1994. Flavonoids of cotton and its relatives. Khim. Prir. Soedin. 1:3-19.

Itai, C. and H. Birnbaum. 1996. Synthesis of plant growth regulators by roots. pp. 273-284. *In:* Y. Waisel, A. Eshel, and U. Kafkafi (eds.). Plant Roots: The Hidden Half. Marcel Dekker,Inc. New York.

Ivanina, L.N. 1991. Possible use of some biochemical indices in the diagnosis of salt resistance in cotton. Konferetsiya biokhimikov respublik Srednei Azii I Kazakhstana, Tashkent. 1991, 182. (Russian).

Ives, P.M., L.T. Wilson, P.O. Cull, W.A. Palmer, C. Haywood, N.J. Thomson, A.B. Hearn, and A.G.L. Wilson. 1984. Field use of Siratac: an Australian computer-based pest management system for cotton. Prot. Ecol. 6:1-21.

Ivleva, L.B. and L.S. Plekhanova. 1992. Influence of salinity on cation-stimulated ATPase activity of the cotton root cell plasmalemma. Fiziologiya i Biokhimiya Kul'turnykh Rastenii 24:499-503.

Ivonvina, L.N. and T.P. Ladonina. 1976. Pathways of radial transport of salts in cotton roots. Biologiya Zhivotnykh I Rastenii Turkmenistana 1976:76-79. (Russian).

Iwasaki, I., H. Arata, H. Kijima, and M. Nishimura. 1992. Two types of channels involved in the malate ion transport across the tonoplast of a crassulacean acid metabolism plant. Plant Physiol. 98:1494-1497.

Iyengar, E.R.R. 1978. Evaluation of cotton varieties to salinity stress. Indian J. Plant Physiol. 21:113-117.

Iyengar, R.L.N. 1941. Variation in the measurable characters of cotton fibers. II. Variation among seeds within a lock. Ind. J. Agric. Sci. 11:703-735.

Jackson, B.S. 1991. Simulating yield development using cotton model COTTAM. pp. 171-180. *In:* T. Hodges (ed.). Predicting Crop Phenology. CRC Press, Boca Raton, FL.

Jackson, B.S., and T.J. Gerik. 1990. Boll shedding and boll load in nitrogen-stressed cotton. Agron. J. 82:483-488.

Jackson, E.B. and P.A. Tilt. 1968. Effects of irrigation intensity and nitrogen level on the performance of eight varieties of upland cotton (*Gossypium hirsutum* L.). Agron. J. 60:13-17.

Jackson, M.B. 1985. Ethylene and response of plants to soil waterlogging and submergence. Annu. Rev. Plant Physiol. 36:145-174.

Jackson, M.B., B. Herman, B., and A. Goodenough. 1982. An examination of the importance of ethanol in causing injury to flooded plants. Plant Cell. Environ. 5:163-172.

Jackson, N.B. 1982. Ethylene as a growth promoting hormone under flooded conditions. Pp. 291-301. *In:* P.F. Wareing (ed.). Plant Growth Substances. Academic Press, NY.

Jadhao, J.K., A.M. Degaonkar, and W.N. Narkhede. 1993. Performance of hybrid cotton (*Gossypium* species) cultivars at different plant densities and nitrogen levels under rainfed conditions. Indian J. Agron. 38:340-341.

Jafri, A.Z. and R. Ahmad. 1994. Plant growth and ionic distribution in cotton (*Gossypium hirsutum* L.) under saline environment. Pakistan J. Bot. 26:105-114.

Jafri, A.Z. and R. Ahmad. 1995. Effect of soil salinity on leaf development, stomatal size, and its distribution in cotton (*Gossypium hirsutum* L.). Pakistan J. Bot. 27:297-303.

Jaglo-Ottosen, K.R., S.J. Gilmour, D.G. Zarka, O. Schabenberger, and M.F. Thomashow. 1998. *Arabidopsis* CBF1 overexpression induces *cor* genes and enhances freezing tolerance. Science 28:104-106.

Jaio, J. and R. Chollet. 1988. Light/dark regulation of maize leaf phosphoenolpyruvate carboxylase by *in vivo* phosphorylation. Arch. Biochem. Biophys. 261:409-417.

Jakobsen, I. 1999. Transport of phosphorus and carbon in arbuscular mycorrhizas. p. 305-332. *In:* A. Varma and B. Hock (eds.). Mycorrhiza: Structure, Function, Molecular Biology and Biotechnology. 2nd edition. Springer-Verlag, Berlin.

Jalaluddin, M. 1993. Effect of VAM fungus (*Glomus intraradices*) on the growth of sorghum, maize, cotton and *Pennisetum* under salt stress. Pakistan J. Bot. 25:215-218.

Jameson, D.A. 1963. Responses of individual plants to harvesting. Bot. Rev. 29:532-594.

Janardhan, K.V., A.S. Murthy, K. Giriraj, and S. Panchaksharaiah. 1976. Salt tolerance of cotton a potential use of saline water for irrigation. Current Sci. 45:334-336.

Janes, L.D., D.M. Oosterhuis, F.M. Bourland, and C.S. Rothrock. 1993. Foliar-applied potassium nitrate effects on cotton genotypes. *In:* W.E. Sabbe (ed.). Arkansas Soil Fertility Studies 1992. Univ. of Arkansas Agri. Exp. Stn. Research Series 425:77-79.

Jaquet, J.-P., A.J. Buchala, and H. Meier. 1982. Changes in the non-structural carbohydrate content of cotton (*Gossypium* spp.) fibres at different stages of development. Planta 156:481-486.

Jaroszewski, J.W., T. Strom-Hansen, S.H. Hansen, and O. Thastrup. 1992. On the botanical distribution of chiral forms of gossypol. Planta Medica 58:454-458.

Jasdanwala, R.T., Y.D. Singh, and J.J. Chinoy. 1977. Auxin metabolism in developing cotton hairs. J. Expt. Bot. 28:1111-1116.

Jasper, D.A., L.K. Abbott, and A.D. Robson. 1989a. Soil disturbance reduces the infectivity of external hyphae of vesicular-arbuscular mycorrhizal fungi. New Phytol. 112: 93-99.

Jasper, D.A., L.K. Abbott, and A.D. Robson. 1989b. Hyphae of a vasicular-arbuscular mycorrhizal fungus maintain infectivity in dry soil, except when soil is disturbed. New Phytol. 112:101-107.

Jasper, D.A., L.K. Abbott, and A.D. Robson. 1991. The effect of soil disturbance on vesicular-arbuscular mycorrhizal fungi in soils from different vegetation types. New Phytol. 118:471-476.

Jasper, D.A., L.K. Abbott, and A.D. Robson. 1992. Soil disturbance in native ecosystems-the decline and recovery of infectivity of VA mycorrhizal fungi. pp. 151-155. *In:* D.J. Read, D.H. Lewis, A.H. Fitter, and I.J. Alexander (eds.). Mycorrhizas in Ecosystems. CAB International, Oxon, UK.

Jayalalitha, K. and A. Narayanan. 1996. Growth and mineral composition of magnesium deficient cotton plants grown in solution culture. Ann. Plant Physiol. 10:11-16.

Jefferson, P.G., D.A. Johnson, and M.D. Rumbaugh. 1988. Genetic analysis of epicuticular wax production in alfalfa. Genome 30:896-899.

Jeffries, P., T. Spyropoulos, and E. Vardavakis. 1988. Vesicular-arbuscular mycorrhizal status of various crops in different agricultural soils of northern Greece. Biol. Fertil. Soils 5:333-337.

Jenkins, J.N. 1994. Host plant resistance in cotton pp. 359-372 *In:* G.C. Constable and N.W. Forrester (eds.). Challenging the Future: Proceedings of the World Cotton Res Conf 1, Brisbane Australia, CSIRO, Melbourne.

Jenkins, J.N., P.A. Hedin, J.C. McCarty, Jr., and W.L. Parrott. 1987. Effects of mepiquat chloride on allelochemicals of cotton. J. Miss. Acad. Sci. 32:73-78.

Jenkins, J.N., J.C. McCarty, Jr., and W.L Parrott. 1990a. Effectiveness of fruiting sites in cotton yield. Crop Sci. 30:365-369.

Jenkins, J.N., J.C. McCarty, Jr., and W.L. Parrott 1990b. Fruiting efficiency in cotton: Boll size and boll set percentage. Crop Sci. 30:857-860.

Jenkins, J.N., W.L. Parrott, and J.C. McCarty. 1991. State of the art of host plant resistance to insects in cotton. Proc. Beltwide Cotton Conf. 2:627-633. National Cotton Council of America, Memphis, Tenn.

Jenkins, J.N., W.L. Parrott, J.C. McCarty Jr., F.E. Callahan, S.A. Berberich, and W.R. Deaton. 1993. Growth and survival of *Heliothis virescens* (Lepidoptera: Noctuidae) on transgenic cotton containing a truncated form of the delta endotoxin gene from *Bacillus thuringiensis*. J. Econ. Entom. 86:181-185.

Jensen, R. 1986. The biochemistry of photosynthesis. pp. 157-182. *In*: J.R. Mauney and J.McD. Stewart (eds.). Cotton Physiology. Cotton Foundation, Memphis.

Jensen, W.A. 1963. Cell development during plant embryogenesis. Brookhaven Symposia in Biology 16:179-200.

Jensen, W.A. 1968. Cotton embryogenesis: the zygote. Planta 79:346-366.

Jeschke, W.D. and J.S. Pate. 1992. Temporal pattern of uptake, flow and utilization of nitrate, reduced nitrogen and carbon in a leaf of salt-treated castor bean (*Ricinis communis* L.). J. Exp. Bot. 43:393-402.

Jianming, S., W. Jingyin, Z. Hanyang, W. Haibo, C. Zhixian, L. Shujun, and Y. Jianxiong. 1991. Effects of nitrogen source and hormones on genesis and development of somatic embryos in protoplast culture of cotton (*Gossypium hirsutum* L.). Jiangsu J. of Agr. Sci. 7(4):20-24.

Jiang , S.H., D.Z. Ye, M. Lin, and Q.Z. Liang. 1986. The available soil boron in cotton growing area in Zhejiang and effects of B fertilizers. Zhejiang Agricultural Science (in Chinese) 1:25-30.

Jing, S.W. and P.D. Coley. 1990. Dioecy and herbivory: the effect of growth rate on plant defense in *Acer negundo*. Oikos 58:369-377.

Joham, H.E. 1986. Effects of nutrient elements on fruiting efficiency. *In:* J.R. Mauney and J. McD. Stewart (eds.). Cotton Physiology, Book 1. The Cotton Foundation, Memphis, pp. 79-89.

Joham, H.E. and J.V. Amin. 1967. The influence of foliar and substrate application of manganese on cotton. Plant Soil 26:369-379.

Joham, H.E. and J.S. Calahan. 1978. The influence of Ca on the salt tolerance of cotton. Proc. Beltwide Cotton Prod. Res. Conf. Dallas. 1978, 56-57.

John, M.E. 1992. Genetic engineering of cotton for fiber modification. pp. 91-106. *In:* C.R. Benedict (ed.). Proc. Cotton Fiber Cellulose: Structure, Function, and Utilization Conference. National Cotton Council, Memphis Tenn.

John, M.E. 1994. Re-engineering Cotton Fibre. Chemistry and Industry 5 September:676-679.

John, M.E. 1995. Characterization of a cotton (*Gossypium hirsutum* L.) fiber mRNA (Fb-B6). Plant Physiol. 107:1477-1478.

John, M.E. 1996. Structural characterization of genes corresponding to cotton fiber mRNA, E6: reduced E6 protein in transgenic plants by antisense gene. Plant Mol. Biol. 30:297-306.

John, M.E. and L.J. Crow. 1992. Gene expression in cotton (*Gossypium hirsutum* L.) fiber: Cloning of the mRNAs. Proc. Natl. Acad. Sci. USA 89:5769-5773.

John, M.E. and G. Keller. 1995. Characterization of mRNA for a proline-rich protein of cotton fiber. Plant Physiol. 108:669-676.

John, M.E. and G. Keller. 1996. Metabolic pathway engineering in cotton: Biosynthesis of polyhydroxy-butyrate in fiber cells. Proc. Natl. Acad. Sci. USA 93:12768-12773.

John, M.E. and M.W. Petersen. 1994. Cotton (*Gossypium hirsutum* L.) pollen-specific polygalacturonase mRNA: Tissue and temporal specificity of its promoter in transgenic tobacco. Plant Mol. Biol. 26:1989-1993.

John, M.E. and J.M. Stewart. 1992. Genes for jeans: biotechnological advances in cotton. Tibtech 10:165-170.

Johnson, A.W., N.A. Minton, T.B. Brenneman, J.W. Todd, G.A. Herzog, G.J. Gascho, S.H. Baker, and K. Bondari. 1998. Peanut-cotton-rye rotations and soil chemical treatment for managing nematodes and thrips. J. Nematology 30:211-225.

Johnson, B.L. 1975. *Gossypium palmeri* and a polyphyletic origin of the New World cottons. Bull. Torrey Bot. Club 102:340-349.

Johnson, B.L., and M.M. Thein. 1970. Assessment of evolutionary affinities in *Gossypium* by protein electrophoresis. Amer. J. Bot. 57:1081-1092.

Johnson, D.A., R.A. Richards, and N.C. Turner. 1983. Yield, water relations, gas exchange, and surface reflectances of near isogenic wheat lines differing in glaucousness. Crop Sci. 23:318-325.

Johnson, D.L. and L.S. Albert. 1967. Effect of selected nitrogen bases and boron on the ribonucleic acid content, elongation, and visible deficiency symptoms of tomato root tips. Plant Physiol. 42:1307-1309.

Johnson, J.T. and F.M. Bourland. 1996. Growth pattern evaluation of contrasting cotton cultivars using COTMAN procedures. p. 628-629. *In:* P. Dugger and D.A Richter (eds.). Proc. Beltwide Cotton Conf., National Cotton Council of America, Memphis, Tenn.

Johnson, K.B. 1987. Defoliation, disease and growth: A reply. Phytopathology 77:1495-1497.

Johnson, N. and F.L. Pfleger. 1992. Vesicular-arbuscular mycorrhizae and cultural stresses. pp. 71-100. *In:* G.J. Bethlenfalvay and R.G. Linderman (eds.). Mycorrhizae in Sustainable Agriculture. ASA Special Publication Number 54, Madison, WI.

Johnson, R.E. and F.T. Addicott. 1967. Boll retention in relation to leaf and boll development in cotton *(Gossypium hirsutum* L.). Crop Sci. 7:571-575.

Johnson, R.M., G.F. Sassenrath-Cole, and J.M. Bradow. 1997. Prediction of cotton fiber maturity from environmental parameters. pp. 1454-1455. *In:* Proc. Beltwide Cotton Conf., National Cotton Council of America, Memphis, Tenn.

Johnson, S.J. 1984. Larvae development, consumption and feeding behavior of the cotton leaf worm, *Alabama argillacea* (Hubner). Southwest. Entomol. 9:1-6.

Jones, C.G. and J.S. Coleman. 1991. Plant stress and insect herbivory: toward an integrated perspective. pp. 249-280. *In:* H.A. Mooney, W.E. Winner, E.J. Pell, and E. Chu (eds.). Response of Plants to Multiple Stresses. Academic Press, New York.

Jones, E.J., G.D. Willis, and J.E. Hanks. 1998. Influence of surfactants, nitrogen salts, and ethephon on cotton defoliation. pp. 1489-1490. *In:* P. Dugger and D.A. Richter (eds.). Proc. Beltwide Cotton Prod. Res. Conf., National Cotton Council of America, Memphis, Tenn.

Jones, H.G. 1992. Plants and microclimate: A quantitative approach to environmental plant physiology. 2nd edition. Cambridge Univ. Press, New York. 428 pp.

Jones, J., Jr., B. Wolf, and H.A. Mills. 1991. Plant analysis handbook. Micro-Macro Publishing Inc., Athens, GA.

Jones, K., T. Kerby, H. Collins, T. Wofford, M. Batis, J. Presley, and J. Burgess. 1996. Performance of NuCOTenn with Bollgard. pp. 46-48. *In:* P. Dugger and D. Richter (eds.). Proc. Beltwide Cotton Conf., National Cotton Council of America, Memphis, Tenn.

Jones, M.A. and C.E. Snipes. 1999. Tolerance of transgenic cotton to topical applications of glyphosate. J. Cotton Science 3:19-26.

Jones, M.A. and R. Wells. 1997. Dry matter allocation and fruiting patterns of cotton grown at two divergent plant populations. Crop Sci. 37:797-802.

Jones, M.A., R. Wells, and D.S. Guthrie. 1996. Cotton response to seasonal patterns of flower removal: I. Yield and fiber quality. Crop Sci. 36:633-638.

Jones, M.N. and N.C. Turner. 1978. Osmotic adjustment in leaves of sorghum in response to water deficit. Plant Physiol. 61:122-126.

Jones, P., L.H. Allen, Jr., and J.W. Jones. 1985. Responses of soybean canopy photosynthesis and transpiration to whole-day temperature changes in different CO_2 environments. Agron. J. 77:242-249.

Jones, R.G., P.J. Bauer, M.E. Roof, and M.A. Langston. 1990. Reduced rates of ethephon on late-season insect oviposition and feeding sites in cotton. J. Entomol. Sci. 25:246-252.

Joost, O. 1993. Early Genetic Events in the Interaction of *Verticillium dahliae* and *Gossypium* species. Ph. D. Thesis, Texas A&M University, College Station, TX. 78 pp.

Joost, O., G. Bianchini, A.A. Bell, C.R. Benedict, and C.W. Magill. 1995. Differential induction of 3-hydroxy-3-methylglutaryl CoA reductase in two cotton species following inoculation with *Verticillium*. Molecular Plant-Microbe Interactions 8:880-885.

Jordan, W.R. 1970. Growth of cotton seedlings in relation to maximum daily plant-water potential. Agron. J. 62:699-701.

Jordan, W.R. 1979. Influence of edaphic parameters on flowering, fruiting, and cutout A. role of plant water deficit. *In:* Proc. Beltwide Cotton Prod. Res. Conf. National Cotton Council, Memphis, Tenn. pp. 297-301.

Jordan, W.R. 1983. Cotton. pp. 213-254. *In:* I.D. Teare and M.M. Peets (eds.). Crop Water Relations. John Wiley & Sons, New York.

Jordan, W.R., R.L. Monk, F.R. Miller, D.T. Rosenow, and L.E. Clark. 1983. Environmental physiology of sorghum. I. Environmental and genetic control of epicuticular wax load drought resistance, adaptation, and heritability. Crop Sci. 23:552-558.

Jordan, W.R. and J.T. Ritchie. 1971. Influence of soil water stress on evaporation, root absorption and internal water stress in cotton. Plant Physiol. 48:783-788.

Jordan, W.R., P.J. Shouse, A. Blum, F.R. Miller, and R.L. Monk. 1984. Environmental physiology of sorghum. II. Epicuticular wax load and cuticular transportation. Crop Sci. 24:1168-1173.

Joshi, P.C. and B.M. Johri. 1972. *In vitro* growth of ovules of *Gossypium hirsutum*. Phytomorphology 22(2):195-209.

Joshi, P.C. and N.S. Pundir. 1966. Growth of ovules in the cross *Gossypium arboreum* x *G. hirsutum in vivo* and *in vitro*. Indian Cotton Journal XX(1):23-29.

Joshi, P.A., J.M. Stewart, and E.T. Graham. 1985. Localization of β-glycerophosphatase activity in cotton fiber during differentiation. Protoplasma 125:75-85.

Joshi, P.A., J.M. Stewart, and E.T. Graham. 1988. Ultrastructural localization of ATPase activity in cotton fiber during elongation. Protoplasma 143:1-10.

Jost, P.H. and J.T. Cothren. 1999. Ultra-narrow row and conventionally-spaced cotton: growth and yield comparisons. p. 559. *In:* P. Dugger and D.A. Richter (eds.). Proc. Beltwide Cotton Conf., National Cotton Council of America, Memphis, Tenn.

Jost, P.H. and J.T. Cothren. 2000. Growth and yield comparison of cotton planted in conventional and ultra-narrow row spacing. Crop Sci. 40:430-435.

Jost, P., T. Cothren, and T.J. Gerik. 1998. Growth and yield of ultra-narrow row and conventionally-spaced cotton. p. 1383. *In:* P. Dugger and D.A. Richter (eds.). Proc. Beltwide Cotton Conf., National Cotton Council of America, Memphis, Tenn.

Joyce, R.J.V. 1958. Effect on the cotton plant in the Sudan Gezira of certain leaf feeding pests. Nature 182:1463-1464.

Juniper, B.E. and C.E. Jefree. 1983. Plant surfaces. Edward Arnold, London.

Jury, W.A., W.R. Gardner, and W.H. Gardner. 1991. Soil Physics. John Wiley & Sons, New York.

Kader, J.C. 1996. Lipid transfer proteins in plants. Annu. Rev. Plant Physiol. Plant Mol. Biol. 47:627-654.

Kadir, Z.B.A. 1976. DNA evolution in the genus *Gossypium*. Chromosoma 56:85-94.

Kafkafi, U. 1990. The functions of plant K in overcoming environmental stress situations. pp.81-93 *In:* Proc. 22nd Colloquium, International Potash Institute, Bern, Switzerland.

Kafkafi, U. 1992. Foliar feeding of potassium nitrate in cotton. Better Crops with Plant Food 76(2)16-17.

Kafkafi, U., N. Valoras, and J. Letey. 1982. Chloride interaction with nitrate and phosphate nutrition in tomato (*Lycopersicon esculentum* L.). J. Plant Nutr. 5:1369-1385.

Kaldorf, M., E. Schmelzer, and H. Bothe. 1998. Expression of maize and fungal nitrate reductase genes in arbuscular mycorrhiza. Mol. Plant-Microbe Interact. 11:439-448.

Kallinis, T.L. and H. Vretta-Kouskoleka. 1967. Molybdenum deficiency in cotton. Soil Sci. Soc. of Am. Proc. 21:507-509.

Kamprath, E.J.; W.L. Nelson, and J.W. Fitts. 1957. Sulfur removed from soils by field crops. Agron. J. 49:289-293.

Kandler, O. and H. Hopf. 1982. Oligosaccharides based on sucrose (sucrosyl oligosaccharides). pp. 348-383. *In:* A. Pirson and M. H. Zimmermann (eds.). Encyclopedia of Plant Physiology, New Series, vol. 13.

Kannan, S. 1986. Foliar absorption and transport of inorganic nutrients. CRC Critical Reviews in Plant Sciences 4:341-375.

Kapazoglou, A., F. Sagliocco, and L. Dure, III. 1995. PSII-T, a new nuclear encoded lumenal protein from photosystem II. J. Biol. Chem. 270:12197-12202.

Kaplan, D.I., D.C. Adriano, C.L. Carlson, and K.S. Sajuran. 1990. Vanadium toxicity and accumulation by beans. Water Air Soil Pollut. 49:81-91.

Kapulnik, Y., N. Yalpani, and I. Raskin. 1992. Salicylic acid induces cyanide-resistant respiration in tobacco cell-suspension cultures. Plant Physiol. 100:1921-1926.

Kapur, M.L. and G.S. Sekhorn. 1985. Rooting pattern, nutrient uptake and yield of pearl millet (*Pennusetum typhoideum* Pers.) and cotton (*Gossypium herbaceum*) as affected by nutrient availability from the surface and subsurface soil layers. Field Crops Res. 10:77-86.

Karami, E., D.R. Krieg, and J.E. Quisenberry. 1980. Water relations and carbon-14 assimilation of cotton with different leaf morphology. Crop Sci. 20:421-426.

Karban, R. 1987. Environmental conditions affecting the strength of induced resistance against mites in cotton. Oecologia 73:414-419.

Karban, R. 1988. Resistance to beet armyworms (*Spodoptera exigua*) induced by exposure to spider mites (*Tetranychus turkestani*) in cotton. Am. Midl. Nat. 119:77-82.

Karban, R., R. Adamchak, and W.C. Schnathorst. 1987. Induced resistance and interspecific competition between spider mites and a vascular wilt fungus. Science 235:678-680.

Karban, R., K.A. Brody, and W.C. Schnathorst. 1989. Crowding and a plants ability to defend itself against herbivores and diseases. American Naturalist 134:740-760.

Karban, R. and J.R. Carey. 1984. Induced resistance of cotton seedlings to mites. Science 225:53-55.

Karban, R. and J.H. Meyers. 1989. Induced plant responses to herbivory. Annu. Rev. Ecol. System. 20:331-348.

Kariev, A. 1981. Growth substances increase the seed oil content. Khlopkovodstvo 1981, 37-38. (Russian).

Karimdzhanov, A.K., N.N. Kuznetsova, and S.A. Dzhataev. 1997. Phenolic compounds of the plant *Gossypium hirsutum* and of callus tissue from its anthers. Chem. Nat. Compd. 33:187-189.

Karlen, D.L., Varvel, G.E., Bullok, D.G., Cruse, R.M. 1994. Crop rotations for the 21st Century. Adv. Agron. 53: 1-45.

Kasimov, N.A., Z.I. Abbasova, and G. Gunduz. 1998. Effects of salt stress of the respiratory components of some plants. Turkish J. Bot. 22:389-396.

Kaspar, T.C. and W.L. Bland. 1992. Soil temperature and root growth. Soil Sci. 145:290-299.

Kasperbauer, M.J. 1971. Spectral distribution of light in a tobacco canopy and effects of end-of-day light quality on growth and development. Plant Physiol. 47:775-778.

Kasperbauer, M.J. 1988. Phytochrome involvement in regulation of the photosynthetic apparatus and plant adaptation. Plant Physiol. Biochem. 26:519-524.

Kasperbauer, M.J. 1994. Cotton plant size and fiber development responses to FR/R ratio reflected from soil surface. Physiol. Plant 91:317-321.

Kasperbauer, M.J., and Hunt, P.G. 1992. Cotton seedling morphogenic responses to FR/R ratio reflected from different colored soils and soil covers. Biochem. Photobiol. 56:579-584.

Katan, T. and J. Katan. 1988. Vegetative compatibility grouping of *Fusarium oxysporum* f.sp. *vasinfectum* from tissue and the rhizosphere of cotton plants. Phytopathology 78:852-855.

Kato, K., M. Abe, K. Ishiguro, and Y. Ueno. 1979. Isolation and characterization of new tetrasaccharides from cottonseed. Agric. Biol. Chem. 43:293-297.

Katterman, F.R.H., M.D. William, and W.F. Clay. 1977. The influence of a strong reducing agent upon the initiation of callus from the germinating seedlings of *Gossypium barbadense*. Physiol. Plant 40:90-100.

Kaufman, P.B., S.J. Cassell, and P.A. Adams. 1965. On the nature of intercalary growth and cellular differentiation in internodes of *Avena sativa*. Bot. Gaz. 126:1-13.

Kaur, R., S. Jund, R.J.K. Anand, U. Kanwar, and S.N. Sanyal. 1995. Effects of gossypol on *in vitro* changes in lipids, lipid peroxidation and enzymes of buffalo spermatozoa. Fitoterapia 66:431-438.

Kawagoe, Y. and D.P. Delmer. 1997. Pathways and genes involved in cellulose biosynthesis. Genetic Engineering 19:63-87.

Keeling C.D., T.P. Whorf ,M. Wahlen, and J. van der Plicht. 1995. Interannual extremes in the rate of rise of atmospheric carbon dioxide since 1980. Nature. 375:660-670.

Keep, E. 1969. Accessory buds in the genus *Rubus* with particular reference to *R. idaeus* L. Ann. Bot. 33:191-204.

Kefu, Z., R. Munns, and R.W. King. 1991. Abscisic acid levels in NaCl-treated barley, cotton, and saltbush. Aust. J. Plant Physiol. 18:17-24.

Keino, J.K., C.A. Beyrouty, and D.M. Oosterhuis. 1994. Relationship between roots and shoots of irrigated and non-irrigated field-grown cotton. P.147.46 Agronomy Abstracts. ASA, Madison, WI.

Keisling, T.C., H.J. Mascagni, R.L. Maples, and K.C. Thompson. 1995. Using cotton petiole nitrate-nitrogen concentration for prediction of cotton nitrogen nutritional status on a clay soil. J. Plant Nutr. 18:35-45.

Kende, H. 1964. Preservation of chlorophyll in leaf sections by substances obtained from root exudate. Science 145:1066-1067.

Kendrick, R.E. and G.H.M. Kronenberg. 1993. Photomorphogenesis in plants. Kluwer Acad. Publ., Norwell, MA. 864 pp.

Kenerley, C.M., T.L. White, M.J. Jeger, and T.J. Gerik. 1998. Sclerotial formation and strand growth of *Phymatotrichopsis omnivora* in minirhizotrons planted with cotton at different soil water potentials. Plant Pathology (Oxford) 47:259-266.

Kennedy, C.W. and R.L. Hutchinson. 1996. Cotton growth and development under different tillage systems. pp. 1234. *In:* Proc. Beltwide Cotton Conf., National Cotton Council of America, Memphis, Tenn.

Kennedy, C.W. and J.E. Jones. 1990. Chemical removal of early squares: Treatment efficacy and effects of reproductive growth of superokra-leaf cotton. p. 53. *In:* J.M. Brown and D.A. Richter (eds.). Proc. Beltwide Cotton Prod. Res. Conf., National Cotton Council of America, Memphis, Tenn.

Kennedy, C.W., W.C. Smith, Jr., and J.E. Jones. 1986. Effect of early season square removal on three leaf types of cotton. Crop Sci. 26:139-145.

Kennedy, C.W., W.C. Smith, Jr., and J.E. Jones. 1991. Chemical efficacy of early square removal and subsequent productivity of superokra-leaf cotton. Crop Sci. 31:791-796.

Kennedy, C.W., W.C. Smith, Jr., and B.A. Schumacher. 1990. Computer programs for recording and analyzing cotton fruit location maps. J. Agron. Educ. 19:167-171.

Kennedy, G.G. and M.E. Whalon. 1995. Managing pest resistance to *Bacillus thuringiensis* endotoxins: constraints and incentives to implementation. J. Econ. Entomol. 88:454-460.

Kenney, D.S. and K.S. Arthur. 1994. Effect of Kodiak® seed treatment on cotton root development. pp. 236-237. *In:* Proc. Beltwide Cotton Conf., National Cotton Council of America, Memphis, Tenn.

Kent, L.M. and A. Läuchli. 1985. Germination and seedling growth of cotton: Salinity-calcium interactions. Plant, Cell Environ. 8:155-159.

Kerby, T.A. 1985. Cotton response to mepiquat chloride. Agron. J. 77:515-518.

Kerby, T.A. 1991. Designing a production system for maximum quality and profit. pp. 42-45. *In:* D.J. Herber (ed.). Proc. Beltwide Cotton Conferences, National Cotton Council of America, Memphis, Tenn.

Kerby, T.A. 1994. Nodes above white flower and cutout. The Australian Cotton Grower. 15(2):10-12.

Kerby, T.A. 1998. UNR cotton production system trial in the Mid-South. *In:* P. Dugger and D. Richter (eds.) Proc. Beltwide Cotton Conf. 5-9 Jan., 1998, San Diego, CA. Natl. Cotton Counc. of Am., Memphis, Tenn. pp. 87-88.

Kerby, T.A. and F. Adams. 1985. Potassium nutrition of cotton. *In:* R.D. Munson (ed.). Potassium in Agriculture. ASA, CSSA, and SSSA, Madison, WI. pp. 843-860.

Kerby, T.A. and D.R. Buxton. 1976. Fruiting in cotton as affected by leaf type and population density. pp. 67-70. *In:* J.M. Brown (ed.). Proc. Beltwide Cotton Prod. Res. Conf., National Cotton Council of America, Memphis, Tenn.

Kerby, T.A. and D.R. Buxton. 1978. Effect of leaf shape and plant population on rate of fruiting position appearance in cotton. Agron. J. 70:535-538.

Kerby, T.A. and D.R. Buxton. 1981. Competition between adjacent fruiting forms in cotton. Agron. J. 73:867-871.

Kerby, T.A., D.R. Buxton, and K. Matsuda. 1980. Carbon source-sink relationships within narrow row cotton canopies. Crop Sci. 20:208-213.

Kerby, T.A., K.G. Cassman, and M. Keeley. 1990a. Genotypes and plant densities for narrow-row cotton systems. I. Height, nodes, earliness, and location of yield. Crop Sci. 30:644-649.

Kerby, T.A., K.G. Cassman, and M. Keeley. 1990b. Genotypes and plant densities for narrow-row cotton systems. II. Leaf area and dry-matter partitioning. Crop Sci. 30:649-653.

Kerby, T.A. and P. Goodell. 1990. Production decisions based on plant data. California Cotton Review 15:1-8.

Kerby, T.A. and K. Hake. 1996. Monitoring Cotton's Growth. p. 335-355. *In:* T.A. Kerby, K. Hake, and S. Hake (eds.). Cotton Production Manual, Publication 3352, Division of Agriculture and Natural Resources, University of California; Oakland, CA.

Kerby, T.A., S.J. Hake, K.D. Hake, L.M. Carter, and R.H. Garber. 1996. Seed quality and planting environment. pp. 203-209. *In:* S.J. Hake, T.A. Kerby, and K.D. Hake (eds.). Cotton Production Manual. ANR Publications, University of California, Oakland, CA.

Kerby, T.A., K. Hake, and M. Keeley. 1986. Cotton fruiting modification with mepiquat chloride. Agron. J. 78:907-912.

Kerby, T.A. and M. Keeley. 1987. Cotton seedlings can withstand some early leaf loss. Calif. Agr. 41(1):18-19.

Kerby, T.A., M. Keeley, and S. Johnson. 1987. Growth and development of Acala cotton. California Agric. Exp. Sta. Bull. 1921.

Kerby, T.A, M. Keeley, and S. Johnson. 1989. Weather and seed quality variables to predict cotton seedling emergence. Agron. J. 81:415-419.

Kerby, T.A., M. Keeley, and M. Watson. 1993. Variation in fiber development as affected by source to link relationships. pp. 1248-1251. *In:* Proc. Beltwide Cotton Confs., National Cotton Council of America, Memphis, Tenn.

Kerby, T.A., J. Presley, J. Thomas, M. Bates, and J. Burgess. 1995. Environment and variety contributions to earliness across the belt. p. 1096-1099. *In:* D.A. Richter and J. Armour (eds.). Proc. Beltwide Cotton Conf., National Cotton Council of America, Memphis, Tenn.

Kerby, T.A., R.E. Plant, and R.D. Horrocks. 1997. Height-to-node ratio as an index of early season cotton growth. J. Prod. Agr. 10:80-83.

Kerby, T.A., R.E. Plant, S. Johnson-Hake, and R.D. Horrocks. 1998. Environmental and cultivar effects on height-to-node ratio and growth rate in Acala cotton. J. Prod. Agric. 11:420-427.

Kerby, T.A., B. Roberts, R. Vargas, and B. Weir. 1992a. Impact of early season growth on subsequent development and yield. California Cotton Review 24:6-7.

Kerby, T.A. and G.F. Ruppenicker. 1989. Node and fruiting branch position effects on fiber and seed quality characteristics. pp. 90-100 *In:* J.M. Brown and D.A. Richter (eds.). 1989 Proc. Beltwide Cotton Conferences. National Cotton Council, Memphis.

Kerby, T.A., J. Supak, J.C. Banks, and C. Snipes 1992b. Timing defoliations using nodes above cracked boll. p. 155-156. *In:* D.J. Herber and D.A. Richter (eds.). Proc. Beltwide Cotton Conf., National Cotton Council of America, Memphis, Tenn.

Kerby, T.A., L. Zelinski, J. Burgess, M. Bates, and J. Presley. 1996. Genetic and environmental contributions to earliness. p.592-594. *In:* P. Dugger and D.A. Richter (eds.). Proc. Beltwide Cotton Conf., National Cotton Council of America, Memphis, Tenn.

Keren, R., A. Meiri, and Y. Kalo. 1983. Plant spacing effect on yield of cotton irrigated with saline waters. Plant and Soil 74:461-465.

Keren, R. and I. Shainberg. 1978. Irrigation with sodic and brackish water and its effect on the soil and on cotton fields. Harrade 58:963-976.

Kerimov, F., Vl. V. Kuznetsov, and Z.B. Shamina. 1993. Organism and cell levels of salt tolerance in two cotton cultivars (133 and Inébr-85. Russian Plant Physiol. 40:128-131.

Kerr, T. 1936. The structure of growth rings in the secondary wall of the cotton hair. Protoplasma 27:229-243.

Ketcheson, J.W. 1968. Effect on controlled air and soil temperature and starter fertilizer on growth and nutrient composition of corn. Soil Sci. Soc. Amer. Proc. 3253 1-534.

Key, J.L. 1989. Modulation of gene expression by auxin. *BioEssays* 11:52-58.

Khaddar, V.K. and N. Ray. 1988. The principles and perspective of paddy cotton intercropping. Advancement of crops and monitoring of environment. Progress in Ecology 10:403-418.

Khalil, M.A., F. Amer, and M.M. Elgabaly. 1967. A salinity-fertility interaction study on corn and cotton. Soil Sci. Soc. Amer. Proc. 31:683-686.

Khalilian, A, M.J. Sullivan, and J.D. Mueller. 1995. Controlled traffic patterns for 30- and 38-inch row cotton in three cropping systems. *In:* M.R. McClelland, T.D. Valco, and R.E. Frans (eds.). Conservation-Tillage Systems for Cotton. University of Arkansas Agric. Expt. Sta. Spec. Rpt. 169:8-13.

Khan, A.N., R.H. Qureshi, N. Ahmad, and A. Rashid. 1995a. Selection of cotton cultivars for salinity tolerance at seedling stage. Sarhad J. of Agric. 11:153-159.

Khan, A.N., R.H. Qureshi, N. Ahmad, and A. Rashid. 1995b. Response of cotton cultivars to salinity at various growth development stages. Sarhad J. Agric. 11:729-731.

Khan, A.N., R.H. Qureshi, and N. Ahmad. 1995c. Performance of cotton cultivars in saline growth media at germination stage. Sarhad J. Agric. 11:643-646.

Khan, A.N., R.H. Qureshi, and N. Ahmad. 1998a. Performance of cotton cultivars as affected by types of salinity. I Growth and yield. Sarhad J. Agric. 14:67-71.

Khan, A.N., R.H. Qureshi, and N. Ahmad. 1998b. Performance of cotton cultivars as affected by types of salinity. II Ionic composition. Sarhad J. Agric. 14:73-77.

Khan, M.A., J.McD. Stewart, and J.B. Murphy. 1999. Evaluation of the *Gossypium* gene pool for foliar terpenoid aldehydes. Crop Sci. 39:253-258.

Kharche, S.G. 1984. Validation of GOSSYM: Effects of irrigation, leaf shape, and plant population on can-opy light interception, growth and yield of cotton *(Gossypium hirsutum* L). Ph.D. Thesis. Mississippi State University.

Khoshkhoo, N., P.A. Hedin, and J.C. McCarty. 1993. Effects of bioregulaton on the terpenoid aldehydes in root-knot nematode infected cotton plants. J. Agric. Food. Chem. 41:2442-2446.

Khoshkhoo, N., P.A. Hedin, and J.C. McCarty, Jr. 1994. Terpenoid aldehydes in root-knot nematode susceptible and resistant cottonseeds as determined by HPLC and aniline methods. J. Agric. Food Chem. 42:804-806.

Kidambi, S.P., J.R. Mahan, and A.G. Matches. 1990. Purification and thermal dependence of glutathione reductase from two forage legume species. Plant Physiol. 92:363-367.

Kijne, J.W. 1998. Yield response to moderately saline irrigation water: Implications for feasibility of management changes in irrigation systems for salinity control. Z. Bewasserungswirtschaft 33:261-277.

Kim, H.L., M.C. Calhoun, and R.D. Stipanovic. 1996. Accumulation of gossypol enantiomers in ovine tissues. Compo Biochem. Physiol. 113B:417-420.

Kimball, B.A. 1983. Carbon-dioxide and agricultural yield: an assemblage and analysis of 430 prior observations. Agron. J. 75:779-788.

Kimball, B.A., R.L. LaMorte, R.S. Seay, P.J. Pinter, Jr., R.R. Rokey, D.J. Hunsaker, W.A. Dugas, M.L. Heuer, J.R. Mauney, G.R. Hendry, K.F. Lewin, and J. Nagy. 1994. Effects of free-air CO_2 enrichment on energy balance and evapotranspiration of cotton. Agric. For. Met. 70:259-278.

Kimball, B.A. and Mauney, J.R. 1993. Response of cotton to varying CO_2, irrigation, and nitrogen: Yield and growth. Agron. J. 85:706-712.

Kimball, B.A., J.R. Mauney, D.H. Akey, D.L. Hendrix, S.G. Allen, S.B. Idso, J.W. Radin, and E.A. Lacatos. 1987. Effects of increasing atmospheric CO_2 on the growth, water relations, and physiology of plants grown under optimal and limited levels of water and nitrogen. Response of Vegetation to Carbon Dioxide, No. 49. U.S.Dept. of Energy, Carbon Dioxide Research Division and the U. S. Dept. of Agriculture, Agricultural Research Service; Washington, D.C. 127 pp.

Kimball, B.A., J.R. Mauney, F.S. Nakayama, and S.B. Idso. 1993. Effects of increasing atmospheric CO_2 on vegetation. Vegetation 104-105:65-75.

Kimball, B.A., J.R. Mauney, J.W. Radin, F.S. Nakayama, S.B. Idso, D.L. Hendrix, D.H. Akey, S.G. Allen, M.G. Anderson, and W. Hartung. 1986. Effects of increasing atmospheric CO_2 on the growth, water relations, and physiology of plants grown under optimal and limiting levels of water and nitrogen. Response of Vegetation to Carbon Dioxide, No. 39.

U.S. Dept. of Energy, Carbon Dioxide Research Division, and U.S. Dept of Agriculture, Agricultural Research Service, Washington, D.C. 125 pp.

Kimball, B.A., P.J. Pinter, Jr., and J.R. Mauney. 1992. Cotton leaf and boll temperatures in the 1989 FACE experiment. Crit Rev. Plant Sci. 11:233-240.

King, C.J. and J.T. Presley. 1942. A root disease of cotton caused by *Thielaviolpsis basciola*. Phytopath. 32:752-761.

King, E.G., R.J. Coleman, D.R. Reed, and J.L. Hayes. 1990. Pre-bloom ethephon application: Effects on cotton yield maturaion date, quality, and the boll weevil. pp. 290-295. *In:* J.M. Brown and D.A. Richter (eds.). Proc. Beltwide Cotton Prod. Res. Conf., National Cotton Council of America, Memphis, Tenn.

King, E.G., L.N. Namken, and R.J. Coleman. 1992. Early square removal with ethephon: Response of cotton fruiting and boll weevil. pp. 756-759. *In:* D.J. Herber and D.A. Richter (eds.). Proc. Beltwide Cotton Conf., National Cotton Council of America, Memphis, Tenn.

Kingsley, M.T., D.W. Gabriel, G.C. Marlow, and P.D. Roberts. 1993. The *ops X* locus of *Xanthomonas campestris* affects host range and biosynthesis of lipopolysaccharide and extracellular polysaccharide. J. Bacteriology 175:5839-5850.

Kiniry, J.R., W.D. Rosental, B.S. Jackson, and G. Hoogenboom. 1991. Predicting leaf development of crop plants. p. 39-42. *In:* T. Hodges (ed.). Predicting Plant Phenology. CRC Press., Inc., Boca Raton, FL.

Kirkham, M.B., W.R. Gardner, and G.C. Gerloff. 1972. Regulation of cell division and cell enlargement by turgor pressure. Plant Physiol. 49:961-962.

Kirkpatrick, T.L., D.M. Oosterhuis, and S.D. Wullschleger. 1991. Interaction of *Meloidogyne incognita* and water stress in two cotton cultivars. J. Nematology 23:462-467.

Kittock, D.L., H.R. Arle, and L.A. Bariola. 1973. Termination of late season cotton fruiting with growth regulators as an insect-control technique. J. Environ. Qual. 2:405-408.

Kittock, D.L., T.J. Henneberry, L.A. Bariola, B.B. Taylor, and W.C. Hofmann. 1983. Cotton boll period response to water stress and pink bollworm. Agron. J. 75: 17-20.

Kittock, D.L., R.A. Selley, C.J. Cain, and B.B. Taylor. 1986. Plant population and plant height effects on Pima cotton lint yield. Agron. J. 78:534-538.

Kittock, D.L., B.B. Taylor, and W.C. Hofmann. 1987. Partitioning yield reduction from early cotton planting. Crop Sci. 27:1011-1015.

Klann, E.M., B. Hall, and A.B. Bennett. 1996. Antisense acid invertase *(TIVI)* gene alters soluble sugar composition and size in transgenic tomato fruit. Plant Physiol. 112:1321-1330.

Kleczkowski, L.A. 1994. Glucose activation and metabolism through UDP-glucose pyrophosphorylase in plants. Phytochem. 37:1507-1515.

Kleifeld, I., T. Blumenfeld, and A. Bargoti. 1978. Weed killers selective for cotton in the soils of the Hulah Valley. Hassadeh 58:610-615.

Klein, I. and S.A. Weinbaum. 1984. Foliar application of urea to olive. Translocation of urea nitrogen as influenced by sink demand and nitrogen deficiency. J. Amer. Soc. Hort. Sci. 109:356-360.

Klepper, B., H.M. Taylor, M.G. Huck, and E.L. Fiscus. 1973. Water relations and growth of cotton in drying soil. Agron J.65:307-310.

Kletter, E. and D. Wallach. 1982. Effects of fruiting form removal on cotton reproductive development. Field Crops Res. 5:69-84.

Kloth, R.H. 1989. Changes in the level of tubulin subunits during development of cotton (*Gossypium hirsutum*) fiber. Physiol. Plant. 76:37-41.

Kloth, R.H. 1992. Variability of malate dehydrogenase among cotton cultivars with differing fiber traits. Crop Sci. 32:617-621.

Kloth, R.H. and R.B. Turley. 1997. Homologue of ribosomal protein RL37a from cotton (*Gossypium hirsutum* L.). Plant Physiol. 120:933.

Knapp, K.C. 1992. Irrigation management and investment under saline, limited drainage conditions. 2. Characterization of optimal decision rules. Water Resour. Res. 28:3091-3097.

Knight, R.L. 1945. The theory and application of the backcross technique in cotton breeding. J. Genet. 47:76-86

Knudson-Butler, L. and T.W. Tibbits. 1979. Stomatal mechanisms determining genetic resistance to ozone in *Phaseolus vulgaris* L. J. Amer. Soc. Hort. Sci. 104:213-216.

Kobayashi, H., H. Fukuda, and H. Shibaoka. 1989. Interrelationship between the spatial deposition of actin filaments and microtubles during the differentiaion of tracheary elements in cultured *Zinnia* cells. Protoplasma 143:29-37.

Koch, K.E. 1996. Carbohydrate-modulated gene expression in plants. Annu. Rev. Plant Physiol. Plant Mol. Biol 47:509-540.

Koch, K.E., Y. Wu, and J. Xu. 1996. Sugar and metabolic regulation of genes of sucrose metabolism: Potential influence of maize sucrose synthase and soluble invertase responses on cabon partitioning and sugar sensing. J. Exp. Bot. 47:1179-1185.

Kochman, J.K., R.D. Davis, N.Y. Moore, S. Bentley, and N.R. Obst. 1996. Characterization of the Fusarium wilt pathogen of cotton in Australia. pp. 651-656. *In*: Proc. Australian Cotton Conference.

Kohel, R.J. and A.A. Bell. 1999. Genetic analysis of two terpenoid variants in cotton (*Gossypium hirsutum* L.). Journal of Heredity 90:249-251.

Kohel, R.J., C.R. Benedict, and G.M. Jividen. 1993. Incorporation of ^{14}C-glucose into crystalline cellulose in aberrant fibers of mutant cotton. Crop Sci. 33:1036-1040.

Kohel, R.J. and S.C. McMichael. 1990. Immature fiber mutant of Upland cotton. Crop Sci. 30:419-421.

Kohel, R.J., J.E. Quisenberry, and C.R. Benedict. 1974. Fiber elongation and dry weight changes in mutant lines of cotton. Crop Sci. 1974:471-474.

Koide, R. 1985. The nature of growth depressions in sunflower caused by vesicular-arbuscular mycorrhizal infection. New Phytol. 99:449-462.

Koide, R.T. and R.P. Schreiner. 1992. Regulation of the vesicular-arbuscular mycorrhizal symbiosis. Annu. Rev. Plant Physiol. Mol. Biol. 43:557-581.

Kolattukudy, P.E. 1970. Plant waxes. Lipids 5:259-275.

Kolattukudy, P.E. 1980. Biopolyester membranes of plants: Cutin and suberin. Science 208:990-1000.

Koli, S.E. and L.G. Morrill. 1976. Effects of narrow row, plant population, and nitrogen application on cotton fiber characteristics. Agron. J. 68:794-797.

Kondo, T. and M. Ishiura. 1999. The circadian clocks of plants and cyanobacteria. Trends Plant Sci. 4:171-176.

Koomeef, M., S.W.M. Dellaert, and J.H. van der Veen. 1982. EMS- and radiation-induced mutation frequencies at individual loci in *Arabidopsis* (L) Heynh. Mutat. Res. 93:109-123.

Koontz, D.A. and J.H. Choi. 1993. Evidence for phosphorylation of tubulin in carrot suspension cells. Physiol. Plant. 87:576-583.

Körner, C.H. 1991. Some often overlooked plant characteristics as determinants of plant growth: a reconsideration. Functional Ecology 5:162-173.

Kornish, K. 1988. Why does photosynthesis decline with leaf age? pp. 50-55. Proc. 1988 Beltwide Prod. Confs., National Cotton Council of America, Memphis, Tenn.

Kosmidou-Dimitropoulou, K. 1986. Hormonal influences in fiber development. pp. 361-374. *In:* J.R. Mauney and J.M. Stewart (eds.). Cotton Physiology. The Cotton Foundation, Memphis.

Kostka-Rick, R. and W.J. Manning. 1993. Dose-response studies with ethylenediurea (EDU) and radish. Environ. Pollut. 79:249-260.

Kostopoulos, S. and A. Chlichlias. 1979. Influence of row spacings and plant population densities on yield, earliness and fiber properties of two Greek cotton cultivars (*Gossypium hirsutum* L.). Agric. Res. 4:343-355.

Koukkari, W.L., S.H. Duke, F. Halberg, and J.K. Lee. 1974. Circadian rhythm leaflet movements: Student exercise in chronobiology. Chronobiologia 1:182-302.

Koukkari, W.L. and S.B. Warde. 1985. Rhythms and their relations to hormones. pp. 37-77. *In:* R.P. Pharis and D.M. Reid (eds.). Encyclopedia of Plant Physiology, New Series, Vol. 11. Oxford Press, Clarendon, UK.

Kovar, J.L., R.L. Hutchinson, and E.R. Funderburg. 1994. Effect of starter fertilizer rate and placement of cotton root growth and lint yield in two tillage systems. pp. 1559-1560. *In:* Proc. Beltwide Cotton Cant., National Cotton Council of America, Memphis, Tenn.

Kramer, P.J. 1969. Plant and soil relationships: A modern synthesis. McGraw-Hill, New York.

Kramer, P.J. and J.S. Boyer. 1995. Water relations of plants and soils. Academic Press, San Diego. pp. 495.

Krieg, D.R. 1988. Leaf age - gas exchange characteristics. *In:* Proc. Beltwide Cotton Prod. Res. Conf. National Cotton Council, Memphis, Tenn. pp. 55-57.

Krieg, D.R. and J.D. Carroll. 1978. Cotton seedling metabolism as influenced by germination temperature, cultivar, and seed physical properties. Agron. J. 70:21-25.

Krieg, D.R. and F.J.M. Sung. 1979. Source-sink relations of cotton as affected by water stress during boll development. *In:* Proc. Beltwide Cotton Prod. Res. Conf. National Cotton Council, Memphis, Tenn. pp. 302-305.

Krieg, D.R., J.L. Hatfield, A.C. Gertsis, and S.A. Staggenborg. 1993. Plant monitoring: water stress and water management. *In:* Proc. Beltwide Cotton Prod. Res. Conf. National Cotton Council, Memphis, Tenn. pp. 1191-1193.

Krizek, D.T. 1986. Photosynthesis, dry matter production and growth in CO_2-enriched atmospheres. pp. 193-225 *In:* J.R. Mauney and J.McD. Stewart (eds.). Cotton Physiology. Cotton Foundation, Memphis.

Krogmann, D.W., A.T. Jagendorf, and M. Avron. 1959. Uncouplers of spinach chloroplast photosynthetic phosphorylation. Plant Physiol. 34:272-277.

Kruger, N.J. 1990. Carbohydrate synthesis and degradation. pp. 59-79. *In:* D.T. Dennis and D.H. Turpin (eds.). Plant Physiology, Biochemistry, and Molecular Biology. Longman Scientific and Technical, Essex UK.

Kudlicka, K., R.M. Brown, Jr.; L. Li, J.H. Lee, H. Shin, and K. Shigenori. 1995. β-Glucan synthesis in the cotton fiber. IV. In vitro assembly of the cellulose I allomorph. Plant Physiol. 107:111-123.

Kulaeva, O.N. 1962. The effect of root on leaf metabolism in relation to the action of kinetin on leaves. Fiziol. Rasteni (Soviet Plant Physiol., Engl. Translation). 9:182-189.

Kuo, T.M., J.V. VanMiddlesworth, and W.J. Wolf. 1988. Content of raffinose oligosaccharides and sucrose in various plant seeds. J. Agric. Food Chem. 36:32-36.

Kurle, J,E. and F.L. Pfleger. 1994. The effects of cultural practices and pesticides on VAM fungi. pp. 101-132. *In:* F.L Pfleger and R.G. Linderman (eds.). Mycorrhizae and Plant Health. APS Press, St Paul, MN.

Kuroiwa, S. 1970. Total photosynthesis of a foliage in relation to inclination of leaves. pp. 79-89. *In:* Prediction and Measurement of Photosynthetic Productivity, Proc. Technical Meeting, Trebon, 14-21 September, 1969.

Kurth, E., G.R. Cramer, A. Läuchli, and E. Epstein. 1986. Effects of NaCl and CaCl$_2$ on cell enlargement and cell production in cotton roots. Plant Physiol. 82:1102-1106.

Kurz, W.A. and M.J. Apps. 1994. The carbon budget of Canadian forests: A sensitivity analysis of changes in disturbance regimes, growth rates, and decomposition rates. Envir. Pollution. 83:55-61.

Kutschera, U. and P. Schopfer. 1986. Effect of auxin and abscisic acid on cell wall extensibility in maize coleoptiles. Planta 167:527-535.

Kuznetsov, V.V., B.T. Khydyrov, B.V. Roschupkin, and N.N. Borisova. 1990. Common systems of resistance to salinity and high temperature in cotton - facts and hypotheses. Soviet Plant Physiol. 37:744-752.

Kuznetsov, V.V., B.T. Khydyrov, N.I. Shevyakova, and V.Y. Rakitin. 1991. Heat shock induction of salt tolerance in cotton - involvement of polyamines, ethylene and proline. Soviet Plant Physiol. 38:877-883.

Kuznetsov, V.V., V.U. Rakitin, N.N. Borisova, and B.V. Roshchupkin. 1993. Why does heat shock increase salt resistance in cotton plants. Plant Physiol. Biochem. 31:181-188.

Kuznetsov, V.V., V.Y. Rakitin, and V.N. Zholkevich. 1999. Effects of preliminary heat-shock treatment on accumulation of osmolytes and drought resistance in cotton plants during water deficiency. Physiologia Plant. 107:399-406.

Kytö, M., P. Nimelä, and S. Larsson. 1996. Insects on trees: population and individual response to fertilisation. Oikos 75:148-159.

Labate, C.A. and R.C. Leegood. 1989. Influence of low temperatures on respiration and contents of phosphorylated intermediates in darkened barley leaves. Plant Phys. 91:905-910.

Ladewig, H. and J.K. Thomas. 1992. A follow-up evaluation of the GOSSYM/COMAX Cotton Program. Texas Agric. Ext. Serv. pp. 1-47.

LaDuke, J.C. and J. Doebley. 1995. A chloroplast DNA based phylogeny of the Malvaceae. Syst. Bot. 20:259-271.

Lambais, M.R. 2000. Regulation of plant defense-related genes in arbuscular mycorrhizae. p. 45-59. *In:* G.K. Podila and D.D. Douds (eds.). Current Advances in Mycorrhizae Research. APS Press, St Paul, MN.

Lambert, J.R. and D.N. Baker. 1984. RHIZOS, A simulator of root growth and soil processes: Model descriptions. South Carolina Agric. Expt. Tech. Bull. 1080.

Lammers, J., D.M. Oosterhuis, F.M. Bourland, and E.D. Vories. 1996. Identification of the last effective flower population using nodes above white flower. p. 628. *In:* P. Dugger and D.A Richter (eds.). Proc. Beltwide Cotton Conf., National Cotton Council of America, Memphis, Tenn.

Lance, C. and P. Rustin. 1984. The central role of malate in plant metabolism. Phsyiol. Veg. 22:625-641.

Lands' End Inc. Catalogue. April 1997 34(4):2-5, 67, 137.

Landivar, J.A. 1991. Physiological characteristics affecting the performance of cotton cultivars in different environments. pp. 97-99 *In:* D.J. Herber and D.A. Richter (eds.). Proc. Beltwide Cotton Conf., National Cotton Council of America, Memphis, Tenn.

Landivar, J.A. 1993. PMAP, a plant map analysis program for cotton. Texas Agric. Exp. Sta. MP 1740.

Landivar, J.A. 1998. The MEPRT method to determine time and rate of mepiquat chloride applications: Use and misuses. *In:* P. Dugger and D. Richter (eds.) Proc. Beltwide Cotton Conf. 5-9 Jan., 1998, San Diego, CA. Natl. Cotton Counc. of Am., Memphis, Tenn. pp. 1414-1416.

Landivar, J.A., D.N. Baker, and H.F. Hodges 1988. Apparent photosynthesis of cotton genotypes as a function of leaf age. p. 60-63. *In:* J.M. Brown (ed). Proc. Beltwide Cotton Prod. Res. Conf., National Cotton Council of America, Memphis, Tenn.

Landivar, J.A., D.N. Baker, and J.N. Jenkins. 1983. Application of GOSSYM to genetic feasibility studies. II. Analyses of increasing photosynthesis, specific leaf weight and longevity of leaves in cotton. Crop Sci. 23:504-510.

Landivar, J.A. and J.H. Benedict. 1996. Monitoring system for the management of cotton growth and fruiting. Bulletin B-2. Texas A&M University Agricultural Research and Extension Center, Corpus Christi, TX.

Landivar, J.A.,J.T. Cothren, and S. Livingston. 1996. Development and evaluation of the average five internode length technique to determine time of mepiquat chloride application. pp. 1153-1156 *In:* P. Dugger and D. Richter (eds.). Proc. Beltwide Cotton Conf., National Cotton Council of America, Memphis, Tenn.

Landivar, J.A., Y. Guo, D. Locke, D. Moseley, and D. Dromgoole. 1994. Crop monitoring system using a weather station network, plant mapping, and crop simulation. p. 1373. *In:* D.J. Herber and D.A. Richter (eds.). Proc. Beltwide Cotton Conf., National Cotton Council of America, Memphis, Tenn.

Landivar, J.A., D. Locke, Z. Cespedes, and D. Moseley. 1995. The effect of estimated plant concentration of Pix on leaf area expansion and main stem elongation rate. pp. 1081-1085. *In:* D. Richter and J. Armour (eds.). Proc. Beltwide Cotton Conf., National Cotton Council of America, Memphis, Tenn.

Landivar, J.A., D. Locke, and D. Moseley. 1994. The effects of sub lethal rates of glyphosate on regrowth control, lint yield, and fiber quality. pp. 1276-1278. *In:* D.J. Herber and D.A. Richter (eds.). Proc. Beltwide Cotton Conf., National Cotton Council of America, Memphis, Tenn.

Landivar, J.A., R. Perkins, A. Johnson, and G. Davidonis. 1997. Texas Coast Plains cotton, motes and fiber qualitiy II. effects of Pix (mepiquat chloride). *In:* Proc. Beltwide Cotton Prod. Res. Conf. National Cotton Council, Memphis, Tenn. pp. 1418-1420.

Landivar, J.A., S. Zupman, D.J. Lawlor, J. Vaske, and C. Crenshaw. 1992. The use of an estimated plant Pix concentration for the determination of timing and rate of application. pp. 1047-1049. *In:* D.J. Herber and D.A. Richter (eds.). Proc. Beltwide Cotton Conf., National Cotton Council of America, Memphis, Tenn.

Lane, H.C. 1959. Simulated hail damage experiments in cotton. Texas Agricultural Expt Station Bulletin 934, 16 pp.

Lane, H.C. and M.F. Schuster. 1981. Condensed tannins of cotton leaves. Phytochemistry 20: 425-427.

Lang, A.R.G. 1973. Leaf orientation of a cotton plant. Agr. Meteorol. 11:37-51.

Lanker, T., T.G. King, S.W. Arnold, and W.H. Flurkey. 1987. Active, inactive, and in vitro synthesized forms of polyphenoloxidase during leaf development. Physiol. Plant. 69:323-329.

Larew, L.A. and D. C. Johnson. 1988. Quantitation of chromatographically separated maltooligosaccharides with a single calibration curve using a postcolumn enzyme reactor and pulsed amperometric detection. Analyt. Chem. 60:1867-1872.

Larkins, J.C., N. Young, M. Prigge, and M.D. Marks. 1996. The control of trichome spacing and number in *Arabidopsis.* Development 122:997-1005.

Larsen, W.E. and M.D. Cannon. 1966. Planting cotton to a stand. Univ.of Arizona Agric. Exp. Stn. Bull. A-46.

Lashin, M.H. and M. Atanasiu. 1972. Studies on the effect of salt concentrations on the formation of dry matter, uptake of mineral nutrients, and mineral composition of cotton plants during the vegetative growth period. Z. Acker- und Pflanzenbau 135:178-186.

Latzko, E. and G.J. Kelly. 1983. The many faceted function of phosphoenolpyruvate carboxylase in C3 plants. Physiol. Veg. 21:805-815.

Läuchli, A. 1999. Salinity-potassium interactions in crop plants. *In:* D.M. Oosterhuis and G.A. Berkowitz (eds.). Frontiers in Potassium Nutrition: New Perspectives on the Effects of Potassium on Physiology of Plants. Potash & Phosphate Institute, Norcross, GA pp. 71-76.

Läuchli, A. and E. Epstein. 1984. Mechanisms of salt tolerance in plants. Calif. Agric. 38:18-20.

Läuchli, A., L.M. Kent, and J.C. Turner. 1981. Physiological responses of cotton genotypes to salinity. *In:* Proc. Beltwide Cotton Prod. Res. Conf. National Cotton Council, Memphis, Tenn. p. 40.

Läuchli, A. and W. Stelter. 1982. Salt tolerance of cotton genotypes in relation to K/Na-selectivity. *In:* A. San Pietro (ed.). Biosaline Research. Plenum Press, New York. pp. 511-514.

Lauter, D.J., A. Meiri, and M. Shuali. 1988. Isoosmotic regulation of cotton and peanut at saline concentrations of K and Na. Plant Physiol. 87:911-916.

Law, R.D. and W.C. Plaxton. 1995. Purification and characterization of a novel phosphoenolpyruvate carboxylase from banana fruit. Biochem. J. 307:807-816.

Lawlor, D.J., J.A. Landivar, C. Crenshaw, and J. Vasek. 1992. Soil water storage and productivity of cotton in conventional vs reduced tillage systems. pp. 1045-1046. *In:* Proc. Beltwide Cotton Conf., National Cotton Council of America, Memphis, Tenn.

Lawlor, D.J., J.A. Landivar, J. Vasek, and C. Crenshaw,. 1991. Cotton root growth in conventional vs reduced tillage systems. pp. 817-819. *In:* Proc. Beltwide Cotton Conf., National Cotton Council of America, Memphis, Tenn.

Leclant, F. and J.P. Deguine. 1994. Aphids (Hemiptera: Aphididae). pp. 285-323 *In*: G.A. Matthews and J.P. Tunstall (eds.). Insects Pests of Cotton. CAB International, Wallingford.

Lee, C., V. Parikh, T. Itsukaichi, K. Bae, and E. Isaac. 1996. Resetting the *Drosophlia* clock by photic regulation of PER and a PER-TIM complex. Sci. 271:1740-1744.

Lee, G. 1993. Non-motor microtubule-associated proteins. Curr. Opin. Cell Biol. 5:88-94.

Lee, J.A. 1981. Genetics of D_3 complementary lethality in *Gossypium hirsutum* and *G. barbadense.* J. Hered. 72:299-300.

Lee, J.A. 1984. Cotton as a world crop. pp. 1-25. *In:* R.J. Kohel and C.F. Lewis (eds.). Cotton. American Soc. Agron., Madison WI.

Lee, J.A. 1986. An early example of a viable hybrid from a cross of *Gossypium barbadense* L. and *G. davidsonii* Kell. J. Hered. 77:56-57.

Lee, J.A. 1987. Induction of adventitious shoots in cotton. Crop Sci. 27:349-350.

Lee, J. 1996. A new spin on naturally colored cottons. Agr. Res. 44:20-21.

Lee, J.I., J.K. Bang, R.K. Park, Y.H. Park, and K.Y. Chung. 1991. Salt resistance and high lint yielding cotton variety "Suwon 3". Res. Rep. Rural Develop. Admin., Upland and Industrial Crops 33:69-73.

Lee, K.Y., J. Townsend, J. Tepperman, M. Black, C.F. Chui, B. Mazur, P. Dunsmuir, and J. Bedbrook. 1988. The molecular basis of sulfonylurea herbicide resistance in tobacco. Embo J. 7:1241-1248.

Lee, L.S., E.J. Conkerton, K.C. Ehrlich, and A. Ciegler. 1983. Reducing sugars and minerals from lint of unopened cotton bolls as a substrate for aflatoxin and kojic acid synthesis by *Aspergillus flavus*. Phytopath. 73:734-736.

Lee, M.S., R.T. Mullen, and R.N. Trelease. 1997. Oilseed isocitrate lyases lacking their essential type I peroxisomal targeting signal are piggybacked to glyoxysomes. Plant Cell 9:185-197.

Lee, S.M., N.A. Garas, and A.C. Waiss, Jr. 1986. High-performance liquid chromatographic determination of sesquiterpenoid stress metabolites in *Verticillium dahliae* infected cotton stele. J. Agri. Food Chem. 34:490-493.

Lee, W.S., B.I. Chevone, and J.R. Seiler. 1990. Growth response and drought susceptibility of red spruce seedlings exposed to simulated acidic rain and ozone. For. Sci. 36:265-275.

Leece, D.R. 1976. Composition and ultrastructure of leaf cuticles from fruit trees relative to differential foliar absorption. Aust. J. Plant Physiol. 3:833.

Leece, D.R. 1978. Foliar absorption in *Prunus domestica* L. 1. Nature and development of the surface wax barrier. Aust. J. Plant Physiol. 5:749-766.

Leffler, H.R. 1983. Plant density affects the development of plant structure and yield. Proc. Beltwide Cotton Prod. Res. Conf. National Cotton Council, Memphis ., pp. 45.

Leffler, H.R. 1986. Mineral compartmentation within the boll. *In:* J.R. Mauney and J.McD. Stewart (eds.). Cotton Physiology. Cotton Foundation, Memphis, Tenn. pp. 301-309.

Leffler, H.R. and B.S. Tubertini. 1976. Development of cotton fruit: II. Accumulation and distribution of mineral nutrients. Agron. J. 68:858-861.

Lefohn, A.S. 1992. The characterization of ambient ozone exposures. *In:* A.S. Lefohn (ed.). Surface Level Ozone Exposures and their Effects on Vegetation. Lewis Publ. Inc., Chelsea, MI. pp. 31-92.

Lefol, E., V. Danielou, H. Darmency, F. Boucher. J. Maillet. and M. Renard. 1995. Gene dispersal from transgenic crops. I. Growth of interspecific hybrids between oilseed rape and the wild hoary mustard. J. Appl. Ecol. 32: 803-808.

Legaspi, B.A.C.Jr, Sterling, W.L., Hartstack, A.W.Jr, and Dean, D.A. 1989. Testing the interactions of pest-predator-plant components of the TEXCIM model. Environ. Entomol. 18: 157-163.

Legé, K.E., M.J. Sullivan, J.T. Walker, and T. Smith. 1996. Evaluation of PGR-IV in South Carolina. pp. 1146-1149. *In:* Proc. Beltwide Cotton Conf., National Cotton Council of America, Memphis, Tenn.

Leger, R.J.S., S.E. Screen, and B. Shams-Pirzodeh. 2000. Lack of host specialization in *Aspergillus flavus*. Applied and Environmental Microbiology 66:320-324.

Legge, A.H., D.J. Savage, and R.B. Walker. 1979. A portable gas-exchange leaf chamber. *In:* W.W. Heck, S.V. Krupa, S.N. Linzon, and E.R. Frederick (eds.). Methodology for the Assessment of Air Pollution Effects on Vegetation. Air Pollut. Control Assoc., Pittsburgh, PA. pp. 16-12 to 16-24.

Leggett, J. E. 1992. Comparison of arthropods sampled from cultivars of upland and Pima cotton with drip and furrow irrigation. Southwest. Entomol. 18: 37-43.

Lehle, F.R, A.M. Zegeer, and O.K. Ahmed. 1991. Ethanolic fermentation in hypoxic cotton seed. Crop Sci. 31:746-750.

Lehnherr, B., F. Machler, A. Grandjean, and J. Fuhrer. 1988. The regulation of photosynthesis in leaves of field-grown spring wheat (*Triticum aestivum* L. cv Albis) at different levels of ozone in ambient air. Plant Physiol. 88:1115-1119.

Leidi, E.O. 1994. Genotypic variation of cotton in response to stress by NaCl or PEG. *In:* M.C Peeters (ed.). Cotton Biotechnology. FAO, Rome. pp. 67-73.

Leidi, E.O. and A. de Castro. 1997. Inclusion de sodio: ¿Principal adaptacion relacionada con tolerancia a salinidad en el algodonero?. *In:*R. Sarmiento-Solis, E.O. Leidi-Montes, and A. Troncoso-de-Arce (eds.). Nutricion Mineral de las Plantas en la Agricultura Sostenible. Consejeria de Agricultura y Pesca, J. Andalucia, Sevilla. pp. 204-210.

Leidi E.O., J.C. Gutierrez, and J. McD. Stewart. 2000. Variability in K and Na uptake in wild and commercial *Gossypium hirsutum* seedlings under saline conditions. *In:* Proceedings World Cotton Conference 2, Athens, Greece 1:567-570.

Leidi, E.O., M. Lopez, J. Gorham, and J.C. Gutierrez. 1999. Variation in carbon isotope discrimination and other traits related to drought tolerance in upland cotton cultivars under dryland conditions. Field Crops Res. 61:109-123.

Leidi, E.O., R. Nogales, and S.H. Lips. 1991. Effect of salinity on cotton plants grown under nitrate or ammonium nutrition at different calcium levels. Field Crops Res. 26:35-44.

Leidi, E.O. and J.F. Saiz. 1997. Is salinity tolerance related to Na accumulation in upland cotton (*Gossypium hirsutum*) seedlings? Plant Soil 190:67-75.

Leidi, E.O., M. Silberbush, M.I.M. Soares, and S.H. Lips. 1992. Salinity and nitrogen nutrition studies on peanut and cotton plants. J. Plant Nutr. 15:591-604.

Leigh, R.A. and R.G. Wyn-Jones. 1986. Cellular compartmentation in plant nutrition: the selective cytoplasm and the promiscuous vacuole. *In:* B. Tinker and A. Läuchli (eds.). Advances in plant nutrition 2:249-279. Paeger Publishers, NY.

Leigh, T.F., D.W. Grimes, W.L.Dickens, and C.E. Jackson. 1974. Planting pattern, plant population, irrigation, and insect interactions in cotton. Environ. Ent. 3:492-496.

Leigh, T.F., D.W. Grimes, H. Yamada, D. Bassett, and J.R. Stockton. 1970. Insects in cotton as affected by irrigation and fertilisation practices. California Agriculture 24:12-14.

Leigh, T.F., R.E. Hunter, and A.H. Hyer. 1968. Spider mite effects on yield and quality of four cotton varieties, California Agric (October): 4-5.

Leigh, T.L., T.A. Kerby, and P.F. Wynholds 1988. Cotton square damage by the plant bug, *Lygus hesperus* (Hemiptera: Heteroptera: Miridae), and abscission rates. J. Econ. Ent. 81:1328-1337.

Lemmon, H. 1986. COMAX: An expert system for cotton crop management. Science 233:29-33.

Lemmon, H., N. Chuk, V. Reddy, B. Acock, Y. Pachepsky, and D. Timlin. 1996. The SIGMA+ cotton model. pp. 529-531. *In:* Proc. Beltwide Cotton Conf., National Cotton Council of America, Memphis, Tenn.

Lemon, R.G., T.A. Hoelewyn, A. Abrameit, and T.J. Gerik. 1999. Evaluation of CGA-248757 (Action) as a harvest-aid in central Texas. pp. 605-606. *In:* P. Dugger and D. Richter (eds.). Proc. Beltwide Cotton Prod. Res. Conf., National Cotton Council of America, Memphis, Tenn.

Lennon, A.M., U.H. Neuenschwander, M. Ribas-Carbo, L. Giles, J.A. Ryals, and J.N. Siedow. 1997. The effects of salicylic acid and tobacco mosaic virus infection on the alternative oxidase of tobacco. Plant Physiol. 115:783-791.

Leon, J.M. and M.J. Bukovac. 1978. Cuticle development and surface morphology of olive leaves with reference to penetration of foliar-applied chemicals. J. Amer. Soc. Hort. Sci. 103:465-472.

Leonard, O.A. 1945. Cotton root development in relation to natural aeration of some Mississippi blackbelt and delta soils. Agron. J. 37:55-71.

Leonard, O.A. and J.A. Pinckard. 1946. Effects of various oxygen and carbon dioxide concentrations on cotton root development. Plant Phys. 21:18-36.

Leonard, R.T. and P.D. Hepler (eds.). 1990. Calcium in Plant Growth and Development. Proc. 13th Ann. Symp. Plant Physiol., Riverside, CA. Am. Soc. Plant Physiol., Rockville, MD.

LePage-Degivry, M.T., P. Barthe, I. Prevost, and B. Boulou. 1989. Regulation of abscisic acid translocation during embryo maturation of *Phaseolus vulgaris.* Physiol. Plant. 77:81-86.

Lepiniec, L., J. Vidal, R. Chollet, P. Gadal, and C. Cretin. 1994. Phosphoenolyruvate carboxylase: Structure, regulation, and evolution. Plant Sci. 99:111-124.

Lety, J., L.H. Stolzy, G.B. Blank, and O.R. Lunt. 1961. Effect of temperature on oxygen-diffusion rates and subsequent shoot growth and mineral content of two plant species. Soil Sci. 92:314-321.

Levan, M.A., J.W. Ycas, and J.W. Hurnmel. 1987. Light leak effects on near-surface soybean rooting observed with minirhizotrons. pp. 89-90. *In:* H.M. Taylor (ed.). Minirhizotron observation tubes: Methods and applications for measuring rhizosphere dynamics. ASA Spec. Publ. 50. ASA, CSSA, and SSSA, Madison, WI.

Levintanus, A. 1992. Saving the Aral Sea. Int. J. Water Res. Develop. 8:60-64.

Levorson, J.P. and C.A. Chlan. 1996. Isolation of a genomic DNA clone from *Gossypium hirsutum* with high similarity to class I endochitinase plant sequences (Accession No. U60197) (PGR96-054). Plant Physiol. 111:1354.

Levorson, J.P. and C.A. Chlan. 1997. Cloning of an ethylene-responsive chitinase from cotton (Accession No. U78888) (PGR97-034). Plant Physiol. 113:665.

Levy, J. and I. Horesh. 1984. Importance of penetration through stomata in the correction of chlorosis with iron salts and low surface tension surfactants. J. Plant Nutr. 7:279-281.

Levy, S., W.S. York, R. Stuike-Prill, B. Meyer, and L.A. Staehelin. 1991. Stimulation of the static and dynamic molecular conformation of xyloglucan. The role of the fucosylated side-chain in surface-specific side-chain folding. Plant J. 1:195-215.

Lewin, K.F., G.R. Hendry, J. Nagy, and R.L. LaMorte. 1994. Design and application of a free-air carbon dioxide enrichment facility. Agric. For. Met. 70:15-29.

Lewis, D.H 1980. Are there inter-relations between the metabolic role of boron, synthesis of phenolic phytoalexins and the germination of pollen? New Phytol. 84:261-270.

Lewis, H.L. 1971. What is narrow-row high population cotton? Ginners Journal & Yearbook. p. 49.

Lewis, H. 1992. Future horizons in cotton research. pp. 5-18. *In:* C.R. Benedict (ed.). Proc. Cotton Fiber Cellulose: Structure, Function, and Utilization Conference. National Cotton Council, Memphis Tenn.

Lewis, H.L. and C.R. Benedict. 1994. Development of cotton fiber strength. pp. 121-124. *In:* G. Jividen and C.R. Benedict (eds.). Proc. Biochemistry of Cotton Workshop. Cotton Incorporated, Raleigh NC.

Ley, T.W., R.L. Elliott, W.C. Bausch, P.W. Brown, D.L. Elwell, and B.D. Tanner. 1994. Review of ASAE standards project X505: Measurement and reporting practices for automatic agricultural weather stations. ASAE paper no. 942086.American Soc. of Agric. Engrs., St. Joseph, MI.

Li, F.G., X.L. Li, and F.L. Li. 1992. A preliminary report on quick salt tolerance identification of cotton explant. China Cottons 1992, 16-17. (Chinese).

Li, F.G., F.L. Li, and X.L. Li. 1994. Effect of salt stress on the activity of protective enzymes in cotton seedlings. Journal of Hebei Agricultural University 17:52-56. (Chinese).

Li, J.M., P. Nagpal, V. Vitart, T.C. McMorris, and J. Chory. 1996. A role for brassinosteroids in light-dependent development of Arabidopsis. Science. 272(5260):398-401.

Li, L. and R.M. Brown, Jr. 1993. β-Glucan synthesis in the cotton fiber. II. Regulation and kinetic properties of β-glucan synthases. Plant Physiol. 101:1143-1148.

Li, L., R.R. Drake, Jr., S. Clement, and R.M. Brown, Jr. 1993. β-Glucan synthesis in the cotton fiber. III. Identification of the UDP-glucose binding subunits of β-glucan synthases by photoaffinity labeling with [β-^{32}P]5'-N$_3$-UDP-glucose. Plant Physiol. 101:1149-1156.

Li, R., D.M. Stelly, and N.L. Trolinder. 1989. Cytogenic abnormalities in cotton (*Gossypium hirsutum* L.) cell cultures. Genome 32:1128-1134.

Li, W.J. H.Z. Dong, Q.Z. Guo, J.Q. Pang, and J. Zhang. 1998. Physiological response of a good upland hybrid and its parent to PEG and NaCl stresses. China Cottons, 25:7-10.

Liakatas, A., D. Roussopoulos, and W.J. Whittington. 1998. Controlled-temperature effects on cotton yield and fibre properties. J. Agric. Sci. 130:463-471.

Liang, X.S., A.J. Rogers, C.L. Webber, T.J. Ormsby, M.E. Tiritan, S.A. Matlin, and C.C. Benz. 1995. Developing gossypol derivatives with enhanced antitumor activity. Invest. New Drugs 13:181-186.

Liebhold, A.M. 1994. Use and abuse of insect and disease models in forest pest management: past, present, and future. pp. 204-210 *In:* W.W. Covington and L.F. DeBano (eds.). Sustainable Ecological Systems: Implementing an Ecological Approach to Land Management. USDA For. Res. Serv. Tech. Rep. RM-247.

Lin, H., S.S. Salus, and K.S. Schumaker. 1997. Salt sensitivity and the activities of the H$^+$-ATPases in cotton seedlings. Crop Sci. 37:190-197.

Lin, H.-J., H.-Y. Cheng, C.-H. Chen, H.-C. Huang, and T.-Y. Feng. 1997. Plant amphipathic proteins delay the hypersensitive response caused by harpin-Pss and *Pseudomonas syringae* pv. *syringae*. Physiological and Molecular Plant Pathology 51:367-376.

Lin, J., Z. Zhu, B. Fan, J.D. Lin, Z.Y. Zhu, and B.X. Fan. 1995. Physiological reaction of cotton varieties under different levels of salt stress. China Cottons 22:16-17.

Lin, T.S., R. Schinazi, B.P. Griffith, E.M. August, B.F.H. Eriksson, D.K. Zheng, L. Huang, and W.H. Prusoff. 1989. Inhibition of human immunodeficiency virus type 1 replication by the (-) but not the (+)- enantiomer of gossypol. Antimicrob. Agents Chemother. 33:2149-2151.

Lincoln, C. and G. Dean. 1976. Yield and blooming of cotton as affected by insecticides. Ark. Farm Res. 25:5.

Linderman, R.G. 1992. Vesicular-arbuscular mycorrhizae and soil microbial interactions. pp. 45-70. *In:* G.J. Bethlenfalvay and R.G. Linderman (eds.). Mycorrhizae in Sustainable Agriculture. ASA Special Publication Number 54, Madison, WI.

Linderman, R.G. 1994. Role of VAM fungi in biocontrol. p. 1-25. *In:* F.L. Pfleger and R.G. Linderman (eds.). Mycorrhizae and Plant Health. APS Press, St Paul, MN.

Lingle, J.C. and O.A. Lorenz. 1969. Potassium nutrition of tomatoes. J. Am. Soc. Hort. Sci. 94:679-683.

Lipe, J.A. and P.W. Morgan. 1973. Location of ethylene production in cotton flowers and fruits. Planta 115:93-96.

Littler, M.M., D.S. Littler, and P.R. Taylor. 1995. Selective herbivore increases biomass of its prey: a chiton coralline reef-building association. Ecology 76:1666-1681.

Liu, B., H.C. Joshi, and B.A. Palevitz. 1995. Experimental manipulation of γ-tubulin distribution in *Arabidopsis* using antimicrotubule drugs. Cell Motil. Cytoskeleton 31:113-129.

Liu, B., J. Marc, H.C. Joshi, and R.A. Palevitz. 1993. A γ-tubulin-related protein associated with the microtubule arrays of higher plants in a cell cycle-dependent manner. J. Cell Sci. 104:1217-1228.

Liu, C.-J., P. Heinstein, and X.-Y. Chen. 1999. Expression pattern of genes encoding farnesyl diphosphate synthase and sesquiterpene cyclase in cotton suspension-cultured cells treated with fungal elicitors. Molecular Plant-Microbe Interactions 12:1095-1104.

Liu, C., Y. Meng, and X.Y. Chen. 1997. Direct submission to database on March 26,1997. (Accession No. Y 12082).

Liu, C.-J., Y.-L. Meng, S.-S. Hou, and X.-Y. Chen. 1998. Cloning and sequencing of a cDNA encoding farnesyl pyrophosphate synthase from *Gossypium arboreum* and its expression pattern in the developing seeds of *Gossypium hirsutum* cv. 'Sumian-6.' Acta Botanica Sinica 40:703-710.

Liu, C.Q., L.M. Lu, and J.D. Liu. 1993. Establishment of the salinity tolerance of cotton germplasm resources. Crop Genetic Res. 1993, 21-22. (Chinese).

Liu, G.,J. Wu, H. Wu, W. Bao, and Y. Hsi. 1985. Somatice embryogenesis of cotton *in vitro* and its cytological observation. J. Agr. Sci. Proc. Sino-Jap Symp. on Biotech. Nanjing, Vol 2(S):36-41.

Liu, J., C.R. Benedict, R.D. Stipanovic, and A.A. Bell. 1999. Purification and characterization of S-adenosyl-L-methionine: Desoxyhemigossypol-6-O-methyl-transferase from cotton plants. An enzyme capable of methylating the defense terpenoids of cotton. Plant Physiology 121:1017-1024.

Liu, J., W. Ye, B. Fan, J.D. Liu, W.W. Ye, and B.X. Fan. 1998. Research on stress resistance in cotton and its utilization in China. China Cottons 25:5-6.

Liu, P.B.W. and J.B. Loy. 1976. Action of gibberellic acid on cell proliferation in the subapical shoot meristem of watermelon seedlings. Am. J. Bot. 63:700-704.

Liu, Q., S. Singh, P. Sharp, A. Green, and D.R. Marshall. 1996. Nucleotide sequence of a cDNA from *Gossypium hirsutum* encoding a stearoyl-acyl carrier protein desaturase (Accession No. X95988) (PGR96-018). Plant Physiol. 110:1436.

Liu, R.J. 1995. Effect of vesicular-arbuscular mycorrhizal fungi on Verticillium wilt of cotton. Mycorrhizia 5:293-297.

Liu, R.-J., H.-F. Li, C.-Y. Shen, and W.-F. Chiu. 1995. Detection of pathogenesis-related proteins in cotton plants. Physiological and Molecular Plant Pathology 47:357-363.

Livingston, S.D. and R.D. Parker. 1994. Lint yield responses to application of PGR-IV and mepiquat chloride applied to five cotton varieties in South Texas. pp. 1263-1266. *In:* D.J. Herber and D.A. Richter (eds.). Proc. Beltwide Cotton Conf., National Cotton Council of America, Memphis, Tenn.

Lloyd, A.M., M. Schena, V. Walbot, and R.W. Davis. 1994. Epidermal cell fate determination in *Arabidopsis:* Patterns defined by a steroid inducible regulator. Science 266:436-439.

Lloyd, R.W. and D.R. Krieg. 1987. Cotton development and yield as affected by insecticides. J. Econ. Entomol. 80:854-858.

Lofland, H.B., Jr. 1950. *In vitro* culture of cotton embryo. Bot. Gazz. 111:307-311.

Loguercio, L.L., H.C. Scott, N.L. Trolinder, and T.A. Wilkins. 1999. Hmg-CoA reductase gene family in cotton (*Gossypium hirsutum* L.): Unique structural features and differential expression of *hmg 2* potentially associated with synthesis of specific isoprenoids in developing embryos. Plant and Cell Physiology 40:750-761.

Lois, L.M., N. Campos, S.R. Putra, K. Danielsen, M. Rohmer, and A. Boronat. 1998. Cloning and characterization of a gene from *Escherichia coli* encoding a transketolase-like enzyme that catalyzes the synthesis of D-1-deoxyxylulose 5-phosphate, a common precursor for isoprenoid, thiamin, and pyridoxol biosynthesis. Proc. Natl. Acad. Sci. USA 95:2105-2110.

Long, D.W., J.B. Graves, B.R. Leonard, E. Burris, L.M. Southwick, and D.C. Rester. 1991. Effects of co-application of urea solutions and insecticides on chemical deposition and biological activity. pp. 762-768. *In:* Proc. Beltwide Cotton Conferences. National Cotton Council, Memphis, Tenn.

Long, S.P. 1991. Modification of the response of photosynthetic productivity to rising temperature by atmospheric CO2 concentrations: Has its importance been underestimated? Plant Cell Environ. 14:729-739.

Longenecker, D.E. 1973. The influence of soil salinity upon fruiting and shedding, boll characteristics, fibre quality, and yield of two cotton species. Soil Sci. 115:294-302.

Longenecker, D.E. 1974. The influence of high sodium in salts upon fruiting and shedding boll characteristics, fibre properties, and yield of two cotton species. Soil Sci. 118:387-396.

Longenecker, N. 1994. Nutrient deficiencies and vegetative growth. *In:* A.S. Basra (ed.). Mechanisms of Plant Growth and Improved Productivity. Marcel Dekker, New York. pp. 137-172.

Longenecker, D.E. and L.J. Erie. 1968. Irrigation water management. *In:* F.C. Elliott et al. (ed.). Advances in production and utilization of quality cotton: Principles and practices. The Iowa State Univ. Press, Ames. pp. 322-345.

Longer, D.E. and D.M. Oosterhuis. 1995. Regrowth of defoliated cotton seedlings in laboratory and field environments. *In:* D.M. Oosterhuis (ed.). Proc. 1995 Cotton Research Meeting and Summaries of Research in Progress. University of Arkansas Agri. Exp. Station Special Report 172:86-89.

Longer, D.E. and D.M. Oosterhuis. 1999. Cotton regrowth and recovery from early season leaf loss. Env. And Exp. Botany 41:67-73.

Longer, D.E., D.M. Oosterhuis, and M. Withrow. 1993. Cotton seedling recovery from partial and complete defoliation. Arkansas Farm Res. 42:10-11.

Longstreth, D.J. and P.S. Nobel. 1979a. Salinity effects on leaf anatomy. Plant Physiol. 63:700-703.

Longstreth, D.J. and P.S. Nobel. 1979b. Nutrient influences on leaf photosynthesis. Effects of nitrogen, phosphorus, and potassium for *Gossypium hirsutum* L. Plant Physiol. 65:541-543.

Loomis, W.D. and R.W. Durst. 1992. Chemistry and biology of boron. BioFactors 3:229-239.

Lord, E. and S.A. Heap. 1988. The origin and assessment of cotton fibre maturity. International Inst. for Cotton, Manchester, UK, pp 1-38.

Loughrin, J.B., A. Manukian, R.R. Heath, T.C.J. Turlings, and J.H. Tumlinson. 1994. Diurnal cycle of emission of induced volatile terpenoids by herbivore-injured cotton plants. Proc. Natl. Acad. Sci. USA 91:11836-11840.

Low, A., J.D. Hesketh, and H. Muramoto. 1969. Some environmental effects on the varietal node number of the first fruiting branch. Cotton Gr. Rev. 46:181-188.

Lu, Z.M., J.W. Chen, R.G. Percy, and E. Zeiger. 1997. Photosynthetic rate, stomatal conductance and leaf area in two cotton species (*Gossypium barbadense* and *Gossypium hirsutum*) and their relation with heat resistance and yield. Aust. J. Plant Physiol. 24:693-700.

Lu, Z., R.G. Percy, C.O. Qualset, and E. Zeiger. 1998. Stomatal conductance predicts yields in irrigated Pima cotton and bread wheat grown at high temperatures. J. Exp. Bot. 49:453-460.

Lucas, M.C., T. Fowler, and D.R. Gossett. 1996. Glutathione S-transferase activity in cotton plants and callus subjected to salt stress. Proc. Beltwide Cotton Conf., Nashville, Tenn, USA, January 9 12, 1996: Vol. 2 1177-1178;

Lucas, M.C., D.R. Gossett, E.P. Millhollon, and M.M. Marney. 1993. Antioxidant scavengers ands salt stress in cotton. Proc. Beltwide Cotton Conf., New Orleans, 1993, 1259-1261.

Lucena, J.J., A. Garate, A.M. Ramon, and M. Manzanares. 1990. Iron nutrition of a hydroponic strawberry culture (*Fragaria vesca* L) supplied with different Fe chelates. Plant Soil 123:9-15.

Ludlow, M.M. and R.C. Muchow. 1988. Critical evaluation of the possibilities modifying crops for high production per unit of precipitation. *In:* F.R. Bidinger and C.A. Johansen (eds.). Drought Research Priorities for the dryland tropics. ICRISAT, Patancheru.

Ludlow, M.M., F.J. Santamareia, and S. Fakai. 1989. Role of osmotic in reducing the loss of grain yield in sorghum due to drought. Australian Soghum Workshop. February, 1989. 25:23-30.

Ludtke, S.J., K. He, W.T. Heller, T.A. Harroun, L. Yang, and H.W. Huang. 1996. Membrane pores induced by magainin. Biochemistry 35:13723-13728.

Ludwig, C.A. 1932. Germination of cottonseed at low temperatures. J. Agr. Res. 44:367-380.

Ludwig, L.J., T. Saeki, and L.T. Evans.1965. Photosynthesis in artificial communities of cotton plants in relation to leaf area. I. Experiments with progressive defoliation of mature plants. Aust. J. Biol. Sci. 18:1103-1118.

Lugo, A.E. and S. Brown. 1992. Tropical forests as sinks for atmospheric carbon. Forest Ecol. Management. 54:239-255.

Lukefahr, M.J., L.W. Noble, and J.E. Houghtaling. 1966. Growth and infestation of bollworms and other insects on glanded and glandless strains of cotton. J. Econ. Entomol. 59:817-820.

Lumsden, P. 1991. Circadian rhythms and phytochrome. Annu. Rev. Plant Physiol. Plant Mol. Biol. 42:351-371.

Lutrick, M.C., H.A. Peacock, and J.A. Cornell. 1986. Nitrate monitoring for cotton lint production on a Typic Paleudult. Agron. J. 78:1041-1046.

Luttge, U. and J.A.C. Smith. 1984. Mechanism of passive malic acid efflux from vacuoles of the CAM plant *Kalanchoe daigremontiana*. J. Memb. Biol. 81:149-158.

Luttrel, R.G. 1994. Cotton pest management. Part 2. A U.S. perspective. Ann. Rev. Entomol. 39:527-542.

Luttrel, R.G., G.P. Fitt, F.S. Ramalho, and E.S. Sugonyaev. 1994. Cotton pest management. Part 1. A worldwide perspective. Ann. Rev. Entomol. 39:517-526.

Lynch, J.; A. Läuchli, and E. Epstein. 1991. Vegetative growth of the common bean in response to phosphorus nutrition. Crop Sci. 31:380-387.

Lyon, B.R., Y.L. Cousins, D.J. Llewellyn, and E.S. Dennis. 1993. Cotton plants transformed with a bacterial degradation gene are protected from accidental spray drift damage by the herbicide 2,4-dichloro-phenoxyacetic acid. Transgenic Res. 2:162-169.

Ma, D.-P., H.-C. Liu, H. Tan, R.G. Creech, J.N. Jenkins, and Y.-F. Chang. 1997. Cloning and characterization of a cotton lipid transfer protein gene specifically expressed in fiber cells. Biochim. Biophys. Acta 1344:111-114.

Ma, D.-P., H. Tan, Y. Si, R.G. Creech, and J.N. Jenkins. 1995. Differential expression of a lipid transfer protein gene in cotton fiber. Biochim. et Biophys. Acta 1257:81-84.

Maas, E.V. 1985. Crop tolerance to saline sprinkling water. Plant and Soil 89:273-284.

Maas, E.V. 1990. Crop salt tolerance. *In:* K.J. Tanji (ed.). Agricultural Salinity Assessment and Management. American Society of Civil Engineers, New York. pp. 262-304.

Maas, E.V., S.R. Grattan, and G. Ogata. 1982. Foliar salt accumulation and injury in crops sprinkled with saline water. Irrig. Sci. 3:157-168.

MacAdam, J.W.; J.J. Volenec, and C.J. Nelson. 1989. Effect of nitrogen on mesophyll cell division and epidermal cell elongation in tall fescue leaf blades. Plant Physiol. 89:549-556.

Mace, M.E. and A.A. Bell. 1981. Flavanol and terpenoid aldehyde synthesis in tumors associated with genetic incompatibility in a *Gossypium hirsutum* x *G. gossypioides* hybrid. Canadian J. Botany 59:951-955.

Mace, M.E., A.A. Bell, and R.D. Stipanovic. 1974. Histochemistry and isolation of gossypol and related terpenoids in roots of cotton seedlings. Phytopathology 64:1297-1302.

Mace, M.E. and R.D. Stipanovic. 1995. Mode of action of the phytoalexin desoxyhemigossypol against the wilt pathogen, *Verticillium dahliae.* Pestic. Biochem. Physiol. 53:205-209.

Mace, M.E., R.D. Stipanovic, and A.A. Bell. 1985. Toxicity and role of terpenoid phytoalexins in Verticillium resistant cotton. Physiol. Plant Pathol. 26:209-218.

Mace, M.E., R.D. Stipanovic, and A.A. Bell. 1987. Toxicity of dihydroxy-cadalene to *Verticillium dahliae.* Phytopathology 77:1692.

Mace, M.E., R.D. Stipanovic, and A.A. Bell. 1989. Histochemical localization of desoxyhemigossypol, a phytoalexin in *Verticillium dahliae*-infected cotton stems. New Phytol. 111:229-232.

Mace, M.E., R.D. Stipanovic, and H.H. Mollenhauer. 1992. Toxicity of terpenoid phytoalexins: Effects on the plasmalemma of *Verticillium dahliae.* p. 196. *In*: Proc. Beltwide Cotton Conf., Volume 1, National Cotton Council of America, Memphis, Tenn.

Machler, F., A. Oberson, A. Crub, and S. Nosberger. 1988. Regulation of photosynthesis in nitrogen deficient wheat seedlings. Plant Physiol. 87:46-49.

Macilwain, C. 1996. Bollworms chew hole in gene-engineered cotton. Nature 382:289.

MacKenzie, A.J. and P.H. van Schaik. 1963. Effect of nitrogen on yield, boll, and fiber properties of four varieties of irrigated cotton. Agron. J. 55:345-347.

MacMillan, J. 2001. Occurrence of gibberellins in vascular plants, fungi and bacteria. J. Plant Growth Regul. 20:387-442.

Macphail, M.K. and E.M. Truswell. 1989. Palynostratigraphy of the central west Murray Basin. J. Aust. Geol. Geophys. 11:301-331.

MacRobbie, E.A.C. 1997. Signaling in guard cells and regulation of ion channel activity. J. Exp. Bot. 48:515-528.

Maeshima, M. 1990. Development of vacuolar membranes during elongation of cells in mung bean hypocotyls. Plant Cell Physiol. 31:311-317.

Maeshima, M. 1992. Characterization of the major integral proteins of vacuolar membrane. Plant Physiol. 98:1248-1254.

Magboul, B.I., H. Khalifa, H.A. El-Fako, and A.W. El Khidir. 1978. Variation of gossypol content in some species of *Gossypium.* Sudan J. Food Sci. Technol. 10:37-44.

Mahall, B.E. and R.M. Callaway. 1991. Root communication among desert shrubs. Proc. Natl. Acad. Sci. 88:874-876.

Mahall, B.E. and R.M. Callaway. 1992. Root communication mechanisms and intracommunity distributions of 2 Mojave desert shrubs. Ecology 73:2145-2151.

Maier-Maercker, U. 1979. Peristomatal transpiration and stomatal movement: A controversial view. I. Additional proof of peristomatal transpiration by photography and a comprehensive discussion in the light of recent results. Z. Pflanzenphysiol. 91:25-43.

Malik, M.N.A., J.P. Evenson, and D.G. Edwards. 1978. The effect of level of nitrogen nutrition on earliness in upland cotton (*Gossypium hirstum* L.). Aust. J. Agric. Res. 29:1213-1221.

Malik, M.N. and M.I. Makhdum. 1987. Salinity tolerance of cotton cultivars (*G. hirsutum* L.) at germination. The Pakistan Cottons 31:171-174.

Malik, R.S., J.S. Dhankar, and N.C. Turner. 1979. Influence of soil water deficits on root growth of cotton seedlings. Plant and Soil 53:109-115.

Maltby, D., N.C. Carpita, D. Montezinos, C. Kulow, and D.P. Delmer. 1979. β-1,3-glucan in developing cotton fibers. Plant Physiol. 63:1158-1164.

Manchandia, A.M., S.W. Banks, D.R. Gossett, B.A. Bellaire, M.C. Lucas, and E.P. Millhollon. 1999. The influence of alpha-amanitin on the NaCl-induced up-regulation of antioxidant enzyme activity in cotton callus tissue. Free Radical Res. 30:429-438.

Manjunath, A. and M. Habte. 1991. Root morphology characteristics of host species having distinct mycorrhizal dependency. Can. J. Bot. 69:671-676.

Manning, W.J. and S.V. Krupa. 1992. Experimental methodology for studying the effects of ozone on crops and trees. *In:* A.S. Lefohn (ed.). Surface Level Ozone Exposures and their Effects on Vegetation. Lewis Publ. Inc., Chelsea, MI. pp. 93-156.

Mantell, A., H. Frenkel, and A. Meiri. 1985. Drip irrigation of cotton with saline-sodic water. Irrig. Sci. 6:95-106.

Maples, R. 1981. Effects of Pix and varying rates of nitrogen on cotton. 1980. Ark. Farm Res. 30:5.

Maples, R.L. and J.L. Keogh. 1971. Cotton fertilization studies on loessial plains of eastern Arkansas. Ark. Agri. Exp. Sta. Rep. Ser. 194.

Maples, R.L. and J.L. Keogh. 1973. Phosphorus fertilization experiments with cotton in delta soils of Arkansas. Ark. Agri. Exp. Sta., Bulletin 781.

Maples, R.L., J.L. Keogh, and W.E. Sabbe. 1977. Nitrate monitoring for cotton production in Loring-Calloway silt loam. Univ. of Arkansas Agri. Exp. Sta. Bulletin 825.

Maples, R.L., W.N. Miley, and T.C. Keisling. 1990. Nitrogen recommendations for cotton and how they were developed in Arkansas. pp 33-39. *In:* W.N. Miley and D.M. Oosterhuis (eds.). Nitrogen nutrition in cotton: Practical issues. Amer. Soc. Agron. Madison, WI.

Maples, R.L., W.N. Miley, T.C. Keisling, S.D. Carroll, J.J. Varvil, and W.H. Baker. 1992. Agronomic criteria and recommendations for cotton nutrient monitoring. Ark. Agri. Exp. Sta. Spec. Report 155.

Maples, R.L., W.E. Sabbe, R.F. Ford, M. Appleberry, and R.J. Mahler. 1987. Response of cotton to nitrogen in problem fields. Ark. Agr. Exp. Sta. Bull. 901.

Maples, R.L., W.R. Thompson, and J. Varvil. 1988. Potassium deficiency in cotton takes on a new look. Better Crops with Plant Food (73):6-9. Potash and Phosphate Institute, Atlanta, Ga.

Marani, A. 1979. Growth rate of cotton bolls and their components. Field Crops Res. 2:169-175.

Marani, A. and A. Amirav. 1971. Effects of soil moisture stress on two varieties of upland cotton in Israel. I. The coastal plain region. Exper. Agric. 7:213-224.

Marani, A., D.N. Baker, V.R. Reddy, and J.M. McKinion. 1985. Effect of water stress on canopy senescence and carbon exchange rates in cotton. Crop Sci. 25:798-802.

Marani, A., G.E. Cardon, and C.J. Phene. 1992a. CALGOS, a version of GOSSYM adapted for irrigated cotton. I. Drip irrigation, soil water transport, and root growth. pp. 1352-1357 *In:* D.J. Herber and D. A. Richter (eds.). Proc. Beltwide Cotton Conf., National Cotton Council of America, Memphis, Tenn.

Marani, A., C.J. Phene, and G.E. Cardon. 1992b. CALGOS, a version of GOSSYM adapted for irrigated cotton. II. Leaf water potential and the effect of water stress. pp. 1358-1360 *In:* D.J. Herber and D. A. Richter (eds.). Proc. Beltwide Cotton Conf., National Cotton Council of America, Memphis, Tenn.

Marani, A., C.J. Phene, and G.E. Cardon. 1992c. CALGOS, a version of GOSSYM adapted for irrigated cotton. III. Leaf an boll growth routines. pp. 1361-1364 *In:* D.J. Herber and D. A. Richter (eds.). Proc. Beltwide Cotton Conf., National Cotton Council of America, Memphis, Tenn.

Marani, A. and J. Ephrath. 1985. Penetration of radiation into cotton crop canopies. Crop Sci. 25:309-313.

Marani, A. and D. Levi.1973. Effect of soil moisture during early stages of development on growth and yield of cotton plants. Agron. J. 65:637-641.

Marani, A., M. Zur, A. Eshel, H. Zimmerman, R. Carmeli, and B. Karaduaid. 1973. Effect of time and rate of application of two growth retardants on growth, flowering, and yield of upland cotton. Crop Sci. 13:429-32.

Marcus-Wyner, L. and D.W. Rains. 1982. Nutritional disorders of cotton. Comm. Soil. Plant. Anal. 13:685-736.

Marks, M.D. and J.J. Esch. 1992. Trichome formation in *Arabidopsis* as a genetic model system for studying cell expansion. Current Topics in Plant Biochem. and Physiol. 11:131-142.

Marquis, R.J. 1988. Intra-crown variation in leaf herbivory and seed production in striped maple, *Acer pennsylvanicum* L. (Aceracea). Oecologia 77:51-55.

Marquis, R.J. and H.M. Alexander. 1992. Evolution of resistance in plant-herbivore and plant-pathogen interactions. Trends Ecol. Evol. 7:126-129.

Marquis R.J. and C. Whelan. 1996. Plant morphology and recruitment of the third thropic level: subtle and little-recognized defences? Oikos 75: 330-333.

Marre, E. 1977. Effects of fusicoccin and hormones on plant cell membrane activities: Observations and hypotheses. *In* E. Marre and O. Ciferri (eds.) .*Regulation of Cell Membrane Activities in Plants*. Elsevier/North-Holland Biomedical Press.

Marroqauin, L.D., D. Elyassnia, J.S. Griffitts, J.S. Feitelson, and R.V. Aroian. 2000. *Bacillus thuringiensis* (Bt) toxin susceptibility and isolation of resistance mutants in the nematode *Caenorhabditis elegans*. Genetics 155:1693-1699.

Marschner, H. 1995. Mineral nutrition of higher plants. Academic Press. New York, NY.

Marschner, H. and I. Cakmak. 1989. High light intensity enhances chlorosis and necrosis in leaves of zinc, potassium, and magnesium deficiency bean (*Phaseolus vulgaris*). J. Plant Physiol. 134:308-315.

Marschner, H., V. Romheld, and M. Kissel. 1986. Different strategies in higer plants to mobilize and increase uptake of iron. J. Plant Nutr. 9:695-713.

Marshall, C. and I.F. Wardlaw. 1973. A comparative study of the distribution and speed of movement of ^{14}C assimilates and foliar-applied ^{32}P labelled phosphate in wheat. Aust. J. Biol. Sci. 26:1-6.

Martin, C. and J. Paz-Ares. 1997. MYB transcription factors in plants. T.I.G. 13:67-73.

Martin, R.D., W.W. Ballard, and D.M. Simpson 1923. Growth of fruiting parts in cotton plants. J. Agr. Res. 25:195-208.

Martin, T.J. and B.E. Juniper. 1970. The cuticles of plants. St. Martin's Press, New York.

Martinez, C., J.L. Montillet, E. Bresson, J.P. Agnel, G.H. Dai. J.F. Daniel, J.P. Geiger, and M. Nicole. 1998. Apoplastic peroxidase generates superoxide anions in cells of cotton cotyledons undergoing the hypersensitive reaction to *Xanthomonas campestris* pv. *malvacearum* race 18. Molecular Plant-Microbe Interactions 11:1038-1047.

Martinez, V. and A. Läuchli. 1991. Phosphorus translocation in salt-stressed cotton. Physiol. Plant. 83:627-632.

Martinez, V. and A. Läuchli. 1994. Salt-induced inhibition of phosphate uptake in plants of cotton. New Phytol. 125:609-614.

Martinoia, E., D.I. Flügge, G. Kaiser, D. Heber, and H.W. Heldt. 1985. Energy-dependent uptake of malate into vacuoles isolated from barley mesophyll protoplasts. Biochim. Biophys. Acta 806:311-319.

Martinoia, E. and D. Rentsch. 1994. Malate compartmentalization- responses to a complex metabolism. Ann. Rev. Plant Physiol. 45:447-468.

Martsinkovskaya, A.I., R.S. Moukhamedov, and A.A. Abdukarimov. 1996. Potential use of PCR-amplified ribosomal intergenic sequences for differentiation of varieties and species of *Gossypium* cotton. Plant Mol. Biol. Rep. 14:44-49.

Marx-Figini, M. and G.V. Schulz. 1966. Uber die kinetik und den Mechanismus der Biosynthese der cellulose in den hoheren pflanzen (nach versuchen an den samenhaaren der baumwolle). Biochem. Biophys. Acta 112:81-101.

Mascagni, J.H., Jr., T.C. Keisling, and R.L. Maples. 1993. Response of fast fruiting cotton cultivars to nitrogen on a clayey soil. J. Prod. Agric. 6:104-108.

Mascagni, H.J., M.E. Terhune, R.L. Maples, and W.H. Miley. 1990. Influence of sulfur on cotton yield. Univ. of Arkansas Coop. Ext. Ser., Little Rock, AR. Fact Sheet 2058.

Maschinski, J. and T.G. Whitham. 1989. The continuum of plant responses to herbivory: the influence of plant association, nutrient availability, and timing. Am. Nat. 134:1-19.

Masciadri, R, W. Angst, and D. Arigoni. 1985. A revised scheme for the biosynthesis of gossypol. J. Chem. Soc., Chem. Commun. pp. 1573-1574.

Mason, T.G. 1922. Growth and abscission in Sea Island cotton. Ann. Bot. 36:457-484.

Mass, E.V. and G.J. Hoffman. 1977. Crop salt tolerance-current assessment. J. Irr. Drain. Div. Am. Soc. Civil Eng. 102(IR2)115-134.

Masucci, J.D., W.G. Rerie, D.R. Foreman, M. Zhang, M.E. Galway, M.D. Marks, and J.W. Schiefelbein. 1996. The homeobox gene *GLABRA2* is required for position-dependent cell differentiation in root epidermis of *Arabidopsis thaliana*. Development 122:1253-1260.

Masucci, J.D. and J.W. Schiefelbein. 1996. Hormones act downstream of *TTG* and *GL2* to promote root hair outgrowth during epidermis development in the *Arabidopsis* root. Plant Cell 8:1505-1517.

Mathews, J.N.A. 1994. The Benefits of overcompensation and herbivory - the difference between coping with herbivores and liking them. Am. Nat. 144:528-533.

Mathur, O.P., S.K. Mathur, and N.R. Talati. 1983. Effect of addition of sand and gypsum to fine-textured salt-affected soils on the yield of cotton and jower (sorghum) under Rajasthan Canal Command Area conditions. Plant and Soil 74:61-65.

Matic-Leigh, R. and D.A. Cauthen. 1994. Determining cotton fiber maturity by image analysis. Part I: Direct measurement of cotton fiber characteristics. Textile Res. J. 64:534-544.

Matlin, S.A., A. Belenguer, R.G. Tyson, and A.N. Brookes. 1987. Resolution of gossypol: analytical and large-scale preparative HPLC on non-chiral phases. J. High Resolut. Chromatogr., Chromatogr. Commun. 10:86-87.

Matlin, S.A., S. Roshdy, Q.B. Cass, L.C.G. Freitas, R.L. Longo, and I. Malvestiti. 1990. Structural investigations of gossypol Schiff's bases. J. Braz. Chem. Soc. 1:128-133.

Matlin, S.A., R. Zhou, G. Bialy, R.P. Blye, R.H. Naqvi, and M.C. Lindberg. 1985. (-)Gossypol: an active male antifertility agent. Contraception 31:141-149.

Matocha, J.E. 1993. Cotton production under conservation tillage in southern Texas. pp. 99-103. *In:* T.D. Valco and M.R. McClelland (eds.). Conservation-Tillage Systems for Cotton. University of Arkansas Agricultural Experiment Station Special Report 169. Fayetteville, AR.

Matocha, J.E., D.L. Coker, and F.L. Hopper. 1994. Potassium fertilization effects on cotton yields and fiber properties. pp. 1597-1600. *In:* Proc. Beltwide Cotton Conf., National Cotton Council of America, Memphis, Tenn.

Matsumoto, T., F. Sakai, and T. Hayashi. 1997. A xyloglucan-specific endo-1,4-β-glucanase isolated from auxin-treated pea stems. Plant physiol. 114:661-667.

Matsuoka, M. and S. Hata. 1987. Comparative studies of phosphoenolpyruvate carboxylase from C_3 and C_4 plants. Plant Physiol. 85:947-651.

Matthews, G.A. 1994a. The effects if insect attack on the yield of cotton. pp. 427-430 *In*: G.A. Matthews and J.P. Tunstall (eds.). Insect Pests of Cotton. CAB International, Wallingford.

Matthews, G.A. 1994b. Insect and mite pests: general introduction. pp. 29-37 *In*. G.A. Matthews and J.P. Tunstall (eds.). Insect Pests of Cotton. CAB International, Wallingford.

Matthews, G.A. and J.P. Tunstall. 1994. Insect pests of cotton. C.A.B. International, University Press, Cambridge, 539 pp.

Matthysse, A.G. and S. McMahan. 1998. Root colonization by *Agrobacterium tumefaciens* is reduced in *cel, att B, att D,* and *att R* mutants. Applied and Environmental Microbiology 64:2341-2345.

Mauderli, A.F., C.A. Madramootoo, and G.T. Dodds. 2000. Water-use efficiency, enhanced farmers participation - means for controlling environmental impacts in irrigation and drainage. Sustainability of irrigated agriculture: Managing environmental changes due to irrigation and drainage Cairo, Egypt, 17 September 1996. 1996, 13-26.

Mauney, J.R. 1961. The culture *in vitro* of immature cotton embryos. Bot. Gaz. 122:205-209.

Mauney, J.R. 1966. Floral initiation of upland cotton *Gossypium hirsutum* L. in response to temperatures. J. Exp. Bot. 17:452-459.

Mauney, J.R. 1984. Anatomy and morphology of the cotton plant. pp. 25-79. *In:* R.J. Kohel and C.F. Lewis (eds.), Cotton: Principles and practice. Iowa State University Press, Ames, IA.

Mauney, J.R. 1984b. Anatomy and morpholoy of cultivated cottons. pp. 59-80. *In:* R.J. Kohel and C.F. Lewis (eds.), Cotton: Principles and practice. Iowa State University Press, Ames, IA.

Mauney, J.R. 1986a. Vegetative growth and development of fruiting sites. pp. 11-28. *In:* J.R. Mauney and J.McD. Stewart (eds.). Cotton Physiology. Cotton Foundation, Memphis.

Mauney, J.R. 1986b. Carbohydrate production and partitioning in the canopy. pp. 183-191. *In:* J.R. Mauney and J.McD. Stewart (eds.). Cotton Physiology. Cotton Foundation, Memphis.

Mauney, J.R., J. Chappell, and B.J. Ward. 1967. Effects of malic acid salts on growth of young cotton embryos *in vitro*. Bot. Gaz. 128:198-200.

Mauney, J.R., G. Guinn, K.E. Fry, and J.D. Hesketh, J.D. 1979. Correlation of photosynthetic carbon dioxide uptake and carbohydrate accumulation in cotton, soybean, sunflower and cotton. Photosynthetica 13:260-266.

Mauney, J.R., G. Guinn, J.D. Hesketh, K.E. Fry, and J.W. Radin. 1976. Inhibition of photosynthesis by leaf starch. p. 60 *In:* Beltwide Cotton Prod. Res. Conf., National Cotton Council of America, Memphis, Tenn.

Mauney, J.R., K.E. Fry, and G. Guinn. 1978. Relationship of photosynthetic rate to growth and fruiting of cotton, soybean, sorghum, and sunflower. Crop Sci. 18:259-263.

Mauney, J.R. and D.L. Hendrix. 1988. Responses of glasshouse grown cotton to irrigation with carbon-dioxide saturated water. Crop Sci. 28:835-838.

Mauney, J.R., B.A. Kimball, P.J. Pinter, Jr., R.L. LaMorte, K.F. Lewin, J. Nagy, and G.R. Hendrey. 1994. Growth and yield in cotton in response to a free-air carbon dioxide enrichment (FACE) environment. Agric. and Forest. Meterol. 70:49-67.

Mauney, J.R., K.F. Lewin, G.R. Hendry, and B.A. Kimball. 1992. Growth and yield of cotton exposed to free-air CO_2 enrichment (FACE). Crit. Rev. Plant Sci. 11:213-222.

Mauney, J.R. and Stewart, J.M. 1986. Cotton Physiology. The Cotton Foundation, Memphis, Tenn.

May, O.L. 1999. Genetic variation in fiber quality. pp. 183-229. *In:* A.S. Basra (ed.) .Cotton Fibers. Food Products Press, NY.

May, O.L. and C.C. Green. 1994. Genetic variation for fiber properties in elite Pee Dee cotton populations. Crop Sci. 34:684-690.

Mayer, R., P. Ross, H. Weinhouse, D. Amikam, G. Volman, P. Ohana, R.D. Calhoon, H.C. Wong, A.W. Emerick, and M. Benziman. 1991. Polypeptide composition of bacterial cyclic diguanylic acid-dependent cellulose synthase and the occurrence of immunologically crosreacting proteins in higher plants. Proc. Natl. Acad. Sci. 88:5472-5476.

Mayeux, J.V. 1992. Cytokin and Pix interaction increases cotton production. pp. 45-46. *In:* D.J. Herber and D.A. Richter (eds.). Proc. Beltwide Cotton Prod. Res. Conf., National Cotton Council of America, Memphis, Tenn.

Mayeux, J.V., V.L. Illum, and R.A. Beach. 1986. Influence of the bioregulator Burst® Yield Booster™ on boll load and yields of cotton. pp. 506-508. *In* T.C. Nelson (ed.). Proc. Beltwide Cotton Prod. Res. Conf., National Cotton Council of America, Memphis, Tenn.

Mayeux, J.V., V.L. Illum, and R.A. Beach. 1987. Effect of Burst® Yield Booster™ on cotton yield components and yields. pp. 76-78. *In:* T.C. Nelson (ed.). Proc. Beltwide Cotton Conf., National Cotton Council of America, Memphis, Tenn.

Mayeux, J.V. and J. Kautz. 1992. Cotton response to foliar application of the fruiting hormone cytokin: A four year study. pp. 1057-1058. *In:* T.C. Nelson (ed.). Proc. Beltwide Cotton Prod. Res. Conf., National Cotton Council of America, Memphis, Tenn.

McAinsh, M.R., H. Clayton, T.A. Mansfield, and A.M. Hetherington. 1996. Changes in stomatal behavior and guard cell cytosolic free calcium in response to oxidative stress. Plant Physiol. 111:1031-1042.

McAllister, C.B., I. Garcia-Romera, A. Godeas, and J.A. Ocampo. 1994. Interactions between *Trichoderma koningii, Fusarium solani* and *Glomus mosseae*: Effects on plant growth, arbuscular mycorrhizas and the saprophyte inoculants. Soil Biology & Biochemistry 26:1369-1374.

McBride, R.G., R.L. Mikkelsen, and K.R. Barker. 1999. Survival and infection of root-knot nematodes added to soil amended with rye at different stages of decomposition and cropped with cotton. Applied Soil Ecology 13:231-235.

McBryde, J.B. and H. Beal. 1896. The chemistry of cotton. *In:* A.C. True (ed.). Its history, botany, chemistry, and uses. USDA Office of Exp. Stn. Bull. 33. U.S. Gov. Print. Office, Washington, D.C. pp. 81-142.

McCabe, D.E. and B. J. Martinell. 1993. Transformation of elite cotton cultivars via particle bombardment of meristems. Biotechnology. 11:596-598.

McCall, P.J., T.C.J. Turlings, J. Loughrin, A.T. Proveaux, and J.H. Tumlinson. 1994. Herbivore-induced volatile emissions from cotton (*Gossypium hirsutum* L.) seedlings. J. Chem. Ecol. 20:3039-3050.

McCann, M.C. and K. Roberts. 1994. Changes in cell wall architecture during cell elongation. J. Exp. Bot. 45:1683-1691.

McCanny, S.J., P.A. Keddy, T.J. Arnason, C.L. Gaudet, D.R.J. Moore, and B. Shipley. 1990. Fertility and the food quality of wetland plants: A test of the resource availability hypothesis. Oikos 59:373-381.

McCarty, J.C., Jr. and P.A. Hedin. 1994. Effects of 1,1-dimethylpiperidinium chloride on the yields, agronomic traits, and allelochemicals of cotton (*Gossypium hirsutum* L.), a nine year study. J. Agric. Food. Chem. 42:2302-2304.

McCarty, J.C. Jr., P.A. Hedin, and R.D. Stipanovic. 1996. Cotton *Gossypium* spp. plant gossypol contents of selected GL2 and GL3 alleles. J. Agric. Food Chem. 44:613-616.

McCarty, W.H. and C. Baskin. 1997. Cotton: Understanding and using results of cottonseed germinations tests. Cooperative Extension Service, Mississippi State University, Mississippi State, Miss. Available at http://www.ces.msstate.edu/pubs/is1364.htm

McClelland, C.K. 1916. On the regularity of blooming in the cotton plant. Science (Washington, DC). 44:578-581.

McClung, A.C., L.M.M. De Freitas, D.S. Mikkelsen, and W.L. Lott. 1961. Cotton fertilization on Campo Cerrado soils, State of Sao Pauol, Brazil. IBEC Research. Institute. Bulletin. 27.

McClung, A.D., A.D. Carroll, and N.H. Battey. 1994. Identification and characterization of ATPase activity associated with maize (*Zea mays*) annexins. Biochem. J. 303:709-712.

McColl, A.L. and R.M. Noble. 1992. Evaluation of a rapid mass-screening technique for measuring antibiosis to *Helicoverpa* spp in cotton cultivars. Aust. J. Exp. Agric. 32:1127-1134.

McConnell, J.S., W.H. Baker, B.S. Frizzell, and J.J. Varvil. 1992. Response of cotton to nitrogen fertilization and early multiple applications of mepiquat chloride. J. Plant Nutr. 45:457-468.

McConnell, J.S., W.H. Baker, B.S. Frizzell, and J.J. Varvil. 1993. Nitrogen fertilization of cotton cultivars of differing maturity. Agron. J. 85:1151-1156.

McConnell, J.S., W.H. Baker, and R.C. Kirst, Jr. 1998. Yield and petiole nitrate concentrations of cotton treated with soil-applied and foliar-applied nitrogen. J. Cotton Sci. 2:143-152.

McConnell, J.S., W.H. Baker, D.M. Miller, B.S. Frizzell, and J.J. Varvil. 1993. Nitrogen fertilization of cotton cultivars of differing maturity. Agron. J. 85:1151-1156.

McConnell, J.S., B.S. Frizzell, R.L. Maples, M.H. Wilkerson, and G.A. Mitchell. 1989. Relationships

of irrigation methods and nitrogen rates in cotton production. Ark. Agri. Exp. Sta. Res. Ser. 310.

McConnell, J.S., B.S. Frizzell, and M.H. Wilkerson. 1989. Effects of soil compaction and subsoil tillage of two alfisols on the growth and yield of cotton. J. Prod. Agric. 2:140-146.

McConnell, J.S., R.E. Glover, E.D. Vories, W.H. Baker, B.S. Frizzell, and F.M. Bourland. 1995. Nitrogen fertilization and plant development of cotton as determined by nodes above white flower. J. Plt. Nutr. 18:1027-1036.

McCool, P.M. 1988. Effect of air pollutants on mycorrhizae. *In:* S. Schulte-Hostede, N.M. Darrall, L.W. Blank, and A.R. Wellburn (eds.). Air Pollutants and Plant Metabolism. Elsevier Appl. Sci. Publ., London. pp. 356-363.

McCool, P.M., J.A. Menge, and O.C. Taylor. 1982. Effect of ozone injury and light stress on response of tomato to infection by the vesicular-arbuscular mycorrhizal fungus, *Glomus fasciculatus*. J. Amer. Soc. Hort. Sci. 107:839-842.

McCully, M.E. 1994. Accumulation of high levels of potassium in the developing xylem elements of roots of soybean and some other dicotyledons. Protoplasma 183:1-4.

McDougall, G.J. and S.C. Fry. 1991. Xyloglucan nonasaccharide, a naturally occurring oligosaccharin, arises *in vivo* by polysaccharide breakdown. Plant. Physiol. 137:332-336.

McGarr, R.L. 1942. Relation of fertilisers to the developement of the cotton aphid. J. Econ. Entomol. 35:482-483.

McGartland, A.M. 1987. The implications of ambient ozone standards for U.S. agriculture: A comment and some further evidence. J. Environ. Manage. 24:139-146.

McGee, P.A., G.S. Pattinson, R.A. Heath, C.A. Newman, and S.J. Allen. 1997. Survival of propagules of arbuscular mycorrhizal fungi in soils in eastern Australia used to grow cotton. New Phytol. 135:773-780.

McGee, P.A., V. Torrisi, and G.S. Pattinson. 1999. The relationship between density of *Glomus mosseae* propagules and the initiation and spread of arbuscular mycorrhizas in cotton roots. Mycorrhiza 9:221-225.

McGonigle, T.P., D.G. Evans, and M.H. Miller. 1990. Effects of degree of soil disturbance on mycorrhizal colonization and phosphorous absorption by maize in growth chamber and field experiments. New Phytol. 116:629-636.

McGonigle, T.P. and M.H. Miller. 2000. The inconsistent effect of soil disturbance on colonization of roots by arbuscular mycorrhizal fungi: a test of the inoculum density hypothesis. Appl. Soil Ecol. 14:147-155.

McGovern, J.M. 1990. Fibers, Vegetable. pp. 412-430. *In:* J.I. Kroschwitz (ed.). Polymers: Fibers and Textiles, A Compendium. John Wiley, NY.

McHargue, J.S. 1926. Mineral constituents of the cotton plant. J. Am. Soc. Agron. 18:1076-1083.

McIntyre, G.I. 1997. The role of nitrate in the osmotic and nutritional control of plant development. Australian J Plant Physiol. 24: 103-118.

McKinion, J.M., D.N. Baker, J.D. Hesketh, and J.W. Jones. 1975. SIMCOT II: A simulation of cotton growth and yield. ARS-S-52, USDA, pp. 27-82.

McKinion, J.M. and D.N. Baker. 1982. Modeling, experimentation, verification, and validation: closing the feedback loop. Trans. ASAE. 25:647-653.

McKinion, J.M., D.N. Baker, F.D. Whisler, and J.R. Lambert. 1989. Application of the GOSSYM/COMAX system to cotton crop management. Agric. Syst. 31:29-33.

McLean, K.S., J.F. Denison, and G.W. Lawrence. 1994. Efficacy of in-furrow fungicides in cotton seedling survival and yield. pp. 254-255. *In:* Proc. Beltwide Cotton Conf., National Cotton Council of America, Memphis, Tenn.

McLaughlin, S.B. and R.K. McConathy. 1983. Effects of SO_2 and O_3 on allocation of ^{14}C-labeled photosynthate in *Phaseolus vulgaris*. Plant Physiol. 73:630-635.

McMichael, B.L. 1979. The influence of plant water stress on flowering and fruiting of cotton. *In:* Proc. Beltwide Cotton Prod. Res. Conf. National Cotton Council, Memphis, Tenn. pp. 301-302.

McMichael, B.L. 1986. Growth of roots. p. 29-38. *In:* J.R. Mauney and J.McD. Stewart (eds.). Cotton Physiology. The Cotton Foundation, Memphis, Tenn.

McMichael, B.L. 1990. Root-shoot relationships in cotton. pp. 232-251 *In:* J.E. Box and L.C. Hammond (eds.). Rhizosphere Dynamics. Westview Press.

McMichael, B.L. and J.J. Burke. 1994. Metabolic activity of cotton roots in response to temperature. Environ. Expl. Bot. 34:201-206.

McMichael, B.L., J.J. Burke, J.D. Berlin, J.L. Hatfield, and J.E. Quisenberry. 1985. Root vascular bundle arrangement among cotton strains and cultivars. Environ. and Exp. Bot. 25:23-30.

McMichael, B.L. and C.D. Elmore. 1977. Proline accumulation in water stressed cotton leaves. Crop Sci. 17:905-908.

McMichael, B.L. and G. Guinn. 1980. The effects of moisture deficits on square shedding. *In:* Proc. Beltwide Cotton Prod. Res. Conf. National Cotton Council, Memphis, Tenn. p. 38.

McMichael, B.L., J. Hacskaylo, and G.W. Cathey. 1974. The effects of nitrate versus ammonium nitrogen on the growth and development of cotton. Proc. Beltwide Cotton Prod. Res. Conf., Memphis, 39.

McMichael, B.L., W.R. Jordan, and R.D. Powell. 1973. Abscission processes in cotton: Induction by plant water deficit. Agron. J. 65:202-204.

McMichael, B.L. and J.E. Quisenberry. 1991. Genetic variation for root-shoot relationships among cotton germplasm. Environ. and Exp. Bot. 31:461-470.

McMichael, B.L. and J.E. Quisenberry. 1993. The impact of the soil environment on the growth of root systems. Environ. and Exp. Bot. 33:53-61.

McMichael, B.L., J.E. Quisenberry, and D.R. Upchurch. 1987. Lateral root development in exotic cottons. Environ. and Exp. Bot. 27:499-502.

McMichael, B.L., D.R. Upchurch, and J.J. Burke. 1996. Soil temperature derived prediction of root density in cotton. Environ. Expt. Botany 36:303-312.

McMillian, K.D. and A. Rikin. 1990. relationships between circadian rhythm of chilling resistance and acclimation to chilling in cotton seedlings. Planta 182:455-460.

McNaughton, S.J. 1983a. Physiological and ecological implications of herbivory. pp. 657-677 *In*: O.L. Lange, P.S. Nobel, C.B. Osmond, and H. Zeigler (eds.). Physiological Plant Ecology Responses to the Chemical and Biological Environment. Springer-Verlag, New York.

McNaughton, S.J. 1983b. Compensatory plant growth as a response to herbivory. Oikos 40:329-336.

McNaughton, S.J. 1986. On plants and herbivores. Am Nat 128:765-770.

McQueen-Mason, S.J. and D.J. Cosgrove. 1995. Expansion mode of action on cell walls. Analysis of wall hydrolysis, stress relaxation, and binding. Plant Physiol. 107:87-100.

McQueen-Mason, S. and D.J. Cosgrove. 1994. Disruption of hydrogen bonding between plant cell wall polymers by proteins that induce wall extension. Proc. Natl. Sci. USA 91:6574-6578.

McQueen-Mason, S., D.M. Durachko, and D.J. Cosgrove. 1992. two endogenous proteins that induce cell wall extension in plants. Plant Cell 4:1425-1433.

Megie, A.C., R.W. Pearson, and A.E. Hitbold. 1967. Toxicity of decomposing crop residues to cotton germination and seedling growth. Agron J. 59:197-199.

Mehetre, S.S. and M.V. Thombre. 1982. Note on the reaction of some promising mutants of upland cotton to the attack of bollworms - *Gossypium* spp., *Heliothis zea, Areas vitella*. Indian J. Agric. Sci. 52:474-476.

Meinert, M.C. and D.P. Delmer. 1977. Changes in biochemical composition of the cell wall of the cotton fiber during development. Plant Physiol. 59:1088-1097.

Meiri, A., H. Frenkel, and A. Mantell. 1992. Cotton response to water and salinity under sprinkler and drip irrigation. Agron. J. 84:44-50.

Meiri, A. and J. Shaihevet. 1973. Crop growth under saline conditions. *In:* B. Yaron, E. Danfors, and Y. Vaadia (eds.). Ecological studies, Vol.5. Arid zone irrigation. Springer-Verlag. Heidelberg. pp. 277-290.

Mellon, J.E. and B.A. Triplett. 1989. De novo synthesis of peroxidase in cotton ovule culture medium. Physiol. Plantarum 77:302-307.

Meng, Y.-L., J.-W. Jia, C.-J. Liu, W.-Q. Liang, P. Heinstein, and X.-Y. Chen. 1999. Coordinated accumulation of (+)-δ-cadinene synthase mRNAs and gossypol in developing seeds of *Gossypium hirsutum* and a new member of the *cad1* family from *G. arboreum*. J. Natural Products 62:248-252.

Mengel, D.B., J.R. Moncrief, and E.E. Schulte. 1992. Fertilizer management. pp.83-87. *In:* Conservation Tillage Systems and Management. MidWest Plan Serv. Pub. MWPS-45. Iowa State Univ., Ames, IA.

Mengel, K. 1985. Potassium movement within plants and its importance in assimilate transport. *In:* R.D. Munson (ed.). Potassium in agriculture. Amer. Soc. Agron. Madison, WI. pp. 397-411.

Mengel, K. and W.W. Arneke. 1982. Effect of potassium on the water potential, the osmotic potential and cell elongation in leaves of *Phaseolus vulgaris*. Physiol. Plant. 54:402-408.

Mengel, K. and E.A. Kirkby. 1987. Principles of plant nutrition, 4th ed. International Potash Institute, Switzerland.

Mengel, K., M.Th. Breininger, and W. Bubl. 1984. Bicarbonate, the most important factor inducing iron chlorosis in vine graps on calcareous soil. Plant Soil 81:333-344.

Menzel, M.Y. and M.S. Brown. 1954. The significance of multivalent formation in three-species *Gossypium* hybrids. Genetics 39:546-557.

Meredith, W.R., Jr. 1984. Influence of leaf morphology on lint yield of cotton - enhancement by the sub okra trait. Crop Sci. 24:855-857.

Meredith, W.R., Jr. 1985. Lint yield genotype x environment interaction in upland cotton as influenced by leaf canopy isolines. Crop Sci. 25:509-512.

Meredith, W.R. Jr. 1990. Yield and fiber-quality potential for second-generation cotton hybrids. Crop Sci. 30:1045-1048.

Meredith, W.R. 1991. Contributions of introductions to cotton improvement. Pp. 127-146 *In:* H.L. Shands and L.E. Wiesner (eds.). Use of Plant Introductions in Cultivar Development, Part 1. Crop Sci. Soc. Amer., Madison.

Meredith, W.R., Jr. 1992a. Fiber quality variation among U.S. cotton growing regions. pp. 105-106. *In:* J.M. Brown (ed.). Beltwide Cotton Production Research Conference. National Cotton Council, Memphis Tenn.

Meredith, W.R., Jr. 1992b. Improving fiber strength through genetics and breeding. pp. 289-302. *In:* C.R. Benedict (ed.). Proc. Cotton Fiber Cellulose: Structure, Function, and Utilization Conference. National Cotton Council, Memphis Tenn.

Meredith, W.R., Jr. and R.R. Bridge. 1972. Heterosis and gene action in cotton, *Gossypium hirsutum* L. Crop Sci. 12:304-310.

Meredith, W.R., Jr., W.T. Pettigrew, and J. J. Heitholt. 1996a. Sub-okra, semi-smoothness, and nectariless effect on cotton lint yield. Crop Sci. 36:22-25.

Meredith, W.R., Jr., P.E. Sasser, and S.T. Rayburn. 1996b. Regional High Quality fiber properties as measured by conventional and AFIS methods. pp. 1681-1684. *In:* Proc. Beltwide Cotton Conf., National Cotton Council of America, Memphis, Tenn.

Meredith, W.R., Jr. and R. Wells. 1989. Potential for increasing cotton yields through enhanced partitioning to reproductive structures. Crop Sci. 29:636-639.

Merkens, W.S.W., G.A. de Zoeten, and G. Gaard. 1972. Observations on ectodesmata and the virus infection process. J. Ultrastruct. Res. 41:397-405.

Merrill, S.D. and D.R. Upchurch. 1994. Converting root numbers observed at minirhiztrons to equivalent root length density. Soil Sci. Soc. Am. J. 58:1061-1067.

Mert, H.H. 1989. Photosynthesis and photorespiration in 2 cultivars of cotton under salt stress. Biol. Plant 31:413-414.

Mert, H.H. 1993. Investigation on the endogenous ABA levels of seeds of cotton cultivars under salt stress conditions. Doga, Turk Botanik Dergisi, 17:201-205. (Turkish).

Merzouki, M., N. Bernet, J.P. Delgenes, R. Moletta, and M. Benlemlih. 1999a. Kinetic behavior of some phosphate-accumulating bacteria isolates in the presence of nitrate and oxygen. Current Microbiology 38:300-308.

Merzouki, M., J.P. Delgenes, N. Bernet, R. Moletta, and M. Benlemlih. 1999b. Polyphosphate-accumulating and denitrifying bacteria isolated from anaerobic-anoxic and anaerobic-aerobic sequencing batch reactors. Current Microbiology 38:9-17.

Meyer, G.A. 1993. A comparisson of the impacts of leaf- and sap-feeding insects on growth and allocation of goldenrod. Ecology 74:1101-1116.

Meyer, G.A. and T.H. Whitlow. 1992. Effects of leaf and sap feeding insects on photosyntehtic rate of goldenrod. Oecologia 92:480-489.

Meyer, J.R. and R.G. Linderman. 1986. Selective influence on populations of rhizosphere or rhizoplane bacteria and actinomycetes by mycorrhizas formed by *Glomus fasciculatum*. Soil Biol. Biochem. 18:191-196.

Meyer, J.R. and V.G. Meyer. 1961. Origin and inheritance of nectariless cotton. Crop Sci. 1:167-169.

Meyer, R.F. and J.S. Boyer. 1972. Sensitivity of cell division and cell elongation to low water potentials in soybean hypocotyls. Planta (Berl.) 108:77-87.

Meyer, R., V. Slater, and I.A. Dubery. 1994. A phytotoxic protein-lipopolysaccharide complex produced by *Verticillium dahliae*. Phytochemistry 35:1449-1453.

Meyer, V.G. 1966. Environmental effects on the differentiation of abnormal cotton flowers. Amer. J. Bot. 53:976-980.

Meyer, V.G. 1969. Some effects of genes, cytoplasm, and environment on male sterility of cotton (*Gossypium*). Crop Sci. 9:237-242.

Meyer, V.G. 1974. Interspecific cotton breeding. Econ. Bot. 28:56-60.

Meyer, V.G. 1975. Male sterility from *Gossypium harknessii*. J. Hered. 66:23-27.

Meyer, V.G. and W.R. Meredith, Jr. 1978. New germplasm from crossing Upland cotton (*Gossypium hirsutum*) with *G. tomentosum*. J. Hered. 69:183-187.

Meyer, W.S. and H.D. Barrs. 1985. Non-destructive measurement of wheat roots in large undisturbed and repacked clay soil cores. Plant and Soil 85:237-247.

Meyer, W.S., D.C. Reicosky, H.D. Barrs, and R.C.G. Smith,1987. Physiological responses of cotton to a single waterlogging at high and low N-levels. Plant and Soil 102:161-170.

Meyers, A.I. and J.J. Willemsen. 1998. An oxazoline based approach to (S)-gossypol. Tetrahedron 1998:10493-10511.

Michaelson, M.J., H.J. Price, J.R. Ellison, and J.S. Johnston. 1991. Comparison of plant DNA contents determined by Feulgen microspectrophotometry and laser flow cytometry. Am. J. Bot. 78:183-188.

Micinski, S., P.D. Colyer, K.T. Nguyen, and K.L. Koonce. 1990. Effects of planting date and early-season pest control on yield in cotton. J. Prod. Agric. 3:597-602.

Micklin, P.P. 1988. Desiccation of the Aral Sea - A water management disaster in the Soviet Union. Science 241:1170-1175.

Micklin, P.P. 1994. The Aral Sea problem. Proc. Inst. Civil Engineers - Civil Engineering 102:114-121.

Middleton, L.J. and J. Sanderson. 1965. The uptake of inorganic ions by plant leaves. J. Exp. Bot. 16:197-15.

Miernyk, J.A. 1989. Glycolysis, the oxidative pentose phosphate pathway and anaerobic respiration. pp. 77-100. *In:* D.T. Dennis and D.H. Turbin (eds.).

Plant Physiology, Biochemistry, and Molecular Biology. Longman Singapore.

Mikkelsen, D.S., L.M.M. de Freitas, and A.C. McClung. 1963. Effects of liming and fertilizing cotton, corn and soybeans on Campo Cerrado soils, State of Sao Paulo, Brazil. IRI Research Institute Bulletin 271. New York.

Miley, W.N., G.W. Hardy, M.B. Sturgis, and J.E. Sedberry. 1969. Influence of boron, nitrogen, and potassium on yield, nutrient uptake, and abnormalities of cotton. Agron. J. 61:9-13.

Miley, W.N. and R.L. Maples. 1988. Cotton nitrate monitoring in Arkansas. *In:* D.M. Oosterhuis (ed.). Proc. Cotton Research Meeting. Univ. of Ark. Agri Exp. Sta. Special Report 132:15-21.

Miley, W.H., H.J. Mascagni, J.S. McConnell, and R.L. Maples. 1990. Sulfur for cotton. Univ. of Arkansas Coop. Ext. Ser., Little Rock, AR. Cotton Comments 1-90.

Miley, W.N. and D.M. Oosterhuis. 1989. Nitrogen and carbohydrate nutrition of cotton. University of Arkansas, Cooperative Extension Service, Little Rock. Fact Sheet. 2045. p 4.

Miley, W.N. and D.M. Oosterhuis. 1994. Three-year comparison of foliar feeding of cotton with five potassium sources. Proc. Beltwide Cotton Production Conf., San Diego, National Cotton Council, Memphis, Tenn. pp. 1534-1536.

Millar, A.A. and E.S. Dennis. 1996. The alcohol dehydrogenase genes of cotton. Plant Mol. Biol. 31:897-904.

Millar, A.A., M.R. Olive, and E.S. Dennis. 1994. The expression and anaerobic induction of alcohol dehydrogenase in cotton. Biochem. Genet. 32:279-300.

Miller, A.J. 1998. Molecular intrigue between phototransduction and the circadian clock. Ann. Bot. 81:581-587.

Miller, J.E. 1988. Effects on photosynthesis, carbon allocation, and plant growth associated with air pollutant stress. *In:* W.W. Heck, O.C. Taylow, and D.T. Tingey (eds.). Assessment of Crop Loss from Air Pollutants. Elsevier Appl. Sci., London. pp. 287-314.

Miller, J.E., R.P. Patterson, W.A. Pursley, A.S. Heagle, and W.W. Heck. 1989. Response of soluble sugars and starch in field-grown cotton to ozone, water stress, and their combination. Environ. Exp. Bot. 29:477-486.

Miller, J.K., D.K. Krieg, and P.E. Peterson. 1996. Relationships between dryland cotton yields and weather parameters on the Southern High Plains. *In:* Proc. Beltwide Cotton Prod. Res. Conf. National Cotton Council, Memphis, Tenn. pp. 1165-1166.

Miller, J.E., R.P. Patterson, A.S. Heagle, W.A. Pursley and W.W. Heck. 1988. Growth of cotton under chronic ozone stress at two levels of soil moisture. J. Environ. Qual. 17:635-643.

Miller, J.E., R.P. Patterson, W.A. Pursley, A.S. Heagle, and W.W. Heck. 1989. Response of soluble sugars and starch in field-grown cotton to ozone, water stress, and their combination. Environ. Exp. Bot. 29:477-486.

Miller, M.E. and P.S. Chourey. 1992. The maize invertase-deficient *minature-1* seed mutation is associated with aberrant pedicel endosperm development. Plant Cell 4:297-305.

Miller, R.H. 1982. Apple fruit cuticle and the occurrence of pores and transcuticular channels. Ann Bot. 50:355-360.

Miller, R.H. 1985. The prevalence of pores and canals in leaf cuticular membranes. Ann Bot. 55:459-471.

Millhollon, E.P., D.R. Gossett, M.C. Lucas, M.M. Marney, and T.M. Moreau. 1993. Varietal response of cotton plants and corresponding cell cultures to NaCl stress. Proc. Beltwide Cotton Conf., New Orleans, 1993, 1281-1282.

Milnera, S.M. 1987. Colorimetric determination of cotton maturity. Canadian Textile J. 104:26-27.

Minaei, S. and C.G. Coble. 1989. Effect of oxygen on germination and respiration of MAR cotton. ASAE Paper No. 89-1558. American Society of Agricultural Engineers, St. Joseph, Mich.

Minaei, S. and C.G. Coble. 1990. Cotton seed germination as affected by carbon dioxide and oxygen. ASAE Paper No. 90-1531. American Society of Agricultural Engineers, St. Joseph, Mich.

Minashina, N.G. 1996. Soil environmental changes and soil reclamation problems in the Aral Sea basin. Eurasian Soil Sci. 28:184-195.

Minton, E.B. 1980. Effects of row spacing and cotton cultivars on seedling diseases, Verticillium wilt, and yield. Crop Sci. 20:347-350.

Minton, E.B. and M.W. Ebelhar. 1991. Potassium and aldicarb-disulfoton effects on verticillium wilt, yield, and quality of cotton. Crop Sci. 31:209-212.

Minton, E.B. and J.R. Supak. 1980. Effects of seed density on stand, Verticillium wilt, and seed and fiber characters of cotton. Crop Sci. 20:345-347

Minyard, J.P., A.C. Thompson, and P.A. Hedin. 1968. Constituents of the cotton bud. VIII. β-Bisabolol, a new sesquiterpene alcohol. J. Org. Chem. 33:909-911.

Minyard, J.P., J.H. Tumlinson, and P.A. Hedin. 1965. Constituents of the cotton bud. Terpene hydrocarbons. J. Agric. Good. Chem. 13:599-602.

Minyard, J.P., J.H. Tumlinson, A.C. Thompson, and P.A. Hedin. 1966. Constituents of the cotton bud. Sesquiterpene hydrocarbons. J. Agric. Food. Chem. 14:332-336.

Minyard, J.P., J.H. Tumlinson, A.C. Thompson, and P.A. Hedin. 1967. Constituents of the cotton bud. The carbonyl compounds. J. Agric. Food Chem. 15:517-525.

Misaghi, I.J., P.J. Cotty, and D.M. Decianne. 1995. Bacterial antagonists of *Aspergillus flavus*. Biocontrol Science & Technology 5:387-392.

Misaghi, I.J., J.E. DeVay, and J.M. Duniway. 1978. Relationships between occlusion of xylem elements and disease symptoms in leaves of cotton plants infected with *Verticillium dahliae*. Canadian J. Botany 56:339-342.

Misra, R.D. 1973. Responses of corn to different sequences of water stress as measured by evapotranspiration deficits. Ph.D. diss. Univ. California, Davis (Diss. Abstr. 73:32288).

Mistric, J.R. 1968. Effects of Nitrogen fertilization on cotton under boll weevil attack in North Carolina. J. Econ. Entomol. 61:282-283.

Mitchell, C.C., F.J. Arriaga, and D.A. Moore. 1995a. Sixty years of continuous fertilization in central Alabama. Proc. Beltwide Cotton Conferences, San Diego, CA. pp. 1340-1344.

Mitchell, C.C., G.L. Mullins, and C. H. Burmester. 1995b. Soil K levels and response to foliar KNO_3 at five locations. Proc. Beltwide Cotton Conf., San Diego, CA. pp. 1321-1324.

Mitten, D.H. 1985. Somatic embryogenesis in *Gossypium hirsutum* L. In Vitro 9:395-398.

Mizuno, K. 1993. Induction of cold stability of microtubules in cultured tobacco cells. Plant Physiol. 100:740-748.

Mizuno, K. 1994. Inhibition of gibberellin-induced elongation, reorientation of cortical microtubules and change of isoform of tubulin in epicotyl segments of azuki bean by protein kinase inhibitors. Plant Cell Physiol. 35:1149-1157.

Mobley, J.B. and D.W. Albers. 1993. Evaluation of cotton growth in ridge till systems in southeast Missouri. pp. 508-509. *In:* Proc. Beltwide Cotton Conf., National Cotton Council of America, Memphis, Tenn.

Mohamed, M.E.S., E.H. El-Haddad, and E.A. Mashaly. 1997. Cotton yield and fiber quality in relation to soil water-table and irrigation frequency. Egyptian J. Soil Sci. 37:251-266.

Mok, M.C., D.W.S. Mok, D.J. Armstrong, K. Shudo, Y. Isogai, and T. Okamoto. 1982. Cytokinin activity of N-phenyl-N-1,2,3,-thiadiazol-5-5yl urea (thidiazuron). Phytochem. 21:1509-1511.

Molin, W.T. and R.A. Khan. 1996. Differential tolerance of cotton (*Gossypium* sp.) cultivars to the herbicide prometryn. Pestic. Biochem. Physiol. 56:1-11.

Molina, R., H. Massicotte, and J.M. Trappe. 1992. Specificity phenomena in mycorrhizal symbioses: community-ecological consequences and practical implications. p. 357-423. *In:* M.F. Allen (ed.).

Mycorrhizal Functioning: an Integrative Plant-Fungal Process. Chapman and Hall, New York.

Momtaz, O.A., M.M. El Baghdady, A.E. Elawady, and M.A. Madkour. 1995. Salt-induced protein expression in some commercial Egyptian cotton varieties. Proc. Beltwide Cotton Conf., San Antonio, TX, USA, January 4 7, 1995: Volume 1 500-503.

Montalvo, J.G., Jr. 1991a. A comparative study of NIR diffuse reflectance of cottons grouped according to fiber cross-section dimensions. Part I. Fundamentals. Appl. Spectroscopy 45:779-794.

Montalvo, J.G., Jr. 1991b. A comparative study of NIR diffuse reflectance of cottons grouped according to fiber cross-section dimensions. Part II. Optical path simulations. Appl. Spectroscopy 45:790-794.

Montalvo, J.G., Jr. 1991c. A comparative study of NIR diffuse reflectance of cottons grouped according to fiber cross-section dimensions. Part III. Experimental. Appl. Spectroscopy 45:795-807.

Montalvo, J.G., Jr., S.E. Faught, and S.M. Buco. 1989. An investigation of the relationship between cotton fineness and light-scattering from thin webs as measured by near-infrared transmission spectroscopy. Applied Spectroscopy 43:1459-1471.

Montalvo, J., S. Faught, and R. Grimball. 1996. Determination of maturity/fineness by FMT and diode-array HVI. Part 1. FMT (Micromat model) procedure optimization. pp. 1276-1278. *In:* Proc. Beltwide Cotton Conf., National Cotton Council of America, Memphis, Tenn.

Montandon, R., J.E. Slosser, and L.E. Clark. 1994. Late-season termination effects on cotton fruiting, yield, and boll weevil (Coleoptera Curculonidae) damage in Texas dryland cotton. J. Econ. Entomol. 87:1647-1652.

Montandon, R., R.D. Stipanovic, H.J. Williams, W.L. Sterling, and S.B. Vinson. 1987. Nutritional indices and excretion of gossypol by *Alabama argillacea* (Hubner) and *Heliothis virescens* (F.) (Lepidoptera: Noctuidae) fed glanded and glandless cotyledonary cotton leaves. J. Econ. Entomol. 80:32-36.

Montheith, J.L. 1977. Climate and efficiency of crop production in Britain. Phil. Trans. Royal Soc. London, Series B 281:277-294.

Monteith, J.L. and M.H. Unsworth. 1990. Principles of environmental physics. Edward Arnold, London. 291 pp.

Montezinos, D. and D.P. Delmer. 1980. Characterization of inhibitors of cellulose synthesis in cotton fibers. Planta 148:305-311.

Mooney, H.A., W.E. Winner, E.J. Pell, and E. Chu. 1991. Response of plants to multiple stresses. Academic Press, New York .

Moore, J.F. 1996. Cotton Classification and Quality. pp. 51-57. *In:* E.H.Glade, Jr., L.A.Meyer, and H.Stults (eds.). The Cotton Industry in the United States. U.S. Dept. of Agriculture, Agricultural Economic Research Service, Agricultural Economic Report No. 739. USDA, ERS, Washington, DC.

Moore, S.H. and G.A. Breitenbeck. 1993. Nitrogen rate effect on yield and nutrient uptake in cotton. Proc. Beltwide Cotton Conf., New Orleans, LA. pp. 1333-1334.

Mor, U. 1997. FCT- a system for defining different levels and profiles of stickiness and its connection to other contaminants such as seed coat fragments. pp. 1639-1642. *In:* Beltwide Cotton Prod. Res. Conf., National Cotton Council of America, Memphis, Tenn.

Moraghan, J.T. 1979. Manganese toxicity in flax growing on certain calcareous soils low in available iron. Soil Sci. Soc. J. 43:1177-1180.

Moraghan, B.J., J. Hesketh, and A. Low. 1968. Effects of temperature and photoperiod among strains of cotton. Cotton Grow. Rev.45:91-100.

Moreno, F., F. Cabera, L. Andreu, R. Vaz, J. Martinaranda, and G. Vachaud. 1995. Water movement and salt leaching in drained and irrigated marsh soils of southwest Spain. Agric. Water Manage. 27:25-44.

Moreno, F., E. Fernandez-Boy, F. Cabrera, J.E. Fernandez, M.J. Palomo, I.F. Giron, and B. Bellido. 1998. Irrigation with saline water in the reclaimed marsh soils of southwest Spain: Impact on soil properties and cotton crop. *In:* R. Ragab (ed.). The use of saline and brackish water for irrigation. Implications for the management of irrigation, drainage and crops. Proc. Intern. Workshop Tenth ICID Afro-Asian Regional Conf. on Irrigation and Drainage. Denpasas, Bali, Indonesia, 19-26 July, 1998. 51-58; Pearce-G. Indonesian National Committee on Irrigation and Drainage (INACID), Directorate General of Water Resources Development, Ministry of Public Works; Jakarta; Indonesia.

Moreno, F., B.E. Fernandez, F. Cabrera, J.E. Fernandez, M.J. Palomo, I.F. Giron, and B. Bellido. 2000. Irrigation with saline water in the reclaimed marsh soils of southwest Spain:impact on soil properties and cotton crop. The use of saline and brackish water for irrigation drainage and crops. Proc. International Workshop at the Tenth ICID Afro-Asian Regional Conference on Irrigation and Drainage, Denpasas, Bali, Indonesia, 19-26 July, 1998. 1998, 51-58; 4 ref., Denpasas, Bali, Indonesia.

Morgan, J.M. 1980. Osmotic adjustment in the spikelets and leaves of wheat. Exp. Bot. 31:655-665.

Morgan, J.M. 1984. Osmoregulation and water stress in higher plants. Ann. Rev. Plant Physiol. 35:299-319.

Morgan, J.M. and Condon. 1986. Water use, grain yield and osmoregulation in wheat. Aust. J. Plant Physiol. 13:523-532.

Morgan, P.M., H.E. Joham, and J. Amin. 1966. Effect of manganese toxicity on the Indoleacetic acid oxidase system of cotton. Plant Physiol. 41:718-724.

Morgan, P.M., D.M. Taylor, and H.E. Joham. 1976. Manipulation of IAA-oxidase activity and auxin deficiency symptoms in intact cotton plants with manganese nutrition. Physiol. Plant. 37:149-156.

Morgan, P.W. 1967. Plant growth control by ethylene. pp. 151-155. 1967 Proc. Beltwide Cotton Prod Res. Confs. National Cotton Council of America, Memphis, Tenn.

Morgan, P.W. and H.W. Gausman. 1966. Effects of ethylene on auxin transport. Plant Physiol. 41:45-52.

Morgan, P.W. and W.C. Hall. 1964. Accelerated release of ethylene by cotton following application of indole-3-acetic acid. Nature 201:99.

Morgan, P.W., C. He, and M.C. Drew. 1992. Intact leaves exhibit a climacteric-like rise in ethylene production before abscission. Plant Physiol. 100:1587-1590.

Moricca, S., A. Ragazzi, T. Kasuga, and K.R. Mitchelson. 1998. Detection of *Fusarium oxysporum* f.sp. *vasinfectum* in cotton tissue by polymerase chain reaction. Plant Pathology (Oxford) 47:486-494.

Morris, D.A. 1965. Photosynthesis by the capsule wall and bracteoles of the cotton plant. Cotton Grow. Rev. 42:49-51.

Morrison, J.E. and F.W. Chichester. 1994. Tillage system effects on soil and plant nutrient distributions on vertisols. J. Prod. Agric. 7:364-373.

Morrison, K.D. and E.G. Reekie. 1995. Pattern of defoliation and its effects on photosynthetic capacity in *Oenothera biennis*. J. Ecol. 83:759-767.

Morrison, T.A., J.R. Kessler, and D.R. Buxton. 1994. Maize internode elongation patterns. Crop Sci. 34:1055-1060.

Morton, J.B. and G.L. Benny. 1990. Revised classification of arbuscular mycorrhizal fungi (Zygomycetes): A new order, Glomales, two new suborders, Glomineae and Gigasporineae, and two new families, Acaulosporaceae and Gigasporaceae, with an emendation of Glomaceae. Mycotaxon 37:471-491.

Moseley, G. 1983. Variations in the epicuticular wax content of white and red clover leaves. Grass Forage Sci. 38:201-204.

Moss, J.M. and C.W. Bednarz. 1999. Compensatory growth after early season fruit removal in cotton. p. 524. *In:* P. Dugger and D.A. Richter (eds.). Proc. Beltwide Cotton Prod. Res. Conf., National Cotton Council of America, Memphis, Tenn.

Moustafa, A.T.A., M.E. Ibrahim, and H.K. Bakhati. 1975. Effect of depth of saline water table on cotton yield and its consumptive use. Agric. Res. Rev. 53:21-24.

Mudd, J.B. 1975. Sulfur dioxide. *In:* J.B. Mudd and T.T. Kozlowski (eds.). Responses of Plants to Air Pollution. Academic Press, N.Y. pp. 9-22.

Mueller, W.C. and C.H. Beckman. 1976. Ultrastructure and development of phenolic-storing cells in cotton roots. Canadian J. Botany 54:2074-2082.

Mueller, W.C. and A.T. Morgham. 1993. Ultrastructure of the vascular response of cotton to *Verticillium dahliae*. Canadian J. Botany 71:32-36.

Mueller, W.C. and C.H. Beckman. 1978. Ultrastructural localization of polyphenoloxidase and peroxidase in roots and hypocotyls of cotton seedlings. Can. J. Bot. 56:1579-1587.

Mueller, W.C., A.T. Morgham, and E.M. Roberts. 1994. Immunocytochemical localization of callose in the vascular tissue of tomato and cotton plants infected with *Fusarium oxysporum*. Canadian J. Botany 72:505-509.

Muhammed, S. M.I. and Makhdum. 1973. Effect of soil salinity on the composition of oil and amino acids and on the oil content of sunflower seed. Pakistan J. Agric. Sci. 10:71-76.

Mühling, K.H. and A. Läuchli, A. 2002. Determination of apoplastic Na in intact leaves of cotton by *in vivo* fluorescence ratio imaging. Functional Plant Biology 29:1491-1499.

Mukhanedkhanova, F.S., M.P. Muminova, and A.A. Abdukarimov. 1991. Effect of salt stress on callus cultures of cotton. 1 S"ezd fiziologov rastenii Uzbekistana, Tashkent, 1991, 128. (Russian).

Mullen, M.B. and D.O. Howard. 1992. Vertical and horizontal distribution of soil C, N, P, K, and pH in continuous no-tillage corn production. pp. 6-10. *In:* Proc. South. Conserv. Tillage Conf. Sust. Agric. Tenn. Agric. Exp. Sta. Spec. Pub. 92-01. Univ. of Tenn., Knoxville, Tenn.

Muller, E. 1968. Getting a stand. pp. 6-7. *In:* Proc. West. Cotton Prod. Res. Conf., Cotton Council, Memphis, Tenn.

Muller, J. 1981. Fossil pollen records of extant angiosperms. Bot. Rev. 47:1-142.

Muller, J. 1984. Significance of fossil pollen for angiosperm history. Ann. Missouri Bot. Gard. 71:419-443.

Muller, L.L. and T.J. Jacks. 1983. Intracellular distribution of free sugars in quiescent cottonseed. Plant Physiol. 71:703-704.

Mullins, G.L. 1993. Cotton root growth as affected by P fertilizer placement. Fert. Res. 34:23-26.

Mullins, G.L. and C.H. Burmester. 1991. Dry matter, nitrogen, phosphorus, and potassium accumulation by four cotton varieties. Agron. J. 82:729-736.

Mullins, G.L. and C.H. Burmester. 1992. Uptake of calcium and magnesium by cotton grown under dryland conditions. Agron. J. 84:564-569.

Mullins, G.L. and C.H. Burmester. 1993a. Accumulation of copper, iron, manganese and zinc by four cotton cultivars. Field Crops Res. 32:129-140.

Mullins, G.L. and C.H. Burmester. 1993b. Uptake of sulfur by four cotton cultivars grown under field conditions. J. Plant Nutr. 16(6):1071-1081.

Mullins, G.L. and C.H. Burmester. 1994.. Cotton response to the source of foliar potassium. pp. 1537-1539. *In:* Proc. Beltwide Cotton Conferences. National Cotton Council, Memphis, Tenn.

Mullins, G.L. and C.H. Burmester. 1995. Response of cotton to the source of foliar potassium. Proc. Beltwide Cotton Conf., San Diego, CA. pp. 1313-1315.

Mullins, G.L. and C.H. Burmester. 2009. Relation of growth and development to mineral nutrition. pp. 98-106. *In:* J.M.Stewart, D.M. Oosterhuis, J.J. Heitholt, and J.R. Mauney (eds.). Physiology of Cotton. National Cotton Council of America, Memphis,Tenn. pp. Springer, London.

Mullins, G.L., C.H. Burmester, and D.W. Reeves. 1992a. Cotton response to surface and deep placement of potassium fertilizer. Proc. Beltwide Cotton Conf., Nashville, Tenn. pp. 77-79.

Mullins, G.L., D.W. Reeves, C.H. Burmester, and H.H. Bryant. 1992b. Effects of subsoiling and the deep placement of K on root growth and soil water depletion by cotton. Proc. Beltwide Cotton Conf., Nashville, Tenn. pp. 1134-1138.

Mullins, G.L.; D.W. Reeves, C.H. Burmester, and H.H. Bryant. 1994. In-row subsoiling and potassium placement effects on root growth and potassium content of cotton. Agron. J. 86:136-139.

Mulrooney, J.E., P.A. Hedin, W.L. Parrott, and J.N. Jenkins. 1985. Effects of PIX, a plant growth regulator, on allelochemical content of cotton and growth of tobacco budworm larvae (Lepidoptera: Noctuidae). J. Econ. Entomol. 78:1100-1104.

Multani, D.S. and B.R. Lyon. 1995. Genetic fingerprinting of Australian cotton cultivars with RAPD markers. Genome 38:1005-1008.

Munk, D.S. and B. Roberts. 1995. Growth and development of Pima and Acala cotton on saline soils. Proc. Beltwide Cotton Conferences, San Antonio. 1995 1:90-92.

Munk, D.S. and J. Wroble. 1995. Irrigation management for increased conjunctive use of a shallow saline water table. Proc. Beltwide Cotton Conf., San Antonio. 1995, 2:1372-1374.

Munns, R. 1993. Physiological processes limiting plant growth in saline soils - Some dogmas and hypotheses. Plant Cell Environ. 16:15-24.

Munns, R. and A. Termaat. 1986. Whole-plant responses to salinity. Aust. Plant Physiol. 13:143-60.

Munoz, S.C. 1994. Nutrient uptake and plant physiology enhancement by PHCA treatments on cotton plants. pp. 1360-1363 *In:* 1994 Proc. Beltwide Cotton Conferences, National Cotton Council of America, Memphis, Tenn.

Munro, J.M. 1971. An analysis of earliness in cotton. Cotton Grow. Rev. 48:28-41.

Munro, J.M. 1987. Cotton. 2nd. Ed. Longman-Wiley, New York.

Munro, J.M. and H.G. Farbrother 1969. Composite plant diagrams in cotton. Cotton Growing Review 46:261-282.

Muramoto, H., J. Hesketh, and M. El-Sharkawy. 1965. Relationships among rate of leaf area development, photosynthetic rate, and rate of dry matter production among American cultivated cottons and other species. Crop Sci. 5:163-166.

Murray, A.K. 1996. The use of glycoconjugate analysis to monitor growth and environmental stress in developing cotton fibers. pp. 1255-1259. *In:* Proc. Beltwide Cotton Conf., National Cotton Council of America, Memphis, Tenn.

Murray, A.K. and J. Brown. 1996. Glycoconjugate analysis of developing cotton fibers from several varieties grown on the same site. pp. 1205-1209. *In:* Proc. Beltwide Cotton Conf., National Cotton Council of America, Memphis, Tenn.

Murray, A.K. and J. Brown. 1997. Glycoconjugate profiles of developing fibers from different fruiting branches on the same plant. pp. 1496-1499. *In:* Proc. Beltwide Cotton Conf., National Cotton Council of America, Memphis, Tenn.

Murray, F.R., D.J. Llewellyn, W.J. Peacock, and E.S. Dennis. 1997. Isolation of the glucose oxidase gene from *Talaromyces flavus* and characterization of its role in the biocontrol of *Verticillium dahliae*. Current Genetics 32:367-375.

Murray, F., D. Llewellyn, H. McFadden, D. Last, E.S. Dennis, and W.J. Peacock. 1999. Expression of the *Talaromyces flavus* glucose oxidase gene in cotton and tobacco reduces fungal infection, but is also phytotoxic. Molecular Breeding 5:219-232.

Murty, P.S., D.N. Ragu, and G.V. Rao. 1976. Effect of plant growth regulators on flower and boll drop in cotton. pp. 9-12. Food Farming Agric.

Musselman, R.C. and W.J. Massman. 1999. Ozone flux to vegetation and its relationship to plant response and ambient air quality standards. Atmos. Environ. 33:65-73.

Musselman, R.C., P.M. McCool, R.J. Oshima, and R.R. Teso. 1986. Field chambers for assessing crop loss to air pollutants. J. Environ. Quality 15:152-157.

Mutchler, C.K., L.L. McDowell, and J.D. Greer. 1985. Soil loss from cotton with conservation tillage. Trans. ASAE 28:160-163.

Muthuchamy, I. and K. Valliappan. 1993. Salt dynamics in the root zone. Madras Agric.J. 80:51-52.

Mutsaers, H.J. 1976. Growth and assimilaate conversion of cotton bolls (*Gossypium hirsutum* L.). I. Growth of fruits and substrate demand. Ann. Bot. 40:300-315.

Mutsaers, H.J.W. 1980. The effect of row orientation, date and latitude on light absorption by row crops. J. Agric. Sci. 95:381-386.

Mutsaers, H.J.W. 1983a. Leaf growth in cotton (*Gossypium hirsutum* L.). 1. Growth in area of mainstem and sympodial leaves. Ann. Bot. 51:503-520.

Mutsaers, H.J.W. 1984. KUTIN: a morphogenetic model for cotton (*Gossypium hirsutum* L.). Agric. Sys. 14:229-257.

Myers, M.P., E. Wager-Smith, A. Rothenfluh-Hilfiker, and M.W. Young. 1996. Light-induced degradation of TIMELESS and entrainment of the *Drosophila* circadian clock. Sci. 271: 1736-1740.

Nable, R.O., G.S. Banuelos, and J.G. Paull. 1997. Boron toxicity. Plant and Soil 193:181-198.

Nable, R.O. and J.F. Loneragan. 1984. Translocation of manganese in subterranean clover (*Trifolium subterraneum* L. cv. Seaton Park). II. Effects of leaf senescence and of restricting supply to part of a split root system. Austral. J. Plant Physiol. 11:113-118.

Nadarajan, N. and S.R.S. Rangasamy. 1988. Inheritance of the fuzzless-lintless character in cotton (*Gossypium hirsutum*). Theor. Appl. Genet. 75:728-730.

Nadian, H., S.E. Smith, A.M. Alston, R.S. Murray, and B.D. Siebert. 1998. Effects of soil compaction on phosphorus uptake and growth of *Trifolium subterraneum* colonised by four species of vesicular-arbuscular mycorrhizal fungi. New Phytol. 139:155-165.

Nagarajah, S. 1975. The relation between photosynthesis and stomatal resistance of each leaf surface in cotton leaves. Physiol Plant 34:62-66.

Nagarajah, S. 1976. The effects of increased illumination and shading on the low-light-induced decline in photosynthesis in cotton leaves. Physiol. Plant. 36:338-342.

Nagarajah, S. 1978. Some differences in the responses of stomata of the two leaf surfaces in cotton. Ann. Bot. 42:1141-1147.

Nagy, J., K.F. Lewin, G.R. Hendry, E. Hassinger, and R.L. LaMorte. 1994. FACE facility CO_2 concentration control and CO_2 use in 1990 and 1991. Agric. For. Met. 70:31-48.

Naidoo, G., J. McD. Stewart, and R.J. Lewis. 1978. Accumulation sites of Al in snap bean and cotton roots. Agron. J. 70:489-492.

Naithani, S.C. 1987. The role of IAA oxidase, peroxidase and polyphenol oxidase in the fiber initiation on the cotton ovules. Beitr. Biol. Pflanzen. 62:79-90.

Naithani, S.C., N. Rama Rao, P.N. Krishnan, and V.D. Singh. 1981. Changes in o-diphenol oxidase during fiber development in cotton. Ann. Bot. 48:379-385.

Naithani, S.C., N. Rama Rao, and Y.D. Singh. 1982. Physiological and biochemical changes associated with cotton fiber development. I. Growth kinetics and auxin content. Physiol. Plant. 54:225-229.

Nakamura, Y., H. Sawada, S. Kobayashi, I. Nakajima, and M. Yoshikawa. 1999. Expression of soybean β-1,3-endoglucanase cDNA and effect on disease tolerance in kiwifruit plants. Plant Cell Reports18:527-532.

Namken, L.N. 1984. Effect of cytokinin on yield components, fiber quality and yield enhancement of cotton. p. 67 *In:* J.M. Brown (ed.). Proc. Beltwide Cotton Prod. Res. Conf., National Cotton Council of America, Memphis, Tenn.

Namken, L.N. and H.W. Gausman. 1978. Practical aspects of chemical regulation of cotton plant growth and fruiting. pp. 23-25. *In:* J.M. Brown (ed.). Proc. Beltwide Cotton Prod. Mech. Conf., National Cotton Council of America, Memphis, Tenn.

Namken, L.N. and E.G. King. 1991. Cotton fruiting response to early season ethephon applications. pp. 1019-1023. *In:* D.J. Herber and D.A. Richter (eds.). Proc. Beltwide Cotton Conf., National Cotton Council of America, Memphis, Tenn.

Nap, J.P., L. Mlynárová, and W.J. Stiekema. 1996. From transgene expression to public acceptance of transgenic plants: a matter of predictibility. Field Crops Res. 45:5-10.

Narayanan, S.S., J. Singh, and P.K. Varma. 1984. Introgressive gene transfer in *Gossypium*. Goals, problems, strategies and achievements. Cot. Fib. Trop. 39:123-135.

Narbuth, E.V. and R.J. Kohel. 1990. Inheritance and linkage analysis of a new fiber mutant in cotton. J. Hered. 81:131-133.

Naresh, R.K., P.S. Minhas, A.K. Goyal, C.P.S. Chauhan, and R.K. Gupta. 1993. Conjunctive use of saline and non-saline waters. II. Field comparisons of cyclic uses and mixing for wheat. Agric. Water Manage. 23:139-148.

Navasero, R.C. and S.B. Ramaswamy. 1991. Morphology of leaf surface trichomes and its influence on egglaying by *Heliothis virescens*. Crop Sci. 31:342-353.

Navrezova, N., M. Agzamova, N.N. Stepanichenko, and B. Makhsudova. 1986. Effect of the flavonoids quercetin 3-rutinoside and kaempferol 3,7-dirharnnoside on the biosynthesis of melanin in *Verticillium dahliae*. Khim. Prir. Soedin. 2:244-245

Nawar, M.T. 1995. Effect of trash content on micronaire reading and color measurements by high speed instrument. pp. 1171-1172. *In:* Proc. Beltwide Cotton Conf., National Cotton Council of America, Memphis, Tenn.

Nawar, M.T., A.M. Zaher, K. El-Sahhar, and S.A. Abdel-Rahim. 1994. Influence of salinity on reproductive growth of three Egyptian cotton cultivars. *In:* G.A. Constable and N.W. Forrester (eds.). Challenging the Future, Proc. World Cotton Res. Conf. Brisbane 1994. CSIRO, Melbourne. pp. 109-114.

Nawar, M.T., A.M. Zaher, K. El Sahhar, and S.A. Abdel-Rahim. 1995. Effect of salinity on botanical characters and fibre maturity of three Egyptian cotton cultivars. 1995 Proc. Beltwide Cotton Conf., San Antonio, TX, USA, January 4 7, 1995:Volume 1 566-569.

Nawaz, A., N. Ahmad, and R.H. Qureshi. 1986. Salt tolerance of cotton. *In:* R. Ahmad and A. San Pietro (eds.). Prospects for Biosaline Research. Karachi University, Pakistan. pp. 285-291.

Nayakekorala, H. and H.M. Taylor. 1990. Phosphorus uptake rates by cotton roots at different growth stages from different soil layers. Plant and Soil 122:105-110.

Nayyar, H., K. Kaur, A.S. Basra, and C.P. Malik. 1989. Hormonal regulation of cotton fiber elongation in *Gossypium arboreum* L. *in vitro* and *in vivo*. Biochem. Physiol. Pflanzen. 185:415-421.

Nazirov, N.N. 1973. Lines for further development in the biology and breeding of cotton. Khlopkovodstvo 1973, 37-39. (Russian).

Nazirov, N.N., N.T. Tashmatov, and A. Vakhabov. 1978. Resistance of cotton radiomutants and their initial forms to salinity. Doklady Vsesoyuznoi Ordena Lenina I Ordena Trudovogo Krasnogo 1978, 9-11. (Russian).

Nazirov, N.N., N.T. Tashmatov, and A. Vakhabov. 1979. Cotton mutants with better salinity tolerance. Mutation Breeding Newsletter, 1979, 1.

Nazirov, N.N., N.T. Tashmatov, A. Vakhabov, and A.G. Nabiev. 1981. Respiration rate and incorporation of ^{32}P in organic phosphorus compounds in radiation mutants of cotton and their original forms under saline conditions. Doklady Vsesoyuznoi Ordena Lenina I Ordena Trudovogo Krasnogo 1981, 18-19. (Russian).

Negi, M., M.S. Sachdev, and K.V.B.R. Tilak. 1990. Influence of soluble phosphorus fertilizer on the interaction between a vesicular-arbuscular mycorrhizal fungus and *Azospirillum brasilense* in barley (*Hordeum vulgare* L.). Biol. Fertil. Soils 10:57-60.

Negi, L.S. and A. Singh. 1956. A preliminary study on the effect of some hormones on yield of cotton. Indian Cotton Grower Rev. 10:153.

Nehl, D.B., S.J. Allen, and J.F. Brown. 1996. Mycorrhizal colonisation, root browning and soil properties associated with a growth disorder of cotton in Australia. Plant Soil 179:171-182.

Nehl, D.B., S.J. Allen, and J.F. Brown. 1997. Deleterious rhizosphere bacteria: an integrating Perspective. Appl. Soil Ecol. 5:1-20.

Nehl, D.B., S.J. Allen, and J.F. Brown. 1998. Slow arbuscular mycorrhizal colonisation of cotton caused by

environmental conditions in the soil. Mycorrhiza 8:159-167.

Nehl, D.B., J.F. Brown, and S.J. Allen. 1994. Mycorrhizas and early season growth disorder: the lazy cotton plant gets into trouble. p. 337-347. *In:* Proc. 7th Australian Cotton Conference. Australian Cotton Growers Research Association, Broadbeach, QLD, Australia.

Nehl, D.B., P.A. McGee, V. Torrisi, G.S. Pattinson, and S.J. Allen. 1999. Patterns of arbuscular mycorrhiza down the profile of a heavy textured soil do not reflect associated colonisation potential. New Phytol. 142:495-504.

Nelson, E.B. 1987. Biological control of Pythium seed rot and pre-emergence damping-off of cotton with *Enterobacter cloacae* and *Erwinia herbicola* applied as seed treatments. Plant Disease 71:140-142.

Nelson W.W. and R.R. Allmaras. 1969. An improved monolith method for excavating and describing roots. Agron. J. 61:751-754.

Nepomuceno, A., D.M. Oosterhuis, and A. Steger. 1997a. Duration of activity of the plant growth regulators PGR-lV and mepiquat chloride. *In:* D.M. Oosterhuis (ed.). Proc. Arkansas Cotton Research Meeting and Summaries of Research in Progress 1996. University of Arkansas Agricultural Experiment Station Special Report 183:136-139.

Nepomuceno, A.L., D.M. Oosterhuis, and J.M. Stewart. 1997b. Physiological characterization of four genotypes representing diverse water stress tolerance. pp.1438-1 439. *In:* P. Dugger and D.A. Richter (eds.). Proc. Beltwide Cotton Conferences. National Cotton Council of America, Memphis, Tenn.

Nester, P.R. 1978. Effect of selected synthetic chemicals on cotton growth patterns and yield. M.S. Thesis, University of Arkansas, Fayetteville, AR.

Neuhaus, H.K., K.-P. Krause, and M. Stitt. 1990. Comparison of pyrophosphate turnover and maximum catalytic activity of pyrophosphate: fructose-6-phosphotransferase in leaves. Phytochem. 29:3411-3415.

Neumann, P.M. 1988. Agrochemicals: Plant physiological and agricultural perspectives. pp. 1-13. *In:* P.M. Neumann (ed.). Plant Growth and Leaf-Applied Chemicals. CRC Press, Boca Raton, FL.

Neumann, P.M. and R. Prinz. 1974. The effects of organosilicone surfactants in foliar nutrient sprays on increased absorption of phosphate and iron salts through stomatal infiltration. Isr. J. Agric. Res. 23:123-128.

Nevell, T.P. and S.H. Zeronian. 1985. Cellulose chemistry fundamentals. pp. 15-29. *In:* T.P. Nevell and S.H. Zeronian (eds.). Cellulose Chemistry and Its

Applications. Ellis Horwood, Ltd., Chichester, UK.

Ni, W. and R.N. Trelease. 1991a. Two genes encode the two subunits of cottonseed catalase. Arch. Biochem. Biophys. 289:237-243.

Ni, W. and R.N. Trelease. 1991b. Post-transcriptional regulation of catalase isozyme expression in cotton seeds. Plant Cell 3:737-744.

Ni, W., R.N. Trelease, and R. Eising. 1990a. Two temporally synthesized charge subunits interact to form the five isoforms of cottonseed (*Gossypium hirsutum*) catalase. Biochem. J. 269:233-238.

Ni, W., R.B. Turley, and R.N. Trelease. 1990b. Characterization of a cDNA encoding cottonseed catalase. Biochim. Biophys. Acta, 1049:219-222.

NIAB. 1987. Fifteen Years of NIAB. Third Five Year Report. Nuclear Institute for Agriculture and Biology, Faisalabad, Pakistan. 202pp.

NIAB. 1992. Twenty Years of NIAB. Fourth Five Year Report. Nuclear Institute for Agriculture and Biology, Faisalabad, Pakistan. 203pp.

Nichiporovich, A.A. 1954. Photosynthesis and the theory of obtaining high crop yields. 15th Timiryazev Lecture 57 pp.

Nicholls, R.L. 1996. Cotton Incorporated News, Vol. 5 (1). First Quarter, 1996.

Nickerson, D. 1950. Color measurements of cotton: Preliminary report on application of new automatic cotton colorimeter. USDA, Production and Marketing Admin., Cotton Branch, Pub. No. PMA 61. USDA, Washington, DC.

Nickerson, D. and F.E. Newton. 1958. Grade and Color Indexes Developed for Evaluating Results of USDA Cotton Finishing Tests. Cotton Div., AMS. No. 245, USDA, 14 pp.

Nickrent, D.L. and D.E. Soltis. 1995. A comparison of angiosperm phylogenies from nuclear 18S rRNA and *rbc*L sequences. Ann. Missouri. Bot. Gard. 82:208-234.

Nida, D.L., S. Patzer, P. Harvey, R. Stipanovic, R. Wood, and R.L. Fuchs. 1996. Glyphosate tolerant cotton: The composition of the cottonseed is equivalent to that of conventional cottonseed. J. Agri. Food Chem. 44:1967-1974.

Nielsen, K.F. 1974. Roots and root temperature. pp 293-335 *In:* E.W. Carson (ed.). The Plant and Its Environment. Univ. Press of Virginia, Charlottesville.

Nielsen, K.F., R.L. Halstead, A.J. Maclean, S.J. Bourget, and R.M. Holmes. 1961. The influence of soil temperature on the growth and mineral composition of corn, bromegrass, and potatoes. Soil Sci. Soc. Amer. Proc. 25:369-372.

Nieman, R.H. and L.L. Poulsen. 1967. Interactive effects of salinity and atmospheric humidity on the growth of bean and cotton plants. Bot. Gaz. 128:69-73.

Niemira, B.A., G.R. Safir, and M.C. Hawes. 1996. Arbuscular mycorrhizal colonization and border cell production: a possible correlation. Phytopathology 86:563-565.

Niles, G.A. and R.F. Bader. 1986. Response of short- and full season cotton cultivars to mepiquat chloride. II. Yield components and fiber properties. pp. 517-520. *In:* T.C. Nelson (ed.). Proc. Beltwide Cotton Prod. Res. Conf., National Cotton Council of America, Memphis, Tenn.

Niles, G.A. and C.V. Feaster. 1984. Breeding. Pp. 201-231 *In:* R.J. Kohel and C.F. Lewis (eds). Cotton. Amer. Soc. Agronomy, Madison.

Nilsson, P., J. Tuomi, and M. Åstrom. 1996. Bud dormancy as a bet-hedging strategy. Am. Nat. 147:269-281.

Nirenberg, H.I. 1996. Evaluation of the systematic position of *Fusarium oxysporum* f.sp. *vasinfectum* – Morphological, phytopathologic, and molecular experiments with 8 races. pp. 10. *In:* Proc. First International Fusarium Biocontrol Workshop, October 28-31, 1996, College Park, MD.

Nishio, J.N., S.E. Taylor, and N. Terry. 1985. Changes in thylakoid galactolipids and proteins during iron nutrition-mediated chloroplast development. Plant Physiology 71:688-91.

Nobel, P.S. 1991. Physicochemical and environmental plant physiology. Academic Press, San Diego, CA. 635 pp.

Noggle, G.R. and G.J. Fritz. 1983a. Assimilation of ammonia. pp. 262-264. *In:* Introductory Plant Physiology, 2nd Edition. Prentice Hall Inc., Engelwood Cliffs, New Jersey.

Noggle, G.R. and G.J. Fritz. 1983b. Plant growth substances: Biosynthesis, analysis, transport, and mechanism of action. pp. 473-476. *In:* Introductory Plant Physiology, 2nd Edition. Prentice-Hall, Inc., Englewood Cliffs, New Jersey.

Nolte, K.D., D.L. Hendrix, J.W. Radin, and K.E. Koch. 1995. Sucrose synthase localization during initiation of seed development and trichome differentiation in cotton ovules. Plant Physiol. 109:1285-1293.

Norfleet, M.L., D.W. Reeves, C.H. Burmester, and C.D. Monks. 1997. Optimal planting dates for cotton in the Tennessee Valley of North Alabama. pp. 644-647. *In:* Proc. Beltwide Cotton Conf., National Cotton Council of America, Memphis, Tenn.

Normanly, J., J.P. Slovin, and J.D. Cohen. 1995. Rethinking auxin biosynthesis and metabolism. Plant Physiol. 107:323-329.

Norris, R.F. 1974. Penetration of 2,4-D in relation to cuticle thickness. Amer. J. Bot. 61:74-79.

Novick, R.G., J.E. Jones, W.S. Anthony, W. Aguillard, and J.I. Dickson. 1991. Genetic trait effects on nonlint trash of cotton. Crop Sci. 31:1029-1034.

O'Kelly, J.C. 1953. The use of ^{14}C in locating growth regions of the cell walls of elongating cotton fibers. Plant Physiol. 28:281-286.

O'Sullivan, A.C. 1997. Cellulose: the structure slowly unravels. Cellulose 4:173-207.

O'Toole, J.C. and R.T. Cruz. 1983. Genotypic variation in epicuticular wax of rice. Crop Sci. 23:392-394.

Oertli, J.J. 1968. The significance of the external water potential and of salt transport to water relations in plants. Intl. Soil science Trans. pp. 95-107.

Oertli, J.L. and J.A. Roth. 1969. Boron supply of sugar beet, cotton, and soybean. Agron. J. 61:191-195.

Ogata, G., and E.V. Maas. 1973. Interactive effects of salinity and ozone on growth and yield of garden beet. J. Environ. Qual. 2:518-520.

Ogborn, J.E.A. and J.H. Proctor. 1962. Bollworm attack and the water status of the cotton crop. Emp. Cott. Grow. Rev. 39:131-135.

Ognyanov, V.I., O.S. Petrov, E.P. Tiholov, and N.M. Mollov. 1989. Synthesis of gossypol analogues. Helvetica Chimica Acta 72:353-360.

Ohana, P., M. Benziman, and D.P. Delmer. 1993. Stimulation of callose synthesis in vivo correlates with changes in intracellular distribution of the callose synthase activator β-furfuryl-β-glucoside. Plant Physiol. 101-187-191.

Ohana, P., D.P. Delmer, G. Volman, J.C. Steffens, D.E. Matthews, and M. Benziman. 1992. β-Furfuryl-β-glucoside: an endogenous activator of higher plant UDP-glucose:(1-3)-β-Glucan synthase. Plant Physiol. 98:708-715.

Ohki, K. 1975. Lower and upper critical zinc levels in relation to cotton growth and development. Physiol. Plant. 35:96-100.

Ohlrogge, J.B. 1987. Biochemistry of Plant Acyl Carrier Proteins. pp. 137-157. *In:* P.K. Stumpf and E.E. Conn (eds.). *The Biochemistry of Plants Vol 9.* Academic Press, New York.

Ohkuma, K., J.L. Lyon, F.T. Addicott, and O.E. Smith. 1963. Abscisin 11, an abscission accelerating substance from young cotton hit. Science 142:1592-1593.

Ohnmeiss, T.E. and I.T. Baldwin. 1994.The allometry of nitrogen allocation to growth and inducible defense under nitrogen-limited growth. Ecology 75: 995-1002.

Oikawa, H., K. Katayama, Y. Suzuki, and A. Ichihara. 1995. Enzymatic activity catalyzing exoselective Diels-Alder reactions in solanapyrone biosynthesis. J. Chem. Soc., Chem. Commun., 13:1321-1322.

Oikawa, H., Y. Murakami, and A. Ichihara. 1992. Biosynthetic study of chaetoglobosin A: origins of the oxygen and hydrogen atoms, and indirect evidence for biological Diels-Alder reaction. J. Chem. Soc. Perkin Trans. I pp. 2955-2959.

Okano, K., O. Ito, G. Takaba, A. Shimizu, and T. Totsuks. 1984. Alteration of ^{13}C-assimilate partitioning in plants of *Phaseolus vulgaris* exposed to ozone. New Phytol. 97:155-163.

Oku, T., Y. Wakasaki, N. Adachi, C.I. Kado, K. Tsuchiya, and T. Hibi. 1998. Pathogenicity, non-host hypersensitivity and host defense non-permissibility regulatory gene *hrp X* is highly conserved in *Xanthomonas* pathovars. J. Phytopathology 146:197-200.

Okuda, K., L. Li, K. Kudlicka, S. Kuga and R.M. Brown, Jr. 1993. β-Glucan synthesis in the cotton fiber. I. Identification of β-1,4- and β-1,3 glucans synthesized *in vitro*. Plant Physiol. 101:1131-1142.

Oliveira, F.A de, T.G.S. da Campos, and B.C. Oliveira. 1998. Effect of saline substrate on germination, vigor, and growth of herbaceous cotton. Engenharia Agricola 18:1-10.

Olsen, R.A. 1957. Absorption of sulfur dioxide from the atmosphere by cotton plants. Soil Science 84:107-111.

Olson, L.C. and R.P. Bledsoe. 1942. The chemical composition of the cotton plant and the uptake of nutrients at different stages of growth. Georgia Agric. Exp. Stn. Bull. 222.

Olsson, P.A. and A. Johansen. 2000. Lipid and fatty acid composition of hyphae and spores of arbuscular mycorrhizal fungi at different growth stages. Mycol. Res. 104:429-434.

Olszyk, D., A. Bytnerowicz, G. Kats, C. Reagan, S. Hake, T. Kerby, D. Millhouse, B. Roberts, C. Anderson, and H. Lee. 1993. Cotton yield losses and abient ozone concentrations in California's San Joaquin valley. J. Environ. Qual. 22:602-611.

Oosterhuis, D.M. 1976. Foliar application of fertilizer. pp. 11-12. *In:* Annual Report Cotton Research Institute 1974-75, Gatooma. Government Printer, Salisbury, Rhodesia.

Oosterhuis, D.M. 1987. A technique to measure components of root water potential. Plant Soil 103:285-288.

Oosterhuis, D.M. 1990. Growth and development of a cotton plant. *In:* W.N. Miley and D.M. Oosterhuis (eds.). Nitrogen nutrition of cotton: Practical issues. ASA, CSSA, and SSSA, Madison, WI. pp. 1-24.

Oosterhuis, D.M. 1992. Foliar feeding with potassium nitrate in cotton. Proc. Beltwide Cotton Conf., Nashville, Tenn. pp. 71-72.

Oosterhuis, D.M. 1994. Effects of PGR-IV on the growth and yield of cotton: A review. *In:* G.A. Constable and N.W. Forrester (eds.) Challenging the future. CSIRO, Australia. pp. 29-39.

Oosterhuis, D.M. 1995a. Effects of PGR-IV on the growth and yield of cotton: A review. pp. 29-39. *In:* G.A. Constable and N.W. Forrester (eds.). *Challenging the Future: Proceedings of the World Cotton Research Conferences*, Brisbane, Australia. 14-17 Feb. 1994. CSIRO, Australia.

Oosterhuis, D.M. 1995b. Potassium nutrition of cotton in the USA, with particular reference to foliar fertilization. *In:* G.A. Constable and N.W. Forrester (eds.). Challenging the Future: Proc. World Cotton Conference-1. Brisbane Australia. CSIRO, Melbourne. pp. 133-146.

Oosterhuis, D.M. 1996. Research on chemical plant growth regulation of cotton at the University of Arkansas. Proc 1996 Cotton Research Meeting and Research Summaries. University of Arkansas, Agri Exp. St., Special Report 178:10-19.

Oosterhuis, D.M. 1997. Physiological aspects of potassium deficiency in cotton. Ark. Agri. Exp. Sta. Spec. Rpt. 183:61-73.

Oosterhuis, D.M. 1999a. The cotton leaf cuticle and absorption of foliar-applied chemicals. pp. 15-26. *In:* P. McMullan (ed.). Proc. Fifth International Symposium on Adjuvants for Agrochemicals. Volume 2. Memphis, Tenn.

Oosterhuis, D.M. 1999b. Foliar fertilization of potassium. *In:* D.M. Oosterhuis and G.A. Berkowitz (eds.). Frontiers in Potassium Nutrition: New Perspectives on the Effects of Potassium on Physiology of Plants. Potash and Phosphate Institute and Crop Science Society of America, Publ. PPI, Atlanta, Georgia. pp. 87-100.

Oosterhuis, D.M., O. Abaye, D.W. Albers, W.H. Baker, C.H. Burmester, J.T. Cothren, M.W. Ebelhar, D.S. Guthrie, M.G. Hickey, S.C. Hodges, D.D. Howard, R.L. Hutchinson, L.D. Janes, G.L. Mullins, B.A. Roberts, J.C. Silvertooth, P.W. Tracy, and B.L. Weir. 1994. A summary of a three-year Beltwide study of soil and foliar fertilization with potassium nitrate in cotton. Proc. Beltwide Cotton Conf., San Diego, CA. pp. 1532-1533.

Oosterhuis, D.M. and B.R. Bondada. 2001. Yield response of cotton to foliar nitrogen as influenced by sink strength, petiole and soil nitrogen. J. Plant Nutr. 24:413-422.

Oosterhuis, D.M., F.M. Bourland, A. Steger, O. Abaye, M. Holman, and C. Bednarz. 1998. NAWF as a signal of physiological cutout. p. 1743. *In:* P. Dugger and D.A Richter (eds.). Proc. Beltwide Cotton Conf., National Cotton Council of America, Memphis, Tenn.

Oosterhuis, D.M., F.M. Bourland, N.P. Tugwell, and M.J. Cochran. 1996. Terminology and concepts related to the COTMAN crop monitoring system. University of Arkansas Agricultural Experiment Station Special Report 174.

Oosterhuis, D.M., J. Chipamaunga, and G.C. Bate. 1983. Nitrogen uptake of field grown cotton. I. Distribution in plant components in relation to fertilization and yield. Expl. Agric. 19:91-101.

Oosterhuis, D.M., J.T. Cothren, and J.A. Landivar. 1997. Late-season applications of PGR-IV to remedi- ate fruit shed and enhance maturity and yield. pp. 1399. *In:* P. Dugger and D.A. Richter (eds.). Proc. Beltwide Cotton Prod. Res. Conf., National Cotton Council of America, Memphis, Tenn.

Oosterhuis, D.M., S. Gomez, and C. Meek. 2000. Effect of CoRoN slow release foliar nitrogen fertilizer on cotton growth and yield. Proc. Beltwide Cotton Conferences, National Cotton Council of America, Memphis, Tenn. Vol. 1:712-714.

Oosterhuis, D.M., R.E. Hampton, and S.D. Wullschleger. 1991a. Water deficit effects on cotton leaf cuticle and the efficiency of defoliants. J. Agron. Prod. 4:260-265.

Oosterhuis, D.M., R.E. Hampton, S.D. Wullschleger, and K.S. Kim. 1991b. Characteristics of the cotton leaf cuticle. Arkansas Farm Research 40(5):12-13.

Oosterhuis, D.M., R.G. Hurren, W.N. Miley, and R.L. Maples. 1991c. Foliar-fertilization of cotton with potassium nitrate. Proc. 1991 Cotton Research Meeting. Univ. Arkansas Agri. Exp. Sta., Special Report 149:21-25.

Oosterhuis, D.M. and J. Jernstedt. 1999. Morphology and anatomy of the cotton plant. Chapter 2.1 pp. 175-206. *In:* W. Smith and J.S Cothren (eds.). Cotton: Origin, History, Technology and Production. John Wiley and Sons, Inc.

Oosterhuis, D.M. and A. Steger. 1999. Evaluation of plant growth regulators as in-furrow applications to enhance cotton growth and yield. *In:* D.M. Oosterhuis (ed.). Proc. Arkansas Cotton Research Meeting and Summaries of Research in Progress. University of Arkansas Agricultural Experiment Station Special Report 193:147-149.

Oosterhuis, D.M. and M.J. Urwiler. 1988. Cotton mainstem leaves in relation to vegetative development and yield. Agron. J. 80:65-67.

Oosterhuis, D.M. and L.A. Venter. 1976. Cotton boron trail. Cotton Research Institute Annual Report. pp. 14-16. Ministry of Agriculture, Salisbury, Rhodesia.

Oosterhuis, D.M. and H.H. Wiebe. 1980. Hydraulic conductivity and osmotic adjustment in drought acclimated cotton. Plant Physiol. 65:5 (suppl).

Oosterhuis, D.M. and H.H. Wiebe. 1986. Water stress preconditioning and cotton root pressure-flux relationships. Plant Soil 95:69-76.

Oosterhuis, D.M. and S.D. Wullschleger. 1987a. Osmotic adjustment in cotton (*Gossypium hirsutum* L.) leaves and roots in response to water stress. Plant Physiol. 84:1154-1157.

Oosterhuis, D.M. and S.D. Wullschleger. 1987b. Water flow through cotton roots in relation to xylem anatomy. J. Exp. Bot. 38:1866-1874.

Oosterhuis, D.M. and S.D. Wullschleger. 1988. Carbon partitioning and photosynthetic efficiency during boll development. pp. 57-60 *In:* D.A. Richter (ed.). 1988 Proc. Beltwide Cotton Conferences. National Cotton Council, Memphis.

Oosterhuis, D.M., S.D. Wullschleger, R.L. Maples, and W.N. Miley. 1990. Foliar-application of potassium nitrate in cotton. Better Crops Summer 1990. pp. 8-9.

Oosterhuis, D.M., S.D. Wullschleger, and J.M. Stewart. 1987. Osmotic adjustment in commercial cultivars and wild types of cotton. Agron. Abstracts p. 97.

Oosterhuis, D.M., S.D. Wullschleger, and J.M. Stewart. 1988. Diversity in cotton root xylem anatomy. Ark. Farm Res. 37:19.

Oosterhuis, D.M., S.D. Wullschleger, and S. Rutherford. 1991. Plant physiological responses to PIX. pp. 47-55. *In:* D.M. Oosterhuis (ed.) Proc. 1991 Arkansas Cotton Research Meeting. University of Arkansas Agric. Expt. Sta. Special Report 149.

Oosterhuis, D.M. and D. Zhao. 1993. Physiological effects of PGR-IV on the growth and yield of cotton. p. 1270. 1993 Proc. Beltwide Cotton Prod. Res. Confs. National Cotton Council of America, Memphis, Tenn.

Oosterhuis, D.M. and D. Zhao. 1994. Increased root length and branching in cotton by soil application of the plant growth regulator PGR-4. Plant and Soil 167:51-56.

Oosterhuis, D.M. and D. Zhao. 1995. Increased root length and branching in cotton by soil application of the plant growth regulator PGR-IV. pp. 107-112. *In:* K. Baluska et al. (eds.). Structure and Function of Roots. Academic Publishers, The Netherlands.

Oosterhuis, D.M., B. Zhu, and S.D. Wullschelger. 1989. Uptake of foliar-applied nitrogen by cotton. *In:* D.M. Oosterhuis (ed.). Proc. Arkansas Cotton Prod. and Res. Meeting. Univ. of Arkansas, Agric. Exp. Stn., Special Report 138:23-26.

Ordentlich, A., R.A. Linzer, and I. Raskin. 1991. Alternative respiration and heat evolution in plants. Plant Physiol. 97:1545-1550.

Orozco-Cardenas, M., B. McGurl, and C.A. Ryan. 1993. Expression of an antisense prosystemin gene in tomato plants reduces resistance toward *Manduca sexta* larvae. Proc. Natl. Acad. Sci. USA 90: 8273-8276.

Orum, T.V., D.M. Bigelow, P.J. Cotty, and M.R. Nelson. 1999. Using predictions based on geostatistics to monitor trends in *Aspergillus flavus* strain composition. Phytopathology 89:761-769.

Oshima, R.J., P.K. Braegelmann, R.B Flagler, and R.R. Teso. 1979. The effects of ozone on the growth, yield, and partitioning of dry matter in cotton. J. Environ. Qual. 8:474-479.

Osusky, M., G.Q. Zhou, L. Osuska, R.E. Hancock, W.W. Kay, and S. Misra. 2000. Transgenic plants expressing cationic peptide chimeras exhibit broad-spectrum resistance to phytopathogens. Nature-Biotechnology 18:1162-1166.

Ouellet, T., R.G. Rutledge, and B.L. Miki. 1992. Members of the acetohydroxyacid synthase multigene family of *Brassica napus* have divergent patterns of expression. Plant J. 2:321-330.

Owen, D.F. 1980. How plants may benefit from the animals that eat them? Oikos 35:230-235.

Owen, D.F. and R.G. Wiegert. 1976. Do consumers maximize plant fitness? Oikos 27:488-492.

Owen, D.F. and R.G. Wiegert. 1981. Mutualism between grasses and grazers: an evolutionary hypothesis. Oikos 36:376-378.

Paige, K.N. and T.G. Whitham. 1987. Overcompensation in response to mammalian hervibory: the advantage of being eaten. Am Nat 129:407-416.

Painter, R.H. 1951. Insect resistance in crop plants. The MacMillan Company, New York.

Pallas, J.E., B.E. Michel, and D.G. Harris. 1967. Photosynthesis, transpiration, leaf temperature and stomatal activity of cotton plants under varying water potentials. Plant Physiol. 42:76-88.

Palmer, J.C., H.B. Cooper, Jr., J.W. Pellow, K.E. McRae, and D.M. Anderson. 1996a. El Dorado, a new high quality Acala cotton for the San Joaquin Valley. pp. 45-46. *In:* Proc. Beltwide Cotton Conf., National Cotton Council of America, Memphis, Tenn.

Palmer, J.C., H.B. Cooper, Jr., J.W. Pellow, K.E. McRae, and D.M. Anderson. 1996b. Fiber properties and large scale processing efficiency of two Acala Cottons. pp. 1686-1687. *In:* Proc. Beltwide Cotton Conf., National Cotton Council of America, Memphis, Tenn.

Palmer, J.D. 1976. An introduction to biological rhythms. Academic Press, New York, N.Y.

Palmer, J.D. 1991. Plastid chromosomes: structure and evolution. Cell Culture and Somatic Cell Genetics of Plants 7A:5-53.

Palmer, J.W. 1989. Canopy manipulation for optimum utilization of light. pp. 245-262. *In:* C.J. Wright (ed.). Manipulation of Fruiting. Butterworths, London.

Pan, Y.B., S. Gong, and X.S. Guan. 1993. Techniques and effects of interplanting *Sesbania* with cotton directly mulched in saline soils. China Cottons No. 1:24.

Panda, D., H.P. Miller, A. Banerjee, R.F. Luduena, and L. Wilson. 1995. Microtubule dynamics *in vitro* are regulated by the tubulin isotype composition. Proc. Natl. Acad. Sci. USA 91:11358-11362.

Pandey, R., M.L. Gupta, H.B. Singh, and S. Kumar. 1999. The influence of vesicular-arbuscular mycorrhizal fungi alone or in combination with *Meloidogyne incognita* on *Hyoscymus niger* L. Bioresource Technology 69:275-278.

Pandey, S.S. and V.V.R. Subrahmanyam. 1986. Rapid disappearance of cyclopropennoid fatty acids (CPFA) during germination of cottonseed. J. Am. Oil Chem. Soc. 63:268.

Papdiwal, P.B. and K.B. Deshpande. 1983. Pectin methylesterase of *Xanthomonas campestris* pv. *malvacearum*. Indian Phytopathology 36:134-135.

Paré, P.W. and J.H. Tumlinson. 1998. Cotton volatiles synthesized and released distal to the site of insect damage. Phytochemistry. 47:521-526.

Paris, Q. 1994. Von Liebig's law of the minimum and low-input technologies. Plant production on the threshold of a New Century, Proc. Int. Conf. on the Ocassion of the 75th Anniversary of the Wageningen Agricultural University, Wageningen, The Netherlands, 28 June-1 July 1993. Kluwer Academic Publishers, Dordrecht pp. 169-177.

Parker, C.J., M.K.V. Carr, N.J. Jarvis, B.O. Puplampu, and V.H. Le. 1991. An evaluation of the minirhizotron technique for estimating root distribution in potatoes. J. Agric. Sci. Camb. 116:341-350.

Parker, L.W. and P.L. Salk. 1990. Foliar Triggrr™ cotton: A comprehensive statistical review. pp. 659-660. *In:* J.M. Brown and D.A. Richter (eds.). Proc. Belt. Cotton Prod. Res. Conf., National Cotton Council of America, Memphis, Tenn.

Parker, P.W., W.H. Baker, J.S. McConnell, R.L. Maples, and J.J. Varvil. 1993. Cotton response to foliar nitrogen applications. Proc. Beltwide Cotton Conf., National Cotton Council, Memphis, Tenn. pp. 1364-1366.

Parks, C.R., W.L. Ezell, D.E. Williams, and D.L. Dreyer. 1975. The application of flavonoid distribution to taxonomic problems in the genus *Gossypium*. Bull. Torrey Bot. Club 102:350-361.

Parrott, W.L., P.A. Hedin, J.N. Jenkins, and J.C. McCarty, Jr. 1987. Feeding and recovery of gossypol and tannin from tobacco budworm larvae. Southwest. Entomol. 12:197.

Parrott, W.L., J.N. Jenkins, and J.C. McCarty, Jr. 1983. Feeding behaviour of first-stage tobacco budworm (Lepidoptera: Noctuidae) on three cotton cultivars. Ann. Entomol. Soc. Am. 76:167-170.

Parrott, W.L., J.N. Jenkins, J.E. Mulrooney, J.C. McCarty, Jr., and R.L. Shepherd. 1989. Relationship between gossypol gland density on cotton squares and resistance to tobacco budworm (Lepidoptera: Noctuidae) larvae. J. Econ. Entomol. 82:589-592.

Parsons, J.E., C.J. Phene, D.N. Baker, J.R. Lambert, and J.M. McKinion. 1979. Soil water stress and photosynthesis in cotton. Physiol. Plant. 47:185-189.

Parvin, D. and R. Atkins. 1997. Three years experiences with a new PGR- *Bacillus cereus* (BC). pp. 1396-1398. *In:* P. Dugger and D. Richter (eds.). Proc. Belt. Cotton Prod. Res. Conf., National Cotton Council of America, Memphis, Tenn.

Passioura, J.B. 1973. Sense and nonsense in crop simulation . J. Aust. Inst. Agric. Sci. 39: 181-183.

Passioura, J.B. 1977. Grain yield, harvest index and water use of wheat. J. Aust. Inst. Agric. Sci. 43:117-120.

Passioura, J.B. 1996. Simulation models: science, snake oil, education, or engineering? Agronomy J. 88:690-694.

Passioura, J.B. and C.B. Tanner. 1985. Oscillations in apparent hydraulic conductance of cotton plants. Aust. J. Plant Physiol. 2:455-461.

Pasternak, D., M. Twersky, and Y. De Malach. 1977. Salt resistance in agricultural crops. *In:* H. Mussell and R.C. Staples (eds.). Stress Physiology in Crop Plants. Wiley, New York. pp. 127-142.

Pasternak, D. and G.L. Wilson. 1973. Illuminance, stomatal opening and photosynthesis in sorghum and cotton. Aust. J. Agric. Res. 24:527-532.

Patel, J.K. 1992. Effect of triacontanol and naphthalene acetic acid on lint yield, fiber quality, and nitrogen, phosphorus and potash uptake in cotton (*Gossypium* species). Indian J. Agron. 37:332-337.

Paterson, A.H., T.-H. Lan, K.P. Reischmann, C. Chang, Y.-R. Lin, S.-C. Liu, M.D. Burow, S.P. Kowalski, C.S. Katsar, T.A. DelMonte, K.A. Feldmann, K.F. Schertz, and J.F. Wendel. 1996. Toward a unified genetic map of higher plants, transcending the moncot-dicot divergence. Nat. Genet. 14:380-382.

Patil, N.B. and M. Singh. 1995. Development of medium-staple high-strength cotton suitable for rotor spinning systems. pp. 264-267. *In:* G.A. Constable and N.W. Forrester (eds.). Challenging the Future. Proc. World Cotton Conference I, CSIRO, Australia.

Patterson, D.T., J.A. Bunce, R.S. Alberte, and E. van Volkenburgh. 1977. Photosynthesis in relation to leaf characteristics of cotton from controlled and field experiments. Plant Physiol. 59:384-387.

Pattinson, G.S. and P.A. McGee. 1997. High densities of arbuscular mycorrhizal fungi maintained during long fallows in soils used to grow cotton except when soil is wetted periodically. New Phytol. 136:571-580.

Pear, J.R., Y. Kawagoe, W.E. Schreckengost, D.P. Delmer, and D.M. Stalker. 1996. Higher plants contain homologs of the bacterial celA genes encoding the catalytic subunit of cellulose synthase. Proc. Natl. Acad. Sci. 93:12637-12642.

Pearce, F. 1994. Neighbours sign deal to save Aral Sea. New Scientist 141:10.

Pearce, F. 1995. Poisoned Waters. New Scientist 148:29-33.

Pearcy, R.W. and O. Björkman. 1983. Physiological effects. pp. 65-105. *In:* E.R. Lemon (ed.). Carbon Dioxide and Plants: The Response of Plants to Rising Levels of Atmospheric Carbon Dioxide. AAAS Selected Symp. 84. Westview Press, Boulder, CO.

Pearson, E.Q. 1958. The insect pests of cotton in Tropical Africa, CAB, London, 455 pp.

Pearson, G.A. 1960. Tolerance of crops to exchangeable sodium. USDA Inf. Bull. 216, Washington D.C.

Pearson, R.W., L.F. Ratliff, and H.M. Taylor. 1 970. Effect of soil temperature, strength and pH on cotton seedling root elongation. Agron. J. 62:243-246.

Peeters, M.C., W. Dillemans, and S. Voets. 1991. Nucleolar activity in differentiating cotton fibers is related to the position of the boll on the plant. J. Exp. Bot. 42:353-357.

Peeters, M.C., S. Voets, G. Dayatilake, and E.D. DeLanghe. 1987. Nucleolar size at early stages of cotton fiber development in relation to final fiber dimension. Physiol. Plant 71:436-440.

Peeters, M.C., S. Voets, J. Wijsmans, and E.D. DeLanghe. Pattern of nucleolar growth in differentiating cotton fibers (*Gossypium hirsutum* L.). Ann. Bot. 62:377-382.

Peeters, M.C., K. Willems, and R. Swennen. 1994. Protoplast-to-plant regeneration in cotton (*Gossypium hirsutum* L. cv. Coker 312) using feeder layers. Plant Cell Reports 13:208-211.

Pell, E.J. and E. Brennan. 1973. Changes in respiration, photosynthesis, adenosine 5'-triphosphate, and total adenylate content of ozonated pinto bean foliage as they relate to symptom expression. Plant Physiol. 51:378-381.

Pell, E.J., N. Eckardt, and A.J. Enyedi. 1992. Timing of ozone stress and resulting status of ribulose bisphosphate carboxylase/oxygenase and associated net photosynthesis. New Phytol. 120:397-405.

Pell, E.J., P.J. Temple, A.L. Friend, H.A. Mooney, and W.E. Winner,. 1994. Compensation as a plant response to ozone and associated stresses: An analysis of the ROPIS experiments. J. Environ. Qual. 23:429-436.

Pellow, J.W., H.B. Cooper, Jr., J.C. Palmer, and K.E. McRae. 1996. Fineness, maturity, micronaire and dyeability of two Acala cottons. pp. 1691-1693. *In:* Proc. Beltwide Cotton Conf., National Cotton Council of America, Memphis, Tenn.

Peng, S. and D.R. Krieg. 1991. Single leaf and canopy photosynthesis response to plant age in cotton. Agron. J. 83:704-708.

Pennington, D.A. and H.C. Pringle, III. 1987. Effect of sprinkler irrigation on open cotton flowers. *In:* Proc. Beltwide Cotton Prod. Res. Conf. National Cotton Council, Memphis, Tenn. pp. 69-71.

Percival, A.E. 1987. The national collection of *Gossypium* germplasm. So. Cooperative Series Bulletin No. 321. Texas Ag. Exp. Sta., Texas A& M University, College Station, Texas.

Percy, R.G. 1986. Effects of environment on ovule abortion in interspecific F$_1$ hybrids and single species cultivars of cotton. Crop Sci. 26:938-942.

Percy, R.G., M.C. Calhoun, and H.L Kim. 1996. Seed gossypol variation within *Gossypium barbadense* L. cotton. Crop Sci. 36:193-197.

Percy, R.G. and J.F. Wendel. 1990. Allozyme evidence for the origin and diversification of *Gossypium barbadense* L. Theor. Appl. Genet. 79:529-542.

Perkins, H.H., Jr. 1991. Cotton stickiness -- A major problem in modern textile processing. pp. 523-524. *In:* Proc. Beltwide Cotton Conf., National Cotton Council of America, Memphis, Tenn.

Perkins, H. H. 1993. A survey of sugar and sticky cotton test methods. pp. 1136-1141 *In:* Proc. Beltwide Cotton Proc. Res. Conf., National Cotton Council of America, Memphis, Tenn.

Perkins, H.H., Jr. and D.E. Brushwood. 1995. Interlaboratory evaluation of the thermodetector cotton stickiness test method. pp. 1189-1191. *In:* Proc. Beltwide Cotton Conf., National Cotton Council of America, Memphis, Tenn.

Perkins, H.H., Jr., D.E. Ethridge, and C.K. Bragg. 1984. Fiber. pp. 437-509. *In:* R.J. Kohel and C.F. Lewis (eds.). Cotton. American Soc. Agron., Madison WI.

Perlak, F.J., R.W. Deaton, T.A. Armstrong, R.L. Fuchs, S.R. Sims, J.T. Greenplate, and D.A. Fischhoff. 1990. Insect resistant cotton plants.: Bio/technology 8: 939-943.

Perry, R.P. 1976. Processing of RNA. Ann. Rev. Biochem. 45:605-629.

Pessarakli, M. 1995. Physiological responses of cotton (*Gossypium hirsutum* L.) to salt stress. *In:* M. Pessarakli (ed.). Handbook of Plant and Crop Physiology. Marcel Dekker, New York. pp. 679-693.

Pessarakli, M. and T.C. Tucker. 1985a. Uptake of nitrogen-15 by cotton under salt stress. Soil Sci. Soc. Am. J. 49:149-152.

Pessarakli, M. and T.C. Tucker. 1985b. Ammonium (^{15}N) metabolism in cotton under salt stress. J. Plant Nutr. 8:1025-1045.

Peterson, N.K. and E.R. Purvis. 1961. Development of molybdenum deficiency symptoms in certain crop plants. Soil Sci. Soc. Am. Proc. 25:111-117.

Petkar, B.M., D.V. Mhadgut, and P.G. Oka. 1986. Distribution of cross-sectional area, perimeter, circularity, and secondary wall thickness of different cotton fibers during growth. Text. Res. J. 56:642-645.

Pettiet, C. 1993a. Evaluation of deep bon placement of a N-K blend: A leaf analysis comparison. Proc. Beltwide Cotton Conf., New Orleans, LA. pp. 1330-1332.

Pettiet, J.V. 1993b. Potassium chloride as a source of foliar fertilizer. pp. 1307-1309. *In:* Proc. Beltwide Cotton Conferences. National Cotton Council, Memphis, Tenn.

Pettigrew, W.T. 1994. Source-to-sink manipulations effects on cotton lint yield and yield components. Agron. J. 86:731-735.

Pettigrew, W.T. 1995. Source-to-sink manipulation effects on cotton fiber quality. Agron. J. 87:947-952.

Pettigrew, W.T. 1996. Low light conditions compromise the quality of fiber produced. pp. 1238-1239. *In:* Proc. Beltwide Cotton Conf., National Cotton Council of America, Memphis, Tenn.

Pettigrew, W.T. 1999a. Potassium deficiency and cotton fiber development. pp. 161-163. *In:* D.M. Oosterhuis and G. Berkowitz (eds.). Frontiers in Potassium Nutrition: New Perspectives on the Effects of Potassium on Physiology of Plants. Potash and Phosphate Institute, Norcross, GA and American Society of Agronomy, Madison, WI.

Pettigrew, W.T. 1999b. Potassium deficiency increases specific leaf weights and leaf glucose levels in field-grown cotton. Agronomy Journal 91:962-968.

Pettigrew, W.T., J.J. Heitholt, and W.R. Meredith, Jr. 1992. Early season floral bud removal and cotton growth, yield, and fiber quality. Agron. J. 84:209-214.

Pettigrew, W.T., J.J. Heitholt, and W.R. Meredith, Jr. 1993. Early season ethephon application effects on cotton photosynthesis. Agron. J. 85:821-825.

Pettigrew, W.T., J.J. Heitholt, and W.R. Meredith, Jr. 1996. Genotypic interactions with potassium and nitrogen in cotton of varied maturity. Agron. J. 88:89-93.

Pettigrew, W.T., J.J. Heitholt, and K.C. Vaughn. 1993. Gas exchange differences and comparative anatomy among cotton leaf-type isolines. Crop Sci. 33:1295-1299.

Pettigrew, W.T. and W.R. Meredith, Jr. 1997. Dry matter produced, nutrient uptake and growth of cotton as affected by potassium fertilization. J. Plant Nutr. 20: 531-548.

Pettigrew, W.T and R.B. Turley. 1998. Variation in photosynthetic components among photosynthetically diverse cotton genotypes. Photo. Res. 56:15-25.

Pfeffer, P.E., D.D. Douds, G. Bécard, and Y. Sharchar-Hill. 1999. Carbon uptake and the metabolism and transport of lipids in an arbuscular mycorrhiza. Plant Physiol. 120:587-598.

Phene, C.J., D.N. Baker, J.R. Lamber, J.E. Parsons, and J.M. McKinion. 1978. SPAR - A soil-plant- atmosphere-research system. Trans. ASAE. 21:924-930.

Phillips, A.L. and A.K. Huttly. 1994. Cloning of two gibberellin-regulated cDNAs from *Arabidopsis thaliana* by subtractive hybridization: Expression of the tonoplast water channel, γ-TIP, is increased by GA₃. Plant Mol. Biol. 24:603-615.

Phillips, J.R., G.A. Herzog, and W.F. Nicholson. 1977. Effect of chlordimeform on fruiting characteristics and yield of cotton. Ark. Farm Res. 26(2):4.

Phillips, L.L. 1961. The cytogenetics of speciation in Asiatic cotton. Genetics 46:77-83.

Phillips, L.L. 1962. Segregation in new aalloploids of *Gossypium*. IV. Segregation in New World x Asiatic and New World x wild American hexaploids. Am. J. Bot. 49:51-57.

Phillips, L.L. 1963. The cytogenetics of *Gossypium* and the origin of New World cottons. Evolution 17:460-469.

Phillips, L.L. 1966. The cytology and phylogenetics of the diploid species of *Gossypium*. Am. J. Bot. 53:328-335.

Phillips, L.L. 1976. Cotton *Gossypium* (Malvaceae). pp. 196-200. *In:* N.W. Simmond (ed.). *Evolution of Crop Plants*.

Phillips, L.L. 1977. Interspecific incompatibility in *Gossypium*. 4. Temperature conditioned lethality in hybrids of *G. klotzschianum*. Amer. J. Bot. 64:914-915.

Phillips, L.L. and D. Clement. 1967. Variation in the diploid *Gossypium* species of Baja California. Madroño 19:137-147.

Phillips, L.L and M.A. Strickland. 1966. The cytology of a hybrid between *Gossypium hirsutum* and *G. longicalyx*. Can. J. Genet. Cytol. 8:91-95.

Phillips, S.A., D.R. Melville, L.G. Rodriguez III, R.H. Bruysbacher, R.L. Rogers, J.S. Roussel, D.L. Robinson, J.L. Bartleson, and R.E. Henderson. 1987. Nitrogen fertilization influences on cotton yields, petiole nitrate concentrations and residual nitrate levels at the Macon Ridge, Northeast, and Red River Research Stations. La. Agr. Exp. Sta. Bull. 779.

Phillips, V.A. and P.A. Hedin. 1990. Spectral techniques for structural analysis of the cotton terpenoid aldehydes gossypol and gossypolone. J. Agri. Food Chem. 38:525-528.

Phillips, V.A., P.A. Hedin, E.G. Alley, and P. Oldham. 1990. Mass spectral techniques for the identification of cotton terpenoid aldehydes. J. Agri. Food Chem. 38:521-524.

Phillis, E. and T.G. Mason. 1933. Studies on the transport of carbohydrates in the cotton plant. III. The polar distribution of sugars in the foliage leaf. Ann. Bot. 47:585-634.

Pierce, M.L., M. Essenberg, and A.J. Mort. 1993. A comparison of the quantities of exopolysaccharide produced by *Xanthomonas campestris* pv. *malvacearum* in susceptible and resistant cotton cotyledons during early stages of infection. Phytopathology 83:344-349.

Pikul, J.L., Jr. and J.F. Zuzel. 1994. Soil crusting and water infiltration affected by long-term tillage and residue management. Soil Sci. Soc. Am. J. 58:1524-1530.

Pilet, P.E. and P.W. Barlow. 1987. The role of abscisic acid in root growth and gravireaction: A critical review. Plant Growth Regul. 6:217-265.

Pillonel, C., A.J. Buchala, and H. Meier. 1980. Glucan synthesis by intact cotton fibers fed with different precursors at the stages of primary and secondary wall formation. Planta 149:306-312.

Pillonel, C. and H. Meier. 1985. Influence of external factors on callose and cellulose synthesis during incubation in vitro of intact cotton fibers with [14C]sucrose. Planta 165:76-84.

Pinkas, L.L.H. 1972. Modification of flowering in Pima cotton with ethephon. Crop Sci. 12:465-466.

Pinkhasov, Y.I. and L.V. Tkachenko. 1981. Competitive relations in assimilate consumption between different fruits of cotton plants. Translated from Fiziologiya Rastenii 28:130-135.

Pinter, P.J., Jr., S.B. Idso, D.L. Hendrix, R.R. Rokey, R.S. Rauschkolb, J.R. Mauney, B.A. Kimball, G.R. Hendry, K.F. Lewin, and J. Nagy. 1994a. Effect of free-air CO_2 enrichment on the chlorophyll content of cotton leaves. Agric. For. Met. 70:163-170.

Pinter, P. J., Jr., B.A. Kimball, J.R. Mauney, G.R. Hendry, K.F. Lewin, and J. Nagy. 1994b. Effects of free-air carbon-dioxide enrichment on the PAR absorption and conversion efficiency by cotton. Agric. For. Met. 70:209-230.

Pissarek, H.P. 1973. Zur Entwicklung der Kalium-Mangelsymptome von Sommerraps. Z. Pflanzenernahr. Bodenk. 136:1-19. (Cited in Mengel and Kirkby, 1987)

Pitman, M.G., U. Luttige, D. Kramer, and E. Ball. 1974. Free space characteristics of barley leaf slices. Aust. J. Plant Physiol. 1:65-75.

Plant, R.E. 1989. An integrated expert decision support system for agricultural management. Agric. Sys. 29:49-66.

Plant, R.E. and L.G. Bernheim. 1996. CottonPro: Software for plant mapping and analysis. p. 1198-1200. *In:* P. Dugger and D.A Richter (eds.). Proc. Beltwide Cotton Conf., National Cotton Council of America, Memphis, Tenn.

Plant, R.E. and T.A. Kerby. 1995. CPM: Software for cotton final plant mapping. Agron. J. 87:1143-1147.

Plant, R.E., T.A. Kerby, L.J. Zelinski, and D.S. Munk. 1998. A qualitative simulation model for cotton growth and development. Computers and Electronics in Agric. 20:165-183.

Plaut, Z. 1989. Response to photosynthesis to water and salt stress - similarities and dissimilarities. *In:* K.H. Kreeb, R.M. Richter, and T.M. Hinckley (eds.). Structural and functional responses to enirionmental stresses: water shortage. SPB Academic Publishing., The Hague. pp. 155-163.

Plaut, Z. and E. Federman. 1991. Acclimation of CO_2 assimilation in cotton leaves to water stress and salinity. Plant Physiol. 97:515-522.

Plaxton, W.C. 1996. The organization and regulation of plant glycolysis. Annu. Rev. Plant Physiol. Plant Mol. Biol. 47:185-214.

Pleban, S., F. Ingel, and I. Chet. 1995. Control of *Rhizoctonia solani* and *Sclerotium rolfsii* in the greenhouse using endophytic *Bacillus spp.* European J. Plant Pathology 101:665-672.

Plenchette, C., J.A. Fortin, and V. Furlan. 1983. Growth responses of several plant species to mycorrhizae in a soil of moderate P-fertility. I. Mycorrhizal dependency under field conditions. Plant Soil 70:199-209.

Plochl, M., T. Lyons, J. Ollerenshaw, and J. Barnes. 2000. Simulating ozone detoxification in the leaf apoplast through the direct reaction with ascorbate. Planta 210:454-467.

Podesta, F.E and W.C. Plaxton. 1994. Regulation of cytosolic carbon metabolism in germinating *Ricinus communis* cotyledons. II. Properties of physphoenolpyruvate carboxylase and cytosolic pyruvate kinase associated with the regulation of glycolysis and nitrogen assimilation. Planta 194:318-387.

Pollard, D.G. 1973. Plant penetration by feeding aphids (Hemiptera, Aphidoidea): a review. Bull. ent. Res. 62:631-714.

Polsky, B., S.J. Segal, P.A. Baron, J.W.M. Gold, H. Ueno, and D. Armstrong. 1989. Inactivation of human immunodeficiency virus in vitro by gossypol. Contraception 39:579-587.

Porter, D.D. 1936. Positions of seeds and motes in locks and lengths of cotton fibers from bolls borne at different positions on plants in Greenville, TX. USDA Technical Bull. 509.

Porter, P.M., M.J. Sullivan, and L.H. Harvey. 1996. Cotton cultivar response to planting date on the southeastern coastal plain. J. Prod. Agric. 9:223-227.

Post-Beittenmiller, D. 1996. Biochemistry and molecular biology of wax production in plants. Annu. Rev. Plant Physiol. Plant Mol. Biol. 47:405-430.

Potikha, T.S., C.C. Collins, D.I. Johnson, D.P. Delmer, and A. Levine. 1999. The involvement of hydrogen peroxide in the differentiation of secondary walls in cotton fibers. Plant Physiol. 119:849-858.

Potikha, T.S. and D.P. Delmer. 1997. cDNA clones for Annexin AnnGh1 (Accession No. U73746) and AnnGh2 (Accession No. U73747) from *Gossypium hirsutum* (cotton). (PGR97-003) Plant Physiol. 113:305.

Potter, J.R. and J.W. Jones.1977. Leaf area partitioning as an important factor in growth. Plant Physiol. 59:10-14.

Powell, R.D. 1969. Effect of temperature on boll set and development of *Gossypium hirsutum*. Cotton Grow. Rev. 46:29-36.

Pozo, M.J., C. Azcón-Aguilar, E. Dumas-Gaudot, and J.M. Barea. 1999. ß-1,3-Glucanase activities in tomato roots inoculated with mycorrhizal fungi and /or *Phytophthora parasitica* and their possible involvement in bioprotection. Plant Sci. 141:149-157.

Pradet, A. 1982. Oxidative phosphorylation in seeds during the initial phases of germination. pp. 347-369. *In:* A.A. Khan (ed.). The Physiology and Biochemistry of Seed Development, Dormancy, and Germination. Elsevier Biomedical.

Precoda, N. 1991. Requiem for the Aral Sea. Ambio 20:109-114.

Pressley, E.H. 1937. A study of the effect of pollen upon the length of cotton fibers. University of Arizona Agric. Exper. Sta. Bulletin No. 70., pp 255-292.

Price, H.J. and R.H. Smith. 1979. Somatic embryogenesis in suspension cultures of *Gossypium klotschiamum* Anderss. Planta 145:305-307.

Price, H.J., R.H. Smith, and R.M. Grumbles. 1977. Callus cultures of six species of cotton (*Gossypium* L.) on defined media. Plant Sci. Lett. 10:115-119.

Price, N.S., R.W. Roncandori, and R.S. Hussey. 1989. Cotton root growth as influences by phosphorous nutrient and vesicular-arbuscular mycorrhizas. New Phytol. 111:61-66.

Price, P.W., C.E. Bouton, P. Gross, B.A. McPheron, J.N. Thompson, and A.E. Weis. 1980. Interaction among three trophic levels: influence of plants on interactions between insect herbivores and natural enemies. Annu. Rev. Ecol. Syst. 11: 14-65.

Prior, S.A., H.H. Rogers, G.B. Runion, and J.R. Mauney. 1994. Effects of free-air CO_2 enrichment on cotton root growth. Agric. For. Meteorol. 70:69-86.

Pritchard, J., R.G. Wyn Jones, and A.D. Tomos. 1991. Turgor, growth and rheological gradients of wheat roots following osmotic stress. J. Exp. Bot. 42:1043-1049.

Prokof'ev, A.A. and D.I. Igamberdieva. 1971. Assimilate content in fugacious and retained fruiting organs of cotton. Soviet Plant Physiol.18:850-854.

Prokof'ev A.A. and S. Rasulov. 1975. Optimal number of fruit elements in cotton. Soviet Plant Physiol.22:624-628.

Pryor, W.A. 1992. How far does ozone penetrate into the pulmonary air/tissue boundary before it reacts? Free Radical Biol. Med. 12:83-88.

Pryor, W.A., J.W. Lightsey, and D.G. Prier. 1982. The production of free radicals in vivo from the action of xenobiotics: The initiation of autooxidation of polyunsaturated fatty acids by NO_2 and O_3. *In:* K. Yagi (ed.). Lipid Peroxidases in Biology and Medicine. Academic Press, N.Y.. pp. 1-22.

Puckhaber, L.S., R.D. Stipanovic, and A.A. Bell. 1998. Kenaf phytoalexin: Toxicity of *o*-hibiscanone and its hydroquinone to the plant pathogens *Verticillium dahliae* and *Fusarium oxysporum* f.sp. *vasinfectum*. J. Agric. and Food Chemistry 46:4744-4747.

Puech-Suanzes, I., T.C. Hsiao, E. Fereres, and D.W. Henderson. 1989. Water-stress effects on the carbon exchange rates of three upland cotton (*Gossypium hirsutum*) cultivars in the field. Field Crops Res. 21:239-256.

Pugh, L.M., R.W. Roncadori, and R.S. Hussey. 1981. Factors affecting vesicular-arbuscular mycorrhizal development and growth of cotton. Mycology 73:869-879.

Puhalla, J.E. and A.A. Bell. 1981. Genetics and biochemistry of wilt pathogens. pp. 145-192. *In*: M.E. Mace, A.A. Bell, and C.H. Beckman (eds.). Fungal Wilt Diseases of Plants. Academic Press, New York.

Puhalla, J.E. and C.R. Howell. 1975. Significance of endopolygalacturonase activity to symptom expression of Verticillium wilt of cotton, assessed by the use of mutants of *Verticillium dahliae* Kleb. Physiological Plant Pathology 7:147-152.

Puhalla, J.E. and M. Hummel. 1983. Vegetative compatibility groups within *Verticillium dahliae*. Phytopathology 73:1305-1308.

Purcell, J.P., J.T. Greenplate, M.G. Jennings, J.S. Ryerse, J.C. Pershing, S.R. Sims, M.J. Prinsen, D.R. Corbin, M. Tran, and R.D. Sammons. 1993. Cholesterol oxidase: a potent insecticidal protein active against boll weevil larvae. Biochem. Biophys. Res. Comm. 196:1406-1413.

Purcell, J.P., J.T. Greenplate, J.S. Ryerse. and D.R. Corbin. 1995. Insecticidal activity of cholesterol oxidase against lepidopteran larvae. FASEB J.: 9:A1375.

Pushnik, J.C. and G.W. Miller. 1989. Iron regulation of chloroplast photosynthetic function: mediation for PS I development. J. Plant. Nutr. 12:407-421.

Pyke, B.A. and E.H. Brown. 1996. The cotton pest and beneficial guide. Cotton Res. Dev. Corp. and Coop. Res. Centre Trop. Pest Manag., Woolloongaba, 51 pp.

Qadir, M. and M. Shams. 1997. Some agronomic and physiological aspects of salt tolerance in cotton (*Gossypium hirsutum* L.). J. Agron. Crop Sci. 179:101-106.

Qi, G.-R., G.-J. Cao, P. Jiang, X.-L. Feng, and X.-R. Gu. 1988. Studies on the sites expressing evolutionary changes in the structure of eukaryotic 5S ribosomal RNA. J. Mol. Evol. 27:336-340.

Quader, H., W. Herth, U. Ryser, and E. Schepf. 1987. Cytoskeletal elements in cotton seed hair development *in vitro*: their possible regulatory role in cell wall organization. Protoplasma 137:56-62.

Quisenberry, J.E. and J.R. Gipson. 1974. Growth and productivity of cotton grown from seed produced under four night temperatures. Crop Sci. 14:300-302.

Quisenberry, J.E., W.R. Jordan, B.A. Roark, and D.W. Fryrear. 1981. Exotic cottons as genetic sources for drought resistance. Crop Sci. 21:889-895.

Quisenberry, J.E. and R.J. Kohel. 1975. Growth and development of fiber and seed in upland cotton. Crop Sci. 15:463-467.

Quisenberry, J.E., L.D. McDonald, and B.L. McMichael. 1994. Responses of photosynthetic rates of genotypic differences in sink-to-source ratios in upland cotton (*Gossypium hyrsutum* L.). Env. Exp. Bot. 34:245-252.

Quisenberry, J.E. and B.L. McMichael. 1985. Potential for using leaf turgidity to select drought tolerance in cotton. Crop Sci. 25:294-299.

Quisenberry, J.E. and B.L. McMichael. 1991. Genetic variation among cotton germplasm for water-use efficiency. Environ. Exp. Bot. 31:453-460.

Quisenberry, J.E. and B.L. McMichael. 1996. Screening cotton germplasm for root growth potential. Environ. and Exp. Bot. 36:333-338.

Quisenberry, J.E., B. Roark, and B.L. McMichael. 1982. Use of transpiration decline curves to identify drought-tolerant cotton germplasm. Crop Sci. 22:918-922.

Quisenberry, J.E. and D.R. Rummel. 1979. Natural resistance to thrips injury in cotton as measured by differential leaf area reduction. Crop Sci. 19:879-881.

Rabin, L.B. and R.S. Pacovsky, 1985. Reduced larva growth of two Lepidoptera (Noctuidae) on excised leaves of soybean infected with a mycorrhizal fungus. J. Econ. Entomol. 78:1358-1363.

Rader, L.F., L.M. White, and C.W. Whittaker. 1943. The salt index: a measure of the effects of fertilizer on the concentration of the soil solution. Soil Science 55:201.

Radin, J.W. 1983 .Control of plant growth by nitrogen: differences between cereals and broadleaf species. Plant, Cell and Environ. 6:65-68.

Radin, J.W. 1990. Response of transpiration and hydraulic conductance to root temperature in nitrogen and phosphorus deficient cotton seedlings. Plant Physiol. 92:855-857.

Radin, J.W. 1992. Reconciling water-use efficiencies of cotton in field and laboratory. Crop Sci. 32:13-18.

Radin, J.W. and R.C. Ackerson. 1981. Water relations of cotton plants under nitrogen deficiency. III. Stomatal conductance. Plant Physiol. 67:115-119.

Radin, J.W. and J.S. Boyer. 1982. Control of leaf expansion by nitrogen nutrition in sunflower plants: Role of hydraulic conductivity and turgor. Plant Physiol. 69:771-775.

Radin, J.W. and M.P. Eidenbock. 1984. Hydraulic conductance as a factor limiting leaf expansion of phosphorus-deficient cotton plants. Plant Physiol. 75:372-377.

Radin, J.W. and M.P. Eidenbock. 1986. P nutrition and hydraulic conductivity. Plant Physiol. 82:869-871.

Radin, J.W., B.A. Kimball, D.L. Hendrix, and J.R. Mauney. 1987. Photosynthesis of cotton plants exposed to elevated levels of carbon dioxide in the field. Photosyn. Res. 12:191-203.

Radin, J.W., Z. Lu, R.G. Percy, and E. Zeiger. 1994. Genetic variability for stomatal conductance in Pima cotton and its relation to improvements of heat adaptation. Proc. Natl. Acad. Sci. USA 91:7217-7221.

Radin, J.W. and M.A. Matthews. 1988. Hydraulic conductivity of roots and root cells of nutrient stressed cotton seedlings. Proc. Beltwide Cotton Conf. 8-8 Jan. 1988. New Orleans, LA. National Cotton Council. pp. 81.

Radin, J.W. and J.R. Mauney. 1986. The nitrogen stress syndrome. *In.* Cotton Physiology, The Cotton Foundation, Memphis, pp. 91-105.

Radin, J. W., J.R. Mauney, and P.C. Kenidge. 1989. Water uptake of cotton roots during fruit filling in relation to irrigation frequency. Crop Sci. 29:1000-1005.

Radin, J.W., J.R. Mauney, and P.C. Kerridge. 1991. Effects of nitrogen fertility on water potential of irrigated cotton. Agron. J. 83:739-743.

Radin, J.W. and L.L. Parker. 1979. Water relation of cotton plants under nitrogen deficiency. I. Dependence upon leaf structure. Plant Physiol. 64:495-498.

Radin, J.W., L.L. Parker, and G. Guinn. 1982. Water relations of cotton plants under nitrogen deficiency. Plant Physiol. 70:1066-1070.

Radin, J.W., L.L. Parker, and C.R. Sell. 1978. Partitioning of sugar between growth and nitrate reduction in cotton roots. Plant Physiol. 62:550-553.

Radin, J.W., L.L. Reaves, J.R. Mauney, and O.F. French. 1992. Yield enhancement by frequent irrigation during fruiting. Agron. J. 84:551-557.

Radin, J.W. and C.R. Sell. 1975. Growth of cotton plants on nitrate and ammonia nitrogen. Crop Sci. 15:707-710.

Radulovich, R.A. 1984. Reproductive behavior and water relations of cotton. Ph.D. Diss., Univ. California, Davis (Diss. Abstr. 84-25012).

Rains, D.W., S. Goyal, R. Weyrauch, and A. Läuchli. 1987. Saline drainage water reuse in a cotton rotation system. California Agric. 41:24.

Rajguru, S., S.W. Banks, D.R. Gossett, and C. Lord. 1996. Antioxidant response to salt stress in cotton ovule cultures. Plant Physiol. 111:No.2SS, 149.

Rajguru, S.N., S.W. Banks, D.R. Gossett, M.C. Lucas, T.E.J. Fowler, and E.P. Millhollon. 1999. Antioxidant response to salt stress during fiber development in cotton ovules. Journal of Cotton Science 3:11-18.

Rama Rao, N. S.C. Naithani, R.T. Jasdanwala, and Y.D. Singh. 1982a. changes in indoleacetic acid oxidase and peroxidase activities during cotton fibre development. Z. Pflanzenphysiol. 106:157-165.

Rama Rao, N., S.C. Naithani, and Y.D. Singh. 1982b. Physiological and biochemical changes associated with cotton fibre development. II. Auxin oxidising system. Physiol. Plant. 55:204-208.

Ramalho, F.S. 1994. Cotton pest management. Part 4. A Brazilian perspective.Ann. Rev. Entomol. 39:563-578.

Ramey, H.H. 1966. Historical review of cotton variety development. Proceedings of the 18th Cotton Improvement Conference. Memphis, Tenn

Ramey, H.H. 1982. The meaning and assessment of cotton fibre fineness. International Inst. for Cotton, Manchester UK, pp. 1-19.

Ramey, H.H. 1986. Stress influences on fiber development. *In:* J.R. Mauney and J. McD. Stewart. (eds.). Cotton Physiology. The Cotton Foundation. Memphis, Tenn. pp. 351-359.

Ramsey, J.C. and J.D. Berlin. 1976. Ultrastructure of early stages of cotton fiber differentiation. Bot. Gaz. 137:11-19.

Ramsey, P. 1996. Hormonal regulation of PEP carboxylase during fiber initiation. M.S. Thesis. University of Arizona, Tucson, Ariz.

Ramsey, P., S. Menke, and C. Zeiher. 1996. Stimulation of PEP caboxylase by auxin and gibberellin in cultured cotton ovules. Plant Physiol. Suppl. 111:114(470).

Rana, R.S., K.N. Singh, and P.S. Ahuja. 1980. Chromosomal variation and plant tolerance to sodic and saline soils. Proc. Int. Symp. Salt-affected Soils, Karnal, 1980. pp. 487-493. CCSRI, Karnal.

Randel, R.D., C.C. Chase, Jr., and S.J. Wyse. 1992. Effects of gossypol and cottonseed products on reproduction of mammals. J. of Anim. Sci. 70:1628-1638.

Rangan, T.S., T. Zavala, and A. Ip. 1984. Somatic embryogenesis in tissue cultures of *Gossypium hirsutum* L. In Vitro 20:256.

Rani, A. and S.S. Bhojwani. 1976. Establishment of tissue cultures of cotton. Plant Sci. Lett. 7:163-169.

Rao, I.M. and N. Terry. 1989. Leaf phosphate status, photosynthesis, and carbon partitioning in sugar beet. I. Changes in growth, gas exchange, and Calvin cycle enzymes. Plant Physiol. 90:814-819.

Rao, M.J. and J.B. Weaver, Jr. 1976. Effect of leaf shape on response of cotton to plant population, N rate, and irrigation. Agron. J. 68:599-601.

Rao, N., P.J. Prabbakara-Rao, and M. Amarish. 1980. Research note on effect of Cycocel on M.C.U.-5. Cotton Division. Bombay, Directorate of Cotton Development. pp. 23-24.

Rao, V.S. and J.A. Inamdar. 1985. Leaf architecture in cultivars of cotton. Phyton: Annales Rei Botanicae 25:65-72.

Raschke, K. 1979. Movements of stomata. *In:* W. Haupt and E. Feinleib (eds.). Encyclopedia of Plant Physiology. 7: Physiology of Movements. Springer-Verlag, Berlin.

Rashkes, A.M., N.K. Khidyrova, M.M. Kiktev, U.A. Abdullaev, and Kh.M. Shakhidoyatov. 1997. Investigation of the dynamics of the accumulation of some secondary metabolites of cotton plant leaves. Chem. Nat. Compd. 33(1):89-92.

Rashkes, Y.V., A.M. Rashkes, U.A. Abdullaev, M.M. Kiktev, N.K. Khadyrova, V.N. Plugar, and K.M. Shakhidoyatov. 1994. Composition and growth-inhibiting activity of secondary metabolites from cotton leaves. Khim. Prir. Soedin. 2:278-285.

Raskin, I. 1995. Salicylic acid. pp. 188-205. *In:* Davies, P.J. (ed.). *Plant Hormones: Physiology, Biochemistry, and Molecular Biology* 2nd edition. Kluwer Academics Publishers, Boston, MA.

Rathert, G. 1982a. Influence of extreme K/Na ratios and high substrate salinity on plant metabolism of crops differing in salt tolerance. 6. Mineral distribution variability among different salt tolerant cotton varieties. J. Plant Nutr. 5:183-193.

Rathert, G. 1982b. Influence of extreme K/Na ratios and high substrate salinity on plant metabolism of crops differing in salt tolerance. 7. Relations between carbohydrates and degradative enzymes in salt tolerant and salt sensitive cotton genotypes during initial salinity stress. J. Plant Nutr. 5:1401-1413.

Rathert, G. 1983. Effects of high salinity stress on mineral and carbohydrate metabolism of two cotton varieties. Plant and Soil. 73:247-256.

Rauf, A., A. Razzaq, S. Muhammad, and J. Akhtar. 1990. Improvement in salt tolerance of cotton. Pakistan J. Agric. Sci. 27:193-197.

Raven, J.A. and F.A. Smith. 1976. Nitrogen assimilation and transport in vascular land plants in relation to intercellular pH regulation. New Phytol. 76:415-431.

Rawluk, C.D.L., G.J. Racz, and C.A. Grant. 2000. Uptake of foliar or soil application of 15N-labelled urea solution at anthesis and its affect on wheat grain yield and protein. Can. J. Plant Sci. 80:331-334.

Rawson, H.M. and G.A. Constable. 1980a. Carbon production of sunflower cultivars in field and controlled environments: photosynthesis and transpiration of leaves, stems and heads. Aust. J. Plant Physiol. 7:555-573.

Rawson, H.M. and G.A. Constable. 1980b. Gas exchange of pigeonpea: a comparison with other crops and a model of carbon production and its distribution within the plant. Proc. Workshop in Pigeonpeas, Hyderabad.

Rawson, H.M. and C. Hackett. 1974. An exploration of the carbon economy of the tobacco plant. III. Gas exchange of leaves in relation to position on the stem, ontogeny and nitrogen content. Aust. J. Plant Physiol. 1:551-560.

Rawson, H.M., J.H. Hindmarsh, R.A. Fischer, and Y.M. Stockman. 1983. Changes in leaf photosynthesis with plant ontogeny and relationships with yield per ear in wheat cultivars and 120 progeny. Aust. J. Plant Physiol. 10:503-514.

Ray, N. and V.K. Khaddar. 1983. Formation of adventitious and floating roots in cotton under waterlogged conditions. Current Sci. 52:826-828.

Ray, N. and V.K. Khaddar. 1993. Effect of water stagnation at various growth stages of cotton plants grown under salt affected soil conditions. Adv. Plant Sci. 6:125-136.

Ray, P.M. 1987. Principles of plant cell growth. pp. 1-17. Physiology of cell expansion during plant growth. American Soc. Plant Physiologists, Rockville, MD.

Raybould, A.F. and A.J. Gray. 1994. Will hybrids of genetically modified crops invade natural communities? Trends Ecol. Evol. 9: 85-89.

Rayle, D.L. and R.E. Cleland. 1992. The acid growth theory of auxin-induced cell elongation is alive and well. Plant Physiol. 99:1271-1274.

Razzouk, S. and W.J. Whittington. 1991. Effects of salinity on cotton yield and quality. Field Crops Res. 26:305-314.

Rea, P.A. and R.J. Poole. 1993. Vacuolar H+-translocating pyrophosphatase. Annu. Rev. Plant Physiol. Plant Mol. Biol. 44:147-180.

Read, D.J. 1992. The Mycorrhizal Mycelium. pp. 102-133. *In:* M. Allen (ed.). Mycorrhizal functioning: An integrative plant–fungal process. Chapman and Hall, New York.

Rebenfield, L. 1990. Fibers. pp. 219-305. *In:* J.I. Kroschwitz (ed.) Polymers: Fibers and Textiles, A Compendium. John Wiley, NY.

Reddy, A.R., K.R. Reddy, and H.F. Hodges. 1996a. Mepiquat chloride (Pix)-induced changes in photosynthesis and growth of cotton. Plant Growth Regul. 20:179-183.

Reddy, A.R., K.R. Reddy, R. Padjung, and H.F. Hodges. 1996b. Nitrogen nutrition and photosynthesis in leaves of Pima cotton. J. Plant Nutrition 19:755-770.

Reddy, A.S., B. Ranganathan, R.M. Haisler, and M.A. Swize. 1996. A cDNA encoding acyl-CoA-binding protein from cotton. Plant Physiol. 111:348.

Reddy, K.R., M.L. Boone, A.R. Reddy, H.F. Hodges, S. Turner, and J.M. McKinion. 1995a. Developing and validating a model for a plant growth regulator. Agron. J. 87:1100-1105.

Reddy, K.R., G.H. Davidonis, A.S. Johnson, and B.T. Vinyard. 1999. Temperature regime and carbon dioxide enrichment alter cotton boll development and fiber properties. Agron. J. 91:851-858.

Reddy, K.R., H.F. Hodges, and V.R. Reddy. 1992a. Temperature effects on cotton fruit retention. Agron. J. 84:26-30.

Reddy, K.R., H.F. Hodges, and J.M. McKinion. 1993a. Temperature effects on pima cotton leaf growth and development. Agron. J. 85:681-686.

Reddy, K.R., H.F. Hodges, and J.M. McKinion. 1993b. A temperature model for cotton phenology. Biotronics 22:47-59.

Reddy, K R., H.F. Hodges, and J.M. McKinion. 1995b. Carbon dioxide and temperature effects on pima cotton development. Agron J. 87:820-826.

Reddy, K.R., H.F. Hodges, and J.M. McKinion. 1995c. Cotton crop responses to a changing environment. pp. 3-30. *In:* C. Rosenzweig, J.T. Ritchie, J.W. Jones, G.Y. Tsuji, and P. Hildebrand (eds.). Climate Change and Agriculture: Analysis of Potential International Impacts. ASA Special Publication No. 59. Madison, WI.

Reddy, K.R., H.F. Hodges, and J.M. McKinion. 1996a. Can cotton crops be sustained in future climates? pp. 1189-1196. *In:* 1996 Beltwide Cotton Conferences, National Cotton Council of America, Memphis, Tenn.

Reddy, K.R., H.F. Hodges, and J.M. McKinion. 1996b. Weather and cotton growth: Present and future. Bulletin 1061: Miss. Agri. and For. Exp. Sta. 23 pp.

Reddy, K.R., H.F. Hodges, and J.M. McKinion. 1997a. A comparison of scenarios for the effect of global climate change on cotton growth and yield. Aust. J. Plant Physiol. 24:707-713.

Reddy, K.R., H.F. Hodges, and J.M. McKinion. 1997b. Crop modeling and applications: a cotton example. Adv. Agron. 59:225-290.

Reddy, K.R., H.F. Hodges, and J.M. McKinion. 1997c. Modeling temperature effects on cotton internode and leaf growth. Crop Sci. 37:503-509.

Reddy, K.R., H.F. Hodges, J.M. McKinion, and G.W. Wall. 1992b. Temperature effects on pima cotton growth and development. Agron. J. 84:237-243.

Reddy, K.R., V.R. Reddy, and H.F. Hodges. 1992d. Temperature effects on early season cotton growth and development. Agron. J. 84:229-237.

Reddy, K.R., R.R. Robana, H.F. Hodges, X.J. Liu, and J.M. McKinion. 1998. Influence of atmospheric CO_2 and temperature on cotton growth and stomatal characteristics. Environ. Expt. Bot. 39:117-129.

Reddy, V.R. 1993. Modeling mepiquat chloride-temperature interactions in cotton: the model. Computers and Electronics in Agr. 8:227-236.

Reddy, V.R. 1994. Modeling cotton growth and phenology in response to temperature. Computers and Electronics in Agr. 10:63-73.

Reddy, V.R. 1995. Modeling ethephon and temperature interaction in cotton. Computers Elec. Agric. 13:27-35.

Reddy, V.R., B. Acock, D.N. Baker, and M. Acock. 1989a. Seasonal leaf area-leaf weight relationships in the cotton canopy. Agron. J. 81:1-4.

Reddy, V.R., D.N. Baker, and H.F. Hodges. 1990. Temperature and mepiquat chloride effects on cotton canopy architecture. Agron. J. 82:190-195.

Reddy, V.R., D.N. Baker, and H.F. Hodges. 1991. Temperature effects on cotton canopy growth, photosynthesis, and respiration. Agron. J. 83:699-704.

Reddy, V.R., D.N. Baker, and J.M. McKinion. 1989b. Analysis of effects of atmospheric carbon dioxide and ozone on cotton yield trends. J. Environ. Qual. 18:427-432.

Reddy, V.R., K.R. Reddy, and B. Acock. 1994. Carbon dioxide and temperature effects on cotton leaf initiation and development. Biotronics 23:58-74.

Reddy, V.R., K.R. Reddy, and B. Acock. 1995. Carbon dioxide and temperature interactions on stem extension, node initiation, and fruiting in cotton. Agriculture, Ecosys. & Eviron. 55:17-28.

Reddy, V.R., K.R. Reddy, and D.N. Baker. 1991. Temperature effect on growth and development of cotton during the fruiting period. Agron. J. 83:211-217.

Reddy, V.R., K.R. Reddy, and H.F. Hodges. 1994. Carbon dioxide enrichment and temperature effects on cotton photosynthesis, transpiration, and water use efficiency. in review. Field Crops Res. 41:13-23.

Reddy, V.R., K.R. Reddy, R. Padjung, and H.F. Hodges. 1996. Nitrogen nutrition and photosynthesis in leaves of Pima cotton. J. Plant Nutr. 19:755-770.

Reddy, V.R., A. Trent, and B. Acock, 1992. Mepiquat chloride and irrigation versus cotton growth and development. Agron. J. 84:930-933.

Reese, J.C., B.G. Chan, and A.C. Waiss, Jr. 1982. Effects of cotton condensed tannin, maysin (corn) and pinitol (soybeans) on *Heliothis zea* growth and development. J. Chem. Ecol. 8:1429-1436.

Reeves, D.W., C.H. Burmester, R.L. Raper, and E.C. Burt. 1996. Developing conservation tillage systems for the Tennessee Valley region in Alabama. pp. 1401-1403. *In:* Proc. Beltwide Cotton Conf., National Cotton Council of America, Memphis, Tenn.

Reeves, D.W., C. Mitchell, G. Mullins, and J. Touchton. 1993. Nutrient management for conservation-tillage cotton in the southeast. pp. 23-28. *In:* T.D. Valco and M.R. McClelland (eds.). Conservation-Tillage Systems for Cotton. University of Arkansas

Agricultural Experiment Station Special Report 169. Fayetteville, AR.

Reeves, D.W. and G.L. Mullins. 1995. Subsoiling and potassium effects on water relations and yield of cotton. Agron. J. 87:847-852.

Reeves, R.G. and J.O. Beasley. 1935. The development of the cotton embryo. J. Agric. Res. 51:935-944.

Rehab, F.I. and A. Wallace. 1979. Sodium chloride on growth, mineral composition, and gas exchange characteristics of three cultivars of cotton grown in soil in a glasshouse. Alexandria J. Agric. Res. 27:237-245.

Reicosky, D.C., W.S. Meyer, N.L. Schaefer, and R.D. Sides. 1985. Cotton response to short-term waterlogging imposed with a water-table gradient facility. Agric. Water Manage. 10:127-   143.

Reid, M.S. 1985. Ethylene in plant growth, development, and senescence. pp. 257-279. *In:* P.J. Davies (ed.). Plant Hormones and Their Role in Plant Growth and Development. Martinus Nijhoff Publishers, Dordercht.

Reiling, K. and A.W. Davison. 1992. The response of native, herbaceous species to ozone: Growth and fluorescence screening. New Phytol. 120:29-37.

Reine, A.B. and S.J. Stegink. 1988. Qualitative and quantitative thin-layer chromatographic analysis of polar lipids from leaves of glanded and glandless cotton genotypes. J. Chromatography 437:211-219.

Reinhardt, D.H. and T.L. Rost. 1995a. Salinity accelerates endodermal development and induces an exodermis in cotton seedling roots. Environ. Exp. Bot. 35:563-574.

Reinhardt, D.H. and T.L. Rost. 1995b. On the correlation of primary root growth and tracheary element size and distance from the tip in cotton seedlings grown under salinity. Environ. Exp. Bot. 35:575-588.

Reinhardt, D.H. and T.L. Rost. 1995c. Primary and lateral root development of dark-grown and light-grown cotton seedlings under salinity stress. Botanica Acta 108:457-465.

Reinhardt, D.H. and T.L. Rost. 1995d. Developmental changes of cotton root primary tissues induced by salinity. Int. J. Plant Sci. 156:505-513.

Reinisch, A.J., J. Dong, C.L. Brubaker, D.M. Stelly, J.F. Wendel, and A.H. Paterson. 1994. A detailed RFLP map of cotton, *Gossypium hirsutum* x *G. barbadense*: chromosome organization and evolution in a disomic polyploid genome. Genetics 138:829-847.

Reiser, L. and R.L. Fischer. 1993. The ovule and the embryo sac. Plant Cell 5:1291-1301.

Rengel, Z. 1992. Role of calcium in aluminum toxicity. New Phytol. 121:499-514.

Renou, A. and J. Aspirot. 1984 . Reflections on the use of pyrethroids for cotton protection in Chad. Cot. Fib. Trop. 39:109-116.

Renu, M., C.L. Goswami, and R. Munjal. 1995. Response of chloroplastic pigments to NaCl and GA3 during cotton cotyledonary leaf growth and maturity. Agric. Sci. Digest Karnal, 15:146-150.

Reuter, D.J., A.D. Robson, J.F. Loneragan, and D.J. Tranthim-Fryer. 1981. Copper nutrition of subterranean clover (*Trifolium subterraneum* L. cv. Seaton Park). II. Effects of copper supply on distribution of copper and the diagnosis of copper deficiency by plant analysis. Aust. J. Agric. Res. 32:267-282.

Rhoades, J.D. 1987. Use of saline water for irrigation. Water Quality Bulletin 12:14-20.

Ribas-Carbo, M., R. Aroca, M.A. Gonzalez-Meler, J.J. Irigoyen, and M. Sanchez-Diaz. 2000. The electron partitioning between the cytochrome and alternative respiratory pathways during chilling recovery in two cultivars of maize differing in chilling sensitivity. Plant Physiology. 122:199-204.

Rich, J.R. and G.W. Bird. 1974. Association of early-season vesicular-arbuscular mycorrhizae with increased growth and development of cotton. Phytopathology 64:1421-1425.

Richards, R.A., H.M. Rawson, and D.A. Johnson. 1986. Glaucousness in wheat: its development and effect on water-use efficiency, gas exchange and photosynthetic tissue temperatures. Aust. J. Plant Physiol. 13:465-473.

Richards, R.A., and T.F. Townley-Smith. 1987. Variation in leaf area development and its effect on water use, yield and harvest index of droughted wheat. Aust. J. Agric. Res. 38:983-922.

Richmond, P.A. 1983. Patterns of cellulose microfibril deposition and rearrangement in *Nitella: in vivo* analysis by a birefringence index. Appl. Polym. Symp. 37:107-122.

Rickerl, D.H., W.B. Gordon, E.A. Curl, and J.T. Touchton. 1988. Winter legume and tillage effects on cotton growth and soil ecology. Soil Tillage Res. 11:63-71.

Rickerl, D.H., W.B. Gordon, and J.T. Touchton. 1986. Effect of tillage and legume N on no-till cotton stands and soil organisms. pp. 455-456. *In:* Proc. Beltwide Cotton Conf., National Cotton Council of America, Memphis, Tenn.

Rickerl, D.H., J.T. Touchton, and W.B. Gordon. 1984. Tillage and cover crop effects on cotton survival and yield. pp. 330-331. *In:* Proc. Beltwide Cotton Prod. Res. Conf., National Cotton Council of America, Memphis, Tenn.

Rieseberg, L.H. and J.F. Wendel. 1993. Introgression and its consequences in plants. Pp 70-109 *In:* R. Harrison (ed.). *Hybrid Zones and the Evolutionary Process.* Oxford University Press, NY.

Rikin, A. 1991. Temperature-induced phase shifting of circadian rhythms in cotton seedlings as affected by variation in chilling resistance. Planta 185:407-414.

Rikin, A. 1992. Temporal organization of chilling resistance in cotton seedlings: Effects of low temperature and relative humidity. Planta 187:517-522.

Rikin, A., E. Chalutz, and J.D. Anderson. 1984a. Rhythmicity in ethylene production in cotton seedlings. Plant Physiol. 75:493-495.

Rikin, A., E. Chalutz, and J.D. Anderson. 1985. Rhythmicity in cotton seedlings: Rhythmic ethylene production as affected by silver ions and as related to other rhythmic processes. Planta 163:227-231.

Rikin, A., J.W. Dillwith, and D.K. Bergman. 1993. Correlation between the circadian rhythm of resistance to extreme temperatures and changes in fatty acid composition in cotton seedlings. Plant Physiol. 101:31-36.

Rikin, A., J.B. St John, W.P. Wergen, and J.D. Anderson. 1984b Rhythmical changes in the sensitivity of cotton seedlings to herbicides. Plant Physiol. 76:297-300.

Rimon, D. 1989. Functions of potassium in improving fiber quality of cotton. pp. 319-323. *In:* Methods of Potassium Research in Plants. Proc. 21st Colloquium, Potash and Phosphate Institute,

Rinehart, J.A., M.W. Petersen, and M.E. John. 1996. Tissue-specific and developmental regulation of cotton gene FbL2A. Plant Physiol. 112:1331-1341.

Rios, M.A. and R.W. Pearson. 1964. The Effect of Some Chemical Environmental Factors on Cotton Root Behavior. Soil Science Society Proceedings. pp. 232-235.

Ritchie, J.T. 1972. Model for predicting evaporation from a row crop with incomplete cover. Water Resour. Res. 8:1204-1213.

Ritchie, J.T. 1983. Efficient water use in crop production: Discussion on the generality of relations between biomass production and evapotranspiration. *In:* H.M. Taylor, W.R. Jordan, and T.R. Sinclair (eds.). Limitations to efficient water use in crop production. Amer. Soc. Agron., Madison. pp. 29-44.

Ritchie, J.T. and D.S. NeSmith. 1991. Temperature and crop development. pp. 5-29. *In:* J. Hanks and J.T. Ritchie (eds.). Modeling Plant and Soil Systems. ASA Monograph no. 31, Madison, WI.

Ritter, D., R.D. Allen, N. Trolinder, D.W. Hughes, and G.A. Galau. 1993. Cotton cotyledon cDNA encoding a peroxidase. Plant Physiol. 102:1351.

Roark, B., T.R. Pfrimmer, and M.E. Merkle. 1963. Effects of some formulations of methyl parathion, toxaphene, and DDT on the cotton plant. Crop Sci. 3:338-341.

Roberts, B.A. 1992. Surface applications of potassium. Proc. Beltwide Cotton Conf., Nashville, Tenn. pp. 80-82.

Roberts, B.A., T.A. Kerby, and B.L. Weir. 1993. Foliar fertilization of cotton in California. pp.64-76. *In:* L.S.

Murphey (ed.). Foliar Fertilization of Soybeans and Cotton. Special publication 1993-1. Potash & Phosphate Institute and Foundation for Agronomic Research, Norcross, GA.

Roberts, E.M., N. Rama Rao, J.Y. Huang, N.L. Trolinder, and C.H. Haigler. 1992. Effects of cycling temperatures on fiber metabolism in cultured cotton ovules. Plant Physiology 100:979-986.

Robertson, W.C. and J.T. Cothren. 1991. Plant growth and yield response of cotton to row spacing, mepiquat chloride, population, and fertility. Agron. Absts. 83:159.

Robinson, A.F. 1999. Chapter 3.5 – Cotton nematodes. pp. 595-615. *In:* C.W. Smith and J.T. Cothren (eds.). Cotton: Origin, History, Technology, and Production. John Wiley & Sons, Inc., New York, NY.

Robinson, A.F., C.G. Cook, and A.E. Percival. 1997. Resistance to *Rotylenchulus reniformis* and *Meloidogyne incognita* race 3 in the major cotton cultivars planted since 1950. Crop Science 39:850-858.

Robinson, A.F. and C.M. Heald. 1991. Carbon dioxide and temperature gradients in Baermann funnel extraction of *Rotylenchulus reniformis*. J. Nematology 23:28-38.

Robinson, J.M. and S.J. Britz. 2000. Tolerance of a field grown soybean cultivar to elevated ozone level correlates with higher leaflet ascorbic acid level and higher ascorbate:dehydroascorbate redox status and long term photosynthetic productivity. Photosynthesis Research 64:77-87.

Robson, A.D., L.K. Abbott, and N. Malajczuk. 1994. Management of Mycorrhizas in Agriculture, Horticulture, and Forestry. Kluwer Academic Publishers, Dordecht.

Robson, M.J. and M.J. Deacon. 1978. Nitrogen deficiency in small closed communities of S24 ryegrass. II. Changes in the weight and chemical composition of single leaves during their growth and death. Ann. Bot. 42:1199-1213.

Rocklin, R.D. and C.A. Pohl. 1983. Determination of carbohydrates by anion exchange chromatography with pulsed amperometric detection. J. Liq. Chromatog. 6:1577-1590.

Rodgers, J.P. 1981. Cotton fruit development and abscission: variations in the level of anxins. Trop. Agr. (Trin.) 58:63-72.

Rodkiewicz, B. 1970. Callose in cell walls during megasporogenesis in angiosperms. Planta 93:39-47.

Rodriguez-Garay, B. and J.R. Barrow. 1988. Pollen selection for heat tolerance in cotton. Crop Sci. 28:857-859.

Rodriguez-Navarro, A. 2000. Potassium transport in fungi and plants. Bioch. Biophy. Acta 1469:1-30.

Roelofsen, P.A. 1951. Orientation of cellose fibrils in the cell wall of growing cotton hairs and its bearing on the physiology of cell wall growth. Biochem. Biophys. Acta 7:43-53.

Rogers, C.H. 1937. The effect of three and four year rotations on cotton root rot in the central Texas Blacklands. Agron. J. 29:668-680.

Rogers, H.H., H.E. Jeffries, E.P. Stachel, W.W. Heck, L.A. Ripperton, and A.M. Witherspoon. 1977. Measuring air pollutant uptake by plants: a direct kinetic technique. J. Air Pollut. Control. Assoc. 27:1192-1197.

Rogers, H.H., S.A. Prior, and E.G. O'Neill. 1992. Cotton root and rhizophere responses to free-air CO_2 enrichment. Crit. Rev. Plant Sci. 11:251-263.

Rogers, J.P. 1981. Cotton fruit development and abscission: the role of gibberellin-like components. South African J. of Sci. 77:3563-366.

Rohmer, M. 1999. The discovery of a mevalonate-independent pathway for isoprenoid biosynthesis in bacteria, algae and higher plants. Natural Products Report 16:565-574.

Rohrbach, M.S., T.J. Kreofsky, Z. Vuk-Pavlovic, and D. Lauque. 1992. Cotton condensed tannin: A potent modulator of alveolar macrophage host-defense function. pp.803-824. *In:* R.W. Hemingway and R.E. Lake (eds.). Plant Polyphenols. Plenum Press, New York.

Roitsch, T. 1999. Source-sink regulation by sugar and stress. Curr. Opin. Plant Biol. 2:198-206.

Roitsch, T. and W. Tanner. 1996. Cell wall invertase: bridging the gap. Bot. Acta 109:90-93.

Rojas, M.G., R.D. Stipanovic, H.J. Williams, and S.B. Vinson. 1989. A method for the preparation of labelled gossypol by the incorporation of ^{14}C acetate. J. Labelled Compd. & Radiopharm. XXVII:995-998.

Rollins, M.L. 1945. Applications of nitrogen dioxide treatment to the microscopy of fiber cell wall structure. Text. Res. J. 15:65-77.]\

Rollit, J. and G.A. Maclachlan. 1974. Synthesis of wall glucan from sucrose by enzyme preparations from *Pisum sativum*. Phytochemistry 13:367-374.

Romero, C., J.M. Belles, J.L. Vaya, R. Serrano, and F.A. Culianez-Macia. 1997. Expression of the yeast trehalose-6-phosphate synthase gene in transgenic tobacco. Pleiotropic phenotypes include drought tolerance. Planta 201:293-297.

Romheld, V. and M.M. El-Fouly. 1999. Foliar nutrient application: Challenges and limits in crop production. pp. 1-32. *In:* A. Suwanarit (ed.). Proc. Second International Workshop on Foliar fertilization. Bangkok, Thailand. Thailand Soil Fertilizer Society, Thailand.

Room, P.M. 1979a. Parasites and predatores of *Heliothis* spp. (Lepidoptera: Noctuidae) in cotton in the

Namoi Valley, New South Wales. J. Aust. Ent. Soc. 18:223-228.

Room, P.M. 1979b. A prototype "On-line" system for management of cotton pests in the Namoi Valley, New South Wales. Prot. Ecol. 1:245-264.

Room, P.M., J.S. Hanan, and P. Prusinkiewicz. 1996. Virtual plants: New perspectives for ecologists, pathologists and agricultural scientists. Trends Plant Sci. 1:33-38.

Room, P.M., L. Maillette, and J.S. Hanan. 1994. Module and metamer dynamics and virtual plants. Adv. Ecol. Res. 25:105-157.

Röse, U.S.R., A. Manukian, R.R. Heath, and J.H. Tumlinson. 1996. Volatile semiochemicals released from undamaged cotton leaves. Plant Physiol. 111:487-495.

Rosenbaum, B.J., T.C. Strickland, and M.K. McDowell. 1994. Mapping critical levels of ozone, sulfur dioxide, and nitrogen dioxide for crops, forests, and natural vegetation in the United States. Water, Air, Soil Pollut. 74:307-319.

Rosenow, D.T., J.E. Quisenberry, C.W. Wendt, and L.E. Clark. 1983. Drought tolerant sorghum and cotton germplasm. Agric. Water Man. 7:207-222.

Rosenthal, J.P. and P.M. Kotanen. 1994. Terrestrial plant tolerance to herbivory. Trends Ecol. Evol. 9:145-148.

Rosenthal, W.D. and T.J. Gerik. 1991. Radiation use efficiency among cotton cultivars. Agron. Jour. 83:655-658.

Rosolem, C.A. and G.B. Bastos. 1997. Mineral deficiency symptoms in cotton cultivar IAC 22. Bragantia 56:377-387.

Rosolem, C.A. and A. Costa. 1999. Boron nutrition and growth of cotton as a function of temporary boron deficiency. Anain II Congresso Brasileiro de Algodao: O algodao no seculo XX, perspectivas para a seculo XXI, Ribeirao Preto, SP, Brasil, 5-10 Setembro 1999, 403-406.

Rosolem, C.A. and D.S. Mikkelsen. 1989. Nitrogen source-sink relationship in cotton. J. Plant Nutr. 12(12):1417-1433.

Rosolem, C.A. and D.S. Mikkelsen. 1991. Potassium absorption and partitioning in cotton as affected by periods of potassium deficiency. J. Plant Nutr. 14(9):1001-1016.

Rosolem, C.A., J.P.T. Witacker, S. Vanzolini, and V.J. Ramos. 1999. The significance of root growth on cotton nutrition in an acidic low-P soil. Plant and Soil 212:185-190.

Rossing, W.A.H. and L.A.J.M. van de Wiel. 1990. Simulation of the effects of honeydew in winter wheat caused by the grain aphid *Sitobion avenae*. I. Quantification of the effects of honeydew on gas exchange of leaves and aphid population of different size on crop growth. Netherlands J. Plant Pathol. 96:343-364.

Rossing, W.A.H., M. van Oijen, W. van der Werf, L. Bastiaans, and R. Rabbinge. 1992. Modelling the effects of foliar pests and pathogens on light interception, photosynthesis, growth rate and yield of field crops. pp. 161-180. *In*: P.G. Ayres (ed.). Pest and Pathogens: Plant Responses to Foliar Attack. Bios Scientific Publishers, Oxford.

Rothrock, C.S. 1992. Influence of soil temperature, water and texture on *Thielaviopsis basicola* and black root rot of cotton. Phytopath. 82:1202-1206.

Rothrock, C. 1996. Impact of seedling pathogens on cotton growth and development. pp. 241-242. *In*: Proc. Beltwide Cotton Conf., National Cotton Council of America, Memphis, Tenn.

Rothwell, A., J.W. Brydon, H. Knight, and B.J. Coxe. 1967. Boron deficiency of cotton in Zambia. Cotton Growing Review. 44:23-28.

Rotty, R.M. and G. Marland. 1986. Fossil fuel combustion: Recent amounts, patterns, and trends of CO_2. pp. 474-490. *In*: J.R. Trabalka and D.E. Reichle (eds.). The Changing Carbon Cycle a Global Analysis. Springer-Verlag.

Roush, R.T. 1994a. Managing pests and their resistance to *Bacillus thuringiensis*: can transgenic crops be better than sprays? Biocontrol Sci. Tecnol. 4:501-516

Roush, R.T. 1994b. Can we slow adaptation by pests to insect-resistant transgenic crops? *In*: G. Persley and R. MacIntyre (eds.). Biotechnology for Integrated Pest Management, CAB International, London.

Roussel, J.S., J.C. Weber, L.D. Newsom, and C.E. Smith. 1951. The effects of infestation by the spider mites *Septanychus tumidus* on growth and yield of cotton. J. Econ. Entomol. 44: 523-527.

Rowland, S.P. and N.R. Bertoniere. 1985. Chemical methods of studying supramolecular structure. pp. 112-137. *In*: T.P. Nevell and S.H. Zeronian (eds.). Cellulose Chemistry and its Applications. Ellis Horwood, Chichester UK.

Rowland, S.P., P.S. Howley, and W.S. Anthony. 1984. Specific and direct measurement of the β-1,3-glucan in developing cotton fiber. Planta 161:281-287.

Royer, R.E., L.M. Deck, N.M. Campos, L.A. Hunsaker, and D.L. Vander Jagt. 1986. Biologically active derivatives of gossypol: Synthesis and antimalarial activities of peri-acylated gossylic nitriles. J. Med. Chem. 29:1799-1801.

Royer, R.E., L.M. Deck, T.J. Vander Jagt, F.J. Martinez, R.G. Mills, S.A. Young, and D.L. Vander Jagt. 1995. Synthesis and anti-HIV activity of 1,1'-dideoxygossypol and related compounds. J. Med. Chem. 38:2427-2432.

Ruan, Y.-L. and P.S. Chourey. 1998. A fiberless seed mutation in cotton is associated with lack of fiber cell initiation in ovule epidermis and alterations in su-crose synthase expression and carbon partitioning in developing seeds. Plant Physiol. 118:399-406.

Ruan, Y.-L., P.S. Chourey, D.P. Delmer, and L. Perez-Grau. 1997. The differential expression of sucrose syn-thase in relation to diverse patterns of carbon par-titioning in developing cotton seeds. Plant Physiol. 115:375-385.

Ruan, Y.-L., D.P. Delmer, L. Perez-Grau, and P.S. Chourey. 1996. Biochemical and molecular level expres-sion of sucrose synthase in cotton. Plant Physiol. Suppl. 111:123.

Ruano, A., J. Barcelo, and Ch. Poschenrieder. 1987. Zinc-toxicity induced variation of mineral elements composition in hydroponically grown bush bean plants. J. Plant Nutr. 10:373-384.

Rufty, T.W., S.C. Huber, and R.J. Volk. 1988a. Alteration in leaf carbohydrate metabolism in response to nitro-gen stress. Physiol. Plant. 89:457-463.

Rummel, D.R., M.D. Arnold, J.R. Gannaway, D.F. Owen, S.C. Carroll, and W.R. Deaton. 1994. Evaluation of Bt cottons resistant to injury from bollworm: Implications for pest management in the Texas southern high plains. Southwest. Entom. 19:199-207.

Rummel, D.R. and J.E. Quisenberry. 1979. Influence of thrips injury on leaf development and yield of var-ious cotton genotypes. J. Econ. Entomol. 72:706-709.

Runeckles, V.C. and B.I. Chevone. 1992. Crop responses to ozone. *In:* A.S. Lefohn (ed.). Surface Level Ozone Exposures and their Effects on Vegetation. Lewis Publ. Inc., Chelsea, MI. pp. 189-270.

Runion, G.B., E.A. Curl, H.H. Rogers, P.A. Backman, R. Rodriguez-Kabana, and B.E. Helms. 1994. Effects of free-air CO_2 enrichment on microbial popula-tions in the rhizosphere and phyllosphere of cot-ton. Agri. For. Met. 70:117-130.

Russell, D.A., S.M. Radwan, N.S. Irving, K.A. Jones, and M.C.A. Downham. 1993. Experimental assess-ment of the impact of defoliation by *Spodoptera littoralis* on the growth and yield of Giza 75 cot-ton. Crop Protection 12:303-309.

Russo, D. and D. Bakker. 1987. Crop water production func-tions for sweet corn and cotton irrigated with saline waters. Soil Sci. Soc. Am. J. 51:1554-1562.

Rustamov, K. 1976. Importance of copper in nutrition, me-tabolism and productivity of cotton. Field Crops Abst. 29:791.

Ruttledge, T.R. and E.B. Nelson. 1997. Extracted fatty acids from *Gossypium hirsutum* stimulatory to the seed-rotting fungus, *Pythium ultimum*. Phytochemistry 46:77-82.

Ryals, J.A., U.H. Neuenschwander, M.G. Willits, A. Molina, H.Y. Steiner, and M.D. Hunt. 1996. Systemic ac-quired resistance. Plant Cell 8:11809-1819.

Ryser, U. 1977. Cell wall growth in elongating cotton fi-bers: An autoradiographic study. Cytobiologie 15:78-84.

Ryser, U. 1979. Cotton fiber differentiation; occurrence and distribution of coated and smooth vesicles during primary and secondary wall formation. Protoplasma 98:223-239.

Ryser, U. 1985. Cell wall biosynthesis in differentiating cotton fibers. Eur. J. Cell Biol. 39:236-256.

Ryser, U. 1992. Ultrastructure of the epidermis of develop-ing cotton (*Gossypium*) seeds: Suberin, pits, plas-modesmata, and their implications for assimilate transport into cotton fibers. Am. J. Bot. 79:14-22.

Ryser, U. and P.J. Holloway. 1985. Ultrastructure and chem-istry of soluble and polymeric lipids in cell walls from seed coats and fibres of *Gossypium* species. Planta 163:151-163.

Ryser, U., H. Meier, and P.J. Holloway. 1983. Identification and localization of suberin in the cell walls of green cotton fibers (*Gossypium hirsutum* L. var. green lint). Protoplasma 117:196-205.

Sabbe, W.E. and S.C. Hodges. 2009. Interpretation of plant mineral status. pp. 266-272. *In:* J.M.Stewart, D.M. Oosterhuis, J.J. Heitholt, and J.R. Mauney (eds.). *Physiology of Cotton*. National Cotton Council of America, Memphis,Tenn. pp. Springer, London.

Sabbe, W.E. and A.J. MacKenzie. 1973. Plant analysis as an aid to cotton fertilization. *In:* L.M. Walsh, and J.D. Beaton (eds.). Soil testing and plant analysis. Madison (WI): Soil Science Society of America, Inc. pp. 299-313.

Sabbe, W.E. and L.J. Zelinski. 1990. Plant analysis as an aid in fertilizing cotton. *In:* R.L. Westerman (ed.). Soil Testing and Plant Analysis, 3rd Ed. Soil Science Society of America, Inc., Madison, WI.

Sabino, N.P. 1975. Efeitos da aplicacao de calcario, fosforo e potassio, na qualidade da fibra do alfodoeira cul-tivado em latossolo roxo. Bragantia 34:153-161. (Quoted by Steward, 1986)

Sachs, R.M. 1965. Stem elongation. Annu. Rev. Plant Physiol. 16:73-97.

Sachs, T., A. Novoplansky, and D. Cohen. 1993. Plants as competing populations of redundant organs. Plant Cell Environ. 16:765-770.

Saden, D. 1992. Irrigation of field crops with saline and so-dic water. Water Irrig. Rev. 12:4-6.

Sadras, V.O. 1995. Compensatory growth in cotton af-ter loss of reproductive organs. Field Crops Res. 40:1-18.

Sadras, V.O. 1996a. Cotton compensatory growth after loss of reproductive organs as affected by availabil-ity of resources and duration of recovery period. Oecologia 106:432-439.

Sadras, V.O. 1996b. Cotton responses to simulated insect damage: Radiation-use efficiency, canopy architecture and leaf nitrogen content as affected by loss of reproductive organs. Field Crops Res. 48:199-208.

Sadras, V.O. 1996c. Population-level compensation after loss of vegetative buds: Interactions among damaged and undamaged neighbours. Oecologia 106:417-423.

Sadras, V.O. 1997a. Effects of simulated insect damage and weed interference on cotton growth and reproduction. Annals of Applied Biology. 130:271-281

Sadras, V.O. 1997b. Interference among cotton neighbours after differential reproductive damage. Oecologia 109:427-432.

Sadras, V.O. 1998. Herbivory tolerance of cotton expressing insecticidal proteins from *Bacillus thuringiensis*: responses to damage caused by *Helicoverpa* spp. and to manual bud removal. Field Crops Res. 56:287-299.

Sadras, V.O. and D.J. Connor. 1991. Physiological basis of the response of harvest index to the fraction of water transpired after anthesis: a simple model to estimate harvest index for determinate species. Field Crops Res 26: 227-239.

Sadras, V.O., and G.P. Fitt. 1997a. Apical dominance - Variability among Gossypium genotypes and its association with resistance to insect herbivory. Env. Exp. Bot. (in press)/

Sadras, V.O. and G.P. Fitt. 1997b. Resistance to insect herbivory of cotton lines: Quantification of recovery capacity after damage. Field Crops Research 52: 129-136.

Sadras, V.O. and N. Trápani. 1997 Leaf expansion and phenologic development: key determinants of sunflower plasticity, growth and yield. pp. 205-233 *In:* C. Hamel and D.L. Smith (eds.). Physiological Control of Growth and Yield in Field Crops. Springer-Verlag, Berlin.

Sadras, V. O. and L.J. Wilson. 1996. Effects of timing and intensity of spider mite infestation on the oil yield of cotton crops. Aust. J. Exp. Agric. 36: 577-580.

Sadras, V.O. and L.J. Wilson. 1997a. Growth analysis of spider mite infested cotton crops. I. Light interception and light-use efficiency. Crop Sci. 37: 481-491.

Sadras, V. O. and L.J. Wilson. 1997b. Growth analysis of spider mite infested cotton crops. II. Partitioning of dry matter. Crop Sci. 37:492-497.

Sadras, V.O. and L.J. Wilson. 1997c. Nitrogen accumulation and partitioning in shoots of cotton plants infested with two-spotted spider mites. Austr. J. Agric. Res. 48: 525-533.

Sadras V.O. and L.J. Wilson. 1998. Recovery of cotton crops after early season damage by thrips (Thysanoptera). Crop Sci. 38:399-409.

Sadras, V.O., L.J. Wilson, and D.A. Lally. 1998. Water deficit enhanced cotton resistance to spider mite herbivory. Ann. Bot. 81:273-286.

Saeki, T. 1963. Light relations in plant communities. pp. 79-94. *In:* L.T. Evans (ed.). Environmental Control of Plant Growth. Academic Press, New York and London.

Sage, L.C. 1992. The Chromophore. In. Pigment of the imagination: A history of phytochrome research. Acad. Press, New York. pp.395-409.

Sagers, C.L. 1992. Manipulation of host plant quality: herbivores keep leaves in the dark. Funct. Ecol 6:741-743.

Sagliocco, F., A. Kapazoglou, and L. Dure, III. 1991. Sequence of an *rbcS* gene from cotton. Plant Mol. Biol. 17:1275-1276.

Sagliocco, F., A. Kapazoglou, and L. Dure, III. 1992. Sequence of *cab*-151, a gene encoding a photosystem II type II chlorophyll *a/b*-binding protein in cotton. Plant Mol. Biol. 18:841-842.

Saha, S. and D.M. Stelly. 1994. Chromosomal location of *Phosphoglucomutase7* locus in *Gossypium hirsutum*. J. Hered. 85:35-39.

St. John, J.B. and M.N. Christiansen. 1976. Inhibition of linolenic acid synthesis and modification of chilling resistance in cotton seedlings. Plant Physiology 57:257-259.

Saka, K., F.R. Katterman, and J.C. Thomas. 1987. Cell regeneration and sustained division of protoplasts from cotton (*Gossypium hirsutum* L.). Plant Cell Reports 6:470-472.

Sakurai, A. and S. Fujioka. 1993. The current status of physiology and biochemistry of brassinosteroids. Plant Growth Regul. 13:147-159.

Saleh, H. and R.A. Sikora. 1984. Relationship between *Glomus fasciculatum* root colonization of cotton and its effect on *Meloidogyne incognita*. Nematologica 30:230-237.

Salem, C.E. and J.T. Cothren. 1990. A rapid, inexpensive buffer extraction for sucrose, glucose, and fructose determination in cotton leaf tissue. J. Chromatog. Sci. 28:250-253.

Salerno, G.L., J.L. Ianito, J.A. Tognetti, M.D. Crespi, and H.G Pontis. 1989. Differential induction of sucrose metabolizing enzymes in wheat (*Triticum aestivum* cv. San Agustin) leaf sections. J. Plant Physiol. 134:214-217.

Salih, H.M. and R.K. Abdul-Halim. 1985. Effects of levels of two dominant salt types in Iraq on some components of cotton yield (*Gossypium hirsutum* L.). J. Agric. Water Resour. Res. 4:1-14.

Salisbury, F.B. and C.W. Ross. 1992. Plant Physiology. Fourth Edition. Wadsworth, Belmont, CA. 682 pp.

Salzer, P. and T. Boller. 2000. Elicitor-induced reactions in mycorrhizae and their suppression. p. 1-10. *In:* G.K. Podila and D.D. Douds (eds.). Current

Advances in Mycorrhizae Research. APS Press, St Paul, MN.

Sambo, E.Y., J. Moorby, and F.L. Milthorpe. 1977. Photosynthesis and respiration of developing soybean pods. Aust. J. Plant Physiol. 4:713-721.

Sammis, T.W. 1981. Yield of alfalfa and cotton as influenced by irrigation. Agron. J. 73:323-329.

Samora, P.J., D.M. Stelly, and R.J. Kohel. 1994. Localization and mapping of the *Le₁* and *Gl₂* loci of cotton (*Gossypium hirsutum* L.). J. Hered. 85:152-157.

Sampath, D.S. and P. Balaram. 1986. A rapid procedure for the resolution of racemic gossypol. J. Chem. Soc., Chem. Commun. pp. 649-650.

Sanchez, R.A., J.J. Casal, C.L. Ballare, and A.L. Scopel. 1993. Plant responses to canopy density mediated by photomorphogenic process. pp. 779-786. *In:* Buxton et al. (eds.). International Crop Science I. Crop Science Society of America, Madison, WI.

Sanders, F.E., P.B. Tinker, R.L.B. Black, and S.M. Palmerley. 1977. The development of endomycorrhizal root systems: I. Spread of infection and growth-promoting effects with four species of vesicular-arbuscular endophyte. New Phytol. 78:257-268.

Sanders, J.L. and D.A. Brown. 1978. A new fiber optic technique for measuring root growth of soybeans under field conditions. Agron. J. 70:1073-1076.

Sandstedt, R. 1975. habituated callus cultures from two cotton species. Proc. Beltwide Cotton Prod. Res. Conf. 36:52-53. National Cotton Council of America, Memphis, Tenn.

Santoni, V., C. Bellini, and M. Caboche. 1994. Use of two dimensional protein-pattern analysis for the characterization of *Arabidopsis thaliana* mutants. Planta 192:557-566.

Sargant, J.A. 1965. The penetration of growth regulators into leaves. Ann. Rev. Plant Physiol. 16:1-12.

Saroha, M.K., P. Sridhar, and V.S. Malik. 1998. Glyphosate-tolerant crops: Genes and enzymes. J. Plant Biochemistry and Biotechnology 7:65-72.

Sasek, T.W., E.H. Delucia, and B.R. Strain. 1985. Reversibility of photosynthetic inhibition in cotton after long-term exposure to elevated CO_2 concentrations. Pl. Phys. 78:619-622.

Sassenrath-Cole, G.F. 1995. Dependence of canopy light distribution on leaf and canopy structure for two cotton (*Gossypium*) species. Agric. Forest. Meteorol. 77:55-72.

Sassenrath-Cole, G.F. and P.A. Hedin. 1996. Cotton fiber development: Growth and energy content of developing cotton fruits. pp. 1247-1251. *In:* Proc. Beltwide Cotton Conf., National Cotton Council of America, Memphis, Tenn.

Sassenrath-Cole, G.F. and J.J. Heitholt. 1996. Limitations to optimal carbon uptake within a cotton canopy. pp. 1239-1240. *In:* Proc. Beltwide Cotton Conf., National Cotton Council of America, Memphis, Tenn.

Sassenrath-Cole, G.F., G. Lu, H.F. Hodges, and J.M. McKinion. 1996. Photon flux density versus leaf senescence in determining photosynthetic efficiency and capacity of *Gossypium hirsutum* leaves. Environ. Exper. Bot. 36:439-446.

Sasser, P.E. 1991. The future of cotton quality: Trends and implications -- Fiber quality measurements. pp. 68-70. *In:* Proc. Beltwide Cotton Conferences. National Cotton Council, Memphis, Tenn.

Sasser, P.E. 1992. The physics of fiber strength. pp. 19-28. *In:* C.R. Benedict (ed.). Proc. Cotton Fiber Cellulose: Structure, Function, and Utilization Conference. National Cotton Council, Memphis Tenn.

Sasser, P. and J.L. Shane. 1996. Crop quality -- A decade of improvement. pp. 9-12. *In:* Proc. Beltwide Cotton Conf., National Cotton Council of America, Memphis, Tenn.

Sastri, D.C., K. Hilu, R. Appels, E.S. Lagudah, J. Playford, and B.R. Baum. 1992. An overview of evolution in plant 5S DNA. Pl. Syst. Evol. 183:169-181.

Sattelmacher, B. and H. Marschner. 1978. Nitrogen nutrition and cytokinin in *Solanurn trubersosum.* Physiol. Plant. 42:185-189.

Sauer, J. 1967. Geographic reconnaissance of seashore vegetation along the Mexican Gulf coast. Coastal Studies Series Number 21. Louisiana State University Press, Baton Rouge, LA.

Saunders, J.H. 1961. *The Wild Species of Gossypium and Their Evolutionary History.* Oxford University Press, London.

Sauter, J.J. and S. Kloth. 1986. Plasmodesmatal frequency and radial translocation rates in ray cells of poplar (*populus x canadensis* Moench 'robusta'). Planta 168:377-380.

Sawan, Z.M. and R.A. Sakr. 1990. Response of Egyptian cotton (*Gossypium barbadense*) yield to 1,1-dimethyl piperidinium chloride (PIX). J. Agric. Sci. 114:335-338.

Sawan, Z.M., R.A. Sakr, and M.A. El-Kady. 1984. The effect of ethrel treatment on the yield components and fiber properties of the Egyptian cotton. J. Agron. and Crop Sci. 153:72-78.

Saxena, I.M., R.J. Brown, Jr., M. Fevre, R.A. Geremia, and B. Henrissat. 1995. Multidomain architecture of β-glucosyltransferases: Implications for mechanism of action. J. Bacteriol. 177:1419-1424.

Saxena, I.M., K. Kudlicka, K. Okuda, and R.M. Brown, Jr. 1994. Characterization of genes in the cellulose synthesizing operon (*acs* operon) of *Acetobacter xylinum*: Implications for cellulose crystallization. J. Bacteriol. 176:5735-5752.

Scarsbrook, C.E.; O.L. Bennett, and R.W. Pearson. 1959. The interaction of nitrogen and moisture on cotton yields and other characteristics. Agron. J. 51:718-721.

Schellenbaum, P., M. Vantard, C. Peter, A. Fellous, and A.-M. Lambert. 1993. Co-assembly properties of higher plant microtubule-associated proteins with purified brain and plant tubulins. Plant J. 3:253-260.

Schenachal, N. 1994. Genetic analysis of somatic embryogenesis in *Gossypium hirsutum*. Ph.D. Dissertation. Texas A&M University. College Station, Texas.

Schenk, R.J. and A.C. Hildebrandt. 1972. Medium and techniques for induction and growth of monocotyledonous and dicotyledonous plant cell cultures. Can. J. Bot. 50:199-204.

Schenone, G. 1993. Impact of air pollutants on plants in hot, dry climates. *In:* M.B. Jackson and C.R. Black (eds.). Interacting Stresses on Plants in a Changing Climate. NATO ASI Series, Vol. I 16. Springer-Verlag, Berlin. pp. 139-152.

Schmidt, J.H. 1989. An improved high performance liquid chromatographic method for detecting and quantifying terpenoids and heliocides in cottonseed. LC-GC 7:964-968.

Schmidt, J.R. and R. Wells. 1986. Recovery of soluble proteins from glanded cotton tissues with amines. Anal. Biochem. 154:224-229.

Schmidt, J.R. and R. Wells. 1990. Evidence for the presence of gossypol in malvaceous plants other than those in the "cotton tribe." J. Agric. Food Chem. 38:505-508.

Schmutz, A., A.J. Buchala, and U. Ryser. 1996. Changing the dimensions of suberin lamellae of green cotton fibers with a specific inhibitor of the endoplasmic reticulum-associated fatty acid elongases. Plant Physiology 110:403-411.

Schmutz, A., A. Buchala, U. Ryser, and T. Jenny. 1994a. The phenols in the wax and in the suberin polymer of green cotton fibres and their functions. Acta Hortica 381:269-275.

Schmutz, A., T. Jenny, N. Amrhein, and U. Ryser. 1993. Caffeic acid and glycerol are constituents of the suberin layers in the green cotton fibers. Planta 189:453-460.

Schmutz, A., T. Jenny, and U. Ryser. 1994b. A caffeoyl-fatty acid-glycerol ester from wax associated with green cotton fibre suberin. Phytochemistry 36:1343-1346.

Schnabl, H. and C. Kottmeier. 1984. Properties of phosphoenolpyruvate carboxylase in desalted extracts from isolated guard cell protoplasts. Planta 162:220-225.

Schnathorst, W.C. and D.C. Mathre. 1966. Cross protection with a fungus – *Verticillium albo-atrum*. Phytopathology 56:1204-1208.

Schonbeck, F. and H.W. Dehne. 1977. Damage to mycorrhizal and non-mycorrhizal cotton seedlings by *Thielaviopsis basicola*. Plant Dis. Report. 61:266-267.

Schonherr, J. 1976. Water permeability of isolated cuticular membranes: Effects of cuticular waxes on diffusion of water. Planta 131:159-164.

Schonherr, J. and Bauer, H. 1992. Analysis of effects of surfactants on permeability of plant cuticles. pp. 17-35. *In:* C.L. Foy (ed.). Adjuvants for Agrichemicals. CRC Press, Boca Raton.

Schonherr, J. and M.J. Bukovac. 1970. Preferential polar pathways in the cuticle and their relationship to ectodesmata. Planta 92:189-201.

Schonherr, J. and M.J. Bukovac. 1972. Penetration of stomates by liquids. Dependence on surface tension, wettability, and stomatal morphology. Plant Physiol. 49:813-819.

Schroeder, J.I. and R. Hedrich. 1989. Involvement of ion channels and active transport in osmoregulation and signaling of higher plant cells. Trends Biochem. Sci. 14:187-192.

Schubert, A.M., C.R. Benedict, J.D. Berlin, and R.J. Kohel. 1973. Cotton fiber development-kinetics of cell elongation and secondary wall thickening. Crop Sci. 13:704-709.

Schubert, A.M., C.R. Benedict, C.E. Gates, and R.J. Kohel. 1976. Growth and development of the lint fibers of Pima S-4 cotton. Crop Sci. 16:539-543.

Schubert, A.M., C.R. Benedict, and R.J. Kohel 1986. Carbohydrate distribution in bolls. pp. 311-324. *In*: J.R. Mauney and J. McD. Stewart (eds). Cotton Physiology. The Cotton Foundation, Memphis, Tenn.

Schuller, K.A., D.H. Turpin, and W.C. Plaxton. 1990. Metabolite regulation of partially purified soybean nodule phosphoenolpyruvate carboxylase. Plant Physiol. 94:1429-1435.

Schuller, K.A. and D. Werner. 1993. Phosphorylation of soybean (*Glycine max* L.) nodule phosphoenolpyruvate carboxylase in vitro decreases sensitivity to inhibition by L-malate. Plant Physiol. 101:1267-1273.

Schulz, M., T. Klockenbring, C. Hunte, and H. Schnabl. 1993. Involvement of ubiquitin in phosphoenolpyruvate carboxylase degradation. Bot Acta 106:143-145.

Schulz, P. and W.A. Jensen. 1971. Capsella embryogenesis: The chalazal proliferating tissue. J. Cell Science 8:201-227.

Schulze, D., N. Hopper, N. Hopper, J. Gannaway, and G. Jividen. 1997. Evaluation of chilling tolerance in cotton genotypes. pp. 1383-1385. *In:* Proc. Beltwide Cotton Conf., National Cotton Council of America, Memphis, Tenn.

Schuster, M.F., M.P. Gibbs, and W.C. Smith. 1990. High condensed tannin and *Heliothis* sp. resistance; biological effect on larvae damage and larvae effects on yield. pp. 84-85. *In:* J.M. Brown and D.A. Richter (eds). Proc. Beltwide Cotton Res. Conf.,

National Cotton Council of America, Memphis, Tenn.

Schuster, M.F., F.G. Maxwell, and J.N. Jenkins. 1972. Resistance to twospotted spider mite in certain *Gossypium hirsutum* races, *Gossypium* species, and glanded-glandless counterpart cottons. J. Econ. Entomol. 65:1108-1110.

Schwab, G.J. 1996. Soil potassium in a Typic Paleudult receiving surface applications of potassium fertilizer and nutrient influx in cotton (*Gossypium hursutum*). M.S. Thesis. Auburn University, AL.

Schwab, G.J., G.L. Mullins, and C.H. Burmester. 2000. Growth and nutrient uptake by cotton roots under field conditions. Commun. Soil Sci. Plant Anal. 31(1&2):149-164.

Schweiger, P.F., A.D. Robson, and N.J. Barrow. 1995. Root hair length determines beneficial effect of a *Glomus* species on shoot growth of some pasture species. New Phytol. 131:247-254.

Scott, N.S. and J.V. Possingham. 1980. Chloroplast DNA in expanding spinach leaves. J. Exp. Bot. 31:1081-1092.

Scott, W.P. 1990. Evaluation of aldicarb and ethephon in cotton production. pp. 278-280. *In:* J.M. Brown and D.A. Richter (eds.). Proc. Beltwide Cotton Prod. Res. Conf., National Cotton Council of America, Memphis, Tenn.

Seagull, R.W. 1986. Changes in microtubule organization and wall microfibril orientation during in vitro cotton fiber development: an immunofluorescent study. Can. J. Bot. 64:1373-1381.

Seagull, R.W. 1989a. The plant cytoskeleton. Crit. Rev. Plant Sci. 8:131-167.

Seagull, R.W. 1989b. The role of the cytoskeleton during oriented microfibril deposition. II. Microfibril disposition in cells with disrupted cytoskeletons. pp. 811-825. *In:* C. Schuerch (ed.). Cellulose and Wood, Chemistry and Technology. Wiley Intersciences Press, New York.

Seagull, R.W. 1990a. The effects of microtubule and microfilament disrupting agents on cytoskeletal arrays and wall deposition in developing cotton fibers. Protoplasma 159:44-59.

Seagull, R.W. 1990b. Tip growth and transition to secondary wall synthesis in developing cotton hairs. pp. 261-284. *In:* I.B. Heath (ed.). Tip Growth in Plant and Fungal Cells. Academic Press, New York.

Seagull, R.W. 1991. Role of the cytoskeletal elements in organized wall microfibril deposition. pp. 143-164. *In:* C.H. Haigler and P.J. Weimer (eds.). Biosynthesis and Biodegradation of Cellulose. Marcel Dekker, Inc., New York.

Seagull, R.W. 1992a. Cytoskeletal involvement in microfibril organization during cotton fiber development. pp. 171-192. *In:* C.R. Benedict (ed.). Proc. Cotton Fiber Cellulose: Structure, Function, and

Utilization Conference. National Cotton Council, Memphis Tenn.

Seagull, R.W. 1992b. A quantitative electron microscopic study of changes in microtubule arrays and wall microfibril orientation during *in vitro* cotton fiber development. J. Cell Sci. 101:561-577.

Seagull, R.W. 1993. Cytoskeletal involvement in cotton fiber growth and development. Micron 24:643-660.

Seagull, R.W. 1995. Cotton fiber growth and development: evidence for tip synthesis and intercalary growth in young fibers. Plant Physiol. (Life Sci. Adv.) 14:27-38.

Seagull, R.W. 1998. Cytoskeletal stability affects cotton fiber initiation. Int. J. Plant Sci. 159:590-598.

Seagull, R.W. and J.D. Timpa. 1990. The relationship between reversal frequency and fiber strength. p. 626. *In:* Proc. Beltwide Cotton Conf., National Cotton Council of America, Memphis, Tenn.

Seelanan, T., A. Schnabel, and J.F. Wendel. 1997. Congruence and consensus in the cotton tribe. Syst. Bot. 22:259-290.

Segurra, E., J.W. Keeling, and J.R. Abernathy. 1991. Tillage and cropping system effects on cotton yield and profitability on the Texas southern high plains. J. Prod. Agric. 4:566-571.

Sestak, Z. 1977a. Photosynthetic characteristics during ontogenesis of leaves. 1. Chlorophylls. Photosynthetica 11:367-448.

Sestak, Z. 1977b. Photosynthetic characteristics during ontogenesis of leaves. 2. Photosystems, components of electron transport chain, and photosynthesis. Photosynthetica 11:449-474.

Sestak, Z. 1978. Photosynthetic characteristics during ontogenesis of leaves. 3. Carotenoids. Photosynthetica 12:89-109.

Sexton, P.D. and C.J. Gerard. 1982. Emergence force of cotton seedlings as influenced by salinity. Agron. J. 74:699-702.

Shafer, W.E. and D.W. Reed. 1986. The foliar absorption of potassium from organic and inorganic potassium carriers. J. Plant Nutr. 9:143-157.

Shakhidoyatov, Kh.M., A.M. Rashkes, and N.K. Khidyrova. Components of cotton plant leaves, their functional role and biological activity. Chem. Natl. Compd. 33:605-616.

Shalhevet, J. 1993. Plants under salt and water stress. *In:* L. Fowden, T. Mansfield, and J. Stoddart (eds.). Plant Adaptation to Environmental Stress. Chapman and Hall, London. pp. 133-154.

Shalhevet, J. and T.C. Hsiao. 1986. Salinity and drought. A comparison of their effects on osmotic adjustment, assimilation, transpiration and growth. Irrig. Sci. 7:249-264.

Shanmugham, K. 1992. Seed soaking and foliar application of growth-regulants and anti-transpirant chemicals for increasing drought resistance in rainfed upland cotton (*Gossypium hirsutum*). Indian J. Agric. Sci. 62:744-750.

Shannon, M.C. and L.E. Francois. 1977. Influence of seed pretreatments on salt tolerance of cotton during germination. Agron. J. 69:619-622.

Shantz, H.L. 1927. Drought resistance and soil moisture. Ecology 8:145-157.

Shao, F.M. and C.D. Foy. 1982. Interaction of soil manganese and reaction of cotton to Verticillium wilt and Rhizoctonia root rot. Comm. Soil Sci. Plant Anal. 13:21-38.

Shappley, Z.W., J.N. Jenkins, C.E. Watson Jr., A.L. Kahler, and W.R. Meredith. 1996. Establishment of molecular markers and linkage groups in two F_2 populations of Upland cotton. Theor. Appl. Genet. 92:915-919.

Sharma, P.K., M. Singh, and A.K. Dhawan. 1989. Management of bollworms in *Gossypium arboreum* through nutrient and irrigation in Punjab. J. Res. Punjab Agric. Univ. 26:204-205.

Sharp, R.E. and W.J. Davies. 1979. Solute regulation and growth by roots and shoots of water-stressed maize plants. Plant 147:43-49.

Shaw, P.J. and E.G. Jordan. 1995. The nucleolus. Ann. Rev. Cell Dev. Biol. 11:93-121.

Shen, C.-Y. 1985. Integrated management of Fusarium and Verticillium wilts of cotton in China. Crop Protection 4:337-345.

Shen, F., C. Yin, Y. Yu, F. Lu, Li Chen, F.F. Shen, C.Y. Yin, Y.J. Yu, F.Z. Lu, and L. Chen. 1997. Screening of whole plants and pollen grains of cotton for salt tolerance. Acta Agronomica Sinica 23:620-625.

Sheng, C.F., K.R. Hopper, W.R. Meredith, E.G. King, and S.J. Ma. 1988. Cotton development, yield, and quality after early square removal with ethephon. pp. 121-124. *In:* J.M. Brown (ed.). Proc. Beltwide Cotton Prod. Res. Conf., National Cotton Council of America, Memphis, Tenn.

Shennan, C., S.R. Grattan, D.M. May, C.J. Hillhouse, D.P. Schachtman, M. Wander, B. Roberts, S. Tafoya, R.G. Burau, C. McNeish, and L. Zelinski. 1995. Feasibility of cyclic reuse of saline drainage in a tomato-cotton rotation. J. Environ. Qual. 24:476-486.

Shepherd, R.L. 1974. Breeding root-knot resistant *Gossypium hirsutum* L. using a resistant wild *G. barbadense* L. Crop Sci.14:687-691.

Sherrick, S.L., H.A. Holt, and F.D. Hess. 1986. Effects of adjuvants and environment during plant development on glyphosate absorption and translocation in field bindweed (*Convulvulus arvensis*). Weed Science 34:811-816.

Shevyakova, N.I., V.Y. Rakitin, L.M. Muzychko, and V.V. Kuznetzov. 1998. Stress-induced accumulation of proline in relation to salt tolerance of intact plants and isolated cells. App. Biochem. Microbiol. 34:291-295.

Shi, J., W.C. Mueller, and C.H. Beckman. 1991a. Ultrastructure and histochemistry of lipoidal droplets in vessel contact cells and adjacent parenchyma cells in cotton plants infected by *Fusarium oxysporum* f.sp. *vasinfectum*. Physiological and Molecular Plant Pathology 39:201-211.

Shi, J., W.C. Mueller, and C.H. Beckman. 1991b. Ultrastructural responses of vessel contact cells in cotton plants resistant or susceptible to infection by *Fusarium oxysporum* f.sp. *vasinfectum*. Physiological and Molecular Plant Pathology 38:211-222.

Shi, J., W.C. Mueller, and C.H. Beckman. 1992. Vessel occlusion and secretory activities of vessel contact cells in resistant or susceptible cotton plants infected with *Fusarium oxysporum* f.sp. *vasinfectum*. Physiological and Molecular Plant Pathology 40:133-147.

Shimizu, Y., S. Aotsuka, O. Hasegawa, T. Kawada, T. Sakuno, F. Sakai, and T. Hayashi. 1997. Changes in levels of mRNAs for cell wall-related enzymes in growing cotton fiber cells. Plant Cell Physiol. 38:375-378.

Shimose, N. and J. Sekiya. 1991. Salt tolerance of higher plants and uptake of inorganic nutrients. Sci. Rep. Faculty of Agriculture, Okayama University 77:21-29.

Shimshi, D. and A. Marani. 1971. Effects of soil moisture stress on two varieties of upland cotton in Israel. II. The northern Negev region. Exper. Agric. 7:225-239.

Shin, I.V., I.S. Kasparova, L.G. Gorenkova, Y.E. Giller, and Y.S. Nasyrov. 1991. A photosynthetic apparatus with elevated activity of the PEP system in cotton. Soviet Plant Physiol. 38:16-21.

Shiroya, T. 1963. Metabolism of raffinose in cotton seeds. Phytochem. 2:33-46.

Shishido, Y., M. Seyama, S. Imada, and Y. Hori. 1990. Effect of the photosynthetic light period on the carbon budget of young tomato leaves. Ann. Bot. 66:729-735.

Shoemaker, R.C., L.J. Couche, and D.W. Galbraith. 1986. Characterization of somatic embryogenesis and plant regeneration in cotton (*Gossypium hirsutum* L.). Plant Cell Rep. 3:178-181.

Shorrocks, V. 1997. The occurrence and correction of boron deficiency. Plant Soil. 193:121-148.

Sieverding, E., S. Toro, and O. Mosquera. 1989. Biomass production and nutrient concentrations in spores of VA mycorrhizal fungi. Soil Biol. Biochem. 21:69-72.

Sikora, R.A. and K. Sitaramaiah. 1980. Antagonistic interaction between the endotrophic mycorrhizal fungus *Glomus mosseae* and *Rotylenchulus reniformis* on cotton. Nematropica 10:72-73.

Silberbush, M. and J. Ben-Asher. 1987. The effect of salinity on parameters of potassium and nitrate uptake of cotton. Communications in Soil Sci. and Plant Anal. 18:65-81.

Silow, R.A. 1944. The genetics of species development in the Old World cottons. J. Genet. 46:62-77.

Silva Fernandes, A.M., E.A. Baker, and J.T. Martin. 1964. Studies on plant cuticle. VI. The isolation and fractionation of cuticular waxes. Ann. Appl. Bot. 53:43-58.

Silvertooth, J.C., E.R. Norton, and S.E. Ozuna. 1998. Foliar fertilizer evaluation on Upland cotton, 1997. Cotton: A College of Agriculture Report 1998. Publication AZ1006. The University of Arizona, Tucson, AZ.

Silverthooth, J.C., A.F. Wrona, D.S. Guthrie, K. Hake, T.A. Kerby, and J.A. Landivar. 1996. Vigor indices for cotton management. Cotton Physiology Today 7(3):9-12. National Cotton Council, Memphis, Tenn.

Simon, E.W. 1978. The symptoms of calcium deficiency in plants. New Phytol. 80:1-15.

Simon, E.W. 1979. Seed germination at low temperatures. pp. 37-45. *In:* J.M. Lyons, D. Graham, and J.K. Raison (eds.). Low Temperature Stress in Crop Plants: The Role of the Membrane, Academic Press, New York.

Simon, E.W. 1984. Early events in germination. pp. 77-115. *In:* D.R. Murray (ed.). Seed Physiology II, Germination and Reserve Mobilization. Academic Press Australia, Sydney.

Simpson, M.E. and L.R. Batra. 1983. Ecological relations in respect to a boll rot of cotton caused by *Aspergillus flavus*. pp. 24-32 *In:* H. Kurata and Y. Ueno (eds.). Toxigenic Fungi-Their Toxins and Health Hazard. Elsevier, Tokyo.

Sinclair, T.R. 1992. Mineral nutrition and plant growth response to climate change. J. Exp. Bot. 43:1141-1146.

Sinclair, T.R. and M.M. Ludlow. 1985. Who taught plants thermodynamics? The unfulfilled potential of plant water potential. Aust. J. Plant Physiol. 12:213-217.

Sinclair, T.R. and W.I. Park. 1993. Inadequacy of the Liebig limiting-factor paradigm for explaining varying crop yields. Agron. J. 85:742-746.

Singh, C.S. 1992. Mass inoculum production of vesicular-arbuscular (VA) mycorrhizae: I. Selection of host in the presence of *Azospirillum brasilense*. Zentralblatt fur Mikrobiologie 147:447-453.

Singh, D., M.S. Brar, and A.S. Brar. 1992. Critical concentrations of potassium in cotton (*Gossypium hirsutum*). J. Agri. Sci. 118:71-75.

Singh, J.S., L. Singh, and C.B. Pandey. 1991. Savannization of dry tropical forest increases carbon flux relative to storage. Current Sci. 61:477-480.

Singh, R. and J. Singh. 1996. Irrigation planning in cotton through simulation modelling. Irrig. Sci. 17:31-36.

Singh, R.P. and S. Bhan. 1993. Yield, quality, and economics of summer cotton (*Gossypium* species) as influenced by frequency of irrigation and moisture-conservation practices. Indian J. Agron. 38:439-442.

Singh, S. 1970. Revolution in cotton yield with CCC. Indian Farming. 20:5-6.

Singh, S. 1975. Studies on the effects of soil moisture stress on the yield of cotton. Indian J. Plant Physiol. 18:49-50.

Sirkar, S. and J.V. Amin. 1974. The manganese toxicity of cotton. Plant Physiol. 54:539-543.

Sitaram, M.S. and E.S. Abraham. 1973. Note on effect of gibberellic acid on quality of Laxmi cotton. Cotton Grower Rev. 50:150-151.

Skinner, R.H. and J.W. Radin. 1994. The effect of phosphorus nutrition on water flow through apoplastic bypass on cotton roots. J Exp. Bot. 45:423-428.

Slama, F. 1991. Transport of Na^+ in leaves and sensitivity of plants to NaCl. Assessment of a trap effect at the level of the stems. Agronomie 11:275-281.

Slaymaker, P.H., N.P. Tugwell, C.E. Watson, Jr., M.J. Cochran, F.M. Bourland, and D.M. Oosterhuis. 1995. Documentation of the SQUARMAP procedure and software for mapping squaring nodes. pp. 483-484. Proc. Beltwide Cotton Conf., National Cotton Council of America, Memphis, Tenn.

Smalley, S.A. and E.J. Bicknell. 1982. Gossypol toxicity in dairy cattle. Comp. on Cont. Educ. for the Practicing Veterinarian 4:5376-5381.

Smart, L.B., F. Vojdani, C.-Y. Wan, M. Maeshima, and T.A. Wilkins. 1996. Regulation of genes involved in cell elongation during cotton (*Gossypium hirsutum*) fiber development. Plant Physiol. Suppl. 111:144.

Smets, S.M.P., M. Kuper, J.C. van Dam, and R.A. Feddes. 1997. Salinization and crop transpiration of irrigated fields in Pakistan's Punjab. Agric. Water Manage. 35:43-60.

Smit, F. and I.A. Dubery. 1997. Cell wall reinforcement in cotton hypocotyls in response to a *Verticillium dahliae* elicitor. Phytochemistry 44:811-815.

Smith, B. 1991. A review of the relationship of cotton maturity and dyeability. Textile Res. J. 61:137-145.

Smith, C.E. and S.G. Stephens. 1971. Critical identification of Mexican archaeological cotton remains. Econ. Bot. 25:160-168.

Smith, C. W. 1992. History and status of host plant resistance in cotton to insects in the United States. Adv. Agron. 48:251-296.

Smith, C.W., J.T. Cothren, and J.J. Varvil. 1986. Yield and fiber quality of cotton following application of 2-chloroethyl phosphonic acid. Agron. J. 78:814-818.

Smith, C.W. and G.K. Golladay. 1994. Association of fiber quality parameters and within-boll components of lint yield. pp. 47-50. *In:* G. Jividen and C. Benedict (eds.). Proc. Biochemistry of Cotton Workshop. Cotton Incorporated, Raleigh NC.

Smith, C.W., J.C. McCarty, T.P. Altamarino, K.E. Lege, M.F. Schuster, J.R. Phillips, and J.D. Lopez. 1992. Condensed tannins in cotton and bollworm-budworm (Lepidoptera: Noctuidae) resistance. J. Econ. Entomol. 85:2211-2217.

Smith, C.W., B.A. Waddle, and H.H. Ramey, Jr. 1979. Plant spacings with irrigated cotton. Agron. J. 71:858-860.

Smith, C.W. and J.J. Varvil. 1984. Standard and cool germination tests compared with field emergence in Upland cotton. 76:587-589.

Smith, F., N. Malm, and C. Roberts. 1987. Timing and rates for foliar nitrogen application of cotton. pp. 61-64 *In:* Proc. Beltwide Cotton Prod. Res. Conf., National Cotton Council, Memphis, Tenn.

Smith, G.S. and R.W. Roncadori. 1986. Responses of three vesicular-arbuscular mycorrhizal fungi at four soil temperatures and their effects on cotton growth. New Phytol. 104:89-95.

Smith, G.S., R.W. Roncadori, and R.S. Hussey. 1986. Interaction of endomycorrhizal fungi, superphosphate, and *Meloidogyne incognita* on cotton in microplot and field studies. J. Nematol. 18:208-216.

Smith, H., J.J. Casal, and G.M. Jackson. 1990. Reflection signals and the perception by phytochrome of the proximity of neighbouring vegetation. Plant Cell Environ.13:73-78.

Smith, L.A. 1992. Response of cotton to deep tillage on Tunica clay. pp. 505-506. *In:* Proc. Beltwide Cotton Conf., National Cotton Council of America, Memphis, Tenn.

Smith, R.G., S.K. Hicks, R.W. Lloyd, J.R. Gannaway, and J. Zak. 1989. VA mycorrhizal infection and its implication in the "fallow syndrome" associated with cotton grown in previously fallowed soils. p. 375. *In:* J.M. Brown (ed.). Proc. Beltwide Cotton Production Research Conf., National Cotton Council, Memphis, Tenn.

Smith, R.H., H.J. Price, and J.B. Thaston. 1977. Defined conditions for the initiation and growth of cotton callus *in vitro*. I. *Gossypium arboreum*. In Vitro 13:320-334.

Smith, R.H., M. Schubert, and C.R. Benedict. 1974. The development of isocitric lyase activity in germinating cotton seed. Plant Physiol. 54:197-200.

Smith, S.E. and D.J. Read. 1997. Mycorrhizal Symbiosis, 2nd Edition. Academic Press, London.

Smith, W.H. 1990. Air Pollution and Forests, Second Ed. Springer-Verlag, N.Y.

Smithson, J.B. 1972. Differential sensitivity to boron in cotton in the northern states of Nigeria. Cotton Growing Review. 49:350-353.

Snek, B., L. Burppie, and A. Ogoshi. 1991. Identification of *Rhizoctonia* Species. American Phytopathological Society, St. Paul, MN. 135 pp.

Snipes, C.E. 1996. Weed control in ultra narrow row cotton possible strategies assuming a worst case scenario. pp. 66-67. *In:* P. Dugger and D.A. Richter (eds.). Proc. Beltwide Cotton Conf., National Cotton Council of America, Memphis, Tenn.

Snipes, C.E. and C.C. Baskin. 1994. Influence of early defoliation on cotton yield, seed quality, and fiber properties. Field Crops Res. 37:137-143.

Snow, J.P., S.H. Crawford, G.T. Berggren, and J.G. Marshall. 1981. Growth regulator tested for cotton boll rot control. Louisiana Agric., Louisiana Agric. Exp. Stn. 24:3-24.

Snyder, C.S., G.M. Lorenz, W.H. Baker, D. Vangilder. E. Ebelhar, and P. Ballantyne. 1995. Late season foliar fertilization of cotton with potassium nitrate in Jefferson County: Two-year results. Ark. Agri. Exp. Sta. Res. Ser. 443:83-90.

Snyder, C.S., W.N. Miley, D.M. Oosterhuis, and R.L. Maples. 1991. Foliar application of potassium to cotton. University of Arkansas, Cooperative Extension Service. Cotton Comments. 4-91.

Snyder, C., D.M. Oosterhuis, D.D. Howard, and J.S. McConnell. 1998. Foliar nitrogen and potassium fertilization of cotton. Phosphate and Potash Institute, News & Views Newsletter, June 1998. 4 pp.

Snyder, W.C. and H.N. Hansen. 1940. The species concept in *Fusarium*. American Journal of Botany 27:64-67.

Soil Science Society of America. 1997. Glossary of Soil Science Terms. Soil Sci. Soc. of America, Madison, Wis.

Sokolova, T.B., A.A. Tomakkov, Yu.V. Shlyakhov, Z.A. Karimov, Kh.V. Khegai, and A.Kh. Mutalov, A.Kh. 1991. Polymorphism for salt tolerance and mineral level of the tissues in cotton plants of *Gossypium barbadense* L. 1 S"ezd fiziologov rastenii Uzbekistana, Tashkent, 1991, 139. (Russian).

Solarova, J. and J. Pospisilova. 1983. Photosynthetic characteristics during ontogenesis of leaves. 8. Stomatal diffusive conductance and stomata reactivity. Photosynthetica 17:101-151.

Soliman, M.F., M.A. Farah, and I.M Anter. 1980. Seed germination and root growth of corn and cotton seedlings as affected by soil texture and salinity of irrigation water. Agrochimica 24:113-120.

Solomon, K.H. 1985. Water-salinity-production functions. Trans. ASAE. 28:1975-1980.

Solovova, G.K., O.V. Kalaptur, and M.I. Chumakov. 1999. Analysis of the attachment of agrobacteria to wheat and rice plants. Microbiology 68:63-68.

Soltis, D.E., P.S. Soltis, D.L. Nickrent, L.A. Johnson, W.J. Hahn, S.B. Hoot, J.A. Sweere, R.K. Kuzoff, K.A. Kron, M.W. Chase, S.M. Swensen, E.A. Zimmer, S.-M. Chaw, L.J. Gillespie, W.J. Kress, and K.J. Sytsma. 1997. Angiosperm phylogeny inferred from 18S ribosomal DNA sequences. Ann. Missouri Bot. Gard. 84:1-49.

Somero, G.N. and P.S. Low. 1976. Temperature: A "shaping force" in protein evolution. Biochem. Soc. Symp. 41:33-42.

Song, F. and Z. Zheng. 1997. Involvement of gossypol and tannin in the resistance of cotton seedlings to Fusarium wilt. Journal of Zhejiang Agricultural University 23:529-532.

Song, F.-M. and Z. Zheng. 1998. The correlation between inhibition of ethylene production and trifluralin-induced resistance of cotton seedlings against *Fusarium oxysporum* f.sp. *vasinfectum*. Acta Phytophysiologica Sinica 24:111-118.

Song, P. and R.D. Allen. 1997. Identification of a cotton fiber-specific acyl carrier protein cDNA by differential display. Biochim. et Biophys. Acta 1351:305-312.

Song, P., P. Dang, J. Heinen, and R. Allen. 1996. Isolation and characterization of a cotton fiber-specific promoter. Plant Physiol. Suppl. 111:55.

Sonobe, S. 1990. ATP-dependent depolymerization of cortical microtubules by an extract in tobacco BY-2 cells. Plant Cell Physiol. 31:1147-1153.

Soomro, A.W., and S.A. Waring. 1987. Effect of temporary flooding on cotton growth and nitrogen nutrition in soils with different organic matter levels. Aust. J. Agric. Res. 38:91-99.

Souza, J.G., A.C.Q.T. Barros, and J.V. da Silva 1983. Reservas de hidratos de carbono e resistência do algodoeiro à seca. Pesq. Agropec. Brasileira 18:269-273.

Souza, J.G. and J.V. da Silva, 1987. Protein and carbohydrate changes in cotton leaves linked with age and development. Trop. Agric. 64:46-48.

Souza, J.G., F.A. Neto, and J.V. da Silva 1990. Physiological studies on the productivity of cotton (*Gossypium hirsutum* L.). The partitioning of carbohydrate and the role of leaves. Bull. Soc. Bot. Fr. 137:139-147.

Sowell, W.F., R.D. Rouse, and J.I. Wear. 1957. Copper toxicity of the cotton plant in solution cultures. Agron. J. 49:206-207

Spears, J.F. 1997. Seed quality and seedling emergence. *In:* 1997 North Carolina Cotton Production Guide, Center for IPM, North Carolina State University, Raleigh, N.C. Available at http://ipmwww.ncsu.edu/Production_Guides/cotton/chptr6.html

Specks, U., T.J. Kreofsky, A.H. Limper, P.J. Bates, W.M. Brutinel, and M.S. Rohrbach. 1995. Comparison of neutrophil chemotactic factor release by human and rabbit alveolar macrophages in response to tannin exposure. Journal of Laboratory & Clinical Medicine 125:237-246.

Speer, M. and W.M. Kaiser. 1991. Ion relations of symplastic and apoplastic space in leaves from *Spinacia oleracea* L. and *Pisum sativum* L. under salinity. Plant Physiol. 97:900-997.

Spieth, A.M. 1 933. Anatomy of the transition region in *Gossypium.* Bot. Gaz. 95:338-347.

Spooner, A.E., C.E. Caviness, and W.I. Spurgeon. 1958. Influence of timing of irrigation on yield, quality, and fruiting of Upland cotton. Agron. J. 50:74-77.

Spoor, M. 1998. The Aral Sea Basin crisis: Transition and environment in former Soviet Central Asia. Development and Change 29:409-435.

Sprenger, G.A., U. Schorken, T. Wiegert, S. Grolle, A.A. de Graaf, S.V. Taylor, T.P. Begley, S. Bringer-Meyer, and H. Sahm. 1997. Identification of a thiamin-dependent synthase in *Escherichia coli* required for the formation of the 1-deoxy-D-xylulose 5-phosphate precursor to isoprenoids, thiamin, and pyridoxol. Proc. natl. Acad. Sci. USA. 94:12857-12862.

Spruegel, D.G., T.M. Hinckley, and W. Schaap. 1991. The theory and practice of branch autonomy. Annu. Rev. Ecol. Syst. 22:309-334.

Staggenborg, S.A. and D.R. Krieg. 1993. Fruit production and retention as by plant density and water supply. pp. 1244-1247. *In:* D.J. Herber and D.A. Richter (eds.). Proc. Beltwide Cotton Conf., National Cotton Council of America, Memphis, Tenn.

Staggenborg, S.A. and D.R. Krieg. 1994. Fruiting site production and fruit size responses of cotton to water supply per plant. pp. 135 1-1354. *In:* D.J. Herber and D.A. Richter. (eds.). Proc. Beltwide Cotton Conf., National Cotton Council of America, Memphis, Tenn.

Standley, P.C. 1930. Flora of Yucatan. Field Museum of Natural History Publications, Botanical Series, vol. 3, 157-492. Chicago, IL.

Stanford, G. and H.V. Jordan. 1966. Sulfur requirements of sugar, fiber, and oil crops. Soil Sci. 101:258-266.

Stanton, M.A., J. McD. Stewart, A.E. Percival, and J.F. Wendel. 1994. Morphological diversity and relationships in the A-genome cottons, *Gossypium arboreum* and *G. herbaceum*. Crop Science 34: 519-527.

Stapel, J.O., A.M. Cortesero, C.M. De Moraes, J.H. Tumlinson, and W.J. Lewis. 1997. Extrafloral nectar, honeydew, and sucrose effects on searching behavior and efficiency of *Microplitis croceipes*

(Hymenoptera: Braconidae) in cotton. Environ. Entomol. 26:617-623.

Stapleton, H.N., D.R. Buxton, F.L. Watson, D.J. Nolting, and D.N. Baker. 1973. Cotton: A computer simulation of cotton growth. Arizona Agr. Exp. Sta. Tech. Bull., Tucson, AZ. 123 pp.

Stark, C. 1991. Osmotic adjustment and growth of salt-stressed cotton as improved by a bioregulator. J. Agron. Crop Sci. 167:326-334.

Stark, C. and R. Schmidt. 1991. Behaviour of ^{22}Na in salt-stressed crops as affected by the growth regulator MCBuTTB. Beitrage zur Tropischen Landwirtschaft und Veterinarmedizin 29:435-443.

Starr, J.L. 1998. Cotton. pp. 359-379. *In*: K.R. Barker, G.A. Pederson, and G.L. Windham (eds.). Plant and Nematode Interactions, Agronomy Monograph No. 36, American Society of Agronomy, Madison, WI.

Starr, J.L. and J.A. Veech. 1986. Comparison of development, reproduction, and aggressiveness of *Meloidogyne incognita* race 3 and 4 on cotton. Journal of Nematology 18:413-415.

Steer, M.W. and J.M. Steer. 1989. Pollen tube tip growth. New Phytol. 111:323-358.

Steiner, J.J. and T.A. Jacobsen. 1992. Time of planting and diurnal soil temperature effects of cotton seedling field emergence and rate of development. Crop Sci. 32:238-244.

Stelly, D.M. 1992. Important reproductive angiosperm mutants, and a detailed discussion of the semigamy mutant of cotton. Apomixis Workshop, Atlanta, GA. 11-12 Feb. 1992. USDA-ARS Publ. 104:53-57.

Stelly, D.M., D.W. Altman, R.J. Kobel, T.S. Rangan, and E. Commiskey. 1989. Cytogenetic abnormalities of cotton somaclones from callus cultures. Genome 32:762-770.

Stelly, D.M., K.C. Kautz. and W.L. Rooney. 1990. Pollen fertility of some simple and compound translocations of cotton. Crop Sci. 30:952-955.

Stelter, W., A. Läuchli, and M.G. Pitman. 1979. K/Na selectivity of cotton plants in relation to salt tolerance. Plant Physiol. 63 Supplement:116.

Stephens, S.G. 1946. The genetics of "corky". I. The New World alleles and their possible role as an interspecific isolating mechanism. J. Genet. 47:150-161.

Stephens, S.G. 1949. The cytogenetics of speciation in *Gossypium*. I. Selective elimination of the donor parent genotype in interspecific backcrosses. Genetics 34:627-637.

Stephens, S.G. 1950a. The genetics of "corky". II. Further studies on its genetic basis in relation to the general problem of interspecific isolating mechanisms. J. Genet. 50:9-20.

Stephens, S.G. 1950b. The internal mechanism of speciation in *Gossypium*. Bot. Rev. 16:115-149.

Stephens, S.G. 1958. Salt water tolerance of seeds of *Gossypium* species as a possible factor in seed dispersal. Amer. Nat. 92:83-92.

Stephens, S.G. 1966. The potentiality for long range oceanic dispersal of cotton seeds. Amer. Nat. 100:199-210.

Stephens, S.G. 1967. Evolution under domestication of the New World cottons (*Gossypium* spp.). Ciencia e Cultura 19:118-134.

Stephens, S.G. 1974. Geographic and taxonomic distribution of anthocyanin genes in New World cottons. J. Genetics 61:128-141.

Stephens, S.G. 1975. Some observations on photoperiodism and the development of annual forms of domesticated cotton. Econ. Bot. 30:409-418.

Stephens, S.G. and M.E. Moseley. 1974. Early domesticated cottons from archaeological sites in central coastal Peru. American Antiquity 39:109-122.

Stephens, S.G. and L.L. Phillips. 1972. The history and geographical distribution of a polymorphic system in New World cottons. Biotropica 4:49-60.

Stevens, G., B. Phipps, and J. Mobley. 1996. Managing cotton for reduced wind damage with ridge till systems. pp. 1403-1405. *In:* Proc. Beltwide Cotton Conf., National Cotton Council of America, Memphis, Tenn.

Stevens, P.J.G., J.A. Zabkiewicz, J.H. Barran, K.R. Klitscher, and F. Ede. 1992. Spray formulation with Silwet organosilicone surfactant. pp. 399-403. *In:* C.L. Foy (ed.). Adjuvants for Agrichemicals. CRC Press, Boca Raton.

Stevens, W.E., J.R. Johnson, J.J. Varco, and J. Parkman. 1992. Tillage and winter cover management effects on fruiting and yield of cotton. J. Prod. Agric. 5:570-575.

Stewart, A.M., K.L. Edmisten, and R. Wells. 1998. Use of Prep, Starfire, Cottonquik, and Finish for boll opening. pp. 1382-1383. *In:* P. Dugger and D. Richter (eds.). Proc. Beltwide Cotton Prod. Res. Conf., National Cotton Council of America, Memphis, Tenn.

Stewart, J.I. and R.M. Hagan. 1973. Functions to predict effects of crop water deficits. J. Irrig. Drain. Div. Am. Soc. Civ. Eng. 100:179-199.

Stewart, J.McD. 1975. Fiber initiation on the cotton ovule *(Gossypium hirsutum)*. Amer. J. Bot. 62:723-730.

Stewart, J.McD. 1986. Integrated events in the flower and fruit. pp. 261-297. *In:* J.R. Mauney and J. McD. Stewart (eds.). Cotton Physiology. Cotton Foundation, Memphis, Tenn.

Stewart, J.McD. 1994. Potential for crop improvement with exotic germplasm and genetic engineering. Pp 313-327 *In:* G.A. Constable and N.W. Forrester (eds.). *Challenging the Future: Proceedings of the World Cotton Research Conference-1*. CSIRO, Melbourne, Australia.

Stewart, J. McD., P.A. Fryxell, and L.A. Craven. 1987. The recognition and geographic distribution of *Gossypium nelsonii* (Malvaceae). Brunonia 10:215-218.

Stewart, J.McD. and R.T. Robbins. 1996. Enhancement of cotton germplasm resistant to reniform nematode. Pp. 20-22. *In*: D.M. Oosterhuis (ed.). Proc. 1996 Cotton Research Meeting and 1996 Summaries of Cotton Research in Progress. Univ. of Ark. Agric. Exp. Sta. Special Report 193.

Sticher, L., B. Mauch-Mani, and J.P. Metraux. 1997. Systemic acquired resistance. Annu. Rev. Phytopathol. 35:235-270.

Stichler, C.R. 1996. Seedling health from an agronomic point of view. pp. 240-241. *In:* Proc. Beltwide Cotton Conf., National Cotton Council of America, Memphis, Tenn.

Stipanovic, R.D. 1992. Natural product biosynthesis via the Diels-Alder reaction. pp. 319-328. *In:* R.J. Petroski and S.P. McCormick (eds). Secondary-Metabolite Biosynthesis and Metabolism. Plenum Press, New York.

Stipanovich, R.D., D.W. Altman, D.L. Begin, G.A. Greenblatt, and J.H. Benedict. 1988. Terpenoid aldehydes in upland cottons: analysis by aniline and HPLC methods. J. Agric. Food Chem. 36:509-515.

Stipanovic, R.D., A.A. Bell, and C.R. Benedict. 1999. Cotton pest resistance: the role of pigment gland constituents. pp. 211-220. *In*: H. Cutler and S. Cutler (eds.). Natural Products: Agrochemicals and Pharmaceuticals. CRC Press, Cleveland, Ohio.

Stipanovic, R.D., A.A. Bell, and M.J. Lukefahr. 1977. Natural insecticides from cotton (*Gossypium*). pp. 197-214. *In:* Host Plant Resistance to Pests. ACS Symposium Series, No. 62. American Chemical Society, Washington, D.C.

Stipanovic, R.D., A.A. Bell, M.E. Mace, and C.R. Howell. 1975. Antimicrobial terpenoids of *Gossypium*: 6-methyoxygossypol and 6,6'-dimethoxygossypol. Phytochemistry 14:1077-1081.

Stipanovic, R.D., A.A. Bell, D.H. O'Brien, and M.J. Lukefahr. 1977. Heliocide H$_2$: An insecticidal sesterterpenoid from cotton (*Gossypium*). Tetra. Lett. 6:567-570.

Stipanovic, R.D., A.A. Bell, D.H. O'Brien, and M.J. Lukefahr. 1978. Heliocide H$_1$: A new insecticidal C$_{25}$ terpenoid from cotton (*Gossypium hirsutum*). Ag. Food Chem. 26:115-118.

Stipanovic, R.D., A.A. Bell, and D.H. O'Brien. 1980. Raimondal, a new sesquiterpenoid from pigment glands of *Gossypium raimondii*. Phytochemistry 19:1735-1738.

Stipanovic, R.D., H.L. Kim, D.W. Altman, A.A. Bell, and R.J. Kohel. 1994. Raimondalone, a sesqui-

terpenene from a cotton interspecific hybrid. Phytochemistry 36:953-956.

Stipanovic, R.D., M.E. Mace, D.W. Altman, and A.A. Bell. 1988. Chemical and anatomical response in *Gossypium* spp. challenged by *Verticillium dahliae*. pp. 262-272. *In*: H.G. Cutler (ed.). Biologically Active Natural Products: Potential Use in Agriculture. ACS Symposium Series 380, American Chemical Society, Washington, D.C.

Stipanovic, R.D., M.E. Mace, A.A. Bell, and R.C. Beier. 1992. The role of free radicals in the decomposition of the phytoalexin desoxyhemigossypol. J. Chemical Society, Perkin Transaction I, pp. 3189-3192.

Stipanovic, R.D., M.E. Mace, M.H. Elissalde, and A.A. Bell. 1991. Desoxyhemigossypol, a cotton phytoalexin: Structure-activity relationship. pp. 336-351. *In*: P.A. Hedin (ed.). Naturally Occurring Pest Bioregulators. ACS Symposium Series 449, American Chemical Society, Washington, D.C.

Stipanovic, R.D., L.S. Puckhaber, and A.A. Bell. 1998. Biological activity and synthesis of a kenaf phytoalexin highly active against fungal wilt pathogens. pp. 318-324. *In*: D.R. Baker, J.G. Fenyes, G.S. Basarab, and D.A. Hunt (eds.). Synthesis and Chemistry of Agrochemicals V. ACS Symposium Series 686, American Chemical Society, Washington, D.C.

Stipanovic, R.D., A. Stoessl, J.B. Stothers, D.W. Altman, A.A. Bell, and P. Heinstein. 1986. The stereochemistry of the biosynthetic precursor of gossypol. J. Chem. Soc., Chem. Commun. pp. 100-102.

Stipanovic, R.D., H.J. Williams, D.P. Muehleisen, and F.W. Plapp, Jr. 1987. Synthesis of deuterated and tritated gossypol. J. of Labelled Compounds and Radiopharm. XXIV(6):741-743.

Stipanovic, R.D., H.J. Williams, I. Sattler, A.I. Scott, S.B. Vinson, and J. Liu. 1997. Preparation of two stereochemically defined isomers of deuterium labelled d-cadinene. J. Labelled Comp. Radiopharm. 39:223-230.

Stitt, M., R. Gerhardt, B. Kürzel, and H.W. Heldt. 1983. A role for fructose 2,6-bisphosphate in the regulation of sucrose synthesis in spinach leaves. Plant Physiol. 72:1139-1141.

Stock, J.B., A.M. Stock, and J.M. Mottonen. 1990. Signal transduction in bacteria. Nature 344:395-400.

Stolzy, L.H. 1974. Soil atmosphere. pp. 335-363. *In:* E.W. Carson (ed.). The Plant Root and its Environment. Univ. Press of Virginia, Charlottesville.

Stomberg, L.K. 1960 Need for potassium fertilizer on cotton determined by leaf and soil analysis. Calif. Agric. 14:4-5.

Strand, L.L. and H. Mussell. 1975. Solubilization of peroxidase activity from cotton cell walls by endopolygalacturonases. Phytopathology 65:830-831.

Strogonov, B.P. 1964. Physiological Basis of Salt Tolerance in Plants. I.P.S.T. Jerusalem O.L. Bourne Press, London.

Strunnikova, O.K., N.A. Vishnevskaya, N.M. Labutova, T.F. Zubenko, A.A. Batyrov, and G.S. Muromtsev. 1997. Evaluation of the development of the *Verticillium dahliae* Kleb. depending on various conditions of cotton cultivation. Mikologiya i Fitopatologiya 31:46-53.

Stuart, B.L., C.W. Isbell, C.W. Wendt, and J.R. Abernathy. 1984. Modification of water relations and growth with mepiquat chloride. Agron. J. 76:651-655.

Subbaiah, G.V., P.S. Rao, and C.S. Rao. 1995. Effect of graded levels of salinity of N, P, and K uptake and K/Na ratio in eleven cotton cultivars. J. Potassium Res. 11:176-184.

Subba Rao, N.S.S., K.V.B.R. Tilak, and C.S. Singh. 1985. Synergistic effect of vesicular-arbuscular mycorrhizas and *Azospirillum brasilense* on the growth of barley in pots. Biol. Fertil. Soils 17:119-121.

Subbarao, K.V., A. Chassot, T.R. Gordon, J.C. Hubbard, P. Bonello, R. Mullin, D. Okamoto, R.M. Davis, and S.T. Koike. 1995. Genetic relationships and cross pathogenicities of *Verticillium dahliae* isolates from cauliflower and other crops. Phytopathology 85:1105-1112.

Subbiah, K.K., and A. Mariakulandia. 1972. Application of gibberellic acid and naphthalene acetic acid in preventing bud and boll shedding Cambodia cotton. Madras Agric. J. 59:350-352.

Suelter, C.H. 1985. Role of potassium in enzyme catalysis. *In:* R.D. Munson (ed.). Potassium in agriculture. Amer. Soc. Agron. Madison, WI. pp. 337-350.

Sugiyama, J. 1985. Lattice images from ultrathin sections of cellulose microfibrils in the cell wall of *Valonia macrophysa* Kütz. Planta 166:161-168.

Sugonyaev, E.S. 1994. Cotton pest management. Part 5. A Commonwealth of Independent States perspective.Ann. Rev. Entomol. 39:579-592.

Suh, M.W., X. Cui, and P.E. Sasser. 1996. Small bundle tensile properties of cotton related to MANTIS and HVI data -- A road to yarn strength prediction. pp. 1296-1300. *In:* Proc. Beltwide Cotton Conf., National Cotton Council of America, Memphis, Tenn.

Summy, K.R. and E.G. King. 1992. Cultural control of cotton insect pests in the United States. Crop Protection 11:307-319.

Sumner, D.R. 1995. Seedling diseases of cotton in rotation with peanut. p. 205. *In:* Proc. Beltwide Cotton Conf., National Cotton Council of America, Memphis, Tenn.

Sumner, M.E. 1977. Preliminary NPK foliar diagnostic norms for wheat. Comm. Soil Sci. Plant Anal. 8:149-167.

Sun, T.J., M. Essenberg, and U. Melcher. 1989. Photoactivated DNA nicking, enzyme inactivation, and bacterial inhibition by sesquiterpenoid phytoalexins from cotton. Molecular Plant-Microbe Interactions 2:139-147.

Sun, T.J., U. Melcher, and M. Essenberg. 1988. Inactivation of cauliflower mosaic virus by a photoactivatable cotton phytoalexin. Physiological and Molecular Plant Pathology 33:115-126.

Sun, X., L.H. Rao, Y.S. Zhang, Q.H. Ying, C.X. Tang, and L.X. Qin. 1989. Effect of potassium fertilizer application on physiological parameters, and yield of cotton grown on a potassium deficient soil. Zeit. Pflanz. Boden. 152:269-272.

Sun, Z.Y. and C.J. Xu. 1986. Soil available B in the south of Hebei province and application of Boron to cotton. J. Soil Sci. (in Chinese). 17:130-132.

Sundararajan, K.S. 1980. Circadian rhythms in the cotton plant *Gossypium hirsutum* L. cv. Lakshmi diss. Madurai Kamaraj University, Madurai, India, 1980.

Sundararajan, K.S., R. Subbaraj, M.K. Chandrashekaran,and S. nmugasundaram. 1978. Influence of fusaric acid on circadian leaf movements in the cotton plant, *Gossypium hirsutum*. Planta 144:111-112.

Supack, J.R., T.A. Kerby, J.C. Banks, and C.E. Snipes. 1993. Use of plant monitoring to schedule chemical crop termination. p. 1194-1196. *In:* D.J. Herber and D.A. Richter (eds.). Proc. Beltwide Cotton Conf., National Cotton Council of America, Memphis, Tenn.

Sutton, R. and I.P. Ting. 1977. Evidence for the repair of ozone induced membrane injury. Am. J. Bot. 64:404-411.

Sweeney, B.M. 1969. Rhythmic phenomena in plants. Academic Press, London, New York.

Sweeney, B.M. 1974. The temporal mechanisms in plant morphogenesis, hourglass and oscillator. Brookhaven Symp. Biol. 25: 95-110.

Swietlick, D. and M. Faust. 1984. Foliar nutrition of fruit crops. Hort. Rev. 6:287-355.

Szabolcs, I. 1989. Salt-affected soils. CRC Press Inc., Boca Raton, Florida.

Sze, H. 1985. H+-translocating ATPases: advances using membrane vesicles. Ann. Rev. Plant Physiol. 36:175-208.

Tabashnik, B.E. 1994a. Evolution of resistance to *Bacillus thuringiensis*. Annu. Rev. Entomol. 39: 47-79.

Tabashnik, B.E. 1994b. Delaying insect adaptation to transgenic crops: seed mixtures and refugia reconsidered. Proc. Royal Soc. London, Series B 255:7-12.

Tabashnik, B.E., N. Finson, and M.W. Johnson. 1991. Managing resistance to *Bacillus thuringiensis*: lessons from the diamondback moth (Lepidoptera: Plutellidae). J. Econ. Entomol. 84: 49-55.

Taha, M.A., M.N.A. Malik, F.I. Chaudhry, and I. Makhdum. 1981. Heat induced sterility in cotton sown during early April in West Punjab. Exp Agric 17:189-194.

Taiz, L. 1994. Expansins: Proteins that promote cell wall loosening in plants. Proc. Natl. Acad. Sci. USA 91:7387-7389.

Takahashi, S., T. Kuzuyama, H. Watanabe, and H. Seto. 1998. A 1-deoxy-D-xylulose 5-phosphate reductiosomerase catalyzing th eformation of 2-*C*-methyl-D-erythritol 4-phosphate in an alternative nonmevalonate pathway for terpenoid biosynthesis. Proc. Natl. Acad Sci. USA. 95:9879-9884.

Takeoka, Y., K. Kondo, and P.B. Kaufman. 1983. Leaf surface in rice plants cultured under shade and non-shade conditions. Jpn. J. Crop Sci. 52:534-543.

Talbot, L.D. and P.M. Ray. 1992. Molecular size and separability features of the pea cell wall polysaccharides. Implications for models of primary wall structure. Plant Physiol. 98:357-368.

Tan, Z. and W.F. Boss. 1991. Association of phosphatidylinositol kinase, phosphatidylinositol monophosphate kinase wth the cytoskeleton and F-actin fractions of carrot *(Daucus carota* L.) cells grown in suspension culture. Response to cell wall degrading enzymes. Plant Physiol. 100:2116-2120.

Tang, B., J.N. Jenkins, C.E. Watson, J.C. McCarty, and R.G. Creech. 1996. Evaluation of genetic variances, heritabilities, and correlations for yield and fiber traits among cotton F_2 hybrid populations. Euphytica 91:315-322.

Tanji, K.K. 1990. Nature and extent of agricultural salinity. *In:* K.K. Tanji (ed.). Agricultural Salinity Assessment and Management. Amer. Soc. Civil Engineers, New York, pp. 1-17.

Tanji, K.K., A. Läuchli, and J. Meyer. 1986. Selenium in the San Joaquin Valley. Environment 28:34-39.

Tarczynski, M.C., D.N. Byrne, and W.B. Miller. 1992. High performance liquid chromatography analysis of carbohydrates of cotton-phloem sap and of honeydew produced by *Bemisia tabaci* feeding on cotton. Plant Physiol. 98:753-756.

Tarczynski, M.C., R.G. Jensen, and H.J. Bohnert. 1993. Stress protection of transgenic tobacco by production of the osmolyte mannitol. Science 259:508-510.

Taylor, A. and D.J. Cosgrove. 1989. Gibberellic acid stimulation of cucumber hypocotyl elongation. Effects on growth, turgor, osmotic pressure, and cell wall properties. Plant Physiol. 90:1335-1340.

Taylor, G.J. and C.D. Foy. 1985. Differential uptake and toxicity of ionic and chelated copper in *Triticum aestivum*. Can. J. Bot. 63:1271-1275.

Taylor, O.C. and D.C. MacLean. 1970. Nitrogen oxides and the peroxyacyl nitrates. *In:* J.S. Jacobson and A.C. Hill (eds.). Recognition of Air Pollution Injury to Vegetation: a Pictorial Atlas. Air Pollut. Control. Assoc., Pittsburgh, PA. pp. E1-E14.

Taylor, O.C. and J.D. Mersereau. 1963. Smog damage to cotton. Calif. Agric., Nov. 1963, pp. 2-3.

Taylor, D.M., P.M. Morgan, H.E. Joham, and J.V. Amin. 1968. Influence of substrate and tissue manganese on the IAA-oxidase system in cotton. Plant Phyisol. 43:243-247.

Taylor, H.M. 1983. Managing root systems for efficient water use: An overview. pp. 87-113. *In :*W.R. Jordan and T.R. Sinclair (eds.). Limitations to Efficient Water Use in Crop Production. Amer. Soc. Agron., Madison.

Taylor, H.M. and H.R. Gardner. 1963. Penetration of cotton seedling taproots as influenced by bulk density, moisture content, and strength of soil. Soil Sci. 96:153-156.

Taylor, H.M., M.G. Huck, and B. Klepper. 1972. Root development in relation to soil physical conditions. pp. 57-77 *In:* D.I. Hillel (ed.). Optimizing the soil physical environment toward greater crop yields. Academic Press, New York.

Taylor, H.M. and B. Klepper. 1971. Water uptake by cotton roots during an irrigation cycle. Aust. J. Biol. Sci. 24:853-859.

Taylor, H.M. and B. Klepper. 1974. Water relations of cotton. I. Root growth and water use as related to top growth and soil water content. Agron. J. 66:584-588.

Taylor, H.M. and B. Klepper. 1975. Water uptake by cotton root systems: An examination of assumptions in the single root model. Soil Sci. 1 2057-2067.

Taylor, H.M. and B. Klepper. 1978. The role of rooting characteristics in the supply of water to plants. Adv. Agron. 30:99-128.

Taylor, H.M. and L.F. Ratliff.. 1969. Root elongation rates of cotton and peanuts as a function of soil strength and soil water content. Soil Sci. 108:113-1 19.

Taylor, R.A. 1994. High speed measurements of strength and elongation. pp. 268-273. *In:* G.A. Constable and N.W. Forrester (eds.). Challenging the Future. Proc. World Cotton Conference I, CSIRO, Australia.

Taylor, W.K. 1981. DROPP: Thidiazuron experimental cotton defoliant. p. 70. *In:* J.M. Brown (ed.). Proc. Beltwide Cotton Prod. Mech. Conf., National Cotton Council of America, Memphis, Tenn.

Teeri, J.A. and M.M. Peet. 1978. Adaptation of malate dehydrogenase to environmental temperature variability in two populations of *Potentilla glandulosa* Lindl. Oecologia 34:133-141.

Temple, P.J. 1986. Stomatal conductance and transpirational responses of field-grown cotton to ozone. Plant, Cell and Environ. 9:315-321.

Temple, P.J. 1990a. Water relations of differentially irrigated cotton exposed to ozone. Agron. J. 82:800-805.

Temple, P.J. 1990b. Growth form and yield responses of four cotton cultivars to ozone. Agron. J. 82:1045-1050.

Temple, P.J. 1991. Variations in responses of dry bean (*Phaseolus vulgaris*) cultivars to ozone. Agric. Ecosys. Environ. 36:1-11.

Temple, P.J. and Grantz. 2009. Plant response to gaseous pollutants. pp. 163-174. *In:* J.M.Stewart, D.M. Oosterhuis, J.J. Heitholt, and J.R. Mauney (eds.). *Physiology of Cotton*. National Cotton Council of America, Memphis,Tenn. pp. Springer, London.

Temple, P.J., R.S. Kupper, R.W. Lennox, and K. Rohr. 1988a. Injury and yield responses of differentially irrigated cotton to ozone. Agron. J. 80:751-755.

Temple, P.J., R.S. Kupper, R.L. Lennox, and K. Rohr. 1988b. Physiological and growth responses of differentially irrigated cotton to ozone. Environ. Pollut. 53:255-263.

Temple, P.J., O.C. Taylor, and L.F. Benoit. 1985. Cotton yield responses to ozone as mediated by soil moisture and evapotranspiration. J. Environ. Qual. 14:55-60.

Tennant, D. 1975. A test of a modified line intersect method of estimating root length. J. Appl. Ecol. 63:995-1001.

Terry, L.I. 1992. Effect of early season insecticide use and square removal on fruiting patterns and fiber quality of cotton. J. Econ. Entomol. 85:1402-1412.

Terry, M.E., R.L. Jones, and B. Boner. 1981. Soluble cell wall polysaccharides released by pea stems by centrifugation. I. Effects of auxin. Plant Physiol. 68:531-537.

Terry, N. 1970. Developmental physiology of sugar-beet. II. Effects of temperature and nitrogen supply on the growth, soluble carbohydrate content and nitrogen content of leaves and roots. J. Exp. Bot. 21:477-498.

Terry, N. and I.M. Rao. 1991. Nutrients and photosynthesis: iron and phosphorus as case studies. *In:* J.R. Porter and D.W. Lalor (eds.). Plant Growth: interactions with nutrition and environment. Cambridge University Press, Cambridge, Great Britain. pp. 55-80.

Terry, N., L.J. Waldron, and A. Ulrich. 1971. Effects of moisture stress on the multiplication and expansion of cells in leaves of sugar beet. Planta (Berl.) 97:281-289.

Tester. M. 1990. Plant ion channels: Whole-cell and single channel studies. New Phytol. 114:305-340.

Teubner, F.G., S.H. Wittwer, W.G. Long, and H.B. Tukey. 1957. Some factors affecting absorption and translocation of foliar-applied nutrients as revealed by radioactive isotopes. Quart. Bull. Mich. Agric. Exp. Stn. 39:398-415.

Tewolde, H., C.J. Fernandez, and D. Foss. 1993. Growth, yield, and maturity of nitrogen and phosphorus deficient pima cotton. pp. 1184-1190. *In:* D.J. Herber and D.A. Ritcher (eds.). Proc. Beltwide Cotton Conf., National Cotton Council of America, Memphis, Tenn.

Tewolde, H., C.J. Fernandez, and D.C. Foss. 1994. Maturity and lint yield of nitrogen- and phosphorus-deficit Pima cotton. Agron. J. 86:303-309.

Thaker, V.S., S. Saroop, and Y.D. Singh. 1986a. Physiological and biochemical changes associated with cotton fibre development. III. Indolyl-3-acetaldehyde dehydrogenase. Biochem. Physiol. Pflanzen 181:339-345.

Thaker, V.S., S. Saroop, and Y.D. Singh. 1987. Physiological and biochemical changes associated with cotton fibre development. IV. Glycosidases and β-1,3-glucanase activities. Ann. Bot. 60:579-585.

Thaker, V.S., S. Saroop, P.P. Vaishnav, and Y.D. Singh. 1986b. Role of peroxidase and esterase activity during cotton fiber development. J. Plant Growth Regul. 5:17-27.

Thaker, V.S., S. Saroop, P.P. Vaishnav, and Y.D. Singh. 1989. Genotypic variations and influence of diurnal temperature on cotton fiber development. Field Crops 22:1-13.

Tharp, W.H. 1960. The cotton plant. U.S. Department of Agriculture Agricultural Handbook No. 178.

Tharp, W.H., J.J. Skinner, J.H. Turner, R.P. Bledsoe, and H.B. Brown. 1949. Yield and composition of cottonseed as influenced by fertilization and other environmental factors. USDA Tech. Bull. No. 974. pp. 153.

Theodorou, M.E., F.A. Cornel, S.M.G. Duff, and W.C. Plaxton. 1992. Phosphate starvation inducible synthesis of the ex-subunit of pyrophosphatedependent phosphofructokinase in black mustard suspension cells. J. Biol. Chem. 267:21901-21905.

Theodorou, M.E. and W.C. Plaxton. 1996. Purification and characterization of pyrophosphate-dependent phosphofructokinase from phosphate-starved *Brassica nigra* suspension cells. Plant Physiol. 112:343-351.

Thimann, K.V. 1969. The auxins: pp. 1-45 *In:* M.B. Wilkins (ed.). *The Physiology of Plant Growth and Development.* McGGraw-Hill, New York, N..

Thoma, S., Y. Kaneto, and C.R. Somerville. 1993. A nonspecific lipid transfer protein from *Arabidopsis* is a cell wall protein. Plant J. 3:427-436.

Thomas, H. 1994. Resource rejection by higher plants. pp. 375-385. *In:* J.L. Monteith, R.K. Scott, and M.H.

Unsworth (eds.). Resource Capature by Crops. Proc 52nd Easter School, Univ of Nottingham, School of Agriculture, Nottingham Univ Press, Nottingham.

Thomas, J.R. 1980. Osmotic and specific salt effects on growth of cotton. Agron. J. 72:407-412.

Thomas, J.C., K.W. Brown, and W.R. Jordan. 1973. Stomatal response to leaf water potential as affected by preconditioning water stress in the field. Agron. J. 68:706-708.

Thomas, J.C., E.F. Saleh, and A.M. Akroush. 1996. Plant synthesized compounds for self defense against insects. Abstract. Annual American Society of Plant Physiology Meeting, San Antonio, TX.

Thomas, J.C., M. Sepshi, B. Arendall, and H.J. Bohnert. 1995. Enhancement of seed germination in high salinity by engineering mannitol expression in *Arabidopsis thaliana*. Plant, Cell and Environment 18:801-806.

Thomas, M.D., R.H. Hendricks, T.R. Collier, and G.R. Hill. 1943. The utilization of sulfate and sulfur dioxide for the nutrition of alfalfa. Plant Physiol. 18:345-371.

Thomas R.B. and B.R. Strain. 1991. Root restriction as a factor in photosynthetic acclimation of cotton seedlings grown in elevated carbon dioxide. Plant Physiol. 96:627-634.

Thomas, R.O. 1975. Cotton flowering and fruiting response to application timing of chemical growth retardants. Crop Sci. 15:87-90.

Thomas, R.O. and M.N. Christiansen. 1971. Seed hydration-chilling treatment effects on germination and subsequent growth and fruiting of cotton. Crop Science 11:454-456.

Thomas, R.O., T.C. Cleveland, and G.W. Cathey. 1979. Chemical plant growth suppressants for reducing late-season cotton bollworm-budworm feeding sites. Crop Sci. 19:861-863.

Thomasson, J.A. 1993. Foreign matter effects on cotton color measurement: Determination and correction. Trans. ASAE 36:663-669.

Thomasson, J.A., M.P. Mengüç, and S.A. Shearer. 1995. Radiative transfer model for relating near-infrared micronaire measurements of cotton fibers. Trans. ASAE 38:367-377.

Thomasson, J.A. and R.A. Taylor. 1995. Color relationships between lint and seed cotton. Trans. ASAE 38:13-22.

Thompson, A.C., H.C. Lane, J.W. Jones, and J.D. Hesketh. 1975. Soluble and insoluble sugars in cotton leaves, squares and bolls. pp. 59-63 *In:* Proc. Beltwide Prod. Res. Conf., National Cotton Council of America, Memphis, Tenn.

Thompson, A.C., H.C. Lane, J.W. Jones, and J.D. Hesketh. 1976. Nitrogen concentrations of cotton leaves, buds, and bolls in relation to age and nitrogen fertilization. Agron. J. 68:617-621.

Thompson, W.F. and M.G. Murray. 1981. The nuclear genome: structure and function. pp. 1-81. *In:* A. Marcus (ed.). *The Biochemistry of Plants Vol 6.* Academic Press, New York.

Thomson, N.J. 1987. Host plant resistance in cotton. J. Aust. Inst. Agric. Sci. 53: 262-270.

Thomson, N.J. 1994. Commercial utilisation of the okra leaf mutant of cotton - the Australian experience. pp. 393-401 *In:* G.C. Constable and N.W. Forrester (eds.). Challenging the Future: Proceedings of the World Cotton Res Conf 1, Brisbane Australia, Feb 14-17 1994, CSIRO, Melbourne.

Thomson, N.J. and J.A. Lee. 1980. Insect resistance in cotton: A review and prospectus for Australia. J. Aust. Inst. Agric. Sci. 46: 75-86.

Thomson, N.J., P.E. Reid, and E.R. Williams. 1987. Effects of the okra leaf, nectariless, frego bract and glabrous conditions on yield and quality of cotton lines. Euphytica 36:545-553.

Thorne, L. and G.P. Hanson. 1976. Relations between genetically controlled ozone sensitivity and gas exchange rate in *Petunia hybrida* Vilm. J. Am. Soc. Hort. Sci. 101:60-63.

Tiben, A, M.J. Pearce, T.G. Wood, M.A. Kambal, and R.H. Cowie. 1990. Damage to crops by *Microtermes najdensis* (Isoptera, Macrotermitinae) in irrigated semi-desert areas of the Red Sea Coast. 2. Cotton in the Tokar Delta region of Sudan. Tropical Pest Management 36:296-304.

Ticha, I. 1982. Photosynthetic characteristics during ontogenesis of leaves. 7. Stomata density and size. Photosynthetica 16:375-471.

Ticha, I. and J. Catsky. 1981. Photosynthetic characteristics during ontogenesis of leaves. 5. Carbon dioxide compensation concentration. Photosynthetica 15:401-428.

Tilman, D. 1990. Constraints and trade-offs: toward a predictive theory of competition and succession. Oikos 58: 3-15.

Timpa, J.D. 1991. Application of universal calibration in gel permeation chromatography for molecular weight determinations of plant cell wall polymers: cotton fiber. J. Agric. Food Chem. 39:270-275.

Timpa, J.D. 1992. Molecular chain length distributions of cotton fiber: developmental, varietal, and environmental influences. pp. 199-210. *In:* C.R. Benedict (ed.). Proc. Cotton Fiber Cellulose: Structure, Function, and Utilization Conference. National Cotton Council, Memphis Tenn.

Timpa, J.D., J.J. Burke, J.E. Quisenberry, and C.W. Wendt. 1986. Effects of water stress on the organic acid and carbohydrate compositions of cotton plants. Plant Physiol. 82:724-728.

Timpa, J.D. and H.H. Ramey. 1989. Molecular characterization of three cotton varieties. Text. Res. J. 59:661-667.

Timpa, J.D., A.M. Striegel, A.L. Abellanosa, and B.A. Triplett. 1995. Rapid, inexpensive method for determining glucose concentrations in cotton bolls and other plant tissues. Crop Sci. 35:274-278.

Timpa, J.D. and B.A. Triplett. 1993. Analysis of cell-wall polymers during cotton fiber development. Planta 189:101-108.

Timpa, J.D. and D.F. Wanjura. 1989. Environmental stress responses in molecular parameters of cotton cellulose. p. 1145-1156. *In*: C. Schuerch (ed.). Cellulose and Wood: Chemistry and Technology. John Wiley and Sons, New York.

Ting, I.P. and W.M. Dugger, Jr. 1968. Factors affecting ozone sensitivity and susceptibility of cotton plants. J. Air Pollut. Control Assoc. 18:810-813.

Tinker, P.B., D.M. Durall, and M.D. Jones. 1994. Carbon use efficiency in mycorrhizas: theory and sample calculations. New Phytol. 128:115-122.

Tisdall, J.M. and J.M. Oades. 1979. Stabilization of soil aggregates by the root systems of ryegrass. Austr. J. Soil Res. 17:429-441.

Tiwari, R.J. 1994. Response of gypsum on morpho-physiochemical properties of cotton cultivars under salt affected vertisols of Madhya Pradesh. Crop Res. Hisar 7:197-200.

Tiwari, R.J., K. Dwivedi, and S.K. Verma. 1993. Effect of gypsum on leaf-water potential of cottons (*Gossypium hirsutum*, *G. herbaceum*, and *G. arboreum*) varieties grown in salt-affected Vertisol of Madhya Pradesh. Indian J. Agric. Sci. 63:734-736.

Tiwari, R.J., K. Dwivedi, and S.K. Verma. 1994a. Effect of gypsum application on the yield of cotton in salt affected clay soils. Crop Res. Hisar 7:30-33.

Tiwari, R.J., K. Dwivedi, and S.K. Verma. 1994b. Effect of gypsum application on nutrient content in cotton leaves grown on a sodic vertisol. Crop Res. Hisar 7:193-196.

Tiwari, S.C. and V.S. Polito. 1988. Organization of the cytoskeleton in pollen tubes of *pyrus communis*: A study employing conventional and freeze-substitution electron microscopy, immunofluorescence, and rhodamine-phallodin. Protoplasma 147:100-112.

Tiwari, S.C. and T.A. Wilkins. 1995. Cotton (*Gossypium hirsutum*) seed trichomes expand via diffuse growing mechanism. Can. J. Bot. 73:746-757.

Tobar, R., R. Azcón, and J.M. Barea. 1994. Improved nitrogen uptake and transport from ^{15}N-labelled nitrate by external hyphae of arbuscular mycorrhiza under water-stressed conditions. New Phytol. 126:119-122.

Tognetti, J.A., P.L. Calderon, and H.G. Pontis. 1989. Fructan metabolism: Reversal of cold acclimation. J. Plant Physiol. 134:232-236.

Tolley-Henry, L. and C.D. Raper. 1986. Expansion and photosynthetic rate of leaves of soybean plants during onset of and recovery from nitrogen stress. Bot. Gaz. 147:400-406.

Tolliver, J, B.R. Savoy, and E.A. Drummond. 1997. Cool germination test on cotton -- variability between seed-testing laboratories. pp. 442-443. *In:* Proc. Beltwide Cotton Conf., National Cotton Council of America, Memphis, Tenn.

Tompkins, F.D., J.F. Bradley, and M.S. Kearney. 1990. Cotton performance under five conservation-tillage production systems. pp. 108-112. *In:* Proc. Beltwide Cotton Prod. Res. Conf., National Cotton Council of America, Memphis, Tenn.

Töpperwein, H. 1993. Relationships in the apical region of angiosperms. Angew. Bot. 67:22-30.

Topping, E. and R. M. Broughton, Jr. 1997. A modified Clinitest procedure for estimation of honeydew contamination in cotton. pp. 1637-1639. *In:* Beltwide Cotton Prod. Res. Conf., National Cotton Council of America, Memphis, Tenn.

Torrisi, V., G.S. Pattinson, and P.A. McGee. 1999. Localised elongation of roots of cotton follows establishment of arbuscular mycorrhizas. New Phytol. 142:103-112.

Torsethaugen, G., E.J. Pell, and S.M. Assmann. 1999. Ozone inhibits guard cell K[+] channels implicated in stomatal opening. Proc. National Acad. Sciences 96:13577-13582.

Torssel, K.B.G. 1983. Natural Product Chemistry–A mechanistic and biosynthetic approach to secondary metabolism. Chapter 5. John Wiley & Sons, Ltd.

Tort, N. 1996. Effects of light, different growth media, temperature, and salt concentrations on germination of cotton seeds (*Gossypium hirsutum* L. cv. Naxilli-87). J. Agron. Crop Sci. 176:217-221.

Touchton, J.T. and W.L. Hargrove. 1983. Grain sorghum response to starter fertilizer. Better Crops with Plant Food. 68:3-5.

Touchton, J.T. and D.W. Reeves. 1988. A beltwide look at conservation tillage for cotton. pp. 36-41. *In:* Proc. Beltwide Cotton Prod. Conf., National Cotton Council of America, Memphis, Tenn.

Touchton, J.T., D.H. Rickerl, C.H. Burmester, and D.W. Reeves. 1986. Starter fertilizer combinations and placement for conventional and no-tillage cotton. J. Fert. Issues. 3:91-98.

Trabalka, J.R., J.A. Edrnonds, J.M., Reilly, R.H. Gardner, and D.E. Reichle. 1986. Atmospheric CO_2 projections with globally averaged carbon cycle models. pp. 534-560. *In:* J.R. Trabalka and D.E. Reichle (eds.). The Changing Carbon Cycle a Global Analysis. Springer-Verlag.

Tracy, P.W. and W.P. Sappenfield. 1992. Integrated management influence on short-season cotton yield in the northern Mississippi Delta. J. Prod. Agric. 5:551-555.

Tran, V.N. and A.K. Cavanagh. 1984. Structural aspects of dormancy. pp. 1-44. *In:* D.R Murray (ed.). Seed Physiology II, Germination and Reserve Mobilization. Academic Press Australia, Sydney.

Trappe, J.M. 1987. Phylogenetic and ecologic aspects of mycotrophy in the Angiosperms from an evolutionary standpoint. p. 5-25. *In:* G.R. Safir (ed.). Ecophysiology of VA Mycorrhizal Plants. CRC Press, Boca Raton, FL.

Trelease, R.N. 1984. Biogenesis of glyoxysomes. Ann. Rev. Plant Physiol. 35:321-347.

Trelease, R.N. and D.C. Doman. 1984. Mobilization of oil and wax reserves. pp. 201-245. *In:* D.R Murray (ed.). Seed Physiology II, Germination and Reserve Mobilization. Academic Press Australia, Sydney.

Trelease, R.N., J.A. Miernyk, J.S. Choinski, Jr., and S.J. Bortman. 1986. Synthesis and compartmentation of enzymes during cottonseed maturation. pp. 441-462. *In:* J.R Mauney and J.McD. Stewart (eds.). Cotton Physiology. The Cotton Foundation, Memphis, Tenn.

Trewavas, A. 1981. How do plant growth substances work? Plant Cell Environ. 4:203-228.

Trewavas, A.J. 1985. A pivotal role for nitrate and leaf growth in plant development. pp. 77-91 *In:* N.R. Backer, W.J. Davies, and C.K. Ong (eds.). Control of Leaf Growth. Cambridge Univ Press, Cambridge.

Trewavas, A.J. 1986. Understanding the control of plant development and the role of growth substances. Aust. J. Plant Physiol. 13:447-457.

Trewavas, A.J. and R. Malhó. 1997. Signal perception and transduction: The origin of the phenotype. Plant Cell 9:1181-1195.

Triplett, B.A. 1989. Ovule and suspension culture of a cotton fiber development mutant. In Vitro Cell. Dev. Biol. 25:197-200.

Triplett, B.A. 1990. Evaluation of fiber and yarn from three cotton fiber mutant lines. Textile Res. J. 60:143-148

Triplett, B.A. 1992. Strategies for improving cotton fiber quality. pp. 107-114. *In:* C.R. Benedict (ed.). Proc. Cotton Fiber Cellulose: Structure, Function, and Utilization Conference. National Cotton Council, Memphis Tenn.

Triplett, B.A. 1993. Using biotechnology to improve cotton fiber quality: progress and perspectives. pp. 135-140. *In:* J.F. Kennedy, G.O. Phillips, and P.A. Williams (eds.). Cellulosics: Pulp, Fibre, and Environmental Aspects. Ellis Horwood Ltd., Chichester UK.

Triplett, G.B. and S.M. Dabney. 1995. Long-term crop response to conservation tillage. pp. 93-96. *In:* Proc. 1995. South. Conserv. Tillage Conf. Sust. Agric. MAFES Spec. Bull. 88-7. Miss. State Univ., Mississippi State, MS.

Triplett, G.B., S.M. Dabney, and J.H. Siefker. 1996. Tillage systems for cotton on silty upland soils. Agron. J. 88:507-512.

Triplett, B.A. and R.S. Quatrano. 1982. Timing, localization and control of wheat germ agglutinin synthesis in developing wheat embryos. Dev. Biol. 91:491-496.

Trolinder, N.L. and J.R. Goodin. 1985. Somatic embryogenesis in cell suspension cultures of *Gossypium hirsutum.* p. 46. *In:* Proc. Beltwide Cotton Conf., National Cotton Council of America, Memphis, Tenn.

Trolinder, N.L. and J.R. Goodin. 1987. Somatic embryogenesis and plant regeneration in cotton (*Gossypium hirsutum* L.). Plant Cell Reports 6:231-234.

Trolinder, N.L. and J.R. Goodin. 1988a. Somatic embryogenesis in cotton (*Gossypium*). I. Effects of source of explant and hormone regime. Plant Cell, Tissue, and Organ Culture 12:31-42.

Trolinder, N.L. and J.R. Goodin. 1988b. Somatic embryogenesis in cotton (*Gossypium)*. II. Requirements for embryo development and plant regeneration. Plant Cell, Tissue, and Organ Culture 12:43-53.

Trolinder, N.L. and X. Shang. 1991. *In vitro* selection and regeneration of cotton resistant to high temperature stress. Plant Cell Reports 8:133-136.

Trolinder, N.L. and C. Xhixian. 1989. Genotype specificity of the somatic embryogenesis response in cotton. Plant Cell Reports 8:133-136.

Trumble, J.T., DM. Kolodny-Hirsh, and I.P. Ting. 1993. Plant compensation for arthropod herbivory. Annu. Rev. Entomol. 38:93-119.

Tsuji, W., T. Nakao, A. Hirai, and F. Horii. 1992. Properties and structure of never-dried cotton fibers. III. Cotton fibers from bolls in early stages of growth. J. Appl. Polym. Sci. 45:299-307.

Tuomi, J, 1992. Toward integration of plant defense theories. Trends Ecol. Evol. 7:365-367.

Tuomi, J., P. Nilsson, and M. Astrom. 1994. Plant compensatory responses: Bud dormancy as an adaptation to herbivory. Ecology 75:1429-1436.

Tupper, G.R. 1992. Technologies to solve K deficiency - deep placement. Proc. Beltwide Cotton Conf., Nashville, Tenn. pp. 73-76.

Tupper, G.R. 1995. A new design for reduced soil surface disturbance and power requirement low-till parabolic subsoiler. *In:* M.R. McClelland, T.D. Valco, and R.E. Frans (eds.). Conservation-Tillage Systems for Cotton, University of Arkansas Agric. Expt. Sta. Spec. Rpt. 169:50-53. Fayetteville, Ark.

Tupper, G.R., M.W. Ebelhar, and H.C. Pringle, III. 1992. The effects of subsoiling and deep bonded potassium on non-irrigated DES 199 cotton. Proc. Beltwide Cotton Conf., Nashville, Tenn. pp. 1130-1133.

Tupper, G.R., M.W. Ebelhar, and H.C. Pringle, III. 1993. Correcting subsoil nutrient problems with deep bonding day material application - summary of results. Proc. Beltwide Cotton Conf., New Orleans, LA. pp. 1326-1329.

Tupper, G.R., J.G. Hamill, and H.C. Pringle III. 1989. Cotton response to subsoiling frequency. pp. 523-525. *In:* Proc. Beltwide Cotton Prod. Res. Conf., National Cotton Council of America, Memphis, Tenn.

Tupper, G.R., H.C. Pringle, and W.E. Ebelhar. 1988. Cotton response to deep banding dry fertilizer in the subsoil. Proc. Beltwide Cotton Conferences. National Cotton Council. Memphis, Tenn. pp. 498-501.

Turley, R.B., S.M. Choe, W. Ni, and R.N. Trelease. 1990a. Nucleotide sequence of cottonseed malate synthase. Nucl. Acids Res. 18:3643.

Turley, R.B., S.M. Choe, and R.N. Trelease. 1990b. Characterization of a cDNA clone encoding the complete amino acid sequence of cotton isocitrate lyase. Biochim. Biophys. Acta 1049:223-226.

Turley, R.B. and D.L. Ferguson. 1996. Changes of ovule proteins during early fiber development in a normal and a fiberless line of cotton *(Gossypium hirsutum* L.). J. Plant Physiol. 149:695-702.

Turley, R.B., D.L. Ferguson, and W.R. Meredith, Jr. 1994a. Isolation and characterization of a cDNA encoding ribosomal protein S16 from cotton *(Gossypium hirsutum* L.). Plant Physiol. 105:1219-1220.

Turley, R.B., D.L. Ferguson, and W.R. Meredith, Jr. 1994b. Isolation and characterization of a cDNA encoding ribosomal protein L41 from cotton *(Gossypium hirsutum* L.). Plant Physiol. 105:1449-1450.

Turley, R.B., D.L. Ferguson, and W.R. Meredith, Jr. 1995. A cDNA encoding ribosomal protein S4e from cotton *(Gossypium hirsutum* L.). Plant Physiol. 108:431-432.

Turlings, T.C.J., J.H. Loughrin, P.J. McCall, U.S.R. Rose, W.J. Lewis, and J.H. Tumlinson. 1995. How caterpillar-damaged plants protect themselves by attracting parasitic wasps. Proc. Natl. Acad. Sci. USA. 92:4169-4174.

Turlings, T.C.J. and J.H. Tumlinson. 1991. Do parasitoids use herbivore-induced plant chemical definses to locate hosts? Flor. Entomol 7:42.

Turlings, T.C.J. and J.H. Tumlinson.1992. Systemic release of chemical signals by herbivore-injured corn, Proc. Natl. Acad. Sci. USA 89: 8399-8402.

Turlings, T.C.J., J.H. Tumlinson, R.R. Heath, A.T. Proveaux, and R.E. Doolittle. 1991. Isolation and identification of allelochemicals that attract the larval para-sitoid, *Cotesia margindueutris* (Cresson), to the microhabitat of one of its hosts. J. Chem. Ecol. 17:2235.

Turner, J.H., J. McD. Stewart, P.E. Hoskinson, and H.H. Ramey. 1977. Seed setting efficiency in eight cultivars of upland cotton. Crop Sci. 17:769-772.

Turner, J.H., Jr., S. Worley, Jr., H.H. Ramey, Jr., P.E. Hoskins, and J.M. Stewart. 1979. Relationship of week of flowering and parameters of boll yield in cotton. Agron. J. 71:248-251.

Turner, N.C. 1986. Adaptation to water deficits: a changing perspective. Aust. J. Plant Physiol. 13:175-190.

Turner, N.C., A.B. Hearn, J.E. Begg, and G.A. Constable. 1986. Cotton *(Gossypium hirsutum* L.): Physiological and morphological responses to water deficits and their relationship to yield. Field Crops Res. 14:153-170.

Turner, N.C. and M.M. Jones. 1980. Turgor maintenance by osmotic adjustment: A review and evaluation. pp. 87–103. *In:* N.C. Turner and P.J. Kramer (eds.). Adaptation of Plants to Water and High Temperature Stress. Wiley Interscience, New York.

Turner, N.C., S. Rich, and H. Tomlinson. 1972. Stomatal conductance, fleck injury, and growth of tobacco cultivars varying in ozone tolerance. Phytopathology 62:63-67.

Twersky, M. and D. Pasternak. 1972. Utilization of saline water for irrigation of salt tolerant crops. Negev Institute for Arid Zone Research: Israel, Report for 1971-1972, 61-62.

Tyler, D.O., W. Halfman, H.P. Denton, and P.W. Tracy. 1994. Trafficability and rooting depth comparisons between no-till and tilled soybeans. pp. 137-143. *In:* Proc. South. Conserv. Tillage Cant. Sust. Agric., Clemson Univ., Columbia, SC.

Tyler, F.J. 1908. The nectaries of cotton. U.S. Dept. Agr. Plant Industry Bull. No. 131, Part V., 45-54.

Tyree, M.T. and J.S. Sperry. 1989. Vulnerability of xylem to cavitation and embolism. Annu. Rev. Plant Physiol. Mol. Biol. 40:19-38.

Tyree, M.T., C.A. Tabor, and C.R. Wescott. 1990. Movement of cations through cuticles of *Citrus aurantium* and *Acer saccharum* – diffusion potentials in mixed salt solutions. Plant Physiology 94:120-126.

Ueda, J., K. Miyamoto, and Y. Momotani. 1996. Jasmonates promote abscission in bean petiople explants: Its relationship to the metabolism of cell wall polysaccharides and cellulase activity. J. Plant Growth Regul. 15:189-195.

Ueda, J., K. Miyamoto, T. Sato, and Y. Momotani. 1991. Identification of jasmonic acid from *Euglena gracilis* Z as a plant growth regulator. Agric. Biol. Chem. 55:275-276.

Ulmasov, T.N., M.K. Gulov, K.A. Aliev, V.N. Andrianov, and E.S. Piruzian. 1990. Nucleotide sequences of the chloroplast *psb*A and *trn*H genes from cotton *Gossypium hirsutum*. Nucl. Acids Res. 18:186.

Uma, M.S. and B.C. Patil. 1996. Inter species variation in the performance of cotton under soil salinity stress. Karnataka J. Agric. Sci. 9:73-77.

Umbeck, P., G. Johnson, K. Barton, and W. Swain. 1987. Genetically transformed cotton (*Gossypium hirsutum* L.) plants. Biotech. 5:263-266.

USDA. 1980. The Classification of Cotton. USDA, AMS, Agric. Handbook No. 594, USDA, Washington, DC.

USDA. 1989. Agricultural statistics 1989. U.S. Gov. Print. Office. Washington, DC.

Underbrink, S.M., J.A. Landivar, and J.T. Cothren. 1998. Pix management strategies for Bt cultivars in the Coastal Plains of Texas. pp. 1452-1454. *In:* P. Dugger and D. Richter (eds.). Proc. Beltwide Cotton Conf., National Cotton Council of America, Memphis, Tenn.

Underbrink, S.M., J.A. Landivar, and J.T. Cothren. 1999. Agronomic differences in growth and yield between *Bt* and conventional cotton. pp. 521-523. *In:* P. Dugger and D. Richter (eds.). Proc. Beltwide Cotton Conf., National Cotton Council of America, Memphis, Tenn.

Ungar, E.D., E. Kletter, and A. Genizi. 1992. Conservative response and stress-damage interactions in cotton reproductive development. Agron. J. 84:382-386.

Ungar, E.D., D. Wallach, and E. Kletter. 1987. Cotton responses to bud and boll removal. Agron. J. 79:491-497.

Unger, P.W. and T.C. Kaspar. 1994. Soil compaction and root growth: A review. Agron. J. 86:759-766.

Unruh, B.L. and J.C. Silvertooth. 1996a. Comparisons between an upland and a pima cotton cultivar: I. Growth and yield. Agron. J. 88:583-589.

Unruh, B.L. and J.C. Silvertooth. 1996b. Comparisons between an upland and a pima cotton cultivar: II. Nutrient uptake and partitioning. Agron. J. 88:589-595.

Upchurch, D.R. and J.T. Ritchie. 1984. Battery-operated color video camera for root observations in mini-rhizotrons. Agron. J. 1015-1017.

Upchurch, D.R. and H.M. Taylor. 1990. Tools for studying rhizosphere dynamics. pp. 83-115. *In:* J.E. Box and L.C. Hammond (eds.). Rhizosphere Dynamics. AAA Selected Symposium.

Urwiler, M.J. 1981. Effect of mepiquat chloride on cotton seed germination and subsequent growth at minimal and optimal temperatures. M.S. thesis, University of Arkansas, Fayetteville, AR.

Urwiler, M.J. and D.M. Oosterhuis. 1986. The effect of the growth regulators Pix and IBA on cotton root growth. Ark. Farm Res. 36:(6)5.

Urwiler, M.J. and C.A. Stutte. 1988. Influence of PPG-1721 and PGR-IV of field-grown cotton (*Gossypium hir-*

sutum). p. 68. *In:* J.M. Brown (ed.). Proc. Beltwide Cotton Prod. Res. Conf., National Cotton Council of America, Memphis, Tenn.

Urwiler, M., C. Stutte, S. Jourdan, and T. Clark. 1987. Bioregulant field evaluations on agronomic crops. University of Arkansas Agricultural Experiment Station Research Series 358:1-30.

Vail, S.G. 1992. Selection for overcompensatory plant responses to herbivory: a mechanism for the evolution of plant-herbivore mutualism. Am. Nat. 139:1-8.

Vail, S.G. 1994. Overcompensation, plant-herbivore mutualism, and mutalistic coevolution - A reply to Mathews. Am. Nat. 144:534-536.

Vaissayre, M. 1994. Ecological attributes of major cotton pests: implications for management. *In:* Challenging the future: Proceedings of the World Cotton Res Conf 1, Brisbane Australia, Feb 14-17 1994, G.C. Constable and N.W. Forrester (eds.), pp. 499-510. CSIRO, Melbourne.

Valco, T.D. and M.R. McClelland. 1995. Conservation-tillage systems for cotton. *In:* T.D. Valco and M.R. McClelland (eds.). Conservation-tillage Systems for Cotton. University of Arkansas Agricultural Experiment Station Special Report 69:1-2. Fayetteville, Ark.

Valiček, P. 1978. Wild and cultivated cottons. Cot. Fib. Trop. 33:363-387.

Van Assche, F. and H. Clijsters. 1986a. Inhibition of photosynthesis in *Phaseolus vulgaris* by treatment with toxic concentrations of zinc: Effects on electron transport and photophosphorylation. Physiol. Plant. 66:717-721.

Van Assche, F. and H. Clijsters. 1986b. Inhibition of photosynthesis in *Phaseolus vulgaris* by treatment with toxic concentrations of zinc: Effect on ribulose-1,5-biphosphate carboxylate/oxygenase. J. Plant Physiol. 125:355-360.

Van Atta, D. 1993. The current state of agrarian reform in Uzbekistan. Post-Soviet Geography 34:598-606.

van Buuren, M.L., I.E. Maldonado-Mendoza, A.T. Trieu, L.A. Blaylock, and M.J. Harrison. 1999. Novel genes induced during arbuscular mycorrhizal (AM) symbiosis formed between *Medicago truncula* and *Glomus versiforme*. Mol. Plant-Microbe Interact. 12:171-181.

van der Heijden, M.G.A., J.N. Klironomos, M. Ursic, P. Moutoglis, R. Streitwolf-Engel, T. Boller, A. Wiemken, and I.R. Sanders. 1998. Mycorrhizal fungal diversity determines plant biodiversity, ecosystem variability and productivity. Nature 396:69-72.

van der Meijden, E. 1990. Herbivory as a trigger for growth. Funct. Ecol. 4: 597-598.

van der Meijden, E., M. Wijn, and H.J. Verkaar. 1988. Defense and regrowth, alternative plant strategies in the struggle against herbivores. Oikos 51: 355-363.

VanderWiel, P.L., D.F. Voytas, and J.F. Wendel. 1993. *Copia*-like retrotransposable element evolution in diploid and polyploid cotton (*Gossypium* L.). J. Mol. Evol. 36:429-447.

van Emden, H.F. and P. Hadley. 1994. The application of the concepts of resource capture to the effect of pest incidence on crops. pp. 149-165 *In:* J.L. Monteith, R.K. Scott, and M.H. Unsworth (eds.). Resource Capture by Crops. Proc 52nd Easter School, Univ of Nottingham, School of Agriculture, Nottingham Univ. Press, Nottingham.

Van Hofsten P., I. Faye, K. Kockum, J.Y. Lee, K.G. Xanthopoulos, I.A. Boman, H.G. Boman, A. Engstrom, D. Andreu, and R.B. Merrifield. 1985. Molecular cloning, cDNA sequencing, and chemical synthesis of cecropin B from *Hyalophora cecropia*. Proc. National Academy of Sciences USA 82:2240-2243.

van Iersel, M.W., W.M. Harris, and D.M. Oosterhuis. 1995. Phloem in developing cotton fruits: 6(5)carboxyfluorescein as a tracer for functional phloem. J. Exp. Bot. 46:321-328.

van Loon, L.C. 1977. Induction by 2-chloroethylphosphonic acid of viral-like lesions, associated proteins, and systemic resistance in tobacco. Virology 80:417-420.

van Loon, L.C., P.A.H.M. Bakker, and C.M.J. Pieterse. 1998. Systemic resistance induced by rhizosphere bacteria. Annu. Rev. Phytopathol. 36:453-483.

van Overbeek, J. 1956. Absorption and translocation of plant growth regulators. Annu. Rev. Plant Physiol. 1:355-372.

Van Volkenburgh, E. and W.J. Davies. 1977. Leaf anatomy and water relations of plants grown in controlled environments and in the field. Crop Sci. 17:53-358.

Vangronsveld, J. and H. Clijsters. 1994. Toxic effects of metals. *In:* M. Farago (ed.). Plants and the chemical elements: Biochemistry, uptake, tolerance, and toxicity. VCH, New York. pp. 149-177.

Vantard, M., C. Peter, A. Fellous, P. Schellenbaum, and A.-M. Lambert. 1994. Characterization of a 100-dDa heat-stable microtubule-associated protein from higher plants. Eur. J. Biochem. 220:847-853.

Vantard, M., P. Schellenbaum, A. Fellous, and A.-M. Lambert. 1991. Characterization of maize microtubule-associated proteins, one of which is immunologically related to tau. Biochemistry 30:9334-9340.

Varco, J.J. 1993. Fertilizer nitrogen and cover crop management for no-tillage cotton production. pp. 63-65. *In:* T.D. Valco and M.R. McClelland (eds.). Conservation-Tillage Systems for Cotton.

University of Arkansas Agricultural Experiment Station Special Report 169. Fayetteville, AR.

Varco, J.J. 1997. Proximity effects of a calcium nitrate starter fertilizer solution on cotton. 1997 Beltwide Cotton Conference, New Orleans. National Cotton Council of America, Memphis, Tenn.

Varghese, S., K.V. Patel, M.D. Gohil, P.H. Bhatt, and U.G. Patel. 1995. Response of G-COT-11 levant cotton (*Gossypium herbaceum*) to salinity at germination stage. Indian J. Agric. Sci. 65:823-825.

Vasilas, B.L., J.O. Legg, and D.C. Wolf. 1980. Foliar fertilization of soybeans: absorption and translocation of ^{15}N-labeled urea. Agron. J. 72:271-275.

Vaughn, K.C., A.R. Lax, and S.O. Duke. 1988. Polyphenol oxidase: The chloroplast oxidase with no established function. Physiol. Plant. 72:659-665.

Vaux, H.R. and W.O. Pruitt. 1983. Crop-water production functions. Advances in Irrigation 2:61-97.

Veech, J.A. 1979. Histochemical localization and nematoxicity of terpenoid aldehydes in cotton. J. Nematology 11:240-246.

Veech, J.A. 1982. Phytoalexins and their role in the resistance of plants to nematodes. J. Nematology 14:2-9.

Veech, J.A. 1984. Cotton protection practices in the USA and world. Section C: Nematodes. pp. 309-330. *In*: R.J. Kohel and C.F. Lewis (eds.). Cotton. Agronomy Monograph No. 24, American Society of Agronomy, Inc., Madison, WI.

Veech, J.A., R.D. Stipanovic, and A.A. Bell. 1976. Peroxidative conversion of hemgossypol to gossypol. A revised structure for isohemigossypol. J. Chem. Soc., Chem. Commun. pp. 144-145.

Velten, J., J.E. Quisenberry, and G. Cartwright. 1998. Cotton variety and bacterial strain interactions during Agrobacteria-based genetic transformation. pp. 1392-1393. *In:* Proc. Beltwide Cotton Conf., National Cotton Council of America, Memphis, Tenn.

Venere, R.J. 1980. Role of peroxidase in cotton resistance to bacterial blight. Plant Science Letters 20:47-56.

Venere, R.J. and L. Brinkerhoff. 1974. The role of pectinolytic enzymes in the production of a necrotic lesion in cotton bacterial blight. Proc. American Phytopathological Society 1:78.

Venere, R.J., L.A. Brinkerhoff, and R.K. Gholson. 1984. Pectic enzyme: An elicitor of necrosis in cotton inoculated with bacteria. Proc. Oklahoma Academy of Science 64:1-7.

Vepraskas, M.J., G.S. Miner, and G.F. Peedin. 1986. Relationships of dense tillage pans, soil properties, and subsoiling to tobacco root growth. Soil Sci. Soc. Amer. J. 50:1541-1546.

Verkaar, H.J. 1988. Are defoliators beneficial for their host plants in terrestrial ecosystems - a review? Acta Bot. Neerl. 37:137-152.

Verschraege, L. 1989. Cotton fibre impurities. Neps, motes, and seed coat fragments. ICAC Review Article on Cotton Production Research No. 1, CAB International, Wallingford UK.

Vetter, H. and W. Teichmann. 1968. Feldversuche mit gestaffleten Kupfer- und Stickstoff-Dungergaben in Weser-Ems. Z. Pflanzenernahr. Bodenk. 121:97-111.

Viator, R.P., P.H. Jost, and J.T. Cothren. 1999. Do cotton varieties respond differently to plant growth regulators? p. 603. *In:* P. Dugger and D. Richter (eds.). Proc. Beltwide Cotton Conf., National Cotton Council of America, Memphis, Tenn.

Vilkova, N.A., T.L. Kuznetsova, A.L. Ismailov, and S.Yu. Islambekov. 1989. Effect of cotton cultivars with high content of gossypol on development of the cotton-boll worm, *Heliothis armigera* (Hbn.) (Lepidoptera, Noctuidae). Entomol. Rev. 68:129-137.

Vincke, H., E. DeLanghe, T. Fransen, and L. Verschraege. 1985. Cotton fibres are uniform in length under natural conditions. pp. 2-5. *In:* Cotton Fibres: Their Development and Properties" a technical monograph from the Belgian Cotton Research Group. The International Institute for Cotton, Manchester UK.

Vinson, S.B., G.W. Elzen, and H.J. Williams. 1987. The influence of volatile plant allelochemics on the third tropic level (parasitoids) and their herbivorous hosts. pp. 109-114. *In:* V. Labeyrie, G. Fabres, and D. Lachaise (eds.). Plants. Dr. W. Junk Publishers, Dordrecht, Netherlands.

Vinson, S.B. and H.J. Williams. 1991. Host selection behaviour of *Campoletis sonorensis*: A model system. Biol. Control 1:107-117.

Viswanathan, N. and R. Subbaraj. 1983. Action of gibberellic acid in effecting phase shifts of the circidian leaf-movement rhythm of a cotton plant, *Gossypium hirsutum*. Can. J. Bot. 61:2527-2529.

Vollesen, K. 1987. The native species of *Gossypium* (Malvaceae) in Africa, Arabia and Pakistan. Kew Bull. 42:337-349.

Volk, R. and C. McAuliffe. 1954. Factors affecting the absorption of ^{15}N-labeled urea by tobacco. Soil Sci. Soc. Amer. Proc. 18:308-312.

Voytas, D.F., M.P. Cummings, A. Konieczny, F.M. Ausubel, and S.R. Rodermel. 1992. *Copia*-like retrotransposons are ubiquitous among plants. Proc. Natl. Acad. Sci. USA 89:7124-7128.

Vretta-Kouskoleka, H. and T.L. Kallinis. 1968. Boron deficiency in cotton in relation to growth and nutrient balance. Proc. Soil Sci. Soc. Amer. 32: 253.

Vulkan-Levy, R., I. Ravina, A. Mantell, and H. Frenkel. 1998. Effect of water supply and salinity on pima cotton. Agric. Water Manage. 37:121-132.

Vunkova-Radeva, R., J. Schiemann, R. Mendel, G. Salcheva, and D. Georgieva. 1988. Stress and activity of molybdenum-containing complex (molybdenum cofactor) in winter wheat seeds. Plant Physiol. 87:533-535.

Waddle, B.A. 1974. Using mid-summer measurement as earliness indicators. pp. 78-79. *In:* J.M. Brown (ed.). Proc., Beltwide Cotton Prod. Res. Conf., National Cotton Council of America, Memphis, Tenn.

Waddle, B.A. 1984. Crop growing practices. pp. 233-263. *In:* R.J. Kohel and C.F. Lewis (eds.). Cotton. American Soc. Agron., Madison, WI

Waddle, B.A. 1985. Risk management in getting and keeping a stand -- environment. pp. 3-4. *In:* Proc. 1985 Beltwide Cotton Prod. Conf., National Cotton Council, Memphis, Tenn.

Wadleigh, C.H. 1944. Growth status of the cotton plant as influenced by the supply of nitrogen. Ark. Agr. Exp. Sta. Bull. 446.

Wadleigh, C.H., H.G. Gauch, and D.C. Strong. 1947. Root penetration and moisture extraction in saline soil by crops plants. Soil Sci. 63:341-349.

Wäfler, D. and H. Meier. 1994. Enzyme activities in developing cotton fibres. Plant Physiol. Biochem. 32:697702.

Wakeman, R.J., B.L. Weir, E.J. Paplomatas, and J.E. DeVay. 1993. Association of foliar symptoms of potassium deficiency in cotton (*Gossypium hirsutum*) with infection by *Verticillium dahliae*. pp. 213-215. *In:* Proc. Beltwide Cotton Conferences. National Cotton Council, Memphis, Tenn.

Walbot, V. and L.S. Dure, III. 1976. Developmental biochemistry of cotton seed and germination. VII. Characterization of the cotton genome. J. Mol. Biol. 101:503-536.

Walhood, V.T. and F.T. Addicott. 1968. Harvest-aid programs: Principles and practices. pp. 407-31. *In:* F.C. Elliot, M. Hoover, and W.K. Porter, Jr. (eds.). *Advances in Production and Utilization of Quality Cotton: Principles and Practices.* The Iowa State University Press, Ames, IA.

Walker, J.K. and G.A. Niles. 1971. Population dynamics of the boll weevil and modified cotton types. Tex. Agric. Exp. Stat. Bull. 1109.

Walker, M. and O.O. Brooks. 1964. Broadcast application versus row placement of mixed fertilizer. Georgia Agr. Exp. Sta. Mimeograph Series.

Walker, N.R., T.L. Kirpatrick, and C.S. Rothrock. 1998. Interaction between *Meloidogyne incognita* and *Thielaviopsis basicola* on cotton (*Gossypium hirsutum*). J. Nematology 30:415-422.

Wall, G.W., J.S. Amthor, and B.A. Kimball. 1994. COTCO2: a cotton growth simulation model for global change. Agric. For. Meteorol. 70:289-342.

Wallace, T.P. and K.M. El-Zik. 1989. Inheritance of resistance in three cotton cultivars to the HV1 isolate of bacterial blight. Crop Science 29:1114-1119.

Wallach, D. 1978. A simple model of cotton yield development. Field Crops Res. 1:269-281.

Waller, D.P., N. Bunyapyrahatsara, A. Martin, C.J. Vournazos, M.S. Ahmed, D.D. Soejarto, G.A. Cordell, H.H.S. Fong, L.D. Russel, and J.P. Malone. 1983. Effects of (+)-gossypol on fertility in male hamsters. J. Androl. 4:276-279.

Waller, G.D. and A.N. Mamood. 1991. Upland and Pima cotton as pollen donors for male-sterile upland seed parents. Crop Sci. 31:265-266.

Walter, H., H.W. Gausman, F.R. Rittig, L.N. Namken, D.E. Escobar, and Rodriguez. 1980. Effect of mepiquat chloride on cotton plant, leaf, and canopy structure and dry weights of its components. Beltwide Cotton Prod. Res. Conf. pp. 32-35.

Walter, M.H., T. Fester, and D. Strack. 2000. Arbuscular mycorrhizal fungi induce the non-mevalonate methylerythritol phosphate pathway of isoprenoid biosynthesis correlated with accumulation of the 'yellow pigment' and other apocarotenoids. Plant J. 21:571-578.

Wan, C. and T.A. Wilkins. 1994. Isolation of multiple cDNAs encoding the vacuolar H$^+$-ATPase subunit B from developing cotton (*Gossypium hirsutum* L.) ovules. Plant Physiol. 106:393-394.

Wang, B., M.L. Dale, J.K. Kochman, and N.R. Obst. 1999. Effects of plant residue, soil characteristics, cotton cultivars, and other crops on Fusarium wilt of cotton in Australia. Australian Journal of Experimental Agriculture 39:203-209.

Wang, G.-L., J.-M. Dong, and A.H. Paterson. 1995. The distribution of *Gossypium hirsutum* chromatin in *G. barbadense* germ plasm: molecular analysis of introgressive plant breeding. Theor. Appl. Genet. 91:1153-1161.

Wang, G.-L., R.A. Wing, and A.H. Paterson. 1993. PCR amplification from single seeds, facilitating DNA marker-assisted breeding. Nucl. Acids Res. 21:2527.

Wang, M. 1987. Analysis of gossypol by high performance liquid chromatography. J. Ethnopharmacol. 20:1-11.

Wang, N., L. Zhou, M. Guan, and H. Lei. 1987. Effect of (-)- and (+)-gossypol on fertility in male rats. J. Ethnopharmacol. 20:21-24.

Wang, S.-Y.C. and J.A. Pinckard. 1972. Some biochemical factors associated with the infection of cotton fruit by *Diplodia gossypina*. Phytopathology 62:460-465.

Wang, W.H., Z.H. Yu, M. Cai, and X.M. Xu. 1990. Antifertility actions of gossypol derivatives and analogues. Acta Pharamacologica Sinica 11:268-271.

Wang, Y. and Y. Xi. 1986. Plantlet regeneration from shoot apex of *Gossypium hirsutum* L. Chinese Journal of Agric. Sci. 2(2):16-19.

Wang, Z.Z., R.X. Chen, and Y.H. Wang. 1991a. Effects of vitamin B6 on salt resistance in cotton. China Cottons 1991, 30-35.

Wang, Z.Z., Z.G. Shen, Y. Ding, Z.N. Ye, K.Q. Li, L.F. Hao, D.Z. Guo, and D.L. Zhang. 1991b. A preliminary report of research on the salt tolerance response and screening of cotton explants. China Cottons 1991, 18. (Chinese).

Wanjura, D.F. 1973. Effect of soil physical properties on cotton emergence. USDA Tech. Bull. 1481. U.S. Gov. Print. Office, Washington, D.C.

Wanjura, D.F. 1986. Field environment and stand establishment. pp. 551-554. *In:* J.R Mauney and J.McD. Stewart (eds.). Cotton Physiology. The Cotton Foundation, Memphis, Tenn.

Wanjura, D.F. and D.R. Buxton. 1972a. Hypocotyl and radicle elongation of cotton as affected by soil environment. Agron. J. 64:431-434.

Wanjura, D.F. and D.R. Buxton. 1972b. Water uptake and radicle emergence of cottonseed as affected by soil moisture and temperature. Agron. J. 64:427-431.

Wanjura, D.F., E.B. Hudspeth, Jr., and J.R. Bilbro, Jr. 1967. Temperature-emergence relations of cottonseed under natural diurnal fluctuation. Agron J. 59:217-219.

Wanjura, D.F., E.B. Hudspeth, Jr., and J.D. Bilbro, Jr. 1969. Emergence time, seed quality, and planting depth effects on yield and survival of cotton (*Gossypium hirsutum* L.). Agron. J. 61:63-65.

Wanjura, D.F., J.R. Mahan, and D.R. Upchurch. 1996a. Irrigation starting time effects on cotton under high-frequency irrigation. Agron. J. 88:561-566.

Wanjura, D.F., J.R. Mahan, and D.R. Upchurch. 1996b. Early season irrigation and influence on soil temperature and cotton yield. pp. 517-522. *In:* Proc. Beltwide Cotton Conf., National Cotton Council of America, Memphis, Tenn.

Wanjura, D.F. and E.B. Minton. 1981. Delayed emergence and temperature influences on cotton seedling vigor. Agron J. 73:594-597.

Wanjura, D.F., D.R. Upchurch, J.L. Hatfield, J.J. Burke, and J.R. Mahan. 1988. Cotton irrigation using the thermal kinetic window criteria. pp. 183-185. *In:* J.M. Brown and D.A. Richter (eds.). Proc. Beltwide Cotton Prod. Res. Conf., National Cotton Council of America, Memphis, Tenn.

Wanjura, D.F., D.R. Upchurch, and J.R. Mahan. 1992. Automated irrigation based on threshold canopy temperature. Trans. ASAE 35:153-159.

Wanjura, D.F., D.R. Upchurch, and J.R. Mahan. 1989. Application of TKW principles to cotton production management. pp. 52-56. *In:* J.M. Brown and

D.A. Richter (eds.). Proc. Beltwide Cotton Prod. Res. Conf., National Cotton Council of America, Memphis, Tenn.

Ward, E.R., S.J. Uknes, S.C. Williams, S.S. Dincher, D.L. Widerhold, D.C. Alexander, P. Ahl-Goy, J.-P. Metraux, and J.A. Ryals. 1991. Coordinate gene activity in response to agents that induce systemic acquired resistance. Plant Cell 3:1085-1094.

Ward, J.M. and J.I. Schroeder. 1994. Calcium activated K⁺ channels and calcium-induced calcium release by slow vacuolar ion channels in guard cell vacuoles implicated in the control of stomatal closure. Plant Cell 6:669-683.

Wardlaw, I.F. and C.W. Wrigley. 1994. Heat tolerance in temperate cereals: an overview. Aust. J. Plant Phsyiol. 21:695-703.

Ware, J.O. 1951. Origin, rise and development of American Upland cotton varieties and their status at present. University of Arkansas, Fayetteville, AR

Waring, G.L. and N.S. Cobb. 1992. The impact of plant stress on herbivore population dynamics. Insect-Plant Interactions 4:167-226.

Warner, D.A. and J.J. Burke. 1993. Cool night temperatures alter leaf starch and photosystem II: Chlorophyll fluorescence in cotton. Agron. J. 85:836-840.

Warner, D.A., A.S. Holaday, and J.J. Burke. 1995. Acclimation of carbon metabolism to night temperature in cotton. Agron. J. 87:1193-1197.

Warner, H.L. and A.C. Leopold. 1969. Ethylene evolution from 2-chloroethyl phosphonic acid. Plant Phyiol. 44:156-158.

Wartelle, L.H., J.M. Bradow, O. Hinojasa, A.B. Pepperman, G.F. Sassenrath-Cole, and P. Dastoor. 1995. Quantitative cotton fiber maturity measurements by X-ray fluorescence spectroscopy and AFIS. J. Agric. Food Chem. 43:1219-1223.

Waterkyn, L. 1981. Cytochemical localization and function of the 3-linked glucan callose in the developing cotton fibre cell wall. Protoplasma 106:49-67.

Watson, D.J. 1947. Comparative physiological studies on the growth of field crops. I. Variation in net assimilation rate and leaf area between species and varieties, and within and between years. Ann. Bot. N.S. 11:41-76.

Watson, R.T., H. Rodhe, H. Oeschger, and U. Seigenthaler. 1990. Greenhouse gasses and aerosols. pp. 1-40. In: J.T. Houghton, G.J. Jenkins, and J.J. Ephraums (eds.). Climate Change. The IPCC Scientific Assessment, Cambridge University Press, Cambridge.

Watson, T.F., F.M. Carasso, D.T. Langston, E.B. Jackson, and D.G. Fullerton. 1978. Pink bollworm suppression through crop termination. J. Econ. Entomol. 71:638-641.

Watt, G. 1907. *The Wild and Cultivated Cotton Plants of the World*. Longmans, Green and Co., London.

Watts, J.G. 1937. Reduction of cotton yield by thrips. J. Econ. Entomol. 30:860-863.

Weatherley, P.E. 1950. Studies in the water relations of the cotton plant. I. The field measurements of water deficits in leaves. New Phytol. 49:81-97.

Weathersbee, A.A., D.D. Hardee, and W.R. Meredith. 1994. Effects of cotton genotype on seasonal abundance of cotton Aphid (Homoptera, Aphididae). J. Agr. Entomol. 11:29-37.

Weaver, J.B., Jr. and H.L. Bhardwaj. 1985. Growth regulating effect of chlordimeform in cotton. pp. 288-293. In: J.M. Brown (ed.). Proc. Beltwide Cotton Prod. Res. Conf., National Cotton Council of America, Memphis, Tenn.

Weaver, J.E. 1926. Root Development of Field Crops. McGraw Hill, New York.

Wedding, R.T. 1989. Malic enzymes of higher plants. Plant Physiol. 90:367-371.

Weete, J.D., G.L. Leek, C.M. Peterson, H.E. Currie, and W.D. Branch. 1978. Lipid and surface wax synthesis in water-stressed cotton leaves. Plant Physiol. 62:675-677.

Wei, Y.-A., D.L. Hendrix, and R. Nieman. 1996. Isolation of a novel tetrasaccharide, bemisiotetrose, and glycine betaine from silverleaf whitefly honeydew. J. Agric. Food Chem. 44:3214-3218.

Wei, Y.-A., D.L. Hendrix, and R. Nieman. 1997. Diglucomelezitose, a novel pentasaccharide in silverleaf whitefly honeydew. J. Agric. Food Chem. 45:3481-3486.

Weinbaum, S.A. 1988. Foliar nutrition in fruit trees. pp. 8-100. In: P.M. Neumann (ed.). Plant Growth and Leaf-Applied Chemicals. CRC Press, Boca Raton, FL.

Weir, B.L. and J.M. Gaggero. 1982. Ethephon may hasten cotton boll opening, increase yield. California Agric., California Agric. Ext. Stn. 36:28-29.

Weir, B.L., T.A. Kerby, K.D. Hake, B.A. Roberts, and L.J. Zelinski. 1996. Cotton Fertility. In: S.J. Hake, T.A. Kerby, and K.D. Hake (eds.). Cotton Production Manual. Univ. of Calif. Oakland. pp. 210-227.

Weir, B., R. Miller, R. Vargas, D. Muck, B. Roberts, S. Wright, D. Munier, and M. Keeley. 1995. Evaluation of cotton soil potassium in California. Proc. Beltwide Cotton Conf., San Diego, CA. p. 1353.

Weir, B.L., D. Munk, S. Wright, and B. Roberts. 1994. Responses to PGR-IV of upland and pima cottons in the San Joaquin Valley of California. pp. 1267-1268. In: D.J. Herber and D.A. Richter (eds.). Proc. Beltwide Cotton Conf., National Cotton Council of America, Memphis, Tenn.

Weir, B.L. and B.A. Roberts. 1993. Nitrogen and potassium fertilization. California Cotton Review, May 1993. pp. 12-13.

Weir, B.L., B.A. Roberts, and T.A. Kerby. 1992. Effects of foliar N and K on cotton petiole levels and lint yields. pp. 1162-1163. *In:* Proc. Beltwide Cotton Conferences. National Cotton Council, Memphis, Tenn.

Weis, K.G., K.R. Jacobsen, and J.A. Jernstedt. 1999. Cytochemistry of developing cotton fibers: A hypothesized relationship between motes and non-dyeing fibers. Field Crops Res. 62:107-117.

Wells, R. 1988. Response of leaf ontogeny and photosynthetic activity to reproductive growth in cotton. Plant Physiol. 87:274-279.

Wells, R. and Edmisten, K. 1998. Two years of growth, canopy photosynthesis and yield in response to different formulations of MepPlus. pp. 1424-1425. *In:* P. Dugger and D. Richter (eds.). Proc. Beltwide Cotton Conf., National Cotton Council of America, Memphis, Tenn.

Wells, R. and W.R. Meredith, Jr. 1984a. Comparative growth of obsolete and modern cotton cultivars. I. Vegetative dry matter partitioning. Crop Sci. 24:858-862.

Wells, R. and W.R. Meredith. 1984b. Comparative growth of obsolete and modern cotton cultivars. II. Reproductive dry matter partitioning. Crop Sci. 24:863-868.

Wells, R. and W.R. Meredith, Jr. 1984c. Comparative growth of obsolete and modern cotton cultivars. III. Relationship of yield to observed growth characteristics. Crop Sci. 24:868-872.

Wells, R. and W.R. Meredith, Jr. 1986a. Heterosis in upland cotton. I. Growth and leaf area partitioning. Crop Sci. 26:1119-1123.

Wells, R. and W.R. Meredith, Jr.1986b. Okra leaf vs. Normal leaf interactions. II. Analysis of vegetative and reproductive growth. Crop Sci. 219-228.

Wells, R., W.R. Meredith, Jr., and J.R. Williford. 1986. Canopy photosynthesis and its relationship to plant productivity in near-isogenic cotton lines differing in leaf morphology. Plant Physiol. 82:635-640.

Wells, R. W.R. Meredith, Jr., and J.R. Williford. 1988. Heterosis in upland cotton. II. Relationship of leaf area to plant photosynthesis. Crop Sci. 28:522-525.

Welter, S.C.1989. Arthropod impact on plant gas exchange. pp. 135-150. *In:* E.A. Bernays (ed.). Insect-Plant Interactions Vol I. CRC Press Boca Raton, Florida.

Wendel, J.F. 1989. New World tetraploid cottons contain Old World cytoplasm. Proc. Nat. Acad. Sci. U.S.A. 86:4132-4136.

Wendel, J.F. and V.A Albert. 1992. Phylogenetics of the cotton genus (*Gossypium*): Character-state weighted parsimony analysis of chloroplast DNA restriction site data and its systematic and biogeographic implications. Syst. Bot. 17:115-143.

Wendel, J.F., C.L. Brubaker, and A.E. Percival. 1992. Genetic diversity in *Gossypium hirsutum* and the origin of Upland cotton. Amer. J. Bot. 79:1291-1310.

Wendel, J.F., P.D. Olson, and J. McD. Stewart. 1989. Genetic diversity, introgression, and independent domestication of Old World cultivated cottons. Amer. J. Bot. 76:1795-1806.

Wendel, J.F. and A.E. Percival. 1990. Molecular divergence in the Galapagos Island-Baja California species pair, *Gossypium klotzschianum* Anderss. and *G. davidsonii* Kell. Pl. Syst. Evol. 171:99-115.

Wendel, J.F. and R.G. Percy. 1990. Allozyme diversity and introgression in the Galapagos Islands endemic *Gossypium darwinii* and its relationship to continental *G. barbadense*. Biochem. Syst. Ecol. 18:517-528.

Wendel, J.F., R. Rowley, and J. Stewart. 1994. Genetic diversity in and phylogenetic relationships of the Brazilian endemic cotton, *Gossypium mustelinum* (Malvaceae). Pl. Syst. Evol. 192:49-59.

Wendel, J.F., A. Schnabel, and T. Seelanan. 1995a. Bidirectional interlocus concerted evolution following allopolyploid speciation in cotton (*Gossypium*). Proc. Natl. Acad. Sci. (USA) 92:280-284.

Wendel, J.F., A. Schnabel, and T. Seelanan. 1995b. An unusual ribosomal DNA sequence from *Gossypium gossypioides* reveals ancient, cryptic, intergenomic introgression. Mol. Phyl. Evol. 4:298-313.

Wendel, J.F., J. McD. Stewart, and J.H. Rettig. 1991. Molecular evidence for homoploid reticulate evolution in Australian species of *Gossypium*. Evolution 45:694-711.

Went, F.W. 1963. The concept of the phytotron. *In:* L.T. Evans (ed.). Environmental Control of Plant Growth. Academic Press. New York & London.

Went, F.W. 1974. Reflections and speculations. Ann. Rev. Plant. Physiol.25:1-26.

Westermann, R.L. (ed.). 1990. Soil Testing and Plant Analysis. No. 3. Soil Sci. Soc. Amer., Madison, WI.

Whalon, M.E. and W.H. McGaughey.1993. Insect resistance to *Bacillus thuringiensis*. pp. 215-232 *In:* L. Kim (ed.). Advanced Engineered Pesticide. Marcell Dekker Inc. New York.

Wheeler, M.H., L.S. Puckhaber, A.A. Bell, R.A. Baker, and R.D. Stipanovic. 2000. Discovery and possible significance of nectriafurone, 5-O-methyl-javanicin, and related compounds produced by *Fusarium oxysporum* f.sp. *vasinfectum*. p. 136. *In*: Proc. Beltwide Cotton Conf., National Cotton Council of America, Memphis, Tenn

Wheeler, M.H., R.D. Stipanovic, and L.S. Puckhaber. 1999. Phytotoxicity of equisetin and epi-equisetin isolated from *Fusarium equiseti* and *F. pallidoroseum*. Mycological Research 103:967-973.

Whisler, F.D., B. Acock, D.N. Baker, R.E., Fye, H.F. Hodges, J.R., Larnbert, H.E. Lemmon, J.M. McKinion, and V.R. Reddy. 1986. Crop simulation models in agronomic systems. Adv. Agron. 40:141 -208.

White, H.C. 1914. The feeding of cotton. Georgia Agric. Exp. Stn. Bull. 108.

White, H.C. 1915. The feeding of cotton. II. Georgia Agric. Exp. Stn. Bull. 114.

White, J. 1979. The plant as a metapopulation. Ann. Rev. Ecol. Syst. 10:109-145.

White, P.J. and J.A. Smith. 1989. Proton and anion transport at the tonoplast in crassulacean acid metabolism plants: specificity of the malate-influx system in *Klanchoe daigremontiana.* Planta 179:265-274.

White, R.F. 1979. Acetylsalicylic acid (aspirin) induces resistance to tobacco mosaic virus in tobacco. Virology 99:410-412.

White, R.H., D. Worsham, and U. Blum. 1989. Allelopathic potential of legume debris and aqueous extracts. Weed Sci. 37:674-679.

White, T.C.R. 1993. The Inadequate Environment. Springer-Verlag, Berlin, pp. 425.

Whitfield, D.M. and D.J. Connor. 1980. Penetration of photosynthetically active radiation into tobacco crops. Aust. J. Plant Physiol. 7:449-461.

Whitham, T.G., J. Maschinski, K.C. Larson, and K.N. Paige. 1991. Plant responses to herbivory: the continuum from negative to positive and underlying physiologycal mechanisms. pp. 227-256 *In:* P.W. Price, T.M. Lewinson, G. W. Fernandes, and W.W. Benson (eds.). Plant-Animal Interactions: Evolutionary Ecology in Tropical and Temperate Regions. John Wiley & Sons.

Whitney, J.B. 1941. Effects of the composition of the soil atmosphere the absorption of water by plants. Amer. J. Bot. 28:14.

Wibbe, M.L., M.M. Blanke, and F. Lenz. 1993. Effect of fruiting on carbon budgets of apple tree canopies. Trees: Structure and Function. 8:56-60.

Wiegand, C.L., J.H. Everitt, and A.J. Richardson. 1992. Comparison of multispectral video and SPOT-1 HRV observations for cotton affected by soil salinity. Int. J. Remote Sensing 13:1511-1525.

Wiegand, C.L., J.D. Rhoades, D.E. Escobar, and J.H. Everitt. 1994. Photographic and videographic observations for determining and mapping the response of cotton to soil salinity. Remote Sensing of Environ. 49:212-223.

Wiersum, L.K. 1958. Density of root branching as affected by substrate and separate ions. Acta Botanica Neerlandica 7:174-190.

Wiese, M.V. and J.E. DeVay. 1970. Growth regulator changes in cotton associated with defoliation

caused by *Verticillium dahliae*. Plant Physiology 45:304-309.

Wilcox, L.V. 1960. Boron injury to plants. USDA Inf. Bull. 211., Washington D.C.

Wiles, A.B. 1959. Calcium deficiency in cotton seedlings. Plant Dis. Rep. 43:365-367.

Wilkins, T.A. 1992. Role of proton pumps in cotton fiber development. pp. 141-152. *In:* C.R. Benedict (ed.). Proc. Cotton Fiber Cellulose: Structure, Function, and Utilization Conference. National Cotton Council, Memphis Tenn.

Wilkins, T.A. 1993. Vacuolar H^+-ZTPase 69 - kilodalton catalytic subunit cDNA from developing cotton (*Gossypium hirsutum* L.) ovules. Plant Physiol. 102:679-680.

Wilkins, T.A. and J.A. Jernstedt. 1999. Molecular genetics of developing cotton fibers. pp. 231-267. *In:* A.S. Basra (ed.). *Cotton Fibers.* Hawthorne Press, New York, N.Y., USA.

Wilkins, T.A. and S.C. Tiwari. 1994. Cotton fiber morphogenesis: A subcellular odyssey. pp. 89-94. *In:* G.M. Jividen and C.R. Benedict (eds.). Proc. Biochemistry of Cotton Workshop. Cotton Incorporated, Raleigh NC.

Wilkins, T.A., C.-Y. Wan, and C.-C. Lu. 1994. Ancient origin of the vacuolar H^+-ATPase 69-kilodalton catalytic subunit superfamily. Theor. Appl. Genet. 89:514-524.

Wilkinson, R.E. and M.J. Kasperbauer. 1980. Epicuticular alkane content of tobacco as influenced by photoperiod, temperature and leaf age. Phytochemistry 11:2439-2442.

Williams, G.F. and J.M. Yankey. 1996. New developments in single fiber fineness and maturity measurements. pp. 1284-1289. *In:* Proc. Beltwide Cotton Conf., National Cotton Council of America, Memphis, Tenn.

Williams, H.J., G.W. Elzen, and S.B. Vinson. 1988. Parasitoides-Host-Plant interactions emphasizing cotton (*Gossypium*). pp. 171-200. *In:* P. Barbosa and D. Letourneau (eds.). Novel Aspects of Insect-Plant Interactions. John Wiley & Sons, New York.

Williams, H.J., G. Moyna, S.B. Vinson, A.I. Scott, A.A. Bell, and R.D. Stipanovic 1996. β-Caryophyllene derivatives from the wild cottons *Gossypium armourianum, Gossypium harknessii,* and *Gossypium turneri.* J. Agric. Food Chem. (submitted???)

Williams, H.J., I. Sattler, G. Moyna, A.I. Scott, A.A. Bell, and S.B. Vinson. 1995. Diversity in cyclic sesquiterpene production by *Gossypium hirsutum*. Phytochem. 40:1633-1636.

Williams, H.J., R.D. Stipanovic, L.A. Smith, S.B. Vinson, P.O. Darnell, R. Montandon, D.L. Begin, G.W. Elzen, H. Guanasena, and A.A. Bell. 1987. Effects of gossypol and other cotton terpenoids on

Heliothis virescens development. Rev. Latinoamer. Quim. 18:119-131.

Williford, J.R. 1992. Influence of harvest factors on cotton yield and quality. 1992. Trans. ASAE 35:1103-1107.

Williford, J.R., W.R. Meredith, Jr., and W.S. Anthony. 1986. Factors influencing yield and quality of Delta cotton. Papers of ASAE, No. 86-1561.

Williford, J.R., S.T. Rayburn, and W.R. Meredith, Jr. 1986. Evaluation of a 76-cm row for cotton production. Trans. ASAE 29:1544-1548.

Willison, J.H.M. 1983. The morphology of supposed cellulose-synthesizing structures in higher plants. J. Appl. Polym. Sci: Appl. Polym. Symp. 37:91-105.

Willison, J.H.M. and R.M. Brown, Jr. 1977. An examination of the developing cotton fiber: Wall and plasmalemma. Protoplasma 92:21-41.

Wilson, A.G.L. 1981. *Heliothis* damage to cotton and concomitant action levels in the Namoi Valley, New South Wales. Prot. Ecol.3:311-325.

Wilson, A., F.B. Pickett, J.C. Turner, and M. Estelle. 1990. A dominant mutation in *Arabidopsis* confers resistance to auxin, ethylene, and abscisic acid. Mol. Gen. Genet. 222:377-383.

Wilson, F.D. 1987. Pink bollworm resistance, lint yield, and earliness of cotton isolines in a resistant genetic background, Crop Sci. 27:957-960.

Wilson, F.D. 1991. Twenty years of host plant resistance: progress, problems, prognostications. Proc. Beltwide Cotton Conferences 1:542-544. National Cotton Council of America, Mempis, Tenn.

Wilson, F.D., H.M. Flint, W.R. Deaton, and R.E. Buehler. 1994. Yield, yield components and fiber properties of insect-resistant cotton lines containing a *Bacillus thuringiensis* toxin gene. Crop Sci. 34:38-41.

Wilson, F.D., H.M. Flint, W.R. Deaton, D.A. Fischhoff, F.J. Perlak, T.A. Armstrong, R.L. Fuchs, S.A. Berberich, N.J. Parks, and B.R. Stapp. 1992. Resistance of cotton lines containing a *Bacillus thuringiensis* toxin to pink bollworm (Lepidoptera: Gelechiidae) and other insects.J. Econ. Entomol. 85: 1516-1521.

Wilson, F.D., H.M. Flint, B.R. Stapp, and N.J. Parks.1993. Evaluation of cultivars, germplasm lines, and species of *Gossypium* for resistance to biotype "B" of sweetpotato whitefly, *Bemisia tabaci* (Homoptera: Aleyrodidae).J. Econ. Entomol. 86:1857-1862.

Wilson, F.D. and B.W. George. 1982. Effects of okra-leaf, frego-bract, and smooth-leaf mutants on pink bollworm damage and agronomic properties of cotton. Crop Sci. 22:798-801.

Wilson, F.D., J.A. Lee, and R.R. Bridge. 1968. Genetics of cytology of cotton 1956-67. South. Coop. Ser. Bull. 139.

Wilson, L.J. 1993. Spider mites (Acari: Tetranychidae) affect yield and fiber quality of cotton. J. Econ. Entomol. 86: 566-585.

Wilson, L.J. 1994. Resistance of okra-leaf cotton genotypes to twospotted spider mites (Acari: Tetranychidae). J. Econ. Entomol. 87: 1726-1735.

Wilson, L.J. 1995. Habitats of twospotted spider mites (Acari: Tetranychidae) during winter and spring in a cotton-producing region of Australia. Environ. Entomol. 24:332-340.

Wilson, L.J., V.O. Sadras, and L. Bauer. 1994. Thrips and cotton: the results so far. The Australian Cottongrower (October 1994) pp. 25-29.

Winter, K. and M. Koniger. 1991. Dry matter production and photosynthetic capacity in *Gossypium hirsutum* L. under conditions of slightly sub-optimal leaf temperatures and high levels of irradiance. Oecologia 87:190-197.

Wittwer, S.H., M.J. Bukovac, and H.B. Tukey. 1963. Advances in foliar feeding of plant nutrients. pp. 429-455 *In:* M.H. McVickar, G.L. Bridger, and L.B. Nelson (eds.). Fertilizer Technology and Use. Amer. Soc. Agron., Madison, WI.

Wong, H.C., A.L. Fear, R.D. Calhoon, G.H. Eichinger, R. Mayer, D. Amikam, M. Benziman, D.H. Gelfand, J.H. Meade, A.W. Emerick, R. Bruner, A. Ben-Bassat, and R. Tal. 1990. Genetic organization of the cellulose synthase operon in *Acetobacter xylinum*. Proc. Natl. Acad. Sci. 87:8130-8134.

Wong, S.-C. 1990. Elevated atmospheric partial pressure of CO_2 and plant growth. II. Non-structural carbohydrate concent in cotton plants and its effect on growth parameters. Photosyn. Res. 23:171-180.

Wood, B.W., W.L. Tedders, and C.C. Reilly. 1988. Sooty mould fungus on pecan foliage suppresses light penetration and net photosynthesis. Hort. Sci. 33:851-853.

Wood, C.W., P.W. Tracy, D.W. Reeves, and K.L. Edmiston. 1992. Determination of cotton nitrogen with a hand-held chlorophyll meter. J. Plt. Nutr. 14:1435-1448.

Wood, R. 1986. Comparison of the cyclopropene fatty acid content of cotton seed varieties, glanded and glandless seeds and various seed structures. Biochem. Arch. 2:73-80.

Woodward, R.G. and H.M. Rawson. 1976. Photosynthesis and transpiration in dicotyledonous plants. II. Expanding and senescing leaves of soybean. Aust. J. Plant Physiol. 3:257-267.

Woolhouse, H.W. 1983. Toxicity and tolerance in response of plants to metals. *In:* O.L. Lang (ed.). Encyclopedia of Plant Physiology, New Series. Vol 12C. Springer-Verlag, Berlin. pp. 245-300.

Wright, P.R. 1999. Premature senescence of cotton (*Gossypium hirsutum* L.) – Predominately a potassium deficiency disorder caused by an imbalance of source and sink. Plant and Soil 211:231-239.

Wright, R.J., P.M. Thaxton, K.M. El-Zik, and A.H. Paterson. 1998. D-subgenome bias of *Xcm* resistance genes in tetraploid *Gossypium* (cotton) suggest that polyploid formation has created novel avenues for evolution. Genetics 149:1987-1996.

Wright, S.F. and A. Upadhyaya. 1998. A survey of soils for aggregate stability and glomalin, a glycoprotein produced by hyphae of arbuscular mycorrhizal fungi. Plant Soil 198:97-107.

Wrona, A.F., R.K. Bowman, R.B. Hutmacher, M.A. Jones, and G.B. Padgett. 1999. 1998 Year in Review: weather extremes dock crop. pp. 3-12. *In:* Proc. Beltwide Cotton Conf., National Cotton Council of America, Memphis, Tenn.

Wrona, A.F., J.R. Bradley, R. Carter, R. Deaton, K. Edmisten, B. Finney, K. Gully, B. Guthrie, D.S. Guthrie, T. Kerby, L. Martin, W. McCarty, B. McLendon, R. Rayner, J. Silvertooth, and R. Smith. 1997. Bt cotton requires vigilant management. Cotton Physiol. Today 8(3):25-36.

Wullschleger, S.D. and D.M. Oosterhuis. 1987a. Carbon partitioning within a sympodial branch during cotton boll development. Ark. Farm Res. 36:4.

Wullschleger, S.D. and D.M. Oosterhuis. 1989a. The occurrence of an internal cuticle in cotton (*Gossypium hirsutum* L.) leaf stomates. Environ. Exp. Bot. 29:229-235.

Wullschleger, S.D. and D.M. Oosterhuis. 1987b. Electron microscope study of cuticle abrasion on cotton leaves in relation to water potential measurement. J. Exp. Bot. 38:660-667.

Wullschleger, S.D. and D.M. Oosterhuis. 1989b. Water use efficiency as a function of leaf age and position within the cotton canopy. Plant and Soil. 120:79-86.

Wullschleger, S.D. and D.M. Oosterhuis. 1990a. Photosynthesis of individual field-grown cotton leaves during ontogeny. Photosynthesis Res. 23:163-170.

Wullschleger, S.D. and D.M. Oosterhuis. 1990b. Canopy development and photosynthesis as influenced by nitrogen nutrition. J. Plant Nut. 13:1141-1154.

Wullschleger, S.D. and D.M. Oosterhuis. 1990c. Photosynthetic and respiratory activity of fruiting forms within the cotton canopy. Plant Physiol. 92:463-469.

Wullschleger, S.D. and D.M. Oosterhuis. 1990d. Photosynthetic carbon production and use by developing cotton leaves and bolls. Crop Sci. 30:1259-1264.

Wullschleger, S.D. and D.M. Oosterhuis. 1992. Canopy leaf area development and age-class dynamics in cotton. Crop Sci. 32:451-456.

Wullschleger, S.D, D.M. Oosterhuis, R.G. Hurren, and P.J. Hanson. 1991. Evidence for light-dependent recycling of respired carbon dioxide by the cotton fruit. Plant Physiol. 97:574-579.

Wyn Jones, R.G., and J. Gorham. 1983. Osmoregulation. pp. 35-38. *In:* O.L. Lange, P.S. Nobel, C.B. Osmond, and H. Ziegler (ed.). Encyclopedia of Plant Physiology, New Series 12, Physiological Plant Ecology III. Springer Verlag, Berlin.

Xi, S., R. Lihua, Z. Yongsong, T. Qizhao, and Q. Lianxiang. 1989. Effect of potassium fertilizer application on physiological parameters and yield of cotton grown on a potassium deficient soil. Z. Pflanzenenerndhr. Bodenk. 152:269-272.

Xia, Z.J., P.N. Achar, and B.K. Gu. 1998. Vegetative compatibility groupings of *Verticillium dahliae* from cotton in mainland China. European J. Plant Pathology 104:871-876.

Xiang, S. and W. Yang. 1993. Gossypol and its enantiomers in the seeds of cotton *Gossypium*. Zhonggou Nongye Kexue 26:31-35.

Xie, Q., W.X. Wei, and Y.H. Wang. 1992. Studies on absorption, translocation and distribution of boron in cotton (*Gossypium hirsutum* L.). Acta Agonomica Sinica (in Chinese). 18:31-37.

Xie, W., N.L. Trolinder, and C.H. Haigler. 1993. The effects of cool temperatures on cotton fiber initiation and elongation clarified by analysis of in vitro cultures. Crop Sci. 33:1258-1264.

Xu, D.P., S.J. Sung, T. Loboda, P.P. Kormanik, and C.C. Black. 1990. Characterization of sucryolysis in the uridine diphosphate and pyrophosphate-dependent sucrose synthase pathway. Plant Physiol. 90:635-642.

Xu, P., C.W. Lloyd, C.J. Staiger, and B.K. Drobak. 1992. Association of phophatidylinositol phophate kinase with the plant cytoskeleton. Plant Cell 4:941-9451.

Xu, X.Q. and S.Q. Pan. 2000. An *Agrobacterium* catalase is a virulence factor involved in tumorigenesis. Molecular Microbiology 35:407-414.

Xu, X. and H.M. Taylor. 1992. Increase in drought resistance of cotton seedlings treated with mepiquat chloride. Agron. J. 84:569-574.

Xue, H.Z., Z.M. Guo, A.H. Kong, and G.P. Wu. 1992. The synthesis of unsymmetric analogs of gossypol, 6-O-methylgossypol. Chinese Chemical Letters 3:165-166.

Yamada, Y. 1962. Studies on foliar absorption of nutrients by using radioisotopes. Ph.D. thesis, Kyoto University, Kyoto, Japan.

Yamada, Y., S.H. Wittwer, and M.J. Bukovac. 1964. Penetration of ions through isolated cuticles. Plant Physiol. 39:28-36.

Yamamura, N. and N. Tsuji. 1995. Optimal strategy of plant antiherbivore defense: implications for apparency and resource availability theories. Ecological Research 10:19-30.

Yamauchi, A., Taylor, H.M., Upchurch, D.R. and McMichael, B.R. 1995. Axial resistance to water flow of intact cotton taproots. Agron. J. 87:439-445.

Yang, S. and D.A. Grantz. 1996. Root hydraulic conductance in Pima cotton: Comparison of reverse flow, transpiration, and root pressure. Crop Sci. 36:1580-1589.

Yang, S. and M.T. Tyree. 1993. Hydraulic resistance in *Acer saccharum* shoots and its influence on leaf water potential and transpiration. Tree Physiol. 12:231-242.

Yang, Y., R. De Feyter, and D.W. Gabriel. 1994. Host-specific symptoms and increased release of *Xanthomonas citri* and *X. campestris* pv. *malvacearum* from leaves are determined by the 102-bp tandem repeats of *pth A* and *avrb$_6$*, respectively. Molecular Plant-Microbe Interactions 7:345-355.

Yang, Y. and D.W. Gabriel. 1992. Functional analysis of avirulence genes in *Xanthomonas campestris* pv. *malvacearum*. Phytopathology 82:1167.

Yankey, J.M. and P.C. Jones. 1993. Multidata AFIS. pp. 1129-1131. *In:* Proc. Beltwide Cotton Conf., National Cotton Council of America, Memphis, Tenn.

Yano, K., A. Yamauchi, M. Iijima, and Y. Kono. 1998. Arbuscular mycorrhizal formation in undisturbed soil counteracts compacted soil stress for pigeon pea. Appl. Soil Ecol. 10:95-102.

Yao, K., V. DeLuca, and N. Brisson. 1995. Creation of a metabolic sink for tryptophan alters the phenylpropanoid pathway and the susceptibility of potato to Phytophtora infestans. Plant Cell 7:1787-1799.

Yao, K., Q. Gu, and H. Lei. 1987. Effects of (+/-)-, (+)-, and (-)-gossypol on the lactate dehydrogenase-X activity of rat testis. J. Ethnopharmacol. 20:25-29.

Yasseen, A.I.H., A.Y. Negm, and A.A. Hosny. 1990. Effect of increasing population density and nitrogen on growth and yield of Giza 75 cotton variety. Annals Agric. Sci., Fac. Agric., Ain Shams Univ., Cairo, Egypt 35:751-760.

Yatsu, L.Y. 1983. Morphological and physical effects of colchicine treatment on cotton (*Gossypium hirsutum* L.) fibers. Text. Res. J. 53:515-519.

Yatsu, L.Y., K.E. Espelie, and P.E. Kolattukudy. 1983. Ultrastructural and chemical evidence that the cell wall of green cotton fiber is suberized. Plant Physiology 73:521-524.

Yatsu, L.Y. and T.J. Jacks. 1981. An ultrastructural study of the relationship between microtubules and microfibrils in cotton (*Gossypium hirsutum* L.) cell wall reversals. Amer. J. Bot. 68:771-777.

Ye, W.W. and J.D. Liu. 1994. The effect of NaCl and table salt on the germination of cotton seed. China Cottons 21:14-15.

Yik, C.P. and W. Birchfield. 1984. Resistant germplasm in *Gossypium* species and related plants to *Rotylenchulus reniformis*. J. Nematology 16:146-153.

Yokoyama, V.Y. and B.E. Mackey. 1987. Protein and tannin in upper, middle, and lower cotton plant strata and cigarette beetle (coleoptera: Anobiidae) growth on the foliage. J. Con. Entomol. 80:843.

Yokoyama, V.Y., B.E. Mackey, and T.F. Leigh. 1987. Relation of cotton protein and tannin to methyl parathion treatment and cigarette beetle (Coleoptera: Anobiidae) growth on foliage. J. Con. Entomol. 80:834.

York, A.C. 1982. Interaction of nitrogen rates, plant populations, and varieties with Pix. p. 58. *In:* J.M. Brown (ed.). Proc. Beltwide Cotton Prod. Res. Conf., National Cotton Council of America, Memphis, Tenn.

York, A.C.1983a. Cotton cultivar response to mepiquat chloride. Agron. J. 75:663-667.

York, A.C. 1983b. Response of cotton to mepiquat chloride with varying N rates and plant populations. Agron. J. 75:667-672.

York, W.S., M. McNeil, A.G. Darvill, and P. Albersheim. 1980. Beta-2-linked glucans secreted by fast-growing species of *Rhizobium*. J. Bacteriology 142:243-248.

Young, E.F., Jr., R.M. Taylor, and H.D. Petersen.1980. Day-degree units and time in relation to vegetative development and fruiting for three cotton cultivars. Crop Sci.20:370-374.

Youngman, R.R., T.F. Leigh, T.A. Kerby, N.C. Toscano, and C.E. Jackson. 1990. Pesticides and cotton: Effect on photosynthesis, growth, and fruiting. J. Econ. Entomol. 83:1549-1557.

Yu, F., T.N. Barry, P.J. Moughan, and G.F. Wilson. 1993. Condensed tannin and gossypol concentrations in cottonseed and in processed cottonseed meal. J. Science Food Agriculture 63:7-15.

Yu, Y.W. 1987. Probing into the mechanism of action, metabolism and toxicity of gossypol by studying its (+)- and (-)-stereoisomers. J. Ethnopharmacol. 20:65-70.

Zahn, H. 1988. Latest findings on the microstructure of cotton. Presented at the 19th International Cotton Conference. Bremen Fibre Institute, Bremen, Germany.

Zak, J.C., B. McMichael, S. Dhillion, and C. Friese. 1998. Arbuscular-mycorrhizal colonization dynamics of cotton (*Gossypium hirsutum* L.) growing under several production systems on the Southern High Plains, Texas. Agric. Ecosyst. Environ. 68:245-254.

Zaki, K., I.J. Mishagi, and M.N. Shatla. 1998. Control of cotton seedling damping-off in the field by

Burkholderia (*Pseudomonas*) *cepacia*. Plant Disease 82:291-293.

Zangerl, A.R. and F.A. Bazzaz. 1992. Theory and pattern in plant defense allocation. pp. 363-391 *In:* R.S. Fritz and E.L. Simms (eds.). Plant Resistance to Herbivores and Pathogens: Ecology, Evolution and Genetics, University of Chicago Press, Chicago.

Zasloff, M. 1987. Magainins, a class of antimicrobial peptides from X*enopus* skin - isolation, characterization of 2 active forms, and partial cDNA sequence of a precursor. Proc. of the National Academy of Sciences USA 84:5449-5453.

Zeevaart, J.A.D. and R.A. Creelman. 1988. Metabolism and physiology of abscisic acid. Annu. Rev. Plant Physiol. Plant Mol. Biol. 39:439-473.

Zelinski, L.J. 1996. Interaction of water and nitrogen stress on the growth and development of cotton. Ph.D. Diss., Univ. California, Davis. (UMI Dissertation Services #9608870).

Zeringue, H.J., Jr, 1987. Changes in cotton leaf chemistry induced by volatile elicitors. Pytochemistry 26:1357-1453.

Zeringue, H.J., Jr. 1988. Production of carpel wall phytoalexins in the developing cotton boll. Phytochemistry 27:3429-3431.

Zeringue, H.J., Jr. 1990. Stress effects on cotton leaf phytoalexins elicited by cell-free-mycelia extracts of *Aspergillus flavus*. Phytochemistry 29:1789-1791.

Zeringue, H.J., Jr. 1996. Possible involvement of lipoxygenase in a defense response in aflatoxicgenic *Aspergillus* - cotton plant interactions. Can. J. Botany 74:98-102.

Zeringue, H.J. Jr. and S.P. McCormick. 1989. Relationships between cotton leaf-derived volatiles and growth of *Aspergillus flavus*. JAOCS 66:581-585.

Zhang, B., X. Li, F. Li, B.H. Zhang, X.L. Li, and F.L. Li. 1995. Selection of NaCl tolerant embryogenic cell line and plant regeneration of cotton (*Gossypium hirsutum* L.) in vitro. Scientia Agricultura Sinica 28:33-38.

Zhang, B.H., X.L. Li, and F.L. Li. 1993. Selection and plant regeneration of NaCl-tolerant variant of cotton callus in vitro. Plant Physiol. Communications 29:423-425. (Chinese).

Zhang, F., V. Romheld, and H. Marschner. 1991. Release of zinc mobilizing root exudates in different plant species as affected by zinc nutritional status. J. Plant Nutr. 14:675-686.

Zhang, H.L., A. Nagatsu, H. Okuyama, H. Mizukami, and J. Sakakibara. 1998. Sesquiterpene glycosides from cotton cake oil. Phytochemistry 48:665-668.

Zhang, J. 1995. Identification and Pathogenicity of *Fusarium* Species Associated with Cotton Seedling Roots, and Their Interactions with Biocontrol Agents and Other Soil-borne Pathogens. Ph. D. Thesis, Texas A&M University, College Station. 99 pp.

Zhang, J., C.R. Howell, and J.L. Starr. 1996. Suppression of *Fusarium* colonization of cotton roots and Fusarium wilt by seed treatments with *Gliocladium virens* and *Bacillus subtilis*. Biocontrol Science and Technology 6:175-187.

Zhang, J., C.R. Howell, J.L. Starr, and M.H. Wheeler. 1995. Frequency of isolation and the pathogenicity of *Fusarium* species associated with roots of healthy cotton seedlings. Mycological Research 100:747-752.

Zhang, J., M.E. Mace, R.D. Stipanovic, and A.A. Bell. 1993. Production and fungitoxicity of the terpenoid phytoalexins in cotton inoculated with *Fusarium oxysporum* f. sp. *vasinfectum*. J. Phytopathol. 139:247-252.

Zhang, J.P., N.P. Tugwell, M.J. Cochran, F.M. Bourland, D.M. Oosterhuis, and C.D. Klein. 1994. COTMAN: A computer-aided cotton management system for late-season practices. p. 1286-1287. *In:* D.J. Herber and D.A. Richter (eds). Proc. Beltwide Cotton Conf., National Cotton Council of America, Memphis, Tenn.

Zhang, S., J.T. Cothren, and E.J. Lorenz. 1990. Mepiquat chloride seed treatment and germination temperature effects on cotton growth, nutrient partitioning, and water use efficiency. J. Plant Growth Reg. 9:195-199.

Zhang, X.C. and W.P. Miller. 1996. Physical and chemical crusting process affecting runoff and erosion in furrows. Soil Sci. Soc. Am. J. 60:860-865.

Zhao, D. and D.M. Oosterhuis. 1994. Physiological responses of cotton plants to PGR-IV application under water stress. p. 1373. *In:* D.J. Herber and D.A. Richter (ed.). Proc. Beltwide Cotton Conf., National Cotton Council of America, Memphis, Tenn.

Zhao, D. and D. Oosterhuis. 1995. Effects of PGR-IV on the growth and yield of environmentally stressed cotton. p. 1150. *In:* D. A. Richter and J. Armour (eds.). Proc. Beltwide Cotton Prod. Res. Conf., National Cotton Council of America, Memphis, Tenn.

Zhao, D.L. and D.M. Oosterhuis. 1997. Physiological response of growth chamber-grown cotton plants to the plant growth regulato PGR-IV under water-stress. Env. Expl. Bot. 38:7-14.

Zhao, D. and D. Oosterhuis. 1998a. Physiological and yield responses of shaded cotton to the plant growth regulator PGR-IV. J. Plant Growth Regul. 17:47-52.

Zhao, D. and D.M. Oosterhuis. 1998b. Responses of field-grown cotton to shade: an overview. pp. 1503-1507. *In:* Proc. Beltwide Cotton Prod. Res. Conf., National Cotton Council of America, Memphis, Tenn.

Zhao, D. and D.M. Oosterhuis. 1999. Photosynthetic capacity and carbon contribution of leaves and bracts to developing floral buds in cotton. Photosynthetica 36:279-290.

Zhao, D. and D.M. Oosterhuis. 2000. Pix Plus and Pix mepiquat chloride effects on physiology, growth and yield of field-grown cotton. J. Plant Growth Regulation, 19:415-422.

Zhao, D. and D.M. Oosterhuis, 2003. Growth and physiological responses of cotton to boron deficiency. J. Plant Nutr. 26:855-867.

Zhao, J.Z., Y.L. Fan, and X.P. Shi. 1997. Gene pyramiding: an effective strategy of resistance management for *Helicoverpa armigera* and *Bacillus thuringiensis.* Resistant Pest Management 9:19-21.

Zhao, X., R.A. Wing, and A.H. Paterson. 1995. Cloning and characterization of the majority of repetitive DNA in cotton (*Gossypium* L.). Genome 38:1177-1188.

Zheng, D.K., Si.Y., Kang, M.J. Ke, Z. Jin, and H. Liang. 1985. Resolution of racemic gossypol. J. Chem. Soc., Chem. Commun. pp. 168-169.

Zhong, H. and A. Läuchli. 1988. Incorporation of [^{14}C]glucose into cell wall polysaccharides of cotton roots: Effects of NaCl and CaCl$_2$. Plant Physiol. 88:511-514.

Zhong, H. and A. Läuchli. 1993a. Spatial and temporal aspects of growth in the primary root of cotton seedlings: Effects of NaCl and CaCl$_2$. J. Exp. Bot. 44:763-771.

Zhong, H. and A. Läuchli. 1993b. Changes of cell wall composition and polymer size in primary roots of cotton seedlings under high salinity. J. Exp. Bot. 44:773-778.

Zhong, H. and A. Läuchli. 1994. Spatial distribution of solutes, K, Na, Ca and their deposition rates in the growth zone of primary cotton roots: Effects of NaCl and CaCl$_2$. Planta 194:34-41.

Zhu, B. 1989. The absorption and translocation of foliar-applied nitrogen in cotton. Ph.D. Dissertation. University of Arkansas, Fayetteville.

Zhu, B. and D.M. Oosterhuis. 1992. Nitrogen distribution within a sympodial branch of cotton. J. Plant Nutr. 15:1-14.

Zhu, G.I. and J.S. Boyer. 1992. Enlargement in Chara studied with a trugor clamp. Plant Physiol. 100:2071-2080.

Zhu, S.J. and B.L. Li. 1993. Studies of introgression of the glandless seeds glandless plant trait from *Gossypium bickii* into cultivated upland (*G. hirsutum*) cotton. Cot. Fib. Trop. 48:195-200.

Ziegler, H. 1975. Nature of Transported Substances. pp. 59-100. *In:* M.H. Zimmerman and J.A. Milburn (eds.). Encyclopedia of Plant Physiology, New Series, vol. I. Springer-Verlag, New York.

Zima, J. and Z. Sestak. 1979. Photosynthetic characteristics during ontogenesis of leaves. 4. Carbon fixation pathways, their enzymes and products. Photosynthetica 13:83-106.

Zohary, D. and M. Hopf. 1988. *Domestication of plants in the Old World.* Clarendon Press, Oxford.

Zummo, G.R., J.H. Benedict, and J.C. Segers. 1983. No-choice study of plant-insect interactions for *Heliothis zea* (Boddie) (Lepidoptera: Noctuidae) on selected cottons. Environ. Entomol. 12:1833.

Zummo, G.R., J.H. Benedict, and J.C. Segers. 1984. Effect of the plant growth regulator mepiquat chloride on host plant resistance in cotton to bollworm (Lepidoptera: Noctuidae). J. Econ. Entomol. 77:922-924.

Zuniga, G.E. and L.J. Corcuera. 1987. Glycine-betaine accumulation influences susceptibility of water-stressed barley to the aphid *Schizaphis graminum.* Phytochemistry 26:367-369.

Index

O